SPACECRAFT ATTITUDE DETERMINATION AND CONTROL

ASTROPHYSICS AND SPACE SCIENCE LIBRARY

A SERIES OF BOOKS ON THE RECENT DEVELOPMENTS
OF SPACE SCIENCE AND OF GENERAL GEOPHYSICS AND ASTROPHYSICS
PUBLISHED IN CONNECTION WITH THE JOURNAL
SPACE SCIENCE REVIEWS

VOLUME 73

SPACECRAFT ATTITUDE DETERMINATION AND CONTROL

Edited by

JAMES R. WERTZ

Computer Sciences Corporation

Written by

Members of the Technical Staff
Attitude Systems Operation
· *Computer Sciences Corporation*

Preparation of this material was supported by the Attitude Determination and Control Section, Goddard Space Flight Center, National Aeronautics and Space Administration under Contract No. NAS 5-11999 and by the System Sciences Division, Computer Sciences Corporation.

D. REIDEL PUBLISHING COMPANY

DORDRECHT : HOLLAND / BOSTON : U.S.A.

LONDON : ENGLAND

Library of Congress Cataloging in Publication Data

Computer Sciences Corporation. Attitude Systems Operation.
    Spacecraft attitude determination and control.

    (Astrophysics and space science library; v. 73)
    'Contract no. NAS 5-11999.'
    Bibliography: p.
    Includes index.
    1. Space vehicles–Attitude control systems. 2. Space vehicles–Guidance
systems. I. Wertz, James R. II. Title. III. Series.
TL3260.C65   1978   629.47'42   78-23657
ISBN 90-277-0959-9
ISBN 90-277-1204-2 (pbk.)

Published by D. Reidel Publishing Company,
P.O. Box 17, 3300 AA Dordrecht, Holland.

Sold and distributed in the U.S.A. and Canada
by Kluwer Academic Publishers,
190 Old Derby Street, Hingham, MA 02043, U.S.A.

In all other countries, sold and distributed
by Kluwer Academic Publishers Group,
P.O. Box 322, 3300 AH Dordrecht, Holland.

Preparation of this material was supported by the Attitude Determination and Control Section,
Goddard Space Flight Center, National Aeronautics and Space Administration under Contract No.
NAS 5-11999 and by the System Sciences Division, Computer Sciences Corporation.

(Reprinted 1980, 1984)

Printed in The Netherlands

LIST OF CONTRIBUTING AUTHORS

All the authors are members of the technical staff in the Attitude Systems Operation, System Sciences Division, Computer Sciences Corporation. Sections written by each author are in brackets.

John Aiello—B.S. (Astronomy), Villanova University [5.2, Appendix G]

Jawaid Bashir—Ph.D. (Aerospace Engineering), M.S. (Electrical Engineering), University of Colorado; B.S. (Electrical Engineering), Karachi University, Pakistan [18.1]

Robert M. Beard—M.S. (Mathematics), B.S. (Physics), Auburn University [16.3]

Bruce T. Blaylock—M.S. (Chemistry), University of Virginia; B.S. (Chemistry), Eastern Montana College [6.3]

Lily C. Chen—Ph.D. (Physics), University of Wisconsin, Madison; M.S. (Physics), University of Cincinnati; B.S. (Physics), National Taiwan University [7.1, 11.3, 11.4, 11.5, Chapter 10]

Roger M. Davis—M.S. (Mechanical Engineering), Northeastern University; B.S. (Mechanical Engineering), University of Connecticut [16.4]

Demosthenes Dialetis—Ph.D. (Physics), University of Rochester; B.Sc. (Physics), University of Athens, Greece [16.4]

Lawrence Fallon, III—Ph.D., M.S. (Materials Science), University of Virginia; B.S. (Engineering Physics), Loyola College [6.4, 6.5, 7.6, 7.8, 13.4, 13.5, 21.3, Appendix D]

B. L. Gambhir—Ph.D. (Physics), University of Maryland; M.Sc. (Physics), B.Sc. (Physics, Mathematics, English), Punjab University, India [6.7, 19.1]

David M. Gottlieb—Ph.D. (Astronomy), University of Maryland; B.A. (Mathematics), Johns Hopkins University [5.3, 5.6, 7.7]

Mihaly G. Grell—M.S. (Physics), University of Sciences, Budapest [19.2]

Dale Headrick—Ph.D., M.S. (Physics), Yale University; B.S. (Physics) Louisiana State University [6.6, 18.2, 19.4]

Steven G. Hotovy—Ph.D. (Mathematics), University of Colorado; B.S. (Mathematics), University of Notre Dame [7.2, 13.2, 13.3]

James S. Legg, Jr.—M.S. (Physics), University of North Carolina; A.B. (Physics, Mathematics), Washington and Lee University [8.1, 8.3, 8.4, 9.1]

Gerald M. Lerner—Ph.D. (Physics), University of Maryland; B.A. (Physics), Johns Hopkins University [6.1, 6.2, 6.9, 7.1, 7.5, 9.2, 9.3, 12.2, 12.3, 18.1, 18.3, 19.5, Appendix F]

Menachem Levitas—Ph.D. (Physics), University of Virginia; B.S. (Physics), University of Portland [7.3, 17.3]

K. Liu—B.S. (Physics), National Taiwan University [4.3]

F. L. Markley—Ph.D. (Physics), University of California, Berkeley; B.E.P. (Engineering Physics), Cornell University [7.4, 7.9: 12.1, 15.2, 16.1, 16.2, 17.1, Appendix C]

Prafulla K. Misra—Ph.D., M.S. (Electrical Engineering), University of Maryland; B.Tech. (Electrical Engineering), Indian Institute of Technology, India [6.9]

Janet Niblack—M.A. (Mathematics), University of Texas; B.A. (Mathematics), Florida State University [8.2]

Michael Plett—Ph.D. (Physics), University of Virginia; B.S. (Physics), University of Cincinnati [5.1, 16.3, Appendix H]

Paul V. Rigterink—Ph.D. (Astronomy), University of Pennsylvania; B.A. (Mathematics), Carleton College [13.4]

John N. Rowe—Ph.D., M.S. (Electrical Engineering), Pennsylvania State University; M.A. (Physics), Western Michigan University; B.A. (Physics), Oakland University, Michigan [4.4, 5.4, 5.5]

Ashok K. Saxena—M.S. (Aerospace Engineering), Virginia Polytechnic Institute and State University; B.E. (Mechanical Engineering), Jadavpur University, India [18.4, Appendix I]

Myron A. Shear—M.S. (Physics), University of Illinois; B.A. (Physics, Chemistry), Harvard University [20.1, 20.3, 21.1]

Malcolm D. Shuster—Ph.D. (Physics), University of Maryland; S.B. (Physics), Massachusetts Institute of Technology [19.2]

Peter M. Smith—Ph.D. (Chemistry), Georgetown University; M.Sc. (Spectroscopy), B.Sc. (Chemistry), Manchester University, England [11.1, 11.2, 21.4]

Des R. Sood—D. Eng. Sc. (Mechanical Engineering), Columbia University; M.S. (Mechanical Engineering), Roorkee University, India; B.S. (Mechanical Engineering), Delhi University, India [6.7, 19.1]

C. B. Spence, Jr.—Ph.D., M.S. (Physics), College of William and Mary; B.S. (Physics), University of Richmond [17.1, 17.2]

Conrad R. Sturch—Ph.D. (Astronomy), University of California, Berkeley; M.S., B.A. (Physics), Miami University, Ohio [Appendix J]

Gyanendra K. Tandon—Ph.D. (Physics), Yale University; M.Sc. (Electronics), B.Sc. (Physics, Mathematics), Allahabad University, India [17.4, 21.2, Appendix E]

Vincent H. Tate—M.S., B.S. (Aerospace Engineering), Pennsylvania State University [15.3]

James R. Wertz—Ph.D. (Physics), University of Texas, Austin; S.B. (Physics), Massachusetts Institute of Technology [4.1, 4.2, 9.4, 11.3, 11.4, 11.5, 13.1, 15.1; Chapters 1, 2, 3, 10, 14, 22; Appendices A, B, K, L, M]

Robert S. Williams—Ph.D. (Physics), University of Maryland; B.S. (Physics), California Institute of Technology [6.8, 7.10, 19.3]

Kay Yong—Ph.D., M.S. (Mechanical Engineering), Rensselaer Polytechnic Institute; B.S. (Mechanical Engineering), National Cheng-Kung University, Taiwan. [5.2]

FOREWORD

Roger D. Werking
Head, Attitude Determination and Control Section
National Aeronautics and Space Administration/
Goddard Space Flight Center

Extensive work has been done for many years in the areas of attitude determination, attitude prediction, and attitude control. During this time, it has been difficult to obtain reference material that provided a comprehensive overview of attitude support activities. This lack of reference material has made it difficult for those not intimately involved in attitude functions to become acquainted with the ideas and activities which are essential to understanding the various aspects of spacecraft attitude support. As a result, I felt the need for a document which could be used by a variety of persons to obtain an understanding of the work which has been done in support of spacecraft attitude objectives. It is believed that this book, prepared by the Computer Sciences Corporation under the able direction of Dr. James Wertz, provides this type of reference.

This book can serve as a reference for individuals involved in mission planning, attitude determination, and attitude dynamics; an introductory textbook for students and professionals starting in this field; an information source for experimenters or others involved in spacecraft-related work who need information on spacecraft orientation and how it is determined, but who have neither the time nor the resources to pursue the varied literature on this subject; and a tool for encouraging those who could expand this discipline to do so, because much remains to be done to satisfy future needs.

The primary purpose of this book is to provide short descriptions of various aspects of attitude determination, prediction, and control with emphasis on the ground support which presently must be provided. The initial chapters provide the necessary background and describe environment models and spacecraft attitude hardware. The authors then present the fundamentals that are essential to a basic understanding of the activities in this area as well as flight-proven concepts which can be used as a basis for operational state-of-the-art activities or as a stepping stone to improved processes. In a limited fashion, Chapter 22 presents future activities which affect or are a part of spacecraft attitude support. It is not the intention of this book to advance the state of the art but rather to call attention to the work that has been done in the successful support of spacecraft attitude requirements and to stimulate future thinking.

PREFACE

The purpose of this book is to summarize the ideas, data, and analytic techniques needed for spacecraft attitude determination and control in a form that is readable to someone with little or no previous background in this specific area. It has been prepared for those who have a physics or engineering background and therefore are familiar with the elementary aspects of Newtonian mechanics, vector algebra, and calculus. Summaries of pertinent facts in other areas are presented without proof.

This material has been prepared by 35 members of the technical staff of the Attitude Systems Operation of the System Sciences Division of Computer Sciences Corporation (CSC) for the Attitude Determination and Control Section of NASA's Goddard Space Flight Center. It necessarily reflects our experience in this area and therefore is concerned primarily with unmanned, Earth-orbiting spacecraft. Nonetheless, the basic principles are sufficiently broad to be applicable to nearly any spacecraft.

Chapters 1, 2, 3, 10, and 15 provide introductory material at a more qualitative level than that of the other chapters. The suggested order of reading depends on the background and interest of the reader:

1. Those who are primarily concerned with mission planning and analysis and who would like a general overview should read Chapters 1, 2, 3, 10, 15, and 22.

2. Those who are primarily interested in attitude determination should read Chapters 1 and 2, Sections 3.1 through 3.3, Chapter 10, Appendices A and B, and Chapters 11 through 14.

3. Those who are primarily interested in attitude dynamics and control should read Chapters 1 and 2, Sections 3.1 through 3.3 and 12.1, Chapters 15 through 19, and Appendices C through H.

4. Those who are primarily interested in the space environment, attitude hardware, and data acquisition should read Chapters 1 through 9 and Appendices G through J.

5. Those who are primarily interested in the development of mission related software should read Chapters 1 and 2, Sections 3.1 through 3.3, Chapters 20 and 21, Chapters 8 and 9, Sections 11.1 and 11.2, and Chapters 12, 4, 5, and 7.

The International System of Units is used throughout the book and a detailed list of conversion factors is given in Appendix K. Because nearly all numerical work is now done with computers or hand calculators, all constants are given to essentially their full available accuracy. Acronyms have generally been avoided, except for spacecraft names. The full spacecraft names are listed in Appendix I, which also provides a cross-referenced list of the attitude hardware used on various spacecraft, including all those used as examples throughout the text.

Because much of the material presented here has not appeared in the open literature, many of the references are to corporate or government documents of limited circulation. To improve the exchange of information, Computer Sciences

Corporation reports referenced herein are available for interlibrary loan through
your librarian by writing to:

Head Librarian
Technical Information Center
Computer Sciences Corporation
8728 Colesville Road
Silver Spring, MD 20910

Standard computer subroutines for attitude analysis cited throughout the book are
available from:

COSMIC
Barrow Hall
University of Georgia
Athens, GA 30601

by asking for Program *Number GSC12421, Attitude Determination and Control
Utilities.* Each of these subroutines is briefly described in Section 20.3.

The preparation of this book was a cooperative effort on the part of many
individuals. It is a pleasure to acknowledge the help of Robert Coady, Roger
Werking, and Richard Nankervis of NASA's Goddard Space Flight Center, who
initiated and supported this project. At Computer Sciences Corporation, direction,
support, and help were provided by Richard Taylor, David Stewart, Michael Plett,
and Gerald Lerner. In addition to the authors, who provided extensive review of
each other's sections, particularly helpful reviews were provided by Peter Batay-
Csorba, Stanley Brown, Charles Gray, Lawrence Gunshol, William Hogan, Whit-
tak Huang, James Keat, Anne Long, J. A. Massart, Donald Novak, Franklin
VanLandingham, Donna Walter, and Chao Yang.* Considerable assistance in
obtaining reference material was supplied by Gloria Urban and the staff of the
CSC Technical Information Center. Jo Border and the CSC Publications Department supplied a consistently high quality of support in editing, composition, and
graphics. Jerry Greeson and the Graphics Department staff prepared nearly all of
the 450 illustrations in the book. Figures 17-4, 18-19, and 19-15 are reproduced by
permission of the American Institute of Aeronautics and Astronautics. Anne Smith
edited the final version of the manuscript and Julie Langston, the publications
editor for the manuscript, did an outstanding and professional job of translating
multiple early drafts into grammatical English, handling the numerous details of
producing a finished manuscript, and preparing the final layout.

Silver Spring, Maryland   **James R. Wertz**
July 1978

---

*The editor would appreciate that any residual errors be brought to his attention at the following
address: Computer Science Corporation, 8728 Colesville Road, Silver Spring, Maryland 20910.

STANDARD NOTATION

Standard notation developed in the first three chapters and used throughout the remainder of this book is given below. Unfortunately, notation, coordinate systems, and even definitions are frequently used differently for different spacecraft. The definition of attitude and the orientation of the roll, pitch, and yaw axes vary more than most quantities.

Alphabets and Type Styles

All arc lengths are lowercase Greek, θ

All rotation angles are uppercase Greek, Λ

These are not exclusive because common usage may require Greek characters for some nonangular measure.

All n-vectors are bold face, $\mathbf{E}$ or $\mathbf{x}$. All quaternions are boldface italics, $\boldsymbol{q}$. Boldface is used exclusively for vectors, quaternions, or the identity matrix, $\mathbf{1}$.

Points on the sky are labeled with uppercase italic Roman, E. Antipodal points (the *antipode* is the point 180 deg away from a given point) have a -1 superscript, E^{-1}.

Vectors and Matrices

The treatment of vectors is illustrated by the Sun vector:

S	point on the celestial sphere in the direction of the Sun
S^{-1}	point on the celestial sphere opposite the direction of the Sun
$\mathbf{S}$	vector from spacecraft to Sun
$\hat{\mathbf{S}}$	unit vector from the spacecraft toward the Sun
$\|\mathbf{S}\|$ or S	magnitude of the Sun vector. S is used if there is no possibility of ambiguity. Otherwise $\|\mathbf{S}\|$ is used
S_i	either the ith component of the Sun vector or an arbitrary component of the Sun vector

Both uppercase and lowercase letters will be used for vectors. Matrices will be represented by uppercase Roman letters with the following notation:

M or $[M_{ij}]$	matrix or second-rank tensor
$\det M$ or $\|M_{ij}\|$	determinant of M
M^{-1}	inverse of M
M^{T}	transpose of M
M_{ij}	either an arbitrary component of M or the component in the ith row, jth column
$\mathbf{1}$	identity matrix
I	moment of inertia tensor

Coordinate Systems

Spacecraft-Centered Celestial Coordinates:

α right ascension
δ declination

Body-Fixed Coordinates for Spinning Spacecraft:

ϕ azimuth
λ elevation
or θ coelevation (measured from the spin axis)

No standard definitions apply for body-fixed coordinates for three-axis stabilized spacecraft.

Roll, Pitch, Yaw:

Yaw axis, **Y**, toward the nadir
Pitch axis, **P**, toward negative orbit normal
Roll axis, **R**, such that $\hat{\mathbf{R}} = \hat{\mathbf{P}} \times \hat{\mathbf{Y}}$
Unfortunately, the roll, pitch, and yaw axes do not have generally accepted meanings in spaceflight. Usage depends on the context.

ξ_r roll angle, measured about **R**
ξ_p pitch angle, measured about **P**
ξ_y yaw angle, measured about **Y**

Positive roll, pitch, and yaw are right-handed rotations about their respective axes.

Standard Symbols

The letter "t" in all forms (Roman, Greek, uppercase, and lowercase) is used only for time or time intervals, except that superscript T is used to indicate the transpose of a matrix.

Orbital elements:

a semimajor axis
e eccentricity
i inclination
ω argument of perigee
Ω right ascension of the ascending node
M_0 mean anomaly at epoch
T_0 epoch time
P orbital period

Vectors:

L   angular momentum
N   torque
B   magnetic field
E   nadir vector
S   Sun vector
A   attitude vector
H   horizon crossing vector; $\mathbf{H}_I$ and $\mathbf{H}_O$ for in-crossing and out-crossing
ω angular velocity vector
x m-dimensional state vector
y n-dimensional observation vector

Angles:

β Sun angle $\equiv$ Sun/attitude angular separation
η Nadir angle $\equiv$ Earth center/attitude angular separation
Φ Rotation angle from the Sun to the center of the Earth about the attiutde
ψ Sun/Earth center angular separation

Astronomical Symbols:

$\oplus$ Earth
$\odot$ Sun
Υ Vernal Equinox

Additional astronomical symbols are defined in Fig. 3-10.

Miscellaneous

Δ indicates an arbitrary interval, as $\Delta t = t_2 - t_1$
δ indicates an infinitesimal interval in which first-order approximations
        may be used, as $\mathbf{L} = \mathbf{L}_0 + \mathbf{N}\delta t$

A dot over any symbol indicates dffferentiation with respect to time, i.e., $\dot{x} \equiv dx/dt$.

The Kronecker Delta, δ_j^i, is defined as

$$\delta_j^i \equiv 1 \text{ for } i = j$$

$$\delta_j^i \equiv 0 \text{ for } i \neq j$$

The Dirac delta function, $\delta_D(x - x_0)$, is defined by

$$\delta_D(x - x_0) \equiv 0 \text{ for } x \neq x_0$$

$$\int_{x_1 < x_0}^{x_2 > x_0} \delta_D(x - x_0)dx \equiv 1$$

TABLE OF CONTENTS

List of Contributing Authors   v
Foreword   vii
Preface   viii
Standard Notation   x

PART I—BACKGROUND

1. INTRODUCTION   1
   1.1 *Representative Mission Profile*   3
   1.2 *Representative Examples of Attitude Determination and Control*   10
   1.3 *Methods of Attitude Determination and Control*   16
   1.4 *Time Measurements*   18

2. ATTITUDE GEOMETRY   22
   2.1 *The Spacecraft-Centered Celestial Sphere*   22
   2.2 *Coordinate Systems*   24
   2.3 *Elementary Spherical Geometry*   31

3. SUMMARY OF ORBIT PROPERTIES AND TERMINOLOGY   36
   3.1 *Keplerian Orbits*   36
   3.2 *Planetary and Lunar Orbits*   48
   3.3 *Spacecraft Orbits*   52
   3.4 *Orbit Perturbations*   62
   3.5 *Viewing and Lighting Conditions*   71

4. MODELING THE EARTH   82
   4.1 *Appearance of the Earth at Visual Wavelengths*   83
   4.2 *Appearance of the Earth at Infrared Wavelengths*   90
   4.3 *Earth Oblateness Modeling*   98
   4.4 *Modeling the Structure of the Upper Atmosphere*   106

5. MODELING THE SPACE ENVIRONMENT   113
   5.1 *The Earth's Magnetic Field*   113
   5.2 *The Earth's Gravitational Field*   123
   5.3 *Solar Radiation and the Solar Wind*   129
   5.4 *Modeling the Position of the Spacecraft*   132
   5.5 *Modeling the Positions of the Sun, Moon, and Planets*   138
   5.6 *Modeling Stellar Positions and Characteristics*   143

PART II—ATTITUDE HARDWARE AND DATA ACQUISITION

6. ATTITUDE HARDWARE   155
   6.1 *Sun Sensors*   155
   6.2 *Horizon Sensors*   166
   6.3 *Magnetometers*   180
   6.4 *Star Sensors*   184

	6.5	*Gyroscopes*	196
	6.6	*Momentum and Reaction Wheels*	201
	6.7	*Magnetic Coils*	204
	6.8	*Gas Jets*	206
	6.9	*Onboard Computers*	210

7.	MATHEMATICAL MODELS OF ATTITUDE HARDWARE		217
	7.1	*Sun Sensor Models*	218
	7.2	*Horizon Sensor Models*	230
	7.3	*Sun Sensor / Horizon Sensor Rotation Angle Models*	237
	7.4	*Modeling Sensor Electronics*	242
	7.5	*Magnetometer Models*	249
	7.6	*Star Sensor Models*	254
	7.7	*Star Identification Techniques*	259
	7.8	*Gyroscope Models*	266
	7.9	*Reaction Wheel Models*	270
	7.10	*Modeling Gas Jet Control Systems*	272

8.	DATA TRANSMISSION AND PREPROCESSING		278
	8.1	*Data Transmission*	278
	8.2	*Spacecraft Telemetry*	293
	8.3	*Time Tagging*	298
	8.4	*Telemetry Processors*	304

9.	DATA VALIDATION AND ADJUSTMENT		310
	9.1	*Validation of Discrete Telemetry Data*	312
	9.2	*Data Validation and Smoothing*	315
	9.3	*Scalar Checking*	328
	9.4	*Data Selection Requiring Attitude Information*	334

PART III—ATTITUDE DETERMINATION

10.	GEOMETRICAL BASIS OF ATTITUDE DETERMINATION		343
	10.1	*Single-Axis Attitude*	344
	10.2	*Arc-Length Measurements*	346
	10.3	*Rotation Angle Measurements*	349
	10.4	*Correlation Angles*	353
	10.5	*Compound Measurements—Sun to Earth Horizon Crossing Rotation Angle*	357
	10.6	*Three-Axis Attitude*	359

11.	SINGLE-AXIS ATTITUDE DETERMINATION METHODS		362
	11.1	*Methods for Spinning Spacecraft*	363
	11.2	*Solution Averaging*	370
	11.3	*Single-Axis Attitude Determination Accuracy*	373
	11.4	*Geometrical Limitations on Single-Axis Attitude Accuracy*	389
	11.5	*Attitude Uncertainty Due to Systematic Errors*	402

12. THREE-AXIS ATTITUDE DETERMINATION METHODS   410
     12.1 *Parameterization of the Attitude*   410
     12.2 *Three-Axis Attitude Determination*   420
     12.3 *Covariance Analysis*   429

13. STATE ESTIMATION ATTITUDE DETERMINATION METHODS   436
     13.1 *Deterministic Versus State Estimation Attitude Methods*   436
     13.2 *State Vectors*   438
     13.3 *Observation Models*   443
     13.4 *Introduction to Estimation Theory*   447
     13.5 *Recursive Least-Squares Estimators and Kalman Filters*   459

14. EVALUATION AND USE OF STATE ESTIMATORS   471
     14.1 *Prelaunch Evaluation of State Estimators*   471
     14.2 *Operational Bias Determination*   473
     14.3 *Limitations on State Vector Observability*   476

PART IV—ATTITUDE DYNAMICS AND CONTROL

15. INTRODUCTION TO ATTITUDE DYNAMICS AND CONTROL   487
     15.1 *Torque-Free Motion*   487
     15.2 *Response to Torques*   498
     15.3 *Introduction to Attitude Control*   502

16. ATTITUDE DYNAMICS   510
     16.1 *Equations of Motion*   510
     16.2 *Motion of a Rigid Spacecraft*   523
     16.3 *Spacecraft Nutation*   534
     16.4 *Flexible Spacecraft Dynamics*   548

17. ATTITUDE PREDICTION   558
     17.1 *Attitude Propagation*   558
     17.2 *Environmental Torques*   566
     17.3 *Modeling Internal Torques*   576
     17.4 *Modeling Torques Due to Orbit Maneuvers*   580

18. ATTITUDE STABILIZATION   588
     18.1 *Automatic Feedback Control*   588
     18.2 *Momentum and Reaction Wheels*   600
     18.3 *Autonomous Attitude Stabilization Systems*   604
     18.4 *Nutation and Libration Damping*   625

19. ATTITUDE MANEUVER CONTROL   636
     19.1 *Spin Axis Magnetic Coil Maneuvers*   636
     19.2 *Spin Plane Magnetic Coil Maneuvers*   642
     19.3 *Gas Jet Maneuvers*   649
     19.4 *Inertial Guidance Maneuvers*   655
     19.5 *Attitude Acquisition*   661

PART V—MISSION SUPPORT

20. SOFTWARE SYSTEM DEVELOPMENT   681
    20.1 *Safeguards Appropriate for Mission Support Software*   681
    20.2 *Use of Graphic Support Systems*   686
    20.3 *Utility Subroutines*   690

21. SOFTWARE SYSTEM STRUCTURE   696
    21.1 *General Structure for Attitude Software Systems*   696
    21.2 *Communications Technology Satellite Attitude Support System* 700
    21.3 *Star Sensor Attitude Determination System*   703
    21.4 *Attitude Data Simulators*   709

22. DISCUSSION   714

PART VI—APPENDICES

APPENDIX A—SPHERICAL GEOMETRY   727
APPENDIX B—CONSTRUCTION OF GLOBAL GEOMETRY PLOTS   737
APPENDIX C—MATRIX AND VECTOR ALGEBRA   744
APPENDIX D—QUATERNIONS   758
APPENDIX E—COORDINATE TRANSFORMATIONS   760
APPENDIX F—THE LAPLACE TRANSFORM   767
APPENDIX G—SPHERICAL HARMONICS   775
APPENDIX H—MAGNETIC FIELD MODELS   779
APPENDIX I—SPACECRAFT ATTITUDE DETERMINATION
              AND CONTROL SYSTEMS   787
APPENDIX J—TIME MEASUREMENT SYSTEMS   798
APPENDIX K—METRIC CONVERSION FACTORS   807
APPENDIX L—SOLAR SYSTEM CONSTANTS   814
APPENDIX M—FUNDAMENTAL PHYSICAL CONSTANTS   826
Index   830

PART I

BACKGROUND

CONTENTS

PART I

BACKGROUND

Chapter

1   Introduction   1

2   Attitude Geometry   22

3   Summary of Orbit Properties and Terminology   36

4   Modeling the Earth   82

5   Modeling the Space Environment   113

CHAPTER 1

INTRODUCTION

James R. Wertz

1.1 Representative Mission Profile
1.2 Representative Examples of Attitude Determination and Control
   Spin Stabilized Spacecraft, Three-Axis Stabilized Spacecraft, Attitude Maneuver Using Gas Jets
1.3 Methods of Attitude Determination and Control
1.4 Time Measurements

The *attitude* of a spacecraft is its orientation in space. This book is concerned with all aspects of spacecraft attitude—how it is determined, how it is controlled, and how its future motion is predicted. We describe simple procedures for estimating the attitude and sophisticated methods used to obtain the maximum accuracy from a given set of data. In this chapter, we introduce the basic terminology and provide an overview of the attitude determination and control processes and their place in the overall space mission.

The motion of a rigid spacecraft is specified by its position, velocity, attitude, and attitude motion. The first two quantities describe the translational motion *of* the center of mass of the spacecraft and are the subject of what is variously called *celestial mechanics*, *orbit determination*, or *space navigation*, depending on the aspect of the problem that is emphasized. The latter two quantities describe the rotational motion of the body of the spacecraft *about* the center of mass and are the subject of this book. Although knowledge of the spacecraft orbit frequently is required to perform attitude determination and control functions, the process of orbit determination or orbit maneuver analysis is outside the scope of this book. In general, orbit and attitude are interdependent. For example, in a low altitude Earth orbit, the attitude will affect the atmospheric drag which will affect the orbit; the orbit determines the spacecraft position which determines both the atmospheric density and the magnetic field strength which will, in turn, affect the attitude. However, we will normally ignore this dynamical coupling and assume that the time history of the spacecraft position is known and has been supplied by some process external to the attitude determination and control system. A brief summary of spacecraft orbit properties and terminology is given in Chapter 3.

One distinction between orbit and attitude problems is in their historical development. Predicting the orbital motion of celestial objects is one of the oldest sciences and was the initial motivation for much of Newton's work. Thus, although the space age has brought with it vast new areas of orbit analysis, a large body of theory directly related to celestial mechanics has existed for several centuries. In contrast, although some of the techniques are old, most of the advances in attitude determination and control have occurred since the launch of Sputnik on October 4, 1957.* The result of this is that relatively little information is recorded in books or other coordinated, comprehensive reference sources. The language of attitude determination and control is still evolving and many of the technical terms do not have universally accepted meanings. One purpose of this book is to codify the definitions of terms as they are commonly used and to eliminate some inconsistencies in their use.

Attitude analysis may be divided into determination, prediction, and control. *Attitude determination* is the process of computing the orientation of the spacecraft relative to either an inertial reference or some object of interest, such as the Earth. This typically involves several types of sensors on each spacecraft and sophisticated data processing procedures. The accuracy limit is usually determined by a combination of processing procedures and spacecraft hardware.

Attitude prediction is the process of forecasting the future orientation of the spacecraft by using dynamical models to extrapolate the attitude history. Here the limiting features are the knowledge of the applied and environmental torques and the accuracy of the mathematical model of spacecraft dynamics and hardware.

Attitude control is the process of orienting the spacecraft in a specified, predetermined direction. It consists of two areas—*attitude stabilization*, which is the process of maintaining an existing orientation, and *attitude maneuver control*, which is the process of controlling the reorientation of the spacecraft from one attitude to another. The two areas are not totally distinct, however. For example, we speak of stabilizing a spacecraft with one axis toward the Earth, which implies a continuous change in its inertial orientation. The limiting factor for attitude control is typically the performance of the maneuver hardware and the control electronics, although with autonomous control systems, it may be the accuracy of orbit or attitude information.

Some form of attitude determination and control is required for nearly all spacecraft. For engineering or flight-related functions, attitude determination is required only to provide a reference for control. Attitude control is required to avoid solar or atmospheric damage to sensitive components, to control heat dissipation, to point directional antennas and solar panels (for power generation), and to orient rockets used for orbit maneuvers. Typically, the attitude control accuracy necessary for engineering functions is on the order of 1 deg. Attitude requirements for the spacecraft payload are more varied and often more stringent than the engineering requirements. Payload requirements, such as telescope or antenna orientations, may involve attitude determination, attitude control, or both. Attitude constraints are most severe when they are the limiting factor in experimental accuracy or when it is desired to reduce the attitude uncertainty to a level such that it is not a factor in payload operation. These requirements may demand accuracy down to a fraction of an arc-second (1 arc-second equals 1/3600 deg).

---

*No science is without antecedents. Much background in this area comes from attitude determination and control for earlier airplanes and rockets. In addition, it has been understood since Newton's time that the attitude of the Moon was probably "gravity-gradient stabilized," i.e., locked into its rotation period by the torque resulting from the tidal forces of the Earth on the Moon.

A convenient method for categorizing spacecraft is the procedure by which they are stabilized. The simplest procedure is to spin the spacecraft. The angular momentum of a *spin-stabilized* spacecraft will remain approximately fixed in inertial space for extended periods, because external torques which affect it are extremely small in most cases. However, the rotational orientation of the spacecraft about the spin axis is not controlled in such a system. If the orientation of three mutually perpendicular spacecraft axes must be controlled, then the spacecraft is *three-axis stabilized*. In this case, some form of active control is usually required because environmental torques, although small, will normally cause the spacecraft orientation to drift slowly. (However, environmental torques can be stabilizing in some circumstances.) Three-axis stabilized spacecraft may be either *nonspinning* (fixed in inertial space) or fixed relative to a possibly rotating reference frame, as occurs for an Earth satellite which maintains one face toward the Earth and therefore is spinning at one rotation per orbit. Many missions consist of some phases in which the spacecraft is spin stabilized and some phases in which it is three-axis stabilized. Some spacecraft have multiple components, some of which are spinning and some of which are three-axis stabilized.

1.1 Representative Mission Profile

In this section we describe the profile of a typical space mission to illustrate the attitude determination and control process and to relate this process to the overall mission requirements. There is no single profile characteristic of all space missions. However, most missions have in common three more or less distinct phases: (1) *launch*, consisting of the activities from lift-off until the end of powered flight in a preliminary Earth orbit; (2) *acquisition*, consisting of orbit and attitude maneuvers and hardware checkout; and (3) *mission operations*, consisting of carrying out the normal activities for which the flight is intended.

Launch is the most distinct and well-defined phase and is normally carried out and controlled primarily by personnel concerned with the rocket launch vehicle and who will not be involved in subsequent mission operations. A limited number of launch vehicles are in use. For the United States, these range from the four-stage Scout, which is capable of orbiting a payload of about 100 kg to the three-stage Saturn V, capable of putting 100,000 kg in low-Earth orbit. (See, for example, Glasstone [1965], vonBraun and Ordway [1975], or the excellent nontechnical book by Clarke [1968].) One of the most common launch vehicles for Earth-orbiting spacecraft is the Delta, which has launched over 100 spacecraft and has low-orbit payload capabilities of 240 kg to 1900 kg, depending on the rocket configuration. For future launches, the Space Shuttle will be the most common, though not exclusive, vehicle.

Launch sites are similarly limited. For the United States, most launches for equatorial orbits occur from the Eastern Test Range at Cape Canaveral, Florida, or for polar orbits from the Western Test Range at Vandenberg Air Force Base, California. Some tests and launches of very small vehicles are conducted at White

Sands, New Mexico, and Wallops Island, Virginia.* For the Soviet Union, the launch sites are Tyuratan, 370 km southwest of Baykonur for major manned and planetary flights; Plesetsk, north of Moscow for military and some operational and scientific launches; and Kapustin Yar, near the Caspian Sea for small launches and tests. (For discussions of Soviet space programs, see most issues of the British journal *Spaceflight* or the comprehensive Senate report for the 92nd Congress [1971].)

Once powered flight has ended and the spacecraft has separated from most of the launch vehicle, the acquisition phase of maneuvers and testing begins. (The final launch stage may be left attached to the spacecraft for later maneuvers using the final stage control hardware and fuel. An Agena upper stage, a rocket which may be restarted in space, is frequently used for such maneuvers.) The acquisition phase can last from a few minutes to several months and may be defined differently depending on the particular aspect of the mission that is involved. For example, for someone concerned primarily with the operation of communications hardware, testing and maneuvering may last only a brief period, with "normal" operations beginning well before experiments or operational equipment have been thoroughly tested. Normally, the major portion of the attitude analysis for any mission is concerned with various aspects of the acquisition phase, as described in the example below.

Finally, once the proper orbit and attitude have been obtained and the hardware has been tested, the mission operations phase, in which the spacecraft carries out its basic purpose, is initiated. At this stage, attitude determination and control becomes, or should become, a routine process. On complex missions, such as lunar or planetary explorations, the acquisition phase may be repeated at various intervals as new hardware or new conditions are introduced.

We will illustrate the above phases and the role of the attitude determination and control process by describing the flight of the *Communications Technology Satellite, CTS,* launched aboard a Delta 2914 launch vehicle (Fig. 1-1) from the Eastern Test Range at 23:28 UT (18:28 EST) January 17, 1976.† (Evening or night launches are preferred to avoid evaporation of rocket fuel while sitting on the launch pad.) CTS was a joint project of the United States and Canada in which the Canadian Department of Communications built and operated the spacecraft and the United States National Aeronautics and Space Administration, NASA, provided the launch vehicle, launch facilities, and operational support through the acquisition phase of the mission. The purpose of the mission was to conduct various communications experiments, primarily as a high-power television relay from portable transmitters operating at a frequency of 14 GHz to low-cost 12-GHz receivers. The spacecraft was placed in synchronous orbit, i.e., at an altitude corresponding to an orbital period of 24 hours, so as to remain approximately

---

*Occasionally, U.S. spacecraft are launched from non-American sites. For example, the Small Astronomy Satellite-1 (SAS-1), also called *Uhuru* after the Swahili word for freedom, was launched from the Italian *San Marco Platform,* similar to an oil exploration platform, off the coast of Kenya. This was done to avoid the van Allen radiation belts.

†See Section 1.4 for an explanation of "UT," or Universal Time.

Fig. 1-1. Lift-Off of a Delta 2914 Launch Vehicle From the Eastern Test Range at Cape Canaveral, Florida. (Conditions for the launch of the Synchronous Meteorological Satellite shown here were essentially identical with those for the CTS launch which occurred after dark.)

stationary over the equator at 114° West longitude. This location permitted television transmissions to remote regions of both Canada and Alaska.

The mass of the CTS spacecraft was 676 kg at lift-off, of which approximately 340 kg was in the weight of a rocket motor, called the *apogee boost motor*, required for an orbit maneuver during the acquisition phase. As shown in Fig. 1-2, the spacecraft is approximately cylindrical, 1.88 m high and 1.83 m in diameter. The main operating power is supplied by two extendable solar arrays, each 6.20 m long and 1.30 m wide, with a mass of 15 kg and a power output of 600 watts per array. CTS has a total of 11 attitude sensors*: 4 used exclusively during the acquisition phase when the spacecraft was spin stabilized, 2 used during the operations phase when the spacecraft was three-axis stabilized, and 5 used for the transition from spinning to nonspinning. In addition to the large apogee boost motor used during the acquisition phase, the spacecraft includes 18 small rocket motors (2 "high thrust" and 16 "low thrust") for orbit and attitude maneuvers.

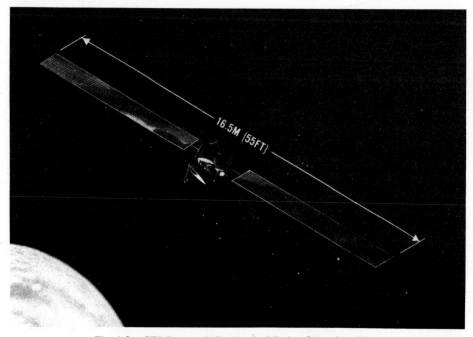

Fig. 1-2.   CTS Spacecraft During the Mission Operations Phase

As shown in Fig. 1-3, CTS was launched into an initial 185-km *parking orbit*. It maintained this orbit for approximately 15 minutes until the spacecraft was over the equator, at which time the third stage was reignited and the spacecraft was *injected* into an elliptical *transfer orbit* which had its low point, or *perigee*, at the parking orbit altitude and its high point, or *apogee*, just above the synchronous

---

*In addition to the 11 sensors, 3 gyroscopes were used to sense the rate of change of the attitude.

altitude of 35,860 km. (See Chapter 3 for a general discussion of orbits and orbit terminology.) Before injection, which marked the end of the launch phase, the spacecraft was controlled by the launch control team at the launch site. Subsequent to injection, the acquisition phase began and control of the spacecraft was transferred to the Operations Control Center at NASA's Goddard Space Flight Center in Greenbelt, Maryland.

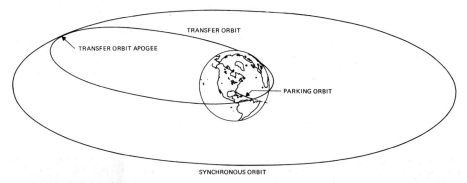

Fig. 1-3. CTS Orbit Maneuvers. (Orbit and Earth drawn to same scale; see text for explanation.)

The purposes of the transfer orbit were to move the spacecraft to synchronous altitude so that the large apogee boost motor could be fired to change the orbit to an approximately circular, synchronous one, as shown in Fig. 1-3, and to control the timing of this apogee motor firing. The smaller motors could then be used for various orbit and attitude refinements. Because the apogee motor was fixed in the spacecraft, it was necessary to reorient the entire spacecraft such that the motor firing would provide the proper orbit change. Thus, the principal activity during the transfer orbit was to determine the attitude, test and calibrate (if needed) the attitude sensors, reorient the spacecraft to the proper apogee motor firing attitude, and make fine adjustments and measurements as needed to ensure that the proper attitude had been obtained. To carry out this sequence and to provide for proper positioning in the synchronous orbit, the spacecraft remained in the transfer orbit for 6-1/2 orbits, or about 3 days.

While in the transfer orbit, the spacecraft was spin stabilized at approximately 60 rpm. Two Sun sensors and two Earth horizon sensors were used for attitude determination with an accuracy requirement of 1 deg for apogee motor firing. The two high-thrust rocket motors were used to reorient the spacecraft spin axis by about 225 deg from its initial orientation (as it left the third stage of the launch vehicle) to the apogee motor firing attitude. A maneuver of more than 180 deg was required to avoid lowering perigee (the minimum altitude point) in the transfer orbit due to the translational thrust of the maneuver jets.

The apogee motor firing changed the orbit to nearly circular and changed the period to approximately 23 hours 15 minutes so that the spacecraft would drift slowly westward relative to the Earth's surface. A series of orbit adjustments made the period nearly identical with the Earth's rotation period when the spacecraft was over the desired longitude. During this station acquisition phase, the principal attitude requirement was to bring the attitude to *orbit normal* (i.e., perpendicular to the orbit plane) after the apogee motor firing and to maintain it there for orbit maneuvering using the two high-thrust jets. The station acquisition sequence required a total of five progressively smaller orbit maneuvers carried out over a period of 9 days after apogee motor firing.

After station acquisition, control of the spacecraft was transferred to the Canadian Research Council, which conducted a major attitude maneuver sequence to transform the spacecraft from a spin-stabilized mode to a three-axis-stabilized, Earth-pointing mode [Bassett, 1976]. This was the most complex attitude maneuver conducted during the mission and consisted of a 2-day sequence of operations divided into 39 specific events. During this phase, attitude determination input was changed from the four sensors previously mentioned to five Sun sensors and two Earth sensors designed for use during the acquisition phase. The set of 16 low-thrust jets was used to despin the spacecraft and to control it in the nonspinning mode. Additional control was supplied by a spinning flywheel, which was used to reorient the spacecraft about the wheel axis by changing the relative angular momentum of the wheel and the spacecraft body. The major events in the maneuver sequence were: despin of the spacecraft, maneuver of the spacecraft to bring the Sun to its desired position in the control system field of view, deployment of the solar arrays, spinup of the flywheel, rotation of the spacecraft about the line to the Sun to orient the flywheel axis perpendicular to the orbit plane, and a series of rotations to achieve the final three-axis attitude with the Earth in the center of the Earth sensor field of view.

Completion of the attitude maneuvers ended the acquisition phase of the mission. After further hardware checks, normal mission operations were initiated. During the planned 2-year life of the spacecraft, the attitude control system will be used to maintain the attitude within 0.1 deg of its nominal orientation. The major factor in mission lifetime is the consumption of fuel for attitude stabilization, although orbit drift, possible mechanical failure, and power loss due to radiation damage to the solar cells may also affect the useful life of CTS.

Three Major Changes Which Will Affect Future Mission Profiles. During the decade of the 1980s, three major spaceflight changes are anticipated which will affect the representative mission profile just described: (1) launch and, for some spacecraft, recovery via the Space Shuttle; (2) increased use of onboard processing for attitude determination and control; and (3) major advances in tracking and spacecraft-to-ground communications via the *Tracking and Data Relay Satellite System, TDRSS,* and the *Global Positioning System, GPS.* The effect of these and unforeseen developments on attitude determination and control hardware and procedures cannot be predicted with precision. It is nevertheless important to consider the probable or possible direction of future developments.*

The principal effect of the Space Shuttle relative to attitude determination and control will be to decrease the cost of near-Earth payloads and increase their numbers. Increased shuttle capacity, relative to expendable launch vehicles, will

---

*For an extended discussion of long-term space developments, see the Future Space Programs report of the 94th Congress [1975] and the NASA Outlook for Space series [1976a, 1976b].

allow heavier and more extensive hardware than previously used. The potential cargo mass for various shuttle orbits is shown in Fig. 1-4. As transportation costs decrease and the number of active payloads increases, new methods will be needed to reduce the cost of attitude determination and control. At present, the most likely procedures to achieve this are (1) increased autonomy with up to 3 days of automatic control without ground support; (2) decreased hardware redundancy, since recoverable payloads will shift cost effectiveness; and (3) standardization of attitude hardware and, possibly, supporting software. Procedures for handling the increased data volume from the greater number of spacecraft will also be important. The development of routine processing procedures which, unlike most present systems, do not require operator intervention, will probably be required.

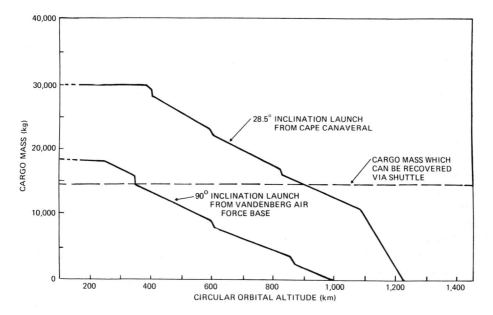

Fig. 1-4. Cargo Mass as a Function of Orbit Altitude and Inclination for the Space Shuttle. See Chapter 3 for definition of orbit parameters. (Data courtesy NASA Public Affairs Office.)

The increased use of processing on board the spacecraft will not substantially affect the analysis described in this volume, since the analytic techniques do not generally depend on the physical location of the processor. However, elimination of complex communication links will eliminate a major source of data irregularities. In addition, the more limited size of onboard computers will place greater emphasis on reducing storage and computational requirements for attitude analysis. This suggests a possible dichotomy of functions in which sophisticated ground-based software will carry out inflight calibration and bias determination and send the resulting critical parameters back to the spacecraft for onboard processing of normal operating data.

The Global Positioning System, also known as *NAVSTAR*, will use a network of 24 satellites for terrestrial, missile, and satellite navigation. Each GPS satellite will broadcast its ephemeris and the time, which will allow a GPS receiver to determine its own position from simultaneous observations of any four GPS satellites. The objective of the system is to provide all users with a positional accuracy of 20 m (either on the ground or in space), a velocity accuracy of 0.06 m/s, and a time accurate to 10 ns. Users with appropriate decoding equipment can obtain twice the accuracy in position and velocity. Finally, the Tracking and Data Relay Satellite System, by which operational satellites will communicate with the ground via geostationary satellites, has the potential for providing increased data coverage over that presently available. (See Section 8.1 for a more detailed discussion of TDRSS.) However, the use of TDRSS will require attitude prediction several days in advance to schedule intersatellite communications. The implications for attitude research and development of these potential changes in future mission profiles are discussed in more detail in Chapter 22.

1.2 Representative Examples of Attitude Determination and Control

The goal of attitude determination is to determine the orientation of the spacecraft relative to either an inertial reference frame or some specific object of interest, such as the Earth. To do this, we must have available one or more *reference vectors*, i.e., unit vectors in known directions relative to the spacecraft. (Recall that for attitude determination we are interested only in the orientation of the spacecraft and not its position; therefore, the magnitude of the reference vector is of no interest except as a possible check on our calculation.) Commonly used reference vectors are the Earth's magnetic field, and unit vectors in the direction of the Sun, a known star, or the center of the Earth.

Given a reference vector, an *attitude sensor* measures the orientation of that vector (or some function of the vector) in the frame of reference of the spacecraft. Having done this for two or more vectors, we may compute the orientation of the spacecraft relative to these vectors, with some possible ambiguity. To clarify this process, we will give two specific examples of attitude determination processes and one example of the use of reference vectors in attitude control.

Although we will present only a single attitude determination method for each situation, most real spacecraft, such as the CTS example, have redundant sensors that can be used in various combinations in the case of sensor or electronic failure. In addition, although the examples are representative of several series of spacecraft, the actual hardware used, and the procedures by which it is used, are normally designed to meet the specific requirements of each individual mission and differ from mission to mission. However, individual missions within a series—such as the Synchronous Meteorological Satellite-1 and -2 or the Atmosphere Explorer-3, -4, and -5 spacecraft—will frequently have identical hardware.

1.2.1 Spin Stabilized Spacecraft

Our first example of the attitude determination process is the case of the CTS spacecraft in its transfer orbit as described above. Because the purpose of attitude determination in the transfer orbit is to support an orbit maneuver and the nozzle of the apogee boost motor is aligned with the spacecraft spin axis, we are interested in the orientation of the spin axis in inertial space.

As shown in Fig. 1-5, two types of attitude sensors are available. (Two sensors of each type were used, primarily for redundancy, but also to provide attitude information at different times.) The *digital Sun sensor* uses a narrow slit and a pattern of photosensitive rectangles to measure the *Sun angle, β,* or the angle between the spin axis and the Sun. The Sun angle is a known function of the rectangles within the instrument on which sunlight falls. The second attitude sensor is an Earth horizon telescope. The telescope has a narrow field of view; as the spacecraft spins, this field of view sweeps out a cone in the sky, as shown in Fig. 1-6. When the sensor scans from space onto the illuminated disk of the Earth it senses a rapid change in the light intensity and produces a pulse. A second pulse is produced when the sensor leaves the Earth. These pulses, produced as the sensor crosses the Earth horizons, are called *in-triggering* and *out-triggering*, or *Earth-in* and *Earth-out*, respectively.

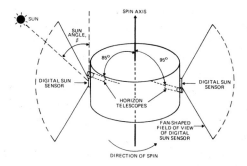

Fig. 1-5. Schematic of CTS Spacecraft Showing Spinning Mode Attitude Determination Hardware

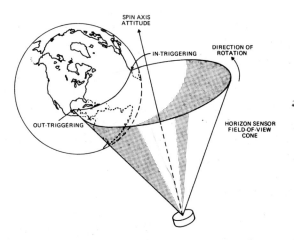

Fig. 1-6. Detection of the Earth by a Horizon Scanner

The time between the in- and out-triggerings, together with the spin period of the satellite and the known size of the Earth, indicate how far above or below the center of the Earth the sensor is scanning. This permits computation of the *nadir angle,* or the angle between the spin axis attitude and the vector from the spacecraft to the center of the Earth, called the *nadir vector.*

We know the vector from the spacecraft to the Sun and have measured the angle between the Sun and the spin axis. Therefore, in inertial space, the spin axis must lie somewhere on a cone centered on the Sun with a radius equal to the measured Sun angle. This cone about the Sun in inertial space is called the *Sun cone.* By a similar argument, the nadir angle measurement implies that the spin axis must lie somewhere on the *nadir cone,* or the cone in inertial space centered on the Earth's center with a radius equal to the nadir angle. The Sun and nadir cones are illustrated in Fig. 1-7. Because the spin axis must lie on both cones, it must lie at one of the two intersections. The choice of which intersection may be based on a third measurement or on a previous estimate as to the probable orientation of the spacecraft. The latter method is commonly used because, as a practical matter, the orientation of the spacecraft is almost never totally unknown.

Figure 1-7 also indicates the problems that are characteristic of attitude determination and which will be discussed throughout this book. For example, the attitude will be poorly determined if the Sun vector and nadir vector are both in the same direction or in opposite directions. It is also possible that because of unavoidable measurement errors, the two cones will not intersect. Even if they intersect but are very nearly tangent to each other, a small error in either measurement means a large shift in the position of the attitude and, therefore, the uncertainty in the calculated position of the attitude is large.

It is convenient to think of the attitude and attitude measurements in terms of cones and cone intersections. However, some fundamental types of measurements are *not* equivalent to cone angles. Although we will always be interested in the intersection of cone-like figures, (i.e., figures that come to a point at the spacecraft), these figures will not always have the simple, circular cross section of the Sun cone or the nadir cone. (See Section 10.3.)

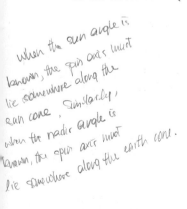
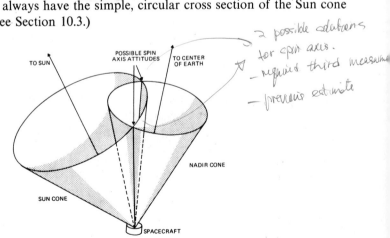

Fig. 1-7.   Attitude Solutions at Intersection of Sun and Nadir Cones

1.2.2 Three-Axis Stabilized Spacecraft

Both the Sun sensor and the horizon sensor described above depend on the rotation of the spacecraft to scan the sky to find the Sun and the Earth. On three-axis stabilized spacecraft, this scanning motion is not available without the addition of moving parts, which are subject to wear and mechanical failure. Thus, we would like to have detectors that could sense the orientation of reference vectors over large portions of the sky without the sensor itself moving.

One such sensor is the *two-axis Sun sensor*, or *solid angle Sun sensor*, which is equivalent to two of the spinning Sun sensors described above mounted perpendicular to each other, as shown in Fig. 1-8. The two Sun angle measurements from the two axes fix the orientation of the Sun in the spacecraft frame of reference. However, this does not fix the orientation of the spacecraft in inertial space, since the spacecraft is free to rotate about the vector to the Sun. That is, no information exists in the Sun measurements about how the spacecraft is oriented "around" the Sun vector, Thus, one other measurement is required to unambiguously specify the spacecraft orientation.

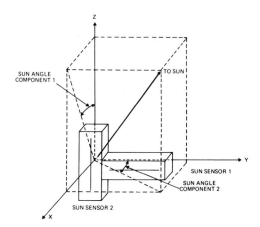

Fig. 1-8.   Two-Axis Sun Sensor as Combination of Two One-Axis Sensors

A second reference vector commonly used on three-axis stabilized spacecraft is the Earth's magnetic field. Three mutually perpendicular magnetometers measure the three components of the Earth's magnetic field vector.* These three measurements may be combined to give the two components of the direction of the magnetic field in the spacecraft reference frame and the magnitude of the field, which can then be used as a check on the measurement accuracy. As with the Sun measurement, the magnetic field measurement does not determine the orientation of the spacecraft "around" the magnetic field vector.

---

*In its attitude acquisition phase, the CTS spacecraft used two-axis Sun sensors but not magnetometers because at synchronous altitudes the Earth's magnetic field was too weak and too poorly known to provide a good attitude reference.

Neither the Sun measurement nor the magnetic field measurement is alone sufficient to determine the inertial attitude of the spacecraft. However, as long as the Sun vector and the magnetic field vector are not parallel, the two pairs of measurements may be combined to determine the spacecraft orientation. Actually, the problem is overdetermined, since there are four independent measurements (two components each of the Sun vector and the magnetic field vector) and only three measurements are required to fix the orientation of the spacecraft in inertial space. For example, we could use two components to fix the orientation of the Sun in spacecraft coordinates and then use a third measurement to specify the rotation of the spacecraft about the Sun vector, thus fixing the spacecraft's orientation completely. We may see explicitly the redundant information by noting that measuring the position of both the Sun vector and the magnetic field vector allows us to calculate the angle between these two vectors in the spacecraft frame of reference. However, this angle is already known because the position of both vectors in inertial space is known and the angle between them does not depend on the frame of reference that is used.

1.2.3 Attitude Maneuver Using Gas Jets

Having seen representative examples of how the attitude is determined, we will present one example of how attitude control maneuvers are performed to reorient the spacecraft from one attitude to another. Specifically, we will use the Sun vector as an attitude reference and gas jets to provide the torque to reorient the spin axis of a spin-stabilized spacecraft.

The reorientation control hardware is shown in Fig. 1-9. There are four Sun sensors, each with a fan-shaped field of view, which produce a pulse as the Sun crosses their field of view. There are two gas jets that point in opposite directions, which are located on the opposite sides of the spacecraft and fire simultaneously. Because the two jets provide an equal force in opposite directions, there is no net effect on the motion of the center of mass of the spacecraft, i.e., its orbit remains unchanged. However, the jets will provide a torque in the direction shown which will change the direction, or *precess*, the angular momentum vector. Note that this definition of precession as the change in direction of the angular momentum vector differs from that normally used in physics.

If the two jets fire continuously, the net torque integrated over a spin period is zero and the spacecraft axis will simply wobble about a position near its initial position. Therefore, the Sun sensors are used to pulse the jets, or to turn them on and off so that they only fire during one-quarter of each spin period. The choice of Sun sensors which turn the jets on and off determines the direction of the precession relative to the Sun. If sensors A and B turn the jet on and off, respectively (abbreviated A/B), then, as shown in Fig. 1-10, jet 1 will be 45 deg past the Sun when it starts and 135 deg past the Sun when it stops. The average position of the jet during firing will be 90 deg past the Sun and the average motion of the spacecraft angular momentum vector will be directly toward the Sun. The four combinations A/B, B/C, C/D, and D/A will precess the spacecraft spin axis in four different directions relative to the Sun cone, as shown in Fig. 1-11.

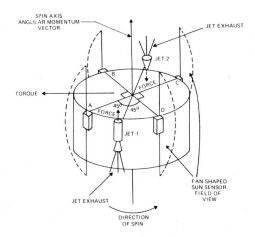

Fig. 1-9.   Representative Control Hardware Orientation. (Note that the exhaust from jet 1 is downward and from jet 2 is upward.)

Combinations A/B and C/D precess the spin axis perpendicular to the Sun cone either directly toward or away from the Sun. Combinations B/C and D/A precess the spin axis along the Sun cone either to the left or to the right relative to the Sun.

We can get from any initial position to any final position by combining one maneuver toward or away from the Sun and one maneuver to the right or the left. However, not all spin axis paths across the sky are equally acceptable. Most missions have a variety of *mission attitude constraints*, which require that the orientation of the spacecraft always meet specific conditions. For example, it may be required that solar panels remain pointed within 30 deg of the Sun, that an antenna remain pointed within 40 deg of the Earth, that accurate attitude determination be possible throughout the maneuver, or that sensitive equipment

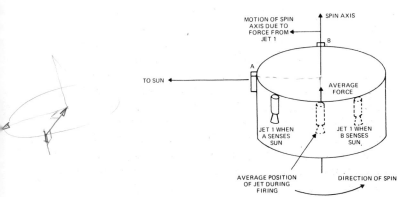

Fig. 1-10.   Attitude Maneuver Geometry for Sensors A and B and Jet 1 of Fig. 1-9

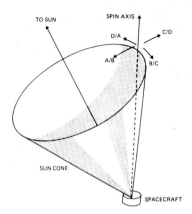

Fig. 1-11. Direction of Motion of the Spin Axis Relative to the Sun Cone for Triggering by the Four Possible Sun Sensor Combinations of Fig. 1-9

never be pointed such that it might look directly at the Sun. A *control strategy* is a procedure by which desired maneuvers can be carried out without violating mission constraints. A significant part of attitude control analysis is devoted to trying to find the best control strategy for a particular attitude maneuver.

1.3 Methods of Attitude Determination and Control

Having seen how the attitude determination and control process works, we summarize in this section the commonly used reference vectors and control torques and the disturbance torques which cause the spacecraft to drift. Table 1-1 lists the various reference sources commonly used for attitude determination. Basically, two alternatives exist: we may either measure the attitude with respect to some external reference vector (the first four items in the table) or we may measure the centrifugal acceleration (the last item listed) to determine the change in the orientation. The latter is referred to as *inertial guidance* and is done by gyroscopes or accelerometers. For space flight, the main problem with inertial guidance is that it depends on integrating small changes in the attitude to propagate the orientation in inertial space from some known initial value. Therefore, small errors accumulate, and periodic updates based on some external reference source are required. One such combination is the use of gyroscopes and star sensors. The efficient use of star sensors typically requires a fairly accurate initial attitude estimate which is then refined by the star sensor data. When a spacecraft undergoes a maneuver from one orientation to another, the gyroscopes provide an accurate measure of the change in orientation and a good initial estimate of the new attitude. This estimate is then refined using the star sensor data. The refined attitude is used both as a measure of the true spacecraft orientation and to update the gyroscopes to eliminate accumulated errors since the previous update.

In general, the strain and torque on spacecraft are several orders of magnitude less in the space environment than they would be at the Earth's surface. Neverthe-less, torques that perturb the attitude do exist. The major environmental torques

Table 1-1.   Attitude Determination Reference Sources

REFERENCE	ADVANTAGES	DISADVANTAGES
SUN	BRIGHT; UNAMBIGUOUS; LOW POWER AND WEIGHT; USUALLY MUST BE KNOWN FOR SOLAR CELLS AND EQUIPMENT PROTECTION	MAY NOT BE VISIBLE DURING PARTS OF ORBIT AROUND LARGE CENTRAL BODY; 1/2 DEG ANGULAR DIAMETER (VIEWED FROM EARTH) LIMITS ACCURACY TO ~1 ARC MINUTE
EARTH, OR OTHER CENTRAL BODY	ALWAYS AVAILABLE FOR NEARBY SPACECRAFT; BRIGHT; LARGELY UNAMBIGUOUS (MAY BE MOON INTERFERENCE); NECESSARY FOR MANY TYPES OF SENSOR AND ANTENNA COVERAGE; ANALYSIS RELATIVELY EASY	TYPICALLY REQUIRES SCAN MOTION TO SENSE HORIZON; SENSORS MUST BE PROTECTED FROM SUN; RESOLUTION LIMITED TO ~ 0.1 DEG BECAUSE OF HORIZON DEFINITION; ORBIT AND ATTITUDE STRONGLY COUPLED
MAGNETIC FIELD	ECONOMICAL; LOW POWER REQUIREMENTS; ALWAYS AVAILABLE FOR LOW-ALTITUDE SPACECRAFT	POOR RESOLUTION (> 0.5 DEG); GOOD ONLY NEAR EARTH—LIMITED BY FIELD STRENGTH AND MODELING ACCURACY; ORBIT AND ATTITUDE STRONGLY COUPLED; SPACECRAFT MUST BE MAGNETICALLY CLEAN (OR INFLIGHT CALIBRATION REQUIRED); SENSITIVE TO BIASES
STARS (INCLUDING DISTANT PLANETS)	HIGH ACCURACY (~ 10^{-3} DEG); AVAILABLE ANYWHERE IN SKY; ESSENTIALLY ORBIT INDEPENDENT, EXCEPT FOR VELOCITY ABERRATION	SENSORS HEAVY, COMPLEX, AND EXPENSIVE; IDENTIFICATION OF STARS FOR MULTIPLE TARGET SENSORS IS COMPLEX AND TIME CONSUMING; SENSORS NEED PROTECTION FROM SUN; DOUBLE AND MULTIPLE STARS CAUSE PROBLEMS; USUALLY REQUIRE SECOND ATTITUDE SYSTEM FOR INITIAL ATTITUDE ESTIMATES
INERTIAL SPACE (GYROSCOPES; ACCELEROMETERS)	REQUIRES NO EXTERNAL SENSORS, ORBIT INDEPENDENT; HIGH ACCURACY FOR LIMITED TIME INTERVALS; EASILY DONE ONBOARD	SENSES CHANGE IN ORIENTATION ONLY—NO ABSOLUTE MEASUREMENT; SUBJECT TO DRIFT; SENSORS HAVE RAPIDLY MOVING PARTS SUBJECT TO WEAR AND FRICTION; RELATIVELY HIGH POWER AND LARGE MASS

that affect the attitude, as listed in Table 1-2, are aerodynamic torque caused by the rapid spacecraft motion through the tenuous upper atmosphere; gravity-gradient torque due to the small difference in gravitational attraction from one end of the spacecraft to the other (the same differential force which produces tides); magnetic torque due to the interaction between the spacecraft magnetic field (including induced magnetism from the surrounding field) and the Earth's magnetic field; and solar radiation torque due to both the electromagnetic radiation and particles radiating outward from the Sun. Early investigators felt that micrometeorites might also supply a significant source of torque. However, this has been found negligible relative to the other torques, except perhaps in some unexplored regions of the solar system, such as inside the rings of Saturn.

The relative strengths of the various torques will depend on both the spacecraft environment and the form and structure of the spacecraft itself. Nonetheless, because of the form of the distance dependence of the environmental torques, the space environment for any specific spacecraft may be divided into

Table 1-2.   Environmental Disturbance Torques

SOURCE	DEPENDENCE ON DISTANCE FROM EARTH	REGION OF SPACE WHERE DOMINANT*
AERODYNAMIC	e^{-ar}	ALTITUDES BELOW ~ 500 km
MAGNETIC	$1/r^3$	~ 500 km TO ~ 35,000 km;
GRAVITY GRADIENT	$1/r^3$	(I.E., OUT TO ABOUT SYNCHRONOUS ALTITUDE)
SOLAR RADIATION	INDEPENDENT	INTERPLANETARY SPACE ABOVE SYNCHRONOUS ALTITUDE
MICROMETEORITES	LARGELY INDEPENDENT; HIGH CONCENTRATION IN SOME REGIONS OF THE SOLAR SYSTEM	NORMALLY NEGLIGIBLE; MAY BE IMPORTANT IN SOME SMALL REGIONS (INTERIOR OF SATURN'S RINGS)

*ALTITUDES LISTED ARE ONLY REPRESENTATIVE; THE SPECIFIC ALTITUDES AT WHICH VARIOUS TORQUES DOMINATE ARE HIGHLY SPACECRAFT DEPENDENT.

three regions where different forces dominate. Close to the Earth, aerodynamic torque will always be the largest. Because this falls off exponentially with distance from the Earth, magnetic and gravity-gradient torques will eventually become more important. Because both of these have the same functional dependence on distance, the relative strength between the two will depend on the structure of the individual spacecraft; either may be dominant. Finally, solar radiation torque, due to both radiation pressure and differential heating, will dominate throughout the interplanetary medium.

Internal torques may also affect the attitude of the spacecraft. This may include seemingly small items such as fuel redistribution or tape recorders turning on and off. The internal torques can become very important, and even dominate the attitude motion, when the spacecraft structure itself is flexible. The role of *flexible spacecraft dynamics* is much more important in spacecraft than would be the case on Earth where flexible structures tend to be torn apart by the strong environmental torques. For example, the small environmental torques permit spacecraft to have wire booms over 100 m long, which causes the flexibility of the booms to dominate the attitude dynamics.

Because torques exist throughout the spacecraft environment, some procedure is necessary for attitude stabilization and control. Spacecraft may be stabilized by either (1) the spacecraft's angular momentum (spin stabilized); (2) its response to environmental torques, such as gravity-gradient stabilization; or (3) active control, using hardware such as gas jets, reaction wheels, or electromagnets. Table 1-3 lists the methods of *passive stabilization* which require no power consumption or external control. Table 1-4 lists the commonly used methods of *active control* which may be used for either maneuver control or active stabilization. In general, active methods of control are more accurate, faster, more flexible, and can be adjusted to meet the needs of the mission. However, active control typically requires a power source and complex logic and may require ground control and the use of spacecraft *consumables* (materials, such as jet fuel, which are brought from the ground and which cannot be replaced once they have been used). For those systems which use consumables, a major constraint on attitude control strategies is to use them as efficiently as possible.

1.4 Time Measurements

Fundamental to both attitude and orbit calculations is the measurement of time and time intervals. Unfortunately, a variety of time systems are in use and sorting them out can cause considerable confusion. A technical discussion of time systems needed for precise computational work is given in Appendix J. In this section, we summarize aspects that are essential to the interpretation of attitude data.

Two basic types of time measurements are used in attitude work: (1) *time intervals* between two events, such as the spacecraft spin period or the length of time a sensor sees the Earth; and (2) *absolute times*, or *calendar times*, of specific events, such as the time associated with some particular spacecraft sensing. Of course, calendar time is simply a time interval for which the beginning event is an agreed standard.

Table 1-3. Passive Stabilization Methods (i.e., those requiring no consumption of spacecraft power or generation of commands)

METHOD	ADVANTAGES	DISADVANTAGES
SPIN STABILIZED	SIMPLE; EFFECTIVE NEARLY ANYWHERE IN ANY ORIENTATION; MAINTAINS ORIENTATION IN INERTIAL SPACE	CENTRIFUGAL FORCE REQUIRES STRUCTURAL STABILITY AND SOME RIGIDITY; SENSORS AND ANTENNAS CANNOT GENERALLY REMAIN POINTED AT A SPECIFIC INERTIAL TARGET; WOBBLE (NUTATION) IF NOT PROPERLY BALANCED; DRIFT DUE TO ENVIRONMENTAL TORQUES
GRAVITY-GRADIENT STABILIZED	MAINTAINS STABLE ORIENTATION RELATIVE TO CENTRAL BODY; NOT SUBJECT TO DECAY OR DRIFT DUE TO ENVIRONMENTAL TORQUES UNLESS ENVIRONMENT CHANGES	LIMITED TO 1 OR 2 POSSIBLE ORIENTATIONS; EFFECTIVE ONLY NEAR MASSIVE CENTRAL BODY (E.G., EARTH, MOON, ETC.); REQUIRES LONG BOOMS OR ELONGATED MASS DISTRIBUTION; SUBJECT TO WOBBLE (LIBRATION); CONTROL LIMITED TO ~1 DEG—PROBLEM OF THERMAL GRADIENTS ACROSS BOOM
SOLAR RADIATION STABILIZED	CONVENIENT FOR POWER GENERATION BY SOLAR CELLS OR SOLAR STUDIES	LIMITED TO HIGH-ALTITUDE OR INTERPLANETARY ORBITS; LIMITED ORIENTATIONS ALLOWED
AERODYNAMIC STABILIZED MAGNETIC STABILIZED WITH PERMANENT MAGNET	{ SPECIAL-PURPOSE METHODS – HIGHLY MISSION AND STRUCTURE DEPENDENT IN ALL OF THEIR CHARACTERISTICS }	

Table 1-4. Active Methods of Stabilization and Control

METHOD	ADVANTAGES	DISADVANTAGES
GAS JETS	FLEXIBLE AND FAST; USED IN ANY ENVIRONMENT; POWERFUL	USES CONSUMABLE (FUEL) WITH LIMITED SUPPLY AVAILABLE DUE TO FUEL WEIGHT; TOO POWERFUL FOR SOME APPLICATIONS (i.e., RELATIVELY COARSE CONTROL); COMPLEX AND EXPENSIVE PLUMBING SUBJECT TO FAILURE
MAGNETIC (ELECTROMAGNETS)	USUALLY LOW POWER; MAY BE DONE WITHOUT USING CONSUMABLES BY USE OF SOLAR POWER	SLOW; NEAR EARTH ONLY; APPLICABILITY LIMITED BY DIRECTION OF THE EXTERNAL MAGNETIC FIELD; COARSE CONTROL ONLY (BECAUSE OF MAGNETIC FIELD MODEL UNCERTAINTIES AND LONG TIME CONSTANTS)
REACTION WHEELS*	PARTICULARLY GOOD FOR VARIABLE SPIN RATE CONTROL; FAST; FLEXIBLE; PRECISE ATTITUDE CONTROL AND/OR STABILIZATION	REQUIRES RAPIDLY MOVING PARTS WHICH IMPLIES PROBLEMS OF SUPPORT AND FRICTION; MAY NEED SECOND CONTROL SYSTEM TO CONTROL OVERALL ANGULAR MOMENTUM ("MOMENTUM DUMPING") IN RESPONSE TO CUMULATIVE CHANGES BY ENVIRONMENTAL TORQUES; EXPENSIVE
ALTERNATIVE THRUSTERS: ION OR ELECTRIC ACTIVE SOLAR, AERODYNAMIC, OR GRAVITY GRADIENT	{ PRIMARILY SPECIAL PURPOSE—LESS EXPERIENCE WITH THESE THAN WITH THOSE LISTED ABOVE; CHARACTERISTICS ARE HIGHLY MISSION DEPENDENT; SOME SYSTEMS MAY SEE MORE USE IN THE FUTURE AS FURTHER EXPERIENCE IS GAINED }	

*REFERS TO ANY DEVICE THAT MAY BE USED IN A PROCESS TO EXCHANGE ANGULAR MOMENTUM WITH THE SPACECRAFT BODY.

Calendar time in the usual form of date and time is used only for input and output, since arithmetic is cumbersome in months, days, hours, minutes, and seconds. Nonetheless, this is used for most human interaction with attitude systems because it is the system with which we are most familiar. Problems arise even with date and time systems, since time zones are different throughout the world and spacecraft operations involve a worldwide network. The uniformly adopted solution to this problem is to use the local time corresponding to 0 deg longitude as the standard absolute time for events anywhere in the world or in space. This is referred to as *Universal Time (UT), Greenwich Mean Time (GMT),* or *Zulu (Z),* all of which are equivalent for the practical purposes of data interpretation. The name

Greenwich Mean Time is used because 0 deg longitude is defined as going through the site of the former Royal Greenwich Observatory in metropolitan London. Eastern Standard Time in the United States is obtained by subtracting 5 hours from the Universal Time.

Because calendar time is inconvenient for computations, we would like an absolute time that is a continuous count of time units from some arbitrary reference. The time interval between any two events may then be found by simply subtracting the absolute time of the first event from that of the second event. The universally adopted solution for astronomical problems is the *Julian day*, a continuous count of the number of days since noon (12:00 UT) on January 1, 4713 BC. This strange starting point was suggested by an Italian scholar of Greek and Hebrew, Joseph Scaliger, in 1582, as the beginning of the current *Julian period* of 7980 years. This period is the product of three numbers: the solar cycle, or the interval at which all dates recur on the same days of the week (28 years); the lunar cycle containing an integral number of lunar months (19 years); and the indiction, or the tax period introduced by the emperor Constantine in 313 AD (15 years). The last time that these started together was 4713 BC and the next time will be 3267 AD. Scaliger was interested in reducing the astronomical dating problems associated with calendar reforms of his time and his proposal had the convenient selling point that it predated the ecclesiastically approved date of creation, October 4, 4004 BC. The Julian day was named after Scaliger's father, Julius Caeser Scaliger, and was not associated with the Julian calendar that had been in use for some centuries.

Tabulations of the current Julian date may be found in nearly any astronomical ephemeris or almanac. A particularly clever procedure for finding the Julian date, or *JD*, associated with any current year (*I*), month (*J*), and the day of the month (*K*), is given by Fliegel and Van Flandern [1968] as a FORTRAN arithmetic statement using FORTRAN integer arithmetic:

$$JD = K - 32075 + 1461 * (I + 4800 + (J - 14)/12)/4$$

$$+ 367 * (J - 2 - (J - 14)/12 * 12)/12$$

$$- 3 * ((I + 4900 + (J - 14)/12)/100)/4$$

For example, December 25, 1981 (*I* = 1981, *J* = 12, *K* = 25) is *JD* 2,444,964.*

The Julian date presents minor problems for space applications. It begins at noon Universal Time, rather than 0 hours Universal Time and the extra digits required for the early starting date limit the precision of times in computer storage. However, no generally accepted substitute exists, and the Julian day remains the only unambiguous continuous time measurement for general use.

For internal computer calculations, the problem of ambiguity does not arise and several systems are used. The *Julian Day for Space*, or *JDS*, is defined as JD − 2,436,099.5. This system starts at 0 hours UT (rather than noon), September 17, 1957, which is the first Julian day divisible by 100 prior to the launch of the first

---

* Note that in FORTRAN integer arithmetic, multiplication and division are performed left to right, the magnitude of the result is truncated to the lower integer value after each operation, and −2/12 is truncated to 0.

manmade satellite by the Soviet Union on October 4, 1957. This system is used internally in NASA orbit programs, with time measured in seconds rather than days. (Measuring all times and time intervals in seconds is convenient for computer use because the large numbers involved do not pose a problem.) The European Space Operations Center uses the *Modified Julian Day*, which starts at 0 hours UT, January 1, 1950. Attitude determination programs at NASA's Goddard Space Flight Center measure time intervals in seconds from 0 hours UT, September 1, 1957. Unfortunately, the origin of this particular system appears to have been lost in antiquity.

References

1. Bassett, D. A., "Ground Controlled Conversion of Communications Technology Satellite (CTS) from Spinning to Three-Axis Stabilized Mode," AIAA paper no. 76-1928 presented at AIAA Guidance and Control Conference, San Diego, CA., Aug. 1976.
2. Clarke, Arthur C., *The Promise of Space*. New York: Harper & Row, 1968.
3. Fliegel, Henry F. and Thomas C. Van Flandern, "A Machine Algorithm for Processing Calendar Dates," *Communications of the ACM*, Vol. 11, p. 657, 1968.
4. *Future Space Programs 1975* (3 volumes), Report prepared for the Subcommittee on Space Science and Applications of the Committee on Science and Technology, U.S. House of Representatives, 94th Congress (Stock numbers 052-070-02889-1, 052-070-02890-4, 052-070-02891-2), July–Sept. 1975.
5. Glasstone, Samuel, *Sourcebook on the Space Sciences*. Princeton, N.J.: D. Van Nostrand Company Inc., 1965.
6. NASA, *Outlook for Space*. NASA SP-386, 1976a.
7. NASA, *A Forecast of Space Technology*. NASA SP-387, 1976b.
8. *Soviet Space Programs, 1966–70*, Staff Report for the Committee on Aeronautical and Space Sciences United States Senate, 92nd Congress, Document No. 92-51, Dec. 9, 1971.
9. von Braun, Wernher, and Frederick I. Ordway, *History of Rocketry and Space Travel*. New York: Crowell, 1975.

CHAPTER 2

ATTITUDE GEOMETRY

James R. Wertz

2.1   The Spacecraft-Centered Celestial Sphere
2.2   Coordinate Systems
      Properties of Coordinate Systems on the Celestial Sphere,
      Spacecraft-Centered Coordinate Systems,
      Nonspacecraft-Centered Coordinate Systems, Parallax
2.3   Elementary Spherical Geometry

This chapter introduces the idea of attitude determination and control as a geometrical problem on the two-dimensional celestial sphere, describes the most common attitude coordinate systems, and summarizes geometry on the celestial sphere.

2.1 The Spacecraft-Centered Celestial Sphere

Recall that the spacecraft attitude is its orientation relative to the Sun, the Earth, or the stars regardless of the distances to these various objects. To think in terms of direction only, it is convenient to form a mental construct of a sphere of unit radius centered on the spacecraft, called the *spacecraft-centered celestial sphere*, which is illustrated in Fig. 2-1. A point on the sphere represents a direction in space. For example, in Fig. 2-1, the points S, E, and A are the directions of the center of the Sun, the center of the disk of the Earth (called the *nadir* or *subsatellite point*), and the spacecraft attitude, respectively, as viewed from the spacecraft. The points E and S on the sphere are both a unit distance from the spacecraft, although the real distance to the Sun and the Earth is vastly different in most cases. Point A corresponds to the direction of a specific spacecraft axis which has no distance associated with it.

As standard notation throughout this book, we represent points on the celestial sphere by uppercase *italic* Roman letters. Points on the sky diametrically opposite a given direction have a "-1" superscript and are called the *antipoint* when speaking of a point on the sphere or the *negative axis* when speaking of the direction of an axis or vector. Thus, S^{-1} is the *antisolar point* and A^{-1} is the *negative attitude axis*. The antinadir, E^{-1}, or direction opposite the center of the Earth, is called the *zenith*. After Fig. 2-2, the spacecraft at the center of the sphere and the lines from the center to the surface of the sphere will be omitted and we will speak of geometry *on the celestial sphere*.

A *great circle* on the celestial sphere is any circle that divides the sphere into two equal hemispheres. Any other circle on the celestial sphere is called a *small circle*. A portion of a great circle is called an *arc*, or an *arc segment*. The arc segments connecting points A, S, and E on Fig. 2-1 form a *spherical triangle* on the

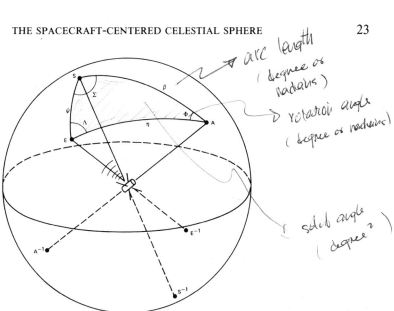

[handwritten annotations: arc length (degree or radians); rotation angle (degree or radians); solid angle (degree²)]

Fig. 2-1.　Spacecraft-Centered Celestial Sphere and Standard Notation for the Sun, Earth, and Attitude and the Angles Between Them

celestial sphere. We will discuss the properties of great circles and spherical triangles in more detail in Section 2.3.

Linear measure, such as metres, has no meaning on the celestial sphere. In general, there are only three types of measurements on the celestial sphere—arc length, rotation angle, and solid angle—and it is important to recognize the distinction between them. The lengths of the sides of a spherical triangle (β, η, and ψ in Fig. 2-1) are *arc length* measurements (or *angular separations*), measured in degrees or radians. The terms *Sun angle* and *nadir angle* (and symbols β and η) are used for the arc lengths from the attitude to the Sun and to the center of the Earth, respectively. The angle at which two arc segments intersect (Φ, Σ, or Λ in the spherical triangle of Fig. 2-1) is called a *rotation angle* and is also measured in degrees or radians. Although arc lengths and rotation angles are measured in the same units, they are different types of quantities and are not interchangeable. Throughout this book, arc lengths will be represented by lowercase Greek letters and rotation angles by uppercase Greek letters. The area of the spherical triangle measured on the curved surface of the unit sphere is an example of the third type of measurement, the *solid angle*. Solid angles are measured in square degrees or steradians and will be represented by uppercase Greek letters. Appendix A gives area formulas for most common shapes on the celestial sphere.

An alternative procedure for thinking of angular measure is in terms of lines and planes from the spacecraft to distant objects. Thus the term *cone angle* is frequently used for arc length, since the angular radius of a cone about a central axis is an arc length measurement. Similarly, the term *dihedral angle* (the angle between two planes) is frequently used for rotation angle, since, for example, the rotation angle Φ is also the angle between the spacecraft-Sun-attitude plane and the spacecraft-nadir-attitude plane.

Finally, it is important to recognize that arc length on the celestial sphere measures the angular separation between two objects as seen from the spacecraft. Thus, the arc length from the Sun to the Earth, ψ, is the angular separation in the sky between these two objects. For convenience in thinking of angular separation, note that 1 deg (1°) is approximately the angular diameter of a dime at a distance of 1 m; 1 minute of arc ($1' = 1/60$th of $1°$) is the angular diameter of a dime at 60 m; and 1 second of arc ($1'' = 1/60$th of $1'$) is the angular diameter of a dime at 3.6 km. The angular diameter of both the Moon and the Sun as seen from the Earth is about 1/2 deg.

2.2 Coordinate Systems

To make measurements on the celestial sphere, it is convenient to use a spherical coordinate system. The general properties of any such system are discussed in Section 2.2.1, and specific examples of the most common spherical and rectangular systems used in practice are discussed in Sections 2.2.2 and 2.2.3. The spherical coordinates are defined in terms of the unit celestial sphere. All of these may be transformed into three-dimensional coordinates by the addition of a third variable, r, the distance from the center of the coordinate system to the point in question.

2.2.1 Properties of Coordinate Systems on the Celestial Sphere

The spherical coordinate systems normally used for spacecraft have a number of properties in common. These are illustrated in Fig. 2-2 for the spacecraft-centered celestial sphere described above. Each spherical coordinate system has two *poles* diametrically opposite each other on the celestial sphere and an *equator*, or great circle, halfway between the poles. The great circles through the poles and perpendicular to the equator are called *meridians* and the small circles a fixed distance above or below the equator are called *parallels*.

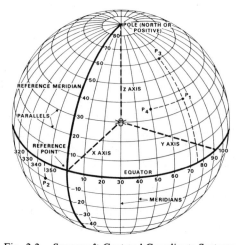

Fig. 2-2.   Spacecraft-Centered Coordinate Systems

The position of any point on the sphere is given in terms of two components equivalent to latitude and longitude on the surface of the Earth. The arc length distance above or below the equator is called the *latitude* or *elevation* component. The angular distance around the equator between the meridian passing through a particular point and an arbitrary *reference meridian*, or *prime meridian*, is known as the *longitude* or *azimuth* component. For example, the reference meridian for longitude on the surface of the Earth is the one passing through the center of the former Royal Greenwich Observatory in London. Thus, we may define the positions of the points P_1 and P_2 in terms of azimuth, ϕ, and elevation, λ, as:

$$P_1 = (\phi_1, \lambda_1) = (75°, 35°)$$

$$P_2 = (\phi_2, \lambda_2) = (345°, -10°)$$

Note that in most spherical coordinate systems, the azimuth coordinate is measured from 0 deg to 360 deg and the elevation component is measured from $+90$ deg to -90 deg. The intersection of the reference meridian and the equator in any system is called the *reference point* and has coordinates $(0°, 0°)$.

Several properties of spherical coordinate systems are shown in Fig. 2-2. For example, a degree of elevation is a degree of arc length in that the angular separation between two points on the same meridian is just the difference between the elevation of the two points. Thus, P_3 at $(75°, 60°)$ is $25°$ from P_1. However, 1-deg separation in azimuth will be less than 1 deg of arc, except along the equator. Point P_4 at $(50°, 35°)$ is less than 25 deg in arc length from P_1. Specific equations for the angular distance along a parallel or between two arbitrary points are given in Appendix A (Eqs. (A-1), (A-4), and (A-5)). In using these equations, a parallel at elevation λ is a small circle of angular radius $90° - |\lambda|$. The distortion in the azimuth component becomes particularly strong near the pole of any coordinate system. At either pole, the azimuth is undefined.

An alternative procedure for specifying the position of a point on the celestial sphere involves three components of a vector of unit length from the center of the sphere to the point on the surface of the sphere. Ordinarily, the x, y, and z axes of such a rectangular coordinate system are defined such that the z axis is toward the $+90$-deg pole of the spherical coordinate system, the x axis is toward the reference point, and the y axis is chosen perpendicular to x and z such that the coordinate system is right handed (i.e., for unit vectors along the x, y, and z axes, $\hat{z} = \hat{x} \times \hat{y}$). A summary of vector notation used in this book is given in the Preface and the coordinate transformations between spherical and rectangular coordinate systems are given in Appendix E. Of course, only two of the three components of the unit vector are independent, since the length (but not the sign) of the third component is determined by requiring that the magnitude of the vector be 1. This constraint on the magnitude is a convenient check that any unit vector has been correctly calculated.

In principle, either spherical or rectangular coordinates can be used for any application. In practice, however, each system has advantages in specific circumstances. In general, computer calculations in long programs should be done in rectangular coordinates because there are fewer trigonometric functions to evaluate and these are relatively time consuming for the computer. Carrying the third component around is conveniently done in computer arrays. However, most input and output and most data intended for people to read are in spherical coordinates, since the geometrical relationships are usually clearer when visualized in terms of a coordinate picture similar to Fig. 2-2. For calculations external to the computer or in short computer runs, spherical coordinates give less likelihood of error because the quantities involved are more easily visualized. Many geometrical theorems, such as those in Chapter 11, are more easily done in spherical geometry than in vector geometry, although either system may be better for any specific problem.

Any spherical coordinate system (or its rectangular equivalent) is fully specified by indicating the positive pole and the choice of either reference meridian or reference point at the intersection of the reference meridian and the equator. On the surface of a sphere, the choice of poles and prime meridian is arbitrary, and any point on the sphere may be used as the pole for a spherical coordinate system. For the Earth, a system defined by the Earth's rotation axis is the most convenient. However, for the sky as viewed by the spacecraft, a variety of alternative coordinate systems are convenient for various uses, as described below.

2.2.2 Spacecraft-Centered Coordinate Systems

The three basic types of coordinate systems centered on the spacecraft are those fixed relative to the body of the spacecraft, those fixed in inertial space, and those defined relative to the orbit and not fixed relative to either the spacecraft or inertial space.

Spacecraft-Fixed Coordinates. Coordinate systems fixed in the spacecraft are used to define the orientation of attitude determination and control hardware and are the systems in which attitude measurements are made. Throughout this book, spacecraft-fixed spherical coordinates will use ϕ for the azimuth component and λ for the elevation. Alternatively, θ will be used for the coelevation; that is, $\theta \equiv 90° - \lambda$. For spinning spacecraft, the positive pole of the coordinate system will be the positive spin vector, unless otherwise specified. The reference meridian is taken as passing through an arbitrary reference point on the spin plane which is the equator of the coordinate system. The three components of a rectangular spacecraft fixed coordinate system will be represented by x, y, and z, with the relation between spherical and rectangular coordinates as defined in Section 2.2.1. For three-axis stabilized (nonspinning) spacecraft, no standard orientation is defined. For attitude-sensing hardware, it is the orientation of the field of view of the hardware in the spacecraft system that is important, *not* the location of the hardware within the spacecraft.

Inertial Coordinates. The most common inertial coordinate system is the system of *celestial coordinates* defined relative to the rotation axis of the Earth, as shown in Fig. 2-3. Recall that the spacecraft is at the center of the sphere in Fig. 2-3. The axis of the spacecraft-centered celestial coordinate system joining the north and south celestial poles is defined as parallel to the rotation axis of the

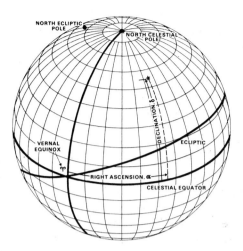

Fig. 2-3. Celestial Coordinates

Earth. Thus, the north celestial pole is approximately 1 deg from the bright star Polaris, the Pole Star. To fully define the coordinate system, we must also define the reference meridian or reference point. The point on the celestial equator chosen as the reference is the point where the *ecliptic*, or plane of the Earth's orbit about the Sun (see Chapter 3), crosses the equator going from south to north, known as the *vernal equinox*. This is the direction parallel to the line from the center of the Earth to the Sun on the first day of spring.

Unfortunately, the celestial coordinate system is not truly inertial in that it is not fixed relative to the mean positions of the stars in the vicinity of the Sun. The gravitational force of the Moon and the Sun on the Earth's equatorial bulge causes a torque which results in the slow rotation of the Earth's spin axis about the ecliptic pole, taking 26,000 years for one complete period for the motion of the axis. This phenomenon is known as the *precession of the equinoxes*, since it results in the vernal equinox sliding along the ecliptic relative to the fixed stars at the rate of approximately 50 sec of arc per year. When the zodiacal constellations were given their present names several thousand years ago, the vernal equinox was in the constellation of Aries, the Ram. Thus, the zodiacal symbol for the Ram, ♈, is used astronomically for the vernal equinox, which is also called the *First Point of Aries*. Since that time, the vernal equinox has moved through the constellation of Pisces and is now entering Aquarius, bringing the dawn of the zodiacal "Age of Aquarius." The other intersection of the ecliptic and the equator is called the *autumnal equinox* and is represented by the zodiacal symbol for Libra. (See Fig. 3-10.)

The importance of the precession of the equinoxes is that it causes a slow change in the celestial coordinates of "fixed" objects, such as stars, which must be taken into account for accurate determination of orientation. Thus, celestial coordinates require that a date be attached in order to accurately define the position of the vernal equinox. The most commonly used systems are *1950 coordinates*, *2000 coordinates*, and *true of date*, or *TOD*. The latter coordinates are defined at the epoch time of the orbit and are commonly used in spacecraft work, where the small corrections required to maintain TOD coordinates are conveniently done with standard computer subroutines.* (See subroutine EQUIN in Section 20.3.)

The elevation or latitude component of celestial coordinates is universally known as the *declination*, δ. Similarly, the azimuth component is known as *right ascension*, α. Although right ascension is measured in degrees in all spacecraft work, in most astronomical tables it is measured in *hours, minutes, and seconds* where 1 hour = 15 deg, 1 min = 1/60th hour = 1/4 deg, and 1 sec = 1/60th min = 0.0041666...deg. Each of these measurements corresponds to the amount of rotation of the Earth in that period of time. *Note that minutes and seconds of right ascension are **not** equivalent to minutes and seconds of arc, even along the equator.*

Although celestial coordinates are the most widely used, several other inertial coordinate systems are used for special purposes. These systems are summarized in Table 2-1.

Orbit-Defined Coordinates. The *l,b,n system* of coordinates is a system for which the plane of the spacecraft orbit is the equatorial plane of the coordinate system. The *l* axis is parallel to the line from the center of the Earth to the ascending node[†] of the spacecraft orbit, the *n* axis is parallel to the orbit normal, i.e., perpendicular to the orbit plane, and the *b* axis is such that for unit vectors along the axes, $\hat{b} = \hat{n} \times \hat{l}$. The *l,b,n* system would be inertial if the spacecraft orbit were fixed in inertial space. In fact, perturbations on the orbit due to nonsphericity of the central body, gravitational attractions of other bodies, etc., cause the orbit to rotate slowly as described in Section 3.4, so the *l,b,n* system is not absolutely inertial.

Lastly, we define a system of coordinates that maintain their orientation relative to the Earth as the spacecraft moves in its orbit. These coordinates are

Table 2-1. Common Inertial Coordinate Systems

SYSTEM	EQUATORIAL PLANE	REFERENCE POINT	AZIMUTH COORDINATE	ELEVATION COORDINATE
CELESTIAL (OR EQUATOR)	CELESTIAL EQUATOR (PARALLEL TO EARTH'S EQUATOR)	VERNAL EQUINOX (T)	RIGHT ASCENSION (α)	DECLINATION (δ)
ECLIPTIC	ECLIPTIC (PLANE OF THE EARTH'S ORBIT)	VERNAL EQUINOX (T)	CELESTIAL LONGITUDE	CELESTIAL LATITUDE
GALACTIC	ADOPTED PLANE OF THE GALAXY*	ADOPTED DIRECTION OF THE GALACTIC CENTER*	GALACTIC LONGITUDE	GALACTIC LATITUDE

*THE PRESENT SYSTEM OF GALACTIC COORDINATES, ADOPTED IN 1958, DEFINES THE NORTH GALACTIC POLE AS BEING AT $\alpha = 192.25^\circ$, $\delta = +27.4^\circ$ AND THE GALACTIC CENTER AS AT $\alpha = 266.6^\circ$, $\delta = -28.917^\circ$.

---

*In addition to the mean motion due to precession, the Earth's true spin axis wobbles with an amplitude of 9.2 arc seconds (0.0026 deg) and a period of 19 years due to the changing inertial orientation of the Moon's orbit. TOD coordinates are updated to the epoch time using the true precessional motion. If the coordinates are updated using the mean precessional motion, they are referred to as *Mean of Date*, or *MOD*.

†For a definition of ascending node and other orbit parameters, see Chapter 3.

known as *roll, pitch, and yaw* or *RPY*, and are illustrated in Fig. 2-4. In this system, the *yaw axis* is directed toward the nadir (i.e., toward the center of the Earth), the *pitch axis* is directed toward the negative orbit normal, and the *roll axis* is perpendicular to the other two such that unit vectors along the three axes have the relation $\hat{\mathbf{R}} = \hat{\mathbf{P}} \times \hat{\mathbf{Y}}$. Thus, in a circular orbit, the roll axis will be along the velocity vector. The *roll, pitch,* and *yaw angles* (ξ_r, ξ_p, and ξ_y) are defined as right-handed rotations about their respective axes. Therefore, for a spacecraft in a circular orbit and an observer on the spacecraft facing in the direction of motion with the Earth below, a positive pitch rotation brings the nose of the spacecraft upward, positive yaw moves it to the right, and positive roll rotates the spacecraft clockwise. The *RPY* system is most commonly used for Earth-oriented spacecraft. **Caution:** *The preceding definition will be used throughout this book. However, individual spacecraft ordinarily define RPY systems unique to that spacecraft and may even define them as spacecraft-fixed coordinates rather than orbit-defined coordinates. Therefore, when reading other material, it is important to know precisely how roll, pitch, and yaw are being defined.*

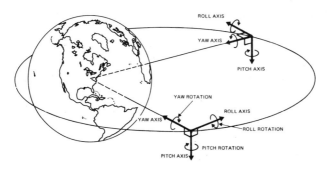

Fig. 2-4.   Roll, Pitch, and Yaw (RPY) Coordinates

2.2.3 Nonspacecraft-Centered Coordinate Systems

For attitude work, the most important coordinate systems are all centered on the spacecraft. However, occasionally the use of nonspacecraft-centered coordinates is convenient, primarily as a means of obtaining reference vectors such as the magnetic field vector or position vectors to objects seen by the spacecraft. Thus, orbit work is ordinarily done in *geocentric inertial coordinates*, equivalent to the celestial coordinates defined above, except that the center of the coordinate system is at the center of the Earth. The position vector of the Earth in spacecraft-centered celestial coordinates is just the negative of the position vector of the spacecraft in geocentric inertial coordinates. Similarly, the positions of the planets within the solar system are ordinarily calculated in *heliocentric coordinates*, or coordinates centered on the Sun. Heliocentric longitude and latitude are defined relative to the ecliptic plane and the vernal equinox as references. *Selenocentric coordinates* are used for spacecraft in lunar orbit and are the same as celestial coordinates except that they are centered on the Moon; that is, the vernal equinox and the celestial equator are used as references.

In some cases, such as analysis of the Earth's magnetic or gravitational field, we may wish to associate a vector with each point in a spherical coordinate system. To do this it is convenient to define at each point in space an orthogonal coordinate system whose three axes are each parallel to the change in one of the three spherical coordinates, as illustrated in Fig. 2-5. Such systems are called *local horizontal coordinates* or *local tangent coordinates*, since the reference plane at any point is always tangent to the sphere centered on the origin of the system and passing through the point in question. If the components of the global spherical coordinate system are r, λ, and ϕ (the radius, elevation, and azimuth, respectively), then at any point in space, the three reference axes of the local horizontal system are: *north axis* (N) in the direction of increasing λ, *east axis* (E) in the direction of increasing ϕ, and *zenith axis* (Z) in the direction of increasing r. *South* (S), *west* (W), and *nadir* (n) are used for the negatives of the three axes. Thus, the names for the axes would correspond to the usual definitions of the four directions, the zenith, and nadir on the surface of a spherical Earth. Within the local horizontal coordinate system, the reference plane is the N-S-E-W plane and the reference direction of 0 azimuth is north. *Elevation* is used for the angular height above the reference plane (i.e., toward the zenith), and *azimuth* is used for the rotation angle in the reference plane measured from north toward east.

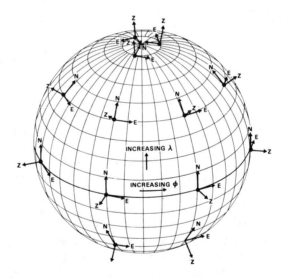

Fig. 2-5. Local Horizontal Coordinates

2.2.4 Parallax

Parallax is the shift in the direction of a nearby object when the observer moves, as illustrated in Fig. 2-6 for the case of the Sun shifting its position relative to the background of the fixed stars. (The amount of parallax for a normal spacecraft orbit is exaggerated in Fig. 2-6.) There is no parallax due to the shift of the center of a coordinate system from one place to another, because the axes are moved parallel to themselves and are defined as maintaining fixed directions in space rather than pointing toward a real object, with the exception of the roll, pitch, yaw coordinates. For example, the pole-to-pole axis of spacecraft-centered celestial coordinates is parallel to the Earth's axis.

In principle, there is a very small parallax because attitude sensors are not mounted precisely at the center of the spacecraft. In practice, this shift is totally negligible. For example, the shift in the position of the Earth's horizon 200 km away for an instrument offset of 1 m is 1 arc sec or 3×10^{-4} deg. Therefore, for attitude-sensing hardware, the orientation of the field of view of the hardware in the spacecraft system is normally important, but the location of the hardware within the spacecraft is not.

Two types of parallax can be important in some circumstances. *Solar parallax* is the shift in the direction of the Sun as an Earth satellite moves in its orbit. (This definition differs from that used in Earth-based astronomy.) The amount of shift as the satellite moves the full diameter of its orbit perpendicular to the direction of the Sun (as illustrated in Fig. 2-6) is 0.005 deg in low Earth orbit, 0.032 deg at

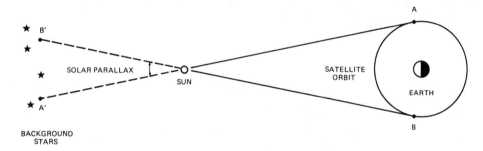

Fig. 2-6. Solar parallax. A' and B' are the apparent positions of the Sun among the background stars when the satellite is at A and B, respectively. The amount of shift is greatly exaggerated.

synchronous altitude, and 0.29 deg at the distance of the Moon. Of course, these shifts are superimposed on the approximately 1-deg-per-day apparent motion of the Sun due to the motion of the Earth in its orbit. Finally, *stellar parallax* is the shift in the direction of very nearby stars (other than the Sun) due to the orbital motion of the Earth about the Sun. **Note:** Stellar parallaxes are quoted as the shift in the direction of a star due to the motion of the observer (perpendicular to the direction of the star) of 1 AU, or the *radius* of the Earth's orbit. Therefore, the maximum shift in stellar direction for an Earth satellite for 1 year will be twice the stellar parallax. The largest stellar parallax is $0.76'' = 2 \times 10^{-4}$ deg for the Sun's nearest neighbor, Alpha Centauri. Thus, stellar parallaxes are only of interest for very accurate star sensor work.

2.3 Elementary Spherical Geometry

This section gives a brief introduction to geometry on the surface of a sphere. Appendix A contains a more complete collection of formulas than is presented here, including a discussion of differential spherical geometry, which is useful in analytic work.

Recall that a *great circle* on a sphere is any circle which divides the sphere into two equal hemispheres. All other circles are known as *small circles*. Great circles are analogous to straight lines in plane geometry in that they have three fundamental properties on a spherical surface: (1) the shortest distance between two points is an arc of a great circle; (2) a great circle is uniquely determined by any two points not 180 deg apart; and (3) great circles are produced by parallel propagation on the surface of a sphere. Parallel propagation means that if we take a straight line on any infinitesimally small region of the sphere and continually extend it parallel to itself in small steps, we will generate a great circle. Spherical triangles and other spherical polygons are constructed from the arcs of great circles.

A fundamental difference between plane and spherical geometry is that on the sphere all great circles eventually intersect. In particular, if we are given a great circle and construct two great circles perpendicular to it (at any distance apart), then these two great circles will intersect exactly 90 deg from the first circle. This is easily seen from Fig. 2-7 by thinking of the first circle as the equator and the second two as any two meridians. Of course, any great circle may be taken as the equator of a coordinate system.

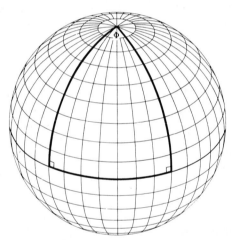

Fig. 2-7.   Intersection of Meridians To Form a Spherical Triangle

A second fundamental difference between spherical and plane geometry involves the sum of the angles of a triangle. The sum of the angles of a plane triangle always equals 180 deg. However, for *any* spherical triangle, the sum, Σ, of the rotation angles formed by the intersecting arcs is always greater than 180 deg. Specifically, the quantity $(\Sigma - 180°)$, known as the *spherical excess*, is directly proportional to the area of a spherical triangle. For example, consider the spherical triangle formed by the equator and two meridians in Fig. 2-7. Because the two angles at the equator are both right angles, the rotation angle at the pole, Φ, between the two meridians is equal to the spherical excess. Clearly, the area of the spherical triangle formed is directly proportional to the rotation angle at the pole. Although more difficult to demonstrate, this theorem is true for any spherical triangle.

Because the surface of a sphere is uniformly curved, we may choose any great circle as the equator of a coordinate system for carrying out geometrical calculations. For the same reason, all the rules of symmetry apply to figures drawn on the sphere. For example, any figure may be reflected about a great circle to produce an identical figure on the other side, except that the reflection will have "right" and "left" interchanged just as in geometrical reflections about a straight line on a plane or physical reflections in a mirror. In addition to symmetry, all the rules of plane geometry hold for any *infinitesimally* small region on the surface of a sphere. For example, if two great circles intersect, the sum of any two adjacent rotation angles at the intersection is 180 deg, as it is for straight lines intersecting in a plane. (Appendix A discusses infinitesimal spherical triangles in detail.)

The fundamental figure for computations in spherical geometry is the spherical triangle, illustrated in Fig. 2-8. Recall from Section 2.1 that lowercase Greek letters are used for the arc length sides of a spherical triangle and uppercase Greek letters are used for the rotation angles. In spherical geometry, any three components, including the three rotation angles, determine a spherical triangle, although in some cases two triangle solutions may exist. Given any three components, Appendix A presents formulas for determining the remaining components.

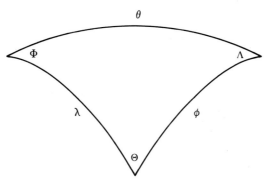

Fig. 2-8.   Notation for a General Spherical Triangle

The relations between sides and angles in plane trigonometry do not hold in spherical trigonometry. (In any infinitesimal region of the sphere, plane geometry is applicable.) However, three fundamental relations do hold among the components of any spherical triangle. Using the notation of Fig. 2-8, these are:

The law of sines,

$$\frac{\sin \theta}{\sin \Theta} = \frac{\sin \lambda}{\sin \Lambda} = \frac{\sin \phi}{\sin \Phi} \qquad (2\text{-}1)$$

The law of cosines for sides,

$$\cos \theta = \cos \lambda \cos \phi + \sin \lambda \sin \phi \cos \Theta \qquad (2\text{-}2)$$

The law of cosines for angles,

$$\cos \Theta = -\cos \Lambda \cos \Phi + \sin \Lambda \sin \Phi \cos \theta \qquad (2\text{-}3)$$

The rules of spherical trigonometry are considerably simplified in the special cases of right and quadrantal triangles. A *right spherical triangle* is one in which one of the rotation angles is 90 deg. A *quadrantal spherical triangle* is one in which one of the sides is 90 deg. In both cases, the relations between the five remaining sides and angles are given by a set of rules developed by John Napier, a 16th-century Scottish mathematician. These are presented in Appendix A. *Napier's Rules and the laws of sines and cosines are particularly useful in attitude analysis and frequently provide simpler, exact analytic expressions than are available from plane geometry approximations.* Napier's Rules may be derived from the law of cosines or may be used to derive it.

As an illustration of the application of spherical geometry, consider the problem illustrated in Fig. 2-9. Here, five solid-angle Sun sensors (also called two-axis sensors; see Section 6.1), *A-E*, are arranged with the optical axis of one toward each pole of a suitably defined coordinate system and the three others equally spaced around the equator. The problem is to determine the maximum angle that a point on the sphere can be from the axis of the closest Sun sensor. By symmetry, one such farthest point must lie along the meridian half way between *C* and *D*. Further, it must lie along the meridian at a point, *P*, such that $\gamma_1 = \gamma_2 = \gamma_3$. (If, for example, γ_1 were greater than γ_2, the point could be moved away from *A* and the distance from the nearest sensor would be increased.) Because $\gamma_1 = \gamma_2 \equiv \gamma$, the triangle *APD* is isosceles and, therefore, $\Gamma_1 = \Gamma_2 \equiv \Gamma$. Because the sensors are symmetrically placed about *A*, $\Gamma_1 = 360°/6 = 60°$. Because ψ is 90 deg, we now know three components of spherical triangle *APD* and the quantity of interest, γ, can be found by several methods. With no further analysis, we may go directly to Table A-1 in Appendix A for a triangle with two angles and the included side known to obtain:

$$\tan \gamma_1 = \frac{\tan \psi \sin \Omega}{\sin(\Gamma_2 + \Omega)}$$

where Ω is defined by

$$\tan \Omega = \tan \Gamma_1 \cos \psi \qquad (2\text{-}4)$$

In this case, the equation cannot be evaluated directly, since $\cos \psi = 0$ and $\tan \psi = +\infty$.

Alternatively, we may apply the law of cosines for sides to *APD* to obtain

$$\cos \gamma_1 = \cos \gamma_2 \cos \psi + \sin \gamma_2 \sin \psi \cos \Gamma_1 \qquad (2\text{-}5)$$

Because $\gamma_1 = \gamma_2$ and $\psi = 90°$, this reduces to

$$\cos \gamma = \sin \gamma \cos \Gamma$$

$$\cot \gamma = \cos \Gamma \qquad (2\text{-}6)$$

We may also create a right spherical triangle by constructing the perpendicular bisector of arc *AD* which will pass through *P*, since *APD* is an isosceles triangle

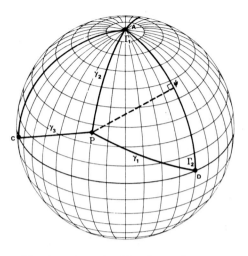

Fig. 2-9. Example of Sun Sensor Geometry

and must be symmetric. Then, from Napier's rules for right spherical triangles (Appendix A), we obtain for the hypotenuse, γ_2

$$\sin(90° - \Gamma_1) = \tan(\psi/2)\tan(90° - \gamma_2) \tag{2-7}$$

or, since $(\psi/2) = 45°$,

$$\cot\gamma = \cos\Gamma$$

Finally, we may also solve the problem by recognizing immediately that APD is a quadrantal triangle and using Napier's rules for quadrantal triangles (Appendix A) to obtain directly

$$\sin\Gamma_1 = \tan\Gamma_2\tan(90° - \gamma_2)$$

$$\cot\gamma = \cos\Gamma \tag{2-8}$$

Because $\Gamma = 60°$, $\gamma = \text{arc}\,\cot(0.5) = 63.42°$. (This suggests why the field of view of solid angle Sun sensors is often approximately a circle of 64-deg radius.) This analysis remains valid for a Sun sensor at each pole and any number of sensors (greater than 1) uniformly distributed about the equator. Thus, with a total of six sensors, four are distributed along the equator, $\Gamma = 45°$, and $\gamma = \text{arc}\,\cot\sqrt{2} = 54.73°$.

CHAPTER 3

SUMMARY OF ORBIT PROPERTIES AND TERMINOLOGY

James R. Wertz

3.1 Keplerian Orbits
3.2 Planetary and Lunar Orbits
3.3 Spacecraft Orbits
3.4 Orbit Perturbations
3.5 Viewing and Lighting Conditions

This chapter provides background information and defines the terminology used in orbit analysis and mission planning. Approximate expressions, appropriate for hand calculation, are provided and motivated so far as possible. Many general technical works are widely available in the areas of celestial mechanics and orbit analysis; among the more popular are Baker [1967], Baker and Makemson [1967], Battin [1964], Escobal [1965], Herrick [1971], Kaplan [1976], McCuskey [1963], Roy [1965], Ruppe [1966], and Thomson [1963].

In contrast with the approximate formulae presented in this chapter, the precise specification of the past or future position of spacecraft or celestial objects is done by means of a numerical table, or *ephemeris*, listing the position at regular intervals. Ephemerides of solar system objects in the form of printed tables are provided in annual editions of the *American Ephemeris and Nautical Almanac* and its British equivalent, the *Astronomical Ephemeris*; versions for computer use are provided on magnetic tape by the Jet Propulsion Laboratory. Both of these sources of definitive information are described in Section 5.5. Definitive spacecraft orbits are normally provided only on magnetic tape or disk and are discussed in Section 5.4.

3.1 Keplerian Orbits

Predicting the motion of the Sun, Moon, and planets was a major part of the scientific revolution of the Sixteenth and Seventeenth centuries. Galileo's discovery of the satellites of Jupiter in 1610 provided a break with Aristotelian science and a strong argument for Copernicus' heliocentric theory. Danish astronomer Tycho Brahe determined the positions of the planets to about 1 minute of arc (1/60 deg) and the length of the year to about 1 second with the unaided eye. Tycho's German assistant, Johannes Kepler, used these precise observations to derive empirically the rules of planetary motion which would later be justified by Newton.

It was the search for the underlying cause of the motion of celestial objects that motivated much of Newton's development of mechanics. In 1665, he determined that if gravity were an inverse square force it could account both for objects falling at the Earth's surface and for the motion of the Moon. However, the detailed theory was not published by Newton until 1687 (*Philosophiae Naturalis Principia Mathematica*). A major cause of this 22-year delay was Newton's inability to show that spherically symmetric objects (e.g., the Earth) behave gravitationally as though all the mass were concentrated at the center, the proof of which required the development of the calculus. Thus, the nearly spherically symmetric planets followed very closely Newton's law of universal gravitation:

$$\mathbf{F} = -(GMm/r^2)\hat{\mathbf{r}} \tag{3-1}$$

where **F** is the force between two objects of mass m and M, **r** is the vector between them, and G is Newton's constant of gravitation. Accurate orbit work includes the effect of the nonspherical symmetry of the Earth, perturbations due to third bodies, and nongravitational forces, but nearly all the basic foundations of orbit theory are direct extrapolations of Newton's work as foreseen by Newton himself.* When gravity is the only force, the orbit defined by two interacting objects is completely determined by their relative position and velocity. In addition, two spherically symmetric masses interacting gravitationally must remain in the plane defined by their relative velocity and position because the forces are central and there is no force to move them out of this plane.

Using gravitational theory and his laws of mechanics, Newton was able to derive Kepler's three laws of planetary motion. These laws apply to any two point masses (or, equivalently, spherically symmetric objects) moving under their mutual gravitational attraction. Kepler's laws, in the form derived by Newton, are as follows:

Kepler's First Law: *If two objects in space interact gravitationally, each will describe an orbit that is a conic section with the center of mass at one focus. If the bodies are permanently associated, their orbits will be ellipses; if they are not permanently associated, their orbits will be hyperbolas.*

Kepler's Second Law: *If two objects in space interact gravitationally (whether or not they move in closed elliptical orbits), a line joining them sweeps out equal areas in equal intervals of time.*

Kepler's Third Law: *If two objects in space revolve around each other due to their mutual gravitational attraction, the sum of their masses multiplied by the square of their period of mutual revolution is proportional to the cube of the mean distance between them; that is,*

$$(m + M)P^2 = \frac{4\pi^2}{G}a^3 \tag{3-2}$$

where P is their mutual period of revolution, a is the mean distance between them, m and M are the two masses, and G is Newton's gravitational constant.

---

*The theory of relativity plays a very minor role in the orbits of planets (principally Mercury) and spacecraft designed specifically to test the theory. For practical purposes, relativistic effects may be totally ignored.

The more massive of the two objects, M, is called the *primary* and the other object is called the *secondary*, or *satellite*. The *barycenter* is the location of the center of mass between the two objects. Kepler's empirical relations presented in two works in 1609 and 1619 were essentially the same except that the constant of proportionality in the third law was obtained empirically and the shape of the orbits specified in the first law was an ellipse (one of the four possible conic sections) because his experience was limited to the planets.

In 1673, Christian Huygens introduced the quantity $\frac{1}{2}mV^2$, which he called the *vis viva* or "living force," to explain the motion of the compound pendulum. The concept was further developed by Gottfried Leibnitz in terms of "living" and "dead" (i.e., static) forces. Application of this kinetic energy theory to celestial mechanics leads to the fourth fundamental relationship for two objects rotating under their mutual gravitational attraction, the *vis viva equation*:

$$V^2 = G(m + M)\left(\frac{2}{r} - \frac{1}{a}\right) \tag{3-3}$$

where r is the instantaneous separation of the objects and V is the magnitude of their relative velocity. For some time there was a bitter controversy between the followers of Huygens-Leibnitz, who believed that $F\Delta x = \Delta(\frac{1}{2}mV^2)$ was the correct measure of the effect of a force, F, and the followers of Galileo-Newton, who believed that $F\Delta t = \Delta(mV)$ was the proper measure. The controversy was resolved in 1743 when Jean D'Alembert published his *Traité de Dynamique*, which showed that both measures were correct and that they were not equivalent. (For a discussion of this controversy see, for example, Girvin [1948] or Dugas [1955].)

Kepler's First Law. Kepler's First Law states that the orbits of celestial objects are *conic sections* i.e., figures produced by the intersection of a plane and a cone (Fig. 3-1), or any quadratic function in a plane. If the objects are permanently associated, this figure will be an *ellipse*, as shown in Fig. 3-2. Geometrically, an ellipse is defined by two points known as *foci*; the ellipse is then the locus of all points such that the sum of the distances from each point on the ellipse to the two foci is $2a$, where a is called the *semimajor axis* and is half the long axis of the ellipse. The semimajor axis is also the *mean distance* between the focus and the boundary of the ellipse and is often listed this way in tables of orbit parameters. In Fig. 3-2, the quantity c is half the distance between the foci, and the *semiminor axis*, b, is half the short axis of the ellipse. One of the foci is the barycenter of the two objects; the other focus is of only geometric interest and is an empty point in space.

The shape of an ellipse is uniquely specified by a single parameter, such as the ratio of the semimajor and semiminor axes. The parameter normally used to specify this shape is the *eccentricity*, e, defined as the ratio $c/a = (a^2 - b^2)^{1/2}/a$. The eccentricity also serves as a convenient ratio to define and parameterize all the conic sections. Specifically, $e = 0$ for a circle; $0 < e < 1$ for an ellipse; $e = 1$ for a *parabola*; and $e > 1$ for a *hyperbola*. In the last case, the nearest point on the curve to the focus is between the focus and the center of the two branches of the

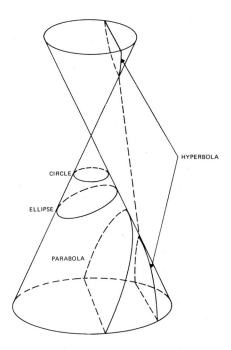

Fig. 3-1.   The four conic sections result from the intersection of a plane and a right circular cone. Two special cases occur when the angle between the plane and the axis of the cone is either 90 deg (resulting in a circle) or equal to the angular radius of the cone (resulting in a parabola).

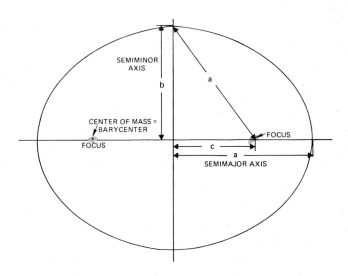

Fig. 3-2.   Geometry of an Ellipse With Eccentricity $e = c/a = 0.6$

hyperbola. These four classes of curves are illustrated in Fig. 3-3, and their properties are summarized at the end of this section.*

Both the circle and parabola represent special cases of the infinite range of possible eccentricities and therefore will never occur in nature. Orbits of objects which are gravitationally bound will be elliptical and orbits of objects which are not bound will be hyperbolic. Thus, an object approaching a planet from "infinity," such as a spacecraft approaching Mars, must necessarily travel on a hyperbolic trajectory relative to the planet and will swing past the planet and recede to infinity, unless some nongravitational force (a rocket firing or a collision with the planet) intervenes. Similarly, a rocket with insufficient energy to escape a planet must travel in an elliptical orbit in the absence of nongravitational forces. Because the ellipse is a closed curve, the rocket will eventually return to the point in space at which the engine last fired.

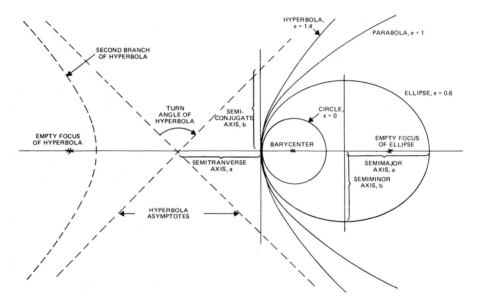

Fig. 3-3. Four Possible Conic Sections. The circle and parabola have uniquely defined shapes, but there is a continuous range of shapes (determined by the eccentricity) for the ellipse and hyperbola.

Kepler's Second Law. As shown in Fig. 3-4, Kepler's Second Law is a restatement of the conservation of angular momentum. The angular momentum is proportional to the magnitude of the radius vector, r, multiplied by the perpendicular component of the velocity, $V_\perp$. In any infinitesimal time interval, δt, the area swept out by a line joining the barycenter and the satellite will be $\frac{1}{2} V_\perp r \delta t$.

---

*Alternatively, we may define a conic section as the locus of all points which maintain a fixed ratio between the distance to the focus and the perpendicular distance to a fixed line called the *directrix*. The directrix is perpendicular to the major axis of any conic section. The ratio of the distance to the focus and to the directrix is the eccentricity, e.

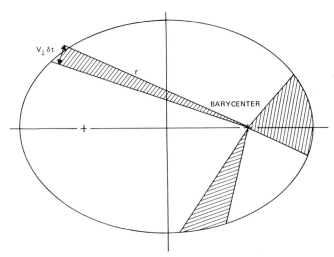

Fig. 3-4. Kepler's Second Law. Because the three shaded areas are equal, the times required for the satellite to cross each area are equal. The area swept out is directly proportional to both the time interval and the angular momentum.

Hence, the area swept out per unit time is proportional to the angular momentum per unit mass which is a constant.

Kepler's Third Law. Kepler's Third Law applies only to elliptical orbits and relates the orbital period to the semimajor axis. In the case of Earth satellites, and very nearly in the case of the planets orbiting the Sun, we may ignore the mass of the secondary and write:

$$a^3 = \left[G(M+m)/4\pi^2 \right] P^2 \approx (GM/4\pi^2) P^2$$

$$\equiv (\mu/4\pi^2) P^2 \tag{3-4}$$

The values of $\mu \equiv GM$ for the major objects in the solar system are given in Appendix L. Note that μ can be measured with considerable precision by astronomical observations. However, the values of M are limited by the accuracy of G to about 0.06%. (This is the most poorly known of the fundamental physical constants.) Therefore, the use of G is normally avoided and calculations are best done in terms of μ and the ratio of the masses of solar system objects.

As long as the mass of the secondary is small, such that Eq. (3-4) holds, then the constant of proportionality in Kepler's Third Law may be evaluated directly from existing orbiting objects. For example, the *astronomical unit*, or *AU*, is a unit of length equal to the semimajor axis of the Earth's orbit about the Sun; thus, in units of years and AU, $\mu/4\pi^2 = 1$ for the Sun and, therefore, $a^3 = P^2$ in units of astronomical units and years for the planets and any other satellites of the Sun.

Kepler's laws and the vis viva equation are independent of the properties of small orbiting objects. This is very different from the case of electrical or magnetic forces and is one of the foundations of the theory of general relativity. On a less subtle level, this independence is also responsible for the phenomenon of "weightlessness" in space. Thus, if two objects, such as an astronaut and his spacecraft, are initially in the same orbit, they will remain in the same orbit and stay adjacent even though they have very different physical properties and are unconnected.

Vis Viva Equation. If we again assume that the mass of the secondary is small, the vis viva equation may be rewritten as

$$V^2 = GM\left(\frac{2}{r} - \frac{1}{a}\right) = \frac{2\mu}{r} - \frac{\mu}{a}$$

$$\frac{1}{2}V^2 - \frac{GM}{r} = -\frac{\mu}{2a} \equiv E \qquad V = \sqrt{\frac{2\mu}{r} - \frac{\mu}{a}} \qquad (3\text{-}5)$$

where E is the total energy per unit mass (kinetic plus potential) of the orbiting object. Thus, the semimajor axis is a function *only* of the total energy. Because the potential energy is a function only of position, the semimajor axis for a satellite launched at any point in space will be a function only of the launch velocity and not the direction of launch. (The shape and orientation of the orbit will, of course, depend on the launch direction.) The orbit is hyperbolic if $E > 0$ and elliptical if $E < 0$. The special case between these, zero total energy, is an orbit with infinite semimajor axis; the velocity in this case is called the *parabolic velocity*, or *velocity of escape*, V_e. At any distance, R, from the center of a spherically symmetric object we have

$$V_e \equiv \sqrt{2\mu/R} \equiv \sqrt{2GM/R} \qquad (3\text{-}6)$$

A satellite launched with this velocity in any direction will not return, assuming that there are no other forces.

The vis viva equation may be used to obtain two other velocities of particular interest. If $R = a$, then

$$V_c \equiv \sqrt{\mu/R} \qquad (3\text{-}7)$$

is the *circular velocity*, or the velocity needed for a circular orbit of radius R. Finally, in a hyperbolic orbit,

$$V_h \equiv \sqrt{2E} = \sqrt{V^2 - 2\mu/R} \qquad (3\text{-}8)$$

is the *hyperbolic velocity*, or the velocity of an object infinitely far away from the primary. Here V is the instantaneous velocity in the hyperbolic orbit at arbitrary distance R from the center of the massive object.

Orbit Terminology. For either hyperbolic or elliptical orbits, the *perifocus* is the point on the orbit where the secondary is closest to the barycenter. As shown in Fig. 3-5, the *perifocal distance*, or the linear separation between the barycenter and the perifocus, is $(a - c) = a(1 - e)$ for an elliptical orbit of semimajor axis, a, and eccentricity, e. Unfortunately, the terminology here is both well established and awkward because different words are used for the point of closest approach to different primaries. Thus, we have *perihelion* (closest approach to the Sun), *perigee* (closest approach to the Earth), *pericynthiane* or *perilune* (closest approach to the Moon), and even *periastron* (closest approach of the two stars in a binary pair).

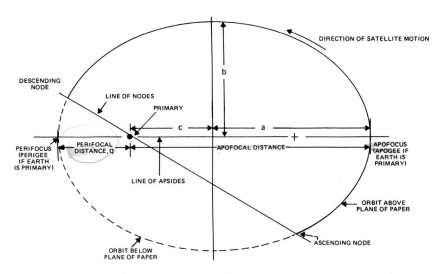

Fig. 3-5. Orbit Terminology for an Elliptical Orbit. The orbit is tilted with respect to the plane of the paper such that the dashed segment is below the paper; the plane of the paper is the reference plane.

Perihelion and perifocus are measured from the center of mass, but *perigee height*, frequently shortened to "perigee" in common usage, is measured from the surface of the Earth. (See Fig. 3-6.) This terminology arises because we are interested primarily in the height above the surface for low-altitude spacecraft. The most unambiguous procedure is to use *perigee height* or *perigee altitude* whenever the distance is being measured from the surface;* however, this is not always done.

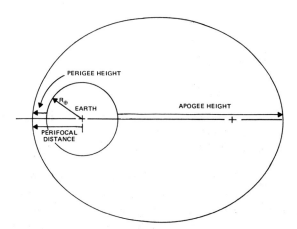

Fig. 3-6. Definition of Perigee Height and Apogee Height

---

*Throughout this book when discussing distances relative to the Earth, we use "height" exclusively for distances measured from the Earth's surface; e.g., "apogee height," "perigee height," or "height of the atmosphere."

In elliptical orbits, the most distant point from the primary is called the *apofocus*; the *apofocal distance* is $c + a$. Again, the words *aphelion, apolune,* and *apogee* or *apogee height* are used, the latter being measured from the Earth's surface. The straight line connecting apogee, perigee, and the two foci is called the *line of apsides*. If h_P, h_A, and $R_\oplus$ are the perigee height, apogee height, and radius of the Earth, respectively, then for an Earth satellite,

$$a = R_\oplus + (h_P + h_A)/2 \tag{3-9}$$

To define an orbit fully we need to specify not only its size and shape, but also the orientation of the orbit plane in space. (See Fig. 3-7.) The *inclination, i,* is the angle between the orbit plane and a *reference plane*, which also contains the barycenter. The most commonly used reference planes are the equatorial plane (the plane of the Earth's equator) for Earth satellites and the plane of the Earth's orbit about the Sun, called the *ecliptic*.

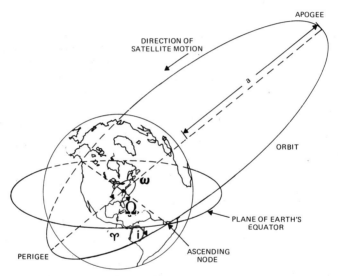

Fig. 3-7. Keplerian Orbital Elements. ♈ marks the direction of the Vernal Equinox. Ω is measured in the plane of the Earth's equator, and ω is measured in the orbit plane.

The intersection of the orbit plane and the reference plane through the center of mass is the *line of nodes*. For an Earth satellite, the *ascending node* is the point in its orbit where a satellite crosses the equatorial plane going from south to north. The *descending node* is the point where it crosses the equatorial plane from north to south. Thus, in. Fig. 3-5, if the plane of the paper is the reference plane and the dashed part of the orbit is below the paper, then the nodes are as illustrated.

To define fully the orbital plane of a satellite we need to specify both its inclination and its eastwest orientation. The latter is defined by the *right ascension of the ascending node, Ω,* or the angle in the equatorial plane measured eastward from the vernal equinox to the ascending node of the orbit. (The vernal equinox is the ascending node of the Earth's orbit about the Sun.) The rotation of the orbit within the orbital plane is defined by the *argument of perigee*, ω, or the angle at the barycenter measured in the orbital plane in the direction of the satellite's motion from the ascending node to perigee.

Finally, we need some method to specify where a satellite is in its orbit. The *true anomaly*, ν, is the angle measured at the barycenter between the perigee point and the satellite. The *mean anomaly*, M, is $360 \cdot (\Delta t / P)$ deg, where P is the orbital period and Δt is the time since perigee passage of the satellite; thus, $M = \nu$ for a satellite in a circular orbit. The mean anomaly at any time is a trivial calculation and of no physical interest. The quantity of real interest is the true anomaly, which is difficult to calculate. The *eccentric anomaly*, E, is introduced as an intermediate variable relating the other two. E is the angle measured at the center of the orbit between perigee and the projection (as shown in Fig. 3-8) of the satellite onto a circular orbit with the same semimajor axis.

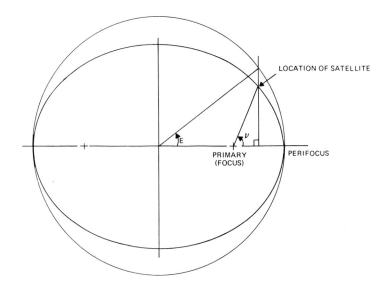

Fig. 3-8.　Definition of True Anomaly, ν, and Eccentric Anomaly, E. The outer figure is a circle with radius equal to the semimajor axis.

The mean and eccentric anomalies are related by *Kepler's equation** (not related to Kepler's Laws):

$$M = E - e \sin E \qquad (3\text{-}10a)$$

where e is the eccentricity. E is then related to ν by *Gauss' Equation*:

$$\tan\left(\frac{\nu}{2}\right) = \left(\frac{1+e}{1-e}\right)^{1/2} \tan\left(\frac{E}{2}\right) \qquad (3\text{-}10b)$$

---

*According to Watson [1958], Joseph Legrange showed in 1770 that the solution to Kepler's Equation could be written as a series expansion in Bessel functions, although the modern name and notation was not applied until after the detailed explanations of Friedrich Wilhelm Bessel in 1824.

$(E/2)$ and $(\nu/2)$ are used because these quantities are always in the same quadrant. For small eccentricities, ν may be expressed directly as a function of M by expanding in a power series in e to yield:

$$\nu \approx M + 2e\sin M + \frac{5}{4}e^2\sin 2M + \mathcal{O}(e^3) \tag{3-11}$$

Ruppe [1966] gives several convenient series expansions for anomalies and other orbit parameters.

The *elements* of an orbit are the parameters needed to fully specify the motion of a satellite. There are several alternative ways in which this can be done. Table 3-1 and Fig. 3-7 show the *classical elements* or *Keplerian elements* for an Earth satellite. (Planetary elements are slightly different and are defined in Section 3.2.) Elements are defined at some reference time or *epoch* because they are slowly changing with time as described in Section 3.4. The semimajor axis and eccentricity define the size and shape of the orbit. The inclination and right ascension of the ascending node define the orbit plane. The rotation of the orbit within the plane (i.e., the orientation of the line of apsides) is defined by the argument of perigee. Finally, the mean anomaly specifies (via Kepler's and Gauss' equations) the position of the satellite in its orbit at the epoch time.

Table 3-1.   Classical Elements for an Earth Satellite

QUANTITY	SPECIFIED BY	SYMBOL	DEFINITION	SPECIAL CASES
REFERENCE TIME	EPOCH	T_0	TIME FOR WHICH ELEMENTS ARE SPECIFIED	
SIZE	SEMIMAJOR AXIS	a	HALF THE LONG AXIS OF THE ELLIPSE	
SHAPE	ECCENTRICITY	e	DISTANCE FROM CENTER OF ELLIPSE TO FOCUS DIVIDED BY SEMIMAJOR AXIS; DIMENSIONLESS RATIO	$e = 0$ FOR A CIRCULAR ORBIT
ORBIT {	INCLINATION	i	ANGLE BETWEEN ORBIT PLANE AND EARTH'S EQUATORIAL PLANE	$i = 0$ FOR EQUATORIAL ORBIT $i = 90$ DEG FOR POLAR ORBIT
PLANE {	RIGHT ASCENSION OF THE ASCENDING NODE	Ω	ANGLE MEASURED AT THE CENTER OF THE EARTH IN THE EQUATORIAL PLANE FROM THE VERNAL EQUINOX EASTWARD TO THE ASCENDING NODE (I.E., THE POINT AS WHICH THE SATELLITE CROSSES THE EQUATOR GOING FROM SOUTH TO NORTH)	$\Omega = 0$ WHEN THE ASCENDING NODE IS AT THE VERNAL EQUINOX
ORIENTATION OF ORBIT IN THE PLANE	ARGUMENT OF PERIGEE	ω	ANGLE MEASURED AT THE CENTER OF THE EARTH IN THE ORBIT PLANE FROM THE ASCENDING NODE TO PERIGEE (CLOSEST APPROACH OF THE SATELLITE TO THE EARTH) MEASURED IN THE DIRECTION OF THE SATELLITE'S MOTION	$\omega = 0$ OR 180 DEG WHEN PERIGEE IS OVER THE EQUATOR
LOCATION OF THE SATELLITE IN ITS ORBIT	MEAN ANOMALY AT EPOCH	M_0	$360° \times \Delta t/P$ WHERE P IS THE PERIOD AND Δt IS THE TIME DIFFERENCE BETWEEN THE EPOCH AND THE LAST PERIGEE PASSAGE BEFORE EPOCH	$M_0 = 0$ WHEN THE EPOCH TIME IS PERIGEE

The Keplerian elements are not always the best choice. For a hyperbolic orbit, the semimajor axis is undefined and is replaced by the *areal velocity*, or the area per unit time swept out by a line joining the satellite and the planet (i.e., the angular momentum per unit mass). In addition, perigee (and, therefore, ω and M) is poorly defined for nearly circular orbits and the line of nodes (and Ω) is poorly defined for orbits with near zero inclination. In these cases, alternative parameters and calculation procedures are sometimes used for specifying orbits.

Finally, *mean orbital elements* given in most general-purpose tables define the average motion over some span of time. For precise calculations, it is preferable to use *osculating elements*, which are the elements of a true Keplerian orbit instantaneously tangent to the real orbit. Thus, the osculating elements will fluctuate continuously as various forces (e.g., aerodynamic drag, multiple body interactions, or nonspherically symmetric mass distributions) alter the classical shape defined by Kepler's Laws.

Summary. Keplerian orbits are not sufficiently accurate for spacecraft ephemerides or for attitude calculations which require a precise knowledge of the position of the spacecraft; however, they are accurate enough to estimate the overall mission characteristics (such as period or altitude) in most regions of space. Table 3-2 summarizes the properties of Keplerian orbits. The normal procedure for adding additional detail to orbit analysis is to treat orbits as Keplerian with additional perturbations produced by any of the various interactions which may be important. Approximations for the most important perturbations are discussed in Section 3.4. Ephemerides based on detailed orbit analysis are described in Sections 5.4 and 5.5.

Table 3-2. Properties of Keplerian Orbits

QUANTITY	CIRCLE	ELLIPSE	PARABOLA	HYPERBOLA
DEFINING PARAMETERS	a = SEMIMAJOR AXIS = RADIUS	a = SEMIMAJOR AXIS b = SEMIMINOR AXIS	q = PERIFOCAL DISTANCE	a = SEMITRANSVERSE AXIS b = SEMICONJUGATE AXIS
PARAMETRIC EQUATION	$x^2 + y^2 = a^2$	$\dfrac{x^2}{a^2} + \dfrac{y^2}{b^2} = 1$	$x^2 = 4qy$	$\dfrac{x^2}{a^2} - \dfrac{y^2}{b^2} = 1$
ECCENTRICITY, e	$e = 0$	$e = \sqrt{a^2 - b^2}/a \quad 0 < e < 1$	$e = 1$	$e = \sqrt{a^2 + b^2}/a \quad e > 1$
PERIFOCAL DISTANCE, q	$q = a$	$q = a(1 - e)$	$q = q$	$q = a(e - 1)$
VELOCITY, V, AT DISTANCE FROM FOCUS, r	$v^2 = \mu/a$	$v^2 = \mu\left(\dfrac{2}{r} - \dfrac{1}{a}\right)$	$v^2 = 2\mu/r$	$v^2 = \mu\left(\dfrac{2}{r} + \dfrac{1}{a}\right)$
TOTAL ENERGY PER UNIT MASS, T.E.	$\text{T.E.} = -\dfrac{\mu}{2a} < 0$	$\text{T.E.} = -\dfrac{\mu}{2a} < 0$	$\text{T.E.} = 0$	$\text{T.E.} = \dfrac{\mu}{2a} > 0$
MEAN ANGULAR MOTION, n	$n = \sqrt{\mu/a^3}$	$n = \sqrt{\mu/a^3}$	$n = \sqrt{\mu}$	$n = \sqrt{\mu/a^3}$
PERIOD, P	$P = 2\pi/n$	$P = 2\pi/n$	$P = \infty$	$P = \infty$
ANOMALY	UNDEFINED	ECCENTRIC ANOMALY, E $\text{TAN}\left(\dfrac{\nu}{2}\right) = \left(\dfrac{1+e}{1-e}\right)^{1/2} \text{TAN}\left(\dfrac{E}{2}\right)$	PARABOLIC ANOMALY, D $\text{TAN}\left(\dfrac{\nu}{2}\right) = D/\sqrt{2q}$	HYPERBOLIC ANOMALY, F $\text{TAN}\left(\dfrac{\nu}{2}\right) = \left(\dfrac{e+1}{e-1}\right)^{1/2} \text{TANH}\left(\dfrac{F}{2}\right)$
MEAN ANOMALY, $M_i \equiv n(t - T_0)$ MODULO 360	UNDEFINED	$M = E - e \sin E$	$M = qD + \dfrac{1}{6}D^3$	$M = e \sinh F - F$
DISTANCE FROM FOCUS, $r = q\left(\dfrac{1+e}{1+e\cos\nu}\right)$	$r = a$	$r = a(1 - e \cos E)$	$r = q + \dfrac{1}{2}D^2$	$r = a(e \cosh F - 1)$
$r \, dr/dt \equiv r\dot{r}$	0	$r\dot{r} = e\sqrt{a\mu} \sin E$	$r\dot{r} = \sqrt{\mu}\, D$	$r\dot{r} = e\sqrt{a\mu} \sinh F$
AREAL VELOCITY, $\dfrac{dA}{dt} = \dfrac{1}{2}r^2\dfrac{d\nu}{dt} = \dfrac{1}{2}r\dot{r}$	$\dfrac{dA}{dt} = \dfrac{1}{2}\sqrt{a\mu}$	$\dfrac{dA}{dt} = \dfrac{1}{2}\sqrt{a\mu(1 - e^2)}$	$\dfrac{dA}{dt} = \sqrt{\mu q/2}$	$\dfrac{dA}{dt} = \dfrac{1}{2}\sqrt{a\mu(e^2 - 1)}$

*μ = GM IS THE GRAVITATIONAL CONSTANT OF THE CENTRAL BODY, ν IS THE TRUE ANOMALY, AND M $\equiv$ n (t–T) IS THE MEAN ANOMALY, WHERE t IS THE TIME OF OBSERVATION, T IS THE TIME OF PERIFOCAL PASSAGE, AND n IS THE MEAN ANGULAR MOTION. SEE HERRICK [1971] FOR ADDITIONAL FORMULAS AND A DISCUSSION AND LISTING OF TERMINOLOGY AND NOTATION. SEE FIG. 3–3 FOR DEFINITIONS OF a AND b.

3.2 Planetary and Lunar Orbits

Orbits of the natural objects in the solar system exhibit considerable regularity; with the exception of cometary orbits, they are nearly circular, coplanar, and regularly spaced.* Although perturbations due to third bodies must be taken into account for computing definitive planetary ephemerides (Section 5.5), the description of the orbits in terms of Keplerian elements adequately describes most orbital characteristics. The position of the center of mass of the solar system relative to the Sun depends on the relative orientation of the planets, but is typically one solar radius from the center of the Sun in the general direction of Jupiter. Thus, for most purposes we regard the Sun as the center of mass of the solar system and, therefore, as at one focus for all planetary orbits.

To define the Keplerian elements for the orbits of either planets or interplanetary spacecraft, we must establish a reference plane through the Sun. The standard plane chosen for this is the *ecliptic*, or the plane of the Earth's orbit about the Sun. This plane is inclined to the Earth's equatorial plane by about 23.44 deg, an angle known as the *obliquity of the ecliptic*. The intersection of the plane of the ecliptic and the plane of the Earth's equator define two opposite directions in space known as the *vernal* and *autumnal equinoxes*, represented by the symbols ♈ and ♎, respectively. The vernal equinox, the direction of the Sun (viewed from the center of the Earth) as it crosses the equatorial plane from south to north, serves as the reference direction for coordinate systems using either the equatorial or ecliptic plane. Perturbative forces on the Earth cause the rotational axis of the Earth to move in a cone of 23.44-deg radius about a vector perpendicular to the ecliptic plane; this *precession of the equinoxes* has a period of about 25,700 years. The effect of this precession on coordinate systems is described in Section 2.2.2.

Because the Earth's orbit is not perfectly Keplerian, and because of the drift of the vernal equinox, the orbital period of the Earth about the Sun depends on how it is measured. The *sidereal year*, about 365.26 days, is the period of revolution of the Earth relative to the fixed stars. The *tropical year* is the Earth's period relative to the vernal equinox and is about 20 minutes shorter than the sidereal year; this is the basis of the civil calendar, since, for calendar purposes, we are interested in the seasons which are determined by the position of the Sun relative to the Earth's equatorial plane. Finally, the *anomalistic year*, 5 minutes longer than the sidereal year, is the period of the Earth relative to perihelion.[†] Recall that *perihelion* is the perifocal point when the Sun is the primary. This shift in the inertial position of perihelion is due to perturbative forces of the other planets.

The orbital elements of the planets and other solar system objects are analogous to the Earth satellite elements defined in Table 3-1, with the Earth's equatorial plane replaced by the ecliptic plane. Thus, the semimajor axis and eccentricity retain the same definitions. The inclination of planetary orbits is

---

*This section describes qualitative characteristics of planetary orbits. For detailed numerical information, see Appendix L.

[†]For a more extended discussion of time measurement systems and precise numerical values, see Appendix J.

measured relative to the ecliptic. The *longitude of the ascending node*, Ω, is the angle from the vernal equinox to the ascending node of the planet's orbit measured eastward along the ecliptic plane. The *argument of perihelion*, ω, is the angle from the ascending node to perihelion measured along the planet's orbit in the direction of its motion. In some tables, ω is replaced by the *longitude* of *perihelion*, $\tilde{\omega} \equiv \omega + \Omega$; note that this is not a true angular measure because ω and Ω are measured in different planes. Finally, the mean anomaly of satellite orbits is replaced by the *time of perihelion passage*, T, which is one of the times (usually the most recent) when the planet was at perihelion. Numerical values for the planetary orbital elements are given in Appendix L.

Planetary orbits within the Solar System are fairly uniform in both shape and orientation; with the exception of Pluto and Mercury, the orbital inclinations are all less than 3.5 deg and the eccentricities are less than 0.1. The semimajor axes of the planetary orbits are also nearly regular and are approximately given by an empirical relation known as *Bode's Law*, in which the semimajor axes of the planets and asteroid belt in AU are approximately 0.4, 0.7, 1.0, 1.6, 2.8, 5.2, etc.

For both interplanetary spacecraft and for determining the visual brightness of the planets as seen by Earth-orbiting spacecraft, we are interested in the orientation of the planets relative to the Earth as well as their orientation relative to the fixed stars. The various geometrical orientations of the planets relative to the observer and the Sun are called *planetary configurations* or *aspects* and are defined in Fig. 3-9. An *inferior* planet is one with an orbit closer to the Sun than the observer and a *superior* planet is one farther from the Sun than the observer. *Conjunction* and *opposition* occur when the planet and the Sun are in the same and opposite directions, respectively;* "the same direction" throughout this discussion is in terms of the relative planet-observer orientation around the ecliptic, regardless of the distance above or below the ecliptic plane. Thus, a full moon occurs when the Moon is at opposition. Conjunction and opposition may also be applied to two

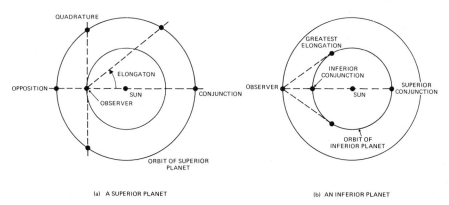

(a) A SUPERIOR PLANET   (b) AN INFERIOR PLANET

Fig. 3-9.   Planetary Configurations. (a) for a superior planet, (b) for an inferior planet.

---

* *Syzygy*, an astronomical contribution to crossword puzzles, refers to either conjunction or opposition, i.e., when the planet, the Sun, and the observer lie on a straight line.

planets; for example, Mars and Jupiter are in conjunction for a particular observer if both planets are in the same direction from the observer. (If only one planet is mentioned, the implied second object is the Sun; "Mars is at opposition" means that Mars and the Sun are in opposition.) For an inferior planet, *inferior conjunction* occurs when the planet is between the Earth and the Sun and *superior conjunction* occurs when the Sun is between the Earth and the planet. *Elongation* is the angular separation between a planet and the Sun measured in the plane of the ecliptic. A superior planet will be brightest near opposition and an inferior planet will be brightest near *greatest elongation*, when it is at the farthest angular separation from the Sun. *Quadrature* occurs when the Sun/observer/planet angle is 90 deg. In astronomical tables, the standard symbols defined in Fig. 3-10 are frequently used for the various aspects of the planets. For example, $\sigma \diagdown \varphi \ \hbar$ is read "Venus and Saturn in conjunction." A discussion of visual magnitude and other optical aspects of the planets relevant to attitude work is presented in Section 3.5.

Fig. 3-10. Standard Astronomical Symbols

The interval between successive oppositions of a superior planet or successive inferior conjunctions of an inferior planet is known as the *synodic period, S*. The relation between the synodic period and the *sidereal period, P*, relative to the fixed stars is shown in Fig. 3-11. If P_1 and P_2 are the periods of the inner and outer planets respectively, then, in general:

$$S^{-1} = P_1^{-1} - P_2^{-1} \tag{3-12}$$

For an observer on the Earth, if S and P are both expressed in years, then for a superior planet the average synodic period is given by

$$S^{-1} = 1 - P^{-1} \tag{3-13}$$

and for an inferior planet by

$$S^{-1} = P^{-1} - 1 \tag{3-14}$$

Thus, for Mars, opposition will occur approximately every 780 days. Because planetary orbits are not circular, the actual synodic periods vary by several weeks. Recent and future oppositions of Mars are listed in Table 3-3. Note that the

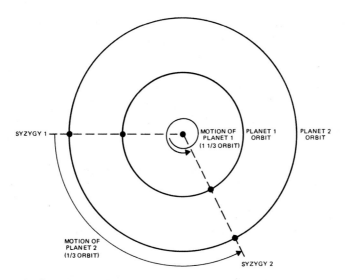

Fig. 3-11.   Determining the Synodic Period (see text for explanation)

synodic period is longest for Mars and Venus (780 and 584 days), shortest for Mercury (116 days), and approaches 1 year for the other planets.

Planetary configurations are important for interplanetary flight as well as for observations because they define the opportunities for planetary travel. For example, as shown in Section 3.3, trips to Mars along a minimum energy trajectory will leave the Earth about 97 days before opposition and arrive at Mars about 162 days after opposition, although various factors may cause the actual flight times, particularly the arrival time, to vary by several weeks. Because an opposition of Mars occurred on December 15, 1975, we would expect flights to leave Earth on about September 10, 1975, and arrive at Mars on about May 27, 1976. The two spacecraft flown during this launch opportunity, Viking I and II, were launched on August 20 and September 9, 1975, and arrived at Mars on June 19 and August 7, 1976.

The orbits of the natural satellites of the solar system are generally less uniform than the orbits of the planets, primarily because perturbations cause substantial variations in satellite orbits. For example, the perigee location for the Moon makes one complete revolution about the Moon's orbit in only 8.85 years and the line of nodes rotates fully around the orbit in 18.6 years. Thus, in analogy

Table 3-3.   Oppositions of Mars

DATE	VISUAL MAG.	DATE	VISUAL MAG.	DATE	VISUAL MAG.
MAY 31, 1969	−2.0	FEBRUARY 25, 1980	−0.8	NOVEMBER 27, 1990	−1.8
AUGUST 10, 1971	−2.7	MARCH 31, 1982	−1.1	JANUARY 7, 1993	−1.2
OCTOBER 25, 1973	−2.2	MAY 11, 1984	−1.7	FEBRUARY 12, 1995	−1.0
DECEMBER 15, 1975	−1.5	JULY 10, 1986	−2.4	MARCH 17, 1997	−1.1
JANUARY 22, 1978	−1.1	SEPTEMBER 28, 1988	−2.5	APRIL 24, 1999	−1.4

with the various types of years, the month is defined in several ways, depending on the measurement reference. For most purposes, the fundamental intervals are: the *sidereal month* relative to the fixed stars, about 27.32 days; the *synodic month* from new moon to new moon, about 29.53 days; and the *nodical* or *draconic* month from node to node, about 27.21 days. Other periods are listed in Appendix L. Section 5.5 gives an algebraic approximation which may be used to model the position of the Moon, as seen from the Earth, to within about 0.25 deg.

3.3 Spacecraft Orbits

Vehicles or objects launched into space are categorized by their orbit. *Ballistic missiles* and *sounding rockets* travel in elliptical orbits which intersect the surface of the Earth; this is frequently called a *ballistic trajectory* because it is also the path of a bullet or cannon shell. The ballistic missile and sounding rocket are distinguished by their function—a missile is used to strike some specific target, whereas a sounding rocket is used to make measurements in or above the Earth's atmosphere. A sounding rocket may either impact the surface, burn up in the atmosphere, or be recovered by parachute.

Any object which travels in an elliptical orbit around a planet is called a *satellite* of that planet. The semimajor axis of a satellite orbit must be at least as large as the radius of the planet, whereas the semimajor axis for a sounding rocket may be as small as approximately half the radius of the planet.* Because the total energy of a spacecraft depends only on the semimajor axis, the energy of a sounding rocket is normally, though not necessarily, much less than that of an Earth satellite. However, sounding rockets and ballistic missiles frequently reach altitudes well above those of low-Earth satellites because they travel in very elongated elliptical orbits.

If the velocity of an object is greater than the escape velocity of a planet, it will be an *interplanetary probe* traveling in a hyperbolic trajectory relative to the planet and, after it has left the vicinity of the planet, traveling in an elliptical orbit about the Sun. Finally, if an object attains a velocity greater than the Sun's escape velocity, it will be an *interstellar probe*. Pioneer 10, swinging past Jupiter in December 1973, gained sufficient energy in the encounter (as described later in this section) to become the first manmade interstellar probe.

All known satellites or probes are assigned an *international designation* by the World Warning Agency on behalf of the Committee on Space Research, COSPAR, of the United Nations. These designations are of the form 1983-14D, where the first number is the year of launch, the second number is a sequential numbering of launches in that year, and the letter identifies each of the separate objects launched by a single vehicle. Thus, the docking module for the first Apollo-Soyuz flight, object 1975-66C, was the third component of the 66th launch in 1975.

In addition to the international designation, most satellites are assigned a

---

*Assume all the mass of the Earth is concentrated at its center and a high platform is built to the former location of New York City. An object dropped from the platform will *not* go all the way to the former location of Australia, but will swing very rapidly around the central mass (with perigee essentially at the center) and return to apogee at the platform tip. (Use Eq. (3-3) with $V=0$.) Therefore, the semimajor axis will be about half the radius of the Earth and the total energy will be a factor of two less than that for a circular orbit at the Earth's surface.

name by the launching agency. For NASA, spacecraft in a series are given a letter designation prior to launch, which is changed to a number after a successful launch. Thus, the second Synchronous Meteorological Satellite was SMS-B prior to launch and SMS-2 after being successfully placed in orbit. Because of launch failures or out-of-sequence launches, the lettering and numbering schemes do not always follow the pattern A = 1, B = 2, etc. For example, in the Interplanetary Monitoring Platform series, IMP-B failed on launch, IMP-E was put into a lunar orbit and given another name, and IMP-H and -I were launched in reverse order; thus, IMP-I became IMP-6 and IMP-H became IMP-7. Satellites may also be assigned numbers in different series; IMP-6 was also Explorer-43 and IMP-7 was Explorer-47.

The *trajectory* of a spacecraft is its path through space. If this path is closed (i.e., elliptical) then the trajectory is an *orbit*. Thus, correct usage would refer to a *satellite* in an elliptical *orbit* or a spacecraft or *probe* on a hyperbolic *trajectory*. Although this distinction is maintained at times, *orbit* and *trajectory* are often used interchangeably. For satellites it is frequently convenient to number the orbits, so that one may refer to "a maneuver on the sixth orbit." In standard NASA practice, that portion of the orbit preceding the first ascending node is referred to as *orbit 0* or *revolution 0*; *orbit 1* or *revolution 1* goes from the first ascending node to the second ascending node, etc. Note that *revolution* refers to one object moving about another in an orbit, whereas *rotation* refers to an object spinning about an axis.

When spacecraft are launched by a multistage vehicle, the initial stages will be fired and subsequently jettisoned; however, the final stage may remain inactive and attached to the spacecraft during a *coasting phase* or *parking orbit*. The final stage is then ignited or reignited to *inject* or place the spacecraft into its proper orbit.* As defined in Chapter 1, a *mission orbit* is one in which a satellite will be conducting normal operations. A *transfer orbit* is one which is used to maneuver a spacecraft from one orbit to another, as in the case of the Apollo transfer orbit to the Moon or the Apollo Command Module transfer orbit back to Earth.

A satellite which revolves about the Earth in the same direction that the planet rotates on its axis is in a *prograde* or *direct* orbit; if it revolves in a direction opposite to the rotation, the orbit is *retrograde*. As shown in Fig. 3-12, the inclination of an Earth-satellite orbit is measured from east toward north; therefore, the inclination of a prograde satellite is less than 90 deg and the inclination of a retrograde satellite is greater than 90 deg. In a *polar orbit*, $i = 90$ deg. Most satellites are launched in a prograde direction because the rotational velocity of the Earth provides a part of the orbital velocity. Although this effect is not large (0.46 km/sec for the Earth's rotational velocity at the equator, compared with a circular velocity of 7.91 km/sec), the available energy is typically the limiting feature in a space mission. Thus, all factors which change the energy which must be supplied by rockets are important.

---

*Only a limited number of upper stages are capable of being reignited. Among these are the Agena used for unmanned spacecraft and the Service Propulsion System from the Apollo program, both of which use *hypergolic* fuel which ignites when the two fuel components come in contact. The largest reignitable American rocket currently planned is the Space Shuttle, which has three engines with a total vacuum thrust of about 6.3×10^6 N (20% of that of the Saturn V first stage) from a hydrogen/oxygen system.

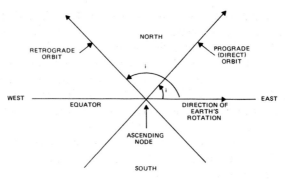

Fig. 3-12.   Inclination, i, for Prograde and Retrograde Orbits. Satellite is moving from south to north.

From Kepler's Third Law, the period, P, of an Earth satellite depends only on the mean altitude (see Eqs. (3-4) and (3-9)); that is,

$$P^2 = (4\pi^2/\mu)(R_\oplus + h_p/2 + h_A/2)^3$$

$$= 2.75118 \times 10^{-8}[6378.14 + (h_p + h_A)/2]^3 \tag{3-15a}$$

or, equivalently,

$$P = (4\pi^2 a^3/\mu)^{1/2}$$

$$= 1.658\,668 \times 10^{-4} a^{3/2} \tag{3-15b}$$

where h_p, h_A, and a are the perigee and apogee heights, and semimajor axis, respectively, in kilometres and P is in minutes. The velocity, V, at any point depends on both the instantaneous altitude and the semimajor axis:

$$V^2 = \mu\left(\frac{2}{R_\oplus + h} - \frac{1}{a}\right)$$

$$= 398,600.5\left[\frac{2}{6378.14 + h} - \frac{1}{a}\right] \tag{3-16}$$

where h is the instantaneous altitude, a is the semimajor axis in kilometres, and V is the velocity in kilometres per second. We may also use the vis viva equation to determine the energy per unit mass which must be supplied to reach a given altitude, starting at rest on the Earth's equator. For a circular orbit at altitude h, Eq. (3-5) may be reformulated to yield:

$$\Delta E = -\frac{\mu}{2(R_\oplus + h)} - \frac{V_\oplus^2}{2} + \frac{\mu}{R_\oplus}$$

$$= -\frac{1.993003 \times 10^{11}}{(6378.14 + h)} + 6.23866 \times 10^7 \tag{3-17}$$

where $V_\oplus$ is the rotational velocity at the Earth's surface, h is in kilometres, and ΔE is the energy increase required in Joules per kilogram. Appendix M tabulates the period, velocity, energy required, and size of the Earth's disk as a function of altitude for Earth-orbiting satellites.

At an altitude of 35,786 km, the period of a satellite equals the sidereal rotation period of the Earth. (The *sidereal period*, or period relative to the stars, is 4 minutes less than the mean period of rotation with respect to the Sun (24 hours) because in 1 day the Sun has moved about 1 deg farther along the ecliptic and the Earth must rotate slightly more than 360 deg to follow the Sun.) Spacecraft at this mean altitude are called *synchronous satellites* because a 0-deg inclination satellite at this altitude will remain over the same location on the Earth's equator. A synchronous satellite in a circular orbit at nonzero inclination travels in a figure "8" relative to the surface of the Earth.

Thus far, we have been concerned with two-body Keplerian orbits. However, there is a simple class of three-body orbits, known as *Lagrange point orbits*,* which is of particular interest to spaceflight. As shown in Fig. 3-13, the *Lagrange points*, or *libration points*, for two celestial bodies in mutual revolution, such as the Earth and the Moon, are the five points such that an object placed at one of them will remain there. The three Lagrange points on the Earth-Moon line are positions of unstable equilibrium; i.e., any small change causes the object to drift away. However, L_4 and L_5, which form equilateral triangles with the Earth and the Moon in the plane of the Moon's orbit, are positions of stable equilibrium. A satellite placed near the Lagrange point (with an appropriate velocity) remains in essentially the same position relative to the Earth and the Moon. Because of the stability of Lagrange point orbits about L_4 and L_5, they have been proposed as one possible location for permanent colonies in space [O'Neill, 1975]. As a natural example of this phenomenon, the *Trojan asteroids* are a group of 14 known asteroids which have collected at the stable Lagrange points of the Jupiter-Sun system.

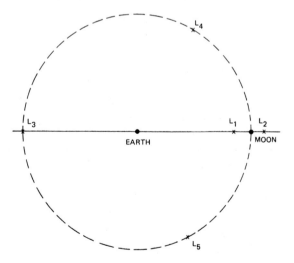

Fig. 3-13. Lagrange Points of the Earth-Moon System

---

*Named after the 18th-Century French mathematician and astronomer Joseph Lagrange.

Orbit Maneuvers.   A significant portion of spaceflight analysis is concerned with orbit maneuvers and optimal methods of changing orbits. Normally we wish to minimize the energy which must be supplied or, equivalently, the velocity change, ΔV, required to go from one orbit to another. There are two basic types of maneuvers: *in-plane maneuvers*, which do not affect the plane of the spacecraft orbit and *out-of-plane* or *plane change maneuvers*, which change the orbital plane. Because the spacecraft velocity vector is in the orbit plane, any component of the rocket thrust normal to the plane of the orbit will have only a small effect on the magnitude of the velocity and, therefore, on the total energy and semimajor axis. If we wish to obtain a specified semimajor axis with a minimum expenditure of energy, then plane changes should be minimized.

The principal orbit characteristic relevant for in-plane maneuvers is that the semimajor axis depends only on the total energy. Therefore, the most efficient way to change the semimajor axis and raise or lower the spacecraft is to change only the magnitude of the velocity by firing the rocket either parallel or antiparallel to the velocity vector. If we wish to transfer between two circular orbits (for example, to travel from the Earth to Mars or from low Earth to synchronous orbit), then we start at the lower orbit and fire the rocket so that the propellent is expelled in the direction opposite the velocity vector, as illustrated in Fig. 3-14. For a minimum energy expenditure, the semimajor axis of the transfer orbit should be such that its apofocus is at the radius of the larger orbit. Such an elliptical orbit with perifocus at the smaller orbit and apofocus at the larger orbit is called a *Hohmann transfer ellipse* and is the minimum energy path between the orbits, either from the smaller to the larger or the larger to the smaller. The Hohmann transfer ellipse is tangent to both the inner and outer orbits.

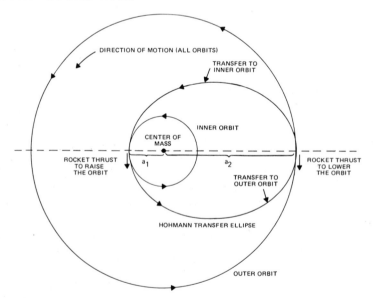

Fig. 3-14.   Hohmann Transfer Ellipse. Note that direction of thrust is opposite the direction propellant is expelled.

Figure 3-14 shows that minimum energy maneuvers between orbits start 180 deg away from the desired end location relative to the central body. If subscripts 1, 2, and H denote the two circular orbits and the Hohmann transfer ellipse, respectively, then, from Fig. 3-14, the semimajor axes are related by:

$$a_H = \frac{1}{2}(a_1 + a_2) \qquad (3\text{-}18)$$

from which the period, P_H, and the transfer time, $T = 0.5P_H$, may be calculated from Kepler's Third Law. Alternatively, if $P_1 \approx P_2$, then we may use the approximation:

$$T^2 = \left(\frac{1}{2}P_H\right)^2 \approx \frac{1}{4}\left[\left(\frac{P_1 + P_2}{2}\right)^2 - \frac{1}{12}(P_1 - P_2)^2\right]$$

$$= \frac{1}{24}\left[(P_1 + P_2)^2 + 2P_1P_2\right] \qquad (3\text{-}19)$$

From the first form above, it is clear that the transfer time is somewhat less than half the mean period of the two orbits. The Hohmann transfer time from low-Earth orbit to synchronous altitude is 5.23 hours and to the orbit of the Moon it is 4.97 days; for these large differences in period, Eq. (3-19) overestimates the time by 4% and 11%, respectively.

Also of interest is the velocity change, ΔV, required to change from a circular orbit to a Hohmann transfer ellipse, or vice versa. This may be obtained directly from the vis viva equation as:

$$|\Delta V| = \sqrt{\mu}\left|\sqrt{\frac{2}{a_i} - \frac{1}{a_H}} - \sqrt{\frac{1}{a_i}}\right| \qquad (3\text{-}20)$$

where a_i may be either a_1 or a_2. (In the case of a Hohmann transfer leaving or approaching a planet or other massive object, $|\Delta V|$ is the hyperbolic velocity of the spacecraft relative to the planet.) Values of the transfer time and velocity change required for trips between the orbits of the planets are given in Table L-10.

As an example of a Hohmann transfer, we consider a flight from the Earth to Mars for which the Hohmann transfer time is 259 days. (The approximation of Eq. (3-19) is in error by less than 0.1%.) As shown in Fig. 3-15, the spacecraft will move 180 deg in true anomaly in 259 days to meet Mars at point C'. The Earth will have moved (259/365) of an orbit or 255 deg in true anomaly. Thus, the Earth will be $255 - 180 = 75$ deg ahead of Mars when the spacecraft arrives. From Table L-1, we find that the mean daily motion in true anomaly of the Earth is 0.462 deg/day faster than that of Mars. Thus, opposition occurred 162 days ($= 75$ deg/(0.462 deg/day)) before arrival when both planets and the Sun were on a straight line. This fixes the mission timing relative to the synodic period of the Earth and Mars. We conclude that flights to Mars should leave Earth $259 - 162 = 97$ days before an opposition of Mars and arrive at their destination 162 days after opposition. Actual flight times will differ from this estimate by several weeks due to the noncoplanar, noncircular orbits of the Earth and Mars.

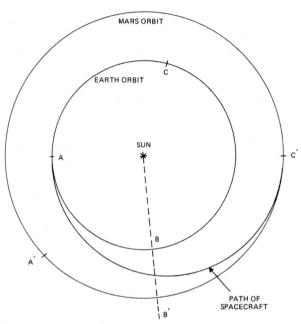

Fig. 3-15.   Hohmann Orbit to Mars. Letters indicate the positions of the Earth and Mars at launch (A, A'), at opposition 97 days after launch (B, B'), and at arrival 259 days after launch (C, C').

To calculate the energy and velocity changes required for a Mars trip, we must further specify the initial conditions because it will be necessary to account for the gravitational attraction of the planets themselves. We assume that we wish to leave from the surface of the Earth and enter a circular orbit just above the surface of Mars. From Tables L-11 and L-10 (or Eqs. (3-6) and (3-20)), the escape velocity of the Earth is 11.18 km/sec and the hyperbolic velocity necessary to achieve a Hohmann transfer orbit to Mars is 2.9 km/sec. Therefore, a Mars probe requires an initial energy change of $0.5 \times 11.18^2 + 0.5 \times 2.9^2 = 66.70$ MJ/kg. (This may be supplied, for example, by either one velocity change of 11.55 km/sec ($= (2 \times 66.7)^{1/2}$) or, less efficiently, by an initial velocity change of 11.18 km/sec and a second change of 2.9 km/sec after the spacecraft has left the vicinity of the Earth.) From Table L-10 or Eq. (3-20), we find that the spacecraft approaches Mars with a hyperbolic velocity of 2.6 km/sec or a total energy per unit mass of +3.38 MJ/kg relative to Mars. From energy conservation, the velocity as the spacecraft passes near the surface will be $(2E + 2\mu/R)^{1/2} = (2E + 2V_c^2)^{1/2}$, where E is the total energy per unit mass, V_c is the circular velocity of Mars, or 3.55 km/sec from Table L-11, and μ and R are the gravitational constant and radius of Mars. Thus, the velocity near the surface of Mars will be $(2 \times 3.38 + 2 \times 3.55^2)^{1/2} = 5.65$ km/sec and a velocity change of $5.65 - 3.55 = 2.10$ km/sec will put the spacecraft into a circular orbit just above the surface of Mars. Without this velocity change, the spacecraft would follow a hyperbolic trajectory away from Mars and return to an elliptical orbit about the Sun. Similarly, a velocity change of 2.10 km/sec is required to return the spacecraft to a Hohmann trajectory back to Earth.

An analysis similar to that described above applies to elliptical orbits as well as circular orbits. Thus, raising or lowering the apogee height is done most efficiently by firing the propellent antiparallel or parallel to the direction of motion when the spacecraft is at perigee. Similarly, perigee height is adjusted by maneuvers at apogee.

In general, plane change maneuvers are more complex than transfer maneuvers within the orbital plane. However, some general characteristics of plane change maneuvers may be determined from the orbital properties described in Section 3.1. Any change in the orbital plane requires a velocity change perpendicular to the initial orbital plane. Because the orbital plane is defined by the velocity vector and the radius vector from the primary to the spacecraft, the general effect of any instantaneous plane change maneuver is to rotate the orbital plane about the radius vector from the primary to the spacecraft. The amount of this rotation is defined in Fig. 3-16. From the figure, we see that a given velocity change provides the maximum orbital rotation when the component, $V_{i\perp}$, of the initial spacecraft velocity perpendicular to the radius vector is a minimum. By Kepler's Second Law (Conservation of Angular Momentum), this occurs at apogee. Of course, the initial and final orbital planes must contain the radius vector to the spacecraft at the time of the maneuver. Therefore, a single impulse maneuver (i.e., one which is nearly instantaneous) must occur whenever the spacecraft lies at the intersection of the initial and final orbital planes.

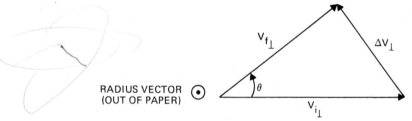

RADIUS VECTOR
(OUT OF PAPER) ⊙

$V_{f\perp}$ $\Delta V_{\perp}$

θ

$V_{i\perp}$

Fig. 3-16. Rotation, θ, of the Orbit Due to Velocity Change ΔV. θ is measured in the plane of the paper which is perpendicular to the radius vector, V_i=initial velocity, V_f=final velocity. $\perp$ indicates components perpendicular to the radius vector.

The most straightforward plane change is a change in the inclination of an orbit. A plane change maneuver performed when the spacecraft is at ascending or descending node will rotate the orbit about the line of nodes, thus changing the inclination without changing the right ascension of the ascending node (except for possibly interchanging the ascending and descending nodes). It is possible for a maneuver at this location to change only the inclination and not affect any of the other orbital elements. It can be shown that inclination changes are most efficiently done when the spacecraft is on the line of nodes. (See, for example, Ruppe [1966].) Maneuvers that change the ascending node but not the inclination or other parameters must be made at one of the two locations which the initial and final orbits have in common.

Most orbit maneuvers require the expenditure of rocket fuel or other spacecraft consumables to make significant modifications in the spacecraft orbit.

However, in *gravity assist trajectories* or *flyby trajectories*, the orbital velocity of a planet is used to change the spacecraft orbit. On a hyperbolic trajectory, a spacecraft will approach and leave the planet at the same speed *relative to the planet* but in a different direction. Thus, the interaction with the planet is an elastic collision analogous to a baseball and bat and can be used to provide additional energy to a spacecraft. If a spacecraft approaches a planet at 5 km/sec relative to the planet and leaves at 5 km/sec at an angle of 90 deg relative to incoming velocity vector, it will undergo a velocity change of $5\sqrt{2} \approx 7.07$ km/sec.

The magnitude of the velocity change, ΔV, that will be produced in a flyby trajectory is a function of the *turn angle*, ψ, between the two asymptotes of the hyperbola (see Fig. 3-3): that is,

$$|\Delta V| = 2 V_h \sin(\psi/2) \qquad (3\text{-}21)$$

where V_h is the hyperbolic velocity. By manipulating the equations for a hyperbola (see Table 3-2), we find

$$\sin(\psi/2) = 1/e \qquad (3\text{-}22)$$

where e is the eccentricity. This may also be expressed in a form more convenient for analysis as:

$$\sin(\psi/2) = \left(1 + \frac{V_h^2 q}{\mu}\right)^{-1} \qquad (3\text{-}23)$$

where q is the perifocal distance or distance of closest approach to the planet and $\mu \equiv GM$. Table 3-4 lists the values of the turn angle and velocity change assuming that the hyperbolic velocity is the velocity of approach for a spacecraft in a Hohmann transfer orbit as given in Table L-2. A Venus gravity assist trajectory was used by Mariner 10 to explore Mercury. However, Jupiter is the most efficient planet in providing large velocity changes. Jupiter flybys can be used in several ways—to give a spacecraft sufficient added energy to become an interstellar probe (Pioneer 10), to reduce the transit time to the onter planets (Pioneer 11 and the 1977 Jupiter-Saturn mission), to obtain a large velocity component perpendicular to the plane of the ecliptic in order to explore the space "above" or "below" the solar system (also Pioneer 11), or to reduce the spacecraft velocity perpendicular to the direction of the Sun (and, therefore, the spacecraft's angular momentum) in order to approach the Sun.

Injection Conditions. The Keplerian orbit of a spacecraft is determined entirely by its position and velocity at any one time. This is of relatively little interest for practical orbit determination because position and velocity are not normally observed directly. However, determining the Keplerian elements from position and velocity is convenient for a variety of analytic studies involving the effect of injection conditions or orbit maneuvers or the determination of approximate elements from data in ephemerides. Thus, we will assume that we are given the radius vector, **R**, from the center of the Earth to the spacecraft and the velocity vector, **V**, at some time, t. We wish to determine the Keplerian elements, a, e, i, ω, Ω, and M. These computations are performed by subroutine ELEM described in Section 20.3.

Table 3-4.   Representative Turn Angle, ψ, and Velocity Change, ΔV, for Flyby Missions to the Planets. The assumed hyperbolic velocity, V_h, is velocity of approach for a Hohmann transfer orbit (from Table L-2). R_P is the radius of the planet and q is the perifocal distance or distance of closest approach.

PLANET	ASSUMED V_h (km/sec)	$q = R_P$ (GRAZING)		$q = 2 R_P$		$q = 10 R_P$	
		ψ (deg)	$\|\Delta V\|$ (km/sec)	ψ (deg)	$\|\Delta V\|$ (km/sec)	ψ (deg)	$\|\Delta V\|$ (km/sec)
MERCURY	9.6	10.2	1.7	5.4	0.9	1.1	0.2
VENUS	2.7	123	4.8	104	4.2	50	2.3
MARS	2.6	81	3.4	58	2.5	18	0.8
JUPITER	5.6	159	11.0	150	10.8	116	9.5
SATURN	5.4	146	10.3	133	9.9	86	7.4
URANUS	4.7	132	8.6	114	7.9	61	4.8
NEPTUNE	4.1	141	7.7	127	7.3	78	5.1
PLUTO	3.7	65	4.0	48	3.0	12	0.8

The orbit plane is that defined by **R** and **V**. Thus, the orbit normal vector, $\hat{\mathbf{N}}$, is

$$\hat{\mathbf{N}} = (\mathbf{R} \times \mathbf{V})/|\mathbf{R} \times \mathbf{V}| \qquad (3\text{-}24)$$

Let the inertial coordinate axes be defined by $\hat{\mathbf{x}}$ in the direction of the vernal equinox, $\hat{\mathbf{z}}$ in the direction normal to the Earth's equator, and $\hat{\mathbf{y}}$ such that the coordinate system is orthonormal and right-handed. Then, the inclination is given by

$$i = \arccos(\hat{\mathbf{N}} \cdot \hat{\mathbf{z}}) \qquad (3\text{-}25)$$

and the right ascension of the ascending node by

$$\Omega = \arctan(-N_x / N_y) \qquad (3\text{-}26)$$

The vector in the direction of the ascending node is $\hat{\mathbf{O}} = \hat{\mathbf{z}} \times \hat{\mathbf{N}}$. The semi-major axis is determined by the vis viva equation as:

$$a = \left(\frac{2}{R} - \frac{V^2}{\mu} \right)^{-1} \qquad (3\text{-}27)$$

where $\mu \equiv GM$ and a negative value of a corresponds to a hyperbolic trajectory. The period for elliptical orbits is then obtained from Kepler's Third Law as:

$$P = \sqrt{4\pi^2 a^3 / \mu} \qquad (3\text{-}28)$$

If we define the *heading* or *flight path angle*, β, by

$$\beta \equiv 90° - \arccos \hat{\mathbf{R}} \cdot \hat{\mathbf{V}} \qquad (3\text{-}29)$$

then the eccentricity, e, and the true anomaly, ν, are given by (see, for example, Thomson [1963]):

$$e^2 = \left(\frac{RV^2}{\mu} - 1\right)^2 \cos^2\beta + \sin^2\beta \tag{3-30}$$

$$\tan\nu = \frac{(RV^2/\mu)\sin\beta\cos\beta}{(RV^2/\mu)\cos^2\beta - 1} \tag{3-31}$$

Note that the heading is the angle by which the velocity vector deviates from being perpendicular to the radius vector. From Eq. (3-31), or the properties of an ellipse, a heading of 0 deg implies that the spacecraft is at either apogee or perigee. The argument of perigee, ω, may be obtained from the true anomaly by

$$\omega = \arccos(\hat{O} \cdot \hat{R}) - \nu \tag{3-32}$$

where the first term on the right is in the range 0 deg to 180 deg if $(\hat{O} \times \hat{R}) \cdot \hat{N} > 0$ and 180 deg to 360 deg if $(\hat{O} \times \hat{R}) \cdot \hat{N} < 0$. Finally, the mean anomaly may be obtained directly from Kepler's and Gauss' equations

$$M = E - e\sin E \tag{3-33}$$

where

$$\tan\left(\frac{E}{2}\right) = \left(\frac{1-e}{1+e}\right)^{1/2} \tan\left(\frac{\nu}{2}\right) \tag{3-34}$$

It is clear from the foregoing principles that injection involving a substantial change in the semimajor axis will normally occur at apogee or perigee ($\beta = 0$ deg) and that changes in the inclination will normally occur at either the ascending or the descending node.

3.4 Orbit Perturbations

Real orbits never follow Kepler's Laws precisely, although at times they may come very close. The Keplerian elements of an orbit provide a convenient analytic approximation to the true orbit; in contrast, a *definitive orbit* is the best estimate that can be obtained with all of the available data on the actual path of a satellite. Because closed-form analytic solutions are almost never available for real orbit problems with multiple forces, definitive orbits are generated numerically based on both orbit theory and observations of the spacecraft. Thus, definitive orbits are only generated for times that have past, although the information from a definitive orbit is frequently extrapolated into the future to produce a predicted orbit.

A *reference orbit* is a relatively simple, precisely defined orbit (usually, though not necessarily, Keplerian) which is used as an initial approximation for the spacecraft motion. *Orbit perturbations* are the deviations of the true orbit from the reference orbit and may be classified according to specific causes, e.g., perturbations due to the Earth's oblateness, atmospheric drag, or the gravitational force of the Moon. In this section, we list the possible causes of orbit perturbations, describe qualitatively the effects of the various perturbations, and, where possible, provide formulae to determine the approximate effect of specific perturbations. Methods for the numerical treatment of perturbations may be found in the references at the end of the chapter.

Effects which modify simple Keplerian orbits may be divided into four classes: nongravitational forces, third-body interactions, nonspherical mass distributions, and relativistic mechanics. The first two effects may dominate the motion of a spacecraft, as in satellite reentry into the atmosphere or motion about one of the stable Lagrange points. Although the effects of nonspherical mass distributions never dominate spacecraft motion, they provide the major perturbation, relative to Keplerian orbits, for most intermediate altitude satellites, i.e., those above where the atmosphere plays an important role and below where effects due to the Moon and the Sun become important. As indicated previously, relativistic mechanics may be completely neglected in most applications. The largest orbit perturbation in the solar system due to general relativity is the rotation of the perihelion of Mercury's orbit in the orbit plane by about 0.012 deg/century or 3×10^{-5} deg/orbit. Although a shift of this amount is measurable, it is well below the magnitude of the other effects which we will consider.

Although the relative importance of the three significant groups of perturbations will depend on the construction of the spacecraft, the details of its orbit, and even the level of solar activity*, the general effect of perturbing forces is clear. Atmospheric effects dominate the perturbing forces at altitudes below about 100 km and produce significant long-term perturbations on satellite orbits up to altitudes of about 1000 km. The major effect resulting from the nonspherical symmetry of the Earth is due to the Earth's oblateness, which changes the gravitational potential by about 0.1% in the vicinity of the Earth. The ratio of the gravitational potential of the Moon to that of the Earth is 0.02% near the Earth's surface. As the satellite's altitude increases, the effect of oblateness decreases and the effect of the Moon increases; the magnitude of their effect on the gravitational potential is the same at about 8000 km altitude. Lunar and solar perturbations are generally negligible at altitudes below about 700 km. (See Section 5.2.)

Nongravitational Forces. For near-Earth spacecraft, the principal nongravitational force is *aerodynamic drag*. Drag is a retarding force due to atmospheric friction and is in the direction opposite the spacecraft velocity vector. (If there is any component of the force perpendicular to the velocity vector, it is called *lift*.) In an elliptical orbit, drag is most important at perigee because the density of the Earth's atmosphere, to which the drag is proportional, decreases exponentially with altitude. (See Section 4.4 for a discussion of atmosphere models.) Because drag forces are tangent to the orbit, opposite to the velocity, and applied near perigee, the qualitative effects of drag are similar to an impulsive in-plane transfer maneuver performed at perigee, as discussed in Section 3.3. As the drag slows the spacecraft at each perigee passage, the apogee height and, consequently, the semimajor axis and eccentricity are reduced. The perigee height and argument of perigee will remain approximately the same. In addition, neither the node nor the inclination will be affected, because the force is within the orbit plane (ignoring the small effect due to the rotation of the atmosphere).

---

*The level of solar activity significantly affects both the atmospheric density at spacecraft altitudes and the structure of the geomagnetic field. See Sections 4.4 and 5.1.

The general process of reducing the total energy and lowering apogee by atmospheric drag is called *orbital decay*. The orbital lifetime of a satellite is the time from launch until it penetrates deeply into the atmosphere such that the spacecraft either burns up or falls to the surface. Figure 3-17 gives the approximate lifetime of a satellite as a function of perigee height, h_p, and eccentricity, e. The ordinate in the figure is the estimated number, N, of orbit revolutions in the satellite lifetime divided by the *ballistic coefficient*, $m/(C_D A)$, where m is the mass of the satellite and A is its cross-sectional area perpendicular to the velocity vector. The drag coefficient, C_D, is a dimensionless number, usually between 1 and 2*. The ballistic coefficient is a measure of the ability of the spacecraft to overcome air resistance. For typical spacecraft, it ranges from 25 to 100 kg/m². Given the value of N from Fig. 3-17, the satellite lifetime L, can be calculated directly from the period, P, or from the perigee height, h_p, and eccentricity, e, by:

$$L = NP = N \left[\frac{4\pi^2}{\mu_\oplus} \left(\frac{h_p + R_\oplus}{1 - e} \right)^3 \right]^{1/2}$$

$$\approx 1.15185 \times 10^{-7} \times N \times \left(\frac{6378.14 + h_p}{1 - e} \right)^{3/2} \qquad (3\text{-}35)$$

where h_p is in kilometres and L is in days. For example, a satellite with a ballistic coefficient of 80 kg/m² in a circular orbit at an altitude of 500 km ($e = 0$, $h_p = 500$) has a value of $N/80$ from Fig. 3-17 of approximately 140; this gives an estimated lifetime of 11,000 revolutions ($= 80 \times 140$) or about 720 days. All formulas or graphs for spacecraft lifetimes are approximations, since the atmospheric density fluctuates considerably. For example, at 800 km, the density can fluctuate by a factor of 3 to 7 due to solar activity [Roy, 1965]. Most lifetime estimates are in error by at least 10%. Simplified relations, such as that of Fig. 3-17, may be in error by 50%.

In addition to atmospheric drag, other nongravitational forces acting on a spacecraft are (1) drag due to induced eddy currents in the spacecraft interacting with the Earth's magnetic field, (2) drag due to the solar wind (Section 5.3) and micrometeoroids (interplanetary dust particles), and (3) solar radiation pressure. The first two effects are very small and are normally ignored. Solar radiation pressure can be important for some satellites, particularly those with large solar panels. For most satellites, the direction of the solar radiation force will be nearly radial away from the Sun[†], although it is theoretically possible to build a solar "sail" which can tack in much the same fashion as a sailboat. The magnitude of the force, $\mathbf{F}_R$, is given by:

$$|\mathbf{F}_R| = KAP \qquad (3\text{-}36)$$

---

* If no measured drag coefficient is available, $C_D = 2$ is a good estimate for satellites whose dimensions are large relative to the mean free path of atmospheric molecules.
† For a spherical object, there will be a slight preferential scattering of light in the direction of the object's motion, which will produce an effective drag. While this *Poynting-Robertson effect* is unimportant for spacecraft, it causes interplanetary meteoroids up to 1 cm in diameter to spiral into the Sun within about 20 million years from an initial distance of 1 AU.

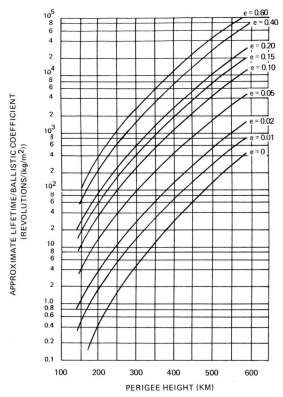

Fig. 3-17. Approximate Lifetimes for Earth Satellites. See text for explanation (adapted from Kendrick [1967]).

where P is the momentum flux from the Sun (4.4×10^{-6} kg·m^{-1}·s^{-2}, as discussed in Section 5.3), A is the cross-sectional area of the spacecraft perpendicular to the sunline, and K is a dimensionless constant in the range $0 \leqslant K \leqslant 2$; $K < 1$ for translucent material, $K = 1$ for perfectly absorbent material (black body), and $K = 2$ for material reflecting all light directly back toward the Sun. Echo I, a 60-kg, 30-m-diameter balloon satellite, was launched into a circular orbit at an altitude of 1600 km; however, solar radiation pressure reduced perigee to 1000 km at times [Glasstone, 1965].

Nonspherical Mass Distribution. For near-Earth satellites above several hundred kilometres, the major source of perturbations is the nonspherical shape of the Earth. This shape approximates an oblate spheroid which would be formed by a rotating fluid. Thus, the major correction for the nonspherical Earth is for the oblateness with much smaller corrections for the deviations from an oblate shape. (See Sections 4.3 and 5.2.)

The oblateness corrections for the Earth are grouped into three catagories. All of the instantaneous, or osculating, orbital elements undergo *short period variations* in which they fluctuate with true anomaly as the spacecraft moves in its orbit. For three of the six elements (a, e, and i), these short period variations average to zero over an orbit; the other three elements undergo cumulative secular variations in which the average value of the parameters changes monotonically. It is the secular variations that are of primary interest in following the gross motion of the satellite.

Because the gravitational force is conservative, the total energy and the mean values of the semimajor axis, the apogee and perigee heights, and, consequently, the eccentricity do not change due to oblateness. To understand the physical cause of the secular variations which occur in other elements, it is convenient to think of the Earth as a point mass and a ring of uniform density in the equatorial plane representing the Earth's equatorial bulge. The easiest perturbation to visualize is the rotation of the orbit plane. Figure 3-18 shows the direction of the orbital angular momentum for a satellite in a prograde orbit. When the satellite is south of the equator, the average torque, $\mathbf{N} = \mathbf{r} \times \mathbf{F}$, due to the pull, $\mathbf{F}$, of the Earth's equatorial bulge produces a westward torque. The average perturbing force north of the equator also produces a westward torque, causing the angular momentum vector and the orbit plane to rotate westward without changing the inclination. This motion of the line of nodes opposite the direction of rotation of the satellite is called *regression of the nodes*.

The other major oblateness perturbation is a motion of the line of apsides. Consider first the case of a satellite with zero inclination. In the equatorial plane, the gravitational force from a ring of mass (i.e., the equatorial bulge) is larger than if all of the mass were concentrated at the center. As shown in Fig. 3-19(a), this added force causes the orbit to curve more strongly, that is, the angular velocity of the satellite about the Earth will be increased. Thus, as shown in Fig. 3-19(b), each successive apogee and perigee will occur farther around than formerly and the line of apsides rotates in the direction of the satellite's motion.

For a satellite over either of the Earth's poles, the distance to the Earth's equator is greater than the distance to the center of the Earth. Therefore, the gravitational force due to the equatorial bulge is less than it would be if the mass of the bulge were at the Earth's center. The smaller force causes the orbit to curve

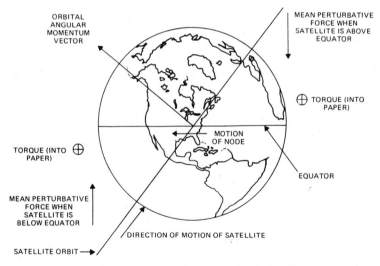

Fig. 3-18.   Regression of Nodes Due to the Earth's Oblateness

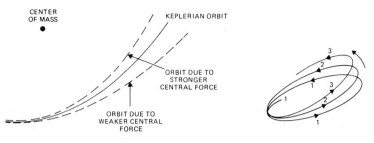

(a) EFFECT OF NONPOINT MASS FORCES   (b) PERIGEE ADVANCE FROM AN EQUATORIAL ORBIT

Fig. 3-19. Motion of the Lines of Apsides Due to the Earth's Oblateness

less, which causes the line of apsides to rotate opposite the direction of motion while the satellite is over the pole. In a polar orbit ($i = 90$ deg), a satellite spends part of the time over the equator where the gravitational force is larger than the mean and part of the time over the poles where it is smaller. It can be shown that for polar orbits the net effect is rotation of the line of apsides opposite the direction of motion, although the rate of rotation is less than for equatorial orbits.

As described in Section 5.2 and Appendix G, the oblateness of the Earth is treated analytically by expanding the gravitational potential in a series of spherical harmonics. The first term in the expansion provides the force resulting from a point mass. The second term, called J_2 *perturbations* (from the second or J_2 coefficient in the expansion), represents the modification to the mean force due to the oblateness of the Earth. Higher order terms represent effects due to the deviation of the shape of the Earth from a simple ellipsoid. Values of the principal coefficients are given in Appendix L; additional coefficients are given by Allen [1973].

The rate of change of the orbital elements arising from the J_2 perturbation is conveniently expressed in terms of the *mean angular motion*, n, equal to the rate of change of mean anomaly (see, for example, Roy [1965]):

$$n \equiv \frac{dM}{dt} = n_0 \left[1 + \frac{3}{2} J_2 \left(\frac{R_\oplus}{a} \right)^2 (1 - e^2)^{-3/2} \left(1 - \frac{3}{2} \sin^2 i \right) \right] \qquad (3\text{-}37)$$

where the nominal mean angular motion, n_0, is given from Table 3-2 by

$$n_0 = \sqrt{\mu / a^3} \qquad (3\text{-}38)$$

The anomalistic period (i.e., perigee to perigee) is

$$P_A = 2\pi / n \qquad (3\text{-}39)$$

and the mean anomaly at time t is

$$M = M_0 + n(t - t_0) \qquad (3\text{-}40)$$

where M_0 is the mean anomaly at the epoch t_0. Note that the mean anomaly is measured relative to the moving perigee. Each of these equations reduces to the nonperturbed form when $J_2 = 0$.

When only the secular variations due to the J_2 term are considered, the right ascension of the ascending node at time t is given by:

$$\Omega = \Omega_0 + \frac{d\Omega}{dt}(t - t_0) \tag{3-41a}$$

$$= \Omega_0 - \frac{3}{2}J_2 n\left(\frac{R_\oplus}{a}\right)^2 (1 - e^2)^{-2}(\cos i)(t - t_0) \tag{3-41b}$$

$$\approx \Omega_0 - \frac{3}{2}J_2\sqrt{\mu}\, R_\oplus^2 a^{-7/2}(1 - e^2)^{-2}(\cos i)(t - t_0) + \mathcal{O}\left(J_2^2\right) \tag{3-41c}$$

$$= \Omega_0 - 2.06474 \times 10^{14} a^{-7/2}(1 - e^2)^{-2}(\cos i)(t - t_0) + \mathcal{O}\left(J_2^2\right) \tag{3-41d}$$

where $(t - t_0)$ is in days, Ω is in degrees, a is the semimajor axis in kilometres, e is the eccentricity, i is the inclination, and Ω_0 is Ω at the epoch time t_0. In the final two forms above, n is approximated by n_0, which may produce an error as large as 0.1% in the $d\Omega/dt$ term.

If the product of the three terms in a, e, and i in Eq. (3-41d) has the value -4.7737×10^{-15} km$^{-7/2}$, the rotation rate of the node will be $= 0.9856$ deg/day or 1 rotation per year. Such an orbit is called *Sun synchronous* because the orientation of the orbit plane will remain nearly fixed relative to the Sun as the Earth moves in its orbit. Thus, the spacecraft will continuously view the surface of the Earth at the same local time at any given latitude. For any satellite in a circular orbit, the local mean time*, T, at which the satellite is over latitude λ is given by (see Appendix J and Eq. (A-6)):

$$T = \frac{1}{15}\left[-\alpha_s + \Omega + \arcsin\left(\frac{\tan\lambda}{\tan i}\right)\right] + 12$$

$$\approx \frac{1}{15}\left[-\frac{360}{365.24}\Delta D + \Omega + \arcsin\left(\frac{\tan\lambda}{\tan i}\right)\right] + 12 \tag{3-42}$$

where α_s is the right ascension of the mean Sun*, Ω is the right ascension of the ascending node of the orbit, i is the orbital inclination, ΔD is the number of days from the vernal equinox to the time Ω is evaluated, the angular quantities in square brackets are in degrees, and T is in hours. For Sun-synchronous orbits, $\alpha_s - \Omega$ is constant. For the orbit to be Sun synchronous, $d\Omega/dt$ must be positive and i must be greater than 90 deg; that is, it must be a retrograde orbit. Such orbits are particularly convenient for surveying the Earth's surface or looking for changes in surface features because the lighting conditions will be nearly the same each time a region is surveyed; however, the lighting conditions change slowly because of the seasonal north-south motion of the Sun. Table 3-5 lists the critical properties of Sun-synchronous orbits.

Finally, the secular variation in the argument of perigee, ω, when the Earth's oblateness is taken into account is given by:

$$\omega = \omega_0 + \frac{d\omega}{dt}(t - t_0) \tag{3-43a}$$

---

*See Appendix J for discussion of time systems and the definition of "mean Sun."

Table 3-5. Properties of Sun Synchronous Orbits (e = eccentricity, i = inclination, h_p = perigee height, h_A = apogee height, $i > 90$ deg indicates retrograde orbit.)

MEAN ALTITUDE (KM)	e = 0	e = 0.1		
	i (DEG)	i (DEG)	h_p (KM)	h_A (KM)
0	95.68			
200	96.33			
400	97.03			
600	97.79			
800	98.60	98.43	82	1518
1000	99.48	99.29	262	1738
2000	104.89	104.59	1162	2838
3000	112.41	111.94	2062	3938
4000	122.93	122.19	2962	5038
5000	138.60	137.32	3862	6138
5974	180.00	168.55	4738	7209

$$= \omega_0 + \frac{3}{2} J_2 n \left(\frac{R_\oplus}{\alpha} \right)^2 (1 - e^2)^{-2} \left(2 - \frac{5}{2} \sin^2 i \right)(t - t_0) \tag{3-43b}$$

$$\approx \omega_0 + \frac{3}{2} J_2 \sqrt{\mu} \, R_\oplus^2 a^{-7/2} (1 - e^2)^{-2} \left(2 - \frac{5}{2} \sin^2 i \right)(t - t_0) + \mathcal{O}\left(J_2^2 \right) \tag{3-43c}$$

$$= \omega_0 + 2.06474 \times 10^{14} a^{-7/2} (1 - e^2)^{-2} \left(2 - \frac{5}{2} \sin^2 i \right)(t - t_0) + \mathcal{O}\left(J_2^2 \right) \tag{3-43d}$$

where the units and variables are the same as those in Eq. (3-41) and ω_0 is the value of ω at epoch t_0. For an equatorial orbit, the line of apsides will rotate in the direction of motion and for a polar orbit it will rotate opposite the direction of motion. At $i = 63.435$ deg, $2 - \frac{5}{2} \sin^2 i = 0$, and the line of apsides does not rotate.

Third-Body Interactions. Perturbations due to the oblateness of the Earth become less important with increasing distance from the Earth. However, as the distance from the Earth increases, perturbations from the gravitational force of the Moon and the Sun become more important. Such third-body interactions are the source of the major orbital perturbations for interplanetary flight and, as is clear from the discussion of Lagrange point orbits in Section 3.3, dominate the motion entirely in some circumstances. Unfortunately, the problem of three interacting gravitational objects is intractable; even series expansions for small perturbations from Keplerian orbits generally have very small radii of convergence. A wide variety of special cases and approximation methods have been studied and are discussed in the references for this chapter. In practice, most work involving significant third-body interactions is done by numerical integration of the equations of motion.

To determine when simple two-body solutions are appropriate, it is convenient to divide space into approximate regions, called *spheres of influence*, in which various orbital solutions are nearly valid. Specifically, consider the case shown in

Fig. 3-20 of a spacecraft of negligible mass moving in the vicinity of two masses, m and M, where $M \gg m$. There exists an approximately spherical Region I of radius R_1 about m such that within this region the perturbing force due to M is less than ϵ times the force due to m, where $\epsilon \ll 1$ is a parameter chosen to reflect the desired accuracy. Within Region I, the motion will be approximately that of a satellite in a Keplerian orbit about m. Similarly, there exists a Region III outside a sphere of radius R_2 centered on m, such that outside this sphere the perturbing force due to m is less than ϵ times the force due to M. Thus, in Region III the motion will be approximately that of a satellite in a Keplerian orbit about M. Within Region II, between R_1 and R_2, the gravitational force from both objects is significant.

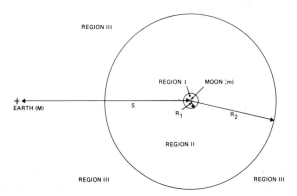

Fig. 3-20.   Spheres of Influence About the Moon for $\epsilon = 0.01$. In Region I, orbits are approximately Keplerian about the Moon; in Region III, they are approximately Keplerian about the Earth.

Simplifications such as the sphere of influence are not precise because the boundaries of the regions are rather arbitrarily defined and are not exactly spherical, and the magnitude of the perturbations within any of the regions is difficult to estimate. Nonetheless, it is a convenient concept for estimating where Keplerian orbits are valid. Approximate formulas for the radii R_1 and R_2 are given by (see, for example, Roy [1965]):

1.   Region I/II Boundary:

$$\epsilon\mu = R_1^2 \left[\frac{1}{(S - R_1)^2} - \frac{1}{S^2} \right]$$

$$R_1 \approx (\epsilon\mu/2)^{1/3} S \qquad \text{for } (\epsilon\mu)^{1/3} \ll 1 \qquad (3\text{-}44)$$

2.   Region II/III Boundary:

$$\epsilon/\mu = (S - R_2)^2 \left(\frac{1}{R_2^2} - \frac{1}{S^2} \right)$$

$$R_2 \approx \left(\frac{-2 + \sqrt{1 + \epsilon/\mu}}{\epsilon/\mu} \right) S \qquad \text{for } \epsilon/\mu \gg 1 \qquad (3\text{-}45)$$

where $\mu = m/M \ll 1$, S is the separation between m and M, and $\epsilon \ll 1$ is the ratio of the perturbing force to the central force. Values of R_1 and R_2 for the various Sun-planet systems for $\epsilon = 0.01$ are given in Table 3-6. For the Earth-Moon system, the radii about the Moon for $\epsilon = 0.01$ are $R_1 = 14{,}900$ km and $R_2 = 189{,}000$ km, as shown in Fig. 3-20.

Table 3-6.   Spheres of Influence for the Planets Relative to the Sun for $\epsilon = 0.01$. (See text for explanation.)

PLANET	R_1 (10^6 KM)	R_2 (10^6 KM)	PLANET	R_1 (10^6 KM)	R_2 (10^6 KM)
MERCURY	0.0544	0.234	SATURN	16	160
VENUS	0.249	1.64	URANUS	17	170
EARTH	0.371	2.52	NEPTUNE	29	280
MARS	0.267	1.28	PLUTO	7	30
JUPITER	13.1	103			

3.5 Viewing and Lighting Conditions

The previous sections have been concerned with orbit kinematics and dynamics and with the general problem of determining the relative position of the spacecraft and other celestial objects. In this section, we assume that these quantities are known and consider the viewing and lighting conditions for planets and natural and artificial satellites. We also discuss the apparent brightness of objects as observed from space.

The geometry of viewing and lighting conditions for either natural or artificial satellites is shown in Fig. 3-21. *Transit* and *occultation* refer to the relative orientations of a planet, a satellite, and an observer. *Transit* occurs when a satellite

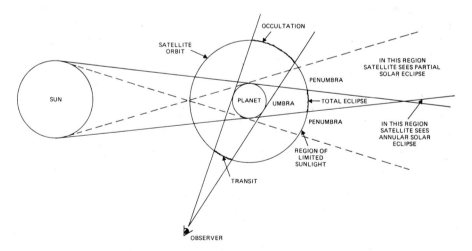

Fig. 3-21. Definition of Viewing Conditions for Satellite. Sun, planet, observer, and orbit are all in the plane of the paper.

passes in front of the disk of a planet as seen by the observer. *Occultation* occurs when the satellite passes behind the disk of the planet and is *occulted* or blocked from the observer's view.

Eclipses are the phenomena of transits and occultations relative to the Sun. An *eclipse of the Sun* is a transit of an object in front of the Sun, blocking all or a significant part of the Sun's radiation from the observer. An eclipse of any other object (e.g., an eclipse of the Moon) is an occultation of that object by another object relative to the Sun. Because the Sun is the largest object in the solar system, the shadows of all the planets and natural satellites are shaped as shown in Fig. 3-21. The *umbra*, or *shadow cone*, is the conical region opposite the direction of the Sun in which the disk of the Sun is completely blocked from view by the disk of the planet. Outside the umbra is the *penumbra*, where a portion of the disk of the Sun is blocked from view and, therefore, where the illumination on objects is reduced.

Unfortunately, the terminology of eclipses depends on whether the observer is thought of as being on the object which is entering the shadow or viewing the event from elsewhere. If the observer enters the shadow, the Sun is partially or wholly blocked from view and the event is referred to as an *eclipse of the Sun* or *solar eclipse* (Fig. 3-22). A *total solar eclipse*, which is frequently shortened to just *eclipse*

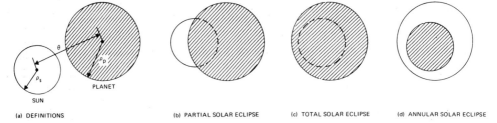

(a) DEFINITIONS   (b) PARTIAL SOLAR ECLIPSE   (c) TOTAL SOLAR ECLIPSE   (d) ANNULAR SOLAR ECLIPSE

Fig. 3-22.   Solar Eclipse Geometry

in spaceflight applications, occurs when the observer enters the umbra. If the observer is farther from the planet than the length of the shadow cone and enters the cone formed by the extension of the shadow cone through its apex, the observer will see an *annular eclipse* in which an annulus or ring of the bright solar disk is visible surrounding the disk of the planet. If the observer is within the penumbra, but outside the umbra, the observer will see a portion of the Sun's disk blocked by the planet and a *partial eclipse of the Sun* occurs.

If the observer is viewing the event from somewhere other than on the satellite being eclipsed, the event is called an *eclipse of the satellite*. A *total eclipse* of the satellite occurs when the entire satellite enters the umbra and a *partial eclipse* occurs when part of an extended satellite (such as the Moon) enters the umbra. If the satellite enters the penumbra only, the event is referred to as a *penumbral eclipse*. Thus, a total eclipse of the Moon to an observer on the Earth is a total solar eclipse to a lunar observer and a penumbral eclipse of a satellite to an observer on the Earth is a partial eclipse of the Sun to an observer on the satellite.

Conditions for Transit and Occultation.   Let **X** be a vector from the observer to the satellite, **P** be a vector from the observer to the center of the planet, and R_p

be the radius of the planet. The angular radius, ρ_p, of the planet as seen by the observer is

$$\rho_p = \arcsin(R_p/P) \qquad (3\text{-}46)$$

The satellite will be in transit, that is, in front of the disk of the planet, whenever:

$$\left. \begin{array}{l} \text{and} \qquad \rho_p > \arccos(\hat{\mathbf{P}} \cdot \hat{\mathbf{X}}) \\[2ex] \qquad\qquad X < P \end{array} \right\} \text{Transit Conditions} \qquad (3\text{-}47)$$

The satellite will be in occultation, that is, behind the disk of the planet, whenever

$$\left. \begin{array}{l} \text{and} \qquad \rho_p > \arccos(\hat{\mathbf{P}} \cdot \hat{\mathbf{X}}) \\[2ex] \qquad\qquad X > P \end{array} \right\} \text{Occultation Conditions} \qquad (3\text{-}48)$$

To apply Eqs. (3-47) and (3-48) to the entire orbit of a satellite, we must determine the satellite position vector and test it against the transit and occultation equations at many places around the orbit. Therefore, it is convenient to determine from general orbital parameters those orbits for which transit and occultation necessarily occur and those for which they cannot occur. We assume that the position of the observer remains fixed. If either transit or occultation occurs, it will be in progress when the angular separation between the spacecraft and the planet is a minimum. (See Fig. 3-23.) Therefore, the minimum angular separation between the planet and the spacecraft determines whether transit or occultation will occur.

In general, it is possible for either transit or occultation to occur in an orbit without the other. However, if the orbit is circular and transit occurs, then occultation must also. This is clear from Fig. 3-23, which shows the general appearance of a circular orbit viewed from nearby and out of the orbit plane. Point A is the closest point on the orbit to the observer and point B is the farthest point from the observer. If point A is in front of the disk, then point B is necessarily behind the disk.

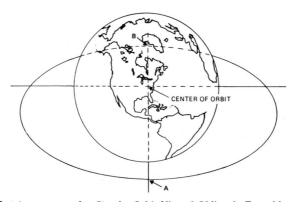

Fig. 3-23.   Appearance of a *Circular* Orbit Viewed Obliquely From Near Point A

For a noncircular orbit, the smallest minimum angular separation between the spacecraft and the planet will occur when perifocus is at B. As shown in Fig. 3-24, let D_p be the perifocal distance and β be the angular separation between the spacecraft and the planet when perifocus is at B. Then:

$$\tan \beta = \left(\frac{D_p}{P + D_p \cos i} \right) \sin i \tag{3-49}$$

where i is the angle between the orbit plane and $\mathbf{P}$, the vector from the observer to the center of the planet*, i may be determined from

$$\cos i = |\hat{\mathbf{P}} \times \hat{\mathbf{N}}| \tag{3-50}$$

where $\hat{\mathbf{N}}$ is the unit vector normal to the orbit plane. Neither transit nor occultation will occur if $\beta > \rho_P$ (**condition 1**).

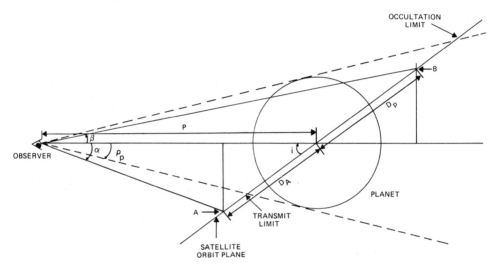

Fig. 3-24.   Geometry for Calculation of Transit and Occultation Limits

The largest minimum angular separation for a given i occurs when apofocus is at B. If β' is the angular separation when apofocus is at B, and D_A is the apofocal distance, then

$$\tan \beta' = \left(\frac{D_A}{P + D_A \cos i} \right) \sin i \tag{3-51}$$

If $\beta' < \rho_P$ (**condition 2**), then occultation of the satellite will occur, but transit may or may not occur. Finally, let α be the angular separation between the center of the planet and the spacecraft at A when apofocus is at A. Then α is given by

$$\tan \alpha = \left(\frac{D_A}{P - D_A \cos i} \right) \sin i \tag{3-52}$$

---

*In astronomical usage, the inclination, i, is normally defined as the complement of the angle defined here.

Both transit and occultation must occur during the orbit if $\alpha < \rho_p$ and $P > D_A \cos i$ (**condition 3**). If none of the above three sets of conditions are fulfilled, then whether transit or occultation occurs depends on both the eccentricity of the orbit and the orientation of the observer relative to the line of apsides.

Eclipse Conditions. To determine the conditions under which eclipses occur, we first determine the length, C, and angular radius, ρ_c, of the shadow cone for any of the planets or natural satellites. (See Fig. 3-25.) Let S be the distance from the planet to the Sun, R_p be the radius of the planet, R_s be the radius of the photosphere (i.e., the visible surface) of the Sun, and C be measured from the center of the planet to the apex of the shadow cone. Then,

$$C = \frac{R_p S}{(R_s - R_p)} \tag{3-53}$$

and

$$\rho_c = \arcsin\left(\frac{R_s - R_p}{S}\right)$$

$$\approx \arcsin\left(\frac{R_s}{S}\right) \tag{3-54}$$

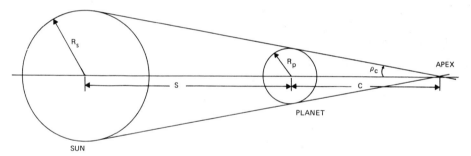

Fig. 3-25. Variables for Eclipse Geometry

For the Earth, the size of the shadow cone for its mean distance from the Sun is $C = 1.385 \times 10^6$ km and $\rho_c = 0.264°$. For the Moon, the mean size is $C = 3.75 \times 10^5$ km and $\rho_c = 0.266°$. The length of the shadow cone for the Moon is just less than the semimajor axis of the Moon's orbit of 3.84×10^5 km. Therefore, eclipses of the Sun seen on the Earth are frequently annular eclipses, and when they are total eclipses they are seen over a very narrow band on the Earth because the maximum radius of the Moon's shadow cone at the distance of the Earth's surface is 135 km.

The presence of an atmosphere on some planets and the non-neglible radius of the natural satellites may be taken into account by adjusting the radius of the planet, as will be discussed later. Initially, we will assume that there are no atmospheric effects and that we are concerned with eclipses seen by objects of neglible size, such as spacecraft. The conditions for the satellite to see a total eclipse of the Sun are exactly those for a transit of the satellite as viewed from the apex of the shadow cone. Similarly, the conditions for spacecraft to see a partial eclipse are nearly the same as those for occultation of the spacecraft viewed from a point in the direction of the Sun equidistant from the planet as the apex of the shadow cone.

To develop specific eclipse conditions, let $\mathbf{D}_s$ be the vector from the spacecraft to the Sun and let $\mathbf{D}_p$ be the vector from the spacecraft to the center of the planet. Three quantities of interest are the angular radius of the Sun, ρ_s, the angular radius of the planet, ρ_p, and the angular separation, θ, between the Sun and planet as viewed by the spacecraft, as shown in Fig. 3-22(a). These are given by:

$$\rho_s = \arcsin(R_s/D_s) \tag{3-55}$$

$$\rho_p = \arcsin(R_p/D_p) \tag{3-56}$$

$$\theta = \arccos(\hat{\mathbf{D}}_s \cdot \hat{\mathbf{D}}_p) \tag{3-57}$$

The necessary and sufficient eclipse conditions are
1. Partial Eclipse:

$$D_s > S \text{ and } \rho_p + \rho_s > \theta > |\rho_p - \rho_s| \tag{3-56}$$

2. Total Eclipse:

$$S < D_s < S + C \text{ and } \rho_p - \rho_s > \theta \tag{3-57}$$

3. Annular Eclipse:

$$S + C < D_s \text{ and } \rho_p - \rho_s > \theta \tag{3-58}$$

These eclipses are illustrated in Fig. 3-22.

The surface brightness of the Sun is nearly uniform over the surface of the disk. Therefore, the intensity, I, of the illumination on the spacecraft during a partial or an annular eclipse is directly proportional to the area of the solar disk which can be seen by the spacecraft. These relations may be obtained directly from Appendix A as:

1. Partial Eclipse

$$I_0 - I = \frac{I_0}{\pi(1 - \cos\rho_s)} \left[\pi - \cos\rho_s \arccos\left(\frac{\cos\rho_p - \cos\rho_s\cos\theta}{\sin\rho_s\sin\theta} \right) \right.$$

$$- \cos\rho_p \arccos\left(\frac{\cos\rho_s - \cos\rho_p\cos\theta}{\sin\rho_p\sin\theta} \right)$$

$$\left. - \arccos\left(\frac{\cos\theta - \cos\rho_s\cos\rho_p}{\sin\rho_s\sin\rho_p} \right) \right] \tag{3-59}$$

2. Annular Eclipse

$$I_0 - I = I_0 \left(\frac{1 - \cos\rho_p}{1 - \cos\rho_s} \right) \tag{3-60}$$

where I_0 is the fully illuminated intensity, and the inverse trigonometric functions in Eq. (3-59) are expressed in radians.

The effect of a planetary atmosphere is difficult to compute analytically because the atmosphere absorbs light, scatters it in all directions, and refracts it into the shadow cone. Close to the surface of the Earth, only a small fraction of the incident light is transmitted entirely through the atmosphere. Thus, the major effects are an increase in the size of the shadow and a general lightening of the entire umbra due to scattering. The scattering becomes very apparent in some eclipses of the Moon, as seen from Earth when the Moon takes on a dull copper color due to refracted and scattered light. The darkness of individual lunar eclipses is noticeably affected by cloud patterns and weather conditions along the boundary of the Earth where light is being scattered into the umbra. The atmosphere of the Earth increases the size of the Earth's shadow by about 2% at the distance of the Moon over the size the shadow would be expected to have from purely geometrical considerations. (See *Supplement to the Astronautical Ephemeris and the American Ephemeris and Nautical Almanac* [1961].) Some ambiguity exists in such measurements because the boundary of the umbra is diffuse rather than sharp. If the entire 2% at the Moon's distance is attributed to an increase in the effective linear radius of the Earth, this increase corresponds to about 90 km.

In considering the general appearance of the solar system as seen by a spacecraft, we may be interested in eclipses of the natural satellites as well as eclipses of spacecraft. In the case of natural satellites, the large diameter of the satellite will have a considerable effect on the occurrence of eclipses. This may be taken into account easily by changing the effective linear diameter of the planet. Let R_p be the radius of the planet, R_m be the radius of the natural satellite, and define the effective planetary radii $R_{e1} \equiv R_p + R_m$ and $R_{e2} \equiv R_p - R_m$. Then, when the center of the satellite is within the shadow formed by an object of radius R_{e1}, at least part of the real satellite is within the real shadow cone. Similarly, when the center of the satellite is within the shadow cone defined by an object of radius R_{e2}, then all of the real satellite is within the real shadow cone; this is referred to as a total eclipse of the satellite when seen from another location. This procedure of using effective radii ignores a correction term comparable to the angular radius of the satellite at the distance of the Sun.

We may use Eqs. (3-49) through (3-52) to determine the conditions on a satellite orbit such that eclipses will always occur or never occur. Let D_p be the perifocal distance, D_A be the apofocal distance, i be the angle between the vector to the Sun and the satellite orbit plane, and C and ρ_c be defined by Eqs. (3-53) and (3-54). We define γ and δ by:

$$\tan \gamma = \left(\frac{D_p}{C - D_p \cos i} \right) \sin i \tag{3-61}$$

$$\tan \delta = \left(\frac{D_A}{C - D_A \cos i} \right) \sin i \tag{3-62}$$

An eclipse will not occur in any orbit for which $\gamma > \rho_c$. An eclipse will always occur in an orbit for which $\delta < \rho_c$.

Planetary and Satellite Magnitudes. The *magnitude*, m, of an object is a logarithmic measure of its brightness or flux density, F, defined by $m \equiv m_0 - 2.5 \log$

F, where m_0 is a scale constant. Two objects of magnitude difference Δm differ in intensity by a factor of $(\sqrt[5]{100})^{\Delta m} \approx 2.51^{\Delta m}$ with smaller numbers corresponding to brighter objects; e.g., a star of magnitude -1 is 100 times brighter than a star of magnitude $+4$. As discussed in detail in Section 5.6, the magnitude of an object depends on the spectral region over which the intensity is measured. In this section, we are concerned only with the visual magnitude, V, which has its peak sensitivity at about 0.55 μm.

Let S be the distance of an object from the Sun in Astronomical Units (AU), r be the distance of the object from the observer in AU, ξ be the *phase angle* at the object between the Sun and the observer, and $P(\xi)$ be the ratio of the brightness of the object at phase ξ to its brightness at zero phase (i.e., fully illuminated). Because the brightness falls off as S^{-2} and r^{-2}, the visual magnitude as a function of ξ and r times S is given by:

$$V(rS, \xi) = V(1,0) + 5\log(rS) - 2.5\log P(\xi)$$

$$= V(1,0) + 5\log r + 5\log S - 2.5\log P(\xi) \tag{3-63}$$

where $V(1,0)$ is the visual magnitude at opposition relative to the observer* (i.e., $\xi = 0$) and at a distance such that $rS = 1$. Note that $P(\xi)$ is independent of distance only as long as the observer is sufficiently far from the object that he is seeing nearly half of the object at any one time; for example, for a low-Earth satellite, the illuminated fraction of the area seen by the satellite depends both on the phase and the satellite altitude.

If the mean visual magnitude, V_0, at opposition to the Earth is the known quantity, then

$$V(1,0) = V_0 - 5\log\left[D(D-1)\right] \tag{3-64}$$

where D is the mean distance of the object from the Sun in AU. Values of V_0 and $V(1,0)$ for the Moon and planets are tabulated in Table L-3.

For the planets, or other objects for which V_0 or $V(1,0)$ is known, the major difficulty is in determining the *phase law*, $P(\xi)$. Unfortunately, there is no theoretical model which is thought to predict $P(\xi)$ accurately for the various phases of the planets. Thus, the best phase law information is empirically determined and, for the superior planets, only a limited range of phases around $\xi = 0$ are observed from the Earth. Although no method is completely satisfactory, the three most convenient methods for predicting the phase law for an object are: (1) assume that the intensity is proportional to the observed illuminated area, that is, $P(\xi) = 0.5(1 + \cos\xi)$; (2) for objects similar in structure to the Moon, assume that the Moon's phase law, which is tabulated numerically in Table L-9, holds; or (3) for the planets, assume that the phase dependence of the magnitude for small ξ is of the form $V = V_0 + a_1\xi$, where the empirical coefficients a_1 are given in Table L-3. For Saturn, the magnitude depends strongly on the orientation of the observer relative

---

*Equation (3-63) holds only for objects which shine by reflected sunlight. Additional terms are needed if lighting is generated internally or by planetary reflections. See Section 5.6 for a discussion of stellar magnitudes.

to the ring system. Because the ring system is inclined to the ecliptic, the orientation of the rings relative to the Earth changes cyclically with a period equal to the period of revolution of Saturn, or about 30 years (Allen [1973]). Additional information on planetary photometry and eclipses is given by Kuiper and Middlehurst [1961] and Link [1969].

For objects for which no a priori magnitude is known, but which shine by reflected sunlight, we may estimate $V(1,0)$ from the relation:

$$V(1,0) = V_\odot - 5\log R_p - 2.5\log g$$
$$= -26.74 - 5\log R_p - 2.5\log g \tag{3-65}$$

where $V_\odot$ is the visual magnitude of the Sun at 1 AU, R_p is the radius of the object in AU, and g is the *geometric albedo* or the ratio of the brightness of the object to that of a perfectly diffusing disk of the same apparent size at $\xi = 0$. For the planets, g ranges from 0.10 for Mercury to 0.57 for Uranus; it is about 0.37 for the Earth, although it is a function of both weather and season. Table L-3 lists the *Bond albedo*, A, of the planets, which is the ratio of total light reflected from an object to the total light incident on it. The Bond and geometric albedos are related by

$$A = gq \tag{3-66}$$

where

$$q \equiv 2\int_0^\pi P(\xi)\sin\xi\,d\xi \tag{3-67}$$

where $P(\xi)$ is the phase law. The quantity q represents the reflection of the object at different phase angles and has the following values for simple objects: $q = 1.00$ for a perfectly diffusing disk; $q = 1.50$ for a perfectly diffusing (Lambert) sphere; $q = 2.00$ for an object for which the magnitude is proportional to the illuminated area; and $q = 4.00$ for a metallic reflecting sphere. For the planets, q ranges from 0.58 for Mercury to about 1.6 for Jupiter, Saturn, Uranus, and Neptune.

As an example of the computation of magnitudes, we calculate the visual magnitude as seen from Earth of the S-IVB (the third stage of the Saturn V rocket) during the first manned flight to the Moon, Apollo 8. The S-IVB which orbited the Moon with the Command and Service modules and several miscellaneous panels, was a white-painted cylinder approximately 7 m in diamter and 18 m long. We assume that the overall Bond albedo was 0.8 because it was nearly all white paint, that $q = 1.5$ corresponding to a diffuse sphere, and that $R_p = 6$ m $= 4\times10^{-11}$ AU, corresponding to the radius of a sphere of the same cross section as the S-IVB viewed from the side. Therefore, the geometric albedo is $0.8/1.5 \approx 0.5$. From Eq. (3-65), we calculate $V(1,0)$ as $V(1,0) = -26.7 + 52.0 + 0.7 = +26.0$.

During the time of the Apollo 8 flight, the angle at the Earth between the spacecraft and the Sun was about 60 deg; therefore, $\xi \approx 120$ deg. If we assume that the intensity is proportional to the illuminated area, then $P = 0.5(1 + \cos 120°) = 0.25$. Setting $S = 1$ AU and $r = 100,000$ km $= 6.7\times10^{-4}$ AU for observations made en route, we find from Eq. (3-63) that the visual magnitude will be approximately $V = +26.0 - 15.9 + 1.5 = \approx +12$. Thus, the S-IVB should be about magnitude $+12$ at 100,000 km, dropping to magnitude $+14.5$ at the distance of the Moon.

The observed magnitudes are in general agreement with this [Liemohn, 1969], although in practice the actual brightness fluctuates by several magnitudes because of the changing cross section seen by the observers, bright specular reflections from windows or other shiny surfaces, and light scattered by exhaust gases during orbit maneuvers.

The visibility of both natural and artificial satellites is a function of both the magnitude of the object itself and its contrast with its surroundings. As illustrated in Fig. 3-26, spacecraft which are orbiting planets are most easily seen when the subsatellite region is in darkness but the spacecraft itself is still in sunlight. Thus, Earth satellites are best seen just after sunset or just before sunrise. Spacecraft orbiting the Moon have the greatest opportunity of being seen when they are not over the disk of the Moon or when they are near the terminator (the boundary between the illuminated and unilluminated portions) above the dark surface of the Moon as seen by the observer.

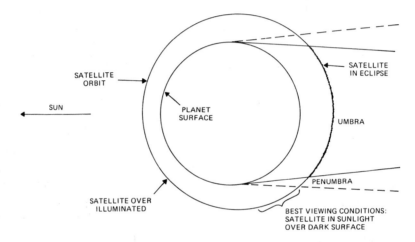

Fig. 3-26.   Best viewing conditions for satellite (either from planet's surface or space) occur when the satellite is in sunlight over unilluminated surface.

References

1. Allen, C. W., *Astrophysical Quantities*, 3rd edition. London: The Athlone Press, 1973.
2. Baker, Robert M. L., Jr., *Astrodynamics, Applications and Advanced Topics*. New York: Academic Press, 1967.
3. ———— and Maud W. Makemson, *An Introduction to Astrodynamics*. New York: Academic Press, 1967.
4. Battin, Richard H., *Astronautical Guidance*. New York: McGraw-Hill, Inc., 1964.
5. Dugas, René, *A History of Mechanics*, translated into English by J. R. Maddox. Neuchatel, Switzerland: Éditions du Griffon, 1955.
6. Escobal, Pedro Ramon, *Methods of Orbit Determination*. New York: John Wiley and Sons, Inc., 1976.

7. Girvin, Harvey F., *A Historical Appraisal of Mechanics*. Scranton, PA.: International Textbook Co., 1948.

8. Glasstone, Samuel, *Sourcebook on the Space Sciences*. Princeton, N.J.: D. Van Nostrand Company, Inc., 1965.

9. Herrick, Samuel, *Astrodynamics* (2 volumes). London: Van Nostrand Reinhold Company, 1971.

10. H. M. Nautical Almanac Office, *Explanatory Supplement to the Astronomical Ephemeris and the American Ephemeris and Nautical Almanac*. London: Her Majesty's Stationery Office, 1961.

11. Kaplan, Marshall H., *Modern Spacecraft Dynamics and Control*. New York: John Wiley and Sons, Inc., 1976.

12. Kendrick, J. B., editor, *TRW Space Data*, TRW Systems Group, Redondo Beach, CA, 1967.

13. Kuiper, G. P. and B. M. Middlehurst, editors, *Planets and Satellites, Vol. III of The Solar System*. Chicago: University of Chicago Press, 1961.

14. Liemohn, Harold B., "Optical Observations of Apollo 8," *Sky and Telescope*, Vol. 37, p. 156–160, 1969.

15. Link, F., *Eclipse Phenomena*. New York: Springer-Verlag, 1969.

16. McCuskey, S. W., *Introduction to Celestial Mechanics*. Reading, MA: Addison-Wesley Publishing Company, Inc., 1963.

17. O'Neill, Gerard K., testimony in *Hearings Before the Subcommittee on Space Science and Applications of the Committee on Science and Technology*, U.S. House of Representatives, U.S.G.P.O., Washington, D.C., 1975.

18. Roy, Archie E., *The Foundations of Astrodynamics*. New York: The Macmillan Company, 1965.

19. Ruppe, Harry O., *Introduction to Astronautics* (2 volumes). New York: Academic Press, 1966.

20. Thomson, William Tyrrell, *Introduction to Space Dynamics*. New York: John Wiley and Sons, Inc., 1963.

21. Watson, G. N., *A Treatise on the Theory of Bessel Functions*. Cambridge: Cambridge University Press, 1958.

CHAPTER 4

MODELING THE EARTH

4.1   Appearance of the Earth at Visual Wavelengths
4.2   Appearance of the Earth at Infrared Wavelengths
4.3   Earth Oblateness Modeling
4.4   Modeling the Structure of the Upper Atmosphere
      *Summary of the Upper Atmosphere Structure, Models of
      the Upper Atmosphere*

The two most commonly used attitude reference sources are the Earth and the Sun. For the purpose of attitude determination, the Sun is normally taken as a point source of light or as a uniformly illuminated disk. In contrast, the Earth as seen from nearby space has a relatively complex appearance, at least some aspects of which must be modeled for accurate attitude determination.

The surface of the Earth is in thermodynamic equilibrium with its surroundings in that the total energy received from the Sun approximately equals the total energy which the Earth radiates into space.* If this were not the case, the Earth would either heat up or cool down until the radiated energy balanced the energy input. Table 4-1 shows the global average radiation for the Earth from meteorological satellite measurements.

Table 4-1.   Radiation Balance of the Earth-Atmosphere System (Data from Lyle, *et al.*, [1971].)

RADIATION	GLOBAL AVERAGE				
	DEC.–FEB.	MAR.–MAY	JUN.–AUG.	SEP.–NOV.	ANNUAL AVERAGE
INCIDENT SOLAR RADIATION (W/M^2)	356	349	342	349	349
ABSORBED SOLAR RADIATION (W/M^2)	244	244	258	251	249
REFLECTED SOLAR RADIATION (W/M^2)	112	105	84	98	100
PLANETARY ALBEDO	0.31	0.30	0.25	0.28	0.29
EMITTED INFRARED RADIATION (W/M^2)	223	230	230	237	230

---

*They are not exactly equal because some energy goes into chemical bonds and some additional energy is supplied by radioactivity and by thermal cooling of the Earth's interior. For the Earth, the heat flow from the interior is approximately 0.004% of the energy received from the Sun.

The *albedo* of an object is the fraction of the incident energy that is reflected back into space. (The word is also used for the reflected radiation itself.) The Earth's albedo is approximately 0.30, although it fluctuates considerably because clouds and ice reflect more light than the land or water surface. The spectral characteristics of the reflected radiation are approximately the same as the incident radiation. (See the solar radiation discussion in Section 5.3.) Thus, the Earth's albedo is most intense in the visual region of the spectrum, i.e., the region to which the human eye is sensitive, from about 0.4 to 0.7 μm wavelength. The appearance of the Earth in the visible spectrum is described in Section 4.1.

Sensors operating in the visible region are called *albedo sensors*, or *visible light sensors*. The principal advantage of this spectral region is that the intensity is greatest here. For attitude sensing, however, a significant disadvantage is the strong variation in albedo—from 0.05 for some soil- and vegetation-covered surfaces to over 0.80 for some types of snow and ice or clouds [Lyle, *et al.*, 1971].

The incident energy which is not reflected from the Earth is transformed into heat and reradiated back into space with a black body spectrum characteristic of the temperature. The Earth's mean surface temperature of approximately 290°K corresponds to a peak intensity of emitted radiation of about 10 μm in the infrared region of the spectrum. Section 4.2 describes the appearance of the Earth in this spectral region. The main advantage of using this *emitted*, or *thermal*, radiation for attitude determination is that the intensity is much more uniformly distributed over the disk of the Earth.

In both Sections 4.1 and 4.2 the Earth is assumed to be spherical. Section 4.3 then describes the oblateness of the Earth and oblateness modeling techniques. Finally, Section 4.4 describes the structure of the upper atmosphere, which is the major source of environmental torque for low-altitude spacecraft.

4.1 Appearance of the Earth at Visual Wavelengths

James R. Wertz

Figure 4-1, taken by the SMS-1 spacecraft in July 1974, shows the appearance of the Earth in the visual region of the spectrum. The location directly beneath the spacecraft is called the *subsatellite point** on the Earth or the *nadir* direction as viewed from the satellite. The subsatellite point for Fig. 4-1 was on the equator at about 50 deg West longitude, on the coast of northern Brazil. Thus, the equator runs through the center of the picture approximately parallel to the lower edge. The outline of the eastern shore of South America is at the center and northern Africa and Spain are visible in the upper right. Cuba and Florida are on either side of a narrow cloud bank in the upper left.

---

*The term *subsatellite point* may be applied to two distinct points on the Earth's surface: (1) the point for which the satellite and the center of the Earth are in opposite directions, or (2) the point from which a line to the satellite is perpendicular to the oblate surface of the Earth. We will use the first definition unless otherwise stated.

Fig. 4-1.   Earth in the Visible Region of the Spectrum. Photograph taken by SMS-1 at 17:40 UT, July
14, 1974, from synchronous altitude. (Courtesy SMS Project Office, NASA. See text for
description.)

At the time that the photograph was taken, the Sun was directly overhead at
the *subsolar point* at approximately 85° West longitude, 20° North latitude, near the
western tip of Cuba. If the satellite is far from the Earth relative to the Earth's size,
and if we think of the Earth as a mirror rather than as a diffuse object, then the
reflection of the Sun would appear at a point midway between the subsatellite
point and the subsolar point. This specular reflection is responsible for the
indistinct bright region (about 3 cm in diameter) along the northern coast of South
America, where the boundaries between land, water, and clouds are difficult to
distinguish. During the course of 24 hours, the subsolar point remains at nearly the
same latitude but rotates through 360 deg in longitude. Thus, the north polar
regions are continuously illuminated and the south polar regions are continuously
dark. The situation is reversed when the Sun crosses the equator in late September.

The *terminator* is the boundary between day and night on a planet or a
planetary satellite and is approximately a great circle 90 deg from the subsolar

point. This is the fuzzy right-hand edge in Fig. 4-1. Because the principal requirement for attitude sensing is to trigger on a well-defined boundary, the very poor definition of the terminator is the best reason for not using visual sensors for attitude determination. However, terminator modeling is required for several purposes, as discussed below.

In contrast to the terminator, the *lit horizon*, i.e., the illuminated edge of the Earth as seen by the spacecraft, provides a sharp boundary which is often used as an attitude reference. In most attitude systems, this boundary is modeled as a step in the intensity at the surface of the Earth. Although atmospheric effects will produce some uncertainty in this boundary, these effects have normally been obscured by other measurement errors and limited sensor resolution. (See, for example, Werking, *et al.*, [1974].)

To determine the brightness of the Earth, a convenient approximation is to ignore variations in the albedo and to think of the Earth as a uniform, diffuse, reflecting, Lambert sphere (i.e., like a white basketball). In this case, the intensity at any point on the surface as viewed by the spacecraft is a function only of the zenith angle of the Sun at that point on the surface of the Earth. The *zenith angle*, θ, is the angle between the sunline and the *zenith*, or the point directly overhead (opposite the direction of the center of the Earth). Thus, the brightness density, d, or the reflected intensity per unit solid angle is

$$d = d_0 \cos\theta \qquad (4\text{-}1)$$

where d_0 is the brightness density at the subsolar point. In this model, the reflected intensity is a maximum at the subsolar point and drops off toward the terminator. Note that d does *not* depend on either the angle at which the surface is viewed or the distance between the viewer and the object. As we approach any planet from space, the intensity per steradian remains constant and the integrated intensity over the planet becomes greater only because the planet subtends a larger solid angle. (This effect is familiar to photographers on Earth. Camera light settings depend on the intensity and position of the light source, not on how close the camera is.)

If we assume that the Earth is spherical, then calculation of the size of the Earth and the portion of the Earth viewed by the spacecraft are as shown in Fig. 4-2. Here, ρ is the angular radius of the Earth as seen by the spacecraft and λ is the angular radius, as seen from the center of the Earth, of the circular segment of the Earth viewed by the spacecraft. By inspection,

$$\sin\rho = \cos\lambda = \frac{R_\oplus}{R_\oplus + h}$$

$$\rho + \lambda = 90° \qquad (4\text{-}2)$$

where h is the height of the spacecraft above the surface and $R_\oplus$ is the radius of the Earth.

We may use the same geometry to find the direction to any point, P, on the surface of the Earth as viewed by the spacecraft. By symmetry, the azimuthal orientation of P about the subsatellite point will be the same whether measured

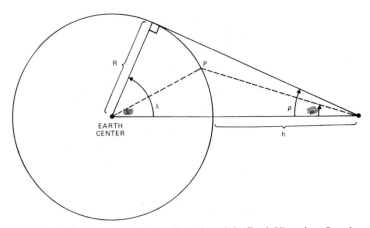

Fig. 4-2. Calculation of the Size of the Earth and Portion of the Earth Viewed, as Seen by a Spacecraft

from the satellite or the surface of the Earth. The coordinates ρ' and λ' are related by:

$$\tan\rho' = \frac{R_\oplus \sin\lambda'}{R_\oplus + h - R_\oplus \cos\lambda'}$$

$$= \frac{\sin\rho \sin\lambda'}{1 - \sin\rho \cos\lambda'} \tag{4-3}$$

Equation (4-3) was used to construct Figs. 4-3 and 4-4, which show the distortion of the Earth as viewed from space and the path of scan lines of three spacecraft sensors, shown as dashed lines in the two figures. Figure 4-3 illustrates a globe of the Earth showing the portion of the surface viewed by a spacecraft at a height, h, of 987 km over the equator at 70° West longitude such that $\lambda = 30°$ and $\rho = 60°$. Figure 4-4 shows the celestial sphere as seen by the spacecraft, including the visible features on the Earth's surface. The spacecraft attitude is toward the north celestial pole with sensors mounted at 40 deg, 60 deg, and 80 deg relative to the attitude; the arrows indicate the direction of scan. Note, particularly, the shape of the ground track of the sensors and the shape of the Earth meridians and parallels of latitude as seen by the spacecraft.

Modeling the Terminator. For the purpose of attitude determination and control, the most important feature of the Earth in the visible region of the spectrum is the terminator. Although the terminator is not normally used as a primary reference for attitude determination, it may be necessary to model the location of the terminator as seen from space for several reasons: (1) to verify coarse attitude or to determine the azimuthal orientation about the nadir (both were done for RAE-2 [Werking, et al., 1974; Lerner, et al., 1975]), (2) to determine the general level of illumination as it affects various attitude sensors (such as star cameras), and (3) to eliminate spurious horizon crossings due to sensor triggerings on the terminator. Section 9.3 includes a detailed discussion of tests for the identification of horizon crossings in attitude data.

Fig. 4-3.   Globe of the Earth Showing Portion of the Earth Viewed by a Spacecraft at $h = 987$ km. See also Fig. 6-29.

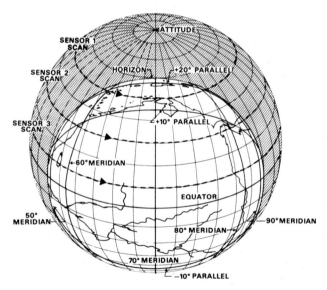

Fig. 4-4.   Celestial Sphere as Viewed by the Spacecraft Showing Distortion in the Appearance of the Portion of the Earth Viewed. Same parameters and horizon as shown in Fig. 4-3. Note the distortion in features and right/left reversal. See text for explanation.

The very poor definition of the terminator is due to three effects: (1) the gradual decrease in overall illumination with increasing solar zenith angle, (2) the extreme albedo variations between clouds and the planetary surface, and (3) the finite angular diameter of the Sun. We define the *dark angle*, ξ, as the angle at the center of a planet from the antisolar point to the point at which a ray from the upper limb of the Sun is tangent to the surface of the planet, as shown in Fig. 4-5. The dark angle, ξ, differs from 90 deg by three small correction terms:

$$\xi = 90° + \rho_E - \rho_S - \sigma \tag{4-4}$$

where ρ_E is the angular radius of the planet as seen from the Sun (i.e., the displacement of the center of the Sun as seen by an observer at the terminator, relative to an observer at the center of the planet), ρ_S is the angular radius of the Sun as seen from the planet, and σ is the atmospheric refraction for an apparent zenith angle of 90 deg. ρ_E is only 0.002 deg for the Earth and ρ_S is 0.267 deg at the

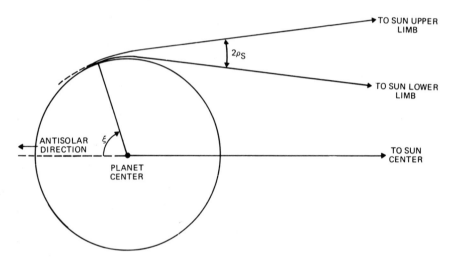

Fig. 4-5.   Dark Angle of a Planet, ξ, Showing the Effects of Atmospheric Refraction and Finite Size of the Disk of the Sun

Earth's mean distance of 1.0 AU. Atmospheric refraction is a strong function of zenith angle near the horizon. (For example, at sunset, the lower limb of the Sun is refracted upward more than the upper limb, producing the appearance of an oblate Sun.) For the Earth, this function is well known. (See, for example, Allen [1973].) At an apparent zenith angle of 90 deg, σ(760 mm Hg, 10 deg C) is 0.590 deg. Thus, for the Earth,

$$\xi_E \approx 90° + 0.002° - 0.267° - 0.590°$$

$$\approx 89.15° \tag{4-5}$$

For the Moon, ρ_M is negligible, ρ_S has the same average value as for the Earth, and $\sigma = 0$. Therefore,

$$\xi_M \approx 90° - 0.267° \approx 89.67° \tag{4-6}$$

The fraction, f_S, of the area of a planet from which at least some portion of the Sun can be seen is given by (see Eq. (A-12))

$$f_S = 0.5(1 + \cos\xi)$$
$$= 0.507 \quad \text{for the Earth}$$
$$= 0.503 \quad \text{for the Moon} \tag{4-7}$$

At best, the expressions for ξ are average values. The correction terms are normally dominated by local effects such as terrain (e.g., valleys where sunset is early and mountains where it is late) and albedo variations depending on both the nature of the surface and the local weather.

In order to model the position of the terminator as seen by the spacecraft, let $\Delta\xi$ be the correction terms in the dark angle, $\Delta\xi \equiv 90° - \xi$; let ρ be the angular radius of the central body as seen by the spacecraft; and let $\psi' = 180° - \alpha + \Delta\xi$, where α is the angle at the center of the planet between the Sun and the spacecraft. ψ' differs from ψ, the angular separation between the planet and the Sun as seen by the spacecraft, by two small correction terms, $\Delta\xi + \epsilon$, where ϵ is the angle between the planet and the spacecraft as viewed from the Sun. Then the following equations hold:

$$\psi' < \rho \qquad\qquad\qquad\qquad\qquad \text{planet fully dark} \tag{4-8}$$

$$\rho < \psi' < 180° - \rho \quad \text{planet partially illuminated, terminator visible} \tag{4-9}$$

$$\psi' > 180° - \rho \qquad\qquad\qquad\qquad \text{planet fully illuminated} \tag{4-10}$$

If the terminator is visible, then Eq. (4-3) may be used to construct the terminator on the celestial sphere, as shown in Fig. 4-6. In general, the terminator as seen by the spacecraft is neither a portion of a small circle nor a portion of an ellipse. The minimum angle from the center of the disk of the planet to the terminator as seen

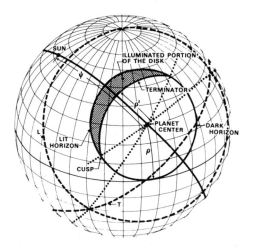

Fig. 4-6. Terminator Geometry. λ' is negative when the planet is less than half lit, as shown.

from the center of the Earth is $\lambda' = \psi' - 90°$, where λ' is positive when the planet is more than half lit and negative when the planet is less than half lit. This angle as seen from the spacecraft, ρ', may be obtained directly from Eq. (4-3) as

$$\rho' = \arctan\left(\frac{\sin\rho\cos\psi'}{1 + \sin\rho\sin\psi'}\right) \tag{4-11}$$

We define the *phase fraction*, P, as the illuminated fraction of the angular diameter of the planet perpendicular to the terminator as seen by the spacecraft. Then

$$P \equiv \frac{\rho + \rho'}{2\rho} \tag{4-12}$$

The orientation of the *cusp*, i.e., the intersection of the terminator and the horizon as seen by the spacecraft, may be constructed geometrically as follows: Construct the small circle, T, of radius ξ centered on the Sun, and the small circle, L, of radius $(90° - \rho)$, centered on the disk of the planet. Then the two great circles connecting the center of the planet to the two intersections of T and L cross the horizon of the disk of the planet at the cusps, with the proper horizon crossing chosen by inspection.

4.2 Appearance of the Earth at Infrared Wavelengths

James R. Wertz

As indicated at the beginning of Chapter 4, infrared radiation from the Earth is thermal radiation from both the surface and the atmosphere resulting from the heat generated by the absorption of sunlight. Figure 4-7, taken by the SMS-1 spacecraft at 14:00 UT, November 18, 1974, shows the Earth in the infrared region of the spectrum from 10.5 μm to 17.6 μm. The subsatellite point is near the equator at approximately 285 deg East longitude, near the point where the borders of Peru, Ecuador, and Colombia meet. Geographic features are difficult to distinguish in the infrared. The western shore of South America runs vertically down the center of the bottom half of the photograph. Northwest Africa is the well-defined region on the right-hand edge and the Great Lakes are clearly visible 1.2 cm from the top of the photograph.

Two general characteristics of the infrared radiation can be seen by comparing Figs. 4-7 and 4-1: (1) the intensity variations are much less than in the visual, and (2) bright and dark areas are reversed relative to the visual. At the time the photograph was taken, the terminator was about 2 cm from the left-hand edge of the photograph; that is, roughly 85% of the visible area is in sunlight and the left-hand 15% is in darkness. The terminator, which dominated the visible photograph, is invisible in the infrared because the Earth's temperature decreases only slightly overnight (relative to absolute zero), with little effect on the infrared thermal radiation. In general, the rapid time fluctuation of the visual radiation is

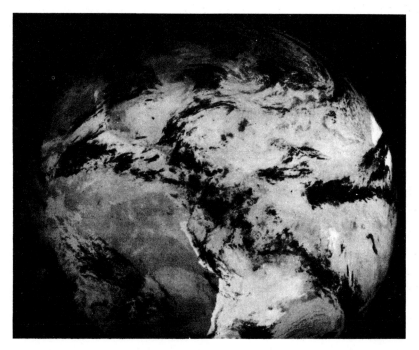

Fig. 4-7. The Earth in the Infrared Region of the Spectrum. Photograph taken by SMS-1 at 14:00 UT, November 18, 1974. (Courtesy SMS Project Office, NASA.)

smoothed by the absorption and gradual reradiation process. Results from the analysis of spacecraft data by Lyle, *et al.*, [1971] indicate variations in the albedo between 0.10 and 0.80 and infrared variations over the more limited range of 105 to 350 W/m^2. Similarly, the maximum diurnal infrared variation is only ±15%. Table 4-1 at the beginning of the chapter shows that global averages for each season vary from the annual mean by 10% in reflected radiation and 3% in emitted radiation.

Regions which reflect the most solar radiation (clouds and ice) are brightest in the visual region. However, because they absorb less energy, they are cooler and radiate less in the infrared. Thus, in contrast to the visual region, the cooler cloud tops and polar caps are dark in the infrared and the warmer, dark-colored, vegetation-covered surfaces are light.

The spectral energy distribution of the emitted infrared radiation is affected by the temperature and, more importantly for the Earth, by the chemical composition of the atmosphere. Figure 4-8 shows the average spectral distribution over midlatitude oceans (solid line) compared with various black body spectra (dashed line). Note the strong absorption bands due to CO_2, O_3 (ozone), and H_2O. In each of these bands radiation is being absorbed and isotropically reemitted by atmospheric molecules. Thus, at these wavelengths it is the atmosphere above the surface which is being viewed.

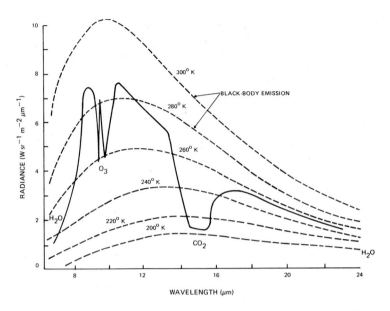

Fig. 4-8. Spectral Distribution of Thermal Emission From the Earth Over Midlatitude Oceans. (Adapted from Lyle [1971].)

For attitude work, we would like to use a spectral region for which the Earth has a uniform intensity. The considerable fluctuations in the 11- to 14-μm window can be seen in Fig. 4-7. Similarly, the H_2O band intensity near 7μm depends on the strongly varying H_2O density in the atmosphere. The CO_2 band provides a more uniform distribution than the H_2O bands [Dodgen and Curfman; 1969]. This spectral region has been used for horizon attitude sensors for a variety of missions, such as SMS/GOES, CTS, AE, and SIRIO, and is proposed for missions requiring precise horizon definition, such as HCMM, SEASAT, DE, and MAGSAT.

The Appearance of the Earth's Horizon at 14.0 to 16.3 μm. To effectively use the infrared radiation from the CO_2 band at 14.0 to 16.3 μm for attitude determination, we need to model the appearance of the Earth's horizon in this spectral region. Although several analytical models of varying complexity have been developed [Bates, et al., 1967; Thomas, et al., 1967a; Thomas, 1967b; Weiss, 1972; Langmaier, 1972; Howard, et al., 1965], the results of only one extended experiment are available in the open literature. These are from Project Scanner, carried out by NASA's Langley Research Center specifically for the study of infrared horizon profiles [McKee, et al., 1968; Whitman, et al., 1968]. Project Scanner consisted of two suborbital rocket flights on August 16 and December 10, 1966, and associated meteorological measurements. Both rockets were launched from Wallops Island, Va., to peak altitudes of 620 and 709 km, respectively. Horizon measurements for both flights covered a latitude range of 10° to 60° North from approximately the northern coast of South America to central Hudson Bay.

Figure 4-9 shows the average radiance profiles for the two flights. The vertical bars are the 1-σ standard deviations, due primarily to latitude variations, which

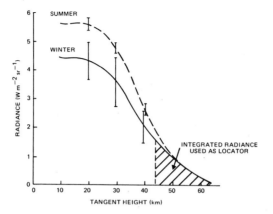

Fig. 4-9.   Average of All Measured Radiance Profiles From Project Scanner in the 14.0-μm to 16.3-μm CO_2 Band. (Adapted from Whitman, *et al.*, [1968].) The sensor triggers after reaching a preset level of integrated radiance, as shown by the shaded area.

were particularly large in the winter flight. The horizontal coordinate is the *tangent height* or the minimum altitude above the surface of the Earth for an unrefracted light ray coming from behind the Earth through the CO_2 layer to the spacecraft, as shown in Fig. 4-10. Thus, the tangent height is the apparent height (at the horizon) from which the radiation is coming. From Fig. 4-9 it is clear that horizon scanners sensitive to the CO_2 band radiation should indicate the presence of the Earth or *trigger* in the general range of 30 to 50 km above the surface.

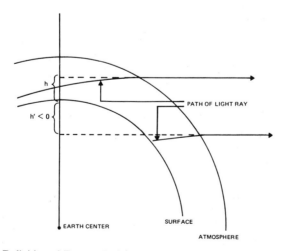

Fig. 4-10.   Definition of Tangent Height, *h*. *h'* is an example of negative tangent height.

To determine the specific altitude at which a given sensor will trigger and thus signal the presence of the horizon, we define the *locator* as the position on the radiance curve at which the sensor will trigger. The choice of a locator depends on both the stability with which it defines a located horizon (i.e., tangent height) and on the electronic processes available to implement the locating procedure. (See Section 6.2.) The field of view of a horizon telescope is typically much wider than the atmospheric band over which the radiance goes from near zero to its peak value. Therefore, the locator is normally defined as a function of the integrated radiance above various tangent heights, as shown in Fig. 4-10. The two most common locators are: (1) a fixed value of the integrated radiance or (2) a fixed percentage of the peak radiance seen by the sensor after it has crossed onto the disk of the Earth. Based on a theoretical analysis, the percentage of peak locator is the more accurate of the two [Dodgen and Curfman, 1969]. However, because of the complex structure of the horizon profiles, the electronic signal processing technique may be as important as the choice of locator. (See, for example, Weiss [1972].)

The radiance profile for the Earth's horizon depends primarily on the effective temperature, the effective pressure, and the optical depth, with temperature fluctuations being the most important factor [Whitman, *et al.*, 1968]. Temperatures at the altitude of the top of the CO_2 layer are governed primarily by latitude, season, and local upper-atmosphere weather conditions. Because of the very limited amount of data, accurate statistics do not exist on the variability of the height of the CO_2 layer or the temperature in the 30- to 50-km altitude range. Figure 4-11 shows the effect of seasonal and latitudinal variations in the Project Scanner data and one example (subfigures (a) and (b)) of longitudinal variations. Note that temperature changes affect the radiance profile most strongly at the peak radiance levels below about 30 km. Thus, in Fig. 4-11, there is greater uniformity in the lower tail than in the peak level.

Because temperature variations appear to be the prime determinant of changes in the radiation intensity in the CO_2 band, it is of interest to examine the degree of nonuniformity in upper atmosphere temperature profiles. Derived temperature profiles for the Project Scanner winter flight are shown in Fig. 4-12 for a vertical cross section covering the latitude range of the horizon scanner data and in Fig. 4-13 for a horizontal cross section at an altitude of 42 km. The approximate boundary of the measured data profiles is also shown in Fig. 4-13. The 42-km profile of Fig. 4-13 goes through the center of a warm pocket over White Sands, New Mexico, and has more horizontal variability than the other altitude profiles which were plotted over the range of 30 to 54 km in 4-km intervals. The temperature profiles for the summer flight were generally more uniform, with less than 5°K variation over the range of the horizon scanner data at an altitude of 40 km.

The most striking feature of Figs. 4-12 and 4-13 is the strong horizontal temperature gradient generally running north/south, but with substantial east/west components in some locations. The large horizontal temperature gradient has two major analytic consequences: (1) it implies a substantial geographical or weather dependence of the radiance profiles and (2) when horizontal temperature

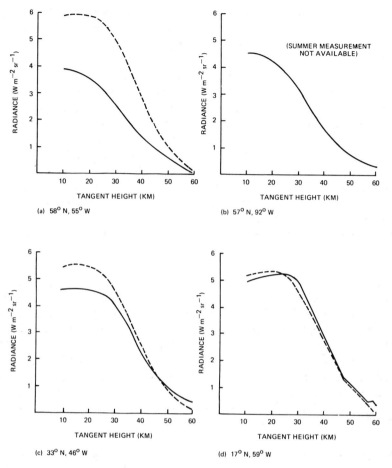

Fig. 4-11. Averaged Radiance Profiles for Several Locations From Project Scanner. Solid line is winter flight; dashed line is summer flight. (Adapted from Whitman, *et al.*, [1968].)

gradients are large, the analytic techniques used to predict radiance profiles for attitude sensing are inadequate [Whitman, *et al.*, 1968]. The analytic techniques used to date employ a shell model for the atmosphere in which the temperature is a function only of the altitude. Because any one scan line from a spacecraft to the horizon covers a wide geographic area, a strong horizontal temperature variation violates a basic assumption of the model. Note that although the temperature strongly affects the height of the CO_2 layer, the top of the CO_2 layer does *not* fall at any specific temperature level.

Figure 4-14 shows the tangent height at which a sensor would trigger based on three different locators and on *analytic* horizon profiles. Note that the locator for Fig. 4-14(a) is a constant radiance level and *not* a constant integrated radiance. On each figure, the solid line is the mean triggering height and the dashed line is the 1-σ standard deviation. The normalized locator of Fig. 4-14(c) is better than either constant radiance locator of Figs. 4-14(a) and (b). However, the analytic modeling

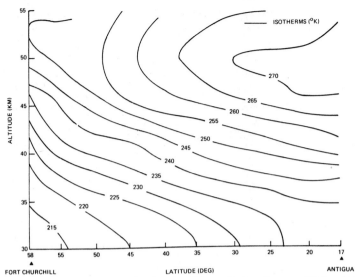

Fig. 4-12. Vertical Temperature Cross Section for Winter Project Scanner Flight. (Adapted from Whitman, *et al.*, [1968].)

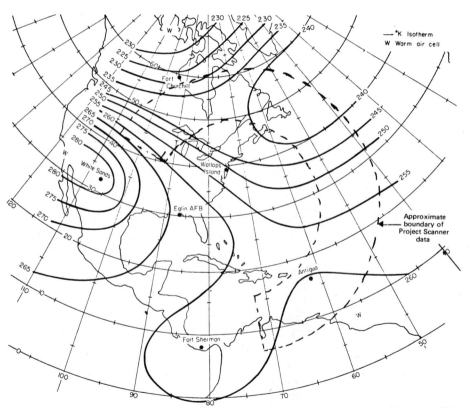

Fig. 4-13. Horizontal Temperature Cross Section at 42 km for Winter Project Scanner Flight. (Adapted from Whitman, *et al.*, [1968].)

procedure does not work well for real conditions with horizontal temperature gradients; these might be expected to have an effect on the peak radiance and, therefore, on the normalization process.

In summary, relatively little real data has been analyzed to determine the appearance of the Earth's horizon in the 14.0- to 16.3-μm CO_2 band. For a fixed radiance level locator, systematic latitudinal variations of 11 km in the triggering height occur during the winter with random 1-σ fluctuations of ± 5.5 km (numerical values in this paragraph are from Dodgen and Curfman, [1969].) Variations appear to be significantly less during the summer. For a fixed integrated radiance locator the winter latitudinal variations were 6 km with random 1-σ fluctuations of 3.5 km. For a locator normalized relative to the peak radiance, the winter latitudinal variation in the mean triggering height was reduced to about 4 km, but

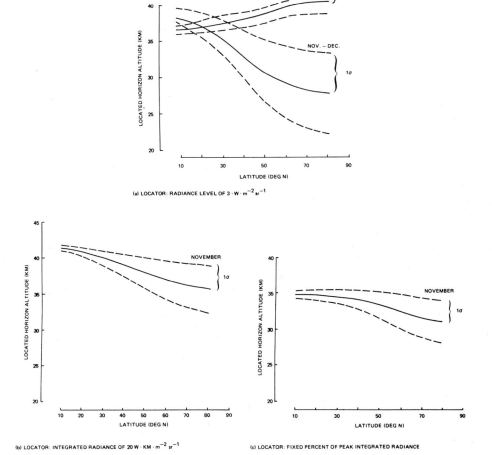

Fig. 4-14. Located Horizon Altitude for Analytic Models of 14.0 μm to 16.3 μm CO_2 Band Radiation (Adapted from Dodgen and Curfman, [1969].)

the 1-σ random fluctuations were only reduced to about 3 km.* These results do not take into account variations due to horizontal temperature gradients. The CO_2 band has been found to be more stable than the H_2O band, but variations at the level of several kilometres are a function of the local meteorology. The need for additional analysis of real data is clearly indicated.

4.3 Earth Oblateness Modeling

K. Liu

In Sections 4.1 and 4.2, we assumed that the Earth was spherical and discussed its appearance primarily from an optical point of view. In this section, we are concerned with the geometrical shape of the Earth. The Earth is basically an oblate spheroid as a result of combined centrifugal and gravitational forces. This is a form generally assumed by a rotating fluid mass in equilibrium. (See Section 5.2 for a description of the Earth's gravitational potential.)

As shown in Table 4-2, the surface of the Earth may be modeled by any of a

Table 4-2.   Comparison of Models of the Shape of the Earth

REFERENCE SURFACE	DEVIATION FROM REFERENCE SPHEROID*
SPHERE OF RADIUS = 6,378.14 km	0 AT EQUATOR TO −21.38 km AT POLE
REFERENCE SPHEROID	——
ELLIPSOID WITH ELLIPTICAL CROSS SECTION ON EQUATOR	AT EQUATOR, $R_{max} - R_{min} \approx 0.10$ km (R_{max} AT 160°, 340° EAST LONGITUDE)
SPHEROID DEFINED BY FOURTH ORDER HARMONICS (SEE SECTION 5.2)	0 AT EQUATOR AND POLE TO -0.005 km AT 45° LATITUDE
GEOID (= MEAN SEA LEVEL)	+0.080 km NEAR NEW GUINEA TO -0.110 km IN INDIAN OCEAN**
TOPOLOGICAL SURFACE (i.e., REAL SURFACE)	+8.8 km (MT. EVEREST) TO -0.4 km (DEAD SEA)
CO_2 LAYER IN ATMOSPHERE	~ +40 km AT EQUATOR TO ~ +30 km AT POLE IN WINTER†

* BASED ON EQUATORIAL RADIUS OF 6,378.140 km AND FLATTENING, f = 1/298.257, AS ADOPTED BY THE IAU [TABLE L-3] IN 1976.

** FOR A DETAILED MAP OF THIS DEVIATION, SEE FIG. 5-8.

† ACTUAL VALUES DEPEND ON HOW THE LAYER IS SENSED AND LOCAL WEATHER. SEE SECTION 4.2.

---

*Studies by Phenneger, *et al.*, [1977a, 1977b] for the SEASAT mission indicate 3-σ random horizon radiance variations for a percentage-of-peak locator of approximately ±0.1 deg. At the SEASAT altitude of 775 km, this corresponds to a triggering level variation of ±7 km.

series of increasingly complex surfaces. Although a simple spherical model is useful for estimation, it is inadequate for most attitude analysis of real spacecraft data. The basic model for most attitude work is the arbitrarily defined *reference spheroid*, which is an ellipse rotated about its minor axis to represent the flattening of the Earth. The ellipse is defined by the Earth's equatorial radius, $R_\oplus \approx 6378.140$ km, and the *ellipticity* or *flattening*,

$$f \equiv \frac{R_\oplus - R_p}{R_\oplus} \approx 0.00335281 = 1/298.257 \qquad (4.13)$$

where R_p is the polar radius of the Earth. These numerical values are those adopted by the International Astronomical Union in 1976 [Muller and Jappel, 1977] and will be used throughout the book, except in cases such as Vanguard units (Appendix K) or geomagnetic field models (Appendix H) where different values are a part of standard numerical models.

At NASA's Goddard Space Flight Center, a common expression used in attitude work for the radius of the Earth at latitude, λ, is:

$$R = R_\oplus(1 - f\sin^2\lambda + k\sin\lambda) + h \qquad (4-14)$$

where the terms in f, h, and k account for the flattening, the height of the atmosphere (for IR sensors which trigger on the atmosphere), and seasonal or other latitudinal variations in the atmosphere height.

A second more complex surface than the reference spheroid is obtained by expanding the Earth's gravitational potential in spherical harmonics and retaining only terms up to fourth order. It can be shown that this and a suitably defined reference spheroid are identical up to the second power of the flattening. A much more complex surface is the equipotential surface of the Earth's gravitational field, known as the *geoid* or *mean sea level*, which has many local irregularities due to the Earth's nonuniform mass distribution. The difference in elevation between the geoid and a reference spheroid is known as the *geoid height* and is shown in Fig. 5-8. Because of its mathematical simplicity, we will use the reference spheroid of Eq. (4-13) as the shape of the Earth throughout the rest of this section.

The Shape of the Earth as Seen From Space. The Earth's shape as viewed from space is defined by the Earth's horizon as seen from the position of the observer. The *horizon* is the point where the observer's line of sight is tangent to the Earth's surface or perpendicular to the surface normal. The spheroidal surface of the Earth is expressed in geocentric coordinates by

$$\frac{x^2 + y^2}{a^2} + \frac{z^2}{c^2} = 1 \qquad (4-15)$$

where a is the equatorial radius and c is the polar radius. This can be rewritten as

$$x^2 + y^2 + \frac{z^2}{(1-f)^2} = a^2 \tag{4-16}$$

$f =$

where f is the flattening. The normal to the surface is given by the gradient of Eq. (4-16), that is,

$$\hat{N} = \left[x^2 + y^2 + \frac{z^2}{(1-f)^4} \right]^{-1/2} \left[x\hat{x} + y\hat{y} + \frac{z}{(1-f)^2}\hat{z} \right] \tag{4-17}$$

If P (u,v,w) and R (x,y,z) represent the location of an observer and a point on the horizon, respectively, the vector from the observer to the horizon or *the horizon vector*, H (Fig. 4-15), is given by

$$H = (x-u)\hat{x} + (y-v)\hat{y} + (z-w)\hat{z} \tag{4-18}$$

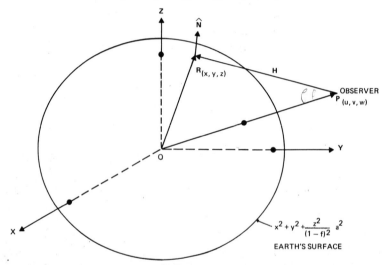

Fig. 4-15. Geometry of the Horizon Vector, H, and Surface Normal, $\hat{N}$, for an Oblate Earth

Because R is a horizon point, H must be perpendicular to $\hat{N}$; that is,

$$\hat{N} \cdot H = 0 \tag{4-19}$$

or

$$x(x-u) + y(y-v) + \frac{z(z-w)}{(1-f)^2} = 0$$

Rearranging terms, this becomes

$$\left(x - \frac{u}{2}\right)^2 + \left(y - \frac{v}{2}\right)^2 + \frac{\left(z - \frac{w}{2}\right)^2}{(1-f)^2} = \left(\frac{u}{2}\right)^2 + \left(\frac{v}{2}\right)^2 + \frac{\left(\frac{w}{2}\right)^2}{(1-f)^2} \tag{4-20}$$

which is the equation for a spheroid of ellipticity f centered at $(u/2, v/2, w/2)$. We call this the *horizon spheroid* or *horizon surface* because it contains all possible horizon points (x, y, z) for an observer at (u, v, w) looking at a spheroidal Earth of ellipticity f and variable sizes. The three principal axes of the horizon spheroid are parallel to those of the Earth spheroid. The intersection of the two surfaces is the Earth's horizon, as shown in Fig. 4-16. By substituting Eq. (4-16) into Eq. (4-20), we obtain

$$ux + vy + \frac{wz}{(1-f)^2} = a^2 \qquad (4\text{-}21)$$

Equation (4-21) defines a plane, called the *horizon plane*, and in this plane the locus of the observed horizon is an ellipse. The normal to the horizon plane is in the direction of $(u, v, w/(1-f)^2)$, or $(\cos\lambda\cos\phi, \cos\lambda\sin\phi, \sin\lambda/(1-f)^2)$, where λ and ϕ are the geocentric latitude and longitude, respectively, of the observer's position.

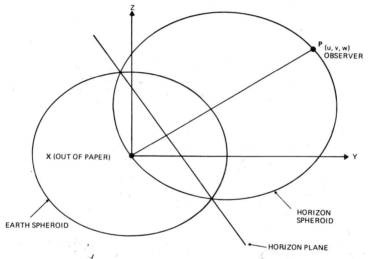

Fig. 4-16.   Meridian Cross Section of the Earth Showing the Horizon Spheroid and the Horizon Plane. (In this figure, the observer is in yz plane.)

   The normal to the horizon plane depends only on the angular position of the observer. Thus, as the observer moves along a fixed nadir* direction, he sees a set of parallel horizon planes. As shown in Fig. 4-17, when the observer is on the Earth's surface (point A), the horizon plane is just the tangent plane at that point. As the observer moves to a distance, d, from the center of the Earth (point B), the parallel horizon plane will intersect the nadir line at a distance

$$D = \frac{R^2}{d} \qquad (4\text{-}22)$$

---

* *Nadir* here means toward the Earth's center.

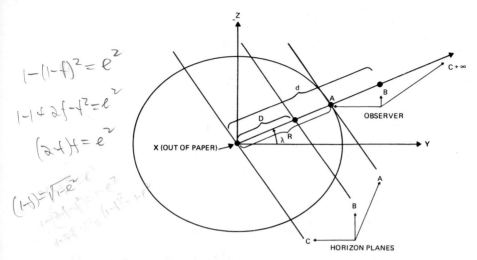

Fig. 4-17. Meridian Cross Section of the Earth Showing Parallel Horizon Planes. (In this figure, the observer is in *yz* plane.)

from the Earth's center, where R is the distance from the Earth's center to point A (the *subobserver* or *subsatellite point*). R is given by:

$$R = \frac{a(1-f)}{\sqrt{1-(2-f)f\cos^2\lambda}} \qquad (4\text{-}23)$$

The horizon plane will approach the center of the Earth as the observer approaches infinity. Note that the nadir line passes through the center of the horizon ellipse.

To find the shape of the horizon ellipse or, equivalently, the shape of the Earth as seen by the observer, it is convenient to solve Eqs. (4-16) and (4-21) in the local tangent coordinate system defined by $\hat{\mathbf{N}}$, $\hat{\mathbf{E}}$, and $\hat{\mathbf{Z}}$ through $\mathbf{P}$, as shown in Fig. 3-6. It can be shown that the angular radius of the Earth or the horizon of the Earth is given by

$$\rho = \text{arc cot}\left\{\left[\frac{(d^2-R^2)}{a^2}\left[1+\frac{(2-f)fR^2\cos^2\lambda}{(1-f)^2a^2}\sin^2\Psi\right]\right]^{1/2}\right.$$

$$\left. + \frac{(2-f)fR^2\sin 2\lambda}{2(1-f)^2a^2}\sin\Psi\right\} \qquad (4\text{-}24)$$

where λ is the geocentric latitude of the observer's position and d and R are the distances from the center of the Earth to the observer and the subobserver point, respectively. As shown in Fig. 4-18, Ψ is the azimuth angle of the horizon vector,

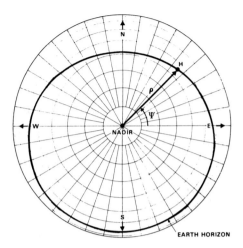

Fig. 4-18. The Shape of the Oblate Earth as Seen From 200 km Above the Earth's Surface at 45 deg Geocentric Latitude. The flattening factor used is 100 times larger than the true value.

H, in local tangent coordinates and ρ is the angle between the nadir vector and the horizon vector. When $f = 0$, that is, when the Earth is spherical, Eq. (4-24) reduces to

$$\rho = \arcsin(a/d) \qquad (4\text{-}25)$$

as expected.

Figure 4-18 shows an example of the shape of the Earth as seen by an observer at 45-deg geocentric latitude and a distance of 200 km above the Earth's surface. To make the oblateness effect noticeable, a flattening factor 100 times larger than the true value was used. Table 4-3 compares the angular radius of the Earth for spheroidal and spherical Earth models.

Table 4-3. Angular Radius of the Earth From the Spheroidal Model at an Altitude of 200 km and Geocentric Latitude of 45 deg. Equation (4-24) and the following parameters were used: $a = 6,378.14$ km, $f = 0.00335281$, $d = a + 200$ km. The angular radius of a spherical Earth of radius a is $\rho_a = 75.8353$ deg.

Ψ (DEG)	ρ (DEG)	$\rho - \rho_a$ (DEG)	Ψ (DEG)	ρ (DEG)	$\rho - \rho_a$ (DEG)
90	75.2786	−0.5567	15	75.5272	−0.3081
75	75.2863	−0.5490	30	75.5664	−0.2689
60	75.3085	−0.5268	45	75.5980	−0.2373
45	75.3430	−0.4923	60	75.6208	−0.2145
30	75.3861	−0.4429	75	75.6346	−0.2007
15	75.4338	−0.4015	90	75.6392	−0.1961
0	75.4821	−0.3532			

Procedure for Finding the Horizon Crossing Vector. A frequent calculation required for attitude analysis is the determination of the horizon crossing vector, **H**, as seen by a sensor with a conical field of view. Any point on the horizon lies at the intersection of the horizon and Earth spheroids. To find the two particular horizon points where a sensor first and last senses the Earth, a third surface is needed. If the Sun angle and the Sun-to-Earth rotation angle are available (Eq. (7-57)), that surface may be provided by

$$\hat{S} \cdot \hat{H} = \cos \psi_H \qquad (4\text{-}26)$$

where $\hat{S}$ is the unit Sun vector and ψ_H is the angle between the Sun vector and the horizon vector. Alternatively, if an iterative procedure is used to find the spacecraft attitude, the attitude vector, $\hat{A}$, combined with the knowledge of the sensor mounting angle, γ, generates the surface of a cone defined by

$$\hat{A} \cdot \hat{H} = \cos \gamma \qquad (4\text{-}27)$$

The horizon-in and -out vectors are obtained by simultaneously solving Eqs. (4-16), (4-20) or (4-21), and (4-26) or (4-27). For a slit horizon sensor, Eqs. (4-26) and (4-27) are replaced by

$$\hat{S} \cdot \hat{N} = \pm \cos \psi_N \qquad (4\text{-}28)$$

and

$$\hat{A} \cdot \hat{N} = \mp \sin \Theta \qquad (4\text{-}29)$$

respectively, where $\hat{N}$ is defined by Eq. (4-17), ψ_N is the angle between the Sun vector and the slit plane normal vector at the horizon crossing, and Θ is the rotation angle between a spacecraft body meridian and the plane of the horizon sensor slit. The upper and lower signs on the right hand side of Eqs. (4-28) and (4-29) are for the horizon-in and -out crossings, respectively. In general, these simultaneous equations cannot be solved analytically and numerical methods are needed.

For example, a linear, iterative method can be used to solve Eqs. (4-16), (4-21), and (4-27). Equation (4-27) can be written as

$$a_1 x + a_2 y + a_3 z = H \cos \gamma + a_1 u + a_2 v + a_3 w \qquad (4\text{-}30)$$

where

$$H = \left[x^2 + y^2 + z^2 - 2(ux + vy + wz) + u^2 + v^2 + w^2 \right]^{1/2}$$

is the length of the horizon vector, H, and a_1, a_2, and a_3 are the rectangular components of $\hat{A}$. Assuming first that H is a constant, then we have two linear and one quadratic equation and (x,y,z) can be determined analytically by expressing them in the form

$$x_i = f_i(H) \qquad (4\text{-}31)$$

$$x_j = f_j(x_i, H) \qquad (4\text{-}32)$$

and

$$x_k = f_k(x_i, H) \tag{4-33}$$

where i, j, and k can be 1, 2, and 3 and $(x_1, x_2, x_3) = (x, y, z)$.

Equations (4-31) through (4-33) can then be used to do the iteration. A good initial estimate of x_i's can be found by assuming a spherical Earth and calculating x_i's using the above three equations with

$$H = \sqrt{u^2 + v^2 + w^2 - a^2}$$

The Effect of Earth Oblateness on Attitude Sensor Data and Solutions. One of the fundamental types of attitude measurements is the rotation angle measurement, such as the Sun-to-Earth horizon crossing or the Earth width (i.e., the rotation angle between two horizon crossings). Clearly, the time and location of the horizon triggering depend on the shape of the Earth as seen from the spacecraft. The effect of the oblateness of the Earth on the rotation angle measurements is a function of the spacecraft position and attitude and the sensor mounting angle.

Table 4-4 gives the difference in Earth width and nadir angles as computed for spheroidal and spherical Earth models. Therefore, this difference is approximately the error that would result from using a spherical Earth to model horizon sensor measurements. In these examples, the effect of oblateness tends to be greater when the spacecraft is at higher geocentric latitudes. However, with the same latitude and sensor mounting angle, the effect is not necessarily smaller when the spacecraft is

Table 4-4. Error in Nadir Angle ($\Delta\eta$) and Earth Width (ΔW) in Degrees Due to Unmodeled Oblateness. Based on circular orbit and attitude at orbit normal.

ALTITUDE	PARAM-ETER (DEG)	INCLINATION, i, AND SUBSATELLITE LATITUDE, λ						INCLINATION, i, AND SUBSATELLITE LATITUDE, λ					
		i = 0°	i = 45°		i = 90°			i = 0°	i = 45°		i = 90°		
		λ = 0°	λ = 0°	λ = 45°	λ = 0°	λ = 45°	λ = 90°	λ = 0°	λ = 0°	λ = 45°	λ = 0°	λ = 45°	λ = 90°
		SENSOR MOUNTING ANGLE = 30 DEG						SENSOR MOUNTING ANGLE = 85 DEG					
h = 200km	ΔW	0.162	0.102	2.38	0.041	1.60	3.11	0.0007	0.046	0.742	0.091	0.756	1.39
	Δη	0.041	0.026	0.595	0.010	0.400	0.775	0.004	0.253	3.76	0.498	3.82	6.61
h = 400km	ΔW	0.266	0.157	2.15	0.048	1.23	2.40	0.001	0.062	0.504	0.122	0.533	0.934
	Δη	0.057	0.033	0.453	0.010	0.260	0.503	0.006	0.328	2.50	0.643	2.63	4.36
h = 800km	ΔW	0.672	0.352	3.06	0.035	1.34	2.67	0.002	0.079	0.325	0.156	0.372	0.584
	Δη	0.077	0.040	0.340	0.004	0.152	0.297	0.008	0.394	1.54	0.766	1.75	2.64
h = 35,800km (SYNCHRONOUS)	ΔW		SENSOR SCAN MISSES EARTH					0.023	0.035	0.019	0.047	0.024	0.002
	Δη							0.017	0.025	0.013	0.033	0.017	0.001
		SENSOR MOUNTING ANGLE = 60 DEG						SENSOR MOUNTING ANGLE = 90 DEG					
h = 200km	ΔW	0.028	0.053	1.06	0.078	0.864	1.62	0	0.046	0.706	0.091	0.753	1.39
	Δη	0.024	0.044	0.874	0.065	0.715	1.34	0	3.22	12.53	4.55	12.93	17.4
h = 400km	ΔW	0.041	0.072	0.789	0.104	0.607	1.10	0	0.061	0.470	0.123	0.531	0.930
	Δη	0.033	0.058	0.626	0.083	0.482	0.869	0	3.13	8.61	4.42	9.15	12.05
h = 800km	ΔW	0.060	0.095	0.592	0.130	0.419	0.704	0	0.079	0.293	0.157	0.371	0.581
	Δη	0.044	0.070	0.432	0.095	0.307	0.514	0	2.95	5.69	4.17	6.40	7.99
h = 35,800km (SYNCHRONOUS)	ΔW		SENSOR SCAN MISSES EARTH					0	0.029	0.0007	0.057	0.029	0.001
	Δη							0	0.502	0.077	0.709	0.507	0.109

at higher altitudes. If the Earth-width and Sun angle measurements are used to calculate the attitude, the resulting discrepancy in attitude using the spherical model will be at least as large as the deviation in the nadir angle, which is significant in certain geometric situations. If the Sun angle is combined with a horizon crossing measurement, the problem becomes more complicated because the relative position of the Sun plays an important role.

4.4 Modeling the Structure of the Upper Atmosphere

John N. Rowe

The effect of the atmosphere on sensor triggerings was discussed in Sections 4.1 and 4.2. In this section, we are primarily concerned with the atmosphere as it affects the spacecraft orbit and attitude. For additional information on the Earth's atmosphere, see, for example, Craig [1965] or Ratcliffe [1960]. An interesting historical reference is Mitra [1952]. Summary atmospheric density tables are given in Appendix L.

The accuracy of upper atmosphere model densities in current atmospheric models is about ±50%, and may be much worse in some regions, such as near an altitude of 120 km where there are few measurements. In addition, the upper atmosphere density is strongly affected by the solar flux in the extreme ultraviolet, and this flux varies with the level of solar activity (see Section 5.3). This level is not entirely predictable, so that predicted densities will have more uncertainty than will historical densities.

4.4.1 Summary of the Upper Atmosphere Structure

The independent variable in describing the mean structure of the atmosphere is the altitude. Figure 4-19 shows the mean temperature distribution between the ground and 500 km. The nomenclature used to describe various regions of the atmosphere is based on the temperature profile, as indicated in the figure. The elevated temperature between the *tropopause* and the *mesopause* is due to the absorption of ultraviolet light at wavelengths from 0.2-0.3 μm by ozone, and the increase in temperature above 100 km is due to absorption of extreme ultraviolet light at wavelength from 0.2 μm down to X-rays by nitrogen and oxygen. Eventually, the heat conductivity becomes so large that an isothermal region called the *exosphere* is formed. The temperature in this region is called the *exospheric temperature*, T_∞.

Above about 1500 km, the ionized component of the atmosphere becomes predominant; this region is called the *magnetosphere*, and its outer boundary is the *magnetopause*. The magnetopause is formed by the interaction of the solar wind with the Earth's magnetic field (see Section 5.1), and lies at a distance of about 10 Earth radii on the day side and at least 80 Earth radii on the night side.

The total atmospheric density is of prime interest for spacecraft. Figure 4-20

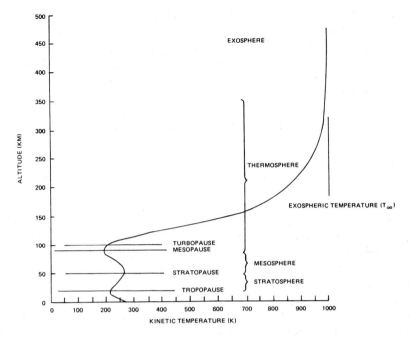

Fig. 4-19. Mean Atmospheric Temperature as a Function of Altitude

shows a profile of the mean density between 25 and 1000 km. The gross exponential behavior of the density is due to hydrostatic equilibrium, in which the pressure and density at any height are determined by the weight of all the air above that height. In this case, both the pressure and density vary as $\exp(-mgz/kT)$ where z

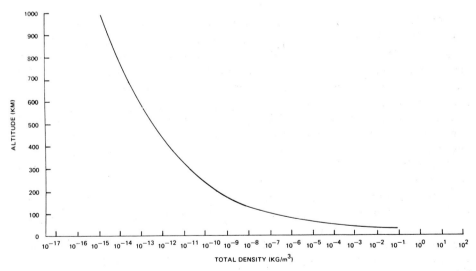

Fig. 4-20. Total Atmospheric Density as a Function of Altitude (See Appendix L for numerical tables.)

is the altitude. The quantity kT/mg is known as the *scale height*, where m is the molecular weight, g is the acceleration due to gravity, T is the temperature, and k is Boltzmann's constant. In the altitude region below the *turbopause* (about 100 km), the atmosphere is dominated by turbulence (sometimes called eddy diffusion), which causes mixing. In a mixed atmosphere, the density of each constituent is a constant fraction of the total density (independent of altitude), and the density falls off with a scale height characteristic of the mean molecular weight of all the component gases. Above the turbopause, turbulence ceases and each constituent diffuses according to its own scale height, resulting in what is known as diffusive separation. The lighter gases, of course, have the larger scale heights and become a larger fraction of the total composition with increasing altitude, resulting in the change in the slope of the total density curve.

The major atmospheric constituents below 1000 km are O_2, N_2, O, and He. Representative minor constituents in this altitude range and above the turbopause are O_3, CO_2, H_2O, NO, electrons, and positive and negative ions. Table 4-5 gives the atmospheric composition in the turbulent region, showing only those constituents which are mixed. The most important minor constituent for attitude determination is CO_2 because of the infrared radiation described in Section 4.2. CO_2 is believed to be mixed [Hays and Olivero, 1970] up to the turbopause. Above the turbopause the CO_2 density is determined by both diffusion and chemical reactions.

Chemistry is a factor in controlling the densities of some atmospheric constituents. Reactions between the various constituents are called *photochemical* because they are generally induced or catalyzed by sunlight. Many minor constituents are controlled by photochemistry rather than by diffusion, and their profiles do not at all resemble diffusive profiles.

The type of flow encountered by a space vehicle, and hence the characteristics of the forces on it (Section 17.2), are controlled to some extent by the *Knudsen number*, which is approximately the ratio of a typical dimension of the spacecraft to the average mean free path, l, of the atmospheric molecules. The mean free path

Table 4-5.   Composition of the Atmosphere Below the Turbopause

CONSTITUENT	FRACTION BY VOLUME
MOLECULAR NITROGEN	0.78084
MOLECULAR OXYGEN	0.209476
ARGON	0.00934
CARBON DIOXIDE	0.000314
HELIUM	0.00000524
KRYPTON	0.00000114

may be estimated from

$$l = \frac{1}{n\sigma}$$

where n is the number density of the atmosphere and σ is a collisional cross section. Figure 4-21 shows a mean free path profile calculated from the mean density profile in Fig. 4-20 and an assumed cross section of 3×10^{-19} m^2.

The subject of structural variations in the atmosphere is complex and not fully understood. Variations in the density may be divided into six types: Diurnal, 27-day, seasonal-latitudinal, semiannual, 11-year, and geomagnetic. Diurnal variations are those related to local time, or, more generally, to the zenith angle of the Sun. The 27-day variation is a result of the rotation of the Sun, and the 11-year variation is a result of the 11-year cycle of solar activity. Geomagnetic variations are due to short-term changes in solar activity (as a result of a flare, for example). The variations in the Earth's magnetic field are used as a measure of this type of solar activity. The semiannual and seasonal-latitudinal variations are only partially solar related and are not well understood. In the thermosphere, the 11-year variation in density is the largest, amounting to order-of-magnitude fluctuations at 350 km. The 27-day and semiannual variations cause density fluctuations by a factor of perhaps 2 or 3, and the other types cause smaller ones. At lower altitudes, the seasonal-latitudinal variations become predominant. Various parameters are used to describe the variations in the atmosphere, in particular, geomagnetic activity indices and solar activity indices. These are tabulated in the Solar-Geophysical Data series published monthly by the National Oceanic and Atmospheric Administration.

4.4.2 Models of the Upper Atmosphere

Models of the atmosphere may be based on either empirical or theoretical work. Real models are a combination of both types, because there generally is

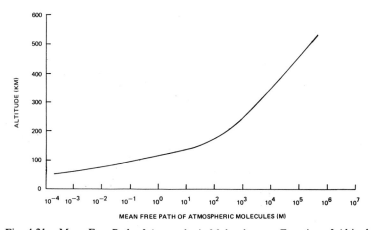

Fig. 4-21.   Mean Free Path of Atmospheric Molecules as a Function of Altitude

insufficient data for a purely empirical model and because the physical processes are not well enough understood to construct an entirely theoretical model. A number of published models are appropriate for attitude use. Those published under the auspices of the Committee on Space Research of the International Council of Scientific Unions (COSPAR) find wide use in atmospheric science and should be considered first. The current version is the *COSPAR International Reference Atmosphere 1972* [1972] (known as CIRA 72). The CIRA 72 model covers the altitude range of 25 km to 2500 km, and includes detailed modeling of the variations mentioned in the previous section. The model below 110 km is the work of Groves [1970], and the model above 90 km, the area of primary interest for spacecraft, is the work of Jacchia [1971].

The Jacchia portion of the model, called J71, is characterized by constant temperature and density at 90 km, analytical temperature profiles (the independent variable being exospheric temperature) and an analytical, fixed, mean molecular weight profile between 90 and 105 km. The density is determined by integration of the static (i.e., time independent) diffusion equations from the lower boundary at 90 km up to 2500 km. The variations in the atmosphere are introduced primarily via the exospheric temperature. This model was constructed to minimize residuals between the density predictions of the model and the densities determined from analysis of the effects of atmospheric drag on the orbits of many satellites.

Earlier Jacchia versions of 1970 [Jacchia, 1970] and 1964 [Jacchia, 1964] have been used as the basis for some analytical models, in particular those of Roberts [1971] and Weidner, *et al.*, [1969]. The latter is known as the NASA Monograph model. Following Walker [1965], Roberts modified Jacchia's temperature profiles so that the diffusion equations would become exact differentials, giving an analytical expression for the density. The resulting densities differ from J70 by less than 5 percent and are thus adequate for attitude work. The NASA monograph model is similar to Roberts model but is based, instead, on J64. The significant difference between J70 and CIRA 72 is in the O/O_2 ratios; there are only small changes in the total density, so that the Roberts model is a reasonable approximation of CIRA 72. Note that Jacchia made some changes in the formulation of the variations, so that the same procedures are not followed between J70 and CIRA 72. In the J64 model, the lower boundary was taken at 120 km, and thus the densities below 150 km and perhaps below 200 km are erroneous.

Other models of interest include CIRA 65 [1965], the U.S. Standard Atmosphere, 1962 [1962] and supplements, 1966 [1966], and the U.S. Standard Atmosphere, 1976 [1976]. CIRA 65 is based on Harris and Priester [1962]. This is a time-dependent diffusion model which is better in principle than a static diffusion model but worse in practice because of the volume of tables necessary to describe the results. In addition, the tables are given for only one latitude, and there is no simple way to account for the variations, as is done in CIRA 72. The U.S. Standard atmospheres 1962 and 1976 are mean atmospheres only, and the supplement of 1966 basically J64, and thus suffer from the defect of too high a lower boundary altitude.

References

1. Allen, C. W., *Astrophysical Quantities* 3rd Edition. London: The Athlone Press, 1973.
2. Bates, Jerry C., David S. Hanson, Fred B. House, Robert O'B. Carpenter, and John C. Gille, *The Synthesis of 15μ Infrared Horizon Radiance Profiles From Meteorological Data Inputs*, NASA CR-724, 1967.
3. COSPAR Working Group IV, *COSPAR International Reference Atmosphere*. Amsterdam: North Holland, 1965.
4. ———, *COSPAR International Reference Atmosphere*. Berlin: Akademie-Verlag, 1972.
5. Craig, R. A., *The Upper Atmosphere-Meteorology and Physics*. New York: Academic Press, 1965.
6. Dodgen, John A., and Howard J. Curfman, Jr., "Accuracy of IR Horizon Sensors as Affected by Atmospheric Considerations," *Proceedings of the Symposium on Spacecraft Attitude Determination, Sept. 30, Oct. 1-2, 1969*, SAMSO-TR-69-417, Vol. 1, p. 161–168, Dec. 1969.
7. Groves, G. V., *Seasonal and Latitudinal Models of Atmospheric Temperature, Pressure, and Density, 25 to 110 km*. Air Force Cambridge Research Laboratories Report 70-0261, 1970.
8. Harris, I. and W. Priester, "Time Dependent Structure of the Upper Atmosphere," *J. of Atmospheric Sc.*, Vol. 19, p. 286–301, 1962.
9. Hays, P. B. and J. J. Olivero, "Carbon Dioxide and Monoxide Above the Troposphere," *Planetary and Space Science*, Vol. 18, p. 1729–1733, 1970.
10. Howard, John N., John S. Garing, and Russell G. Walker, "Transmission and Detection of Infrared Radiation," *Handbook of Geophysics and Space Environments* (S. Valley, ed.). New York: McGraw-Hill, Inc., 1965.
11. Muller, Edith A. and Arnost Jappel, editors, *International Astronomical Union, Proceedings of the Sixteenth General Assembly, Grenoble 1976*. Dordrecht, Holland: D. Reidel Publishing Co., 1977.
12. Jacchia, L. G., *Static Diffusion Models of the Upper Atmosphere With Empirical Temperature Profiles*, SAO Special Report 170, 1964.
13. ———, *New Static Models of the Thermosphere and Exosphere With Empirical Temperature Profiles*, SAO Special Report 313, 1970.
14. ———, *Revised Static Models of the Thermosphere and Exosphere With Empirical Temperature Profiles*, SAO Special Report 332, 1971.
15. Langmaier, J. K., *Techniques for Defining the Position of the Earth's Horizon for Attitude Sensing Purposes*, Ithaco, Inc., Report No. 90370 (File: 10-2565), May 1972.
16. Lerner, G., D. Headrick, and R. Williams, *Lunar RAE Spacecraft Dynamics Verification Study*, Comp. Sc. Corp., CSC/TM-75/6174, Aug. 1975.
17. Lyle, Robert, James Leach, and Lester Shubin, *Earth Albedo and Emitted Radiation*, NASA SP-8067, July 1971.
18. McKee, Thomas B., Ruth I. Whitman, and Richard E. Davis, *Infrared Horizon Profiles for Summer Conditions From Project Scanner*, NASA TN D-4741, Aug. 1968.

19. Mitra, S. K., *The Upper Atmosphere*. Calcutta: The Asiatic Society, First Edition, 1947, Second Edition, 1952.

20. Phenneger, M. C., C. Manders, and C. B. Spence, Jr., *Infrared Horizon Scanner Attitude Data Error Analysis for SEASAT-A*, Comp. Sc. Corp., CSC/TM-77/6064, July 1977a.

21. Phenneger, M. C., C. Manders, C. B. Spense, Jr., M. Levitas, and G. M. Lerner, "Infrared Horizon Scanner Attitude Data Error Analysis for SEA-SAT-A," paper presented at the GSFC Flight Mechanics/Estimation Theory Symposium, Oct. 17 and 18, 1977b.

22. Ratcliffe, J. A. (ed.), *Physics of the Upper Atmosphere*. New York: Academic Press, 1960.

23. Roberts, C. E., "An Analytic Model for Upper Atmosphere Densities Based on Jacchia's 1970 Models," *Celestial Mechanics*, Vol. 4, p. 368–377, 1971.

24. Thomas, John R., Ennis E. Jones, Robert O'B. Carpenter, and George Ohring, *The Analysis of 15μ Infrared Horizon Radiance Profile Variations Over a Range of Meteorological, Geographical, and Seasonal Conditions*, NASA CR-725, April 1967a.

25. Thomas, John R., *Derivation and Statistical Comparison of Various Analytical Techniques Which Define the Location of Reference Horizons in the Earth's Horizon Radiance Profile*, GSFC, NASA CR-726, April 1967b.

26. United States Committee on Extension to the Standard Atmosphere, *U.S. Standard Atmosphere*. Washington: U.S. G.P.O., 1962.

27. ———, *U.S. Standard Atmosphere Supplements*. Washington: U.S. G.P.O., 1966.

28. ———, *U.S. Standard Atmosphere*. Washington: U.S. G.P.O., 1976.

29. Walker, J. C. G., "Analytical Representation of the Upper Atmosphere Density Based on Jacchia's Static Diffusion Models," *J. of Atmospheric Sc.*, Vol. 22, p. 462, 463, 1965.

30. Weidner, D. K., C. L. Hassetine, and R. E. Smith, *Models of Earth's Atmosphere (120 to 1000 km)*, NASA SP-8021, 1969.

31. Weiss, R., "Sensing Accuracy of a Conical Scan CO_2 Horizon Sensor," *J. Spacecraft*, Vol. 9, p. 607–612, 1972.

32. Werking, R. D., R. Berg, K. Brokke, T. Hattox, G. Lerner, D. Stewart, and R. Williams, *Radio Astronomy Explorer-B Postlaunch Attitude Operations Analysis*, GSFC, NASA X-581-74-227, July 1974.

33. Whitman, Ruth I., Thomas B. McKee, and Richard E. Davis, *Infrared Horizon Profiles for Winter Conditions From Project Scanner*, NASA TN D-4905, Dec. 1968.

CHAPTER 5

MODELING THE SPACE ENVIRONMENT

5.1 The Earth's Magnetic Field
5.2 The Earth's Gravitational Field
5.3 Solar Radiation and The Solar Wind
     Solar Radiation, The Solar Wind
5.4 Modeling the Position of the Spacecraft
5.5 Modeling the Positions of the Sun, Moon, and Planets
5.6 Modeling Stellar Positions and Characteristics
     Star Catalog Data Required for Attitude Determination,
     Existing Star Catalogs, Generating a Core Catalog

Chapter 4 described models of the appearance, shape, and atmosphere of the Earth. This chapter is concerned with modeling properties of the spacecraft environment that are relevant to attitude determination and control. Sections 5.1 and 5.2 describe the magnetic and gravitational fields of the Earth, although many of the modeling procedures can be extended to other planets as well. Section 5.3 discusses the interplanetary medium known as the solar wind. The remaining three sections discuss models of the position of various objects needed for attitude determination—the spacecraft itself, the Sun, the Moon, the planets, and the stars.

5.1 The Earth's Magnetic Field

Michael Plett

Although the general characteristics of the Earth's magnetic field have been known for centuries, the first systematic study of the field was initiated by the German mathematician and physicist Karl Gauss* in the early part of the nineteenth century. Since that time, a great deal of data has been accumulated, much of it as a result of spacecraft measurements during the 1960s. Although this body of data has served to increase our ability to accurately describe the field, it has not yet provided the key to the physical processes which produce it or perturb it. Thus, in this section we will describe the observed phenomena and, wherever possible, provide plausible arguments for their existence.

The Main Field. The Earth's magnetic field is predominantly that of a magnetic dipole such as that produced by a sphere of uniform magnetization or a current loop. The strength of the dipole was 7.96×10^{15} Wb·m in 1975. The "south" end of the dipole was in the northern hemisphere at 78.60° N latitude and 289.55° E longitude and drifting westward at about 0.014 deg/year. The dipole strength is decreasing by 0.05%/year. This *secular drift* implies a possible field reversal in several thousand years. There is ambiguous evidence of several reversals

---

*Among his many contributions, Gauss was also the first to apply least-squares analysis to the problem of orbit determination.

in the past with time scales of 70,000 to 100,000 years between reversals [Haymes, 1971].

The plane perpendicular to the Earth-centered dipole is called the *magnetic equator*. The field is weakest there, being about 3×10^4 nT at the surface of the Earth. Figure 5-1 shows the variation in the dipole field strength as a function of altitude at the magnetic equator. The field strength increases by a factor of two as the magnetic latitude increases from 0 deg to 90 deg, as shown in Fig. 5-2. At the geomagnetic equator, the field is horizontal relative to the Earth's surface. At a geomagnetic latitude of about 27 deg, the field is 45 deg down from horizontal.

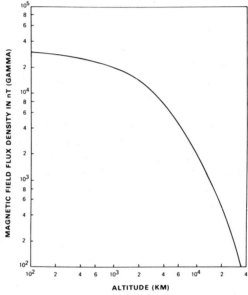

Fig. 5-1. Earth's Magnetic Field Intensity at the Magnetic Equator as a Function of Altitude (Adapted from Schalkowsky and Harris [1969])

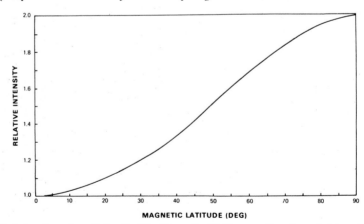

Fig. 5-2. Relative Intensity of the Earth's Magnetic Field as a Function of Magnetic Latitude (Adapted from Schalkowsky and Harris [1969])

Plots of the field strength for various altitudes are given in Figs. 5-3 and 5-4. Note that as the altitude increases, the contours become more regular and begin to resemble a dipole field more closely.

The low in magnetic intensity at about 25°S, 45°W (called the *Brazilian Anomaly*) together with the high at about 10°N, 100°E implies that the center of the magnetic dipole is offset from the Earth's center. In 1975, the eccentric dipole was offset 474.2 km in the direction of 19.5°N, 146.9°E [J. Bartels, 1936]. The eccentric dipole is moving outward at 2.4 km/year, westward at 0.19 deg/year and northward at 0.23 deg/year. The eccentric nature of the dipole can be described mathematically as a quadrupole distribution of magnetization. The maximum deviations of the centered dipole model and the quadrupole model from the actual field of the Earth are shown in Fig. 5-5.

The fact that the field rotates with the Earth is a clear indication that the field originates within the Earth. A coherent dipole field of this nature can be produced either by a uniformly magnetized sphere or by a current loop. However, calculations of the magnetization required lead to values much higher than those observed in the Earth's crust. Magnetization deeper than the crust is unlikely because the *Curie point* (i.e., the temperature at which a magnetized material loses its magnetization) of iron is reached only 20 km below the Earth's surface [Haymes, 1971].

An alternative theory postulates a dynamo effect in the outer core of the Earth driven by thermal convection currents [Garland, 1971]. Basically, a dynamo is a conductor driven in a magnetic field such that it acts to sustain that field. The theory has been refined to include a primary current which produces the dipole,

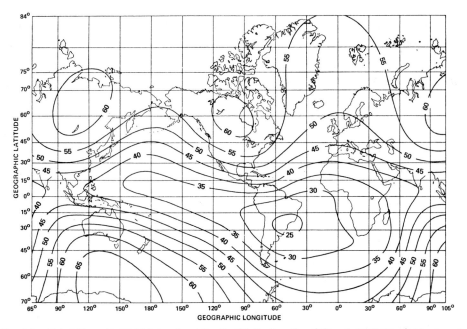

Fig. 5-3.   Total Magnetic Field Intensity at the Earth's Surface (in μT Epoch 1965) (From Harris and Lyle [1969])

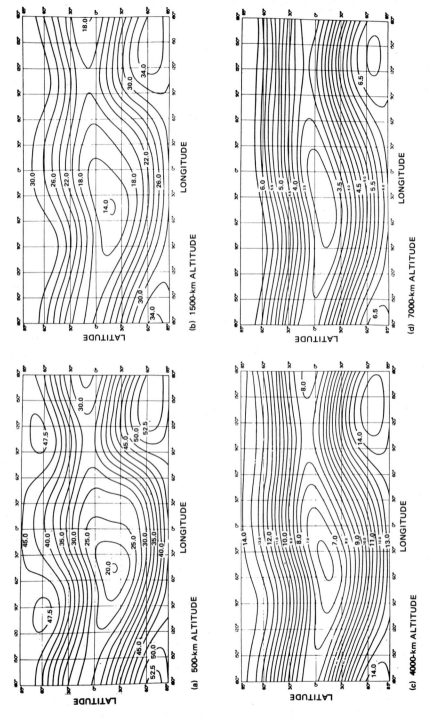

Fig. 5-4.   Magnitude of the Earth's Magnetic Field (in μT Epoch 1960) (From Chernosky, Fougere, Hutchinson [1965])

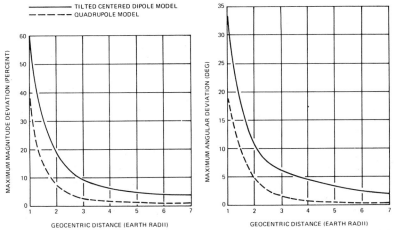

Fig. 5-5. Maximum Deviations of Approximate Models From the Earth's Magnetic Field (From Harris and Lyle [1969])

plus secondary currents or whirlpools near the core-mantle boundary which produce local dipoles. These secondary dipoles are then superimposed to produce the observed multipole nature of the field, as well as local anomalies, which are large surface areas where the magnetic field deviates appreciably from the dipole field. The creation and decay of the whirlpools may cause the secular drift. Another theory of the secular drift is that the core is rotating more slowly than the mantle and crust.

Although the exact nature of the field generator is unknown, the fact that it is internal suggests that the field can be conveniently described as a solution to a boundary value problem. The lack of surface electric currents implies that outside the Earth, the magnetic field, **B**, has zero curl,

$$\nabla \times \mathbf{B} = 0 \qquad (5\text{-}1)$$

which means that the field can be expressed as the gradient of a scalar potential, V

$$\mathbf{B} = -\nabla V \qquad (5\text{-}2)$$

The absence of magnetic monopoles implies

$$\nabla \cdot \mathbf{B} = 0 \qquad (5\text{-}3)$$

Substituting Eq. (5-2) into Eq. (5-3) yields Laplace's equation:

$$\nabla^2 V = 0 \qquad (5\text{-}4)$$

which, because of the spherical nature of the boundary at the Earth's surface, has a solution conveniently expressed in spherical harmonics as

$$V(r,\theta,\phi) = a \sum_{n=1}^{k} \left(\frac{a}{r}\right)^{n+1} \sum_{m=0}^{n} (g_n^m \cos m\phi + h_n^m \sin m\phi) P_n^m(\theta) \qquad (5\text{-}5)$$

where a is the equatorial radius of the Earth; g_n^m and h_n^m are called *Gaussian coefficients*; r, θ, and ϕ are the geocentric distance, coelevation, and east longitude from Greenwich; and $P_n^m(\theta)$ are the associated Legrendre functions. (See Appendix G for a further discussion of spherical harmonics.) The $n = 1$ terms are called dipole; the $n = 2$, quadrupole; the $n = 3$, octupole. The actual calculation of **B** from Eqs. (5-5) and (5-2) is explained in detail in Appendix H.

To use Eq. (5-5) to calculate the field at any point, the Gaussian coefficients must be known. It is the object of theories, such as the core-dynamo theory, to calculate them; however, success has been severely limited. The alternative is to determine the Gaussian coefficients empirically by doing a least-squares fit to magnetic field data using the coefficients as fitting parameters. Data consisting of both magnitude and direction is obtained from a series of magnetic observatories. Unfortunately, these observatories are not distributed uniformly so that the data is sparse in some regions of the Earth. More uniformly distributed data is obtained from field magnitude measurements made by satellites. Although there are some theoretical arguments that obtaining coefficients by simply fitting field magnitudes is an ambiguous process [Stern, 1975], it appears to work quite well in practice [Cain, 1970].

One set of Gaussian coefficients to degree 8 (n in Eq. (5-5)) and order 8 (m in Eq. (5-5)), comprises the *International Geomagnetic Reference Field (IGRF (1975))* [Leaton, 1976] and is given in Appendix H. The field model includes the first-order time derivatives of the coefficients in an attempt to describe the secular variation. Because of the lack of adequate data over a long enough period of time, the accuracy of this (or any) field model will degrade with time. In fact, the IGRF (1975) is an update of IGRF (1965) [Cain and Cain, 1971]. The IGRF (1965) should be used for the period 1955-1975 and the IGRF (1975) should be used for the period 1975-1980. The maximum and root-mean-square (RMS) errors in the field magnitude based on IGRF (1965) are given in Table 5-1 for 1975.

The estimated growth of the errors presented in the table was a factor of two from 1970 to 1975. The errors in direction (i.e., in components of the field) are more difficult to estimate but should not be more than a factor of two greater than the magnitude data shown in Table 5-1. One value in Table 5-1 was verified by GEOS-3 data taken in a polar orbit at an altitude of 840 km [Coriell, 1975].

Substitution of the magnetic field potential in Eq. (5-5) into Eq. (5-2) will show that the strength of the dipole field decreases with the inverse cube of the distance from the center of the dipole, and that the quadrupole decreases with the inverse fourth power. Higher degree multipoles decrease even more rapidly. Thus, at the

Table 5-1.   Errors in the Field Magnitude Derived From the IGRF (1965) for 1975 (From Trombka and Cain [1974])

DISTANCE FROM EARTH CENTER (EARTH RADII)	DIPOLE FIELD* MAGNITUDE (nT)	MAXIMUM ERROR (nT)	RMS ERROR (nT)
1 (SURFACE)	30,800	940	280
1.07 (445 KM ALTITUDE)	25,150	540	180
2	3,850	34	18
3	1,150	8	4
4	480	4	2

*CALCULATED AT THE MAGNETIC EQUATOR.

high altitudes attained by some satellites, it is frequently possible to use a reduced degree of expansion in the field model. Omitting higher multipoles permits reductions in computation time. An estimate of the error resulting from this truncation can be obtained by comparing the full field model with its truncated forms, as shown in Tables 5-2 and 5-3. The choice of degree should be based on the accuracy of the model in Table 5-1 and on the strength of the perturbations of the main field discussed below. For a given altitude, those truncation errors to the left of the heavy line in Table 5-2 exceed the errors in the field model itself. Generally,

Table 5-2. Field Truncation Errors (nT) Using the IGRF (1965) (From Trombka and Cain [1974])

ALTITUDE	DEGREE						
	1 (DIPOLE)	2 (QUADRUPOLE)	3	4	5	6	7
SURFACE (R = 1 EARTH RADIUS):							
MAXIMUM	20255	13905	8125	3452	1819	858	268
RMS	10231	6942	3685	1640	855	364	129
300 KM ABOVE SURFACE:							
MAXIMUM	16367	10844	6110	2440	1244	564	169
RMS	8281	5431	2764	1171	587	240	81
R = 2 EARTH RADII:							
MAXIMUM	945	369	114	22	6	2	.
RMS	507	191	53	11	3	.	.
R = 3 EARTH RADII:							
MAXIMUM	169	47	10	1	.	.	.
RMS	96	25	5	.	.	.	.
R = 4 EARTH RADII:							
MAXIMUM	51	11	2	.	.	.	.
RMS	30	6	.	.	.	.	.

* = LESS THAN ONE nT.

Table 5-3. Angular Errors (deg) Using Truncations of the IGRF (1965) (From Trombka and Cain [1974])

ALTITUDE	DEGREE						
	1 (DIPOLE)	2 (QUADRUPOLE)	3	4	5	6	7
SURFACE (R = 1 EARTH RADIUS):							
MAXIMUM	32	21	8	5	2	1	0.4
RMS	10	7	4	2	1	0.4	0.2
300 KM ABOVE SURFACE:							
MAXIMUM	29	19	7	4	2	0.8	0.3
RMS	10	7	3	1.3	0.8	0.3	0.1
R = 2 EARTH RADII:							
MAXIMUM	10	4	1	0.3	.	.	.
RMS	4	2	0.4	.	.	.	.
R = 3 EARTH RADII:							
MAXIMUM	6	2	0.3	.	.	.	.
RMS	2	0.7	0.1	.	.	.	.
R = 4 EARTH RADII:							
MAXIMUM	4	1	0.1	.	.	.	.
RMS	1.7	0.4	.	.	.	.	.

* = LESS THAN 0.1 DEG.

beyond 4 Earth radii and especially beyond synchronous altitude, 6.6 Earth radii, the perturbations are sufficiently large to render the harmonic expansion model invalid.

Perturbations to the Main Field.   The primary source of geomagnetic field perturbations is the Sun. The Sun constantly emits a neutral plasma called the *solar wind*, described in Section 5.3. The action of the solar wind is to distort the Earth's field at high altitudes (8 to 10 Earth radii) so that the multipole description of Eq. (5-5) is no longer valid at those altitudes. Because the plasma is highly conductive, it will not allow the Earth's field to enter it. Thus, the plasma compresses the field ahead of it until the plasma energy density equals the magnetic field energy density at a distance of about 10 Earth radii [Haymes, 1971]. At that point, the plasma breaks up so that some of the charged particles are trapped in the magnetic field. Other particles slip around the field and drag the field lines along as they pass the Earth, as shown schematically in Fig. 5-6.

A shock front, similar to a sonic boom, occurs where the solar wind first strikes the geomagnetic field because the solar wind is moving faster than the field can respond [Haymes, 1971]. Just beyond the shock front is a region of magnetic turbulence called the *magnetosheath*. It is characterized by rapidly fluctuating field strengths and directions [Harris and Lyle, 1969]. Within the *magnetosphere*, the field is primarily due to geologic causes; outside, the field is due largely to the solar wind and its interaction with the geomagnetic field. The boundary between the magnetosheath and the magnetosphere is called the *magnetopause*. The region behind the Earth relative to the Sun, where the geomagnetic field lines first fail to close because they are being dragged along by the plasma, is called the *cusp region*. It occurs at a distance of 8 to 16 Earth radii and ± 25 deg geomagnetic latitude [Harris and Lyle, 1969]. That part of the geomagnetic field which is carried by the plasma is called the *magnetotail*. Its extent is not known, but it has been observed by Pioneer VII at 1000 Earth radii [Harris and Lyle, 1969]. The plane which

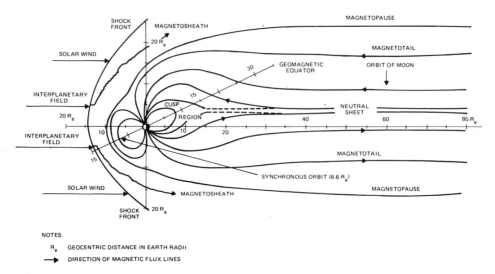

Fig. 5-6.   The Earth's Magnetosphere (From Harris and Lyle [1969])

separates the incoming field lines from the outgoing field lines is called the *neutral sheet*.

Although the solar wind is fairly constant, it is frequently augmented by energetic bursts of plasma emitted by solar flares. When this plasma encounters the geomagnetic field, it compresses the field further giving a rise in field intensity on the surface of the Earth. This rise, called *sudden commencement*, initiates what is referred to as a *magnetic storm* [Haymes, 1971]. The *initial phase* has a typical strength of 50 nT and lasts for about 1 hour. During particularly strong storms, the magnetopause can be compressed to below synchronous altitude. After compressing the field, the plasma burst injects more charged particles into the geomagnetic field. These particles spiral around the field lines in a northsouth direction, reversing direction ("mirroring") at those locations where their velocity is perpendicular to the field line, usually at high latitudes. They will also drift in an eastwest direction, thus developing a ring current at 3 to 5 Earth radii whose magnetic field (up to ∼400 nT) opposes the geomagnetic field. This phenomenon causes the *main phase* of the magnetic storms which lasts for a few hours until the charged particles start to escape from the magnetic entrapment through collisions with the atmosphere. The initial recovery to about 150 nT then takes from 6 hours to 2 days. The field fully recovers in several days.

These phenomena are summarized in Fig. 5-7, which shows the general characteristics of a magnetic storm. The graph, as well as the foregoing discussion, is an oversimplification. Although a storm is observed simultaneously throughout the world, its characteristics will be different for observers at different latitudes. The largest storm effects occur in the auroral zones which are 5 deg either side of 67 deg geomagnetic latitude. At that latitude, the disturbance can exceed 2000 nT. The amplitude decreases rapidly with latitude to about 250 nT at 30 deg latitude and increases to several hundred nT at the equator. At the higher latitudes, the storm is characteristically much more irregular than that shown in Fig. 5-7 [Chernosky, Fougere, and Hutchinson, 1965].

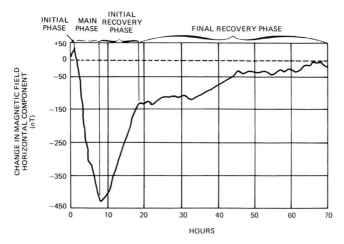

Fig. 5-7. Magnetic Storm Effect on Main Geomagnetic Field Intensity (Observation of a large magnetic storm in Hawaii, February 1958, from Harris and Lyle [1969])

The frequency of the storms is somewhat correlated with Sun-spot activity, since flares are usually associated with Sun spots [Haymes, 1971]. Thus, a storm may recur after 27 days (the length of a solar rotation) and the overall activity tends to follow the 11-year solar cycle. Storms are also more frequent near the equinoxes, possibly because at approximately those times, the position of the Earth is at the highest solar latitude (about 7 deg). Sun spots appear most frequently between the solar latitudes of 5 deg and 40 deg on both sides of the equator. They appear first at the higher solar latitudes at the beginning of a solar cycle.

The geomagnetic field is monitored continuously at a series of stations called *magnetic observatories*. They report observed magnetic activity, such as storms, as an index, K, which is the deviation of the most disturbed component of the field from the average quiet-day value [Chernosky, Fougere, and Hutchinson, 1965]. The K scale is quasi-logarithmic with $K=0$, quiet, and $K=9$, the largest disturbance the station is likely to see. The value of K is averaged and reported for every 3 hours. The values of K for 12 selected stations are corrected for the station's geomagnetic latitude (since activity is latitude dependent) and then averaged to produce the *planetary index*, K_p. The indices are published each month in the *Journal of Geophysical Research*. The value of K_p is a good indicator of the level of magnetic storms and is therefore an indication of the deviation of the geomagnetic field from the model in Eq. (5-5). The 3-hourly planetary index can be roughly (10%) converted to the linear 3-hourly planetary amplitude, a_p, by:

$$a_p = \exp((K_p + 1.6)/1.75) \qquad (5\text{-}6)$$

The value of a_p is scaled such that at 50 deg geomagnetic latitude and a deviation of 500 nT at $K=9$, the field deviation $\Delta B = |B_{\text{disturbed}} - B_{\text{quiet}}|$ is

$$\Delta B \approx 2a_p \qquad (5\text{-}7)$$

For other latitudes, a_p is scaled by dividing the lower limit of ΔB for $K=9$ by 250. Thus, at a higher latitude for which $\Delta B = 1000$ nT corresponds to $K=9$,

$$\Delta B \approx 4a_p \qquad (5\text{-}8)$$

Although K_p is a measure of geomagnetic activity, it is ultimately a measure of solar activity. In fact, it has been found empirically that the velocity of the solar wind can be derived from K_p by:

$$v = 8.44 \sum K_p + 330 \qquad (5\text{-}9)$$

where v is in kilometres per second and $\sum K_p$ is the sum over the eight values of K_p for the day [Haymes, 1971]. Similarly, the interplanetary magnetic field is generated by the Sun, and it has been shown empirically that the interplanetary field is approximately

$$K_p = 0.3B \pm 0.2 \qquad (5\text{-}10)$$

where B is the magnitude of the interplanetary field in nT [Haymes, 1971].

The Sun is also responsible for the diurnal variation of the geomagnetic field. Solar electromagnetic radiation ionizes some atmospheric atoms and molecules at an altitude of roughly 100 km, producing the E-layer of the ionosphere. The Sun's gravitational field then exerts a tidal force causing the ions and electrons to rise.

The interaction of the charged particles with the geomagnetic field produces a rather complex current system which creates a magnetic field. The effect is most pronounced on the day side of the Earth, since it is dependent on the ion density of the E-layer. On solar quiet days, this field causes a deviation from the internal field of 20-40 nT in the middle latitude regions and can cause deviations of 100-200 nT near the magnetic equator [Harris and Lyle, 1969]. At each magnetic observatory, the daily magnetic variations for the five quietest days are averaged together to produce the quiet day solar variation, S_q. This variation is subtracted from the actual variations before generating K_p. The Moon also exerts daily tidal forces which lead to quiet-day variations about 1/30 of that due to the Sun [Harris and Lyle, 1969].

There are two other current systems of some importance: the *polar electrojet* and the *equatorial electrojet*. The polar electrojet is an intense ionospheric current that flows westward at an altitude about 100 km in the auroral zone. Changes in the electrojet can cause negative excursions (called *bays*) as great as 2000 nT and are typically about 1000 to 1500 nT at the Earth's surface. The excursions can last from 0.5 to 2 hours. Like magnetic storms, auroral activity has a 27-day periodicity and reaches a maximum at the equinoxes [Harris and Lyle, 1969]. The equatorial electrojet is an intense west-to-east current in the sunlit ionosphere. It is partly responsible for the high intensity of the magnetic storms. It produces a 220 nT discontinuity in the total field between 96- and 130-km altitude. At 400 km, the field is 30 to 40 nT at longitudes across South America and 10 to 20 nT elsewhere [Zmuda, 1973].

5.2 The Earth's Gravitational Field

John Aiello
Kay Yong

Two point masses, M and m, separated by a vector distance $\mathbf{r}$, attract each other with a force given by Newton's law of gravitation as

$$\mathbf{F} = \frac{GMm}{r^2}\hat{\mathbf{r}} \tag{5-11}$$

where G is the gravitational constant (see Appendix M). If $M_\oplus$ is the mass of the Earth and m is the mass of the body whose motion we wish to follow, then it is convenient to define the *geocentric gravitational constant*, $\mu_\oplus$, and the *Earth gravitational potential*, U, by:

$$\mu_\oplus \equiv GM_\oplus \tag{5-12}$$

$$U \equiv -\frac{GM_\oplus}{r} \tag{5-13}$$

From Eqs. (5-12) and (5-13), Eq. (5-11) may be rewritten as the gradient of a scalar potential:

$$\mathbf{F} = -\frac{\mu_\oplus m}{r^2}\hat{\mathbf{r}}$$

$$= -m\nabla U \tag{5-14}$$

where **r** is the unit vector from the Earth's center to the body (assumed to be a point mass). A gravitational potential satisfying Eq. (5-14) may always be found due to the conservative nature of the gravitational field.

By extending the single point mass, m, to a collection of point masses, the gravitational potential at a point outside a continuous mass distribution over a finite volume can be defined. For example, consider a solid body of density ρ, situated in a rectangular coordinate system with the mass elements coordinates denoted by (ξ, η, ζ) and the point coordinates denoted by (x, y, z). The gravitational potential at the point, $P = (x, y, z)$, due to the body can be written as [Battin, 1963]

$$U(x,y,z) = -G \int_\xi \int_\eta \int_\zeta \frac{\rho(\xi,\eta,\zeta)}{\left[(x-\xi)^2 + (y-\eta)^2 + (z-\zeta)^2\right]^{1/2}} d\xi \, d\eta \, d\zeta \quad (5\text{-}15)$$

Successive applications of Gauss' law and the divergence theorem show that U satisfies Poisson's equation,

$$\nabla^2 U = 4\pi G \rho \quad (5\text{-}16)$$

which, in the region exterior to the body (i.e., where $\rho = 0$), reduces to Laplace's equation:

$$\nabla^2 U = 0 \quad (5\text{-}17)$$

Because of the spherical symmetry of most astronomical objects, it is convenient to write Eq. (5-15) in the spherical coordinate system (r, θ, ϕ). In this case, solutions to Eq. (5-17) may be written in terms of spherical harmonics as described in Appendix G. Specifically, U for the Earth can be expressed in the convenient form

$$U = -\frac{\mu_\oplus}{r} + B(r,\theta,\phi) \quad (5\text{-}18)$$

where $B(r,\theta,\phi)$ is the appropriate spherical harmonic expansion to correct the gravitational potential for the Earth's nonsymmetric mass distribution. $B(r,\theta,\phi)$ may be written explicitly* as [Meirovitch, 1970; Escobal, 1965]

$$B(r,\theta,\phi) = \frac{\mu_\oplus}{r} \left\{ \sum_{n=2}^{\infty} \left[\left(\frac{R_\oplus}{r}\right)^n J_n P_{n0}(\cos\theta) \right.\right.$$

$$\left.\left. + \sum_{m=1}^{n} \left(\frac{R_\oplus}{r}\right)^n (C_{nm}\cos m\phi + S_{nm}\sin m\phi) P_{nm}(\cos\theta) \right] \right\} \quad (5\text{-}19)$$

Here, $R_\oplus$ is the radius of the Earth, J_n are *zonal harmonic coefficients*, $P_{n,m}$ are Legendre polynomials, and C_{nm} and S_{nm} are *tesseral harmonic coefficients* for $n \neq m$ and *sectoral harmonic coefficients* for $n = m$ (see Appendix G).

In Eq. (5-19), we see that the zonal harmonics depend only on latitude, not on longitude. These terms are a consequence of the Earth's oblateness. The tesseral

---

*Note that the $n = 0$ term is written explicitly in Eq. (5-18) as $-\dfrac{\mu_\oplus}{r}$, and that the $n = 1$ term is absent due to the origin of the coordinate system being coincident with the Earth's center of mass.

harmonics represent longitudinal variations in the Earth's shape. Although generally smaller than zonal terms, tesseral components become important in the case of geosynchronous spacecraft because the satellite remains nearly fixed relative to the Earth; consequently, longitudinal variations do not average to zero over a long period of time. For most satellites other than geosynchronous ones, the assumption of axial symmetry of the Earth is usually valid, and only the zonal harmonic corrections are needed. Thus, the expression for the gravitational potential of the Earth can be approximated as

$$U \approx -\frac{GM}{r}\left[1 - \sum_{n=2}^{\infty}\left(\frac{R_\oplus}{r}\right)^n J_n P_{n0}(\cos\theta)\right] \qquad (5\text{-}20)$$

The zonal harmonics are a major cause of perturbations for Earth-orbiting spacecraft, being the primary source of changes in orbital period, longitude of the ascending node, and argument of perigee (Section 3.4).

The gravitational potential of Eq. (5-18), when combined with the potential due to the angular momentum of the Earth's rotation, describes a mathematical model or reference figure for the shape of the Earth, known as the *geoid* or *mean sea level*. The geoid is a surface coincident with the average sea level (i.e., less meteorological and tidal effects) over the globe or in an imaginary channel cut in the continents. A number of measurement techniques have been used [King-Hele, 1976] to map the geoid, including satellite-to-satellite tracking, in which a geosynchronous satellite measures the relative velocity of a lower orbiting satellite; radar altimetry from satellites; and laser ranging to reflectors both on satellites and the Moon. The last method has been the most accurate.

Figure 5-8 illustrates the *geoid height* or deviation of the geoid from a reference

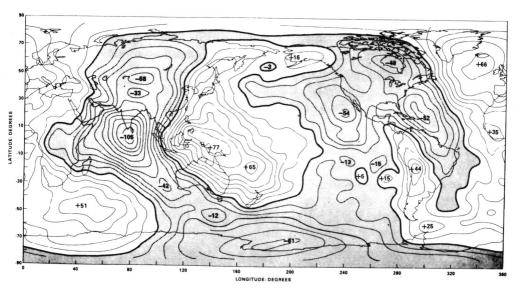

Fig. 5-8.   Geoid Heights From Goddard Earth Model-8 (GEM-8). Contours are at 10-m intervals. (From Wagner, *et al.*, [1976]) .

spheroid of flattening 1/298.255 and semimajor axis of 6378.145 km* as given by
the Goddard Earth Model-8 (GEM-8) [Wagner, *et al.*, 1976]. Particularly noticeable is the variation from 77 m above the geoid near New Guinea to 105 m below
in the Indian Ocean.

The accuracy of the potential function, i.e., the number of terms included in
the infinite spherical harmonic series, has a greater effect on orbital dynamics than
on attitude dynamics. For analysis of gravity-gradient torques on spacecraft,
inclusion of the J_2 term in the harmonic series is normally sufficient because of the
uncertainties in other environmental disturbance torques. For practical purposes,
the point mass potential function, Eq. (5-14), is adequate for spinning satellites or
those with only short appendages. Table 5-4 lists the differential acceleration,
$da/dr = -2\mu/r^3$, experienced by these satellites for various altitudes.

Table 5-4.   Differential Acceleration, $\Delta a/l$ for Point Mass Gravitational Field ($\Delta a \equiv a_2 - a_1$ is the
difference in acceleration between points p_1 and p_2 whose distances from the center of the
massive object are r_1 and r_2 and $l \equiv r_2 - r_1$.)

DISTANCE FROM CENTER OF OBJECT (km)	$\Delta a/l$ (m · sec^{-2}/m)		
	MOON	EARTH	SUN
2,000	1.23×10^{-6}	–	–
6,400	3.74×10^{-8}	3.04×10^{-6}	–
8,000	1.92×10^{-8}	1.56×10^{-6}	–
20,000	1.23×10^{-9}	9.96×10^{-8}	–
40,000	1.53×10^{-10}	1.25×10^{-8}	–
10^5	9.80×10^{-12}	7.97×10^{-10}	–
10^6	9.80×10^{-15}	7.97×10^{-13}	2.65×10^{-7}
10^7	–	7.97×10^{-16}	2.65×10^{-10}
10^8	–	–	2.65×10^{-13}

*$\Delta a = a_2 - a_1$ IS THE DIFFERENCE IN ACCELERATION BETWEEN POINTS p_1 AND p_2 WHOSE
DISTANCES FROM THE CENTER OF THE MASSIVE OBJECT ARE r_1 AND r_2 AND $l \equiv r_2 - r_1$.

For gravity-gradient or three-axis-stabilized satellites with long, flexible appendages, the J_2 effect becomes significant in overall attitude motion. For example,
consider the potential function for the Earth written as

$$U \cong \frac{GM}{r} \left[U_0 + U_{J_2} + U_{J_3} + U_{J_4} + \dots \right] \qquad (5\text{-}21)$$

where

$$U_0 = -1$$

$$U_{J_2} = \left(\frac{a}{r}\right)^2 J_2 \frac{1}{2} (3\cos^2\theta - 1)$$

$$U_{J_3} = \left(\frac{a}{r}\right)^3 J_3 \frac{5}{2} \left(\cos^3\theta - \frac{3}{5}\cos\theta\right)$$

$$U_{J_4} = \left(\frac{a}{r}\right)^4 J_4 \frac{35}{8} \left(\cos^4\theta - \frac{6}{7}\cos^2\theta + \frac{3}{35}\right)$$

---

*See Section 4.3 for definition of various reference surfaces.

The values for each of these terms are compared in Table 5-5 for $\theta = 90$ deg, where it is seen that although U_{J_2} is large compared with U_{J_3} and U_{J_4}, it is only about 0.05% the value of U_0.* Moreover, the importance of zonal harmonic corrections

Table 5-5. Comparison of Various Zonal Harmonic Terms in the Expansion for the Earth's Gravitational Potential Over the Equator

ALTITUDE (KM)	$\dfrac{GM}{r}$ (KM^2/SEC^2)	U_0	U_{J_2}	U_{J_3}	U_{J_4}
0	62.495	1	0.54×10^{-3}	0	-0.59×10^{-6}
200	60.595	1	0.51×10^{-3}	0	-0.52×10^{-6}
500	57.952	1	0.46×10^{-3}	0	-0.44×10^{-6}
1,000	54.025	1	0.40×10^{-3}	0	-0.33×10^{-6}
2,000	47.576	1	0.31×10^{-3}	0	-0.20×10^{-6}
10,000	24.337	1	0.82×10^{-4}	0	-0.45×10^{-8}
36,000*	9.401	1	0.13×10^{-4}	0	-0.30×10^{-11}

*APPROXIMATELY GEOSYNCHRONOUS ORBIT.

becomes less significant at higher altitudes. When the spacecraft is in an almost geosynchronous orbit, the inclusion of lunar and solar attractions may become more important than the spherical harmonic correction, as shown in Table 5-6.

Table 5-6. Acceleration Due to Gravity With Lunar and Solar Corrections for Earth, Moon, and Sun in Syzygy*

HEIGHT (KM)	a_{EARTH}	$\lvert \Delta a_{MOON} \rvert$		$\lvert \Delta a_{SUN} \rvert$	
		+	−	+	−
100	9.498	1.091×10^{-6}	1.15×10^{-6}	5.131×10^{-4}	5.138×10^{-4}
500	8.426	1.16×10^{-6}	1.22×10^{-6}	5.447×10^{-4}	5.454×10^{-4}
2,000	5.678	1.40×10^{-6}	1.49×10^{-6}	6.634×10^{-4}	6.646×10^{-4}
10,000	1.486	2.66×10^{-6}	3.02×10^{-6}	1.296×10^{-3}	1.300×10^{-3}
35,786	0.2242	6.23×10^{-5}	8.68×10^{-6}	3.328×10^{-3}	3.356×10^{-3}
100,000	0.0352	1.28×10^{-5}	3.02×10^{-5}	8.342×10^{-3}	8.522×10^{-3}

*THE ACCELERATION DUE TO GRAVITY WITH LUNAR AND SOLAR CORRECTIONS FOR EARTH, MOON, AND SUN IN SYZYGY IS

$$a_E = \frac{\mu_E}{r^2}$$

WHERE r = HEIGHT (KM)

$$\lvert \Delta a_{MOON} \rvert = \mu_M \left\lvert \left[\frac{1}{(R_{MOON-SAT})^2} - \frac{1}{(R_{EARTH-MOON})^2} \right] \right\rvert \qquad \lvert \Delta a_{SUN} \rvert = \mu_s \left\lvert \left[\frac{1}{(R_{SUN-SAT})^2} - \frac{1}{(R_{EARTH-SUN})^2} \right] \right\rvert$$

---

*Values of the major spherical harmonic coefficients for the Earth are given in Table L-5.

For attitude dynamics, the following gravitational potential function is normally sufficient for the computation of gravity-gradient torques:

$$U_p = -\frac{\mu_\oplus}{r_p} + \frac{3}{2}J_2\frac{\mu_\oplus}{r_p}\left(\frac{R_\oplus}{r_p}\right)^2\left[\left(\frac{r_{3p}}{r_p}\right)^2 - \frac{1}{3}\right] \tag{5-22}$$

where $\mathbf{r}_p$ is the vector from the center of the Earth to the point P, $\mathbf{r}_{3p}$ is the component of $\mathbf{r}_p$ parallel to the Earth's spin axis, and $R_\oplus$ is the equatorial radius of the Earth.

The gravitational potential given by Eq. (5-22) can be expanded in Taylor series about the center of mass of the spacecraft to give

$$U_p = U_c + \nabla U_c\cdot(\mathbf{r}_p - \mathbf{r}_c) + (\mathbf{r}_p - \mathbf{r}_c)\cdot\frac{[G]}{2}\cdot(\mathbf{r}_p - \mathbf{r}_c) \tag{5-23}$$

Therefore, the gravitational acceleration at point $\mathbf{r}_p$ is

$$\mathbf{g}_p = \frac{d^2\mathbf{r}_p}{dt^2} = -\nabla U_p = -\nabla U_c + [G]\cdot(\mathbf{r}_c - \mathbf{r}_p)$$

$$= \mathbf{g}_c + [G]\cdot(\mathbf{r}_c - \mathbf{r}_p) \tag{5-24}$$

where U_c is a constant; $\mathbf{r}_p$ and $\mathbf{r}_c$ are the position vectors of an arbitrary point and the center of mass of the spacecraft, respectively, in geocentric coordinates; and $\mathbf{g}_e \equiv -\nabla U_c$ is the gravitational acceleration of the center of mass. Specifically,

$$\mathbf{g}_c = -\frac{\mu_\oplus}{r_c^3}\mathbf{r}_c + \frac{9}{2}\frac{\mu_\oplus}{r_c^3}J_2\left(\frac{R_\oplus}{r_c}\right)^2\left[\left(\frac{r_{3c}}{r_c}\right)^2 - \frac{1}{3}\right]\mathbf{r}_c + \frac{3}{2}\frac{\mu_\oplus}{r_c^2}J_2\left(\frac{R_\oplus}{r_c}\right)^2\sin^2\delta\,\hat{\delta} \tag{5-25}$$

where $\hat{\delta}$ is a unit vector in the direction of increasing declination, δ is the declination, $\cos^{-1}(r_{3c}/r_c)$, and $[G]$ is the gravity-gradient tensor with components

$$G_{ij} = \frac{\partial U_c}{\partial r_{ic}\partial r_{jc}} \tag{5-26}$$

The gravity-gradient tensor is important for attitude work; explicit expressions for the components are

$$G_{11} = \frac{\mu_\oplus}{r_c^3}\left[1 - 3\left(\frac{r_{1c}}{r_c}\right)^2\right] + \frac{3}{2}J_2\frac{\mu_\oplus}{r_c^3}\left(\frac{R_\oplus}{r_c}\right)^2\left\{1 - 5\left[\left(\frac{r_{1c}}{r_c}\right)^2 + \left(\frac{r_{3c}}{r_c}\right)^2\right]\right.$$

$$\left. + 35\left(\frac{r_{1c}}{r_c}\right)^2\left(\frac{r_{3c}}{r_c}\right)^2\right\} \tag{5-27}$$

$$G_{22} = \frac{\mu_\oplus}{r_c^3}\left[1 - 3\left(\frac{r_{2c}}{r_c}\right)^2\right] + \frac{3}{2}J_2\frac{\mu_\oplus}{r_c^3}\left(\frac{R_\oplus}{r_c}\right)^2\left\{1 - 5\left[\left(\frac{r_{2c}}{r_c}\right)^2 + \left(\frac{r_{3c}}{r_c}\right)^2\right]\right.$$

$$\left. + 35\left(\frac{r_{2c}}{r_c}\right)^2\left(\frac{r_{3c}}{r_c}\right)^2\right\}$$

$$G_{33} = \frac{\mu_\oplus}{r_c^3}\left[1 - 3\left(\frac{r_{3c}}{r_c}\right)^2\right] + \frac{3}{2}J_2\frac{\mu_\oplus}{r_c^3}\left(\frac{R_\oplus}{r_c}\right)^2\left\{3 - 30\left(\frac{r_{3c}}{r_c}\right)^2 + 35\left(\frac{r_{3c}}{r_c}\right)^4\right\}$$

$$G_{12} = G_{21} = -3\frac{\mu_\oplus}{r_c^3}\left(\frac{r_{1c}}{r_c}\right)\left(\frac{r_{2c}}{r_c}\right) + \frac{3}{2}J_2\frac{\mu_\oplus}{r_c^3}\left(\frac{R_\oplus}{r_c}\right)^2\left\{-5\left(\frac{r_{1c}}{r_c}\right)\left(\frac{r_{2c}}{r_c}\right)\right.$$

$$\left. + 35\left(\frac{r_{1c}}{r_c}\right)\left(\frac{r_{2c}}{r_c}\right)\left(\frac{r_{3c}}{r_c}\right)^2\right\}$$

$$G_{13} = G_{31} = -3\frac{\mu_\oplus}{r_c^3}\left(\frac{r_{1c}}{r_c}\right)\left(\frac{r_{3c}}{r_c}\right) + \frac{3}{2}J_2\frac{\mu_\oplus}{r_c^3}\left(\frac{R_\oplus}{r_c}\right)^2\left\{-5\left(\frac{r_{1c}}{r_c}\right)\left(\frac{r_{3c}}{r_c}\right)\right.$$

$$\left. + 35\left(\frac{r_{1c}}{r_c}\right)\left(\frac{r_{3c}}{r_c}\right)\left(\frac{r_{3c}}{r_c}\right)^2\right\}$$

$$G_{23} = G_{32} = -3\frac{\mu_\oplus}{r_c^3}\left(\frac{r_{2c}}{r_c}\right)\left(\frac{r_{3c}}{r_c}\right) + \frac{3}{2}J_2\frac{\mu_\oplus}{r_c^3}\left(\frac{R_\oplus}{r_c}\right)^2\left\{-5\left(\frac{r_{2c}}{r_c}\right)\left(\frac{r_{3c}}{r_c}\right)\right.$$

$$\left. + 35\left(\frac{r_{2c}}{r_c}\right)\left(\frac{r_{3c}}{r_c}\right)\left(\frac{r_{3c}}{r_c}\right)^2\right\}$$

The gravity-gradient matrix becomes particularly useful for spacecraft that have long appendages or that are highly asymmetrical.

For a spacecraft in a high-altitude orbit, the J_2 term may be less important than the lunar and solar perturbations which are included in the equation of motion of a satellite [Escobal, 1965; Battin, 1963] as follows:

$$\frac{d^2\mathbf{r}}{dt^2} = -\nabla U - GM_M\left(\frac{\mathbf{r}_{MV}}{r_{MV}^3} - \frac{\mathbf{r}_{EM}}{r_{EM}^3}\right) - GM_S\left(\frac{\mathbf{r}_{SV}}{r_{SV}^3} - \frac{\mathbf{r}_{ES}}{r_{ES}^3}\right) \qquad (5\text{-}28)$$

where the subscripts M, S, E, and V denote Moon, Sun, Earth, and spacecraft, respectively, and $\mathbf{r}_{MV}$ is the vector from the Moon to the spacecraft.

5.3 Solar Radiation and The Solar Wind

David M. Gottlieb

Solar radiation includes all the electromagnetic waves emitted by the Sun, from X-rays to radio waves. *Solar wind* is the particulate radiation expelled from the Sun and consists mainly of ionized nuclei and electrons. Both solar radiation and solar wind may produce torques which affect the spacecraft attitude. The charged particles and the magnetic field embedded in the solar wind may also affect sensor performance or ground-spacecraft communication.

For most applications, torques due to solar radiation pressure will be much larger than those due to the solar wind. Torque is proportional to the *momentum flux* (momentum per unit area per unit time), and the solar radiation momentum flux is two to three orders of magnitude greater than that of the solar wind. Furthermore, the solar wind does not penetrate the Earth's magnetopause (see Section 5.1) except in the vicinity of the magnetic poles.

5.3.1 Solar Radiation

The mean solar energy flux integrated over all wavelengths is proportional to the inverse square of the distance from the Sun. To within 0.3%, the mean integrated energy flux *at the Earth's position* is given by:

$$F_e = \frac{1358}{1.0004 + 0.0334 \cos D} \, W/m^2 \tag{5-29}$$

where 1358 W/m^2 is the mean flux at 1 AU, and the denominator is a correction for the true Earth distance. *D* is the "phase" of year, measured from July 4, the day of Earth aphelion [Smith and Gottlieb, 1974]. This is equivalent to a mean momentum flux of 4.4×10^{-6} kg·m^{-1}·s^{-2}. Variations in this flux from this formula are always less than 0.5%. Solar radiation is largely emitted in the visible and near-infrared portions of the spectrum, as shown in Fig. 5-9. Note that the three curves coincide for wavelengths longer than 14 nm.

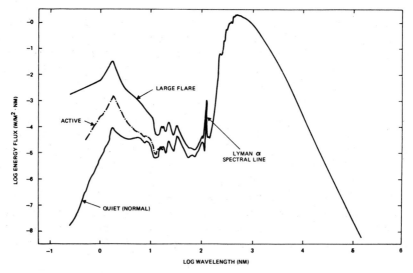

Fig. 5-9. Solar Energy Flux at 1 AU in the Ecliptic Plane (From Smith and Gottlieb [1974])

5.3.2 The Solar Wind

The solar wind was first postulated to explain the aurorae, geomagnetic disturbances, and the bending of comet tails, and was first observed directly by the Russian Luna 2 spacecraft in 1959 and Explorer 10 in 1961. The solar wind is coronal gas ejected from the Sun by a process that "can be deduced only by true believers, usually with a parental relationship to one of the competing ideas or models" [Hundhausen, 1972]. Its composition is typical of that of the corona, meaning that the relative abundance of elements is essentially solar, with hydrogen dominating, helium being second most common, and all other elements two or more orders of magnitude less abundant. For a table of solar abundances, see Allen [1973]. The solar wind is ionized virtually completely at least to a distance of 5 AU or more, with the ionization state of the elements being those that would be expected from a 1.5×10^6 °K gas (i.e., nearly all electrons with binding energies less than 130 eV are stripped from their nuclei). Table 5-7 lists some properties of the "quiet" solar wind at 1 AU in the ecliptic plane.

Table 5-7. Properties of the Quiet Solar Wind at 1 AU in the Ecliptic Plane (Adapted from Hundhausen and Wolfe [1972])

PROPERTY	VALUE	PROPERTY	VALUE
MEAN VELOCITY IN THE ECLIPTIC	$3.0 - 3.5 \times 10^5$ ms^{-1}	MEAN PROTON TEMPERATURE	4×10^4 °K
TYPICAL NONRADIAL VELOCITY IN THE ECLIPTIC	1.8×10^4 ms^{-1}	TYPICAL MAGNETIC FIELD	5 nT
MEAN VELOCITY PERPEN-DICULAR TO THE ECLIPTIC	1.8×10^4 ms^{-1}	MEAN MOMENTUM FLUX DENSITY IN THE ECLIPTIC	2.3×10^{-9} kg·m^{-1}·s^{-2}
PROTON DENSITY = ELECTRON DENSITY	8.7×10^6 m^{-3}	TYPICAL NONRADIAL COM-PONENT OF MOMENTUM FLUX DENSITY IN THE ECLIPTIC	7.9×10^{-12} kg·m^{-1}·s^{-2}
MEAN ELECTRON TEMPERATURE	1.5×10^5 °K	MEAN MOMENTUM FLUX DENSITY PERPENDICULAR TO THE ECLIPTIC	7.9×10^{-12} kg·m^{-1}·s^{-2}

Variations from the quiet solar wind values occur frequently. Figure 5-10 shows observed solar wind velocity distributions as observed by the Vela 3 spacecraft from 1965 to 1967. Other parameters listed in Table 5-7 probably vary, but the correlation of their variation with velocity is poorly known. One explanation of the variations is the sporadic occurrence of "high velocity streams" in the solar wind. The velocity increases over the period of a day to typically 6.5×10^5 m/s, and then declines over several days. High densities occur for the first day, followed by several days of abnormally low densities. The temperatures vary proportionally to the velocity. The direction of the wind moves west of radial up to about 8 deg near maximum velocity after being east of radial the same amount at the leading edge of the stream [Hundhausen, 1972].

High-velocity streams are associated with energetic solar storms, but the exact relationship is unknown. These must be regarded as unpredictable at this time. The solar wind appears to be split in regions (*sectors*), which may be connected with the high-velocity stream phenomenon. These sectors, each 30 to 180 deg across, are

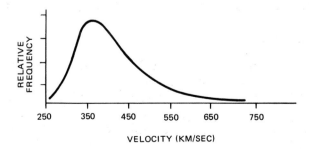

Fig. 5-10. Observed Solar Wind Velocity Distribution as Recorded by Vela 3 Spacecraft

best defined by the alternating direction of the interplanetary magnetic field within them, as shown in Fig. 5-11. The sector structure lasts for several months.

Data on the solar wind at distances other than 1 AU in the ecliptic plane is sparse. Pioneers 10 and 11, which took measurements of solar wind velocity from 1 to 5 AU, found that the mean velocity was essentially constant and that the velocity variation decreased with increasing distance [Collard and Wolfe, 1974]. Nothing is known about the solar wind outside the ecliptic plane.

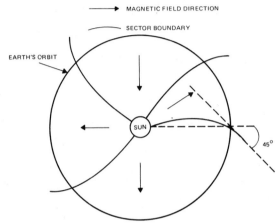

Fig. 5-11. Sector Boundaries and the Direction of the Interplanetary Magnetic Field (Adapted from Harris and Lyle [1969])

5.4 Modeling the Position of the Spacecraft

John N. Rowe

To determine attitude reference vectors for nearby celestial objects such as the Sun, the Earth, or other planets, it is necessary to have an accurate model of the position of the spacecraft itself. In this section, we discuss both definitive orbits as they are generated at NASA's Goddard Space Flight Center and a simple orbit generator using the basic equations presented in Chapter 3. The latter method is satisfactory for most aspects of prelaunch attitude analysis and for generating simulated data. However, the analysis of real spacecraft data generally requires use of the ephemeris files generated by one of the much more sophisticated orbit programs.

The orbit of a spacecraft is determined from observations of its position or its distance and radial velocity at different points in its orbit; distance and radial velocity are the most commonly used and are often referred to as *range* and *range rate*, respectively. Because six elements are to be determined, at least six pieces of information are required. This means pairs of right ascension and declination or pairs of distance and radial velocity at a minimum of three points in the orbit. Usually such data is obtained at more than three points, and a differential correction procedure (see Chapter 13) is used to estimate the elements.

Spacecraft Ephemeris Files. When the position of the spacecraft is needed for attitude determination, it is normally obtained from files generated by numerical integration incorporating all significant forces. This is accomplished at Goddard Space Flight Center using the Goddard Trajectory Determination System (GTDS), a detailed discussion of which is beyond the scope of this book. (See Capellari, *et al.*, [1976].)

GTDS generates two types of files. One type is the multilevel direct access or *ORBIT* file, which contains the spacecraft acceleration from which the position and velocity may be recovered. This file is read with the standard utility routine GETHDR (Section 20.3). Table 5-8 shows the contents of the two header records and Table 5-9 shows the contents of the data file. The header information is returned in arrays HDR and IHDR of GETHDR according to the following scheme: bytes 1 through 608 of header 1 and bytes 1 through 608 of header 2 are returned in that order in HDR; bytes 609 through 1092 of header 1 and bytes 609 through 1092 of header 2 are returned in that order in IHDR.

The second type of GTDS file is the sequential *EPHEM* file, which contains the spacecraft position and velocity at regular time intervals. The position and

Table 5-8. Goddard Space Flight Center ORBIT File Header Records. Contents of bytes marked "internal use" are given by Cappellari, *et al.*, [1976] and Zavaleta, *et al.*, [1975].

BYTES	NAME	DESCRIPTION	BYTES	NAME	DESCRIPTION	BYTES	NAME	DESCRIPTION
FIRST RECORD IN FILE (2204 BYTES)								
1–8	SATNAM	SATELLITE NAME (EBCDIC)	257–368	OBLINT(L)	L = 1, 14 AUXILIARY ORBITAL ELEMENTS AT EPOCH IN COORDINATE SYSTEM OF INTEGRATION	625–1048		(INTERNAL USE)
9–16	AREA	CROSS-SECTIONAL AREA OF SATELLITE (cm^2)				1049–1052	NBROBS	NUMBER OF OBSERVATIONS IN FITTED DATA SPAN FOR ELEMENTS SET
17–24	SCMASS	MASS OF SATELLITE (kg)			L = 1, ECCENTRIC ANOMALY (rad) 2, PERIOD (sec)	1053–1056	NSTATE	NUMBER OF STATE PARTIALS
25–32	CSUBR	SATELLITE REFLECTIVITY CONSTANT			3, TIME DERIVATIVE OF PERIOD 4, MEAN MOTION (rad/sec)	1057–1084		(INTERNAL USE)
33–40	CSUBDZ	DRAG COEFFICIENT			5, TRUE ANOMALY (rad) 6, PERIFOCAL HEIGHT (km)	1085–1088	ICENT	CENTRAL BODY INDICATOR
41–48	YMDOUT	YYMMDD. OF STARTING DATE OF FILE			7, APOFOCAL HEIGHT (km) 8, TIME DERIVATIVE OF ARGUMENT OF PERIGEE			ICENT = 1, EARTH 2, MOON
49–56	HMSOUT	HHMMSS. SSSS OF STARTING DATE OF FILE			(rad/sec) 9, TIME DERIVATIVE OF ASCENDING NODE (rad/sec)			3, SUN 4, MARS 5, JUPITER
57–64	YMDFN	YYMMDD. OF ENDING DATE OF FILE			10, VELOCITY AT APOGEE (km/sec)			6, SATURN 7, URANUS 8, NEPTUNE
65–72	HMSFN	HHMMSS. SSSS OF ENDING DATE OF FILE			11, VELOCITY AT PERIGEE (km/sec) 12, LATITUDE (rad)			9, PLUTO 10, MERCURY
73–80	YMSIC	YYMMDD. OF EPOCH DATE			13, LONGITUDE (rad) 14, HEIGHT (km)	1089–1092	IND(1)	11, VENUS ORBIT GENERATOR INDICATOR
81–88	HMSIC	HHMMSS. SSSS OF EPOCH DATE	369–376	OBSYMD	YYMMDD. OF START OF FITTED DATA SPAN FOR ELEMENT SET			= 1, TIME REGULARIZED COWELL ORBIT GENERATOR
89–96	YMDREF	YYMMDD. OF REFERENCE LINE FOR TIME COORDINATE SYSTEM	377–384	OBSHMS	HHMMSS. SSSS OF START FITTED SPAN FOR ELEMENT SET			= 2, COWELL ORBIT GENERATOR
97–104	EGHA	GREENWICH HOUR ANGLE OF THE VERNAL EQUINOX AT EPOCH (rad)	385–392	OBEYMD	YYMMDD. OF END OF FITTED DATA SPAN FOR ELEMENT	SECOND RECORD IN FILE		
105–112	EJED	JULIAN EPHEMERIS DATE OF EPOCH	393–400	OBEHMS	HHMMSS. SSSS OF END OF FITTED DATA SPAN FOR ELEMENT SET	1–16		(INTERNAL USE)
113–160	AEINT(K)	K = 1, 6 KEPLERIAN ELEMENTS AT EPOCH IN COORDINATE SYSTEM OF INTEGRATION	401–408	WRMS	WEIGHTED RMS OF FIT FOR ELEMENT SET	17–104	GM(I)	I = 1, 11 GRAVITATIONAL CONSTANT TIMES THE MASS OF CENTRAL BODY' (km^3/sec^2)
		K = 1, SEMIMAJOR AXIS (km) 2, ECCENTRICITY	409–576	COVMAT(I)	I = 1, 21 UPPER TRIANGLE OF STATE COVARIANCE MATRIX	105–612		(INTERNAL USE)
		3, INCLINATION (rad) 4, LONGITUDE OF ASCENDING NODE (rad)	577–584	AZERO	DIFFERENCE BETWEEN ATOMIC TIME (TAI) AND UTC (sec)	613–616	IND(41)	DRAG PARTIALS INDICATOR = 1, YES
		5, ARGUMENT OF PERIHELION (rad) 6, MEAN ANOMALY AT EPOCH (rad)	585–592	TZERO	TIME FROM BEGINNING OF YEAR (sec)	617–620	IND(42)	= 2, NO SOLAR RADIATION PARTIALS INDICATOR
161–208	SPINT(K)	K = 1, 6 SPHERICAL ELEMENTS AT EPOCH IN COORDINATE SYSTEM OF INTEGRATION	593–600	DEPOCH	JULIAN DATE OF EPOCH			= 1, YES = 2, NO
			601–608	SPARE	SPARE LOCATION	621–624	IND(43)	POTENTIAL PARTIALS INDICATOR
		K = 1, RIGHT ASCENSION (rad) 2, DECLINATION (rad)	609–612	IDSAT	SATELLITE NUMBER			= 1, YES = 2, NO
		3, FLIGHT PATH ANGLE (rad) 4, AZIMUTH (rad)	613–616	NBRRUN	RUN NUMBER	625–628	IND(44)	THRUST PARTIALS INDICATOR
		5, RADIUS (km) 6, VELOCITY (km/sec)	617–620	NBRELS	ELEMENT SET NUMBER			= 1, YES = 2, NO
209–256	PVINT(K)	K = 1, 6 POSITION AND VELOCITY OF SATELLITE IN COORDINATE SYSTEM OF INTEGRATION (km, km/sec)	621–624	ISO	INERTIAL COORDINATE SYSTEM REFERENCE INDICATOR	629–1092	ISPARE	(INTERNAL USE)
					ISO = 1, MEAN EQUATOR AND EQUINOX OF 1950.0 2, TRUE OF DATE			

velocity for intermediate times are obtained with a six-point interpolation procedure using the standard routine RO1TAP (Section 20.3). Table 5-10 shows the header, data, and trailer records of an EPHEM file. Note that the file is in units of 864 sec and 10^4 km, whereas the output position and velocity from RO1TAP are in units of 6378.166 km and 7.90538916 km/sec (these are known as Vanguard units; see Appendix K).

A similar file is anticipated for payloads flown on the space shuttle. The shuttle orbit information will be given as a series of position and velocity vectors in geocentric celestial coordinates (mean of 1950).

Table 5-9. Goddard Space Flight Center ORBIT File Data Record

BYTES	NAME	DESCRIPTION	BYTES	NAME	DESCRIPTION
1–8	TN	TIME OF nTH (LAST) ACCELERATION IN XDD ARRAY EPHEMERIS SECONDS FROM EPOCH OR s VARIABLE OF LAST ACCELERATION IN XDD FOR TIME REGULARIZED FILE	393–416	SX2(I)	I = 1, 3; SECOND SUM VECTOR OF SATELLITE ACCELERATIONS
			417–5696	XVDD(K,I,J)	K = 1, 11; I = 1, 3; J = 1, 20 ACCELERATION PARTIALS
9–16	H	INTEGRATOR STEPSIZE (sec)			
17–280	XDD(K,I)	K = 1, 11; I = 1, 3 SATELLITE ACCELERATION VECTORS (km/sec^2)	5697–6176	SV1(I,J)	I = 1, 3; J = 1, 20 FIRST SUM MATRICES FOR ACCELERATION PARTIALS
			6177–6656	SV2(I,J)	I = 1, 3; J = 1, 20 SECOND SUM MATRICES FOR ACCELERATION PARTIALS
281–368	TDD(K)	K = 1, 11 TIME REGULARIZATION ARRAY OF TIMES (sec)			
			6657–6660	NBRSEC	SECTION NUMBER
369–392	SX1(I)	I = 1, 3; FIRST SUM VECTOR OF SATELLITE ACCELERATIONS			

NOTE: THIS FORMAT APPLIES TO ALL RECORDS EXCEPT THE FIRST AND THE SECOND. THE BLOCK SIZE FOR A FILE WITH NO PARTIALS IS 1092. THE BLOCK SIZE OF A FILE WITH PARTIALS IS 6660. A DATA RECORD WHEN NO PARTIALS ARE PRESENT IS 420 BYTES LONG; THIS RECORD CONSISTS OF BYTE LOCATIONS 1–416 AND 6657–6660.

Orbit Generators. Orbit generators may be classified as those which use Kepler's equation to determine position and velocity, and those which integrate the equations of motion directly using models of the forces. Only the former are discussed here and only elliptical orbits are considered. Recall from Chapter 3 that Kepler's equation relates, for an eliptic orbit, the mean anomaly,[*] M, at some time, t, to the eccentric anomaly, E, at the same time by

$$M = E - e \sin E \qquad (5\text{-}30)$$

where e is the eccentricity. If t_0 is the epoch time of the elements, then the mean anomaly at time t is found from the mean anomaly at epoch by

$$M = M_0 + n(t - t_0) \qquad (5\text{-}31)$$

where $n \equiv 2\pi/\text{period}$ is the mean motion. The utility routine ORBGEN (Section 20.3) solves Kepler's equation numerically to find E at any time before or after the epoch using an iterative solution to Eq. (5-30) (obtained using Newton's method). Successive estimates of E are given by

$$E_i = E_{i-1} + \frac{M + e \sin(E_{i-1}) - E_{i-1}}{1 - e \cos(E_{i-1})} \qquad (5\text{-}32)$$

_____
[*] Angles are expressed in radians throughout this section.

Table 5-10.   Goddard Space Flight Center EPHEM File Format

BYTES	DESCRIPTION	BYTES	DESCRIPTION	BYTES	DESCRIPTION
	FIRST RECORD (2800 BYTES)	329-336	DRAG PERTURBATION INDICATOR[1]	729-760	(INTERNAL USE)
1-8	TAPE IDENTIFIER - EPHEM	337-344	t_0 - EPOCH TIME OF ELEMENTS: CENTIDAYS SINCE SEPTEMBER 18, 1957 (ATOMIC TIME SYSTEM)	761-1536	(SPARES)
9-16	SATELLITE NUMBER			1537-1544	START TIME OF EPHEMERIS
17-24	TIME SYSTEM INDICATOR	345-352	YEAR OF ELEMENTS EPOCH	1545-1552	END TIME OF EPHEMERIS
	• 1, TAI	353-360	MONTH OF ELEMENTS EPOCH	1553-1560	TIME INTERVAL BETWEEN EPHEMERIS POINTS
	• 2, UTC	361-368	DAY OF ELEMENTS EPOCH	1561-1568	PRECESSION AND NUTATION INDICATOR
25-32	DATE OF START OF EPHEMERIS YYMMDD	369-376	HOUR OF ELEMENTS EPOCH	1569-1576	GREENWICH HOUR ANGLE OF THE VERNAL EQUINOX AT t_0 (rad)
33-40	DAY COUNT OF YEAR FOR START OF EPHEMERIS	377-384	MINUTE OF ELEMENTS EPOCH		
41-48	SECONDS OF DAY FOR START OF EPHEMERIS	385-392	SECONDS OF ELEMENTS EPOCH	1577-1584	(INTERNAL USE)
49-56	DATE OF END TIME OF EPHEMERIS YYMMDD	393-400	SEMIMAJOR AXIS[2] AT t_0 (km)	1585-2344	(SPARES)
57-64	DAY COUNT OF YEAR FOR END TIME OF EPHEMERIS	401-408	ECCENTRICITY[2] AT t_0 (km)	2345-2800	(INTERNAL USE; HARMONIC COEFFICIENTS)
65-72	SECONDS OF DAY FOR END TIME OF EPHEMERIS	409-416	INCLINATION[2] AT t_0 (rad)		SECOND RECORD (2800 BYTES)
73-80	TIME INTERVAL BETWEEN EPHEMERIS POINTS (sec)	417-424	ARGUMENT OF PERIGEE[2] AT t_0 (rad)	1-2800	(INTERNAL USE, HARMONIC COEFFICIENTS)
81-208	REFERENCE DATE FOR DODS YYMMDD 570918	425-432	RIGHT ASCENSION OF THE ASCENDING NODE[2] AT t_0 (rad)	1-8	DATE OF FIRST EPHEMERIS POINT YYMMDD
209-216	RUN IDENTIFIER (UP TO 128 EBCDIC CHARACTERS)	433-440	MEAN ANOMALY[2] AT t_0 (rad)	9-16	DAY COUNT OF YEAR FOR FIRST EPHEMERIS POINT
217-220	COORDINATE SYSTEM TYPE INDICATOR (EBCDIC)	441-448	TRUE ANOMALY AT t_0 (rad)	17-24	SECONDS OF DAY FOR FIRST EPHEMERIS POINT
221-224	COORDINATE SYSTEM TYPE INDICATOR (NUMERIC)	449-456	ARGUMENT OF LATITUDE AT t_0 (rad)	25-32	TIME INTERVAL BETWEEN DATA POINTS (sec)
225-232	ORBIT THEORY USED (EBCDIC)	457-464	FLIGHT PATH ANGLE AT t_0 (rad)	33-56	FIRST POSITION VECTOR (X, Y, Z)
233-248	(SPARES)	456-472	ECCENTRIC ANOMALY AT t_0 (rad)	57-80	FIRST VELOCITY VECTOR (X, Y, Z)
249-256	C_D, COEFFICIENT OF DRAG	473-480	PERIOD AT t_0	81-2432	POSITION AND VELOCITY VECTOR SETS FOR DATA SETS FOR DATA POINTS 2-50
257-264	C_R, SHAPE FACTOR	481-488	PERIGEE HEIGHT AT t_0 (km)	2433-2440	TIME OF FIRST EPHEMERIS DATA POINT
265-272	ATMOSPHERIC DENSITY MODEL ID (EBCDIC)	489-496	APOGEE HEIGHT AT t_0 (km)	2441-2448	MAGNITUDE OF THE INTERVAL BETWEEN DATA POINTS
273-280	CROSS-SECTIONAL AREA OF SATELLITE (cm²)	497-504	MEAN MOTION AT t_0	2449-2800	(SPARES)
281-288	MASS OF SATELLITE (gm)	505-512	RATE OF CHANGE OF THE ARGUMENT OF PERIGEE AT t_0		LAST RECORD (2800 BYTES)
289-296	ZONAL AND TESSERAL HARMONICS INDICATOR[1]	513-520	RATE OF CHANGE OF THE RIGHT ASCENSION OF THE ASCENDING NODE AT t_0	80 BYTES FOLLOWING LAST DATA POINT	TEN (TEN SENTINELS (0.999999999999999 × 10¹⁶)
297-304	BROUWER COMPLEMENTARY PERTURBATIONS INDICATOR[1]	521-544	POSITION VECTOR (X, Y, Z) AT t_0	REMAINDER	(SPARES)
305-312	LUNAR GRAVITATION PERTURBATION INDICATOR[1]	545-568	VELOCITY VECTOR (X, Y, Z) AT t_0		
313-320	SOLAR RADIATION PERTURBATION INDICATOR[1]	569-728	(SPARES)		
321-328	SOLAR GRAVITATION PERTURBATION INDICATOR[1]				

NOTES
TIME IN UNITS OF 946 SECONDS · 0.01 DAY
DISTANCE IN UNITS OF 10⁴ km
THE FIRST DATA POINT FOLLOWING THE LAST VALID DATA POINT CONSISTS OF A SET OF SIX VALUES EQUAL TO 0.999999999999999 × 10¹⁶

where the starting value $E_0 = M$. (Note that if M is identically 0, this method will fail; the solution in this case is trivial, that is, $M = 0$ implies $E = 0$.) In routine ORBGEN, the iteration proceeds until either the correction term is less than 1×10^{-8} or until 25 iterations have occurred. Once the eccentric anomaly is found, the true anomaly, ν, and the distance, r, may be found from (see Chapter 3):

$$\sin(\nu) = \frac{(1 - e^2)^{1/2} \sin E}{1 - e \cos E} \tag{5-33}$$

$$\cos(\nu) = \frac{\cos(E) - e}{1 - e \cos E} \tag{5-34}$$

$$r = a(1 - e \cos E) \tag{5-35}$$

Equations (5-33) through (5-35) thus give the position of the spacecraft in the orbit plane. We need to take into account the orientation of the orbit in space to find the position relative to an inertial system. Use of the spherical triangles shown in Figure 5-12 gives

$$x = r\left[\cos(\omega + \nu)\cos(\Omega) - \sin(\omega + \nu)\sin(\Omega)\cos(i)\right] \tag{5-36}$$

$$y = r\left[\cos(\omega + \nu)\sin(\Omega) + \sin(\omega + \nu)\cos(\Omega)\cos(i)\right] \tag{5-37}$$

$$z = r\left[\sin(\omega + \nu)\sin(i)\right] \tag{5-38}$$

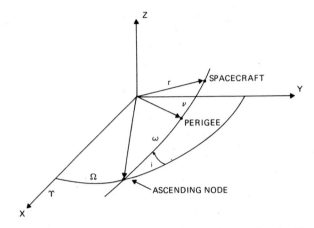

Fig. 5-12.   Defining the Orientation of an Orbit in Space. (See also Fig. 3-7.)

where Ω, ω, and i are the longitude of the ascending node, the argument of perigee, and the inclination; and x, y, and z are resolved in the coordinate system in which the elements are defined. It is possible to use Eqs. (5-33) and (5-34) to remove the explicit true anomaly dependence in Eqs. (5-36) through (5-38) and thus compute the position directly.

The velocity at time t may be found by applying the chain rule to Eqs. (5-36) through (5-38). Thus, for example,

$$\frac{dx}{dt} = \frac{\partial x}{\partial r}\frac{\partial r}{\partial E}\frac{\partial E}{\partial t} + \frac{\partial x}{\partial v}\frac{\partial v}{\partial E}\frac{\partial E}{\partial t} \tag{5-39}$$

where ω, Ω, and i are assumed constant. The quantity dE/dt may be found by differentiating Kepler's equation to obtain

$$\frac{dM}{dt} = \frac{dE}{dt} - e\cos(E)\frac{dE}{dt} \tag{5-40}$$

so

$$\frac{dE}{dt} = \left(\frac{1}{1-e\cos E}\right)\frac{dM}{dt} \tag{5-41}$$

The equations for the velocity are as follows:

$$\frac{dx}{dt} = \frac{na}{r}\left[bl_2\cos E - al_1\sin E\right] \tag{5-42}$$

$$\frac{dy}{dt} = \frac{na}{r}\left[bm_2\cos E - am_1\sin E\right] \tag{5-43}$$

$$\frac{dz}{dt} = \frac{na}{r}\left[bn_2\cos E - an_1\sin E\right] \tag{5-44}$$

where $b = a(1-e^2)^{1/2}$

$$l_1 = \cos\Omega\cos\omega - \sin\Omega\sin\omega\cos i$$

$$m_1 = \sin\Omega\cos\omega + \cos\Omega\sin\omega\cos i$$

$$n_1 = \sin \omega \sin i$$

$$l_2 = -\cos \Omega \sin \omega - \sin \Omega \cos \omega \cos i$$

$$m_2 = -\sin \Omega \sin \omega + \cos \Omega \cos \omega \cos i$$

$$n_2 = \cos \omega \sin i$$

The above procedure is not coordinate-system-dependent; that is, the position and velocity will be in whatever coordinate system the elements of the orbit are defined. For Earth-orbiting spacecraft, the elements are usually given in geocentric inertial coordinates, whereas for interplanetary orbits, the elements are usually given in heliocentric coordinates (see Section 2.2).

The simple two-body orbit generator described above may be modified to take into account noncentral forces (or forces from a third body). The procedure, which is known as the method of general perturbations (Section 5.5), is to obtain series solutions to the equations of motion in the form of perturbations to the orbit elements. These elements then become functions of time, and the method outlined above for solving Kepler's equation is applied with different elements each time the position is to be calculated. (Some simple results of general perturbations, in showing the effect of the oblateness of the Earth on the orbit of an Earth satellite, are given in Section 3.4, Eqs. (3-37), (3-38), (3-42), and (3-43).) Increasingly accurate descriptions of the actual motion of a spacecraft can be obtained by including an increasing number of perturbation terms (both periodic and secular). This forms the basis of the *Brouwer method*, which is a detailed application of the theory of general perturbations to the motion of artificial Earth satellites.

A simpler, more direct approach to the detailed calculation of orbits is to integrate the equations of motion directly, given the initial conditions. This is known as the method of special perturbations (Section 5.5). Models of all forces which are expected to be significant are included. Two commonly used integration schemes are those of Cowell (integration in rectangular coordinates) and Encke (calculation of an osculating conic section for which integration gives the differences between the real coordinates and the coordinates given by the conic section). Methods of this type are used in GTDS.

Utility and Accuracy of Two-Body Orbit Generators. The question arises as to the utility and accuracy of simple two-body orbit generators. Some comparisons, using as a reference an integration model from GTDS incorporating Sun, Moon, Earth harmonics, atmospheric drag, and solar radiation pressure are shown in Table 5-11. The references in Table 5-11 to "error with and without perturbations" refer to the inclusion in the two-body generator of the secular perturbations in the right ascension of the ascending node, the argument of perigee, and the mean motion due to the second-order gravitational harmonics of the Earth. (These are the J_2 perturbations discussed in Chapter 3.)

The near-circular orbits with moderate inclination and altitudes (cases 1 and 2) show improvement by including the perturbation. The highly elliptical orbit (case 4) and the low-altitude, high-inclination orbit (case 3) show no pronounced effect from including the perturbation (slight degradation in case 3 and slight improvement in case 4) and the absolute errors are large. The study also indicated

Table 5-11. Comparison of Simple Two-Body Orbit Generators. (From Shear [1977]; see text for discussion.)

CASE	ORBIT TYPE	SPACECRAFT	SEMIMAJOR AXIS (KM)	ECCENTRICITY	INCLINATION TO EQUATOR (DEG)	APOGEE ALTITUDE (KM)	PERIGEE ALTITUDE (KM)
1	LOW-ALTITUDE CIRCULAR	OSO-8	6,929	0.0004	32.9	553	548
2	HIGH-ALTITUDE CIRCULAR (SYNCHRONOUS)	ATS-6	42,166	0.0001	0.3	35,788	35,788
3	VERY LOW-ALTITUDE CIRCULAR, HIGH INCLINATION	AE-3	6,632	0.0008	68.0	249	259
4	HIGH-ALTITUDE ELIPTIC	GOES-2 (TRANSFER)	24,940	0.737	23.8	36,939	184

CASE	ERROR WITHOUT PERTURBATION				ERROR WITH PERTURBATION			
	AFTER ONE-HALF ORBIT		AFTER ONE ORBIT		AFTER ONE-HALF ORBIT		AFTER ONE ORBIT	
	POSITION (DEG)	DISTANCE (KM)	POSITION (DEG)	DISTANCE (KM)	POSITION (DEG)	DISTANCE (KM)	POSITION (DEG)	DISTANCE (KM)
1	0.2	10.0	0.5	0.3	0.02	10.0	0.02	0.3
2	0.02	3.2	0.02	0.6	0.01	3.2	0.005	0.6
3	0.2	9.0	0.4	0.08	0.3	9.0	0.7	0.09
4	0.01	6.0	0.5	15.0	0.08	3.0	0.4	17.0

that with the perturbation, the position errors are almost entirely in track; without the perturbation, the out-of-plane error becomes significant, especially for cases 1 and 3.

A second study was performed by Legg and Hotovy [1977] for an orbit similar in size and shape to that of case 1 ($e = 0.001$, $a = 6928$ km) but with an inclination of 97.8 deg. This study showed the major perturbations to be due to J_2, but that implementing only the secular component resulted in a significant degradation in accuracy. (The maximum absolute errors for two orbits without perturbations were on the order of 6 deg in position and 8 km in distance.)

The above studies are not sufficient to establish general recommendations for whether or not to include the secular perturbations in simple orbit generators. In view of the overall large error in these generators, there seems to be no point in including the perturbations unless a study indicates they are useful in a particular situation. The use of simple orbit generators is limited mainly to two situations: (1) when the orbit is so poorly known that the error contributed by the orbit generator is not a limitation (such as during an immediate postlaunch period); and (2) for simulations or analysis when the two-body position is in effect the true position. Even in this case, however, excessive extrapolation may result in the generation of an unrealistic position.

5.5 Modeling the Positions of the Sun, Moon, and Planets

John N. Rowe

To use the Sun, the Moon, or the major planets in the solar system as reference vectors for attitude determination, it is necessary to model the changing positions of these objects. The accuracy of such modeling should be such that attitude accuracy is not limited by the ephemeris accuracy; i.e., the uncertainty in the reference vectors should contribute a negligible fraction of the allowable attitude uncertainty. Uncertainty in the reference vectors arises from two sources: uncertainty in the position of the spacecraft as described in the previous section, and uncertainty in the position of the reference objects themselves. The error in modeling the positions of the solar system bodies using the complete theories described below is typically on the order of 0.1 arc-sec (see, for example, [Clemence, 1961]); this quantity is limited mainly by the error in the observations used to compute the model parameters*.

The orbits of the major planets and the Moon are nearly circular. They are characterized by small eccentricity (the largest being about 0.2 for Mercury) and small inclination to the ecliptic (the largest being about 7 deg, again for Mercury). General characteristics of planetary orbits are discussed in Section 3.2 and tables of orbital elements are given in Appendix L.

The basic problem to be solved is the same as that discussed in Section 5.4 for the orbital motion of spacecraft, except that in the present case the body in question may not necessarily be treated as a point of negligible mass. Two approaches have been adopted to the modeling of the motions of bodies in the solar system. These are the *method of general perturbations* and the *method of special perturbations*† [Danby, 1962]. In the former, the motion of the body is obtained from series solutions to the differential equations of motion. These solutions are expressed in closed form and typically involve power series in the time to give the mean motion, along with trigonometric series which provide corrections to the mean motion. The arguments of these correction terms are linear combinations of quantities relating to the mean motion, with amplitudes that are either constant or are slowly varying functions of time. The method of special perturbations uses a direct numerical solution of the equations of motion. The result is a series of state vectors at different times; there is no closed-form expression that can be used to compute directly the locations of the bodies.

Both of the above methods are used in the computation of ephemeris information. The locations of the inner planets (Mercury, Venus, Earth, and Mars) are usually computed using general perturbations because, for these objects, the Sun-planet interaction is closely described by two-body solutions. The locations of the five outer planets are computed using special perturbations because the two-body approximation is less valid. Historically, the motion of the Moon has been computed by general perturbations yielding a quite complex solution, involving over 1600 periodic terms. However, a preliminary theory of special perturbations for the Moon has been published [Garthwaite, et al., 1970].

The primary sources for background information on ephemerides of solar system bodies and for tabular ephemerides are the *American Ephemeris and Nautical Almanac*‡ and the corresponding *Explanatory Supplement to the Ephemeris*

---

* In some cases, spacecraft orbital data has provided positions much better than this. The mean residual in the position of the center of mass of Mars is currently on the order of 100 m, or about 10^{-4} arc sec at closest approach to the Earth [Standish, 1975]. Similar work has been done for the Moon [Garthwaite, et al., 1970].
† An illuminating nonmathematical discussion of both methods is given by Clemence, et al., [1960].
‡ Published annually in the United States by the U.S. Government Printing Office and in the United Kingdom by H. M. Stationery Office under the title *The Astronomical Ephemeris*.

[H. M. Nautical Almanac Office, 1961]. The ephemerides in the *American Ephemeris* are based on the work of Newcomb [1898] for the inner planets, on Eckert, *et al.*, [1951] for the outer planets, and on Brown [1919] for the Moon. Details on these sources, modifications to them, and further references may be found in the *Explanatory Supplement*.

The information in the *American Ephemeris* is given in the form of printed tables and therefore is not well suited to computer use. This defect is remedied by the *JPL magnetic tapes* (see, for example, Devine [1967]) produced and periodically updated by the Jet Propulsion Laboratory of Pasadena, California, primarily for the support of deep space and planetary probes. The JPL tapes give planetary ephemeris information in a form suitable for computer use. The information is derived from numerical integration of the equations of motion between various epochs. These epochs are chosen to minimize the least-squares deviation between the calculated positions and "source positions"; these source positions are determined from the theories used in the *American Ephemeris*, except for the lunar ephemeris, which is computed directly from the theory of general perturbations. The JPL ephemeris data consists of the position and velocity in rectangular, heliocentric coordinates for Mercury, Venus, the Moon, the Earth-Moon barycenter, Mars, Jupiter, Saturn, Uranus, Neptune, and Pluto; these are referred to the mean equator and equinox of 1950.0. A routine for reading these tapes to obtain Sun and Moon positions (RJPLT) is described in Section 20.3.

The JPL ephemeris tapes are still not especially convenient because of the high computer input/output time required to extract the needed information. For this reason, an adaptation of the JPL tapes on disk storage is used at Goddard Space Flight Center [Armstrong, *et al.*, 1973]; these are referred to as the *Solar-Lunar-Planetary* (SLP) files and are accessed with routine SUNRD (Section 20.3). The information on the SLP files is in the form of Chebyshev polynomial coefficients valid for intervals of time. The size of these intervals is a function of the speed of the body. The SLP files also contain coefficients allowing transformations between mean equator and equinox of 1950.0 and true equator and equinox of date. SUNRD, however, returns only the solar, lunar, and planetary position information.

Algebraic Approximations.   The accuracy of the complete ephemeris solutions is not always necessary in attitude analysis, and tapes and files are not always available. In fact, fairly simple closed-form expressions may be obtained for the motion of the solar system bodies. The simplest approximation is to consider the mean motion only. This is equivalent to taking two-body solutions with circular orbits. This approximation will give, in most cases, errors in excess of 1 deg.

A more accurate approximation is to consider two-body solutions with eccentric orbits. The small eccentricities of the orbits under consideration allow the use of a series solution to Kepler's equation, known as the *equation of the center*. In this approach, the true anomaly, v, is expressed in terms of the mean anomaly, M, by

$$v = \sum_i A_i \cos(iM) \qquad i = 1, 2, 3, \ldots \qquad (5\text{-}45)$$

Alternatively, Keplerian orbital elements (Section 3.1) may be used with an orbit generator, such as ORBGEN (Section 20.3). This approach will give errors well below 1 deg for the inner planets and the Sun (over a period of years), but in excess of 1 deg for the Moon. For the outer planets, osculating elements at particular epochs may be used with an orbit generator to obtain accuracy better than 1 deg, even for periods of several years on either side of the epoch. Additional accuracy may be obtained by including selected terms from the method of general perturbations. In the case of the Moon, the use of about 20 periodic terms will result in errors below 0.25 deg. The remainder of this section outlines the algorithms that can be used. Routines which implement these algorithms (SUN1X, SMPOS, and PLANET) are described in Section 20.3.

The mean motion of the Sun is given in the *American Ephemeris* as:

$$L = 279.°696678 + 0.9856473354(d) + 2.267 \times 10^{-13}(d^2)$$

$$M_\odot = 358.°475845 + 0.985600267(d) - 1.12 \times 10^{-13}(d^2) - 7 \times 10^{-20}(d^3)$$

$$e = 0.016751 \tag{5-46}$$

where L is the mean longitude of the Sun, measured in the ecliptic from the mean equinox of date; $M_\odot$ is the mean anomaly of the Sun; e is the eccentricity of the Earth's orbit; and d is the number of ephemeris days since 1900 January 0, 12^h Ephemeris Time* (Julian date 2,415,020).

For most attitude work, the number of ephemeris days may be assumed equal to the number of Julian days, and the reduction of universal time to Ephemeris Time may be omitted. (See Appendix J for a discussion of time systems.) In addition, the terms in d^2 and d^3 may be omitted. A correction δL is applied to the mean longitude to find the true longitude, and to the mean anomaly to find the true anomaly. The first two terms in the series for δL, as given by Newcomb [1898], are

$$\delta L = 1.°918 \sin(M_\odot) + 0.°02 \sin(2M_\odot) \tag{5-47}$$

The above is used in SMPOS and SUN1X. The distance, R, from the Earth to the Sun may be found from the following relationship between the distance and the true anomaly, ν:

$$R = \frac{1.495 \times 10^8 (1 - e^2)}{1 + e \cos \nu} \, \text{km} \tag{5-48}$$

The mean motion of the Moon, as described in the *American Ephemeris*, is given by

$$L_m = 270°.434164 + 13.1763965268(d) - 8.5 \times 10^{-13}(d^2) + 3.9 \times 10^{-20}(d^3)$$

$$\Gamma' = 334°.329356 + .1114040803(d) - 7.739 \times 10^{-12}(d^2) - 2.6 \times 10^{-19}(d^3)$$

$$\Omega = 259°.183275 - .0529539222(d) + 1.557 \times 10^{-12}(d^2) + 5 \times 10^{-20}(d^3)$$

$$D = 350°.737486 + 12.1907491914(d) - 1.076 \times 10^{-12}(d^2) + 3.9 \times 10^{-20}(d^3)$$

$$i = 5°.145396374 \tag{5-49}$$

---

*That is, 12^h Ephemeris Time on December 31, 1899.

where L_m is the mean longitude of the Moon, measured in the ecliptic from the mean equinox of date to the mean ascending node, and then along the orbit; Γ' is the mean longitude of the Moon's perigee, measured as above; Ω is the longitude of the mean ascending node of the lunar orbit, measured in the ecliptic from the mean equinox of date; $D \equiv L_m - L$ is the mean elongation of the Moon from the Sun; and i is the inclination of the lunar orbit to the ecliptic. Again, the terms in d^2 and d^3 may be neglected.

The corrections, δL_{m_i}, to the mean longitude are given by Brown [1919] in the form:

$$\delta L_{m_i} = A_i \sin \left[f_i(L_m - \Gamma') + f_i'(M_0) + g_i(L_m - \Omega) + g_i'(D) \right] \qquad (5\text{-}50)$$

where f_i, f_i', g_i, and g_i' are integer constants. The true longitude is obtained by adding the sum of the δL_{m_i} to L_m. The constants in Eq. (5-50) are given in

Table 5-12. Periodic Terms for Calculating the Longitude of the Moon

i	A_i (DEG)	f_i	f_i'	g_i	g_i'
1	+6.289	1	0	0	0
2	−1.274	1	0	0	−2
3	+0.658	0	0	0	2
4	+0.213	2	0	0	0
5	−0.185	0	1	0	0
6	−0.114	0	0	2	0
7	−0.059	2	0	0	−2
8	−0.057	1	1	0	−2
9	+0.053	1	0	0	+2
10	−0.046	0	1	0	−2
11	+0.041	1	−1	0	0
12	−0.035	0	0	0	1

Table 5-12 for the 12 terms used in SMPOS. The distance R from the Earth to the Moon is determined from the lunar parallax, P, as

$$R = \frac{6378.388}{\sin P} \text{ km} \qquad (5\text{-}51)$$

Brown gives a cosine series for P with the same arguments as for δL_{m_i}. The terms used in SMPOS are given in Table 5-13.

The positions of the planets may be calculated using mean elements for the inner planets and osculating elements for the outer planets. Subroutine PLANET uses elements for December 19, 1974, as given in Appendix L. The positions from PLANET are within 0.02 deg for times within 2 years of the epoch and are within 0.1 deg for times within 6 or 7 years of the epoch. These elements should be updated periodically by consulting a current *American Ephemeris*.

Table 5-13.   Periodic Terms for Calculating the Lunar Parallax

i	A_i (ARC-SEC)	f_i	f_i'	g_i	g_i'
1	3422.7	0	0	0	0
2	186.5398	1	0	0	0
3	34.3117	1	0	0	−2
4	28.2373	0	0	0	2
5	10.1657	2	0	0	0
6	3.0861	1	0	0	2
7	1.9178	0	1	0	−2
8	1.4437	1	1	0	−2
9	1.1528	1	−1	0	0

5.6 Modeling Stellar Positions and Characteristics

David M. Gottlieb

The value of using star observations for attitude determination lies in the high degree of accuracy that can be obtained. This accuracy derives from the point source nature of stars. Identifying observations with catalogued stars is, however, difficult (see Section 7.7). To alleviate star identification problems, and to obtain as much precision as possible from available star data, it is crucial to have an accurate and complete star catalog.

5.6.1 Star Catalog Data Required for Attitude Determination

Each star in any catalog used for attitude determination should have an identifying number to facilitate checkout of computer software and to aid the investigation of anomalous results. Unfortunately, there are many identification systems in use and few catalogs cross-reference more than one or two of them. Four major systems are in common use:

BD/CD/CPD. The most widely used and extensive system, generated from three positional catalogs: the Bonner Durchmusterung (BD) [Argelander, 1859–1862 and Schönfeld, 1886], the Cordoba Durchmusterung (CD) [Thome, 1892–1914 and Perrine, 1932], and the Cape Photographic Durchmusterung (CPD) [Gill and Kapteyn, 1896–1900]. Unfortunately, regions of the sky covered by these systems overlap, resulting in nonunique numbers.

HD. The Henry Draper number [Cannon, 1918–1924], also widely used. Because the catalog is virtually complete to eighth visual magnitude, most stars visible to present star sensors have HD numbers.

HR. Number from the Catalog of Bright Stars [Hoffleit, 1964], frequently cross-referenced in the literature. However, few stars dimmer than sixth visual magnitude have HR numbers.

SAO. The Smithsonian Astrophysical Observatory number [Smithsonian Institute, 1971], relatively new and used principally in the SAO catalog itself. It covers approximately as many stars as the HD, but it is seldom cross-referenced.

All catalogs also contain star positions, given as right ascension and declination at some epoch. The accuracy of the stated catalog position depends on the accuracy of the original observation and the time between the epoch of observation and the epoch of the catalog position. Star positions reported in the SAO or AGK-3 [Astronomisches Rechen Institut, 1975] catalogs are accurate to approximately 1 arc-sec. For about 2% of the stars brighter than eighth magnitude, and 15% from eighth to ninth magnitude, only the nineteenth-century HD positions exist with typical inaccuracies of 1 arc-min (one standard deviation) in both right ascension and declination.

Because star catalogs give positions at an epoch (typically 1900.0 to 1950.0) that differs from the time of the spacecraft observations, the star positions must be updated to the observation time. Corrections are usually required for the precession of the equinoxes (Section 2.2) and the *proper motion*, or space motion, of each individual star. Proper motion can be applied linearly for periods of several hundred years when the rates are available in the star catalog. For 95% of the stars brighter than ninth magnitude, proper motion is less than 10 arc-sec per century, and for 99.9%, it is less than 1 arc-min.

An additional correction may be required for *aberration*—the apparent shift in the position of a star caused by the motion of the spacecraft. The original observation of aberration by Astronomer Royal Bradley in 1728 was one of the first confirmations of Roemer's postulate that the speed of light was finite. For Earth-orbiting spacecraft, the motion of the Earth around the Sun causes a maximum aberration of about 20 arc-sec; the motion of the spacecraft about the Earth accounts for less than 5 arc-sec of additional aberration. The aberration, $\Delta\theta$, may be computed from the spacecraft velocity relative to the Sun, $\mathbf{v}$, by:

$$\Delta\theta = \frac{|\mathbf{v}|}{c}\sin\theta \qquad (5\text{-}52)$$

where c is the speed of light and θ is the angular separation between $\mathbf{v}$ and the star vector, $\mathbf{s}$. The star appears shifted toward $\mathbf{v}$ in the $\mathbf{v}$-$\mathbf{s}$ plane.

Star intensity is included in most catalogs and is measured by *magnitude*, a logarithmic quantity defined by $m = -2.5 \log(F) + m_0$, where m_0 is constant and F is the brightness or flux density. Note that brightness *decreases* as magnitude increases. Magnitudes are usually reported in one of two systems. The UBV (Ultraviolet, Blue, Visual) system of Johnson and Morgan [1953] is the more modern and accurate of the two. Commonly, only the V magnitude and sometimes the B magnitude are available. Figure 5-13 defines these magnitudes in terms of sensitivity versus wavelength. Some catalogs list V and the difference, $B - V$. The second system is the photographic-photovisual magnitude used in the Henry Draper Catalog [1918-1924]. These are more frequently available, but are far less accurate than the UBV system. No sensitivity-wavelength plots exist for them. If only photographic and photovisual magnitudes are available for a star, these can be used analogously for B and V, respectively.

Observed B and V magnitudes have errors of about 0.02 magnitude (one standard deviation). Only about 20% of the stars brighter than eighth magnitude have observed B and V magnitudes, and very few fainter than eighth magnitude. Photographic and photovisual magnitudes are uncertain to about 0.3 and 0.2

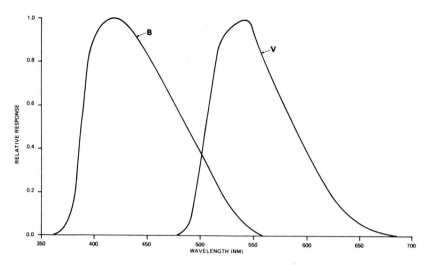

Fig. 5-13. Curves Defining B and V Magnitude Scales

magnitude, respectively. Conversion of these to B and V adds an additional error of 0.1 magnitude. Because sensor responses do not normally coincide with the wavelength sensitivities of either B or V magnitudes, some combination of these will be required to accurately represent star magnitudes on an instrumental scale (see Section 7.6).

Because star sensors detect light from the entire segment of the sky covered by their apertures, an additional requirement for modeling stellar magnitudes is the integrated intensity of faint background stars. Table 5-14 summarizes mean star densities and background level for the entire sky and for regions near the galactic plane. The background level is the integrated contribution of all stars fainter than the limiting magnitude, expressed in terms of stars of brightness equal to the limiting magnitude per square degree. Only 0.5% of the stars in the sky brighter than ninth magnitude have magnitudes known to vary with time by more than 0.1 magnitude. Some catalogs flag these stars and give values of the maximum and minimum magnitude to be expected. Brighter stars are more likely to be known variables, because dimmer ones are not observed as frequently.

Finally, those components of a multiple star system which are separated by about 1 to 5 arc-min may cause misidentifications and position errors. Many star

Table 5-14. Star Densities (Adapted from Allen [1973])

LIMITING VISUAL MAGNITUDE	AVERAGE REGION NEAR THE GALACTIC POLE		AVERAGE REGION NEAR THE GALACTIC PLANE		WHOLE SKY
	MEAN NUMBER OF STARS PER SQUARE DEG	BACKGROUND LEVEL*	MEAN NUMBER OF STARS PER SQUARE DEG	BACKGROUND LEVEL*	TOTAL NUMBER OF STARS
3.0	0.002	0.10	0.009	0.84	187
4.0	0.006	0.23	0.028	2.05	556
5.0	0.020	0.52	0.083	5.0	1660
6.0	0.063	1.20	0.25	12.9	5146
7.0	0.18	2.58	0.69	28.8	15095
8.0	0.50	5.40	1.95	68.0	44700

*VALUES ARE STARS OF THE LIMITING MAGNITUDE PER SQUARE DEG.

catalogs identify these multiple stars and give their separations. *Optical doubles* which appear close together in the sky but are not physically associated cause the same difficulty. About 50% of stars brighter than ninth magnitude have another star brighter than ninth magnitude within 0.2 deg, and about 90% have one within 0.35 deg.

5.6.2 Existing Star Catalogs

The discussion here is limited to the three catalogs that are most useful for attitude determination: The Catalog of Bright Stars [Hoffleit, 1964], the Smithsonian Astrophysical Observatory Catalog [5.6-8], and the SKYMAP Catalog [Gottlieb, 5.6-17]. Each exists on magnetic tape and in printed versions.

The Catalog of Bright Stars. The Catalog of Bright Stars contains approximately 9100 stars to visual magnitude 7.0 and is complete to visual magnitude 6.0. The HR number (the sequential index for the catalog), the HD number, the BD/CD/CPD number, and the star name (number or Greek letter and constellation) are given for each star when available. State-of-the-art* right ascensions and declinations are given in epochs 1900.0 and 2000.0. Position errors are not available. Proper motion and precession are given per hundred years.

The Catalog of Bright Stars gives V and B−V magnitudes, which are slightly out of date, with about 50% of the star magnitudes quoted actually being the old and inaccurate photovisual magnitudes instead of V. Those stars having photovisual magnitudes are flagged. The B−V values are accurate, but additional magnitudes are now available which are not included in the catalog. Some spectral types are given, and these can be used to compute a B magnitude if only V is given as described by Gottlieb [1969].

The following multiple star data are given where applicable: separation, difference in magnitude between the brightest and second brightest component, and the number of components. An indication of whether each star is variable or not is given, but no other variability data is available on the tape versions. An appendix to the printed version lists the type of variable and the period.

Smithsonian Astrophysical Observatory Catalog (SAO). The SAO contains almost 260,000 stars down to about tenth visual magnitude, and is over 98% complete to 8.0 visual magnitude. The SAO was created by merging a number of existing positional catalogs. It was designed to have at least four stars per square degree everywhere in the sky regardless of magnitude; therefore, the effective limiting magnitude of the catalog varies across the sky. The catalog gives SAO and BD/CD/CPD numbers. The absence of HD numbers is a serious limitation to the user who wishes to cross-reference SAO stars to other catalogs.

The SAO gives state-of-the-art right ascensions and declinations, epoch 1950.0. Errors in the position are quoted for each star. These errors average about 0.5 arc-sec at epoch 1950.0 [Smithsonian Institute, 1971]. Proper motion per year is listed. Precession is not given. Only photographic-photovisual magnitudes are

---

*"State-of-the-art" means values which are as accurate as possible given the current data. Some current catalogs are compiled with older data.

cited; these are accurate to about 0.5 magnitude. Some HD spectral types are quoted; however, like the magnitudes, the quality is poor. *The SAO is not primarily a magnitude catalog and should not be used as one.* No multiple star or variable star data are available.

SKYMAP Catalog. The SKYMAP Catalog was prepared in 1975 specifically for attitude determination purposes. It contains approximately 255,000 stars down to 10.0 visual magnitude and is 90% to 100% complete to 9.0 magnitude, V or B, whichever is the fainter. It is impossible to establish the completeness level more accurately than this without extensive observational surveys. The catalog contains HR, HD, SAO, BD/CD/CPD numbers and star names. SKYMAP numbers are also assigned.

State-of-the-art right ascensions and declinations are given at epoch 2000.0. Errors in position are quoted. Positions and errors in position were taken from the SAO catalog or the AGK-3 [Astronomisches Rechen Institut, 1975] when available (accuracy, about 1 arc-sec), and from the HD (accuracy, 35 arc-sec) for most of the remainder. Proper motion (also from the SAO) is given per year, and the sum of proper motion and precession is quoted per hundred years. Nearest neighbor computations, epoch 2000.0, including both multiple stars and optical double stars, are given for a variety of limiting magnitudes and magnitude differences.

State-of-the-art values of V and B were taken from Blanco, *et al.*, [1968] and Mermilliod [1973] (accuracy, 0.02 magnitude) or converted from photographic and photovisual magnitudes (accuracy, 0.15 magnitude) [Gottlieb, 1978 (in press)]. State-of-the-art spectral types were taken from Jaschek, *et al.*, [1964] or converted from HD spectral types [Gottlieb]. For multiple stars, the separation between the brightest and second brightest component, the difference in magnitude, and the year of observation are given. Variable star data include the type of variable, the magnitude range, the epoch, and the period. Other data include the reddening index and the U (ultraviolet) magnitude.

5.6.3 Generating a Core Catalog

For many automated computational functions, it is convenient to maintain a rapid access *core catalog* consisting of only those portions of the star catalog that may be required during a single program run. To reduce the time needed to create such a core catalog, it is appropriate to presort the whole sky *master catalog* into smaller regions, or *zones*, so that only a limited number of zones must be searched to generate any one core catalog.

A technique that divides the sky into zones that overlap in right ascension and declination by 50% (Fig. 5-13) was used for SAS-3 and HEAO-1, and is planned for HEAO-B and MAGSAT [Gottlieb, 1978 (in press)]. The use of overlapping zones has the advantage that the entire sensor field of view will always lie entirely in a single zone, provided the zone size has been chosen to be at least twice the diameter of the sensor field of view. This simplifies specification of the zones required to generate the core catalog. Zone overlapping has the disadvantage that a single star appears in up to four zones. This redundancy increases the size of the presorted catalog with an attendant increase in read time and storage requirements.

In the system illustrated in Fig. 5-14, zones adjacent in right ascension have the same declination limit and overlap in right ascension in the manner: 0 to 10 deg, 5 to 15 deg, 10 to 20 deg, 15 to 25 deg, etc. To prevent some of the redundant storage of data, stars can be stored in "half zones" (0 to 5 deg, 5 to 10 deg, 10 to 15 deg), which can then be merged computationally to simulate an original zone. Using this technique, no overlap in right ascension is required. A similar procedure does not work for declination overlap because the right ascension boundaries of two zones adjacent in declination will generally not align.

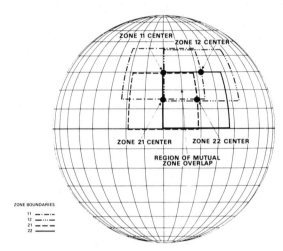

Fig. 5-14. Zone Overlap

Once the sky has been divided into zones, it is a simple matter to sort the whole-sky data base into these zones. The choice of zones to be read to create a specific core catalog is then determined by the directions to which the optical axis of the sensor will point during the program run. The optical axis pointings can be defined in several ways:

1. For three-axis stabilized spacecraft, a series of expected pointings will be known *a priori* or the analysis program must obtain a rough attitude each time a new pointing is reached.

2. For spinning spacecraft, the path the optical axis takes during one spacecraft rotation can be represented as a sequence of discrete pointings, ϵ degrees apart, as given by:

$$\mathbf{O_i} = [A]\mathbf{O_i'}$$

$$\mathbf{O_i'} = (\cos\varphi_i \sin\gamma, \sin\varphi_i \sin\gamma, \cos\gamma)$$

$$\varphi_i = \frac{(i-1)\epsilon}{\sin\gamma} \quad i = 1, \ldots, \frac{360°\sin\gamma}{\epsilon}$$

where $\mathbf{O_i}$ is the optical axis in inertial coordinates, $\mathbf{O_i'}$ is the optical axis in spacecraft coordinates, $[A]$ is the coordinate transformation matrix or attitude matrix (see Section 11.3), and γ is the angle between the spacecraft spin axis and the sensor optical axis.

3. For slowly spinning spacecraft, the interval of analysis may be less than one spacecraft rotation, and a portion of the optical path defined above may suffice.

Once the optical axis pointings have been defined, the zone center nearest to each O_i can be computed. This procedure yields a list of the zones that must be read to generate the core catalog. However, not every star in every selected zone need go into the core catalog because some will lie outside the sensor field of view. To store a minimum of stars in the core catalog, an *augmented field-of-view* size may be specified such that only those stars falling within this field will appear in the core catalog. The augmented field-of-view radius, ρ, should be the sum of (1) the radius of the smallest circle that can be circumscribed about the field of view; (2) the maximum anticipated error in the attitude relative to the sensor optical axis; (3) the maximum expected precession and nutation amplitudes; and (4) the maximum expected secular motion of the optical axis during the interval of analysis for a pointed spacecraft, or the maximum expected secular motion of the spin axis for a spinning spacecraft. To build a core catalog, each star in each selected zone is examined and included in the catalog if it lies within the augmented field of view of an optical axis pointing. For nonspinning spacecraft, a star is included if

$$\cos^{-1}(|\mathbf{S} \cdot \mathbf{O_i}|) \leqslant \rho \quad \text{for some } i = 1, 2, \ldots$$

where $\mathbf{S}$ is the star unit vector. For a spinning spacecraft, a star is included in the core catalog if

$$\gamma - \rho \leqslant \cos^{-1}(|\mathbf{S} \cdot \mathbf{Z}|) \leqslant \gamma + \rho$$

where $\mathbf{Z}$ is the spacecraft spin axis unit vector.

Although the core catalog is already limited to only those stars that might be required, it may still be too large to read each time an observation is to be identified. A rapid way of finding the desired star in the catalog is to compute catalog star longitudes, Ψ_c, defined in Fig. 5-15. The longitudes of all stars in the core catalog can be computed (see Appendix C or subroutine VPHASE in Section 20.3) and the core catalog sorted in order of increasing longitude. A cross-reference table can then be created such that the ith position in the table refers to the first star with longitude greater than $C \cdot i$ deg, where C is any desired constant. Therefore, to identify a specific observation it is sufficient to search longitudes in the range

$$\Psi_o - \epsilon_\Psi < \Psi_c < \Psi_o + \epsilon_\Psi$$

where Ψ_o is the estimated longitude of the observation and ϵ_Ψ is the maximum expected error in the longitude.

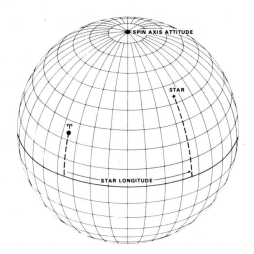

Fig. 5-15. Inertial Coordinate System for Defining Star Longitudes (For nonspinning spacecraft, a suitable reference axis must be selected to take the place of the spin axis.)

References

1. Allen, C. W., *Astrophysical Quantities*, 3rd edition. London: The Athlone Press, 1973.
2. Argelander, F. W. A., "Bonner Sternverzeichniss, Sections 1–3," *Astron. Beob.* Sternwarte Konigl. Rhein. Friedrich-Wilhelms-Univ., Bonn, 1859–1862, Vols. 3–5.
3. Armstrong, M. G., *Generation and Use of the Goddard Trajectory Determination System, SLP Ephemeris Files*, NASA TM-X-66185, GSFC, 1973.
4. Astronomisches Rechen-Institut, *Dritter Katalog der Astronomischen Gesellschaft*. Heidelberg, 1975.
5. Bartels, J., "The Eccentric Dipole Approximating the Earth's Magnetic Field," *Terrestial Magnetism and Atmosphere Electricity*, Vol. 41, no. 3, p. 225–250, Sept. 1936.
6. Battin, Richard H., *Astronautical Guidance*. New York: McGraw-Hill, Inc., 1964.
7. Blanco, V. M., S. Demers, G. G. Douglass, and M. P. Fitzgerald, *Publ. U.S. Naval Obs.*, 2nd series, Vol. 21, 1968.
8. Boss, B., *General Catalog of 33,342 Stars for the Epoch 1950*, Vols. 1–5. Washington, D.C.: Carnegie Institute of Washington, 1937.
9. Brown, E. W., *Tables of the Motion of the Moon*. New Haven: Yale University Press, 1919.
10. Cain, Joseph C., *Geomagnetic Models From Satellite Surveys*, NASA X-645-70-263, GSFC, July 1970.
11. ——— and Shirley J. Cain, *Derivation of the International Geomagnetic Reference Field (IGRF 10/68)*, NASA TN O-6237, Aug. 1971.
12. Cannon, A. J. and E. C. Pickering, *Harvard Ann.*, 1918–1924, Vols. 91–99.
13. Cappellari, J. O., C. E. Velez, and A. J. Fuchs, *Mathematical Theory of the Goddard Trajectory Determination System*, NASA X-582-76-77, April 1976.
14. Chernosky, Edwin J., Paul F. Fougere, and Robert O. Hutchinson, "The

Geomagnetic Field," *Handbook of Geophysics and Space Environments* (S. Valley, ed.). New York: McGraw-Hill, Inc., 1965.

15. Clemence, G. M., *Astronomical Papers of the American Ephemeris*, Vol. 16, part II, 1961.

16. ———, D. Brouwer, and W. J. Eckert, "Planetary Motions and the Electronic Calculator," *Source Book in Astronomy 1900–1950*. Cambridge: Harvard University Press, 1960.

17. Collard, H. R. and J. H. Wolfe, in *Solar Wind Three* (C. T. Russel, ed.). University of California at Los Angeles Press, 1974.

18. Coriell, K., *Geodynamics Experimental Ocean Satellite-3 Postlaunch Attitude Determination and Control Performance*, Comp. Sc. Corp., CSC/TM-75/6149, Aug. 1975.

19. Danby, J. M. A., *Fundamentals of Celestial Mechanics*. New York: MacMillan and Co., Ltd., 1962.

20. Devine, C. J., JPL Tech. Report No. 32-1181, 1967.

21. Eckert, W. J., D. Brouwer, and G. M. Clemence, *Astronomical Papers of the American Ephemeris*, Vol. 12, 1951.

22. Escobal, Pedro Raman, *Methods of Orbit Determination*. New York: John Wiley & Sons, Inc., 1976.

23. Fricke, W. and A. Kopff, *Fourth Fundamental Catalog*. Heidelberg: Veröff. Astron. Rechen-Inst., no. 10, 1963.

24. Garland, George D., *Introduction to Geophysics*. Philadelphia: W. B. Saunders Co., 1971.

25. Garthwaite, K., D. B. Holdridge, and J. D. Mulholland, "A Preliminary Special Perturbation Theory for the Lunar Motion," *Astron. J.*, Vol. 75, p. 1133, 1970.

26. Gill, D. and J. C. Kapteyn, "Cape Photographic Durchmusterung, Parts I–III," *Ann. Cape Obs.*, 1896–1900, Vols. 3–5.

27. Goldstein, Herbert, *Classical Mechanics*. Reading, MA: Addison-Wesley Publishing Company, Inc., 1950.

28. Gottlieb, D. M., *Astrophys. J. Supplements* (in press).

29. ——— and W. L. Upson III, *Astrophys. J.*, Vol. 157, p. 611, 1969.

30. Harris, M. and R. Lyle, *Magnetic Fields—Earth and Extraterrestrial*. NASA SP-8017, March 1969.

31. Haymes, Robert C., *Introduction to Space Science*. New York: John Wiley & Sons, Inc., 1971.

32. H.M. Nautical Almanac Office, *Explanatory Supplement to the Astronomical Ephemeris and the American Ephemeris and Nautical Almanac*. London: Her Majesty's Stationery Office, 1961.

33. Hoffleit, D., *Catalog of Bright Stars*. New Haven, CT: Yale University Obs., 1964.

34. Hundhausen, A. J., *Coronal Expansion and Solar Wind*. New York: Springer-Verlag, 1972.

35. Jaschek, C., H. Conde, and A. C. deSierra, *Publ. La Plata Obs. Ser. Astron.*, Vol. 28, 1964.

36. Johnson, H. L. and W. W. Morgan, *Astrophys. J.*, Vol. 117, p. 313, 1953.

37. King-Hele, D., "The Shape of the Earth", *Science*, Vol. 192, no. 4246, June 1976.

38. Kopff, A., *Dritter Fundamental-Katalog des Berliner Astronomisches Jahrbuchs, Part I*. Berlin: Veröff. Astron. Rechen-Inst., no. 54, 1937.
39. ——, *Dritter Fundamental-Katalog des Berliner Astronomisches Jahrbuchs, Part II*. Abh. Preus. Akad. Wiss., Phys-math-Kl, no. 3, 1938.
40. Leaton, B. R., "International Geomagnetic Reference Field 1975," *Transactions of the American Geophysical Union (E ⊕ S)*, Vol. 57, no. 3, p. 120, 1976.
41. Legg, J. S. and S. G. Hotovy, Private Communication, 1977.
42. Meirovitch, L., *Methods of Analytical Dynamics*. New York: McGraw-Hill, Inc., 1970.
43. Mermilliod, J., *Bulletin D'Information du Centre des Données Stellaires de Strasbourg*, Jan. 1973.
44. Morgan, H. R., "Catalog of 5268 Standard Stars," *Astronomical Papers of the American Ephemeris*, Vol. 13, part 3, 1950.
45. Newcomb, S., *Astronomical Papers of the American Ephemeris*, Vol. 6, 1898.
46. Perrine, C. D., "Cordoba Durchmusterung, Part V," *Resultados Obs. Nacional Argentino*, 1932, Vol. 21.
47. Schalkowsky, S., and M. Harris, *Spacecraft Magnetic Torques*, NASA SP-8018, March 1969.
48. Schonfeld, E., "Bonner Sternverzeichniss, Section 4," *Astron. Beob.* Sternwarte Konigl. Rhein. Friedrich-Wilhelms-Univ., Vol. 8, Bonn, 1886.
49. Schorr, R. and Kohlschütter, *Zweiter Katalog der Astr. Gesellschaft*, Vols. 1–15, 1951–1953.
50. Shear, M., Private Communication, 1977.
51. Smith, E. v. P. and D. M. Gottlieb, *Possible Relationships Between Solar Activity and Meteorological Phenomena*, NASA X-901-74-156, GSFC 1974.
52. Smithsonian Institute Staff, *Smithsonian Astrophysical Observatory Star Catalog, Parts I–IV*. Washington, D.C.: Smithsonian Institute, 1971.
53. Standish, E. M., *JPL Planetary Ephemeris Development*. Presented at the Fight Mechanical Estimation Theory Symposium, October 29–30, 1975, NASA CP-2003, GSFC, 1976.
54. Stern, David P., *Representations of Magnetic Fields in Space*, GSFC-X-602-75-57, March 1975.
55. Thome, J. M., "Cordoba Durchmusterung, Parts I–IV," *Resultados Obs. Nacional Argentino*, 1892–1914, Vols. 16–19.
56. Trombka, B. T. and J. C. Cain, *Computation of the IGRF I. Spherical Expansions*, NASA X-922-74-303, GSFC, Aug. 1974.
57. Wagner, C. A., F. J. Lerch, J. E. Brownd, and J. A. Richardson, *Improvement in the Geopotential Derived From Satellite and Surface Data (GEM 7 and 8)*, NASA X-921-76-20, GSFC, Jan. 1976.
58. Wolfe, John H., "The Large Scale Structure of the Solar Wind," in *Solar Wind*. Elmsford, N.Y.: Pergamon, 1972.
59. Zavaleta, E. L., E. J. Smith, C. Berry, J. Carlson, C. Chang, J. Fein, B. Green, G. Hibdon, A. Kapoor, A. Long, R. Luczak, M. O'Neill, *Goddard Trajectory Determination System User's Guide*, Comp. Sc. Corp., CSC/SD-75/6005, April 1975.
60. Zmuda, A. J., "The Geomagnetic Field and Its Harmonic Description," *Geomagnetism and Aeronomy*, Vol. 13, no. 6, p. 9, 1973.

PART II

ATTITUDE HARDWARE
AND
DATA ACQUISITION

CONTENTS

PART II

ATTITUDE HARDWARE AND
DATA ACQUISITION

Chapter

6   Attitude Hardware   155

7   Mathematical Models of Attitude Hardware   217

8   Data Transmission and Preprocessing   278

9   Data Validation and Adjustment   310

CHAPTER 6

ATTITUDE HARDWARE

6.1 Sun Sensors
   Analog Sensors, Sun Presence Detectors, Digital Sensors,
   Fine Sun Sensors
6.2 Horizon Sensors
   Sensor Components, Horizon Sensor Systems
6.3 Magnetometers
6.4 Star Sensors
   Overview of Star Sensor Hardware, BBRC CS-103 V-Slit
   Star Scanner for OSO-8,
   BBRC CT-401 Fixed-Head Star Tracker
6.5 Gyroscopes
   Rate Gyros, Rate-Integrating Gyros, Control Moment Gyros
6.6 Momentum and Reaction Wheels
6.7 Magnetic Coils
6.8 Gas Jets
6.9 Onboard Computers

In this chapter we describe representative examples of spacecraft hardware used for both attitude determination and attitude control. Extensive hardware experimentation has taken place over the 20-year history of spaceflight. Although this experimentation and development is still continuing, a variety of basic functional types of attitude hardware have emerged. This chapter describes the physical characteristics and operating principles of a variety of sensors. *The mathematical models associated with these sensors are presented in Chapter 7.* Additional summaries of attitude hardware are given by Fontana, *et al.*, [1974], Hatcher [1967], and Schmidtbauer, *et al.*, [1973]. A summary of attitude hardware for specific spacecraft is given in Appendix I.

6.1 Sun Sensors

Gerald M. Lerner

Sun sensors are the most widely used sensor type; one or more varieties have flown on nearly every satellite. The Sun sensor owes its versatility to several factors. Unlike the Earth, the angular radius of the Sun is nearly orbit independent and sufficiently small (0.267 deg at 1 AU) that for most applications a point-source approximation is valid. This simplifies both sensor design and attitude determination algorithms. The Sun is sufficiently bright to permit the use of simple, reliable equipment without discriminating among sources and with minimal power requirements. Many missions have solar experiments, most have Sun-related thermal constraints, and nearly all require the Sun for power.* Consequently, missions are concerned with the orientation and time evolution of the Sun vector in body coordinates. Attitude control systems are frequently based on the use of a Sun reference pulse for thruster firings, or, more generally, whenever phase-angle information is required. Sun sensors are also used to protect sensitive equipment such as star trackers, to provide a reference for onboard attitude control, and to position solar power arrays.

The wide range of Sun sensor applications has led to the development of numerous sensor types with fields of view (FOV) ranging from several square arc-minutes (10^{-7} sr) to 128 by 128 deg (approximately π sr) and resolutions of several degrees to less than an arc-second. The three basic classes of Sun sensors are *analog sensors*, which have an output signal that is a continuous function of the Sun angle and is usually monotonic; *Sun presence sensors*, which provide a constant output signal whenever the Sun is in the FOV; and *digital sensors*, which provide an encoded, discrete output which is a function of the Sun angle. A summary of sensor types manufactured by the Adcole Corporation is presented in Table 6-1.

6.1.1 Analog Sensors

Analog sensors are frequently called cosine detectors because a common type is based on the sinusoidal variation of the output current of a silicon solar cell with Sun angle as shown in Fig. 6-1. Specifically, the energy flux, E, through a surface of area dA with unit normal $\hat{n}$ is

$$E = \mathbf{P} \cdot \hat{n}\, dA \qquad (6\text{-}1)$$

where $\mathbf{P}$ is the *Poynting vector*, which gives the direction and magnitude of energy flow for electromagnetic radiation. Thus, the energy deposited in a photocell and, consequently, the output current, I, is proportional to the cosine of the angle of incidence of the solar radiation.

$$I(\theta) = I(0)\cos\theta \qquad (6\text{-}2)$$

Small transmission losses due to Fresnel reflection, the effective photocell area, and angle-dependent reflection at the air-cell interface are omitted from the simple model given by Eq. (6-2).

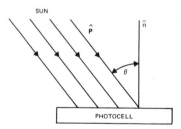

Fig. 6-1.   Cosine Detector Sun Sensor

---

*Spacecraft that do not use solar power include the Pioneer missions, which use nuclear power because of the $1/r^2$ decrease in solar flux with distance from the Sun.

Table 6-1. Representative Sun Sensors Produced by Adcole Corporation. (Data Courtesy Adcole.)

SENSOR	ADCOLE MODEL NUMBER	FIELD OF VIEW	MAXIMUM NUMBER OF SENSORS[2]	LEAST SIGNIFICANT BIT	ACCURACY[4]	TRANSFER FUNCTION[5]	OUTPUT[10]	DIMENSIONS — ELECTRONICS	DIMENSIONS — SENSOR	WEIGHT (gm) — ELECTRONICS	WEIGHT (gm) — SENSOR	INPUT POWER	MISSIONS FLOWN
DIGITAL SENSOR FOR SPINNING VEHICLES	17083	180°[1]	1	1.0°	0.50°	(6)	9-BIT SERIAL GRAY CODE	84 x 51 x 89	51 x 51 x 31	445	141	−24.5 vdc 12 ma	AE-3, 4, 5, RAE-2
	15761	5.6°[1]	1	0.02°	0.1°	(7)	8-BIT SERIAL BINARY CODE	102 x 89 x 56	33 x 33 x 33	540	109	+5 vdc 80 ma / +8 vdc 4 ma / −8 vdc 4 ma	AEROS-1 AEROS-2
		180°[1]	1	1.0°	0.5°	(6)	8-BIT SERIAL GRAY CODE		46 x 43 x 20		109		
TWO-AXIS DIGITAL SENSORS	18273	64°[1]	2	0.25°	0.1°	(6)	8-BIT SERIAL GRAY CODE	193 x 114 x 64	66 x 38 x 25	1043	109	+28 vdc 600 mw	CTS
	15486	128° x 128°	5	1.0°[3]	0.5°	(8)	7-BIT/AXIS SERIAL GRAY CODE	76 x 76 x 51	84 x 41 x 18	358	73	+15 vdc 19.5 ma	ATS-6
	17115	128° x 128°	1	1.0°[3]	0.5°	(8)	7-BIT PARALLEL GRAY CODE	165 x 114 x 64	84 x 41 x 18	957	82	5.0 ±0.2 vdc 12 ma	ATS-6, GEOS-3 SAS-3 RAE-2
		128° x 128°	3	0.5°[3]	0.5°	(8)	7-BIT/AXIS PARALLEL GRAY CODE		84 x 41 x 18		113		
	16764	128° x 128°	5	0.5°[3]	0.25°	(8)	8-BIT/AXIS PARALLEL 3-BIT IDENTITY GRAY CODE	89 x 114 x 31	81 x 81 x 20	295	259	+12 ±0.36 vdc 8 ma / −12 ±0.36 vdc 2 ma	AEM-A, B[11]
	17032	64° x 64°	4	0.125°[3]	0.1°	(8)	9-BIT/AXIS SERIAL GRAY CODE	197 x 114 x 60	97 x 71 x 23	1148	277	−24.5 vdc 560 mw	NIMBUS-6
	15381	64° x 64°	1	0.004°[3]	0.017°	(9)	14-BIT/AXIS PARALLEL BINARY CODE	198 x 114 x 64	97 x 104 x 25	1361	372	+28 vdc 61.5 ma	OAO-3
	18960	64° x 64°	2	0.004°[3]	0.017°	(9)	15-BIT/AXIS SERIAL GRAY AND BINARY CODE	206 x 157 x 30	84 x 110 x 25	455	341	+28 vdc 1.8 w	IUE SEASAT[11]
TWO-AXIS ANALOG SYSTEM	12202	30° CONE	N/A	N/A	1' AT NULL	±1° LINEAR	:4 ma	N/A	64 x 30 x 33	NONE	55	NONE	OAO-2, 3, ATS-6
	18394	FULL HEMISPHERE	1	N/A	2° AT NULL	±30° LINEAR	:0.1 ma PEAK	N/A	48 x 48 x 33	NONE	82	NONE	OAO-2, 3, ATS-6
SINGLE-AXIS ANALOG SYSTEM	17470	40° x 60°	1	N/A	6' AT NULL	±1° LINEAR	0-5v 2.5 V AT NULL	69 x 51 x 28		118		±15 ±0.6 vdc 100 mw	CTS
COSINE-LAW ANALOG	11866	160° CONE	N/A	N/A	2 μ	COSINE OF ANGLE OF INCIDENCE	0.1 ma PEAK	N/A	23 diam x 10	NONE	4.6	NONE	OAO-2, 3 ATS-6

[1] THE FIELD OF VIEW IS FAN SHAPED. AN OUTPUT PULSE IS PROVIDED WHEN THE FAN CROSSES THE SUN AND THE DIGITAL SUN ANGLE IS READ AT THIS TIME AND STORED. THE SENSOR SHOULD BE MOUNTED SO THAT THE PLANE OF THE FAN IS PARALLEL TO THE SPIN AXIS

[2] SUPPORTED BY ELECTRONICS

[3] THE LEAST SIGNIFICANT BIT SIZE IS AN AVERAGE OF THE ONE-AXIS STEP SIZES OVER THE FIELD OF VIEW.

[4] FOR A DIGITAL SENSOR, THE ERROR IS DEFINED AS THE ABSOLUTE VALUE OF THE DIFFERENCE BETWEEN THE SUN ANGLE CALCULATED FROM THE TRANSFER FUNCTION AND THE MEASURED ANGLE AT A STEP.

[5] SEE SECTION 7.1 AND FIG. 7-8 FOR A DERIVATION OF TRANSFER FUNCTIONS AND SENSOR ANGLES.

[6] EQS. (7.13)
$J \neq 0$
$a = k_0 + k_1 N + 0$ (N[3])
$J = 0$

[7] $a = 0.0108 + 0.0216 N$

[8] EQS. (7.17) THROUGH (7.21).

[9] EQS. (7.38).

[10] PARALLEL OUTPUT USES A WIRE FOR EACH BIT, SERIAL OUTPUT USES A BUFFER STORAGE REGISTER.

[11] PROPOSED.

Apertures are used to limit the FOV of an analog sensor, and the cosine detectors used to position solar angle generally have conical FOVs. A group of cosine detectors, or *eyes*, each with a limited FOV, can provide intermediate accuracy over a wide angular range, as shown in Figs. 6-2 and 6-3.

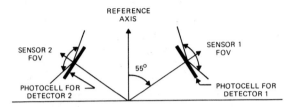

Fig. 6-2. Orientation of Two Cosine Detectors To Provide Sun Angle Measurements Over a Wide Angular Range

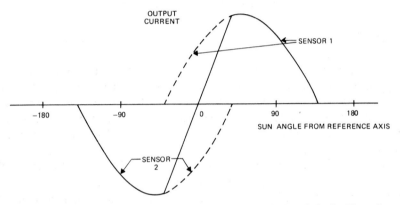

Fig. 6-3. Summed Output From the Two Cosine Detectors When the Sun Is in the Plane Containing the Reference Axis and the Normal to the Detectors. The dashed lines give the output from each sensor; the solid line is the summed output.

A second analog sensor type uses a bar or mask to shadow a portion of one or more photocells. Different configurations can yield a one-axis sensor (Fig. 6-4) or a two-axis sensor (Fig. 6-5) with varying FOVs and resolution. The two-axis sensor shown is similar to that flown on HEAO-1 [Gray, *et al.*, 1976].

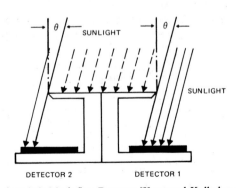

Fig. 6-4. One-Axis Mask Sun Detector [Koso and Kollodge, 1969]

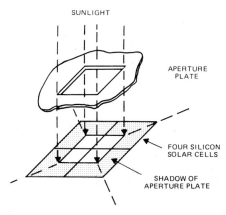

Fig. 6-5. Two-Axis Mask Sun Detector [Schmidtbauer, *et al.*, 1973]. The sunline is normal to the
aperture plate if the output of all four solar cells is equal.

6.1.2 Sun Presence Detectors

Sun presence detectors are used to protect instrumentation, to activate hardware, and to position the spacecraft or experiments. Ideally, Sun presence detectors
provide a step function response that indicates when the Sun is within the FOV of
the detector. For example, the shadow bar detector shown in Fig. 6-6 has a steep
output slope and, consequently, a limited FOV and a 1-arc-minute accuracy. The
sensor mass is less than 200 g.

The critical angle prism illustrated in Fig. 6-7(a) is based on Snell's law,
$n \sin\theta = \sin\theta'$. Consider radiation incident normal to the base of an isosceles
triangular prism with index of refraction n, and base angle γ, such that $n \sin\gamma = 1$.
The angle of the refracted radiation is $\theta' = 90°$, and the total output current from
the photocells will be zero. Non-normal incidence will yield current in the detector
for which $\theta' < 90°$. Figure 6-7(b) illustrates the total transmission for near normal
incidence.

Another type of highly accurate null detector is illustrated in Fig. 6-8. The
sensor optics are such that a null Sun angle will center the Sun image at the top of
the wedge and mirrors on the side of the wedge will reflect radiation to yield a
current balance in the photocells.

Spinning spacecraft frequently employ one or more Sun-presence detectors
composed of two slits and a photocell, as shown in Fig. 6-9. Whenever the Sun lies

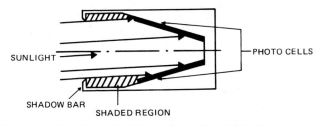

Fig. 6-6. Shadow Bar Sun Sensor With Steep Output Slope [Schmidtbauer, *et al.*, 1973]

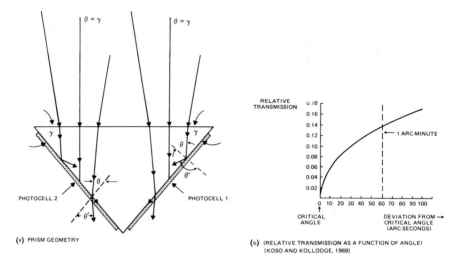

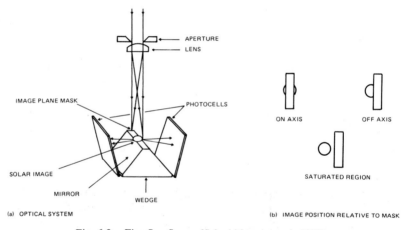

Fig. 6-7.   Critical Angle Prism Sun Sensor

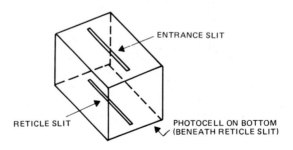

Fig. 6-8.   Fine Sun Sensor [Schmidtbauer, *et al.*, 1973]

ENTRANCE SLIT

RETICLE SLIT

PHOTOCELL ON BOTTOM
(BENEATH RETICLE SLIT)

Fig. 6-9.   Two-Slit Sun Presence Detector

in the plane formed by the entrance and reticle slits and makes an angle with the normal to the sensor face of less than a specified limit (typically 32 or 64 deg), the photocell will indicate Sun presence. When two such sensors are placed in a **V** configuration, usually with one sensor entrance slit parallel to the spin axis, the time between Sun pulses is a measure of the Sun angle, as illustrated in Fig. 6-10 (see also Section 7.1).

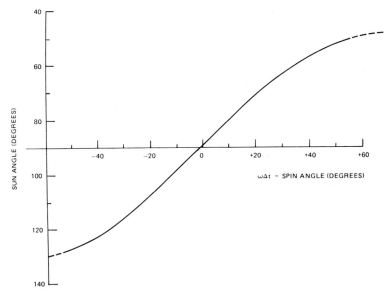

Fig. 6-10.   Sun Angle as a Function of Spin Angle for Typical Solar V-Beam Sensor With 45-Deg Tilt Angle Between Slits

6.1.3 Digital Sensors

A common digital Sun sensor for spinning spacecraft consists of the two basic components, command and measurement, as illustrated in Fig. 6-11.* The command component is the same as the Sun presence detector shown in Fig. 6-9.

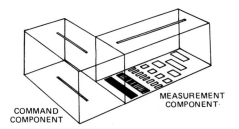

Fig. 6-11.   Basic One-Axis Digital Sensor Components

---

*The discussion in this section is based on sensors manufactured by the Adcole Corporation and flown on a variety of spacecraft. Table 6-1 summarizes the physical data for these sensors.

Because the nominal FOV for Adcole sensors is limited to ± 64 deg, full 180-deg coverage is accomplished by mounting two or more sensor units with overlapping FOVs as shown in Fig. 6-12.

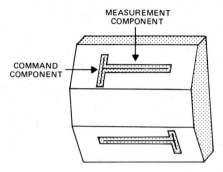

Fig. 6-12.   Two One-Axis Sun Sensors for Spinning Spacecraft With 180-Deg FOV (Adcole Model 17083.)

The measurement component generates an output which is a digital representation of the angle between the sunline and the normal to the sensor face when the Sun is in the FOV of the command component, as shown in Fig. 6-13. The measurement component illustrated in Fig. 6-14 is a composite (similar to that flown on Nimbus-6, Adcole model 17032) that shows most of the features of interest. The Sun image is refracted by a material of index of refraction, n, which may be unity, and illuminates a pattern of slits. The slits are divided into a series of rows with a photocell beneath each row. Four classes of rows are illustrated: (1) an automatic threshold adjust (ATA), (2) a sign bit, (3) encoded bits (Gray code, described below, is shown), and (4) fine bits.

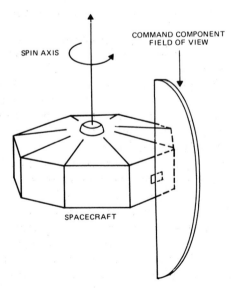

Fig. 6-13.   Sun Sensor Command Component Field of View for Spinning Spacecraft

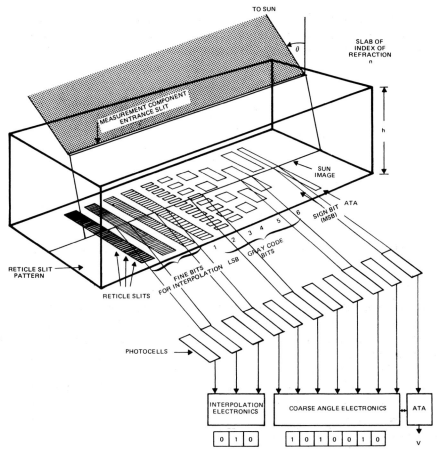

Fig. 6-14.   Detail of Sun Sensor Measurement Component

Because the photocell voltage is proportional to cos θ (θ = Sun angle), a fixed threshold is inadequate for determining the voltage at which a bit is turned on. This is compensated for by use of the ATA slit, which is half the width of the other slits. Consequently, the ATA photocell output is half that from any other fully lit photocell independent of θ as long as the Sun image is narrower than any reticle slit. A bit is turned "on" if its photocell voltage is greater than the ATA photocell voltage and, consequently, "on" denotes that a reticle slit is more than half illuminated (independent of the Sun angle).

The sign bit or most significant bit determines which side of the sensor the Sun is on. The encoded bits provide a discrete measure of the linear displacement of the Sun image relative to the sensor center line or *null*. Several codes are used in Adcole sensors, including V-brush and Gray [Susskind, 1958]. *Gray code*, named after the inventor, is the most widely used and is compared with a binary code in Table 6-2 and Fig. 6-15. The advantage of a Gray code may be seen by comparing the binary and Gray codes for a Sun angle near −16 deg. As the Sun angle decreases across the transition, the binary code changes from −001111 to −010000 and the Gray code from −101000 to −111000. Thus, five binary bits change but

Table 6-2. Gray-to-Binary Conversion. The most significant bit is the same in either binary or Gray code. Each succeeding binary bit is the complement of the corresponding Gray bit if the preceding binary bit is 1 or is the same if the preceding binary bit is 0. (See Section 8.4 for conversion algorithm.)

DECIMAL	BINARY	GRAY	DECIMAL	BINARY	GRAY
0	0	0	11	1011	1110
1	1	1	12	1100	1010
2	10	11	13	1101	1011
3	11	10	14	1110	1001
4	100	110	15	1111	1000
5	101	111	16	10000	11000
6	110	101	17	10001	11001
7	111	100	18	10010	11011
8	1000	1100	19	10011	11010
9	1001	1101	20	10100	11110
10	1010	1111	21	10101	11111

only one Gray bit changes. By inspection, the Gray code is an equidistant code. That is, one and only one bit changes for each unit distance whereas one or more binary bits change for the same unit distance. Because some imperfection in the reticle pattern is inevitable and a transition may occur while the photocell is being interrogated for transmission, the possible decoded angles for a binary code near -16 could range from -0 to -16, whereas for a Gray code, only -16 or -15 is

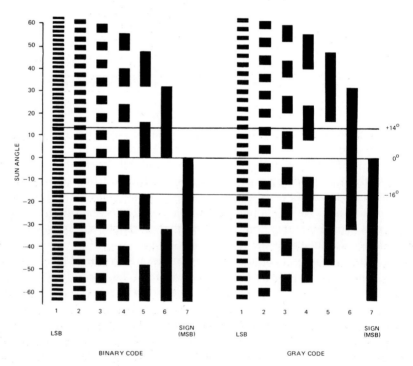

Fig. 6-15. Gray and Binary Coded Reticle Patterns for a ± 64-Deg FOV Digital Sun Sensor With a 1-Deg Least Significant Bit

possible. Algorithms for converting between Gray and binary codes are given in Section 8.4.

The calibration of the encoded bits is verified by plotting the output from each photocell versus Sun angle, as shown in Fig. 6-16 for the two least significant bits (LSBs). Note that the envelope of the sinusoidal output of both bits is roughly proportional to $\cos\theta$ and the ATA output follows the envelope with half the amplitude. A characteristic of the Gray code is that the peak output of one bit corresponds to alternate minima of the next lesser bit. The angular error at a bit on-off transition is typically half the LSB.

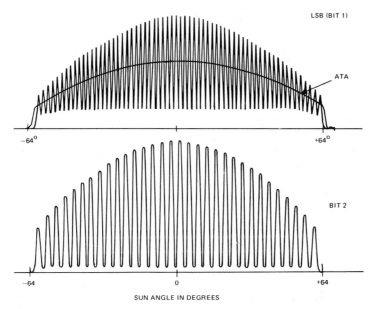

Fig. 6-16. Plot of the Output From Representative Photocells Versus Sun Angle for Adcole Digital Sun Sensors

The fine bits in Fig. 6-14 are used by an interpolation circuit to provide increased resolution. Straightforward addition of encoded rows to the pattern is not possible because the 0.53-deg angular diameter of the Sun from near the Earth would blur the output from adjacent bits. This effectively limits Gray code transitions to a 1/2-deg LSB. By combining the output of 2 or 3 offset LSB patterns in an interpolation circuit, 1/4- or 1/8-deg transitions are obtained.

Two-axis sensors consist of two measurement components mounted at right angles, yielding a 64- by 64-deg or 128- by 128-deg FOV as shown in Fig. 6-17. Full 4π sr coverage for the two-axis sensors is obtained by use of five or more 128- by 128-deg sensors. (See Sections 2.1 and 7.1.) Onboard logic for selecting and telemetering data from the illuminated Sun sensor in multisensor configurations is based on monitoring the output of the ATA photocell and selecting the sensor with the highest output signal (effectively the smallest angle relative to the optical null or *boresight angle*).

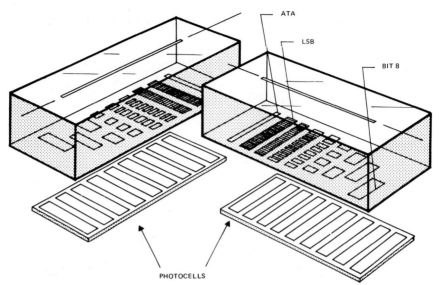

Fig. 6-17. Reticle and Photocell Assemblies for Two-Axis Sun Sensor. Illustration represents a 1/2-Deg LSB, Adcole Model 16764. (Information courtesy of Adcole Corporation.)

6.1.4 Fine Sun Sensors

Increasingly stringent attitude accuracy requirements, such as for IUE, MAG-SAT, or SMM, imply Sun sensor *absolute* accuracies of several arc-minutes to 5 arc-seconds and even better *relative* accuracies. Resolutions of less than 1/8-deg LSB, the practical limit of the device shown in Fig. 6-14, to an LSB of 0.1 arc-second can be achieved by electronically combining the output current from four offset photocells beneath a reticle pattern as shown in Fig. 6-18. The SMM fine Sun sensor, shown here, consists of an entrance slit composed of 72 pairs of alternately opaque and transparent rectangles of equal width (0.064 mm); a 1.5-cm spacer; an exit slit composed of four offset reticle patterns, each with 68 alternately opaque-transparent rectangle pairs; and four photocells, one beneath each pattern [Adcole, 1977]. The entrance and exit slits are separated by a vacuum to reduce the effect of spectral dispersion on the accuracy.

The Sun sensor output is periodic, with a period which can be adjusted to meet the accuracy and FOV requirements, e.g., a 2-deg period for SEASAT with a ±32-deg FOV and a 1-deg period for SMM with a ±2-deg FOV. The fine Sun sensor is combined with a digital Sun sensor to resolve ambiguities in the output angle. Two sensors are mounted perpendicular to one another for two-axis output. The sensor operation is described more fully in Section 7.1.

6.2 Horizon Sensors

Gerald M. Lerner

The orientation of spacecraft relative to the Earth is of obvious importance to space navigation and to communications, weather, and Earth resources satellite

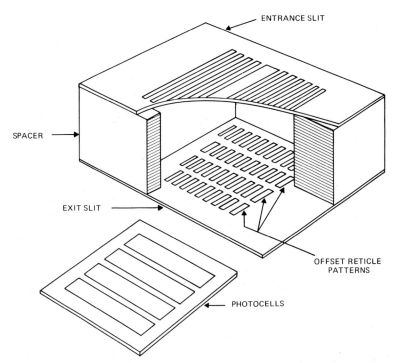

ENTRANCE SLIT

SPACER

EXIT SLIT

OFFSET RETICLE PATTERNS

PHOTOCELLS

Fig. 6-18. Solar Maximum Mission Experimental Sun Sensor. [Adcole, 1977].

payloads. To a near-Earth satellite, the Earth is the second brightest celestial object and covers up to 40% of the sky. The Earth presents an extended target to a sensor (3.9 sr at a 500-km altitude) compared with the generally valid point source approximations employed for the Sun (7×10^{-5} sr) and stars.* Consequently, detecting only the presence of the Earth is normally insufficient for even crude attitude determination and nearly all sensors are designed to locate the Earth's horizon. (The detection of the presence of small planetary bodies, such as the Moon from a near-Earth orbit, is, however, sufficient for coarse attitude determination.) Horizon sensors are the principal means for directly determining the orientation of the spacecraft with respect to the Earth. They have been employed on aircraft and were used on the first U.S. manned flights in the Mercury and Gemini programs [Hatcher, 1967]. In this section we describe the requirements imposed on horizon sensors, outline the characteristics of several generic types, and describe the operating principles of horizon sensor systems in common use.

As described in Chapter 4, the location of the horizon is poorly defined for a body possessing an atmosphere because of the gradual decrease in radiated intensity away from the *true* or *hard horizon* of the solid surface. However, even a body possessing no atmosphere, such as the Moon, poses a horizon sensor design problem due to variations in the radiated intensity. To illustrate an extreme case, a detector triggering on the lunar horizon in the 14- to 35-μm infrared spectral region

---

*Betelgeuse, the star with the largest angular radius, subtends 6×10^{-14} sr.

will experience fiftyfold variations in radiance (120°K to 390°K) between illuminated and unilluminated horizons. As illustrated in Fig. 6-19, if the radiation integrated over half the sensor field of view (FOV) is just above threshold at a cold Moon, the horizon location error at a hot Moon is half the sensor FOV because the sensor would then trigger at the edge of the FOV. Lowering the threshold or decreasing the sensor FOV may not be possible because of the low intensity of emitted radiation relative to noise for practical detectors. Thus, for the lunar horizon, a different choice of spectral region (the visible) is frequently employed to provide a sufficient radiation intensity with a small FOV.

Earth resources, communications, and weather satellites typically require pointing accuracies of 0.05 deg to less than a minute of arc, which is beyond the state of the art for horizon sensors. However, Earth-oriented spacecraft frequently employ autonomous attitude control systems based on error signals from horizon sensors with accuracy requirements of 0.5 to 1 deg. Thus, although payload requirements may not be met by current horizon detectors, control requirements are easily met, and significant cost savings may be realized by increasing the accuracy of horizon sensors to meet payload and control accuracy requirements simultaneously and thereby avoid the necessity of flying star sensors.

In Chapter 4, we showed that the position of the Earth's horizon is least ambiguous in the spectral region near $15\mu m$ in the infrared. Most horizon sensors now exploit the narrow 14- to 16-μm CO_2 band. Use of the infrared spectral band avoids the large attitude errors encountered on Mercury, Gemini, and OGO due to spurious triggerings of visible light (albedo) horizon sensors off high-altitude clouds [Hatcher, 1967]. In addition, the operation of an infrared horizon sensor is unaffected by night or by the presence of the terminator. Infrared detectors are less susceptible to sunlight reflected by the spacecraft than are visible light detectors

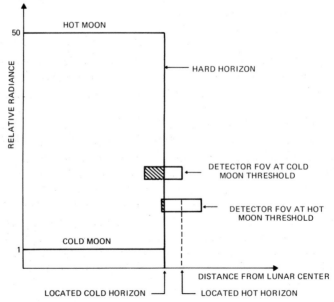

Fig. 6-19. Horizon Location Errors Due to Radiance Variations

and therefore avoid reflection problems such as those encountered on RAE-2 [Werking, *et al.*, 1974]. Sun interference problems are also reduced in the infrared where the solar intensity is only 400 times that of the Earth, compared with 30,000 in the visible [Trudeau, *et al.*, 1970]. However, albedo sensors have some advantages over the infrared sensors, including lower cost, faster response time (microseconds for the photodiode employed by visible sensors versus milliseconds for the thermistor employed by many infrared sensors), and higher signal-to-noise ratio because the radiated intensity is highest in the visible (see Section 4.1).

Attitude acquisition (see Section 19.5) frequently requires horizon detection far from the nominal mission orbit and attitude. Consequently, sensor versatility is a common design requirement. Wide FOV detectors, such as the two-axis Sun sensors described in Section 6.1, cannot accurately define the horizon of the large, relatively dim Earth. Consequently, horizon sensors frequently employ some means to scan the celestial sphere with a small, typically 2- by 2-deg FOV.

Finally, Sun rejection capability is important for horizon sensors, particularly for those used for onboard control. Redundant sensors or optical systems are used to provide Sun rejection by comparing the output of spatially displaced optical systems. Alternatively, Sun rejection may be based on a priori knowledge of the Sun position, the sensor output pulsewidth or intensity, or the output of special-purpose Sun sensors.

6.2.1 Sensor Components

Most horizon sensors consist of four basic components: a *scanning mechanism*, an *optical system*, a *radiance detector*, and *signal processing electronics*. They are normally categorized by the *scanning mechanism* or the method used to search the celestial sphere. Several methods are employed, the simplest of which is to rigidly attach the sensor to the body of a spinning spacecraft. For such *body-mounted horizon sensors* fixed at a selected angle relative to the spin axis, the FOV is typically a small circle or square of about 2-deg diameter, although a sensor consisting of two fan-shaped slits, 1- by 120-deg, was flown on the COS-B and ISEE-2 [Massart, 1974; Wetmore, *et al.*, 1976]. *Wheel-mounted horizon sensors* are similar to body-mounted sensors except that they are attached to the momentum wheel of a spacecraft and the wheel, rather than the spacecraft, provides the scanning motion.

In contrast to the wheel-mounted horizon sensor, *Scanwheels*, a registered trademark of Ithaco, Inc., are integrated systems consisting of a momentum wheel, a horizon sensor, and electronics which may be used for both attitude determination and control (these are discussed further in Section 6.2.2). For spacecraft for which the angular momentum of a wheel-mounted sensor or Scanwheel is undesirable, designs employing a slowly rotating turret, such as the *panoramic attitude sensor* (PAS) flown on IUE and the ISEE-1; a rotating mirror, such as the *nonspinning Earth sensor assembly* flown on CTS;* or counterrotating scanwheels or wheel-mounted horizon sensors with zero net angular momentum may be used. For all horizon sensors, the sensor mounting angle, γ, is defined as the angle between the spin and optical axes.

---

*The original PAS flown on RAE-2 also employed a rotating mirror.

The *optical system* of a horizon sensor consists of a filter to limit the observed spectral band and a lens to focus the target image on the radiance detector. Optical system components depend greatly on sensor design. In many cases, rotating mirrors or prisms are incorporated into the optical system to provide the scanning mechanism. The spectral sensitivity characteristics of the proposed SEASAT infrared Scanwheels built by Ithaco are illustrated in Fig. 6-20.

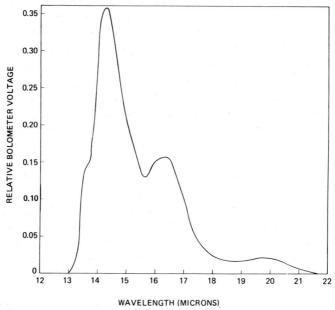

Fig. 6-20. SEASAT-A Scanwheel Optics Spectral Sensitivity. Optical system includes a window, an optical wedge, and a bolometer. (Courtesy of Ithaco Corporation.)

Radiance detectors used to detect the presence of a horizon may be classified by their region of spectral sensitivity. A *photodiode*, illustrated in Fig. 6-21, consists of a P-N junction operated under reverse bias. Light falling on the photodiode increases the number of electrons and holes in the junction region, thereby increasing the leakage current, i_F [Ryder, 1967]. Photodiodes have a peak sensitivity in the near infrared at about 1.2 μm for germanium and 0.8 μm for silicon. (The visible spectrum extends from about 0.4 to 0.7 μm.)

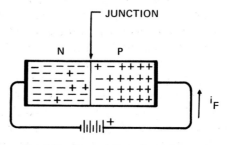

Fig. 6-21. Typical Photodiode Schematic

Detectors which respond to the longer wavelengths of infrared blackbody radiation are based on the operating principles of the thermistor, thermocouple, or pyroelectric crystal [Barnes Engineering Co., 1976]. A *bolometer* is a very sensitive resistance thermometer, or *thermistor*, used to detect infrared radiation. Thermistors consist of fused conglomerates, or *sinters*, of manganese, cobalt, and nickel oxide formed into *flakes*, typically 0.5 mm by 0.5 mm by 10 μm thick, bonded to a heat-dissipating substrate or *heat sink*. Impinging radiation heats the flake and alters the resistance, typically by 3.5%/°K, which is sensed by conversion to a voltage and amplification. When radiation is removed from the flake, its temperature returns to that of the heat sink with a time constant depending on the thermal conductance of the flake-substrate bond. A typical time constant is 3 ms. Bolometers are able to sense temperature changes of 0.001°K due to radiation despite ambient temperature changes four orders of magnitude greater [Astheimer, 1976]. The minute temperature change is observed by modulating the incoming radiation by, for example, scanning across the target, and thereby removing the effect of ambient temperature changes on the output voltage by capacitance coupling to the amplifier.

A bolometer may have either one or two flakes in the focal plane of the optical system. The two flakes of a *dual-flake* system detect radiance originating from different regions of the celestial sphere. Consequently, the two output signals may be combined in an electronic AND circuit to provide Sun rejection if the separation between the flakes is such that the Sun cannot be seen by both flakes simultaneously. Thermistors are often *immersed* in or surrounded by a germanium lens (transparent to infrared radiation) to increase the intensity of radiation at the thermistor.

A *thermopile* consists of a string of thermocouple junctions connected in series. Each thermocouple consists of a hot junction and a cold junction. The hot junctions are insulated from a heat sink and coated with a blackening agent to reduce reflection. The cold junctions are connected directly to a heat sink. When exposed to impinging infrared radiation, the hot junction is heated and yields a measurable output voltage. Thermocouple junctions commonly use bismuth and antimony. Thermopile detectors are simple, requiring minimal electronics and no moving parts; however, they suffer from a slow response time and are used only in nonscanning systems.

Pyroelectric detectors consist of a thin crystal slab, such as triglycine sulfate, sandwiched between two electrodes. Impinging radiation raises the temperature of the crystal, causes spontaneous charge polarization of the crystal material, and yields a measurable potential difference across the electrodes. Pyroelectric detectors may be used in scanning systems because they are fast and have a high signal-to-noise ratio with no low-frequency noise.

The output from a scanning horizon sensor is a measure of the time between the sensing of a reference direction and the electronic pulse generated when the radiance detector output reaches or falls below a selected threshold. The reference direction for a body-mounted sensor is generally a Sun pulse from a separate sensor, whereas wheel-mounted sensors typically use a magnetic pickoff fixed in the body. If the detector output is increasing across the threshold, the pulse corresponds to a dark-to-light transition or *acquisition of signal (AOS)*. If the detector output is decreasing across the threshold, the pulse corresponds to a light-to-dark transition or *loss of signal (LOS)*. The AOS and LOS pulses are also referred to as *in-crossings* and *out-crossings*, or *in-triggering* and *out-triggering*, respectively.

Various electronic systems provide the *reference to AOS time* ($t_I \equiv t_{AOS} - t_{REF}$), the *reference to LOS time* ($t_O \equiv t_{LOS} - t_{REF}$), the *Earthwidth* ($t_W \equiv t_{LOS} - t_{AOS}$), and the *reference to midscan time** ($t_M \equiv ((t_{LOS} + t_{AOS})/2 - t_{REF}$). The percentage of the scan period that the radiance is above threshold is the *duty cycle*. Figure 6-22 illustrates the various possible outputs. Knowledge of the scan rate or duty cycle allows the conversion from time to angle either onboard or on the ground. As described in Sections 4.1, 7.2, and 7.3, the horizon crossing times depend on the sensor field of view, the radiance profile of the scanned body, the *transfer function*, and the *locator*. The transfer function relates the radiation pulse incident on the detector to the electronic output of the horizon sensor. The transfer function includes the thermal response time of the detector and time constants associated with pulse amplification and shaping. Typically, sensors are designed and calibrated such that the system output may be used directly for attitude control and determination within a specified accuracy under normal conditions. The electronic technique used to define the threshold for horizon detection, called the locator, can significantly affect the overall attitude accuracy of the system. Many locators have been studied [Thomas, 1967] and two are widely used: the *fixed threshold* locator specifies the observed detector output which defines the horizon; the *fixed percentage of maximum output* or *normalized* locator redefines the threshold for each scan period as a fixed percentage of the maximum output encountered by an earlier scan. Better results are obtained with the normalized threshold locator because it is less sensitive to seasonal or geographical variations in radiance (see Section 4.2 for specific radiance profiles of the Earth). A slightly modified locator has been proposed for SEASAT which continuously resets the threshold to 40% of the mean detector output observed on the Earth between 5 and 11 deg from the located horizon. The thresholds for AOS and LOS are determined independently.

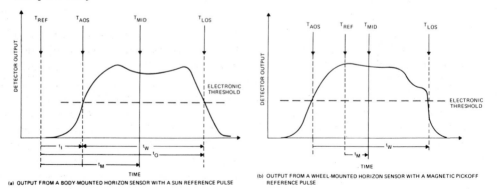

Fig. 6-22. Scanning Horizon Sensor Output

---

*The negative of the reference to midscan time is the *split-to-index time*.

6.2.2 Horizon Sensor Systems

The simplest horizon sensor system is a body-mounted horizon sensor sensitive to visible light. Such a system consists of an aperture and lens to define the field of view and a photodiode to indicate the presence of a lit body. Body-mounted sensors are cheap and reliable and have been used on IMP; slightly more complex versions, sensitive to the infrared spectrum, have been used on AE, SMS/GOES, CTS, and SIRIO. A body-mounted infrared sensor is shown in Fig. 6-23. Body-mounted sensors are suitable only for spinning spacecraft and their fixed mounting angle makes target acquisition a substantial problem for many missions. In this subsection we describe the operating principles of several more versatile sensor systems that have been used operationally.

Fig. 6-23.   Body Mounted Horizon Sensor Used on the CTS Spacecraft. (Photo courtesy of Barnes Engineering Co.)

Panoramic Attitude Sensor (PAS).   The PAS, flown on IUE and ISEE-1 and planned for ISEE-C, is manufactured by the Ball Brothers Research Corporation and is a modification of the original design flown on RAE-2. The PAS is among the most versatile of all horizon scanners because of its ability to use either an internal scanning motion or the spacecraft rotation with a variable sensor mounting angle which may be controlled by ground command. Thus, the PAS may detect both the Earth and the Moon for virtually any attitude and central-body geometry. A slit Sun sensor and a visual wavelength telescope, both employing photodiode detectors, are included in the PAS assembly, as illustrated in Figs. 6-24 and 6-25. The telescope has an 0.71-deg FOV diameter, and its optical axis may assume any of 512 discrete positions with a specified positional accuracy of 0.1 deg. The PAS functions as a variable-angle body-mounted sensor when the spacecraft is spinning about an axis parallel to the X axis in Fig. 6-24. Outputs from the system in this

Fig. 6-24. Panoramic Attitude Sensor. (Photo courtesy of Ball Brothers Research Corporation.)

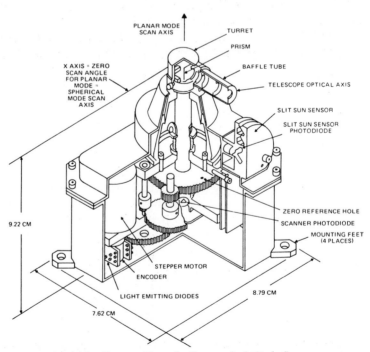

Fig. 6-25. Cutaway View of a Panoramic Attitude Sensor

spherical mode are the times from the Sun pulse to AOS and LOS. The threshold for detection is specified as 0.1 times the maximum lunar radiance, which corresponds to a first- or third-quarter Moon as viewed from the vicinity of the Earth.

When the spacecraft is despun, the scanning motion in the *planar mode* is accomplished by rapidly stepping the turret. Various commands are available to control the operation of the PAS in both the spherical and the planar modes. The turret can be commanded to step continuously in either direction, to reverse directions at specified limit angles, or to inhibit stepping altogether. The detector records and stores, in a series of registers, the encoded steps at each dark-to-light or light-to-dark transition. The telescope is baffled to prevent detection of the Sun at separation angles between the telescope axis and the Sun of 12 deg or more.

Nonspinning Earth Sensor Assembly (NESA). The NESA, built by TRW for the synchronous orbit CTS and the ATS-6 spacecraft, consists of two independent infrared sensors that scan across the Earth, measuring rotation angles to define the spacecraft attitude relative to the Earth from synchronous altitude. The detector senses radiation in the 13.5- to 25- μm spectral band and uses a fixed percentage of maximum output locator. A small Sun detector is located near the infrared telescope to identify intrusion of the Sun in the FOV. This sensor consists of a small mirror, two fixed mechanical apertures, and a silicon photodiode detector. The Sun detector FOV is concentric with, but larger than, the infrared detector FOV.

The sensor geometry near the mission attitude (pitch = roll = 0) is shown in Fig. 6-26 with the spacecraft Z axis in the nadir direction. A scanning field of view, approximately in the spacecraft X-Z or Y-Z plane, is created by oscillating a beryllium mirror at 4.4 Hz. The scan plane is tilted 3.5 deg so that, for the mission attitude, the scan paths are slightly offset from the Earth center. The sensor geometry is chosen such that either the north-south (NS) or east-west (EW) scanner

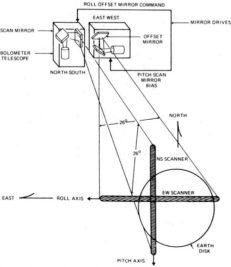

Fig. 6-26. Nonspinning Earth Sensor Assembly. View from Earth toward spacecraft at synchronous altitude.

provides the angular error about both the pitch and roll axes. For the EW scanner, pitch is measured using a binary up-down counter to accumulate encoder pulses from the scan mirror. Encoder pulses are counted during the time an Earth radiance signal is present with the direction of count reversed when the scan crosses the spacecraft pitch axis. When the scan is centered, the up count and the down count are equal and a zero count or *null pointing angle* is obtained. Roll is measured by comparing the total number of encoder pulses with a nominal value corresponding to the expected Earth width at zero roll angle and synchronous altitude. For the NS scanner, the pitch and roll angle computations are reversed. Thus, the system provides redundant output over a ±2.82-deg linear range. The specified sensor accuracy is 0.05 deg with a 0.01-deg least significant bit. In the mission mode, the NESA provides error signals to an autonomous control system. During the CTS attitude acquisition phase (Section 19.5), the NESA data were available over an approximate 26-deg by 26-deg field of view, although most of the data provided pitch and roll quadrant information only.

Scanwheels. Integral horizon scanner/momentum wheel systems, similar to the SAS-3 design manufactured by the Ithaco Corporation and illustrated in Figs. 6-27 and 6-28, have flown on numerous spacecraft including ERTS, NIMBUS, and SAS-3 and are proposed for HCMM, SEASAT, SAGE and MAGSAT. The flywheel and attached prism rotate at a variable rate, generally near 1600 rpm, providing both angular momentum and a conical scan about the flywheel axis. Radiation originating at an angle defined by the prism passes through a germanium

Fig. 6-27.   Ithaco Scanwheel Showing Rotating Components With Cover and Germanium Window Removed. Compare with Fig. 6-28. (Courtesy of Ithaco Corporation.)

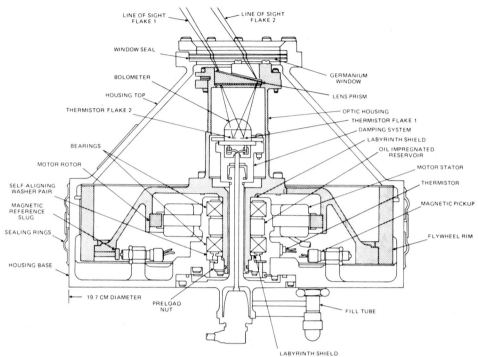

Fig. 6-28. SAS-3 Type B Scanwheel. Shading indicates rotating components consisting of flywheel and prism assembly. (Courtesy of Ithaco Corporation.)

window (to define the optical passband) and is focused on a thermistor fixed in the spacecraft body. A magnetic reference slug, located on the flywheel, provides a reference pulse when it is sensed by a body-mounted magnetic pickup each revolution. The Scanwheel electronics measure the duty cycle before and after the reference pulse. Scanwheel and similar systems have been the subject of detailed analysis by Wertz, *et al.*, [1975], Hotovy, *et al.*, [1976], and Nutt, *et al.*, [1978]. Table 6-3 summarizes the physical characteristics of scanwheel systems.

Two Scanwheel configurations are commonly flown. Use of a single scanwheel on SAS-3 and HCMM yields a dual-spin spacecraft with the horizon sensor scanning at a fixed mounting angle relative to the wheel axis. Dual Scanwheel systems, illustrated in Fig. 6-29, can provide momentum about two axes; their

Table 6-3. Scanwheel and Wheel-Mounted Horizon Sensor Systems

MISSION	DETECTOR*	SPECTRAL BAND (μm)	LOCATOR	FIELD OF VIEW	RANGE (DEG)	NOMINAL RPM	WHEEL INERTIA (kg·m²)	MASS (kg)	ROTOR DIAMETER (cm)	TORQUE (J)	POWER (W)
SAS-3†	DUAL-FLAKE	13.5–16	FIXED THRESHOLD	2° X 2° SQUARE	± 10	1400–1600	0.0271	6.713	18.3	0.025	5.4 @ 1500 RPM
AE-3‡	SINGLE-FLAKE	14–16	25% OF MAXIMUM RADIANCE	2.75° DIAMETER	§	200–360	3.64		130	1.13	
NIMBUS 4-6, G, ERTS†	SINGLE-FLAKE	12.5–18	FIXED THRESHOLD	2° X 2° SQUARE	ABOUT ±60°	600	0.0060	3.719	15.3	0.025	2.25
SEASAT-A†	SINGLE-FLAKE	14–16	40% OF INTEGRATED RADIANCE 5 TO 11 DEG FROM LO-CATED HORIZON	2° X 2° SQUARE	±10	900	0.0271	6.713	18.3	0.025	5.4
HCMM†	DUAL-FLAKE	14–16	FIXED THRESHOLD	2° X 2° SQUARE	±20	1940	0.0271	6.713	18.3	0.025	5.4

*ALL EMPLOY A THERMISTOR BOLOMETER.   †WHEEL-MOUNTED HORIZON SENSOR.
†SCAN WHEEL.   §FULL RANGE, EARTH WIDTH IS SUPPLIED IN SECONDS.

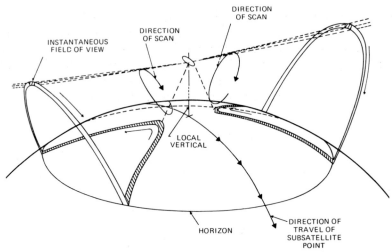

Fig. 6-29.   SEASAT Scanwheel Configuration. (Compare with Fig. 4-3.)

attitude control properties are described in Section 18.3. Attitude accuracy and reliability is increased by dual scanner systems; in particular, and attitude accuracy about the velocity vector, or roll axis, is improved by using the difference between the two observed duty cycles, which is less sensitive to attitude variations than the individual scanner duty cycle measurements.

   Slit Sensors.   Slit sensors of a much different design than the narrow field-of-view detectors described earlier in this section were flown on the spinning spacecraft COS-B, ISEE-2, and SIRIO. The unit used on COS-B and ISEE-2 is called an *attitude sensor unit*, or ASU, and consists of two 120-deg by 1-deg slits [Massart, 1974]. The complete sensor system consists of two ASUs, mounted on COS-B as shown in Figure 6-30. On ISEE-2, both ASUs were mounted with the same orientation, similar to that of the unit on the right-hand side of Fig. 6-30 [Wetmore, *et al.*, 1976]. A cylindrical lens focuses radiation on a photodiode*,

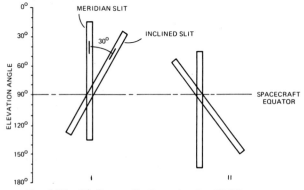

Fig. 6-30.   Slit Sensor Configuration for COS-B

---

*A thermistor bolometer with a 14- to 16-μm optical passband was used for SIRIO.

yielding a pulse which is identified by the sensor electronics. The sensor discriminates between Sun and Earth or Moon pulses and measures the time differences, or rotation angles, between transits of the Sun and Earth by each slit as illustrated in Fig. 6-31. COS-B and ISEE-2 did not fly separate Sun angle sensors, because the Sun angle can be computed from the time difference between the Sun transit of the meridian and inclined slits, as described in Section 7.1.

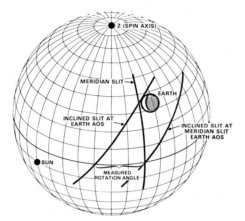

Fig. 6-31.   Sun and Earth Measurement Geometry for COS-B Slit Horizon Sensors

The primary advantage of the ASU is that wide angle coverage can be obtained by a single sensor with no moving parts and no complex commanding logic as is required for the PAS. In addition, a single ASU provides Sun angle, nadir angle, and multiple Sun/Earth rotation angle measurements. However, the ASU has a lower sensitivity to the Earth and the Moon and more complex signal processing electronics than does the PAS. Section 22.1 contains a further discussion of the relative advantages of slit sensors and other sensor types.

Other Systems.   Many systems have been designed for specific mission conditions and thereby achieve increased accuracy and simplicity at the cost of reduced

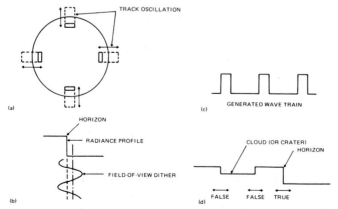

Fig. 6-32.   Moving Edge Tracker Horizon Sensor [Schwartz, 1966]

versatility. These systems operate over a narrow range of orbits and attitudes and include moving and static edge trackers and radiometric balance systems. The *moving edge tracker*, illustrated in Fig. 6-32, has been used on OGO and Gemini. Four oscillating detectors dither across the horizon edge and generate a train of nearly rectangular pulses, as shown in Fig. 6-32(b) and (c). The spacing and width of the pulses vary depending on the null error or the position of the detector relative to the horizon. As will be shown in Section 6.3, the second harmonic of the detector output is related to the null error.

The moving edge tracker uses a feedback system to null the second harmonic or attitude error. For earth trackers, irregularities in the atmosphere composition or temperature can generate false structure, as shown in Fig. 6-32(d), and an erroneous second harmonic in the tracker output [Schwartz, 1966]. The Gemini V-edge trackers experienced track loss at sunrise and sunset [Hatcher, 1967].

The static Earth sensor flown on Symphonie at synchronous altitude used an array of thermopiles configured as shown in Fig. 6-33(a) [Ebel, 1975]. The difference in the output between opposite thermopiles provided a measure of the attitude error. Eight thermopiles, rather than four, were used to provide redundancy and Sun discrimination. Radiometric balance systems, illustrated in Fig. 6-33(b), are similar to the static edge tracker except that a wide field-of-view sensor is employed. Such systems work well if the target radiance is uniform. An accurate radiometric balance sensor, manufactured by Quantic, is used on ERTS/LANDSAT to provide 0.1-deg accuracy [General Electric Space Systems, 1971].

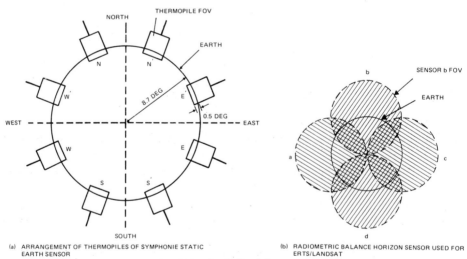

(a) ARRANGEMENT OF THERMOPILES OF SYMPHONIE STATIC
     EARTH SENSOR

(b) RADIOMETRIC BALANCE HORIZON SENSOR USED FOR
     ERTS/LANDSAT

Fig. 6-33.   Static Earth Sensors

6.3 Magnetometers

Bruce T. Blaylock

Magnetometers are widely used as spacecraft attitude sensors for a variety of reasons: they are vector sensors, providing both the direction and magnitude of the magnetic field; they are reliable, lightweight, and have low power requirements;

they operate over a wide temperature range; and they have no moving parts. However, magnetometers are not accurate inertial attitude sensors because the magnetic field is not completely known and the models used to predict the magnetic field direction and magnitude at the spacecraft's position are subject to relatively substantial errors, as discussed in Section 5.1. Furthermore, because the Earth's magnetic field strength decreases with distance from the Earth as $1/r^3$, residual spacecraft magnetic biases eventually dominate the total magnetic field measurement, generally limiting the use of magnetometers to spacecraft below 1000 km; however, attitude magnetometers were flown successfully on RAE-1 at an altitude of 5875 km.

As illustrated in Fig. 6-34, magnetometers consist of two parts: *a magnetic sensor* and an *electronics unit* that transforms the sensor measurement into a usable format. Magnetic field sensors are divided into two main categories: *quantum magnetometers*, which utilize fundamental atomic properties such as Zeeman splitting or nuclear magnetic resonance; and *induction magnetometers*, which are based

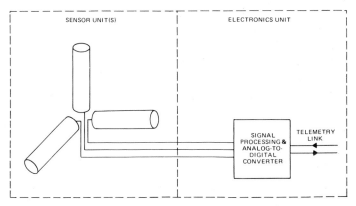

Fig. 6-34.　Generalized Magnetometer Block Diagram

on Faraday's Law of Magnetic Inductance. Faraday's law is the observation that an electromotive force (EMF), **E**, is induced in a conducting coil placed in a time-varying magnetic flux, $d\Phi_B/dt$, such that the line integral of **E** along the coil is

$$V = \oint \mathbf{E} \cdot d\mathbf{l} = -\frac{d\Phi_B}{dt} \tag{6-3}$$

The two types of induction magnetometers are search-coil and fluxgate magnetometers. In a *search-coil magnetometer*, a solenoidal coil of N turns surrounds a ferromagnetic core with magnetic permeability μ, and cross-sectional area A. The EMF induced in the coil when placed in a magnetic field produces a voltage, V, given by

$$V = \oint \mathbf{E} \cdot d\mathbf{l} = -AN\mu(dB_\perp/dt) \tag{6-4}$$

where $B_\perp$ is the field component along the solenoid axis. The output voltage is clearly time dependent and can be rewritten for a coil rotating at a fixed frequency, $f = \omega/2\pi$, about an inertially fixed axis normal to a constant field $\mathbf{B}_0$ as

$$V(t) = - A N \mu B_0 \cos \omega t \tag{6-5}$$

Search-coil magnetometers based on the above principle are used mainly on spin-stabilized spacecraft to provide precise phase information. Because the search coil is sensitive only to variations in the component of the field along the solenoid axis, any spacecraft precession or nutation will greatly complicate the interpretation of data [Sonett, 1963].

The second type of magnetic induction device is the *fluxgate magnetometer*, illustrated in Fig. 6-35. The primary coil with leads P_1 and P_2 is used to alternately drive the two saturable cores SC_1 and SC_2 to states of opposite saturation. The presence of any ambient magnetic field may then be observed as the second harmonic of the current induced in the secondary coil with leads S_1 and S_2. The purpose of the two saturable cores wound in opposite directions is to cause the secondary coil to be insensitive to the primary frequency. Other geometries used to achieve primary and secondary decoupling utilize helical and toroidal cores.

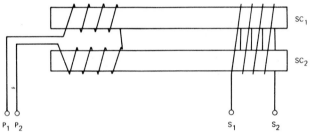

Fig. 6-35.   Dual-Core Fluxgate Magnetometers With Primary and Secondary Induction Coils. (Adapted From Geyger [1964].)

The functional operation of a fluxgate magnetometer is illustrated in Fig. 6-36. If the voltage across the primary coil has a triangular waveform of frequency $2\pi/T$ and the amplitude of the resultant magnetic intensity is H_D, then the core elements saturate at a flux density of $\pm B_S$, when the magnetic intensity reaches $\pm H_C$. The net magnetic intensity is displaced from zero by the ambient magnetic intensity, ΔH. The secondary coil will experience an induced EMF, V_S, while the core elements are being switched or the magnetic flux density is being gated from one saturated state to the other (hence the name *fluxgate*). V_S consists of a train of pulses of width $K_1 T$, separated by time intervals $K_2 T$ or $(1 - K_2)T$ where

$$K_1 = \frac{H_c}{4 H_D} \qquad K_2 = \frac{1}{2}\left(1 - \frac{\Delta H}{H_D}\right) \tag{6-6}$$

The ambient magnetic intensity may then be derived from the pulse spacing in the fourth graph of Fig. 6-36 as

$$\Delta H = (1 - 2 K_2) H_D \tag{6-7}$$

To model the response of the magnetometer electronics, V_S is expressed in a Fourier series as

$$V_S = A \sum_{n=1}^{\infty} \left[1 - \exp(-i2\pi n K_2) \right] \frac{\sin K_1 n\pi}{n\pi} \cos\left(2\pi n \frac{t}{T} \right) \qquad (6\text{-}8)$$

$$A = \frac{2B_S}{K_1 T}$$

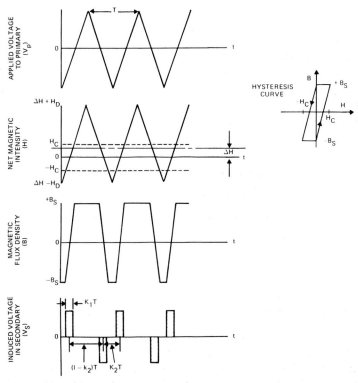

Fig. 6-36.   Operating Principles of Fluxgate Magnetometers

In the absence of an external magnetic field (i.e., $\Delta H = 0$), then $K_2 = 1/2$ and the bracketed term in Eq. (6-8) becomes

$$[1 - \cos(n\pi)] = \left\{ \begin{matrix} +2 \\ 0 \end{matrix} \right\} \quad \text{for} \ \left\{ \begin{matrix} n = 1, 3, 5, \ldots \\ n = 0, 2, 4, \ldots \end{matrix} \right\} \qquad (6\text{-}9)$$

Equations (6-8) and (6-9) imply that even harmonics of the primary frequency can occur only in the presence of an external magnetic field. The ratio of the second harmonic to the first is

$$r = \left\{ \frac{1 - \exp\left[-i2\pi(1 - \Delta H / H_D) \right]}{1 - \exp\left[-i\pi(1 - \Delta H / H_D) \right]} \times \left[\frac{\sin 2\pi K_1}{2\sin \pi K_1} \right] \right\} \qquad (6\text{-}10)$$

For $\Delta H \ll H_D$ and $H_C \ll H_D$, then $K_1 \ll 1$ and

$$r \approx \frac{1-1+i\sin(2\pi\Delta H/H_D)}{1+1-i\sin(\pi\Delta H/H_D)} = i\frac{\Delta H}{H_D}\pi \tag{6-11}$$

This means that the second harmonic is ± 90 deg out of phase with the primary. The sign of the second harmonic gives the sense of ΔH relative to the core axis and the amplitude is proportional to $\Delta H / H_D$. The fluxgate external magnetic field measurement may be degraded if the sensor electronics cannot produce a primary waveform free of the second harmonic or if residual spacecraft biases are present. A list of operating specifications for several fluxgate magnetometers is given in Table 6-4.

Magnetometers in the second broad category are termed "quantum" devices because they utilize fundamental atomic properties in the measurement of magnetic field direction and magnitude. Quantum sensors have been used for experimental field measurements onboard several spacecraft. However, because of weight and power requirements they are not appropriate as attitude sensors on small spacecraft.

The simplest of the quantum devices is the proton precession magnetometer. If a hydrogenous sample is placed in a strong magnetic field, it will exhibit a weak magnetic field after the strong field is removed. Further, the induced magnetic field will precess about any external field, $\mathbf{H}$, with the Larmor frequency, $\gamma_\rho|\mathbf{H}|$, where γ_ρ is the gyromagnetic ratio [Grivet and Malner, 1967]. Measurement of the resulting precessional frequency then gives a precise measure of the magnitude of the external magnetic field; however, because the magnetic field direction is unobservable, proton precession magnetometers are not used as attitude sensors.

A second type of quantum magnetometer relies on a process called optical pumping, which was first reported by Dehmelt [1957]. Magnetometers based on optical pumping have a light source producing an intense collimated beam of resonance radiation, a circular polarizer, an absorption cell containing the vapor to be optically pumped, a radio frequency coil to produce resonance in the pumped vapor, and a photocell to monitor the transmission of light [Bloom, 1962]. These optically pumped magnetometers measure ambient magnetic fields as a complicated function of the vapor transparency. Rubidium, cesium, and helium have been used as the optically pumped gas [Slocum and Reilly, 1963]. Optically pumped magnetometers provide both magnetic field direction and magnitude and are generally used as research magnetometers. A rubidium vapor magnetometer with a range of 15000 nT to 64000 nT and a sensitivity of ± 2 nT was flown onboard OGO-II. It weighed 4.4 kg and required 8 W of power for operation, of which 6 W were required for the lamp alone [Farthing and Folz, 1967]. The high weight and power requirements normally prohibit the use of these magnetometers as attitude instruments.

6.4 Star Sensors

Lawrence Fallon, III

Star sensors measure star coordinates in the spacecraft frame and provide attitude information when these observed coordinates are compared with known

Table 6-4. Operating Specifications of Fluxgate Magnetometers Manufactured by Schonstedt Instrument Company [1976]

MODEL NUMBER	FIELD RANGE	NUMBER OF AXES	ZERO FIELD OUTPUT	SENSITIVITY	ORTHOGONALITY	INPUT VOLTAGE	CURRENT	SIZE	WEIGHT
SAM-72C-HR	±60,000 nT	2	2.5 V ±10 mV	41.7 mV/100 nT ±1%	±1° AXIS TO AXIS AND AXIS TO REFERENCE SURFACE	14V TO 34V	20 mA	70.5 CM³	71 g
SAM-73C	±60,000 nT	3	2.5 V ±10 mV	4.17 mV/100 nT ±1%	±1° AXIS TO AXIS AND AXIS TO REFERENCE SURFACE	14V TO 34V	35 mA	88.0 CM³	142 g
SAM-42C-2	±10,000 nT	2 MINIMUM	2.5 V ±25 mV	1.0 mV/3 nT ±1%		24V TO 33V	14 mA	60.3 CM³/SENSOR 459 CM³/ELEC	63.8 g/SENSOR 539 g/ELEC
SPM-43B-4	±20 nT	3	2.5 V	2.5 V		12V, 0V, −12V	POWER < 600 mW	688 CM³/SENSOR 1180 CM³/ELEC	408 g/SENSOR 678 g/ELEC
SPM-43B-10	±128 nT AND ±32 nT	3	2.5 V ±2%	2.5 V/32 nT 2.5 V/128 nT	WITHIN ±15 MINUTES OF ARC	12V ±1%	37.5 mA	688 CM³/SENSOR 1020 CM³/ELEC	350 g/SENSOR 850 g/ELEC
SPM-61B-1	±200 nT	1		2.5 V/200 nT ±1%		12V ±1%	POWER < 500 mW	35.2 CM³/SENSOR 538 CM³/ELEC	60 g/SENSOR 500 g/ELEC
SAM-63B-1	±40,000 nT AND ±10,000 nT	1 OR MORE		−40,000 nT OVER 4.5 V TO 0.5 V		14V TO 18V	40 mA	15.01 CM³/SENSOR 287 CM³/ELEC	< 90.7 g/SENSOR < 453 g/ELEC
SPM-63B-1	±320 nT AND ±3200 nT	3	2.5 V ±1%	2.5 V/320 nT 2.5 V/3,200 nT	±6 MINUTES OF ARC	12V ±1%	55 mA	688 CM³/SENSOR 1180 CM³/ELEC	460 g/SENSOR 950 g/ELEC
SPM-63B-7	±600 nT	3		2.5 V/2,500 nT 2.5 V/60,000 nT	±0.25°	±10V AND ±15V ±1%	75 mA	603 CM³/SENSOR 660 CM³/ELEC	< 453 g/SENSOR < 272 g/ELEC
SPM-63B-2	±200 nT	3	2.5 V ±1%	40 V/200 nT	±15 MINUTES OF ARC	12V ±1%	37.5 mA	770 CM³/SENSOR 1020 CM³/ELEC	225 g/SENSOR 850 g/ELEC

star directions obtained from a star catalog. In general, star sensors are the most accurate of attitude sensors, achieving accuracies to the arc-second range. This impressive capability is not provided without considerable cost, however. Star sensors are heavy, expensive, and require more power than most other attitude sensors. In addition, computer software requirements are extensive, because measurements must be preprocessed and identified before attitudes can be calculated (see Sections 7.7 and 21.3). Star sensors also suffer from both occultation and interference from the Sun, the Earth, and other bright sources. In spite of these disadvantages, the accuracy and versatility of star sensors have led to applications in a variety of different spacecraft attitude environments. This section presents an overview of the operation and construction of star sensors and detailed descriptions of two representative sensors: the **V** slit Star Scanner used on the OSO-8 mission and the Fixed Head Star Tracker used on the SAS-3 and HEAO-1 missions and planned for HEAO-C and MAGSAT.

6.4.1 Overview of Star Sensor Hardware

Star sensing and tracking devices can be divided into three major classes: *star scanners*, which use the spacecraft rotation to provide the searching and sensing function; *gimbaled star trackers*, which search out and acquire stars using mechanical action; and *fixed head star trackers*, which have electronic searching and tracking capabilities over a limited field of view. Sensors in each of these classes usually consist of the following components, as illustrated in Fig. 6-37 for a **V** slit star scanner: a Sun shade; an optical system; an image definition device which defines the region of the field of view that is visible to the detector; the detector; and an electronics assembly. In addition, gimbaled star trackers have gimbal mounts for angular positioning.

Stray light is a major problem with star sensors. Thus, an effective *Sun shade* is critical to star sensor performance. Carefully designed light baffles are usually employed to minimize exposure of the optical system to sunlight and light scattered by dust particles, jet exhaust, and portions of the spacecraft itself. Even with a well-designed Sun shade, star sensors are typically inoperable within 30 to 60 deg of the Sun.

The star sensor *optical system* consists of a lens which projects an image of the star field onto the focal plane. The *image definition device* selects a portion of the star field image in the sensor's *field of view* (FOV) which will be visible to the

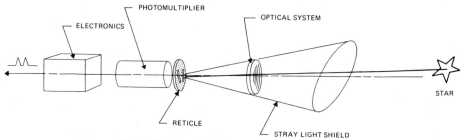

Fig. 6-37.   Simplified Diagram of a **V** Slit Star Sensor

detector. This portion is known as the *instantaneous field of view* (IFOV). The image definition device may be either a reticle consisting of one or more transparent slits etched on an otherwise opaque plate, or an image dissector tube in which the IFOV electronically scans the FOV. The *detector* transforms the optical signal (i.e., whatever light is not blocked by the image definition device) into an electrical signal. The most frequently used detector is a photomultiplier. Solid-state detectors are also commonly employed, but they are usually noisier than photomultipliers. Finally, the *electronics assembly* filters the amplified signal received from the photomultiplier and performs many functions specific to the particular star sensor.

Star scanners used on spinning spacecraft are the simplest of all star sensors because they have no moving parts. The image definition device employed by this type of sensor consists of a slit configuration, such as the **V** slit arrangement shown in Fig. 6-37. The spacecraft rotation causes the sensor to scan the celestial sphere. As the star image on the focal plane passes a slit, the star is sensed by the detector. If the amplified optical signal passed from the detector to the electronics assembly is above a threshold value, then a pulse is generated by the electronics signifying the star's presence. The accuracy of this sensor is related to the width of the slits and is typically on the order of 0.5 to 30 arc-minutes, although more accurate models exist. Star scanners have been used successfully on several missions, including the OSO and SAS series. Table 6-5 lists the characteristics of several typical star scanners. The OSO-8 star scanner is further described in Section 6.4.2.

The interpretation of star scanner measurements becomes increasingly difficult as spacecraft motion deviates from a non-nutating, uniformly spinning rigid body. For example, data from the SAS-3 star scanner is useful only during the constant spin rate portions of the mission. The nominal spin rate at 1 rpo (approximately 0.07 deg/sec) is at the lower range for successful interpretation of star scanner data. Problems of noise and the generation of false star crossing signals are greater at this spin rate than at 2 or 3 rpo. Interpretation of the SAS-3 star scanner data is virtually impossible during the portion of the mission when the spin rate changes rapidly.

Gimbaled star trackers, illustrated in Fig. 6-38, are commonly used when the spacecraft must operate at a variety of attitudes. This type of tracker has a very

Table 6-5.   Parameters for Representative Star Scanners

SENSOR	DETECTOR	CONFIGURATION	SENSITIVITY (VISUAL MAGNITUDE)	FOV (DEG)	CALIBRATED ACCURACY-1σ	REFERENCE
APPLIED PHYSICS LABORATORY STAR SENSOR FOR SAS 1, 2, AND 3	PHOTO-MULTIPLIER	N	BRIGHTER THAN +4	5 BY 10	±1 ARC-MIN	1
BBRC[5] CS–103 STAR SCANNER FOR OSO–8	PHOTO-MULTIPLIER	V	+3.5 TO –2.0	5 BY 10	±0.1 DEG	2 3
BBRC[5] CS–201 STAR SCANNER	SOLID STATE (SILICON)	V	+1.4 TO –1.4	EACH SLIT 25.0 BY 0.41	±0.5 DEG	–
HONEYWELL SPARS STAR SENSOR	SOLID STATE (SILICON)	6–SLIT	BRIGHTER THAN +3.15	8.8 WIDE	±2 ARC-SEC	4
HONEYWELL BLOCK 5D/DMSP STRAPDOWN STAR SCANNER	SOLID STATE (SILICON)	6–SLIT	BRIGHTER THAN +3.7	10.0 WIDE	±2 ARC-SEC	–

1 FOUNTAIN, 1972

2 WETMORE, et al., 1974

3 RCA, 1975

4 SCOTT AND CARROLL, 1969

5 BBRC = BALL BROTHERS RESEARCH CORPORATION

small optical FOV (usually less than 1 deg). The gimbal mounts, however, give the sensor a much larger effective FOV. The coarse-alignment star trackers on OAO, for example, were gimbaled to cover an area with a 43-deg half-cone angle [NASA, 1970]. Gimbaled star trackers normally operate on a relatively small number of target stars (e.g., 38 for the OAO trackers).

Many different kinds of image definition devices are used in gimbaled star trackers to determine the position of the star with respect to the center, or *null*, position in the small FOV. The electronics assembly causes the gimbals to move so that the star image remains centered in the small FOV. The star's position is then given by the gimbal angle readout positions. Some image-definition devices employ an optical or electronic scan of the small FOV to provide star position information. For example, small FOV image dissector tubes may perform this function. Another type of scanning device is an optical wedge-slit system. A rotating optical wedge causes the star image to be deflected past an L-shaped slit. As the wedge rotates, the image of the star follows a circular path over the L slit. The optical wedge is designed so that the radius of the circle grows as the star image diverges from the null position. The electronics assembly determines the position of the star with

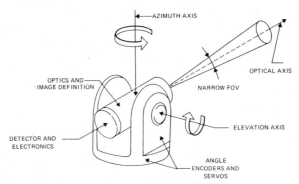

Fig. 6-38.   Gimbaled Star Tracker

respect to null by comparing the time difference between slit crossings and the rotation phase of the optical wedge.

One type of gimbaled star tracker does not use an image definition device at all, but rather reflects a defocused star image onto four photomultipliers in a square array. The star's position is determined by comparing the output signals of the four photomultipliers. This system has the advantage of simplicity. However, it suffers from disadvantages: temperature variations and changes in photomultiplier characteristics due to aging may introduce systematic biases; nonuniform background light or the presence of a second star within the small FOV causes serious errors.

Errors in determining the star position with respect to null and gimbal angle readout errors affect the overall gimbaled star tracker's accuracy. Typical accuracies range from 1 to 60 arc-seconds, excluding tracker misalignment. A major disadvantage of gimbaled star trackers is that the mechanical action of the gimbals reduces their long-term reliability. In addition, the gimbal mount assembly is frequently large and heavy.

Spacecraft which maintain an inertially fixed direction commonly employ gimbaled star trackers which have a unique target star. The positions of Polaris and Canopus near the north celestial and south ecliptic poles, respectively, make these two stars particularly useful. The location of Canopus makes it especially useful as a reference direction for determining the rotation about the sunline. A Sun/Canopus attitude reference system has been used for Mariner, Surveyor, and Lunar Orbiter [NASA, 1970]. The Polaris tracker used on ATS-6 [Moore and Prensky, 1974] was adapted from a Canopus tracker. A serious disadvantage of unique star trackers is that they may occasionally track either the wrong star or particles scattering stray light, such as paint chips from the spacecraft.

Fixed-head star trackers use an electronic scan to search their field of view and acquire stars. They are generally smaller and lighter than gimbaled star trackers and have no moving parts. The image definition device used by fixed-head star trackers is usually an image dissector, although vidicons have been used and recently developed image detecting *charge coupled devices* (CCD) are showing promise. A *charge coupled device star tracker* is essentially an optical system combined with a digitally scanned array of photosensitive elements whose output is fed to a microprocessor. Such a tracker operates by integrating a charge pattern corresponding to the image of the star field on the focal plane of the optical system. The charge pattern is then read out serially line by line to an analog-to-digital converter and then to a microprocessor. Star trackers incorporating this technology have been built by the Jet Propulsion Laboratory [Salmon and Goss, 1976].

A typical fixed-head star tracker using an image dissector tube is shown in Fig. 6-39. The photocathode contains the star-field image created by the optical system. An electron replica of this image is deflected past a fixed receiving aperture by the magnetic deflecting coils. This aperture defines a small IFOV (usually in the arc-minute range) on the photocathode and hence on the star-field image. Although this aperture does not move in the dissector tube, the IFOV scans across the fixed image as the current through the deflection coils is varied. An image dissector searches its FOV for stars by moving the IFOV in a *search pattern*, such as a right-to-left staircase-retrace pattern as described in Section 6.4.3, or a center-to-edge rosette pattern. When the detector finds a visual signal above a threshold value, the electronics assembly engages a *track pattern*. The IFOV then moves in a small figure-eight or square pattern so that the electronics assembly can

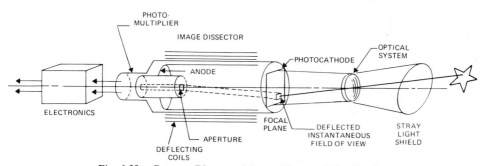

Fig. 6-39.   Cutaway Diagram of Image Dissector Tube Star Sensor

locate the center of the star image. The IFOV will either remain in the track pattern until the star is lost, or it will automatically resume searching after a predetermined time interval (depending on mission requirements and sensor electronics).

If a photomultiplier is placed after the receiving aperture of the image dissector in a fixed-head star tracker, the instrument is referred to as an *analog image dissector*. Alternatively, if a photoelectron counter is used, the instrument is called a *photon counting image dissector*. The characteristics of several fixed-head star trackers using mage dissectors are listed in Table 6-6.

Table 6-6. Fixed-Head Star Trackers Using Image Dissectors

SENSOR	DETECTOR TYPE	SENSITIVITY (VISUAL MAGNITUDE)	FOV (DEG)	CALIBRATED ACCURACY-1σ (ARC-SEC)	REFER-ENCE
BBRC FINE ERROR SENSOR (IUE)	PHOTO-MULTIPLIER	+14 TO +7	VARIABLE WITHIN 16 ARC-MIN CIRCLE	±2.25	1
BBRC CT401 FIXED HEAD STAR TRACKER (SAS−3, HEAO−1)	PHOTO-MULTIPLIER	+6.5 OR BRIGHTER	8 BY 8	±10	2
BBRC CT411 LARGE FOV STAR TRACKER (SPACE SHUTTLE ORBITER)	PHOTO-MULTIPLIER	+3 TO −7	10 BY 10	±60	−
HONEYWELL PHOTON COUNTING STAR TRACKER (HEAO−B)	PHOTO-ELECTRON COUNTER	+9 OR BRIGHTER	2 BY 2	±1.5	3
TRW PADS TRACKER	PHOTO-MULTIPLIER	+10 OR BRIGHTER	1 BY 1	±1.5	4

1 ADAMS, 1974   3 TSAO AND WOLLMAN, 1976

2 CLEAVINGER AND MAYER, 1976   4 GATES AND McALOON, 1976

Image dissectors are subject to errors from stray electric and magnetic fields. Electric and transverse magnetic field effects can be reduced by shielding. However, it is more difficult to shield against axial magnetic fields. Errors due to these effects become significant in the outer regions of a large FOV image dissector. Correction procedures to remove these effects as well as temperature effects are described in Section 7.7. Image dissectors have the advantages of high sensitivity, low noise, and relative mechanical ruggedness.

The choices of field-of-view size and star magnitude sensitivity for any star sensor generally depend on the attitude accuracy requirements. A small FOV tracker can provide more accurate star positions than can a larger FOV tracker with comparable components. However, a small FOV tracker must be sensitive to dimmer stars to ensure that enough stars are visible to it. Use of a larger FOV demands extensive prelaunch ground calibration for temperature, distortion, and magnetic effects, as well as postlaunch preprocessing of data to correct for these effects.

6.4.2 BBRC CS-103 V-Slit Star Scanner for OSO-8

As an example of a star scanning sensor we will describe the CS-103 V-slit star scanner built for OSO-8 by the Ball Brothers Research Corporation (BBRC). It is designed to provide spacecraft attitudes accurate to ±0.1 deg at a nomimal spin rate of 6 ± 1 rpm. The star scanner, shown in Fig. 6-40, is oriented such that as the spacecraft rotates, the scanner's FOV sweeps a 10-deg band in the sky with a half-cone angle of 53 deg about the spin axis. The scanner generates two pulses each time its FOV sweeps past a star that is brighter than the preselected level.

Fig. 6-40.   CS-103 Star Scanner. (Photo courtesy of Ball Brothers Research Corporation.)

Thus, during spacecraft rotation, the sensor generates a series of pulse pairs corresponding to the bright stars that pass through its field of view. Characteristics of the CS-103 are summarized in Table 6-7.

Table 6-7.   BBRC CS-103 Star Scanner Characteristics

CHARACTERISTIC	VALUE
VERTICAL FIELD OF VIEW	±5 DEG
HORIZONTAL FIELD OF VIEW	±2.5 DEG
WIDTH OF EACH SLIT LEG	0.036 DEG
DETECTABLE STAR RANGE	−2.0 TO +3.5 MAGNITUDE (SELECTABLE GAIN THRESHOLDS +1.75 TO +3.5 IN 0.25-MAGNITUDE STEPS)
MAXIMUM EQUIVALENT BACKGROUND	+3.5 MAGNITUDE
TOTAL POWER CONSUMPTION	1.89 W
ACCURACY (TWO AXES)	±0.1 DEG (3σ)

The sensor lens assembly focuses the light from stars within its field on an opaque quartz reticle with a **V**-shaped slit etched in its surface (see Fig. 6-41). As the lens assembly sweeps past a star, the photomultiplier produces a pulse at the crossing of each leg of the **V** slit. The crossing time of the first leg (which is vertical) is proportional to the star's azimuth angle. The elapsed time between the crossing of the first leg and the second slanted leg is a function of the star's elevation in the spacecraft coordinate system. This procedure is essentially the same as that used in the **V**-slit Sun and Earth sensors described in Sections 6.1 and 6.2.

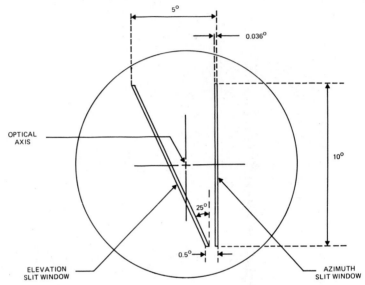

Fig. 6-41.   V-Slit Sensor Reticle Configuration Showing Field of View With Respect to Optical Axis

The star scanner level detector receives the star pulses and excludes those whose magnitudes are dimmer than the selected threshold. This prevents overloading of the data handling system due to clusters of dim stars or background noise. The level detector may be commanded to any of eight detecting thresholds (from $+3.5\ m_v$ to $+1.75\ m_v$ in 0.25-m_v increments).

The electronic processor generates a 24-bit word for each star encountered. This 24-bit word consists of 14 azimuth bits denoting the time at which the leading slit was transited, 1 flag bit indicating if three or more pulses were encountered within the maximum azimuth time of 250 ms, and 9 bits representing the time between the pulses from the vertical and slanted slits. At 6 rpm, the 250-ms time interval corresponds to an angular separation of approximately 9 deg between the two pulses. The transit times are counted with a 1600-Hz clock which is periodically resynchronized to the spacecraft clock.

The star sensor electronics generate a limited number of *false sightings* which do not correspond to valid star transits. These false sightings arise from photomultiplier and electronics noise. Because they occur in a random fashion, discrimination from valid star transmits may be readily accomplished.

6.4.3 BBRC CT-401 Fixed-Head Star Tracker

As an example of the fixed-nead star tracker, we will describe the CT-401, manufactured by BBRC and shown in Fig. 6-42. The CT-401 has flown on the SAS-3 and HEAO-1 missions and is scheduled for HEAO-C and MAGSAT. Schematically, the tracker is similar to the instrument shown in Fig. 6-39. The specified accuracy of the tracker over its 8- by 8-deg field of view is ±3 arc-minutes without calibration, or ±10 arc-seconds after applying corrections for electro-optical distortion, temperature, ambient magnetic field, and star intensity obtained from preflight calibration. Because this instrument is capable of observing several stars within a relatively short period of time, it is frequently called a *star camera*.

Fig. 6-42.   BBRC CT-401 Fixed-Head Star Tracker. (Photo courtesy of Ball Brothers Research Corporation.)

The tracker has four commandable thresholds corresponding to selection of stars brighter than approximately +3.0, +4.0, +5.0, and +6.5 m_v. These settings add to the flexibility of the instrument, because it can be used both for coarse attitude determination or attitude acquisition when set at +3.0 m_v or for fine attitude determination when set at +6.5 m_v. (As discussed in Section 7.7, many fewer stars are measured at +3.0 m_v than at 6.5 m_v, which permits star identification with a coarse a priori attitude.) A 9- by 9-arc-minute receiving aperture (IFOV) scans the 8- by 8-deg FOV using the search pattern shown in Fig. 6-43. The

FOV scan continues until a star is found which exceeds the threshold level. When this occurs, the current scan line is completed, and the track pattern shown in Fig. 6-44 begins. A feedback system in the electronics assembly centers the track pattern on the star image. Two-axis star position signals are determined by the electronics assembly from the rising and falling edges of the star image. These signals, u and v, determine the tangent plane coordinates of the star, as explained in Section 7.6. As a star moves in the FOV due to attitude changes, the track pattern follows and remains centered on the star image. Tracking of the same star continues until the tracker is commanded to return to the search mode, the star signal drops below the threshold, or the star leaves the FOV.

Figure 6-44 shows the electronic sampling procedure used in the track mode. As the IFOV is deflected past the star image, a video signal is produced. When this signal exceeds and then falls below the threshold level, two pulses for each axis are generated by the electronics. These pulses correspond to the position of the IFOV during the leading and trailing edge crossings. The electronics combines these signals to determine the position of the star image in the FOV. These star position signals are then used to keep the track pattern centered over the star image. Star coordinates on both axes are updated every 100 ms. The resulting data samples are filtered through an RC filter with a time constant of approximately 450 ms. If significant motion of the star image with respect to the previously sampled position occurs, major tracking problems will result. For example, the CT-401 may fail to track stars moving faster than approximately 0.6 deg/sec in the FOV.

When the search mode resumes, the v coordinate for the beginning of the new search line will be the v coordinate of the star which was last tracked, plus a small increment (0.4 deg) to avoid retracking the same star. If this would place the aperture beyond the edge of the FOV, the search pattern returns to the start position. The starting u coordinate is at the beginning of a new line.

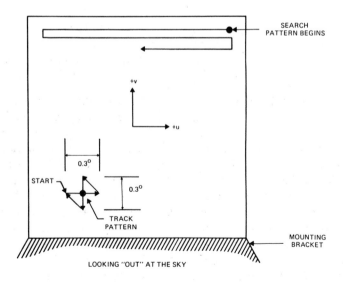

Fig. 6-43.   BBRC CT-401 Star Tracker Search and Track Mode Patterns

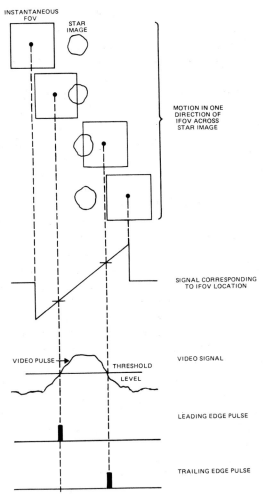

INSTANTANEOUS FOV

STAR IMAGE

MOTION IN ONE DIRECTION OF IFOV ACROSS STAR IMAGE

SIGNAL CORRESPONDING TO IFOV LOCATION

VIDEO PULSE

THRESHOLD LEVEL

VIDEO SIGNAL

LEADING EDGE PULSE

TRAILING EDGE PULSE

Fig. 6-44. BBRC CT-401 Star Tracker Sampling Sequence in the Track Mode

The initiate search command can be used to vary the style of operation. If the tracker is frequently commanded to resume search, it operates more like a star "mapper" than a "tracker." If it is commanded only infrequently, it operates more like a "tracker." A bright object sensor and shutter mechanism protect the image dissector from excess energy from the Sun, the Moon, or the lit Earth. The sensor will close the tracker's shutter when a bright object approaches the FOV. This occurs for the CT-401 when the Sun lies within 42 deg of the FOV optical axis. Additional output signals from the tracker include indication for search or track mode, a bright-object sensor ON or OFF, a high-voltage monitor, and a temperature monitor. The temperature monitor can be used to correct the star position coordinates for temperature effects, as described in Section 7.6. Additional details concerning the CT-401 tracker are given by Cleavinger and Mayer [1976] and Gottlieb, et al., [1976].

6.5 Gyroscopes

Lawrence Fallon, III

A gyroscope, or gyro, is any instrument which uses a rapidly spinning mass to sense and respond to changes in the inertial orientation of its spin axis. Three basic types of gyroscopes are used on spacecraft: *rate gyros (RGs)* and *rate-integrating gyros (RIGs)* are attitude sensors used to measure changes in the spacecraft orientation; *control moment gyros (CMGs)* are used to generate control torques to change and maintain the spacecraft's orientation.

Rate gyros measure spacecraft angular rates and are frequently part of a feedback system for either spin rate control or attitude stabilization. The angular rate outputs from RGs may also be integrated by an onboard computer to provide an estimate of spacecraft attitude displacement from some initial reference. *Rate-integrating gyros* measure spacecraft angular displacements directly. In some applications, the RIG output consists of the total spacecraft rotation from an inertial reference. In other cases, output consists of the amount of incremental rotation during small time intervals. An accurate measure of the total attitude displacement may then be obtained by integrating the average angular rates constructed from the incremental displacements. Average angular rates constructed in this manner may also be used for spin rate control or stabilization via a feedback system. RIGs are generally more accurate than RGs when used for either of these procedures. They are usually much more expensive, however.

Control moment gyros are not attitude sensors like RGs or RIGs, but are used to generate attitude control torques in response to onboard or ground command. They operate much like reaction wheels (Section 6.6) except that their spin axis is gimbaled. Torques are generated by commanding a gimbal rotation and thereby changing the spin axis orientation. CMGs may be used in conjunction with RGs or RIGs and an onboard computer as components of an attitude determination and control system. Because of their expense and weight, CMGs are used only on large spacecraft.

All gyros have the basic construction geometry shown in Fig. 6-45. The angular momentum vector of an RG or an RIG is fixed in magnitude and parallel to the gyro's *spin axis*. Because this vector maintains its inertial orientation in the absence of applied torques, spacecraft motion about the gyro's *input axis* causes the *gimbal* supporting the spin axis to precess about the *output axis*, or *gimbal rotation axis*. The output of an RG or an RIG is obtained from the motion of the gimbal. A CMG operates essentially in the reverse manner. A commanded displacement of the gimbal and the resultant change in the angular momentum vector causes a control torque about the gyro's input axis.

The example shown in Fig. 6-45 is a *single-degree-of-freedom*, or *SDOF*, gyro because the spin axis is supported by only one gimbal, and the gyro is thus sensitive in only one direction. In many applications the spin axis is also supported by a second gimbal, resulting in a *2-degree-of-freedom*, or *TDOF*, gyro. Two or more TDOF gyros or three or more SDOF gyros may be used to provide sensing or control about all three axes. For example, a configuration of four RIGs, referred to

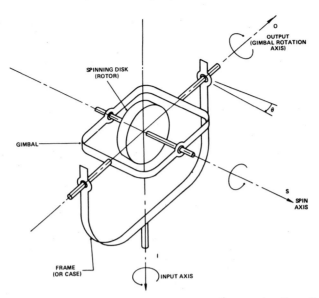

Fig. 6-45.　Single-Degree-of-Freedom Gyroscope Construction Geometry

as an *Inertial Reference Assembly*, or *IRA*, is used for HEAO-1 attitude determination and spin rate control. The input axes of these gyros are oriented so that any combination of three gyros will provide complete three-axis information. The extra gyro is included for redundancy. The IRA proposed for SMM will consist of three TDOF RIGs. A configuration of three TDOF CMGs, shown in Fig. 6-46, was used in the Skylab attitude control system.

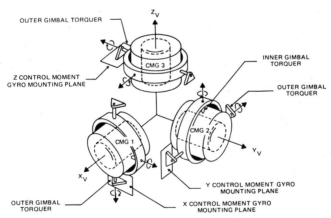

Fig. 6-46.　Configuration of Skylab's Control Moment Gyros

6.5.1　Rate Gyros

The output of a rate gyro is obtained by measuring the rotation of the gimbal about the output axis. The excursion of the rate gyro's gimbal is inhibited by viscous damping and a spring restraint, where the spring constant is chosen to be large compared with the damping effects. The relationship between the rate about the input axis and the angular displacement, θ, about the output axis may be derived by examining the total angular momentum, $\mathbf{H}$, of the gyro system:

$$\mathbf{H} = \mathbf{L} + I_0 \dot{\theta} \hat{\mathbf{O}} \qquad (6\text{-}12)$$

where $\mathbf{L} \equiv L \hat{\mathbf{S}}$ is the angular momentum of the rotor, I_O is the moment of inertia of the gimbal system about the output axis, $\hat{\mathbf{O}}$ is a unit vector in the direction of the gyro's output axis, and $\hat{\mathbf{S}}$ is a unit vector in the direction of the gyro's spin axis. Newton's laws applied to the gyro system, whose angular velocity relative to an inertial system is $\boldsymbol{\omega} \equiv \omega_I \hat{\mathbf{I}} + \omega_O \hat{\mathbf{O}} + \omega_S \hat{\mathbf{S}}$, yields the following (see Section 16.1):

$$\sum \text{Torques} = \left(\frac{d\mathbf{H}}{dt} \right)_{\text{Inertial}} = \left(\frac{d\mathbf{H}}{dt} \right)_{\text{Gyro}} + \boldsymbol{\omega} \times \mathbf{H} \qquad (6\text{-}13)$$

where ω_I, ω_O, and ω_S are angular velocity components along the gyro's input, output, and spin axes, respectively, and $\hat{\mathbf{I}}$ is a unit vector in the direction of the gyro's input axis.

The torque on the single-degree-of-freedom gyro is the sum of restoring and viscous damping terms,

$$\sum \text{Torques} = -(K\theta + D\dot{\theta})\hat{\mathbf{O}} \qquad (6\text{-}14)$$

Substitution of Eqs. (6-12) and (6-14) into (6-13) yields

$$I_0 \ddot{\theta} + D\dot{\theta} + K\theta - \omega_I L = 0 \qquad (6\text{-}15)$$

for the component along the $\hat{\mathbf{O}}$ axis, where ω_I is the angular velocity component along the gyro's input axis. The steady-state solution to Eq. (6-15), (i.e., $\dot{\theta} = \ddot{\theta} = 0$) is

$$\theta = \frac{\omega_I L}{K} \qquad (6\text{-}16)$$

The output of an RG is thus proportional to the spacecraft angular rate about the input axis.

Rate gyros are the simplest and the least expensive gyros. Their accuracy is generally suitable for spin rate control in a feedback system, but their integrated output requires frequent correction for precise attitude determination using other sensors such as Sun sensors or star trackers (Section 21.3). Errors in RG output are generally caused by nonlinearity, drift, and hysteresis. In addition, input accelerations may affect their accuracy if the gimbal is not perfectly balanced.

An improvement over the conventional rate gyro is the *closed-loop rate gyro*, in which an electromagnetic torque rebalance system reduces gimbal angular excursions by about three orders of magnitude. The gyro output is then derived from the current required to maintain the gimbal at the null (i.e., zero deflection) position. The restricted gimbal deflection improves linearity and reduces drift rate instability. Some characteristics of representative closed-loop rate gyros manufactured by the Bendix Corporation are listed in Table 6-8. Descriptions of several other types of rate gyros are given by Schimdtbauer, et al., [1973]. Additional details concerning the operation of rate gyros are given by Greensite [1970] and Thomson [1963].

Table 6-8.   Characteristics of Representative Closed-Loop Rate Gyros (Source: Bendix Corporation)

CHARACTERISTIC	VALUE
SIZE	~ 7.8 x 3.0 x 4.8 CM
WEIGHT	0.34 KG
ANGULAR MOMENTUM	15,000, 30,000 OR 60,000 GM–CM^2/SEC
MAXIMUM GIMBAL DISPLACEMENT	±0.6 DEG
INPUT RATE RANGE (FULL SCALE)	5 TO 1,000 DEG/SEC
GYRO OUTPUT (FULL SCALE)	±10 VOLTS
TEMPERATURE SENSITIVITY	< 0.02%/°K
LINEARITY	0.5% FULL SCALE TO ½ SCALE 2% FULL SCALE FROM ½ TO FULL SCALE
RESOLUTION	< 0.01 DEG/SEC
HYSTERESIS	< 0.01% FULL SCALE
LINEAR ACCELERATION SENSITIVITY	< 0.03 DEG/SEC/G

6.5.2   Rate-Integrating Gyros

Because of its high accuracy and low drift, the rate-integrating gyro is the type most often used in spacecraft attitude sensing. The gimbal is mounted so that its motion is essentially frictionless and without spring restraint. It is usually a sealed cylinder which is immersed or floated in a fluid. The spin axis in the cylinder is generally supported by either gas or ball bearings. Because the viscous damping and spring constants are both small, the steady-state solution to Eq. (6-15) indicates that an RIG's output (i.e., the rotation of the gimbal about the gyro's output axis) is proportional to the spacecraft's angular displacement about its input axis.

In practical applications, gimbal motion is usually limited to a few degrees. Two different procedures are frequently used to measure larger angles and to improve accuracy when measuring smaller angles. In the first method, the gyro is mounted on a platform which is rotated in a closed-loop system using the gimbal motion signal to maintain the gimbal position near the zero point. The gyro's output is then proportional to the rotation of the platform which, in turn, is proportional to the rotation of the spacecraft about the input axis. Alternatively, the gyro can be fixed in the spacecraft with the gimbal torqued magnetically using a closed-loop system to maintain its deflection near null. As is done for similar RGs, the gyro's output is derived from the torque current, which is proportional to the spacecraft rotation. Such a gyro is referred to as a *strapdown torque rebalanced RIG*. The torque current may be either analog or pulsed. Pulsed torquing has gained in popularity because of its utility in computer applications. The torque current from either type of torque rebalanced gyro may be differenced after small time intervals before being output, so that the resultant gyro output during any one of these small time intervals is proportional to the differential spacecraft rotation and, thus, to the average spacecraft velocity during the interval. An RIG operating in this way is referred to as a *rate-integrating gyro in the rate mode*.

The principal source of error in an RIG is drift rate instability. The systematic errors of drift, input axis misalignment, and scale factor error can be modeled and corrected for as described in Section 7.7. In torque rebalanced RIGs with floated gimbals, the component of drift instability caused by thermal effects is minimized by automatically controlled heaters. Most of the residual drift instability normally results from random null shifts in the torque rebalance control loop. A short-term component of this instability, referred to as *random drift*, can be related to float torque noise (i.e., noise in the torque applied to the floated gimbal). Similarly, a random walk component, referred to as *drift rate ramp*, can be related to float torque derivative noise. The effects of both of these noise sources on the uncertainty in gyro outputs can be modeled so that spacecraft attitude error can be predicted (Section 7.7). Occasionally, however, fluctuations in the spacecraft voltage or changes in the magnetic environment cause systematic null shifts, which are difficult or impossible to model. In many cases, use of a regulated gyro power supply reduces the voltage fluctuation effects, although the cost of the gyro package is considerably increased.

Additional information concerning RIGs is given by Schimdtbauer, *et al.*, [1973]; Greensite [1970]; Thomson [1963]; and Scott and Carroll [1969]. Table 6-9 lists the characteristics of typical RIGs manufactured by Bendix and Honeywell.

Table 6-9. Characteristics of Representative Rate-Integrating Gyros (Source: Bendix Corporation and Honeywell, Inc.)

GYRO	VOLUME DIAMETER, LENGTH (CM)	WEIGHT (KG)	OPERATING POWER (W)	RANDOM DRIFT 1σ (DEG/HR)	INPUT RANGE (DEG/SEC)	ANGULAR MOMENTUM (GM CM^2/SEC)
HONEYWELL GG 334 RIG (SDOF)	5.89 11.94	0.77	17 MAX	0.003	± 5.6	185,000
BENDIX 64 RIG (SDOF) (FOR IUE AND HEAO—1)	6.35 27.94	0.77	8—16	0.006	± 2.5	430,000

6.5.3 Control Moment Gyros

A control moment gyro's angular momentum is due to the rotor which is spinning about the spin axis with a constant angular rate. Because the spin axis is gimbaled, a commanded gimbal rotation causes the direction of the angular momentum vector to change, thus creating a control torque parallel to the output axis. The magnitude of this torque depends upon the speed of the rotor and the gimbal rotation rate. Because gimbal excursion is often limited by position stops and gimbal rotation rates must not exceed specified maximum values, a partitioning of torque components among several CMGs is often required [Chubb, *et al.*, 1975; Coon and Irby 1976]. Occasionally, however, undesirable momentum configurations will result, and momentum dumping using an auxiliary control system (such as gas thrusters) becomes necessary (Section 6.6). The characteristics of several Bendix CMGs are listed in Table 6-10.

Table 6-10.   Characteristics of Representative Control Moment Gyros (Source: Bendix Corporation)

MODEL	WEIGHT (KG)	ROTOR SPEED (RPM)	ANGULAR MOMENTUM (KG · M^2/S)	MAXIMUM OUTPUT TORQUE (N · M)	GIMBAL FREEDOM (DEG)	MAXIMUM GIMBAL RATE (DEG/SEC)	APPROXIMATE SIZE
BENDIX DOUBLE GIMBAL MA−2000	253	4,000 TO 12,000	1400−4100	237	UNLIMITED	5 30	1.1 M DIA. SPHERE
BENDIX DOUBLE GIMBAL MA−2300 FOR SKYLAB	190	9,000	3100	165	± 80 ± 175	4 7	1.0 M DIA. SPHERE
BENDIX SINGLE GIMBAL MA−500 AC	66	7,850	340−1000	680	± 170	57.3	CYLINDER 0.51 M DIAM X 0.81 M LONG
BENDIX SINGLE GIMBAL MA−5−100−1	17	8,000	7	140	UNLIMITED	1146	CYLINDER 0.25 M DIAM X 0.25 M LONG

6.6 Momentum and Reaction Wheels

Dale Headrick

Devices for the storage of angular momentum, sometimes called simply *momentum* in attitude work, are used on spacecraft for several purposes: to add stability against disturbance torques, to provide a variable momentum to allow operation at 1 rpo for Earth-oriented missions, to absorb cyclic torques, and to transfer momentum to the satellite body for the execution of slewing maneuvers. These devices depend on the momentum of a spinning wheel, $\mathbf{h} = I\omega$, where I is the moment of inertia about the rotation axis and ω is the angular velocity. (See Sections 11.1 and 16.1.) Unfortunately, the terminology of momentum wheels in the literature is not uniform. We adopt the following:

Flywheel, or *inertia wheel*, is any rotating wheel or disk used to store or transfer momentum. It refers to the wheel itself, exclusive of electronics or other associated devices.

Momentum wheel is a flywheel designed to operate at a *biased*, or nonzero, momentum. It provides a variable-momentum storage capability about its rotation axis, which is usually fixed in the vehicle.

Reaction wheel is a flywheel with a vehicle-fixed axis designed to operate at zero bias.

Momentum wheel assembly consists of the flywheel and its associated parts: bearings, torque motors, tachometers, other sensing devices, caging devices for launch, and control electronics.

Control moment gyro (CMG), or *gyrotorquer*, consists of a single- or a double-gimbaled wheel spinning at a constant rate. The gimbal rings allow control of the direction of the flywheel momentum vector in the spacecraft body. The CMG is discussed in Section 6.5.

Single Momentum Wheel.   The capacity of a typical momentum wheel varies from 0.4 to 40 kg·m^2/s. Because the same momentum can be achieved with a small, high-speed flywheel as with a large low-speed one, design tradeoffs generally favor the smaller wheel because of size and weight. The high-speed wheel has the disadvantage of greater wear on the bearings, which may shorten its lifetime. As described in Section 6.2, horizon scanners have been incorporated as an integral part of the momentum wheel assembly on several spacecraft. A momentum wheel-horizon scanner combination is shown in Figs. 6-27 and 6-28. Typical values of momentum wheel parameters are given in Table 6-11.

Table 6-11.   Typical Values of Momentum Wheel Parameters

MANUFAC-TURER	SPACECRAFT	MASS (KG)	MOMENT OF INERTIA (KG · M²)	SPEED RANGE (RPM)	ANGULAR MOMENTUM (KG · M²/SEC)
APL	GEOS-3 } SAS-1 }	3.18	0.0115	2000	2.41 @ 2000 RPM
BENDIX	ATS NIMBUS OAO SERIES	8.84 2.36 5.13	0.0880 0.0034 0.0297	±1450 ±1400 ± 900	¡1.52 @ 1250 RPM 0.447 @ 1250 RPM 2.79 @ 900 RPM
ITHACO	NIMBUS −4, −5, −6 } LANDSAT −1, −2 } SAS-3 HCMM } SEASAT }	3.72 6.71	0.0060 0.0272	600−2000 1000−2000	1.49 @ 2000 RPM 5.69 @ 2000 RPM
RCA	AE SERIES ITOS SERIES	18.66 14.43	3.4604	95−392 120−160	128.03 @ 353.32 RPM
SPERRY	HEAO−B	13.38	0.1913	±2000	40.071 @ 2000 RPM

Torque motors, used to transfer momentum between the wheel and the spacecraft body, may be of two types: an AC two-phase induction motor or a DC brushless motor. Because the AC motor requires no brushes or sliprings, it has high reliability and a long lifetime, but also low efficiency, low torque, and a high operating speed. The high-speed motor requires use of a gearing system, with associated friction and backlash problems. By comparison, DC motors are efficient and provide high torque at low speed, thus allowing direct drive without gearing. The conventional brush commutators are normally replaced with electronic or brushless commutation.

Because of evaporation, bearings have a lubrication problem when the seal is exposed to the space environment. Low vapor pressure oils and labyrinthine seals have been used, for example, on the Atmosphere Explorer series. Also, dry lubricants, such as Teflon® compounds, have been used on spacecraft, such as SAS-1, although Teflon may deform under impact during launch. With small momentum wheels, completely sealed systems can be used. With reaction wheels, which may go through the zero-speed region, special care must be taken to minimize static friction (often called *stiction*). Recent work has been done on prototype magnetic suspension systems which have the potential of avoiding wear altogether [Sabnis, *et al.*, 1974].

Tachometers, which measure the wheel speed, often consist of a wheel-mounted magnet and a fixed sensor, such as a simple pickoff coil. The pulse train can be converted to a DC voltage for use as a controlling error signal for either a constant speed or a commanded variable-speed mode. Another type, the DC tachometer, uses the back electromotive force (emf) generated by the armature winding to produce an analog voltage proportional to the rotational speed. Optical encoders are also used with light-emitting diodes.

Dual-Spin Spacecraft.   A dual-spin spacecraft is one which has two sections with different spin rates. It usually consists of a despun section and a flywheel. The dual-spin OSO spacecraft are somewhat different, however, with a despun "sail" section containing Sun-pointed instruments, while a "wheel" section, rotating from 6 to 30 rpm, provides angular momentum and the requisite stability. Because the wheel section contains experiments and requires three-axis attitude determination, it is considerably more elaborate than a typical momentum wheel.

Momentum wheels may be operated at either a constant or a variable speed and are used to control the spin rate and attitude about the wheel axis. The former application is less common and generally is used only on satellites such as GEOS-3, where a large gravity-gradient restoring torque (see Sections 17.2, 18.5, and 19.5) is available about the wheel axis.

A *momentum bias* design is common for dual-spin Earth-orbiting spacecraft, in which a momentum wheel is mounted along the pitch axis, which is controlled to orbit normal. This allows the instruments to scan over the Earth. For example, the AE series was designed to operate with a nominal angular momentum of 125 $kg \cdot m^2/s$ with a wheel capacity which allows operation of the body either despun at 1 rpo or spinning at 4 rpm. An integral wheel horizon scanner provides information for closed-loop pitch control and open-loop roll/yaw control using magnetic coils. The SAS-3 spacecraft uses its momentum wheel in several different operational modes: spin rate control mode using gyro rate sensing, Earth-oriented mode using horizon scanner pitch data, and a three-axis stabilized mode using star camera data for pitch control [Mobley, *et al.*, 1974].

Practical problems which should be considered in the design of momentum wheel systems include bearing noise, quantization, jitter, variation of the bearing friction with temperature, offset of the wheel axis from the body principal axis, and nutation. Difficulties have been experienced on the AE series in dissipating nutation with a distributive damper. It is suspected that a mechanism such as standing wave patterns reduced the effectiveness of the fluid-loop damper. Control system problems may occur, especially in switching from the spinning to the despun mode, where there may be difficulty in achieving pitch lock if the body rate is too high.

Multiple Reaction Wheels.   Because reaction wheels are operated with nominally zero momentum, they are used primarily for absorbing cyclic torques and temporarily storing momentum from the body during *slew*, or reorientation, maneuvers. However, the secular disturbance torques, which are about the same magnitude as the cyclic terms, would eventually saturate the momentum storage capacity. Therefore, provision is made for periodic *momentum dumping* through *external* torques produced by gas jets or magnetic coils. (These are differentiated from *internal* torques due to sources such as torque motors and bearings which do not change the total angular momentum of the system.)

Normally, three reaction wheels are used to control a vehicle, with the wheel axes aligned with the body principal axes; a redundant fourth wheel is also common. A redundant fourth skewed wheel has been flown on IUE and the MMS series as a provision against failure of one of the orthogonal wheels [NASA, 1974 and 1975]. Also, a nonorthogonal four-wheel configuration has been designed for HEAO-B [Rose and Hoffman, 1976].

6.7 Magnetic Coils

B. L. Gambhir
Des R. Sood

Magnetic coils, or electromagnets, are used to generate magnetic dipole moments for attitude and angular momentum control. They are also used to compensate for residual spacecraft biases and to counteract attitude drift due to environmental disturbance torques.

Consider a single, plane, wire loop enclosing an area, A, through which a current, I, is flowing (see Fig. 6-47). Then the *magnetic moment*, **m**, is given by

$$\mathbf{m} = IA\mathbf{n} \qquad (6\text{-}17)$$

where **n** is a unit vector normal to the plane of the loop. The positive sense of the magnetic moment is determined by the right-hand rule; that is, the direction of the magnetic moment is the direction of the thumb of the right hand when the fingers of the right hand are cupped in the direction of the electric current in the loop. For a coil of N turns, the principle of superposition gives

$$\mathbf{m} = NIA\mathbf{n} \qquad (6\text{-}18)$$

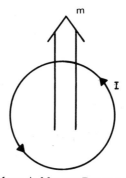

Fig. 6-47.   Magnetic Moment Due to a Current Loop

The *magnetic dipole moment* depends on the material enclosed by the current-carrying coil and is given by

$$\mathbf{d} = \mu\mathbf{m} \qquad (6\text{-}19)$$

where μ is the permeability of the core material. In SI units, the permeability of free space, μ_0, has the value $4\pi \times 10^{-7}$ N/A^2 (see Appendix K). Thus, for a coil or an electromagnet enclosing a plane area, A, the magnetic dipole moment is given by

$$\mathbf{d} = \mu(NI)A\mathbf{n} \qquad (6\text{-}20)$$

It is apparent from Eq. (6-20) that to generate a requisite amount of dipole, parameters such as core material, μ; coil configuration, N and A; and the current level, I, must be appropriately selected. The selection is dictated by mission requirements and is influenced by considerations such as weight, power consumption, and bulk.

The choice of the core material is the most important design parameter. Ferromagnetic materials, such as Permalloy (78% nickel, 22% iron), and Permendur (50% cobalt, 50% iron) have very high permeabilities and, when used as core materials, lead to a substantial reduction in power consumption as well as bulk. However, ferromagnetic materials have magnetization curves which saturate at relatively low values of applied magnetic field intensity and exhibit both nonlinearity and hysteresis. Moreover, in ferromagnetic materials, permeability is a function of the magnitude of the magnetic field intensity. (See, for example, Jackson [1965].) Consequently, with ferromagnetic cores, it is difficult to predict accurately the magnetic dipole moment and, hence, they have been very infrequently used. Magnetic coils on most satellites have "air" cores.

The material of the current-carrying element is chosen on the basis of weight and ability to dissipate the heat generated by the current without an adverse impact on the electrical properties. For example, SAS-3 used coils wound with no.18 aluminum magnet wire of 1.02-mm diameter. Table 6-12 summarizes pertinent information concerning the spin axis magnetic coils flown on some representative missions.

Table 6-12.   Characteristics of Spin Axis Magnetic Coils on Representative Missions. See Appendix I for mission details.

SPACECRAFT	SPACECRAFT ANGULAR MOMENTUM $(kg \cdot m^2 \cdot s^{-1})$	MAXIMUM DIPOLE $(W \cdot m)$	MAXIMUM PRECESSION RATE IN A FIELD OF 24 A/m NORMAL TO THE SPIN AXIS (deg/hour)	REMARKS
SAS–3	4.465	$6.28 \cdot 10^{-5}$	69.3	COIL CONSISTS OF 260 TURNS. MAXIMUM CURRENT IS 0.6 A AND MAXIMUM POWER CONSUMPTION IS 10 W
OSO–8	342.6	$5.33 \cdot 10^{-5}$	0.75	COIL CONSISTS OF 360 TURNS. MAXIMUM CURRENT IS 0.075 A
AE–3	127.7	$2.94 \cdot 10^{-4}$	11.3	TWO COILS; EACH HAS 500 TURNS. MAXIMUM POWER CONSUMPTION IS 12 W

Accurate prediction of the magnetic control torques requires that the coils be supplied with a constant current. Control of the coil current is necessary for two reasons: the supply voltage may fluctuate considerably ($\pm 30\%$ from nominal for some missions), and the resistance of the current-carrying element changes with temperature.

Figure 6-48 shows the electronic system used to drive the spin axis coil of the SAS-3 spacecraft [Mobley, et al., 1974]. The current in the coil is controlled in closed-loop fashion by sensing the voltage drop across the feedback resistor, R_{fb}. Bidirectional operation is achieved by coil current reversal through the use of a remote operated latching-type "sense" switching relay. On SAS-3, the spin axis coil served a dual purpose: when the power to the constant current source was turned off, the coil was automatically switched to the "trim" system to generate a small magnetic dipole to counteract the spin axis attitude drift.

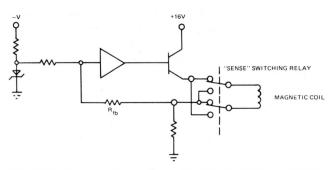

Fig. 6-48.   Constant Current Source Used for SAS-3 Magnetic Coil

6.8 Gas Jets

Robert S. Williams

All *jets* or *thrusters* produce thrust by expelling propellant in the opposite direction. The resultant torques and forces are used for five principal spacecraft functions: (1) to control attitude, (2) to control spin rate, (3) to control nutation, (4) to control the speed of momentum wheels, and (5) to adjust orbits. *Gas jets* produce thrust by a collective acceleration of propellant molecules, with the energy coming from either a chemical reaction or thermodynamic expansion, whereas *ion jets* accelerate individual ionized molecules electrodynamically, with the energy ultimately coming from solar cells or self-contained electric generators. Gas jets are widely used, whereas ion jets are not yet developed enough for spacecraft use. Schmidtbauer, *et al.*, [1973] provide a survey of all types; Junge and Sprengel [1973], Pye [1973], LeGrives and Labbe [1973] and Vondra and Thomassen [1974] describe work on ion thrusters which may lead to flight-qualified units. A hybrid flight-qualified unit in which solid Teflon® is vaporized by a high-voltage electric discharge is described by Au and Baumgarth [1974]. Gas jet hardware and applications to attitude control are discussed here, mathematical models in Sections 7.10 and 17.4, and control laws in Section 19.3.

Gas jets are classified as *hot gas* when the energy is derived from a chemical reaction or *cold gas* when it is derived from the latent heat of a phase change, or from the work of compression if no phase change is involved. Hot-gas jets generally produce a higher thrust level (> 5 N) and a greater total *impulse* or time integral of the force. Cold-gas systems operate more consistently, particularly when the system is operated in a pulsed mode, because there is no chemical reaction which must reach steady state. The lower thrust levels ($\lesssim 1$ N) of cold-gas systems may facilitate more precise control than would be available with a high-thrust system.

Hot-gas systems may be either *bipropellant* or *monopropellant*. Fuel and oxidizer are stored separately in a bipropellant system; very high thrust levels ($\gtrsim 500$ N) can be obtained, but the complexity of a two-component system is justified only when these thrust levels are required. Monopropellant systems use a catalyst or, less frequently, high temperature to promote decomposition of a single component, which is commonly hydrazine (N_2H_4) or hydrogen peroxide (H_2O_2). Hydrazine with catalytic decomposition is the most frequently used hot-gas monopropellant system on spacecraft supported by Goddard Space Flight Center. The problem of consistency, mentioned above, manifests itself in two ways. First, the thrust is below nominal for the initial few seconds of firing because the reaction rate is below the steady-state value until the catalyst bed reaches operating temperature. Second, the *thrust profile*, or time dependence of thrust, changes as a function of total thruster firing time; this is significant when a long series of short pulses is executed, because the thrust profile for the later pulses will differ from that for the earlier pulses. The latter problem has been ascribed by Holcomb, *et al.*, [1976] to aniline impurities in the grade of hydrazine usually used as fuel. Much, *et al.*, [1976], Pugmire and O'Connor [1976], and Grabbi and Murch [1976] describe the development of electrothermal thrusters in which decomposition of the hydrazine is initiated at a heated surface within the thruster; these thrusters reportedly function consistently, with a well-defined thrust profile, over a wide range of pulsewidths and total burn times. Variable thrust profiles can be modeled as described in Section 19.3, but the models are more complicated and are probably less accurate than those for consistently reproducible thrust profiles.

In near-Earth orbits, either jets or magnetic coils (Section 6.7) can be used for many of the same purposes. The control laws for jets (Section 19.3) are simpler than those for coils (Sections 19.1 and 19.2), primarily because jets produce larger torques. The magnetic torque produced by a coil depends on the local magnetic field, which varies as the spacecraft moves in its orbit; a coil command frequently must extend over a large fraction of an orbit or over several orbits to achieve the desired results. The propellant supply required for jets is the major limitation on their use; a *fuel budget* is an important part of mission planning for any system using jets. Other considerations are the overall weight of the system and the need to position thrusters where the exhaust will not impinge on the spacecraft. The latter consideration is especially important when hydrazine is used, because the exhaust contains ammonia, which is corrosive. The only magnetic fields associated with gas-jet systems are those generated by the solenoid valves; these will generally be smaller than those associated with magnetic control coils, but may be significant in some cases if experiments on the spacecraft are adversely affected by stray fields.

In more distant orbits (certainly beyond geosynchronous altitude), jets are the only practical means of interchanging momentum with the environment. High-thrust or total impulse requirements may indicate a hot-gas system. Otherwise, the cold-gas system may be favored because hydrazine freezes at about 0° C and may require heaters if lower temperatures will be encountered during the mission. Specific components may affect the relative system reliability; for example, hydrazine systems use tank diaphragms to separate the propellant from the pressurizing agent and also require a catalyst or heater to initiate decomposition; cold-gas systems may have a pressure regulator between the tank and the thruster.

IUE Hydrazine System. As a representative hot-gas attitude control system, we describe the Hydrazine Auxiliary Propulsion System manufactured by the Hamilton-Standard division of United Technologies for the International Ultraviolet Explorer (IUE) spacecraft [Sansevero and Simmons 1975]. The IUE spacecraft will perform measurements of ultraviolet spectra of stars from a geo-synchronous orbit. The hydrazine system will be used for attitude, spin rate, and nutation control in the transfer orbit and for orbit adjustments and momentum-wheel speed control thereafter. A hydrazine system is needed to meet total impulse requirements over the 3- to 5- year mission lifetime. The *plume*, or envelope of the thruster exhaust, has been analyzed to determine whether ammonia is likely to condense on the telescope optics; this was found not to be a problem.

Figure 6-49 shows the IUE hydrazine system, which was designed for as-sembly as a complete unit to be attached subsequently to the spacecraft. The octagonal framework is approximately 137 cm between opposite faces. Four thrusters are mounted on the octagonal faces; two of these can be seen on the leftmost face in Fig. 6-49. Eight additional thrusters are mounted in two clusters of four thrusters each; one cluster is suspended from the octagonal face closest to the camera, the other from the opposite face. Each cluster contains two large thrusters, each generating about 20 N, which are used for attitude and nutation control and orbit adjustments. The two small thrusters in each cluster and the four body-mounted thrusters each generate about 0.4 N for spin rate control and momentum unloading.

Fig. 6-49. Hydrazine System for the IUE Spacecraft. (Photo courtesy of Hamilton-Standard Division of United Technologies Corp.)

Spherical fuel tanks are mounted in six of the eight bays. A diaphragm in each tank separates the nitrogen pressurizing gas from the hydrazine fuel. Opposite tanks are connected in pairs to minimize imbalance as fuel is consumed. Fuel flows through one filter and two latch valves between the tanks and thruster assemblies. Pressure transducers located between the filters and latch valves allow the amount of fuel remaining in each pair of tanks to be estimated.

As shown in Fig. 6-50, all lines are interconnected between latch valves to minimize the effect of a valve failure. Fuel lines from the spacecraft body to the suspended clusters are heated to prevent freezing. The thrusters are also provided with heaters to maintain the proper operating temperature, which is measured by thermocouples on each thruster. An additional filter and solenoid valve is associated with each thruster. In operation, latch valves are open and the system is controlled with the solenoid valves on selected thrusters.

The system as built weighs almost 25 kg without fuel. The fuel budget for the mission is about 12 kg. During the mission, thrusters will be operated with firing times as short as 0.1 sec during attitude maneuvers and as long as several minutes during spin rate changes or orbit maneuvers.

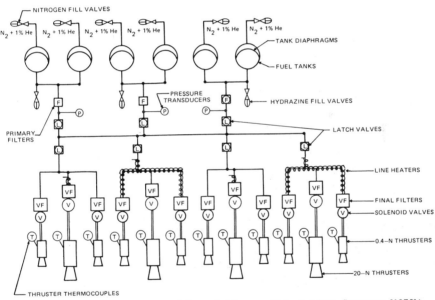

Fig. 6-50. IUE Hydrazine System Schematic Diagram. (Adapted from Sansevero [1975].)

Other Representative Systems. Most cold-gas systems are functionally similar to the IUE system. The major differences are that the propellant is stored as a liquid above the critical pressure and is self-pressurizing. Heaters are not required because the propellant is a gas at low pressure at any temperature likely to be encountered in operation. Pressure regulators are usually used to control propellant flow rate.

The OSO series of Earth-orbiting spacecraft combines a cold-gas system with magnetic coils. The coils are usually used for attitude control. The gas jets are used for occasional rapid maneuvers which cannot be performed with the coils and for correcting secular angular momentum changes caused by gravity-gradient and residual magnetic torques.

The RAE-2 spacecraft, placed in lunar orbit in 1973, carried both a hydrazine hot-gas system for orbit corrections and a Freon® cold-gas system for attitude and spin rate control. Although single systems can be designed to perform all three functions, the RAE-2 mission required the ejection of the orbit-correction system before all attitude and spin rate control functions were completed. The impulse potential of a hot-gas system was required for the orbit changes, but a simpler cold-gas system sufficed for the other requirements.

The ISEE-1 and -2 spacecraft were placed simultaneously in an orbit with an apogee of about 22 Earth radii on October 22, 1977, to study the interaction of the solar wind with the magnetosphere. Attitude control is used to maintain the spin axes at the North Ecliptic Pole; spin rate control is needed to maintain a constant spin rate; and orbit maneuvers are performed to maintain the desired distance between the two spacecraft. The estimated total impulse requirements for all three functions over the 3-year mission lifetime can be met with a Freon® cold-gas system.

6.9 Onboard Computers

Gerald M. Lerner
Prafulla K. Misra

In general, onboard attitude control is obtained by combining onboard sensors and torquers through a *control law* (Chapter 18), or control strategy, which is implemented via analog logic or a digital computer. Because attitude control systems are normally chosen for reliability and cost, control laws which are easily implemented through analog logic have been widely used. Sensors such as analog Sun sensors (Section 6.1), and wheel-mounted horizon sensors (Section 6.2), are well suited for such applications because the sensor output is simply related to an angle which is to be controlled. Reaction wheels, momentum wheels (Section 6.6), or jets (Section 6.8) are preferred torquing devices because in many applications there is a simple relationship between attitude errors and the appropriate torque commands. In addition, magnetic torquers (Section 6.7) are often used in conjunction with a magnetometer.

Increasingly stringent spacecraft attitude control and autonomy requirements (Chapter 22) have resulted in the need for *onboard computers* (*OBCs*) or digital processors. Digital processors afford several advantages over analog systems [Schmidtbauer *et al.*, 1973], including the capability of processing complex types of data—such as star tracker, gyroscope, or digital Sun sensor data—and of modifying programmed control laws via ground command.

In an attempt to standardize flight hardware, NASA's Goddard Space Flight Center is developing the *NASA Standard Spacecraft Computer* NSSC, which was derived from the OAO-3 onboard computer and is similar to that of IUE. The NSSC-I will be flown on the Solar Maximum Mission (SMM) and on subsequent flights in the Multi-mission Modular Spacecraft series. A second, larger version, the NSSC-II, will be used for Spacelab payloads and the Space Telescope. The specifications for the NSSC-I, NSSC-II, and HEAO digital processor are shown in Table 6-13. Timing estimates for the NSSC-I and NSSC-II are given in Table 6-14.

Table 6-13. Specifications for the NASA Standard Spacecraft Computers NSSC-I and NSSC-II and the HEAO Digital Processor

PARAMETER	NSSC–I*	NSSC–II	HEAO DIGITAL PROCESSOR
POWER (W)	38 MAXIMUM (6 STANDBY)	130 TO 242[†]	15
MASS (Kg)	6.4	8.3 (32K BYTE)	4.5
VOLUME (LITERS)	9.4	6.4	2.4
WORD LENGTH (BITS)	18	8, 16, 32, OR 64	16
NUMBER OF INSTRUCTIONS	55	171	42
MEMORY	8K WORD MODULES TO 64K WORDS (32K NOMINAL)**	16K BYTES[††] (EXPANDABLE BY 16K INCREMENTS)	8K WORDS

*WITH 32K WORDS OF MEMORY.
**EACH 8K WORD MODULE IS DIVIDED INTO TWO 4K WORD LOGICAL BANKS.
†DEPENDING ON CONFIGURATION.
††8-BIT BYTES.

Table 6-14. Timing Estimates for NSSC-I and NSSC-II

OPERATION	SINGLE PRECISION (μs)		DOUBLE PRECISION (μs)	
	NSSC–I*	NSSC–II**	NSSC–I*	NSSC–II**
ADD/SUBTRACT	18	1.7	63/83	2
MULTIPLY	57	8.3	233	33.5
DIVIDE	85	16.9	2500	54.9
SINE/COSINE	375	–	1600	–
SQUARE ROOT	540 to 840	–	3530 to 4920	–

*ESTIMATES INCLUDE A LOAD AND STORE WHICH REQUIRE ABOUT 13 μs; MULTIPLY OR DIVIDE BY POWERS OF 2 TAKES APPROXIMATELY ONE-TENTH THE NOMINAL MULTIPLICATION TIME.

**ESTIMATES ARE FOR REGISTER-TO-REGISTER ONLY.

NSSC-I. This computer uses 18-bit words and fixed point, two's complement arithmetic. A 55-instruction set is available with a basic cycle time of 1.25 μs and a 5-μs requirement for an add operation. (These values are still uncertain and may be revised in subsequent versions of the computer.) A detailed description of the instruction set is given by Merwarth [1976].

A set of mathematical subroutines for the NSSC-I has been designed to provide elementary 18- and 36-bit operations [DeMott, 1976]. Timing estimates for these are given in Table 6-14. The efficiency of coding the NSSC-I is limited because it has only three registers: an accumulator and an extended accumulator (which are combined into a double-length register for products and dividends in multiplication and division) and one index register. A further complication is introduced by the small word size which allows only 12 bits for operand addresses in instructions. The NSSC-I therefore uses a page register to specify the *logical* *bank,* or $2^{12} = 4096$ word* region, from which the operand is to be retrieved. Loading, reloading, and (especially) saving and restoring the page register is cumbersome, so NSSC-I programs can address directly only 4096 words of data and only data defined within the independently assembled module which addresses it.

An interrupt system provides 16 hardware interrupts and one programmable interrupt. Input and output are provided from 16 devices [Merwarth, 1976]. The onboard computer transmits data to the ground at the rate of one word per telemetry frame (see Chapter 8) and receives commands at the rate of 2000 bits per second (a 48-bit command every 24 ms). Memory dump, via S-band, is available at 32,000 bits per second.

Memory for the NSSC-I is expandable in 8192 word modules to a maximum of 8 modules. Hardware protection against changing data or instructions within selected address limits is provided. The proposed memory layout for MMS is shown in Fig. 6-51. A flight executive is used to schedule the various tasks of the onboard computer. These tasks include high-priority attitude control operations (probably every 128 ms for SMM) in addition to low-priority housekeeping functions. The latter include performing functions normally provided by analog devices such as thermostats and other spacecraft hardware.

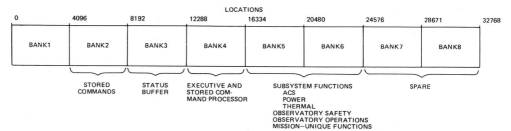

Fig. 6-51.   NSSC-I Memory Layout for the MMS Spacecraft. The MMS attitude control system core requirements are 10 K to 16 K words of program and data storage; 10 to 1000 stars at 4 words per star; and 100 to 3000 words for 72 hours of ephemeris data

NSSC-II. This computer is a microprogrammed general-purpose computer that is compatible with the standard instruction set of IBM S/360 ground-based computer systems [NASA, 1977]. The machine microcode implements a total of 171 instructions including 16-, 32-, and 64-bit fixed-point and 32-bit floating point instructions. In addition, the design accommodates 512 words of microcode memory capacity for special instructions or routines programmed or specified by the user. The semiconductor memory is expandable in 16 K-byte (8-bit byte) increments.

The NSSC-II uses 8-, 16-, 32- or 64-bit fixed-point data words at the user's option. The basic cycle time for the machine is 440 ns. The system has 16 general registers. The word size allows 20-bit operand addresses in instructions; thus, NSSC-II programs can address directly up to 1 M-byte words of data.

HEAO Computer. The digital processor employed for the HEAO spacecraft will process gyroscope data and compute jet commands (every 320 ms) for

---

* Eighteen-bit words.

HEAO-1 and -C and will process both gyroscope and star tracker data to compute jet commands for HEAO-B [Hoffman, 1976]. The specifications for the HEAO digital processor are given in Table 6-13.

References

1. Adams, D. J., *Hardware Technical Summary for IUE Fine Error Sensor*, Ball Brothers Research Corp., TN74-51, Oct. 1974.
2. Adcole Corp., *Sun Angle Sensor Systems Short Form Catalog*, Feb. 1975.
3. Adcole Corp., *Design Review Data Package Fine Pointing Sun Sensor for Solar Maximum Mission*, Oct. 1977.
4. Astheimer, Robert W., "Instrumentation for Infrared Horizon Sensing," *Proceedings of the Symposium on Spacecraft Attitude Determination, Sept. 30, Oct. 1–2, 1969*, El Segundo, CA; Air Force Report No. SAMSO-TR-69-417, Vol. I; Aerospace Corp. Report No. TR-0066(5306)-12, Vol. I, 1969.
5. Au, G. F. and S. F. J. Baumgarth, "Ion Thruster ESKA 8 for North-South Stationkeeping of Synchronous Satellites," *J. Spacecraft*, Vol. 11, p. 618–620, 1974.
6. Barnes Engineering Co., *Infrared Detectors, Thermal and Photon*, Barnes Engineering Bulletin 2-350A, 1976.
7. Bloom, A. L., "Principles of Operation of the Rubidium Vapor Magnetometer," *Applied Optics*, Vol. 1, p. 61–68, 1962.
8. Chubb, W. B., H. F. Kennel, C. C. Rupp and S. M. Seltzer, "Flight Performance of Skylab Attitude and Pointing Control System," *J. Spacecraft*, Vol. 12, p. 220–227, 1975.
9. Cleavinger, R. L., and W. F. Mayer, *Attitude Determination Sensor for Explorer 53*, AIAA Paper No. 76-114, AIAA 14th Aerospace Sciences Meeting, Wash. DC, Jan. 1976.
10. Coon, T. R., and J. E. Irby, "Skylab Attitude Control System," *IBM Journal of Research and Development*, Jan. 1976.
11. Dehmelt, H. G., "Modulation of a Light Beam by Precessing Absorbing Atoms," *Phys. Rev. 2nd Series*, Vol. 105, p. 1924–1925, 1957.
12. DeMott, A., *Preliminary Study of Onboard Attitude Control for the Multi-Mission Modular Spacecraft*. Comp. Sc. Corp., Feb. 1976.
13. Ebel, B., *In Flight Performance of the French German Three-Axis Stabilized Telecommunications Satellite SYMPHONIE*, AIAA Paper No. 75-099, AAS/AIAA Astrodynamics Specialist Conference, Nassau, Bahamas, July 1975.
14. Farthing, W. H. and W. C. Folz, "Rubidium Vapor Magnetometer for Near Earth Orbiting Spacecraft," *Rev. Sci Instr.*, Vol. 38, p. 1023–1030, 1967.
15. Fontana, R., R. Baldassini, and G. Simoncini, *Attitude Sensors Review and General Applications*, Vol. 2 of *Study of Detection and Estimation Techniques Applied to Attitude Measurements of Satellites*, ESRO, ESRO-CR(P)-551, April 1974.
16. Fountain, G. H., *SAS-B Star Sensor Telemetry Data*, Applied Physics Laboratory, S2P-2-499, Feb. 1972.
17. Gates, R. F., and K. J. McAloon, *A Precision Star Tracker Utilizing Advanced Techniques*, AIAA Paper No. 76-113, AIAA 14th Aerospace Sciences Meeting, Wash., DC, Jan. 1976.
18. General Electric Space Systems, *Earth Resources Technology Satellite Image*

Annotation Processing (IAP) Software Description, Document 71SD5216, Valley Forge Space Center, Oct. 1971.

19. Geyger, W. A., *Non-linear Magnetic Control Devices*. New York: McGraw-Hill, Inc., Chapters 13 and 14, 1964.
20. Gottlieb, D. M., C. M. Gray, and L. Fallon, *High Energy Astronomy Observatory-A (HEAO-A) Star Tracker Assembly Description*, Comp. Sc. Corp., CSC/TM-75/6203, June 1976.
21. Grabbi, R. and C. K. Murch, "High Performance Electrothermal Hydrazine Thruster (Hi PEHT) Development, " AIAA Paper No. 76-656, AIAA/SAE Twelfth Propulsion Conference, Palo Alto, CA, July 1976.
22. Gray, C. M., L. Fallon, D. M. Gottlieb, M. A. Holdip, G. F. Meyers, J. A. Niblack, and M. Rubinson, *High Energy Astronomy Observatory-A (HEAO-A) Attitude Determination System Specifications and Requirements*, Comp. Sc. Corp., CSC/SD-76/6001, Feb. 1976.
23. Greensite, A. L., *Control Theory: Volume II, Analysis and Design of Space Vehicle Flight Control Systems*. New York: Spartan Books, 1970.
24. Grivet, P. A. and L. Malner, "Measurement of Weak Magnetic Fields by Magnetic Resonance," *Advances in Electronics and Electron Physics*. New York: Academic Press, p. 39–151, 1967.
25. Hatcher, Norman M., *A Survey of Attitude Sensors for Spacecraft*, NASA SP-145, 1967.
26. Hoffman, D. P., "HEAO Attitude Control Subsystem—A Multimode/-Multimission Design," *Proceedings AIAA Guidance and Control Conference*, San Diego, CA, Aug. 1976.
27. Holcomb, L., L. Mattson, and R. Oshiro, "The Effects of Aniline Impurities on Monopropellant Hydrazine Thruster Performance," AIAA Paper No. 76-659, AIAA/SAE Twelfth Propulsion Conference, Palo Alto, CA, July 1976.
28. Hotovy, S. G., M. G. Grell, and G. M. Lerner, *Evaluation of the Small Astronomy Satellite-3 (SAS-3) Scanwheel Attitude Determination Performance*, Comp. Sc. Corp., CSC/TR-76/6012, July 1976.
29. Jackson, John David, *Classical Electrodynamics*. New York: John Wiley & Sons, Inc., 1965.
30. Junge, Hinrich J., and Uwe W. Sprengel, "Direct Thrust Measurements and Beam Diagnostics on an 18-cm Kaufman Ion Thruster," *J. Spacecraft*, Vol. 10, p. 101–105, 1973.
31. Koso, D. A. and J. C. Kollodge, "Solar Attitude Reference Sensors," *Proceedings of the Symposium on Spacecraft Attitude Determination, Sept. 30, Oct. 1–2, 1969*, El Segundo, CA; Air Force Report No. SAMSO-TR-69-417, Vol. I; Aerospace Corp. Report No. TR-0066(5306)-12, Vol. I, 1969.
32. LeGrives, E. and J. Labbe, "French Research on Cesium Contact Ion Sources," *J. Spacecraft.*, Vol. 10, p. 113–118, 1973.
33. Massart, J. A., *A Survey of Attitude Related Problems for a Spin-Stabilized Satellite on a Highly Eccentric Orbit*, ESOC Internal Note 152, Aug. 1974.
34. Merwarth, A., *Multimission Modular Spacecraft (MMS) Onboard Computer (OBC) Flight Executive Definition*, NASA S-700-55, March 1976.

35. Mobley, F. F., Konigsberg, K., and Fountain, G. H., *Attitude Control System of the SAS-C Satellite*, AIAA Paper No. 74-901; AIAA Mechanics and Control of Flight Conference, Anaheim, CA., Aug. 1974.

36. Moore, W., and W. Prensky, *Applications Technology Satellite, ATS-6, Experiment Check-out and Continuing Spacecraft Evaluation Report*, NASA X-460-74-340, Dec. 1974.

37. Murch, C. K., R. L. Sackheim, J. D. Kuenzly, and R. A. Callens, "Noncatalytic Hydrazine Thruster Development, 0.050 to 5.0 Pounds Thrust," AIAA Paper No. 76-658, AIAA/SAE Twelfth Propulsion Conference, Palo Alto, CA, July 1976.

38. NASA, *NASA Standard Spacecraft Computer -II (NSSC-II)*.CAT. NO. 4.006, Standard Equipment Announcement, Revision 1, Aug. 1, 1977.

39. NASA, *Spacecraft Star Trackers*, NASA SP-8026, July 1970.

40. NASA, *System Design Report for International Ultraviolet Explorer (IUE)*, GSFC, Greenbelt, MD, April 1974.

41. Nutt, W. T., M. C. Phenniger, G. M. Lerner, C. F. Manders, F. E. Baginski, M. Rubinson, and G. F. Meyers, *SEASAT-A Attitude Analysis and Support Plan*, NASA X-XXX-78-XXX, April 1978.

42. Pugmire, T. K., and T. J. O'Connor, "5 Pound Thrust Non-Catalytic Hydrazine Engine," AIAA Paper No. 76-660, AIAA/SAE Twelfth Propulsion Conference, Palo Alto, CA, July 1976.

43. Pye, J. W., "Component Development for a 10-cm Mercury Ion Thruster," *J. Spacecraft*, Vol. 10, p. 106–112, 1973.

44. Pyle, E. J., Jr., *Solar Aspect System for the Radio Astronomy Explorer*, NASA X-711-68-349, Sept. 1968.

45. Quasius, G., and F. McCanless, *Star Trackers and Systems Design*. Wash., DC: Spartan Books, 1966.

46. RCA Service Company, *OSO-I Spacecraft Subsystems Description Document*, for GSFC, POB-3SCP/0175, May 1975.

47. Rose, R. E., and D. P. Hoffman, *HEAO-B Attitude Control and Determination Subsystem Critical Design Review*, TRW Systems Group, Redondo Beach, CA, Oct. 19, 1976.

48. Ryder, J. D., *Engineering Electronics*. New York: McGraw-Hill, Inc., 1967.

49. Sabnis, A. V., J. B. Dendy and F. M. Schmitt, *Magnetically Suspended Large Momentum Wheels*, AIAA Paper No. 74-899, AIAA Mechanics and Control of Flight Conference, Anaheim, CA, Aug. 1974.

50. Salmon, P. M. and W. C. Goss, *A Microprocessor-Controlled CCD Star Tracker*, AIAA Paper No. 76-116, AIAA 14th Aerospace Sciences Meeting, Wash., DC, Jan. 1976.

51. Sansevero, V. J., Jr., and R. A. Simmons, *International Ultraviolet Explorer Hydrazine Auxiliary Propulsion System Supplied Under Contract NAS 5-20658*, Hamilton Standard Division of United Technologies Corporation, Windsor Locks, CT, Oct. 1975.

52. Schmidtbauer, B., Hans Samuelsson, and Arne Carlsson, *Satellite Attitude Control and Stabilisation Using On-Board Computers*, ESRO, ESRO-CR-100, July 1973.

53. Schonstedt Instrument Company, Reston, Virginia, Private Communication, 1976.

54. Schwarz, Frank, and Thomas Falk, "High Accuracy, High Reliability Infrared Sensors for Earth, Lunar, and Planetary Use," *Navigation*, Vol. 13, p. 246–259, 1966.

55. Scott, R. T., and J. E. Carroll, "Development and Test of Advanced Strapdown Components for SPARS," *Proceedings of the Symposium on Spacecraft Attitude Determination Sept. 30, Oct. 1–2, 1969*, El Segundo, CA; Air Force Report No. SAMSO-TR-69-417, Vol. I; Aerospace Corp. Report No. TR-0066(5306)-12, Vol. I, 1969.

56. Slocum, R. E. and F. N. Reilly, "Low Field Helium Magnetometer," *IEEE Transactions on Nuclear Science*, Vol. NS-10, p. 165–171, 1963.

57. Smith, B. S., *Hardware Technical Summary Fine (Digital) Sun Sensor System (FSS) (IUE)*, Adcole Corp., QD10153, Jan. 1975.

58. Sonett, C. P., "The Distant Geomagnetic Field II, Modulation of a Spinning Coil EMF by Magnetic Signals," *J. Geophys. Res.* Vol. 68, p. 1229–1232, 1963.

59. Spetter, D. R., *Coarse Detector Output Model*, TRW Systems Group, HEAO-74-460-204, Dec. 1974.

60. Susskind, Alfred K., *Notes on Analog-Digital Conversion Techniques*. The Technology Press of MIT, Cambridge, MA, 1958.

61. Thomas, J. R., *Derivation and Statistical Comparison of Various Analytical Techniques Which Define the Location of Reference Horizons in the Earth's Horizon Radiance Profile*, NASA CR-726, April 1967.

62. Thomson, William Tyrrell, *Introduction to Space Dynamics*. New York: John Wiley & Sons, Inc., 1963.

63. Trudeau, N. R., F. W. Sarles, Jr. and B. Howland, *Visible Light Sensors for Circular Near Equatorial Orbits*, AIAA Paper 70-477, Third Communications Satellite Systems Conference, Los Angeles, CA, 1970.

64. Tsao, H. H., and H. B. Wollman, *Photon Counting Techniques Applied to a Modular Star Tracker Design*, AIAA Paper No. 76-115, AIAA 14th Aerospace Sciences Meeting, Wash., DC, Jan. 1976.

65. Vondra, R. J. and K. I. Thomassen, "Flight Qualified Pulsed Electric Thruster for Satellite Control," *J. Spacecraft*, Vol. 11, p. 613–617, 1974.

66. Werking, R. D., R. Berg, T. Hattox, G. Lerner, D. Stewart, and R. Williams, *Radio Astronomy Explorer-B Postlaunch Attitude Operations Analysis*, NASA X-581-74-227, July 1974.

67. Wertz, J. R., C. F. Gartell, K. S. Liu, and M. E. Plett, *Horizon Sensor Behavior of the Atmosphere Explorer-C Spacecraft*, Comp. Sc. Corp., CSC/TM-75/6004, May 1975.

68. Wetmore, R., S. Cheuvront, K. Tang, R. Bevacqua, S. Dunker, E. Thompson, C. Miller, and C. Manders, *OSO-I Attitude Support System Specification and Requirements*, Comp. Sc. Corp., 3000-26900-01TR, Aug. 1974.

69. Wetmore, R., J. N. Rowe, G. K. Tandon, V. H. Tate, D. L. Walter, R. S. Williams, and G. D. Repass, *International Sun-Earth Explorer-B (ISEE-B) Attitude System Functional Specifications and Requirements*, Comp. Sc. Corp., CSC/SD-76/6091, Sept. 1976.

CHAPTER 7

MATHEMATICAL MODELS OF ATTITUDE HARDWARE

7.1   Sun Sensor Models
      V-Slit Sensors, Digital Sensors
7.2   Horizon Sensor Models
      *Horizon Sensor Geometry, Nadir Vector Projection Model
      for Body-Mounted Sensor, Central Body Width Model,
      Split Angle Model for Wheel-Horizon Scanner, Biases*
7.3   Sun Sensor/Horizon Sensor Rotation Angle Models
7.4   Modeling Sensor Electronics
      Theory, Example: IR Horizon Sensor
7.5   Magnetometer Models
      *Calibration of Vector Magnetometers, Magnetometer
      Biases*
7.6   Star Sensor Models
      *Star Scanner Measurements, Image Dissector Tube Star
      Measurements, Modeling Sensor Intensity Response*
7.7   Star Identification Techniques
      *Direct Match Technique, Angular Separation Match
      Technique, Phase Match Technique, Discrete Attitude
      Variation Technique*
7.8   Gyroscope Models
      *Gyro Measurements, Model for Measured Spacecraft
      Angular Velocity, Calculation of Estimated Angular Ve-
      locity From the Gyro Measurements, Modeling Gyro Noise
      Effects*
7.9   Reaction Wheel Models
7.10 Modeling Gas-Jet Control Systems

Chapter 6 described the physical properties of representative examples of attitude hardware. However, to use sensor output or to predict control performance we need specific mathematical models of the hardware and its output. Various models encountered at NASA's Goddard Space Flight Center are presented in this chapter.

The hardware and its mathematical model should be thought of as distinct entities. It is possible for the hardware to be refined or modified without requiring a change in the mathematical formulation. Similarly, it is possible to refine or improve the mathematical model even though the hardware has not changed. For example, the horizon sensor models in Section 7.2 implicitly assume that the sensor responds instantaneously to a change in intensity as the sensor scans the sky. This concept was used for spacecraft supported at Goddard Space Flight Center prior to the launch of SMS-2 in February 1975. With the rather good data from SMS-2 it became apparent that the anomalous behavior of the data when the scan crossed only a small segment of the Earth could most easily be explained in terms of the finite response time of the sensor electronics. The mathematical model of the sensor electronics presented in Section 7.3 was subsequently developed and greatly improved our understanding of the data. Thus, the continuing development of both hardware and mathematical models can proceed at least somewhat independently.

7.1 Sun Sensor Models

Lily C. Chen
Gerald M. Lerner

In this section, we will derive general expressions for data reduction and simulation for two classes of Sun sensors: slit sensors for which the measurement is the fraction of the spin period required for the Sun image to traverse a slit pattern, and digital sensors for which the measurement is the linear deflection of the image of a narrow slit upon traversing a refractive medium.

7.1.1 V-Slit Sensors

A V-slit Sun sensor used for spinning spacecraft normally contains two plane field (PF) sensors making an angle θ_0 with respect to each other. Each PF sensor has a planar field of view (FOV). Thus, the projection of the FOV onto the celestial sphere is a segment of a great circle. The sensor provides an event pulse whenever the FOV crosses the Sun. Therefore, the Sun angle, β, can be obtained directly from the measurements of ω, the spin rate, and Δt, the time interval between the two Sun-sighting events from the two PF sensors.

Nominal Case. In the nominal case, one of the PF sensors (PF-1) is parallel to the spacecraft spin axis and the other (PF-2) is inclined at an angle θ_0 to PF-1, as shown in Fig. 7-1. The two sensor FOVs nominally intersect the spin equator at the same point. In Fig. 7-1, A is the spin axis and S is the Sun. The great circle SB is

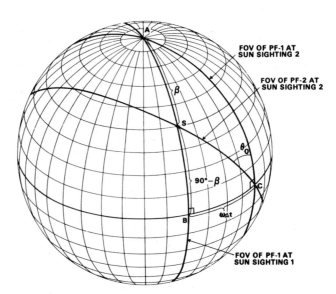

Fig. 7-1. V-Slit Sun Sensor Nominal Geometry

the FOV of PF-1 when it senses the Sun, and the great circles AC and SC are the FOVs of PF-1 and PF-2, respectively, when PF-2 senses the Sun. The arc length, $\omega\Delta t$, between B and C is the rotation angle between the two Sun-sighting events, where ω is the spin rate and Δt is the time interval. By a direct application of Napier's rules (Appendix A) to the right spherical triangle SBC, we obtain

$$\tan\beta = \frac{\tan\theta_0}{\sin\omega\Delta t} \tag{7-1}$$

For data simulation, the inverse expression for Δt is

$$\Delta t = \frac{1}{\omega}\arcsin(\cot\beta\tan\theta_0) \tag{7-2}$$

Misalignment Considerations. Three kinds of sensor misalignment are possible. A *separation misalignment* is an error in the angular separation such that $\theta = \theta_0 + \Delta\theta$. For this type of error, both Eqs. (7-1) and (7-2) hold by simply replacing θ_0 with $\theta_0 + \Delta\theta$.

An *elevation misalignment* occurs when PF-1 is not parallel to the spacecraft spin axis but rather makes an angle, ϵ, with the spin axis, as shown in Fig. 7-2. Note that the great circle SB no longer passes through A but rather makes an angle ϵ with great circle AB. θ_0 is still the angle between the two PF sensors; therefore SC makes an angle $\theta_0 + \epsilon$ with AC. The arc length, ϕ, between B and D is the angular shift of the Sun-sighting events due to the elevation misalignment, ϵ. By applying relation (7-1) to the two spherical triangles SDB and SDC, we obtain

$$\tan\beta_\epsilon = \frac{\tan\epsilon}{\sin\phi} = \frac{\tan(\theta_0 + \epsilon)}{\sin(\phi + \omega\Delta t)} \tag{7-3}$$

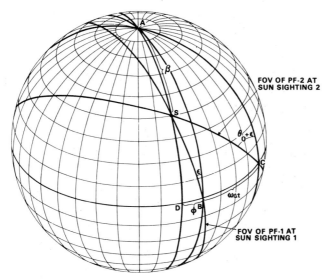

Fig. 7-2.   V-Slit Sun Sensor Geometry With Elevation Misalignment

Eliminating ϕ from Eq. (7-3), we have

$$\tan^2\beta_\epsilon = \left[\frac{\tan(\theta_0+\epsilon)-\tan\epsilon\cos\omega\Delta t}{\sin\omega\Delta t} \right]^2 + \tan^2\epsilon \qquad (7\text{-}4)$$

For small ϵ, we may keep only the first-order terms in ϵ, so that

$$\tan\beta_{\epsilon\to0} = \frac{\tan(\theta_0+\epsilon)-\epsilon\cos\omega\Delta t}{\sin\omega\Delta t} \qquad (7\text{-}5)$$

Finally, an *azimuth misalignment* occurs when the two FOV intersections with the spin equator are separated by an angle δ in the spin plane, as shown in Fig. 7-3. Due to the azimuth misalignment, δ, the actual rotation angle between the two Sun-sighting events is BD rather than BC. Comparing Fig. 7-3 with Figs. 7-1 and 7-2, it is clear that all of the previously derived equations are still valid if $\omega\Delta t$ is replaced with $\omega\Delta t - \delta$. Thus, from Eq. (7-1), with only the azimuth misalignment we have

$$\tan\beta_\delta = \frac{\tan\theta_0}{\sin(\omega\Delta t-\delta)} \qquad (7\text{-}6)$$

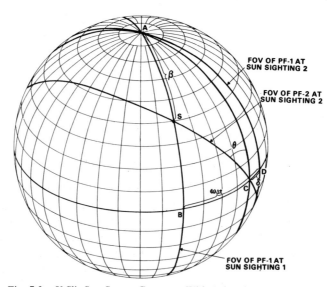

Fig. 7-3. V-Slit Sun Sensor Geometry With Azimuth Misalignment

With all possible misalignments, the general expression for the Sun angle can be obtained by replacing θ_0 with $\theta_0+\Delta\theta$ and $\omega\Delta t$ with $\omega\Delta t - \delta$ in Eq. (7-4). That is,

$$\tan^2\beta_{\Delta\theta,\epsilon,\delta} = \left[\frac{\tan(\theta_0+\Delta\theta+\epsilon)-\tan\epsilon\cos(\omega\Delta t-\delta)}{\sin(\omega\Delta t-\delta)} \right]^2 + \tan^2\epsilon \qquad (7\text{-}7)$$

For simulation, the inverse expression for Δt as a function of β and the misalignment angles is

$$\cos(\omega\Delta t - \delta) = \frac{1}{a}\left[\, b + \sqrt{b^2 - ac}\,\,\right]$$

where $a = \tan^2\beta$

$b = \tan\epsilon \tan(\theta_0 + \Delta\theta + \epsilon)$

$c = \tan^2(\theta_0 + \Delta\theta + \epsilon) + \tan^2\epsilon - \tan^2\beta$

7.1.2 Digital Sensors

As indicated in Section 6.1, one- and two-axis digital sensors are closely related, the former consisting of a command component (A) and a measurement component (B) and the latter consisting of two Gray-coded measurement components (A and B) as shown schematically in Fig. 7-4.

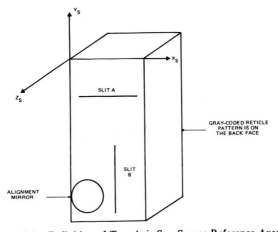

Fig. 7-4.   Definition of Two-Axis Sun Sensor Reference Axes

Alignment of Digital Sensors. The alignment of digital sensors consists of two distinct processes. Internal alignment is performed by the sensor manufacturer to ensure that the sensor slits, the Gray-coded reticle patterns, and the alignment mirror form a self-consistent unit. External alignment of the sensor unit relative to the spacecraft attitude reference axes is performed by the spacecraft manufacturer. In this section, we will model only the external alignment and assume that there are no errors in the internal alignment.

The alignment mirror is used to orient the sensor boresight. The remaining alignment parameter is the rotation of the sensor about the boresight axis. For single-axis sensors, the command component entrance slit is generally parallel to the spacecraft spin axis. In most cases, two-axis sensors are mounted such that either the A or the B measurement slit is parallel to the spacecraft X-Y plane (see Fig. 7-5).

We define the sensor Z axis, Z_S, as the outward normal of the plane containing the alignment mirror and the entrance slits of both components. The

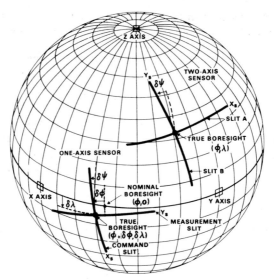

Fig. 7-5.   Orientation of Digital Sun Sensors

Z_S-axis is the sensor boresight and is the optical null of both the A and B components. The X_S and Y_S sensor axes are perpendicular to Z_S as defined in Fig. 7-4 and Table 7-1. Note that some of the internal alignment parameters could be modeled by treating two-axis sensors as two independently aligned one-axis sensors, although we will not use that model here.

The orientation of a one-axis sensor with boresight located at $(\phi' = \phi + \delta\phi, \delta\lambda)$, misaligned slightly from the nominal location of $(\phi, 0)$, and rotated through the angle $\delta\psi$ about the boresight, is shown in Fig. 7-5. The transformation which rotates a vector from sensor to spacecraft coordinates may be expressed as the transpose of a 3-2-3 Euler rotation with angles $\theta_1 = \phi'$, $\theta_2 = 90° - \delta\lambda$, and $\theta_3 = \delta\psi$,* where $\delta\lambda$, $\delta\phi$, and $\delta\psi$ are small misalignment angles and the 0-deg nominal value of

Table 7-1.   Definition of Reference Axes for Digital Sensors

AXIS	ONE-AXIS MODELS	TWO-AXIS MODELS
Z_S	NORMAL TO THE PLANE CONTAINING COMMAND AND MEASUREMENT SLITS. IT IS ALSO THE OPTICAL NULL OF BOTH COMPONENTS	NORMAL TO THE PLANE CONTAINING THE A AND B SLITS. IT IS ALSO THE OPTICAL NULL OF BOTH COMPONENTS
X_S	PARALLEL TO COMMAND SLIT. POSITIVE SENSE DEFINED BY THE OUTPUT OF THE DETECTOR BENEATH THE GRAY-CODED RETICLE. SUN ANGLE IS MEASURED ALONG THE X_S AXIS	PARALLEL TO THE MEASUREMENT SLIT OF COMPONENT A. POSITIVE SENSE DEFINED BY OUTPUT OF COMPONENT B
Y_S	PARALLEL TO THE MEASUREMENT SLIT	PARALLEL TO THE MEASUREMENT SLIT OF COMPONENT B. POSITIVE SENSE DEFINED TO COMPLETE RIGHT-HANDED ORTHOGONAL SYSTEM AND CONSISTENT WITH OUTPUT OF COMPONENT A. NOTE THAT THE LATTER REQUIREMENT DEPENDS ON THE INTERNAL SENSOR ALIGNMENT AND MAY REQUIRE INVERTING THE SIGN OF THE A COMPONENT OUTPUT

*$\delta\lambda$ is positive if the boresight is above the spacecraft X-Y plane; $\delta\psi$ is a positive rotation of the sensor about the boresight.

θ_3 is chosen so that the sensor output is positive toward the $+Z$-axis. Using Table E-1, we obtain the small angle approximation for the rotation matrix,

$$A_{SS1} = \begin{pmatrix} \delta\lambda\cos\phi' - \delta\psi\sin\phi' & -\sin\phi' & \cos\phi' \\ \delta\lambda\sin\phi' + \delta\psi\cos\phi' & \cos\phi' & \sin\phi' \\ -1 & \delta\psi & \delta\lambda \end{pmatrix} \qquad (7\text{-}8)$$

The orientation of a two-axis sensor with boresight located at (ϕ,λ) may be expressed similarly as the transpose of a 3-2-3 Euler rotation with angles $\theta_1 = \phi$, $\theta_2 = 90° - \lambda$, and $\theta_3 = 90° + \delta\psi$, where $\delta\psi$ is a small misalignment angle about the boresight. Using Table E-1 we obtain the small angle approximation for the rotation matrix,

$$A_{SS2} = \begin{pmatrix} -\sin\phi - \delta\psi\sin\lambda\cos\phi & \delta\psi\sin\phi - \sin\lambda\cos\phi & \cos\phi\cos\lambda \\ \cos\phi - \delta\psi\sin\lambda\sin\phi & -\delta\psi\cos\phi - \sin\lambda\sin\phi & \sin\phi\cos\lambda \\ \delta\psi\cos\lambda & \cos\lambda & \sin\lambda \end{pmatrix} \quad (7\text{-}9)$$

Note that in the example shown, slit A is nominally parallel to the spacecraft X-Y plane and the spacecraft Z-axis is in the sensor Y-Z plane.

One-Axis Digital Sensor. The geometry of a ray incident on a block of material with index of refraction n is illustrated in Fig. 7-6. Snell's law relating the angle of incidence, θ, and the angle of refraction, θ', is

$$n\sin\theta' = \sin\theta \qquad (7\text{-}10)$$

where the index of refraction of space is unity. The detectors beneath the reticle pattern of the sensor yield a signed, digitized output, N, proportional to the deflection, x, such that

$$x = kN \qquad (7\text{-}11)$$

where k is the reticle step size. From Eq. (7-10) and simple trigonometry, we have

$$\sin\theta = n\sin\theta' = \frac{nkN}{\left[(kN)^2 + h^2\right]^{1/2}} \qquad (7\text{-}12)$$

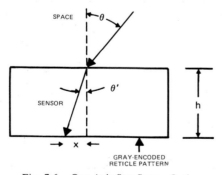

Fig. 7-6. One-Axis Sun Sensor Optics

Expanding in a trigonometric series in $\epsilon = x/h = kN/h \ll 1$ and retaining terms through ϵ^5, we obtain

$$\theta \approx n\epsilon - n(3 - n^2)\epsilon^3/6 + n(15 - 10n^2 + 3n^4)\epsilon^5/40 \tag{7-13}$$

where θ is in radians. A design goal of digital sensors is a linear relation between the sensor output, N, and the measured angle. For a material with $n \approx \sqrt{3}$, the term dependent on ϵ^3 becomes negligible, yielding the approximate result

$$\theta \approx nkN/h \tag{7-14}$$

A further useful simplification results if the reticle geometry is chosen such that $180nk/(\pi h) = 1$. In this case, $\theta \approx N$ where θ is now expressed in degrees.

When the Sun angle measurement, θ, is made, the Sun vector in sensor reference coordinates is $(-\sin\theta, 0, \cos\theta)^T$. In spacecraft coordinates, the Sun vector is

$$\hat{V}_B = A_{SS1} \begin{bmatrix} -\sin\theta \\ 0 \\ \cos\theta \end{bmatrix} \tag{7-15}$$

from which the azimuth and elevation of the Sun in spacecraft coordinates may be computed.

Two-Axis Digital Sensors.   The derivation of the data reduction equations for two-axis Adcole sensors is analogous to that for the single-axis sensors [Adcole, 1975]. The geometry is shown in Fig. 7-7. Note that OZ_S is the optical null (or boresight) of both the A and B sensors. The refracted ray (OP') is deflected by the slab with index of refraction n and strikes the Gray-coded rear reticle at P' with coordinates (b, a). Application of Snell's law yields

$$\sin\theta = n\sin\theta'$$
$$\phi = \phi' \tag{7-16}$$

By analogy with Eq. (7-11), the output of the A and B components denoted by NA and NB, respectively, is converted to a displacement by

$$a = k_m(NA - 2^{m-1} + 0.5)$$
$$b = k_m(NB - 2^{m-1} + 0.5) \tag{7-17}$$

NA and NB are unsigned decimal equivalents of the m-bit Gray-coded sensor output and k_m is a sensor constant. (See Table 7-2 for representative values of the sensor constants.) The form of Eq. (7-17), particularly the addition of 0.5 to NA and NB, is a consequence of the Adcole alignment and calibration procedure.

Right triangles $OO'P'$ and $O'Q'P'$ yield the relations

$$\phi = \phi' = \text{ATAN2}\,(a, b) \tag{7-18}$$

$$\tan\theta' = \frac{d}{h} = \frac{(a^2 + b^2)^{1/2}}{h} \tag{7-19}$$

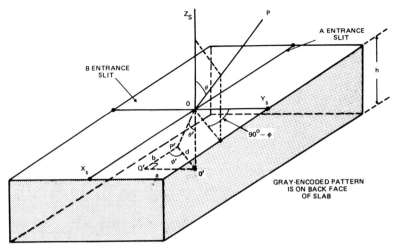

Fig. 7-7. Two-Axis Sun Sensor Optics

where the FORTRAN function ATAN2 is used in Eq. (7-18) to resolve the quadrant ambiguity. If we substitute Eq. (7-16) into (7-19), and rearrange terms, we get

$$\theta = \arctan\left\{ \frac{n(a^2+b^2)^{1/2}}{\left[h^2 - (n^2-1)(a^2+b^2) \right]^{1/2}} \right\} < 90° \qquad (7\text{-}20)$$

The angles ϕ and θ are the azimuth and coelevation, respectively, of the Sun vector in sensor coordinates which have the positive pole along the sensor boresight and the reference meridian along the $+X_s$ axis. The Sun vector may be transformed into spacecraft coordinates by using Eq. (7-9). Two-axis digital Sun sensor data are commonly reparameterized in terms of the angles between the projections of the sunline on the Y_s-Z_s and X_s-Z_s planes and the Z_s-axis, as illustrated in Figs. 7-8 and 7-9. The angles α and β are rotations about the $-X_s$ and Y_s axes, respectively, given by

$$\tan\beta = \tan\theta \cos\phi = nb/R$$
$$\tan\alpha = \tan\theta \sin\phi = na/R \qquad (7\text{-}21)$$

where

$$R^2 = h^2 - (n^2-1)(a^2+b^2) \qquad (7\text{-}22)$$

The specified field of view of the Adcole two-axis sensor is "square" as illustrated in Fig. 7-9 for a 128- by 128-deg sensor. The effective FOV is often considered circular with radius 64 deg because this is the maximum angle of incidence which guarantees valid sensor data (i.e., sufficient intensity) independent of ϕ. For Sun angles near the "corners" of the FOV, $\phi = \pm 45$ or ± 135, valid sensor

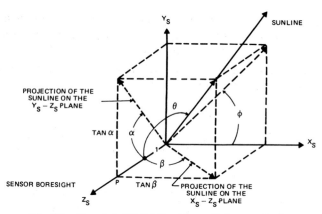

Fig. 7-8.   Two-Axis Digital Sun Sensor Reference Angles

data are obtained for θ up to 71 deg.* In Section 2.3, we proved that five two-angle sensors may be dispersed to provide 4π sr coverage with a maximum θ angle of 63.5 deg. We now see that this result is valid for two-axis digital sensors independent of sensor alignment about the boresight. For an 8-bit sensor with $n = 1.4553$, the coordinates of various *grid points* within the FOV, expressed as (NA, NB), are shown in Fig. 7-9. For $n \neq 1$, lines of constant α or β are *not* lines of constant NA or NB and, in particular, the grid point corresponding to $[\alpha, \beta] = [64°, 0°]$ is (255, 127.5) and $[64°, 64°]$ is (226, 226). The boresight is at the center of the four grid points (127, 127), (127, 128), (128, 127) and (128, 128). Because of the refractive sensor medium, a ray normal to the boresight at $\theta = 90$ deg and $\phi = 45$ deg will

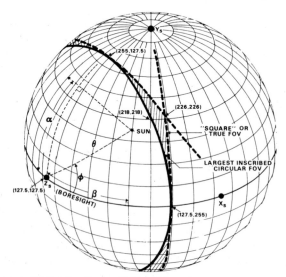

Fig. 7-9.   Two-Axis Digital Sun Sensor Field of View. See text for explanation of coordinates.

---

* From Eq. (7-21), we have $\phi = 45$ deg and $\beta = 64$ deg; hence, $\tan\theta = \tan\beta/\cos 45° = \tan 64°/\cos 45°$ or $\theta = 70.97$ deg.

reach the reticle pattern (with zero intensity) and fall at (236, 236). Sensor data with grid points corresponding to $\theta > 90$ deg are necessarily anomalous and an application of Eqs. (7-17) and (7-22) to such data would yield $R^2 < 0$.

To simulate data, we must solve for NA, NB, and the selected sensor in terms of the Sun vector in sensor coordinates,

$$\hat{v}_{SS} = A_{SS2}^T \hat{v}_B = \begin{bmatrix} X_S \\ Y_S \\ Z_S \end{bmatrix} = \frac{1}{(\tan^2\alpha + \tan^2\beta + 1)^{1/2}} \begin{bmatrix} \tan\beta \\ \tan\alpha \\ 1 \end{bmatrix} \qquad (7\text{-}23)$$

where $\hat{v}_B$ and $\hat{v}_{SS}$ are the Sun vector in spacecraft and sensor coordinates, respectively. Using Eqs. (7-21), (7-22) and Fig. 7-8, we obtain the result,

$$a = Y_S \gamma^{1/2}$$

$$b = X_S \gamma^{1/2} \qquad (7\text{-}24)$$

where

$$\gamma = \left[h^2 / (n^2 - X_S^2 - Y_S^2) \right] \geqslant 0 \qquad (7\text{-}25)$$

Finally, the sensor output is

$$NA = \text{INT}(a/k_m + 2^{m-1})$$

$$NB = \text{INT}(b/k_m + 2^{m-1}) \qquad (7\text{-}26)$$

where $\text{INT}(x)$ is the integral part of x and NA and NB are Gray coded by the reticle pattern. The Sun is visible to a specific sensor (although the intensity may be below the ATA threshold) if both γ and Z_S are positive. The selected sensor for multisensor configurations is determined by the ATA output, i.e., the sensor with the largest (positive) Z_S.

For state estimation, the digital sensor angular outputs may be computed using Eqs. (7-26) but the sensor identification for multisensor configurations cannot be reliably predicted. The actual sensor selected is a function of the precise threshold settings whenever the Sun is near the Earth's horizon or is between the fields of view of adjacent sensors. Sensor identification should be used merely to validate sensor data for state estimation.

Fine Sun Sensors. The operation of the fine Sun sensor described in Section 6.1 is illustrated in Fig. 7-10 (compare with Fig. 6-9). In the figure, the horizontal axis has been expanded to illustrate the effect of the 32-arc-minute angular diameter of the Sun (from near the Earth), which requires the use of an analog sensor rather than a finely gridded digital sensor. Incident sunlight falling on the entrance slits with spacing s produces the photocell current shown schematically in Fig. 7-10(d). The nearly sinusoidal output signal is a consequence of the Sun's finite size. If four reticle patterns are offset by $s/4$ the photocell current, I, beneath each pattern may be written as a function of $x = 2\pi w/s$ where $w = t\tan\alpha$, t is the distance between the two reticle patterns, and α is the Sun angle. I is given by

$$I_1 = f(2\pi w/s)$$
$$I_2 = f(2\pi w/s + \pi/2)$$
$$I_3 = f(2\pi w/s + \pi)$$
$$I_4 = f(2\pi w/s + 3\pi/2) \tag{7-27}$$

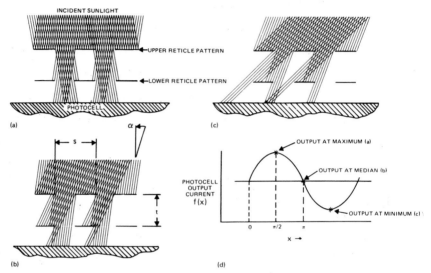

Fig. 7-10.   Schematic Representation of Fine Sun Sensor Photocell Output Current. Rays coming from different directions represent light from opposite sides of the Sun. (The angular spread of these rays is greatly exaggerated.)

where angles are measured in radians. The fine Sun sensor electronics forms the quantity $\arctan y = \arctan[(I_1 - I_3)/(I_2 - I_4)]$, which is related to the Sun angle by

$$\arctan y = \frac{2\pi t}{s}\tan\alpha + \text{small error term} \tag{7-28}$$

Equation (7-28) may be derived as follows. The function $f(x)$ is periodic with period s and has a maximum at $x = \pi/2$. Because $f(x)$ is symmetric about $x = \pi/2$, it may be expanded in a Fourier cosine series as [Markley, 1977]:

$$f(x) = a_0 + a_1\cos(x - \pi/2) + a_2\cos(2x - \pi) + a_3\cos(3x - 3\pi/2) + \cdots$$
$$= a_0 + a_1\sin x - a_2\cos 2x - a_3\sin 3x + \cdots \tag{7-29}$$

The fine Sun sensor electronics forms the quantities $I_1 - I_3, I_2 - I_4, y = (I_1 - I_3)/(I_2 - I_4)$, and $\arctan y$, which are approximated as follows:

$$I_1 - I_3 = a_1\big[\sin x - \sin(x + \pi)\big] - a_2\big[\cos 2x - \cos(2x + 2\pi)\big]$$
$$\qquad - a_3\big[\sin 3x - \sin(3x + 3\pi)\big] + \cdots$$
$$= 2a_1\sin x\big[1 - a_3(4\cos^2 x - 1)/a_1\big] + \cdots \tag{7-30a}$$

$$I_2 - I_4 = 2a_1 \cos x \left[1 + a_3(4\cos^2 x - 3)/a_1 \right] + \cdots \tag{7-30b}$$

$$y \approx \tan x \left[1 - 4a_3(2\cos^2 x - 1)/a_1 \right] = \tan x \left[1 - 4a_3 \cos 2x/a_1 \right] \tag{7-30c}$$

Equation (7-30c) can be rewritten in the more convenient form

$$\arctan y = x - \arctan \epsilon \tag{7-30d}$$

where ϵ is a small error term, by taking the tangent of both sides of the above equation, and using the trigonometric identity

$$\tan(a + b) = (\tan a \pm \tan b)/(1 \mp \tan a \tan b) \tag{7-31a}$$

to obtain

$$(\tan x - \epsilon)/(1 + \epsilon \tan x) = \tan x - \epsilon(1 + \tan^2 x) + \mathcal{O}(\epsilon^2) \tag{7-31b}$$

Comparing this result with Eq. (7-30c) to obtain ϵ, we have

$$\arctan y \approx x - \arctan(a_3 \sin 4x/a_1) \approx x - a_3 \sin 4x/a_1 \tag{7-32}$$

For small $w = t \tan \alpha$, we obtain

$$\arctan y = \frac{2\pi t}{s} \tan \alpha - \frac{a_3}{a_1} \sin\left(\frac{8\pi t \tan \alpha}{s} \right) \tag{7-33}$$

which is the desired result.

Thus, if the photocell output is adequately represented by the first three terms of a Fourier cosine series, the output of the fine Sun sensor electronics, $\arctan y$, is given by a term proportional to the tangent of the incident angle, α, plus a sinusoidal error term.

In practice, the inverse of Eq. (7-33) is required for sensor data processing. The digital sensor output, NA, is related to the analog output by

$$\arctan y = k_1(NA) + k_2 \tag{7-34}$$

where k_1 and k_2 are sensor constants. Equation (7-33) can be rewritten as

$$\tan \alpha = \frac{s}{2\pi t}(k_1 NA + k_2) + \frac{a_3}{a_1} \frac{s}{2\pi t} \sin\left(\frac{8\pi t}{s} \tan \alpha \right) \tag{7-35}$$

Defining the sensor constants

$$A_1 = sk_2/2\pi t \tag{7-36a}$$

$$A_2 = sk_1/2\pi t \tag{7-36b}$$

$$A_3 = sa_3/2\pi t a_1 \tag{7-36c}$$

then successive approximations, $\alpha^{(n)}$, to α are given by

$$\tan \alpha^{(0)} = A_1 + A_2 NA \tag{7-37a}$$

$$\tan \alpha^{(n+1)} = \tan \alpha^{(n)} + A_3 \sin\left(\frac{8\pi t}{s} \tan \alpha^{(n)} \right) \tag{7-37b}$$

or, to the same order as Eq. (7-33),

$$\tan \alpha \approx A_1 + A_2 NA + A_3 \sin(A_4 NA + A_5) \tag{7-37c}$$

For IUE, the sensor output is encoded into 14-bit (0–16,383) words and the $\pm$ 32-deg field of view is measured with a 14 arc-second least significant bit. The transfer function is as follows [Adcole, 1977]:

$$\alpha = \alpha_0 + \arctan\left[A_1 + A_2 NA + A_3 \sin(A_4 NA + A_5) + A_6 \sin(A_7 NA + A_8)\right]$$

$$\beta = \beta_0 + \arctan\left[B_1 + B_2 NB + B_3 \sin(B_4 NB + B_5) + B_6 \sin(B_7 NB + B_8)\right] \quad (7\text{-}38)$$

where NA and NB denote the digitized sensor output; the parameters A_i, B_i, α_0, and β_0 are obtained by ground calibration; and α and β are defined in Fig. 7-8.

The parameters defining the slab thickness, index of refraction, alignment, and resolution vary depending on the sensor model and specific calibration. Table 7-2 lists values which are representative and convenient for simulation and error analysis. (See also Table 6-1.)

Table 7-2.   Representative Constants for Digital Sun Sensors Manufactured by the Adcole Corporation

PROPERTY	SYMBOL	VALUE
TRANSFER FUNCTIONS 6 AND 8*		
INDEX OF REFRACTION	n	1.4553
SLAB THICKNESS	h	0.56896 CM
RESOLUTION (7-BIT MODEL)	k	0.006985 CM/UNIT
RESOLUTION (8-BIT MODEL)	k	0.0034925 CM/UNIT
MODEL 18960* (IUE) (14-BIT OUTPUT)		
CALIBRATION CONSTANTS	A_1, B_1	−0.624869
	A_2, B_2	7.6278×10^{-5}
	A_3, A_6, B_3, B_6	$<10^{-4}$
	A_4, B_4	0.703125[†] DEG/COUNT
	A_7, B_7	1.40625[††] DEG/COUNT
	A_5, A_8, B_5, B_8	ARBITRARY, 0 TO 360 DEG
ALIGNMENT ANGLES	α_0, β_0	<0.1 DEG

*SEE TABLE 6–1
[†]CORRESPONDS TO 32 OSCILLATION PERIODS OVER THE ±32–DEG FIELD OF VIEW.
[††]CORRESPONDS TO 64 OSCILLATION PERIODS OVER THE ±32–DEG FIELD OF VIEW.

7.2 Horizon Sensor Models

Steven G. Hotovy

In this section, we provide several observation models for any sensor which scans the celestial sphere in a small circle and is sensitive to the presence of electromagnetic radiation from a body in its field of view. Such sensors, described in Section 6.2, may be divided into three categories:

1. *Body-Mounted Sensor (BHS)*—a visual or infrared telescope fixed on the body of a spinning spacecraft

2. *Panoramic Scanner (PS)*—a visual scanner operating in the scan mode on a despun spacecraft

3. *Wheel-Mounted Sensor (WHS)*—an infrared scanner consisting of a bolometer attached to the body of the spacecraft into which the field of view of a lens or mirror mounted on a rapidly spinning wheel is reflected

7.2.1   Horizon Sensor Geometry

Figure 7-11 depicts the movement of the optical axis of a BHS sensor as the spacecraft spins. $\hat{\mathbf{A}}$ is the spin axis attitude, $\hat{\mathbf{X}}$ is a reference point in the spacecraft body, Φ_P is the azimuth, and γ_N is the nominal coelevation of the sensor optical axis in body coordinates. As the optical axis, $\hat{\mathbf{P}}$, sweeps through the sky, the sensor detects an *in-crossing* (entrance of the central body into the sensor field of view) at time t_I at the point $\hat{\mathbf{H}}_I \equiv \hat{\mathbf{P}}(t_I)$. At some later time, t_O, it will detect an *out-crossing* (departure of the central body from the sensor field of view) at the point $\hat{\mathbf{H}}_O \equiv \hat{\mathbf{P}}(t_O)$.

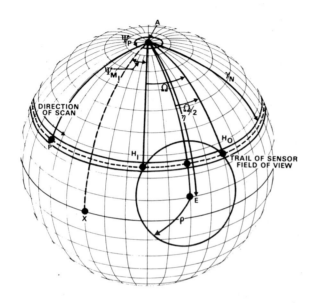

Fig. 7-11.   Horizon Sensor Geometry

The same geometry applies to a wheel-mounted scanner, with some change in interpretation. In this case, $\hat{\mathbf{A}}$ is the spin axis of the wheel and $\hat{\mathbf{X}}$ is some reference vector in the spacecraft body which lies in the plane of the wheel. As the mirror rotates, the sensor detects an in-crossing and an out-crossing as before. However, for a wheel-mounted sensor, a magnetic pickup is mounted on the body of the spacecraft at some index point and a magnet is mounted on the wheel. These are used to measure the time of one complete revolution of the wheel. As a result, a wheel-mounted sensor can measure the *central body width*, Ω, equal to the rotation angle about $\hat{\mathbf{A}}$ from $\hat{\mathbf{H}}_I$ to $\hat{\mathbf{H}}_O$. In addition, it can measure the *split-to-index time*, t_{SI}, the time between the detection of the midscan of the central object, $0.5(t_O + t_I)$, and the detection of the magnet by the magnetic pickup, t_{index}; that is

$$t_{SI} \equiv t_{index} - 0.5(t_O + t_I)$$

7.2.2 Nadir Vector Projection Model for Body-Mounted Sensor

The nadir vector projection model for a BHS is

$$\hat{\mathbf{E}} \cdot \hat{\mathbf{P}} - \cos \rho = 0 \qquad (7\text{-}39)$$

where $\hat{\mathbf{P}}$ is the unit vector along the line of sight of the sensor, $\hat{\mathbf{E}}$ is the nadir vector to the central body, and ρ is the apparent angular radius of the central body as seen from the satellite. The value of $\hat{\mathbf{E}} \cdot \hat{\mathbf{P}}$ will oscillate sinusoidally approximately once per spacecraft rotation as $\hat{\mathbf{P}}$ sweeps through the sky. The value of $\hat{\mathbf{E}} \cdot \hat{\mathbf{P}} - \cos \rho$ will be zero when the angle between $\hat{\mathbf{E}}$ and $\hat{\mathbf{P}}$ is equal to the apparent angular radius of the central body, i.e., at horizon crossing times when $\hat{\mathbf{P}} = \hat{\mathbf{H}}_I$ or $\hat{\mathbf{H}}_O$.

The values of $\hat{\mathbf{E}}$ and ρ may be determined from an ephemeris, which provides the spacecraft-to-central body vector, $\mathbf{E}$. If we assume that the central body is a sphere, then ρ satisfies

$$\cos \rho = \left(E^2 - R_E^2 \right)^{1/2} / E \qquad (7\text{-}40)$$

where R_E is the radius of the central body. If the central body is the Earth, oblateness may be considered (procedures for modeling an oblate central body are discussed in Section 4.3), in which case R_E in Eq. (7-40) is latitude dependent.

To evaluate Eq. (7-39), it is necessary to express $\hat{\mathbf{P}}$ in inertial coordinates. This expression is given in terms of the sensor location in the spacecraft frame and the spacecraft orientation in inertial space. For this model, we assume that the spacecraft has an inertially fixed spin axis and is spinning at a constant rate (i.e., nutation, coning, precession, and spin rate variations are assumed to be negligible). The pertinent parameters are the initial phase of the spacecraft Φ_0 at time t_0; the spin rate, ω; and the spin axis vector, $\hat{\mathbf{A}}$. The phase at a time t is, then,

$$\Phi_t = \Phi_0 + \omega(t - t_0) \qquad (7\text{-}41)$$

The position of the center line of sight in spacecraft coordinates is

$$\hat{\mathbf{P}}_{SC} = \begin{bmatrix} \sin \gamma \cos \Phi_P \\ \sin \gamma \sin \Phi_P \\ \cos \gamma \end{bmatrix} \qquad (7\text{-}42)$$

where $\gamma = \gamma_N + \Delta\gamma$ is the true mounting angle.

The attitude matrix, $B(t)$, for a spinning spacecraft at time t is given by (see Section 12.2 and Appendix E)

$$B(t) = \frac{1}{\left(A_1^2 + A_2^2 \right)^{1/2}}$$

$$\times \begin{bmatrix} A_1 A_3 \cos \Phi_t - A_2 \sin \Phi_t & A_2 A_3 \cos \Phi_t + A_1 \sin \Phi_t & -\left(A_1^2 + A_2^2 \right) \cos \Phi_t \\ -A_1 A_3 \sin \Phi_t - A_2 \cos \Phi_t & -A_2 A_3 \sin \Phi_t + A_1 \cos \Phi_t & \left(A_1^2 + A_2^2 \right) \sin \Phi_t \\ A_1 \left(A_1^2 + A_2^2 \right)^{1/2} & A_2 \left(A_1^2 + A_2^2 \right)^{1/2} & A_3 \left(A_1^2 + A_2^2 \right)^{1/2} \end{bmatrix} \qquad (7\text{-}43)$$

where $\hat{A} = (A_1, A_2, A_3)^T$, the spin axis unit vector, is now expressed in inertial coordinates and Φ_t is as in Eq. (7-41). Thus, the location of the line of sight of the sensor at time t in inertial coordinates is

$$\hat{P}(t) = B^T(t)\hat{P}_{SC} \tag{7-44}$$

This is then substituted into Eq. (7-39). This model is not valid in the case of terminator crossings for a visible light sensor; thus, terminator rejection is required.

7.2.3 Central Body Width Model

In the case of valid in- and out-crossings from a BHS or a PS, we may develop a model incorporating both crossing times t_I and t_O. This model is

$$t_O - t_I = (\Omega + 360° n)/\omega \tag{7-45}$$

where ω is the body rate (again assumed to be constant), n is the number of complete spacecraft rotations between t_I and t_O, and Ω is the central body width (in degrees), which can be calculated as follows.

Applying the law of cosines to spherical triangle AEH in Fig. 7-11, we obtain

$$\cos\rho = \cos\gamma\cos\eta + \sin\gamma\sin\eta\cos\left(\frac{\Omega}{2}\right) \tag{7-46}$$

which becomes, upon solving for Ω,

$$\Omega = 2\arccos\left(\frac{\cos\rho - \cos\gamma\cos\eta}{\sin\gamma\sin\eta}\right) \tag{7-47}$$

Here $\gamma = \gamma_N + \Delta\gamma$, where $\Delta\gamma$ is a fixed mounting angle bias. A fixed bias can similarly be included in ρ.

When other effects (such as oblateness or height of the CO_2 layer) are considered, the expression for Ω becomes

$$\Omega = \arccos\left(\frac{\cos\rho_I - \cos\gamma\cos\eta}{\sin\gamma\sin\eta}\right) + \arccos\left(\frac{\cos\rho_O - \cos\gamma\cos\eta}{\sin\gamma\sin\eta}\right) \tag{7-48}$$

where ρ_I and ρ_O are the effective scan-in and scan-out radii of the central body, including all correction factors to the nominal radius.

For a WHS, the central body width can often be obtained directly from telemetry data. The scanners aboard SMS-1 and -2 and AE-3, -4, and -5, for example, provided the Earth-in and -out times, t_I and t_O, and the wheel speed, ω. From Eq. (7-45), we have

$$\Omega = (t_O - t_I)\omega \tag{7-49}$$

On other spacecraft (SAS-3, for example), the telemetry data consisted of a voltage which was converted to an Earth width, Ω, from a calibration curve.

The assumptions and limitations for the nadir vector projection model hold true for the central body width model as well. In addition, we must assume that the orbital motion of the spacecraft is negligible between in- and out-crossings. This effect is more troublesome for BHSs and PSs than for WHSs because wheel rates are generally much faster than spacecraft body rates.

Knowledge of Ω permits the calculation of the nadir angle, η. Equation (7-46) leads to a quadratic equation in $\cos\eta$ with solutions

$$\cos\eta = \frac{\cos\gamma\cos\rho \pm k\left(\cos^2\gamma + k^2 - \cos^2\rho\right)^{1/2}}{\cos^2\gamma + k^2}$$

$$k = \sin\gamma\cos(\Omega/2) \tag{7-50}$$

Because both solutions are geometrically meaningful, more information, such as an a priori attitude estimate, is needed to resolve the ambiguity. Once it has been resolved, however, we know that the spin axis of the spacecraft (or of the wheel in the case of a WHS) lies on the cone in inertial space centered on $\hat{\mathbf{E}}$ and of radius η.

7.2.4 Split Angle Model for Wheel-Horizon Scanner

As mentioned previously, a wheel-mounted scanner provides two readings that are not available from a body-mounted sensor: the wheel rate, ω_W, and the split-to-index time, t_{SI}. These can be combined to determine the azimuth, α, of the magnetic pick-off relative to the midscan of the central body. As shown in Fig. 7-12, we have

$$\alpha = \omega_W t_{SI} + \Delta\alpha \tag{7-51}$$

where $\Delta\alpha$ is the azimuthal misalignment of the pickoff from its nominal value. This can be combined with the spin axis attitude to determine the three-axis attitude of the spacecraft, since the spin angle model specifies the azimuthal orientation of the spacecraft body about the wheel spin axis.

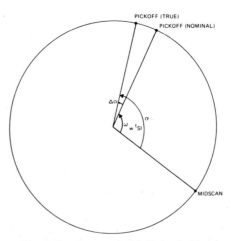

Fig. 7-12.   Geometry of Split Angle Model

7.2.5 Biases

The model developed above may not accurately explain sensor behavior because of the presence of additional sensor biases.* For example, there may be an azimuthal mounting angle bias, $\Delta\Phi$, due to either a mounting misalignment or incorrectly calibrated sensor electronics. (See Section 7.4.) This bias can be added to the nadir vector projection model by replacing Φ_p with $\Phi_p + \Delta\Phi$ in Eqs. (7-42) and (7-44). If this bias is due to sensor electronics, it may be appropriate to use separate in- and out-crossing biases, $\Delta\Phi_I$ and $\Delta\Phi_O$, since the electronic response may be different in these two cases. This may be incorporated into the central body-width model by changing Eq. (7-45) to

$$t_O - t_I = (\Omega - \Delta\Phi_O + \Delta\Phi_I + 360°\,n)/\omega \qquad (7\text{-}52)$$

Another possible bias is a systematic variation, $\Delta\rho$, in the angular radius of the central body. This may be caused by a genuine uncertainty in the size of the effective triggering radius of the central body itself, or, more likely, may reflect the sensor triggering performance as shown in Fig. 7-13. Under nominal circumstances, we assume that the FOV of the sensor is circular and that the sensor will register an

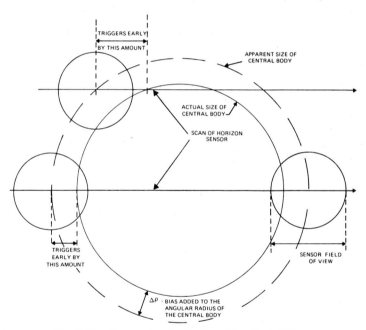

Fig. 7-13.   Bias on Angular Radius of the Central Body

in- or out-crossing when the central body occupies 50 percent of the FOV. However, if the sensor triggers at some value other than 50 percent, the effective size of the central body changes. In Fig. 7-13, the horizon sensor triggers when the central body occupies only about 10 percent of the FOV. This means that the apparent size of the central body is greater than the actual size. Note that $\Delta\rho$ is *independent* of the path of the sensor across the central body, although the

---

*Each of the biases described here has been found to have a significant effect on real data for some missions.

difference in triggering times will vary with the path. This effect can be added to the nadir vector projection and central body width models by replacing ρ with $\rho + \Delta\rho$ in Eqs. (7-39), (7-46) through (7-48), and (7-50).

Finally, for a WHS, the optical axis of the bolometer (see Section 6.2) mounted on the body may be misaligned relative to the spin axis of the wheel. This results in a sinusoidal oscillation of the central body width data with a frequency equal to the body spin rate relative to the central body. This phenomenon was first observed on the AE-3 spacecraft [Wertz, et al., 1975]. The phase and amplitude of the oscillation will depend on the phase and amplitude of the bolometer misalignment, as shown in Fig. 7-14. Here, S is the spin axis of the wheel; B_1 and B_2 are the positions of the bolometer optical axis at times t_1 and t_2; H_{I_j} and H_{O_j} are the in- and out-crossings of the bolometer at time t_j; and M_{I_j} and M_{O_j} are the positions of the mirror normal at these times. Figure 7-14 shows that the bolometer 2 Earth width, which is the rotation angle about the spin axis from M_{I_2} to M_{O_2}, is greater

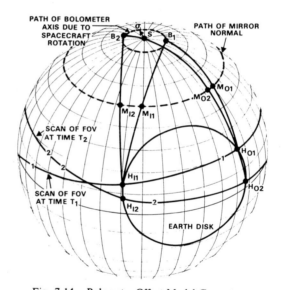

Fig. 7-14. Bolometer Offset Model Geometry

than that from bolometer 1. The nadir angle/Earth width model, Eq. (7-50), can be changed to reflect a bolometer offset, although the derivation of this new model is not straightforward [Wertz, et al., 1975; Liu and Wertz, 1974]. The model is

$$\cos\rho = \cos\sigma(\cos\gamma\cos\eta + \sin\gamma\sin\eta\cos L_I)$$
$$+ \sin\sigma\left\{\left[\sin\gamma\cos\eta + (1-\cos\gamma)\sin\eta\cos L_I\right]\cos(B-L_I) - \sin\eta\cos B\right\} \quad (7\text{-}53a)$$

$$\cos\rho = \cos\sigma(\cos\gamma\cos\eta + \sin\gamma\sin\eta\cos L_O)$$
$$+ \sin\sigma\left\{\left[\sin\gamma\cos\eta + (1-\cos\gamma)\sin\eta\cos L_O\right]\cos(B+L_O) - \sin\eta\cos B\right\} \quad (7\text{-}53b)$$

where σ is the offset angle between the bolometer and the spin axes, B is the rotation angle about the spin axis from the bolometer axis to the nadir vector, L_I is the rotation angle about the spin axis from the Earth-in to the nadir vector, and L_O is the rotation angle about the spin axis from the nadir vector to the Earth-out. Thus,

$$L_I + L_O = \Omega \tag{7-54}$$

and

$$B + (\Omega/2 - L_I) + t_{SI}(\omega_W - \omega_S) + \Phi_B = 360° \text{ or } 720° \tag{7-55}$$

where ω_W is the wheel rate, ω_S is the body spin rate, and Φ_B is the phase of the bolometer offset. For a fixed bolometer position, the four unknowns in these equations are η, L_I, L_O, and B. Equations (7-54) and (7-55) determine B and L_O in terms of L_I, and these are substituted into Eq. (7-53), whereupon η and L_I are solved for, usually in an iterative fashion. Alternatively, for a fixed spacecraft attitude, the observed WHS data may be used to compute the bolometer offset parameters σ and Φ_B (Liu and Wertz [1974]).

7.3 Sun Sensor/Horizon Sensor Rotation Angle Models

Menachem Levitas

In this section, we describe observation models for the following Sun sensor/horizon sensor rotation angle measurements: Sun-to-Earth-in, Sun-to-Earth-out, and Sun-to-Earth-midscan. Related azimuth biases are discussed for body-mounted horizon sensors and panoramic scanners (Section 6.2). For additional modeling procedures, see Joseph, *et al.*, [1975]. In every case, the observable quantity is a time difference, Δt. For the Sun-to-Earth-in model, $\Delta t = t_I - t_S$, where t_I is the horizon-in crossing time and t_S is the Sun sighting time. (Note that these times are measured by different sensors at different orientations in the spacecraft.) For the Sun-to-Earth-out and the Sun-to-Earth-midscan models, t_I is replaced by the horizon-out crossing time, t_O, and the midscan crossing time, $t_m = 1/2(t_I + t_O)$, respectively.

The relevant geometry for the Sun-to-Earth-in model is shown in Fig. 7-15. We assume that the Earth is spherical; that the spin rate, ω, is constant; and that there is no nutation. Therefore, the total rotation angle change between t_S and t_I is $\omega(t_I - t_S) = \omega \cdot \Delta t_I$ and the observation model is

$$\Delta t_I = \frac{1}{\omega}(\Phi_I - \Phi_H + 360° n) \tag{7-56}$$

Here Φ_I is the rotation angle from the Sun, S, to the horizon in-crossing, H_I; Φ_H is the azimuthal mounting angle between the Sun sensor and the horizon sensor onboard the spacecraft; and $n = \pm 1$, or 0. Φ_I can be calculated from

$$\Phi_I = \arctan\left[\frac{\hat{\mathbf{A}} \cdot (\hat{\mathbf{S}} \times \hat{\mathbf{H}}_I)}{\hat{\mathbf{S}} \cdot \hat{\mathbf{H}}_I - (\hat{\mathbf{S}} \cdot \hat{\mathbf{A}})(\hat{\mathbf{H}}_I \cdot \hat{\mathbf{A}})}\right] \tag{7-57a}$$

where $\hat{\mathbf{A}}$ is the spin axis attitude, $\hat{\mathbf{S}}$ is the Sun unit vector, and $\hat{\mathbf{H}}_I$ is a unit vector along the horizon sensor line of sight at the time $t = t_I$. Here $\hat{\mathbf{A}}$ is assumed known, $\hat{\mathbf{S}}$

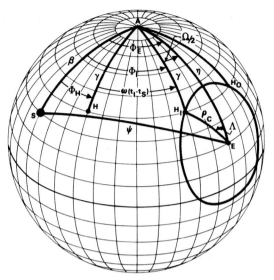

Fig. 7-15. Geometry for Sun Sensor/Horizon Sensor Rotation Angle Model

is provided by an ephemeris (Section 5.5) evaluated at $t = t_S$, and $\hat{\mathbf{H}}_I$ is calculated below.

Equation (7-57a) is derived as follows: Let $\hat{\mathbf{S}}_P$ and $\hat{\mathbf{H}}_P$ be the normalized components of $\hat{\mathbf{S}}$ and $\hat{\mathbf{H}}_I$ in the spin plane, i.e., the plane whose normal is $\hat{\mathbf{A}}$. Then

$$\hat{\mathbf{S}}_P = [\hat{\mathbf{S}} - (\hat{\mathbf{S}} \cdot \hat{\mathbf{A}})\hat{\mathbf{A}}]/|\hat{\mathbf{S}} - (\hat{\mathbf{S}} \cdot \hat{\mathbf{A}})\hat{\mathbf{A}}|$$

and

$$\hat{\mathbf{H}}_P = \left[\hat{\mathbf{H}}_I - (\hat{\mathbf{H}}_I \cdot \hat{\mathbf{A}})\hat{\mathbf{A}}\right]/|\hat{\mathbf{H}}_I - (\hat{\mathbf{H}}_I \cdot \hat{\mathbf{A}})\hat{\mathbf{A}}|$$

Performing the dot product of $\hat{\mathbf{H}}_P$ and $\hat{\mathbf{S}}_P$, we obtain the following expression for $\cos\Phi_I$:

$$\cos\Phi_I = \hat{\mathbf{H}}_P \cdot \hat{\mathbf{S}}_P = \left[\hat{\mathbf{S}} \cdot \hat{\mathbf{H}}_I - (\hat{\mathbf{S}} \cdot \hat{\mathbf{A}})(\hat{\mathbf{H}}_I \cdot \hat{\mathbf{A}})\right]/(D_S D_H) \qquad (7\text{-}57\text{b})$$

where D_S and D_H are the denominators in the expressions for $\hat{\mathbf{S}}_P$ and $\hat{\mathbf{H}}_P$, respectively. Using a different manipulation of $\hat{\mathbf{H}}_P$, $\hat{\mathbf{S}}_P$, and $\hat{\mathbf{A}}$, we obtain the following expression for $\sin\Phi_I$:

$$\sin\Phi_I = \hat{\mathbf{A}} \times \hat{\mathbf{S}}_P \cdot \hat{\mathbf{H}}_P = \hat{\mathbf{A}} \cdot \hat{\mathbf{S}}_P \times \hat{\mathbf{H}}_P = \hat{\mathbf{A}} \cdot (\hat{\mathbf{S}} \times \hat{\mathbf{H}}_I)/(D_S D_H) \qquad (7\text{-}57\text{c})$$

Equation (7-57a) is then obtained by dividing Eq. (7-57c) by Eq. (7-57b).

The unit vector $\hat{\mathbf{H}}_I$ is calculated as follows: Let $\hat{\mathbf{M}}$ be a unit vector perpendicular to both $\hat{\mathbf{A}}$ and $\hat{\mathbf{E}}$. Then

$$\hat{\mathbf{M}} = \hat{\mathbf{A}} \times \hat{\mathbf{E}}/\sin\eta$$

where η is the angle between $\hat{\mathbf{A}}$ and $\hat{\mathbf{E}}$ (the nadir angle). Let $\hat{\mathbf{N}}$ be a unit vector perpendicular to both $\hat{\mathbf{E}}$ and $\hat{\mathbf{M}}$. Then $\hat{\mathbf{E}}$, $\hat{\mathbf{M}}$, and $\hat{\mathbf{N}}$ form an orthonormal triad such that $\hat{\mathbf{N}} = \hat{\mathbf{E}} \times \hat{\mathbf{M}}$. Because $\hat{\mathbf{E}} \cdot \hat{\mathbf{H}}_I = \cos\rho$, where ρ is the angular radius of the Earth, $\hat{\mathbf{H}}_I$ can be written as

$$\hat{\mathbf{H}}_I = \cos\rho\,\hat{\mathbf{E}} + \sin\rho\,(\hat{\mathbf{M}}\sin\Lambda + \hat{\mathbf{N}}\cos\Lambda)$$

where Λ is a phase angle which can be determined from the dot product between $\hat{\mathbf{A}}$ and $\hat{\mathbf{H}}_I$. This is done as follows: If γ is the horizon sensor mounting angle, then

$$\hat{\mathbf{H}}_I \cdot \hat{\mathbf{A}} = \cos\gamma = \cos\rho\,\hat{\mathbf{E}}\cdot\hat{\mathbf{A}} + \sin\rho\,\hat{\mathbf{N}}\cdot\hat{\mathbf{A}}\cos\Lambda$$

which simplifies to

$$\cos\gamma = \cos\rho\cos\eta + \sin\rho\sin\eta\cos\Lambda \qquad (7\text{-}58a)$$

or

$$\cos\Lambda = \frac{\cos\gamma - \cos\rho\cos\eta}{\sin\rho\sin\eta} \qquad (7\text{-}58b)$$

and

$$\sin\Lambda = \pm(1 - \cos^2\Lambda)^{1/2} \qquad (7\text{-}58c)$$

Because Eq. (7-58a) is the law of cosines applied to the spherical triangle AEH_I in Fig. 7-15, the phase angle Λ must be the rotation angle about E between A and H_I. Due to our choice of the unit vectors $\hat{\mathbf{M}}$ and $\hat{\mathbf{N}}$, the negative sign in Eq. (7-58c) is associated with $\hat{\mathbf{H}}_I$ and the positive sign with $\hat{\mathbf{H}}_O$. The nadir vector $\hat{\mathbf{E}}$ is determined from the spacecraft ephemeris.

For the Sun-to-Earth-out and Sun-to-Earth-midscan models, the procedure is identical with that for the Sun to Earth-in model, except that the quantities Φ_I, t_I, and $\hat{\mathbf{H}}_I$ are replaced everywhere by Φ_O, t_O, and $\hat{\mathbf{H}}_O$ or by Φ_m, t_m, and $\hat{\mathbf{H}}_m$. Here, $\hat{\mathbf{H}}_m$ is a unit vector in the direction of $\mathbf{H}_I + \hat{\mathbf{H}}_O$. Φ_I can also be calculated directly, using the following relation (see Fig. 7-15):

$$\Phi_I = \Phi_m - \Omega/2 \qquad (7\text{-}59a)$$

An expression for Φ_m is obtained by applying the law of cosines to triangle SAE in Fig. 7-5, yielding $\cos\psi = \cos\eta\cos\beta + \sin\eta\sin\beta\cos\Phi_m$ which becomes, upon solving for Φ_E,

$$\Phi_m = \arccos\left[\frac{\cos\psi - \cos\eta\cos\beta}{\sin\eta\sin\beta}\right] \qquad (7\text{-}59b)$$

An expression for the Earth width, Ω, is obtained analogously from triangle $H_I AE$:

$$\Omega = 2\arccos\left[\frac{\cos\rho - \cos\eta\cos\gamma}{\sin\eta\sin\gamma}\right] \qquad (7\text{-}59c)$$

The quantities η, β, and ψ are computed from ephemerides evaluated at the proper times.

Figure 7-16 shows the relevant geometry when biases in the orientation of both sensors are included and the horizon sensor is assumed to have a fixed mounting angle, γ. In Fig. 7-16, β_M is the measured Sun angle, β is the true Sun angle, and ϵ_S and $\Delta\beta$ are the inclination and elevation biases which cause the difference between β_M and β. $\Delta\Phi_S$ is the resulting rotation angle bias. Similarly, γ_N is the nominal mounting angle of the horizon sensor line of sight, relative to the spin axis; $\Delta\gamma$ is the difference, $\gamma - \gamma_N$, between γ_N and the true mounting angle, γ; $\Delta\Phi_H$ is a constant bias on Φ_H, the nominal azimuthal mounting angle difference between the sensors; ρ_C is the computed angular radius of the Earth; and $\Delta\rho$ is a fixed angular

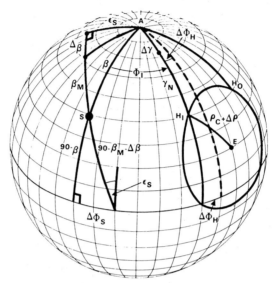

Fig. 7-16.   Effect of Sun Sensor Biases on Geometry for Sun Sensor/Horizon Sensor Rotation Angle Model

bias on ρ_C, resulting primarily from a constant bias on the triggering threshold of the horizon sensor (see Sections 6.2 and 7.2).

In terms of the above quantities, the observation model becomes

$$\Delta t = \frac{1}{\omega}(\Phi - \Delta\Phi_S - \Delta\Phi_H - \Phi_H + 360°n) \tag{7-60}$$

where

$$\Phi = \Phi(\hat{\mathbf{A}}, \hat{\mathbf{S}}, \hat{\mathbf{E}}, \rho_C, \Delta\rho, \Delta\gamma)$$

$$\Delta\Phi_S = \Delta\Phi_S(\beta_M, \Delta\beta, \epsilon_S) = \Delta\Phi_S(\beta, \epsilon_S)$$

and $\Delta\Phi_H$ and Φ_H are constants. $\hat{\mathbf{E}}$ is the unit nadir vector.

It remains to express Φ and $\Delta\Phi_S$ in terms of their arguments. For the case of the Sun-to-Earth-in model, $\Phi = \Phi_I$ and is again calculated from Eq. (7-57) or Eq. (7-59), in which ρ and γ are replaced everywhere by $\rho_C + \Delta\rho$ and $\gamma N + \Delta\gamma$, respectively. $\Delta\Phi_S$ is calculated by applying Napier's rules for right spherical triangles to the lower triangle associated with the Sun sensor in Fig. 7-16, yielding:

$$\sin\Delta\Phi_S = \tan(90° - \beta)\tan[90° - (90° - \epsilon_S)]$$

which simplifies to

$$\Delta\Phi_S = \arcsin(\cot\beta\tan\epsilon_S) \tag{7-61}$$

where β is computed from

$$\beta = \arccos(\hat{\mathbf{A}} \cdot \hat{\mathbf{S}}) \tag{7-62}$$

Here, $\hat{\mathbf{A}}$ is the known attitude, and $\hat{\mathbf{S}}$ is determined from an ephemeris and is evaluated at $t = t_s$. Note that this description of horizon sensor biases is valid only for horizon sensors with fixed mounting angles, γ.

In the case of panoramic scanners, the nominal mounting angle, γ_N, varies by fixed increments (Section 6.2) in a plane inclined at an angle ϵ_H to its nominal orientation, as shown in Fig. 7-17 for the case of the Earth-in models. Thus, $\Delta\gamma$ and γ, the true mounting angle, are related to the other quantities as follows:

$$\cos\gamma = \cos\epsilon_H \cos(\gamma_N + \Delta\gamma) \tag{7-63}$$

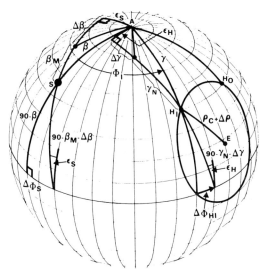

Fig. 7-17.   Effect of Horizon Sensor Biases on Geometry for Sun Sensor/Horizon Sensor Rotation Angle Model

The observation model is again described by Eq. (7-60), in which Φ and Δt are replaced by Φ_I and Δt_I. $\Delta\Phi_S$ and Φ_H are as before, but now Φ_I depends on ϵ_H (through γ) in addition to the other biases, and $\Delta\Phi_H$ is defined by

$$\Delta\Phi_H = \Delta\Phi_{HM} - \Delta\Phi_{HR} \tag{7-64}$$

where $\Delta\Phi_{HM}$ is a constant azimuth bias on the horizon sensor mounting angle, and $\Delta\Phi_{HR}$ is an additional horizon sensor rotation angle bias caused by ϵ_H and $\Delta\gamma$ as shown in Fig. 7-17.

$\Delta\Phi_{HR}$ is calculated by applying Napier's rules for right spherical triangles to the lower triangle associated with the horizon sensor in Fig. 7-17 to obtain

$$\sin\left[90° - (90° - \epsilon_H)\right] = \tan(\Delta\Phi_H)\tan\left[90° - (90° - \gamma_N - \Delta\gamma)\right]$$

which simplifies to

$$\Delta\Phi_{HR} = \arctan\left[\frac{\sin\epsilon_H}{\tan(\gamma_N + \Delta\gamma)}\right] \qquad (7\text{-}65)$$

All of the above applies also to the Sun-to-Earth-out and Sun-to-Earth-midscan models, where Φ_I, t_I, and $\hat{H}_I$ are replaced everywhere by Φ_O, t_O, and $\hat{H}_O$ or Φ_m, t_m, and $\hat{H}_m$, as before.

From Fig. 7-17 we see that the geometrical relationship between $\Delta\Phi_S$, $\Delta\beta$, and ϵ_S is identical with that between $\Delta\Phi_H$, $\Delta\gamma$, and ϵ_H. The practical difference is that β can be found directly from Eq. (7-62), whereas γ cannot and therefore must be expressed in terms of γ_N, $\Delta\gamma$, and ϵ_H. It is the independent knowledge of β which makes it possible to eliminate $\Delta\beta$ from the expressions for Φ and $\Delta\Phi_S$. Once ϵ_S is found, $\Delta\beta$ can be computed from β_m, β, and ϵ_S.

The expression on the right side of Eq. (7-60) is a complicated function of the following biases: ϵ_S, ϵ_H, $\Delta\gamma$, $\Delta\rho$, and $\Delta\Phi_{HM}$. The values of the various coefficients in that expression depend on the numerical values of the attitude $\hat{A}$ and the time. To determine the above biases, at least five independent equations are necessary, although the numerical solutions of such a system would generally not be unique. Such equations can be obtained by taking measurements at various attitudes and times. When reasonable initial estimates are available, ambiguities can generally be resolved and satisfactory solutions obtained.

7.4 Modeling Sensor Electronics

F. L. Markley

7.4.1 Theory

In the previous three sections, ideal mathematical models have been constructed for Sun sensor and horizon sensor systems. Effects of electronics signal processing on the sensor outputs have been considered only in an *ad hoc* fashion (e.g., the azimuth biases and central body angular radius biases introduced in Section 7.2). In this section we consider the electronics signal processing systems from a more fundamental viewpoint. For a large class of such systems, which we shall assume to include all cases of interest to us, the output signal, $S_O(t)$, is related to the input, $S_I(t)$, by

$$S_O(t) = \int_{-\infty}^{\infty} h(t,t')S_I(t')dt' \qquad (7\text{-}66)$$

where $h(t,t')$ is called the *impulse response function* of the system. A system that obeys Eq. (7-66) is called a *linear system*, because the response of such a system to the linear input combination $S_I = S_{I1} + S_{I2}$ is the output $S_O = S_{O1} + S_{O2}$, where S_{O1} is related to S_{I1} by Eq. (7-66) and, similarly, for S_{O2} and S_{I2}. Extensive literature exists on the subject of linear systems. See, for example, Schwarz and Friedland [1965], Kaplan [1962], or Hale [1973].

If the input to a linear system is the *unit impulse function* or the *Dirac delta function* (see the Preface),

$$S_I(t) = \delta_D(t - t_0)$$

then the output is

$$S_O(t) = h(t, t_0)$$

which is why h is called the *impulse response function*. A linear system is *time invariant* if the impulse response depends only on the time elapsed since the application of the input impulse, and not otherwise on the input time. That is,*

$$h(t, t') = h(t - t') \tag{7-67}$$

For a time-invariant linear system, the input/output relation resulting from combining Eqs. (7-66) and (7-67) is

$$S_O(t) = \int_{-\infty}^{\infty} h(t - t') S_I(t') dt' \tag{7-68}$$

Any integral of the form

$$\int_{-\infty}^{\infty} f(\zeta) g(x - \zeta) d\zeta$$

is called a *convolution integral* [Schwarz and Friedland, 1965; Kaplan, 1962; Hale, 1973; Churchill, 1972]. Comparison with Eq. (7-68) shows that the output signal of a linear, time-invariant system is given by the convolution integral of the input signal and the impulse response function. This property will be used shortly.

It is often convenient to work in the frequency domain rather than the time domain. The *Fourier transform*, $\tilde{X}(\omega)$, of a function $X(t)$ is defined by

$$\tilde{X}(\omega) \equiv \int_{-\infty}^{\infty} X(t) e^{-i\omega t} dt \tag{7-69}$$

The inverse transformation is given by [Churchill, 1972]

$$X(t) = \frac{1}{2\pi} \int_{-\infty}^{\infty} \tilde{X}(\omega) e^{i\omega t} d\omega \tag{7-70}$$

The *convolution theorem* [Churchill, 1972] states that the Fourier transforms of functions obeying the convolution relation Eq. (7-68) obey the product relation

$$\tilde{S}_O(\omega) = \tilde{h}(\omega) \tilde{S}_I(\omega) \tag{7-71}$$

---

*This equation means that the function $h(t, t')$ depends only on the difference $t - t'$, and can be written as a function of that single variable. Clearly, the two functions in Eq. (7-67) must be different mathematically, but confusion should not result from using the same symbol for them.

A linear system is *causal* if the output at time t depends only on the input at times $t' \leqslant t$; that is, if

$$h(t, t') = 0 \text{ for all } t' > t$$

All the systems we consider (in particular, that defined by the transfer function of Eq. (7-78)) are causal, and the infinite upper limit of all t' integrals can be replaced by t.

The simplicity of Eq. (7-71) as compared with Eq. (7-68) explains the usefulness of analysis in terms of frequency dependence rather than time dependence. Finally, the *transfer function*

$$H(i\omega) \equiv \tilde{h}(\omega) \tag{7-72}$$

is often used in place of $\tilde{h}(\omega)$ to specify response characteristics of a linear system.

7.4.2 Example: IR Horizon Sensor

As an example of the application of sensor electronics modeling, we shall consider the performance of the infrared horizon sensors on the Synchronous Meteorological Satellite-2, SMS-2, launched in February 1975 [Philco-Ford, 1971; Chen and Wertz, 1975]. The input to the sensor electronics system is the intensity of infrared radiation in the 14–16 μm wavelength range falling on the sensor. If we assume that the sensor has uniform sensitivity over its field of view and that the Earth is a uniformly bright disk (a good approximation for this wavelength range, as discussed in Section 4.2), then the input signal is proportional to the overlap area on the celestial sphere between the sensor field of view and the Earth disk.

The sensor field of view is nominally square, 1.1 deg on a side [Philco-Ford, 1971], but for simplicity we model it as a circle with an angular radius of $\epsilon = 0.62$ deg to give the same sensor area. We ignore the oblateness of the Earth (shown in Section 4.3 to be a reasonable approximation) and treat the Earth disk as a circle of radius $\rho = 8.6$ deg, the appropriate value for the SMS-2 drift orbit [Chen and Wertz, 1975]. Then the input signal is given by a constant, K, times the overlap area between two small circles on the celestial sphere as shown in Fig. 7-18. Using Eq. (A-14) for the area and the notation of Fig. 7-18, we have

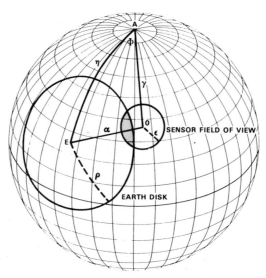

Fig. 7-18. Earth Sensor Geometry

$$S_I = 0, \qquad \alpha > \rho + \epsilon$$

$$= 2K\left[\pi - \cos\rho \arccos\left(\frac{\cos\epsilon - \cos\rho\cos\alpha}{\sin\rho\sin\alpha} \right) - \cos\epsilon \arccos\left(\frac{\cos\rho - \cos\epsilon\cos\alpha}{\sin\epsilon\sin\alpha} \right) \right.$$

$$\left. - \arccos\left(\frac{\cos\alpha - \cos\epsilon\cos\rho}{\sin\epsilon\sin\rho} \right) \right], \qquad |\rho - \epsilon| < \alpha < \rho + \epsilon$$

$$= \text{Min}\left[2\pi(1 - \cos\rho), 2\pi(1 - \cos\epsilon) \right], \qquad \alpha < |\rho - \epsilon| \tag{7.73}$$

where Min denotes the lesser of the two function values in the brackets. This function is rather intractable mathematically, so we prefer to work with its derivative:

$$\frac{dS_I}{dt} = -2K(1 - \cos^2\alpha - \cos^2\rho - \cos^2\epsilon + 2\cos\alpha\cos\rho\cos\epsilon)^{1/2}\frac{1}{\sin\alpha}\frac{d\alpha}{dt},$$

$$|\rho - \epsilon| < \alpha < \rho + \epsilon \tag{7-74a}$$

$$= 0, \qquad \text{otherwise} \tag{7-74b}$$

Because the angular radii of the sensor field of view and of the Earth disk, ϵ and ρ, are constant, the only time dependence in S_I is through the time dependence of $\alpha(t)$, the arc-length distance between the centers of the small circles. The horizon sensors on SMS-2 are rigidly mounted on the spacecraft; the motion of their fields of view is due to the spacecraft's spin. Let A in Fig. 7-18 be the spacecraft spin axis, and let the sensor mounting angle and nadir angle be denoted by γ and η, respectively, Then the law of cosines applied to spherical triangle AEO gives

$$\cos\alpha(t) = \cos\eta\cos\gamma + \sin\eta\sin\gamma\cos\Phi(t) \tag{7-75}$$

Differentiating Eq. (7-75) gives

$$\frac{1}{\sin\alpha}\frac{d\alpha}{dt} = \frac{\sin\eta\sin\gamma\sin\Phi}{1 - \cos^2\alpha}\frac{d\Phi}{dt} \tag{7-76}$$

Substituting Eqs. (7-75) and (7-76) into Eqs. (7-73) and (7-74) gives S_I and dS_I/dt as functions of the rotation angle, Φ. These functions are plotted in Figs. 7-19(a, b) and 7-20(a, b) for $\eta = 81$ deg and $\eta = 78$ deg, respectively, and for $\gamma = 86$ deg, the mounting angle for the SMS-2 primary Earth sensor [Chen and Wertz, 1975]. The points where the center of the sensor field of view crosses the edge of the Earth disk are indicated by Φ_I and Φ_O on the figures.

The calculation of the output signal requires a numerical integration of Eq. (7-68) or its equivalent. Substituting Eq. (7-70) with $X = h$, and Eq. (7-72) into Eq. (7-70), and then integrating by parts [so we can use Eq. (7-74) rather than Eq. (7-73)] gives*

---

*A horizon scanner actually makes repeated scans of the Earth, so dS_I/dt is a rather complicated function. We shall include only one Earth scan in the integral; this is a good approximation if the transfer function is such that the output signal from one Earth scan has decreased to a negligible value before the next scan, as is the case for this example. With this approximation, the integrated part of the integration by parts vanishes at infinite time. The quantity in brackets in Eq. (7-77b) is that integral of the quantity in brackets in Eq. (7-77a) that is finite at $\omega = 0$.

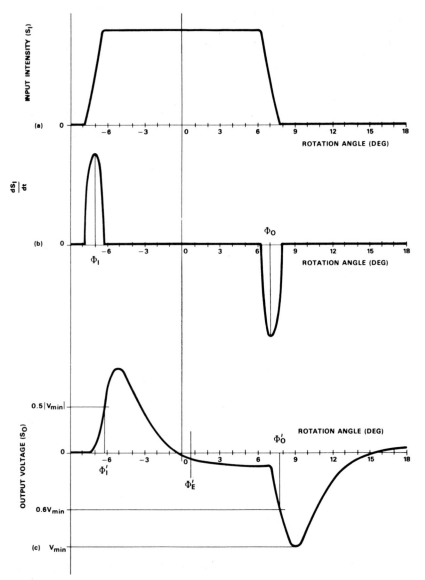

Fig. 7-19. Model of SMS-2 Earth Sensor Electronics Response for 81-Deg Nadir Angle

$$S_O(t) = \int_{-\infty}^{\infty} S_I(t') \left[\frac{1}{2\pi} \int_{-\infty}^{\infty} H(i\omega) e^{i\omega(t-t')} d\omega \right] dt' \qquad (7\text{-}77a)$$

$$= \int_{-\infty}^{\infty} \frac{dS_I}{dt'} \left[\int_{-\infty}^{\infty} H(i\omega) \frac{e^{i\omega(t-t')} - 1}{2\pi i\omega} d\omega \right] dt' \qquad (7\text{-}77b)$$

The spin rate, $d\Phi/dt$, of SMS-2 was taken to be 600 deg/sec which is close to the measured value [Chen and Wertz, 1975].

The SMS-2 Earth sensor transfer function is [Philco-Ford, 1971]

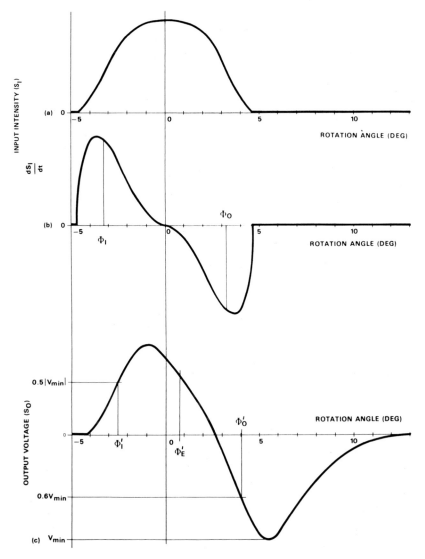

Fig. 7-20. Model of SMS-2 Earth Sensor Electronics Response for 78-Deg Nadir Angle

$$H(i\omega) = \left(\frac{1}{i\omega T_b + 1}\right)\left(\frac{i\omega T_1}{i\omega T_1 + 1}\right)\left(\frac{1}{i\omega T_2 + 1}\right)\left(\frac{i\omega T_3}{i\omega T_3 + 1}\right)\left(\frac{1}{i\omega T_4 + 1}\right)\left(\frac{i\omega T_5}{i\omega T_5 + 1}\right) \quad (7\text{-}78)$$

where $T_b =$ 1.8 ms = detector cutoff

$T_1 =$ 80 ms = preamplifier lower cutoff

$T_2 = 0.238$ ms = preamplifier upper cutoff

$T_3 =$ 2.66 ms = main amplifier lower cutoff

$T_4 = 0.560$ ms = main amplifier upper cutoff

$T_5 =$ 80 ms = output transformer lower cutoff

For this transfer function, the quantity inside the brackets in Eq. (7-77b) can be evaluated in closed form by the method of residues [Churchill, 1972]. The t' integral in Eq. (7-77b) is then evaluated numerically with dS_I/dt given by Eqs. (7-74) through (7-76). The output signals, S_O, for $\eta = 81$ deg and 78 deg are plotted in Fig. 7-19(c) and 7-20(c), respectively. They resemble the curves of dS_I/dt to some extent, but the peaks are broadened and time delayed, the positive and negative peaks have unequal height, and S_O does not return to zero between the peaks in the cases where dS_I/dt does. Thus, the electronics acts something like a differentiating circuit, although its response is quite a bit more difficult to characterize completely.

The telemetry signal from the SMS-2 horizon sensor is not the output signal, S_O, of the sensor electronics, but rather the time intervals from Sun sightings to Earth-in and -out crossings. The latter times are determined by onboard threshold detection logic [Philco-Ford, 1971]. A negative edge peak detector measures the amplitude of the negative peak of S_O and holds it in the form of a direct-current voltage. The Earth-in crossing is specified as the point where the positive pulse reaches 50 $\pm 5\%$ of the magnitude of the saved peak voltage, and the Earth-out crossing is specified to be where the negative peak voltage is 60 $\pm 5\%$ of the peak. These points are indicated by Φ'_I and Φ'_O on Figs. 7-19(c) and 7-20(c); and the apparent Earth center, defined as the midpoint between Φ'_I and Φ'_O, is indicated by Φ'_E. Note that Φ'_E is displaced from the true Earth center, $\Phi = 0$. Figure 7-21(a) shows Φ'_I and Φ'_O as a function of nadir angle, η, plotted at 0.1-deg intervals. This figure also includes a curve showing the rotation angles at which the center of the sensor field of view crosses the Earth's horizon, corrected by a constant offset so that the points fall on this curve for large Earth scan widths (the offset is equal to the value of Φ'_E at large Earth widths). The deviation of the two curves at the left of the figure indicates that modeling IR sensor electronic effects as a fixed bias on the angular radius of the Earth, as discussed in Section 7.2, fails at small Earth widths. Figure 7-21(b) is similar to Fig. 7-21(a) except that it was calculated with 15% and 25% threshold levels for Earth-in and Earth-out times, respectively. The deviations at small Earth widths are exaggerated at these threshold levels, as compared with the nominal levels shown in Fig. 7-21(a). Figure 7-22 shows actual Earth-in and

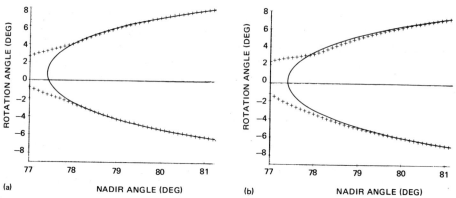

Fig. 7-21. Simulated Earth-In and -Out Data as a Function of Nadir Angle. (a) Nominal SMS-2 Triggering Levels; (b) Decreased Triggering Levels. (See text for explanation.)

-out data from SMS-2, which is further described in Section 9.4. The slope of the ellipse is due to orbital motion effects and is excluded from this section because it is not important for our purposes. What is important is the deviation of the theoretical and experimental points at small Earth widths, called the *pagoda effect*

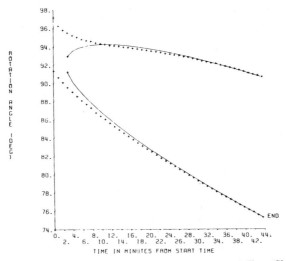

Fig. 7-22.   Actual Earth-In and -Out Data as a Function of Time (SMS-2 Data)

because of the characteristic shape of the Earth-out curve [Chen and Wertz, 1975]. The similarity between Figs. 7-21(b) and 7-22 indicates that a more sophisticated treatment of sensor electronics than customarily used in attitude systems may lead to an understanding of the pagoda effect and other related anomalies.

7.5 Magnetometer Models

Gerald M. Lerner

This section develops the models used to decode data from fluxgate magnetometers described in Section 6.3 for use in attitude determination algorithms. Equations are included to encode magnetometer data for simulation.

The basic measurement provided by a single-axis fluxgate magnetometer is a voltage, V, related to the component of the local field, $\mathbf{H}$, along the input axis, $\hat{\mathbf{n}}$, by

$$V = a(\hat{\mathbf{n}} \cdot \mathbf{H}) + V_0 \qquad (7\text{-}79)$$

where a is the magnetometer scale factor, V_0 is the magnetometer bias, and $\mathbf{H}$ is the net local magnetic intensity in body coordinates. The output voltage passes through an analog-to-digital converter for transmission, yielding a discrete output

$$N_V = \text{Int}\{ c[a(\hat{\mathbf{n}} \cdot \mathbf{H}) + V_0] + 0.5 \} \qquad (7\text{-}80)$$

where $\text{Int}(x)$ is the integral part of x and c is the analog-to-digital scale factor.

An alternative system provides a *zero-crossing* measurement in which a telemetry flag is set when the magnetometer output voltage changes sign. This type of output is used by spinning spacecraft to provide phase information for either attitude determination or control. The flag generally is set and time-tagged in the first telemetry frame following a change in sign.

Vector magnetometer systems consist of three mutually orthogonal, single-axis fluxgate magnetometers. The system can be packaged as a single unit mounted within the spacecraft or attached to a boom, or as separate units dispersed about the spacecraft (for example, on extendable paddles).

The remainder of this section is concerned primarily with vector magnetometers. The models developed are independent of the specific geometry; however, probable biases are highly dependent on both packaging and location within the spacecraft. A magnetometer located at the end of a long boom is unlikely to be exposed to internal magnetic fields but may be misaligned relative to the spacecraft. The opposite is likely to be true for a magnetometer located within the spacecraft interior. In a system consisting of three separate units (particularly units dispersed on extendable hardware) individual units may not be mutually orthogonal and units may be misaligned relative to the spacecraft reference axes.

7.5.1 Calibration of Vector Magnetometers

By analogy with Eq. (7-79), the output of a vector magnetometer system is

$$\mathbf{V} = A\mathbf{H} + \mathbf{V}_0 \qquad (7\text{-}81)$$

where the components of $\mathbf{V}$ are the outputs of the three units, A is a 3-by-3 matrix, including both scale factor and alignment data, and $\mathbf{V}_0$ is the magnetometer bias voltage. Magnetic testing is performed by placing the spacecraft in a Helmholtz coil and measuring the magnetometer response, $\mathbf{V}$, to a systematically varied external field, $\mathbf{H}$. A least-squares fit of the data to Eq. (7-81) yields the 12 parameters, A and $\mathbf{V}_0$, which define the magnetometer calibration. The analog output, $\mathbf{V}$, is passed through an analog-to-digital converter to provide the digitized output

$$\mathbf{N}_V = \begin{bmatrix} \text{Int}(c_1 V_1 + 0.5) \\ \text{Int}(c_2 V_2 + 0.5) \\ \text{Int}(c_3 V_3 + 0.5) \end{bmatrix} \qquad (7\text{-}82)$$

where c_1, c_2, and c_3 are the analog-to-digital conversion factors.

The matrix A in Eq. (7-81) is diagonal if the three magnetometer input axes are colinear with the spacecraft reference axes and crosstalk is absent. *Crosstalk* refers to induced magnetic fields normal to an applied field caused by ferromagnetic material or currents in the magnetometer and associated electronics. If crosstalk is absent and the three magnetometer units are mutually orthogonal, then A can be diagonalized by a similarity transformation (see Appendix C). This would imply a coherent misalignment of an orthogonal magnetometer package relative to the spacecraft reference axes. In practice, crosstalk, internal misalignment, and external misalignment cannot be separated and for the remainder of this section we will assume that A is not diagonal.

Equation (7-81) may be inverted and combined with Eq. (7-82) to yield the best estimate of the external magnetic intensity in body coordinates, $\overline{\mathbf{H}}$,

$$\overline{\mathbf{H}} = A^{-1}(\mathbf{V} - \mathbf{V}_0) \tag{7-83}$$

$$\overline{\mathbf{H}} = \frac{1}{c} B^{-1} \begin{bmatrix} a_1 N_{V1} - b_1 \\ a_2 N_{V2} - b_2 \\ a_3 N_{V3} - b_3 \end{bmatrix} \tag{7-84}$$

where $c = (c_1 + c_2 + c_3)/3$ is the mean analog-to-digital scale factor. The components of the matrix B are

$$B_{ij} = A_{ij}/d_i \tag{7-85}$$

where

$$d_i \equiv \left\{ \sum_{j=1}^{3} (A_{ij})^2 \right\}^{1/2} \tag{7-86}$$

$$a_i \equiv c/(c_i d_i) \tag{7-87}$$

$$b_i \equiv c V_{0i}/d_i \tag{7-88}$$

and i is 1, 2, or 3.

Note that Eq. (7-84) is a *linear* function of the magnetometer output and is thus analogous to the gyroscope model described in Section 7.8. Equation (7-84) assumes that matrix B has a unique inverse (see Appendix C) and requires that no two magnetometer input axes be collinear.

Matrix B defines the effective (not necessarily physical) orientation of the single-axis magnetometers relative to the spacecraft reference axes.* Thus, the effective coelevation, θ_i, and azimuth, ϕ_i, of the ith magnetometer are

$$\theta_i = \arccos(B_{i3})$$

$$\phi_i = \arctan(B_{i2}/B_{i1}) \tag{7-89}$$

7.5.2 Magnetometer Biases†

Sources of the bias term, $\mathbf{V}_0$, in Eq. (7-83) include magnetic fields generated by spacecraft electronics and electromagnetic torquing coils, and residual magnetic fields caused by, for example, permanent magnets induced in ferromagnetic spacecraft components. It is important to distinguish between the sources of magnetometer bias and of magnetic dipole torque on the spacecraft. Although both are manifestations of uncompensated spacecraft magnetism, a unique relation between the two cannot be derived. The magnetic induction, $\mathbf{B}^c(\mathbf{x})$, due to all the localized

---

*This is a heuristic definition which ignores crosstalk and assumes that misalignment is the source of off-diagonal terms in A.

†Much of this development follows the formulation of Jackson [1963], where more complete derivations can be found.

current distributions, $\mathbf{J}(\mathbf{x})$, contained within the spacecraft can be expressed as the curl of a vector potential, $\mathbf{B}^c(\mathbf{x}) = n \times \mathbf{A}(\mathbf{x})$, where the vector potential, $\mathbf{A}(\mathbf{x})$, is [Jackson, 1963],

$$\mathbf{A}(\mathbf{x}) = \frac{\mu_0}{4\pi} \int \frac{\mathbf{J}(\mathbf{x}')\,d^3x'}{|\mathbf{x}-\mathbf{x}'|} \tag{7-90}$$

Equation (7-90) may be expressed as a multipole expansion using

$$\frac{1}{|\mathbf{x}-\mathbf{x}'|} = \frac{1}{x} + \frac{\mathbf{x}\cdot\mathbf{x}'}{x^3} + \cdots \tag{7-91}$$

to yield the components of $\mathbf{A}(\mathbf{x})$,

$$A_i(\mathbf{x}) = \frac{\mu_0}{4\pi}\left[\frac{1}{x}\int J_i(\mathbf{x}')\,d^3x' + \frac{1}{x^3}\mathbf{x}\cdot\int J_i(\mathbf{x}')\mathbf{x}'\,d^3x' + \cdots \right] \tag{7-92}$$

For a localized steady-state current distribution, the volume integral of $\mathbf{J}$ is zero because $n \cdot \mathbf{J} = 0$. Therefore, the first term, which is analogous to the monopole term in electrostatics, is also zero. Manipulation of the lowest order (in $1/x$) nonvanishing term in Eq. (7-92) can be shown [Jackson, 1963] to yield

$$\mathbf{A}(\mathbf{x}) = \frac{\mu_0}{4\pi}\left[\mathbf{m}\times\mathbf{x}/x^3 + \mathcal{O}\,(x^{-3}) \right] \tag{7-93a}$$

and therefore

$$\mathbf{B}^c(\mathbf{x}) = \frac{\mu_0}{4\pi}\left[3\hat{\mathbf{x}}(\hat{\mathbf{x}}\cdot\mathbf{m}) - \mathbf{m} \right]/x^3 + \mathcal{O}\,(x^{-4}) \tag{7-93b}$$

where

$$\mathbf{m} = \frac{1}{2}\int \mathbf{x}'\times\mathbf{J}(\mathbf{x}')\,d^3x' \tag{7-94}$$

is the magnetic moment of the current distribution $\mathbf{J}$.

The total force on a current distribution, $\mathbf{J}$, in an external field, $\mathbf{B}$, is

$$\mathbf{F} = \int \mathbf{J}(\mathbf{x})\times\mathbf{B}(\mathbf{x})\,d^3x \tag{7-95}$$

and the total torque is

$$\mathbf{N} = \int \mathbf{x}\times(\mathbf{J}(\mathbf{x})\times\mathbf{B}(\mathbf{x}))\,d^3x \tag{7-96}$$

For $\mathbf{B}$ constant over the dimensions of the current distribution the net force, $\mathbf{F}$, vanishes and Eq. (7-96) may be reformulated as

$$\mathbf{N} = \mathbf{m}\times\mathbf{B}(0) \tag{7-97}$$

The difference between the magnetometer bias and residual spacecraft dipole torque can now be seen from Eqs. (7-93b) and (7-97). A *magnetometer bias* is a measure of $\mathbf{B}^c(\mathbf{x})$ in the near field and terms of all order in $\mathbf{x}$ contribute because the magnetometer may be in close proximity to magnetic material. However, the *residual dipole torque* results only from the interaction of the dipole term with the environment because the higher order multipoles do not contribute to the torque.

Magnetometer biases will be induced by magnetic coils used for spacecraft attitude control. These biases may be lessened by winding small coils near the magnetometer in series with the larger control coils to produce a near zero net field independent of the coil current.

To simulate magnetometer biases for prelaunch analysis, or to remove magnetometer biases for postlaunch processing, the field of an electromagnet (see Section 6.7) may be computed as follows. The magnetic induction of a coil with dipole m and radius a is given by Jackson [1963] as

$$B_r = \frac{1}{r\sin\theta} \frac{\partial}{\partial\theta}(\sin\theta A_\phi)$$

$$B_\theta = -\frac{1}{r} \frac{\partial}{\partial r}(rA_\phi)$$

$$B_\phi = 0 \tag{7-98}$$

where

$$A_\phi(r,\theta) = \frac{m\mu_0}{\pi^2 a(a^2 + r^2 + 2ar\sin\theta)^{1/2}} \left[\frac{(2-k^2)K(k) - 2E(k)}{k^2} \right] \tag{7-99}$$

K and E are complete elliptic integrals of the first and second kinds with argument

$$k = \left[\frac{4ar\sin\theta}{a^2 + r^2 + 2ar\sin\theta} \right]^{1/2} \tag{7-100}$$

Figure 7-23 defines the relevant geometry. For small k^2, Eq. (7-99) becomes

$$A_\phi(r,\theta) = \frac{\mu_0 mr\sin\theta}{4\pi(a^2 + r^2 + 2ar\sin\theta)^{3/2}} \tag{7-101}$$

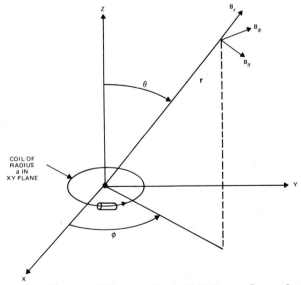

Fig. 7-23. Geometry of Magnetic Dipole Field From a Current Loop

and the field is

$$B_r = \frac{\mu_0 m}{4\pi}\cos\theta \frac{(2a^2 + 2r^2 + ar\sin\theta)}{(a^2 + r^2 + 2ar\sin\theta)^{5/2}} \qquad (7\text{-}102)$$

$$\lim_{(a/r)\to 0} B_r = \mu_0 m\cos\theta/(2\pi r^3) \qquad (7\text{-}103)$$

and

$$B_\theta = -\frac{\mu_0 m}{4\pi}\sin\theta \frac{(2a^2 - r^2 + ar\sin\theta)}{(a^2 + r^2 + 2ar\sin\theta)^{5/2}} \qquad (7\text{-}104)$$

$$\lim_{(a/r)\to 0} B_\theta = \mu_0 m\sin\theta/(4\pi r^3) \qquad (7\text{-}105)$$

The elliptic integrals may be computed analytically [Abramowitz and Stegun, 1964] or numerically using subroutines CEL1 and CEL2 in the *IBM Scientific Subroutine Package* [1968].

7.6 Star Sensor Models

Lawrence Fallon, III

This section describes the mathematical relationships between star sensor measurements and catalog star positions (Section 5.6) for the slit- and image dissector-type sensors described in Section 6.4. Sensor response to star magnitudes is also discussed. Because of the close interaction between the interpretation of star sensor measurements, spacecraft dynamics models, and attitude determination techniques, we will refer to material in Sections 16.2 and 17.1.

7.6.1 Star Scanner Measurements

Star scanners, or slit star sensors, use a photomultiplier and electronic assembly to detect stars crossing a slit configuration. The exact form of the scanner measurements will depend on the particular type of instrument being used (Section 6.4). In general, however, scanner output will consist of a series of times corresponding to star crossings with detected intensity greater than a specified threshold, or a series of detected intensities from which crossing times may be deduced. The mathematical model for star scanner measurements presented in this section follows the analysis of Grosch, *et al.*, [1969] and Paulson, *et al.*, [1969].

Consider a transparent slit etched on an otherwise opaque plate in the focal plane of a star scanner optical system. If a slit is a straight line segment and the optical system is free of distortion, then a plane, known as the *slit plane*, is defined which contains the slit and the optical center of the lens. A distant bright point source, e.g., a star, will be sensed by a detector behind the slit, if and only if it lies in the slit plane. The instant the star crosses this plane is called the *transit time*.

At any star transit time,

$$\hat{\mathbf{n}}\cdot\hat{\mathbf{S}} = 0 \qquad (7\text{-}106)$$

where $\hat{\mathbf{n}}$ is the unit vector normal to the slit plane and $\hat{\mathbf{S}}$ is the unit vector in the direction of the star. This equation holds for each star encountered as the spacecraft scans the celestial sphere. This results in the set of conditions

$$\hat{\mathbf{n}}(t_i)\cdot\hat{\mathbf{S}}_i = 0 \tag{7-107}$$

The spacecraft three-axis attitude at any one instant is defined by three independent angles. Equation (7-107), however, provides only one condition at each star transit time. To provide the additional information, the equations of motion of the spacecraft may be used to obtain a time-dependent characterization of the attitude, which involves just a few parameters. Alternatively, an attitude time history may be provided by a system of gyros. This attitude model or history, described further in Section 17.1, may then be used to internally couple the conditions in Eq. (7-107). In addition, if more than one slit is employed, each star will yield two spatially independent measurements. The additional information per star which is gained from a multislit system may be exploited to reduce the number of required stellar targets or to increase the data sampling rate.

The normal vector of the jth slit, $\mathbf{n}_j$, is fixed in spacecraft body coordinates. The star vector in body coordinates, $\hat{\mathbf{S}}$, is related by the spacecraft attitude matrix, A, to the star vector, $\mathbf{S}^I$, fixed in inertial space by

$$\hat{\mathbf{S}} = A(t)\mathbf{S}^I \tag{7-108}$$

Therefore, Eq. (7-107) may be rewritten

$$\hat{\mathbf{n}}_j \cdot A(t_i)\hat{\mathbf{S}}_i^I = 0 \tag{7-109}$$

This set of equations may be used with the attitude model to identify observed stars, as described in Section 7.7. After star identification, these equations may be used to refine attitude model parameters as described in Chapter 3.

In some cases, the relationship between slit geometry and the attitude model allows considerable simplification of Eq. (7-109). For example, consider the "N" slit sensor, as shown in Fig. 7-24, mounted on a uniformly spinning satellite in a torque-free environment (Section 16.2). As the satellite spins, transit pulses will be generated at times t_1, t_2, and t_3 by a star passing the three slits.

If the satellite's spin rate is assumed constant and nutation is neglected between times t_1 and t_3, the star's elevation, λ, in the spacecraft frame is constant between t_1 and t_3. Using Napier's Rules, this angle may be calculated as

$$\tan(\lambda - \delta) = \tan\Gamma\sin\left[\zeta\left(-0.5 + \frac{t_2 - t_1}{t_3 - t_1}\right)\right] \tag{7-110}$$

where ζ, Γ, and δ are defined in Fig. 7-24. If ϕ is the azimuth of the first slit in the spacecraft body frame, the observed star unit vector at time t_1 in the spacecraft body frame is given by

$$\hat{\mathbf{S}}(t_1) = \begin{bmatrix} \cos\phi\cos\lambda \\ \sin\phi\cos\lambda \\ \sin\lambda \end{bmatrix} \tag{7-111}$$

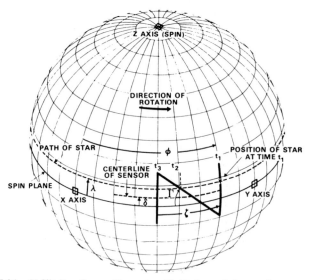

Fig. 7-24.   N-Slit Star Sensor Geometry. N slit shown is larger than actual size.

At $t_i = t_1$, Eq. (7-108) may be replaced by

$$\hat{S}(t_1) = A(t_1)\hat{S}^I \tag{7-112}$$

This equation may be used for observation identification and attitude model refinement instead of Eq. (7-109).

7.6.2 Image Dissector Tube Star Measurements

Image dissector tube star sensors, such as the Ball Brothers CT-401 Fixed Head Star Tracker used on the SAS-3 and HEAO-1 missions (Section 6.4), measure two coordinates U and V, as shown in Fig. 7-25, which are ideally proportional to the position of the observed star's image on the sensor's focal plane.

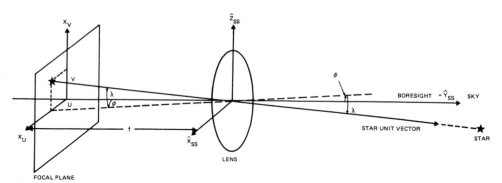

Fig. 7-25.   Image Dissector-Type Star Sensor Geometry

The line of sight of the star's image on the focal plane is located using the angles ϕ and λ, which are defined with respect to the sensor's reference frame, as shown in Fig. 7-25. λ is the elevation of the image from the X_{ss}-Y_{ss} plane and ϕ is the angle between the negative Y_{ss} axis and the projection of the image line of sight onto the X_{ss}-Y_{ss} plane. Figure 7-26 illustrates the use of these angles to locate the unit vector, $\hat{S}_{ss}$, of the star corresponding to the image on the focal plane. The components of $\hat{S}_{ss}$ in the sensor's reference frame in terms of ϕ and λ are

$$\hat{S}_{ss} = \begin{bmatrix} -\sin\phi\cos\lambda \\ \cos\phi\cos\lambda \\ -\sin\lambda \end{bmatrix} \qquad (7\text{-}113)$$

From Fig. 7-25, the relationship between ϕ and λ and the coordinates U and V is

$$\tan\phi = U/f$$
$$\tan\lambda = (V/f)\cos\phi \qquad (7\text{-}114)$$

where f is the focal length of the lens.

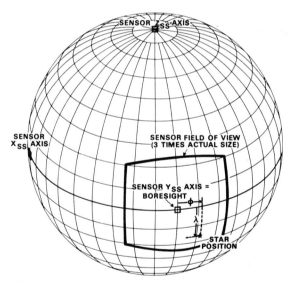

Fig. 7-26. Star Position on Spacecraft-Centered Celestial Sphere Showing Positive Sense of ϕ and λ Measurements

Because image dissector tube sensors are subject to optical and electronic distortion, temperature, magnetic, and star intensity effects, the simple relationships in Eq. (7-114) are not precise. Gates and McAloon [1976], Cleavinger and Mayer [1976] and Gray, et al., [1976] describe the calculation of ϕ and λ from U and V using an empirical model based on laboratory calibrations. The following series for the computation of ϕ and λ takes into account optical and electronic distortion and temperature effects.

$$\phi = C_0 + C_1 u + C_2 v + C_3 u^2 + C_4 uv + C_5 v^2 + C_6 u^3$$
$$+ C_7 u^2 v + C_8 uv^2 + C_9 v^3 \qquad (7\text{-}115)$$

$$\lambda = D_0 + D_1 u + D_2 v + D_3 u^2 + D_4 uv + D_5 v^2 + D_6 u^3$$
$$+ D_7 u^2 v + D_8 uv^2 + D_9 v^3$$

where $u = U/f$, $v = V/f$, and the coefficients C and D are temperature dependent. The magnitudes of magnetic and star intensity effects vary depending on the particular sensor being used. For the Ball Brothers star trackers used on HEAO-1, magnetic effects are approximately 0 to 20 arc-sec, depending on the magnetic field strength and the star position in the field of view. Intensity effects are approximately 0 to 45 arc-sec, depending on the star magnitude and the position in the field of view. Algorithms for the calculation of corrections to ϕ and λ due to these effects are discussed by Gray *et al.*, [1976].

The observed star unit vector $\hat{\mathbf{S}}_{ss}$, calculated from Eq. (7-113), is related to a unit vector, $\hat{\mathbf{S}}^I$, in the rectangular celestial frame by

$$\hat{\mathbf{S}}_{ss} = MA\hat{\mathbf{S}}^I \qquad (7\text{-}116)$$

where A is the spacecraft attitude matrix and M is the transformation matrix from the spacecraft body frame to the star sensor reference frame.

7.6.3 Modeling Sensor Intensity Response

In general, a star sensor's spectral response is such that neither the visual, V, nor the blue, B, star magnitudes defined in Section 5.6 accurately corresponds to the magnitude measured by the sensor. *Instrumental star magnitudes*, that is, magnitudes which take into account the spectral response characteristics of the sensor, must be calculated for each sensor to create a star catalog which contains a minimum of stars but includes all of those that the sensor is likely to observe. Instrumental magnitudes are also necessary for modeling the output of sensors which provide intensity measurements.

A system for computing instrumental magnitudes has been proposed by Gottlieb [1977] for the SAS-3 and HEAO-1 star trackers. The instrumental magnitude, m_I, is modeled as a linear combination of the V and B magnitudes

$$m_I = hV + (1-h)B \qquad (7\text{-}117)$$

where h is a constant between 0 and 1. This constant may be determined by comparing the laboratory-measured wavelength response of the sensor and the wavelength sensitivities of the B and V magnitudes given in Section 5.6. An experimental value for h may be obtained by varying h until the observed distribution of instrumental magnitudes best matches a theoretical distribution of stellar magnitudes or until a sharp sensor magnitude limit is obtained.

Several slit and image dissector sensors provide an output signal which is related to the intensity of the detected star. This measurement, I, is related to the star's instrumental magnitude by

$$\log_{10} I = C_I + D_I m_I$$

where C_I and D_I are sensor-dependent coefficients obtained from laboratory calibration. See Section 5.6 for a further discussion of stellar magnitudes.

7.7 Star Identification Techniques

David M. Gottlieb

Star identification refers to the process used to associate sensor observations with stars in a star catalog. Because the sensor observations can be related to a spacecraft reference frame, and the star catalog gives star positions in an inertial frame of celestial coordinates (CC), this identification allows computation of the spacecraft attitude.

The process of star identification usually begins with the transformation of the sensor observations to a frame that is as close as possible to celestial coordinates, called the *estimated CC frame*. This permits the identification algorithm to operate with the smallest possible error window, thereby reducing misidentifications and ambiguities. If the initial attitude estimate is poor or if an inaccurate model of spacecraft motion is used, the estimated CC frame may be very far from the true CC frame; this would normally be the case, for example, during attitude acquisition. When the model of the spacecraft motion is poor, the estimated frame may also be seriously *distorted*; in other words, the angular distance between observations in the estimated frame may differ significantly from the angular distance between the corresponding catalog stars in an undistorted frame. This greatly complicates the star identification process, and may even make it impossible.

In this section we discuss four types of star identification algorithms: direct match, angular separation match, phase match, and discrete attitude variation. The *direct match* technique matches each observation with a catalog star lying within a specified tolerance of its position. This requires that the estimated coordinate frame be very close to the true coordinate frame. Onboard processors use this technique whenever a good guess of the attitude is available (e.g., HEAO-1; see Gray, *et al.*, [1976]. The *angular separation* technique matches angular distances between observations with angular distances between catalog stars and is used when the observations are in a frame that is only slightly distorted but the initial attitude estimate is not sufficiently accurate to permit the use of the direct match technique. SAS-3, where the motion model is sometimes inaccurate and the initial attitude poorly known, uses this technique [Berg, *et al.*, 1974]. The *phase match* technique is a one-dimensional version of the angular separation match. Known star *azimuths** are compared with observed star longitudes as the phase between the estimated and true frames is stepped through 360 deg. This may be used when the observations are in a frame that is only slightly distorted and when the spin axis of the spacecraft is well known but the phase angle about the spin axis is poorly known. HEAO-1 used a phase match for its original attitude acquisition [Gray, *et al.*, 1976]. Finally, the *discrete attitude variation* technique, appropriate when everything else fails, uses the direct match or angular separation technique, as the initial attitude and motion parameters are stepped through various values in their possible range.

---

*Azimuth will be used to mean the longitude of the observation in the estimated CC frame.

In discussing the above methods, we will use the following definitions and assumptions.

Distortion.   If the estimated CC frame is severely distorted, star identification will be difficult or impossible because the distortion causes the need for large error windows, which causes an excessive number of *field stars* (random catalog stars located close to the observed star) to invade the windows. Sometimes the interval of analysis can be shortened to alleviate this problem.

Scores.   The result of any attempted identification procedure can be reduced to a numerical *score*, such as the total number of unambiguous matches attained by the direct match technique. The identification is accepted if the score is sufficiently high. Alternatively, a number of attempts, exhausting a complete set of possible values of some parameter(s), can be made, and the one yielding the highest score accepted.

Coordinate Frame.   We assume here that the observations are transformed to an estimated CC frame, but the star catalog could be transformed into an estimated sensor frame instead. Normally, the technique that requires fewer transformations would be the most advantageous.

Multidimensional Matches.   If a sensor also observes something other than star position (brightness, for example), a multidimensional match on position and the other observed variable can be performed. This increases the power of any technique.

Related Problems.   The problem of star identification is quite distinct from the apparently related problems of pattern matching such as those encountered by Earth resources satellites or by character scanning. For these, the search is for some specific set of patterns. In contrast, star identification presents an ever-changing set of search patterns, distinct for each point on the sky.

7.7.1 Direct Match Technique

The direct match technique matches an observation in the estimated CC frame with catalog stars lying sufficiently close to it. For this to succeed, the identification window must be small enough to avoid the incursion of too many field stars. This means that the initial attitude estimate and model of spacecraft motion must be accurate. An observation is matched with a catalog star if

$$d(\hat{\mathbf{O}}', \hat{\mathbf{S}}) < \epsilon \qquad (7\text{-}118)$$

where $d(\hat{\mathbf{O}}', \hat{\mathbf{S}})$ is the angular distance between $\hat{\mathbf{O}}'$, the observation unit vector in the estimated CC frame, and $\hat{\mathbf{S}}$, the catalog star unit vector in the true CC frame; and ϵ is the error window radius.

After checking an observation against all possible catalog stars, one of three outcomes is possible: *no identification* (no catalog star within the error window), an *ambiguous identification* (two or more catalog stars within the error window), or a *unique identification* of the observation with exactly one catalog star. In the last case, the identification is hopefully a *correct identification*; if, however, the observation is not in the star catalog or lies outside the error window, it is a *misidentification*. Misidentifications, even in small numbers, can cause some attitude solution techniques to diverge. Therefore, care must be exercised to keep the proportion of misidentifications small enough for the attitude solution technique employed. The score for a direct match technique can be either the number of stars identified or the percentage identified. In either case, a match is successful if the score is sufficiently high.

Because of its simplicity, the direct match technique may be statistically analyzed using a Poisson distribution. The fraction of correct identifications is

$$qh \exp(-E)$$

where

$$E = 1 + 2\pi\rho(1 - \cos\epsilon) \qquad (7\text{-}119)$$

where q is the probability that the correct star is in the star catalog, and is equal to the catalog completeness fraction for objects detectable by the sensor, as discussed in Section 5.6; ρ is the density of detectable stars in the region of observation brighter than the sensor detection limits, expressed in stars per steradian; and h is the probability that an observation lies inside the window. The probability, h, is a function of both window size and the error distribution function. Because this last must include sensor errors, data processing inaccuracies and star catalog position errors, it is very sensor dependent. Frequently, errors follow a Gaussian distribution with some known or estimatable standard deviation. Note that E in Eq. (7-119) is one plus the expectation value for the number of stars to be found within the window. (See Eq. (A-12), Appendix A.)

The fraction of no identifications is:

$$(1 - qh)\exp(-E) \qquad (7\text{-}120)$$

The fraction of misidentifications is

$$(1 - qh)E \exp(-E) \qquad (7\text{-}121)$$

The fraction of ambiguous identifications is the sum of the fraction of cases where the correct star and one or more field stars are in the window, and the fraction of cases where the correct star is not in the window but two or more field stars are

$$[1 - \exp(-E)]qh + [1 - (1 + E)\exp(-E)](1 - qh) \qquad (7\text{-}122)$$

The four probabilities given above can be used to optimize the choice of the window size, ϵ. To do so, note that both nonidentifications and ambiguous identifications are normally dropped from consideration by identification algorithms. Misidentifications, even in small numbers, may cause erroneous results and should be minimized. However, too small a choice for ϵ may result in an insufficient number of correct identifications. The optimum value of the window size is the one that produces the desired tradeoff between the number of correct identifications and the number of misidentifications. To illustrate this point, consider the following example, which is similar to the performance of HEAO-1.

We assume that the limiting magnitude of the sensor is 7.5 visual (V). At this limit, the star catalog is estimated to be 98% complete. Hence, $q = 0.98$. Near the galactic plane, the star density to 7.5 V is approximately 1 star/deg^2 (from Table 5-14). Hence, $\rho = 3.3 \times 10^3$ stars/sr. We model h as a hyperbolic function of the expected error, α, such that at $\alpha = 0$, $h = 1$; at $\alpha = \infty$, $h = 0$; and at $\alpha = \langle\alpha\rangle$ (the mean expected error), $h = 0.5$. Thus, $h = \epsilon/(\epsilon + \langle \alpha \rangle)$ for a window of radius ϵ. Assigning a value of 0.05 deg to $\langle \alpha \rangle$, we use Eqs. (7-119) to (7-122) to compute the probability of each direct match outcome as a function of ϵ. Table 7-3 presents the results. Note that as ϵ is increased, the probability of a correct identification rises steadily until $\epsilon \approx 2\langle \alpha \rangle$. The increase then slows, reaching a maximum at $\epsilon \approx 4\langle \alpha \rangle$, after which the probability of a correct identification declines due to an increase in the probability of an ambiguous identification. The probability of a misidentification also rises with increasing ϵ. Therefore, the optimum choice of ϵ would be $\epsilon \lesssim 4\langle \alpha \rangle$. The more important it is to avoid misidentifications, or the smaller the fraction of observations that must be identified for the proper functioning of the rest of the algorithm, the smaller ϵ should be.

Table 7-3.   Probability of Direct Match Outcomes for Limiting Visual Magnitude 7.5, 98% Catalog Completeness, and Stars in the Galactic Plane. (See text for explanation.)

ϵ (DEG)	h (DEG)	PROBABILITY OF			
		CORRECT IDENTIFICATION	NO IDENTIFICATION	MISIDENTIFICATION	AMBIGUOUS IDENTIFICATION
0.005	0.091	0.089	0.911	< 0.0001	< 0.0001
0.020	0.286	0.280	0.719	0.009	0.0003
0.040	0.444	0.433	0.562	0.0027	0.0021
0.060	0.546	0.529	0.460	0.0050	0.0059
0.080	0.615	0.591	0.389	0.0075	0.0118
0.100	0.667	0.634	0.336	0.0100	0.0198
0.150	0.750	0.687	0.248	0.0162	0.0493
0.200	0.800	0.697	0.192	0.0214	0.0901
0.250	0.833	0.683	0.153	0.0251	0.1384
0.350	0.875	0.619	0.103	0.0286	0.2493

7.7.2 Angular Separation Match Technique

The *angular separation match* technique matches angular distances between observations with angular distances between catalog stars. For this technique to work, the estimated CC frame must be sufficiently undistorted. That is, the spacecraft motion must be sufficiently well modeled such that

$$|d(\hat{\mathbf{O}}'_1, \hat{\mathbf{O}}'_2) - d(\hat{\mathbf{O}}_1, \hat{\mathbf{O}}_2)| \ll \mu \qquad (7\text{-}123)$$

holds for enough observations, where $\hat{\mathbf{O}}_1$ and $\hat{\mathbf{O}}_2$ are unit vectors for two stars in the true CC frame and $\hat{\mathbf{O}}'_1$ and $\hat{\mathbf{O}}'_2$ are the corresponding unit vectors in the estimate CC frame. μ is an error allowance for the inaccuracies of observation and the distortion of the estimated CC frame.

In addition, the number of candidates, or catalog stars which could possibly be identified with an observation, must be manageable. Because candidates are chosen for their proximity to the observation, the initial attitude estimates must be sufficiently accurate so that the number of catalog stars within the *candidate search window* is not so large as to impose storage or processing time problems; i.e.,

$$d(\hat{\mathbf{O}}', \hat{\mathbf{S}}) < \epsilon \qquad (7\text{-}124)$$

holds for sufficiently few catalog stars, where ϵ is the radius of the candidate search window. Experience with the SAS-3 spacecraft suggests that an average of two or three candidates per observation is satisfactory. However, if this number exceeds five to ten, ambiguities and misidentifications may lead to an insufficient number of correct unambiguous identifications.

The simplest angular separation match is a pairwise match between just two observations. This is done by picking two observations with unit vectors $\hat{\mathbf{O}}_1'$ and $\hat{\mathbf{O}}_2'$ in the estimated CC frame. The candidate catalog stars for $\hat{\mathbf{O}}_1'$ are those that meet the requirement

$$d(\hat{\mathbf{O}}_1', \hat{\mathbf{S}}) < \epsilon$$

The candidates for $\hat{\mathbf{O}}_2'$ are similarly selected. A match exists and $\hat{\mathbf{O}}_1'$ and $\hat{\mathbf{O}}_2'$ are associated with $\hat{\mathbf{S}}_1$ and $\hat{\mathbf{S}}_2$, respectively, provided that

$$|d(\hat{\mathbf{O}}_1', \hat{\mathbf{O}}_2') - d(\hat{\mathbf{S}}_1, \hat{\mathbf{S}}_2)| < \mu \qquad (7\text{-}125)$$

If this condition is met by more than one pair of catalog stars, the match is ambiguous.

Polygon matches can help resolve ambiguities and generally increase the reliability of the star identification. This technique consists of selecting a set of N observations ($N > 2$). Each pair of observations can be matched with catalog stars as above. The polygon match is considered successful when each pairwise distance match is successful *and* when the catalog star associated with each observation is the same for all pairwise distance matches involving that observation. An alternative approach is to form m vectors containing the distances between all pairs of observations, where $m = \binom{N}{2} = N \cdot (N-1)/2$. For example, for $N = 4$, the m vector for observations 1, 2, 3, and 4 would be

$$\left[d(\hat{\mathbf{O}}_1', \hat{\mathbf{O}}_2'), d(\hat{\mathbf{O}}_1', \hat{\mathbf{O}}_3'), d(\hat{\mathbf{O}}_1', \hat{\mathbf{O}}_4'), d(\hat{\mathbf{O}}_2', \hat{\mathbf{O}}_3'), d(\hat{\mathbf{O}}_2', \hat{\mathbf{O}}_4'), d(\hat{\mathbf{O}}_3', \hat{\mathbf{O}}_4') \right]^T$$

A successful match occurs when the observation m vector is sufficiently close to the m vector of distances between four catalog stars. The disadvantages of the polygon match are that it requires more data than a pairwise distance match and that the computation time is longer. In the most efficient models, the computation time will increase approximately as N^2.

The angular separation match technique is difficult to analyze statistically; this makes the choice of ϵ and μ more difficult than for the direct match method. ϵ must be large enough to allow for the error in the initial attitude estimate plus the error caused by inaccuracies in the motion model. For SAS-3, $\epsilon = 5$ deg gave satisfactory results. μ need only be large enough to allow for the distortion caused by the motion model inaccuracy. It should be set to the maximum anticipated error in attitude at the end of the interval of analysis, assuming that the initial attitude was perfect. If μ is too large, ambiguous identifications and misidentifications will arise; if it is too small, no identification will be possible. The analyst must choose μ on the basis of the data accuracy and previous experience with the particular algorithm.

7.7.3 Phase Match Technique

The phase match technique computes a phase angle about a known spin axis by matching observation longitudes and catalog star azimuths about that spin axis. To use this technique, the frame of the sensor observations must be nearly undistorted; i.e.,

$$|\Phi(\hat{O}'_1, \hat{O}_1) - \Phi(\hat{O}'_2, \hat{O}_2)| < \delta_1 \qquad (7\text{-}126)$$

where Φ is the phase or azimuth difference and δ_1 is an error tolerance to allow for distortion. (For the HEAO-1 attitude acquisition algorithm, $\delta_1 = 1$ deg.) For the phase match technique to work, the spin axis of the spacecraft must be known to an accuracy substantially better than δ_1. The phase about the spin axis need not be known, however.

To implement a phase match, compute the phases of all observations in an estimated CC frame with an arbitrary zero phase. Next, extract from a star catalog all stars, $\hat{S}$, meeting the requirement

$$\cos(\theta + \zeta + \delta_2) \leqslant \hat{Z} \cdot \hat{S} \leqslant \cos(\theta - \zeta - \delta_2) \qquad (7\text{-}127)$$

where $\hat{Z}$ is the spin axis unit vector; θ is the angle between the spin axis and the sensor optical axis; ζ is the radius of the sensor field of view; and δ_2 is the maximum anticipated error in the spin axis position. Compute the longitude of each catalog star as discussed in Section 5.6. Divide the entire azimuth circle (0 to 360 deg) into bins of equal width, δ, such that

$$\delta \geqslant \delta_1 + \delta_2 \sin \zeta \qquad (7\text{-}128)$$

where the second term allows for errors in catalog star longitudes caused by errors in the spin axis position. The score, R, is given by

$$R = \sum_{i=1}^{B} \sum_{j=1}^{B} N_i M_j \qquad (7\text{-}129)$$

where $N_i = 0$ if there are no observations in the ith azimuth bin and $N_i = 1$ otherwise; $M_j = 0$ if there are no catalog stars in the jth longitude bin and $M_j = 1$ otherwise; and B is the number of bins.

Rotate the observation frame by δ by adding δ to each observation azimuth and compute a score for the new configuration. Repeat this process for a complete 360-deg circuit. The highest score corresponds to the correct phase for the observations. Alternatively, the process can be stopped when a score is attained which the analyst feels is sufficiently large to ensure that the correct phase has been found.

A major limitation of the phase match technique is that it fails if either the N_i or M_j values in Eq. (7-129) are mostly 1.

To ensure that $M_j = 0$ often enough, the mean number of catalog stars per longitude bin, $\langle S \rangle$, must be $\lesssim 2$; $\langle S \rangle$ is given by

$$\langle S \rangle = 2\delta(\zeta + \delta_2)\rho \qquad (7\text{-}130)$$

where ρ is the density of stars brighter than the limiting magnitude of the sensor (see Section 5.6). If $\langle S \rangle \gtrsim 1$, those bins where there are no stars become the important ones; if the star catalog is complete, these "holes" will never contain observations when the correct phase is found.

Because the catalog must contain all or nearly all stars to the limiting magnitude of the sensor, the only way to control $\langle S \rangle$ is by adjusting the sensor sensitivity. If the threshold is sufficiently high (i.e., only fairly bright stars are detected), the star density, ρ, will be low; thus, $\langle S \rangle$ will be low and M_j will be zero sufficiently often. This procedure also ensures that N_i will be zero often enough because the sensor cannot observe more stars than are in the catalog, if the catalog is nearly complete.

Several refinements to this technique are possible. If the star catalog is complete, we can assume that a single selected observation will match some star in the catalog. By matching the observation with each catalog star in turn, a set of possible phase angles is generated. Because the number of catalog stars is normally far less than the number of bins, this reduces the number of scores which must be calculated.

A second refinement makes use of elevation and azimuth information. Because the maximum elevation error for an observation is δ_2, the elevation information is useful if

$$\delta_2 \ll \delta$$

To include elevation information, redefine the score given in Eq. (7-129) as

$$R = \sum_{i=1}^{B} \sum_{j=1}^{B} N_i M_j E_{i,j} \qquad (7\text{-}131)$$

where $E_{i,j} = 1$ if the elevation of one of the observations in the ith bin is within δ_2 of the elevation of one of the catalog stars in the jth bin, and is 0 otherwise.

7.7.4 Discrete Attitude Variation Technique

This technique should be used only as a last resort because it involves the repeated use of one of the other matching techniques and is therefore very costly in computation time. No knowledge of the initial attitude is required, but any information available can be used to limit the number of attitude guesses that must be tried. However, the observation frame must be undistorted to the extent required by the identification technique that will be used. To implement the discrete attitude variation technique, create an array of trial attitudes such that no possible attitude is more than ϵ angular distance from one of the discrete trial attitudes in the array. For each trial attitude, apply any of the other identification techniques described above. The attitude that gives the highest score is taken to be correct.

Because the number of possible attitudes may be very large (e.g., if $\epsilon = 1$ deg, there are over 40,000 of them), refinements to the technique which cut down the number of guesses are critical. For one such refinement, assume that no information is available concerning the initial attitude and that the number of catalog stars is much less than the number of discrete attitude guesses that are possible. Provided that any given observation is very likely to correspond to one of the catalog stars, then a substantial savings in computer time can be realized by assuming that the observation matches each catalog star in turn. For any given catalog star, this determines two of the attitude axes; the third is discretely estimated as above. Each catalog star and third-axis attitude is tried until a sufficiently high score is attained. This refinement is very powerful if it is possible to narrow the field of candidates in some way for any one of the observations. For example, if some brightness information is available, the brightest observation might be used. The star catalog candidates for this observation are then limited to only the brightest stars, thus proportionately decreasing the number of attitude guesses that must be tried.

7.8 Gyroscope Models

Lawrence Fallon, III

As described in Section 6.5, gyroscopes form the major component of inertial guidance systems, which are used extensively for attitude propagation and control. This section describes mathematical models for the estimation of spacecraft angular rates from gyro measurements, the simulation of gyro outputs from true spacecraft angular rates, and the modeling of noise in gyro outputs. We are concerned primarily with torque rebalanced single-degree-of-freedom gyros and draw largely on analysis performed by the TRW Systems Group for the High Energy Astronomy Observatory (HEAO) missions [McElroy, 1974]. The notation of Section 6.5 is used throughout.

7.8.1 Gyro Measurements

The gyro output, θ, represents a voltage proportional to the torque current in an analog rebalanced gyro or the number of rebalance pulses in a pulse rebalanced gyro (Section 6.5). The relationship between θ and ω_i, the angular rate component in the direction of the gyro's input axis, depends on the type of gyro in use. For example, rate gyros supply an angular displacement, θ_R, which is ideally proportional to ω_i. Thus

$$\omega_i{}^M = K_R \theta_R \qquad (7\text{-}132)$$

where $\omega_i{}^M$ is the gyro's measurement of ω_i, and K_R is the rate gyro scale factor. Rate-integrating gyros operating in the rate mode (Section 6.5) provide an output θ_I which is ideally proportional to the integral of ω_i over a sampling interval δt_I. Thus, the gyro's measurement of the average angular velocity over the interval is obtained from

$$\omega_i{}^M = \frac{K_I \theta_I}{\delta t_I} \qquad (7\text{-}133)$$

where K_I is the rate-integrating scale factor. The interval, δt_I, typically in the 200- to 500-ms range, must be chosen such that θ_I remains small. In high angular velocity environments, such a choice may not be practical, and errors in the computation of ω_i^M result if the gyro's input axis moves significantly within δt_I. A more sophisticated algorithm given by Paulson, et al., [1969] reduces errors due to this effect.

7.8.2 Model for Measured Spacecraft Angular Velocity

The spacecraft's angular velocity ω_i in the direction of the gyro's input axis is related to the gyro's measurement of this quantity ω_i^M by the following model from Iwens and Farrenkopf [1971].

$$\omega_i^M = (1 + k_i)\omega_i + b_i + n_i \qquad (7\text{-}134)$$

where k_i is a small correction to the nominal scale factor in Eq. (7-132) or Eq. (7-133), because K_R and K_I are not precisely known; b_i is the drift rate; and n_i is white noise on the gyro output. In torque rebalanced gyros, b_i represents a null shift in the torque rebalance control loop which generally is not constant but may be influenced by gyro noises and systematic effects. Modeling of gyro noise sources is discussed in more detail in Section 7.8.4.

If the direction of the gyro's input axis is given by a unit vector, $\hat{U}_i$, in the spacecraft coordinate frame, Eq. (7-134) becomes

$$\omega_i^M = (1 + k_i)\hat{U}_i \cdot \omega + b_i + n_i \qquad (7\text{-}135)$$

where ω is the true spacecraft angular velocity vector.

Consider a configuration of N single-degree-of-freedom gyros with input axes oriented to measure the three components of ω (Section 6.5). To account for the N gyros, the following vectors are constructed:

$$\omega_g^M = \begin{bmatrix} \omega_{i1}^M \\ \vdots \\ \omega_{iN}^M \end{bmatrix}; \quad \mathbf{b}_g = \begin{bmatrix} b_{i1} \\ \vdots \\ b_{iN} \end{bmatrix}; \quad \mathbf{n}_g = \begin{bmatrix} n_{i1} \\ \vdots \\ n_{iN} \end{bmatrix} \qquad (7\text{-}136)$$

Gyro geometry and scale factor error matrices are similarly constructed:

$$U = \begin{bmatrix} \mathbf{U}_{i1}^T \\ \vdots \\ \mathbf{U}_{iN}^T \end{bmatrix} \qquad K = \begin{bmatrix} 1 + k_{i1} & & 0 \\ & \ddots & \\ 0 & & 1 + k_{iN} \end{bmatrix} \qquad (7\text{-}137)$$

It follows that the vector ω_g^M, representing the collective output of the gyro configuration, is given by

$$\omega_g^M = KU\omega + \mathbf{b}_g + \mathbf{n}_g \qquad (7\text{-}138)$$

7.8.3 Calculation of Estimated Angular Velocity From the Gyro Measurements

An expression for the true spacecraft angular velocity is found by solving Eq. (7-138) for ω,

$$\omega = C\left(\omega_g^M - \mathbf{b}_g - \mathbf{n}_g\right) \qquad (7\text{-}139)$$

where

$$C = [U^T K^T K U]^{-1} U^T K^T$$

In the case of three gyros, this expression reduces to

$$C = [KU]^{-1}$$

Because the expected value of $\mathbf{n}_g$ is zero, an estimate of the spacecraft angular velocity, $\langle \omega \rangle$, becomes

$$\langle \omega \rangle = C \omega_g^M - \mathbf{b} \tag{7-140}$$

where $\mathbf{b} = C\mathbf{b}_g$ is the effective gyro drift rate vector in the spacecraft frame.

In practice, the scale factor corrections and input axis orientations are not known precisely and will vary slightly with time. The matrix C is calculated before launch based upon ground calibrated values for K and U and then remains invariant. To take time variations into account, a misalignment/scale factor correction matrix, G, is introduced such that

$$\langle \omega \rangle = (I - G)(C \omega_g^M - \mathbf{b}) \tag{7-141}$$

where input estimates for $\mathbf{b}$ and G are used. Equation (7-141) is a convenient algorithm for the calculation of spacecraft angular velocities from gyro measurements for use in attitude propagation (Section 17.1) or in attitude control (Section 19.4).

Because each gyro contributes a scale factor uncertainty and a 2-deg-of-freedom input axis alignment error, the elements of G will in general be independent for $N \geqslant 3$. G may be initialized at zero if C contains all scale factor and alignment information after gyro calibration and spacecraft assembly. After launch, it may be necessary to refine G occasionally due to small scale factor and alignment changes. Estimates for the gyro drift are available after gyro calibration but must be redetermined frequently after launch. Procedures for the refinement of $\mathbf{b}$ and G are presented in Section 13.4 and by Gray, et al., [1976].

7.8.4 Modeling Gyro Noise Effects

Gyro noise may seriously degrade the accuracy of the calculated spacecraft angular velocities and of attitude estimates based on these angular velocities. For torque rebalanced gyros, it is convenient to model gyro noise as being composed of electronic noise, float torque noise, and float torque derivative noise, as introduced in Section 6.5. The models given here for these three noise sources follow the formulation of Farrenkopf [1974] and McElroy [1975] for rate integrating gyros in the rate mode.

Electronic noise is modeled as a time-correlated colored noise* of standard deviation σ_e on the gyro output. At the kth readout time interval, the electronic noise, $n_e(k)$, is

$$n_e(k) = e^{-\delta t_I/\tau}\left[n_e(k-1) - \sigma_e\zeta_e(k-1)\right] + \left(1 - \sqrt{1 - e^{-2\delta t_I/\tau}}\right)\sigma_e\zeta_e(k) \quad (7\text{-}142)$$

where $\zeta_e(k)$ is a normally distributed random number with zero mean and unit standard deviation, τ is the torque rebalance loop time constant (Section 6.5), and δt_I is the gyro readout time interval. If τ is much less than δt_I, then Eq. (7-142) is simply

$$n_e(k) = \sigma_e\zeta_e(k)$$

Float torque noise is assumed to be white Gaussian noise of standard deviation σ_v on the gyro drift rate. It is modeled as a noise, $n_v(k)$, on the gyro output corresponding to the kth readout interval given by

$$n_v(k) = \sigma_v\sqrt{\delta t_I}\; \zeta_v(k)$$

where $\zeta_v(k)$ is a normally distributed random number with zero mean and unit standard deviation independent from $\zeta_e(k)$.

Float torque derivative noise is integrated white noise of standard deviation σ_u, and is modeled as a noise, $n_u(k)$, on the drift rate at the kth readout interval. Thus,

$$n_u(k) = n_u(k-1) + \sigma_u\sqrt{\delta t_I}\; \zeta_u(k)$$

where $\zeta_u(k)$ is a random number analogous to but independent of both $\zeta_e(k)$ and $\zeta_v(k)$.

Gyro noise effects cause an uncertainty in the angular rates calculated from Eqs. (7-132) and (7-141), which then cause cumulative uncertainties in attitudes determined using these angular rates. If the spacecraft attitude and drift rate are known exactly at time t_1, then at time $t_2 = t_1 + \Delta t$ the attitude uncertainty will follow a Gaussian probability distribution with standard deviation

$$\sigma_\theta = \left(\sigma_e^2 + \sigma_v^2\Delta t + \frac{1}{3}\sigma_u^2(\Delta t)^3\right)^{1/2} \quad (7\text{-}143)$$

As an example of noise levels, the HEAO-1 gyros have specified values of $\sigma_e = 0.5$ arc-sec, $\sigma_v = 0.22$ arc-sec/sec$^{1/2}$, and $\sigma_u = 4.7 \times 10^{-5}$ arc-sec/sec$^{3/2}$. A plot of Eq. (7-143) using these parameters is shown in Fig. 7-27. At time t_1 (0.32 sec) on this figure, attitude uncertainty at the 1 σ level is 0.5 arc-sec; at t_2 (32 min) it is 9.6 arc-sec; and at t_3 (24 hours), 690 arc-sec. The frequency at which the attitude reference and drift rate must be redetermined depends largely on the noise characteristics of the particular gyros in use.

---

*If the value of a noise at one time influences its value at some other time, then it is a *colored noise*. The value of *white noise* at one time gives no information regarding its value at any other time.

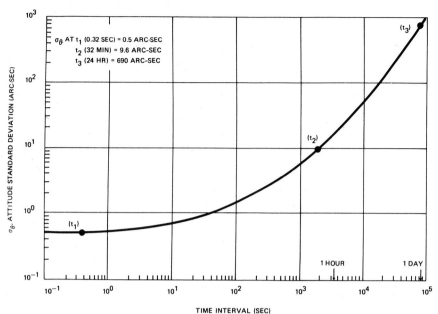

Fig. 7-27. Model of Attitude Determination Uncertainty Due to Gyro Noise Effects Using Parameters Specified for the High Energy Astronomy Observatory-1 Mission. (Adapted from McElroy, [1975].)

7.9 Reaction Wheel Models

F. L. Markley

If active spacecraft control is to be modeled in a simulation system, mathematical models of the control system are needed. Specifically, models of the reaction wheel torque and friction characteristics are needed to model reaction wheel control systems. As discussed in Section 6.6, reaction wheel characteristics differ widely among spacecraft; therefore, we choose as a single illustrative example the reaction wheel proposed for the IUE spacecraft [Welch, 1976]. This wheel is equivalent to that used on the yaw axis of Nimbus and has a moment of inertia of 0.00338 kg-m^2 and a *synch speed* of 1500 rpm.*

The wheel torque is provided by an AC two-phase induction motor, which is driven by square pulses provided by a reaction wheel drive electronics package. The torque level is controlled by varying the *duty cycle*, or fraction of each half-cycle in which the applied square-wave voltage is nonzero. The duty cycle, X_{dc}, is varied between $+1$ and -1 by a control voltage, V, as shown in Fig. 7-28.†

---

*The *synch speed*, or synchronous speed, is the speed of the wheel at which the electromagnetic torque N_{em}, defined below, is zero.

† The torque applied to an induction motor is proportional to the square of the applied voltage. The drive electronics includes a square root circuit so that the applied torque is proportional to the control voltage. The duty cycle as used in Eq. (7-144) and Fig. 7-28 is actually the squared input to the wheel. The nonlinearity of the function graphed in Fig. 7-28 near $V=0$ is due to the fact that the mathematical square foot function has infinite slope at the origin, which can be modeled only approximately in the wheel drive electronics package. For a negative duty cycle, the phase relation between the signals applied to the two phases of the wheel is reversed.

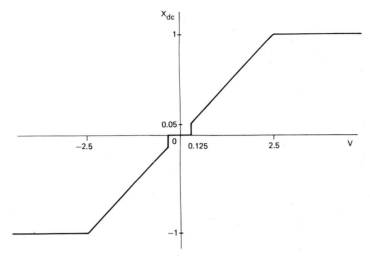

Fig. 7-28.   Duty Cycle, X_{dc}, as a Function of Control Voltage, V

The net torque on the wheel is given by

$$N = X_{dc} N_{em} - N_{friction} \qquad (7\text{-}144)$$

where both N_{em}, the applied electromagnetic torque when the duty cycle is unity, and $N_{friction}$, the bearing friction torque, depend on the wheel speed, s. The dependence of N_{em} on s is shown in Fig. 7-29. For accurate simulation, a table of values and an interpolation scheme should be used for N_{em}. For less precise calculations, the following approximation is adequate:

$$N_{em} = 2N_0 \alpha r (\alpha^2 + r^2)^{-1} \qquad (7\text{-}145)$$

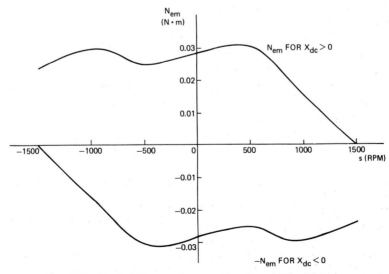

Fig. 7-29.   Applied Torque, N_{em}, as a Function of Wheel Speed, s

where $r = 1 - s/s_{max}$ for $X_{dc} > 0$, $r = 1 + s/s_{max}$ for $X_{dc} < 0$, s_{max} is the synch speed, α is the value of r for which N_{em} has maximum magnitude, and N_0 is the maximum magnitude of N_{em}. The friction torque is most simply modeled as the sum of Coulomb and viscous terms

$$N_{friction} = N_c \text{sgn}(s) + fs \qquad (7\text{-}146)$$

For the IUE wheel, the Coulomb friction coefficient is $N_c = 7.06 \times 10^{-4}$ N·m, and the viscous friction coefficient is $f = 1.21 \times 10^{-6}$ N·m/rpm.

A considerably more sophisticated friction model developed by Dahl [1968] can be used where the simple model described above is inadequate. The Dahl model is a statistical model of friction as the random making and breaking of bonds between solid surfaces. It includes "*stiction*," the increased friction found when the relative velocity between the sliding surfaces is zero.

7.10 Modeling Gas-Jet Control Systems

Robert S. Williams

A mathematical model of a gas-jet control system is used to predict the spacecraft response when a given set of commands is input to the control system. This prediction may be used for simulation, for refinement of initial estimates when computing commands, or for comparison with the actual spacecraft response during or after the execution of a command.

The main factors modeled in a gas-jet system are (1) the *thrust profile*, or time-dependence of the jet thrust relative to the commanded on and off times, and (2) the alignment of the thruster in the spacecraft body coordinate system. Both factors are ordinarily measured before launch. However, the measured thrust profile may be erroneous if the thrust vector does not lie along the thruster symmetry axis or if launch vibration affects the alignment. Consequently, inflight calibration, discussed in Section 19.3, may be desirable if several maneuvers must be performed and the fuel budget is tight.

Several additional effects may be considered, although in most cases they will be negligible. These are the change in spin rate resulting from the conservation of angular momentum as propellant flows from storage tanks to thrusters, and the change in center of mass and moments of inertia as propellant is consumed. These effects can be easily estimated given the geometry of the tanks and thrusters and the propellant flow rate. The major uncertainty is in the distribution of the propellant within the tanks.

Thrust Profile. A hypothetical thrust profile is shown in Fig. 7-30. The commanded start time is t_0; the thrust begins buildup at t_1 and reaches a steady state at t_2; the commanded stop time is t_3; the thrust begins to decay at t_4 and reaches zero at t_5. The delays between t_0 and t_1 and between t_3 and t_4 are due to electrical and mechanical delays in the valve circuits and to the time for the propellant to flow from the valves to the thrusters. The intervals from t_1 to t_2 and from t_4 to t_5 are the *rise* and *fall times*, respectively; these are nonzero because of the time required to establish steady-state propellant flow and (for hot-gas systems) reaction rate in the thruster. The exact shape of the buildup and decay does not follow any well-established law; Fig. 7-30 is descriptive rather than quantitative. The thrust may not even reach steady state if very short pulses are commanded.

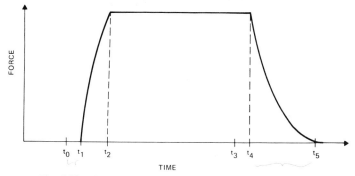

Fig. 7-30.   General Thrust Profile for Gas Jet Control Systems

Typical values for delays and rise and fall times range from a few milliseconds to a few hundred milliseconds. For example, Werking, *et al.*, [1974] cite a fall time of 300 ms for the cold-gas thrusters on RAE-2, whereas a rise time of 10 ms can be inferred from measurements on 0.4 N thrusters for the hot-gas IUE system [Sansevero and Simmons 1975]. Sansevero also reports a delay time of 5 to 15 ms for opening and closing solenoid valves in the latter system.

In both hot-gas and cold-gas systems, the peak force increases with increasing propellant flow rate. In cold-gas systems, a pressure regulator is ordinarily used to maintain a constant flow rate and, hence, peak force as long as the propellant supply pressure remains above the regulator output pressure. If the pressure is not regulated, the flow rate will depend on individual system characteristics, but will drop as propellant is consumed and the supply pressure drops. In hydrazine-fueled systems, the flow rate of the propellant, which is a liquid, is not regulated. In these systems, the peak force is measured at various supply pressures so that thruster performance may be predicted over the entire range of pressures which will be encountered during the mission. The pressure dependence of the force is sometimes described in a parametric form suitable for the thruster model. Otherwise, the model must use interpolation between calibration points to predict thruster performance.

Hydrazine thrusters may not reach their full rated performance until the catalyst heats up. According to measurements on the IUE system [Sansevero and Simmons 1975], the initial thrust from a cold thruster may be as little as 50% of the rated value, rising to 90% after 3 sec of operation and to essentially 100% after 30 sec. Hydrazine thrusters using electrothermal rather than catalyzed decomposition [Murch, *et al.*, 1976; Pugmire and O'Connor, 1976; Grabbi and Murch, 1976] at least partially alleviate this problem because the reaction chamber is heated electrically before propellant is fed in. As with pressure dependence of the force, the time dependence is sometimes available directly from the calibration data in a suitable form; otherwise, the initial thrust and the firing time required to reach full thrust can be used to create a simple piecewise linear model of the increase in force.

The response of the spacecraft to control torques is proportional to the time integral of the torque. When the thruster firing time is long compared with turn-on and turn-off delays, rise and fall times, and warm-up times, response will depend only on the peak force and the time. If the response of the spacecraft is a rate change about an axis about which the rate is directly measured, a detailed model may be unnecessary even for short thruster firing times, if commands can be sent until the measured rate equals the desired rate. As an example of a case in which neither of the above simplifications ordinarily applies, consider precession of a spinning spacecraft, in which the thruster is fired for a series of short intervals, each a fraction of a spin period. (See Section 1.3.) The direction of the applied torque changes with time, so that an average direction and magnitude must be computed, and rise and fall times can be expected to be significant.

The geometry for the computation of the average torque is shown in Fig. 7-31. Here, **L** is the angular momentum vector; **r** is the radius vector from the center of mass to the thruster; **F**(t) is the thrust, assumed to lie in the **L**/**r** plane; and **N** = **r** × **F** is the resultant torque. **N**$_2$ and **N**$_4$, corresponding to the forces **F**(t_2) and **F**(t_4), are shown relative to **N**$_c$, which is the direction of the *average torque*. The

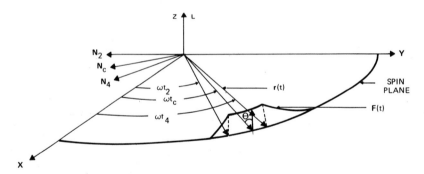

Fig. 7-31.   Geometry of Precession Jet Firing. The torque vector, **N**, is assumed to lie in the spin plane.

centroid, or time (or equivalent angle) at which the instantaneous torque is parallel to the average torque, is computed by requiring that the integral of the torque component, **N**$_\perp$, perpendicular to **N**$_C$ vanish:

$$0 = \int N_\perp(t)\,dt = r\sin\theta \int F(t)\sin\omega(t - {'t_c})\,dt \qquad (7\text{-}147a)$$

where ω is the spin rate, θ is the angle between **r** and **F**, and the integral is computed over one spin period. The *effective torque* or *impulse*, I_c, is then calculated as the time integral of the torque component, $N_{\|}$, parallel to **N**$_c$:

$$I_c \equiv \int N_\|(t)\,dt = r\sin\theta \int F(t)\cos\omega(t - t_c)\,dt \qquad (7\text{-}147b)$$

The required integrals can be computed numerically from the thrust profile. If a trapezoidal approximation, as shown in Fig. 7-32, is sufficiently accurate, the integrals can be performed analytically, with the results that

$$\tan \omega t_c = b/a \tag{7-148}$$

and

$$\frac{\omega I_c}{r \sin \theta} = a \cos \omega t_c + b \sin \omega t_c \tag{7-149}$$

where

$$a = \frac{\cos \omega t_2 - \cos \omega t_1}{\omega(t_2 - t_1)} - \frac{\cos \omega t_5 - \cos \omega t_4}{\omega(t_5 - t_4)}$$

$$b = \frac{\sin \omega t_2 - \sin \omega t_1}{\omega(t_2 - t_1)} - \frac{\sin \omega t_5 - \sin \omega t_4}{\omega(t_5 - t_4)} \tag{7-150}$$

(Time points are labeled to correspond to Fig. 7-30.) The trapezoid model is commonly used for modeling gas-jet thrust profiles at Goddard Space Flight Center.

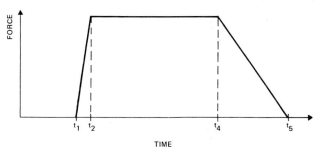

Fig. 7-32.   Trapezoidal Approximation for Gas Jet Thrust Profile

References

1. Abramowitz, M., and I. A. Stegun, *Handbook of Mathematical Functions*, National Bureau of Standards Applied Mathematics Series 55, June 1964.
2. Adcole Corporation, *Digital Solar Aspect Systems*, Waltham, MA, Feb. 1975.
3. ———, *Functional Test Report Five Sun Sensor Head (FSS) Model 19020*, March 1977.
4. Berg, R. A., B. L. Blaylock, W. A. Fisher, B. L. Gambhir, K. E. Larsen, P. V. Rigterink, C. B. Spence, and G. F. Meyers, *Small Astronomy Satellite-C (SAS-C) Attitude Support System Specification and Requirements*, Comp. Sc. Corp., 3000-05700-01TN, July 1974.
5. Chen, L. C., and J. R. Wertz, *Analysis of SMS-2 Attitude Sensor Behavior Including OABIAS Results*, Comp. Sc. Corp., CSC/TM-75/6003, April 1975.
6. Churchill, Ruel V., *Operational Mathematics*, Third Edition, New York: McGraw-Hill, Inc., 1972.

7. Cleavinger, R. L., and W. F. Mayer, *Attitude Determination Sensor for Explorer 53*, Paper No. 76-114, AIAA 14th Aerospace Sciences Meeting, Wash., DC, Jan. 1976.

8. Dahl, P. R., *A Solid Friction Model*, Aerospace Corporation, TOR-0158 (3107-18)-1, El Segundo, CA, May 1968.

9. Farrenkopf, R. L., *Generalized Results for Precision Attitude Reference Systems Using Gyros*, AIAA Paper No. 74-903, AIAA Mechanics and Control of Flight Conference, Anaheim, CA, Aug. 1974.

10. Gates, R. F. and K. J. McAloon, *A Precision Star Tracker Utilizing Advanced Techniques*, Paper No. 76-113, AIAA 14th Aerospace Sciences Meeting, Wash., DC, Jan. 1976.

11. Gottlieb, D. M., *Small Astronomy Satellite-3 (SAS-3) Y-Axis Star Tracker Limiting Magnitude and Instrumental Response*, Comp. Sc. Corp., CSC/TM-76/6047, Jan. 1977.

12. Grabbi, R. and C. K. Murch, "High Performance Electrothermal Hydrazine Thruster (Hi PEHT) Development," AIAA Paper No. 76-656, AIAA/SAE Twelfth Propulsion Conference, Palo Alto, CA, July 1976.

13. Gray, C. M., L. Fallon, D. M. Gottlieb, M. A. Holdip, G. F. Meyers, J. A. Niblack, M. Rubinson, *High Energy Astronomy Observatory-A (HEAO-A) Attitude Determination System Specifications and Requirements*, Comp. Sc. Corp., CSC/SD-76/6001, March 1976.

14. Grosch, C. B., A. E. LaBonte, and B. D. Vannelli, "The SCNS Attitude Determination Experiment on ATS-III," *Proceedings of the Symposium on Spacecraft Attitude Determination, Sept. 30, Oct. 1–2, 1969*, El Segundo, CA; Air Force Report No. SAMSO-TR-69-417, Vol. I; Aerospace Corp. Report No. TR-0066(5306)-12, Vol. I, 1969.

15. Hale, Francis J., *Introduction to Control System Analysis and Design*. Englewood Cliffs, NJ: Prentice-Hall, Inc., 1973.

16. IBM Corporation, *System/360 Scientific Subroutine Package (360A-CM-03X), Version III Programmers Manual*, Document No. H20-0205-3, 1968.

17. Iwens, R. P., and R. L. Farrenkopf, *Performance Evaluation of a Precision Attitude Determination System (PADS)*, AIAA Paper No. 71-964, AIAA Guidance, Control, and Flight Mechanics Conference, Hempstead, NY, Aug. 1971.

18. Jackson, John David, *Classical Electrodynamics*. New York: John Wiley and Sons, Inc., 1963.

19. Joseph, M., J. E. Keat, K. S. Liu, M. E. Plett, M. E. Shear, T. Shinohara, and J. R. Wertz, *Multisatellite Attitude Determination/Optical Aspect Bias Determination (MSAD/OABIAS) System Description and Operations Guide*, Comp. Sc. Corp., CSC/TR-75/6001, April 1975.

20. Kaplan, Wilfred, *Operational Methods for Linear Systems*. Reading, MA: Addison-Wesley Publishing Company, Inc., 1962.

21. Liu, K. S., and J. R. Wertz, *A Bolometer Offset Model for Atmosphere Explorer-C*, Comp. Sc. Corp. 3000-25600-01TM, Dec. 1974.

22. Markley, F. L., private communication, 1977.

23. McElroy, T. T., *Reference Gyro Assembly Model*, TRW Systems Group, HEAO-74-460-085, Oct. 1974.

24. McElroy, T. T., *Gyro Noise Model—Parameter Considerations*, TRW Systems Group, HEAO-75-460-055, Jan. 1975.

25. Murch, C. K., R. L. Sackheim, J. D. Kuenzly, and R. A. Callens, "Non-Catalytic Hydrazine Thruster Development—0.050 to 5.0 Pounds Thrust," AIAA Paper No. 76-658, AIAA/SAE Twelfth Propulsion Conference, Palo Alto, CA, July 1976.

26. Paulson, D. C., D. B. Jackson, and C. D. Brown, "SPARS Algorithms and Simulation Results," *Proceedings of the Symposium on Spacecraft Attitude Determination, Sept. 30, Oct. 1–2, 1969*, El Segundo, CA; Air Force Report No. SAMSO-TR-69-417, Vol. I; Aerospace Corp. Report No. TR-0066(5306)-12, Vol. I, 1969.

27. Philco-Ford, *Synchronous Meteorological Satellite Phase C Design Report*, WDL-TR4545, June 1971.

28. Pugmire, T. K., and T. J. O'Connor, "5 Pound Thrust Non-Catalytic Hydrazine Engine," AIAA Paper No. 76-660, AIAA/SAE Twelfth Propulsion Conference, Palo Alto, CA, July 1976.

29. Repass, G. D., G. M. Lerner, K. P. Coriell, and J. S. Legg, Jr., *Geodynamics Experimental Ocean Satellite-C (GEOS-C) Prelaunch Report*, NASA X-580-75-23, GSFC, Feb. 1975.

30. Sansevero, V. J., Jr., and R. A. Simmons, *International Ultraviolet Explorer Hydrazine Auxiliary Propulsion System Supplied Under Contract NAS 5-20658*, Hamilton Standard Division of United Technologies Corp., Windsor Locks, CT, Oct. 1975.

31. Schwarz, Ralph J., and Bernard Friedland, *Linear Systems*. New York: McGraw-Hill, 1965.

32. Welch, Raymond V., *Control System Model for the IUE Spacecraft*, Stabilization and Control Branch Report 269, GSFC, July 1976.

33. Werking, R. D., R. Berg, K. Brokke, T. Hattox, G. Lerner, D. Stewart, and R. Williams, *Radio Astronomy Explorer-B Postlaunch Attitude Operations Analysis*, NASA-X-581-74-227, GSFC, July 1974.

34. Wertz, J. R., C. F. Gartrell, K. S. Liu, and M. E. Plett, *Horizon Sensor Behavior of the Atmosphere Explorer-C Spacecraft*, Comp. Sc. Corp., CSC/TM-75/6004, May 1975.

CHAPTER 8

DATA TRANSMISSION AND PREPROCESSING

8.1 Data Transmission
  Generation of Data and Insertion into the Telemetry Stream, Tracking Stations, Receiving Stations, Transmission from the Receiving Station to Attitude Determination Computers, Transmission of Attitude Results and Spacecraft Commands
8.2 Spacecraft Telemetry
8.3 Time Tagging
  Spacecraft Clock Time Tagging, Ground-Based Time Tagging
8.4 Telemetry Processors

This chapter describes the process by which data are transmitted from sensors onboard a spacecraft to the point at which these data are used for attitude determination by a software system. Section 8.1 provides an overview of the data transmission process from the spacecraft to the attitude determination software system, and the command transmission process from the ground to the spacecraft. Sections 8.2 and 8.3 provide a detailed view of two particular aspects of interest— the content and form of the telemetry data, and the process used to associate a time with the telemetered data, or *time tagging*. Section 8.4 describes the part of an attitude software system which transforms telemetry data into engineering data.

8.1 Data Transmission

James S. Legg, Jr.

In this section we describe methods by which data are obtained and transmitted from sensors onboard a spacecraft to a ground data base, or *downlinked*, and the methods by which commands are transmitted from the ground to the spacecraft, or *uplinked*. We will follow the flow of data from the spacecraft sensors to the telemetry transmission antenna, from the antenna to a ground tracking station, from the tracking station to a receiving station (e.g., an Operations Control Center at Goddard Space Flight Center), and from the receiving station to a telemetry data base accessible to an attitude determination processing computer. After data analysis, commands to the spacecraft may be generated, which follow the reverse process to the spacecraft. An overview of this two-way transmission of spacecraft telemetry data and commands is shown in Fig. 8-1 (for additional detail, see Gunshol and Chapman [1976]).

8.1.1 Generation of Data and Insertion Into the Telemetry Stream

Measurements of many different physical properties are performed automatically or on ground command onboard a spacecraft, and the results of these measurements are used by other spacecraft components, telemetered to the ground,

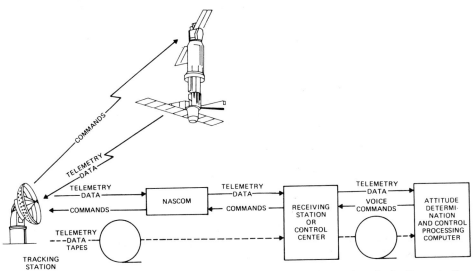

Fig. 8-1.　Telemetry and Command Data Flow Between Spacecraft and Attitude Determination Computer

or tape recorded for later transmission. Two or more of these functions may be performed simultaneously. Several types of measurement require *analog-to-digital conversion* (*ADC*) of the value of a physical quantity before storage or transmission. Analog measurements of voltages are often digitized by measuring the time required for a ramp voltage of known slope (i.e., linearly changing with time) to equal the voltage being measured. The time required is proportional to the measured voltage and is stored in binary form in memory chips in *parallel format*; i.e., each memory chip contains a bit which is either 0 or 1. The stored value is later transmitted via telemetry in *serial format*; i.e., the bits are transmitted one at a time, usually the most significant bit first.

Some measurements, such as digital sun sensor data, are intrinsically digital and need not be converted before transmission. Such measurements frequently occur in a Gray rather than a binary code, as described in Section 6.1. Other types of hardware which provide digital data include optical or magnetic shaft encoders, gyroscopes, and pendulum dampers.

A third common measurement is a time interval between events, e.g., the time between successive Sun pulses, the time between a slit Sun sensor pulse and acquisition of the Earth's horizon by an Earth sensor, or the time between acquisition and loss of an Earth presence signal. These measurements are generally made by a crystal-controlled oscillator circuit which counts the number of vibrations of a piezoelectric crystal between the two events.

Measurements are sampled from sensors in a cyclic order, the sampling rate being determined by a spacecraft clock (a crystal oscillator). Sensors often produce signals which are not directly suitable for telemetering. In these cases, the sensor output is applied to the input of a *signal conditioner*, which adapts the signal to suit the input of the telemetry transmission system. This process includes signal amplification and, when necessary, analog-to-digital conversion. Sensors typically requiring signal conditioning include thermocouples, strain-gages, variable reluctance devices, and small-change variable resistance devices. Sometimes a single signal conditioner may be used with more than one sensor, allowing several signals to be *multiplexed* into the same conditioner; i.e., the same signal conditioner is used on a timesharing basis. Sensors which do not generally require signal conditioning are potentiometer pressure gages, accelerometers, bimetallic thermometers, gyroscopes, displacement gages, and angle-of-attack meters.

Figure 8-2 summarizes the process of sampling the sensors and inserting the sampled data into the telemetry stream together with other information, such as the synchronization (sync) pattern. The multiplexer, driven at a rate determined by the spacecraft clock, samples various sensors via the signal conditioners. These data are fed into the encoder, which generates digital data corresponding to the analog output from the multiplexer. The encoder output then goes to the signal mixing gates, which transmit one signal at a time, in a time-ordered sequence. The signal going to the telemetry transmitter is determined by the mixing gates, based on the spacecraft clock. The spacecraft clock supplies time interval information to the sync generator, which supplies the sync pattern characteristic of the spacecraft. Nonmultiplexed data (such as time history codes, output from an onboard computer, and output signals from sensors producing digital data) are interleaved with the multiplexed data. The bit rate at which data are transmitted is an integral power-of-two multiple or submultiple of the spacecraft clock rate.

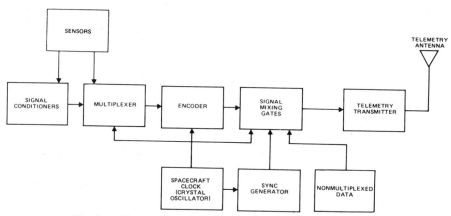

Fig. 8-2. Transfer of Data From Sensors to the Telemetry Stream

The signal transmitted from the spacecraft is generally transmitted on a frequency modulated (FM) carrier wave which is modulated by one or more subcarrier oscillators, which are, in turn, frequency modulated by signals containing the information to be transmitted. This type of telemetry signal is referred to as FM/FM. There are three basic ways, one digital and two analog, of superimposing the data pulses on the carrier signal. The most widely used is the digital method of *pulse code modulation*, or *PCM*, in which the sensor data are transformed to binary numbers and transmitted serially. In this case the noise is minimized because the signal consists of only two voltages, corresponding to 1 and 0, each pulse being the same width. When superimposed on an FM/FM signal, the composite is called *PCM/FM/FM*.

The manner in which the PCM pulses are generated depends on how transitions from 1 to 0 are treated by the transmitter. The three most common methods are illustrated in Fig. 8-3. In *nonreturn-to-zero level* (*NRZL*), ones have a specific assigned voltage and zeros have another. The signal remains constant during the entire bit period in both cases. In *nonreturn-to-zero mark* (*NRZM*), there are two voltage levels, but neither corresponds exclusively to ones or zeros; the voltage level changes value whenever a 1 occurs. Finally, in *return-to-zero* (*RZ*), a 1 is represented by a pulse for one-half the bit period and a 0 is represented by no pulse at all.

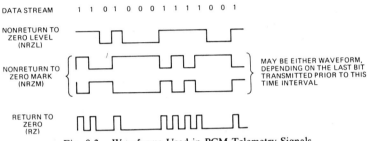

Fig. 8-3.   Waveforms Used in PCM Telemetry Signals

Two methods of superimposing analog data on the FM carrier wave that do not require digitization prior to transmission are *pulse amplitude modulation* (*PAM*) and *pulse duration modulation* (*PDM*). PAM consists of generating a high-frequency signal whose amplitude depends on the value of the data to be transmitted. PDM generates a constant frequency signal whose pulsewidth is proportional to the value of the transmitted data. The latter method of modulation improves the signal-to-noise ratio significantly over the PAM method because sharp spikes caused by the multiplexer switching from one sensor to another affect the amplitude of the signal but not the width of the transmitted pulses.

When transmitting telemetry data, it is necessary for the telemetry transmitter to identify the data item corresponding to each segment of the bit stream. This is done by inserting a synchronization bit string into the telemetry stream on a regular basis, usually at the beginning or end of each repeating telemetry (multiplexer) cycle. This generally corresponds to the beginning or end of a minor or major frame of telemetry data (see Section 8.2). As described in Section 8.2, the sync pattern is usually 24 bits long and is unique to each satellite. The method of selecting the format of the sync bit pattern varies from one spacecraft to another. The important feature of all sync patterns is that they be recognizable; i.e., the correct position of each bit must be recoverable in the event of an occasional loss of one or more bits or the insertion of extraneous bits. This is critical because data items can be identified only by their position relative to the sync pattern. There are generalized sync patterns which provide optimal correlation properties, and display relative immunity to phase displacement by random pulses occurring immediately adjacent to the pattern [Stiltz, 1961; Jackson, 1953].

International standards have been adopted regarding transmitting frequencies, bandwidth, and other characteristics which are applicable to commercial, scientific, and military spacecraft of participating nations. The board responsible for these decisions is the *Inter-Range Instrumentation Group* (*IRIG*), which annually publishes standards for telemetry designers and users. Samples of such guidelines are [Stiltz, 1961]:

> The number of bits per frame shall not exceed 2048, including those used for frame synchronization. The frame length selected for a particular mission shall be kept constant. Word length for a given channel can range from 6 to 64 bits but shall be kept constant for a given channel for a particular mission....
> Frames shall be identified by a unique frame synchronization word which shall be limited to a maximum length of 33 bits....

When the telemetry bit error rate is expected to be greater than 1 bit in a million, NASA usually specifies that parity bits be included in the telemetry stream, typically doubling the number of bits to be transmitted but affording an opportunity to detect bit errors. This process is called *convolutional encoding*. A *convolutional decoder* is required at the receiving station to detect and correct random bit errors. This reduces the raw bit error rate by orders of magnitude and effectively increases the signal strength by approximately 6 dB, or a factor of four in power. Convolutional encoders have been flown on far-Earth spacecraft such as RAE-2 (lunar orbit), IUE, ISEE, and IMP-6, 7, and 8.

Finally, the type and shape of the transmitting antenna depends on the transmitting frequencies desired and the pointing accuracy required. Most spacecraft have at least one parabolic or turnstile antenna for transmitting in the very high frequency (VHF) range and one for the S-band range. The VHF antenna transmits in the 30- to 300-MHz frequency range and is used for transmitting telemetry data, for tracking, and for receiving commands. It is generally used during both launch and mission modes. The S-band antenna is normally designed in a logarithmic spiral configuration, transmits in the 1.55- to 5.20-GHz range, and is used during the mission mode. The bandwidth of an S-band antenna is typically subdivided into 13 subbands, each of which conveys information independently; thus, a higher data rate can be achieved with S-band than with VHF by sending more than one stream of information simultaneously. The subbands are listed in Table 8-1, with the letters which characterize them. The letters are used as

Table 8-1. Subband Frequency Ranges in the Microwave S-Band

SUB-BAND	FREQUENCY (GHZ)	WAVELENGTH (CM)	SUB-BAND	FREQUENCY (GHZ)	WAVELENGTH (CM)
e	1.55 − 1.65	19.3 − 18.3	s	2.90 − 3.10	10.3 − 9.67
f	1.65 − 1.85	18.3 − 16.2	a	3.10 − 3.40	9.67 − 8.32
t	1.85 − 2.00	16.2 − 15.0	w	3.40 − 3.70	8.32 − 8.10
c	2.00 − 2.40	15.0 − 12.5	h	3.70 − 3.90	8.10 − 7.69
q	2.40 − 2.60	12.5 − 11.5	z	3.90 − 4.20	7.69 − 7.14
y	2.60 − 2.70	11.5 − 11.1	d	4.20 − 5.20	7.14 − 5.77
g	2.70 − 2.90	11.1 − 10.3			

subscripts to denote the subband; e.g., a signal at 1.60 GHz is designated S_e. Radio frequencies have been designated for various uses within the western hemisphere in the Radio Regulations of the International Telecommunication Union, which meets in Geneva, Switzerland. A sample of these regulations appears in Table 8-2 for the frequency range 450 MHz to 6.425 GHz, illustrating the proportion of allocations devoted to aerospace use [*Reference Data for Radio Engineers*, 1968].

Table 8-2. Designations of Radio Frequencies Between 450 MHz and 6.425 GHz; Aerospace Frequencies Are Underlined. (From *Reference Data for Radio Engineers* [1968].)

MEGAHERTZ	SERVICE	MEGAHERTZ	SERVICE
450.0–460.0	FIXED MOBILE	2550–2690	FIXED MOBILE
460.0–470.0	FIXED MOBILE METEOROLOGICAL–SATELLITE	2690–2700	RADIO ASTRONOMY
470.0–890.0	BROADCASTING	2700–2900	AERONAUTICAL RADIO NAVIGATION RADIO LOCATION
890.0–942.0	FIXED RADIO LOCATION	2900–3100	RADIO NAVIGATION (GROUND-BASED RADARS) RADIO LOCATION
942.0–960	FIXED		
960.0–1215	AERONAUTICAL RADIO NAVIGATION	**GIGAHERTZ**	**SERVICE**
1215–1300	RADIO LOCATION AMATEUR	3.100–3.300	RADIO LOCATION
1300–1350	AERONAUTICAL RADIO NAVIGATION RADIO LOCATION	3.300–3.400	RADIO LOCATION AMATEUR
1350–1400	RADIO LOCATION	3.400–3.500	RADIO LOCATION COMMUNICATION–SATELLITE (SATELLITE TO EARTH) AMATEUR
1400–1427	RADIO ASTRONOMY	3.500–3.700	FIXED MOBILE RADIO LOCATION COMMUNICATION–SATELLITE (SATELLITE TO EARTH)
1427–1429	FIXED MOBILE EXCEPT AERONAUTICAL SPACE (TELECOMMAND)		
1429–1435	FIXED MOBILE	3.700–4.200	FIXED MOBILE COMMUNICATION–SATELLITE (SATELLITE TO EARTH)
1435–1525	MOBILE FIXED	4.200–4.400	AERONAUTICAL RADIO NAVIGATION
1525–1535	SPACE (TELEMETERING) FIXED MOBILE	4.400–4.700	FIXED MOBILE COMMUNICATION–SATELLITE (EARTH TO SATELLITE)
1535–1540	SPACE (TELEMETERING)	4.700–4.990	FIXED MOBILE
1540–1660	AERONAUTICAL RADIO NAVIGATION	4.990–5.000	RADIO ASTRONOMY
1660–1664.4	METEOROLOGICAL AIDS METEOROLOGICAL–SATELLITE	5.000–5.250	AERONAUTICAL RADIO NAVIGATION RADIO LOCATION SPACE RESEARCH
1664.4–1668.4	METEOROLOGICAL–SATELLITE METEOROLOGICAL–SATELLITE RADIO ASTRONOMY	5.250–5.255	RADIO LOCATION
1668.4–1670	METEOROLOGICAL AIDS METEOROLOGICAL–SATELLITE	5.255–5.350	RADIO LOCATION
1670–1690	METEOROLOGICAL AIDS FIXED MOBILE EXCEPT AERONAUTICAL	5.350–5.460	AERONAUTICAL RADIO NAVIGATION RADIO LOCATION
1690–1700	METEOROLOGICAL AIDS METEOROLOGICAL–SATELLITE	5.460–5.470	RADIO NAVIGATION RADIO LOCATION
1700–1710	SPACE RESEARCH (TELEMETERING AND TRACKING)	5.470–5.650	MARITIME RADIO NAVIGATION RADIO LOCATION
1710–1770	FIXED MOBILE	5.650–5.670	RADIO LOCATION AMATEUR
1770–1790	FIXED MOBILE METEOROLOGICAL–SATELLITE	5.670–5.725	RADIO LOCATION AMATEUR SPACE RESEARCH (DEEP SPACE)
1790–2290	FIXED MOBILE	5.725–5.925	RADIO LOCATION AMATEUR
2290–2300	SPACE RESEARCH (TELEMETERING AND TRACKING IN DEEP SPACE)	5.925–6.425	FIXED MOBILE COMMUNICATION–SATELLITE (EARTH TO SATELLITE)
2300–2450	RADIO LOCATION AMATEUR FIXED MOBILE		
2450–2550	FIXED MOBILE RADIO LOCATION		

8.1.2 Tracking Stations

Telemetry data from NASA-supported spacecraft are received by a worldwide network of tracking stations called the *Spaceflight Tracking and Data Network* (*STDN*). The northernmost station is located in Fairbanks, Alaska, at 65° North latitude; the southernmost is in Orroral Valley, Australia, at 35° South latitude. The locations of fixed STDN stations are given in Fig. 8-4 and detailed locations are listed in Table 8-3. In the second column of this table, the abbreviations USB and GRARR indicate *unified S-band* and *Ground Range and Range Rate*, respec-

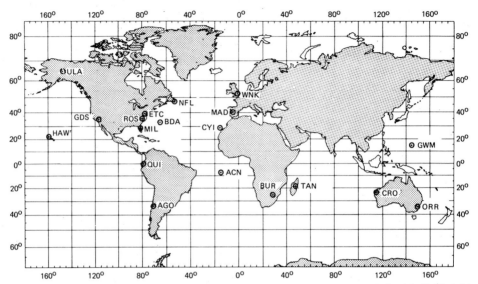

Fig. 8-4.   Locations of STDN Stations. (The three-letter station designators are defined in Table 8-3.)

tively. In addition to the permanent tracking stations, the network includes portable land-based stations (vans), a ship (the *USNS Vanguard*), and several specially equipped *Advanced Range Instrumented Aircraft* (*ARIA*). Figures 8-5 through 8-8 show examples of STDN facilities. This network of receiving stations supports NASA's Earth-orbiting scientific and applications satellites, interplanetary missions, and manned space flight.*

One of the larger STDN stations is located at Rosman, North Carolina, and is shown in Figs. 8-8(a) to 8-8(d). The aerial view in Fig. 8-8(a) shows seven of the station's tracking and data acquisition antennas. The largest ones are two 26-m telemetry antennas at either end of the large clearing just above the center of the photograph. (Closeups are shown in Figs. 8-5 and 8-8(b)). In the clearing above and to the left of the large antenna in the center are two *Satellite Automatic Tracking Antennas* (*SATAN*). The two antennas in the clearing in the lower left corner of the aerial view are range and range-rate antennas and are shown in Fig. 8-8(d). Finally, the VHF *Satellite Command Antenna on Medium Pedestal* (*SCAMP*) antenna of Fig. 8-8(c) is in a small foreground clearing along the righthand edge of the aerial view.

Operational control and scheduling of the network is provided by the *Network Operations Control Center* (*NOCC*) located at *Goddard Space Flight Center* (*GSFC*)

---

*The *Jet Propulsion Laboratory* (*JPL*) *Deep Space Network* (*DSN*) handles communications for interplanetary missions. Because of the long distances involved, this system uses the 26-m antennas in Madrid, Spain, and Goldstone, California.

Table 8-3. Geodetic Coordinates of STDN Stations Tracking System, Referenced to Fisher '60' Ellipsoid, Semimajor Axis = 6378166 m and 1/flattening = 298.3. (See Appendix L for the transformation to geocentric coordinates.) System locations are subject to minor changes as refinements in positional accuracies are made.

STATION	SYSTEM	LATITUDE[1]	LONGITUDE (E)	HEIGHT ABOVE ELLIPSOID (METERS)
ASCENSION ISLAND (ACN)	9m USB	− 7°57'17.37"	345°40'22.57"	528
SANTIAGO, CHILE (AGO)	9m USB	−33°09'03.58"	289°20'01.08"	706
	VHF GRARR	−33°09'06.06"	289°20'01.07"	706
	INTERFEROMETER	−33°08'58.10"	289°19'54.20"	694
BERMUDA (BDA)	9m USB	32°21'05.00"	295°20'31.94"	− 33
	FPQ-6 RADAR	32°20'53.05"	295°20'47.90"	− 35
GRAND CANARY ISLAND (CYI)	9m USB	27°45'51.61"	344°21'57.88"	167
ENGINEERING TRAINING CENTER, MARYLAND (ETC)	9rr. USB	38°59'54.84"	283°09'26.23"	− 1
	9m USB (ERTS)	38°59'54.08"	283°09'29.21"	4
	INTERFEROMETER	38°59'57.25"	283°09'38.71"	− 5
GOLDSTONE, CALIFORNIA (GDS)	26m USB	35°20'29.66"	243°07'35.06"	919
	9m USB (ERTS)	35°20'29.64"	243°07'37.45"	913
GUAM (GWM)	9m USB	13°18'38.25"	144°44'12.53"	116
HAWAII (HAW)	9m USB	22°07'34.46"	200°20'05.43"	1139
	FPS-16 RADAR	22°07'24.37"	200°20'04.02"	1143
MADRID, SPAIN (MAD)	26m USB	40°27'19.67"	355°49'53.59"	808
MERRITT ISLAND, FLORIDA (MIL)	9m USB NO. 1	28°30'29.79"	279°18'23.85"	− 55
	9m USB NO. 2	28°30'27.91"	279°18'23.85"	− 55
ORRORAL VALLEY, AUSTRALIA (ORR)	INTERFEROMETER	−35°37'32.19"	148°57'15.15"	926
QUITO, ECUADOR (QUI)	INTERFEROMETER	−00°37'22.04"	281°25'16.10"	3546
ROSMAN, NORTH CAROLINA (ROS)	4.3m USB	35°11'45.99"	277°07'26.96"	810
	VHF GRARR	35°11'42.02"	277°07'26.97"	810
TANANARIVE, MALAGASY REPUBLIC (TAN)	4.3m USB	−19°01'13.87"	47°18'11.87"	1368
	VHF GRARR	−19°01'16.34"	47°18'11.83"	1368
	INTERFEROMETER	−19°00'31.66"	47°17'59.75"	1347
	FPS-16 RADAR	−19°00'05.52"	47°18'53.46"	1307
FAIRBANKS, ALASKA (ULA)	9m USB	64°58'19.20"	212°29'13.39"	339
	VHF GRARR	64°58'17.50"	212°29'19.12"	339
	INTERFEROMETER	64°58'36.91"	212°28'31.89"	282
WINKFIELD, ENGLAND (WNK)	INTERFEROMETER	51°26'46.12"	359°18'09.13"	87

[1] A MINUS SIGN (−) INDICATES SOUTH LATITUDE.

in Greenbelt, Maryland. Selection of which station tracks a given satellite at a given time is made by the NOCC based on requests from the *Project Operations Control Centers* (*POCC*) for unmanned spacecraft, and from the *Mission Control Center* (*MCC*) at the *Lyndon B. Johnson Space Center* (*JSC*) in Houston, Texas, for manned spacecraft.

Telemetry data received by STDN stations are either transmitted in near real time to GSFC, as discussed in Section 8.1.3, or are recorded on magnetic tapes and mailed to the receiving station. Range (position) and range-rate (velocity) data from the spacecraft are also acquired by radar or laser techniques at the tracking stations and relayed for use in orbit determination. Spacecraft command data are transmitted to the spacecraft in near real time or stored at the station for later transmission.

Computer facilities are located at each STDN station for processing spacecraft-associated data and performing local equipment test and control func-

Fig. 8-5.   NASA Multiband Telemetry Antenna (26-m Diameter) at Rosman, North Carolina

Fig. 8-6. USNS Vanguard used for Spacecraft Tracking

Fig. 8-7. Advanced Range Instrumented Aircraft

tions. Data processing capabilities range from simple header generation to relatively sophisticated data compression operations [Scott, 1974]. One of the processing functions provided by STDN stations is time tagging, or attaching the *Greenwich Mean Time* (*GMT*) to processed data (see Section 8.3).

It is anticipated that in the future the ground-based communications network will be enhanced by satellite relay systems. For example, the *Tracking and Data Relay Satellite System* (*TDRSS*), scheduled to become operational in 1979, consists of two communications satellites in geostationary orbits which can relay telemetry

Fig. 8-8(a). NASA Tracking Stations at Rosman, North Carolina. Aerial view of tracking station showing seven antennas. See text for description.

data in real-time from other spacecraft which are not within the line of sight of any STDN station and which can also relay real-time commands from the tracking stations to the spacecraft. The two TDRSS satellites will be approximately 130 deg apart, at 41° and 171° West longitude. The inclination of their orbits will be between 2 and 7 deg. A ground tracking station located within the continental United States (presently planned for White Sands, New Mexico) will remain in constant contact with the *Tracking and Data Relay Satellites* (*TDRS*) providing telecommunication for orbital tracking data, telemetry data, and, in the case of manned spaceflight, voice communication (Fig. 8-9). This network will provide coverage of at least 85% of all orbits below 5000 km. For orbits above this altitude, the remaining STDN stations will provide coverage. To ensure reliability, a redundant TDRS will be placed in orbit midway between the two operational satellites and a fourth will be maintained on the ground for rapid replacement launch, if required. Redundant antenna systems will also be provided at the

Fig. 8-8(b). 26-m tracking antenna showing the two-wheel tracking assembly.

8-8(c). Satellite Command Antenna on Medium Pedestal (SCAMP).

Fig. 8-8(d). Range and range rate antennas.

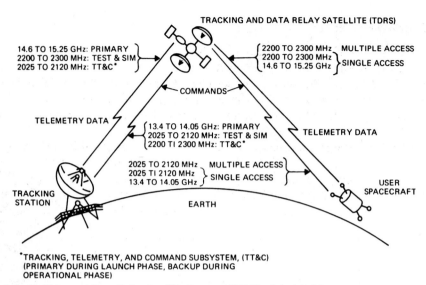

Fig. 8-9. Tracking and Data Relay Satellite System (TDRSS), Scheduled for Implementation in 1979

primary STDN tracking site. The number of worldwide full-time tracking stations will be reduced to approximately five when the TDRSS is fully operational.

Two modes of operation are being considered for the TDRSS. The first, *multiple access* (*MA*), allows each TDRS to transmit telemetry and commands for as many as 20 spacecraft simultaneously. The disadvantage of MA is that the probability of transmission errors increases for spacecraft with altitudes greater than 5000 km. The *single access* (*SA*) method allows each TDRS to transmit telemetry and commands to only two spacecraft at a time. Its advantage is its transmission efficiency for spacecraft with altitudes up to 12,000 km.

Tracking and data acquisition in the Soviet space program differs in several respects from NASA's program; detailed information on the Soviet program is both limited and somewhat dated. (See the U.S. Senate Report [1971] for a comprehensive discussion.) Although the United States has developed an extensive network of tracking stations in foreign countries, the Soviet Union relies primarily on stations within its own territory and on sea-based support. Because of the larger land area, stations within the Soviet Union can provide greater contact time than could a similar set of stations spread throughout the United States. Soviet references have been made to tracking stations in the United Arab Republic, Mali, Guinea, Cuba, and Chad.

At least 10 ships have been identified as working for the Soviet Academy of Sciences, the majority of which are involved in some phase of space operations. Among the most advanced of these are the *Kosmonaut Vladimir Komarov* and the *Akademik Sergey Korolev*. The latter is a space satellite control ship which was launched in 1971 and is described as the largest scientific research ship in the world, 182 m long and displacing 21,250 metric tons. The ships maintain contact with the Soviet Union via the *Molniya* communications satellites.

8.1.3 Receiving Stations

NASA's STDN tracking stations are linked with each other and with GSFC and JSC by the *NASA Communications Network (NASCOM)*. This system provides voice, data, teletype, television from selected stations, and other wideband communication. The network uses land lines, submarine cables, and microwave links. Redundant, geographically diverse routes are provided so that communication will not be lost if a primary route fails.

NASCOM leases full-time voice circuits (2-kHz bandwidth) to nearly all stations and control centers in its network. Most communication is routed through the GSFC *Switching, Conferencing and Monitoring Arrangement (SCAMA)*. When these circuits are used for data transmission, the data format in Fig. 8-10 is used. The length of the data block may be any multiple of 12 bits, but the use of a

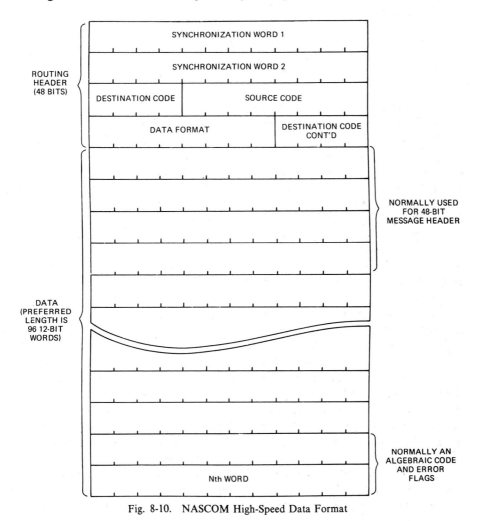

Fig. 8-10. NASCOM High-Speed Data Format

1200-bit block is encouraged so that it will be compatible with planned future STDN data-handling systems. A 48-bit message header normally follows the first 48-bit routing header, and the last 24 bits normally include a 22-bit algebraic code and 2 bits for flagging detected errors.

8.1.4. Transmission From the Receiving Station to Attitude Determination Computers

When telemetry data arrive at the receiving station, whether by NASCOM or mailed tapes, they are processed by a control center computer before delivery to an attitude determination processing computer. At GSFC, this function is performed by an *Operations Control Center* (*OCC*) for near-real-time data and/or by the *Information Processing Division* (*IPD*) for playback (tape recorded) data.

The processing performed by the OCC is minimal, since it is performed in near real time, and consists of stripping out data to be relayed to several destinations, one of which is the attitude determination computer. The sync pattern is examined and a quality flag is attached to the data, based on the number of incorrect bits in the sync pattern (Section 9.1). Sometimes the current GMT is attached to the data as well. The current date and the name of the tracking station which received the data are also inserted. The data are then transmitted to the attitude determination computer via a communication line controlled by a software package called the *Attitude Data Link* (*ADL*).

Processing performed by the IPD is more extensive, since the data need not be relayed immediately. Data are collected from tracking stations for periods of a day or more, and are then time ordered before transmission to the attitude determination computer. Segments of data which were incorrectly time tagged by the tracking station are detected and corrected. Other functions performed by the OCC are also performed by the IPD. The data are then transmitted to the attitude determination computer via a communication line under control of the ADL.

8.1.5 Transmission of Attitude Results and Spacecraft Commands

After the attitude determination computer processes the attitude data, it generates a definitive attitude history file, which is relayed to the IPD computers via the ADL and processed by a software package called the *Telemetry On-Line Processing System* (*TELOPS*). The data are then available for processing by experimenters. (For more detail on TELOPS, IPD, and their role in the data transmission process, see Gunshol and Chapman [1976].)

Commands may be uplinked to the spacecraft based on analysis of data on the attitude determination computer. Command requests, in engineering units, may be relayed from the attitude computer area to the OCC by voice (telephone lines). These requests are translated into coded commands by the OCC and transmitted to the tracking station via NASCOM. The tracking station then stores the command for later transmission or relays it to the spacecraft immediately in near real time. Sometimes the relayed commands are stored in a computer onboard the spacecraft for later execution. These are referred to as *delayed commands*.

8.2 Spacecraft Telemetry

Janet Niblack

Telemetry is a sequence of measurements being transmitted from one location to another.* The data are usually a continuous stream of binary digits (or pulses representing them). A single stream of digits is normally used for the transmission of many different measurements. One way of doing this is to sequentially sample various data sources in a repetitive manner. This process is called *commutation*, and the device which accomplishes the sequential switching is a *commutator*. The commutator may be either a mechanical or electronic device or a program in an onboard computer.

A *minor frame* of telemetry data contains measurements resulting from one complete cycle of the main commutator. Each frame consists of a fixed number of bit segments called *telemetry words*. Each word in a frame is a *commutator channel*. If the telemetry word contained in a main commutator channel is supplied by another commutator (called a subcommutator), data appearing in that channel are said to be *subcommutated*. If a single data source is sampled more than once within a minor frame, the data item is said to be *supercommutated*. The level of commutation for a particular data item determines the relative frequency at which it is transmitted. Whether a data item should be commutated, subcommutated, or supercommutated depends on how the measurement will be used and at what rate the value will change.

A *major frame* (sometimes called a *master frame*) contains the minimum number of minor frames required to obtain one complete cycle of all subcommutators, or an integral multiple of this number. (Because not all spacecraft telemetry systems use subcommutators, the major frame concept is not always relevant.) A *minor frame counter* or *minor frame ID* is often telemetered to identify the position of a minor frame within a major frame. This counter is particularly useful when minor frames are lost in transmission, since minor frame location determines what type of data a subcommutator channel will contain. Figure 8-11 shows a simple eight-channel main commutator with two subcommutators. Table 8-4 gives the sequence of telemetry words which would be generated by this commutator for one major frame. Note that the relative frequency at which a subcommutated data item appears depends on the number of channels in the subcommutator.

Because commutation involves time-dependent functions, some method of establishing and maintaining exact sychronization of data sampling is necessary. Spacecraft clocks provide the signals for synchronization. A *frame synchronization signal*, described in Section 8.1, is a series of pulses which marks the start of a minor frame period. These pulses are transmitted as part of each main commutator cycle and are used in identifying individual frames when the data are received on the ground.

The assignment of specific data items to commutator and subcommutator channels defines the *telemetry format*. Commutator or subcommutator channels are allocated to experimental data, to attitude determination and control data, and to

---

*For an extended discussion of spacecraft telemetry, see Stiltz [1961].

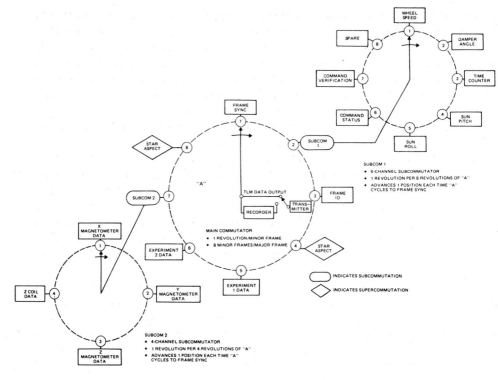

Fig. 8-11.   Levels of Commutation. See text for description.

Table 8-4.   Contents of a Typical Major Frame of Telemetry Data. Major frame words 2 and 7 are subcommutated. Words 4 and 8 are supercommutated.

MINOR FRAME WORD / MINOR FRAME NUMBER	1	2	3	4	5	6	7	8
1	FRAME SYNC	WHEEL SPEED	FRAME ID	STAR ASPECT	EXPERIMENT 1 DATA	EXPERIMENT 2 DATA	X MAGNETOMETER DATA	STAR ASPECT
2	FRAME SYNC	DAMPER ANGLE	FRAME ID	STAR ASPECT	EXPERIMENT 1 DATA	EXPERIMENT 2 DATA	Y MAGNETOMETER DATA	STAR ASPECT
3	FRAME SYNC	TIME COUNTER	FRAME ID	STAR ASPECT	EXPERIMENT 1 DATA	EXPERIMENT 2 DATA	Z MAGNETOMETER DATA	STAR ASPECT
4	FRAME SYNC	SUN PITCH	FRAME ID	STAR ASPECT	EXPERIMENT 1 DATA	EXPERIMENT 2 DATA	Z COIL DATA	STAR ASPECT
5	FRAME SYNC	SUN ROLL	FRAME ID	STAR ASPECT	EXPERIMENT 1 DATA	EXPERIMENT 2 DATA	X MAGNETOMETER DATA	STAR ASPECT
6	FRAME SYNC	COMMAND STATUS	FRAME ID	STAR ASPECT	EXPERIMENT 1 DATA	EXPERIMENT 2 DATA	Y MAGNETOMETER DATA	STAR ASPECT
7	FRAME SYNC	COMMAND VERIFICA- TION	FRAME ID	STAR ASPECT	EXPERIMENT 1 DATA	EXPERIMENT 2 DATA	Z MAGNETOMETER DATA	STAR ASPECT
8	FRAME SYNC	SPARE	FRAME ID	STAR ASPECT	EXPERIMENT 1 DATA	EXPERIMENT 2 DATA	Z COIL DATA	STAR ASPECT

general spacecraft maintenance (or *housekeeping*) data. A telemetry system may have either a fixed format or several formats which correspond to various operating modes. For example, immediately following launch, attitude data, power supply data, and other data related to the "health" of the spacecraft are needed at a high frequency. Later, these items can be telemetered at a reduced rate and the amount of experimental data can be increased. Failure conditions, such as an undervoltage condition, may cause resumption of a launch-type format.

All spacecraft telemetry systems have one or more telemetry formats established before launch. However, it is often difficult to predict which data items will be the most useful. Telemetry systems with *programmable formats* allow the formats to be changed by remote command during flight. For example, in the SAS-3 telemetry system, several telemetry formats were defined at the time of launch. Within a year, due to the permanent failure of several spacecraft instruments, use of any of these formats resulted in the transmission of a significant amount of useless data. Two new formats were added (by remote command) to allow additional magnetometer and experimental data to be telemetered in place of the useless data.

Digital Codes. Although a measurement is telemetered as a series of binary digits, the value of this measurement need not be represented in the natural binary code. Although the binary code is frequently used, other digital codes are more convenient and reliable for certain applications. One of the problems with the natural binary code is that a change of one unit may require a change in several binary digits (e.g., $127_{10} = 01111111_2$ to $128_{10} = 10000000_2$). Thus, if the value is sampled when the bits are changing, it is possible for a gross error to occur. To circumvent this problem, codes have been developed in which only one bit changes per unit change in value. The *Gray code*, also called the *reflected binary code*, is the most widely used code of this type and is described in Sections 6.1 and 8.4.

Binary codes in which a 1 represents the presence of a pulse and a 0 represents the absence can be generated using simple hardware circuitry. However, errors in such a code arising from minor imperfections in the telemetry system, such as transmission noise or the momentary failure of a relay contact are common. To increase data reliability, error-checking codes are used, the simplest type being a *parity code*. In parity codes, one bit is added to the original code and the extra bit is set so that the number of bits with value one is always even (*even parity*) or odd (*odd parity*). The addition of parity bits to a natural binary code is shown in Table 8-5. With a single parity bit, it is possible to detect bit errors, but *not* to determine which bit is in error.

Table 8-5.   Even and Odd Parity Codes

DECIMAL NUMBER	EVEN PARITY		ODD PARITY		DECIMAL NUMBER	EVEN PARITY		ODD PARITY	
	BINARY NUMBER	PARITY BIT	BINARY NUMBER	PARITY BIT		BINARY NUMBER	PARITY BIT	BINARY NUMBER	PARITY BIT
0	0000	0	0000	1	5	0101	0	0101	1
1	0001	1	0001	0	6	0110	0	0110	1
2	0010	1	0010	0	7	0111	1	0111	0
3	0011	0	0011	1	8	1000	1	1000	0
4	0100	1	0100	0	9	1001	0	1001	1

More intricate codes have been designed which not only detect bit errors but also correct them. An example of a simple error-correcting code is an 8-bit *Hamming code* which consists of 4 message bits and 4 parity bits. This code is capable of correcting an error in any one bit and detecting errors in any two bits. We define the Hamming matrix, H, as

$$H \equiv \begin{bmatrix} 0 & 0 & 0 & 0 & 1 & 1 & 1 & 1 \\ 0 & 0 & 1 & 1 & 0 & 0 & 1 & 1 \\ 0 & 1 & 0 & 1 & 0 & 1 & 0 & 1 \\ 1 & 1 & 1 & 1 & 1 & 1 & 1 & 1 \end{bmatrix} \tag{8-1}$$

Note that the first three entries of each column form the binary numbers 0 to 7. The *message row vector*, $\mathbf{M}$, is defined as

$$\mathbf{M} \equiv (p_1, p_2, p_3, a, p_4, b, c, d) \tag{8-2}$$

where a, b, c, and d are the 4 message bits and the p's are parity bits to be defined. A received message vector is tested by forming the *syndrome vector*, $\mathbf{S}$, defined by

$$\mathbf{S} = H\mathbf{M}^{\mathrm{T}} = (S_1, S_2, S_3, S_4)^{\mathrm{T}} \tag{8-3}$$

To code a message, we set the parity bits such that $\mathbf{S}$ is identically $\mathbf{0}$ (mod 2); that is

$$\begin{aligned} p_4 + b + c + d &= 0 \\ p_3 + a + c + d &= 0 \\ p_2 + a + b + d &= 0 \\ p_1 + p_2 + p_3 + a + p_4 + b + c + d &= 0 \end{aligned} \tag{8-4}$$

(The parity bits are located as indicated in Eq. (8-2) so that only one new parity bit is involved in each component of Eq. (8-4) and the message is easy to code.) When a message vector is received, the syndrome vector is calculated. If $\mathbf{S}=0$, we assume that no error occurred. If $S_4 = 1$, we assume that a single error occurred and the binary number $S_1 S_2 S_3$ gives the number of the component of $\mathbf{M}$ which is in error. If $S_4 = 0$ and one or more of the other syndrome bits is 1, then two bits of $\mathbf{M}$ are incorrect, and the error is detectable but uncorrectable. It is possible for errors in more than two bits to be undetected or incorrectly corrected.

To illustrate the above Hamming code, assume that we wish to send the message 1100. From Eq. (8-4) we choose the parity bits such that the 8-component message vector is 00111100. If an error occurs in bit 5 (counting the left-most as bit 0), then the received message is 00111000. The syndrome of the received message is 1011. Because $S_4 = 1$, we assume there is a single error in bit $101 = 5$ and the corrected message vector is 00111100, from which we extract our original message (1100) from bits 3, 5, 6, and 7.

If the probability of an error in any bit is 1%, then the probability of at least one error in a 4-bit message is approximately 4%. If a single parity bit is added, errors will occur in approximately 5% of the messages, which must then be discarded, and undetected errors in two message bits will occur in approximately 0.25% of the messages. If the 8-component Hamming code is used, errors in one bit which are then corrected will occur in approximately 8% of the messages. Errors in two of the message bits for which **M** must be discarded will occur in approximately 0.64% of the messages, and undetected errors in three or more bits will occur in approximately 0.05% of the messages. Thus, even simple error-correcting codes improve the amount of information correctly transmitted and reduce the probability of undetected errors.

Error-checking and self-correcting codes can be costly in terms of the number of bits required. Therefore, their use is justified only when the possibility of error is large. Because errors increase with transmission distance, these additional bits are frequently used for lunar and interplanetary missions. For an extended discussion of error-correcting codes, see Peterson and Weldon [1972] or Ryder [1967].

Often the natural binary code is used to represent a particular range of positive values, with a sign bit provided to allow representation of negative numbers. Normally, the sign bit is set to 0 for the positive values, although occasionally 1 is used. In either case, a negative number can be represented in natural binary or as either the one's or two's complement of the corresponding positive number. The *one's complement* is obtained by inverting every bit, i.e., changing each original 0 to 1 and each original 1 to 0. The *two's complement* is obtained by inverting every bit and adding 1 to the result. Computers often use the two's complement form for negative numbers because, with fixed-length arithmetic of n bits, the two's complement of x is $2^n - x$. Thus, the two's complement behaves much like the negative of the number; for example, the sum of a binary number and its two's complement is zero. Table 8-6 shows four methods of representing positive and negative binary numbers, using four bits. Note than when two's complement is used, only one representation of zero is possible, and the largest magnitude of a negative number is one greater than the largest positive value.

A major factor in the choice of a digital code is whether the sensor is digital or analog. Using a code such as those previously described, digital sensors generate

Table 8-6.   Alternative Representations of Positive and Negative Binary Numbers

DECIMAL EQUIVALENT	ONE'S COMPLEMENT SIGN BIT = 0 FOR +	TWO'S COMPLEMENT SIGN BIT = 0 FOR +	ONE'S COMPLEMENT SIGN BIT = 1 FOR +	NATURAL BINARY SIGN BIT = 0 FOR +
7	0111	0111	1111	0111
6	0110	0110	1110	0110
5	0101	0101	1101	0101
4	0100	0100	1100	0100
3	0011	0011	1011	0011
2	0010	0010	1010	0010
1	0001	0001	1001	0001
0	0000	0000	1000	0000
−0	1111	−	0111	1000
−1	1110	1111	0110	1001
−2	1101	1110	0101	1010
−3	1100	1101	0100	1011
−4	1011	1100	0011	1100
−5	1010	1011	0010	1101
−6	1001	1010	0001	1110
−7	1000	1001	0000	1111
−8	−	1000	−	−

binary digits which can be inserted directly into the telemetry stream. As an example, sensors which measure position often generate Gray-coded output by using a pattern of conducting and nonconducting surfaces with contacting brushes or, as in the case of the digital Sun sensor described in Section 6.1, a patterned mask or reticle and photocells.

Analog sensors generate signals which vary continuously with the magnitude of the measured quantity. For telemetering, output from analog sensors must be converted from an analog to a digital form. This is accomplished by an *analog-to-digital converter* (*ADC*), sometimes simply called an *encoder*. The ADC generates a series of bits describing the magnitude of the analog sample being encoded. The format of the bits is determined by the characteristics of the ADC. Generally, an ADC generates one of the four signed binary codes shown in Table 8-6.

8.3 Time Tagging

James S. Legg, Jr.

Telemetry data transmitted from a spacecraft generally have little significance without knowledge of when the data were measured. Slowly varying quantities, such as information describing the mode of operation of the spacecraft or the telemetry format, do not need to be accurately time tagged; however, most attitude data change continuously with time as the spacecraft position and attitude change. Consequently it is important to accurately correlate telemetry data items with the time at which they entered the telemetry stream, i.e., the time at which they were transmitted in the case of real-time data, or the time at which they were recorded onboard the spacecraft for tape recorded playback data. Two methods of providing accurate timing are used: (1) "clocks" onboard the spacecraft, and (2) time tagging at a tracking or a receiving station.

8.3.1 Spacecraft Clock Time Tagging

Spacecraft clocks measure time intervals, rather than absolute time. They normally consist of a piezoelectric crystal to which a known voltage is applied, causing oscillation at a constant frequency. Electronic circuits count the number of oscillations between two events and, hence, the elapsed time between them. In this sense, the crystal and its associated circuitry constitute a "clock." These mechanisms are used both to measure time intervals and to control the timing of spacecraft events. By using divider circuits, the effective output frequency of the oscillator can be decreased by successive factors of two to drive electronic components at lower frequencies. For example, this output is normally used to determine the rate at which telemetry data are sampled and transmitted. The spacecraft clock is generally activated shortly before launch and continues running indefinitely thereafter. Typically once per major or minor frame of telemetry data, the count of the clock is transmitted. After a time on Earth* has been associated with a spacecraft clock count, data received at other times can be correlated with Earth time by using the current spacecraft clock count and the known clock update frequency.

---

* Times attached on Earth are *Coordinated Universal Time*, or *UTC*, as broadcast by radio time stations. This is also referred to as *Greenwhich Mean Time* (*GMT*) or *Zulu* (*Z*). See Appendix J.

Crystal oscillator frequencies drift due to aging and environmental effects such as temperature; for example, spacecraft clocks often run slightly faster when in sunlight than in shadow. To minimize such effects, the crystal is cut along a particular crystallographic axis and heaters stabilize its temperature. Other effects, such as magnetic fields and relativistic effects, are negligible for most applications. Spacecraft clocks are typically stable to 1 part in 10^{10} per orbit [Fang, 1975]. The count of the spacecraft clock can be altered by several occurrences, depending on the clock. The SEASAT clock can be reset to zero and the oscillator adjusted to meet and maintain synchronization with GMT to within 100 μsec.

The register onboard the spacecraft which contains the clock count generally contains enough bits to allow the clock to run from several days to a year before it returns to zero, or *rolls over*. If the clock counts are in milliseconds, this requires 36 bits. The SEASAT clock contains 40 bits, providing time steps of 30 μsec.

The count of the clock read into telemetry obviously gives only one time—usually the time of the beginning of a minor frame of data. If more accurate timing is required, ground software can use *subblock time tagging* to account for the time interval between the beginning of the minor frame and the time the critical data item was measured within the minor frame. This is often unnecessary for attitude data because times accurate to the nearest second are generally adequate.

Both systematic and nonsystematic errors occur in the spacecraft clock reading. Systematic errors are caused by clock rollover or resetting. Nonsystematic errors, such as noise in the telemetry signal, can cause one or more bits in the clock count to be received incorrectly by a tracking station. In this case, randomly distributed times in the telemetry stream assume random values. This can generally be detected and corrected by ground software, as described in Section 9.1.

8.3.2 Ground-Based Time Tagging

Data received at ground tracking stations can be tagged with the local *Greenwich Mean Time (GMT)* at the time they are received by the station, or (for NASA-supported spacecraft) they can be transmitted in near real-time to an *Operations Control Center (OCC)* at Goddard Space Flight Center in Greenbelt, Maryland, where they are time tagged when they are received. In either case, the attached GMT consists of milliseconds of year at the time the data were received, or day of year and milliseconds of day. When the *Information Processing Division (IPD)* (see Section 8.1) at Goddard Space Flight Center processes data, the time tag always consists of day of year and milliseconds of day. Time tagging to this accuracy is made feasible by the *Long Range Navigation-C (Loran-C)* timing network supported by the U.S. Coast Guard, the National Bureau of Standards, and the U.S. Naval Observatory in Washington, D.C. Coarse clock synchronization (± 1 sec) is accomplished via frequency and time signals transmitted by radio stations *WWV* and *WWVH* by the National Bureau of Standards, and fine synchronization is accomplished via signals transmitted by the U.S. Naval Observatory over the Loran-C network. These signals provide timing accuracy to 25 μsec, although improvement to ± 2.5 μsec is anticipated in the late 1970s.

The Loran-C network is important for spacecraft because it is the primary source of timekeeping for all of the *Spaceflight Tracking and Data Network (STDN)* tracking stations. In addition to STDN time tagging, Loran-C is used for navigation and aviation (its primary function), precise timing and frequency standards for industrial purposes such as crystal manufacturing, network synchronization for power companies, and scientific measurements such as very long baseline interferometry, pulsar frequencies, and Lunar laser ranging. Eight stations located around the world, called *master stations*, receive extremely accurate 100-kHz timing signals from the U.S. Naval Observatory. Each of the master stations then transmits a 100-kHz signal to two or more Loran-C *slave stations*. Each master and slave group is called a *chain*. The eight chains in use as of 1972 are shown in Fig. 8-12 [Hefley, 1972]. The master station, M, in each chain transmits a group of eight pulses separated by 1 msec, followed by a 2-msec delay and a ninth pulse. The ninth pulse distinguishes the signal transmitted by the master station from those of the slaves (denoted X, Y, and Z, or W, X, Y, and Z) which contain only the initial eight pulses.

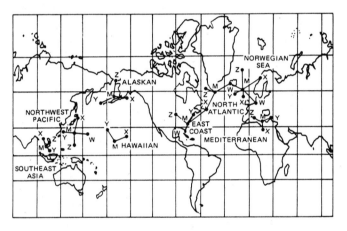

Fig. 8-12.   Worldwide Loran-C Chains as of 1972

When a slave station receives the wave train from the master station, it delays a preset time interval and transmits an eight-pulse signal similar to that received from the master. The preset delay is generated by on-site atomic clocks.

A STDN tracking station receiving the pulse train from either a master or a slave station can determine time and time intervals to an accuracy of approximately 25 μsec. The pulse trains transmitted by master and slave stations are shown in Fig. 8-13. The slave station delay times are such that a receiver within direct radio distance will always receive all of the slave transmissions before the next master transmission, and will receive slave transmissions in the same order: W first, followed by X, Y, and Z. Because the geocentric coordinates of the master or slave station being received are known, the time delay from the transmitter to the receiver can be calculated.

Some timing errors are computed by the receiving station and corrected for when the data are *not* processed in near real time. These include known errors in the daily timing signals transmitted by the U.S. Naval Observatory, the propagation delay between the master or slave transmitter and the tracking station receiver, the time delay within the receiver electronics, and (when appropriate) the delay

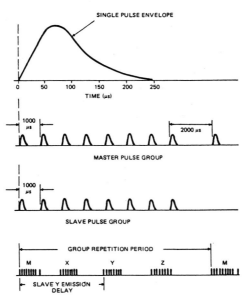

Fig. 8-13.   Typical Loran-C Pulse Trains

time within the slave station. When data are processed in near real time, these errors are ignored. When portable STDN stations are out of range of direct master or slave signals, delayed signals reflected off the Earth's ionosphere are used. These reflections, called *skywaves*, are less reliable because their arrival time depends on local atmospheric conditions between the transmitter and the receiver. The pulsed transmission described above facilitates distinguishing between direct signals and skywaves, which typically lag direct signals by 30 to 40 μsec. Loop antennas are used at STDN receivers to directionalize reception; therefore, signals from only one Loran-C station are received at a time.

Time tagging errors other than those in the timing scheme itself include (1) propagation time from the spacecraft to the tracking station, (2) electronic hardware and software delays during the time-tagging process, (3) uncertainties in the position of the spacecraft, and (4) uncertainties in the position of the tracking station. These errors are typically small compared with the timing accuracies required by attitude determination and control software and are usually neglected. Their magnitudes are on the order of milliseconds. (The propagation delay from the Moon is ~1.3 sec.)

When required, all NASA STDN tracking stations can also employ a cesium beam frequency standard as the primary source for time and time interval measurements, with rubidium atomic frequency standards as a first backup. At some tracking stations, a highly stable oven-controlled quartz crystal frequency standard provides a secondary backup. All stations provide automatic switchover from primary to secondary timing source in the event of low signal amplitude from the primary source. Secondary timing sources are phase-locked to the primary, eliminating frequency and time jumps during switchover [Scott, 1974].

For precise attitude determination and experimental data processing, the actual time of measurement of each data type is determined from the time tag. In this determination, considerations are made for time delays caused by pulse shaping within the electronics, the location of the data within the minor frame, the time delay between the time the sensor measurement was made and the time the sensor was sampled, and so forth (see Section 8.1).

Time Tagging of Near-Real-Time Data. Data received from the spacecraft in near real time and transmitted from the tracking station to a receiving station via NASCOM (Section 8.1) are usually time tagged by the tracking station, but can be tagged by the receiving station. In either case, the process is as described above. Attitude determination software must handle random erroneous times in this case, since the tracking or receiving station software does not have time to detect and correct incorrect times. Limit checks on the times are typically sufficient.

Time Tagging of Playback Data. When a spacecraft does not have continuous tracking station coverage, which is generally the case for low-Earth orbits, and continuous attitude information is required, the data are stored on tape recorders onboard the spacecraft and played back while over a tracking station. Time tagging of data in this case is done either by the tracking station or the receiving station at Goddard Space Flight Center and is generally accomplished by correlating playback data with near-real-time data. During the period of the orbit when a tracking station is not available, all data are recorded for later *playback*. When a tracking station is available, the tape recorder continues recording, and telemetry data are simultaneously transmitted to the ground in real time. This continues for a fraction of the station pass, after which real-time data are neither transmitted nor recorded, unless there is a second tape recorder available. The recorded data are *dumped*, or played back at high speed (usually a factor of at least five faster than they were recorded), and are transmitted to the ground. After the recorder has been dumped, it resumes recording and the process is subsequently repeated.

The tracking or receiving station time tags the tape recorded data by searching for the data segment which was recorded simultaneously with the transmitted real-time data. After a match or *correlation* is found, the data segment can be tagged with ground time, since the real-time data were tagged with ground time as they were received. Once this segment of recorded data has been tagged with ground time, the rest of the recorded data can be tagged by working backward from the known segment. Periods of missing data, or *data dropout*, can be detected by examining the spacecraft clock count in the recorded data. This process will be illustrated by the time tagging scheme used for the RAE-2 spacecraft.

The Radio Astronomy Explorer-2 (RAE-2) spacecraft is in a lunar orbit. It is unable to transmit data while on the far side of the Moon, so it contains two tape recorders to record and then transmit data when the spacecraft is in view of tracking stations. The spacecraft clock count register is updated by one count every 20 minutes, and is read into the telemetry stream during 1 of 10 *calibration frames* occurring at 20-minute intervals along with other spacecraft *housekeeping* information, such as battery temperatures. The sequence of events during transmission to tracking stations is depicted in Fig. 8-14. When *acquisition of signal* at a tracking station occurs, the tape recorder in use continues to record data while the station receives real-time data. When a calibration sequence is received, a command is transmitted to the spacecraft which causes (1) the tape recorder to begin dumping its recorded data, (2) the other tape recorder to begin recording data, and (3) real-time data transmission to cease. The tape recorder dump requires 45 minutes, after which tape recording and real-time transmission resume. A ground command then turns the tape recorder off again, while real-time transmission continues until *loss of signal* [Grant and Comberiate 1973; Ferris, 1973].

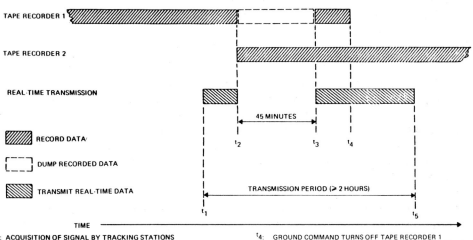

Fig. 8-14. Sequence of Events During Transmission of Data From RAE-2 to Ground Tracking Stations

The data are dumped at a rate five times the recording speed, so that data from the 225-minute orbit can be dumped in 45 minutes. For this reason, the process is used even when continual station coverage is available. The data segment recorded between times t_1 and t_2 in Fig. 8-14 is matched with the real-time segment received and time tagged during the same interval. The rest of the recorded data are then time tagged based on the tags during this segment and the values of the spacecraft clock contained in the remainder of the recorded data. During the next data transmission, the roles of the tape recorders are reversed.

Data Processing at the Receiving Station. Data are processed by the IPD at Goddard Space Flight Center at two major levels. The first consists of analysis by the *input processing* computer, which includes calculations to account for short-term (one-orbit) drift in the spacecraft clock; this step produces attached times of sufficient accuracy for rough calculations. The second step consists of analysis by the *intermediate processing* computer. During this process, calculations include the change in orbital position during the tracking station pass and the corresponding time-dependent spacecraft-to-Earth transmission delay, the tracking station-dependent hardware delay time (measured onsite at each tracking station), and the long-term (several orbits) drift in the spacecraft clock. Data are then processed by IPD software to validate the attached times. Incorrectly time-tagged data are detected and corrected. Time-tagged data from the IPD are hence more reliable than near-real-time data from the Multisatellite Operations Control Center. Time-ordered data are then transmitted to an attitude determination computer via a communication data link or on magnetic tape.

Summary. Attitude-related data are typically time tagged to the nearest millisecond, though timing capabilities exist at NASA STDN tracking and receiving stations to 25 μsec, with accuracies of ± 2.5 μsec expected by 1980. Time tags in definitive data are processed by the IPD, and incorrect tags are detected and corrected. Near-real-time data are time tagged by tracking stations or by the receiving station, and tagging errors are detected and corrected by attitude determination software.

8.4 Telemetry Processors

James S. Legg, Jr.

After telemetry data have been received by an attitude determination computer as described in Section 8.1, they are analyzed by an attitude determination software system. The first subsystem involved in this procedure is the *telemetry processor*. The functions performed by telemetry processors vary from mission to mission, but routinely include the following:

1.   *Reading* telemetry records from a permanent telemetry disk data set or from a telemetry tape

2.   *Unpacking* selected data items, i.e., placing telemetered values into arrays in core

3.   *Converting* the data to engineering units

4.   *Validating* the data (see Section 9.1)

5.   *Correcting* invalid data

6.   *Time-checking* the attached times and/or spacecraft clock count (see Section 9.1) and

7.   *Generating segments of valid data*, usually corresponding to minor or major frames of telemetry data

Functions 1, 2, and 7 are always performed; functions 3, 4, 5, and 6 are generally available, and are performed as necessary.

Reading and Unpacking Telemetry Data. Data are read and processed from the telemetry data set one record at a time; a record may contain several major frames of data (GEOS-3 records contain 3 major frames), or may contain only a portion of a major frame (SAS-3 records contain 8 minor frames; 32 records are required to complete a major frame of 256 minor frames). After each telemetry record is read into core, selected data items are extracted and placed into arrays for subsequent processing. Sometimes this process requires extracting and examining the values of one or more data items before the extraction of other items; e.g., a flag in the data may indicate which of several formats the data appear in, and the location of other data items within the record depends on this format. The method of reading data records depends on whether they are being read singly in *near real time* or in large groups in the *batch processing* mode. In the near-real-time mode, data from the spacecraft are received by a tracking station as they are being measured and transmitted. The data are relayed immediately to a receiving station and made available to attitude determination software on an as-available basis.

This means that the amount of data available for processing increases steadily with time, record by record. The telemetry processor must read each new record, process the data, pass control to the attitude determination system for further processing, and upon receiving control again, read the next record and repeat the process. If the read attempt occurs before the next record is received, an *end-of-file* condition occurs. When this happens, the telemetry processor generally waits a brief interval (typically ~1 sec), and attempts to read the record again. If the record is still not available, the process is repeated until a specified limit on the number of attempts is reached, at which time the telemetry processor displays an appropriate message and waits for operator action.

In the batch processing mode, all data to be processed have already been received. The telemetry processor can read all the data desired, process them, and pass all results to the attitude determination system at one time. The amount of data to be read and processed is limited only by the size of the arrays to be filled or by the amount of telemetry data available.

Preread or *quicklook* features are often provided in the batch processing mode to read and unpack selected data items for display purposes for rapid determination of whether the data are suitable for processing. Several types of data items are examined in such a mode; an example of a quicklook display for SAS-3 is shown in Fig. 8-15, in which data are normalized to arbitrary units for common display. In

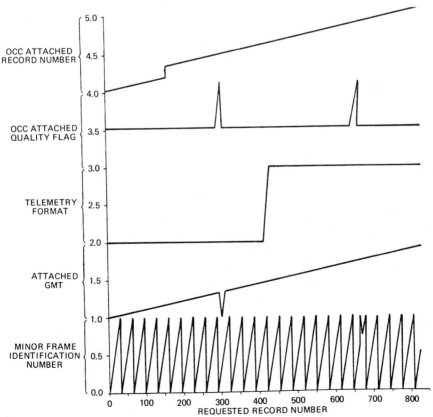

Fig. 8-15.   Sample Quicklook Display for SAS-3

the figure, a bad GMT attached time occurred at record number 300 and an out-of-sequence minor frame number occurred at approximately record number 660. A spurious event apparently happened at these times, because the OCC attached quality flag is also bad for these two records. The telemetry format changed at approximately record 425. Record dropout occurred at approximately record 160.

Converting Telemetry Data. Data items telemetered to the ground or attached by ground software frequently require conversion to engineering units prior to processing by attitude determination software. For example, the time attached to the data samples frequently consists of milliseconds of year, or day of year and milliseconds of day, both of which are typically converted to seconds since 0 hour UT Sept. 1, 1957.* (See Section 1.4.) A second type of conversion is required when the bits representing the magnitude of a data item are inverted when the number is negative. This is frequently the case, for example, with magnetometer or other analog data. In this case, the first bit in the data represents its polarity, and is often assigned a value (0 for negative and 1 for positive) opposite to the sign convention on standard computers. The sign bit must be extracted and examined; if it implies a negative number, the remaining bits must be inverted and a negative sign inserted which the processing computer will recognize (see Section 8.2).

A third example of conversion is the application of a linear multiplicative scale factor, or an additive constant, to telemetered values. In this case the converted value, y, is related to the telemetered value, x, by

$$y = ax + b \qquad (8-5)$$

where the constants a and b are based on measurements performed prior to launch.

If the relation between the telemetered value and the converted value is nonlinear, some form of table lookup may be required. Examples of this type of conversion include infrared scanner pitch angle data, solar panel position data, and damper angle data for SAS-3 and Sun angle data for the SMS/GOES series.† One common nonlinear relation is the conversion of a Gray code (see Section 6.1) to engineering units. Telemetry processors convert from Gray to binary code, and telemetry data simulators convert from binary to Gray code; algorithms for both processes are presented below.

To convert from Gray to binary code:
1.   Invert the left-most bit or retain it as is, depending on the sensor.
2.   Invert the next bit to the right if the left-most bit is now a 1.
3.   Treat the remaining bits in a similar left-to-right pairwise manner, inverting each new bit if the preceding bit is now a 1.

To convert from binary to Gray code:
1.   Perform a logical shift to the right on the binary bit string (i.e., delete the right-most bit, move each of the remaining bits one place to the right, and insert a 0 as the new left-most bit: 11101011 becomes 01110101).
2.   Perform an *exclusive or* of the resulting bit string with the original binary

---

* These conversions are performed by subroutines TCON40 and TCON20. See Section 20.3.2.
† See Section 22.1 for a discussion of linear and nonlinear calibration.

number. The result of the *exclusive or* operation is the Gray-coded representation of the original binary number.

These conversions are illustrated in Fig. 8-16 (see also Section 6.1).

Values obtained from the above conversions may require further conversion before they are suitable for processing by attitude determination software. For example, data obtained from Sun or magnetometer sensors may need to be transformed by a suitable Euler transformation from sensor coordinates to spacecraft body coordinates. (See, for example, Section 7.1.) As another example, Sun sensor data after being converted from Gray to binary code can result in a bit

GRAY-TO-BINARY CONVERSION		BINARY-TO-GRAY CONVERSION	
GRAY CODED VALUE	1 0 0 1 1 1 1 0	BINARY VALUE (235$_{10}$)	1 1 1 0 1 0 1 1
LEAVE LEFT-MOST BIT UNALTERED	1	LOGICAL SHIFT TO RIGHT	0 1 1 1 0 1 0 1
FIRST BIT IS 1 : INVERT NEXT BIT	1 1	RESULT OF EXCLUSIVE OR	1 0 0 1 1 1 1 0
SECOND BIT IS 1 : INVERT NEXT BIT	1 1 1		
THIRD BIT IS 1 : INVERT NEXT BIT	1 1 1 0		
FOURTH BIT IS 0 : DO NOT INVERT NEXT BIT	1 1 1 0 1		
FIFTH BIT IS 1 : INVERT NEXT BIT	1 1 1 0 1 0		
SIXTH BIT IS 0 : DO NOT INVERT NEXT BIT	1 1 1 0 1 0 1		
SEVENTH BIT IS 1 : INVERT NEXT BIT	1 1 1 0 1 0 1 1		
THE RESULT IS 11101011 BINARY, OR 235$_{10}$			

Fig. 8-16. Conversion Between Gray and Binary Codes. The left-most (sign) bit may or may not require inversion, depending on the sensor convention.

pattern which results in a value of -0; i.e., the sign bit indicates negative but the magnitude is zero. Because this value will be converted to $+0$ by most computers, a legitimate sensor reading (-0) will be converted to an erroneous value $(+0)$. Consequently, such data are usually converted to a range of positive numbers so that each value remains unique. Thus, the range of legitimate values

$$-63, -62, \ldots, -1, -0, +0, +1, \ldots, +62, +63$$

may be converted to

$$+0, +1, \ldots, +62, +63, +64, +65, \ldots, +126, +127$$

by a judicious choice of the Gray-to-digital conversion scheme.

In addition, data is recorded in *buckets*, or integral steps, and may require that half a stepsize be added to or subtracted from the transmitted value, so that the converted value corresponds to the most probable value of the quantity measured.

Validating and Correcting Telemetry Data. Validation of telemetry data within the telemetry processor is done on a discrete point-by-point basis and is discussed in Section 9.1. In some instances, invalid data can be corrected based on other data, but corrections in the telemetry processor are usually minimal and highly spacecraft dependent. Invalid data can be deleted or flagged as incorrect, but it is generally left to attitude determination software to attempt corrections. If the data contain any parity bits, they can be checked by the telemetry processor or by attitude determination software. (See Section 8.2 for a discussion of parity bits.)

Time-Checking Telemetry Data. Time-checking of telemetry data usually consists of comparing the times associated with different data points for self-consistency. The values checked are usually the attached time, the telemetered count of the spacecraft clock, and/or the minor or major frame number. If one of the three values checked is in error, it can frequently be corrected by using one of the other two. Many algorithms have been developed to accomplish time checking; there is no general agreement as to the best type to use.* Upper and lower limit checking is also often performed (see Section 9.1).

Generating Segments of Valid Attitude Data. The telemetry processor, having performed all conversions and validation, generates segments of valid (or, in some cases, flagged) data which are passed to the attitude determination system for further processing. The segments generated may or may not contain data found to be invalid, depending on the option chosen by the operator. If he elects not to accept invalid data, the data segments must be generated with gaps at the beginning, middle, or end, depending on which data are invalid. Often gaps are left in segments of one type of data because another type of data was invalid during the gap and the first type is useless without the second.

Data segments are usually generated on the basis of an integral number of minor or major telemetry frames. For example, in the case of SAS-3, all major frames are initially flagged as containing invalid data, and as valid data are identified, they replace the flagged values. As a result, a major frame at the beginning of a pass may contain flag values at the beginning of the frame, and a major frame at the end may contain flags at the end, as shown in Fig. 8-17.

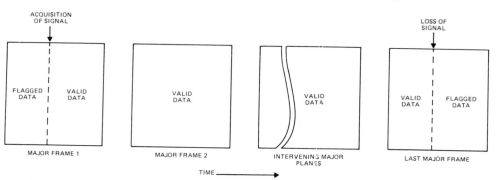

Fig. 8-17.   SAS-3 Major Frames of Data Showing Flagged Data at the Beginning and End of a Data Pass

When more data are available than required by the attitude determination software, data segments are often filled with every nth available data point, thereby reducing the size of the arrays which must be allocated in core for processing. Alternatively, when the volume of data is great enough to prohibit retaining all results in core, the telemetry processor can write its results to an intermediate storage device so the attitude determination software can read and process only the data needed for a particular function (e.g., star camera data, Sun and magnetometer data, slit star sensor data, or infrared Earth horizon sensor data). This method was used for SAS-3 and HEAO-1.

---

*For examples of time-checking schemes, see Lerner, *et al.*, [1974], Williams, *et al.*, [1974], or Cheuvront and Eiserike [1975].

References

1. Cheuvront, S. E., and S. S. Eiserike, *OSO-I Attitude Determination System User's Guide*, Comp. Sc. Corp., CSC/SD-75/6039, June 1975.

2. Fang, A. C., *An Analysis of Spacecraft Data Time Tagging Errors*, NASA TN D-8073, 1975.

3. Ferris, A. G., *NASA-GSFC Mission Opplan 1-73 for the Radio Astronomy Explorer (RAE-B)*, NASA X-513-73-10 GSFC 1973.

4. Grant, M. M., and M. A. Comberiate, *Operating the Radio Astronomy Explorer-B Spacecraft (RAE-B)*, NASA X-714-73-152, GSFC, 1973.

5. Gunshol, L. P., and G. A. Chapman, *Flight Dynamics Resource Requirements for the TDRSS/STS—Shuttle-Era Payload Support*, Comp. Sc. Corp., CSC/TM-76/6105, May 1976.

6. Hefley, G., *The Development of Loran-C Navigation and Timing*, National Bureau of Standards Monograph 129, 1972.

7. Jackson, Willis, *Communication Theory*. London: Academic Press, Inc., 1953.

8. Lerner, G. M., J. S. Legg, Jr., R. W. Nelson, J. A. Niblack, and J. F. Todd, *MAPS/GEOS-C Operating Guide and System Description*, Comp. Sc. Corp., 3000-17400-01TR, March 1974.

9. Peterson, W. Wesley, and E. S. Weldon, Jr., *Error Correcting Codes*. Cambridge, MA: The MIT Press, 1972.

10. *Reference Data for Radio Engineers*, Fifth Edition. Wash., D.C., Judd & Detweiler, Inc., 1968.

11. Ryder, J. D., *Engineering Electronics*. New York: McGraw-Hill, 1967.

12. Scott, J. N., *STDN User's Guide Baseline Document (Revision 2)*, STDN No. 101.1, GSFC, 1974.

13. Stiltz, H. L., *Aerospace Telemetry*. Englewood Cliffs, N.J.: Prentice-Hall, 1961.

14. U.S. Senate, *Soviet Space Programs, 1966–70*, Staff Report for the Committee on Aeronautical and Space Sciences, 92nd Congress, Document No. 92-51, Dec. 1971.

15. Williams, R., D. Stewart, C. Cowherd, T. Hattox, D. Kramer, and W. Palmer, *Radio Astronomy Explorer-B (RAE-B) Attitude Support System Program Description and Operating Guide—Volume I*, Comp. Sc. Corp., 3000-05300-02TR, May 1974.

CHAPTER 9

DATA VALIDATION AND ADJUSTMENT

9.1 Validation of Discrete Telemetry Data
Checking Data Flags and Sensor Identification, Validation of Discrete Data Points, Handling Invalid Data
9.2 Data Validation and Smoothing
9.3 Scalar Checking
Representative Scalars, Applications of Scalar Checking, Central Body and Horizon / Terminator Identification
9.4 Data Selection Requiring Attitude Information

Prelaunch anticipation and postlaunch analysis of erroneous data are commonly the most time-consuming aspects of attitude analysis. However, careful software design can permit automatic detection and/or correction of many types of data errors and mitigate time-consuming and costly manual data correction. For real-time operation, automatic correction or deletion of bad data is essential because of the time required for manual processing.

Bad data may be categorized in several ways:

1. According to the *processing stage* at which the erroneous data can be recognized and corrected, such as errors which may be identified in the telemetry processor and those which cannot be identified until an initial estimate of the attitude is obtained;

2. According to the *source* or cause of the erroneous data, such as transmission, hardware, software, or operator errors;

3. According to the *result* or manifestation of the erroneous data, such as biased output, incorrect sensor identification, or random bit errors.

The sections in this chapter are organized according to the first category. Section 9.1 describes tests which may be performed on a single frame of telemetry data and Section 9.2 describes tests appropriate to larger segments of data which may be performed in the telemetry processor or the attitude determination system and which do not require additional information such as ephemeris data or an initial estimate of the attitude. Section 9.3 describes tests requiring ephemeris information which may be done in the early stages of attitude processing before an attitude estimate is available. Section 9.4 describes tests requiring some estimate of the attitude before the tests can be conducted.

Errors which commonly occur in attitude data are summarized in Table 9-1, which is a representative sample rather than a complete list. All of the items listed have been observed in real data. Complete hardware failures, calibration errors, biases, and misalignments have not been included.

There are four general sources of error encountered in spacecraft data: data transmission, operator error, hardware or software malfunction, and "non-nominal" operating conditions. Data transmission problems, caused by a weak signal or electronic interference, increase the probability of random errors in the transmitted bit stream. No telemetry signal, or a signal below the receipt threshold, will result in *data dropout*, or no data. Although transmission problems may be critical to mission performance, they do not present a significant processing

Table 9-1.   Representative Examples of Telemetry Data Errors (including hardware failure, calibration errors, biases, and sensor misalignments)

	DATA TYPE	ERROR	PROBABLE CAUSE	SPACECRAFT	SECTION WHERE DISCUSSED
1.	ALL DATA	RANDOM BIT ERROR ("BIT SLIPPAGE")	HARDWARE/COMMUNICATIONS	ALL	9.1
2.	ALL DATA	DATA DROPOUT	COMMUNICATIONS (ALL PHASES) NORMALLY LOSS OF SYNCHRONIZATION	NEARLY ALL	9.1
3.	QUALITY FLAGS	INCORRECT FLAGGING	GROUND SOFTWARE	SSS-1, SAS-3	9.1
4.	SPACECRAFT CLOCK	TIME JUMPS	SPURIOUS RESET OF SPACECRAFT CLOCK	SAS-3, SSS-1	8.3
5.	TIME TAGGING	TIME JITTER (JUMPS OR GAPS)	GROUND-DUPLICATE TRANSMISSION, OPERATOR ERROR	SAS-1, 2, 3	8.3
6.	TIME TAGGING	RELATIVE TIMES CORRECT ABSOLUTE TIMES INCORRECT	OPERATOR ERROR IN ATTACHING TIMES TO TAPE-RECORDED DATA	SAS-3, SAS-1, NIMBUS-6	9.1
7.	TIME TAGGING	CORRECT TIME, INCORRECT DATE	OPERATOR	NEARLY ALL	9.1
8.	MULTIPLE SENSORS (ANY TYPE)	INCORRECT SENSOR ID	ELECTRONICS/GROUND SOFTWARE	NEARLY ALL	9.1
9.	MAGNETOMETERS	TIME-DEPENDENT BIAS	FIELD DUE TO EQUIPMENT ON IN SPACE-CRAFT	AE-3, SAS-2, 3, OSO-8	6.3
10.	VISUAL HORIZON SCANNERS	FREQUENT SPURIOUS EVENTS	REFLECTIONS OFF SPACECRAFT	RAE-2	9.4
11.	VISUAL HORIZON SCANNERS	EXTRA SPIN PERIODS ADDED TO EARTH	ELECTRONICS	IMP-7	
12.	IR HORIZON SCANNER	INVALID EARTH WIDTHS BELOW 12°	SENSOR ELECTRONICS	SMS-1, 2; GOES-1; CTS	9.4
13.	IR HORIZON SCANNER	CORRECT DATA ATTACHED TO INSERTED SPURIOUS TIMES	UNKNOWN (TRANSMISSION?)	AE-C	9.1
14.	IR HORIZON SCANNER	SPURIOUS SIGNALS WHEN SENSOR MISSES EARTH	ELECTRONICS	CTS	
15.	IR HORIZON SCANNER	MOON INTERFERENCE	–	GOES-1	
16.	IR HORIZON SCANNER	SYSTEMATIC 2-1/2° ERROR IN EARTH-OUT FOR 10-MINUTE INTERVALS	LOOSE WIRE AT GROUND STATION	GOES-1	
17.	IR HORIZON SCANNER	NOISY DATA	"CROSSTALK" DUE TO STAR TRACKER	ATS-6	
18.	WHEEL-MOUNTED HORIZON SENSORS	REDUCED EARTH WIDTHS	LOGICAL "AND" IN ELECTRONICS FOR SUN REJECTION	ATS-6, AE-3, SAS-3	
19.	WHEEL-MOUNTED HORIZON SENSORS	SINUSOIDAL OSCILLATION IN OUTPUT DATA AT BODY RATE	MISALIGNED BOLOMETER	AE-3, 4, 5	7.2
20.	DIGITAL SUN SENSORS	FAILURE OF 1 BIT	HARDWARE	AE-3, IMP-8	7.1
21.	POLARIS TRACKER	LOSES POLARIS	FOLLOWS DUST PARTICLES	ATS-6	6.4
22.	STAR SENSOR	HIGH NOISE LEVEL	CHARGED PARTICLES IN SOUTH ATLANTIC ANOMALY/REFLECTIONS/SENSOR FAILURE	OSO-6, 7, 8; SSS-1; SAS-1	6.4
23.	STAR SENSOR	SENSITIVITY DROPS WITH AGE	UNKNOWN	SAS-2	
24.	SLIT STAR SCANNER	FREQUENT REPEATED SPURIOUS EVENTS	MULTIPLE TRIGGERINGS BY 1 STAR WHEN SLOWLY SPINNING/IMPROPER THRESHOLD/ MULTIPLE REFLECTIONS	SAS-3, SSS, OSO	6.4

problem because they are easily recognized by elementary tests so that the affected data may be removed at an early level. In contrast to transmission problems, operator errors are frequently the most difficult to recognize because they do not occur with any regular pattern and normally no indication exists within the data stream itself as to which data were attached manually at some stage of the data transmission process.

The detectability of hardware or software malfunctions depends on the type of malfunction. The best method for identifying subtle malfunctions (i.e., biases which shift output values by a small amount) is the use of independent, redundant attitude hardware and processing techniques. Non-nominal operating conditions may also produce subtle errors that are difficult to detect. For example, spacecraft in synchronous orbits may have Earth horizon sensors which have been thoroughly analyzed and tested for normal mission conditions, but which are essentially untested for conditions arising during attitude maneuvers or transfer from low Earth orbit to synchronous altitude. (See, for example, the "pagoda effect" described in Section 9.4.) Each of these possible sources of bad data should be considered in preparation for mission support.

9.1 Validation of Discrete Telemetry Data

James S. Legg, Jr.

Validation of discrete telemetry data consists of checking individual data items. The two principal methods of validation are (1) checking the actual value of data items, such as quality flags and sensor identification numbers, to determine if associated data are valid, and (2) checking that values of selected data items fall within specified limits.

In describing errors in raw telemetry, it is pertinent to distinguish between systematic and random errors. *Systematic errors*, or those which occur over a non-negligible segment of telemetry data, often are more troublesome to detect and

REC. NO.	TIMES HH.MM.SS	FELDBX	FELDBY	FELDBZ	ID	NA	NB
49	17.16.52	47.	-67.	-343.	3	20	72
49	17.16.56	47.	-67.	-343.	3	20	0
50	17.17.07	47.	-67.	-343.	3	20	71
50	17.17.09	16.	-67.	-343.	3	20	70
50	17.17.13	55.	-67.	-343.	3	20	70
51	17.17.15	-500.	-500.	-500.	3	19	70
51	17.17.16	-496.	-480.	-465.	3	19	70
51	17.17.20	-500.	-500.	-500.	3	19	89
52	17.17.22	51.	-67.	-339.	3	19	70
52	17.17.25	51.	-67.	-339.	3	19	69
52	17.17.27	-500.	-500.	-500.	3	19	69
53	17.17.30	-500.	-500.	-500.	3	19	69
53	17.17.32	51.	-67.	-339.	0	127	90
53	17.17.41	51.	-67.	-339.	3	19	69
54	17.17.42	51.	-67.	-339.	3	19	69
54	17.17.44	51.	-67.	-370.	3	19	69
54	17.17.46	51.	429.	-339.	3	19	68
55	17.17.58	-461.	-71.	-276.	3	18	68
55	17.18.08	358.	-47.	-472.	3	18	67
55	17.18.28	55.	-71.	-335.	3	18	66
56	17.18.29	55.	-71.	-335.	3	18	66
56	17.18.37	39.	-71.	-366.	3	17	65
56	17.18.41	-500.	-500.	-500.	3	14	94
57	17.18.49	-500.	-500.	-500.	0	64	32
57	17.18.50	59.	-71.	-335.	3	16	64
57	17.18.55	165.	0.	-484.	3	17	64
58	17.18.58	59.	-71.	-335.	3	17	64
58	17.19.04	311.	-71.	-83.	3	17	63
58	17.19.12	461.	217.	-461.	3	16	64
59	17.19.20	59.	-71.	-343.	3	16	63
59	17.19.22	71.	-71.	-319.	3	16	33
59	17.19.34	63.	-71.	-335.	3	16	62
60	17.19.36	63.	-71.	-335.	3	16	62
60	17.19.38	0.	-75.	-457.	1	85	15
60	17.19.49	-323.	406.	441.	3	16	61
61	17.19.57	480.	472.	-496.	3	15	60
61	17.20.02	-500.	-500.	-500.	3	85	85
61	17.20.05	67.	-75.	-331.	3	15	59
62	17.20.11	67.	-75.	-331.	3	15	59
62	17.20.13	67.	-12.	-331.	3	15	60
62	17.20.15	67.	-75.	-331.	3	84	0
63	17.20.30	67.	-75.	-331.	3	15	58
63	17.20.32	-500.	-500.	-500.	3	85	85

Fig. 9-1.   Random Errors in Sun and Magnetometer Data From GEOS-3. Columns 3 through 5 list the *x*, *y*, and *z* components of the measured magnetic field vector. The last three columns list the Sun sensor identification and the two Sun angles. Underlined values are spurious.

correct than *random errors*, which occur at isolated points within the data. Examples of these types of error in data from the GEOS-3 spacecraft are presented in Figs. 9-1 and 9-12 (Section 9.3). These examples include random hit errors in Sun and magnetometer data (Fig. 9-1) and systematic errors in Sun data (Fig. 9-12). Figure 9-2 illustrates rotation angle data from the AE-3 spacecraft which contained so much random noise that automatic data validation was impossible. Operator intervention and iterative processing were necessary to identify valid data (at rotation angles of about 450 deg).

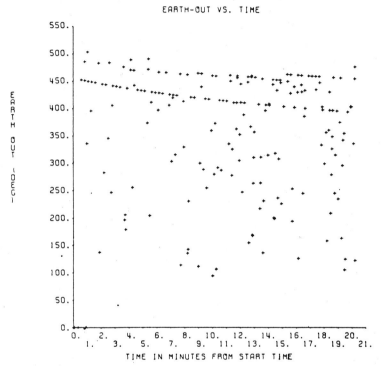

Fig. 9-2.   Earth-Out Horizon Scanner Data From AE-3. Large quantity of spurious data makes identification of valid data difficult.

9.1.1 Checking Data Flags and Sensor Identification

The first method of validation is concerned with the type of data being analyzed, rather than whether the values of these data are acceptable. For example, there may be tell-tales or flags indicating whether the data were received in real time from the spacecraft (real-time data) or were recorded on a tape recorder aboard the spacecraft and transmitted later while over a tracking station (playback data). There may be one or more flags used to determine in which of several formats the data were transmitted. Flags may also describe the operating mode of the spacecraft at the time of transmission and what attitude determination sensors were operating at that time.

These flags are normally evaluated before attempting to read other data because they determine what types of data are present and where and how often these data occur in the telemetry. An example of the need for this form of validation is GEOS-3, which has two telemetry formats, one containing a single data sample from the two-axis digital Sun sensors in each major frame of data and the other containing four Sun sensor data sample per major frame. A flag byte, included in the raw telemetry data, is examined to determine the number of Sun data items present before extracting them from the telemetry frame.

9.1.2 Validation of Discrete Data Points

The most common method of validation for discrete data points is upper- and lower-limit checking; that is, the value of the data must fall within specified limits to be acceptable. These limits can be constant (a maximum sensor voltage) or time varying (the proper day for an attached time). If the value of a data point lies outside the prescribed limits, it is invalid and may be corrected, flagged, or deleted. This method of validation is often performed on data types such as the attached times and the spacecraft clock. Sometimes limit checking is useful even when the data will not be used in further attitude calculations. For example, if the data are to be plotted automatically, outlying data points may adversely affect the limits of plot axes, causing valid data to lose significance. Limit checking is not useful when all the values a data item may assume are acceptable. In these cases, a discrete data item cannot be classified as erroneous without examining its value relative to other data, as discussed in Sections 9.2 through 9.4.

Another method of discrete data point validation is examination of the quality flag attached to the data by previous ground software processing. In data processed at Goddard Space Flight Center, this flag is set by an Operations Control Center or the Information Processing Division. The quality flag denotes whether a minimum number of bits in the telemetry synchronization (sync) pattern for each major or minor frame are incorrect, and hence indicates the likelihood of remaining bits in the data segment being bad. (The number of incorrect bits in the sync pattern which causes the quality flag to indicate bad data can vary from satellite to satellite; it is generally 9 bits out of 24.) A quality flag indicating bad data does not necessarily imply that bad data are present, but rather that there is a greater probability of bad data, since the sync pattern itself is in error. This flag can be validated as a discrete data point and the remaining data in the major or minor frame flagged or deleted accordingly.

Data may also be validated on a discrete point-by-point basis by comparing one type of data with another. For example, one can compare the value of the spacecraft clock for a given data sample with the time attached to that sample, or either of these might be compared with the minor or major frame number. Another example is comparison of the selected two-axis Sun sensor identification number with the analog output of the ATA photocell for each sensor (see Section 6.1 for a description of ATA) to determine if the Sun sensor ID corresponds to the Sun sensor most intensely illuminated. A third example is determination that star tracker data are valid by checking the values of associated flags, which indicate whether the tracker is in the track mode and whether the intensity of the object being tracked is within acceptable limits.

Validation may also be performed on information contained in the header provided by the receiving station. Information such as the location of the tracking station that received the data, the date the data were received, the start time of the data, and the spacecraft ID may be validated if desired.

9.1.3 Handling Invalid Data

Data which have been determined to be invalid can sometimes be corrected. For example, if the attached time is invalid but the spacecraft clock reading is valid and a known attached time corresponds to a known spacecraft clock reading, a current attached time may be computed on the basis of the current spacecraft clock time. The minor and major frame numbers might be used in a similar manner. Another example is correcting the two-axis Sun sensor identification based on the largest of the ATA readings.

When data have been examined and found to be invalid and no method exists to correct them on a discrete point-by-point basis, we must decide what to do with the bad data. In some cases, an invalid data point is useless and renders other data gathered at the same time useless as well. In these cases, all the data in question can be deleted and not processed by attitude determination software. In other cases, although a particular data value is useless, related data may be useful and should be retained. Sometimes the invalid data itself may be worth examining in further analysis. In these cases, the data are retained and used in further attitude determination calculations or corrected as discussed in the following sections. Data so treated are often flagged so that subsequent software can readily identify questionable data and correct or ignore them. The two most common methods of flagging data are *internal flagging* (changing the value of the data to a flag value, such as 99999) and *external flagging* (setting the value of a corresponding flag variable to a flag value). The latter method has the advantage of retaining incorrect data values for further analysis and the disadvantage of requiring extra computer storage for flag variables; extra core is generally required even when no data are flagged.

Similar manipulation can be done manually when data are viewed in interactive mode on a graphic display device. This enables the operator to evaluate the data and selectively process those considered acceptable. As seen in Fig. 9-2, it is often impossible to foresee all the ways in which the data will be bad and to provide fully automatic validation checks in the software; consequently an interactive processing capability is included in most software systems to permit manual data validation and manipulation.

After data validation and processing, it may become apparent from attitude solutions that telemetry data should have been selected, validated, or processed in a different manner. In this case, the entire procedure may be repeated using different discrete data validation criteria. Iterative procedures of this type are discussed in Section 9.4.

9.2 Data Validation and Smoothing

Gerald M. Lerner

Data validation is a procedure by which we either accept or reject measurements but do not otherwise alter them. Rigorously, validating data by rejecting measurements which are "obviously" incorrect alters the statistical characteristics of the data. For example, data with Gaussian noise will have, on the average, one measurement in 1.7 million with an error of 5σ or more. In a practical sense, however, rejecting such data is justified because all spacecraft data are subject to random bit errors (see Section 9.1), which typically occur much more frequently than 5σ Gaussian noise errors.

Data smoothing is a technique which is widely used both to preprocess and validate data before attitude determination and to postprocess computed attitude solutions, primarily to reduce random noise or to derive attitude rates. Data smoothing is the only processing required for some data types which are used or displayed directly, such as boom length, accelerometer, or spin rate data.

Data smoothing is one method used to obtain an expected value for a measurement which is then used for validation. In using smoothing as a validation technique, one assumes that the telemetered data frequency is high compared with the frequencies characteristic of the data type and that similar measurements made at nearby times are reliable. In this section, we describe techniques used to "smooth" or to obtain an expected value for either measured or processed data. The expected value may be used either for subsequent processing or just for validation.

In addition to validating, data smoothing may be used to:

1.   *Remove high-frequency noise.* The effects of sensor data digitization and noise may be reduced by the use of an algorithm which attenuates high-frequency components in the data.

2.   *Reduce data volume.* If telemetry data rates are sized for a particular data type or operating mode, a large fraction of the telemetered data may be redundant and can be discarded to reduce the data processing load significantly without degrading attitude solutions.

3.   *Interpolate.* For postprocessing, short periods of data dropout may be bridged. For preprocessing, interpolation is useful for data display or for providing estimated data at times other than those measured.

4.   *Improve accuracy.* For some data types, the intrinsic accuracy of the sensor exceeds the telemetered least significant bit (LSB). For example, digital Sun sensor errors are typically less than half the LSB at transitions. Processing techniques which emphasize data when the LSB changes can, in principle, improve the accuracy of computed attitudes.*

5.   *Compute Rates.* Attitude rates are required for some applications such as the initialization of data predictors and verification of control system performance. Magnetometer rate data is required for some attitude control systems and is usually obtained by analog differentiation; however, backup ground support may require numerical differentiation.

6.   *Filter Data.* Some data types, such as AE-3 accelerometer data [Dennis, 1974], are used directly and may be enhanced by filtering, which can remove high-frequency noise.

---

*The practical worth of this scheme is doubtful because the reduced data volume may nullify the increased data accuracy. This procedure was implemented by Pettus [1973] with two-axis digital Sun sensor data. Pettus concluded that it was not useful because of the reduced data volume.

7. *Display Data.* A smooth function through noisy data can improve the intelligibility of graphic displays.

Four basic techniques for data smoothing are filtering, curve fitting, sifting, and preaveraging. *Filtering* is a data weighting scheme which is applied symmetrically to each measurement, y_i, to produce a filtered measurement

$$\bar{y}_i = \sum_{k=0}^{N} a_k(y_{i-k} + y_{i+k}) \tag{9-1}$$

Choice of the range, N, and weights, a_k, permits the selective attenuation of high-frequency components in the data and significantly alters the statistical characteristics of the data.* Removal of high-frequency noise highlights the actual frequency characteristics of the data. The digital filter, Eq. (9-1), is derived from analog filters used for electronic signal processing.

Curve fitting is a technique which assumes a functional form for the data over a time interval and computes a set of coefficients which represent the data over the given interval. The function selected to fit the data may be (1) a linear combination of orthogonal polynomials, or (2) a nonlinear combination of functions chosen to represent the probable characteristics of the data. Curve fitting techniques treat the data asymmetrically because end points tend to influence the coefficients less than midrange points; thus they are most appropriate for batch processing (the end points are the most important for recursive or real-time processing). Data represented by coefficients are difficult to treat statistically and should not be used for many types of subsequent processing. Curve fitting is most frequently used for data display and interpolation.

Sifting is a technique which subdivides an interval into discrete bins and replaces the data in each bin, independent of other bins, with a randomly or systematically selected data point within the bin. *Preaveraging* is similar to sifting except that the data within the bin are replaced with the arithmetic mean. Either sifting or preaveraging must be combined with another method for data validation, such as a comparison with data from adjacent bins. Sifting and preaveraging are most appropriate for reducing the quantity of data to be processed. They are the preferred choices for preprocessing data which are subsequently used in a differential corrector or any algorithm that depends on the statistical characteristics of the data. Note that any data preprocessing method alters or destroys some statistical properties inherent in the measurement, including the systematic process employed by the spacecraft to insert sensor data into the telemetry stream. The advantage of data sifting (and, to a lesser extent, preaveraging) is that it is a less destructive method of preprocessing, and sifted data more closely conform to the requirements of the attitude determination algorithms described in Chapter 13.

The guidelines presented above should not replace the careful consideration of the processing requirements and the characteristics of each data type before selecting a preprocessing method. The most important data characteristic is its implicit or explicit frequencies. Smoothing is most useful for measurement frequencies, ω_m, such that $\omega_m = 2\pi/\Delta t \gg \omega_s$, where Δt is the telemetered data rate

---

*End points, e.g., $\bar{y}_i$ for $i < N+1$, must be treated separately. One approach is to assume $\bar{y}_i = y_i$ for $i < N+1$.

and ω_s is any real frequency associated with the data which is to be retained. Dominant low frequencies of interest are related to the orbital rate, which is $\omega_o \approx 2\pi/100$ minutes $= 10^{-3}$ rad per sec for near-Earth spacecraft. The orbital rate affects the thermal profile and solar and aerodynamic torques. The dominant gravity-gradient frequency for a pencil-shaped spacecraft is $\sqrt{3}\,\omega_o$, and for a polar orbit the magnetic torque frequency is approximately $2\omega_o$. High frequencies of interest are related to the spin period, onboard control, flexible components, and rastering instruments and are typically 0.1 to 50 rad per sec, which is also the frequency range of telemetered data. Thus, telemetry data rates are often a limiting factor in the extraction of high-frequency information.

To summarize, the tradeoff between preprocessing sensor data before attitude determination and postprocessing computed attitudes must be established for each spacecraft. In general, it is better to preprocess only for the purpose of data validation and postprocess to reduce random (or high frequency) noise, primarily because attitudes have a time dependence which is simpler than sensor data and preprocessing may destroy important statistical properties used in some attitude determination algorithms. For postprocessing, curve fitting may use low-order polynomials or well-established functional forms.

Curve Fitting. Curve-fitting techniques require a data model which may be either purely phenomenological, such as a linear combination of orthogonal functions, or a nonlinear function chosen to approximate the assumed dynamics characteristics of the data. Fitting techniques, as described in Section 13.4, may be either sequential or batch. A sequential method (see subroutine RECUR in Section 20.3) has been used successfully on the AE mission to postprocess computed nadir angles with a nonlinear model of the following form [Grell, 1976]:

$$y(t) = A_1\sin(\omega_1 t + \phi_1) + A_2\sin(\omega_2 t + \phi_2) \qquad (9\text{-}2)$$

The state parameters, A_1, A_2, ω_1, ω_2, ϕ_1, and ϕ_2, are updated sequentially with the covariance matrix controlled to track or smooth the measurements to allow for large model deficiencies. Curve fitting was used on AE to validate computed nadir angles and extract approximate nutation and coning frequencies, phases, and amplitudes. Nonlinear models, such as Eq. (9-2), generally require special techniques to obtain an initial estimate of the model parameters. For AE, a frequency analysis based on a fast Fourier transform [Gold, 1969] was used to obtain ω_1 and ω_2.

Linear models are preferred for curve fitting because of the ease of solution. Power series, spherical harmonics, and Chebyshev polynomials are used frequently, although any set of orthogonal polynomials may be used. Care must be taken to ensure that the correct degree of the representation is selected. If n data points are to be fitted with representation of degree r, clearly r must be less than n. However, if r is either too small or too large for a given n, a poor compromise between minimizing truncation error and reducing random noise will be obtained.

One procedure to automatically select the degree of the representation is to monitor the *goodness of fit* or *chi-squared function*,

$$\chi^2(r) = \frac{S_r}{n-r} = \frac{1}{n-r}\sum_{i=1}^{n}\left(y_i - \sum_{k=0}^{r-1}C_k g_k(t_i)\right)^2 \Big/ \sigma_i^2 \qquad (9\text{-}3a)$$

where $g_k(t_i)$ is the kth basis polynomial evaluated at the ith value of the independent variable, y_i is the measured data, σ_i is the standard deviation of y_i, and the parameters C_k are selected by a linear least-squares algorithm to minimize $\chi^2(r)$. $\chi^2(r)$ decreases rapidly with increasing degree. The degree may be chosen to be the lowest such that either absolute or relative convergence is obtained; i.e.,

$$\chi^2(r) < \epsilon_1 \approx 10 \tag{9-3b}$$

$$|[\chi^2(r) - \chi^2(r-1)]/\chi^2(r)| < \epsilon_2 \approx 0.1 \tag{9-3c}$$

Assuming the model is adequate, χ^2 should range from 1 to 10 for a correct r; $\chi^2 < 1$ is indicative of too high a degree, r, or an overestimate of the standard deviations, σ_i.

As an example of curve fitting, we consider the use of Chebyshev polynomials. We wish to smooth the data, y_i, measured at discrete times, t_i. Let $g_k(x)$ be a sequence of orthogonal polynomials defined for $x = (-1, 1)$. As described in Section 13.4, the problem is to determine the coefficients, C_k, to minimize the quantity

$$S_r = \sum_{i=1}^{n} w_i \left[y_i - \sum_{k=0}^{r-1} C_k g_k(x_i) \right]^2 \tag{9-4}$$

where

$$x_i = (2t_i - t_{max} - t_{min})/(t_{max} - t_{min}) \tag{9-5}$$

where the weight of the ith measurement is $w_i \equiv 1/\sigma_i^2$ and t_{max} and t_{min} are the maximum and minimum values of t_i, the independent variable. The mapping function, Eq. (9-5), limits the range of the independent variable to that permitted for the orthogonal polynomials which satisfy the relation

$$\int_{-1}^{1} g_i(x) g_j(x) dx = \delta_{ij} \tag{9-6}$$

The solution for the coefficients, C_k, requires inverting the $r \times r$ positive definite matrix (see Chapter 13)

$$A = GWG^T \tag{9-7}$$

to compute

$$C = (GWG^T)^{-1} GWY \tag{9-8}$$

where

$$C = \begin{bmatrix} C_0 \\ C_1 \\ \vdots \\ C_{r-1} \end{bmatrix}; \quad Y = \begin{bmatrix} y_1 \\ y_2 \\ \vdots \\ y_n \end{bmatrix}; \quad W = \begin{bmatrix} w_1 & & & \\ & w_2 & & \\ & & \ddots & \\ & & & w_n \end{bmatrix} \tag{9-9}$$

and

$$G = \begin{bmatrix} g_0(x_1) & g_0(x_2) & \cdots & g_0(x_n) \\ g_1(x_1) & & & \\ \vdots & & & \\ g_{r-1}(x_1) & & \cdots & g_{r-1}(x_n) \end{bmatrix}$$

(9-10)

In practice, the matrix A is ill conditioned (i.e., difficult to invert in practice) for a power series, $g_k(x) = x^k$, and power series representations are not practicable for $r > 4$ because of the greatly varying magnitude of the elements of A. However, an alternative representation using *Chebyshev polynomials* will greatly improve the condition of the matrix A for most applications. Chebyshev polynomials are solutions to the differential equation

$$(1 - x^2)\frac{d^2 g_k}{dx^2} - x\frac{d g_k}{dx} + k^2 g_k = 0$$

(9-11)

and satisfy the recursion relation

$$g_{k+1}(x) = 2x g_k(x) - g_{k-1}(x)$$

(9-12)

with the starting polynomials

$$g_0(x) = 1$$
$$g_1(x) = x$$
$$g_2(x) = 2x^2 - 1$$
$$g_3(x) = 4x^3 - 3x$$

(9-13)

Subroutines are available to set up (APCH) and solve (APFS) the normal equations, Eq. (9-8), using the Chebyshev polynomials [IBM, 1968]. APFS selects the polynomial degree by computing S_r until the equation

$$S_r < \epsilon S_0$$

(9-14)

is satisfied, where ϵ is an input parameter. Note that ϵ must be greater than approximately 10^{-6} for single-precision arithmetic on IBM System/360 computers.
Given the coefficients, C_k, the smoothed value of y_i is

$$\bar{y}_i = \sum_{k=0}^{r-1} C_k g_k(x_i)$$

(9-15)

A *residual edit* may be performed by discarding data, y_i, for which

$$|y_i - \bar{y}_i| > n_\sigma \sigma_i$$

(9-16)

where n_σ is a tolerance parameter. The data are processed iteratively, first obtaining the coefficients by solving Eq. (9-8), then editing using Eq. (9-16) until no additional data are discarded and the process converges. Convergence requires a high ratio of valid to invalid data, typically 10 to 1 or greater. If the data are very noisy, or substantial data dropout is present, automatic processing will reject all data and manual intervention will be required. (See, for example, Fig. 9-2.) Note that the use of Eq. (9-16) for data validation does *not* depend on the method used to obtain $\bar{y}_i$ and, consequently, the preceding caveat applies to any validation algorithm employing data smoothing.

An estimate of the derivative, $d\bar{y}_i/dx$, is obtained by differentiation of Eq. (9-15),

$$\frac{d\bar{y}_i}{dx} = \sum_{k=0}^{r-1} C_k \frac{dg_k(x)}{dx}\bigg|_{x=x_i} \tag{9-17}$$

The derivative of the Chebyshev polynomials satisfies the recursion relation [Abramowitz and Stegun 1964]

$$g_k'(x) \equiv \frac{dg_k(x)}{dx} = \frac{-k}{(1-x^2)}(xg_k(x) - g_{k-1}(x)) \tag{9-18}$$

for $|x| < 1$ and $g_k'(\pm 1)$ equals k^2 for k odd and $\pm k^2$ for k even.

Figures 9-3 through 9-5 illustrate the use of Chebyshev polynomials for data smoothing. Figure 9-3 shows GEOS-3 magnetometer data for an early orbit. Despite the highly nonlinear data, a 20th-degree Chebyshev polynomial produces a satisfactory qualitative fit, which is useful for display, for determining crude attitude rates, and for data validation. The quantitative fit is poor because the telemetered data rate is too low relative to the attitude rate; therefore, a higher degree or nonlinear representation should be used. Figure 9-4 illustrates the use of low-degree Chebyshev polynomials to fit deterministic attitude solutions. The noise on the solutions is dominated by sensor resolution and the Sun-magnetic field

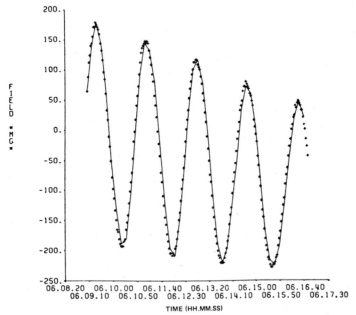

Fig. 9-3. Curve Fitting for GEOS-3 Magnetometer Data Using a Twentieth-Degree Chebyshev Polynomial

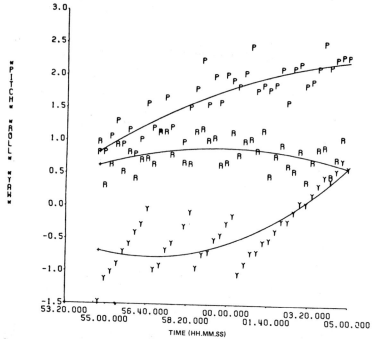

Fig. 9-4. Curve Fitting for GEOS-3 Attitude Data (P = pitch, R = roll, Y = yaw) Using a Third-Degree Chebyshev Polynomial

geometry. The sensor data was validated but not otherwise preprocessed. Figure 9-5 illustrates a difficulty with preprocessed sensor data. In this case, Sun sensor and magnetometer data were preprocessed and the smoothed value used for deterministic attitude solutions. Note that the observed structure in the attitude data is artificial and the apparent high accuracy of the attitude solution is misleading.

A major problem with batch process curve fitting is the asymmetric treatment of the data and the difficulty of obtaining a satisfactory compromise between a polynomial of degree high enough to avoid truncation error and low enough to reduce random noise.

Filtering. With data filters, some of the problems of curve fitting can be avoided by first fixing the degree and then selecting a data interval about each measurement. This approach is often called a *moving arc filter* because it is sequentially centered on the measurement to be smoothed and is therefore symmetric.

A useful filter is the *least-squares quadratic* filter, which has the following expressions for the smoothed function and its derivative [Budurka, 1967]:

$$\bar{y}_i = \sum_{k=i-m_1}^{i+m_2} C_k y_i \qquad (9\text{-}19)$$

$$\dot{\bar{y}}_i = \sum_{k=i-m_1}^{i+m_2} D_k y_i \qquad (9\text{-}20)$$

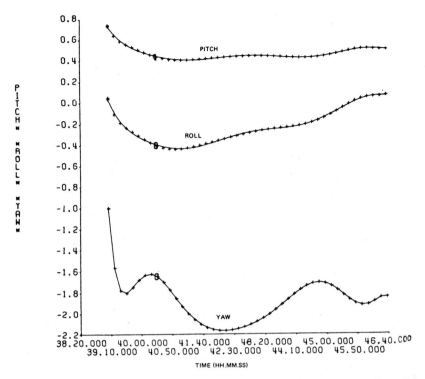

Fig. 9-5. GEOS-3 Attitude Solutions Using Smoothed Sun Sensor and Magnetometer Data. Solutions appear unrealistically accurate as discussed in the text. The resolution of the GEOS-3 attitude sensors is approximately 0.5 to 1 deg, and the only frequencies which should be observed in the data are related to the orbital period. The observed data span covers only 8% of an orbital period.

where the filter coefficients are

$$C_k = \left(P - Qy_k + Ry_k^2\right)/D \tag{9-21a}$$

$$D_k = \left(-Q + Ty_k - Sy_k^2\right)/D \tag{9-21b}$$

$$P = \sum y_k^2 \sum y_k^4 - \left(\sum y_k^3\right)^2 \tag{9-21c}$$

$$Q = \sum y_k \sum y_k^4 - \sum y_k^2 \sum y_k^3 \tag{9-21d}$$

$$R = \sum y_k \sum y_k^3 - \left(\sum y_k^2\right)^2 \tag{9-21e}$$

$$S = (m_1 + m_2 + 1) \sum y_k^3 - \sum y_k \sum y_k^2 \tag{9-21f}$$

$$T = (m_1 + m_2 + 1) \sum y_k^4 - \left(\sum y_k^2\right)^2 \tag{9-21g}$$

$$D = (m_1 + m_2 + 1)P - Q \sum y_k + R \sum y_k^2 \tag{9-21h}$$

and the sums in Eqs. (9-21) are over the range $i-m_1 \leqslant k \leqslant i+m_2$. If the data points are equally spaced in time, the simplified expressions for $m_1=m_2=m$ are as follows:

$$\bar{y}_i = C_0 y_i + \sum_{k=1}^{m} C_k (y_{i-k} + y_{i+k}) \tag{9-22}$$

$$\dot{y}_i = \sum_{i=1}^{m} D_k (y_{i-k} - y_{i+k}) \tag{9-23}$$

where

$$C_k = \frac{3(3m^2+3m-1)-15k^2}{(2m+1)(3(3m^2+3m-1)-5m(m+1))} \tag{9-24a}$$

$$D_k = \frac{3k}{m(m+1)(2m+1)} \tag{9-24b}$$

Filters may be described by their effect on various frequency components in the data. Figure 9-6 illustrates the relative attenuation of frequencies for a least-squares quadratic filter. For 25 data points, attenuation is substantial for $\omega > 0.07\omega_m$ and negligible for $\omega < 0.05\omega_m$. The quantity $\omega_m = 2\pi/\Delta t$ is the measurement frequency and Δt is the time interval between measurements. The number of data points must be carefully selected to avoid removing desired information from the data.

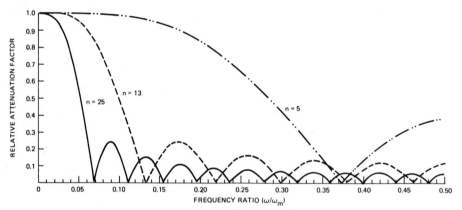

Fig. 9-6.   Frequency Attenuation for a Least-Squares Quadratic Filter [Budurka, 1967]

For some applications, the poor frequency cutoff characteristics of the quadratic least-squares filter (manifested by the persistent sinusoidal oscillation at high frequencies) are undesirable. The *Butterworth filter* [Dennis, 1974; Budurka, 1967; Rabiner and Gold, 1975; Stanley, 1975] has a much sharper cutoff, as shown in Fig. 9-7. The coefficients depend on both the order and the cutoff frequency, ω_c. The difference equation for the fifth-order Butterworth is

$$\bar{y}_i = \sum_{k=0}^{5} G_0 A_{-k} y_{i-k} + \sum_{k=1}^{5} B_{-k} \bar{y}_{i-k} \tag{9-25}$$

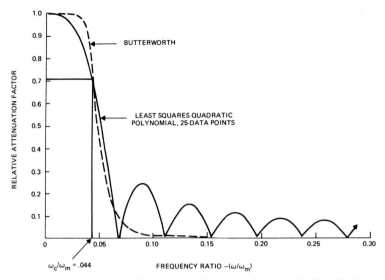

Fig. 9-7.   Magnitude-Frequency Function for Fifth-Order Butterworth Filter [Budurka, 1967]

where the coefficients for $\omega_c = 0.044\omega_m$ are given in Table 9-2 for equally spaced data. The recursive nature of Eq. (9-25) implies an infinite memory; that is, the improved bandwidth characteristic (e.g., selective frequency attenuation) is achieved by linking together all the measurements. The infinite memory causes an initial transient response in the filter output. The Butterworth filter is particularly

Table 9-2.   Coefficients for Fifth-Order Butterworth Filter [Budurka, 1967]

A_0 = 1.000000	A_{-3} = 10.000000	G_0 = 3.20567×10^{-5}	B_{-3} = 5.724077
A_{-1} = 5.000000	A_{-4} = 5.000000	B_{-1} = 4.113261	B_{-4} = −2.415025
A_{-2} = 10.000000	A_{-5} = 1.000000	B_{-2} = −6.833588	B_{-5} = 0.410249

well suited for real-time applications because it depends only on previous measurements to obtain the filtered value. Figures 9-8 through 9-11 illustrate the use of various data filters on simulated attitude data which has been contaminated with Gaussian noise. In Fig. 9-8, a constant input signal plus noise is processed by a Butterworth (order = 50, $\omega_c = 2\pi/50$ sec^{-1}), least-squares quadratic or LSQ ($m_1 = m_2 = m = 25$), and averaging* filter. In the figure, points denote the noisy input data, and the dotted, dashed, and solid lines denote the data after processing with a Butterworth, LSQ, and averaging filter, respectively. Except for the initial transient in the Butterworth filter's response, all three filters effectively attenuate the noise.

   In Fig. 9-9, a sinusoidal input signal, $V = 1 + \cos\omega t$ ($\omega = 2\pi/50$ sec^{-1}), has been contaminated as before. The averaging filter removes both the noise and the signal, whereas the Butterworth and LSQ filters remove the noise and only

---

*Each data sample is replaced with the arithmetic mean of the 25 preceding and subsequent samples.

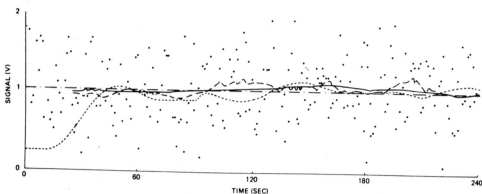

Fig. 9-8. Response of Butterworth (Dotted Line), Least-Squares Quadratic (Dashed Line), and Averaging (Solid Line) Filters to Gaussian Noise (Mean = 1 V, Standard Deviation = 0.5 V)

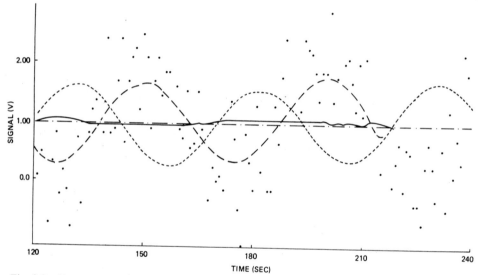

Fig. 9-9. Response of Butterworth (Dotted Line), Least-Squares Quadratic (Dashed Line), and Averaging (Solid Line) Filters. $(V(t) = 1 + \cos \omega t + \nu$, where $\omega_m = 0.13$ sec^{-1}, $E(\nu) = 0.25$, $v(\nu) = 0.25$ and $\omega_c = 0.13$ sec^{-1}.)

attenuate the signal. The predicted attenuation factor is 0.5 for the Butterworth for which $\omega_c = \omega$. Note the phase lag in the response of the Butterworth filter.

Figures 9-10 and 9-11 illustrate the use of the frequency response of the Butterworth and LSQ filters to obtain a desired output frequency spectrum. In Fig. 9-10, the Butterworth filter cutoff, $\omega_c = 2\pi/100$, is chosen to attenuate the input frequency, whereas in Fig. 9-11 the cutoff, $\omega_c = 2\pi/12.5$, is chosen to pass the input frequency and only attenuate the noise. The frequency dependence of the Butterworth filter's phase lag is apparent by comparing Figs. 9-9 and 9-11.

The frequency response of the LSQ filter is not as easily controlled as that of the Butterworth. In Fig. 9-10, with $m = 50$, there is some attenuation of the input frequency, whereas in Fig. 9-11, with $m = 5$, the signal attenuation is negligible but the noise is not removed completely.

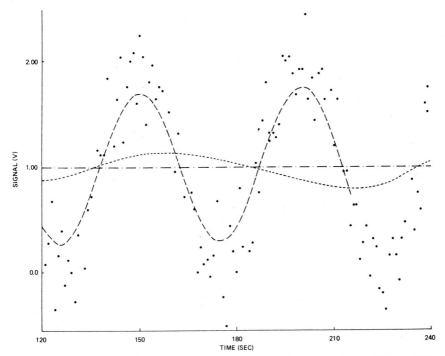

Fig. 9-10.   Response of Butterworth (Dotted Line) and Least-Squares Quadratic Filters. ($V(t) = 1 +$ $\cos \omega t + \nu$, where $E(\nu) = 0$, $\nu(\nu) = 1/16$, $\omega_m = 0.13$ sec^{-1}, and $m = 50$.)

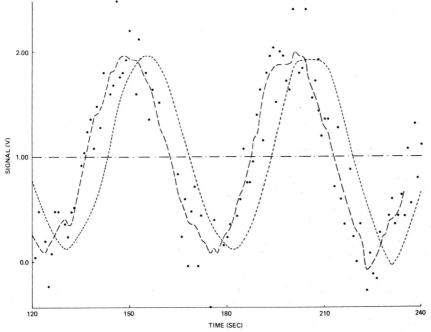

Fig. 9-11.   Response of Butterworth (Dotted Line) and Least-Squares Quadratic Filters. ($V(t) = 1 +$ $\cos \omega t + \nu$, where $E(\nu) = 0$, $\nu(\nu) = 1/16$, $\omega_m = 0.13$ sec^{-1}, $\omega_c = 0.525$ sec^{-1}, and $m = 5$.)

9.3 Scalar Checking

Gerald M. Lerner

Data validation based on scalar checking occupies an intermediate position in the data validation hierarchy; data must be time tagged and ephemeris information computed, but an attitude estimate is not required. Scalar checking tests the self-consistency of attitude data and is used to remove or correct spurious data prior to the actual attitude computation.

Scalar checking is based on the elementary principle that scalars, such as the magnitude of a vector or the angle between two vectors, do not depend on the coordinate system in which they are evaluated. In particular, a scalar computed from measurements in the body frame must equal that computed in any convenient reference frame.

9.3.1 Representative Scalars

The scalar which is validated most frequently in attitude determination systems is the measured magnitude of the Earth's magnetic field, $B_M = |\mathbf{B}_M|$.* Although attitude determination algorithms generally require only the measured field direction, the measured magnitude may be compared with the calculated magnitude, $B_C = |\mathbf{B}_C|$, computed from the spacecraft ephemeris and a model for the Earth's magnetic field (Section 5.1). Measured data is rejected if

$$|B_M - B_C| > \epsilon_B \qquad (9\text{-}26)$$

where ϵ_B is a tolerance parameter based on the magnetometer resolution and unmodeled magnetometer biases.

Comparison of B_M and B_C for a data segment is particularly useful for identifying errors in time tagging or in the spacecraft ephemeris. The former error is manifested by a systematic phase difference between $B_M(t)$ and $B_C(t)$ such that $B_M(t) \approx B_C(t + t_0)$, and the latter by a qualitative difference in both amplitude and phase. The root-mean-square of the quantity $\Delta B(t_i) = B_M(t_i) - B_C(t_i)$ is a measure of the fidelity of the field model or an indicator of the presence of systematic magnetometer biases. Assuming a magnetometer quantization size of X_B, the mean square residual [Coriell, 1975]

$$\langle (\Delta B)^2 \rangle \equiv \frac{1}{N} \sum_i (\Delta B(t_i))^2 \qquad (9\text{-}27)$$

must be greater than $X_B^2/12$. The residual error, $\delta B = (\langle (\Delta B)^2 \rangle - X_B^2/12)^{1/2}$, in the IGRF (1968) magnetic field model has been shown to be less than 200 nT (Section 5.1) for intermediate altitudes of 700 to 800 km. Computed residual errors in excess of this value are indicative of unmodeled biases.

For missions which fly magnetometers, Sun sensor data may be validated by comparing the measured angle between the Sun and the magnetic field vectors with

---

*Rigorously, the measurement $\mathbf{B}_M$ is a true vector only if the magnetometer triad is orthogonal; otherwise, $\mathbf{B}_M$ denotes three ordered measurements which are treated algebraically as a vector.

that computed from the spacecraft ephemeris and a field model. Assuming that $\mathbf{B}_M$ satisfies Eq. (9-26), Sun data is flagged if

$$|\cos^{-1}(\hat{\mathbf{B}}_M \cdot \hat{\mathbf{S}}_M) - \cos^{-1}(\hat{\mathbf{B}}_C \cdot \hat{\mathbf{S}}_C)| > \epsilon_\theta \qquad (9\text{-}28)$$

where $\hat{\mathbf{S}}_M$ and $\hat{\mathbf{S}}_C$ denote the measured and the calculated Sun vectors and ϵ_θ is a tolerance based on the magnetometer and Sun sensor resolution, the accuracy of the field model, and unmodeled biases. An equation analogous to Eq. (9-27) may be used to obtain a measure of the relative Sun sensor and magnetometer alignment and the error in the model field direction. Expected root-mean-square (rms) residuals, due to Sun sensor and magnetometer resolution, will contribute a residual to Eq. (9-28) analogous to the $X_B^2/12$ term in the discussion following Eq. (9-27). However, this residual is highly orbit dependent and can best be established via simulation [Coriell, 1975]. Unmodeled rms residual angular errors in the IGRF reference field are of the order 0.3 to 0.5 degree (Section 5.1) at 700 to 800 km.

Earth horizon scanners and similar devices measure the Earth nadir vector in body coordinates, $\hat{\mathbf{E}}_M$. This vector must satisfy the condition

$$|\cos^{-1}(\hat{\mathbf{E}}_M \cdot \hat{\mathbf{B}}_M) - \cos^{-1}(\hat{\mathbf{E}}_C \cdot \hat{\mathbf{B}}_C)| < \epsilon_M \qquad (9\text{-}29)$$

when used with magnetometer measurements, and

$$|\cos^{-1}(\hat{\mathbf{E}}_M \cdot \hat{\mathbf{S}}_M) - \cos^{-1}(\hat{\mathbf{E}}_C \cdot \hat{\mathbf{S}}_C)| < \epsilon_S \qquad (9\text{-}30)$$

when used with Sun sensor measurements. $\hat{\mathbf{E}}_C$ is the nadir vector in inertial coordinates and ϵ_M and ϵ_S are tolerances associated with the magnetometer and Sun sensor accuracies, respectively.

Clearly, mean and root-mean-square residuals of scalar quantities are useful for assessing the magnitude of unmodeled errors in sensor data. Displays of predicted-versus-observed scalars are useful in identifying time-tagging or other systematic errors in the data, particularly before mission mode when tests based on an a priori attitude are not available.

9.3.2 Applications of Scalar Checking

In addition to its use in validating and assessing sensor data, scalar checking has been used in star identification (Section 7.7), magnetometer bias estimation [Gambhir, 1975], and Sun sensor data reduction. For example, a procedure to periodically compensate for magnetometer biases is based on the assumption that the biases are constant for some appropriate time interval (typically an hour or more). Neglecting noise, we may write

$$\mathbf{B}_U(t_i) = \mathbf{B}_M(t_i) - \mathbf{b} \qquad (9\text{-}31)$$

where $\mathbf{B}_U(t_i)$ is the unbiased measured magnetic field at time t_i, $\mathbf{B}_M(t_i)$ is the biased measured magnetic field at time t_i, and $\mathbf{b}$ is the magnetometer bias which is assumed constant in time. Although the components of the vector $\mathbf{B}_U(t_i)$ are attitude dependent, the magnitude $B_U(t_i)$ is not. Assuming that the magnetometer triad is orthogonal, we may equate $B_U(t_i)$ to the model field magnitude, $B_C(t_i)$,

$$B_C^2(t_i) = B_U^2(t_i) = |\mathbf{B}_M(t_i) - \mathbf{b}|^2 = B_M^2 + b^2 - 2\mathbf{b} \cdot \mathbf{B}_M \qquad (9\text{-}32)$$

or

$$B_C^2 - B_M^2 \equiv Y(t_i) = b^2 - 2\mathbf{b} \cdot \mathbf{B}_M \tag{9-33}$$

where the explicit time dependence of $\mathbf{B}_C$ and $\mathbf{B}_M$ has been suppressed for convenience.

The vectors $\mathbf{B}_M$ are known from measurements, and the corresponding values of $\mathbf{B}_C$ can be calculated using spacecraft ephemerides and geomagnetic field models. Therefore, the values of $Y(t_i)$ corresponding to each value of $\mathbf{B}_M$ can be calculated, and a least-squares fit of the data to Eq. (9-33) (see Chapter 13 and Gambhir, [1975]) can be made to obtain the best estimates of the three components of $\mathbf{b}$.

As a second application, a scalar test may be applied to correct anomalous two-axis Sun sensor data. Figure 9-12 illustrates a problem encountered with the

REC. NO.	$ F	TIMES HH.MM.SS	F M	FELDBX	FELDBY	FELDBZ	F S	I D	NA	NB	
119		02.15.44		165.	-94.	-205.					
120		02.15.46		165.	-94.	-205.	3	15	15		
120		02.15.48		165.	-94.	-205.	3	15	15		
120		02.15.5?		165.	-94.	-205.	3	15	15		
121		02.15.52		165.	-94.	-205.	3	15	14		
121		02.15.54		165.	-94.	-205.	3	15	14		VALID DATA,
121		02.15.56		165.	-94.	-205.	3	15	14		SUN JUST ABOVE
122		02.15.58		165.	-94.	-205.	3	15	14		HORIZON
122		02.16.00		165.	-94.	-201.	3	15	14		
122		02.16.02		165.	-94.	-201.	3	15	14		
123		02.16.04		165.	-94.	-201.	3	15	14		
123		02.16.07		165.	-94.	-201.	0	15	14		
123		02.16.09		155.	-94.	-201.	0	15	13		
124		02.16.11		165.	-94.	-201.	0	15	13		
124		02.16.13		165.	-94.	-197.	0	15	13		
124		02.16.15		165.	-94.	-197.	0	15	13		
125		02.16.17		155.	-94.	-197.	0	15	13		OUTPUT BELOW
125		02.16.19		165.	-94.	-197.	0	12	13		SENSOR SELECT
125		02.16.21		159.	-94.	-197.	0	12	13		THRESHOLD
126		02.16.23		169.	-94.	-197.	0	12	13		
126		02.16.25		169.	-94.	-197.	0	12	13		
126		02.16.27		169.	-94.	-193.	0	12	13		
127		02.16.29		169.	-94.	-193.	0	19	13		
127		02.16.31		169.	-94.	-193.	0	21	85		
127		02.16.33		169.	-94.	-193.	0	85	85		
128		02.16.35		169.	-94.	-193.	0	85	85		
128		02.16.37		169.	-94.	-193.	0	85	85		
128		02.16.39		169.	-94.	-193.	0	85	85		
129		02.16.41		169.	-94.	-189.	0	85	85		
129		02.16.43		169.	-94.	-189.	0	85	85		
129		02.16.45		169.	-94.	-189.	0	85	85		
130		02.16.47		169.	-98.	-189.	0	85	85		
130		02.16.50		169.	-98.	-189.	0	85	85		
130		02.16.52		169.	-98.	-189.	0	85	85		SUN BELOW
131		02.16.54		169.	-98.	-189.	0	85	85		HORIZON
131		02.16.56		169.	-98.	-185.	0	85	85		
131		02.16.58		173.	-98.	-185.	0	85	85		
132		02.17.00		173.	-98.	-185.	0	85	85		
132		02.17.02		173.	-98.	-185.	0	85	85		
132		02.17.04		173.	-98.	-185.	0	85	85		
133		02.17.06		173.	-98.	-181.	0	85	85		
133		02.17.08		173.	-98.	-181.	0	85	85		
133		02.17.10		173.	-98.	-181.	0	85	85		

Fig. 9-12. Observed Sun Sensor Data for GEOS-3 in Mission Mode, June 28, 1975

GEOS-3 Sun sensors in mission mode. Correct angular measurements (NA and NB) were telemetered but the sensor head (ID) selected by onboard electronics was incorrect when the Sun was near the Earth horizon. (Similar problems were observed when the Sun traversed the field of view of two sensors.) Let S_{Mi} denote the measured Sun vector in body coordinates, assuming that the measurement corresponds to the ith sensor. Then the correct sensor selection minimizes the quantity

$$f_i = |\mathbf{S}_{Mi} \cdot \mathbf{X}_M - \mathbf{S}_C \cdot \mathbf{X}_C| \tag{9-34}$$

where $\mathbf{X}_M$ and $\mathbf{X}_C$ are the measured and calculated values for any inertial reference vectors (e.g., nadir or magnetic field) and S_C is the calculated inertial Sun vector. Note that this procedure can fail for certain geometries. For example, if the Sun sensor boresights are dispersed at a half cone angle θ with respect to the spacecraft Z-axis, Eq. (9-34) is independent of i when X_M is colinear with the Z-axis.

9.3.3 Central Body and Horizon/Terminator Identification

Data from horizon scanners, either those sensitive to the visible or the infrared portion of the electromagnetic spectrum, must be validated to reject spurious triggerings caused by the Sun, the Moon, or reflections from spacecraft hardware. For visible light sensors, a further test is required to distinguish between horizon and terminator crossings (Section 4.1).

Although most spurious triggerings are relatively simple to identify (Section 8.1), terminator crossings escape most preprocessing tests and normally are eliminated after the attitude computation by a data regeneration test [Joseph, 1972] or solution averaging (Section 11.2). However, a simple scalar test based on the arc length separation, α, between the Sun, $\hat{S}$, and the triggering event, $\hat{X}$, will suffice for both central body identification and terminator rejection for all cases for which the data regeneration test will succeed [Williams, 1972].

For any triggering, the angle α may be computed by applying the law of cosines to the spherical triangle shown in Fig. 9-13. Thus,

$$\cos \alpha = \cos \beta \cos \gamma + \sin \beta \sin \gamma \cos \Phi \qquad (9\text{-}35)$$

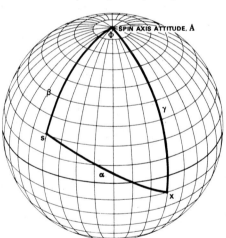

Fig. 9-13.   Spherical Geometry for a Sensor Event at **X**

Figure 9-14 illustrates the terminator geometry for a central body, **R**, of angular radius ρ, less than half-lit (*crescent*). Alternatively, Fig. 9-14 illustrates the case for the central body more than half-lit (*gibbous*) if the Sun is at the opposite pole. Since the Sun is at a pole of the coordinate grid, the latitude lines are lines of constant α. For infrared sensors, the only restriction on α is that

$$\alpha_0 \equiv \psi - \rho \leqslant \alpha \leqslant \psi + \rho \equiv \alpha_3 \qquad (9\text{-}36)$$

where $\cos \psi = \hat{\mathbf{R}} \cdot \hat{\mathbf{S}}$.

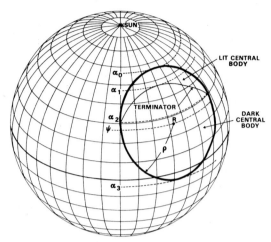

Fig. 9-14.   Definition of Sun to Central Body Angles Which Define Event Classification. Light and dark portions of the central body are interchanged for the Sun at the opposite pole. α_i and ψ are measured from the Sun and are defined in Eqs. (9-36) through (9-43).

The requirements for visible light data are more restrictive. First, consider the crescent geometry of Fig. 9-14. Note that the angles α_i and ψ on the figure are measured from the Sun. The small circle of constant α_1 is tangent to the terminator on the **SR** great circle and the small circle of constant α_2 passes through the cusps or the points where the horizon and terminator intersect. Clearly, a triggering at latitudes $\alpha_0 \leqslant \alpha \leqslant \alpha_1$ can result only from the central body horizon. Triggerings at latitudes $\alpha_1 < \alpha \leqslant \alpha_2$ can result from *either* the horizon or the terminator.* This is defined as the indeterminate case. Triggerings at latitudes $\alpha > \alpha_2$ are necessarily spurious.

The angles α_1 and α_2 may be calculated with the aid of the upper half of Fig. 9-15. The plane of the figure contains the vectors $\hat{\mathbf{S}}$ and $\mathbf{R}$ with the spacecraft at the origin and the Sun along the $+Y$ axis. By symmetry, this plane also contains $\mathbf{X}$ at the angle $\alpha = \alpha_1$. D_c (89.15 deg) is the dark angle defined in Section 4.1.

Let $\mathbf{R}_\oplus$ be the vector from the center of the central body to the terminator crossing. Taking components of $\mathbf{X}$ and $\mathbf{R}_\oplus$, perpendicular and parallel to the sunline, we obtain

$$X \sin\alpha_1 + R_\oplus \sin D_c = R \sin\psi \qquad (9\text{-}37)$$

$$X \cos\alpha_1 + R_\oplus \cos D_c = R \cos\psi \qquad (9\text{-}38)$$

with the result

$$\alpha_1 = \arctan\left[\left(\sin\psi - \frac{R_\oplus}{R}\sin D_c\right)\bigg/\left(\cos\psi - \frac{R_\oplus}{R}\cos D_c\right)\right] \qquad (9\text{-}39)$$

To compute α_2, we note that when $\alpha = \alpha_2$, $\mathbf{X}$ is located on both the horizon and the terminator. Points on the terminator are formed by rotating $\mathbf{R}_\oplus$ about the

---

*Data regeneration procedures will attribute this data unambiguously (and incorrectly) to a horizon crossing.

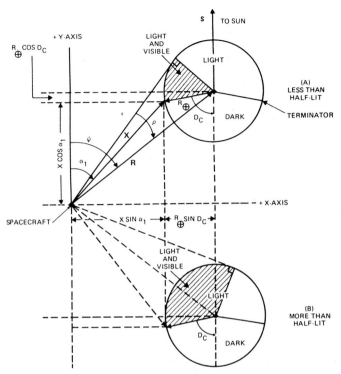

Fig. 9-15. Computation of α_1, the angular separation between the Sun and the closest point on the terminator to the Sun. (See Fig. 9-14.)

sunline. Clearly, all points on the terminator satisfy Eq. (9-38) for components parallel to the sunline,

$$X \cos\alpha_2 + R_\oplus \cos D_c = R \cos\psi \tag{9-40}$$

The condition that $\mathbf{X}$ is on the horizon is simply

$$X^2 + R_\oplus^2 = R^2 \tag{9-41}$$

Therefore, we have the result

$$\cos\alpha_2 = \frac{\cos\psi - \dfrac{R_\oplus}{R}\cos D_c}{\left(1 - R_\oplus^2/R^2\right)^{1/2}} \tag{9-42}$$

As seen from Fig. 9-15, as ψ increases beyond 90 deg, α_1 will exceed α_2 and there exists a value of ψ such that

$$\psi \equiv \psi_c = \arccos(R_\oplus \cos D_c / R) \tag{9-43}$$

and

$$\alpha_1 = \alpha_2 = 90 \text{ deg}$$

This is the condition for which the central body is half lit.

The conditions for the gibbous central body may be obtained by inspection. Note that for $\alpha_1 \geqslant \alpha > \alpha_2$, a triggering must occur at the terminator (the corresponding horizon is dark) and the indeterminate case is absent. Table 9-3 summarizes the results for terminator identification.

Table 9-3. Definition of Triggering Event Classification in Terms of α

CENTRAL BODY LIGHTING	ψ	RANGE OF α			
		HORIZON	TERMINATOR	INDETERMINATE	SPURIOUS
FULLY LIT	$180° \geqslant \psi \geqslant 90° - \rho + D_c$	$\alpha_3 \geqslant \alpha \geqslant \alpha_0$	–	–	$\alpha < \alpha_0, \alpha > \alpha_3$
GIBBOUS	$90° - \rho + D_c > \psi > \psi_c$	$\alpha_2 \geqslant \alpha \geqslant \alpha_0$	$\alpha_1 \geqslant \alpha \geqslant \alpha_2$	–	$\alpha < \alpha_0, \alpha > \alpha_1$
HALF-LIT	$\psi = \psi_c$	$90° > \alpha \geqslant \alpha_0$	–	–	$\alpha > 90°, \alpha < \alpha_0$
CRESCENT	$\psi_c > \psi > \rho + D_c - 90°$	$\alpha_1 \geqslant \alpha \geqslant \alpha_0$	–	$\alpha_2 \geqslant \alpha \geqslant \alpha_1$	$\alpha < \alpha_0, \alpha > \alpha_2$
DARK	$\rho + D_c - 90° > \psi \geqslant 0$	–	–	–	$\alpha < 180°$

9.4 Data Selection Requiring Attitude Information

James R. Wertz

Previous sections have described data selection or validation procedures which do not require any knowledge of the attitude and, therefore, can be performed at an early stage of attitude processing. In contrast, data selection requiring attitude information must be a part of the attitude determination process itself and may require an iterative procedure to determine if the data selection process is consistent with the computed attitude and to reselect the data if it is not.

The most straightforward data selection of this type is that which requires only an *a priori attitude estimate*, that is, an attitude estimated or assumed before any processing is done. In practice, some a priori knowledge is usually available and this is generally sufficient to resolve quadrant ambiguities or to choose the correct attitude solution from the two possible solutions generated by intersecting cones. The latter procedure is described in more detail in Section 11.2. For spacecraft using automatic control, the intended or *null* attitude may be used to validate data used for a definitive attitude solution; however, any such test may also effectively hide a failure of the control system, since data that is inconsistent with the intended attitude would automatically be rejected.

When data selection requires attitude information, it may become the most complex and time-consuming aspect of the attitude determination process. For example, in Fig. 9-2, an attitude estimate and manual data editing were required to determine which of the two groupings of data at the top of the figure was valid and which was anomalous.

On RAE-2 a horizon sensor of a new design (the panoramic attitude sensor described in Section 6.2.2) was flown. During the translunar portion of the flight, much of the data was spurious. Figures 9-16(a) and 9-16(b) illustrate displays which were used to manually distinguish valid lunar sightings from spurious data due to the Sun, spacecraft reflections, or noise. The ordinate of Fig. 9-16(a) is the angle, γ, from the spin axis to the scanner line of sight and the abscissa is the rotation angle,

Φ, from the Sun to a sighting event. The panoramic scanner is a variable mounting angle sensor; for RAE-2 the angle γ changed by 0.707 deg approximately every 15 sec. The crosses in Fig. 9-16(a) mark observed light to dark (LOS) or dark to light (AOS) transitions and the ovals mark the expected location of solar and lunar data for the a priori attitude 5 hours before insertion into lunar orbit. The data observed near $\Phi = 0$ deg or $\Phi = 180$ deg and 60 deg $> \gamma > 0$ deg are clearly spurious and are believed to have been caused by reflected sunlight [Werking, 1974]. The relatively small amount of data near $\Phi = 65$ deg and $\gamma = 30$ deg are valid AOS or LOS events from the lunar horizon or terminator. The expanded view of these data in Fig. 9-16(b) shows that most of the valid data were LOSs at the terminator. Closer

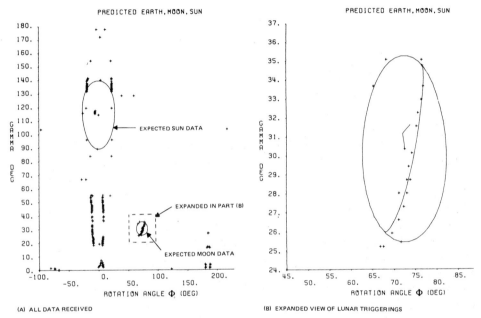

(A) ALL DATA RECEIVED   (B) EXPANDED VIEW OF LUNAR TRIGGERINGS

Fig. 9-16.   RAE-2 Data Selection Based on an A Priori Attitude

inspection of Fig. 9-16(a) suggests the nature of the anomaly. Nearly all of the AOSs are spurious and only the lunar presence at 25 deg $< \gamma < 35$ deg resulted in valid LOSs at the lunar terminator. For RAE-2, these displays, which utilized an a priori attitude and interactive graphics, were essential for attitude determination and maneuver planning during the early portion of the mission.

The attitude determination process is more complex than described above whenever data selection based on an a priori attitude is not sufficiently accurate. This occurs, for example, in the presence of smoothly varying systematic anomalies in which some of the data are clearly invalid but presumably "valid" and "invalid" data run smoothly together. Attitude determination in the presence of such errors requires iterative processing to obtain successive attitude estimates. The general procedure for this is as follows:

1.   Discard "obviously" bad data (in addition to the rejection of random errors as described in Sections 9.2 and 9.3).

2.   Use the remaining data to estimate the attitude *as accurately as possible*.

3.   Use the new attitude estimate to reject additional data (or recover previously rejected data) as appropriate.

4.   Iterate until a self-consistent solution has been obtained, i.e., when step 3 makes no change in the set of selected data.

This procedure does *not* establish that the final attitude estimate is correct, or that the data selection has been correct. It is also possible that the iterative procedure will not converge—it may eventually reject all the data or oscillate between two distinct data sets. This method can at best obtain an attitude solution which is consistent with the data selection process. Therefore, whenever problems of this type are encountered, it is important to attempt to find the physical cause or a mathematical model of the data anomaly to provide an independent test of whether the data selection is correct.

The central problem of the above iteration procedure is the data rejection in step 3. Operator judgment is the main criterion used, both because general mathematical tests are unavailable and because the anomaly is usually unanticipated. (Otherwise it would have been incorporated as part of the attitude determination model.) Tables of data are of little or no use for operator identification of systematic anomalies; therefore, data plots are normally required. Four types of data plots are commonly used for this purpose:

1.   Plots of raw data

2.   Plots of deterministic attitude solutions obtained from individual pairs of points within the data

3.   Plots of residuals between the observed data and predictions from a least-squares or similar processing method based on the entire collection of data

4.   Plots comparing directly the observed data and predicted data based on the most recent attitude estimate

In practice, the author has found the fourth type of plot to be the most useful in defining the boundary between valid and invalid data. To illustrate the use of various plot types and the process by which anomalous data are identified, we describe the data selection process which was used to eliminate the "pagoda effect" identified in SMS-2 data.

The SMS-2 Pagoda Effect. The Synchronous Meteorological Satellite-2, launched from the Eastern Test Range on Feb. 6, 1975 (Fig. 1-1), was the second test satellite for the Geostationary Operational Environmental Satellite series used by the U.S. National Oceanic and Atmospheric Administration to provide daily meteorological photographs of the western hemisphere and other data. During the transfer orbit to synchronous altitude, attitude data were supplied by two Sun sensors and five body-mounted, infrared horizon sensors. As illustrated in Fig. 1-6, each horizon sensor sweeps out a conical field of view or scan cone. Because the spin axis was nearly fixed in inertial space, the scan cone of a single Earth sensor encounters the Earth during one or two segments of the spacecraft orbit and moves across the disk of the Earth as the spacecraft moves. As shown in Fig. 9-17(a), a major anomaly, called the *pagoda effect*, occurred in the Earth data [Chen and Wertz, 1975]; this is most easily seen in the sharp upturn of Earth-out data as the

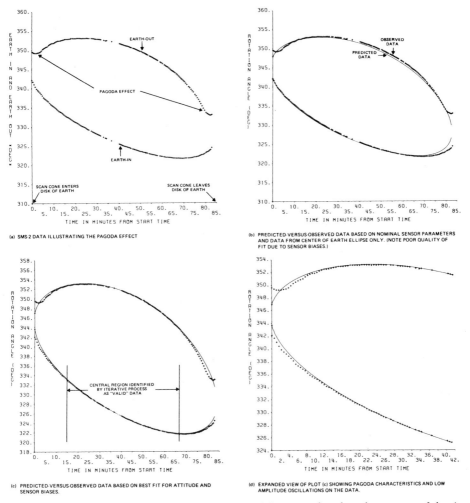

Fig. 9-17.   Raw Data from SMS-2 horizon sensors. (All figures are based on the same set of data.)

scan cone of the sensor enters or leaves the disk of the Earth. The effect is common to all of the SMS-2 horizon sensors.

It is clear that the data at the ends of the Earth are invalid. It is hoped that the data in the center are valid; otherwise, there is little potential for successful attitude determination. Therefore, the main question is how far into the Earth the bad data extend; i.e., what subset of the data should be used to provide the best attitude estimate. Figure 9-18 shows a plot of the spin axis declination determined from the Sun angle and the midscan rotation angle (i.e., rotation angle from the Sun to the midpoint between Earth-in and -out) from the data in Fig. 9-17(a). The data in the central region give at least somewhat consistent solutions, but the 20 to 30 frames of data at both ends are clearly part of the systematic anomaly and should be discarded.

Data selection is performed according to the procedure described above. Both

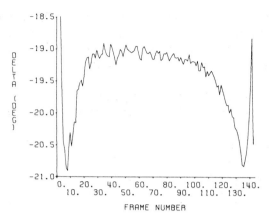

Fig. 9-18. Declination Versus Frame Number as Determined From Data in Fig. 9-17 Using Nominal Sensor Parameters. (Slope of solutions in central region is caused by biases in the sensor parameters.)

ends of the data sample are eliminated and the best available solution for the central portion of the data is obtained. Predicted-versus-observed data plots are then used to refine the selection process in an iterative manner and other plots are used as needed to check the consistency of the results. The need for an accurate attitude solution in step 2 of the process is shown in Fig. 9-17(b),* which compares the observed data with predicted data based on nominal sensor parameters and the data from the central portion of the pass. Particularly on the right side of the figure, it is clear that the anomaly involves both Earth-in and -out and that it extends at least somewhat beyond the end of the "pagoda." However, it is impossible to precisely determine the invalid data from Fig. 9-17(b) because of the poor overall fit, even in the central region. (A solution based on all the data yields an even worse fit.)

Figures 9-17(c) (showing the fit to the central portion of the data) and 9-17(d) (showing the pagoda characteristics) compare the observed data with the predicted data based on results from the central portion of the data pass using attitude and sensor bias parameters obtained from a bias determination subsystem similar to that described in Section 21.2. Once an accurate fit to the data has been obtained, the general character of the pagoda effect becomes clear. Both Earth-in and -out begin varying systematically from predicted values when the *Earth width,* or the difference between Earth-in and -out, drops below about 20 deg. At an Earth width of 12 deg, the Earth-out data turn sharply upward.[†] (The small ripple most noticeable in the Earth-out data on the left of Fig. 9-17(d) is not a plotting artifact; although the cause is unknown, it may result from variations in the height of the Earth's atmosphere in the infrared.)

---

*The attitude solution is based on data from the central region only. Using this attitude, data are predicted for the full data pass (including the end regions) to provide a visual comparison in order to identify the data anomaly. This procedure was used for Figs. 9-17(b) through 9-17(d).
[†]The differences quoted and those shown in the figures are in terms of rotation angle. The arc-length separation between Earth-in and -out is about 14 deg when the effect begins and 7.5 deg when the upturn occurs.

At the time of the above analysis, the cause of the pagoda effect was unknown. Subsequent investigation indicated that it is probably due to delays inherent in the sensor electronics, as described in Section 7.4. The results of that analysis, shown in Fig. 7-21, indicate that the data selection described above is at least approximately correct.

The value of predicted-versus-observed data plots as part of the data validation procedure is shown in Fig. 9-19, which illustrates a data pass where the sensor scan cone does not drop off the Earth before reversing direction and moving toward the Earth's center. In Fig. 9-19(a), there is no visually detectable anomaly, although the previous example suggests that there might be a problem at small Earth widths. This is confirmed by the predicted-versus-observed display of Fig. 9-19(b), which reveals the pagoda effect data which must be eliminated to obtain accurate attitude solutions.

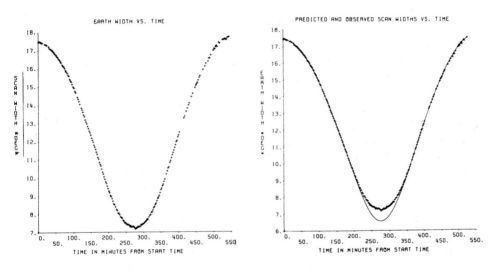

(a) PAGODA EFFECT DATA WHICH IS NOT VISUALLY DETECTABLE

(b) PREDICTED VERSUS OBSERVED DATA, AS FOR PART (a), SHOWING PAGODA EFFECT AT CENTER

Fig. 9-19. Earth Width Data for the SMS-2 Spacecraft Illustrating the Pagoda Effect

References

1. Abramowitz, M., and I. A. Stegun, *Handbook of Mathematical Functions*, NBS Applied Mathematics Series 55, 1964.
2. Budurka, W. J. *Data Smoothing and Differentiation in Program TRIAD*, IBM Report No. TM-67-27, Dec., 1967.
3. Chen, L. C., and J. R. Wertz, *Analysis of SMS-2 Attitude Sensor Behavior Including OABIAS Results*, Comp. Sc. Corp., CSC/TM-76/6003, April 1975.
4. Coriell, K., *Geodynamics Experimental Ocean Satellite-3 Postlaunch Attitude Determination and Control Performance*, Comp. Sc. Corp., CSC/TM-75/6149, Aug. 1975.

5. Dennis, A. R., *Digital Filter Design and Development for the Atmospheric Explorer-C (AE-C) Miniature Electrically Suspended Accelerometer (MESA)*, Comp. Sc. Corp., CSC/TR-74/6028, Dec. 1974.

6. Gambhir, B., *Determination of Magnetometer Biases Using Module RESIDG*, Comp. Sc. Corp., 3000-32700-01TN, March 1975.

7. Gold, B., and C. M. Rader, *Digital Processing of Signals*, McGraw-Hill, 1969.

8. Grell, M. G., and B. Chapman, *Multisatellite Attitude Determination/Atmospheric Explorer (MSAD/AE) System Description*, Comp. Sc. Corp., CSC/SD-76/6031, May 1976.

9. IBM, *System/360 Scientific Subroutine Package (360A-CM-03X) Version 3 Programmers Manual*, International Business Machines, IBM H20-0205-3, 1968.

10. Joseph, M., and M. Shear, *Optical Aspect Data Prediction (ODAP) Program System Description and Operating Guide*, Comp. Sc. Corp., 9101-13300-02TR, Oct. 1972.

11. Pettus, W., *Attitude Determination Accuracy for GEOS-C*, Comp. Sc. Corp. Internal Memo., Jan. 1973.

12. Rabiner, L., and B. Gold, *Theory and Application of Digital Signal Processing*. Englewood Cliffs, NJ: Prentice Hall, 1975.

13. Stanley, W. D., *Digital Signal Processing*. Englewood Cliffs, NJ: Prentice Hall, 1975.

14. Werking, R. D., R. Berg, K. Brokke, T. Hattox, G. Lerner, D. Stewart, and R. Williams, *Radio Astronomy Explorer-B Postlaunch Attitude Operations Analysis*, NASA X-581-74-227, GSFC, July 1974.

15. Williams, R., private communication, July 1972.

PART III

ATTITUDE DETERMINATION

CONTENTS

PART III

ATTITUDE DETERMINATION

Chapter

10   Geometrical Basis of Attitude Determination   343

11   Single-Axis Attitude Determination Methods   362

12   Three-Axis Attitude Determination Methods   410

13   State Estimation Attitude Determination
       Methods   436

14   Evaluation and Use of State Estimators   471

CHAPTER 10

GEOMETRICAL BASIS OF ATTITUDE DETERMINATION

Lily C. Chen
James R. Wertz

10.1   Single-Axis Attitude
10.2   Arc-Length Measurements
10.3   Rotation Angle Measurements
10.4   Correlation Angles
10.5   Compound Measurements—Sun to Earth
        Horizon Crossing Rotation Angle
10.6   Three-Axis Attitude

In Part II we described both the hardware and the process by which attitude data are gathered, transmitted to the attitude software system, and assembled in a manner appropriate for attitude determination. In Part III we describe the procedures by which these data are processed to determine the spacecraft attitude. Chapter 10 introduces the types of attitude measurements and the geometrical meaning of these measurements. Chapters 11 and 12 describe methods for combining as many measurements as there are observables (usually two or three) to produce a single, possibly multivalued determination of the attitude. Chapters 13 and 14 then describe filtering methods to provide optimum estimates of the attitude, given many data points.

As discussed in Chapter 1, there are two types of attitude. *Single-axis attitude* is the specification of the orientation of a single spacecraft axis in inertial space. Ordinarily, this single axis is the spin axis of a spin-stabilized spacecraft. However, it could be any axis in either a spinning or a three-axis stabilized spacecraft. Single-axis attitude requires two independent numbers for its specification, such as the right ascension and declination of the spin axis. The attitude of a single axis may be expressed either as a unit vector in inertial space or as a geometrical point on the unit celestial sphere centered on the spacecraft. (See Section 2.1.) Generally, we will use the vector representation of the attitude for numerical or computer calculations, and the geometrical representation for analytical work and physical arguments. However, because of the direct correspondence between the two representations, we will often move back and forth between them as convenient for the particular problem.

If the orientation of a single axis is specified, the complete spacecraft orientation is not fixed because the rotation of the spacecraft about the specified axis is still undetermined. A third independent attitude component, such as the azimuth about the spin axis of a point on the spacecraft relative to some object in inertial space, completely fixes the inertial orientation of a rigid spacecraft. Such a three-component attitude is commonly called *three-axis attitude* because it fixes the orientation of the three orthogonal spacecraft axes in inertial space.

Throughout Part III we will frequently ignore the distinction between single- and three-axis attitude. If we refer to the attitude as the orientation of a single axis, this may be taken as either single-axis attitude or one axis of a three-axis system. Specifically, in Sections 10.1 through 10.5 we will discuss the types of single-axis measurements. Measurements concerned specifically with determining the third component in three-axis systems will then be discussed in Section 10.6.

10.1 Single-Axis Attitude

Specifying the orientation of a single axis in space requires two independent attitude measurements. Therefore, if only one of these measurements is known, an infinite set of possible single-axis attitude orientations exists which maps out a curve, or *locus*, on the celestial sphere. This is illustrated in Fig. 10-1 for the *Sun angle measurement*, β, which is the arc-length separation between the attitude and the Sun. *Any two attitude measurements are equivalent if and only if they correspond to the same locus of possible attitudes on the celestial sphere.*

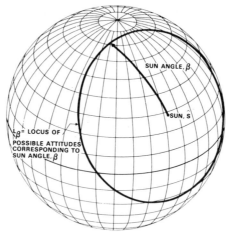

Fig. 10-1. Locus of Attitudes Corresponding to Measured Sun Angle, β (arbitrary inertial coordinates)

Given both independent attitude measurements, each having a distinct locus of possible values, the attitude must lie at their intersection. In general, there may be multiple intersections resulting in ambiguous attitude solutions. Because no measurement is exact, the possible attitudes corresponding to any real measurement lie in a band on the celestial sphere about the corresponding locus with the width of the band determined by the uncertainty in the measurement, as illustrated in Fig. 10-2. The Sun angle measurement, β, with uncertainty U_β, implies that the attitude must lie somewhere in a band centered on L_β of width ΔL_β. Similarly, the *nadir angle measurement*, η (i.e., the arc-length separation between the attitude and the center of the Earth), with uncertainty U_η, implies that the attitude lies in the band defined by L_η and ΔL_η. Clearly, the attitude must lie in one of the two parallelograms formed by the intersection of the two bands. We will assume that the region of intersection is sufficiently small that we may use plane geometry to describe these parallelograms. The correct parallelogram may be chosen and the attitude ambiguity resolved either from an a priori estimate of the attitude, or, if the attitude is constant in time, by processing many measurements from different times and selecting that solution which remains approximately constant. (See Section 11.2.)

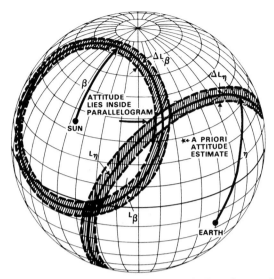

Fig. 10-2. Determination of Single-Axis Attitude From Intersecting Loci

Figure 10-3 shows an expanded view of the parallelogram of intersection formed by the two bands on the celestial sphere. The center of the parallelogram, where the two measurement loci intersect, is the measured value or estimate of the attitude. The size of the parallelogram is the uncertainty in the attitude result. For any measurement, m (either β or η in Figs. 10-2 and 10-3), the width, ΔL_m, of the attitude uncertainty band on the celestial sphere is determined by the measurement uncertainty, U_m, and the *measurement density*, d_m, which is the change in measurement per unit arc-length change between adjacent loci, measured perpendicular to the loci. Thus,

$$d_m = U_m / \Delta L_m$$

To obtain a more formal definition, let m_1 and m_2 be two values of the measurement m (e.g., β_1 and β_2), and let σ_{m_1, m_2} be the arc-length separation between L_{m_1} and L_{m_2} measured

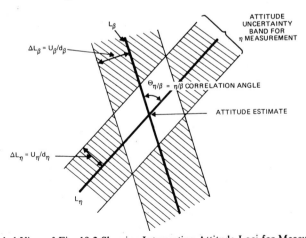

Fig. 10-3. Expanded View of Fig. 10-2 Showing Intersecting Attitude Loci for Measurements β and η

perpendicular to the loci. Then the *measurement density*, d_m, is the two-dimensional gradient of m on the celestial sphere for a fixed position of the reference vectors (i.e., Sun vector, nadir vector, etc.). That is,

$$d_m \equiv |\nabla m| \qquad \text{for fixed reference vectors}$$

$$\equiv \lim_{\sigma_{m_1, m_2} \to 0} |m_2 - m_1| / \sigma_{m_1, m_2} \qquad (10\text{-}1)$$

If we let m_1 and m_2 be the limits of uncertainty in measurement m, then σ_{m_1, m_2} corresponds to the width, ΔL_m, of the attitude uncertainty band on the celestial sphere. Thus,

$$\Delta L_m = U_m / d_m \qquad (10\text{-}2)$$

If the measurement density is low (i.e., if the spacing between loci is large), a small measurement error will result in a large shift in the measured attitude and a small measurement uncertainty will produce a wide attitude uncertainty band, ΔL_m, on the celestial sphere.

In addition to the width of the two attitude uncertainty bands, the size of the parallelogram of intersection is determined by the angle at which the two loci intersect, called the *correlation angle*, Θ. (A more formal definition of Θ will be given in Section 10.4.) Thus, for any two measurements, e.g., β and η, the attitude uncertainty corresponding to these measurements is determined by three factors: (1) the measurement uncertainties, U_β and U_η, (2) the measurement densities, d_β and d_η, and (3) the correlation angle between the loci, $\Theta_{\eta/\beta}$. For given measurement uncertainties, the measurement densities determine the widths of the attitude uncertainty bands and the correlation angle determines how these bands will combine to produce an overall attitude uncertainty. Thus, attitude accuracy analysis for pairwise measurement combinations may be reduced to determining the various measurement densities and correlation angles. Specific formulas for transforming these parameters into measures of the attitude uncertainty (i.e., the size of the error parallelogram) are given in Section 11.3.

Although there are many types of attitude sensors (e.g., Sun sensors, horizon scanners, magnetometers), the analysis of attitude measurements can be greatly simplified by classifying them according to the shape of the corresponding loci of possible attitudes. Thus, we will say that two attitude *measurements* are of the same *type* if and only if the attitude loci corresponding to the two measurements have the same shape, i.e., if both loci satisfy parametric equations of the same form. Although the number of attitude sensors and measurements is large, these measurements correspond to only a few basic types.

The two most fundamental types of attitude measurements are: (1) *arc-length measurements* from a known reference vector, such as the Sun angle measurement of Fig. 10-1, and (2) *rotation angle measurements* about the attitude between two known reference vectors, discussed further in Section 10.3. In addition, there are some compound measurement types (such as the rotation angle from the Sun to the Earth's horizon, as described in Section 10.5) that are not as well understood.

10.2 Arc-Length Measurements

The *arc-length measurement,* represented by the Sun angle, β, as shown in Fig. 10-1, is the simplest measurement type. For this type, the locus of possible attitudes is a small circle centered on the known reference vector with an angular radius equal to the measured arc length. If the arc length is measured directly, as in the case of β, the loci are uniformly distributed over the celestial sphere. That is, if L_{β_1}

and L_{β_2} are the loci corresponding to Sun angle measurements β_1 and β_2, then they are concentric small circles and the perpendicular separation, σ_{β_1,β_2}, between them is constant along the two curves and equal to the difference in Sun angle, i.e., $\sigma_{\beta_1,\beta_2} = |\beta_2 - \beta_1|$. As defined in Eq. (10-1), the *Sun angle density*, d_β, is the magnitude of the two-dimensional gradient on the surface of the celestial sphere of the family of attitude loci, L_β, for a fixed position of the Sun, $\hat{\mathbf{S}}$. That is,

$$d_\beta \equiv |\nabla \beta| \equiv \lim_{\sigma_{\beta_1,\beta_2} \to 0} |\beta_2 - \beta_1| / \sigma_{\beta_1,\beta_2} = 1 \qquad (10\text{-}3)$$

A second example of an arc-length measurement is the *Earth width*, Ω, or the rotation angle about the attitude between the two Earth horizon crossings for either a rotating sensor or a fixed sensor mounted on a spinning spacecraft. A given Earth width implies that the *nadir angle*, η, between the attitude and the center of the Earth must have one of two possible values, as shown in Fig. 10-4. *Thus, although the Earth-width measurement is a rotation angle, it is classified as an arc-length measurement because the resulting attitude loci are small circles.*

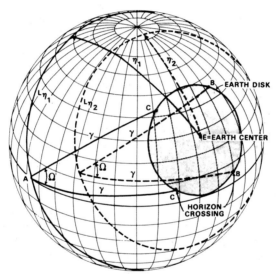

Fig. 10-4.   A given Earth width, Ω, results in two possible nadir angles, η_1 and η_2. η_1 corresponds to the scanner scanning "above" the center of the Earth $(C - C)$ and η_2 corresponds to the sensor scanning "below" the center of the Earth $(B - B)$. This gives two sets of loci, L_{η_1} and L_{η_2}, for the possible position of the attitude. γ is the fixed angle between the horizon sensor and the spin axis.

The Earth-width measurement is more complex than the Sun angle measurement in two ways. First, as mentioned above, a given Earth width corresponds to two possible discrete nadir angles. As illustrated in Fig. 10-4, a given Earth width, Ω, can correspond to the two horizon crossing points, C. As the points move about the perimeter of the Earth's disk, point A traces out the locus, L_{η_1}, of possible attitudes at a fixed nadir angle, η_1, from the Earth's center, E. However, the same value of Ω can also correspond to horizon crossing points at B. In this case, the locus of possible attitudes is L_{η_2}, with nadir angle $\eta_2 < \eta_1$.

The second complexity of the Earth-width measurements is that although the nadir angles have unit density ($d_\eta = 1$) over the entire celestial sphere, the Earth-width measurements do not. Figure 10-5 shows a plot of nadir angle as a function of Earth width for the geometry of Fig. 10-4. From this figure, it is clear that $\partial\Omega/\partial\eta$ varies from infinity to zero. Thus, because Ω is a function only of η (for a spherical Earth), we may write

$$d_\Omega \equiv |\nabla\Omega| \equiv |\nabla\Omega(\alpha,\delta)| \qquad \hat{E}, \rho, \text{ and } \gamma \text{ fixed}$$
$$= (\partial\Omega/\partial\eta)d_\eta$$
$$= (\partial\Omega/\partial\eta) \qquad \hat{E}, \rho, \text{ and } \gamma \text{ fixed} \tag{10-4}$$

where $\hat{E}$ and ρ are the position and angular radius of the Earth, and γ is the mounting angle between the sensor and the attitude.

Figure 10-6 shows a plot of attitude loci corresponding to Earth widths from Fig. 10-5 of approximately 5, 10, 15, 20,... deg. The loci do not cover the sky because the nadir angle must lie between $\gamma + \rho$ and $\gamma - \rho$, where γ is the mounting

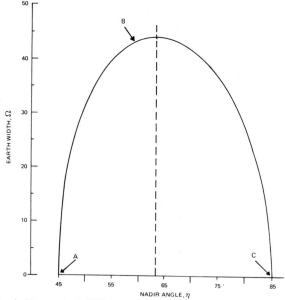

Fig. 10-5. Nadir Angle Versus Earth Width for Geometry of Fig. 10-4 ($\gamma = 65$ deg, angular radius of Earth $= 20$ deg). Such curves are symmetric about $\eta = \gamma$ only for $\gamma = 90$ deg. Loci resulting from nadir angles at A, B, and C are labeled on Fig. 10-6.

angle from the attitude to the horizon sensor and ρ is the angular radius of the Earth's disk. The physical interpretation of the measurement density as the density of loci on the celestial sphere is clear from the figure. When the attitude is near the A or C loci, the measurement density, d_Ω, is high. (Compare with Eq. (10-4) and Fig. 10-5.) Here, an uncertainty in Ω of 5 deg corresponds to only a small uncertainty in the attitude, ΔL_Ω, and the attitude uncertainty band will be narrow. In contrast, when the attitude is near B, the Earth-width measurement density is low and a shift in Ω of 5 deg corresponds to a large uncertainty in the attitude. The numerical form of the curve plotted in Fig. 10-5 and of the density $\partial\Omega/\partial\eta$ are given in Section 11.3.

Many standard attitude observations are arc-length measurements. The shape of attitude loci, *not* the method by which the attitude data are processed, determines the type of measurement. For example, the elevation of an identified star

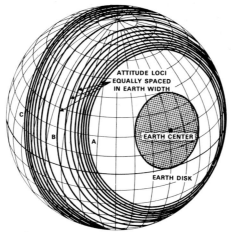

Fig. 10-6. Attitude loci equally spaced in Earth width for conditions of Fig. 10-5. *A*, *B*, and *C* correspond to the similarly lettered points on Fig. 10-5. The measurement density, d_Ω, is high near *A* and *C* and low near *B*.

above the spin plane is equivalent to an arc-length measurement of the attitude relative to the star. A single magnetometer reading, in a known magnetic field, measures the arc-length distance between the magnetometer axis and the magnetic field vector. Most observations involving a single, known reference vector are arc-length measurements.

10.3 Rotation Angle Measurements

The second fundamental type of attitude measurement is a measured rotation angle about the attitude between two *known* reference vectors, as illustrated in Fig. 10-7. For concreteness, we will assume throughout this section that the two

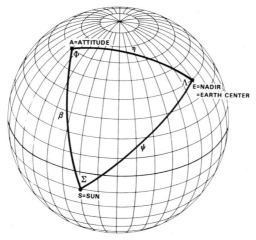

Fig. 10-7. Standard Notation for Attitude Angles. Φ is the rotation angle about the attitude between the Sun and the Earth.

reference vectors are the Sun and the center of the Earth,* although they could equally well be any two points of known orientation in the sky.

The geometry here is more complex than in the case of the constant arc-length measurements. Specifically, a fixed rotation angle, Φ, about the attitude between the Sun and the center of the Earth implies that the single-axis attitude lies on a curve with two discrete segments on the celestial sphere. Representative plots of these curves of constant Φ are shown in Figs. 10-8 and 10-9 for an Earth-Sun separation, ψ, of 30 deg. The curves overlying the coordinate grid are the lines of constant Φ. For example, the curve labeled "40°" covers all possible orientations of the spacecraft attitude such that the rotation angle from the Sun to the Earth (about the attitude) is 40 deg. Thus, the set of constant Φ curves has the same relation to the rotation angle measurement as the set of all small circles centered on the Sun has to the Sun angle measurement.

The five views in Figs. 10-8 and 10-9 are centered at varying Sun angles, β, and azimuthal angles relative to the Earth-Sun great circle. The rotation angle curves are plotted in 10-deg intervals, except that curves between the Earth and Sun for rotation angles between 120 deg and 240 deg have been omitted because of the high measurement density in that region. In Fig. 10-9(c), 25- and 35-deg rotation angle curves have been added as dotted lines to show the shape of the curves in the region of the *null*, or Sun vector/nadir vector cross product. The null will prove to be an important reference vector for many aspects of Sun-Earth-attitude geometry.

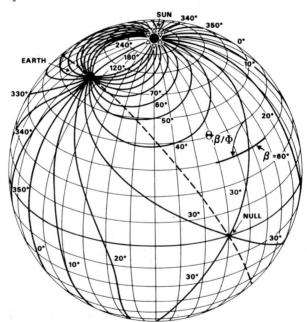

Fig. 10-8.   Rotation Angle Geometry for a 30-Deg Sun-Earth Angular Separation. View centered at $\beta = 60$ deg, azimuth = 60 deg from Earth-Sun great circle.

---

*The center of the Earth is a known reference vector, but a horizon crossing is *not*. This is discussed in detail in Section 10.5.

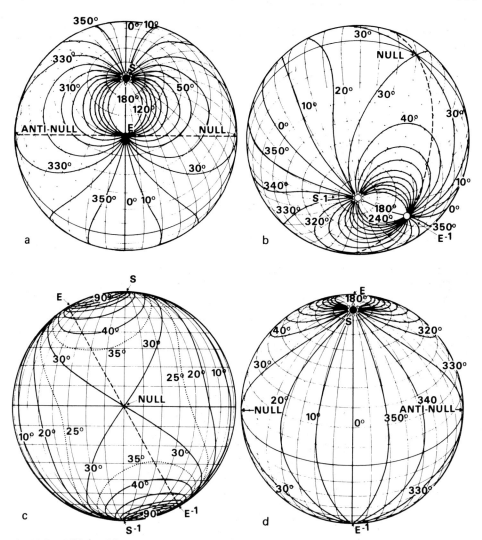

Fig. 10-9. Different Views of Rotation Angle Geometry for a 30-Deg Sun-Earth Angular Separation. Views centered at β, azimuth coordinates of: (a) 30 deg, 0 deg; (b) 150 deg, 60 deg; (c) 90 deg, 90 deg; and (d) 60 deg, 180 deg.

The general character of the rotation angle curves is evident from the plots. As can be seen most clearly in Fig. 10-9(a), the great circle containing the Earth and the Sun divides the celestial sphere into two hemispheres. All rotation angle curves between 0 deg and 180 deg are in one hemisphere and all curves between 180 deg and 360 deg are in the other. In addition, the 30- and 330-deg rotation angle curves (i.e., the curves with $\Phi = \psi$) divide each hemisphere into four quadrants, as can be seen in Fig. 10-9(c). Each rotation angle curve (except those of 30 deg and 330 deg) consists of *two nonintersecting segments* in opposite quadrants of one hemisphere (Fig. 10-9(c)). All segments start and end on the Earth, the Sun, the *zenith* (E^{-1}), or the *antisolar point* (S^{-1}).

In contrast to the uniformly distributed small circles of β and η (i.e., $d_\beta = d_\eta = 1$), the rotation angle curves are characterized by their greatly varying density. For $\psi < 90$ deg, the rotation angle density ($d_\Phi \equiv |\nabla\Phi(\alpha, \delta)|$, $\hat{S}$ and $\hat{E}$ fixed) is greatest between the Earth and the Sun (and between the zenith and the antisolar point) and approaches zero as a limit in the region of the null or the antinull. Recall that a low rotation angle density means that a small change in rotation angle corresponds to a large change in attitude. Figure 10-9(c) shows that a change of only 5 deg in rotation angle from 30 deg at the null to either 25 deg or 35 deg corresponds to a shift in attitude from the null to a point over 30 deg of arc away. *Thus, the region around the null or the antinull will yield poor attitude solutions based on the rotation angle measurement, because a small uncertainty in rotation angle corresponds to a large uncertainty in the attitude.* Similarly, for the geometry of Figs. 10-8 and 10-9, the area between the Earth and the Sun (or between the zenith and the antisolar point) will result in particularly good attitude solutions from rotation angle data. Expressions for the rotation angle density, which may be used to quantitatively evaluate the attitude accuracy, are given in Section 11.3.

Figure 10-10 shows the rotation angle curves for an Earth-Sun separation of 90 deg. In general, the rotation angle curves have become more uniformly distributed over the celestial sphere, although a large region of low density remains in the vicinity of the null and the antinull.

When the angular separation between the Earth and the Sun is greater than 90 deg, the geometry is equivalent to that of separations of less than 90 deg with the location of the Earth and the zenith interchanged. Thus, the geometry for the 30-deg angular separation shown in Figs. 10-8 and 10-9 is equivalent to the geometry for a 150-deg separation with the zenith and Earth interchanged.

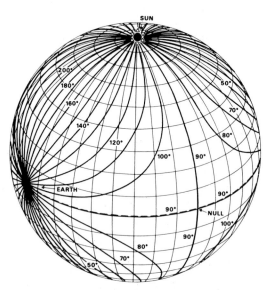

Fig. 10-10.   Rotation Angle Geometry for a 90-Deg Sun-Earth Angular Separation. Note that the measurement density is more nearly uniform than in Fig. 10-8, but is still low in the vicinity of the null.

10.4 Correlation Angles

We have seen that the attitude uncertainty depends on the measurement uncertainties, the measurement densities, and the correlation angle (or angle of intersection of the attitude loci). In this section, we give a formal definition of the correlation angle, expressions for the correlation angles among the arc-length and rotation angle measurements described above, and an example of the application of correlation angles and measurement densities to determine the accuracy of the Sun position in a two-axis Sun sensor.

To specify the angle of intersection between two loci, several choices are available. Geometrically, it is convenient to define the correlation angle as the acute angle between the tangents to the loci as was done in Section 10.1. However, for computer work or algebraic manipulation, this involves continuous tests on the range of an angle and adjustments when it falls outside the range 0 to 90 deg. Thus, for algebraic use, it is more convenient to define a unique correlation angle covering the range 0 to 360 deg. Given two arbitrary loci, L_i and L_j, we formally define the correlation angle between them, $\Theta_{i/j}$, as the rotation angle at the intersection of the loci from the positive gradient of L_i counterclockwise (as viewed from infinity toward the spacecraft) to the positive gradient of L_j, as illustrated in Fig. 10-11. This is equivalent in its effect on attitude uncertainties to defining $\Theta_{i/j}$ as the acute angle between the tangents to L_i and L_j. Note that from the formal definition, we have

$$\Theta_{j/i} = 360° - \Theta_{i/j} \tag{10-5}$$

As an example of the correlation angle for two arc length measurements, consider the Sun angle/nadir angle correlation angle, $\Theta_{\beta/\eta}$, shown in Fig. 10-11.

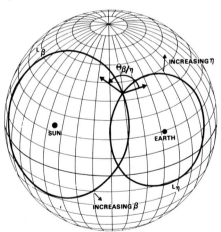

Fig. 10-11. Definition of the Correlation Angle, $\Theta_{\beta/\eta}$

$\Theta_{\beta/\eta}$ equals the angle between the radii of the two small circles at their intersection; however, this is just the Sun-Earth rotation angle, Φ, defined in Fig. 10-7. Thus,

$$\Theta_{\beta/\eta} = \Phi \tag{10-6}$$

When the correlation angle is 0 deg or 180 deg, the two small circles are tangent and the two measurements give essentially the same information about the attitude. Thus, when the correlation angle is small (or near 180 deg), the attitude uncertainty is largest, because the component of the attitude tangent to the two circles is essentially unknown. In contrast, when the correlation angle is near 90 deg or 270 deg, the two measurements are independent and the attitude uncertainty is smallest.

As an example of the correlation angle for arc length/rotation angle intersections, we consider the Sun angle and Sun-Earth rotation angle measurements. Figures 10-8 through 10-10 are convenient for studying the general character of these loci intersections. The latitude lines on the underlying coordinate grid in these figures are curves along which the Sun angle is constant, because a fixed Sun angle, β, implies that the attitude lies on a small circle centered on the Sun. The $\beta = 60$ deg locus is marked on Fig. 10-8. Thus, in Figs. 10-8 through 10-10, $\Theta_{\beta/\Phi}$ is simply the angle between the constant Φ curves and the constant β curves as indicated in Fig. 10-8 between the $\beta = 60$ deg and $\Phi = 30$ deg loci. The value of the angle at any point on the celestial sphere is derived in Section 11.3 as

$$\Theta_{\beta/\Phi} = \arc\tan\left(\frac{\tan \eta}{\tan \beta \sin \Phi} - \cot \Phi \right) \tag{10-7}$$

$\Theta_{\beta/\Phi} = 0$ implies that the constant β and constant Φ curves are tangent. As is most easily seen in Figs. 10-8 and 10-9 (c), this occurs when the attitude lies on the great circle containing the Earth and the null* (shown as a dashed line on the figures). Consequently, along this great circle, no information is available on the component of the attitude tangent to the constant β and Φ curves.

Because the Sun angle and the nadir angle are the same type of measurement, a similar relationship must hold for the nadir angle/rotation angle correlation angle, $\Theta_{\eta/\Phi}$:

$$\Theta_{\eta/\Phi} = \arc\tan\left(\frac{\tan \beta}{\tan \eta \sin \Phi} - \cot \Phi \right) \tag{10-8}$$

Also by symmetry with β/Φ, $\Theta_{\eta/\Phi} = 0$ when the attitude lies on the Sun-null great circle.

The set of all possible correlation angles relating any set of attitude measurements satisfies an addition theorem. For example, if $\hat{G}_\beta$, $\hat{G}_\eta$, and $\hat{G}_\Phi$ are the directions of the gradients of the constant β, η, and Φ curves, respectively, then we can see from Fig. 10-12 that

$$\Theta_{\beta/\eta} + \Theta_{\eta/\Phi} + \Theta_{\Phi/\beta} = (360 \cdot n)\text{deg} \qquad n = 1 \text{ or } 2 \tag{10-9}$$

where $n = 1$ if the vectors are in the order $\hat{G}_\beta$, $\hat{G}_\eta$, $\hat{G}_\Phi$ and $n = 2$ if they are in the order $\hat{G}_\beta$, $\hat{G}_\Phi$, $\hat{G}_\eta$. Equation (10-9) is particularly useful for the approximate evaluation of correlation angles, because frequently one or two of them are easy to estimate.

---

*This is true everywhere along the Earth-null great circle except at the Earth, Zenith, null, and antinull, where the tangent to the constant Φ curve is undefined.

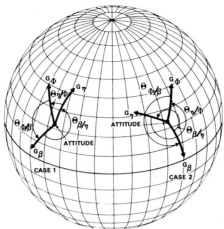

Fig. 10-12.   Addition Theorem for Correlation Angles. The sum of the correlation angles for any set of measurements must sum to a multiple of 360 deg.

Correlation angles will be used extensively in Chapter 11 to determine the attitude accuracy from various measurement types. To illustrate the versatility of the correlation angle and measurement density concepts, we analyze here the internal accuracy of the solid angle Sun sensors described in Section 6.1. Specifically, given a sensor with a circular field of view 128 deg in diameter, and a uniform reticle pattern with an 0.5-deg step size (or least significant bit) on both axes *at the boresight*, we wish to determine the maximum inaccuracy in the measured position of the Sun, assuming that there is no error in the sensor measurements. Our procedure will be first to determine the type of measurement made by the Sun sensor and then to determine the measurement densities and the correlation angle between the two sensor measurements.

Figure 10-13 shows the locus of Sun positions corresponding to given output angles Γ and Λ. (Compare Fig. 10-13 with Fig. 7-9.) Clearly, the loci of Sun positions corresponding to a given Γ or Λ output signal are great circles at a constant rotation angle *about the X and Y axes*, respectively, from the sensor boresight which defines the center of the field of view of the sensor.* (These rotation angle loci are different from those of Figs. 10-8 through 10-10 because the rotation angle is being measured about the sensor axis rather than about the position of the Sun.)

To determine the measurement density on the celestial sphere, we note that the separation between the sensor input slit and the reticle pattern on the back of the sensor is a constant. Therefore, equal steps along the reticle pattern correspond to equal steps in the tangent of the angle from the boresight to the Sun along the two axes, i.e., $\tan\Gamma$ and $\tan\Lambda$. Therefore, the density of the step boundaries on the celestial sphere is the derivative of the tangent of the measurement angles. (Compare with Eq. (10-4).) Thus, $d_\Gamma = 1/\cos^2\Gamma$, $d_\Lambda = 1/\cos^2\Lambda$, and the measurement step

---

*The locus of Sun positions for constant Γ or Λ is a great circle only if the index of refraction, n, of the material inside the sensor is 1. If $n \neq 1$, then the loci will deviate slightly from great circles and Λ and Γ will not be independent. In this example, we will assume $n = 1$, as is commonly true for high resolution sensors.

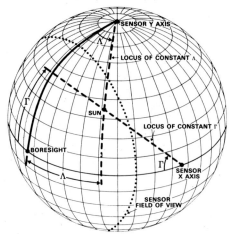

Fig. 10-13. Correlation Angle and Measurement Density Geometry for a Solid Angle Sun Sensor

size on the celestial sphere is $0.5°/d$. At the boresight the measurement density is 1 and the Sun angle is being measured in 0.5-deg steps; along an axis at the edge of the sensor (for example, at $\Gamma = 0$, $\Lambda = 64$ deg) the density is $1/\cos^2 64° = 5.20$, and the steps are $0.5°/5.20 = 0.096$ deg. Thus, ignoring problems of diffraction, reduced intensity, and manufacturing imperfections, all of which tend to be worse at the edge, the resolution at the edge of the sensor is approximately five time better than at the center.

To evaluate the uncertainty in the position of the Sun, we still need to determine the correlation angle, $\Theta_{\Lambda/\Gamma}$, between the two loci. This can be obtained by inspecting Fig. 10-13. Specifically, $\Theta_{\Lambda/\Gamma}$ equals the rotation angle about the Sun from the sensor $+Y$ axis to the $-X$ axis. The angular separation between the X and Y axes is 90 deg; therefore, the correlation angle at any point in the sensor field of view may be evaluated using Fig. 10-10 with the Sun, Earth, and null replaced by the sensor $+Y$ axis, the $-X$ axis, and the boresight, respectively, as shown in Fig. 10-14 with the center of the view shifted to the boresight axis.

Figure 10-14 shows that $\Theta_{\Lambda/\Gamma}$ is near 90 deg in the vicinity of the boresight and along the X and Y axes. For $\Theta_{\Lambda/\Gamma} = 90$ deg, the uncertainty in the position of the Sun, U_S, is smallest and is equal to half the length of the diagonal of a rectangle whose sides are the angular step size. At the boresight, $U_S = 0.5 \times 0.5° \times \sqrt{2} = 0.354$ deg. At the sensor boundary along one of the axes, $U_S = 0.5 \times (0.5^2 + 0.096^2)^{\frac{1}{2}} = 0.255$ deg. Along a line midway between the X and Y axes, steps in Λ and Γ are of equal size, but the measurement loci do not intersect at right angles. Along this line we may use Napier's Rules (Appendix A) to obtain

$$\tan \Lambda = \tan \Gamma = \cos 45° \tan \beta \qquad (10\text{-}10a)$$

and

$$\tan(0.5\Theta_{\Lambda/\Gamma}) = \cos \beta \qquad (10\text{-}10b)$$

where β is the angle from the boresight to the Sun. At the sensor boundary, $\beta = 64°$ and, therefore, $\Theta_{\Lambda/\Gamma} = 47.34°$, $\Lambda = \Gamma = 55.40°$, $d_\Lambda = d_\Gamma = 3.10$, and the step size is

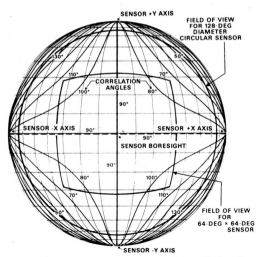

Fig. 10-14.   Correlation Angles for Solid Angle Sun Sensor

$0.5°/3.10 = 0.161°$. This may be further evaluated using the upper form of Eq. (11-14) to give $U_S = 0.201$ deg. (If the loci at the boundary midway between the two axes formed a rectangle on the celestial sphere rather than a parallelogram, the attitude uncertainty there would be 0.114 deg.) Thus, although the single-axis resolution varies by a factor of 5 over the full range, the attitude uncertainty fluctuates by only about 50%. A similar analysis for a Sun sensor with a 32-deg "square" field of view (Fig. 10-14) and other properties as above gives $U_S = 0.354$ deg at the boresight, 0.308 deg at the center of each edge, and 0.300 deg at the corners.

10.5 Compound Measurements—Sun-to-Earth Horizon Crossing Rotation Angle

Sections 10.1 through 10.4 described measurements involving one or two reference vectors whose orientation in inertial space is known. The technique that we have used is to examine the locus of possible attitudes for any given measurement to classify that measurement. However, some common attitude measurements do not fall into the basic categories that we have established thus far. One example of such *compound measurements* is the rotation angle about the attitude from the Sun to the Earth's horizon, Φ_H. The horizon sensor which produces this measurement is assumed to have a field of view which is a point on the celestial sphere, which thus sweeps out a small circle as the spacecraft rotates and provides an output pulse upon crossing the Earth's horizon.

The Sun–to-Earth horizon crossing rotation angle differs from other rotation angle measurements in that the location of the horizon crossing on the celestial sphere is unknown. We know only that the horizon crossing is a given arc-length distance from the nadir vector. Thus, Φ_H is neither a rotation angle measurement nor an arc-length measurement and the attitude loci corresponding to constant values of Φ_H do not have the same form as the loci corresponding to β or Φ measurements.

Let ψ, γ, and ρ be the Sun-nadir separation, the sensor mounting angle (relative to the attitude), and the angular radius of the Earth, respectively. Figures 10-15 and 10-16 show the shape of several constant Φ_H curves for $\psi > \gamma + \rho$ and $\psi < \gamma + \rho$, respectively. The solid curves are the attitude loci for a constant Sun to Earth-in horizon crossing angle and the dashed curves are the attitude loci for a constant Sun to Earth-out horizon crossing angle. *Earth-in*, or *in-triggering*, denotes

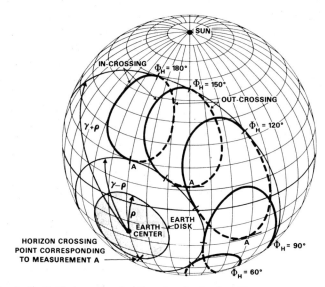

Fig. 10-15. Sun-to-Earth Horizon Crossing Rotation Angle, Φ_H, Geometry for $\psi > \gamma + \rho$. The disk of the Earth is shaded.

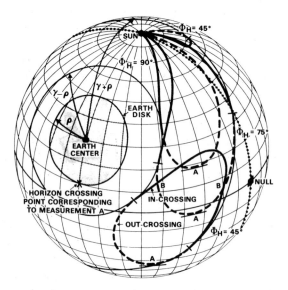

Fig. 10-16. Sun-to-Earth Horizon Crossing Rotation Angle, Φ_H, Geometry for $\psi < \gamma + \rho$. The disk of the Earth is shaded.

a sensor crossing from space onto the disk of the Earth and *Earth-out*, or *out-triggering*, denotes a crossing from the disk of the Earth to space.

An examination of the figures shows several characteristics of the Φ_H loci. The loci are entirely contained between small circles of radii $\gamma + \rho$ and $\gamma - \rho$, because these are the only conditions under which the sensor will cross the Earth. In the limit $\rho \to 0$, the set of all Φ_H loci lie on a small circle of radius γ about the Earth; that is, $\eta = \gamma$. The constant Φ_H loci are neither small circles nor constant rotation angle curves, but a third, distinct measurement type. Note that in Fig. 10-16 the 45-deg locus consists of *two*, discrete, nonintersecting, closed curves. In this case, it is possible to have four discrete, ambiguous attitude solutions when combining Φ_H with the nadir angle measurement. Unfortunately, the formulae for the various correlation angles involving Φ_H and other attitude measurements are inconvenient to use. However, the location of several of the correlation angle singularities can be identified from the figures. The Φ_H measurement may be combined with either the Sun angle or the nadir angle measurement to determine the attitude. Attitude singularities will occur whenever the Φ_H loci are *tangent to* small circles centered on the Sun (L_β) or the Earth (L_η), respectively. Reference to Figs. 10-15 and 10-16 shows that for the horizon angle/nadir angle method, the constant Φ_H loci are tangent to small circles centered on the Earth at the transitions from solid to dashed lines. Thus, an η/Φ_H singularity occurs when there is a transition from Earth-in crossing to Earth-out crossing. Equivalently, $\Theta_{\eta/\Phi_H} = 0$ whenever the sensor field-of-view small circle is tangent to the Earth. Similarly, $\Theta_{\beta/\Phi_H} = 0$ whenever the constant Φ_H loci are tangent to small circles centered on the Sun. Again, these L_β curves are the latitude lines of the underlying grid. Representative points where $\Theta_{\beta/\Phi_H} = 0$ have been marked by the letter A on Figs. 10-15 and 10-16. This occurs when the Sun vector, the nadir vector, and the horizon crossing vector are coplanar.

Finally, we may determine the attitude by using two horizon crossing measurements, an Earth-in crossing, and an Earth-out crossing. (For example, an Earth-in rotation of 45 deg and an Earth-out rotation angle of 75 deg implies that the attitude must be at one of the two points marked B on Fig. 10-16.) For a spherical Earth, this gives us the same information as an Earth-width measurement plus a Sun-to-nadir vector rotation angle measurement. As discussed in Section 10.3, a singularity occurs in the Earth-width/rotation angle method whenever the attitude lies on the Sun-null great circle shown as a dotted line in Fig. 10-16. At any point along this line, the Φ_H loci passing through that point are mutually tangent. Although this cannot be clearly established from the figure, it is at least consistent with the shape of the attitude loci along the Sun-null great circle.

10.6 Three-Axis Attitude

Thus far we have described procedures for using two independent measurements to determine the orientation of a single spacecraft axis. For single-axis attitude, this is all of the information that is desired. However, to completely determine the orientation of a rigid spacecraft, three parameters must be determined and, therefore, an additional measurement is required. For three-axis-stabilized spacecraft, these three parameters are frequently chosen to be three angles, known as Euler angles, which define how the spacecraft-fixed coordinates are related to inertial coordinates. This procedure is described in Section 12.1.

An alternative procedure frequently used for spinning spacecraft is to define the orientation in space of a single spacecraft axis (such as the spin axis) and then to define the rotational orientation of the spacecraft about this axis. This rotation angle, also called the *azimuth* or *phase angle,* may be specified as the azimuth of some arbitrary point in the spacecraft relative to some reference direction in inertial space, as illustrated in Fig. 10-17. In this figure, the underlying coordinate grid is fixed in inertial space.

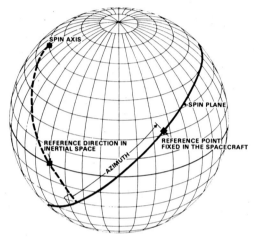

Fig. 10-17. Defining the Three-Axis Orientation of the Spacecraft by Defining the Spin Axis and Azimuth in Inertial Coordinates. The underlying coordinate grid is an inertial coordinate system.

To determine the three-axis attitude specified by the spin axis direction and azimuth, we first determine the orientation in inertial space of the spacecraft spin axis using any of the methods described in Sections 10.1 through 10.5. The one remaining attitude component is then measured by measuring the rotation angle about the attitude between some fixed direction in inertial space and an arbitrarily defined reference direction fixed in the spacecraft. For example, we might record the time at which a slit Sun sensor parallel to the spin axis sees the Sun and assume that the spacecraft is rotating uniformly to determine its relative azimuth at any other time. Alternatively, if we are using a wheel-mounted horizon scanner (Section 6.2), we could measure the relative azimuth between the center of the disk of the Earth (midway between the telescope Earth-in and -out crossings for a spherical Earth) and some reference mark fixed in the body of the spacecraft, as has been done for the AE series of spacecraft. This *pitch angle* may be used directly for Earth-oriented satellites or, with ephemeris data, transformed into an inertial azimuthal measurement. The reference point for the inertial azimuth is arbitrary. However, the perpendicular projection of the vernal equinox onto the spin plane is commonly used.

Another alternative procedure, frequently used on three-axis stabilized spacecraft, is to determine the attitude by measuring the orientation in spacecraft coordinates of two reference vectors fixed in inertial space. For example, three orthogonal magnetometers may be used to measure the orientation of the Earth's magnetic field in spacecraft coordinates. Similarly, a two-axis Sun sensor can provide the coordinates of the Sun vector in spacecraft coordinates. The specification of these two vectors in spacecraft coordinates fixes the orientation of the spacecraft in inertial space.

When using two reference vectors, the attitude problem is overdetermined because we have measured four parameters (two orientation parameters for each reference vector) but have only three independent variables. This is clearly shown by the Sun sensor/magnetometer example. The Sun sensor output defines a spacecraft axis which is pointing toward the Sun. It remains only to determine an azimuth about this axis. However, specifying the direction of the magnetic field vector in spacecraft coordinates determines both the azimuth of the spacecraft axis parallel to the magnetic field and also the angular separation of the magnetic field vector and the Sun vector. (See Fig. 10-18.) The latter quantity is not an independent parameter because it is fixed by knowing the direction in inertial space of both the Sun vector and the magnetic field vector. (See Sections 9.3 and 12.2 for a discussion of using this fourth parameter as a test for invalid data.)

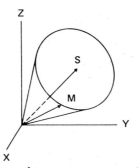

Fig. 10-18.   Sun ($\hat{S}$) and Magnetic Field ($\hat{M}$) Geometry

If we are determining three-axis attitude by determining the orientation in spacecraft coordinates of two reference vectors, then all the analysis of Sections 10.1 through 10.5 can be applied directly to determining the orientation of each reference vector. For example, the theory of correlation angles was applied to the output of a two-axis Sun sensor at the end of Section 10.4. Similarly, a single magnetometer measurement in a known magnetic field is an arc-length measurement specifying the angle between the external magnetic field and the magnetometer axis.* The same analytic procedures can be applied to other types of sensors as well.

---

*Ordinarily, magnetometer measurements are obtained from three mutually perpendicular magnetometers. The sum of the squares of the readings determines the overall field strength. Any two of the measurements may then be taken as the remaining independent numbers. These are two arc-length measurements which together determine the orientation of the magnetic field in spacecraft coordinates to within a discrete ambiguity which may be resolved by the sign of the third measurement.

CHAPTER 11

SINGLE-AXIS ATTITUDE DETERMINATION METHODS

11.1 Methods for Spinning Spacecraft
General Requirements, Specific Solution Methods, Nonintersecting Loci

11.2 Solution Averaging

11.3 Single-Axis Attitude Determination Accuracy
Attitude Accuracy for Measurements With Uncorrelated Uncertainties, Attitude Accuracy for Measurements With Correlated Uncertainties, Measurement Densities and Correlation Angles

11.4 Geometrical Limitations on Single-Axis Attitude Accuracy
Limitations on the Attitude Direction Due to Attitude Accuracy Requirements, Limitations on Reference Vector Direction, Applications

11.5 Attitude Uncertainty Due to Systematic Errors
Behavior of Single-Frame Solutions, Identification of Singularities, State Vector Formulation

This chapter describes standard procedures for determining the orientation in space of any single spacecraft axis. For illustration we will normally assume that this is the spin axis of a spin-stabilized spacecraft. However, this axis could equally well be that of an attitude sensor, such as the rotation axis of a scanning horizon sensor, or any axis in a three-axis stabilized spacecraft.

The methods presented here are all deterministic in that they use the same number of observations as variables (normally the two parameters required to specify the orientation of a single axis). The models presented have all been used for the operational support of a variety of spacecraft. The directions to the Sun and to the center of the Earth or to a point on the Earth's horizon are used as reference directions for illustration; however, the techniques presented may equally well be applied to any known reference vectors. All of the models given involve different observations made at the same time. However, if the attitude is assumed constant or if a dynamic model for attitude motion is available, these methods may be applied to observations made at different times.

Section 11.1 describes the basic, deterministic single-axis methods and the problem of nonintersecting loci. Section 11.2 describes the resolution of solution ambiguities, data weighting, and solution averaging. Sections 11.3 and 11.4 then provide analytic expressions for single-axis uncertainties, limitations on solution accuracy due to the relative geometry of reference vectors, and application of this information to mission analysis. Finally, Section 11.5 describes the behavior of single-axis solutions in the presence of systematic biases, identifies the specific singularity conditions for each of the models in Section 11.1, and introduces the need for state estimation procedures to resolve the biases characteristic of real spacecraft data.

11.1 Methods for Spinning Spacecraft

Peter M. Smith

Determining the attitude of a spin-stabilized spacecraft in the absence of nutation is equivalent to fixing the orientation of the unit spin vector axis with respect to some inertial coordinate system. The most common system used is that of the celestial coordinates right ascension and declination, described in Section 2.2. In general, ambiguous solutions for the attitude are obtained, due to multiple intersections of the attitude loci, and must be resolved either by comparison with an a priori attitude or by using the method of block averaging described in Section 11.2

11.1.1 General Requirements

For a deterministic, two-component attitude solution, we require two reference vectors with their origin at the spacecraft and either (a) an arc-length measurement from the spin vector to each reference vector, or (b) one arc-length measurement and a rotation angle measurement about the spin axis between the reference vectors.

As shown in Fig. 11-1, each arc-length measurement for case (a) defines a cone* about each reference vector; the intersections of these cones are possible attitude solutions. For concreteness, let us assume that the two known reference

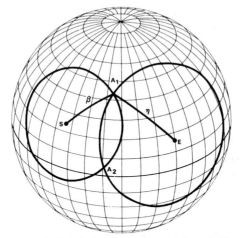

Fig. 11-1.   Single-Axis Attitude Solution Using Two Arc-Length Measurements, Case (a)

---

*Recall from Chapter 2 that we may define single-axis attitude either by three components of a unit vector, $\hat{A}$, or by the coordinates (α, δ) of the point at which that vector intersects the unit celestial sphere. In the former case, we think of an arc-length measurement as determining a cone about the reference vector; in the latter case, as determining a small circle (the intersection of the cone with the celestial sphere) on the celestial sphere about the reference point. Because the two representations are equivalent, we will use them interchangeably as convenient. See Section 2.2.1 for a discussion of the relative merits of the spherical and rectangular coordinate systems.

vectors are the Sun and nadir vectors, $\hat{\mathbf{S}}$ and $\hat{\mathbf{E}}$. The cone about $\hat{\mathbf{S}}$ has a half angle, β, equal to the angular separation of this vector and the unknown attitude vector, $\hat{\mathbf{A}}$; similarly, the cone about $\hat{\mathbf{E}}$ has a half angle, η, equal to the angular separation between $\hat{\mathbf{E}}$ and $\hat{\mathbf{A}}$. The possible solutions for the attitude are $\hat{\mathbf{A}}_1$ and $\hat{\mathbf{A}}_2$.

Analytically, this geometrical problem is specified by three simultaneous equations in three unknowns, A_i, A_j, A_k:

$$\hat{\mathbf{A}} \cdot \hat{\mathbf{S}} = \cos \beta \tag{11-1}$$

$$\hat{\mathbf{A}} \cdot \hat{\mathbf{E}} = \cos \eta \tag{11-2}$$

$$\hat{\mathbf{A}} \cdot \hat{\mathbf{A}} = 1 \tag{11-3}$$

These three equations may be solved using the following technique due to Grubin [1977]. Let

$$x \equiv \frac{\cos \beta - \hat{\mathbf{E}} \cdot \hat{\mathbf{S}} \cos \eta}{1 - (\hat{\mathbf{E}} \cdot \hat{\mathbf{S}})^2} \tag{11-3a}$$

$$y = \frac{\cos \eta - \hat{\mathbf{E}} \cdot \hat{\mathbf{S}} \cos \beta}{1 - (\hat{\mathbf{E}} \cdot \hat{\mathbf{S}})^2} \tag{11-3b}$$

$$z = \pm \sqrt{\frac{1 - x \cos \beta - y \cos \eta}{1 - (\hat{\mathbf{E}} \cdot \hat{\mathbf{S}})^2}} \tag{11-3c}$$

[handwritten annotation: $-ve \Rightarrow$ no solution]

$$\mathbf{C} = \hat{\mathbf{S}} \times \hat{\mathbf{E}} \tag{11-3d}$$

Then, the solutions for $\hat{\mathbf{A}}$ are given by

$$\hat{\mathbf{A}} = x\hat{\mathbf{S}} + y\hat{\mathbf{E}} + z\mathbf{C} \tag{11-3e}$$

Equation (11-3e) gives the two possible ambiguous attitude solutions. If the radicand in Eq. (11-3c) is negative, then no real solution exists; i.e., the cones do not intersect. Utility subroutine CONES8, described in Section 20.3, may also be used to solve for the intersection of two cones.

Figure 11-2 shows case (b), in which an arc-length measurement and a rotation angle measurement are combined to solve for the attitude. The arc-length measurement, say β, constrains the attitude to lie on a small circle of radius β centered at $\hat{\mathbf{S}}$. This small circle is called the *Sun cone*, where the reference vector is the Sun vector. In addition, the rotation angle measurement, Φ, requires that the attitude, $\hat{\mathbf{A}}$, lie at the intersection of the great circles defined by $(\hat{\mathbf{A}}, \hat{\mathbf{S}})$ and $(\hat{\mathbf{A}}, \hat{\mathbf{E}})$, where $\hat{\mathbf{S}}$ and $\hat{\mathbf{E}}$ are known reference vectors. To solve for the attitude, we first solve for η. Using the law of cosines for the sides of spherical triangle AES (Appendix A), it follows that

$$\cos \psi = \cos \beta \cos \eta + \sin \beta \sin \eta \cos \Phi \tag{11-4}$$

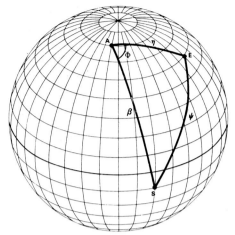

Fig. 11-2. Single-Axis Attitude Solution Using an Arc-Length and a Rotation Angle Measurement, Case (b)

where ψ is the arc length between $\hat{\mathbf{E}}$ and $\hat{\mathbf{S}}$. Solving for η gives two possible solutions:

$$\cos\eta = \frac{\cos\beta\cos\psi \pm \sin\beta\cos\Phi\sqrt{\sin^2\beta\cos^2\Phi + \cos^2\beta - \cos^2\psi}}{\sin^2\beta\cos^2\Phi + \cos^2\beta} \tag{11-5}$$

Each arc length, η, defines a small circle about $\hat{\mathbf{E}}$. When $\hat{\mathbf{E}}$ represents the nadir vector, this small circle is called the *nadir cone*. The evaluation of $\hat{\mathbf{A}}$ now reduces to case (a) as the attitude is constrained to lie at the intersections of the small circles with radii equal to arc-length measurements β and η. Because of the twofold ambiguity in η, a maximum of four possible solutions can be obtained for $\hat{\mathbf{A}}$. However, of the two possible attitudes computed for each value of η, only one member of each pair will be consistent with the original rotation angle, Φ; hence, the fourfold ambiguity is reduced to a twofold ambiguity as in case (a).

Adjustments to the data are required for certain sensor types, before deterministic solutions can be computed. For visible-light Earth sensors, terminator crossings must be differentiated from horizon crossings and removed from further processing. The problem of identification of terminator crossings is described in Section 9.3. Attitude determination methods which use the angular radius of the Earth must obtain a value based on an oblate Earth. Methods for modeling the Earth's oblateness are described in Section 4.3. Both Earth oblateness and spacecraft orbital motion cause distortion in the observed Earth width. This may be corrected by constructing a fictitious Earth width—i.e., that which would have been observed if the spacecraft were stationary—as shown in Fig. 11-3. The two disks represent the position of the Earth at horizon-in and -out crossing times, t_I and t_O. Each disk has a radius (ρ_I, ρ_O) equal to the angular Earth radius for the appropriate horizon crossing. $\mathbf{E}_I$ and $\mathbf{E}_O$ are the nadir vectors evaluated at t_I and t_O. The open dots represent the horizon crossing events for an Earth fixed at its position at time t_I, and the solid dots represent the horizon crossing events for an Earth fixed at its position at t_O. The observed Earth width, Ω, is corrected to the

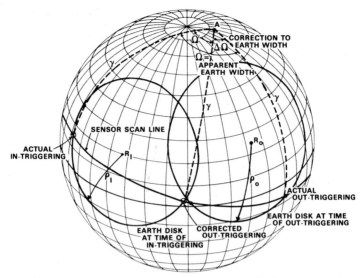

Fig. 11-3.   Earth-Width Corrections Caused by Orbital Motion of the Spacecraft Between Times of In-crossing Observation and Out-crossing Observation. (Motion is greatly exaggerated.)

fictitious value, $\Omega' = \Omega - \Delta\Omega$. Because the horizon vectors for each of the four crossing events in Fig. 11-3 may be calculated (given an a priori attitude and Earth model), utility routine PHASED (Section 20.3) may be used to compute $\Delta\Omega$. Ω' is the width that would have been observed for a spherical Earth of radius ρ_I fixed at its position at time t_I and may be used to obtain an accurate attitude solution. For a spin-stabilized spacecraft, the spin axis attitude is completely described by its right ascension and declination. However, for some spacecraft, a knowledge of the pointing direction of a particular body fixed vector, **X**, may be required. The azimuth angle measures the rotation of this body vector about the spin axis. For spinning spacecraft with Sun or Earth sensors, this azimuth angle is effectively measured with every sensor triggering. Thus, the calculation of an azimuth angle is trivial.

11.1.2 Specific Solution Methods

The attitude determination methods described below all use the procedures for case (a), case (b), or some multiple-step combination of these. The reference vectors are arbitrary, but for convenience, we will use the Sun vector and the nadir vector throughout. For a spacecraft equipped with one Sun sensor and two Earth sensors mounted at different angles from the spin axis and capable of simultaneous operation, six attitude determination methods may be used:
1. Earth-width/Sun angle method
2. Dual Earth-width/Sun angle method
3. Earth midscan rotation angle/Sun angle method
4. Earth-width/Earth midscan rotation angle method
5. Dual Earth-width/Earth midscan rotation angle method
6. Single horizon rotation angle/Sun angle method

Each of these is described below. Figure 11-4 summarizes the geometry and

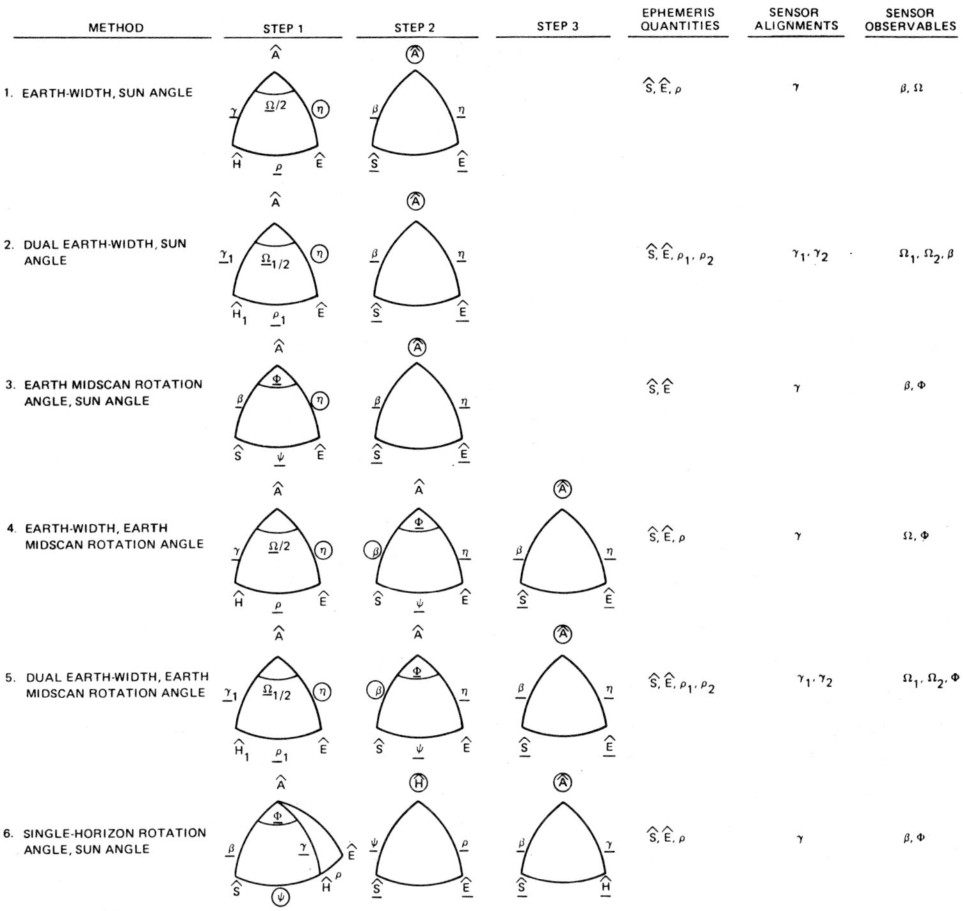

KEY: A = ATTITUDE VECTOR, S = SUN VECTOR, E = NADIR VECTOR, β = SUN ANGLE, γ = SENSOR MOUNTING ANGLE
      H = HORIZON CROSSING VECTOR
      ρ = ANGULAR EARTH RADIUS, η = NADIR ANGLE, ψ = ARC LENGTH ANGLE BETWEEN EARTH AND SUN,
      $\Omega/2$ = EARTH HALF WIDTH, Φ = ROTATION ANGLE

Fig. 11-4. Single-Axis Attitude Determination Methods. In each step, the independent variables are underlined and the variable being solved for is circled.

notation for the various methods. The steps used in evaluating the attitude for each method are given with the known observables used in the computations. The circled variables represent the parameters being solved for in each of the steps. For example, in the Earth-width/Sun angle method, spherical triangle AHE is used in step 1 to compute the nadir angle, η, with twofold ambiguity. The computed values of η are then used in step 2 to calculate the attitude with fourfold ambiguity. For each method, at least a twofold ambiguity may remain, and this may be resolved either by comparison with an a priori attitude or by the block averaging process described in the next section.

In principle, any of the six methods can be used alone to determine the spacecraft attitude. In practice, however, all of the applicable methods are normally used and the final attitude is taken as some weighted average of the results. The principal reasons for this redundancy are: (1) to ensure that a solution is obtained when some types of data are not available because of hardware or software malfunctions, (2) to reduce the effect of possible biases or other systematic errors by using methods which vary in their dependence on the observables, and (3) to aid in the identification of biases and estimation of the size of systematic errors by comparing the various solutions throughout the data pass. (See Section 11.5 for a further discussion of the latter procedure.) The principal disadvantage of using multiple solutions is that it makes the correct statistical treatment of the variables more difficult because a single measurement, such as a Sun angle, enters several calculations and is formally treated as an independent measurement each time it is used.

11.1.2.1 Two Arc-Length Methods

1. Earth-Width/Sun Angle Method. The attitude observables used are $\hat{S}$, $\hat{E}$, β, γ, Ω, and ρ. Using the law of cosines for sides on the spherical triangle shown in step 1 for the method, we may solve for the nadir angle, η:

$$\cos\eta = \frac{\cos\gamma\cos\rho \pm \sin\gamma\cos(\Omega/2)\sqrt{\sin^2\gamma\cos^2(\Omega/2)+\cos^2\gamma-\cos^2\rho}}{\sin^2\gamma\cos^2(\Omega/2)+\cos^2\gamma} \qquad (11\text{-}6)$$

The nadir angle is determined with twofold ambiguity. The intersections of the nadir cone and Sun cone shown in step 2 are then used to compute a maximum of four solutions for the attitude using utility routine CONES8 (Section 20.3) or the analytic procedure described above.

2. Dual Earth-Width/Sun Angle Method. The attitude observables used are $\hat{S}$, $\hat{E}$, β, γ_1, γ_2, ρ_1, ρ_2, Ω_1, and Ω_2, where the suffixes 1 and 2 refer to separate Earth sensors 1 and 2. This method requires simultaneous Earth coverage by both Earth sensors. In step 1 of Fig. 11-4, the attitude geometry is shown only for sensor 1; however, a similar spherical triangle exists for sensor 2. The law of cosines of the sides may be applied to both triangles to give

$$\cos\rho_1 = \cos\gamma_1\cos\eta_1 + \sin\gamma_1\cos(\Omega_1/2)\sin\eta_1 \qquad (11\text{-}7)$$

$$\cos\rho_2 = \cos\gamma_2\cos\eta_2 + \sin\gamma_2\cos(\Omega_2/2)\sin\eta_2 \qquad (11\text{-}8)$$

If noise and bias differences between the two sensors are ignored, then $\eta_2 = \eta_1 = \eta$ and the two equations may be combined to give

$$\eta = \arctan\left[\frac{\cos\gamma_2\cos\rho_1 - \cos\gamma_1\cos\rho_2}{\sin\gamma_1\cos(\Omega_1/2)\cos\rho_2 - \sin\gamma_2\cos(\Omega_2/2)\cos\rho_1}\right] \qquad (11\text{-}9)$$

Unlike the single Earth-width method, this method provides an unambiguous nadir angle, η. The analysis now parallels the single Earth-width method described above. As shown in step 2, $\hat{A}$ is constrained to lie at the intersections of the nadir cone and Sun cone and may be obtained using either CONES8 or the analytic procedure.

In general, $\eta_1 \neq \eta_2$ because the sensors will have different biases and different noise. Thus, Eqs. (11-7) and (11-8) are overspecified and Eq. (11-9) will provide only an approximate solution which fails to account for the differences in uncertainties between the observables. Equation (11-9) generally should not be regarded as the statistically best estimate of η.

11.1.2.2 Arc-Length Rotation Angle Methods

The methods discussed in this subsection make use of the *midscan rotation angle*, Φ, or the rotation angle from the Sun vector to the nadir vector, measured about the spin axis. Typically, this angle is observed by measuring the in-crossing and out-crossing rotation angles separately and averaging them.

3. Earth Midscan Rotation Angle/Sun Angle Method. The attitude observables used are $\hat{\mathbf{S}}$, $\hat{\mathbf{E}}$, β, and Φ. Referring to the attitude geometry shown in step 1 for this method, we see that the law of cosines for sides may be used on spherical triangle ASE to compute the nadir angle with twofold ambiguity. Thus,

$$\cos\eta = \frac{\cos\beta\cos\psi \pm \sin\beta\cos\Phi\sqrt{\sin^2\beta\cos^2\Phi + \cos^2\beta - \cos^2\psi}}{\sin^2\beta\cos^2\Phi + \cos^2\beta} \tag{11-10}$$

Step 2 shows that the attitude is constrained to lie at the intersection of the Sun cone and one of the two possible nadir cones. CONES8 may therefore be used to compute up to a maximum of four solutions, which may be decreased to two solutions by comparing the observed midscan rotation angle and calculated values for this angle using each of the four possible attitude solutions. Only two of the four attitudes will yield the correct value for Φ.

4. Earth-Width/Earth Midscan Rotation Angle Method. The attitude observables are $\hat{\mathbf{S}}$, $\hat{\mathbf{E}}$, Ω, Φ, ρ, and γ. Step 1 shows that the law of cosines for sides may be applied to spherical triangle AHE to compute the nadir angle with twofold ambiguity, as given by Eq. (11-6). The Sun angle, β, is then solved for in step 2 by applying the law of cosines for sides to triangle ASE:

$$\cos\beta = \frac{\cos\eta\cos\psi \pm \sin\eta\cos\Phi\sqrt{\sin^2\eta\cos^2\Phi + \cos^2\eta - \cos^2\psi}}{\sin^2\eta\cos^2\Phi + \cos^2\eta} \tag{11-11}$$

The two possible Sun cones and two nadir cones may be combined, as shown in step 3, to yield a total of eight attitude solutions from CONES8. Comparison of observed and calculated values of Φ reduces the number of solutions to four.

5. Dual Earth-Width/Earth Midscan Rotation Angle Method. This method requires simultaneous Earth coverage by two Earth sensors. The attitude observables are $\hat{\mathbf{S}}$, $\hat{\mathbf{E}}$, Ω_1, Ω_2, ρ_1, ρ_2, γ_1, γ_2, and Φ. The analysis for step 1 of this method parallels the dual Earth-width/Sun angle method and, hence, Eq. (11-9) may be used to calculate an unambiguous value for the nadir angle. Step 2 shows how this nadir angle may be combined with the observed midscan angle to compute a Sun angle. This computation parallels that described above in step 2 for the single Earth-width/Earth midscan rotation angle method. Hence, Eq. (11-11) may be used to compute the Sun angle. Using CONES8, the two possible Sun cones and nadir cone are combined to calculate four solutions for the attitude. The fourfold ambiguity is reduced to two by comparing the observed and the calculated Φ angles.

11.1.2.3 Compound Arc-Length Rotation Angle Method

6. Single Horizon Rotation Angle/Sun Angle Method. This method superficially resembles the arc-length/rotation angle class. It is, however, a more complicated multistep process because the inertial location of the horizon crossing vector is unknown. (See Section 10.5 for a discussion of the geometry of this measurement.) The attitude observables are $\hat{S}$, $\hat{E}$, β, γ, Φ, and ρ. The attitude geometry for this method is shown in step 1 of Fig. 11-4. Applying the law of cosines for sides to spherical triangle ASH gives ψ:

$$\cos\psi = \cos\beta\cos\gamma + \sin\beta\sin\gamma\cos\Phi \qquad (11\text{-}12)$$

Step 2 shows that CONES8 may be used to solve for the horizon vector, $\hat{H}$, using the arc-length measurements ψ and ρ. Two solutions are obtained for $\hat{H}$. Step 3 shows the application of CONES8 to triangle ASH using as reference vectors $\hat{S}$ and the two solutions for $\hat{H}$. A maximum of four solutions are obtained for the attitude. Two of these solutions are rejected by comparing observed and calculated values for Φ.

11.1.3 Nonintersecting Loci

Throughout the foregoing discussion we have tacitly assumed that the various cone intersections always produce an analytical solution. However, in the presence of biases and random noise, the possibility arises that the pair of solution cones do not intersect, or that arc cosine and square root functions are undefined. There are two methods for obtaining a solution in this case. One method is to input some predetermined bias, such as a sensor mounting angle bias or Earth angular radius bias. This change may produce the desired intersection and an attitude solution. Alternatively, we may force the cones to intersect by computing a fictitious intersection point midway between the cones at their point of closest approach. If the necessary change in the arc-length measurements which define the half angles of the two cones exceeds a given tolerance, the forced solution is rejected.

11.2 Solution Averaging

Peter M. Smith

Application of the deterministic attitude methods described in Section 11.1 to a span of data results in a set or *block* of attitude solutions consisting of several attitude estimates for each data frame. The number of attitude solutions for a frame depends on the number of valid methods used in processing. For example, the deterministic processing subsystem for the CTS spacecraft [Shear, *et al.*, 1976], can in principle use up to 12 methods and return a maximum of 2 ambiguous solutions for each processing method.

This section describes qualitatively how the method of *block averaging* may be used to resolve the ambiguous solutions and how the resultant block of chosen attitude vectors is averaged to provide the best estimate for the spin vector. Block averaging requires only that the true solution vary more slowly with time than the false solutions in the ambiguous sets. Therefore, it has general applicability and has been used in evaluating the pitch, roll, and yaw angles of the three-axis stabilized RAE-2 spacecraft as well as the spin axis right ascension and declination of spin-stabilized spacecraft such as AE, CTS, SMS, GOES, SIRIO, IUE and ISEE.

Data Weighting. As the first step in the averaging process, a weight, W, is assigned to each individual attitude solution within the block. The weight for any one solution is the inverse square of the arc-length uncertainty for that solution. Several methods for calculating the single-frame arc-length uncertainties are described in Sections 11.3 and 12.3. The forced attitude solutions described in Section 11.1.3 are arbitrarily assigned a small weight to minimize their contribution to the averaging process, except for cases in which only forced solutions are available.

Resolution of Ambiguous Solution and Block Averaging. Several methods are available to select the true solution from a block of data containing ambiguous solutions. The first method is to compare each set of ambiguous solutions with an a priori value for the attitude and to select the solution lying closest to this initial estimate. If no a priori estimate is available, an alternative procedure is to plot all of the attitude solutions in a right-ascension-versus-declination plot, as shown in Figs. 11-5 and 11-6. The set of correct solutions should form a cluster, because the correct attitude is assumed to remain approximately constant; the incorrect solutions from each of the ambiguous sets will usually be more scattered, because the geometry of the incorrect solutions changes as the orbital position of the spacecraft changes during the data pass. Any attitude near the center of the cluster may then be used as an a priori attitude for the subsequent elimination of ambiguous solutions.

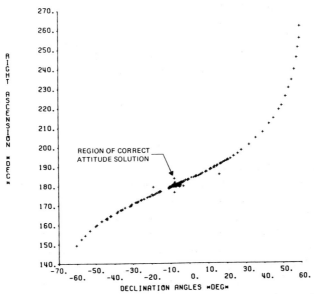

Fig. 11-5. Right Ascension Versus Declination Plot Including Both Solutions From Each Ambiguous Pair. Concentration of points near declination of −8 deg indicates correct solution should be in that region.

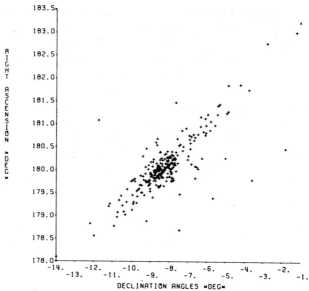

Fig. 11-6.   Right Ascension Versus Declination Plot Keeping Only Selected Solution From Each Ambiguous Pair for the Same Data as in Fig. 11-5. Note the greatly reduced range on the axes relative to Fig. 11-5.

A third method is to use each of the solutions in each ambiguous set as trial solutions and choose the one which provides the best fit to the data. Thus, the first solution in the block is arbitrarily selected to provide the trial attitude and used to resolve the ambiguities. The set of remaining attitude solutions is averaged to obtain an initial attitude estimate. Because erroneous attitudes due to noisy or biased data or unrejected terminator crossings may be present, a residual edit process is performed. The residual, R_i, for each attitude is defined as the angle between the individual attitude vector and the average attitude vector. The R_i values are used to compute a standard deviation, σ, for a set of selected attitudes, according to

$$\sigma = \sqrt{\frac{\sum W_i R_i^2}{\sum W_i}} \qquad (11\text{-}13)$$

where W_i is the weight assigned to each selected attitude solution. The summation is over all the selected attitudes in the block. The selected attitudes are then compared with the average attitude, and any that differ in arc-length separation by more than $N\sigma$ are rejected. (N is normally chosen in the range 3 to 5.) A new averaged attitude is calculated from the edited group of attitudes. This editing procedure is repeated until no additional attitudes are rejected. A goodness-of-fit parameter associated with the averaged spin vector and the M attitudes remaining is defined as σ/M. In a similar manner, other ambiguous solutions present in the original unresolved set of attitudes are selected as trial attitudes and the whole procedure is repeated. The averaged spin vector and the set of attitudes associated with the lowest σ/M value are selected as the true set of attitude solutions.

After the resolution of ambiguous solutions, the processing is the same for all the methods. The a priori attitude, or the selected trial attitude, is used to process the unresolved block of attitudes. The chosen attitudes are then averaged to obtain a new attitude estimate. The chosen attitudes are residually edited in iterative fashion until a self-consistent set of solutions remains. In practice, the laborious search for a trial attitude is conducted over a small subset of data because processing time is proportional to the square of the number of solutions in the original set. In addition, for a trial search to be successful, it is necessary for at least one attitude in the block to lie close to the final averaged attitude and for the true attitude solution to vary more slowly than the false attitude solutions. A time-varying attitude or large systematic or random errors present in constant attitude data would result in wildly fluctuating attitude solutions and can cause the block averaging process to fail.

Reliability of the Averaged Attitude Solution. The quality of the computed average attitude solution may be evaluated using either statistical measures or solution plots. As an example of the latter, the plots of right ascension versus declination shown in Figs. 11-5 and 11-6 immediately reveal if the chosen attitudes are clustering about a constant value. These plots display either the selected and the rejected attitude solutions or only the selected solutions. Plots of attitude solutions versus frame number may also be used to search for systematic variations or incorrect editing. For example, Fig. 9-18 in Section 9.4 shows a plot of declination versus frame number before residual editing for real SMS-2 data. The downward spikes are due to a systematic anomaly in the sensor performance; these data should be removed before further processing. A set of suspect attitude solutions would be revealed by a wide scatter in the plotted data. Similarly, poor quality solutions or a processing method giving inconsistent results can be eliminated. (See Section 11.5.)

A second possible method for evaluating the quality of solutions is to compute statistical indicators. For example, standard deviations can be calculated for the sets of solutions associated with each method separately, for an average of the single-sensor methods for each Earth sensor and for an average over all the attitude methods.

11.3 Single-Axis Attitude Determination Accuracy

Lily C. Chen
James R. Wertz

In this section, we calculate the uncertainty in deterministic single-axis solutions due to both the statistical noise on the data and estimates of any systematic errors which may be present. The purpose of this calculation is both to determine the attitude accuracy available from given measurements and to provide weights for the various data and measurement types as described in Section 11.2.

Attitude uncertainties can be obtained through two different approaches. In the direct calculation procedure discussed in Section 12.3, the attitude uncertainty is obtained directly from the uncertainties of the various observables via the partial derivatives of the attitude parameters with respect to the observables. Although this is adequate for determining the attitude uncertainty for specific values of the observables, it provides little insight into the underlying causes of the attitude uncertainty and does not lend itself to mission or maneuver planning or analytic attitude studies, where a wide range of alternatives are considered.

An alternative method, described in this section and in Section 11.4, is to express the attitude uncertainty in terms of three factors (involving the various partial derivatives) which have well-defined physical and geometrical interpretations. We then use this factorization to develop explicit analytic expressions for the attitude uncertainty and to discuss the geometrical causes of large uncertainties to provide the perspective necessary for prelaunch analysis and mission or maneuver planning.

In this section, the discussion is restricted to single-axis attitude determined from measurements taken at a single time. The attitude, which corresponds to a point on the celestial sphere as discussed in Chapter 10, is defined as the spin axis for a spinning spacecraft or as the direction of some convenient axis fixed in the body for a three-axis stabilized spacecraft.

If the uncertainties in two measurements are due to independent error sources such as random noise or unrelated systematic errors, then the uncertainties are *uncorrelated*. In other words, an error in one measurement does not imply any error in the other measurement, and vice versa. Alternatively, part of the uncertainties in two measurements may come from the same error source. For example, if the attitude is determined by measuring the Earth width with a horizon telescope at two different times, then a misalignment in the sensor mounting angle would cause an error in both measurements and the uncertainties are *correlated*. Measurement uncertainties are also correlated whenever there exists a systematic error which can introduce uncertainties in both measurements.

11.3.1 Attitude Accuracy for Measurements With Uncorrelated Uncertainties

Quantized Measurements. The easiest measurements to interpret physically are *quantized measurements*, i.e., those for which the measurement uncertainty is the result of the step size or bucket size in which the measurements are made or transmitted.* As discussed in Section 10.4, two measurements, m and n, each imply that the attitude lies within a band on the celestial sphere as shown in Fig. 11-7. Here, a plane geometry approximation is made such that the constant measurement curves on the celestial sphere are approximated by straight lines. L_m is the locus of attitudes corresponding to measurement m; $\Delta L_m = U_m / d_m$ is the change in the attitude perpendicular to L_m due to the uncertainty, U_m (1/2 step size); d_m is the measurement density for measurement m as defined in Section 10.1, and $\Theta_{m/n}$ is the correlation angle between measurements m and n, as defined in Section 10.4.

The correct attitude solution may be anywhere inside the *error parallelogram* shown in Fig. 11-7. The probability of the correct attitude being in any small area of the parallelogram is the same regardless of the location within the parallelogram.

---

*In a strict sense, all attitude measurements are quantized by the process of transmitting them as binary numbers. However, we will regard the quantization as important only when the quantization step is sufficiently larger than the noise such that in a continuous string of measurements of an approximately constant observable, only one or two discrete values are reported.

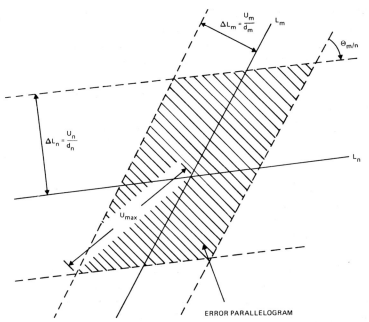

Fig. 11-7. Error Parallelogram for Quantized Measurements. (An example of this error parallelogram for Sun angle and nadir angle measurements is shown in Figs. 10-2 and 10-3.)

For example, the probability of the attitude being in some small area at the tip of the parallelogram is the same as the probability of the attitude being in an equal area at the center of the parallelogram. The *probability density*, or probability per unit area on the celestial sphere, is constant inside the parallelogram and zero outside the parallelogram.

The *attitude uncertainty* for quantized measurements *is* the error parallelogram. To fully specify this uncertainty requires both the size and orientation of the parallelogram, which depend on four independent parameters. For example, we could give the width of both bands and the azimuthal orientation of each band relative to an arbitrary reference direction.

It is frequently convenient to characterize the attitude uncertainty by a single number. Clearly, this cannot be done in any precise sense because no one number completely defines the error parallelogram. We define three convenient error parameters which may be used depending on the nature of the uncertainty requirements. The *component uncertainty* is the distance from the center to the edge of the parallelogram along some specified direction, e.g., right ascension uncertainty. The *maximum uncertainty*, U_{max}, is the semilength of the longest diagonal, or, equivalently, the radius of a circle cimcumscribed about the parallelogram.

$$U_{max} = \frac{1}{|\sin \Theta_{m/n}|} \left[(\Delta L_m)^2 + (\Delta L_n)^2 + 2(\Delta L_m)(\Delta L_n)|\cos \Theta_{m/n}| \right]^{1/2}$$

$$= \frac{1}{|\sin \Theta_{m/n}|} \left[\left(\frac{U_m}{d_m} \right)^2 + \left(\frac{U_n}{d_n} \right)^2 + 2 \left(\frac{U_m}{d_m} \right) \left(\frac{U_n}{d_n} \right) |\cos \Theta_{m/n}| \right]^{1/2} \quad (11\text{-}14)$$

If we have not specified otherwise, U_{max} will be taken as the attitude uncertainty for quantized measurements. Finally, we define the *mean uncertainty*, U_{mean}, as the radius of a circle with area equal to that of the parallelogram:

$$U_{mean} = 2\left[\frac{\Delta L_m \Delta L_n}{\pi |\sin \Theta_{m/n}|}\right]^{1/2}$$

$$= 2\left[\frac{U_m U_n}{\pi d_m d_n |\sin \Theta_{m/n}|}\right]^{1/2} \tag{11-15}$$

Continuous Measurements. If the uncertainty in a transmitted measurement is due to either Gaussian-distributed random noise or any unknown systematic error which is assumed to have a Guassian probability distribution, then the attitude uncertainty corresponds to an *error ellipse* on the celestial sphere. For illustration, we first consider the simplest case in which the two independent measurements, m and n, correspond to attitude loci which are orthogonal on the celestial sphere, as shown in Fig. 11-8. Let x be the attitude component perpendicular to L_m and σ_x be the standard deviation in x resulting from the uncertainty in m; i.e., $\sigma_x \equiv U_m/d_m$, where U_m is now the standard deviation of the measurement m. By the definition of a Gaussian distribution, the probability of the x-component of the attitude lying between x and $x + \delta x$ is given by

$$\rho(x)\delta x = \left(\frac{1}{\sigma_x \sqrt{2\pi}}\right)\exp\left(-x^2/2\sigma_x^2\right)\delta x \tag{11-16}$$

where $\rho(x)$ is called the *probability density* for x. Similarly, if y is the attitude component perpendicular to L_n, then the y-component probability is

$$\rho(y)\delta y = \left(\frac{1}{\sigma_y \sqrt{2\pi}}\right)\exp\left(-y^2/2\sigma_y^2\right)\delta y \tag{11-17}$$

The probability both that the x-component lies between x and $x + \delta x$ *and* that the y-component lies between y and $y + \delta y$ is

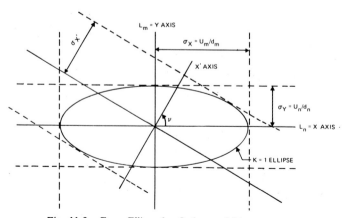

Fig. 11-8.   Error Ellipse for Orthogonal Measurements

$$\rho(x)\delta x\, \rho(y)\delta y = \left(\frac{1}{2\pi\sigma_x\sigma_y}\right)\exp\left[-\frac{1}{2}\left(x^2/\sigma_x^2\right)-\frac{1}{2}\left(y^2/\sigma_y^2\right)\right]\delta x\,\delta y$$

$$\equiv \rho(x,y)\delta x\,\delta y \tag{11-18}$$

$\rho(x,y)$ is the two-dimensional probability density on the celestial sphere. From Eq. (11-18) it is clear that the lines of constant probability density are ellipses defined by $(x^2/\sigma_x^2)+(y^2/\sigma_y^2)=K$, where K is a constant. The standard deviations of x and y are just the semimajor and semiminor axes of the $K=1$ ellipse. As shown in Fig. 11-8, the standard deviation of any arbitrary component, x', is the perpendicular projection of the $K=1$ ellipse onto the x' axis. That is,

$$\sigma_{x'} = \frac{1}{\sqrt{2}}\left[\left(\sigma_x^2+\sigma_y^2\right)+\left(\sigma_x^2-\sigma_y^2\right)\cos 2\nu\right]^{1/2} \tag{11-19}$$

where ν is the angle between the x' axis and the major axis of the ellipse. The same relationship holds for the $K\sigma$ uncertainties in any attitude component.

In general, we would like to consider the independent measurements, m and n, corresponding to nonorthogonal loci, as shown in Fig. 11-9. For computation, we choose an orthogonal coordinate system, x and y, for which the y-component is perpendicular to L_n. Thus, the standard deviation for the y-component is

$$\sigma_y = \frac{U_n}{d_n} \tag{11-20}$$

The standard deviation for the x-component is now more complex. As we will show later (Eq. (11-27b)):

$$\sigma_x = \frac{1}{|\sin\Theta_{m/n}|}\sqrt{\left(\frac{U_n}{d_n}\right)^2\cos^2\Theta_{m/n}+\left(\frac{U_m}{d_m}\right)^2} \tag{11-21}$$

where $\Theta_{m/n}$ is the correlation angle between L_m and L_n. Note that for $\Theta=90$ deg, the previous result is recovered.

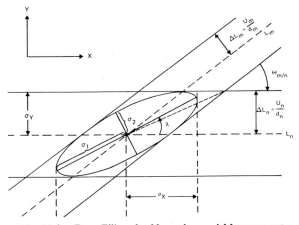

Fig. 11-9. Error Ellipse for Nonorthogonal Measurements

Because L_m and L_n are not orthogonal, the x and y components are not independent. A measure of the degree of their interdependence is the *correlation coefficient*, $\tilde{C}_{xy}$, given by (Eq. (11-27d))

$$\tilde{C}_{xy} = \tilde{C}_{yx} = -\frac{\left(\frac{U_n}{d_n}\right)^2 \cot \Theta_{m/n}}{\sigma_x \sigma_y}$$

$$= -\frac{\frac{U_n}{d_n} \cos \Theta_{m/n}}{\sqrt{\left(\frac{U_m}{d_m}\right)^2 + \left(\frac{U_n}{d_n}\right)^2 \cos^2 \Theta_{m/n}}} \tag{11-22}$$

Note that $\tilde{C}_{xy} = 0$ (measurements m and n are independent) when $\Theta_{m/n} = 90°$ or $270°$, and $|\tilde{C}_{xy}|$ is a maximum when $\Theta_{m/n} = 0°$ or $180°$.

The above results may be established by use of the covariance analysis introduced in Section 12.3. The derivation is summarized here.

The covariance matrix, P, which defines the attitude uncertainty determined from measurements m and n, can be obtained from the following equation:

$$P^{-1} = M^{-1} + G^T U^{-1} G \tag{11-23}$$

where M and U are the initial estimates of the square of the errors in attitude and measurements, respectively, and G is given by

$$G = \begin{bmatrix} \frac{\partial m}{\partial x} & \frac{\partial m}{\partial y} \\ \frac{\partial n}{\partial x} & \frac{\partial n}{\partial y} \end{bmatrix} \tag{11-24}$$

where m and n are the two measurements and x and y are any two orthonormal components on the celestial sphere (such as $\alpha \cos \delta$ and δ where α and δ are the right ascension and the declination of the attitude). Because U_m and U_n are uncorrelated, the uncertainty matrix, U, is diagonal. By definition,

$$U = \begin{bmatrix} U_m^2 & 0 \\ 0 & U_n^2 \end{bmatrix} \tag{11-25}$$

If we assume $M^{-1} = 0$ and let (x,y) be the two perpendicular coordinates shown in Fig. 11-9, then

$$\frac{\partial m}{\partial x} = -d_m \sin \Theta_{m/n}$$

$$\frac{\partial m}{\partial y} = d_m \cos \Theta_{m/n}$$

$$\frac{\partial n}{\partial x} = 0 \tag{11-26}$$

$$\frac{\partial n}{\partial y} = d_n$$

By substituting Eqs. (11-24) through (11-26) into Eq. (11-23), we obtain

$$P = \begin{bmatrix} P_{xx} & P_{xy} \\ P_{yx} & P_{yy} \end{bmatrix} \tag{11-27a}$$

where

$$P_{xx} = \frac{\left(\dfrac{U_n}{d_n}\right)^2 \cos^2\Theta_{m/n} + \left(\dfrac{U_m}{d_m}\right)^2}{\sin^2\Theta_{m/n}} = \sigma_x^2 \qquad (11\text{-}27b)$$

$$P_{yy} = \left(\frac{U_n}{d_n}\right)^2 = \sigma_y^2 \qquad (11\text{-}27c)$$

$$P_{xy} = P_{yx} = -\left(\frac{U_n}{d_n}\right)^2 \cot\Theta_{m/n} = \tilde{C}_{xy}\sigma_x\sigma_y \qquad (11\text{-}27d)$$

The semimajor axis, σ_1; semiminor axis, σ_2; and the orientation, λ, of the error ellipse in Fig. 11-9 can be expressed in terms of σ_x, σ_y and $\tilde{C}_{xy}$ by the following equations (see, for example, Keat and Shear [1973]):

$$\sigma_1^2 = \frac{1}{2}\left[\sigma_x^2 + \sigma_y^2 + \sqrt{(\sigma_x^2 - \sigma_y^2)^2 + 4\tilde{C}_{xy}^2\sigma_x^2\sigma_y^2}\right] \qquad (11\text{-}28a)$$

$$\sigma_2^2 = \frac{1}{2}\left[\sigma_x^2 + \sigma_y^2 - \sqrt{(\sigma_x^2 - \sigma_y^2)^2 + 4\tilde{C}_{xy}^2\sigma_x^2\sigma_y^2}\right] \qquad (11\text{-}28b)$$

$$\tan 2\lambda = -\frac{2\tilde{C}_{xy}\sigma_x\sigma_y}{\sigma_x^2 - \sigma_y^2} \qquad (11\text{-}28c)$$

By substituting Eqs. (11-20) through (11-22) into Eq. (11-28), the following expressions for σ_1, σ_2, and λ in terms of U_m, U_n, d_m, d_n, and $\Theta_{m/n}$ are obtained:

$$\sigma_1^2 = \frac{1}{2}\frac{A+B}{\sin^2\Theta_{m/n}}\left[1 + \sqrt{1 - \frac{4AB}{(A+B)^2}\sin^2\Theta_{m/n}}\right] \qquad (11\text{-}29a)$$

$$\sigma_2^2 = \frac{1}{2}\frac{A+B}{\sin^2\Theta_{m/n}}\left[1 - \sqrt{1 - \frac{4AB}{(A+B)^2}\sin^2\Theta_{m/n}}\right] \qquad (11\text{-}29b)$$

$$\tan 2\lambda = \frac{B\sin 2\Theta_{m/n}}{A + B\cos 2\Theta_{m/n}} \qquad (11\text{-}29c)$$

where

$$A \equiv \frac{U_m^2}{d_m^2}$$

$$\qquad\qquad\qquad\qquad\qquad (11\text{-}29d)$$

$$B \equiv \frac{U_n^2}{d_n^2}$$

Note that the long axis of the error ellipse is *not*, in general, aligned with the long diagonal of the error parallelogram. The uncertainty of the attitude component along any specific direction making an angle ν with respect to the semimajor axis of the error ellipse, or making an angle $\nu + \lambda$ with respect to L_n, can then be obtained by substituting Eq. (11-29) into Eq. (11-19), with σ_x replaced by σ_1 and σ_y replaced by σ_2. That is,

$$\sigma_\nu^2 = \frac{1}{2} \frac{A+B}{\sin^2\Theta_{m/n}} \left[1 + \sqrt{1 - \frac{4AB}{(A+B)^2} \sin^2\Theta_{m/n} \, \cos 2\nu} \right] \qquad (11\text{-}30)$$

where A and B are defined in Eq. (11-29d).

The physical interpretation of the error ellipse in Fig. 11-9 is different in several respects from that of the quantized error parallelogram of Fig. 11-7. As shown in Eq. (11-18) and Fig. 11-10, the probability density is no longer uniform, but is a maximum at the center and falls off continuously away from the center. The boundaries of the error ellipses are lines of constant probability density. The $n\sigma$ uncertainty along any arbitrary axis is given by the perpendicular projection of the $n\sigma$ error ellipse onto that axis. Thus, the 1σ uncertainty along the y' axis is the distance from the origin to the point A in Fig. 11-10; that is, the probability that the y' component of the attitude lies between A and A' is 0.68.

Although the probability of any one component being within the 1σ uncertainty boundary is 0.68, the probability of both attitude components in any orthogonal coordinate system being within the 1σ error ellipse is less than 0.68.* Specifically, the probability of the attitude lying somewhere inside the 1σ error ellipse is 0.39. Table 11-1 gives the probability for the attitude to lie within various error ellipses and for any one component to lie within the boundary of the error ellipse.

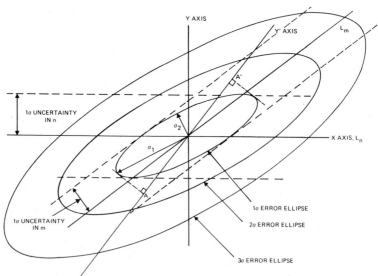

Fig. 11-10.   Probability Interpretation of Error Ellipse From Fig. 11-9

---

* This is easily visualized by considering a two-component error rectangle. If the two components, x and y, have upper limits at the boundary of the rectangle of B_x and B_y, then four possibilities exist: $x \leqslant B_x$ and $y \leqslant B_y$, $x > B_x$ and $y > B_y$, $x \leqslant B_x$ and $y > B_y$, and $x > B_x$ and $y \leqslant B_y$. However, only the first combination results in the point defined by (x,y) being inside the box. The probability of this occurring is clearly less than either the probability of $y \leqslant B_x$ or the probability of $y \leqslant B_y$.

Table 11-1.   Probability for Location of the Attitude with Gaussian Measurement Errors

UNCERTAINTY LEVEL, K	PROBABILITY OR CONFIDENCE LEVEL			CONFIDENCE LEVEL	UNCERTAINTY LEVEL, K		
	SINGLE COMPONENT	TWO COMPONENTS (SINGLE-AXIS ATTITUDE)	THREE COMPONENTS (THREE-AXIS ATTITUDE)		SINGLE COMPONENT	TWO COMPONENTS (SINGLE-AXIS ATTITUDE)	THREE COMPONENTS (THREE-AXIS ATTITUDE)
$1\,\sigma$	0.6827	0.3935	0.1987	0.50	$0.675\,\sigma$	$1.177\,\sigma$	$1.538\,\sigma$
$2\,\sigma$	0.9545	0.8647	0.7385	0.68	$0.994\,\sigma$	$1.510\,\sigma$	$1.872\,\sigma$
$3\,\sigma$	0.9973	0.9889	0.9707	0.90	$1.645\,\sigma$	$2.146\,\sigma$	$2.500\,\sigma$
$4\,\sigma$	0.99994	0.99966	0.9989	0.95	$1.960\,\sigma$	$2.448\,\sigma$	$2.795\,\sigma$
				0.99	$2.576\,\sigma$	$3.035\,\sigma$	$3.368\,\sigma$
				0.995	$2.807\,\sigma$	$3.255\,\sigma$	$3.583\,\sigma$

A precise statement of the attitude uncertainty for Gaussian errors requires the specification of three independent numbers, e.g., the size and eccentricity of the error ellipse and the orientation of the long axis relative to some arbitrary direction. As in the case of quantized measurements, we would like to characterize the uncertainty by a single number. Again there is no precise way to do so, because specifying the ellipse is the only unambiguous procedure. One option for a single accuracy parameter would be to use Eq. (11-29a) to obtain σ_1. This is then the long axis of the error ellipse and corresponds approximately to U_{max} for quantized measurements.

An alternative one-parameter estimate for the attitude uncertainty would be the radius of a small circle on the celestial sphere which had the same integrated probability as the corresponding error ellipse. A numerically convenient approximation to this radius is given by

$$U_A = \sqrt{(\sigma_1^2 + \sigma_2^2)/2}$$

$$= \sqrt{(P_{xx} + P_{yy})/2}$$

$$= \frac{1}{\sqrt{2}\,|\sin\Theta_{m/n}|}\left[\left(\frac{U_m}{d_m}\right)^2 + \left(\frac{U_n}{d_n}\right)^2\right]^{1/2} \qquad (11\text{-}31)$$

This approximation is good for $\sigma_1 \approx \sigma_2$. That is, if U_m and U_n are the 3σ uncertainties in m and n, then the probability that the attitude will lie within U_A of the estimated value, if $\sigma_1 \approx \sigma_2$, is 0.989. If $\sigma_1 \gg \sigma_2$, then the approximation of Eq. (11-31) is less accurate, being a 37% overestimate for the 1σ uncertainty radius and a 16% underestimate for the 3σ uncertainty radius. As σ_1 becomes much larger than σ_2, the error ellipse becomes very elongated and any single number representation becomes less meaningful. In this case, the best choice for the one-parameter attitude uncertainty would be the semimajor axis of the error ellipse, which is approximately

$$U_A' \approx \frac{1}{|\sin\Theta_{m/n}|}\left[\left(\frac{U_m}{d_m}\right)^2 + \left(\frac{U_n}{d_n}\right)^2\right]^{1/2} \qquad (11\text{-}32)$$

as can be obtained from Eq. (11-29) when σ_2 approaches zero. An alternative physical interpretation of U_A, as defined by Eq. (11-31), is to note that U_A is the exact formula for the attitude component inclined at 45 deg to both the semimajor and semiminor axes of the error ellipse (see Eq. (11-19)).

Another option for a single accuracy parameter would be to use the radius of a circle with the same geometrical area as the error ellipse. That is,

$$U_{mean} = \sqrt{\sigma_1 \sigma_2}$$

$$= \left[\frac{U_m U_n}{d_m d_n |\sin \Theta_{m/n}|} \right]^{1/2} \tag{11-33}$$

Equation (11-33) is analogous to Eq. (11-15) for the quantized measurements. Note that this representation also gives a poor estimate of the attitude uncertainty when $\sigma_1 \gg \sigma_2$ because $U_{mean} \to 0$ when $\sigma_2 \to 0$. Again, when $\sigma_1 \gg \sigma_2$, Eq. (11-32) should be used for one-parameter attitude uncertainty. The application of a three-dimensional analog of U_{mean} to three-axis attitude is discussed in Section 12.3. Throughout the rest of this chapter, we will use U_A as defined by Eq. (11-31) as our one-parameter estimate of the attitude uncertainty, unless stated otherwise.

11.3.2 Attitude Accuracy for Measurements With Correlated Uncertainties

Whenever there exists a systematic error which can introduce uncertainties in both measurements m and n, then the measurement uncertainties contain a *correlated component*. For example, a sensor mounting angle bias will produce a correlated uncertainty component when using the Earth-width/Sun-to-Earth-in rotation angle method.

When attitude is determined from two measurements with a correlated un-certainty component, the measurement uncertainty matrix given in Eq. (11-25) will contain off-diagonal terms. That is,

$$U = \begin{bmatrix} U_m^2 & C_{m/n} \\ C_{m/n} & U_n^2 \end{bmatrix} \tag{11-34}$$

where

$$U_m^2 = R_m^2 + \sum_i \left(\frac{\partial m}{\partial S_i} \right)^2 (\Delta S_i)^2$$

$$U_n^2 = R_n^2 + \sum_i \left(\frac{\partial n}{\partial S_i} \right)^2 (\Delta S_i)^2 \tag{11-35}$$

$$C_{m/n} = \sum_i \left(\frac{\partial m}{\partial S_i} \right) \left(\frac{\partial n}{\partial S_i} \right) (\Delta S_i)^2$$

In Eq. (11-35), U_m and U_n are the total uncertainties in measurements m and n; R_m and R_n are the random errors in measurements m and n; ΔS_i is the ith systematic error existing in either measurement; $\partial m/\partial S_i$ and $\partial n/\partial S_i$ are the partial derivatives of m and n with respect to the ith systematic error; and $C_{m/n}$ is the correlated uncertainty component between the two measurements.

In this case, the attitude uncertainty can be obtained from the covariance matrix approach given in Section 11.3.1 with Eq. (11-25) replaced by Eq. (11-34). The result is

$$
U_A = \frac{1}{\sqrt{2}\,|\sin\Theta_{m/n}|}\left[\frac{U_m^2}{d_m^2} + \frac{U_n^2}{d_n^2} - 2\frac{C_{m/n}}{d_m d_n}\cos\Theta_{m/n}\right]^{1/2}
\tag{11-36}
$$

Equation (11-36) gives the general expression for the attitude uncertainty determined by two measurements with total uncertainties U_m and U_n, and correlated uncertainty component $C_{m/n}$. This equation can be applied to any single-axis attitude determination procedure regardless of the type of measurements and attitude determination methods.

Equation (11-36) shows that the attitude accuracy in general is determined by three factors: the measurement uncertainties U_m, U_n, and $C_{m/n}$; the measurement densities, d_m and d_n; and the correlation angle, $\Theta_{m/n}$. Note that the attitude uncertainty goes to infinity (i.e., a singularity occurs) whenever d_m, d_n, or $\sin\Theta_{m/n}$ is zero.

The expressions for the measurement uncertainties are given in Eq. (11-35), and the expressions for d and Θ, which depend on the types of the two measurements, are given in Section 11.3.3 for arc-length and rotation angle measurements.

11.3.3 Measurement Densities and Correlation Angles

Expressions for the measurement density, d_m, and the correlation angle, $\Theta_{m/n}$, depend on the types of measurements. Because arc-length and rotation angles are the most fundamental and most commonly used measurements, we derive explicit expressions for d_m, $\Theta_{m/n}$, and U_A in terms of the geometrical parameters involved. The results are presented in Table 11-2 using the notation defined in Figs. 11-11 and 11-14. The attitude uncertainty U_A, for any deterministic attitude method using arc-length and rotation angle measurements, can be obtained by substituting the expressions from Table 11-2 into Eq. (11-31) or (11-36).

To make the discussion specific, the Sun and the Earth are used as the two reference vectors. However, final expressions are *not* limited to the Sun/Earth system. The results are generally applicable for any single-axis attitude determination procedure using arc-length or rotation angle measurements. We emphasize that the uncertainties presented in Table 11-2 are a result of the observations which are used for a deterministic solution and do *not* depend on the numerical procedure by which the attitude is computed. For example, Section 11.1.2.2 describes a procedure for computing the attitude from the measurements β and Φ. First, β, Φ and the reference vector parameters are used to compute η, and then β and η are used to compute the attitude. The uncertainty for this method may be obtained directly from line 3 of Table 11-2, irrespective of the fact that η was used as a numerically convenient intermediate variable in computing the attitude.

Table 11-2 gives the attitude uncertainty in terms of simple functions of measurement uncertainties and geometrical conditions, which enables one to give quick attitude uncertainty estimates, frequently without computer computations. This is a major advantage of the geometrical approach over other computational techniques, in terms of time, cost, and the need for prompt decisions.

Table 11-2.   Summary of Single-Axis Attitude Accuracy for Arc-Length and Rotation Angle Measurements. (See Section 11.5 for a list of attitude singularities for these methods.)

METHOD	d_m	d_n	$\Theta_{m/n}$
β/η	1	1	Φ
β/Ω	1	$2\left\| \dfrac{\text{COT }\gamma}{\text{SIN }\frac{\Omega}{2}} - \text{COT }\eta\,\text{COT }\dfrac{\Omega}{2} \right\|$	Φ
β/Φ	1	$\left\| \dfrac{\text{SIN }\Phi}{\text{SIN }\beta} \right\| \sqrt{\cos^2\beta + \cot^2\Sigma}$ OR $\left\| \dfrac{\text{SIN }\Phi}{\text{SIN }\eta} \right\| \sqrt{\cos^2\eta + \cot^2\Lambda}$ OR $\left\| \dfrac{\text{SIN }\upsilon\,\text{SIN }\xi}{\text{SIN }\beta\,\text{SIN }\eta} \right\|$	$\text{TAN}^{-1}\left[\dfrac{\text{TAN }\eta}{\text{TAN }\beta\,\text{SIN }\Phi} - \text{COT }\Phi \right]$ OR $\text{TAN}^{-1}\left[\dfrac{\text{COT }\Lambda}{\text{COS }\eta} \right]$
η/Φ	1	SAME AS ABOVE	$\text{TAN}^{-1}\left[\dfrac{\text{TAN }\beta}{\text{TAN }\eta\,\text{SIN }\Phi} - \text{COT }\Phi \right]$ OR $\text{TAN}^{-1}\left[\dfrac{\text{COT }\Sigma}{\text{COS }\beta} \right]$
Ω/Φ	$2\left\| \dfrac{\text{COT }\gamma}{\text{SIN }\frac{\Omega}{2}} - \dfrac{\text{COT }\eta}{\text{TAN }\frac{\Omega}{2}} \right\|$	SAME AS ABOVE	SAME AS ABOVE

NOTE:

$$(U_A)_{m/n} = \frac{1}{\sqrt{2}\,|\text{SIN }\Theta_{m/n}|}\left[\left(\frac{U_m}{d_m}\right)^2 + \left(\frac{U_n}{d_n}\right)^2 - 2\left(\frac{C_{m/n}}{d_m d_n}\right)\cos\Theta_{m/n} \right]^{1/2}$$

WHERE U_m, U_n, AND $C_{m/n}$ ARE AS GIVEN IN EQ. (11–35).

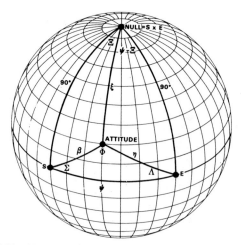

Fig. 11-11.   Notation Used for Attitude Uncertainty Computations

As an example of the application of Table 11-2, we compute attitude uncertainty for the IUE spacecraft with its spin axis attitude oriented toward the north ecliptic pole [Boughton and Chen, 1978]. Figure 11-12 shows the IUE

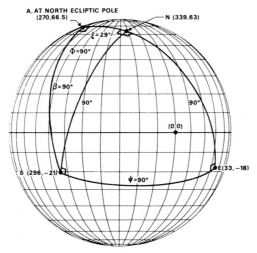

Fig. 11-12. Attitude Determination Geometry for IUE Mission Attitude

mission attitude geometry on the spacecraft-centered celestial sphere. A is the attitude at the north ecliptic pole, S is the direction of the Sun for a January 15 launch, E is the direction of the Earth as seen by the panoramic scanner, and $N = S \times E$ is the direction of the null. E is 90 deg from S.

In this example, $\beta = \psi = 90°$, $\eta = 119°$, $\xi = 29°$, and $\Phi = 90°$. From Table 11-2 or Fig. 11-18 in the next section, we obtain

$$\Theta_{\beta/\eta} = \angle SAE = 90°$$

$$\Theta_{\Phi/\beta} = \angle EAN = 0°$$

$$\Theta_{\Phi/\eta} = \angle SAN = 90°$$

and

$$d_{\Phi} = \frac{\sin\psi \sin\xi}{\sin\beta \sin\eta} = \frac{\sin 29°}{\sin 119°} = 0.55$$

Assuming that $U_{\beta} = 0.1°$, $U_{\eta} = 0.7°$, and $U_{\Phi} = 4.9°$ with no correlated components, we get

$$(U_A)_{\beta/\eta} = \frac{1}{\sin 90°} \sqrt{(0.1°)^2 + (0.7°)^2} = 0.71°$$

$$(U_A)_{\Phi/\beta} = \frac{1}{\sin 0°} \sqrt{(0.1°)^2 + \left(\frac{4.9°}{0.55}\right)^2} \rightarrow \infty$$

$$(U_A)_{\Phi/\eta} = \frac{1}{\sin 90°} \sqrt{(0.1°)^2 + \left(\frac{4.9°}{0.55}\right)^2} = 8.95°$$

The weighted mean of the attitudes obtained from the three methods will thus have an uncertainty of (see, for example, Bevington [1969])

$$U_A = \left[\frac{1}{(U_A)^2_{\beta/\eta}} + \frac{1}{(U_A)^2_{\Phi/\beta}} + \frac{1}{(U_A)^2_{\Phi/\eta}} \right]^{-1/2}$$

$$= 0.71°$$

Thus, in this case, all of the attitude information is coming from the β/η attitude method.

Measurement Density Derivation. For any arc-length measurement such as the Sun angle, β, or the nadir angle, η, an error in the measurement produces the same amount of error in the attitude along the gradient to the constant measurement curve, as shown in Fig. 11-13. Thus, by definition,

$$d_\beta = |\nabla \beta(\alpha, \delta)| \qquad \hat{S} \text{ held fixed}$$

$$= \frac{|\Delta\beta|}{|L_{\beta+\Delta_\beta} - L_\beta|_\perp}$$

$$= 1 \tag{11-37}$$

Similarly,

$$d_\eta = |\nabla \eta(\alpha, \beta)| = 1$$

However, if η is not measured directly but is obtained instead from an Earth-width measurement, Ω, the measurement density for Ω is

$$d_\Omega = |\nabla\Omega(\alpha, \delta)|$$

$$= |\nabla\eta(\alpha, \delta)| \left| \frac{\partial\Omega}{\partial\eta} \right|$$

$$= \left| \frac{\partial\Omega}{\partial\eta} \right| \tag{11-38}$$

Let ρ be the angular radius of the Earth and γ be the sensor mounting angle; then, from Fig. 11-14,

$$\cos\rho = \cos\gamma\cos\eta + \sin\gamma\sin\eta\cos\frac{\Omega}{2} \tag{11-39}$$

By differentiating Eq. (11-39) and substituting it into Eq. (11-38), we get

$$d_\Omega = 2\left| \frac{\cot\gamma}{\sin(\Omega/2)} - \cot\eta\cot(\Omega/2) \right| \tag{11-40}$$

Thus, d_Ω can change rapidly during a data pass. Specifically, $d_\Omega = 0$ when $\cot\gamma = \cot\eta\cos\Omega/2$ or $\angle AEH_I = \angle AEH_O = 90°$. This means that when Earth-width measurements are used for attitude determination, a singularity occurs when the sensor scans near the middle of the disk of the Earth.

The geometry for the rotation angle density, shown in Fig. 11-15, is considerably more complex than for arc-length densities. In the figure, the attitude changes from A to A' along the direction perpendicular to L_Φ due to an infinitesimal change in rotation angle from Φ to $\Phi+\Delta\Phi$. To obtain the arc length AA', let B be the intersection of $L_{\Phi+\Delta\Phi}$ with the extension of EA. Then AB is $\Delta\eta$ along the constant Λ direction due to the change $\Delta\Phi$; that is, $AB = \Delta\eta|_\Lambda$. By definition, the angle $BAA' = \Theta_{\eta/\Phi}$. Therefore,

$$AA' = |L_{\Phi+\Delta\Phi} - L_\Phi|_\perp$$

$$= AB\cos(\angle BAA')$$

$$= \Delta\eta|_\Lambda\cos\Theta_{\eta/\Phi}$$

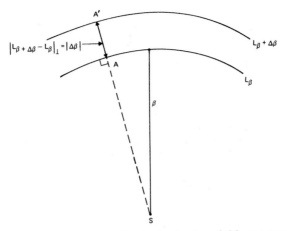

Fig. 11-13. Measurement Density for Arc-Length Measurements

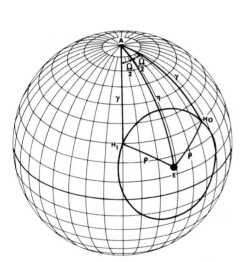

Fig. 11-14. Geometry of Earth-Width Measurements

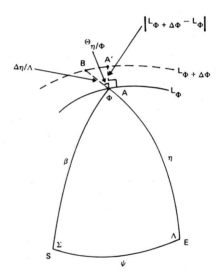

Fig. 11-15. Measurement Density for Rotation Angle Measurements

The rotation angle density, d_Φ, is

$$d_\Phi = |\nabla \Phi(\alpha, \beta)| \qquad \hat{S}, \hat{E} \text{ fixed}$$

$$= \frac{|\Delta\Phi|}{|L_{\Phi+\Delta\Phi} - L_\phi|_\perp}$$

$$= \frac{1}{|\cos \Theta_{\eta/\Phi}|} \left| \frac{\Delta\Phi}{\Delta\eta} |_\Lambda \right| \tag{11-41}$$

From spherical triangle SAE, we have

$$\cot\psi \sin\eta = \cot\Phi \sin\Lambda + \cos\eta \cos\Lambda \tag{11-42}$$

which yields

$$\left|\frac{\Delta\Phi}{\Delta\eta}|_\Lambda\right| = \left|\frac{\sin\Phi}{\tan\beta}\right|$$

From Eq. (11-54) to be derived later, we have

$$|\cos\Theta_{\eta/\Phi}| = \frac{|\cos\beta|}{\sqrt{\cos^2\beta + \cot^2\Sigma}} \tag{11-43}$$

Substituting Eqs. (11-42) and (11-43) into Eq. (11-41), we obtain

$$d_\Phi = \frac{|\sin\Phi|\sqrt{\cos^2\beta + \cot^2\Sigma}}{|\sin\beta|} \tag{11-44}$$

Because of the Sun/Earth symmetry in Fig. 11-15, d_Φ also can be expressed in terms of η and Λ as

$$d_\Phi = \frac{|\sin\Phi|\sqrt{\cos^2\eta + \cot^2\Lambda}}{|\sin\eta|} \tag{11-45}$$

Note that $d_\Phi = 0$ when $\Sigma = \beta = 90°$ or when $\Lambda = \eta = 90°$, i.e., when the attitude is at the null. If we define ξ as the arc-length separation between the attitude and the null (Sun-Earth cross product), then Eqs. (11-44) and (11-45) can be reformulated as

$$d_\Phi = \left|\frac{\sin\psi\sin\xi}{\sin\beta\sin\eta}\right| \tag{11-46}$$

where ψ is the angular separation between the Sun and the Earth and ξ is given by

$$\cos\xi = \frac{\sin\beta\sin\eta\sin\Phi}{\sin\psi} \tag{11-47}$$

Thus, $d_\Phi = 0$ when the attitude is at null or when the Sun and Earth are in the same direction. Conversely, $d_\Phi \to \infty$ when $\beta = 0$ or $\eta = 0$, i.e., when the attitude is close to the Sun or the Earth.

Correlation Angle Derivation. The correlation angle between two arc-length measurements is simply the rotation angle, Φ, between the two reference vectors. That is,

$$\Theta_{\beta/\eta} = \Phi \tag{11-48}$$

where Φ is measured from $\hat{S}$ to $\hat{E}$ about $\hat{A}$.

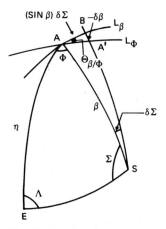

Fig. 11-16.   Derivation of $\Theta_{\beta/\Phi}$

The correlation between an arc-length measurement and a rotation angle measurement can be derived from the infinitesimal spherical triangle shown in Fig. 11-16. L_Φ and L_β are the constant Φ and constant β curves through A. By definition, $\Theta_{\beta/\Phi}$ is the angle between L_β and L_Φ. Let the attitude move an infinitesimal amount from A to A' along L_Φ; then, β will change by $-\delta\beta$ perpendicular to L_β and Σ will change by $\delta\Sigma$, which gives an arc-length change of $\delta\Sigma\sin\beta$ along L_β. From the infinitesimal right triangle ABA', we obtain

$$\tan\Theta_{\beta/\Phi} = \frac{-\delta\beta}{(\sin\beta)\delta\Sigma}\bigg|_\Phi = -\frac{1}{\sin\beta}\frac{\delta\beta}{\delta\Sigma}\bigg|_\Phi \qquad (11\text{-}49)$$

From the spherical triangle AES, we have

$$\cot\psi\sin\beta = \cos\beta\cos\Sigma + \cot\Phi\sin\Sigma \qquad (11\text{-}50)$$

By differentiating Eq. (11-50) and expressing ψ and Σ in terms of β, η and Φ, it can be shown that [Wertz and Chen, 1976]

$$\tan\Theta_{\beta/\Phi} = \left[\frac{\tan\eta}{\tan\beta\sin\Phi} - \cot\Phi\right] \qquad (11\text{-}51)$$

This can be reformulated in terms of Λ and η as

$$\tan\Theta_{\beta/\Phi} = \frac{\cot\Lambda}{\cos\eta} \qquad (11\text{-}52)$$

Again, by symmetry between the Sun and the Earth, the correlation angle between η and Φ measurements can be written in the same form:

$$\tan\Theta_{\eta/\Phi} = \left[\frac{\tan\beta}{\tan\eta\sin\Phi} - \cot\Phi\right] \qquad (11\text{-}53)$$

$$= \frac{\cot\Sigma}{\cos\beta} \qquad (11\text{-}54)$$

11.4 Geometrical Limitations on Single-Axis Attitude Accuracy

Lily C. Chen
James R. Wertz

In Section 11.3 we described how to determine the attitude accuracy for given geometrical conditions. However, for most aspects of mission planning—such as hardware configuration studies, maneuver and attitude planning, contingency analysis, or launch window analysis—the inverse problem is more relevant. Instead of determining the attitude uncertainty for given conditions, we wish to select the geometrical conditions such that the required attitude accuracy can be achieved. Thus, we would like to understand the effect of any change in the mission conditions on the attitude uncertainties.

In this section, we present a graphical method to study the geometrical limitations on attitude accuracy by applying the equations derived in Section 11.3. With this method, we obtain an overview of the attitude determination geometry and an insight into the effect of changes in mission parameters. Specifically, the equations of Section 11.3 will be used to identify "poor" geometry regions on the celestial sphere for either the attitude or one of the two reference vectors [Chen and Wertz, 1977].

Two cases are considered. In Section 11.4.1, the two reference vectors are assumed fixed and the attitude direction is treated as a variable. The poor geometry regions on the celestial sphere are defined such that whenever the attitude is inside one of these regions, one or more of the attitude determination methods of Section 11.1 will not provide the required attitude accuracy. In Section 11.4.2, the attitude and one of the two reference vectors are assumed fixed and the other reference vector is treated as the variable. In this case, the poor geometry regions on the celestial sphere are defined such that whenever the variable reference vector falls inside one of these regions, the attitude uncertainty *evaluated at the attitude* will be high for one or more of the attitude determination methods. Examples of the application of this geometrical study to mission support activities are given in Section 11.4.3. Again, throughout this section, we use for convenience the Sun and the Earth as the two reference vectors. However, the discussion and conclusions can be applied to any pair of known reference vectors. The notation defined in Figs. 2-1 or 11-18 is used throughout.

11.4.1 Limitations on the Attitude Direction Due to Attitude Accuracy Requirements

We wish to determine the regions of single-axis attitude directions on the celestial sphere which give poor attitude accuracy for fixed positions of the Sun and the Earth. As introduced in Sections 10.1 and 11.3, two geometrical factors limit the attitude accuracy: the correlation angle, $\Theta_{m/n}$, and the measurement densities, d_m and d_n. From Eq. (11-36), the attitude uncertainty becomes infinite whenever $\Theta_{m/n} = 0$ or 180 deg, or either d_m or d_n equals zero, That is, poor geometry regions occur when either the correlation between the two measurements is high or the measurement density is low.

Regions of High Correlation. Regions of poor geometry due to high correlations can be defined for each of the three attitude determination methods: β/η, β/Φ, and η/Φ. Although specific attitude accuracy limits are mission dependent, we define a region of "poor geometry" as any region in which the attitude

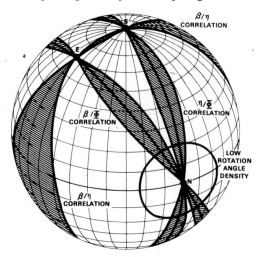

Fig. 11-17. Poor Geometry Regions for the Location of Attitude

uncertainty given in Eq. (11-36) is more than five times greater than the measurement uncertainty, assuming $U_m = U_n$, $C_m = C_n = 0$, and $d_m = d_n = 1$. From Eq. (11-36), this corresponds to a region in which $\Theta_{m/n}$ is in the range $0° \pm 11.5°$ or $180° \pm 11.5°$.

For the β/η method, the analysis is simple because $\Theta_{\beta/\eta} = \Phi$, as given in Table 11-2. Thus, the constant correlation angle curves for the β/η method are the same as the constant Sun-to-Earth rotation angle curves given in Section 10.3. From Fig. 10-9, it is obvious that the singularity occurs when the attitude lies along the Sun/Earth great circle where $\Theta_{\beta/\eta} = 0$ or 180 deg and the poor geometry regions due to high β/η correlation must be regions around this great circle bounded by constant rotation angle curves. This poor geometry region for a Sun/Earth separation of 30 deg is shown as the shaded region labeled "β/η correlation" in Fig. 11-17 (preceding page).

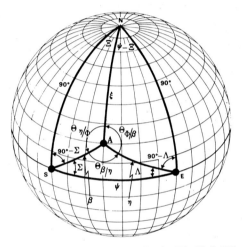

Fig. 11-18.   Relations Among Attitude (A), Sun (S), Earth (E), Null (N), and Correlation Angles ($\Theta_{i/j}$)

For the β/Φ and η/Φ methods, the interpretation of the expressions for $\Theta_{\beta/\Phi}$ and $\Theta_{\eta/\Phi}$ from Table 11-2 is more difficult. However, this interpretation may be simplified by using the *Null, N,* or Sun-Earth cross product, as shown in Fig. 11-18. Applying Napier's rule to the spherical triangles *EAN* and *SAN,* and comparing the results with Eq. (11-52) and (11-54), the rotation angle *NAE* equals the correlation angle $\Theta_{\beta/\Phi}$ and the rotation angle *NAS* equals the correlation angle $\Theta_{\eta/\Phi}$. From Fig. 11-18, it is clear that the constant $\Theta_{\beta/\Phi}$ and $\Theta_{\eta/\Phi}$ curves are the constant rotation angle curves between the Earth and the null and between the Sun and the null, respectively. Because the Earth/null and the Sun/null separations are always 90 deg, the rotation angle curves given in Fig. 10-10 can be used to obtain the regions of high β/Φ or η/Φ correlation. The poor geometry regions for which $\Theta_{\beta/\Phi}$ and $\Theta_{\eta/\Phi}$ lie within 11.5 deg of 0 deg or 180 deg are shown as the shaded regions labeled "β/Φ correlation" and "η/Φ correlation," respectively, in Fig. 11-17. Note that the centers of these regions are the Earth/null and the Sun/null great circles, respectively.

Regions of Low Measurement Density. Poor geometry regions due to a low measurement density (d) occur only for the rotation angle measurements. As discussed in Sections 10.3 and 11.3, the rotation angle density goes to zero, i.e., the attitude uncertainty goes to infinity, when the attitude approaches the null or the antinull. Therefore, poor geometry regions for the attitude due to low rotation angle densities are regions around the null or the antinull bounded by the constant rotation angle density curves. These curves can be obtained by using Eq. (11-46) to obtain a quadratic equation in $(\sin^2\xi)$ in terms of Ξ, ψ, and d:

$$\left[d^2 \cos^2\Xi \cos^2(\psi-\Xi) \right]\sin^4\xi - \left\{ d^2 \left[\cos^2\Xi + \cos^2(\psi-\Xi) \right] + \sin^2\psi \right\}\sin^2\xi + d^2 = 0$$

$$(11\text{-}55)$$

Note that Ξ is defined in Figs. 11-11 and 11-18. The result in Eq. (11-55) for $\psi = 30$ deg and $d = 0.2$ is shown as the unshaded region about the null in Fig. 11-17. That is, whenever the attitude lies inside this region, the attitude component

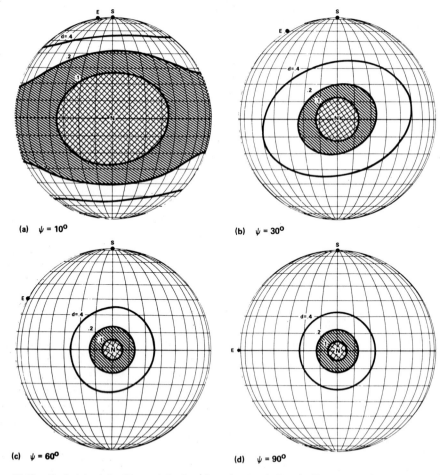

(a) $\psi = 10^o$ (b) $\psi = 30^o$

(c) $\psi = 60^o$ (d) $\psi = 90^o$

Fig. 11-19. Evolution of the Shape of the Low Rotation Angle Density Region for Varying Separation Between the Reference Vectors. Each subfigure is centered on the null.

determined from the rotation angle measurements will have an uncertainty at least five times greater than the rotation angle measurement uncertainty. The evolution of the shape of this region for varying Sun/Earth separations is shown in Fig. 11-19 (preceding page).

Combination of High Correlation and Low Density. In Fig. 11-17, only the β/η subfigure gives the poor geometry region directly because the measurement densities are unity. The β/Φ and η/Φ regions must be obtained by combining the high-correlation effect with the low-density effect. This can be done numerically using Eq. (11-36) and results are shown in Fig. 11-20. This figure shows the poor geometry regions for the attitude such that within the shaded regions, the attitude uncertainty will be five times greater than the measurement uncertainties for the β/Φ or η/Φ method, assuming equal uncertainties in the two measurements.

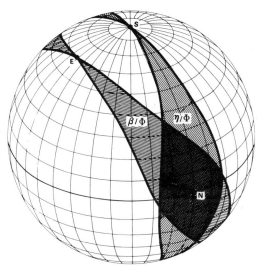

Fig. 11-20.   Poor Geometry Regions for the Attitude for the β/Φ and η/Φ Methods. When the attitude lies inside the shaded region, the attitude uncertainty will be more than five times the measurement uncertainty, based on equal uncertainty in the two measurements. (Compare with Fig. 11-17.)

11.4.2 Limitations on Reference Vector Direction

For many mission support activities, the attitude direction is predetermined while one of the two reference vector directions either remains to be determined or is moving, as in the case of a satellite with an inertially fixed attitude moving around the Earth which is being used as one of the reference vectors. In this case, we wish to obtain the poor geometry regions for the variable reference vectors such that whenever this vector is located inside these regions, one or more of the attitude determination methods will result in poor attitude accuracy.

For convenience, the Earth will be used as the varying reference vector. However, due to the symmetry between the Sun and the Earth, the results can be equally applied if the Sun position is treated as the variable instead, as will be shown below for the launch window analysis.

Regions of High Correlation.   As in Section 11.4.1, poor geometry regions can be defined for each of the three attitude determination methods. For the β/η method, Eq. (11-48) can be used directly. Because $\Theta_{\beta/\eta}=\Phi$, the poor geometry region lies between two great circles which intersect at the attitude at an angle $\Theta_{\eta/\beta}$ on either side of the Sun-attitude great circle, as shown in Fig. 11-21 for $\Theta_{\beta/\eta}=\pm$ 11.5 deg.

For the β/Φ correlation, the poor geometry regions can be obtained from Eqs. (11-51) and (11-52). From Eq. (11-52), Λ must be a right angle when $\Theta_{\beta/\Phi}=0$ or 180 deg. That is, a singularity occurs when the Earth lies on the 90-deg or 270-deg constant rotation angle curve between the Sun and the attitude. This is equivalent to the attitude lying on the Earth/null great circle, as shown in Fig. 11-17. The boundaries of this poor geometry region may be obtained by reformulating Eq. (11-51) into an expression for η in terms of $\Theta_{\beta/\Phi}$, β and Φ, as shown in Fig. 11-21 for $\Theta_{\beta/\Phi}=\pm 11.5$ deg about 0 deg or 180 deg. The evolution of the shape of this region as a function of β is shown in Fig. 11-22. As seen most clearly in Fig. 11-22(c) and (d), this region is *not* symmetric under an interchange of the Sun and the attitude. Except for a Sun angle of 90 deg, the β/Φ correlation region consists of two unconnected areas, one near the attitude and the other near the antiattitude.

For the η/Φ correlation, Eq. (11-54) may be used directly. When $\Theta_{\eta/\Phi}=0$ or 180 deg, $\Sigma=90$ deg or 270 deg and the Earth lies on the great circle through the Sun perpendicular to the Sun-attitude great circle. The poor geometry region

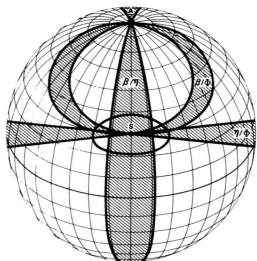

Fig. 11-21.   Poor Geometry Regions for the (Earth) Reference Vector Due to High Correlations or Low Rotation Angle Densities. (Contrast with Fig. 11-17 showing poor geometry regions for the attitude with fixed Earth/Sun positions.)

around this great circle is bounded by two great circles intersecting at the Sun and the antisolar point and making a constant angle with the $\Sigma=90$ deg or 270 deg great circle. The shaded area in Fig. 11-21 labeled "η/Φ" shows this region for $\Theta_{\eta/\Phi}=\pm 11.5$ deg.

Regions of Low Measurement Density.   Finally, in addition to the poor geometry regions due to measurement correlations, Eqs. (11-44) and (11-46) can be

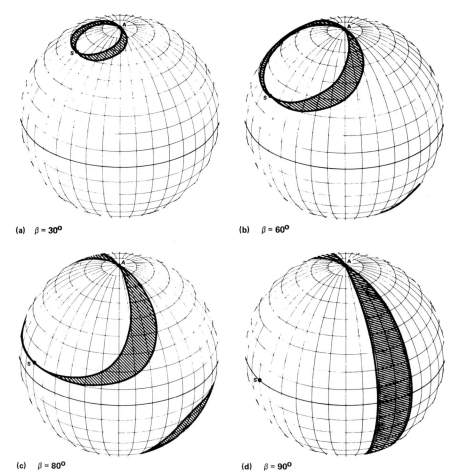

(a) $\beta = 30^{\circ}$ (b) $\beta = 60^{\circ}$

(c) $\beta = 80^{\circ}$ (d) $\beta = 90^{\circ}$

Fig. 11-22.   Evolution of the Shape of Poor Geometry Regions for the Earth Due to β/Φ Correlation

used to obtain the poor geometry regions due to low rotation angle density. Equation (11-46) shows that the rotation angle density approaches zero whenever ψ or ξ is near 0 or 180 deg. $\xi = 0$ deg or 180 deg implies that the attitude is at the null or the antinull and both β and η are equal to 90 deg. For a given β other than 90 deg, the poor geometry region due to low rotation angle density depends strongly on ψ; the rotation angle density becomes low when the Earth is close to the Sun or the antisolar point.

We may determine two regions around the Sun and the antisolar point, such that if the Earth lies inside either region, the rotation angle density, *evaluated at the attitude*, will be less than a specified value. The boundary of this region can be obtained from Eq. (11-44) by substituting Φ in terms of Σ, ψ, and β and reformulating the equation to yield

$$\cot\psi = \frac{1}{\sin\beta}\left[\cos\beta\cos\Sigma \pm \sqrt{\frac{(\sin^2\Sigma\cos^2\beta + \cos^2\Sigma)}{d^2\sin^2\beta}} - \sin^2\Sigma\right] \quad (11\text{-}56)$$

This region around the Sun for $\beta = 30$ deg and $d = 0.2$ is shown in Fig. 11-21 and the evolution of the shape of this region for varying β is shown in Fig. 11-23. Note that when $\beta = 90$ deg, the low rotation angle density regions become a single continuous band bounded by small circles of fixed nadir angle such that

$$\sin^2 \eta = \frac{1}{1 + d^2} \qquad (11\text{-}57)$$

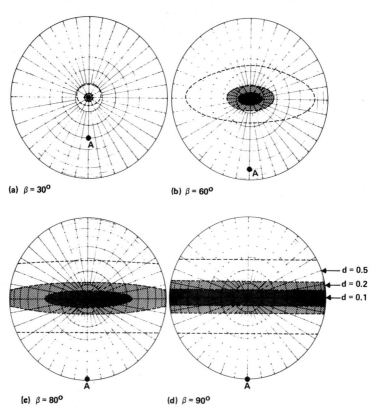

(a) $\beta = 30°$ (b) $\beta = 60°$

(c) $\beta = 80°$ (d) $\beta = 90°$

d = 0.5
d = 0.2
d = 0.1

Fig. 11-23.   Evolution of the Shape of the Low Rotation Angle Density Region for Positions of the Earth. (The Sun is at the center of each plot.) When the Earth lies inside the darkly shaded region, the rotation angle density, d, *at the attitude* is less than 0.1. The lightly shaded and unshaded regions are for $d = 0.2$ and 0.5, respectively.

Combination of High Correlation and Low Density.   Similar to the discussion in Section 11.4.1, among the four regions in Fig. 11-21, only the β/η region provides the poor geometry area directly. For the β/Φ and η/Φ methods, the correlation regions must be combined with the low rotation angle density region to obtain the regions corresponding to a factor of five between the attitude uncertainty and the measurement uncertainties. This can be done by substituting Eqs. (11-45) and (11-52) or Eqs. (11-44) and (11-54) into Eq. (11-36) and expressing η in terms of Φ, β, and f (the ratio between attitude uncertainty and the measurement uncertainty). The results for $\beta = 30$ deg and $f = 5$ are shown in Fig. 11-24.

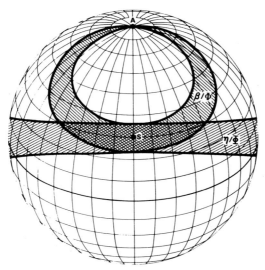

Fig. 11-24.   Poor Geometry Regions for the Earth for β/Φ and η/Φ Methods. When the Earth is in the shaded region, the attitude uncertainty will be more than five times the measurement uncertainty, based on equal uncertainty in the two measurements. (Compare with Fig. 11-21.)

11.4.3 Applications

The geometrical study of the limitations on attitude accuracy described in this section has been applied in both prelaunch and postlaunch analysis for SMS-2; GOES-1, -2, and -3; AE-4, and -5; CTS; and SIRIO and in prelaunch analysis for ISEE-C and IUE [Wertz and Chen, 1975, 1976; Chen and Wertz, 1975; Tandon, *et al.*, 1976; Chen, *et al.*, 1976, 1977; Chen, 1976; Lerner and Wertz, 1976; Rowe, *et al.*, 1978]. To illustrate the procedure, we will discuss the attitude determination accuracy for SMS-2 and the attitude launch window constraints for SIRIO. The profile for both missions is similar to that of CTS, as described in Section 1.1. An alternative formulation is given by Fang [1976].

SMS-2 Attitude Determination.   The Synchronous Meteorological Satellite, SMS-2, was launched into an elliptical transfer orbit on February 6, 1975. Shortly after launch, the attitude was maneuvered to that appropriate for Apogee Motor Firing (AMF). On the second apogee, the AMF put the spacecraft into a circular near-synchronous drift orbit over the equator. Over the next 3 days, the attitude was maneuvered to orbit normal with two intermediate attitudes. The data collected in both the transfer and drift orbits allowed the measurement of 20 attitude bias parameters on five Earth horizon sensors and one Sun sensor. The geometrical methods described here were used extensively in the analysis of SMS-2 attitude and bias determination and contributed substantially to the result obtained [Chen and Wertz, 1975; Wertz and Chen, 1975, 1976.]

As the spacecraft moves in its orbit, the attitude determination geometry changes due to the motion of the position of the Earth (as seen by the spacecraft) relative to the Sun and the attitude. A convenient vehicle for examining this changing geometry is a plot of the celestial sphere as seen by the spacecraft, with the directions of the Sun and the attitude fixed. Figures 11-25 and 11-26 show examples of such plots* for the nominal transfer orbit and apogee motor firing attitude near apogee. The region around perigee was of less interest because the spacecraft was then out of contact with the Earth.

As usual, the spacecraft is at the center of the sphere. The heavy solid line is the orbit of the Earth around the spacecraft as seen from the spacecraft. The Earth is moving toward increasing right ascension, i.e., from left to right on the plots. Tic marks denote the time from apogee in 10-minute intervals. The dotted line surrounding the orbit denotes the envelope of the disk of the Earth as it moves across the sky. AP marks the location of the Earth when the spacecraft is at the apogee. S^{-1}, A, and A^{-1} mark the location of the antisolar point, attitude, and negative attitude axis, respectively.

The small solid circles labeled $ES\,1$ and $ES\,4$ and centered on the A/A^{-1} axis are the fields of view (FOV) for two of the five SMS-2 Earth horizon sensors as the spacecraft spins about the A/A^{-1} axis. Arrowheads on the FOV lines indicate the direction in which the sensors scan the sky. Acquisition of signal (AOS) and loss of signal (LOS) of the Earth by each sensor are marked by arrowheads along the orbit with primed numbers for LOS and unprimed numbers for AOS.

The three dashed curves in Fig. 11-25 are the central lines of the poor geometry regions for the position of the Earth due to strong correlations, and the

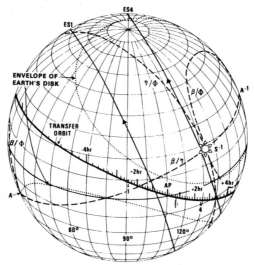

Fig. 11-25. SMS-2 Attitude Determination Geometry for the Transfer Orbit and Apogee Motor Firing Attitude. See text for explanation.

---

*Since their initial use for the SMS-2 mission in 1975, global plots of the sky as seen by the spacecraft, such as Fig. 11-25, have been used by the authors for each of the missions they have supported. These plots have been very convenient for examining sensor fields of view and optimum sensor placement, Sun and Earth coverage, attitude uncertainties, the relative geometry of reference vectors, and other aspects of mission analysis. See Section 20.3 for a description of the subroutines used to generate these plots. With practice, they may also be drawn quickly by hand using the blank grids and methods given in Appendix B.

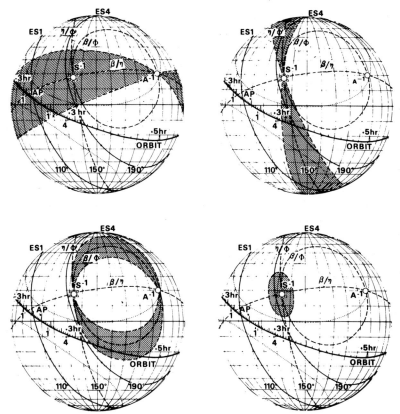

Fig. 11-26.   Poor Geometry Regions for the Earth for the Geometry in Fig. 11-25. (Relative to Fig. 11-25, the center of the plot has been shifted down to the celestial equator and to the right of the antisolar point.) Here the poor geometry regions are bounded by a correlation angle of 23 deg, whereas Fig. 11-21 shows the same regions bounded by correlation angles of 11.5 deg.

four shaded areas shown in Fig. 11-26 are the poor geometry regions for the Earth analogous to those shown in Fig. 11-21. Thus, whenever the Earth moves inside one of these regions, one or more of the attitude determination methods will give poor results. By comparing the Earth coverage regions (from AOS to LOS) for each of the five sensors with the poor geometry regions for the Earth, we can easily choose the preferred attitude determination method for each of the five data passes. For example, $ES1$ sees the Earth in a region of poor geometry due to high correlation between the Sun angle and the nadir angle measurements. Therefore, the attitude determined by the β/η method would yield high uncertainties and the other two methods should be used instead. Similar results can be obtained for the data passes from other sensors. None of the data passes falls inside the low rotation angle density region. Therefore, attitude uncertainties due to low rotation angle density were not a problem during the SMS-2 transfer orbit.

SIRIO Launch Window Constraints.   SIRIO is an Italian satellite launched in August 1977, which uses the Sun angle data and the IR Earth sensor data to determine the spinning spacecraft attitude, similar to SMS and CTS. We briefly describe analysis of the attitude determination constraints on the SIRIO launch window. (For additional details, see Chen [1976] and Chen, *et al.*, [1977].) The purpose of this analysis is to obtain the launch window (in terms of right ascension of the orbit's ascending node versus the launch date) which will give the required attitude accuracy.

Figure 11-27(a) shows the nominal geometry for SIRIO attitude determination in the transfer orbit, in a plot analogous to Fig. 11-25. The position of the antisolar point is plotted for a January 15 launch. As the launch date changes, so does the Sun position and attitude determination geometry. Thus, determining the launch window constraints is equivalent to determining the constraints on the position of the Sun to obtain good attitude determination geometry. Thus, instead of considering the attitude and the Sun to be fixed, as in the previous example, we consider the attitude and the Earth as fixed and treat the Sun position as a variable.

The position of the Earth can be determined by the sensor coverage. Attitude determination is most important before AMF. Therefore, we require that the attitude be determined to within the specified attitude accuracy from a data pass

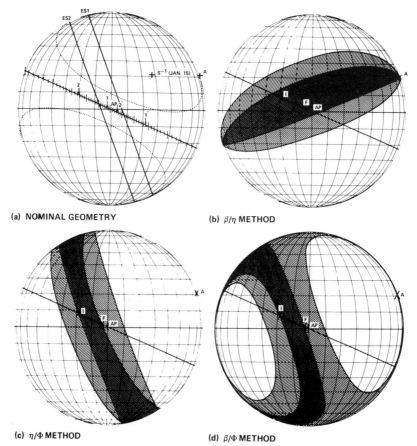

(a) NOMINAL GEOMETRY   (b) β/η METHOD

(c) η/Φ METHOD   (d) β/Φ METHOD

Fig. 11-27.   SIRIO Attitude Determination Geometry in the Transfer Orbit. Shaded areas give Sun locations for which the attitude determination geometry is poor. See text for explanation. (Compare with Figs. 11-21 and 11-24.)

covering the time period of 150 minutes before AMF to 30 minutes before AMF; this is the darkened region between I and F along the orbit in Figs. 11-27(b), (c), and (d).

The method described in Section 11.4.2 can be used to obtain the poor geometry regions for the position of the Sun using the three attitude determination methods, having the Earth located at I and F, respectively. Figures 11-27(b), (c), and (d) show these poor geometry regions for β/η, β/Φ and η/Φ methods, respectively (compare with Figs. 11-21 and 11-24). In each figure, two regions are plotted, corresponding to Earth positions at I and F. Thus, the overlapping regions in Figs. 11-27(b) to (d) give the positions of the Sun (or the antisolar point) such that for all locations of the Earth between I and F, the attitude determined by that particular method will not give the required accuracy.

The poor geometry regions shown in Figs. 11-27(b) through (d) provide the constraints on the position of the Sun relative to the ascending node. As the launch date changes, the Sun position changes, and the ascending node and the attitude are rotated to maintain the relative positions of the Sun and the node. Therefore, the Sun constraints can be transformed into constraints on the right ascension of the ascending node versus launch date, as desired. Figures 11-28(a) through (c) show such results for the three attitude determination methods for a full year, and Fig. 11-28(d) gives the constraints on the launch window where none of the three attitude determination methods would give required attitude accuracy.

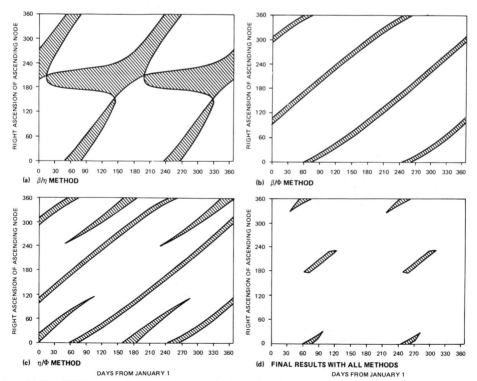

Fig. 11-28. SIRIO Attitude Constraints on the Launch Window for AMF Attitude Determination Accuracy. (See text for explanation.)

Further Applications.   Both of the examples discussed above show the applications of the poor geometry regions for the reference vectors as discussed in Section 11.4.2. However, the poor geometry regions for the attitude discussed in Section 11.4.1 may also be used in mission planning activities, especially in maneuver planning. For example, if we plot an attitude maneuver on the geometry plots given in Fig. 11-17 or 11-20, it is clear where along the maneuver the attitude can be best determined using any attitude determination method. Thus, we can stop the maneuver at the appropriate position for attitude and bias determination. Alternatively, we can change the route of a maneuver to provide attitude accuracy for maneuver monitoring, or we can plan an attitude maneuver purely for the purpose of attitude and sensor bias determination. Activities of this type were used successfully on AE-4 and -5 to evaluate attitude sensor biases. Similar analyses have been performed to provide optimal Sun sensor configurations for SEASAT [Lerner and Wertz, 1976] and to examine Earth and Moon coverages as ISEE-C transfers to the Sun-Earth libration point approximately 6 lunar orbit radii from the Earth [Rowe, *et al.*, 1978].

11.5 Attitude Uncertainty Due to Systematic Errors

Lily C. Chen
James R. Wertz

The causes of single-frame attitude uncertainty may be separated into the two categories of random and systematic errors. A random error is an indefiniteness of the result due to the finite precision of the experiment, or a measure of the fluctuation in the result after repeated experimentation. A systematic error is a reproducible inaccuracy introduced by faulty equipment, calibration, or technique. The attitude uncertainty due to random errors can be reduced by repeated measurement. When a measurement is repeated n times, the mean value of that measurement will have an uncertainty $\sqrt{n}$ times smaller than the uncertainty of each individual measurement. However, this statistical reduction does not apply to systematic errors. Therefore, the attitude uncertainty due to systematic errors is usually much larger than that due to random errors.* Therefore, to reduce the attitude uncertainty, we must identify and measure as many as possible of the systematic errors present in each of the attitude measurements. In this section, we compare the behavior of the single-frame attitude solutions with and without systematic errors, discuss the singularity conditions for various attitude determination methods, and introduce the concept of data filters and state estimation to solve for the systematic errors.

11.5.1 Behavior of Single-Frame Solutions

Although systematic errors cannot be reduced by measurement statistics, they will usually reveal themselves when the same measurements are repeated at different times along the orbit under different geometrical conditions. Thus, a

---

*If the random errors dominate, normally more measurements will be taken until the uncertainty is again dominated by the systematic error.

study of the behavior of the single-frame attitude solutions as a function of time can help reveal the existence of systematic errors.

For an ideal case in which no systematic error exists in any of the attitude measurements or attitude determination models, the behavior of the single-frame attitude solutions may be summarized as follows:

1. For each attitude determination method, the attitude solution should follow a known functional variation with time except for the fluctuations due to random errors. If the attitude is inertially fixed and nutation and coning are small, the attitude solutions should remain constant in time.

2. The attitude solutions obtained from different attitude determination methods should give consistent results. Most spacecraft provide redundant measurements for attitude determination to avoid problems of sensor inaccuracies or failure.* Therefore, more than one attitude determination method is generally available. If no systematic error exists, the same attitude solution should result from all methods at any one time in the orbit, to within the random noise on the data.

3. Near an attitude solution singularity, the attitude solutions will have large fluctuations about a uniform mean value because these uncertainties are due entirely to random errors. An attitude singularity is any condition for which the uncertainty of the attitude solution approaches infinity.

Figure 11-29 shows the behavior of single-frame solutions for a near-ideal case. In the figure, the spin axis declination from one real SMS-2 data pass obtained when the spacecraft was in near-synchronous orbit is plotted against the frame number. In obtaining the plotted results, the biases obtained from a bias determination subsystem (as described in Section 21.2) have been used to compensate for most of the systematic errors present in the data. Consequently, apart from the beginning of the data pass, the solutions obtained from the four different attitude determination methods show nearly constant and consistent results throughout. Also, the solutions near singularities fluctuate about the mean value, as most easily seen from solution 2 near frame 160. The inconsistency in results near the beginning of the data pass and the small deviation in the solutions from a constant value indicate the presence of small residual systematic errors.

The ideal situation normally does not exist in a real mission using nominal parameters. In general, systematic errors are difficult to avoid and contribute most of the uncertainty in single-frame attitude solutions. The systematic errors usually encountered in attitude determination fall into three categories: (1) *sensor and modeling parameter biases*, which include all possible misalignments in the position and orientation of the attitude sensors and erroneous parameter values used in the models; (2) *incorrect or imperfect mathematical models*, which include all possible erroneous assumptions or errors in the mathematical formulation of the attitude determination models, such as the shape of the Earth, the dynamic motion of the attitude, or unmodeled sensor electronic characteristics; and (3) *incorrect reference vector directions*, which include all possible errors in the instantaneous orientation

---

*In some cases, the same sensors may be used to provide attitude solutions based on different targets. For example, the Earth and the Moon provide redundant information for RAE-2 [Werking, *et al.*, 1974] and ISEE-1.

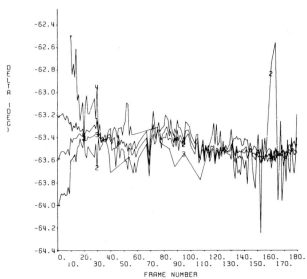

Fig. 11-29.   Behavior of Single-Frame Solutions With Small Systematic Errors for Real SMS-2 Data. Numbers on plots indicate solution method: 1 = Sun angle/Earth-in crossing, 2 = Sun angle/Earth-out crossing, 3 = Sun angle/Earth width, 4 = Sun angle/Earth midscan.

of reference vectors, such as orbit errors; errors in ephemeris information for the Sun, the Moon, the planets, and the stars; time-tagging errors; and the errors in direction of the magnetic or gravitational field.

Because of systematic errors, the real behavior of the single-frame attitude solutions are generally quite different from the ideal situation. Specifically, the following items characterize the behavior of single-frame solutions with significant systematic errors:

1. For each attitude determination method, the attitude solution departs from the known functional variation with time. This behavior is most easily observed for the spin-stabilized spacecraft where, ideally, the attitude should remain constant in time, and in the presence of systematic errors it shows an apparent time variation.

2. The solutions obtained from different attitude determination methods give different attitude results and show relative variations with time.

3. Near attitude determination singularities, attitude solutions tend to diverge drastically from the mean value.

Thus, the analyst can normally identify the existence of systematic errors by examining the time dependence of the attitude solution from each method, the consistency of results from different methods, and the behavior of solutions nearing singularities.

The behavior of single-frame solutions with significant systematic errors is illustrated in Fig. 11-30, which shows the spin axis declination determined from the same data set as that in Fig. 11-29, except that here the systematic errors have not been removed (i.e., nominal parameters for all sensors were used). Note that the vertical scales are different in the two figures and that solutions which are outside the scale are not plotted. Here, solutions vary strongly in time, show substantial inconsistency among different methods, and diverge rapidly near singularities.

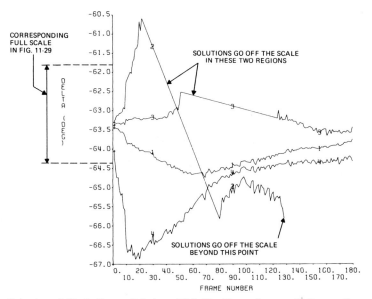

Fig. 11-30. Behavior of Single-Frame Solutions With Significant Systematic Errors. Same data and attitude determination methods as Fig. 11-29.

The singularity conditions for the data pass of Figs. 11-29 and 11-30 can be obtained from the predicted arc-length uncertainty plot shown in Fig. 11-31. Again, points outside the scale range are not plotted. It is seen from this figure that singularities occur near the middle of the data pass for method 3 (β/Ω) and near frame 40 and beyond frame 130 for method 2 (β/Φ_{H_o}); these are also the places where the solutions diverge in Fig. 11-30. (See Section 11.1 for a description of the attitude determination methods.) An analysis of the location of the singularities is given in the next subsection.

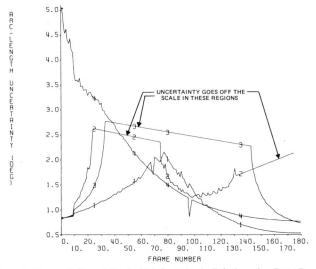

Fig. 11-31. Arc-Length Uncertainties of Single-Frame Attitude Solutions for Data Pass of Figs. 11-29 and 11-30.

11.5.2 Identification of Singularities

Because the existence of systematic errors can be recognized from the behavior of solutions near singularities, it is important to determine the singularity conditions for the commonly used attitude determination methods. The general expression of the attitude uncertainty given by Eq. (11-36) shows that attitude uncertainty approaches infinity only under two conditions: (1) when the measurement density, d_m or d_n, approaches zero, or (2) when the correlation angle, $\Theta_{m/n}$, approaches 0 or 180 deg, i.e., when $\sin\Theta_{m/n} = 0$.

The singularity conditions for the attitude determination methods listed in Table 11-2 can be summarized as follows:

1. *Sun Angle/Nadir Angle* (β/η) *Method.* The only singularity for this method occurs when $\Theta_{\beta/\eta} = 0$ or 180 deg. From Fig. 11-18, this occurs when the attitude lies on the Sun/Earth great circle, that is, when the Sun vector, the nadir vector, and attitude are coplanar. This singularity condition can be generally applied to any method using two arc-length measurements relative to two *known** reference vectors to determine a third unknown vector direction. A singularity always occurs when the unknown vector is coplanar with the two reference vectors.

2. *Sun Angle/Earth-Width* (β/Ω) *Method.* Because the correlation angle here is the same as that for the β/η method, the singularity condition above also applies here. However, an additional singularity exists due to the measurement density of Ω. From Eq. (11-40), $d_\Omega = 0$ when $\cot\gamma = \cot\eta \cos\Omega/2$. This is the condition for which the dihedral angles AEH_I and AEH_O equal 90 deg, that is, at maximum Earth width for constant ρ.

3. *Sun Angle/Rotation Angle* (β/Φ) *Method.* Two types of singularities exist for this method. The singularity due to the correlation angle is most easily obtained from Fig. 11-18, from which, $\Theta_{\beta/\Phi} = 0$ or 180 deg when the attitude lies on the Earth/Null great circle, that is, when the nadir vector, the null, and the attitude are coplanar. From Eq. (11-46), the singularity due to low rotation angle density occurs when either $\sin\psi$ or $\sin\xi$ equals zero, that is, when the nadir vector is parallel or antiparallel to the Sun vector or when the attitude is parallel or antiparallel to the null.

4. *Nadir Angle/Rotation Angle* (η/Φ) *Method.* The singularity conditions for this method are similar to those for the β/Φ method except that here the correlation angle equals 0 or 180 deg when the attitude lies along the Sun/null great circle; that is, when the Sun vector, the null, and the attitude are coplanar.

5. *Earth-Width/Rotation Angle* (Ω/Φ) *Method.* Singularities in this method come from three sources: the high correlation between Ω and Φ or the low measurement density of either Ω or Φ. The singularity condition due to the low measurement density for Ω is given in the β/Ω method, and those due to the other two sources are the same as those given in the η/Φ method.

In addition to these five attitude determination methods, a sixth common method is the Sun angle/Sun-to-horizon crossing rotation angle (β/Φ_H) method, in which the horizon vector, $\hat{\mathbf{H}}$, can be either the Earth-in or -out vector (see Section 11.1.2). As discussed in Section 10.5, the Sun-to-horizon crossing rotation

---

*The critical nature of this condition is discussed below (and in Section 10.4) in terms of the unknown horizon crossing vector as an attitude reference.

angle measurement is a compound data type because the horizon vector is *not* a known vector but rather is determined from the Sun angle and the rotation angle measurements. Therefore, the equations given in Table 11-2 cannot be directly applied to this method to identify the singularities. However, the singularity conditions for this method can be obtained by considering the attitude determination procedures described in detail in Section 11.1.2. Three steps are involved in attitude determination in the β/Φ_H method: the computation of ψ_H from γ, β and Φ_H; the determination of the horizon vector from the arc-length angles ψ_H and ρ relative to the Sun and the Earth, respectively; and the determination of attitude using β and γ. The first step is a direct application of the cosine law for spherical triangles and gives no singularities. The second step causes a singularity when the Sun vector, the nadir vector, and the horizon vector are coplanar as a result of the general rule discussed in the β/η method. However, this rule *cannot* be applied to the third step because the horizon vector is not a known reference vector but is determined from the knowledge of the rotation angle, Φ_H. In other words, the third step not only uses the values of β and γ, but also implicitly uses the value of Φ_H. Because the singularity condition for the β/γ method in general provides good geometry for methods using Φ_H, with all three data items (β, γ and Φ_H) available, step three will not introduce any attitude determination singularity. Thus, the only singularity condition for the β/Φ_H method occurs when the Sun vector, the nadir vector, and the horizon vector are coplanar. This is confirmed by examining Figs. 10-15 and 10-16 and noting that the β/Φ_H correlation angle singularity occurs when small circles centered on the Sun are tangent to the Φ_H attitude loci. This occurs at the points labeled A on Figs. 10-15 and 10-16.

Table 11-3 summarizes the singularity conditions for all methods discussed in this section. Singularity conditions for other attitude determination methods can be obtained in the same manner by analyzing the data types, the measurement densities, and the correlation angles.

11.5.3 State Vector Formulation

Because of the large number of possible systematic errors, single-frame deterministic attitudes normally have large uncertainties relative to what would be expected from the noise alone. To meet the attitude accuracy requirement for most

Table 11-3. Singularity Conditions for Common Attitude Determination Methods. A singularity occurs when any one of the conditions is met.

METHOD	SINGULARITY CONDITIONS
β/η	$\hat{S}, \hat{E}, \hat{A}$ COPLANAR
β/Ω	$\hat{S}, \hat{E}, \hat{A}$ COPLANAR; $\angle AEH_I$ AND $\angle AEH_O = 90°*$
β/Φ	$\hat{E}, \hat{N}, \hat{A}$ COPLANAR; $E = \pm\hat{S}$; $\hat{A} = \pm\hat{N}$
η/Φ	$\hat{S}, \hat{N}, \hat{A}$ COPLANAR; $E = \pm\hat{S}$; $\hat{A} = \pm\hat{N}$
Ω/Φ	$\hat{S}, \hat{N}, \hat{A}$ COPLANAR; $E = \pm\hat{S}$; $\hat{A} = \pm\hat{N}$; $\angle AEH_I$ AND $\angle AEH_O = 90°*$
β/Φ_H	$\hat{S}, \hat{E}, \hat{H}$ COPLANAR

*CONDITION FOR MAXIMUM Ω IF THE ANGULAR RADIUS OF THE EARTH (ρ) IS CONSTANT.

missions, it is necessary to measure as many existing systematic errors as possible. Although the existence of systematic errors can usually be revealed from the behavior of the single-frame attitude solutions, it is our experience that making a quantitative determination of the individual systematic errors solely from this behavior is almost impossible. Hence, in addition to deterministic-type attitude systems, we need techniques and systems which provide the capability of determining systematic errors without relying solely on operator evaluation. Such *bias determination* systems have become a standard part of most attitude systems in use at NASA's Goddard Space Flight Center.

In bias determination systems, the attitude parameters and the most commonly encountered systematic errors are treated as the components of a general state vector. The various *estimators* or *data filters* are built to allow some or all of the components of the state vector to vary to optimize the fit to the attitude data. In this manner, systematic errors which give observable effects on the data can be solved for quantitatively and the results can then be fed back to the attitude determintion systems to compensate for these errors and to obtain more accurate attitude results. A detailed discussion of data filters and state estimation techniques is the topic of Chapters 13 and 14.

References

1. Bevington, Philip R., *Data Reduction and Error Analysis for the Physical Sciences*. New York: McGraw-Hill, Inc., 1969.
2. Boughton, W. L., and L. C. Chen, *Despun-Mode Attitude Determination for IUE*, Comp. Sc. Corp., CSC/TM-78/6013, Jan. 1978.
3. Bryson, Arthur E., Jr., and Yu-Chi Ho, *Applied Optimal Control*. Waltham, Ma: Ginn and Company, 1969.
4. Chen, L. C., and J. R. Wertz, *Analysis of SMS-2 Attitude Sensor Behavior Including OABIAS Results*, Comp. Sc. Corp., CSC/TM-75/6003, April 1975.
5. ——, H. L. Hooper, J. E. Keat, and J. R. Wertz, *SIRIO Attitude Determination and Control Systems Specifications*, Comp. Sc. Corp., CSC/TM-76/6043, Feb. 1976.
6. ——, *Attitude Determination Accuracy Constraints on the SIRIO Launch Window*, Comp. Sc. Corp., CSC/TM-76/6210, Sept. 1976.
7. ——, N. Hauser, L. Hooper, T. McGann, and R. Werking, *SIRIO Attitude Analysis and Support Plan*, NASA X-581-77-169, GSFC, July 1977.
8. ——, and J. R. Wertz, *Single-Axis Attitude Determination Accuracy*, AAS-/AIAA Astrodynamics Conference, Grand Teton Nat. Park, WY, Sept. 7–9, 1977.
9. Fang, Bertrand T., "General Geometric Theory of Attitude Determination from Directional Sensing," *J. Spacecraft*, Vol. 13, p. 322–323+, 1976.
10. Grubin, Carl, "Simple Algorithm For Intersecting Two Conical Surfaces," *J. Spacecraft*, Vol. 14, p. 251–252, 1977.
11. Keat, J., and M. Shear, *GRECRS Test Results*, Comp. Sc. Corp., 3000-06000-02TM, Sept. 1973.
12. Lerner, G. M., and J. R. Wertz, *Sun Sensor Configurations for SEASAT-A*, Comp. Sc. Corp., CSC/TM-76/6147, July 1976.

13. Rowe, J. N., P. Batay-Csorba, S. K. Hoven, G. Repass, *International Sun-Earth Explorer-C (ISEE-C) Attitude Analysis and Support Plan,* Comp. Sc. Corp., CSC/TM-78/6082, June 1978.

14. Shear, M., G. Page, S. Eiserike, V. Brown, K. Tasaki, and J. Todd, *Infrared (IR) Attitude Determination System Communications Technology Satellite Version System Description,* Comp. Sc. Corp., CSC/TM-76/6083, June 1976.

15. Tandon, G. K., M. Joseph, J. Oehlert, G. Page, M. Shear, P. M. Smith, and J. R. Wertz, *Communications Technology Satellite (CTS) Attitude Analysis and Support Plan,* Comp. Sc. Corp. CSC/TM-76/6001, Feb. 1976.

16. Werking, R. D., R. Berg, K. Brokke, T. Hattox, G. Lerner, D. Stewart, and R. Williams, *Radio Astronomy Explorer-B Postlaunch Attitude Operations Analysis,* NASA X-581-74-227, GSFC, July 1974.

17. Wertz, James R., and Lily C. Chen, *Geometrical Procedures for the Analysis of Spacecraft Attitude and Bias Determinability,* Paper No. AAS75-047, AAS/AIAA Astrodynamics Specialist Conference, Nassau, Bahamas, July 28 – 30, 1975.

18. ———, and Lily C. Chen, "Geometrical Limitations on Attitude Determination for Spinning Spacecraft," *J. Spacecraft,* Vol. 13, p. 564–571, 1976.

CHAPTER 12

THREE-AXIS ATTITUDE DETERMINATION METHODS

12.1 Parameterization of the Attitude
12.2 Three-Axis Attitude Determination
       Geometric Method, Algebraic Method, q Method
12.3 Covariance Analysis

Chapter 11 described deterministic procedures for computing the orientation of a single spacecraft axis and estimating the accuracy of this computation. The methods described there may be used either to determine single-axis attitude or the orientation of any single axis on a three-axis stabilized spacecraft. However, when the three-axis attitude of a spacecraft is being computed, some additional formalism is appropriate. The attitude of a single axis can be parameterized either as a three-component unit vector or as a point on the unit celestial sphere, but three-axis attitude is most conveniently thought of as a coordinate transformation which transforms a set of reference axes in inertial space to a set in the spacecraft. The alternative parameterizations for this transformation are described in Section 12.1. Section 12.2 then describes three-axis attitude determination methods, and Section 12.3 introduces the covariance analysis needed to estimate the uncertainty in three-axis attitude.

12.1 Parameterization of the Attitude

F. L. Markley

Let us consider a rigid body in space, either a rigid spacecraft or a single rigid component of a spacecraft with multiple components moving relative to each other. We assume that there exists an orthogonal, right-handed triad $\hat{\mathbf{u}}$, $\hat{\mathbf{v}}$, $\hat{\mathbf{w}}$ of unit vectors fixed in the body, such that

$$\hat{\mathbf{u}} \times \hat{\mathbf{v}} = \hat{\mathbf{w}} \tag{12-1}$$

The basic problem is to specify the orientation of this triad, and hence of the rigid body, relative to some reference coordinate frame, as illustrated in Fig. 12-1.

It is clear that specifying the components of $\hat{\mathbf{u}}$, $\hat{\mathbf{v}}$, and $\hat{\mathbf{w}}$ along the three axes of the coordinate frame will fix the orientation completely. This requires nine parameters, which can be regarded as the elements of a 3×3 matrix, A, called the *attitude matrix*:

$$A \equiv \begin{bmatrix} u_1 & u_2 & u_3 \\ v_1 & v_2 & v_3 \\ w_1 & w_2 & w_3 \end{bmatrix} \tag{12-2}$$

where $\hat{\mathbf{u}} = (u_1, u_2, u_3)^{\mathrm{T}}$, $\hat{\mathbf{v}} = (v_1, v_2, v_3)^{\mathrm{T}}$, and $\hat{\mathbf{w}} = (w_1, w_2, w_3)^{\mathrm{T}}$. Each of these elements is the cosine of the angle between a body unit vector and a reference axis; u_1, for

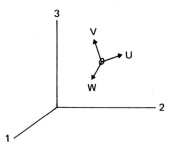

Fig. 12-1. The fundamental problem of three-axis attitude parameterization is to specify the orientation of the spacecraft axes $\hat{u}$, $\hat{v}$, $\hat{w}$ in the reference 1, 2, 3 frame.

example, is the cosine of the angle between $\hat{u}$ and the reference 1 axis. For this reason, A is often referred to as the *direction cosine matrix*. The elements of the direction cosine matrix are not all independent. For example, the fact that $\hat{u}$ is a unit vector requires

$$u_1^2 + u_2^2 + u_3^2 = 1$$

and the orthogonality of $\hat{u}$ and $\hat{v}$ means that

$$u_1 v_1 + u_2 v_2 + u_3 v_3 = 0$$

These relationships can be summarized by the statement that the product of A and its transpose is the identity matrix

$$A A^T = 1 \qquad (12\text{-}3)$$

(See Appendix C for a review of matrix algebra.) This means that A is a *real orthogonal* matrix. Also, the definition of the determinant is equivalent to

$$\det A = \hat{u} \cdot (\hat{v} \times \hat{w})$$

so the fact that $\hat{u}$, $\hat{v}$, $\hat{w}$ form a right-handed triad means that $\det A = 1$. Thus, A is a *proper* real orthogonal matrix.

The direction cosine matrix is a coordinate transformation that maps vectors from the reference frame to the body frame. That is, if $\mathbf{a}$ is a vector with components a_1, a_2, a_3 along the reference axes, then

$$A\mathbf{a} = \begin{bmatrix} u_1 & u_2 & u_3 \\ v_1 & v_2 & v_3 \\ w_1 & w_2 & w_3 \end{bmatrix} \begin{bmatrix} a_1 \\ a_2 \\ a_3 \end{bmatrix} = \begin{bmatrix} \hat{u} \cdot \mathbf{a} \\ \hat{v} \cdot \mathbf{a} \\ \hat{w} \cdot \mathbf{a} \end{bmatrix} \equiv \begin{bmatrix} a_u \\ a_v \\ a_w \end{bmatrix} \qquad (12\text{-}4)$$

The components of $A\mathbf{a}$ are the components of the vector $\mathbf{a}$ along the body triad $\hat{u}$, $\hat{v}$, $\hat{w}$. As shown in Appendix C, a proper real orthogonal matrix transformation preserves the lengths of vectors and the angles between them, and thus represents a *rotation*. The product of two proper real orthogonal matrices $A'' = A'A$ represents the results of successive rotations by A and A', in that order. Because the transpose and inverse of an orthogonal matrix are identical, A^T maps vectors from the body frame to the reference frame.

It is also shown in Appendix C that a proper real orthogonal 3×3 matrix has at least one eigenvector with eigenvalue unity. That is, there exists a unit vector, $\hat{e}$, that is unchanged by A:

$$A\hat{e} = \hat{e} \tag{12-5}$$

The vector $\hat{e}$ has the same components along the body axes and along the reference axes. Thus, $\hat{e}$ is a vector along the axis of rotation. The existence of $\hat{e}$ demonstrates Euler's Theorem: *the most general displacement of a rigid body with one point fixed is a rotation about some axis.*

We regard the direction cosine matrix as the fundamental quantity specifying the orientation of a rigid body. However, other parameterizations, as summarized in Table 12-1 and discussed more fully below, may be more convenient for specific applications. In each case, we will relate the parameters to the elements of the direction cosine matrix. Our treatment follows earlier work by Sabroff, *et al.*, [1965].

Table 12-1.   Alternative Representations of Three-Axis Attitude

PARAMETERIZATION	NOTATION	ADVANTAGES	DISADVANTAGES	COMMON APPLICATIONS
DIRECTION COSINE MATRIX	$A = [A_{ij}]$	NO SINGULARITIES NO TRIGONOMETRIC FUNCTIONS CONVENIENT PRODUCT RULE FOR SUCCESSIVE ROTATIONS	SIX REDUNDANT PARAMETERS	IN ANALYSIS, TO TRANSFORM VECTORS FROM ONE REFERENCE FRAME TO ANOTHER
EULER AXIS/ANGLE	$\hat{e}, \Phi$	CLEAR PHYSICAL INTERPRETATION	ONE REDUNDANT PARAMETER AXIS UNDEFINED WHEN SIN $\Phi = 0$ TRIGONOMETRIC FUNCTIONS	COMMANDING SLEW MANEUVERS
EULER SYMMETRIC PARAMETERS (QUATERNION)	q_1, q_2, q_3, q_4 (q)	NO SINGULARITIES NO TRIGONOMETRIC FUNCTIONS CONVENIENT PRODUCT RULE FOR SUCCESSIVE ROTATIONS	ONE REDUNDANT PARAMETER NO OBVIOUS PHYSICAL INTERPRETATION	ONBOARD INERTIAL NAVIGATION
GIBBS VECTOR	g	NO REDUNDANT PARAMETERS NO TRIGONOMETRIC FUNCTIONS CONVENIENT PRODUCT RULE FOR SUCCESSIVE ROTATIONS	INFINITE FOR 180-DEG ROTATION	ANALYTIC STUDIES

EULER ANGLES	ϕ, θ, ψ	NO REDUNDANT PARAMETERS
		PHYSICAL INTERPRETATION IS
		  CLEAR IN SOME CASES
			TRIGONOMETRIC FUNCTIONS
			SINGULARITY AT SOME θ
			NO CONVENIENT PRODUCT RULE
			  FOR SUCCESSIVE ROTATIONS
				ANALYTIC STUDIES
				INPUT/OUTPUT
				ONBOARD ATTITUDE CONTROL OF
				  3-AXIS STABILIZED SPACECRAFT

Euler Axis/Angle.   A particularly simple rotation is one about the 3 axis by an angle Φ, in the positive sense, as illustrated in Fig. 12-2. The direction cosine matrix for this rotation is denoted by $A_3(\Phi)$; its explicit form is

$$A_3(\Phi) = \begin{bmatrix} \cos\Phi & \sin\Phi & 0 \\ -\sin\Phi & \cos\Phi & 0 \\ 0 & 0 & 1 \end{bmatrix} \tag{12-6a}$$

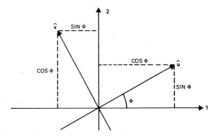

Fig. 12-2.   Rotation About the Three-Axis by the Angle Φ

The direction cosine matrices for rotations by an angle Φ about the 1 or 2 axis, denoted by $A_1(\Phi)$ and $A_2(\Phi)$, respectively, are

$$A_1(\Phi) = \begin{bmatrix} 1 & 0 & 0 \\ 0 & \cos\Phi & \sin\Phi \\ 0 & -\sin\Phi & \cos\Phi \end{bmatrix} \tag{12-6b}$$

$$A_2(\Phi) = \begin{bmatrix} \cos\Phi & 0 & -\sin\Phi \\ 0 & 1 & 0 \\ \sin\Phi & 0 & \cos\Phi \end{bmatrix} \tag{12-6c}$$

The matrices $A_1(\Phi)$, $A_2(\Phi)$, and $A_3(\Phi)$ all have the trace

$$\mathrm{tr}(A(\Phi)) = 1 + 2\cos\Phi \tag{12-6d}$$

The trace of a direction cosine matrix representing a rotation by the angle Φ about an arbitrary axis takes the same value. This result, which will be used without proof below, follows from the observation that the rotation matrices representing rotations by the same angle about different axes can be related by an orthogonal transformation, which leaves the trace invariant (see Appendix C).

In general, the axis of rotation will not be one of the reference axes. In terms of the unit vector along the rotation axis, $\hat{e}$, and angle of rotation, Φ, the most general direction cosine matrix is

$$A = \begin{bmatrix} \cos\Phi + e_1^2(1-\cos\Phi) & e_1 e_2(1-\cos\Phi) + e_3\sin\Phi & e_1 e_3(1-\cos\Phi) - e_2\sin\Phi \\ e_1 e_2(1-\cos\Phi) - e_3\sin\Phi & \cos\Phi + e_2^2(1-\cos\Phi) & e_2 e_3(1-\cos\Phi) + e_1\sin\Phi \\ e_1 e_3(1-\cos\Phi) + e_2\sin\Phi & e_2 e_3(1-\cos\Phi) - e_1\sin\Phi & \cos\Phi + e_3^2(1-\cos\Phi) \end{bmatrix} \tag{12-7a}$$

$$= \cos\Phi \mathbf{1} + (1-\cos\Phi)\hat{e}\hat{e}^T - \sin\Phi E \tag{12-7b}$$

where $\hat{e}\hat{e}^T$ is the outer product (see Appendix C) and E is the skew-symmetric matrix

$$E \equiv \begin{bmatrix} 0 & -e_3 & e_2 \\ e_3 & 0 & -e_1 \\ -e_2 & e_1 & 0 \end{bmatrix} \tag{12-8}$$

This representation of the spacecraft orientation is called the *Euler axis and angle* parameterization. It appears to depend on four parameters, but only three are independent because $|\hat{e}| = 1$. It is a straightforward exercise to show that A defined by Eq. (12-7) is a proper real orthogonal matrix and that $\hat{e}$ is the axis of rotation, that is, $A\hat{e} = \hat{e}$. The rotation angle is known to be Φ because the trace of A satisfies Eq. (12-6d).

It is also easy to see that Eq. (12-7) reduces to the appropriate one of Eqs. (12-6) when $\hat{e}$ lies along one of the reference axes. The Euler rotation angle, Φ, can be expressed in terms of direction cosine matrix elements by

$$\cos\Phi = \frac{1}{2}[\mathrm{tr}(A) - 1] \tag{12-9}$$

If $\sin\Phi \neq 0$, the components of $\hat{\mathbf{e}}$ are given by

$$e_1 = (A_{23} - A_{32})/(2\sin\Phi) \tag{12-10a}$$

$$e_2 = (A_{31} - A_{13})/(2\sin\Phi) \tag{12-10b}$$

$$e_3 = (A_{12} - A_{21})/(2\sin\Phi) \tag{12-10c}$$

Equation (12-9) has two solutions for Φ, which differ only in sign. The two solutions have axis vectors $\hat{\mathbf{e}}$ in opposite directions, according to Eq. (12-10). This expresses the fact that a rotation about $\hat{\mathbf{e}}$ by an angle Φ is equivalent to a rotation about $-\hat{\mathbf{e}}$ by $-\Phi$.

Euler Symmetric Parameters.   A parameterization of the direction cosine matrix in terms of *Euler symmetric parameters* q_1, q_2, q_3, q_4 has proved to be quite useful in spacecraft work. These parameters are not found in many modern dynamics textbooks, although Whittaker [1937] does introduce them and they are discussed by Sabroff, *et al.*, [1965]. They are defined by

$$q_1 \equiv e_1 \sin\frac{\Phi}{2} \tag{12-11a}$$

$$q_2 \equiv e_2 \sin\frac{\Phi}{2} \tag{12-11b}$$

$$q_3 \equiv e_3 \sin\frac{\Phi}{2} \tag{12-11c}$$

$$q_4 \equiv \cos\frac{\Phi}{2} \tag{12-11d}$$

The four Euler symmetric parameters are not independent, but satisfy the constraint equation

$$q_1^2 + q_2^2 + q_3^2 + q_4^2 = 1 \tag{12-12a}$$

These four parameters can be regarded as the components of a quaternion,

$$\boldsymbol{q} \equiv \begin{bmatrix} q_1 \\ q_2 \\ q_3 \\ q_4 \end{bmatrix} \equiv \begin{bmatrix} \mathbf{q} \\ q_4 \end{bmatrix} \tag{12-12b}$$

Quaternions are discussed in more detail in Appendix D. The Euler symmetric parameters are also closely related to the *Cayley-Klein parameters* [Goldstein, 1950].

The direction cosine matrix can be expressed in terms of the Euler symmetric parameters by

$$A(\boldsymbol{q}) = \begin{bmatrix} q_1^2 - q_2^2 - q_3^2 + q_4^2 & 2(q_1q_2 + q_3q_4) & 2(q_1q_3 - q_2q_4) \\ 2(q_1q_2 - q_3q_4) & -q_1^2 + q_2^2 - q_3^2 + q_4^2 & 2(q_2q_3 + q_1q_4) \\ 2(q_1q_3 + q_2q_4) & 2(q_2q_3 - q_1q_4) & -q_1^2 - q_2^2 + q_3^2 + q_4^2 \end{bmatrix} \tag{12-13a}$$

$$= (q_4^2 - \mathbf{q}^2)\mathbf{1} + 2\mathbf{q}\mathbf{q}^{\mathrm{T}} - 2q_4 Q \tag{12-13b}$$

where Q is the skew-symmetric matrix

$$Q \equiv \begin{bmatrix} 0 & -q_3 & q_2 \\ q_3 & 0 & -q_1 \\ -q_2 & q_1 & 0 \end{bmatrix} \qquad \text{(12-13c)}$$

These equations can be verified by substituting Eqs. (12-11) into them, using some trigonometric identities, and comparing them with Eq. (12-7).

The Euler symmetric parameters corresponding to a given direction cosine matrix, A, can be found from

$$q_4 = \pm \frac{1}{2}(1 + A_{11} + A_{22} + A_{33})^{1/2} \qquad \text{(12-14a)}$$

$$q_1 = \frac{1}{4q_4}(A_{23} - A_{32}) \qquad \text{(12-14b)}$$

$$q_2 = \frac{1}{4q_4}(A_{31} - A_{13}) \qquad \text{(12-14c)}$$

$$q_3 = \frac{1}{4q_4}(A_{12} - A_{21}) \qquad \text{(12-14d)}$$

Note that there is a sign ambiguity in the calculation of these parameters. Inspection of Eq. (12-13) shows that changing the signs of all the Euler symmetric parameters simultaneously does not affect the direction cosine matrix. Equations (12-14) express one of four possible ways of computing the Euler symmetric parameters. We could also compute

$$q_1 = \pm \frac{1}{2}(1 + A_{11} - A_{22} - A_{33})^{1/2}$$

$$q_2 = \frac{1}{4q_1}(A_{12} + A_{21})$$

and so forth. All methods are mathematically equivalent, but numerical inaccuracy can be minimized by avoiding calculations in which the Euler symmetric parameter appearing in the denominator is close to zero. Other algorithms for computing Euler symmetric parameters from the direction cosine matrix are given by Klumpp [1976].

Euler symmetric parameters provide a very convenient parameterization of the attitude. They are more compact than the direction cosine matrix, because only four parameters, rather than nine, are needed. They are more convenient than the Euler axis and angle parameterization (and the Euler angle parameterizations to be considered below) because the expression for the direction cosine matrix in terms of Euler symmetric parameters does not involve trigonometric functions, which require time-consuming computer operations. Another advantage of Euler symmetric parameters is the relatively simple form for combining the parameters for two individual rotations to give the parameters for the product of the two rotations. Thus, if

$$A(q'') = A(q')A(q) \qquad \text{(12-15a)}$$

then

$$q'' = \begin{bmatrix} q_4' & q_3' & -q_2' & q_1' \\ -q_3' & q_4' & q_1' & q_2' \\ q_2' & -q_1' & q_4' & q_3' \\ -q_1' & -q_2' & -q_3' & q_4' \end{bmatrix} q \qquad (12\text{-}15b)$$

Equation (12-15b) can be verified by direct substitution of Eq. (12-13) into Eq. (12-15a), but the algebra is exceedingly tedious. The relationship of Eq. (12-15b) to the quaternion product is given in Appendix D. Note that the evaluation of Eq. (12-15b) involves 16 multiplications and the computation of Eq. (12-15a) requires 27; this is another advantage of Euler symmetric parameters.

Gibbs Vector. The direction cosine matrix can also be parameterized by the *Gibbs vector,** which is defined by

$$g_1 \equiv q_1/q_4 = e_1 \tan\frac{\Phi}{2} \qquad (12\text{-}16a)$$

$$g_2 \equiv q_2/q_4 = e_2 \tan\frac{\Phi}{2} \qquad (12\text{-}16b)$$

$$g_3 \equiv q_3/q_4 = e_3 \tan\frac{\Phi}{2} \qquad (12\text{-}16c)$$

The direction cosine matrix is given in terms of the Gibbs vector by

$$A = \frac{1}{1+g_1^2+g_2^2+g_3^2} \begin{bmatrix} 1+g_1^2-g_2^2-g_3^2 & 2(g_1 g_2+g_3) & 2(g_1 g_3-g_2) \\ 2(g_1 g_2-g_3) & 1-g_1^2+g_2^2-g_3^2 & 2(g_2 g_3+g_1) \\ 2(g_1 g_3+g_2) & 2(g_2 g_3-g_1) & 1-g_1^2-g_2^2+g_3^2 \end{bmatrix} \qquad (12\text{-}17a)$$

$$= \frac{(1-g^2)\mathbf{1}+2gg^T-G}{1+g^2} \qquad (12\text{-}17b)$$

where G is the skew-symmetric matrix

$$G \equiv \begin{bmatrix} 0 & -g_3 & g_2 \\ g_3 & 0 & -g_1 \\ -g_2 & g_1 & 0 \end{bmatrix} \qquad (12\text{-}17c)$$

Expressions for the Gibbs vector components in terms of the direction cosine matrix elements can be found by using Eqs. (12-16) and (12-14). Thus,

$$g_1 = \frac{A_{23}-A_{32}}{1+A_{11}+A_{22}+A_{33}} \qquad (12\text{-}18a)$$

*Gibbs [1901, p. 340] named this vector the "vector semitangent of version." Cayley [1899] used the three quantities g_1, g_2, g_3 in 1843 (before the introduction of vector notation), and he credits their discovery to Rodriguez.

$$g_2 = \frac{A_{31} - A_{13}}{1 + A_{11} + A_{22} + A_{33}} \qquad (12\text{-}18b)$$

$$g_3 = \frac{A_{12} - A_{21}}{1 + A_{11} + A_{22} + A_{33}} \qquad (12\text{-}18c)$$

Note that there is no sign ambiguity in the definition of the Gibbs vector and that the components are independent parameters. The product law for Gibbs vectors analogous to Eq. (12-15b) can be found from that equation and Eq. (12-16), and takes the convenient vector form

$$\mathbf{g}'' = \frac{\mathbf{g} + \mathbf{g}' - \mathbf{g}' \times \mathbf{g}}{1 - \mathbf{g} \cdot \mathbf{g}'} \qquad (12\text{-}19)$$

The Gibbs vector has not been widely used because it becomes infinite when the rotation angle is an odd multiple of 180 deg.

Euler Angles. It is clear from the above discussion that three independent parameters are needed to specify the orientation of a rigid body in space. The only parameterization considered so far that has the minimum number of parameters is the Gibbs vector. We now turn to a class of parameterizations in terms of three rotation angles, commonly known as Euler angles. These are not as convenient for numerical computations as the Euler symmetric parameters, but their geometrical significance is more apparent (particularly for small rotations) and they are often used for computer input/output. They are also useful for analysis, especially for finding closed-form solutions to the equations of motion in simple cases. Euler angles are also commonly employed for three-axis stabilized spacecraft for which small angle approximations can be used.

To define the Euler angles precisely, consider four orthogonal triads of unit vectors, which we shall denote by

$$\hat{\mathbf{x}}, \hat{\mathbf{y}}, \hat{\mathbf{z}}$$

$$\hat{\mathbf{x}}', \hat{\mathbf{y}}', \hat{\mathbf{z}}'$$

$$\hat{\mathbf{x}}'', \hat{\mathbf{y}}'', \hat{\mathbf{z}}''$$

$$\hat{\mathbf{u}}, \hat{\mathbf{v}}, \hat{\mathbf{w}}$$

The initial triad $\hat{\mathbf{x}}, \hat{\mathbf{y}}, \hat{\mathbf{z}}$ is parallel to the reference 1,2,3 axes. The triad $\hat{\mathbf{x}}', \hat{\mathbf{y}}', \hat{\mathbf{z}}'$ differs from $\hat{\mathbf{x}}, \hat{\mathbf{y}}, \hat{\mathbf{z}}$ by a rotation about the i axis ($i = 1$, 2, or 3 depending on the particular transformation) through an angle ϕ.* Thus, the orientation of the $\hat{\mathbf{x}}', \hat{\mathbf{y}}', \hat{\mathbf{z}}'$ triad relative to the $\hat{\mathbf{x}}, \hat{\mathbf{y}}, \hat{\mathbf{z}}$ triad is given by $A_i(\phi)$ for $i = 1$, 2, or 3, one of the simple direction cosine matrices given by Eq. (12-6). Similarly, the $\hat{\mathbf{x}}'', \hat{\mathbf{y}}'', \hat{\mathbf{z}}''$ triad orientation relative to the $\hat{\mathbf{x}}', \hat{\mathbf{y}}', \hat{\mathbf{z}}'$ triad is a rotation about a coordinate axis in the $\hat{\mathbf{x}}', \hat{\mathbf{y}}', \hat{\mathbf{z}}'$ system by an angle θ, specified by $A_j(\theta), j = 1$, 2, or 3, $j \neq i$. Finally, the orientation of $\hat{\mathbf{u}}, \hat{\mathbf{v}}, \hat{\mathbf{w}}$ relative to $\hat{\mathbf{x}}'', \hat{\mathbf{y}}'', \hat{\mathbf{z}}''$ is a third rotation, by an angle ψ, with the direction cosine matrix $A_k(\psi), k = 1$, 2, or 3, $k \neq j$. The final $\hat{\mathbf{u}}, \hat{\mathbf{v}}, \hat{\mathbf{w}}$ triad is the body-fixed triad considered previously, so the overall sequence of three rotations specifies the orientation of the body relative to the reference coordinate axes.

---

*Although Euler angles are rotation angles, we follow the usual convention of denoting them by lower-case Greek letters.

A specific example of Euler angle rotations is shown in Fig. 12-3. Here, the first rotation is through an angle ϕ about the $\hat{\mathbf{z}}$ axis, so that the $\hat{\mathbf{z}}$ and $\hat{\mathbf{z}}'$ axes coincide. The second rotation is by θ about the $\hat{\mathbf{x}}'$ axis, which thus is identical with $\hat{\mathbf{x}}''$. The third rotation is by ψ about the $\hat{\mathbf{z}}''$ (or $\hat{\mathbf{w}}$) axis. This sequence of rotations is called a 3-1-3 sequence, because the rotations are about the 3, 1, and 3 axes, in that order. The labeled points in the figure are the locations of the ends of the unit vectors on the unit sphere. The circles containing the numbers 1, 2, and 3 are the first, second, and third rotation axes, respectively. The solid lines are the great circles containing the unit vectors of the reference coordinate system, $\hat{\mathbf{x}}, \hat{\mathbf{y}}, \hat{\mathbf{z}}$. The cross-hatched lines are the great circles containing the unit vectors of the body coordinate system, $\hat{\mathbf{u}}, \hat{\mathbf{v}}, \hat{\mathbf{w}}$. The dotted and dashed lines are the great circles defined by intermediate coordinate systems.

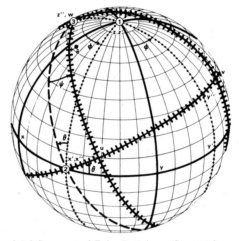

Fig. 12-3.   3-1-3 Sequence of Euler Rotations. (See text for explanation.)

The direction cosine matrix for the overall rotation sequence is the matrix product of the three matrices for the individual rotations, with the first rotation matrix on the right and the last on the left:

$$A_{313}(\phi,\theta,\psi) = A_3(\psi)A_1(\theta)A_3(\phi) =$$

$$\begin{bmatrix} \cos\psi\cos\phi - \cos\theta\sin\psi\sin\phi & \cos\psi\sin\phi + \cos\theta\sin\psi\cos\phi & \sin\theta\sin\psi \\ -\sin\psi\cos\phi - \cos\theta\cos\psi\sin\phi & -\sin\psi\sin\phi + \cos\theta\cos\psi\cos\phi & \sin\theta\cos\psi \\ \sin\theta\sin\phi & -\sin\theta\cos\phi & \cos\theta \end{bmatrix} \quad (12\text{-}20)$$

The Euler axis corresponding to $A_{313}(\phi,\theta,\psi)$ can be found from Eq. (12-10); it is denoted by $\hat{\mathbf{e}}$ in Fig. 12-3.

The 3-1-3 Euler angles can be obtained from the direction cosine matrix elements by

$$\theta = \arccos A_{33} \quad (12\text{-}21a)$$

$$\phi = -\arctan(A_{31}/A_{32}) \quad (12\text{-}21b)$$

$$\psi = \arctan(A_{13}/A_{23}) \quad (12\text{-}21c)$$

Note that Eq. (12-21a) leaves a twofold ambiguity in θ, corresponding to $\sin\theta$ being positive or negative. Once this ambiguity is resolved, ϕ and ψ are determined uniquely (modulo 360 deg) by the signs and magnitudes of A_{13}, A_{23}, A_{31}, and A_{32}, with the exception that when θ is a multiple of 180 deg, only the sum or difference of ϕ and ψ is determined, depending on whether θ is an even or an odd multiple of 180 deg. The origin of this ambiguity is apparent in Fig. 12-3. The usual resolution of this ambiguity is to choose $\sin\theta \geqslant 0$, or $0 \leqslant \theta < 180$ deg.

Other sequences of Euler angle rotations are possible, and several are used. Figure 12-4 illustrates a 3-1-2 sequence: a rotation by ϕ about $\hat{z}$ followed by a rotation by θ about $\hat{x}'$ and then by a rotation by ψ about $\hat{y}''$. This is often referred to as the yaw, roll, pitch sequence, but the meaning of these terms and the order of rotations implied is not standard. The direction cosine matrix illustrated in Fig. 12-4 is

$$A_{312}(\phi,\theta,\psi) = A_2(\psi)A_1(\theta)A_3(\phi) =$$

$$\begin{bmatrix} \cos\psi\cos\phi - \sin\theta\sin\psi\sin\phi & \cos\psi\sin\phi + \sin\theta\sin\psi\cos\phi & -\cos\theta\sin\psi \\ -\cos\theta\sin\phi & \cos\theta\cos\phi & \sin\theta \\ \sin\psi\cos\phi + \sin\theta\cos\psi\sin\phi & \sin\psi\sin\phi - \sin\theta\cos\psi\cos\phi & \cos\theta\cos\psi \end{bmatrix} \quad (12\text{-}22)$$

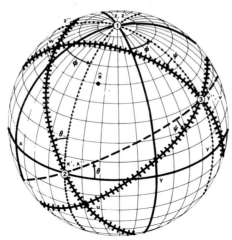

Fig. 12-4.   3-1-2 Sequence of Euler Rotations. (See text for explanation.)

The expressions for the rotation angles in terms of the elements of the direction cosine matrix are

$$\theta = \arcsin A_{23} \quad (12\text{-}23a)$$

$$\phi = -\arctan(A_{21}/A_{22}) \quad (12\text{-}23b)$$

$$\psi = -\arctan(A_{13}/A_{33}) \quad (12\text{-}23c)$$

As in the 3-1-3 case, the angles are determined up to a twofold ambiguity except at certain values of the intermediate angle θ. In this case, the singular values of θ are odd multiples of 90 deg. The usual resolution of the ambiguity is to choose -90 deg $< \theta \leqslant 90$ deg, which gives $\cos\theta \geqslant 0$.

If ϕ, θ, and ψ are all small angles, we can use small-angle approximations to the trigonometric functions, and Eq. (12-22) reduces to

$$A_{312}(\phi,\theta,\psi) \simeq \begin{bmatrix} 1 & \phi & -\psi \\ -\phi & 1 & \theta \\ \psi & -\theta & 1 \end{bmatrix} \qquad (12\text{-}24a)$$

where the angles are measured in radians.

It is not difficult to enumerate all the possible sequences of Euler rotations. We cannot allow two successive rotations about a single axis, because the product of these rotations is equivalent to a single rotation about this axis. Thus, there are only 12 possible axis sequences:

$$313, 212, 121, 323, 232, 131,$$
$$312, 213, 123, 321, 231, 132.$$

Because of the twofold ambiguity in the angle θ mentioned above, there are 24 possible sequences of rotations, counting rotations through different angles as different rotations and ignoring rotations by multiples of 360 deg. The axis sequences divide naturally into two classes, depending on whether the third axis index is the same as or different from the first. Equation (12-20) is an example of the first class, and Eq. (12-22) is an example of the second. It is straightforward, using the techniques of this section, to write down the transformation equations for a given rotation sequence; these equations are collected in Appendix E. In the small-angle approximation, the 123, 132, 213, 231, 312, and 321 rotation sequences all have direction cosine matrices given by Eq. (12-24a) with the proviso that ϕ, θ, and ψ are the rotation angles about the 3, 1, 2 axes, respectively. Comparison with Eq. (12-13) shows that in the small-angle approximation, the Euler symmetric parameters are related to the Euler angles by

$$q_1 \simeq \frac{1}{2}\theta \qquad (12\text{-}24b)$$

$$q_2 \simeq \frac{1}{2}\psi \qquad (12\text{-}24c)$$

$$q_3 \simeq \frac{1}{2}\phi \qquad (12\text{-}24d)$$

$$q_4 \simeq 1 \qquad (12\text{-}24e)$$

12.2 Three-Axis Attitude Determination

Gerald M. Lerner

Three-axis attitude determination, which is equivalent to the complete specification of the attitude matrix, A, is accomplished either by an extension of the geometric techniques described in Chapter 11 or by a direct application of the concept of attitude as a rotation matrix. If the spacecraft has a preferred axis, such as the angular momentum vector of a spinning spacecraft or the boresight of a payload sensor, it is usually convenient to specify three-axis attitude in terms of the attitude of the preferred axis plus a phase angle about that axis. This asymmetric treatment of the attitude angles is usually justified by the attitude sensor configuration and the attitude accuracy requirements, which are generally more severe for the preferred axis. We refer to this method as *geometric* three-axis attitude determination because the phase angle is computed most conveniently using spherical trigonometry. Alternatively, in the *algebraic* method, the attitude matrix is determined directly from two vector observations without resorting to any angular representation. Finally, the *q method* provides a means for computing an optimal three-axis attitude from many vector observations. In this section we describe these methods for the computation of three-axis attitude.

12.2.1 Geometric Method

The geometric method is normally used when there is a body axis—such as the spin axis of a momentum wheel, a wheel-mounted sensor, or the spacecraft itself, about which there is preferential attitude data. Either deterministic techniques, as described in Chapter 11, or differential correction techniques, as will be described in Chapter 13, may be used to compute the attitude of the preferred axis. The phase angle about the preferred axis is then computed from any measurement which provides an angle about that axis.

In many cases, the geometric method is required because the sensor measurements themselves (e.g., spinning Sun sensors or horizon scanners) define a preferred spacecraft axis and provide only poor azimuthal information about that axis.

Figure 12-5 illustrates the geometric method. The reference axes are the celestial coordinates axes, $\hat{\mathbf{X}}_I$, $\hat{\mathbf{Y}}_I$, and $\hat{\mathbf{Z}}_I$. We wish to compute the 3-1-3 Euler angles, ϕ, θ, and ψ, which define the transformation from the celestial to the body

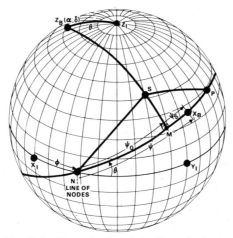

Fig. 12-5. Determination of the Phase Angle, ψ

coordinates, $\hat{\mathbf{X}}_B$, $\hat{\mathbf{Y}}_B$, and $\hat{\mathbf{Z}}_B$. The Euler angles ϕ and θ are related to the attitude (α, δ) of the preferred body axis, $\hat{\mathbf{Z}}_B$, by

$$\phi = 90° + \alpha \qquad (12\text{-}25\text{a})$$

$$\theta = 90° - \delta \qquad (12\text{-}25\text{b})$$

where the right ascension, α, and the declination, δ, are obtained by using any one-axis attitude determination method. ϕ defines the orientation of the node, $\hat{\mathbf{N}}$. The phase angle, ψ, is computed from the azimuth, ψ_S, of the projection of a measured vector, $\hat{\mathbf{S}}$ (e.g., the Sun or magnetic field) on the plane normal to $\hat{\mathbf{Z}}_B$. Let $\hat{\mathbf{M}}$ be the projection of $\hat{\mathbf{S}}$ on the plane normal to $\hat{\mathbf{Z}}_B$ and $\hat{\mathbf{P}} = \hat{\mathbf{Z}}_B \times \hat{\mathbf{N}}$. Application of Napier's rules (Appendix A) to the right spherical triangles SMN and SMP yields

$$\hat{\mathbf{N}} \cdot \hat{\mathbf{S}} = \hat{\mathbf{M}} \cdot \hat{\mathbf{S}} \cos \psi_0 \qquad (12\text{-}26\text{a})$$

$$\hat{\mathbf{P}} \cdot \hat{\mathbf{S}} = \hat{\mathbf{M}} \cdot \hat{\mathbf{S}} \cos(90° - \psi_0) = \hat{\mathbf{M}} \cdot \hat{\mathbf{S}} \sin \psi_0 \qquad (12\text{-}26\text{b})$$

which may be rewritten as*

$$\tan \psi_0 = \hat{\mathbf{P}} \cdot \hat{\mathbf{S}} / \hat{\mathbf{N}} \cdot \hat{\mathbf{S}} \qquad (12\text{-}27)$$

where

$$\hat{\mathbf{N}} = (\cos \phi, \sin \phi, 0)^T \qquad (12\text{-}28\text{a})$$

$$\hat{\mathbf{P}} = (-\cos \theta \sin \phi, \cos \theta \cos \phi, \sin \theta)^T \qquad (12\text{-}28\text{b})$$

The phase angle, ψ, is then given by

$$\psi = \psi_0 + \psi_S \qquad (12\text{-}29)$$

As a more complex example of the geometric technique, we consider the three-axis attitude determination for the CTS spacecraft during attitude acquisition as illustrated in Fig. 12-6. The spacecraft Z axis is along the sunline and the spacecraft Y axis (the spin axis of a momentum wheel) is fixed in inertial space on a great circle 90 deg from the Sun. An infrared Earth horizon sensor has its boresight along the spacecraft Z axis and measures both the rotation angle, Ω_E, from the Sun to the nadir about the spacecraft Y axis and the nadir angle, η, from the spacecraft Y axis to the Earth's center. We wish to compute the rotation angle, Φ_S, about the sunline required to place the spacecraft Y axis into the celestial X-Y plane as a function of the following angles: the Sun declination in celestial coordinates, δ_S; the *clock angle*† or difference between the Earth and Sun azimuth in celestial coordinates, $\Delta\alpha = \alpha_E - \alpha_S$; and either measurement Ω_E or η. As shown in Fig. 12-6, Φ_S is 180 deg minus the sum of three angles:

$$\Phi_S = 180° - (\angle YSR + \Lambda + \Phi_E) \qquad (12\text{-}30)$$

---

*Note that $\hat{\mathbf{M}} \cdot \hat{\mathbf{S}} \geqslant 0$ by the definition of $\hat{\mathbf{M}}$. If $\hat{\mathbf{M}} \cdot \hat{\mathbf{S}} = 0$, ψ_0 is indeterminate because $\hat{\mathbf{S}}$ provides no phase information about $\hat{\mathbf{Z}}_B$. If $\hat{\mathbf{M}} \cdot \hat{\mathbf{S}} > 0$, ψ_0 is obtained unambiguously because the quadrants of both $\sin \psi_0$ and $\cos \psi_0$ are known.

†For the synchronous CTS orbit, the azimuthal difference or clock angle is zero at local midnight and decreases by 15 deg/hour.

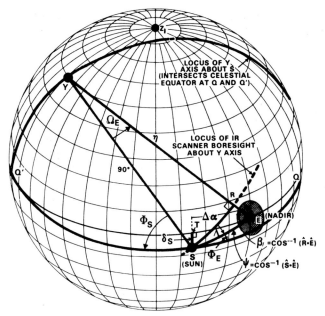

Fig. 12-6.   Attitude Determination Geometry for CTS

Applying Napier's rules to the right spherical triangles *ETS*, *ERS*, and *QTS*, the arc length *SE* is

$$\psi \equiv \arc\cos(\mathbf{\hat{S}} \cdot \mathbf{\hat{E}}) = \arc\cos(\cos\delta_S \cos\Delta\alpha) \tag{12-31}$$

and the rotation angle, Λ, is

$$\Lambda \equiv \angle\,RSE = \arc\sin(\sin\beta_E/\sin\psi) \tag{12-32}$$

where $\beta_E = \eta - 90°$ and the arc length, *TQ*, is 90 deg. Next, the quadrantal spherical triangle, *QES*, is solved for the angle *ESQ*:

$$\Phi_E \equiv \angle\,ESQ = \arc\cos(\cos(90° - \Delta\alpha)/\sin\psi) \tag{12-33}$$

Combining Eqs. (12-30), (12-32), and (12-33) with $\angle\,YSR = 90$ deg gives the result

$$\Phi_S = 90° - \arc\sin(\sin\beta_E/\sin\psi) - \arc\cos(\sin\Delta\alpha/\sin\psi) \tag{12-34}$$

or

$$\Phi_S = \arc\cos(\sin\beta_E/\sin\psi) - \arc\cos(\sin\Delta\alpha/\sin\psi) \tag{12-35}$$

where

$$\sin\psi = \left(\cos^2\Delta\alpha\,\sin^2\delta_S + \sin^2\Delta\alpha\right)^{1/2} \tag{12-36}$$

Finally, Ω_E and β_E are related through the quadrantal spherical triangle, *YSE*, by

$$\beta_E = \arc\cos(\cos\delta_S \cos\Delta\alpha/\cos\Omega_E) \tag{12-37}$$

One problem with the geometric method is apparent from the proliferation of inverse trigonometric functions in Eqs. (12-31) to (12-37), which results in quadrant and consequent attitude ambiguities. Ambiguity is a frequent problem when dealing with inverse trigonometric functions and must be carefully considered in mission analysis. Although from Fig. 12-6, Φ_S and all the rotation angles in Eq. (12-30) are in the first quadrant by inspection, the generalization of Eq. (12-35) for arbitrary angles is not apparent. From the form of Eq. (12-35), it would appear that there is a fourfold ambiguity in Φ_S; however, some of these ambiguities may be resolved by applying the rules for quadrant specification given in Appendix A. There is, however, a true ambiguity in the sign of Λ which may be seen by redrawing Fig. 12-6 for $\Phi_S \approx -70$ deg and noting that, in this case,

$$\Phi_S = 180° - (\angle\, YSR - \Lambda + \Phi_E) \tag{12-38}$$

The ambiguity between Eqs. (12-30) and (12-38) is real if only pitch or roll measurements are available and must be differentiated from apparent ambiguities which may be resolved by proper use of the spherical triangle relations. However, if both Ω_E and η measurements are available, the ambiguity may be resolved by the sign of Ω_E because Ω_E is positive for Eq. (12-30) and negative for Eq. (12-38).

12.2.2 Algebraic Method

The algebraic method is based on the rotation matrix representation of the attitude. Any two vectors, $\mathbf{u}$ and $\mathbf{v}$, define an orthogonal coordinate system with the basis vectors, $\hat{\mathbf{q}}$, $\hat{\mathbf{r}}$, and $\hat{\mathbf{s}}$ given by

$$\hat{\mathbf{q}} = \hat{\mathbf{u}} \tag{12-39a}$$

$$\hat{\mathbf{r}} = \hat{\mathbf{u}} \times \hat{\mathbf{v}} / |\hat{\mathbf{u}} \times \hat{\mathbf{v}}| \tag{12-39b}$$

$$\hat{\mathbf{s}} = \hat{\mathbf{q}} \times \hat{\mathbf{r}} \tag{12-39c}$$

provided that $\hat{\mathbf{u}}$ and $\hat{\mathbf{v}}$ are not parallel, i.e.,

$$|\hat{\mathbf{u}} \cdot \hat{\mathbf{v}}| < 1 \tag{12-40}$$

At a given time, two measured vectors in the spacecraft body coordinates (denoted by the subscript B) $\hat{\mathbf{u}}_B$ and $\hat{\mathbf{v}}_B$, determine the body matrix, M_B:

$$M_B = \left[\hat{\mathbf{q}}_B \,\vdots\, \hat{\mathbf{r}}_B \,\vdots\, \hat{\mathbf{s}}_B\right] \tag{12-41}$$

For example, the measured vectors may be the Sun position from two-axis Sun sensor data, an identified star position from a star tracker, the nadir vector from an infrared horizon scanner, or the Earth's magnetic field vector from a magnetometer. These vectors may also be obtained in an appropriate reference frame (denoted by the subscript R) from an ephemeris, a star catalog, and a magnetic field model. The reference matrix, M_R, is constructed from $\hat{\mathbf{u}}_R$ and $\hat{\mathbf{v}}_R$ by

$$M_R = \left[\hat{\mathbf{q}}_R \,\vdots\, \hat{\mathbf{r}}_R \,\vdots\, \hat{\mathbf{s}}_R\right] \tag{12-42}$$

As defined in Section 12.1, the *attitude matrix*, or *direction cosine matrix*, A, is given by the coordinate transformation,

$$AM_R = M_B \tag{12-43}$$

because it carries the column vectors of M_R into the column vectors of M_B. This equation may be solved for A to give

$$A = M_B M_R^{-1} \qquad (12\text{-}44)$$

Because M_R is orthogonal, $M_R^{-1} = M_R^T$ and, hence (see Appendix C),

$$A = M_B M_R^T \qquad (12\text{-}45)$$

Nothing in the development thus far has limited the choice of the reference frame or the form of the attitude matrix. The only requirement is that M_R possess an inverse, which follows because the vectors $\hat{\mathbf{q}}$, $\hat{\mathbf{r}}$, and $\hat{\mathbf{s}}$ are linearly independent provided that Eq. (12-40) holds. The simplicity of Eq. (12-45) makes it particularly attractive for onboard processing. Note that inverse trigonometric functions are not required; a unique, unambiguous attitude is obtained; and computational requirements are minimal.

The preferential treatment of the vector $\hat{\mathbf{u}}$ over $\hat{\mathbf{v}}$ in Eq. (12-39) suggests that $\hat{\mathbf{u}}$ should be the more accurate measurement;* this ensures that the attitude matrix transforms $\hat{\mathbf{u}}$ from the reference frame to the body frame exactly and $\hat{\mathbf{v}}$ is used only to determine the phase angle about $\hat{\mathbf{u}}$. The four measured angles that are required to specify the two basis vectors are used to compute the attitude matrix which is parameterized by only three independent angles. Thus, some information is implicitly discarded by the algebraic method. The discarded quantity is the measured component of $\hat{\mathbf{v}}$ parallel to $\hat{\mathbf{u}}$, i.e., $\hat{\mathbf{u}}_B \cdot \hat{\mathbf{v}}_B$. This measurement is coordinate independent, equals the known scalar $\hat{\mathbf{u}}_R \cdot \hat{\mathbf{v}}_R$, and is therefore useful for data validation as described in Section 9.3. All of the error in $\hat{\mathbf{u}}_B \cdot \hat{\mathbf{v}}_B$ is assigned to the less accurate measurement $\hat{\mathbf{v}}_B$, which accounts for the lost information.

Three reference coordinate systems are commonly used: celestial, ecliptic, and orbital (see Section 3.2). The *celestial reference system*, M_C, is particularly convenient because it is obtained directly from standard ephemeris and magnetic field model subroutines such as EPHEMX and MAGFLD in Section 20.3. An *ecliptic reference system*, M_E, defined by the Earth-to-Sun vector, $\hat{\mathbf{S}}$, and the ecliptic north pole, $\hat{\mathbf{P}}_E$, is obtained by the transformation

$$M_E = \left[\hat{\mathbf{S}} \vdots \hat{\mathbf{P}}_E \times \hat{\mathbf{S}} \vdots \hat{\mathbf{P}}_E \right]^T M_C \qquad (12\text{-}46)$$

where $\hat{\mathbf{S}}$ and $\hat{\mathbf{P}}_E$ are in celestial coordinates,

$$\hat{\mathbf{P}}_E \approx (0, -\sin\epsilon, \cos\epsilon)^T \qquad (12\text{-}47)$$

and $\epsilon \approx 23.44$ deg is the obliquity of the ecliptic.

An *orbital reference system*, M_O, is defined by the nadir vector, $\hat{\mathbf{E}}$, and the negative orbit normal, $-\hat{\mathbf{n}}$, in celestial coordinates,

$$M_O = \left[-\hat{\mathbf{n}} \times \hat{\mathbf{E}} \vdots -\hat{\mathbf{n}} \vdots \hat{\mathbf{E}} \right]^T M_C \qquad (12\text{-}48)$$

---

*If both measurements are of comparable accuracy, basis vectors constructed from $\hat{\mathbf{u}} + \hat{\mathbf{v}}$ and $\hat{\mathbf{u}} - \hat{\mathbf{v}}$ would provide the advantage of symmetry.

Any convenient representation may be used to parameterize the attitude matrix. Quaternions and various Euler angle sequences are commonly used as described in the previous section.

The construction of vector measurements from sensor data is generally straightforward, particularly for magnetometers (Section 7.5), Sun sensors (Section 7.1), and star sensors (Section 7.6). For Earth-oriented spacecraft using horizon scanners, the nadir vector may be derived from the measured quantities by reference to the orbital coordinate system defined in Fig. 12-7. The Z_O axis is along the nadir vector and the Y_O axis is along the negative orbit normal. The scanner measures both (1) the pitch angle, Ω_E, about the scanner axis (the spacecraft Y axis, $\hat{\mathbf{Y}}_B$) from the spacecraft Z axis, $\mathbf{Z}_B$, to the $Y_B Z_O$ plane, and (2) β_E, the angle from the scanner axis to the nadir minus 90 deg.*

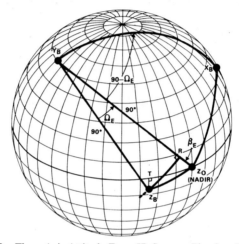

Fig. 12-7.   Three-Axis Attitude From IR Scanner Plus Sun Sensor Data

Solving the quadrantal spherical triangles, $X_B Y_B Z_O$ and $Y_B Z_B Z_O$, gives

$$\hat{\mathbf{Z}}_B \cdot \hat{\mathbf{Z}}_O = \cos\Omega_E \cos\beta_E \qquad (12\text{-}49)$$

$$\hat{\mathbf{X}}_B \cdot \hat{\mathbf{Z}}_O = \sin\Omega_E \cos\beta_E \qquad (12\text{-}50)$$

Hence, the nadir vector in body coordinates is

$$\hat{\mathbf{E}}_B = (\sin\Omega_E \cos\beta_E, \ -\sin\beta_E, \cos\Omega_E \cos\beta_E)^{\mathrm{T}} \qquad (12\text{-}51)$$

12.2.3 q Method

A major disadvantage of the attitude determination methods described thus far is that they are basically ad hoc. That is, the measurements are combined to provide an attitude estimate but the combination is not optimal in any statistical

---

*The angles Ω_E and β_E are analogous to pitch and roll, respectively, as they are defined in Chapter 2. Because standard definitions of pitch, roll, and yaw do not exist, the sign of the quantities here may differ from that used on some spacecraft. (See Section 2.2.)

sense. Furthermore, the methods are not easily applied to star trackers or combinations of sensors which provide many simultaneous vector measurements. Given a set of $n \geqslant 2$ vector measurements, $\hat{\mathbf{u}}_B^i$, in the body system, one choice for an optimal attitude matrix, A, is that which minimizes the loss function

$$J(A) = \sum_{i=1}^{n} w_i |\hat{\mathbf{u}}_B^i - A\hat{\mathbf{u}}_R^i|^2 \tag{12-52}$$

where w_i is the weight of the ith vector measurement and $\hat{\mathbf{u}}_R^i$ is the vector in the reference coordinate system. The loss function is the weighted sum squared of the difference between the measured and transformed vectors.

The attitude matrix may be computed by an elegant algorithm derived by Davenport [1968] and based in part on earlier work by Wahba [1965] and Stuelpnagel [1966]. This algorithm was used for the HEAO-1 attitude determination system [Keat, 1977].

The loss function may be rewritten as

$$J(A) = -2 \sum_{i=1}^{n} \mathbf{W}_i A \mathbf{V}_i + \text{constant terms} \tag{12-53}$$

where the unnormalized vectors $\mathbf{W}_i$ and $\mathbf{V}_i$ are defined as

$$\mathbf{W}_i = \sqrt{w_i}\, \hat{\mathbf{u}}_B^i; \qquad \mathbf{V}_i = \sqrt{w_i}\, \hat{\mathbf{u}}_R^i \tag{12-54}$$

The loss function $J(A)$ is clearly a minimum when

$$J'(A) = \sum_{i=1}^{n} \mathbf{W}_i A \mathbf{V}_i \equiv \text{tr}(W^T A V) \tag{12-55}$$

is a maximum, where the $(3 \times \text{n})$ matrices W and V are defined by

$$W \equiv \left[\mathbf{W}_1 \vdots \mathbf{W}_2 \vdots \cdots \vdots \mathbf{W}_n\right]$$

$$V \equiv \left[\mathbf{V}_1 \vdots \mathbf{V}_2 \vdots \cdots \vdots \mathbf{V}_n\right] \tag{12-56}$$

To find the attitude matrix, A, which maximizes Eq. (12-55), we parameterize A in terms of the quaternion, q, Eq. (12-13b),

$$A(q) = (q_4^2 - \mathbf{q} \cdot \mathbf{q})\mathbf{1} + 2\mathbf{q}\mathbf{q}^T - 2q_4 Q \tag{12-57}$$

where the quaternion has been written in terms of its vector and scalar parts,

$$q = \begin{pmatrix} \mathbf{q} \\ q_4 \end{pmatrix} \tag{12-58}$$

$\mathbf{1}$ is the (3×3) identity matrix, $\mathbf{q}\mathbf{q}^T$ is the (3×3) matrix outer product formed from the vector part of q, and Q is the skew-symmetric matrix

$$Q = \begin{bmatrix} 0 & -q_3 & q_2 \\ q_3 & 0 & -q_1 \\ -q_2 & q_1 & 0 \end{bmatrix} \tag{12-59}$$

Substitution of Eq. (12-57) into (12-55) and considerable matrix algebra [Keat, 1977] yields the following convenient form for the modified loss function:

$$J'(q) = q^T K q \tag{12-60}$$

where the (4×4) matrix K is

$$K = \begin{pmatrix} S - 1\sigma & Z \\ Z^T & \sigma \end{pmatrix} \tag{12-61}$$

and the intermediate (3×3) matrices B and S, the vector $\mathbf{Z}$, and the scalar σ are given by

$$B \equiv W V^T \tag{12-62a}$$

$$S \equiv B^T + B \tag{12-62b}$$

$$\mathbf{Z} \equiv (B_{23} - B_{32}, \; B_{31} - B_{13}, \; B_{12} - B_{21})^T \tag{12-62c}$$

$$\sigma \equiv \mathrm{tr}(B) \tag{12-62d}$$

The extrema of J', subject to the normalization constraint $q^T q = 1$, can be found by the method of Lagrange multipliers [Hildebrand, 1964]. We define a new function

$$g(q) = q^T K q - \lambda q^T q \tag{12-63}$$

where λ is the Lagrange multiplier, $g(q)$ is maximized without constraint, and λ is chosen to satisfy the normalization constraint. Differentiating Eq. (12-63) with respect to q^T and setting the result equal to zero, we obtain the eigenvector equation (see Appendix C)

$$Kq = \lambda q \tag{12-64}$$

Thus, the quaternion which parameterizes the optimal attitude matrix, in the sense of Eq. (12-52), is an eigenvector of K. Substitution of Eq. (12-64) into (12-60) gives

$$J'(q) = q^T K q = q^T \lambda q = \lambda \tag{12-65}$$

Hence, J' is a maximum if the eigenvector corresponding to the largest eigenvalue is chosen. It can be shown that if at least two of the vectors $\mathbf{W}_i$ are not collinear, the eigenvalues of K are distinct [Keat, 1977] and therefore this procedure yields an unambiguous quaternion or, equivalently, three-axis attitude. Any convenient means, e.g., use of the subroutine EIGRS [IMSL, 1975], may be used to find the eigenvectors of K.

A major disadvantage of the method is that it requires constructing vector measurements, which is not always possible, and weighting the entire vector. Alternative, optimal methods which avoid these disadvantages are described in Chapter 13. Variations on the q-method which avoid the necessity for computing eigenvectors are described by Shuster [1978a, 1978b].

12.3 Covariance Analysis

Gerald M. Lerner

   Covariance analysis or the *analysis of variance* is a general statistical procedure for studying the relationship between errors in measurements and error in quantities derived from the measurements. In this section, we first discuss covariance analysis for an arbitrary set of variables and then discuss the interpretation of the covariance parameters for three-axis attitude. For a more extended discussion of covariance analysis, see, for example, Bryson and Ho [1969] or Bevington [1969]. For geometrical procedures for analyzing single-axis attitude error, see Section 11.3.

   We define the *mean*, ξ, and *variance*, v, of the random variable x by

$$\xi \equiv E(x) \tag{12-66a}$$

$$v \equiv E\{(x-\xi)^2\} \equiv E\{(\delta x)^2\} \tag{12-66b}$$

where E denotes the expectation value or ensemble average. The variance is simply the mean square deviation, $\delta x = x - \xi$, of x from the value $x = \xi$. The root-mean-square (rms) deviation, or *standard deviation*, σ, is defined by

$$\sigma \equiv \sqrt{v} \tag{12-67}$$

   The *covariance* of two variables x_1 and x_2, with means ξ_1 and ξ_2 and standard deviations σ_1 and σ_2, is defined by

$$\lambda_{12} = \lambda_{21} \equiv E\{(x_1-\xi_1)(x_2-\xi_2)\} \tag{12-68}$$

and is a measure of the interdependence or correlation between the two variables. The *correlation coefficient* of x_1 and x_2 is the normalized covariance

$$C_{12} = C_{21} \equiv E\left\{\left(\frac{x_1-\xi_1}{\sigma_1}\right)\left(\frac{x_2-\xi_2}{\sigma_2}\right)\right\} = \frac{\lambda_{12}}{\sigma_1\sigma_2} \tag{12-69}$$

which satisfies the inequality

$$|C_{12}| \leqslant 1 \tag{12-70}$$

   For independent variables, $C_{12}=\lambda_{12}=0$, and for totally correlated variables (e.g., $x_1=7x_2$), $|C_{12}|=1$. Covariance analysis relates the presumably known variance and covariance in one set of variables (e.g., measurement errors) to a second set of variables (e.g., computed attitude errors).

   We assume that the n computed quantities, x_i, are functions of the m measurements, y_j, with $m \geqslant n$. Thus,

$$x_i = x_i(y_1,y_2...y_m) \tag{12-71}$$

or, in vector notation, $\mathbf{x}=\mathbf{x}(\mathbf{y})$. In Chapter 11, geometrical techniques were applied to the special case $n=m=2$ and $\mathbf{x}=(\alpha,\delta)^T$. Here, we are primarily concerned with the case where $n = 3$, $m \geqslant 3$, and $\mathbf{x}$ consists of three attitude angles; however, other interpretations and higher dimensions are consistent with the formal development. If $m > 3$, then the problem is overdetermined and the functional form of Eq. (12-71) is not unique. In this case, we assume that a unique function has been chosen.

If the measurement errors, δy_j, are sufficiently small and $\mathbf{x}$ is differentiable, the error in x_i may be estimated by using a first-order Taylor series:*

$$\delta x_i = \sum_{j=1}^{m} \frac{\partial x_i}{\partial y_j} \delta y_j \qquad (12\text{-}72\text{a})$$

or

$$\delta \mathbf{x} = H \, \delta \mathbf{y} \qquad (12\text{-}72\text{b})$$

where H is the $n \times m$ matrix of partial derivatives with the elements $H_{ij} \equiv \partial x_i / \partial y_j$. The expectation value of the outer product of $\delta \mathbf{x}$ with $\delta \mathbf{x}^{\mathrm{T}}$ is

$$\mathrm{E}(\delta \mathbf{x} \, \delta \mathbf{x}^{\mathrm{T}}) = \mathrm{E}(H \, \delta \mathbf{y} \, \delta \mathbf{y}^{\mathrm{T}} H^{\mathrm{T}}) = H \, \mathrm{E}(\delta \mathbf{y} \, \delta \mathbf{y}^{\mathrm{T}}) H^{\mathrm{T}} \qquad (12\text{-}73)$$

which may be rewritten in matrix notation as

$$P_c = H P_m H^{\mathrm{T}} \qquad (12\text{-}74)$$

where the elements of the *covariance matrix*, P_c, and the *measurement covariance matrix*, P_m, are defined by

$$P_{c_{ij}} \equiv \mathrm{E}(\delta x_i \, \delta x_j) \qquad (12\text{-}75\text{a})$$

$$P_{m_{ij}} \equiv \mathrm{E}(\delta y_i \, \delta y_j) \qquad (12\text{-}75\text{b})$$

Thus, the diagonal elements of the $n \times n$ symmetric covariance matrix, P_c, give the variance of the errors in the computed components of $\mathbf{x}$ and the off-diagonal elements give the covariance between the components of $\mathbf{x}$. Similarly, the elements of the $m \times m$ matrix P_m give the variance and covariance of the measurement errors in $\mathbf{y}$.

Equation (12-74) provides the link between the (presumably) known variance and correlation in the measurements, and the desired variance and correlation in the computed quantities. Different algorithms for obtaining $\mathbf{x}$ from $\mathbf{y}$, when $m > n$, will, in general, yield different solutions, different partial derivatives and, consequently, a different computed covariance matrix. Thus, an algorithm, $\mathbf{x}(\mathbf{y})$, might be chosen to avoid undesirable error correlations.

Equation (12-74) relates the variance and covariance in the measurements to the variance and covariance in the computed quantities without implying anything further about the distributions of the errors in either $\mathbf{x}$ or $\mathbf{y}$. However, three specific cases are often used in attitude analysis.

1. If the distribution of errors in $\mathbf{y}$ is Gaussian or normal, then the distribution of errors in $\mathbf{x}$ is also Gaussian.

---

*In general, we are *not* free to pick some appropriately small region about the solution in which the first-order Taylor series is valid. It must be valid over the range of solutions corresponding to the range in measurement errors. Thus, second-order effects, which are ignored here, may become important when using realistic estimates of the measurement errors.

2. If the measurement accuracy is determined by *quantization*, i.e., buckets or steps which return a single discrete value for the measurement or group of measurements, then the variance in the measurement is $S^2/12$, where S is the step size. If all of the measurements are limited by quantization, then the probability density of the attitude is a step function (i.e., uniform within a particular region and 0 outside that region. See Section 11.3.)

3. If there are a large number of uncorrelated measurements, then the *Central Limit Theorem* (see, for example, Korn and Korn [1973]) can be used to infer the distribution of errors in **x**, irrespective of the form of the distribution of the measurement errors. The theorem states that if the m random variables r_j are uncorrelated with mean ξ_j and variance v_j, then as $m \to \infty$, the distribution of the sum

$$\sum_{j=1}^{m} r_j$$

is asymptotically Gaussian [Bevington, 1969] with mean

$$\sum_{j=1}^{m} \xi_j$$

and standard deviation

$$\left(\sum_{j=1}^{m} v_j \right)^{1/2}$$

An application of this theorem to Eq. (12-72a) with

$$r_j = \frac{\partial x_i}{\partial y_j} \delta y_j \qquad (12\text{-}76a)$$

implies that, for m large, the errors in x_i, that is, δx_i, are approximately Gaussian with standard deviation

$$\sigma_{x_i} = \left\{ \sum_{j=1}^{m} \left(\frac{\partial x_i}{\partial y_j} \right)^2 v_{y_j} \right\}^{1/2} \qquad (12\text{-}76b)$$

In practice, the Central Limit Theorem may give reasonable estimates for m as small as 4 or 5, although in such cases the results should be verified by other means. The Central Limit Theorem may also be used to compute the variance and distribution of errors in a measurement which is contaminated by many error sources with (presumably) known variances. To determine the covariance matrix P_c from Eq. (12-74), we need to determine both the measurement covariance matrix, P_m, and the matrix of partial derivatives, H. In practice, P_m is normally assumed to be diagonal; i.e., the measurements are assumed to be uncorrelated. The diagonal elements of P_m are simply the variance of the measurements. If the measurements are correlated but can be written as functions of uncorrelated quantities, then the above analysis may be used to determine P_m. For example, the Sun/Earth-in and Sun/Earth-out rotation angle measurements, Ω_I and Ω_O, described in Section 7.2, are correlated but may be written in terms of the uncorrelated quantities, t_I, the Sun to Earth-in time; t_O, the Sun to Earth-out time; and t_S, the Sun time as

$$\Omega_I = \omega(t_I - t_S) \qquad (12\text{-}77a)$$

$$\Omega_O = \omega(t_O - t_S) \qquad (12\text{-}77b)$$

where ω is the spin rate. Thus,

$$P_m = H \begin{bmatrix} \sigma_{t_S}^2 & 0 & 0 \\ 0 & \sigma_{t_I}^2 & 0 \\ 0 & 0 & \sigma_{t_O}^2 \end{bmatrix} H^T \qquad (12\text{-}78)$$

where σ_{t_S}, σ_{t_I}, and σ_{t_O} are the standard deviations in the (assumed) uncorrelated measurements t_S, t_I, and t_O. The elements of H are

$$H_{11} = \frac{\partial \Omega_I}{\partial t_S} = -\omega; \qquad H_{12} = \frac{\partial \Omega_I}{\partial t_I} = \omega; \qquad H_{13} = \frac{\partial \Omega_I}{\partial t_O} = 0$$

$$H_{21} = \frac{\partial \Omega_O}{\partial t_S} = -\omega; \qquad H_{22} = \frac{\partial \Omega_O}{\partial t_I} = 0; \qquad H_{23} = \frac{\partial \Omega_O}{\partial t_O} = \omega \qquad (12\text{-}79)$$

Substitution of Eqs. (12-79) into Eq. (12-78) yields

$$P_m = \omega^2 \begin{bmatrix} \sigma_{t_S}^2 + \sigma_{t_I}^2 & \sigma_{t_S}^2 \\ \sigma_{t_S}^2 & \sigma_{t_S}^2 + \sigma_{t_O}^2 \end{bmatrix} \qquad (12\text{-}80)$$

The correlation coefficient between the errors in Ω_I and Ω_O is

$$\frac{\sigma_{t_S}^2}{\left[(\sigma_{t_S}^2 + \sigma_{t_I}^2)(\sigma_{t_S}^2 + \sigma_{t_I}^2) \right]^{1/2}} \leqslant 1$$

Given an estimate of P_m, it remains to evaluate the partial derivatives $\partial x_i / \partial y_j$ to obtain H. These partials may be computed either numerically or analytically. Numerical evaluation of partial derivatives is particularly convenient when computer evaluation of the necessary functions is already required for attitude determination, as is normally the case. For example, consider the right ascension of the spin axis attitude, α, as a function of the Sun angle, β; a horizon sensor mounting angle, γ; and other variables [Shear, 1973]: $\alpha = \alpha(\beta, \gamma, \ldots)$. Then, if the variance in β is $v_\beta = \sigma_\beta^2$, and if α is linear over the appropriate range, the partial derivative is approximately

$$\frac{\partial \alpha}{\partial \beta} \approx \frac{\alpha(\beta + \sigma_\beta, \gamma, \ldots) - \alpha(\beta, \gamma, \ldots)}{\sigma_\beta} \qquad (12\text{-}81)$$

Given specific values of β, γ, and the other measurements and their variance and correlation, H can be calculated directly from Eq. (12-81).

This method breaks down if the attitude cannot be computed from the perturbed data, i.e., if $\alpha(\beta + \sigma_\beta, \gamma, \ldots)$ is undefined. (This is clearly an indication that α is nonlinear in this region and therefore Eq. (12-72) is probably invalid.) It is also possible for the perturbed solution to yield only the wrong attitude solution of an ambiguous pair and, therefore, to give absurdly large uncertainties. Numerically, both problems may be resolved by substituting some reduced fraction, e.g., $0.1\sigma_\beta$, for σ_β in Eq. (12-81). The reduced fraction chosen should be small enough to avoid undefined solutions and large enough so that computer round-off error is insignificant.

The alternative to numerical partial derivatives is to find analytic expressions for the partial derivatives. (See, for example, Shear and Smith [1976] for analytic solutions for the partial derivatives for all of the spin-axis methods described in Section 11.1.) This eliminates the major inaccuracy of the numerical computations resulting either from no solution or from an erroneous one. The use of analytic partial derivatives eliminates the problem of undefined solutions or incorrect solution choice at the cost of potentially very complex algebra. The principal advantage of the numerical procedure is that it is simple and direct. In this case, the possibility of algebraic errors in the uncertainties is nearly eliminated because any error in the basic formulas will affect both the perturbed and the unperturbed solutions. Because no additional algebra is required, numerical evaluation of partials can be applied to very complex systems with minimal difficulty.

Interpretation of the Covariance Matrix. The geometrical interpretation of the computed $n \times n$ covariance matrix is generally difficult. As discussed in Section 11.3, no single number adequately represents the "attitude error" nor does the computed variance in each component of $\mathbf{y}$ completely characterize the "error" in that component. Thus, combining attitude solutions obtained by various methods into an "average" solution by weighting according to their variance is frequently misleading.

In practice, the selected set of measurements frequently have uncorrelated errors. Thus, the measurement covariance matrix, P_m, is diagonal, and the diagonal elements of P_c, i.e., the variance in the error of the computed quantities, is given by the simple expression

$$v_{x_i} = \sum_{j=1}^{m} \left(\frac{\partial x_i}{\partial y_j} \right)^2 v_{y_j} \qquad (12\text{-}82)$$

However, even when P_m is diagonal, P_c is diagonal only if there is nearly a 1-to-1 relationship between the measurements and the computed quantities (i.e., if for each x_i there is a y_j such that $|\partial x_i / \partial y_k| \ll |\partial x_i / \partial y_j|$ for all $k \neq j$. This latter condition is rarely satisfied in practice.

Further insight into the significance of the covariance matrix may be obtained by observing that it is positive definite and symmetric by construction and, if $\det (P_c) \neq 0$,* it may be diagonalized by a similarity transformation (see Appendix C). This transformation may be thought of as a rotation of the n correlated errors, $\delta\mathbf{x}$, into a new coordinate system where the transformed errors, $\delta\mathbf{x}'$, are uncorrelated. The covariance matrix for $\delta\mathbf{x}'$ is

$$P_c' = B^T P_c B \qquad (12\text{-}83)$$

---

*Det $(P_c) = 0$ if and only if $|C_{ij}| = 1$ for some i,j. In this case, the phase space of $\mathbf{x}$ should be reduced by one dimension such that either x_i or x_j is eliminated.

where P_c' is diagonal with elements v_{x_i}' and B is the $n \times n$ matrix which diagonalizes P_c. Procedures for computing B are contained in Appendix C.

For $n = 3$, where the elements of $\mathbf{x}$ are the attitude angles, the probability that the transformed attitude error, $\delta\mathbf{x}' = (\delta x_1', \delta x_2', \delta x_3')^T$, is contained in the error ellipsoid

$$\frac{(\delta x_1')^2}{v_{x_1}'} + \frac{(\delta x_2')^2}{v_{x_2}'} + \frac{(\delta x_3')^2}{v_{x_3}'} = K^2 \tag{12-84}$$

is the probability that the chi-square random variable for 3 degrees of freedom is less than K^2 [Abramowitz and Stegun, 1964]. K is commonly called the σ uncertainty level. Thus, the "3σ" attitude error ellipsoid is defined by Eq. (12-84) with $K = 3$. The largest of the v_{x_i}', v_{max}', is the three-dimensional analog of the semimajor axis of the error ellipse described in Section 11.3 and is one measure of the attitude accuracy. An alternative measure of the overall attitude error when none of the v_{x_i}' are much smaller than the others is the radius, ρ, of a sphere whose volume equals that of the error ellipsoid. Thus,

$$\frac{4\pi}{3} K^3 \left(v_{x_1}' v_{x_2}' v_{x_3}' \right)^{1/2} = \frac{4\pi}{3} \rho^3 \tag{12-85}$$

which may be solved for ρ to give (see Appendix C)

$$\rho = K \left(v_{x_1}' v_{x_2}' v_{x_3}' \right)^{1/6} = K (\det P_c')^{1/6} = K (\det P_c)^{1/6} \qquad \left(v_{x_1}' \approx v_{x_2}' \approx v_{x_3}' \right) \tag{12-86}$$

Table 11-1 gives the relationship between K and various confidence levels. As an example, if we wish to assign a 99% confidence level to a three-axis attitude estimate, we obtain from Table 11-1 that $K = 3.37$ and use Eq. (12-83) to determine v_{max}. A conservative measure of the attitude accuracy is then $3.37 v_{max}'$. Alternatively, if the approximation of Eq. (12-86) is valid, we compute the determinent of the covariance matrix and set $\rho = 3.37 \, [\det(P_c)]^{1/6}$.

References

1.   Abramowitz, M., and I. A. Stegun, *Handbook of Mathematical Functions*, National Bureau of Standards Applied Mathematics Series, No. 55, June 1964.

2.   Bevington, Philip R., *Data Reduction and Error Analysis for the Physical Sciences*. New York: McGraw-Hill Book Company, 1969.

3.   Bryson, Arthur E., Jr., and Yu-Chi Ho, *Applied Optimal Control*. Waltham, MA: Ginn and Company, 1969.

4.   Cayley, A., "On the Motion of Rotation of a Solid Body," *Cambridge Mathematical Journal*, Vol. III, No. 1843, p. 224–232. Reprinted in *The Collected Papers of Arthur Cayley*, Vol. I, Cambridge, Cambridge University Press, 1899, p. 28–35.

5.   Davenport, P., Private Communication, 1968.

6.   Gibbs, J. Willard, and E. B. Wilson, *Vector Analysis*, ed. by E. B. Wilson. New York: Scribner, 1901.

7. Goldstein, Herbert, *Classical Mechanics*. Reading, MA: Addison-Wesley Publishing Company, Inc., 1950.

8. Hildebrand, Francis B., *Advanced Calculus for Applications*. Englewood Cliffs, NJ: Prentice-Hall, Inc. 1964.

9. IMSL, *Library 1 Reference Manual*, International Mathematical and Statistical Libraries, Inc. (IMSL), Nov. 1975.

10. Keat, J., *Analysis of Least-Squares Attitude Determination Routine DOAOP*, Comp. Sc. Corp., CSC/TM-77/6034, Feb. 1977.

11. Klumpp, A. R., "Singularity-free Extraction of a Quaternion from a Direction Cosine Matrix," *J. Spacecraft*, Vol. 13, p. 754–755, 1976.

12. Korn, Granino A., and Theresa M. Korn, *Mathematical Handbook for Scientists and Engineers*. New York: McGraw-Hill, Inc., 1968.

13. Sabroff, A., R. Farrenkopf, A. Frew, and M. Gran, *Investigation of the Acquisition Problem in Satellite Attitude Control*, TRW Report AFFDL-TR-65-115, Dec.1965.

14. Shear, M. A., *Optical Aspect Attitude Determination System (OASYS), Version 4.0, System Description and Operating Guide*, Comp. Sc. Corp. 3000-06000-03TM, Dec. 1973.

15. Shear, M. A., and P. M. Smith, *Infrared (IR) Attitude Determination System Communications Technology Satellite Version (CTSADS) Analytical Techniques*, Comp. Sc. Corp., CSC/TM-76/6149, June 1976.

16. Shuster, M. D., *Algorithms for Determining Optimal Attitude Solutions*, Comp. Sc. Corp., CSC/TM-78/6056, April 1978a.

17. ———, *Approximate Algorithms for Fast Optimal Attitude Computation*, paper no. 78-1249, AIAA Guidance and Control Specialist Conference, Palo Alto, CA, Aug. 1978b.

18. Wahba, Grace, "A Least Squares Estimate of Satellite Attitude, Problem 65.1," *SIAM Review*, p. 384–386, July 1966.

19. Whittaker, E. T., *A Treatise on the Analytical Dynamics of Particles and Rigid Bodies*, 4th edition. Cambridge: Cambridge University Press, 1937.

CHAPTER 13

STATE ESTIMATION ATTITUDE DETERMINATION METHODS

13.1   Deterministic Versus State Estimation Attitude Methods
13.2   State Vectors
       State Vector Elements, Choosing State Vector Elements
13.3   Observation Models
13.4   Introduction to Estimation Theory
13.5   Recursive Least-Squares Estimators and Kalman Filters
       Recursive Least-Squares Estimation, Kalman Filters

It became clear at the end of Chapter 11 that some method of dealing with multiple parameters is necessary to obtain accurate attitude estimates. In this chapter, we summarize the basic procedures normally used for handling this problem. *State estimation* methods of attitude determination use the partial derivatives of the observables with respect to various solved-for parameters to correct an a priori estimate of these parameters. The collection of solved-for attitude parameters is called the *state vector*. The process of determining the state vector elements is variously referred to as *state estimation, differential correction,* or *filtering*. Section 13.1 summarizes the state estimation process. Section 13.2 discusses the concept of the state vector and how it should be constructed. Section 13.3 describes how observations are handled in the state vector formulation. Finally, Sections 13.4 and 13.5 summarize the mathematical methods for carrying out the state estimation process.

13.1 Deterministic Versus State Estimation Attitude Methods

James R. Wertz

In Chapters 11 and 12 we have been primarily concerned with *deterministic* attitude methods in which the same number of observations as variables is used to obtain one or more discrete attitude solutions. In contrast, *state estimation* methods of attitude determination correct successive estimates of attitude parameters as illustrated in Fig. 13-1 for an estimator which processes one observation at a time. Here, as introduced in Chapter 10, L_β and L_η are the attitude loci corresponding to the Sun angle measurement, β, and the nadir angle measurement, η, and the state vector, $\mathbf{x}$, consists simply of the attitude, $(\alpha, \delta)^T$.* The initial estimate of $\mathbf{x}$ is $\mathbf{x}_0 = (\alpha_0, \delta_0)^T$. After processing the Sun angle information, β, the state estimate is shifted toward L_β to $\mathbf{x}_1$. The amount of the shift depends on $\partial \mathbf{x}/\partial \beta$ and the uncertainties in $\mathbf{x}_0$ and β. After processing the η measurement, the state vector is shifted toward L_η to $\mathbf{x}_2$. The process can continue with additional measurements.

---

*In Sections 13.1 through 13.3, it is sufficient to regard the state vector as simply a collection of variables. The notation $(\alpha, \delta)^T$ is used because in later sections the state vector will be regarded as a column vector for matrix manipulations.

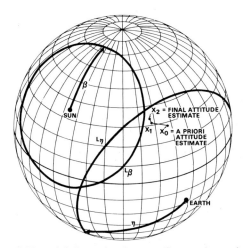

Fig. 13-1. Differential Correction Process. (See text for explanation.)

In state estimation methods, neither the number of solved-for attitude prarmeters nor the number of attitude observations is important as far as the process itself is concerned. (If the number of observations is less than the number of solved-for parameters, some combination of the unknowns will retain their a priori value, or, in some cases, an algebraic singularity will result.) In Fig. 13-1, we can obtain an answer after processing only a single observation, e.g., β, or we can process 1000 observations. Similarly, we can solve for the two-component attitude of Fig. 13-1 or any number, N, of parameters incorporated into an N-dimensional *state vector*. In general, the state vector and the various attitude estimates (i.e., estimates of the values of the N parameters) will be vectors in an N-dimensional phase space. The most common state estimator is the *least-squares filter,** which minimizes the square of the difference between the observations and the calculated results.

In state estimation processes, there are two basic ways to update the state vector. If a new estimate of the state vector is obtained after each observation, the process is called a *sequential estimator*, or *recursive estimator*, as illustrated in Fig. 13-1. If the partial derivatives for all the observations are processed and then combined to produce a single update to the state vector, the process is referred to as a *batch estimator*. Generally, the sequential estimator will be more sensitive to individual data points than will the batch processor; that is, the sequential estimator may converge to a solution more quickly but be less stable than a batch processor. It is also possible to combine batch and sequential methods and update the state vector after some intermediate number of observations has been processed.

Both state estimators (either batch or sequential) and deterministic processors have advantages and disadvantages. The deterministic method nearly always

---

* The word filter is applied to any process which sorts out high-frequency noise from the low-frequency information—for example, the slowly varying nadir or rotation angles. Much of the language of filtering theory comes from electrical engineering, where much of the analysis was initially done.

provides a solution and requires, at most, a very rough a priori estimate of the attitude. The methods and results are easy to interpret physically and geometrically. However, it is both cumbersome and algebraically difficult to model biases, a time-varying attitude, or other attitude-related parameters with deterministic processors. Large quantities of data are difficult to combine with the proper statistical balance in a deterministic processor, which is of particular importance where very accurate attitude solutions are needed.

In contrast to the deterministic processor, state estimation can provide statistically optimal solutions. Expanding the state vector to represent a large range of attitude parameters, such as biases, orbit parameters, or time-varying coefficients, is relatively easy. (We may also use physically meaningless parameters as state vector elements for numerical convenience.) However, state estimators may diverge and provide no solution at all. They may require a dynamic model or a more accurate estimate of the a priori attitude than do deterministic methods, and their increased flexibility and sophistication means that interpreting the physical or geometrical meaning of the results may be very difficult.

In practice, both solution methods are frequently used in a complementary fashion. For spinning spacecraft, a deterministic processor solving only for the attitude is often used to obtain an a priori estimate for a state estimator, which then corrects on an expanded state vector including biases or attitude drift parameters. Results are then confirmed by returning to the deterministic processor to verify that systematic errors have been eliminated. (See, for example, Section 11.5.)

As with deterministic processors, the attitude accuracy for state estimators should be independent of the choice of numerical procedure for handling the data, provided statistically correct ways of combining the data are used. For example, we may use either spherical trigonometry or vector algebra to compute the estimated state vector elements. We can lose information by choice of a particularly poor technique; however, no technique, no matter how clever or sophisticated, can obtain more real information than the statistics of the data will allow. Determining that all of the information content has been obtained from a particular segment of data in a state estimator is not necessarily easy. Generally, this is tested empirically —if different statistical methods and different processing techniques produce essentially the same results, then we assume that we have extracted nearly all of the information in the data. When we discuss fundamental limits to attitude accuracies throughout Part III, we assume that, in general, all processing methods are equivalent in terms of the accuracy obtainable, although the processing efficiency may vary greatly.

13.2 State Vectors

Steven G. Hotovy

As the name implies, a state vector deals with the state, or condition, of some situation; this situation is referred to as the *process*.* For our purpose, the *attitude process* consists of all of the parameters which define or affect the spacecraft

---

*A wide body of literature exists which gives an introduction to state vectors with varying degrees of sophistication. See, for example, Deutsch [1965], Bryson and Ho [1969], and Schmidtbauer, *et al.*, [1973].

attitude and the interrelationships between them. The *state* consists of the values of these parameters at any one time.

The three elements necessary to define the attitude process are described below:

1. The *state vector*, $\mathbf{x}$, is an m-dimensional vector which includes all of the variables necessary to permit accurate attitude determination. It may include such factors as sensor biases and misalignments, attitude propagation parameters (which may include the attitude itself), and orbital parameters. The state vector elements may be constant for the processing interval (e.g., biases and misalignments), or they may be time varying (e.g., the quaternions described in Section 12.1). In the latter case, the propagation of the state vector is given by the differential equation

$$\frac{d\mathbf{x}}{dt} = \mathbf{f}(\mathbf{x}, t) \tag{13-1}$$

2. The *observation vector*, $\mathbf{y}$, is an n-dimensional vector composed of sensor measurements. These measurements may involve direct sensor readouts, such as event times from a Sun sensor, or observations in some processed form, such as Earth-width data obtained from a wheel-mounted horizon scanner.

3. The *observational model vector*, $\mathbf{z}$, is an n-dimensional vector composed of *predicted* values of the observational vector based on estimated values of the state vector elements, i.e.,

$$\mathbf{z} = \mathbf{g}(\mathbf{x}, t) \tag{13-2}$$

The observation model vector is frequently based on the hardware model of the sensor which is providing the corresponding observation.

The way in which these three vectors are used to obtain an estimate of the state vector depends on which state estimation technique is being used, as discussed in Section 13-4. In general, however, for a given estimate $\tilde{\mathbf{x}}_0$ of the state, the observation model vector $\mathbf{z}_0$ is determined, and then compared with the observation vector $\mathbf{y}_0$. A new estimate of the state $\tilde{\mathbf{x}}_1$ is then selected to minimize, in some sense, the difference between $\mathbf{y}_0$ and $\mathbf{z}_0$. The *batch estimator*, introduced in Section 13.1, repeats the cycle with the new estimate $\tilde{\mathbf{x}}_1$ of the state but with the same observation vector, $\mathbf{y}_0$; a *recursive estimator* does the same with a new observation vector $\mathbf{y}_1$, as well as the new state estimate, $\tilde{\mathbf{x}}_1$.

13.2.1 State Vector Elements

The state vector should include those elements which are necessary to allow determination of the spacecraft attitude with sufficient accuracy, using sensor data of varying quality. These elements may be grouped into three main categories: (1) sensor-related parameters, (2) orbital parameters, and (3) attitude propagation parameters (which may include the attitude itself).

Among the most important parameters to include in the state vector are those relating to sensor performance. With each type of sensor is associated a collection of biases which may affect their performance. An extensive list of these biases and misalignments may be obtained from the mathematical sensor models described in Chapter 7. These biases may remain constant for the life of the mission, such as a bolometer offset for a horizon sensor, or may be time varying, such as mag-netometer residual biases and drift parameters for a gyroscope.

Orbital information is necessary for attitude determination whenever the sensor supplying the data produces a measurement which depends on spacecraft position in the orbit (such as horizon scanners and magnetometers). Although any of the orbital parameters could be included in the state vector, the orbital in-track error, which measures how far the spacecraft is behind or ahead of its anticipated position, is the most common because it is frequently the largest source of error and because it is easily modeled by obtaining the spacecraft ephemeris at a time offset from the nominal time.

The third category of state vector elements consists of attitude propagation parameters. The choice of these parameters is based on whether the propagation model being employed includes dynamics. In a *kinematic model*, the elements of the state vector relating to attitude propagation do not include internal or external torque parameters, and the modeled attitude of the spacecraft at any time in the interval of interest, $A(t)$, can be calculated directly from the state vector elements, i.e.,

$$A(t) = h(x, t) \tag{13-3}$$

In a *dynamic model*, some elements of the state vector may include torque-related parameters, and the modeled attitude of the spacecraft at any time in the interval of interest is determined by integrating the equations of motion. In some estimators, the quaternions and body rates are themselves considered part of the vector, and in this case the modeled attitude is determined from the state propagation equation (Eq. (13-1).)

An example of a simple kinematic model is the one used for attitude determination for the SMS-2 spacecraft [Chen and Wertz, 1975]. For this model, it is assumed that the spacecraft spin axis remains inertially fixed and that the spacecraft rotation remains constant over the interval of interest. The state vector elements relating to attitude propagation are

$$\alpha = \text{right ascension of spin axis}$$

$$\delta = \text{declination of spin axis} \tag{13-4}$$

$$\omega = \text{spacecraft rotation rate about the spin axis}$$

This model is appropriate for a spin-stabilized spacecraft which is expected to encounter only small torques (either environmental or control) during the interval of interest. This attitude propagation model was assumed in the development of several of the attitude hardware models of Section 7.

An example of how a state vector can be expanded is provided by examination of the SAS-3 attitude determination system [Rigterink, *et al.*, 1974]. SAS-3 is a spin-stabilized spacecraft for which the constant attitude model described above is appropriate. Therefore, the state vector should minimally contain the right ascension and declination of the spin axis and the spacecraft angular rate. In addition, a magnetometer triad provides data which are used to compute the spacecraft attitude, but the attitude solution accuracy requirements demand compensation for the magnetometer misalignments, residual biases, and errors in the calibration curve. Expanding the state vector, $\mathbf{x}$, to include these elements gives

$$\mathbf{x} = (\alpha, \delta, \omega, \theta, \phi_0, f, b)^{\mathrm{T}} \tag{13-5}$$

where θ is the colatitude of a given component of the magnetometer triad in spacecraft coordinates, ϕ_0 is the corresponding azimuth, f is the slope of the magnetometer calibration curve, and b is the residual bias.

Another kinematics model, useful for three-axis stabilized spacecraft, was developed for the GEOS-3 mission [Repass, *et al.*, 1975]. GEOS-3 is an Earth-oriented, gravity-gradient stabilized spacecraft; thus attitude information is most conveniently expressed in pitch, roll, and yaw angles. It is assumed that each of these angles can be expressed adequately in terms of an initial value and a fixed rate of change for a suitably chosen time interval. The elements of the state vector relating to attitude propagation are as follows:

$$p = \text{initial pitch angle}$$

$$\dot{p} = \text{pitch rate}$$

$$r = \text{initial roll angle}$$

$$\dot{r} = \text{roll rate} \tag{13-6}$$

$$y = \text{initial yaw angle}$$

$$\dot{y} = \text{yaw rate}$$

The pitch, roll, and yaw angles at time t are

$$p(t) = p + \dot{p}\Delta t$$

$$r(t) = r + \dot{r}\Delta t \tag{13-7}$$

$$y(t) = y + \dot{y}\Delta t$$

where Δt is the time since the beginning of the interval of interest. GEOS-3 has a three-axis magnetometer triad which is subject to both misalignment errors and residual biases; the misalignment is expressed by the three angles θ_x, θ_y, and θ_z between the true magnetometer placement and the spacecraft $\hat{\mathbf{x}}$, $\hat{\mathbf{y}}$, and $\hat{\mathbf{z}}$ axes, respectively.

Thus, the complete state vector becomes

$$\mathbf{x} = \left(p, \dot{p}, r, \dot{r}, y, \dot{y}, \theta_x^1, \theta_y^1, \theta_z^1, \theta_x^2, \theta_y^2, \theta_z^2, \theta_x^3, \theta_y^3, \theta_z^3, b_1, b_2, b_3 \right)^{\mathrm{T}} \tag{13-8}$$

where θ^j refers to the jth magnetometer and b_j is the residual bias on the jth magnetometer.

An example of a state vector which incorporates a dynamics model is that of the Nimbus-6 attitude determination system [Lefferts and Markley, 1976]. Attitude determination and control hardware on Nimbus-6 consisted of a horizon scanner, four two-axis Sun sensors, and four reaction wheels. The onboard control system continuously varied the reaction wheel rates (and, hence, the body rates) to maintain the near-nominal attitude of zero pitch, roll, and yaw. For this reason, it was felt that no kinematics model would approximate the spacecraft attitude with sufficient accuracy. The state vector elements relating to attitude propagation were initial estimates of the quaternion, $q = (q_1, q_2, q_3, q_4)^T$, and initial estimates of the spacecraft angular velocity, $\omega = (\omega_x, \omega_y, \omega_z)^T$. The attitude is propagated using the spacecraft equations of motion (Section 16.1). The torque terms in the equations of motion included control system parameters consisting of the moments of inertia of each wheel (m_1, m_2, m_3, m_4) and a constant bias on the speed of each wheel (s_1, s_2, s_3, s_4). The torque terms also include environmental torque parameters, which include constant torques in body coordinates (c_x, c_y, c_z) and the spacecraft magnetic dipole in body coordinates (d_x, d_y, d_z). Thus, the state vector has the following 21 elements:

$$x = (q_1, q_2, q_3, q_4, \omega_x, \omega_y, \omega_z, m_1, m_2, m_3, m_4, s_1, s_2, s_3, s_4, c_x, c_y, c_z, d_x, d_y, d_z)^T \quad (13\text{-}9)$$

Dynamics models are most useful when control system operation or significant external torques make kinematic modeling impossible or when highly accurate attitude solutions are required.

13.2.2 Choosing State Vector Elements

For a complex satellite containing sophisticated attitude determination and control hardware, there are potentially hundreds of state vector elements. Large state vectors are undesirable for several reasons. Two or more elements of the state vector may have nearly the same effect on the data and therefore be redundant and difficult to distinguish. For a given pass of data, only a limited number of state vector elements can be solved for, and the selection of the solved-for parameters is more difficult for large state vectors. Finally, from a computational standpoint, programs involving large state vectors are more difficult to develop and test, more unwieldy to operate, and require more computer time and storage to execute. Therefore, guidelines concerning the selection of state vector elements for a particular application are needed. Specifically, the state vector should include all elements that satisfy the following criteria at some time in the mission:

1. It significantly affects the observation model vector relative to changes in the attitude.
2. It represents a physically real quantity.
3. Its value remains nearly constant over an interval of interest or it is propagated in a dynamics model.

It may be necessary to include additional parameters that do not represent physically real quantities, such as coefficients of polynomial approximations. Therefore, the number of state vector elements can be quite large. However, for a given set of data, the number of vector elements actually solved for can be reduced by constraining the remaining elements to the best estimate of their values. (This topic is addressed in detail in Section 14.3.)

One factor which affects the three criteria for selection of a state vector element is the accuracy requirement of the attitude solution. The "significance" of the effect of a parameter on the observation model vector is relative: what is insignificant when larger attitude uncertainties are permitted may be significant when highly accurate attitudes are required. The "constancy" of the value of a parameter over an interval of interest is also relative: slight variations in the value of a parameter can be tolerated for less accurate attitude solutions or short time intervals, but these variations may be unacceptable for more accurate attitude determination or longer time intervals. This is especially true in attitude propagation modeling.

Another important consideration is the *observability* of a potential state vector element. A state vector element is observable over an interval if an estimate of its value can be made from the observations over that interval. For example, two-axis Sun sensor data provide no attitude information concerning the rotation angle about the sunline. If the sunline and spin axis are collinear, then the spacecraft phase angle about the sun is unobservable from Sun sensor data. Another example is given by a magnetometer triad mounted on the spacecraft pitch, roll, and yaw axes of an Earth-oriented satellite in a polar orbit. Over the magnetic poles, the magnetic field vector is nearly parallel to the yaw axis. For that reason, any misalignment of the pitch and roll magnetometers in the pitch-roll plane is nearly unobservable. If a state vector element is expected to have little observability in the interval of interest, it is best to constrain it to the best estimate of its value.

It may not be possible to separate the effects of several different elements of the state vector on the data. An example of this is provided by a body-mounted horizon sensor on a satellite which has a fixed spin axis attitude and body rate. For a short interval of interest, the geometry of the Earth relative to the spacecraft changes only slightly. Therefore, if the time between observed Earth-in and -out differs from that predicted from the Earth-width model described in Section 7.2, four sensor biases could explain such behavior. These are an in-crossing azimuthal bias, an out-crossing azimuthal bias, a bias on the angular radius of the Earth, and a mounting angle bias. Because the geometry changes only slightly, each parameter alone could be used to correct the predicted time properly, but it would be impossible to solve for more than one parameter because they are so highly correlated. Also, it is impossible to determine which parameter is the true cause of the error. In the case of highly correlated state vector parameters, it is best to constrain all but one of the parameters to the best estimate of their values and solve for the remaining parameter.

Some considerations of state vector formulation are due to the manner in which estimates for these parameters are determined, a topic discussed in detail in Sections 13.4 and 13.5. It is advantageous if an observation varies monotonically with each state vector element influencing the corresponding observation model. This is so because if two values of a particular parameter produce the same value for the observation model, the incorrect value could be selected as the best estimate for the parameter. It is also desirable that the observation model vary nearly linearly as a function of the state vector because the observation model function in Eq. (13-2) is linearized for many state estimation techniques.

13.3 Observation Models

Steven G. Hotovy

In this section we discuss the formation of the observation vector and observation model vector. An *observation* is any quantity which may be computed from a sensor measurement or combination of measurements. It may be a direct sensor measurement, such as the reticle count from a Sun sensor, or a derived quantity, such as the nadir angle from wheel-mounted horizon scanner data. An *observation model* is the predicted value of the observation based on hardware models of the appropriate sensors, sensor measurements, and the value of the state vector. Several hardware and associated observation models are provided in Chapter 7.

The general form of an *n*-dimensional observation model vector is

$$z = g(M, x) \tag{13-10}$$

where x is the *m*-dimensional state vector and M is a *p*-dimensional vector of sensor measurements. The observation and observation model vectors are related by

$$y(M) = z + v = g(M, x) + v \tag{13-11}$$

where $y(M)$ is the true observation and v is the total error due to errors in the measurements and inaccuracies in modeling the observation. Note that we draw a distinction between *observations* and *measurements*. By *measurements* we refer to the data provided directly from a sensor, such as the crossing time and Sun angle from a slit Sun sensor or the Earth width and corresponding time from a wheel-mounted horizon scanner.

The measurements are therefore determined explicitly by the attitude sensor hardware on the satellite. The *observations*, however, are defined by the person developing the estimator. As we will see, it is possible to specify several different observations based on the same measurement or to specify one observation which combines measurements from several different sensors.

Before forming observation models, it is important to analyze carefully the particular application of the estimation process and identify the elements of the state vector. When this has been done, observation models can be selected. The overriding consideration in this selection is the accurate estimation of the state vector elements. Only those models in which at least one of the state vector elements is observable at some time in the mission should be considered. However, there are potentially hundreds of observation models which satisfy this criterion; therefore, some method of selection is necessary.

One requirement is that the observation model be compatible with the state vector. For example, if the bolometer offset of a wheel-mounted horizon scanner is not included in the state vector, the bolometer offset model should not be selected because it is too detailed. On the other hand, if the spacecraft is expected to experience appreciable nutation, the Earth-width model for a horizon scanner, which assumes an inertially fixed spin axis, may be too simplistic.

Observations are preferred which are as close as possible to true sensor measurements. The transformation from true sensor measurements to calculated observations has two undesirable effects [Pettus, 1972]: (1) the transformation may generate statistically correlated observations from uncorrelated measurements and (2) the transformation may involve parameters which are not known accurately and which would therefore lead to inaccuracies in the model.

It is also desirable to select observations which have uncorrelated noise. An example of the effect of correlated noise occurs in Sun sensor/horizon sensor rotation angle models. Suppose that t_S is a vector of Sun sensor crossing times, t_I is a vector of horizon sensor in-crossing times, and $\mathbf{t}_O$ is a vector of horizon sensor out-crossing times, with uncorrelated noise $\Delta\mathbf{t}_S$, $\Delta\mathbf{t}_I$, and $\Delta\mathbf{t}_O$, respectively. We consider two possible pairs of rotation angle model vectors. The first is the Sun-to-Earth-in angle, $\mathbf{\Phi}_I$, and Sun-to-Earth-out angle, $\mathbf{\Phi}_o$:

$$\mathbf{\Phi}_I = (\mathbf{t}_I - \mathbf{t}_S)\omega \tag{13-12a}$$

$$\mathbf{\Phi}_O = (\mathbf{t}_O - \mathbf{t}_S)\omega \tag{13-12b}$$

where ω is the spin rate. The second pair of observation model vectors consists of the Earth width, Ω, and the Sun-to-midscan, $\mathbf{\Phi}_M$, rotation angles:

$$\Omega = (\mathbf{t}_O - \mathbf{t}_I)\omega \tag{13-13a}$$

$$\mathbf{\Phi}_M = [(\mathbf{t}_I + \mathbf{t}_O)/2 - \mathbf{t}_S]\omega \tag{13-13b}$$

Then, the noise ($\Delta\mathbf{\Phi}_I, \Delta\mathbf{\Phi}_O$, $\Delta\Omega$, and $\Delta\mathbf{\Phi}_M$) on the first observation model vectors are related by

$$\Delta\mathbf{\Phi}_I \cdot \Delta\mathbf{\Phi}_O = (\Delta\mathbf{t}_I - \Delta\mathbf{t}_S)\cdot(\Delta\mathbf{t}_O - \Delta\mathbf{t}_S)\omega^2 = |\Delta\mathbf{t}_S|^2\omega^2 \tag{13-14}$$

whereas, for the second pair,

$$\Delta\Omega \cdot \Delta\mathbf{\Phi}_M = (\Delta\mathbf{t}_O - \Delta\mathbf{t}_I)\cdot\left[\tfrac{1}{2}(\Delta\mathbf{t}_I + \Delta\mathbf{t}_O) - \Delta\mathbf{t}_S\right]\omega^2 = \tfrac{1}{2}\left(|\Delta\mathbf{t}_O|^2 - |\Delta\mathbf{t}_I|^2\right)\omega^2 \tag{13-15}$$

Thus, the first pair of rotation angle measurements has correlated noise, whereas the second does not if the magnitudes of the noise for the Earth-in and Earth-out times are the same. The implication of Eq. (13-14) is that noise in the Sun sensor crossing time data affects both the $\mathbf{\Phi}_I$ and $\mathbf{\Phi}_O$ rotation angle models in the same manner. Because it is assumed in the use of many estimation algorithms that the random errors in the observations are uncorrelated, the Earth-width and Sun-to-midscan rotation angles are the preferable observation models.

There are also guidelines concerning the number of observation models to employ for a particular application. There should be enough observation models to make use of all sensor measurements which would be helpful in estimating the elements of the state vector. Also, it is desirable that there be several different models based on the same sensor data so that the user can select the appropriate models for changing mission conditions. Different state vector elements may be more observable in different models and the changing geometry or sensor performance may require changing observation models. Finally, there should be some observation models which depend on a minimum of sensor measurements so that the loss of one or more measurements does not invalidate all observation models. Within these guidelines, the observation model vector should be as small as possible to minimize design and operating complexity. The largest number of *independent* observation models that can be used at any one time is the same as the number of independent measurements.

Examples. The SAS-3 attitude determination problem described in Section 13.2 provides one example of an observation model vector. To successfully obtain spin axis attitude solutions using data from a single-axis induction magnetometer requires that the state vector include both attitude and magnetometer bias parameters, as defined by Eq. (13-5), that is,

$$\mathbf{x} = (\alpha, \delta, \omega, \theta, \phi_0, f, b)^T$$

The measured component, B_M, of the magnetic field parallel to the sensitive axis of the magnetometer is given by

$$B_M = fV_M + V_0 \tag{13-16}$$

where V_M is the voltage measured by the magnetometer and V_0 is the voltage measured by the magnetometer in the absence of a magnetic field.

The predicted measurement, B_P, based on the values of the state vector parameters is calculated as follows. The attitude matrix, $A(t)$, at any time t after the beginning of the interval is

$$A(t) = A_1(t)D \tag{13-17}$$

where

$$A_1(t) = \begin{bmatrix} \cos\omega t & \sin\omega t & 0 \\ -\sin\omega t & \cos\omega t & 0 \\ 0 & 0 & 1 \end{bmatrix}$$

and the D matrix, which transforms a vector from celestial coordinates to a coordinate system whose z axis coincides with the spacecraft spin axis, is

$$D = \begin{bmatrix} -\sin\alpha & \cos\alpha & 0 \\ -\cos\alpha\sin\delta & -\sin\alpha\sin\delta & \cos\delta \\ \cos\alpha\cos\delta & \sin\alpha\cos\delta & \sin\delta \end{bmatrix}$$

In spacecraft coordinates, the position vector, $\hat{\mathbf{r}}$, of the sensitive axis of the magnetometer is

$$\hat{\mathbf{r}}_B = \begin{bmatrix} \cos\phi_0\cos\theta \\ \sin\phi_0\cos\theta \\ \sin\theta \end{bmatrix} \tag{13-18}$$

Thus, in inertial coordinates, the magnetometer position vector, $\hat{\mathbf{r}}_I$, is

$$\hat{\mathbf{r}}_I = A^{\mathrm{T}}(t)\hat{\mathbf{r}}_B = D^{\mathrm{T}} \begin{bmatrix} \cos(\omega t + \phi_0)\cos\theta \\ \sin(\omega t + \phi_0)\cos\theta \\ \sin\theta \end{bmatrix} \tag{13-19}$$

The predicted measurement, B_P, is then

$$B_P = \hat{\mathbf{r}}_I \cdot \mathbf{B}_I + b = \cos\theta\cos(\omega t + \phi_0)\left[\cos\alpha B_y - \sin\alpha B_x\right]$$
$$+ \left[\sin\theta\cos\delta - \cos\theta\sin\delta\sin(\omega t + \phi_0)\right]\left[\cos\alpha B_x + \sin\alpha B_y\right]$$
$$+ B_z(\cos\delta\cos\theta\sin(\omega t + \phi_0) + \sin\theta\sin\delta) + b \tag{13-20}$$

where $\mathbf{B}_I \equiv (B_x, B_y, B_z)^{\mathrm{T}}$ is the magnetic field in inertial coordinates.

The observation is chosen to be the magnetometer voltage, V. Thus, the one-dimensional observation vector, $\mathbf{y}$, is

$$\mathbf{y} = V_M \tag{13-21}$$

while the observation model vector, $\mathbf{z}$, is

$$\mathbf{z} = V_P \equiv (B_P - V_0)/f \tag{13-22}$$

where B_P, which is a function of the state vector, is given by Eq. (13-20).

The Optical Aspect Bias Determination System, OABIAS, provides an example of a system with several observation models [Joseph, *et al.*, 1975]. (A modified version of OABIAS was used as the bias determination subsystem for CTS, as described in Section 21.2.) This system processes Sun sensor and body-mounted horizon scanner data for attitude and bias determination for spin-stabilized spacecraft, and its state vector consists of the spacecraft attitude, the phase of the spacecraft at the start of the processing interval, the satellite spin rate, four alignment angles for the horizon scanner, two alignment angles for the Sun sensor, an error in the central body angular radius, and an orbital timing bias or in-track error. The four sensor measurements are

1. Time of Sun sighting
2. Sun angle at this time
3. Time of horizon in-crossing
4. Time of horizon out-crossing

OABIAS provides the following seven observation models:

1. Sun angle model (Section 7.1)
2. Sun sighting time model
3. Horizon sensor nadir vector projection model (Section 7.2)
4. Horizon sensor crossing time model
5. Horizon sensor Earth-width model (Section 7.2)
6. Sun-to-Earth-in and Sun-to-Earth-out rotation angle models (Section 7.3)
7. Sun-to-Earth-midscan rotation angle model (Section 7.3)

Models 1, 2, and 4 are included because they are the observation models corresponding to the four sensor measurements. Each sensor measurement can be used in at least one observation model. Also, Model 1 requires Sun angle data only, Model 2 requires Sun sighting time data only, and Models 3 and 4 can be used for either Earth-in or -out data, so that even if only one data type were available, there would be at least one valid observation model. Models 3, 4, and 5 require horizon sensor data, but each model uses different state vector elements in its formulation, thus providing flexibility for the user. Models 6 and 7 are more complicated because they require both Sun sensor and horizon sensor data, but are useful because some state vector elements may be more observable in these observation models, and they have a clearer physical interpretation than Models 2 and 4. As described above, Models 5 and 7 have uncorrelated noise, whereas the two models in 6 have correlated noise. This collection of observation models follows the guidelines which have been suggested and has been employed in several missions, as described in Section 20.4

13.4 Introduction to Estimation Theory

Lawrence Fallon, III
Paul V. Rigterink

The purpose of an *estimator* or *data filter* is to calculate a state vector which is optimum by some measure. For example, a least-squares filter determines the state vector which minimizes the square of the difference between the observed data and the expected data computed from an observation model. The contribution of an individual observation in this process may be weighted according to the observation's expected accuracy and importance. Because they provide the best estimate of the state parameters when the uncertainty is a result of Gaussian noise, least-squares filters are by far the most common and are the only type considered here.

There are two major classes of least-squares estimators: batch and sequential. A *batch estimator* updates a state vector at an epoch or reference time using a block of observations taken during a fixed timespan. For example, suppose that a state vector consists of the spacecraft attitude and other model parameters and that it is desired to estimate these parameters at a given epoch. Observations made at any other time can be used to update the epoch state vector if a mathematical model is available to relate the state parameters at each measurement time to their values at the epoch.

In a *sequential estimator*, the state vector is updated after each observation (or a small set of observations) is processed. The two major types of sequential estimators are recursive least-squares estimators and Kalman filters. Like a batch estimator, a *recursive least-squares estimator* corrects the state vector at an epoch time. The recursive least-squares estimator's confidence in the updated state at the epoch time improves as more and more data are processed. Consequently, the sensitivity of this type of estimator to the observations diminishes as time passes. A *Kalman filter* is a sequential estimator with a fading memory. It generally corrects the state vector at the time of each of the observations rather than at an epoch time. After the state is updated using one or more observations, it is propagated or extrapolated by a mathematical model to the time of the next set of observations to provide an initial estimate for the next update*. The filter's confidence in its estimate of the state is allowed to degrade from one update to another using models of noise in the state vector. This causes the influence of earlier data on the current state to fade with time so that the filter does not lose sensitivity to current observations.

Batch Least-Squares Estimation. This subsection describes the mathematical formulation of the *Gauss-Newton least-squares procedure* initially formulated independently by Karl Gauss and Adrien Legendre in the early Nineteenth Century. We begin by considering the m-component state vector, $\mathbf{x}$, which is allowed to vary with time according to the function

$$\mathbf{x}(t) = \mathbf{h}(\mathbf{x}^0, t) \tag{13-23}$$

where $\mathbf{x}^0$ is the state vector at the *epoch* or *reference time*, t_0. The batch least-squares algorithm estimates this epoch state vector, $\mathbf{x}^0$; this estimate is denoted by $\tilde{\mathbf{x}}^0$.[†] The simplest time variation occurs when $\mathbf{x}$ is constant; that is, $\mathbf{x} = \mathbf{x}^0$. However,

---

*What we actually describe here is an extended Kalman filter. The distinction between the basic Kalman filter and the extended Kalman filter will be clarified in Section 13.5.2.

[†]In estimation theory texts, the notation $\hat{\mathbf{x}}$ is often used to denote an estimate. We use $\tilde{\mathbf{x}}$ to avoid confusion with the notation for unit vectors.

if $\mathbf{x}$ contains parameters whose time variation is nonnegligible, propagation of the state, as described in Section 17.1, will be required. If the state undergoes a minor unmodeled variation during the time spanned by the observations, a batch estimator will calculate a weighted "average" value for $\tilde{\mathbf{x}}^0$. In this case, a Kalman filter as described in Section 13.5 may allow better tracking of the state variations than will a batch technique.

Consider a set of n observations,

$$\mathbf{y} \equiv [y_1, \dots, y_n]^{\mathrm{T}} \qquad (13\text{-}24)$$

taken during the timespan of interest, as described in Section 13.2. To determine the state vector, $\mathbf{x}$, we assume that $\mathbf{y}$ equals the *observation model vector*, $\mathbf{g}(\mathbf{x}, t)$, based on the mathematical model of the observations plus additive random noise, $\mathbf{v}$. Thus, for each element of $\mathbf{y}$,

$$y_i = g_i(\mathbf{x}(t_i), t_i) + v_i \qquad (13\text{-}25)$$

Loss Function. We will use Eq. (13-25) to estimate $\mathbf{x}^0$, given an *a priori* estimate $\tilde{\mathbf{x}}_A^0$, the observations $\mathbf{y}$, the functional forms' of $\mathbf{h}(\mathbf{x}^0, t)$ and $\mathbf{g}(\mathbf{x}, t)$, and the statistical properties of $\mathbf{v}$. To accomplish this, we use the least-squares criterion as a measure of "goodness of fit"; the best value of $\mathbf{x}^0$ minimizes the weighted sum of the squares of the residuals between the elements of the observation and observation model vectors. This is done quantitatively by minimizing the *loss function*,

$$J \equiv \tfrac{1}{2} \boldsymbol{\rho}^{\mathrm{T}} W \boldsymbol{\rho} \qquad (13\text{-}26)$$

where the *observation residual vector*, $\boldsymbol{\rho}$, is defined by

$$\boldsymbol{\rho} \equiv \mathbf{y} - \mathbf{g} \qquad (13\text{-}27)$$

W is an $(n \times n)$ symmetric, nonnegative definite matrix chosen to weight the relative contribution of each observation, according to its expected accuracy or importance. In the simplest case, W is the identity matrix indicating that equal weight is given to all observations. Throughout the rest of this section, we assume that W has the form

$$W = \begin{bmatrix} \sigma_1^{-2} & 0 & \cdots & 0 \\ 0 & \sigma_2^{-2} & & \\ \vdots & & & \\ 0 & & & \sigma_n^{-2} \end{bmatrix} \qquad (13\text{-}28)$$

where σ_i $(i = 1, 2, \dots, n)$ is the uncertainty in the ith observation.

An important variation of the loss function given by Eq. (13-26) penalizes any deviation from the a priori estimate in proportion to the inverse of the uncertainty in that estimate; that is

$$J = \tfrac{1}{2} \boldsymbol{\rho}^{\mathrm{T}} W \boldsymbol{\rho} + \tfrac{1}{2} [\mathbf{x}^0 - \tilde{\mathbf{x}}_A^0]^{\mathrm{T}} S_0 [\mathbf{x}^0 - \tilde{\mathbf{x}}_A^0] \qquad (13\text{-}29)$$

where S_0 is the $(m \times m)$ *state weight matrix*. If the elements of S_0 are zero, no weight is assigned to the a priori state estimate, and Eq. (13-29) is equivalent to Eq. (13-26). Commonly, S_0 has the form

$$
S_0 = \begin{bmatrix} \sigma_{x1}^{-2} & 0 & \cdots & 0 \\ 0 & \sigma_{x2}^{-2} & & \\ \vdots & & \ddots & \\ 0 & & & \sigma_{xm}^{-2} \end{bmatrix}
$$

where σ_{xk} $(k=1,2,\ldots,m)$ is the uncertainty in the a priori estimate, $\tilde{x}_A^0$. The use of S_0 is especially valuable when lack of *observability* is a problem. This occurs when a change in one or more state parameters causes little change in the observations, i.e., when the observations do not contain enough information to completely specify the state. (The problem of state observability is discussed from a practical point of view in Chapter 14.)

The loss function given in Eq. (13-29) is particularly useful in the later discussion of sequential estimators. Other criteria for goodness of fit are discussed by Hamming [1962].

Locating the Loss Function Minimum.   For J to be a minimum with respect to x^0, $\partial J/\partial x^0$ must be zero. Therefore, the value of x^0 which minimizes J is a root of the equation

$$
\frac{\partial J}{\partial x^0} = -\rho^T WG + \left[x^0 - \tilde{x}_A^0\right]^T S_0 = 0^T \tag{13-30a}
$$

where G is the $(n \times m)$ matrix

$$
G \equiv \frac{\partial g}{\partial x^0} \equiv \begin{bmatrix} \dfrac{\partial g_1}{\partial x_1^0} & \dfrac{\partial g_1}{\partial x_2^0} & \cdots & \dfrac{\partial g_1}{\partial x_m^0} \\ & \vdots & & \\ \dfrac{\partial g_n}{\partial x_1^0} & \dfrac{\partial g_n}{\partial x_2^0} & \cdots & \dfrac{\partial g_n}{\partial x_m^0} \end{bmatrix} \tag{13-30b}
$$

Values for $\partial g_i/\partial x$ are normally computed analytically from the observation model. Values for $\partial g_i/\partial x^0$ are then calculated from

$$
\frac{\partial g_i}{\partial x^0} = \frac{\partial g_i}{\partial x}(t_i)\frac{\partial x}{\partial x^0}(t_i) = \frac{\partial g_i}{\partial x}(t_i)D(t_i,t_0) \tag{13-31}
$$

where $D(t_i,t_0)$ is the $(m \times m)$ *state transition matrix* consisting of the partial derivatives of the state at t_i with respect to the state at the epoch time, t_0; that is,

$$
D(t_i,t_0) \equiv \frac{\partial x}{\partial x^0}(t_i) \equiv \begin{bmatrix} \dfrac{\partial x_1(t_i)}{\partial x_1^0} & \dfrac{\partial x_1(t_i)}{\partial x_2^0} & \cdots & \dfrac{\partial x_1(t_i)}{\partial x_m^0} \\ & \vdots & & \\ \dfrac{\partial x_m(t_i)}{\partial x_1^0} & \dfrac{\partial x_m(t_i)}{\partial x_2^0} & \cdots & \dfrac{\partial x_m(t_i)}{\partial x_m^0} \end{bmatrix}
$$

The elements of D may be calculated either numerically or analytically, depending on the functional form of $\mathbf{h}(\mathbf{x}^0, t)$. If $\mathbf{x}$ is assumed to be constant, then $D(t_i, t_0)$ is the identity, and

$$\frac{\partial g_i}{\partial \mathbf{x}^0} = \frac{\partial g_i}{\partial \mathbf{x}}$$

The most common method of solving Eq. (13-30) is to linearize $\mathbf{g}$ about a *reference state vector*, $\mathbf{x}_R^0$, and expand each element of $\mathbf{g}$ in a Taylor Series of $\mathbf{x}_R^0$. Note that $\mathbf{x}_R^0$ may be different from $\tilde{\mathbf{x}}_A^0$. If higher order terms are truncated, this yields for each element of $\mathbf{g}$:

$$g_i = g_i(\mathbf{x}_R^0) + \frac{\partial g_i}{\partial \mathbf{x}^0}(\mathbf{x}_R^0)[\mathbf{x}^0 - \mathbf{x}_R^0]$$

In general, each element of $\mathbf{g}$ could be evaluated at a different reference vector. Expressing the above equation in vector form gives

$$\mathbf{g} = \mathbf{g}_R + G_R \mathbf{x}^0 - G_R \mathbf{x}_R^0 \tag{13-32}$$

if the same reference vector is used for each element of $\mathbf{g}$. (The possibility of using distinct reference vectors for different elements of $\mathbf{g}$ will be useful in the later development of a sequential least-squares algorithm.) The subscript R signifies evaluation at $\mathbf{x}^0 = \mathbf{x}_R^0$.

Substituting Eq. (13-32) into Eq. (13-30) yields

$$[S_0 + G_R^T W G_R]\mathbf{x}^0 = S_0 \tilde{\mathbf{x}}_A^0 + G_R^T W[\mathbf{y} - \mathbf{g}_R + G_R \mathbf{x}_R^0] \tag{13-33}$$

We now solve this equation for $\mathbf{x}^0$, and denote the result by $\tilde{\mathbf{x}}^0$,

$$\tilde{\mathbf{x}}^0 = \mathbf{x}_R^0 + [S_0 + G_R^T W G_R]^{-1}[G_R^T W(\mathbf{y} - \mathbf{g}_R) + S_0(\tilde{\mathbf{x}}_A^0 - \mathbf{x}_R^0)] \tag{13-34}$$

If $\mathbf{x}_R^0 = \tilde{\mathbf{x}}_A^0$, and if $\mathbf{g}$ is a linear function in $\mathbf{x}^0$, then this equation will provide the best estimate for $\mathbf{x}_0$. If $\mathbf{g}$ is nonlinear, $\mathbf{x}^0$ will not be corrected exactly by Eq. (13-34) unless $\tilde{\mathbf{x}}_A^0$ is already very close to the optimum value.

If the correction determined from Eq. (13-34) is not small, then an iterative procedure is usually necessary. In this case, $\mathbf{g}$ is first linearized about the a priori estimate, which is then corrected to become $\tilde{\mathbf{x}}_1^0$, as follows:

$$\tilde{\mathbf{x}}_1^0 = \tilde{\mathbf{x}}_A^0 + [S_0 + G_A^T W G_A]^{-1}[G_A^T W(\mathbf{y} - \mathbf{g}_A)] \tag{13-35}$$

The corrected value, $\tilde{\mathbf{x}}_1^0$, then replaces $\tilde{\mathbf{x}}_A^0$ as a reference for the linearization of $\mathbf{g}$ in the next iteration. The $(k+1)$st estimate for $\mathbf{x}^0$ is derived from

$$\tilde{\mathbf{x}}_{k+1}^0 = \tilde{\mathbf{x}}_k^0 + [S_0 + G_k^T W G_k]^{-1}[G_k^T W(\mathbf{y} - \mathbf{g}_k) + S_0(\tilde{\mathbf{x}}_A^0 - \tilde{\mathbf{x}}_k^0)] \tag{13-36}$$

These iterations continue until the *differential correction* (i.e., the difference between $\tilde{\mathbf{x}}_{k+1}^0$ and $\tilde{\mathbf{x}}_k^0$) approaches zero and/or until the loss function no longer decreases. At this time, $\tilde{\mathbf{x}}_{k+1}^0$ has *converged* to its optimum value. If the estimator fails to converge, a new a priori estimate should be attempted. If this is not successful, improved mathematical modeling, additional data, or higher quality data may be necessary. A block diagram of the batch least-squares alorithm is shown in Fig. 13-2.

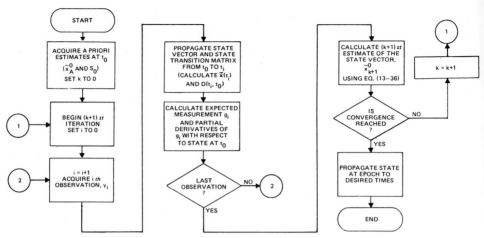

Fig. 13-2.   Block Diagram of Batch Least-Squares Estimator Algorithm

Statistical Information.   For a converged solution, several statistical quantities are useful. The $(m \times m)$ *error covariance matrix* is given by*

$$P = \left[S_0 + G^T W G \right]^{-1} = E(ee^T) \qquad (13\text{-}37)$$

assuming that $E(e) = 0$, where the estimation error vector $e \equiv x^0 - \tilde{x}^0$, and E denotes expected value. Provided the estimation process has converged, uncertainties in the estimated state parameters may be calculated from the diagonal elements of P by

$$\sigma_j = \sqrt{P_{jj}} \qquad (13\text{-}38)$$

These uncertainties are realistic error estimates only if the observations are uncorrelated and contain only random errors. The mathematical models characterizing state propagation and the relationship of the observations to the state are also considered to be known with sufficient accuracy. As discussed in Chapter 14, these assumptions are seldom fulfilled completely in practice. To account for this problem, Bryson and Ho [1968] recommend modifying the uncertainty to be

$$\sigma_j = \sqrt{2 J_0 P_{jj} / (n + m)} \qquad (13\text{-}39)$$

where J_0 is the loss function based on the final estimate for x^0 and $E(J) = \frac{1}{2}(n + m)$ is the expected value of J for n observations and m state vector elements (See Eq. (13-29).)

The off-diagonal elements of P represent the interdependence or *correlation* among errors in state parameters. The *correlation coefficient*,

$$C_{jl} \equiv \frac{P_{jl}}{\sqrt{P_{jj} P_{ll}}} \qquad (13\text{-}40)$$

---

*The properties and physical significance of the error covariance matrix are more fully described in Section 12.3. Eqs. (13-37) and (12-74) are equivalent if $S_0 = 0$, and if the components of the observation vector are uncorrelated so that W is diagonal.

measures the correlation between the jth and lth state parameters. Correlation coefficients range from -1 to $+1$; either extreme value indicates that the two parameters are completely dependent and one may be eliminated from the estimation process.

Another useful quantity is the *weighted root-mean-square (rms) residual*, given by

$$\rho_{rms} \equiv \left[\frac{\sum\limits_{i=1}^{n} W_i (y_i - g_i)^2}{\sum\limits_{i=1}^{n} W_i} \right]^{\frac{1}{2}} \tag{13-41}$$

where W_i is the ith diagonal element of the weight matrix and the units of ρ_{rms} are the same as for the y_i. The *rms* residual must be calculated using only observations of the same units. If y contains observations of different types, then a ρ_{rms} value may be calculated for observations of each data type.*

Because ρ_{rms} is normalized according to the sum of observation weights, it is frequently more useful than the loss function as a relative measure of the degree to which the solution fits the observed data. However, this parameter alone is insufficient for detecting the two major causes of a poor fit—unmodeled biases and a high level of noise in the observations. Some insight into the contributions of these two phenomena in specific cases is gained by writing ρ_{rms} in the form

$$\rho_{rms}^2 = k^2 + \sigma_\rho^2 \tag{13-42}$$

where k is the weighted mean of residuals,

$$k \equiv \frac{\sum W_i \rho_i}{\sum W_i} \tag{13-43}$$

and σ_ρ is the weighted rms deviation of the residuals,

$$\sigma_\rho \equiv \left[\frac{\sum W_i (\rho_i - k)^2}{\sum W_i} \right]^{1/2} \tag{13-44}$$

The mean of the residuals should be near zero, because the ρ_i can be either positive or negative. A large value for k indicates that unmodeled biases are probably present in the observations. A large value of σ_ρ indicates that the observation noise is large.

---

*Alternatively, a fractional rms residual, r_{rms}, may be calculated using all observations, as follows:

$$r_{rms} = \left[\frac{\sum\limits_{i=1}^{n} W_i (y_i - g_i)^2}{\sum\limits_{i=1}^{n} W_i y_i^2} \right]^{\frac{1}{2}}$$

Example of a Simplified Batch Least-Squares Application. Consider a spacecraft that is spinning uniformly about the axis $\hat{\mathbf{A}}$, which may be expressed in rectangular celestial coordinates (Section 2.2) as

$$\hat{\mathbf{A}} = A_x \hat{\mathbf{X}} + A_y \hat{\mathbf{Y}} + A_z \hat{\mathbf{Z}}$$

or in terms of right ascension, α, and declination, δ, as

$$A_x = \cos\alpha \cos\delta$$

$$A_y = \sin\alpha \cos\delta$$

$$A_z = \sin\delta$$

Suppose that $\hat{\mathbf{A}}$ is fixed in inertial space during an interval spanned by a series of observations from an onboard sensor, and that an initial estimate for $\hat{\mathbf{A}}$ is available using attitude solutions from previous intervals. Suppose, also, that the observations consist of arc lengths, θ_i, between $\hat{\mathbf{A}}$ and a known time-varying reference vector, $\hat{\mathbf{U}}(t)$, such as the nadir vector. We choose

$$\mathbf{x} = \begin{bmatrix} \alpha \\ \delta \end{bmatrix}$$

as the state vector. Because $\hat{\mathbf{A}}$ is constant, $\mathbf{x} = \mathbf{x}^0$ and D is the identity matrix. The observation vector consists of n values of θ_i, i.e.,

$$\mathbf{y} = \left[\theta_1, \theta_2, \ldots, \theta_n \right]^{\mathrm{T}}$$

To construct the observation model vector, $\mathbf{g}$, we express the θ_i in terms of the elements of $\hat{\mathbf{A}}$ by

$$\theta_i = \cos^{-1}\left(\hat{\mathbf{U}}_i \cdot \hat{\mathbf{A}}\right) = \cos^{-1}\left(U_{x_i} A_x + U_{y_i} A_y + U_{z_i} A_z\right)$$

where $\hat{\mathbf{U}}_i$ be calculated at the time of the ith observation, i.e., $\hat{\mathbf{U}}_i \equiv \hat{\mathbf{U}}(t_i)$. The elements of $\mathbf{g}$ are then given by

$$g_i = \cos^{-1}\left(U_{x_i}\cos\alpha\cos\delta + U_{y_i}\sin\alpha\cos\delta + U_{z_i}\sin\delta\right)$$

The $(n \times 2)$ matrix of partial derivatives of the observation model with respect to the state vector is

$$G = \begin{bmatrix} \partial g_1/\partial\alpha & \partial g_1/\partial\delta \\ \vdots & \vdots \\ \partial g_n/\partial\alpha & \partial g_n/\partial\delta \end{bmatrix}$$

where

$$\frac{\partial g_i}{\partial\alpha} = \frac{U_{x_i}\sin\alpha\cos\delta - U_{y_i}\cos\alpha\cos\delta}{\sqrt{1 - \left(U_{x_i}\cos\alpha\cos\delta + U_{y_i}\sin\alpha\cos\delta + U_{z_i}\sin\delta\right)^2}}$$

and

$$\frac{\partial g_i}{\partial \delta} = \frac{U_{x_i}\cos\alpha\sin\delta + U_{y_i}\sin\alpha\sin\delta - U_{z_i}\cos\delta}{\sqrt{1 - (U_{x_i}\cos\alpha\cos\delta + U_{y_i}\sin\alpha\cos\delta + U_{z_i}\sin\delta)^2}}$$

If the a priori state vector estimate, $\tilde{\mathbf{x}}_A = [\alpha_A, \delta_A]^T$, is known to an estimated accuracy of $[\sigma_\alpha, \sigma_\delta]$ and all observations are measured with equal estimated accuracies of σ_θ, then Eq. (13-35) gives the following solution for $\tilde{\mathbf{x}}$, assuming that $\tilde{\mathbf{x}}_A$ and $\mathbf{y}$ are weighted according to their expected accuracies:

$$\tilde{\mathbf{x}}_1 = \begin{bmatrix} \alpha_0 \\ \delta_0 \end{bmatrix} + \begin{bmatrix} \begin{bmatrix} \sigma_\alpha^{-2} & 0 \\ 0 & \sigma_\delta^{-2} \end{bmatrix} + \sigma_\theta^{-2}G_A^T G_A \end{bmatrix}^{-1} \begin{bmatrix} \sigma_\theta^{-2}G_A^T(\mathbf{y} - \mathbf{g}_A) \end{bmatrix}$$

This solution may be improved by additional iterations. Equation (13-36) gives the solution in the $(k+1)$st iteration as:

$$\tilde{\mathbf{x}}_{k+1} = \begin{bmatrix} \alpha_k \\ \delta_k \end{bmatrix} + \begin{bmatrix} \begin{bmatrix} \sigma_\alpha^{-2} & 0 \\ 0 & \sigma_\delta^{-2} \end{bmatrix} + \sigma_\theta^{-2}G_k^T G_k \end{bmatrix}^{-1} \begin{bmatrix} \sigma_\theta^{-2}G_A^T(\mathbf{y} - \mathbf{g}_k) + \begin{bmatrix} \sigma_\alpha^{-2} & 0 \\ 0 & \sigma_\delta^{-2} \end{bmatrix} \begin{bmatrix} \alpha_0 - \alpha_k \\ \delta_0 - \delta_k \end{bmatrix} \end{bmatrix}$$

Corrections to $\mathbf{x}$ will continue in this fashion until convergence is achieved.

Convergence and Marquardt's Algorithm. The Gauss-Newton differential correction procedure outlined above may be unsuitable for some nonlinear problems because convergence cannot be guaranteed unless the a priori estimate is close to a minimum in the loss function. Moreover, its rate of convergence can be difficult to control [Melkanoff and Sanada, 1966; Wilde, 1964]. An alternative approach to solving the batch least-squares problem which guarantees convergence is the *gradient search method*, or *method of steepest descent*. With this technique, the state parameters are adjusted so that the resultant direction of travel in state space is along the negative gradient of J, i.e., in the direction of steepest descent of the loss function. Although this method initially converges rapidly, it slows down when the solution approaches the vicinity of the minimum.

To overcome both the difficulties of the Gauss-Newton technique when an accurate initial estimate is not available, and the slow convergence problems of the gradient search approach when the solution is close to the loss function minimum, D. W. Marquardt [1963] proposed an algorithm which performs an optimal interpolation between the two techniques. For simplicity, let $S_0 = 0$, reflecting no confidence in the a priori estimate. Equation (13-29) then shows that

$$\frac{\partial J}{\partial \mathbf{x}^0} = -\rho^T W G \tag{13-45}$$

Correction of the state estimate $\tilde{\mathbf{x}}_k^0$ in the direction of the negative gradient of J yields the following expression for $\tilde{\mathbf{x}}_{k+1}^0$:

$$\tilde{\mathbf{x}}_{k+1}^0 = \tilde{\mathbf{x}}_k^0 + \lambda^{-1}G_k^T \dot{W}[\mathbf{y} - \mathbf{g}_k] \tag{13-46}$$

where λ is a proportionality constant. The Marquardt technique uses an expression of the form

$$\tilde{\mathbf{x}}_{k+1}^0 = \tilde{\mathbf{x}}_k^0 + \left[G_k^T W G_k + \lambda \mathbf{1} \right]^{-1} G_k^T W \left[\mathbf{y} - \mathbf{g}_k \right] \qquad (13\text{-}47)$$

If λ is small, Eq. (13-47) is equivalent to the Gauss-Newton procedure. If λ is large, $\mathbf{x}^0$ is corrected in the direction of the negative gradient of J, but with a magnitude which decreases as λ increases.

An example of the use of Marquardt's algorithm for improved convergence is as follows:

1. Compute the loss function using the a priori state estimate, $\tilde{\mathbf{x}}_A^0$.
2. Apply the first state correction to the state to form $\tilde{\mathbf{x}}_1^0$ using Eq. (13-47) with $\lambda >> G^T W G$.
3. Recompute the loss function at $\tilde{\mathbf{x}}^0 = \tilde{\mathbf{x}}_1^0$. If $J(\tilde{\mathbf{x}}_1^0) \geq J(\tilde{\mathbf{x}}_A^0)$, then $\tilde{\mathbf{x}}_1^0$ is discarded and λ is replaced by λK, where K is a fixed positive constant, usually between 1 and 10. The state estimate $\tilde{\mathbf{x}}_1^0$ is then recomputed uisng the new value of λ in Eq. (13-47). If $J(\tilde{\mathbf{x}}_1^0) < J(\tilde{\mathbf{x}}_A^0)$, then $\tilde{\mathbf{x}}_1^0$ is retained, but λ is replaced by λ / K.
4. After each subsequent iteration, compare $J(\tilde{\mathbf{x}}_{k+1}^0)$ and replace λ by λK or λ / K as in step 3. The state estimate $\tilde{\mathbf{x}}_{k+1}^0$ is retained if J continues to decrease and discarded if J increases.

This procedure continues until the difference in J between two consecutive iterations is small, or until λ reaches a small value. Additional details are given by Marquardt [1963] and Bevington [1969].

Advantages and Disadvantages of Batch Estimators. The major advantage of batch estimators is that they are the simplest to implement; they are also generally less sensitive to bad data points than are the somewhat more sophisticated algorithms described below. Another advantage of batch estimators is that all observation residuals can be seen simultaneously, so that any obviously invalid observations, i.e., those with unusually large residuals, can be removed. An observation is commonly removed if the absolute value of its residual is greater than three times the weighted rms residual.

The computer execution time required for a batch estimator depends on the number of state parameters, the number of observations, the complexity of the state and observation models, and the number of iterations required for convergence. If a large number of iterations is required, a recursive estimator should be considered. The computer storage required to contain the observations for possible future iterations is also a disadvantage of batch estimators; therefore, for applications in which computer storage is limited, recursive estimators or Kalman filters, described in the next section, may be preferred.

Example of a "Single-Frame" Least-Squares Estimator. In the previous example in this section, we wished to determine the constant attitude of a spinning spacecraft based on large number of measurements at different times. In this example we assume that there are several measurements at one time and that we wish to determine the attitude at that time. (This is commonly done when the control and environmental torques are poorly known and it is impossible to predict how the attitude will change between data samples.)

In particular, we wish to compute the three-axis attitude for an Earth-oriented spacecraft with a horizon scanner, a two-axis digital Sun sensor, and a three-axis magnetometer. The attitude is parameterized by pitch (ξ_p), roll (ξ_r), and yaw (ξ_y). Thus, the state vector is

$$\mathbf{x} = (\xi_p, \xi_r, \xi_y)^T$$

with the a priori estimate, $\tilde{\mathbf{x}}_A = (0,0,0)^T$. Note that all observation vector and state vector elements are evaluated at the same time, so that the functional dependence on time is ignored. The seven-component observation vector is

$$\mathbf{y} = (p_m, r_m, H_x, H_y, H_z, NA, NB)^T \tag{13-48}$$

where p_m and r_m are the measured pitch and roll angles from the horizon scanner, $\mathbf{H} = (H_x, H_y, H_z)^T$ is the measured magnetic field in spacecraft body coordinates, and NA and NB are the measured Sun sensor reticle counts (see Section 7.1).

To simplify the construction of the seven-component observation model vector, $\mathbf{g}(\mathbf{x})$, we define yaw, roll, and pitch as the 3-1-2 Euler angle sequence which rotates a vector from orbital to body coordinates (see Section 12.2). The Sun and magnetic field vectors in orbital coordinates, $\hat{\mathbf{S}}_0$ and $\mathbf{H}_0$, are obtained from an ephemeris and magnetic field model. The nadir vector is $\hat{\mathbf{E}}_0 = (0,0,1)^T$ by the definition of the orbital coordinate system. The Sun, magnetic field, and nadir vectors in body coordinates are

$$\hat{\mathbf{S}}_B = A\hat{\mathbf{S}}_0; \qquad \hat{\mathbf{H}}_B = A\mathbf{H}_0; \qquad \hat{\mathbf{E}}_B = A\hat{\mathbf{E}}_0 \tag{13-49}$$

where, from Table E-1, the attitude matrix is

$$A(\xi_p, \xi_r, \xi_y) =$$

$$\begin{bmatrix} \cos\xi_y\cos\xi_p - \sin\xi_y\sin\xi_r\sin\xi_p & \sin\xi_y\cos\xi_p + \cos\xi_y\sin\xi_r\sin\xi_p & -\cos\xi_r\sin\xi_p \\ -\sin\xi_y\cos\xi_r & \cos\xi_y\cos\xi_r & \sin\xi_r \\ \cos\xi_y\sin\xi_p + \sin\xi_y\sin\xi_r\cos\xi_p & \sin\xi_y\sin\xi_p - \cos\xi_y\sin\xi_r\cos\xi_p & \cos\xi_r\cos\xi_p \end{bmatrix} \tag{13-50}$$

Substitution of Eq. (13-50) with $\mathbf{E}_0 = (0,0,1)^T$ into the above expression for $\hat{\mathbf{E}}_B$ and comparison with Eq. (12-51) (see Section 12.2) gives the first two observation model equations as*

$$g_1 = \xi_p; \qquad g_2 = \xi_r \tag{13-51}$$

The reason for choosing the 3-1-2 Euler angle sequence is apparent from the simple form of Eq. (13-51). The observation model equations for $\mathbf{H}$ are given directly from the second part of Eq. (13-49) as

$$(g_3, g_4, g_5)^T = A\mathbf{H}_0 \tag{13-52}$$

Finally, the observation model equations for NA and NB are obtained using Eqs. (7-23) through (7-26), with the result

$$g_6 = \text{INT}(a/k_m + 2^{m-1}); \qquad g_7 = \text{INT}(b/k_m + 2^{m-1}) \tag{13-53}$$

---

*The sign of ξ_p and ξ_r, as defined here, is opposite to Ω_E and β_E defined in Section 12.2.

where

$$a = S_2 \gamma^{1/2}$$

$$b = S_1 \gamma^{1/2}$$

$$\gamma = h^2 / (n^2 - S_1^2 - S_2^2)$$

where

$$(S_1, S_2, S_3)^T \equiv A_{SS}^T \hat{S}_B = A_{SS}^T A \hat{S}_0$$

A_{SS} is the transformation matrix from Sun sensor to body coordinates (see Eq. (7-9), and m, k_m, h, and n are sensor constants defined in Table 7-2.

The partial derivatives of the observation model vector elements, g_i, with respect to the state vector elements, x_j, are

$$\frac{\partial g_1}{\partial x_j} = \delta_{1j}; \qquad \frac{\partial g_2}{\partial x_j} = \delta_{2j} \tag{13-54a}$$

$$\frac{\partial g_i}{\partial x_j} = \sum_{k=1}^{3} \frac{\partial A_{ik}}{\partial x_j} H_{0k}; \qquad i = 3, 4, 5 \tag{13-54b}$$

$$\frac{\partial g_6}{\partial x_j} = \frac{1}{k_m} \left(\frac{\partial S_2}{\partial x_j} \gamma^{1/2} + \frac{S_2}{2\gamma^{1/2}} \frac{\partial \gamma}{\partial x_j} \right) \tag{13-54c}$$

$$\frac{\partial g_7}{\partial x_j} = \frac{1}{k_m} \left(\frac{\partial S_1}{\partial x_j} \gamma^{1/2} + \frac{S_1}{2\gamma^{1/2}} \frac{\partial \gamma}{\partial x_j} \right) \tag{13-54d}$$

where

$$\frac{\partial \gamma}{\partial x_j} = \frac{2\gamma^2}{h^2} \left(S_1 \frac{\partial S_1}{\partial x_j} + S_2 \frac{\partial S_2}{\partial x_j} \right) \tag{13-54e}$$

$$\frac{\partial S_i}{\partial x_j} = \sum_{k,l=1}^{3} \frac{\partial A_{kl}}{\partial x_j} A_{SS_{ki}} S_{0l}; \qquad i = 1, 2, 3 \tag{13-54f}$$

and $j = 1, 2,$ or 3.

The observation weights are taken to be the inverse of the corresponding variance. For pitch and roll data, the weights are

$$w_1 = 1/\sigma_p^2; \qquad W_2 = 1/\sigma_r^2 \tag{13-55}$$

and for the magnetometer data, the three observation weights are assumed to be equal and given by

$$w_3 = w_4 = w_5 = 1/\sigma_{mag}^2 \tag{13-56}$$

The errors in the Sun sensor data are assumed to be dominated by the step size (see Section 12.3); hence,

$$w_6 = w_7 = 12 \tag{13-57}$$

Note that the weights have the same units as the square of the corresponding inverse measurement. The elements of the weighted matrix, W, are given by

$$W_{ij} = w_i \delta_{ij}; \qquad i, j = 1, 2, 3 \tag{13-58}$$

assuming that the measurement errors are uncorrelated.

With the above definitions, we now wish to find the state vector estimate, $\tilde{\mathbf{x}}$, which minimizes the loss function defined by Eq. (13-29) with $S_0 = 0$ (to indicate no confidence in the a priori solution). The solution is given by Eq. (13-36) with the a priori solution $\tilde{\mathbf{x}}_0 \equiv \tilde{\mathbf{x}}_A$ as

$$\tilde{\mathbf{x}}_{k+1} = \tilde{\mathbf{x}}_k + \left(G_k^T W G_k \right)^{-1} G_k^T W \left[\mathbf{y} - \mathbf{g}_k \right]$$

where G_k is given by Eqs. (13-30b) and (13-54) and the subscript k denotes that the partial derivatives are evaluated at $\mathbf{x} = \tilde{\mathbf{x}}_k$. The covariance matrix of the computed state vector (see Section 12.3),

$$P = \left(G_k^T W G_k \right)^{-1} \tag{13-60}$$

is obtained as a byproduct of the differential correction algorithm.

13.5 Recursive Least-Squares Estimators and Kalman Filters

Lawrence Fallon, III

13.5.1 Recursive Least-Squares Estimation

Consider an n-component observation vector, $\mathbf{y}$, which is partitioned into p members; that is

$$\mathbf{y} = \begin{bmatrix} \mathbf{y}_1 \\ \mathbf{y}_2 \\ \vdots \\ \mathbf{y}_p \end{bmatrix} \tag{13-61}$$

Each member contains q observations which are generally measured at nearly the same time. For example, consider an observation vector which contains $n = 100$ star tracker measurements obtained in a 30-minute interval. If $\mathbf{y}$ consists of angular coordinates which are measured two at a time, then it would be partitioned into $p = 50$ members, each containing $q = 2$ components. The observation model vector, $\mathbf{g}$, is also partitioned in the same manner as $\mathbf{y}$.

A batch least-squares estimate of the m-component state vector $\mathbf{x}^0$, determined with observations from only member $\mathbf{y}_1$, will be denoted $\tilde{\mathbf{x}}_1^0$. The state estimate determined using observations from both members $\mathbf{y}_1$ and $\mathbf{y}_2$ will be denoted $\tilde{\mathbf{x}}_2^0$, and so forth. We want an expression for $\tilde{\mathbf{x}}_2^0$ using $\tilde{\mathbf{x}}_1^0$ and observations from member $\mathbf{y}_2$. This will then be extended to form an expression for $\tilde{\mathbf{x}}_k^0$, using $\tilde{\mathbf{x}}_{k-1}^0$ and the observations from member $\mathbf{y}_k$.

The loss function in Eq. (13-29) leads to the following relation for $\tilde{\mathbf{x}}_1^0$:

$$\left[S_0 + G_{1R_1}^T W_1 G_{1R_1} \right] \tilde{\mathbf{x}}_1^0 = S_0 \tilde{\mathbf{x}}_A^0 + G_{1R_1}^T W_1 \left[\mathbf{y}_1 - \mathbf{g}_{1R_1} + G_{1R_1} \mathbf{x}_{R_1}^0 \right] \tag{13-62}$$

which is equivalent to Eq. (13-33), except that W_1, G_{1R_1}, and $\mathbf{g}_{1R_1}$ contain observations from member $\mathbf{y}_1$ only. The subscript R_1 signifies evaluation at reference state vector $\mathbf{x}_{R_1}^0$. (We will use a different reference state vector for each of the p members of $\mathbf{g}$.) Similarly, for $\mathbf{x}_2^0$,

$$\left[S_0 + G_{SR}^T W_S G_{SR} \right] \tilde{\mathbf{x}}_2^0 = S_0 \tilde{\mathbf{x}}_A^0 + G_{SR}^T W_S \left[\mathbf{y}_S - \mathbf{g}_{SR} + \mathbf{b}_{SR} \right] \tag{13-63}$$

where the subscript S means that observations from both members $\mathbf{y}_1$ and $\mathbf{y}_2$ are included. The $2q$-vectors $\mathbf{y}_S$ and $\mathbf{b}_{SR}$ are defined by

$$\mathbf{y}_S = \begin{bmatrix} \mathbf{y}_1 \\ \mathbf{y}_2 \end{bmatrix}$$

$$\mathbf{b}_{SR} = \begin{bmatrix} G_{1R_1}\mathbf{x}^0_{R_1} \\ G_{2R_2}\mathbf{x}^0_{R_2} \end{bmatrix}$$

where $\mathbf{x}^0_{R_2}$ is a reference state vector which is generally different from $\mathbf{x}^0_{R_1}$. The $2q$-vector $\mathbf{g}_{SR}$, the $(2q \times m)$ matrix G_{SR}, and the $(2q \times 2q)$ matrix W_S are analogous to $\mathbf{y}_S$ and $\mathbf{b}_{SR}$.

Using Eq. (13-62) and (13-63), we obtain

$$S_{2R}\tilde{\mathbf{x}}^0_2 = S_{1R}\tilde{\mathbf{x}}^0_1 + G^T_{2R_2}W_2\left[\mathbf{y}_2 - \mathbf{g}_{2R_2} + G_{2R_2}\mathbf{x}^0_{R_2}\right] \qquad (13\text{-}64)$$

where the $(m \times m)$ matrices S_{1R} and S_{2R} are defined by

$$S_{1R} \equiv S_0 + G^T_{1R_1}W_1 G_{1R_1}$$

$$S_{2R} \equiv S_{1R} + G^T_{2R_2}W_2 G_{2R_2}$$

We emphasize that the $(q \times q)$ matrix W_2, the $(q \times m)$ matrix G_{2R_2}, and the q-vector $\mathbf{g}_{2R_2}$ pertain to observations from $\mathbf{y}_2$ only. Solving Eq. (13-64) for $\tilde{\mathbf{x}}^0_2$ yields

$$\tilde{\mathbf{x}}^0_2 = \tilde{\mathbf{x}}^0_1 + P_2 G^T_{2R_2}W_2\left[\mathbf{y}_2 - \mathbf{g}_{2R_2} + G_{2R_2}\left(\mathbf{x}^0_{R_2} - \tilde{\mathbf{x}}^0_1\right)\right] \qquad (13\text{-}65a)$$

where the $(m \times m)$ matrix P_2 is given by

$$P_2 \equiv S^{-1}_{2R} = \left[P^{-1}_1 + G^T_{2R_2}W_2 G_{2R_2}\right]^{-1} \qquad (13\text{-}65b)$$

and the $(m \times m)$ matrix P_1,

$$P_1 \equiv S^{-1}_{1R} = \left[S_0 + G^T_{1R_1}W_1 G^T_{1R_1}\right]^{-1} \qquad (13\text{-}65c)$$

Equation (13-65) is an expression for $\tilde{\mathbf{x}}^0_2$ which depends on $\tilde{\mathbf{x}}^0_1$, P_1 (the covariance of error in $\tilde{\mathbf{x}}^0_1$), and quantities associated with observations from member $\mathbf{y}_2$. Once $\tilde{\mathbf{x}}^0_1$ and P_1 have been calculated, observations from $\mathbf{y}_1$ are no longer necessary for the estimation of $\tilde{\mathbf{x}}_2$.

By analogy, the expression for $\tilde{\mathbf{x}}^0_k$, the state estimate using the first k members of the observation vector, is

$$\tilde{\mathbf{x}}^0_k = \tilde{\mathbf{x}}^0_{k-1} + P_k G^T_{kR_k}W_k\left[\mathbf{y}_k - \mathbf{g}_{kR_k} + G_{kR_k}\left(\mathbf{x}^0_{R_k} - \tilde{\mathbf{x}}^0_{k-1}\right)\right] \qquad (13\text{-}66)$$

In many sequential least-squares applications, the estimate derived from processing the previous observations becomes the reference vector for the current estimation, i.e., $\mathbf{x}^0_{R_k} = \tilde{\mathbf{x}}^0_{k-1}$. In such cases, the state estimation is frequently not iterated in the batch least-squares sense. The state estimate "improves" as additional data are processed. Occasionally, however, the same reference vector will be used for the entire group of data, and iterations may or may not be used.

The algorithm given by Eq. (13-66) requires the inversion of the $(m \times m)$ matrix S_{kR} at each step to compute P_k. This algorithm may be transformed into one which requires inverting a $(q \times q)$ matrix. Recall that q is the dimension of the members of $\mathbf{y}$, which are generally smaller than the m-dimensional state vector. Applying the matrix identity

$$[A^{-1} + BC]^{-1} = A - AB[1 + CAB]^{-1}CA \qquad (13\text{-}67)$$

to Eq. (13-65b) and extending the result to the kth estimate of P yields

$$P_k = P_{k-1} - P_{k-1}G_k^T[R_k + G_k P_{k-1} G_k^T]^{-1} G_k P_{k-1}$$

where R_k, the $(q \times q)$ *measurement covariance matrix*, is the inverse of W_k. Substitution of Eq. (13-67) into Eq. (13-66) and some matrix manipulation yields

$$\tilde{\mathbf{x}}_k^0 = \tilde{\mathbf{x}}_{k-1}^0 + K_k\left[\mathbf{y}_k - \mathbf{g}_k + G_k\left(\tilde{\mathbf{x}}_{R_k}^0 - \tilde{\mathbf{x}}_{k-1}^0\right)\right] \qquad (13\text{-}68)$$

where the $(m \times q)$ *gain matrix*, K_k, is given by

$$K_k = P_{k-1}G_k^T[R_k + G_k P_{k-1} G_k^T]^{-1} \qquad (13\text{-}69)$$

If Eq. (13-69) is now substituted into Eq. (13-67), we obtain the following algorithm for P_k, the error covariance matrix of the state estimate, $\tilde{\mathbf{x}}_k^0$:

$$P_k = [1 - K_k G_k]P_{k-1} \qquad (13\text{-}70)$$

Equations (13-68) to (13-70) are the basic equations used in sequential least-squares estimators. In these equations, the q-vector $\mathbf{g}_k$ and the $(q \times m)$ matrix G_k are the parameters associated with observation k, evaluated at $\mathbf{x}^0 = \mathbf{x}_{R_k}^0$. If $\mathbf{x}_{R_k}^0 = \tilde{\mathbf{x}}_{k-1}^0$, then Eq. (13-68) reduces to

$$\tilde{\mathbf{x}}_k^0 = \tilde{\mathbf{x}}_{k-1}^0 + K_k[\mathbf{y}_k - \mathbf{g}_k] \qquad (13\text{-}71)$$

The $(q \times q)$ matrix to be inverted in Eq. (13-69) thus has the dimensions of $\mathbf{y}_k$. In a common application of this algorithm, the observations are processed one at a time. In this case, $q = 1$ and no matrix inversion is necessary.

Because of computer roundoff errors, P_k can become nonpositive definite and therefore meaningless. An alternative is to use the *Joseph algorithm* for the computation of P_k:

$$P_k = [1 - K_k G_k]P_{k-1}[1 - K_k G_k]^T + K_k W_k^{-1} K_k \qquad (13\text{-}72)$$

This algorithm requires more computation than Eq. (13-70), but ensures that P_k will remain positive definite. Substituting Eq. (13-71) into (13-72) reproduces Eq. (13-70), which indicates that the two methods are analytically equivalent for any K defined by Eq. (13-69). Figure 13-3 summarizes the procedures used by a recursive least-squares algorithm.

Advantages and Disadvantages of Recursive Least-Squares Estimators. The principal computational advantage of recursive least-squares estimators occurs in applications where iterations are not required. In these cases, the estimator converges (i.e., the difference between $\tilde{\mathbf{x}}_k^0$ and $\tilde{\mathbf{x}}_{k-1}^0$ approaches zero) as additional data

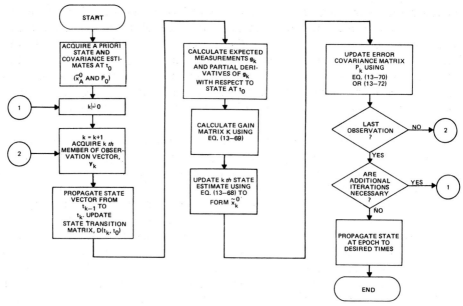

Fig. 13-3.   Block Diagram of Recursive Least-Squares Estimator Algorithm

are processed. This results from the use of $\tilde{x}^0_{k-1}$ as the reference state vector for the estimation of $\tilde{x}^0_k$. Only information pertaining to the kth set of observations must be stored in these cases. Use of a recursive estimator instead of a batch estimator (in which the reference state vector is not replaced until all observations have been processed) will thus result in a reduction of computer storage requirements and a decrease in execution time. The principal disadvantage of the recursive least-squares estimator is that it is more sensitive to bad data, particularly at the beginning of a pass, than is the batch estimator.

If the state undergoes minor unmodeled variation during the time spanned by the observations, the recursive least-squares estimator will calculate a weighted "average" value for $\tilde{x}^0$ which is essentially equivalent to the value estimated by a batch procedure. In contrast, the Kalman filter described below will generally track state variations better than either the recursive or batch least-squares algorithms.

13.5.2 Kalman Filters

To estimate the value of a state vector at an arbitrary time, t_k, the state estimate at t_0, from a batch or recursive algorithm, must be propagated from t_0 to t_k using a model of the system dynamics. The *Kalman filter*, on the other hand, estimates the m-component state vector $\tilde{x}(t_k)$ directly based on all observations up to and including y_k and the dynamics model evaluated between observations.*

---

*This subsection describes a continuous-discrete Kalman filter which assumes that the system dynamics varies continuously with time and that observations are available at discrete time points. The two other classes are *continuous* and *discrete Kalman filters*, in which both the system dynamics and observation availability are either continuous or discrete. The continuous-discrete filter is the most common for spacecraft attitude determination. Much of the development of these filters was done by R. E. Kalman [1960, 1961] in the early 1960s. The basic Kalman filter in each of these classes assumes that the observation models and system dynamics are linear. What we describe is actually an *extended Kalman filter* because nonlinear observation models will be allowed.

Although all filters require a dynamics model (the simplest of which is $\mathbf{x} =$ constant) to propagate the state estimate between observations, the accuracy requirements for this model are normally less severe for the Kalman filter than for batch or recursive estimators because propagation is not performed at one time over the entire block of data. In addition, the Kalman filter compensates for dynamics model inaccuracy by incorporating a noise term which gives the filter a *fading memory*—that is, each observation has a gradually diminishing effect on future state estimates.

Each time a set of q observations, $\mathbf{y}_k$, is obtained, the Kalman filter uses it to update the a priori state vector estimate at t_k, denoted by $\tilde{\mathbf{x}}_{k-1}(t_k)$, to produce an a posteriori estimate $\tilde{\mathbf{x}}_k(t_k)$. It also converts the a priori error covariance matrix estimate, $P_{k-1}(t_k)$, into the a posteriori estimate, $P_k(t_k)$. These a posteriori estimates are then propogated to t_{k+1} to become the a priori estimates $\tilde{\mathbf{x}}_k(t_{k+1})$ and $P_k(t_{k+1})$ for the next observation set, $\mathbf{y}_{k+1}(t_{k+1})$. The subscript k on $\tilde{\mathbf{x}}$ and P indicates that the estimate is based on all observations up to and including the observations in $\mathbf{y}_k$.

The updating equations for the Kalman filter are the same as those for the recursive least-squares estimator except that we are now estimating the state vector and covariance matrix values at the time t_k rather than at a fixed epoch t_0. Thus, the Kalman filter update equations are

$$\tilde{\mathbf{x}}_k(t_k) = \tilde{\mathbf{x}}_{k-1}(t_k) + K_k[\mathbf{y}_k - \mathbf{g}_k] \tag{13-73}$$

with the $(m \times q)$ gain matrix

$$K_k = P_{k-1}(t_k)G_k^T[R_k + G_k P_{k-1}(t_k)G_k^T]^{-1} \tag{13-74}$$

and either

$$P_k(t_k) = [1 - K_k G_k]P_{k-1}(t_k) \tag{13-75}$$

or, alternatively,

$$P_k(t_k) = [1 - K_k G_k]P_{k-1}(t_k)[1 - K_k G_k]^T + K_k R_k K_k^T \tag{13-76}$$

for the $(m \times m)$ error covariance matrix. In these equations, the q-vector $\mathbf{g}_k$ and the $(q \times m)$ matrix G_k are evaluated at $\tilde{\mathbf{x}}_{k-1}(t_k)$.

In some cases it is necessary to iterate the estimate of $\tilde{\mathbf{x}}_k(t_k)$ to reduce the effects of nonlinearities in the observation model. If this occurs, then $\mathbf{g}_k$ and G_k will be evaluated about a reference vector $\mathbf{x}_{R_k}(t_k)$, which may be different from $\tilde{\mathbf{x}}_{k-1}(t_k)$. Equation (13-73) is then replaced by the more general form

$$\tilde{\mathbf{x}}(t_k) = \mathbf{x}_{R_k}(t_k) + K_k[\mathbf{y}_k - \mathbf{g}_k + G_k\{\mathbf{x}_{R_k}(t_k) - \tilde{\mathbf{x}}_{k-1}(t_k)\}] \tag{13-77}$$

Iteration may then be done using Eq. (13-77) with $\mathbf{x}_{R_k}(t_k) = \tilde{\mathbf{x}}_{k-1}(t_k)$ to estimate $\tilde{\mathbf{x}}_k(t_k)$. The operation is cyclically repeated using $\mathbf{x}_{R_k}(t_k) = \tilde{\mathbf{x}}_k(t_k)$ and so on, until the change in the $\tilde{\mathbf{x}}_k(t)$ between successive iterations is negligible. Jazwinski [1970] provides additional information concerning local iteration techniques. In attitude determination, the time between observations is normally short and local iteration is generally not needed. Thus, Eq. (13-73) is the more common expression. Additional techniques for nonlinear problems are discussed by Athans, *et al.*, [1968].

Propagation of $\tilde{x}$ and P Between Observation Times. The Kalman filter assumes that the system dynamics is linear and of the form

$$\frac{d}{dt}x = Fx + Bu + Nn \tag{13-78}$$

where F, B, and N are known matrices which may be time varying and the matrix F has dimensions $(m \times m)$. The dimensions of B and N are such that the terms Bu and Nu are m-vectors.* The vector u is a known, deterministic driving function which does not depend on x. For example, u could consist of attitude-independent environment or control torques. In many attitude determination problems, u is zero and the term Bu may thus be ignored. The vector n is zero mean white noise which is assumed to be uncorrelated with the observation noises, v_n; that is

$$E(n(t)) = 0$$

$$E(n(t)n(\tau)^T) = V(t)\delta_D(t - \tau) \tag{13-79}$$

$$E(n(t)v_k^T) = 0$$

for all t, τ, and k, where V is a known, symmetric, nonnegative definite matrix; $\delta_D(t - \tau)$ is the Dirac delta; and E denotes the expectation value. It is the matrix V which is selected to give the filter its desired fading memory characteristics. White noise processes such as $n(t)$ do not exist in nature. The term Nn in Eq. (13-78) is included more as compensation for imperfections in the dynamics model than as a literal approximation of actual system inputs.

The Kalman filter propagates the state vector estimate via Eq. (13-78) with the noise term omitted; that is,

$$\frac{d}{dt}\tilde{x} = F\tilde{x} + Bu \tag{13-80}$$

The solution to this equation is [Wilberg, 1971]

$$\tilde{x}(t) = D(t, t_k)\tilde{x}(t_k) + r(t, t_k) \tag{13-81}$$

where the m-vector r is given by

$$r(t, t_k) = \int_{t_k}^{t} D(t, \tau)B(\tau)u(\tau)d\tau \tag{13-82}$$

Implementation of these two equations requires that the state transition matrix, D, be determined. When F is time invariant, D is the exponential of $(m \times m)$ matrix F; that is,

$$D(t, \tau) = \exp(F(t - \tau)) \tag{13-83}$$

and when F is time varying, D is obtained by integrating the following matrix equation, either analytically or numerically:

$$\frac{d}{dt}D(t, t_k) = FD(t, t_k) \tag{13-84}$$

with $D(t_k, t_k) = 1$.

---

*The dimensions of vectors u and n will depend upon the nature of the filtering application. They are frequently of different dimensions than the state vector.

In many applications the propagated state vector is updated at each observation set, $\mathbf{y}_k$, and the propagation computations are restarted using the a posteriori estimate $\tilde{\mathbf{x}}_k(t_k)$. Thus, Eq. (13-81) takes the form

$$\tilde{\mathbf{x}}_k(t_{k+1}) = D(t_{k+1}, t_k)\tilde{\mathbf{x}}_k(t_k) + \mathbf{r}(t_{k+1}, t_k) \tag{13-85}$$

If the function $\mathbf{u}$ is zero, then the propagated state vector is given by the simpler equation

$$\tilde{\mathbf{x}}_k(t_{k+1}) = D(t_{k+1}, t_k)\tilde{\mathbf{x}}_k(t_k) \tag{13-86}$$

We now develop an expression for propagating P, which uses D, N, and V. Recall that the $(m \times m)$ matrix P represents the covariance of errors in the state estimate; that is,

$$P(t) = E\big(\mathbf{e}(t)\mathbf{e}(t)^T\big)$$

where $\mathbf{e}(t) = \mathbf{x}(t) - \tilde{\mathbf{x}}(t)$, and it is assumed that $E(\mathbf{e}(t)) = 0$. Differentiating this expression with respect to time, using Eqs. (13-78) and (13-79), and performing some algebra yields the *matrix Riccati equation*,

$$\dot{P} = FP + PF^T + NVN^T \tag{13-87}$$

which has the solution [Wilberg, 1971; Meditch, 1967]

$$P(t) = D(t, t_k)P(t_k)D^T(t, t_k) + Q(t, t_k) \tag{13-88}$$

where the $(m \times m)$ matrix, $Q(t, t_k)$, is defined by

$$Q(t, t_k) \equiv \int_{t_k}^{t} D(t, \tau)N(\tau)V(\tau)N^T(\tau)D^T(t, \tau)d\tau \tag{13-89}$$

A more explicit form of Eq. (13-88) for propagating P between observations $\mathbf{y}_k$ and $\mathbf{y}_{k+1}$ is

$$P_k(t_{k+1}) = D(t_{k+1}, t_k)P_k(t_k)D^T(t_{k+1}, t_k) + Q(t_{k+1}, t_k) \tag{13-90}$$

$Q(t_{k+1}, t_k)$ is called the *state noise covariance matrix*.* A block diagram of the Kalman filter algorithm is given in Fig. 13-4.

Example of $\tilde{\mathbf{x}}$ and P Propagation. As an example of state and error covariance propagation in a Kalman filter, we describe a simplified version of an algorithm employed for attitude determination on the ATS-6 spacecraft. This

---

*When the system dynamics is nonlinear, the state propagation is commonly performed by integrating an equation of the form

$$\frac{d}{dt}\tilde{\mathbf{x}} = \mathbf{f}(\mathbf{x}, t) + B\mathbf{u}$$

with the initial condition $\mathbf{x}(t_k) = \tilde{\mathbf{x}}(t_k)$, where $\mathbf{f}$ is a function which is nonlinear in $\mathbf{x}$. Additional techniques for propagation of attitude parameters are given in Section 17.1. The state transition matrix to be used in the propagation of P is calculated using Eq. (13-84) with

$$F = \left[\frac{\partial \mathbf{f}}{\partial \mathbf{x}}\right]$$

This linear approximation causes the resulting estimate for P to be only a first-order approximation to the true covariance matrix.

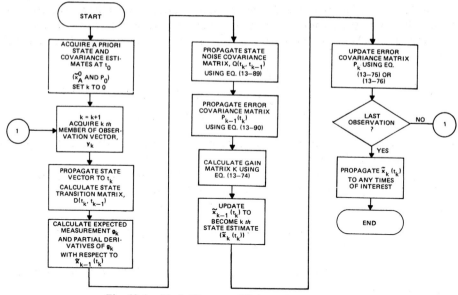

Fig. 13-4. Block Diagram of Kalman Filter Algorithm

spacecraft was Earth-pointing and three-axis stabilized by an active onboard control system. The motion about the three axes was assumed to be uncoupled and each axis was modeled independently. The dynamic model for any one of the three axes is of the form

$$x_1 = \theta$$

$$x_2 = \dot{x}_1 = \dot{\theta}$$

$$x_3 = \dot{x}_2 = \ddot{\theta}$$

$$\dot{x}_3 = n$$

where θ is a small angular deviation from a known reference attitude and n is zero mean white noise. The matrix form of the above equation is

$$\frac{d}{dt}\mathbf{x} = F\mathbf{x} + \mathbf{N}n$$

where $\mathbf{x} = (x_1, x_2, x_3)^T$, $\mathbf{N} = (0, 0, 1)^T$, and

$$F = \begin{bmatrix} 0 & 1 & 0 \\ 0 & 0 & 1 \\ 0 & 0 & 0 \end{bmatrix}$$

Note that the vector $\mathbf{n}$ is actually a scalar in this application. This causes the matrix N to have dimensions (3×1). It is therefore denoted by the 3-vector $\mathbf{N}$. The state transition matrix determined from the above equations is

$$D(t, t_k) = \begin{bmatrix} 1 & \Delta t & 0.5\Delta t^2 \\ 0 & 1 & \Delta t \\ 0 & 0 & 1 \end{bmatrix}$$

where $\Delta t = t - t_k$. Because the system dynamics does not include a deterministic driving function, the term $\mathbf{u}$ in the state vector propagation equation is zero.

The state noise covariance matrix can be established analytically through Eq. (13-89) as

$$Q = \frac{V}{120}\begin{bmatrix} 6\Delta t^5 & \cdot 15\Delta t^4 & 20\Delta t^3 \\ 15\Delta t^4 & 40\Delta t^3 & 60\Delta t^2 \\ 20\Delta t^3 & 60\Delta t^2 & 120\Delta t \end{bmatrix}$$

where V is a scalar selected by experience with real and simulated data to give the filter its desired fading memory characteristics.

Suppose that $\tilde{\mathbf{x}}(t_k)$ and $P(t_k)$ are known and that propagation to t_{k+1} is desired. Then, $\tilde{\mathbf{x}}_k(t_{k+1})$ is calculated using Eq. (13-86); that is,

$$\tilde{\mathbf{x}}_k(t_{k+1}) = \begin{bmatrix} 1 & \Delta t & 0.5\Delta t^2 \\ 0 & 1 & \Delta t \\ 0 & 0 & 1 \end{bmatrix}\tilde{\mathbf{x}}_k(t_k)$$

where, in this case, $\Delta t = t_{k+1} - t_k$. $P_k(t_{k+1})$ is then calculated using Eq. (13-88):

$$P_k(t_{k+1}) =$$

$$\begin{bmatrix} 0 & \Delta t & 0.5\Delta t^2 \\ 0 & 1 & \Delta t \\ 0 & 0 & 1 \end{bmatrix}P_k(t_k)\begin{bmatrix} 1 & 0 & 0 \\ \Delta t & 1 & 0 \\ 0.5\Delta t^2 & \Delta t & 1 \end{bmatrix} + \frac{V}{120}\begin{bmatrix} 6\Delta t^5 & 15\Delta t^4 & 20\Delta t^3 \\ 15\Delta t^4 & 40\Delta t^3 & 60\Delta t^2 \\ 20\Delta t^3 & 60\Delta t^3 & 120\Delta t \end{bmatrix}$$

$\tilde{\mathbf{x}}(t_{k+1})$ and $P_k(t_{k+1})$ may then be updated by Eqs. (13-73) to (13-76) using the observations at t_{k+1} to become $\tilde{\mathbf{x}}_{k+1}(t_{k+1})$ and $P_{k+1}(t_{k+1})$.

Divergence. A Kalman filter achieves a *steady state* when the corrections to the state vector reach a consistent level and when the error covariance matrix is stable. *Divergence* occurs when the estimated state moves away from the true state. This is the most common problem associated with Kalman filters. The most frequent causes of Kalman filter divergence are linearization errors, cumulative roundoff and truncation errors, modeling errors, and unknown noise statistics. Linearization problems can be reduced by local iteration, as described earlier, or more frequent selection of observations. Roundoff and truncation errors may be partially solved by using a Kalman filter variation, called a *square-root filter* [Andrews, 1968], which substitutes the square root of the error covariance matrix for its full value in the filter gain equation. Another useful variation which is as numerically stable as the square root filter but which requires less computation is the UDU^T *filter* discussed by Bierman [1977]. In the *adaptive filter* [Jazwinski, 1970], the state noise covariance matrix, Q, is adjusted using the residuals between actual and computed observations. This variation is intended to reduce the effects of modeling errors.

Problems associated with unknown noise statistics may be solved after extensive testing with both simulated and real data. Proper filter response will only result when the appropriate balance between the state noise and measurement noise covariance matrices is found. A data rejection scheme which removes all observations whose uncertainties are not accurately known is also necessary to prevent divergence. If, for example, state noise has been underestimated with respect to observation noise, the state estimation procedure will become less and less sensitive to the observation residuals. Divergence could then result even though the filter may have reached a steady state. Alternatively, if observation noise has been underestimated, the state estimation procedure may be incorrectly influenced by the observation errors. Reviews of alternative methods to solve filter divergence problems are given by Cappellari, et al., [1976] and Morrison [1969].

Advantages and Disadvantages of Kalman Filters. The Kalman filter is frequently chosen for use in onboard attitude determination and for applications where constant tracking of a changing attitude is required. It is useful for onboard processing because it does not need to recycle through previously observed data and is frequently able to estimate the current state in real time. The execution time required for a Kalman filter depends on the complexity of the calculations required to update the state transition and state noise covariance matrices. In some applications, the advantages resulting from the lack of iteration in the batch least-squares sense are partially offset by the time required to update these quantities.

Sequential Pseudoinverse Estimator. When the observations are much more accurate than the state propagation process, and when correlation among the elements of the state vector can be ignored, the *sequential pseudoinverse estimator* can serve as a useful substitute for the Kalman filter. This type of estimator computes the minimum correction to the state vector such that the difference between the observation vector, $\mathbf{y}_k$, and observation model vector, $\mathbf{g}_k$, becomes zero. This causes the state to match the observations exactly at the time of the update. (This exact match is actually possible only when q, the dimension of $\mathbf{y}_k$ and $\mathbf{g}_k$, is less than or equal to m, the dimension of the state vector.) The sequential pseudoinverse method does not actually provide any filtering of observation noise. Its performance is therefore generally inferior to that of the Kalman filter. It does, however, have the advantage of being relatively simple and computationally fast. The application of a sequential pseudoinverse estimator to onboard attitude determination using gyro and star tracker data is discussed by McElroy and Iwens [1975].

The governing equations for this algorithm are obtained from the Kalman filter by choosing the measurement covariance matrix, R_k, to be a null or zero matrix and setting the error covariance matrix, P_{k-1}, to a multiple of the identity matrix. Thus, from Eqs. (13-73) and (13-74) we obtain for the pseudoinverse estimator

$$\tilde{\mathbf{x}}_k(t_k) = \tilde{\mathbf{x}}_{k-1}(t_k) + K_k[\mathbf{y}_k - \mathbf{g}_k]$$

where

$$K_k = G_k^{\mathrm{T}}(G_k G_k^{\mathrm{T}})^{-1} \qquad (13\text{-}90)$$

In this algorithm, the gain matrix depends only on the current observation set so that the state estimate always corresponds to an exact fit to the most recent observations. Thus, this estimator has a rapidly fading memory; that is, the dependence of the current estimate of the state on the earlier observations diminishes completely as new information is provided by the latest observations.

References

1. Andrews, A., "A Square Root Formulation of the Kalman Covariance Equations," *AIAA Journal*, Vol. 6, p. 1165–1166, 1968.
2. Athans, M., R. P. Wishner, and A. Bertolini, "Suboptimal State Estimation for Continuous-Time Nonlinear Systems from Discrete Measurements," *IEEE Transactions on Automatic Control*, Vol. AC-13, no. 5, p. 504–514, Oct. 1968.
3. Bevington, Philip R., *Data Reduction and Error Analysis for the Physical Sciences.* New York: McGraw-Hill Book Co., 1969.
4. Bierman, G. J., *Factorization Methods for Discrete Sequential Estimation.* New York: Academic Press, Inc., 1977.
5. Bryson, Arthur E., Jr., and Yu-Chi Ho, *Applied Optimal Control.* Waltham, MA: Ginn and Company, 1969.
6. Cappellari, J. O., C. E. Velez, and A. J. Fuchs (editors), *Mathematical Theory of the Goddard Trajectory Determination System—Chapter 8 Estimation*, NASA X-581-76-77, GSFC, 1976.
7. Chen, L. C., and J. R. Wertz, *Analysis of SMS-2 Attitude Sensor Behavior Including OABIAS Results*, Comp. Sc. Corp. CSC/TM-75/6003, April 1975.
8. Deutsch, Ralph, *Estimation Theory.* Englewood Cliffs, NJ: Prentice-Hall, Inc., 1965.
9. Hamming, R. W., *Numerical Methods for Scientists and Engineers.* New York: McGraw-Hill, Inc., 1962.
10. Jazwinski, A. H., *Stochastic Processes and Filtering Theory.* New York: Academic Press, Inc., p. 278–280, 1970.
11. Joseph, M., J. E. Keat, K. S. Liu, M. E. Plett, M. A. Shear, T. Shinohara, and J. R. Wertz, *Multisatellite Attitude Determination/Optical Aspect Bias Determination (MSAD/OABIAS) System Description and Operating Guide*, Comp. Sc. Corp. CSC/TR-75/6001, April 1975.
12. Kalman, R. E., "A New Approach to Linear Filtering and Prediction Problems," *J. Basic Eng.*, Vol. 82, p. 35–45, 1960.
13. ———, and R. S. Bucy, "New Results in Linear Filtering and Prediction Theory," *J. Basic Eng.*, Vol. 83, p. 95–108, 1961.
14. Lefferts, E. J., and F. L. Markley, *Dynamic Modeling for Attitude Determination*, AIAA Paper No. 76-1910, AIAA Guidance and Control Conference, San Diego, CA, Aug. 1976.
15. Marquardt, D. W., "An Algorithm for Least-Squares Estimation of Nonlinear Parameters," *SIAM Journal*, Vol. 11, p. 431, 1963.
16. McElroy, T. T., and R. P. Iwens, "Precision Onboard Reference Using Generalized Inverse," *Proceedings of the 12th Annual Allerton Conference on Circuit and System Theory (Oct. 1974)*, Univ. of Illinois, Urbana, IL., 1975.
17. Meditch, J. S., *Stochastic Optimal Linear Estimation and Control.* New York: McGraw-Hill, Inc., p. 146-147, 1967.

18. Melkanoff, M. A., and T. Sanada "Nuclear Optical Model Calculations," *Methods in Computational Physics*, Vol. 6. New York: Academic Press, Inc., 1966.

19. Morrison, N., *Introduction to Sequential Smoothing and Prediction.* New York: McGraw-Hill, Inc., 1969.

20. Pettus, W., G. Fang, and S. Kikkawa, *Evaluation of Filtering Methods for Optical Aspect and Horizon Sensor Data*, Comp. Sc. Corp., CSC/5035-22300/05TR 1972.

21. Repass, G. D., G. Lerner, K. Coriell, and J. Legg, *Geodynamics Experimental Ocean Satellite-C (GEOS-C) Prelaunch Report*, NASA X-580-75-23, GSFC, 1975.

22. Rigterink, P. V., R. Berg, B. Blaylock, W. Fisher, B. Gambhir, K. Larsen, C. Spence, and G. Meyers, *Small Astronomy Satellite-C (SAS-C) Attitude Support System Specifications and Requirements*, Comp. Sc. Corp. CSC/3000-05700/01TN, June 1974.

23. Schmidtbauer, B., Hans Samuelsson, and Arne Carlson, *Satellite Attitude Control and Stabilization Using On-Board Computers*, ESRO-CR-100, July 1973.

24. Wiberg, D. M., *State Space and Linear Systems.* New York: McGraw-Hill, Inc., p. 109, 1971.

25. Wilde, D. J., *Optimum Seeking Methods.* Englewood Cliffs, NJ: Prentice-Hall, Inc., 1964.

CHAPTER 14

EVALUATION AND USE OF STATE ESTIMATORS

James R. Wertz

14.1 Prelaunch Evaluation of State Estimators
14.2 Operational Bias Determination
14.3 Limitations on State Vector Observability

Chapters 10 through 13 have provided the basic analytic foundations of attitude determination and attitude state vector estimation or data filtering. This chapter discusses practical aspects of state estimation and makes specific suggestions for the evaluation and operational use of estimators. There is no optimum procedure for analyzing attitude data, and much of the discussion is necessarily subjective. This chapter is intended as a practical guide, to be modified as appropriate for the problem under consideration. Section 14.1 describes specific procedures for the prelaunch evaluation and testing of estimators, and Section 14.2 describes the operational use of estimators. Finally, Section 14.3 discusses the limitations on state vector estimation and procedures for determining which state vector elements should be solved for in a particular situation. Although the discussion is as spacecraft independent as possible, the examples used are drawn primarily from systems using horizon sensors and Sun sensors on a spinning spacecraft, such as the CTS spacecraft described in Section 1.1.

14.1 Prelaunch Evaluation of State Estimators

Given a functioning estimator or data filter, how do we determine whether it behaves correctly? The two basic requirements for testing estimators are a data simulator and a detailed test plan. Normally, real data is not sufficient for testing because the correct solution is unknown. However, it may be possible to use real data when the attitude has been determined by more accurate, redundant sensors. For example, attitudes based on SAS-3 star tracker data were used to evaluate infrared horizon sensor data [Hotovy, 1976] and a similar procedure is anticipated for MAGSAT [Levitas, *et al.*, 1978]. When a data simulator is used it must be at least as sophisticated (i.e., incorporate as many effects) as the state estimator and preferably more sophisticated.

A test plan for the evaluation of an estimator should be designed by an individual or group other than those who designed the estimator itself; otherwise, that which was overlooked in the design will also be overlooked in the testing. The test should be designed to test each part of the estimator independently and should also test that all of the parts of the estimator work together. Unfortunately, with a complex state estimation system, it is effectively impossible to test all possible combinations of processing options. Therefore, it is important to attempt to identify both combinations that may present problems (e.g., a change in sensors in the middle of a data pass) and those that will normally be used in practice.

There are three levels of testing that may be performed on any state estimator: (1) testing under nominal and contingency mission conditions, (2) testing under conditions specifically designed to identify and isolate problems, and (3) testing under conditions specifically designed to identify system limits. The first type of test is the least severe and is the minimum requirement for any test procedure. There are two major problems with type (1) tests—it is unknown how the estimator will behave under unanticipated (but possible) mission conditions and it is not really known what to expect from nominal mission conditions. The nominal situation is usually sufficiently complex that analytic solutions for the various parameters are not available. Therefore, the only conclusions that can be drawn from tests under nominal or contingency conditions are that the estimator behaves approximately as one would expect or that it obtains the correct answer on "perfect" data within the limits required by the design and the conditions antici-pated. Both of these conclusions are weak, although they may be adequate for some purposes.

Type (2) tests eliminate much of the ambiguity by choosing tests specifically designed to identify problems or evaluate performance, although the conditions chosen may be unrealistic. The usual procedure here is to choose conditions unrealistically simple such that analytic solutions for the observables are available or at least such that the analyst has an intuitive "feel" for the results. For example, horizon sensor modeling might be tested using a spherical Earth model and a circular, equatorial spacecraft orbit; star sensor calibration algorithms might be tested using an evenly spaced, rectangular grid of stars.

Type (3) tests are the most stringent and are the analog of destructive testing in civil engineering because the intent of the test is to determine the limits of system performance. For example, the attitude might be chosen at the celestial pole to determine how the estimator handles coordinate singularities; or unmodeled biases, such as a deliberate orbit error, might be included in the data to determine how well the estimator behaves in the presence of unmodeled systematic variations. The latter test is particularly valuable because unmodeled systematic errors are the practical accuracy limit for most state estimators.

Finally, we describe three specific tests to examine the operational characteristics of state estimators. These are designed to test the statistical compu-tations and resulting uncertainties, the accuracy of observation models, and the accuracy of computed partial derivatives.

Statistical and Uncertainty Tests. For any state estimator, we want to deter-mine whether uncertainties calculated by the estimator truly reflect the variations in the solutions due to the noise on the data. This can be conveniently done with a data simulator which adds pseudorandom noise to the data based on an algebraic random number generator. (For a discussion of the characteristics of algebraic pseudorandom number generators, see, for example, Carnahan, *et al.*, [1969].) A series of 10 to 20 test runs are made under identical conditions including the addition of Gaussian noise to the data; the only difference between the runs should be the *seed* or starting value used for the algebraic random number generator. The standard deviation of the resulting attitude and bias parameter solutions is then a real measure of the spread in the solutions due to the noise on the data; further, it is fully independent of the statistical computations within the estimator (i.e., the spread is dependent only on the estimator solution and not on its statistics). Therefore, this solution spread can be compared with the uncertainties computed by the estimator to determine the accuracy of these computations. Note that the computed uncertainties should be nearly the same for each test run. This test is particularly effective when the underlying statistical analysis is uncertain, as in the case in which nonlinearities may become important or those in which quantized measurements are assumed continuous.

To evaluate this test we need to know how well the standard deviation, s_i, of the resulting attitude or bias solutions, x_i, measures the actual variance, σ_i^2, in the state vector solution; that is, how closely should the observed s_i agree with the value of σ_i computed by the state estimator. Given n random samples from a normally distributed population, the s_i^2 should have a chi-square distribution (see, for example, Freund [1962]) such that

$$\frac{\nu s_i^2}{\chi_{\alpha,\nu}^2} < \sigma_i^2 < \frac{\nu s_i^2}{\chi_{(1-\alpha),\nu}^2} \qquad (14\text{-}1)$$

where $\nu \equiv n-1$ and $(1-2\alpha)$ is the confidence interval for the results. For example, if we make 10 runs on a simulated data set for which the resulting standard deviation in the 10 values of element i is 0.015 deg, then for a 90% confidence level, we obtain from standard statistical tables $\chi_{0.05,9}^2 = 16.92$ and $\chi_{0.95,9}^2 = 3.325$. Therefore, with 90% confidence, σ_i lies in the interval $(9 \times 0.015^2/16.92)^{1/2} = 0.011$ deg to 0.025 deg; that is, the computed uncertainty from the estimator should lie in this range 90% of the time.

Observation Model Tests. The best procedure for testing the accuracy of individual models is to execute the state estimator on simulated, noisefree data and closely examine the solutions or solution residuals. The noise in the system is then the unavoidable noise due to machine round off. For the estimator to be functional, this must be below that anticipated for the actual data; nevertheless, it clearly demonstrates any existing differences between simulator models and models incorporated into the estimator, such as slightly different ephemerides or sensor models.

Partial Derivative Tests. The partial derivatives in an estimator may be tested for accuracy by the following method. Set all of the state vector elements except one to their correct, known value in a simulated data set. Set the one element being tested off of the correct value by a small amount, such as 0.1 to 0.01 deg. Assuming that the problem is approximately linear over this small range, the estimator should converge to the correct solution in a *single* iteration. If the first iteration gives an answer which is significantly in error, then the partial derivatives are probably computed incorrectly. However, even if the partial derivatives are incorrect, the estimator may still converge slowly provided that the sign of the partials is correct. Of course, the estimator will generally operate more efficiently if the values of the partial derivatives are correct. It is also possible that nonlinearities may be important even in a very small region around the correct solution. If this is suspected, then an independent test of the partial derivatives should be sought.

14.2 Operational Bias Determination

This section is concerned primarily with estimators used for *bias determination*, a process of state estimation or data filtering that is performed only infrequently during the life of a space mission, to determine various calibration parameters or biases; these will then be fixed at their estimated values and subsequently used for routine attitude determination. Therefore, we are concerned here with how to get the most possible information about biases from a given set of data. In practice, most of the time spent in bias determination is devoted to selecting the data to be processed (i.e., selecting the data base to be used and eliminating individual samples) and verifying the solution in as many ways as possible.

Operational bias determination is governed by both mission requirements and time constraints. State estimators are used to solve for attitude and biases simultaneously either to find the best estimate of bias parameters or to find the best attitude estimate to be used in conducting maneuvers. The estimation process may be divided into the following two or three steps: (1) a rapid preliminary state vector estimate to provide backup in case of software failure or unanticipated timeline changes (this step is necessary only in real-time analysis); (2) determination of the best possible attitude and bias parameters; and (3) validation of the results, possible revision of the answer, and evaluation of the uncertainties.

To carry out the above program requires a systematic and predesigned procedure to accomplish each step and to record mission parameters and the results of the various tests performed. Recordkeeping is particularly important in areas such as the spacecraft control environment, in which large quantities of data are ordinarily processed and reaccessing data may be difficult or time consuming.

Selecting Data Sets for Processing. To obtain the best possible estimate of bias parameters, we would like to obtain data bases* with the greatest possible information content. To do this, we use the procedures described in Section 14.3 and Chapter 11 to evaluate the information content and correlations for the various data sets that will be available or could become available. Conclusions are then tested on simulated data to establish a formal operational procedure.

An advantage of the geometrical analyses of Chapter 11 and Section 14.3 is that many potential data sets may be evaluated quickly with minimal computer support. Consequently, it may be possible to evaluate many possible geometrical conditions for bias determination or sensor calibration. With this information, data collection can be planned specifically for sensor evaluation. For example, after firing the apogee boost motor for SMS-2, GOES-1, and CTS to put the spacecraft into a near synchronous orbit, an attitude maneuver was required to bring the spin axis to orbit normal. This large maneuver was accomplished by a series of smaller maneuvers with intermediate attitudes, or *stops*, used to obtain one orbit of data for sensor evaluation [Chen and Wertz, 1975; Wertz and Chen, 1976; Tandon, *et al.*, 1976]. The steps were chosen by evaluating the different available geometries along the maneuver path.

Although making the best use of existing maneuvers or normal geometry is the most efficient procedure for sensor evaluation, it may be appropriate to use the geometrical analyses to plan specific *calibration maneuvers* or maneuvers designed explicitly to calibrate or evaluate hardware performance. Small attitude maneuvers are frequently used with gas jet systems to evaluate the approximate performance

---

* *Data base* in this context is any collection of data that can be processed together through a state estimator.

of the jets before using them for long maneuvers. (See, for example, Werking, *et al.*, [1974].) Attitude sensor calibration maneuvers were used on AE-4 and -5 to measure the sensor biases early in each mission to provide accurate deterministic attitudes and are planned for SMM for the calibration of gyros and Sun and star sensors [Branchflower, *et al.*, 1974].

Sample Operating Procedure. Having obtained data suitable for state vector estimation, we would like to process this data to obtain the most information possible. We describe here a sample procedure for spinning spacecraft used with some variations for launch support for CTS; for spacecraft in the AE, SMS, and GOES series; and anticipated for upcoming missions [Chen and Wertz, 1975; Tandon, *et al.*, 1976; Wertz, *et al.*, 1975]. This procedure is based on a processing system similar to the CTS system described in Section 21.2, incorporating a deterministic processor, a bias determination subsystem (normally, a state estimator), and a means for comparing predicted and observed data. The operating procedure which has proved successful in these cases is as follows:

Obtain a Preliminary Solution:
1.   Manually select data or check the automatic data selection process.
2.   Use the deterministic processor to provide an unbiased initial estimate for the bias determination subsystem.
3.   Use the bias determination subsystem to obtain a preliminary state vector estimate based on prelaunch selection of the parameters to be solved for. (See Section 14.3.)

Obtain the Best Available Solution:
4.   Use an iterative procedure to eliminate data anomalies, as described in Section 9.4.
5.   Again, use the deterministic processor on the final set of selected data to provide an unbiased estimate for the bias subsystem.
6.   Obtain a converged solution with the state estimator. Convergence should be based on the convergence of the state vector elements themselves, rather than on statistical tests which may give unrealistic results if the problem is nonlinear, if some of the measurements are quantized, or if the estimator includes the option of data rejection. To test for convergence, the change in each of the state vector elements may be required to be below a predetermined level (e.g., 0.001 deg for angular parameters) or the change may be required to be some fraction (e.g., 0.1 or less) of the change on the preceding iteration. As discussed in Section 14.3, a solution should be obtained which provides the maximum number of state vector elements which can be solved for.

Test the Solution.   Four classes of tests are available:
7.   Use the bias parameters in the deterministic processor to verify that the systematic errors in the deterministic solutions have been reduced. (See Figs. 11-29, 11-30, and 9-18.)
8.   Determine the consistency of the results by comparing the predicted and the observed data. Note any systematic behavior of the residuals. (See Figs. 9-17 through 9-19.)
9.   Add known biases to the real data and see if they can be recovered by the same procedure which was used to obtain the final answer.

10.   To ensure that the estimator is operating in a "linear" region, try different initial estimates for the state vector parameters and check to see that the estimator converges to the same answer.

Discussion. *Typically, state estimation accuracies are limited by systematic rather than statistical errors* because when statistical errors dominate, more data is normally available that could be processed to further reduce the statistical error. Therefore, it is important to observe and record the level of systematic variations and to attempt to estimate the systematic uncertainties. The uncertainties computed by the state estimator will be unrealistically low because they will account only for the statistical variation.

One procedure for estimating the systematic uncertainties in a given set of bias parameters is to examine the scatter among independent determinations. Here, the estimator is used to analyze data taken with different orbits and different geometries, but from the same sensors. The results from the different data sets may then be statistically combined to obtain improved estimates of attitude parameters and their uncertainties. (See, for example, Chen and Wertz [1975].)

Although the above procedure provides the most unambiguous quantitative results, truly "independent" data sets are rarely, if ever, available. Therefore, the resulting uncertainties may still be unrealistically low. We may test further for systematic uncertainties by using an *a priori* knowledge of unmodeled biases such as errors in the orbit, the Earth model, or unobservable sensor biases. The unmodeled biases are applied to simulated data, the estimator is used to solve for the state vector parameters, and the uncertainties are estimated by the amount that the computed state differs from the known values.

A final procedure for testing the level of systematic errors directly is to apply an unmodeled bias to the data and to adjust the magnitude of the bias to give systematic residuals of the same magnitude as those observed in the real solutions. The amount by which the state vector elements are changed by this systematic error is then an estimate of the parameter uncertainty, due to the unknown systematic error. The "unmodeled" bias here may be conveniently chosen as one of the state vector elements which is not solved for during the test.

None of the above methods for determining a realistic uncertainty estimate is completely satisfactory, and the procedures for obtaining the best attitude state vector estimate cannot be quantified. Thus, there is still a need for sophisticated judgment on the part of the operator of a state estimator. The most effective operational procedure for attitude determination is the use of a deterministic processor for continuous, routine attitude determination with the occasional use of a state estimator to determine biases and evaluate possible systematic errors.

14.3 Limitations on State Vector Observability

In general, a good state estimator should provide more state vector elements than can be solved for simultaneously with any real data pass; if not, there is insufficient flexibility in the estimator. Thus, the first requirement for analysis with any estimator—whether in prelaunch testing or operational use—is to correctly choose the state vector elements to be solved for. In principle, this is a straightforward matter based on the estimator covariance matrix; any set of elements for which the off-diagonal terms in the covariance matrix remain small can be resolved by the estimator. In practice, model inadequacies, nonlinearities, and data anomalies frequently make the covariance matrix difficult to interpret and, therefore, of very limited use in choosing the solved-for parameters.

Before discussing practical procedures for choosing state vector elements, we distinguish three types of state vector elements: (I) the attitude itself, which is the basic parameter to be determined; (II) biases which may be represented by a deviation in one particular measurement type—for example, a bias in the Sun angle, an azimuth bias of the Sun sensor relative to a horizon sensor, or a magnetometer bias; and (III) biases which are not associated with any one particular measurement, for example, a bias on the angular radius of the Earth or any orbit parameter such as an orbital in-track error.

If the state vector element being considered consists only of the attitude and constant type II biases, it is sufficient to determine an accurate attitude one time because if the attitude is known, the biases are determined by the various sensor measurements. In addition, correlations among type II biases are the same as those among their respective measurements. Therefore, the analysis of the information content of measurements also indicates a correlation among biases. For example, a correlation between the Sun angle and the Sun-to-Earth rotation angle measurement, as described in Section 11.4, implies a correlation between the Sun angle bias and azimuth bias. In contrast, type III biases cannot be analyzed by the general procedure for type II biases and must be treated individually.

In general, we would like to solve for as many parameters as possible, so long as a meaningful, converged solution can still be obtained. We would like to know beforehand which combinations of parameters are uncorrelated and therefore can be resolved. As a general hypothesis, we suggest that the greater the variation in the geometry, the larger the number of parameters which can be solved for. Repeated measurements under the same geometrical conditions, for example, measuring the Earth width by a spacecraft in a circular orbit with its spin axis at orbit normal, do not provide new information with which to distinguish various sensor biases, such as deviations in the mounting angle of the sensor, the angular radius of the Earth, or the semimajor axis of the spacecraft orbit.* Repeated measurements under the same geometrical conditions serve only to reduce the statistical noise, but this is of limited practical value because the uncertainties are normally dominated by systematic rather than statistical errors. However, if the geometrical conditions change (for example, if the orbit is noncircular or the attitude is not at orbit normal), each measurement provides new information and more parameters may be solved for. Thus, state vector estimation is best done with data which incorporate the widest possible variety of geometrical relationships among the attitude reference vectors.

To determine explicitly which state vector elements are observable or distinguishable, four procedures have been found to be useful and have been successfully applied to the analysis of real spacecraft data: (1) trial and error with

---

*Although biases may be indistinguishable in the instance cited, a full orbit of data may determine the attitude very precisely. The importance of determining the biases depends on whether it is the attitude at the time of the data pass which is important, or whether the biases themselves are needed so that accurate attitudes may be determined at a later time when less or different data are available.

simulated data, (2) analysis of correlations among different measurement types (for type II state vector elements which are correlated with specific observations), (3) analysis of the information content of a single measurement type, and (4) analysis of geometrical procedures which allow a particular parameter to be resolved.

1. Trial-and-Error Procedures. In a trial-and-error procedure, we use the state estimator to process real or simulated data, trying various combinations of state vector elements to determine which combinations give converged solutions. In the case of simulated data, we may also determine which combinations return approximately the correct answer. This procedure is practically useless as a general analytic technique for more than 5 to 10 parameters because of the many possible combinations of state vector elements and processing options. Also, it is difficult to obtain any general insight from trial-and-error analysis which can be applied to conditions different from those tested. Nevertheless, trial and error provides a procedure for testing analytic results obtained by other methods, and may be the most economical procedure for single applications. It serves as the basic test for other techniques and all analytic conclusions should be subsequently tested, so far as possible, first on simulated data and then on real data. Each of the procedures presented below has been tested on both real and simulated data.

2. Correlations Among Different Measurement Types. If two measurements are correlated over some region, the associated type II biases are also correlated over that region and are difficult to distinguish. Therefore, the analysis of Chapter 11 concerning correlations between measurement types may be applied directly to correlations between type II biases. For example, Fig. 11-26(a) indicates that for the SMS-2 spacecraft in its transfer orbit to geosynchronous altitude, the Sun angle/nadir angle correlation angle was near zero during the entire coverage of the Earth by horizon sensor 2. Therefore, the Sun angle and nadir angle measurements are providing nearly the same information about the attitude, and their associated biases are strongly correlated. Thus, we expect that a Sun angle bias is difficult to distinguish from a sensor mounting angle bias* with these data. This conclusion, and similar conclusions with data from other sensors, was confirmed through the analysis of both real and simulated SMS-2 data [Chen and Wertz, 1975].

Although this procedure can provide information on the correlation of biases quickly and easily, it is limited to type II biases and is also limited to data spans over which the geometry does not change greatly. As the data span becomes longer, such that the correlation is changing, the interpretation in terms of correlated and uncorrelated biases is less clear. Procedure 3 is concerned specifically with the changing correlation over long data spans.

3. Information Content of a Single Measurement Type. Procedure 2 is only applicable to correlations at one point or over a small region; however, large data passes are normally needed for bias determination. Therefore, we would like to examine the information content of any single type of attitude measurement by determining the correlation angle for measurements of the same type at the

---

*Note that although a sensor mounting angle bias and a true nadir angle bias are not identical, the distinction is not critical in this case. Both biases shift the computed attitude in the same direction, but the relative amount of the shift is not linearly related for the two biases.

beginning and the end of (or throughout) a long data pass. If this correlation angle is small and if the possibility of a bias in the measurement exists, there is little information content in the measurement. Conversely, if the correlation angle is large, even in the presence of a constant bias on the measurement, it may be possible to use that measurement to determine both the attitude and the corresponding type II bias.

Figure 14-1 illustrates the qualitative physical basis for this interpretation for the case of a cone angle measurement for which the possible attitude loci are small circles on the celestial sphere. For concreteness, we assume that the measurement is a direct measure of the nadir angle and that the $+$'s are the orientations of the Earth (i.e., the centers of the various nadir cones) at the beginning and end of a data pass. The solid line corresponds to the possible attitudes assuming that there is no bias in the measurement. The dashed line corresponds to the possible attitudes if there is a bias $\Delta\eta$ in the nadir angle measurement. The correlation angle, Θ_{η_1/η_2}, between the nadir angle measurement at the beginning (time t_1) and end (time t_2) of the data pass is just the angle of intersection of the two nadir cones. In Fig. 14-1(a), this correlation angle is small. Therefore, the horizontal component of the attitude is poorly defined. However, if there is the possibility of a nadir angle bias, then the radius of the cone is unknown and the vertical component of the attitude is also poorly defined. In this example we have, in effect, only one measurement and one potential bias and, therefore, this measurement provides no information about the attitude. In practice, the bias is not totally unknown and may normally be expected to fall within some assumed limits. However, in the logic of the state

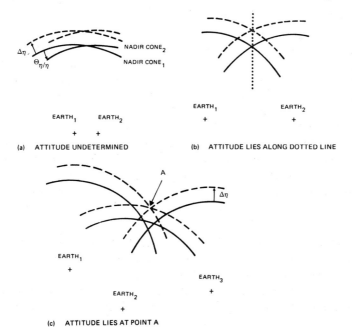

Fig. 14-1. Nadir Angle/Nadir Angle Correlation for Data Passes of Different Lengths. Subscripts 1, 2, and 3 denote the positions of the Earth and nadir cone at times t_1, t_2, and t_3. See text for explanation.

estimator it is usually assumed that the bias is completely unknown and can take on any value.* It is in this sense that a single measurement with a bias of unknown magnitude provides *no* attitude information.

Fig. 14-1(b) illustrates two more widely spaced points corresponding to the ends of a data pass of intermediate length, or to a data pass for which the central data is unavailable. Although we do not know the value of the nadir angle bias, we assume that it is a constant bias for both measurements. Therefore, the attitude must lie along the dotted line. Thus, the horizontal component of the attitude is well determined, but both the vertical attitude component and the nadir angle bias are not determined.

Finally, Fig. 14-1(c) illustrates the information available from a full data pass with a large correlation angle between the measurement at the beginning and at the end. If we assume that the attitude is fixed, then all of the measurements must give the same result if the nadir angle bias has been correctly determined. Because the three solid curves do not intersect in one point, there must be a nadir angle bias. Because the dashed lines do intersect in a point, the attitude must be at that intersection and the nadir angle bias must be equal to $\Delta\eta$. In this case we have used a single measurement type to determine both the attitude and the magnitude of the type II bias associated with that measurement type.

As an example of the above analysis, consider a system similar to the examples of Chapter 11 (such as CTS, SMS/GOES, or SIRIO) consisting of a spinning spacecraft with Sun angle, Sun-to-nadir rotation angle, and nadir angle measurements and possible biases in all three measurements. If data is obtained over a period of less than a day, then the inertial position of the Sun remains essentially fixed. Therefore, if there is a possible Sun angle bias, the Sun angle measurement indicates that the attitude lies on a cone of unknown radius centered on the Sun; that is, there is *no* information in the Sun angle measurement. Adding Sun angle data to a state estimator and including a Sun angle bias in the state vector solved for will affect neither the attitude results nor the values of any of the other state vector elements. Of course, if the attitude is determined from other data, then the Sun angle measurement provides a measure of the Sun angle bias.

To determine the content of the Sun-to-nadir rotation angle measurement, it is convenient to find a general procedure for determining the correlation angle between a measurement at the beginning of a data pass and that same measurement at some other time during the data pass. As shown in Fig. 14-2, the attitude locus, L_β, for a given Sun angle measurement, β, remains nearly fixed on the celestial sphere as the spacecraft moves in its orbit. Therefore, the correlation angle, Θ_{m_1/m_2}, between one measurement, m, at any two positions in the orbit is just the difference between the β/m correlation angles at these two positions. (See Section 10.4 for a discussion of correlation angles.) For example, for $m = \Phi$, where Φ is the Sun-Earth rotation angle:

$$\Theta_{\Phi/\Phi}(\text{time 1 to time 2}) \equiv \Theta_{\Phi_1/\Phi_2} = \Theta_{\beta/\Phi}(\text{time 2}) - \Theta_{\beta/\Phi}(\text{time 1}) \equiv \Theta_{\beta/\Phi_2} - \Theta_{\beta/\Phi_1}$$

$$(14\text{-}2)$$

---

*Assuming that the bias is completely unknown is equivalent to setting the state weight matrix, S_0, in Section 13.4 to zero. If S_0 is nonzero, then a penalty is assigned to deviations of the bias from its nominal value, and a single measurement with a possible bias does constrain the attitude solution.

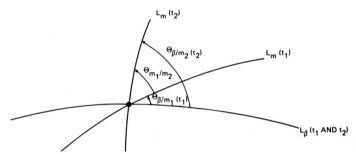

Fig. 14-2.   Computation of Θ_{m_1/m_2} From Correlations, $\Theta_{m/\beta}$, With the Sun Angle. L_β and L_m are the *attitude loci*, or possible positions of the attitude on the celestial sphere, for a given Sun angle, β, and a given measurement value, m. So long as the time interval $t_2 - t_1$ is short enough so that the Sun remains essentially fixed, then $\Theta_{m_1/m_2} = \Theta_{\beta/m_2} - \Theta_{\beta/m_1}$ for any measurement m.

The rotation angle correlation angle, $\Theta_{\Phi/\Phi}$, can be determined from Fig. 14-3, which shows the $\Theta_{\beta/\Phi}$ correlation angle curves at 2-deg intervals over the entire sky for fixed positions of the Sun and attitude and variable positions of the Earth. For example, if the center of the disk of the Earth is at B, then $\Theta_{\beta/\Phi}$ (evaluated at the

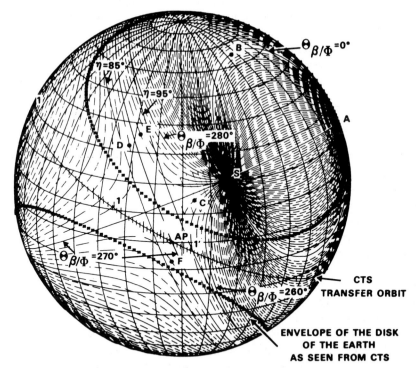

Fig. 14-3.   Sun Angle/Rotation Angle Correlation Angle Curves at 2-Deg Intervals for a Sun Angle of Approximately 65 Deg. See text for explanation. The orbit and Earth envelope illustrated are for the CTS transfer orbit. At 10-deg intervals, the correlation angle curves are solid lines. The lines at ± 2 deg have been omitted to identify the $\Theta_{\beta/\Phi} = 0$ curve. Because the only independent parameter in generating these curves is the Sun angle, β, they may be used for any spacecraft for which $\beta \approx 65$ deg as shown.

attitude) equals zero. Similarly, if the Earth is at C or D, then $\Theta_{\beta/\Phi}=268°$; if the Earth is at E or F, then $\Theta_{\beta/\Phi}=272°$. Figure 14-3 also shows the approximate geometry of the CTS transfer orbit to synchronous altitude. As in Figs. 11-25 and 11-26, the line with vertical tick marks denotes the Earth's orbit about the spacecraft (as seen by the spacecraft), asterisks mark the envelope of the Earth's disk, AP marks the location of apogee, and 1 and 1' mark the interval over which horizon sensor 1 senses the Earth. Thus, horizon sensor 1 picks up the Earth at 1 where $\Theta_{\beta/\Phi}=268°$; $\Theta_{\beta/\Phi}$ then increases to a maximum of approximately 271° just before apogee and drops to about 266° as sensor 1 loses the Earth at 1'. Therefore, the maximum variation in $\Theta_{\beta/\Phi}$ is about 5 deg. From Eq. (14-2), this implies that $\Theta_{\Phi/\Phi}$ has a maximum value of about 5 deg. Thus, the rotation angle correlation angle for the CTS geometry is small and there is very little information content in the rotation angle measurement if the possibility of an unknown bias in the measurement is considered.

The above conclusion about minimal information in the CTS rotation angle measurement is generally applicable under certain common conditions. Note that in the vicinity of the spin plane in Fig. 14-3 (between the lines at nadir angles, η, of 85 deg and 95 deg), $\Theta_{\beta/\Phi}$ is approximately 270 deg and is insensitive to the rotation angle, Φ.* Physically this means that if the attitude is near orbit normal, then as the spacecraft moves through an entire orbit, the rotation angle, Φ, goes from 0 deg to 360 deg, but the loci of possible attitudes remains nearly the same for the various positions of the spacecraft in its orbit. Although the measurement is changing through its full range, the information content as to the possible locations of the attitude is nearly the same for all of these measurements. Therefore, whenever the nadir angle remains near 90 deg for an entire pass (i.e., if either the attitude is at orbit normal *or* the Earth is small and the sensor is mounted near the spin plane as is the case for CTS) and there is the possibility of a rotation angle bias, then there is very little information in the rotation angle measurement.

For the nadir angle measurement, the situation is the opposite of the rotation angle measurement. For an attitude near orbit normal, the measured value of the nadir angle remains approximately fixed, but the corresponding attitude loci rotate through 360 deg as the spacecraft goes around a full orbit. Therefore, the nadir angle measurement contains sufficient information to determine both the attitude and a constant nadir angle bias as illustrated previously in Fig. 14-1. (The nadir angle bias may be a composite of biases in the sensor mounting angle, the angular radius of the Earth, or other parameters.)

In summary, we may determine the information content of any type of measurement in which there may be a constant bias by examining the changing orientation of attitude loci for that measurement. *It is the attitude loci, not the reference vector or measurement values, that is important.* If there is no rotation of the attitude loci (e.g., the Sun angle measurement) and if a bias in the measurement is solved for, then there is no information about the attitude in that measurement. Conversely, if there is a large rotation of the attitude loci (e.g., nadir angle or Earth-width measurements over a full orbit with the attitude near orbit normal),

---

*The same conclusion can be obtained from Eq. (11-52) or Fig. 11-18.

then that information may be used to solve for both the attitude and the magnitude of the constant type II bias in that measurement. These conclusions on the information content of the β, Φ, and η measurements have been verified on both real and simulated data for the GOES-1 and CTS missions [Tandon, *et al.*, 1976].

4. Geometry of Individual Biases.   The procedure described in the preceding paragraphs is only applicable to type II biases. The general procedure for type III biases is to find a region in which the data are very sensitive to the bias in question. Particularly good regions to test in this regard are those where the effect of the bias on the data changes sign or reaches an extremum or where the measurement density is low.

As an illustration of sensitive regions for particular biases, consider the case of a negative bias on the angular radius of the Earth, as illustrated in Fig. 14-4. The solid line is the nominal Earth disk and the dotted line is the sensed or biased Earth disk. If there is a bias, then as the sensor scan moves downward across the disk of the Earth, a measured Earth width corresponding to a scan at A will imply that the scan was crossing at A' where the Earth width for a nominal Earth disk would be the same size as for the biased disk at A. Thus, the computed nadir angle would be significantly larger than the real nadir angle. Similarly, a real scan at B will imply that the scan crossed the nominal Earth at B' and the computed nadir angle will be significantly smaller than the real nadir angle. Thus, as the sensor scans across the diameter of the Earth going from A to B, there will be a large discontinuity in the computed nadir angles if there is an unresolved bias on the angular radius of the Earth. Making the computed attitudes agree (even if the value of the attitude is not particularly well known) as a horizon sensor sweeps across the diameter of the Earth provides a very sensitive measure of the bias on the angular radius of the Earth. This procedure was used on the CTS mission to determine the Earth radius bias to about 0.02 deg on a very short span of data taken as the horizon sensor scan crossed the diameter of the Earth.

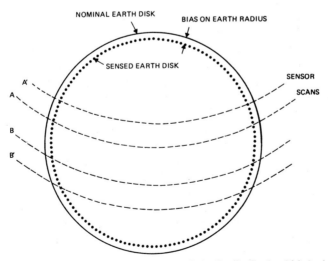

Fig. 14-4.   Sensitivity to Bias on the Angular Radius of the Earth. Earth width is the same on the nominal Earth disk at A' and B' as it is on the sensed Earth disk at A and B.

Although an analysis of this type is necessary for type III biases, such as a bias on the angular radius of the Earth or an orbital in-track error, it may be used for other biases as well. For example, a bias in the mounting angle of an Earth horizon sensor causes a shift in opposite directions on opposite sides of the orbit. Thus, two data passes on opposite sides of the orbit with the spacecraft at a constant attitude were used to successfully determine the sensor mounting angle bias for sensors on the CTS and GOES-1 spacecraft [Tandon and Smith, 1976]. A similar procedure was used for the Panoramic Attitude Scanner on RAE-2 [Werking, *et al.*, 1974] and magnetometer data on the SAS-1 mission [Meyers, *et al.*, 1971].

References

1. Branchflower, G. B., *et al.*, *Solar Maximum Mission (SMM) Systems Definition Study Report*, NASA GSFC, Nov. 1974.
2. Carnahan, Brice, H. A. Luther, and James O. Wilkes, *Applied Numerical Methods*. New York: John Wiley & Sons, Inc., 1969.
3. Chen, L. C. and J. R. Wertz, *Analysis of SMS-2 Attitude Sensor Behavior Including OABIAS Results*, Comp. Sc. Corp., CSC/TM-75/6003, April 1975.
4. Freund, John E., *Mathematical Statistics*, Englewood Cliffs, NJ: Prentice-Hall, Inc., 1962.
5. Hotovy, S. G., M. G. Grell, and G. M. Lerner, *Evaluation of the Small Astronomy Satellite-3 (SAS-3) Scanwheel Attitude Determination Performance*, Comp. Sc. Corp., CSC/TR-76/6012, July 1976.
6. Levitas, M., M. K. Baker, R. Collier, and Y. S. Hoh, *MAPS/MAGSAT Attitude System Functional Specifications and Requirements*, Comp. Sc. Corp., CSC/SD/78-6077, June 1978.
7. Meyers, G. F., M. E. Plett, and D. E. Riggs, *SAS-2 Attitude Data Analysis*, NASA X-542-71-363, GSFC Aug. 1971.
8. Tandon, G. K., M. Joseph, J. Oehlert, G. Page, M. Shear, P. M. Smith, and J. R. Wertz, *Communications Technology Satellite (CTS) Attitude Analysis and Support Plan*, Comp. Sc. Corp., CSC/TM-76/6001, Feb. 1976.
9. Tandon, G. K. and P. M. Smith, *Communications Technology Satellite (CTS) Post Launch Report*, Comp. Sc. Corp., CSC/TM-76/6104, May 1976.
10. Werking, R. D., R. Berg, K. Brokke, T. Hattox, G. Lerner, D. Stewart, and R. Williams, *Radio Astronomy Explorer-B Postlaunch Attitude Operations Analysis*, NASA X-581-74-227, GSFC, July 1974.
11. Wertz, J. R., C. E. Gartrell, K. S. Liu, and M. E. Plett, *Horizon Sensor Behavior of the Atmospheric Explorer-C Spacecraft*, Comp. Sc. Corp., CSC/TM-75/6004, May 1975.
12. Wertz, James R. and Lily C. Chen, "Geometrical Limitations on Attitude Determination for Spinning Spacecraft," *J. Spacecraft*, Vol. 13, p. 564–571, 1976.

PART IV

ATTITUDE DYNAMICS
AND
CONTROL

CONTENTS

PART IV

ATTITUDE DYNAMICS
AND
CONTROL

Chapter

15   Introduction to Attitude Dynamics and
       Control   487

16   Attitude Dynamics   510

17   Attitude Prediction   558

18   Attitude Stabilization   588

19   Attitude Maneuver Control   636

CHAPTER 15

INTRODUCTION TO ATTITUDE DYNAMICS AND CONTROL

15.1 Torque-Free Motion
15.2 Response to Torques
15.3 Introduction to Attitude Control

Dynamics is the study of the relationship between motion and the forces affecting motion. The study of the dynamics of objects in interplanetary or interstellar space is called *astrodynamics* and has two major divisions: celestial mechanics and attitude dynamics. *Celestial mechanics* or *orbit dynamics*, discussed briefly in Chapter 3, is concerned with the motion of the center of mass of objects in space, whereas *attitude dynamics* is concerned with the motion about the center of mass. In Part IV, we deal exclusively with this latter category.

Thus far, we have been concerned primarily with determining the orientation of a spacecraft without consideration of its dynamics, or, at least, with an implicit assumption of a specific and accurate dynamic model. However, knowledge of attitude dynamics is necessary for attitude prediction, interpolation, stabilization, and control. In this chapter, which is less quantitative than the remainder of Part IV, we attempt to provide a physical "feel" for attitude motion and environmental torques affecting the attitude. Chapter 16 then develops the more formal mathematical tools used in the study of attitude dynamics and briefly discusses the effect of nonrigidity in spacecraft structure. *Free-body (i.e., satellite) motion differs in several important respects from the motion of rigid objects, such as a spinning top, supported in a gravitational field. Thus, the reader should be careful to avoid relying on either intuition or previous analytic experience with common rotating objects supported in some way near the surface of the Earth.*

15.1 Torque-Free Motion

James R. Wertz

We consider first the simplest case of the attitude motion of a completely rigid, rotating object in space free of all external forces or torques. In describing this motion, four fundamental axes or sets of axes are important. *Geometrical axes* are arbitrarily defined relative to the structure of the spacecraft itself. Thus, the geometrical z axis may be defined by some mark on the spacecraft or by an engineering drawing giving its position relative to the structure. This is the reference system which defines the orientation of attitude determination and control hardware and experiments.

The three remaining axis systems are defined by the physics of satellite motion. The *angular momentum axis* is the axis through the center of mass parallel to the angular momentum vector. The *instantaneous rotation axis* is the axis about which the spacecraft is rotating at any instant; Euler's Theorem (Section 12.1) establishes the existence of this axis. The angular momentum axis and the instantaneous rotation axis are not necessarily the same. For example, consider the rotation of a symmetric dumbbell, as shown in Fig. 15-1. In elementary mechanics, we define the angular momentum, **L**, of a point mass, m, at position **r** relative to some arbitrary origin as

$$\mathbf{L} = \mathbf{r} \times \mathbf{p} = \mathbf{r} \times m\mathbf{v} \qquad (15\text{-}1)$$

where **p** is the momentum and **v** is the velocity of the particle in question. For a collection of n points,

$$\mathbf{L} = \sum_{i=1}^{n} \mathbf{L}_i = \sum_{i=1}^{n} \mathbf{r}_i \times \mathbf{p}_i \qquad (15\text{-}2)$$

Assume that the dumbbell is rotating with angular velocity ω about an axis through the center of mass and perpendicular to the rod joining the masses (Fig. 15-1(a)). Then, **L** is parallel to ω and the motion is particularly simple because **L** and ω remain parallel as the dumbbell rotates.

However, if the dumbbell is initially rotating about an axis through the center of mass but inclined to the normal to the central rod (Fig. 15-1(b)), **L** is in the plane defined by ω and the two end masses, but **L** is clearly *not* parallel to ω. (Use Eq. (15-2) to calculate the angular momentum about the center of mass.) Now the free-space motion is more complex. Because the conservation of angular momentum requires that **L** remain fixed in inertial space if there are no external torques, the instantaneous axis of rotation, ω, must rotate as the dumbbell rotates. Conversely, if ω is fixed in space by some external supports or axes, a torque must be supplied via the supports to change **L** as the object rotates. (This may be conveniently demonstrated by constructing models of the two dumbbells in Fig. 15-1 out of Tinkertoys.)

Clearly, the motion about the axis in Fig. 15-1(a) is simpler than that in Fig. 15-1(b). Thus, the motion of the dumbbell leads us to define as the third physical axis system, preferred axes about which the motion is particularly simple. Specifically, a *principal axis* is any axis, **P̂**, such that the resulting angular momentum is parallel to **P̂** when the spacecraft rotates about **P̂**. Therefore, for rotation about a principal axis, **L** is parallel to ω, or

$$\mathbf{L} = I_p \omega = I_p \omega \hat{\mathbf{P}} \qquad (15\text{-}3)$$

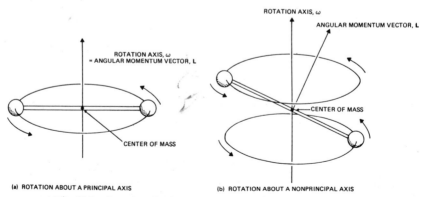

Fig. 15-1.   Rotation of a Symmetric Dumbbell. See text for discussion.

where I_p is a constant of proportionality called the *principal moment of inertia*. Because the magnitude of the angular velocity is defined by $\omega = v/r$, where r is the rotation radius, Eqs. (15-1) through (15-3) imply that *for a principal axis* and a collection of point masses,

$$I_p = \sum_{i=1}^{n} m_i r_i^2 \qquad (15\text{-}4)$$

where r_i is the perpendicular distance of m_i from the principal axis. *For rotation about nonprincipal axes the motion is more complex, and Eq. (15-3) does **not** hold.*

The form of Eq. (15-2) shows that whenever the mass of an object is symmetrically distributed about an axis (i.e., if the mass distribution remains identical after rotating the object $360/N$ deg about the specified axis, where N is any integer greater that 1*), the angular momentum generated by rotation about the symmetry axis will be parallel to that axis. Thus, any axis of symmetry is a principal axis. In addition, we will show in Chapter 16 that any object, no matter how asymmetric, has three mutually perpendicular principal axes defined by Eq. (15-3).

The sets of axes above may be used to define three types of attitude motion called pure rotation, coning, and nutation. *Pure rotation* is the limiting case in which the rotation axis, a principal axis, and a geometrical axis are all parallel or antiparallel, as shown in Fig. 15-2(a). Clearly, *the angular momentum vector* will lie along this same axis. These four axes will remain parallel as the object rotates.

Coning is rotation for which a geometrical axis is not parallel to a principal axis. If the principal and rotation axes are still parallel, the physical motion of the object is precisely the same as pure rotation. However, the "misalignment" of the geometrical axis (which may be intentional) causes this axis to rotate in inertial space about the angular momentum vector, as shown in Fig. 15-2(b). Coning is

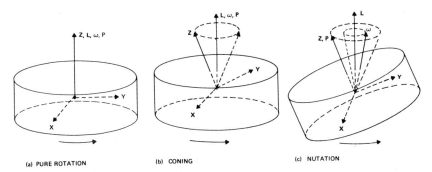

(a) PURE ROTATION   (b) CONING   (c) NUTATION

Fig. 15-2. Types of Rotational Motion. L = angular momentum vector; $\hat{\mathbf{P}}$ = principal axis; ω = instantaneous rotation axis; $\hat{\mathbf{z}}$ = geometrical z axis.

---

*In this case, the mass distribution consists of N symmetrically distributed groups of mass points. If **L** does not lie on the axis of symmetry, then for $N > 2$ it must lie closer to, or farther away from, one group; however, this is impossible because all of the mass points contribute equally to **L**. For $N = 2$, the mass distribution has the form $m(x,y,z) = m(-x, -y, z)$, where z is the symmetry axis. Therefore, any x or y components of **L** cancel when summed and **L** must lie along the z axis.

associated with a coordinate system misalignment rather than a physical misalignment and can be eliminated by a coordinate transformation if the orientation of the principal axes in the body of the spacecraft is known precisely.

Finally, *nutation** is rotational motion for which the instantaneous rotation axis is not aligned with a principal axis, as illustrated in Fig. 15-2(c). In this case, the angular momentum vector, which remains fixed in space, will not be aligned with either of the other physical axes. Both $\hat{P}$ and ω rotate about L. $\hat{P}$ is fixed in the spacecraft because it is defined by the spacecraft mass distribution irrespective of the object's overall orientation. Neither L nor ω is fixed in the spacecraft. ω rotates both in the spacecraft and in inertial space, while L rotates in the spacecraft but is fixed in inertial space. The angle between $\hat{P}$ and L is a measure of the magnitude of the nutation, called the *nutation angle*, θ. Nutation and coning can occur together, in which case none of the four axis systems is parallel or antiparallel.

We now describe the simple case in which two of the three principal moments of inertia are equal; that is, we assume $I_1 = I_2 \neq I_3$. Although this is an idealization for any real spacecraft, it is a good approximation for many spacecraft which possess some degree of cylindrical symmetry. In this case, the angular momentum vector, L, the instantaneous rotation axis, ω, and the $\hat{P}_3$ principal axis are coplanar and the latter two axes rotate uniformly about L. The body rotates at a constant velocity about the principal axis, $\hat{P}_3$, as $\hat{P}_3$ rotates about L and the nutation angle remains constant. (Because $\hat{P}_3$ is a spacecraft-fixed axis and is moving in inertial space, it cannot be the instantaneous rotation axis.)

As shown in Chapter 16, the spacecraft *inertial spin rate*, ω, about the instantaneous rotation axis (when $I_1 = I_2$) can be written in terms of components along $\hat{P}_3$ and $\hat{L}$ as:

$$\omega = \omega_p + \omega_l = \omega_p \hat{P}_3 + \omega_l \hat{L} \tag{15-5}$$

Because $\hat{P}_3$ and L are not orthogonal, the amplitude of ω is given by

$$\omega^2 = \omega_p^2 + \omega_l^2 + 2\omega_p \omega_l \cos\theta \tag{15-6}$$

where the nutation angle, θ, is the angle between $\hat{P}$ and $\hat{L}$; the *inertial nutation rate*, ω_l, is the rotation of $\hat{P}_3$ about L relative to an inertial frame of reference; and the *body nutation rate*, ω_p, is the rotation rate of any point, R, fixed in the body (e.g., a geometrical axis) about $\hat{P}_3$ *relative to the orientation of* L. Figure 15-3 shows a view looking "down" on the motion of the axes when θ is small. Here ω_l and ω are the rotation rates of lines LP_3 and P_3R, respectively, relative to inertial space, and ω_p is the rotation rate of P_3R relative to P_3L. The component angular velocities in Eqs. (15-5) and (15-6) are related by (Eq. 16-68)):

$$\omega_p = \frac{I_1 - I_3}{I_3} \omega_l \cos\theta \tag{15-7}$$

By resolving ω in Eq. (15-5) into components along $\hat{P}_3$ and orthogonal to $\hat{P}_3$ and then using Eq. (15-7), we may obtain an expression for the angle, ζ, between $\hat{P}_3$ and

---

*This definition of nutation is in keeping with common spacecraft usage and differs from that used in classical mechanics for describing, say, the motion of a spinning top. In the latter case, nutation refers to the vertical wobble of the spin axis as it moves slowly around the gravitational field vector.

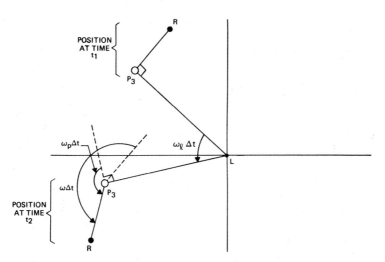

Fig. 15-3.   Position of Principal Axis, $\hat{\mathbf{P}}_3$ and Arbitrary Spacecraft Reference Axis **R**, at Times t_1 and t_2 for a Nutating Spacecraft With Small Values of θ and ζ. $\Delta t = t_2 - t_1$.

ω, as follows:

$$\tan \zeta = \frac{\omega_i \sin \theta}{\omega_p + \omega_i \cos \theta} = \frac{I_3}{I_1} \tan \theta \qquad (15\text{-}8)$$

To obtain a physical feel for the motion described by Eqs. (15-5) through (15-8), we note that in inertial space, ω rotates about **L** on a cone of half-cone angle $(\theta - \zeta)$ called the *space cone*, as illustrated in Fig. 15-4 for $I_1 > I_3$. Similarly, ω maintains a fixed angle, ζ, with $\hat{\mathbf{P}}_3$ and, therefore, rotates about $\hat{\mathbf{P}}_3$ on a cone called the *body cone*. Because ω is the instantaneous rotation axis, the body is instantaneously at rest along the ω axis as ω moves about **L**. Therefore, we may visualize the motion of the spacecraft as the body cone rolling without slipping on the space cone. The space cone is fixed in space and the body cone is fixed in the spacecraft.

Figure 15-4 is correct only for objects, such as a tall cylinder, for which I_1 is greater than I_3. In this case, Eq. (15-7) implies that ω_p and ω_i have the same sign.[*] If I_3 is greater than I_1, as is the case for a thin disk, ω_p and ω_i have opposite signs and the space cone lies inside the body cone, as shown in Fig. 15-5. The sign of ω_p is difficult to visualize, since ω_p is measured relative to the line joining the axes of the two cones. (Refer to Fig. 15-3.) If we look down on the cones from above, in both Figs. 15-4 and 15-5, $\hat{\mathbf{P}}_3$ is moving counterclockwise about **L**. In Fig. 15-4, the dot on the edge of the body cone is moving toward ω and, therefore, is also rotating counterclockwise. In Fig. 15-5, the dot on the edge of the body cone is moving counterclockwise in inertial space, but the ω axis is moving counterclockwise more quickly. Therefore, relative to the $\hat{\mathbf{P}}_3 - \mathbf{L} - \omega$ plane, the dot is moving clockwise and ω_p has the opposite sign of ω. If $I_1 = I_2 = I_3$, the space cone reduces to a line, $\omega_p = 0$, and the spacecraft rotates uniformly about **L**. In this case, any axis is a principal axis.

Figure 15-6 illustrates the motion in inertial space of an arbitrary point, **R**

---

[*]The terms *prolate* and *oblate* are commonly used for $I_1 > I_3$ and $I_3 > I_1$, respectively; these terms refer to the shape of the energy ellipsoid, which is introduced in Section 15.2.

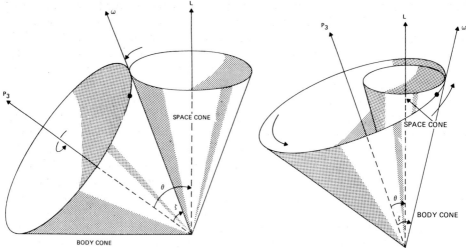

Fig. 15-4. Motion of a Nutating Spacecraft. The body cone rolls on the space cone for $I_1 = I_2 > I_3$.

Fig. 15-5. Motion of a Nutating Spacecraft. The body cone rolls on the space cone for $I_3 > I_1 = I_2$.

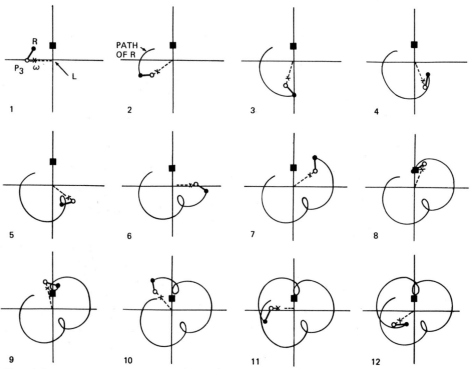

Fig. 15-6. Motion in Inertial Space of a Point, **R**, Fixed on a Nutating Spacecraft With $I_1 = I_2 > I_3$. Origin = **L**, o = $\hat{P}_3$, X = ω, ● = **R** (arbitrary point fixed on the spacecraft); ■ = arbitrary point fixed in inertial space.

(such as the geometrical z axis), fixed in the body of the nutating spacecraft. The coordinate axes in Fig. 15-6 are fixed in inertial space with the angular momentum vector at the origin. The symbols X, o, and ● mark the directions of ω, $\hat{\mathbf{P}}_3$, and $\mathbf{R}$, respectively, on the plane normal to $\mathbf{L}$. Thus, the dashed line is the line connecting the centers of the body cone and the space cone and the X along the dashed line is the point at which the two cones touch. The heavy solid line is a line fixed in the spacecraft joining the principal axis to the arbitrary point, $\mathbf{R}$. The light solid line traces the motion of $\mathbf{R}$ in inertial space as the spacecraft rotates and nutates. The nutation angle, θ, is assumed small. For the case shown, $I_1 = 3.5I_3$. Therefore, $\omega_p = 2.5\omega_l$ and the inertial spin rate, $\omega = 3.5\omega_l$. That is, in one revolution of the dashed line, the heavy solid line rotates 3.5 times in inertial space and 2.5 times relative to the dashed line. In a single frame, the dashed line rotates 36 deg and the heavy solid line rotates 90 deg relative to the dashed line and 126 deg relative to the edge of the page.

Figure 15-7 is identical with frame 12 of Fig. 15-6, except that the point in the body which is followed is farther from the axes and the positions of $\mathbf{R}$ and $\hat{\mathbf{P}}_3$ for each of the 12 frames have been labeled. As seen clearly in this figure, a point in the body at an angle ψ from $\hat{\mathbf{P}}_3$ will always be between $(\psi - \theta)$ and $(\psi + \theta)$ from $\mathbf{L}$.

If the space cone is inside the body cone, then the motion of a general point is as shown in Fig. 15-8. For this example, $I_3 = 3.5I_1$. Thus, assuming θ is small, $\omega_p = -0.714\omega_l$ and $\omega = 0.285\omega_l$. Notice the "backward" rotation of the heavy solid

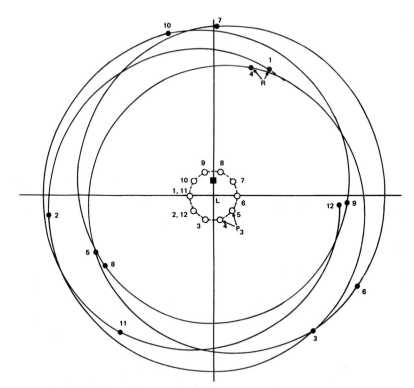

Fig. 15-7.　Frame 12 of Fig. 15-6 for a Point Farther From the Principal Axis

line relative to the dashed line and the very slow rotation of the heavy solid line in inertial space. Also notice that the angular momentum vector is now between the principal axis and the instantaneous rotation axis.

If the values of the two moments of inertia I_1 and I_2 are close but not equal, the motion is physically very similar to that shown in Figs. 15-4 through 15-8, but is considerably more involved mathematically. The space cone and body cone have approximately elliptically shaped cross sections rather than circular ones. Thus, ω, $\mathbf{L}$, and $\hat{\mathbf{P}}$ are no longer coplanar and the nutation angle, θ, is not constant.

Because attitude measurements are made in the spacecraft frame, it is of interest to consider the motion of an object fixed in inertial space as viewed from a frame of reference fixed on the nutating spacecraft. The motions in this frame of reference are just the reverse of those previously discussed. Thus, the body cone remains fixed and the space cone rolls around it carrying the inertial coordinate system. Figure 15-9 illustrates the motion of a point fixed in inertial space as viewed from the spacecraft for the nutation shown in Fig. 15-6. In Fig. 15-9, $\hat{\mathbf{P}}_3$ is fixed at the origin of the coordinate system in each frame and the geometrical point $\mathbf{R}$ from Fig. 15-6 is fixed at the position shown by the ●. The open square marks the position of $\mathbf{L}$. The solid square marks the position of an arbitrary point, $\mathbf{S}$, fixed in inertial space. ($\mathbf{S}$ is shown in Fig. 15-6 as the solid square on the upper axis.) Frames are at the same time intervals in both figures so that the relative orientation of all components is the same for each of the 12 frames in the 2 figures, as is shown most clearly in frame 1.

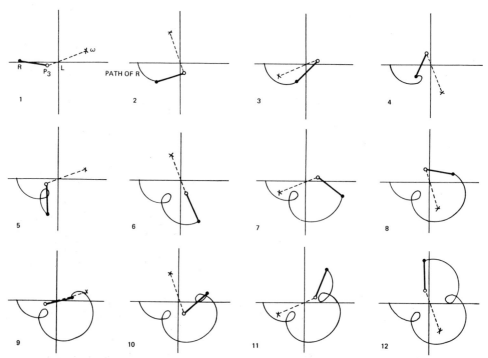

Fig. 15-8.   Motion in Inertial Space of a Point, $\mathbf{R}$, Fixed on a Nutating Spacecraft With $I_3 > I_1 = I_2$. Origin $= \mathbf{L}$, $\circ = \hat{\mathbf{P}}_3$, $\mathsf{X} = \omega$, ● $= \mathbf{R}$.

Example of Real Satellite Motion. Figure 15-10 shows 12 frames taken at equal time intervals (every 1.25 sec) from a motion picture of an actual, small, scientific spacecraft in orbit.* The spacecraft is the Apollo 15 subsatellite, launched into lunar orbit by a spring mechanism from the service module of the Apollo 15 spacecraft at 21:01 UT, August 4, 1971, just before it left lunar orbit for return to Earth [Anderson, *et al.*, 1972]. As the subsatellite moved away from the command module, it was photographed by the astronauts using a hand-held camera operated at 12 frames per second.

The approximate structure and dimensions of the satellite are shown in Fig. 15-11. The satellite was used to measure properties of the magnetic and gravitational fields and the solar plasma in the vicinity of the Moon. A magnetometer on the end of one of the three booms and the wire running along the boom can be identified in Fig. 15-10, frames 4, 6, and 8. This boom has been marked with a white dot. (Tip masses were added to the other two booms for balance.) Following the motion of the white dot (indicated by the solid line and arrow in frame 1) reveals the counterclockwise rotation of the satellite. The frames have been chosen

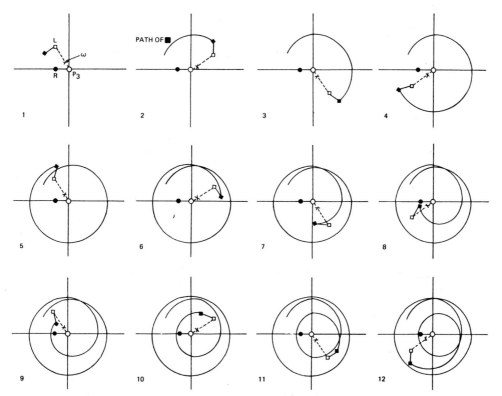

Fig. 15-9. Motion of Point, ■, Fixed in Inertial Space Viewed From a Nutating Spacecraft for Conditions of Fig. 15-6. Origin = $\hat{\mathbf{P}}_3$, X = ω, ● = **R**, □ = **L**. The relative positions of all points are the same as in Fig. 15-6, as seen most easily in frame 1.

---

*Such photographs may become commonplace with shuttle-launched spacecraft. Closeup photographs of orbiting satellites prior to that time are rare. Figure 15-10 is taken from the only existing footage at the time of this writing.

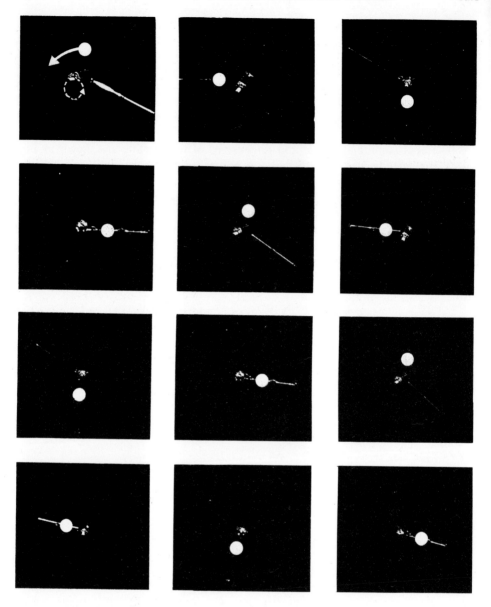

Fig. 15-10. Motion of Apollo 15 Subsatellite (courtesy NASA)

so that the satellite rotates about 360 deg in 4 frames (90 deg/frame), as can be seen by comparing frames 1, 5, and 9, or frames 4, 8, and 12.

By the symmetry of the satellite, we may assume that a principal axis lies along the long axis of the body. Thus, the satellite is nutating, because the principal axis (the body axis) is not fixed in space but is rotating counterclockwise in a small cone whose axis is inclined slightly to the left of center in each frame, as shown by the dashed line in frame 1.

The satellite's inertial spin rate, ω, and inertial nutation rate, ω_I, can be estimated from the figure. From frame 1 to frame 9, the satellite completes slightly *less* than two rotation periods, but slightly *more* than two nutation periods. By measuring the change in orientation of both the booms and the principal axis between frames 1 and 9, we estimate

$$\omega \approx 11.8 \text{ rpm}$$

$$\omega_I \approx 12.4 \text{ rpm}$$

Neglecting the fact that **P** and **L** are not quite collinear, we obtain immediately from Eq. (15-5)

$$\omega_p \approx 11.8 - 12.4 = -0.6 \text{ rpm}$$

and from Eq. (15-7)

$$\frac{I_1}{I_3} \approx 1 + \frac{\omega_p}{\omega_I} = 0.95$$

Thus, $I_3 > I_1$ and the motion of the various axes is as illustrated in Fig. 15-5 and 15-8. We shall see in Section 15.2 that this condition determines that the satellite motion is stable, i.e., that the nutation will not increase and will eventually damp out if there are sufficient dissipative forces. Thus, by examining the photographs we see that the three booms are long enough and heavy enough to provide the satellite with stable, rather than unstable, rotation. Although our quantitative estimates may be in error, our qualitative results depend only on the inequality $\omega_I > \omega$, which is clear from Fig. 15-10.

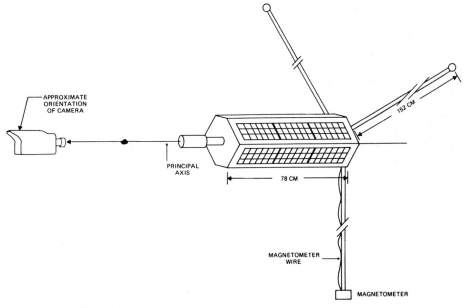

Fig. 15-11. Approximate Size and Shape of Apollo 15 Subsatellite

15.2 Response to Torques

F. L. Markley

We now turn to a qualitative discussion of the effect of an applied force on the motion of a spacecraft about its center of mass. The basic equation of attitude dynamics is obtained from Eq. (15-2), which expresses the angular momentum of the spacecraft as a sum over the masses, m_i, located at positions, r_i, and moving with velocities, v_i, that make up the spacecraft. Differentiation with respect to time gives

$$\frac{d}{dt}\mathbf{L} = \sum_{i=1}^{n} \frac{d}{dt}\mathbf{L}_i = \sum_{i=1}^{n} \frac{d}{dt}(\mathbf{r}_i \times m_i\mathbf{v}_i)$$

$$= \sum_{i=1}^{n} (\mathbf{v}_i \times m_i\mathbf{v}_i + \mathbf{r}_i \times m_i\mathbf{a}_i) = \sum_{i=1}^{n} \mathbf{r}_i \times \mathbf{F}_i$$

$$= \sum_{i=1}^{n} \mathbf{N}_i \equiv \mathbf{N} \tag{15-9}$$

where $\mathbf{a}_i$ and $\mathbf{F}_i$ are the acceleration of m_i and the force applied to it, respectively.

The *torques*, $\mathbf{N}_i$, on the individual points in a rigid body are due to both forces between the points and externally applied forces. Under the very general conditions discussed in Section 16.1, the internal torques sum to zero and the resultant torque, $\mathbf{N}$, is simply the torque due to external forces. The external torques are of two kinds: (1) *disturbance torques* (described in Section 17.2) caused by environmental effects such as aerodynamic drag and solar radiation pressure, and (2) deliberately applied *control torques* from devices such as gas jets or magnetic coils. (Control torques due to reaction wheels do not change the total angular momentum of the spacecraft because they are not external torques. A spacecraft with reaction wheels is not a rigid body; the control torques in this case cause a redistribution of the angular momentum between the wheels and the spacecraft body.) Control torques will be discussed in more detail in Section 15.3.

If a spacecraft is initially spinning about a principal axis, a torque applied parallel or antiparallel to the angular momentum vector, $\mathbf{L}$, will cause an increase or a decrease in the magnitude of $\mathbf{L}$ without affecting its direction. A torque component perpendicular to $\mathbf{L}$, will cause the direction of $\mathbf{L}$ to change without altering its magnitude. The change in direction of the angular momentum vector due to an applied torque is called *precession.** The special case of slow precession due to a small applied torque (such that the magnitude of the integral of the torque over a spin period is much less than $|\mathbf{L}|$) is known as *drift*. Environmental torques are a common source of attitude drift.

Although internal torques do not change the value of the angular momentum in inertial space, they can affect the behavior of $\mathbf{L}$ in spacecraft-fixed coordinates. Additionally, if the internal forces between the components of a spacecraft lead to energy dissipation (through solid or viscous friction or magnetic eddy currents, for

---

*Note that this definition of precession, which has been adopted in spacecraft dynamics, is somewhat different from the meaning usually assumed in physics.

example) the rotational kinetic energy of the spacecraft will decrease. These effects can be qualitatively understood with the aid of two concepts—the *angular momentum sphere* and the *energy ellipsoid*. Consider first the situation in which there is no energy dissipation. In Section 16.1, it is shown that the rotational kinetic energy of a rigid spacecraft is equal to

$$E_k = \frac{1}{2}\left(I_1\omega_1^2 + I_2\omega_2^2 + I_3\omega_3^2\right) = \frac{1}{2}\left(L_1^2/I_1 + L_2^2/I_2 + L_3^2/I_3\right) \qquad (15\text{-}10)$$

where I_1, I_2, and I_3 are the spacecraft principal moments of inertia; ω_1, ω_2, and ω_3 are the angular velocity components about the body-fixed principal axes; and L_1, L_2, and L_3 are the components of the spacecraft angular momentum vector along the principal axes. The equivalence of the two forms of Eq. (15-10) follows from Eq. (15-3). Now consider a representation of the spacecraft angular momentum as a point in a three-dimensional coordinate system, the displacement of the point along the three coordinate axes being proportional to L_1, L_2, L_3, respectively. The locus of points in this *angular momentum space* consistent with a fixed rotational kinetic energy is the set of points satisfying Eq. (15-10). Rewriting this equation as

$$\frac{L_1^2}{2I_1E_k} + \frac{L_2^2}{2I_2E_k} + \frac{L_3^2}{2I_3E_k} = 1 \qquad (15\text{-}11)$$

shows that these points lie on an ellipsoid with semiaxis lengths $\sqrt{2I_1E_k}$, $\sqrt{2I_2E_k}$, and $\sqrt{2I_3E_k}$. This is the *energy ellipsoid* corresponding to rotational kinetic energy E_k.

The components of **L** in inertial space are constant in the absence of torques, but the components of **L** in spacecraft-fixed coordinates are time dependent. The magnitude, $L \equiv |\mathbf{L}|$, of the angular momentum is constant, however. The locus of points in angular momentum space corresponding to a fixed magnitude, L, is just a sphere of radius L, the *angular momentum sphere*. The locus of possible values of **L** *in the spacecraft frame* is the intersection of the angular momentum sphere and the energy ellipsoid. Figure 15-12 shows this intersection for the case $I_1 = I_2 < I_3$, and Fig. 15-13 shows it for $I_1 = I_2 > I_3$. In both cases of axial symmetry, the locus of possible values of the angular momentum consists of two circles about the symmetry axis. The angular momentum vector in the spacecraft frame moves at a constant rate along one of the circles. This motion is nutation, as described in Section 15.1, Fig. 15-12 corresponds to Fig. 15-5 with $\hat{\mathbf{P}}_3$ held fixed, and Fig. 15-13 corresponds to Fig. 15-4.

This pictorial representation of nutation can be used to analyze the effects of energy dissipation. In the presence of dissipative forces, the energy ellipsoid shrinks in size while maintaining its shape, and the angular momentum sphere is unchanged. The angular momentum vector continues to rotate along the intersection of the shrinking energy ellipsoid and the angular momentum sphere. This path has an approximately spiral shape. The shrinking of the energy ellipsoid continues until it lies wholly within, or is tangent to, the angular momentum sphere. For $I_1 = I_2 < I_3$ (Fig. 15-12), this results in **L** being aligned along the positive or negative 3-axis. For $I_1 = I_2 > I_3$ (Fig. 15-13), the limit occurs when **L** lies on the circle

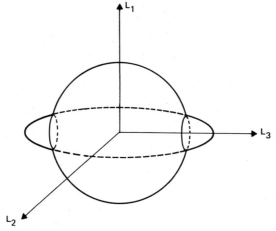

Fig. 15-12. Energy Ellipsoid and Angular Momentum Sphere for $I_1 = I_2 < I_3$. L_1, L_2, and L_3 are the angular momentum components in body principal coordinates.

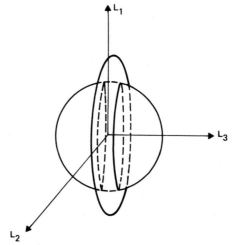

Fig. 15-13. Energy Ellipsoid and Angular Momentum Sphere for $I_1 = I_2 > I_3$. L_1, L_2, and L_3 are the angular momentum components in body principal coordinates.

$L_1^2 + L_2^2 = L^2$, $L_3 = 0$. It will be shown in Chapter 16 that L remains fixed at a point on this circle. Thus, in both examples nutation ceases, and the motion of the body is simple rotation about a fixed axis. Clearly, energy dissipation must also cease (see Section 18.4).

If there is no axis of symmetry, the intersections of the energy ellipsoid and the angular momentum sphere are not circles. A family of intersections for different values of E_k with $I_1 > I_2 > I_3$ is shown in Fig. 15-14. When energy dissipation ceases, the angular momentum vector becomes aligned with the *major principal axis*; i.e. the axis corresponding to the largest principal moment of inertia. It is clear from Eq. (15-10) that when L is constant, the rotational kinetic energy is minimized when rotation is about the major principal axis. If the nominal spacecraft spin axis is the major principal axis, nutation represents excess kinetic energy above that required by the magnitude of the angular momentum. The

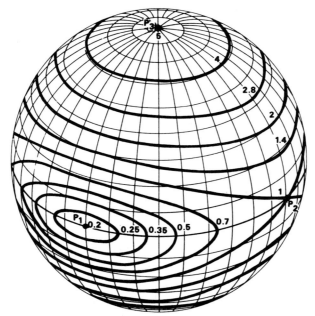

Fig. 15-14. Family of Intersections of Energy Ellipsoid and Angular Momentum Sphere for Various Energies, With $I_1 : I_2 : I_3$ in the Ratio $25:5:1$. Each curve is labeled with its value of $2I_2E/L^2$. P_1, P_2 and P_3 are the principal axes; P_1 is the major principal axis.

reduction of this excess kinetic energy and the corresponding alignment of the rotation axis with the principal axis of largest moment of inertia is known as *nutation damping*. Several mechanisms for nutation damping are discussed in more detail in Section 18.4.

If the nominal spacecraft spin axis is a principal axis other than the major principal axis, energy dissipation will result in an *increase* in nutation. The motion when energy dissipation ceases is pure rotation about an axis perpendicular to the nominal spin axis, a condition known as *flat spin*. A well-known example of this is Explorer 1 (Fig. 15-15), the first U.S. satellite, which was launched on February 1, 1958. It was designed to spin about its longitudinal symmetry axis, which was an

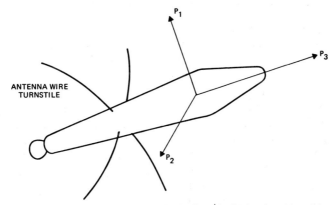

Fig. 15-15. Explorer I. P_3, the nominal spin axis, is the principal axis with minimum moment of inertia.

axis of minimum moment of inertia, but the motion rapidly changed into a flat spin mode. Bracewell and Garriott [1958] explained this as a result of energy dissipation due to vibrational motions of the antennas. This is an example of the case illustrated in Fig. 15-13. In this case, energy dissipation mechanisms will not reduce nutation, and active nutation damping by means of external control torques must be resorted to.

15.3 Introduction to Attitude Control

Vincent H. Tate

Attitude control is the process of achieving and maintaining an orientation in space. An *attitude maneuver* is the process of reorienting the spacecraft from one attitude to another. An attitude maneuver in which the initial attitude is unknown when maneuver planning is being undertaken is known as *attitude acquisition*. *Attitude stabilization* is the process of maintaining an existing attitude relative to some external reference frame. This reference frame may be either inertially fixed or slowly rotating, as in the case of Earth-oriented satellites.

Control System Overview. *Control torques*, such as those produced by gas jets, are generated intentionally to control the attitude. *Disturbance torques* are environmental torques (e.g., aerodynamic drag) or unintended internal torques (e.g., crew motion). Because these can never be totally eliminated, some form of attitude control system is required. An *attitude control system* is both the process and the hardware by which the attitude is controlled. In general, an attitude control system consists of three components: attitude sensors, the control process, and control hardware. An *attitude sensor* locates known reference targets such as the Sun or the Earth to determine the attitude. The *control process* or *control law* determines when control is required, what torques are needed, and how to generate them. The *control hardware* or *actuator* is the mechanism that supplies the control torque.

Control systems can be classified as either open-loop or closed-loop. An *open-loop* system is one in which the control process includes human interaction. For example, attitude data from the attitude sensors is analyzed, and a control analyst occasionally sends commands to the spacecraft to activate the control hardware (e.g., fire the jets). A *closed-loop*, or *feedback*, system is one in which the control process is entirely electrical or computer controlled. For example, attitude sensors send attitude data to an onboard computer which determines the attitude and then activates the control hardware (e.g., fires the jets). Normally, closed-loop systems are more sophisticated and complex but can maintain a much smaller tolerance on the deviation from the desired attitude. Frequently, in a closed-loop system, the attitude sensors are such that the attitude measurement is directly related to the desired orientation. For example, a wheel-mounted horizon sensor with a magnetic index mark on the body is placed so that the mark should be pointed toward the center of the Earth in azimuth. The difference between the time the index is sensed by the wheel and the time of the midscan between the two Earth horizon crossings is a direct measurement of the difference between the real and the desired attitude. This type of attitude measurement is called an *error signal*.

The function of the control system is to maintain the error signal within specified limits.

The example just described is an *active attitude control system* in which continuous decisionmaking and hardware operation is required. The most common sources of torque for active control systems are gas jets, electromagnets, and reaction wheels. In contrast, *passive attitude control* makes use of environmental torques to maintain the spacecraft orientation. Gravity-gradient and solar sails are common passive attitude control methods.

Attitude control systems are highly mission dependent. The decision to use a passive or an active control system or a combination of the two depends on mission pointing and stability requirements, interaction of the control system with onboard experiments or equipment, power requirements, weight restrictions, mission orbital characteristics, and the control system's stability and response time. For example, a near-Earth, spin-stabilized spacecraft could use magnetic coils for attitude maneuvers and for periodic adjustment of the spin rate and attitude. Above synchronous altitudes, gas jets would be required for these functions because the Earth's magnetic field is generally too weak at this altitude for effective magnetic maneuvers. Table 15-1 compares the various types of commonly used control methods.

Table 15-1.   Comparison of Attitude Control Methods

CONTROL METHOD	REGIONS OF SPACE WHERE APPLICABLE	REPRESENTATIVE TORQUE	SATELLITES USING SYSTEM	SECTIONS WHERE DISCUSSED*
GAS THRUSTERS	UNLIMITED	FOR A THRUSTER FORCE OF 0.3 N AND MOMENT ARM OF 2 M; TORQUE = 0.6 N·M	ISEE, OSO, ATS, CTS	6.8 (H); 7.10 (H) 18.3 (S); 19.3 (M)
MAGNETIC COILS	BELOW SYNCHRONOUS ORBIT (<35,000 KM)	FOR A 40,000 POLE-CM ELECTROMAGNET AT AN ALTITUDE OF 550 KM; TORQUE $\simeq$ 0.001 N·M	AEROS, OSO, SAS-3, AE	5.1 (E); 6.7 (H); 18.3 (S); 19.1 (M); 19.2 (M); AP H (E)
GRAVITY GRADIENT	NEAR MASSIVE CENTRAL BODY (EARTH, MOON, ETC.)	FOR AN 800-KM CIRCULAR ORBIT AND AN ELONGATED SATELLITE WITH A TRANSVERSE MOMENT OF INERTIA OF 1000 KG·M^2; TORQUE $\approx$ 5 X 10^{-5} N·M/DEG OF OFFSET FROM NULL ATTITUDE	GEOS, RAE	5.2 (E); 17.2(E); 18.3 (S)
MOMENTUM WHEELS	UNLIMITED	TYPICAL ANGULAR MOMENTUM $\approx$ 10 KG·M^2/S; TYPICAL TORQUE ABOUT THE WHEEL AXIS $\approx$ 0.1 N·M	ATS-6, SAS-3, AE, OAO, GEOS	6.6 (H); 7.9 (H); 18.2 (S); 19.4 (M); 18.3 (S)

*E = ENVIRONMENT MODEL; H = HARDWARE; S = STABILIZATION; M = MANEUVERS.

Passive Attitude Control. The most common passive control techniques are *spin stabilization*, in which the entire spacecraft is rotated so that its angular momentum vector remains approximately fixed in inertial space; *dual-spin stabilization*, in which the spacecraft has a rotating wheel or consists of two rotating components; and *gravity-gradient stabilization*, in which the differential gravitational forces acting on an asymmetric spacecraft force the minor axis (minimum moment of inertia axis) to be perpendicular to the gravitational equipotential. With the exception of gravity-gradient stabilization, passive control normally requires the use of active control systems, such as mass expulsion or magnetic coils, to periodically adjust the spacecraft attitude and spin rate to counteract disturbance torques. They also require some form of nutation damping to eliminate nutation caused by an unbalanced spacecraft or the elasticity of the spacecraft structure.

A spin-stabilized spacecraft normally spins about the major principal axis for stability, as described in Section 15.2. These are sometimes called *single-spin spacecraft* to distinguish them from dual-spin spacecraft discussed below. The basic requirement for spin stabilization is

$$\left|\int \mathbf{N}dt\right| \ll |\mathbf{L}| \qquad (15\text{-}12)$$

where $\mathbf{L}$ is the spacecraft angular momentum, $\mathbf{N}$ is the sum of the disturbance torques discussed in Section 17.2, $\mathbf{N}dt$ is the angular momentum change due to environmental torques over time interval dt, and the integral is carried out over whatever length of time passive stability is required. The integral defines the change in both the spacecraft orientation and the spin rate. If the disturbance torques are cyclic and the maximum attitude change from the torques is less than the mission requirements, no other control technique is required once the mission attitude is achieved. If disturbance torques exhibit either cyclic variations or a secular trend which exceeds the mission attitude constraints, an active control system is required to periodically adjust the attitude and the spin rate. Spin stabilization is a simple and effective technique and requires no moving parts; however, it is limited to spacecraft for which the spin itself does not inhibit the spacecraft function.

The International Sun-Earth Explorer-1, ISEE-1, shown in Fig. 15-16, is an example of a spin-stabilized spacecraft. The spacecraft contains scanning experiments for studying the space environment between the Sun and the Earth. Gas thrusters are used for reorientation maneuvers and for periodic adjustment of the attitude and spin rate to counteract the solar pressure disturbance torque. A passive fluid nutation damper (see Section 18.4) is used to control nutation. A Panoramic Attitude Sensor, which senses the Earth and Moon horizons, and a Sun sensor are used for attitude determination.

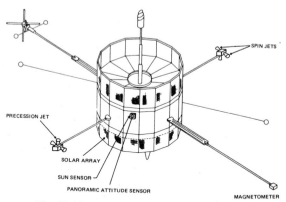

Fig. 15-16.   ISEE-1 Spin-Stabilized Spacecraft

Dual-spin stabilized spacecraft have two components spinning at different rates. Normally, one spacecraft section, such as a wheel, is spinning rapidly and the other section is despun or spinning very slowly to maintain one axis toward the Earth. A dual-spin system operates on the same principle as a single-spin system and usually requires a nutation damper and an active control system as does a single-spin spacecraft. A dual-spin system provides platforms for both scanning and pointing (inertially fixed) instruments. However, with a two-component spacecraft, additional complexity arises because of the need for bearings and support structures separating the two components.

The Orbiting Solar Observatory-8, OSO-8, shown in Fig. 15-17, is an example of a dual-spin spacecraft. The sail and pointing instrument assembly are the despun components and contain the solar power array and Sun pointing experiments. A passive-eddy current nutation damper located in the sail controls the nutation. The wheel is the spinning component and contains scanning instruments. An electromagnetic torquer mounted along the wheel spin axis and a pair of gas jets on the wheel are used for maneuvering the spacecraft and maintaining the spin rate. A wheel-mounted Sun sensor, a magnetometer mounted on one of the ballast arms, and a star scanner are used for attitude determination.

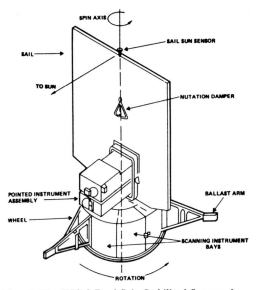

Fig. 15-17. OSO-8 Dual-Spin Stabilized Spacecraft

A gravity-gradient stabilization system interacts with the gravitational field to maintain the spacecraft attitude. Because the gravity-gradient torque decreases as the inverse cube of the distance from the gravitational source, gravity-gradient systems are usually used for near-Earth or -Moon missions requiring one side of the spacecraft to point toward the central body. Due to orbit eccentricity, damper, and thermal heating effects, the potential pointing accuracy is typically 1 to 4 deg. The basic requirement for gravity-gradient stabilization is that the gravity-gradient torque be greater than all other environmental torques. To achieve this, one principal moment of inertia must be smaller than the others, causing the minor axis to align along the nadir vector. To obtain the differential in moments of inertia, booms are often deployed along the minor axis. The gravity-gradient torque causes the spacecraft to oscillate or librate about the pitch axis and a passive damper is generally used to minimize the amplitude of this oscillation. Gravity-gradient systems require no moving parts other than, in some cases, extendable booms or antennas.

Figure 15-18 shows the Radio Astronomy Explorer-2, RAE-2, gravity-gradient stabilized spacecraft placed in orbit about the Moon. During the transfer from the Earth to the Moon, the spacecraft was spin stabilized and used gas jets for attitude control. After achieving a lunar orbit and final boom deployment, it became a

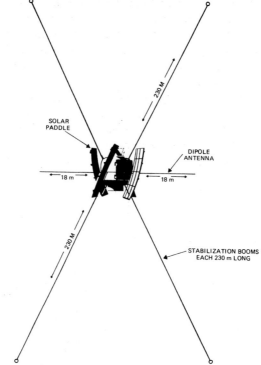

SOLAR PADDLE

230 M

DIPOLE ANTENNA

18 m

18 m

230 M

STABILIZATION BOOMS
EACH 230 m LONG

Fig. 15-18.   RAE-2, Gravity-Gradient Stabilized Spacecraft

nonspinning, gravity-gradient stabilized spacecraft. The libration control system consisted of a passive hysteresis damper and an extendable and retractable tubular boom. Two Panoramic Attitude Sensors were used to sense the Earth and Moon and two single-axis and eight two-axis digital Sun sensors were used for attitude determination.

Active Attitude Control.   The most common active control techniques are mass expulsion devices, such as gas jets or ion thrusters; momentum wheels, which are used to absorb disturbance torques; and electromagnetic coils, which provide a torque by interacting with the Earth's magnetic field.

Mass expulsion control systems used for attitude maneuvering include gas and ion thrusters. Gas thruster systems are efficient in the execution of a maneuver, are simple to operate, and are not limited to a specific environment; however, they are expensive, require complex hardware and plumbing, and are limited in lifetime by the amount of fuel onboard. Gas attitude control systems can also cause orbit changes during a reorientation maneuver. Consequently, the thrusters are usually fired in pairs to minimize translational motion. Gas thruster systems are commonly used with spin-stabilized spacecraft (such as CTS described in Section 1.1) for attitude maneuvering and spin rate control. For this type of spacecraft, a minimum of two reorientation thrusters and two spin rate control thrusters are required. For a three-axis stabilized system, six possible directions ($\pm$ pitch, $\pm$ roll, $\pm$ yaw) are available for maneuvering the spacecraft and a minimum of six thrusters are required.

Momentum wheel control systems can have wheels on 1, 2, or 3 axes and normally require a secondary active control system, such as gas jets, to maintain the wheel and spacecraft momentum in the presence of disturbance torques and friction losses. A dual-spin spacecraft is a single-momentum wheel system. A two-wheel system for an Earth-oriented spacecraft normally has one wheel along the pitch axis for pitch control and another wheel mounted on either the roll or the yaw axis for roll/yaw control. A three-axis system uses momentum wheels along all three axes and may have six or more wheels along nonorthogonal axes. Figure 15-19 illustrates three types of momentum wheel systems with mass expulsion and magnetic coils to control the wheel spin rate. The operation of the momentum wheels is complex and relies on the interaction of mechanical parts, which limits the system lifetime.

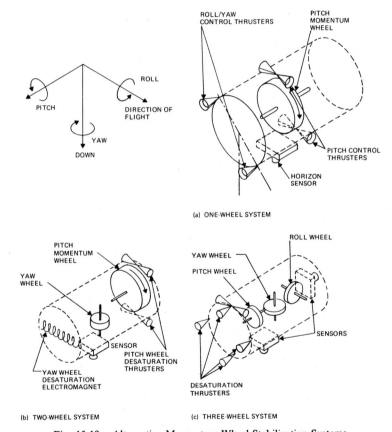

Fig. 15-19. Alternative Momentum Wheel Stabilization Systems

Momentum wheel stabilization systems are used to maintain the attitude by momentum exchange between the spacecraft and the wheel. As a torque acts on the spacecraft along one axis, the momentum wheel reacts, absorbing the torque and maintaining the attitude. As a result, momentum wheels are particularly attractive for attitude control in the presence of cyclic torques or random torques, such as in manned space stations. The wheel spin rate increases or decreases to maintain a constant attitude. Over a full period of a cyclic torque, the wheel speed remains constant. Secular torques acting on the spacecraft cause the momentum wheel speed to either increase or decrease monotonically until the wheel speed moves outside operational constraints. A momentum exchange device (i.e., a gas jet, magnetic coil, or gravity-gradient torque) must then be used to restore the momentum wheel speed to its nominal operating value. The upper operating limit of a momentum wheel is called the *saturation limit*.

Momentum wheel control systems used for maneuvering operate in the same fashion as the stabilization systems. For example, consider a maneuver for an inertially pointed spacecraft. Initially, the spacecraft is motionless and the wheel is spinning with angular momentum **H**. At some time, t, the control system is commanded to maneuver the spacecraft. At this time, a transfer of momentum, Δ**H**, from the wheel to the spacecraft occurs and the spacecraft attitude begins changing. The angular momentum of the wheel becomes **H** − Δ**H** and the angular momentum of the spacecraft becomes Δ**H**. When the spacecraft reaches its desired attitude, the momentum transfer is reversed, the spin rate of the momentum wheel returns to its original value, and the spacecraft body momentum returns to zero. The spacecraft is now pointing at its new attitude with zero angular momentum. The Applications Technology Spacecraft-6, ATS-6, shown in Fig. 15-20, has a three-wheel momentum control system. Momentum wheels are mounted along the pitch, yaw, and roll axes and serve as prime torquers for stabilization and maneuvering.

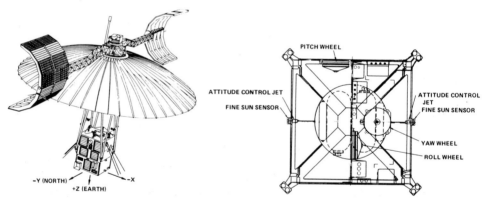

(a) ATS-6 SPACECRAFT

(b)   COMPONENT VIEW OF THE ATTITUDE CONTROL SYSTEM LOCATED AT THE
       BOTTOM OF THE SPACECRAFT AS SEEN LOOKING FROM THE EARTH

Fig. 15-20.   The Three-Axis Momentum Wheel Stabilized ATS-6 Spacecraft

Magnetic coil control systems can be used for maneuvers for virtually all orbits at less than synchronous altitudes (35,000 km). Magnetic control systems are relatively lightweight and require no moving parts, complex hardware, or expendables. This makes magnetic torquing attractive for space applications; however, it requires significant amounts of power, it provides slow maneuvering because of the power constraints, and its operation depends on the magnetic field configuration. Three types of magnetic torquer systems currently being used are permanent magnets, "air-"core torquing coils (i.e., electromagnets), and iron-core torquing coils. Permanent magnets are the heaviest type and are used for limited stabilization. "Air-" and iron-core magnets are used for both stabilization and maneuvering. For a spin-stabilized spacecraft, magnetic coils may be mounted either around

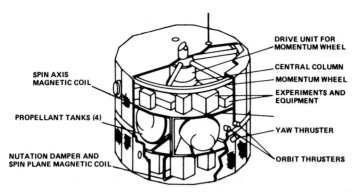

Fig. 15-21. The AE Spacecraft. AE uses a momentum wheel, thrusters, and magnetic coils for attitude control.

or perpendicular to the spin axis. Spin axis coils can be used only for reorientation because torque cannot be applied along the spin axis, whereas a coil with its dipole in the spin plane can provide both reorientation and spin rate control. Electromagnetic control systems vary the control coils' polarity and direction to match the Earth's magnetic field to produce a torque to cause the attitude to change as desired. The Atmospheric Explorer, AE, spacecraft shown in Fig. 15-21, uses magnetic coils both along the spin axis and in the spin plane for attitude stabilization, maneuvering, and momentum wheel control.

References

1. Anderson, K. A., L. M. Chase, R. P. Lin, J. E. McCoy, and R. E. McGuire, "Subsatellite Measurements of Plasmas and Solar Particles," *Apollo 15 Preliminary Science Report*, NASA SP-289, 1972.

2. Bracewell, R. N., and O. K. Garriott, *Nature*, Vol. 182, no. 4638, p. 760–762, 1958.

3. Daugherty, H. J., K. L. Lebsock, and J. J. Rodden, "Attitude Stabilization of Synchronous Communications Satellite Employing Narrow-Beam Antennas," *J. Spacecraft*, Vol. 8, p. 834–841, 1971.

4. Dunker, C., C. Manders, R. Wetmore, and H. Witting, *Orbiting Solar Observatory-I Attitude Prelaunch and Analysis Report*, NASA X-581-75-137, GSFC, June 1975.

5. Grell, M. G., M. A. Firestone, M. C. Phenneger, M. E. Plett, and P. V. Rigterink, *Atmosphere Explorer-D and -E Attitude Determination and Control Prelaunch Report Analysis and Operations Plan*, Comp. Sc. Corp. CSC/TR-75/6018, Oct. 1975.

6. NASA, *Mission Operations Plan 1-73 Radio Astronomy Explorer-B*, NASA X-513-73-110, GSFC, May 1973.

7. ———, *The ATS-F Data Book*, GSFC, May 1974.

8. Repass, G. D., J. N. Rowe, V. H. Tate, D. L. Walter, R. J. Wetmore, and R. S. Williams, *International Sun-Earth Explorer-A (ISEE-A) Attitude System Functional Specifications and Requirements*, Comp. Sc. Corp. CSC/SD-76/6057, Aug. 1975.

CHAPTER 16

ATTITUDE DYNAMICS

16.1 Equations of Motion
  Kinematic Equations of Motion; Rate of Change of Vectors in Rotating Frames; Angular Momentum, Kinetic Energy, and Moment of Inertia Tensor; Dynamic Equations of Motion
16.2 Motion of a Rigid Spacecraft
  Torque-Free Motion—Dynamic Equations, Torque-Free Motion—Kinematic Equations, Variation-of-Parameters Formulation
16.3 Spacecraft Nutation
  Dynamic Motion of a Symmetric Dual-Spin Spacecraft, Nutation Monitoring With Digital Sun Sensor
16.4 Flexible Spacecraft Dynamics
  Flexibility Effects on Spacecraft Attitude Dynamics, Modified Equations of Motion, Characteristics of Various Flexible Spacecraft

This chapter describes the mathematical formulation of attitude dynamics. Alternative descriptions are available in many standard references, such as Goldstein [1950]; Kibble [1966]; Synge and Griffith [1959]; MacMillan [1936]; and Whittaker [1937]; and in more recent books emphasizing spacecraft applications, such as Thomson [1963] and Kaplan [1976]. Section 16.1 is concerned with equations of motion of attitude dynamics, using the notation defined in Section 12.1. Section 16.2 considers the solutions of these equations for torque-free rigid body motion and the use of these solutions in the variation-of-parameters formulation of rigid body dynamics. Section 16.3 discusses dynamics approximations appropriate for determining nutation parameters from attitude sensor data. Finally, the effects of flexible components on spacecraft dynamics are discussed in Section 16.4.

16.1 Equations of Motion

F. L. Markley

The equations of motion of attitude dynamics can be divided into two sets: the *kinematic equations of motion* and the *dynamic equations of motion*. *Kinematics* is the study of motion irrespective of the forces that bring about that motion. The kinematic equations of motion are a set of first-order differential equations specifying the time evolution of the attitude parameters introduced in Section 12.1. These equations, which contain the instantaneous angular velocity vector ω, are considered in Section 16.1.1. Section 16.1.2 presents the relation between the rate of change of a vector in an inertial reference frame and its rate of change in a reference frame rotating with angular velocity ω. In Section 16.1.3, the *angular momentum*, *kinetic energy*, and *moment of inertia tensor* are precisely defined and the relations between them presented. Finally, the dynamic equations of motion, which express the time dependence of ω, are derived in Section 16.1.4. These are needed for dynamic simulations and for attitude prediction whenever gyroscopic measurements of ω are unavailable.

16.1.1 Kinematic Equations of Motion

Several parameterizations for the attitude have been presented in Section 12.1. Each parameterization has an associated set of kinematic equations of motion. Since the Euler symmetric parameter or quaternion parameterization has proved most useful for spacecraft kinematics analysis, we consider this case first.

The time dependence of the Euler symmetric parameters can be derived from the product relation, Eq. (12-15). Let the quaternion q represent the orientation of the rigid body with respect to the reference system at time t, and q'' represent the orientation with respect to the reference system at time $t+\Delta t$. We shall denote these by $q(t)$ and $q(t+\Delta t)$, respectively. Then q' specifies the orientation of the $\hat{u}, \hat{v}, \hat{w}$ triad (Fig. 12-1) at time $t+\Delta t$ relative to the position that it occupied at time t. Equation (12-11) gives

$$q_1' = e_u \sin\frac{\Delta\Phi}{2}$$

$$q_2' = e_v \sin\frac{\Delta\Phi}{2}$$

$$q_3' = e_w \sin\frac{\Delta\Phi}{2}$$

$$q_4' = \cos\frac{\Delta\Phi}{2}$$

where e_u, e_v, e_w are the components of the rotation axis unit vector along the $\hat{u}$, $\hat{v}$, $\hat{w}$ triad at time t (because this is the reference system for q') and $\Delta\Phi$ is the rotation in time Δt. Thus,

$$q(t+\Delta t) = \left\{ \cos\frac{\Delta\Phi}{2}\mathbf{1} + \sin\frac{\Delta\Phi}{2} \begin{bmatrix} 0 & e_w & -e_v & e_u \\ -e_w & 0 & e_u & e_v \\ e_v & -e_u & 0 & e_w \\ -e_u & -e_v & -e_w & 0 \end{bmatrix} \right\} q(t) \quad (16\text{-}1)$$

where $\mathbf{1}$ is the 4×4 identity matrix. Equation (16-1) is particularly useful if the axis of rotation does not change over the time interval Δt, and is often used in inertial navigation. This is discussed more fully in Section 17.1, as are the errors resulting from the use of Eq. (16-1) when the axis is not strictly constant.

For the case of general attitude motion, it is convenient to convert Eq. (16-1) to a differential equation. In this case, Δt is infinitesimal and $\Delta\Phi = \omega\Delta t$, where ω is the magnitude of the instantaneous angular velocity of the rigid body. We use the small angle approximations

$$\cos\frac{\Delta\Phi}{2} \approx 1, \qquad \sin\frac{\Delta\Phi}{2} \approx \frac{1}{2}\omega\Delta t$$

to obtain

$$q(t+\Delta t) \approx [1+\tfrac{1}{2}\Omega\Delta t]q(t) \tag{16-2a}$$

where Ω is the skew-symmetric matrix

$$\Omega = \begin{bmatrix} 0 & \omega_w & -\omega_v & \omega_u \\ -\omega_w & 0 & \omega_u & \omega_v \\ \omega_v & -\omega_u & 0 & \omega_w \\ -\omega_u & -\omega_v & -\omega_w & 0 \end{bmatrix} \tag{16-2b}$$

and $\omega = \omega\hat{e}$ is the angular velocity vector. Then

$$\frac{dq}{dt} \equiv \lim_{\Delta t \to 0} \frac{q(t+\Delta t)-q(t)}{\Delta t} = \tfrac{1}{2}\Omega q \tag{16-3}$$

If Ω is constant, we can formally integrate Eq. (16-3) to obtain

$$q(t) = \exp{(\Omega t/2)}\, q(0) \tag{16-4a}$$

In the weaker case that $\hat{e}$ is constant but ω varies, the integration can still be carried out to yield.

$$q(t) = \exp\left(\tfrac{1}{2}\int_0^t \Omega(t')\,dt'\right) q(0) \tag{16-4b}$$

The meaning of exponential functions of matrices and the relation of Eq. (16-4) to Eq. (16-1) are discussed in Appendix C.

The time dependence of the direction cosine matrix, A, can be similarly derived. We have

$$A(t+\Delta t) = A'A(t) \tag{16-5}$$

where A' is given by Eq. (12-7) with rotation angle $\Delta\Phi$ and with e_1, e_2, e_3 replaced by e_u, e_v, e_w, as discussed above. If Δt is infinitesimal, small-angle approximations can be used for $\cos\Delta\Phi$ and $\sin\Delta\Phi$, yielding

$$A' = 1 + \Omega'\,\Delta t \tag{16-6a}$$

where 1 is the 3×3 identity matrix and

$$\Omega' = \begin{bmatrix} 0 & \omega_w & -\omega_v \\ -\omega_w & 0 & \omega_u \\ \omega_v & -\omega_u & 0 \end{bmatrix} \tag{16-6b}$$

Thus

$$\frac{dA}{dt} \equiv \lim_{\Delta t \to 0} \frac{A(t+\Delta t)-A(t)}{\Delta t} = \Omega'A \tag{16-7}$$

Exponential solutions of this equation similar to Eq. (16-4) can be written, but are not used as frequently.

The kinematic equations of motion for the Gibbs vector, g, can be derived from Eq. (12-19). For infinitesimal Δt we have, from Eq. (12-16),

$$\mathbf{g}' = \hat{\mathbf{e}} \tan \frac{\Delta \Phi}{2} \approx \tfrac{1}{2} \boldsymbol{\omega} \, \Delta t$$

where $\mathbf{g}'$ is the Gibbs vector representing the infinitesimal rotation between times t and $t + \Delta t$, so that

$$\frac{d\mathbf{g}}{dt} = \tfrac{1}{2}\left[\boldsymbol{\omega} - \boldsymbol{\omega} \times \mathbf{g} + (\boldsymbol{\omega} \cdot \mathbf{g})\,\mathbf{g}\right] \tag{16-8}$$

The kinematic equations of motion for the Euler angles (ϕ, θ, ψ) can be derived by a different technique. Consider the 3-1-3 sequence of rotations as illustrated in Fig. 12-3 as an example. The rotations involved are ϕ about $\hat{\mathbf{z}}$, θ about $\hat{\mathbf{x}}'$, and ψ about $\hat{\mathbf{w}}$. If ϕ were the only angle changing, the angular velocity would be $\dot{\phi}\hat{\mathbf{z}}$. Similarly, if only θ or only ψ were changing, the angular velocity would be $\dot{\theta}\hat{\mathbf{x}}'$ or $\dot{\psi}\hat{\mathbf{w}}$, respectively. When all three angles are changing, the angular velocity is the vector sum of these three contributions:

$$\boldsymbol{\omega} = \dot{\phi}\hat{\mathbf{z}} + \dot{\theta}\hat{\mathbf{x}}' + \dot{\psi}\hat{\mathbf{w}} \tag{16-9}$$

Taking components of $\boldsymbol{\omega}$ along the body axes $\hat{\mathbf{u}}, \hat{\mathbf{v}}, \hat{\mathbf{w}}$ gives

$$\omega_u = \dot{\phi}\hat{\mathbf{z}} \cdot \hat{\mathbf{u}} + \dot{\theta}\hat{\mathbf{x}}' \cdot \hat{\mathbf{u}} \tag{16-10a}$$

$$\omega_v = \dot{\phi}\hat{\mathbf{z}} \cdot \hat{\mathbf{v}} + \dot{\theta}\hat{\mathbf{x}}' \cdot \hat{\mathbf{v}} \tag{16-10b}$$

$$\omega_w = \dot{\phi}\hat{\mathbf{z}} \cdot \hat{\mathbf{w}} + \dot{\theta}\hat{\mathbf{x}}' \cdot \hat{\mathbf{w}} + \dot{\psi} \tag{16-10c}$$

Comparison with Eqs. (12-2) and (12-20) gives

$$\hat{\mathbf{z}} \cdot \hat{\mathbf{u}} = A_{13} = \sin\theta \sin\psi$$

$$\hat{\mathbf{z}} \cdot \hat{\mathbf{v}} = A_{23} = \sin\theta \cos\psi$$

$$\hat{\mathbf{z}} \cdot \hat{\mathbf{w}} = A_{33} = \cos\theta$$

The inner products of $\hat{\mathbf{x}}'$ with the body axes are elements of the matrix giving the orientation of the $\hat{\mathbf{u}}, \hat{\mathbf{v}}, \hat{\mathbf{w}}$ triad relative to the $\hat{\mathbf{x}}', \hat{\mathbf{y}}', \hat{\mathbf{z}}'$ triad:

$$A' \equiv A_3(\psi)A_1(\theta) = \begin{bmatrix} \cos\psi & \cos\theta \sin\psi & \sin\theta \sin\psi \\ -\sin\psi & \cos\theta \cos\psi & \sin\theta \cos\psi \\ 0 & -\sin\theta & \cos\theta \end{bmatrix}$$

Thus,

$$\hat{\mathbf{x}}' \cdot \hat{\mathbf{u}} = A'_{11} = \cos\psi$$

$$\hat{\mathbf{x}}' \cdot \hat{\mathbf{v}} = A'_{21} = -\sin\psi$$

$$\hat{\mathbf{x}}' \cdot \hat{\mathbf{w}} = A'_{31} = 0$$

Combining these results gives

$$\omega_u = \dot{\theta}\cos\psi + \dot{\phi}\sin\theta \sin\psi \tag{16-11a}$$

$$\omega_v = -\dot{\theta}\sin\psi + \dot{\phi}\sin\theta \cos\psi \tag{16-11b}$$

$$\omega_w = \dot{\psi} + \dot{\phi}\cos\theta \tag{16-11c}$$

Equation (16-11) can now be solved for $\dot\theta$, $\dot\phi$, and $\dot\psi$ to yield the kinematic equations of motion for the 3-1-3 Euler angle sequence:

$$\dot\theta = \omega_u \cos\psi - \omega_v \sin\psi \tag{16-12a}$$

$$\dot\phi = (\omega_u \sin\psi + \omega_v \cos\psi)/\sin\theta \tag{16-12b}$$

$$\dot\psi = \omega_w - (\omega_u \sin\psi + \omega_v \cos\psi)\cot\theta \tag{16-12c}$$

The lack of uniqueness in the specification of ϕ and ψ when θ is a multiple of 180 deg shows up as a singularity in the kinematic equations of motion, Eqs. (16-12b) and (16-12c), when $\sin\theta = 0$. This is a serious disadvantage of Euler angle formulations for numerical integration of the equations of motion.

For many applications, it is convenient to have expressions for the components of the angular velocity vector, ω, along the reference axes as functions of the Euler angle rates. These are given by

$$\begin{bmatrix} \omega_1 \\ \omega_2 \\ \omega_3 \end{bmatrix} = A_{313}^T(\phi,\theta,\psi) \begin{bmatrix} \omega_u \\ \omega_v \\ \omega_w \end{bmatrix} \tag{16-13}$$

where $A_{313}^T(\phi,\theta,\psi)$, the transpose of the matrix of Eq. (12-20), is the matrix that transforms vector components from the body frame to the reference frame. The result of this matrix multiplication is

$$\omega_1 = \dot\theta\cos\phi + \dot\psi\sin\theta\sin\phi \tag{16-14a}$$

$$\omega_2 = \dot\theta\sin\phi - \dot\psi\sin\theta\cos\phi \tag{16-14b}$$

$$\omega_3 = \dot\psi\cos\theta + \dot\phi \tag{16-14c}$$

Either Eq. (16-11) or Eq. (16-14) can be used to show that

$$\omega^2 = \dot\theta^2 + \dot\psi^2 + \dot\phi^2 + 2\dot\phi\dot\psi\cos\theta \tag{16-15}$$

Similar relations can be derived for other Euler axis sequences and are collected in Appendix E. When the third axis is identical with the first, the kinematic equations are singular when θ is a multiple of 180 deg; when the first and third axes are different, the equations are singular when θ is an odd multiple of 90 deg, as expected.

16.1.2 Rate of Change of Vectors in Rotating Frames

We have resolved vectors into components along coordinate axes in several coordinate systems. We shall now derive the relationship between the time derivatives of an arbitrary vector resolved along the coordinate axes of one system and the derivatives of the components in a different system. For definiteness, we consider the geomagnetic field vector in the body system and reference system, previously introduced. We wish to compare the time derivatives of the field measured by magnetometers fixed in a rotating spacecraft with the derivatives measured by a (possibly fictitious) set of magnetometers traveling with the spacecraft but with a fixed orientation relative to the reference frame. If we denote the components of the vector in the reference system by $\mathbf{a}' = (a_1, a_2, a_3)^T$ and the components in the body by $\mathbf{a} = (a_u, a_v, a_w)^T$, then, according to Eq. (12-4),

$$\mathbf{a} = A\mathbf{a}' \tag{16-16}$$

The time variation of the components of $\mathbf{a}$ is due to the time variation of both A and $\mathbf{a}'$. The former represents the variation due to the change of the relative orientation of the two reference systems. The product rule for differentiation gives

$$\frac{d\mathbf{a}}{dt} = \frac{dA}{dt}\mathbf{a}' + A\frac{d\mathbf{a}'}{dt} \tag{16-17}$$

The first term on the right can be written, using Eq. (16-7), as

$$\frac{dA}{dt}\mathbf{a}' = \Omega' A\mathbf{a}' = \Omega'\mathbf{a} = -\boldsymbol{\omega}\times\mathbf{a}$$

where the last equality follows from the explicit form of Ω'. The second term consists of the components in the body frame of the vector $d\mathbf{a}'/dt$, where the time derivatives are evaluated in the reference frame. If we denote this vector by $(d\mathbf{a}'/dt)_b$, we have

$$\frac{d\mathbf{a}}{dt} = -\boldsymbol{\omega}\times\mathbf{a} + \left(\frac{d\mathbf{a}'}{dt}\right)_b \tag{16-18}$$

If the components of $\mathbf{a}$ along the body axes, a_u, a_v, a_w, are constant, then $d\mathbf{a}/dt = 0$ and

$$\left(\frac{d\mathbf{a}'}{dt}\right)_b = \boldsymbol{\omega}\times\mathbf{a} \tag{16-19}$$

This expression gives the derivatives of $\mathbf{a}$ in the reference coordinate system, but with the vector components resolved along the body coordinate axes. Because it is a vector equation, $\mathbf{a}$ and $\boldsymbol{\omega}$ can be resolved into components along any set of coordinate axes, including the reference axes; therefore, the prime and the subscript b will be omitted in future applications where the distinction is clear from the context. An alternative, geometric derivation can be given which derives Eq. (16-19) directly in the reference coordinate system. Figure 16-1 shows the vector $\mathbf{a}$ at times t and $t + \Delta t$. The motion of $\mathbf{a}$ is in a cone with $\boldsymbol{\omega}$ as the axis, with fixed cone angle η. In the time between t and $t + \Delta t$, the rotation angle is $\omega\Delta t$, and the magnitude of $\Delta\mathbf{a}$, the change in $\mathbf{a}$, is

$$\Delta a = 2(a\sin\eta)\sin\tfrac{1}{2}\omega\Delta t$$

where η is the angle between $\mathbf{a}$ and $\boldsymbol{\omega}$. Then

$$\frac{da}{dt} = \lim_{\Delta t\to 0}\frac{\Delta a}{\Delta t} = a\,\omega\sin\eta = |\boldsymbol{\omega}\times\mathbf{a}|$$

In the limit $\Delta t\to 0$, the direction of $\Delta\mathbf{a}$ is tangent to the circle, perpendicular to the plane containing $\mathbf{a}$ and $\boldsymbol{\omega}$. Thus

$$\frac{d\mathbf{a}}{dt} = \boldsymbol{\omega}\times\mathbf{a}$$

which is Eq. (16-19) in the reference coordinate system.

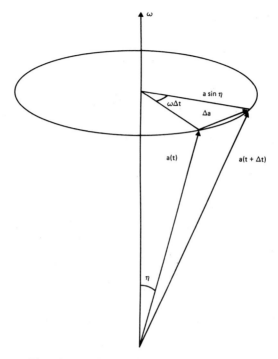

Fig. 16-1. Rate of Change of a Rotating Vector

16.1.3 Angular Momentum, Kinetic Energy, and Moment of Inertia Tensor

The fundamental quantity in rotational mechanics is the angular momentum, $\mathbf{L}$, as discussed in Section 15.1. For a collection of n point masses, the angular momentum is given by

$$\mathbf{L}_{total} \equiv \sum_{i=1}^{n} \mathbf{r}_i \times m_i \mathbf{v}_i \tag{16-20}$$

where m_i, $\mathbf{r}_i$, and $\mathbf{v}_i$ are the mass, position, and velocity, respectively, of the ith point mass. Newton's laws of motion, which are valid only in an inertial coordinate system, will be used to derive an equation of motion for $\mathbf{L}$, so it is important to assume for the present that $\mathbf{r}_i$ and $\mathbf{v}_i$ are the position and velocity in an inertial reference frame. It is convenient to write $\mathbf{r}_i$ as the sum of two terms

$$\mathbf{r}_i = \mathbf{R} + \boldsymbol{\rho}_i \tag{16-21}$$

where $\mathbf{R}$ is the position of a fixed reference point, O', in the rigid body, and $\boldsymbol{\rho}_i$ is the position of the ith mass relative to O', as shown in Fig. 16-2. Differentiating Eq. (16-21) with respect to time gives

$$\mathbf{v}_i = \mathbf{V} + \frac{d\boldsymbol{\rho}_i}{dt} \tag{16-22}$$

where $\mathbf{V}$ is the velocity of O' in the inertial frame. Substituting Eqs. (16-21) and

(16-22) into Eq. (16-20) yields

$$\mathbf{L}_{total} = M\mathbf{R} \times \mathbf{V} + \mathbf{R} \times \frac{d}{dt}\left[\sum_{i=1}^{n} m_i \boldsymbol{\rho}_i\right]$$

$$+ \left[\sum_{i=1}^{n} m_i \boldsymbol{\rho}_i\right] \times \mathbf{V} + \sum_{i=1}^{n} m_i \boldsymbol{\rho}_i \times \frac{d\boldsymbol{\rho}_i}{dt} \tag{16-23}$$

where $M \equiv \sum_{i=1}^{n} m_i$ is the total mass of the body. If O' is taken to be the center of mass of the body,

$$\sum_{i=1}^{n} m_i \boldsymbol{\rho}_i = 0 \tag{16-24}$$

by definition, so the second and third terms on the right side of Eq. (16-23) vanish identically. We will always choose this reference point for rigid body dynamics, giving

$$\mathbf{L}_{total} = M\mathbf{R} \times \mathbf{V} + \mathbf{L} \tag{16-25}$$

where the first term on the right side represents the angular momentum of the total mass considered as a point located at the center of mass, and the second term,

$$\mathbf{L} \equiv \sum_{i=1}^{n} m_i \boldsymbol{\rho}_i \times \frac{d\boldsymbol{\rho}_i}{dt} \tag{16-26}$$

is the contribution of the motion of the n mass points relative to the center of mass.

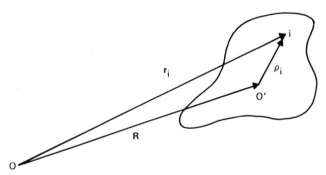

Fig. 16-2.   The position of mass point, i, relative to the origin, O, of an inertial reference frame is the vector sum of its position relative to a reference point, O', fixed in the body and the vector $\mathbf{R}$ from O to O'.

A similar separation between center-of-mass motion and motion relative to the center of mass occurs for the kinetic energy

$$E_{k\,total} \equiv \frac{1}{2} \sum_{i=1}^{n} m_i v_i^2$$

$$= \frac{1}{2} M V^2 + \mathbf{V} \cdot \frac{d}{dt}\left[\sum_{i=1}^{n} m_i \boldsymbol{\rho}_i\right] + \frac{1}{2} \sum_{i=1}^{n} m_i \left(\frac{d\boldsymbol{\rho}_i}{dt}\right)^2 \tag{16-27}$$

The middle term in Eq. (16-27) vanishes identically if O' is the center of mass of the body, so

$$E_{k\ total} = \tfrac{1}{2}MV^2 + E_k \qquad (16\text{-}28)$$

where

$$E_k \equiv \tfrac{1}{2}\sum_{i=1}^{n} m_i \left(\frac{d\rho_i}{dt}\right)^2 \qquad (16\text{-}29)$$

is the kinetic energy of motion relative to the center of mass.

Spacecraft rigidity has not been assumed to this point. If the body is not rigid, a reference point other than the center of mass is often used, in which case all the terms in Eqs. (16-23) and (16-27) must be retained. If we now assume that the spacecraft is *rigid*, i.e., that all the vectors ρ_i are constant in a reference frame fixed in the spacecraft, then all the vectors may conveniently be resolved into components along a spacecraft reference system. In this section, the subscripts 1, 2, 3 will be used for components along spacecraft-fixed axes. This should be distinguished from the notation in Section 12.1, where the subscripts 1, 2, 3 referred to an arbitrary reference system, and u, v, w to the spacecraft reference system. The attitude dynamics problem is only concerned with motion relative to the center of mass, and thus only with the angular momentum, $\mathbf{L}$, and kinetic energy, E_k, defined by Eqs. (16-26) and (16-29), in the rigid body case.

Although the components of ρ_i in the spacecraft frame are constant, the components of $d\rho_i/dt$ are not zero if the spacecraft is rotating with instantaneous angular velocity ω, because the vector $d\rho_i/dt$ is the rate of change of ρ_i *relative to inertial coordinates, resolved along spacecraft-fixed axes.* All time derivatives must be evaluated in an inertial reference frame if Newton's laws of motion are to be applied directly. Equation (16-19) with $\mathbf{a} = \rho_i$ gives

$$\frac{d\rho_i}{dt} = \omega \times \rho_i \qquad (16\text{-}30)$$

where ρ_i and ω are understood to be resolved into components along spacecraft-fixed axes.

Substituting Eq. (16-30) into Eq. (16-26) yields

$$\mathbf{L} = \sum_{i=1}^{n} m_i \rho_i \times (\omega \times \rho_i) = \sum_{i=1}^{n} m_i \left[\rho_i^2 \omega - (\rho_i \cdot \omega)\rho_i \right] \qquad (16\text{-}31)$$

We define the symmetric 3×3 *moment of inertia tensor*, I, by

$$I_{11} \equiv \sum_{i=1}^{n} m_i \left(\rho_{i2}^2 + \rho_{i3}^2\right) \qquad (16\text{-}32a)$$

$$I_{22} \equiv \sum_{i=1}^{n} m_i \left(\rho_{i3}^2 + \rho_{i1}^2\right) \qquad (16\text{-}32b)$$

$$I_{33} \equiv \sum_{i=1}^{n} m_i \left(\rho_{i1}^2 + \rho_{i2}^2\right) \qquad (16\text{-}32c)$$

$$I_{12} = I_{21} \equiv - \sum_{i=1}^{n} m_i \rho_{i1} \rho_{i2} \tag{16-33a}$$

$$I_{23} = I_{32} \equiv - \sum_{i=1}^{n} m_i \rho_{i2} \rho_{i3} \tag{16-33b}$$

$$I_{31} = I_{13} \equiv - \sum_{i=1}^{n} m_i \rho_{i3} \rho_{i1} \tag{16-33c}$$

Then, Eq. (16-31) can be written in matrix form as

$$\mathbf{L} = I \boldsymbol{\omega} \tag{16-34}$$

Substituting Eq. (16-30) into Eq. (16-29) and using Eq. (16-31) for $\mathbf{L}$ yields

$$E_k = \frac{1}{2} \sum_{i=1}^{n} m_i (\boldsymbol{\omega} \times \boldsymbol{\rho}_i) \cdot (\boldsymbol{\omega} \times \boldsymbol{\rho}_i) = \frac{1}{2} \sum_{i=1}^{n} m_i \boldsymbol{\omega} \cdot \left[\boldsymbol{\rho}_i \times (\boldsymbol{\omega} \times \boldsymbol{\rho}_i) \right]$$

$$= \tfrac{1}{2} \boldsymbol{\omega} \cdot \mathbf{L} = \tfrac{1}{2} \boldsymbol{\omega}^T I \boldsymbol{\omega} \tag{16-35}$$

Thus, both the angular momentum and the kinetic energy can be expressed in terms of I and $\boldsymbol{\omega}$.

Some authors (e.g., Whittaker [1937]) define the negatives of the off-diagonal elements

$$P_{jk} \equiv - I_{jk} \qquad j \neq k$$

as *products of inertia*; but other authors (e.g., Goldstein [1950] and Kibble [1966]) define the elements I_{jk}, without the minus sign, as the products of inertia. Still other authors (e.g., Thomson [1963] and Kaplan [1976]) define the products of inertia as Whittaker does, but denote them by I_{jk}, so that the off-diagonal elements of the moment of inertia tensor are $-I_{jk}$. The quantity I is called a *tensor* because it has specific transformation properties under a real orthogonal transformation (see, for example, Goldstein [1950] or Synge and Schild [1964].) It is sufficient for our purposes to think of the moment of inertia tensor as a real, symmetric 3×3 matrix. Because the moment of inertia tensor is a real, symmetric matrix, it has three real orthogonal eigenvectors and three real eigenvalues (see Appendix C) satisfying the equation

$$I \hat{\mathbf{P}}_i = I_i \hat{\mathbf{P}}_i \qquad i = 1, 2, 3 \tag{16-36}$$

The scalars I_1, I_2, and I_3 are the *principal moments of inertia*, and the unit vectors $\hat{\mathbf{P}}_1$, $\hat{\mathbf{P}}_2$, and $\hat{\mathbf{P}}_3$ are the *principal axes*. These quantities were introduced in a more intuitive manner in Section 15.1. If we use the principal axes as the coordinate axes of a spacecraft reference frame, the moment of inertia tensor takes the diagonal form

$$I = \begin{bmatrix} I_1 & 0 & 0 \\ 0 & I_2 & 0 \\ 0 & 0 & I_3 \end{bmatrix} \tag{16-37}$$

In this coordinate frame (and only in this frame), Eqs. (16-34) and (16-35) can be expressed as

$$L_1 = I_1 \omega_1 \tag{16-38a}$$

$$L_2 = I_2 \omega_2 \tag{16-38b}$$

$$L_3 = I_3 \omega_3 \tag{16-38c}$$

$$E_k = \tfrac{1}{2}\left(I_1 \omega_1^2 + I_2 \omega_2^2 + I_3 \omega_3^2\right) \tag{16-39}$$

Combining Eqs. (16-32), (16-33), and (16-37) gives

$$I_1 = \sum_{i=1}^{n} m_i \left(\rho_{i2}^2 + \rho_{i3}^2\right) \tag{16-40a}$$

$$I_2 = \sum_{i=1}^{n} m_i \left(\rho_{i3}^2 + \rho_{i1}^2\right) \tag{16-40b}$$

$$I_3 = \sum_{i=1}^{n} m_i \left(\rho_{i1}^2 + \rho_{i2}^2\right) \tag{16-40c}$$

$$0 = \sum_{i=1}^{n} m_i \rho_{i1} \rho_{i2} \tag{16-41a}$$

$$0 = \sum_{i=1}^{n} m_i \rho_{i2} \rho_{i3} \tag{16-41b}$$

$$0 = \sum_{i=1}^{n} m_i \rho_{i3} \rho_{i1} \tag{16-41c}$$

where the vectors ρ_i are resolved into components along the principal axes in Eqs. (16-40) and (16-41). Equations (16-41) must have balancing positive and negative contributions in the sums on the right-hand sides. Thus, principal axes can be thought of intuitively as axes around which the mass is symmetrically distributed. In particular, any axis of rotational symmetry of the mass distribution is a principal axis. Equation (16-40) shows that I_1, I_2, and I_3 are all nonnegative, and thus $\det I = I_1 I_2 I_3 \geqslant 0$. The determinant is zero only if all the ρ_i are collinear, i.e., if all the mass is along a mathematical straight line. Thus, for real objects $\det I \geqslant 0$, and, because the determinant is invariant under a change of coordinate system (see Appendix C), this holds in all coordinate systems. Consequently, in any coordinate system, the moment of inertia tensor has an inverse, I^{-1}, which is also a 3×3 matrix*. We can thus write Eqs. (16-34) and (16-35) as

$$\omega = I^{-1} \mathbf{L} \tag{16-42}$$

and

$$E_k = \tfrac{1}{2} \mathbf{L}^T I^{-1} \mathbf{L} \tag{16-43}$$

respectively. According to Eq. (16-43), a surface of constant energy is an ellipsoid in L_1, L_2, L_3 space, as was discussed in Section 15.2. Note that Eq. (16-35) similarly

---

*The moment-of-inertia tensor is not an orthogonal matrix, so I^{-1} is not equal to I^T.

defines an energy ellipsoid in ω_1, ω_2, ω_3 space, with semiaxis lengths $\sqrt{2E_k/I_1}$, $\sqrt{2E_k/I_2}$, and $\sqrt{2E_k/I_3}$. This ellipsoid may be used for qualitative discussions of rotational motion, but it will not be considered in this work. According to Whittaker [1937, page 124], "The existence of principal axes was discovered by Euler, *Mem. de Berl.*, 1750, 1758, and by J. A. Segner, *Specimen Th. Turbinem*, 1755. The momental ellipsoid was introduced by Cauchy in 1827, *Exerc. de math.* I, p. 93."

16.1.4 Dynamic Equations of Motion

The basic equation of attitude dynamics relates the time derivative of the angular momentum vector, dL/dt, to the applied torque, N. This relation was introduced in Section 15.2, and Eq. (15-9) gives dL/dt in inertial coordinates. In this section, we consider the time derivatives of the components of L along spacecraft-fixed axes, because the moment of inertia tensor of a rigid body is most conveniently expressed along these axes. Combining Eqs. (15-9), (16-34), and (16-18) gives*

$$\frac{dL}{dt} = N - \omega \times L = I\frac{d\omega}{dt} \qquad (16\text{-}44)$$

where the torque vector, N, is defined as

$$N \equiv \sum_{i=1}^{n} \mathbf{r}_i \times \mathbf{F}_i \qquad (16\text{-}45)$$

and ω is the instantaneous angular velocity vector discussed in Section 16.1. The force, $\mathbf{F}_i$, on the ith mass consists of two parts: an externally applied force, $\mathbf{F}_i^{ext}$, and an internal force consisting of the sum of the forces, $\mathbf{f}_{ij}$, exerted by the other masses (the cohesive forces of the rigid body):

$$\mathbf{F}_i = \mathbf{F}_i^{ext} + \sum_{\substack{j=1 \\ j \neq i}}^{n} \mathbf{f}_{ij} \qquad (16\text{-}46)$$

Thus,

$$N = \sum_{i=1}^{n} \mathbf{r}_i \times \mathbf{F}_i^{ext} + \sum_{i=1}^{n} \sum_{\substack{j=1 \\ j \neq 1}}^{n} \mathbf{r}_i \times \mathbf{f}_{ij} \qquad (16\text{-}47)$$

Each pair of masses contributes two terms to the second sum, $\mathbf{r}_i \times \mathbf{f}_{ij}$ and $\mathbf{r}_j \times \mathbf{f}_{ji}$. By Newton's third law of motion, $\mathbf{f}_{ji} = -\mathbf{f}_{ij}$, so the contribution to the sum of each pair of masses is $(\mathbf{r}_i - \mathbf{r}_j) \times \mathbf{f}_{ij}$. If the line of action of the force between each pair of masses is parallel to the vector between the masses, $\mathbf{r}_i - \mathbf{r}_j$, the cross product vanishes, and the net torque, N, is equal to the torque due to external forces alone. This is always assumed to be the case in spacecraft applications. Some forces, most notably magnetic forces between moving charges, violate this condition, so that the

---

*The moment-of-inertia tensor of a rigid body is constant. This is not the case when flexibility effects (Section 16.4) or fuel expenditure (Section 17.4) are considered.

rate of change of mechanical angular momentum is not equal to the external torque. In the case of electromagnetic forces, this difference can be ascribed to the angular momentum of the electromagnetic field, but this is negligible for spacecraft dynamics problems.

Equation (16-44) is the fundamental equation of rigid body dynamics. The presence of the $\omega \times L$ term on the right side means that L, and hence ω, is not constant in the spacecraft frame, even if $N = 0$. The resulting motion is called *nutation*, and is discussed qualitatively in Section 15.1. Rotational motion without nutation occurs only if ω and L are parallel, that is, only if the rotation is about a principal axis of the rigid body.

Substituting Eq. (16-34) or Eq. (16-42) into Eq. (16-44) gives

$$I \frac{d\omega}{dt} = N - \omega \times (I\omega) \tag{16-48}$$

or

$$\frac{dL}{dt} = N - (I^{-1}L) \times L \tag{16-49}$$

respectively. These equations can be written out in component form, but no insight is gained by it, except when the vector quantities are referred to the principal axis coordinate system. In the principal axis system, Eq. (16-48) has the components:

$$I_1 \frac{d\omega_1}{dt} = N_1 + (I_2 - I_3)\omega_2\omega_3 \tag{16-50a}$$

$$I_2 \frac{d\omega_2}{dt} = N_2 + (I_3 - I_1)\omega_3\omega_1 \tag{16-50b}$$

$$I_3 \frac{d\omega_3}{dt} = N_3 + (I_1 - I_2)\omega_1\omega_2 \tag{16-50c}$$

and Eq. (16-49) has the components:

$$\frac{dL_1}{dt} = N_1 + (1/I_2 - 1/I_3)L_2L_3 \tag{16-51a}$$

$$\frac{dL_2}{dt} = N_2 + (1/I_3 - 1/I_1)L_3L_1 \tag{16-51b}$$

$$\frac{dL_3}{dt} = N_3 + (1/I_1 - 1/I_2)L_1L_2 \tag{16-51c}$$

Equations (16-44), (16-48), and (16-49) and their component forms Eqs. (16-50) and (16-51) are alternative formulations of *Euler's equations of motion*.

A spacecraft equipped with reaction or momentum wheels is not a rigid body, but the dynamic equations derived above can still be used, with one minor modification. When wheels are present, the total angular momentum of the spacecraft, including the wheels, is

$$L = I\omega + h \tag{16-52}$$

where the moment of inertia tensor I includes the mass of the wheels and the vector h is the net angular momentum due to the rotation of the wheels *relative to the spacecraft*. The inverse of Eq. (16-52) is

$$\omega = I^{-1}(L - h) \tag{16-53}$$

Substituting Eq. (16-52) or (16-53) into Eq. (16-44) gives

$$I\frac{d\omega}{dt} = \mathbf{N} - \frac{d\mathbf{h}}{dt} - \omega \times (I\omega + \mathbf{h}) \tag{16-54}$$

or

$$\frac{d\mathbf{L}}{dt} = \mathbf{N} - [I^{-1}(\mathbf{L}-\mathbf{h})] \times \mathbf{L} \tag{16-55}$$

respectively. For numerical calculations the second form is sometimes preferable because it does not involve the derivatives of the wheel angular momenta. The derivative term in Eq. (16-54) has a natural physical interpretation, however. The quantity $d\mathbf{h}/dt$ is the net torque applied to the wheels by the spacecraft body; so, by Newton's third law of motion, $-d\mathbf{h}/dt$ is the torque applied to the spacecraft body by the wheels. Writing Eqs. (16-54) and (16-55) in component form in the principal axis system yields equations similar to Eqs. (16-50) and (16-51).

Euler's equations of motion can be used to discuss the stability of rotation about a principal axis of a rigid spacecraft. Let $\hat{\mathbf{P}}_3$ be the nominal spin axis, so that ω_1 and ω_2 are much smaller than ω_3. Let us also assume that the applied torques are negligible. Then the right side of Eq. (16-50c) is approximately zero, and ω_3 is approximately constant. Taking the time derivative of Eq. (16-50a), multiplying by I_2, and substituting Eq. (16-50b) gives

$$I_1 I_2 \frac{d^2\omega_1}{dt^2} \simeq (I_2 - I_3)I_2 \frac{d\omega_2}{dt}\omega_3$$

$$\simeq (I_2 - I_3)(I_3 - I_1)\omega_3^2\omega_1 \tag{16-56}$$

If $(I_2 - I_3)(I_3 - I_1) < 0$, then ω_1 will be bounded and have sinusoidal time dependence with frequency $\sqrt{(I_2 - I_3)(I_1 - I_3)/(I_1 I_2)}\,\omega_3$; however, if $(I_2 - I_3)(I_3 - I_1) > 0$, then ω_1 will increase exponentially. Thus, the motion is stable if I_3 is either the largest or the smallest of the principal moments of inertia, and unstable if I_3 is the intermediate moment of inertia. This can be seen in the form of the paths of the angular momentum vector in the body shown in Fig. 15-14; the loci in the neighborhood of the principal axes of largest and smallest moment of inertia are elliptical closed curves, but the loci passing near the third principal axis go completely around the angular momentum sphere. Equation (16-56) only establishes the stability over short time intervals; over longer time intervals, energy dissipation effects cause rotational motion about the axis of smallest moment of inertia to be unstable, too, as discussed in Sections 15.2, 17.3, and 18.4.

16.2 Motion of a Rigid Spacecraft

F. L. Markley

We now turn to a discussion of the solutions of the kinematic and dynamic equations of motion presented in the previous section. These equations must be solved simultaneously because, in general, the torque $\mathbf{N}$ depends on the spacecraft attitude. Numerical integration methods and approximate closed-form solution methods for the general case are discussed in Section 17.1.

If **N** is independent of the attitude, the dynamic equations can be solved separately for the instantaneous angular velocity ω, which can then be used to solve the kinematic equations. A special case for which analytic solutions are available is the $N=0$ case, which is treated in Sections 16.2.1 and 16.2.2. These solutions are intrinsically interesting and furnish a useful approximation for the motion when the torques are small. Sections 16.2.1 and 16.2.2 provide an analytic counterpart to the qualitative discussion of attitude motion in Section 15.1. They also provide the starting point for the variation-of-parameters formulation of attitude dynamics presented in Section 16.2.3.

16.2.1 Torque-Free Motion—Dynamic Equations

The vector quantities in this section will be resolved along spacecraft principal axes to simplify the equations of motion. If two of the principal moments of inertia are equal, we shall take these to be I_1 and I_2; this is referred to as the *axial symmetry* case. If no two moments of inertia are equal, we shall denote the intermediate moment by I_2. In this case, the labeling of I_1 and I_3 will be fixed by the following convention: if $L^2 < 2I_2E_k$, we take $I_3 < I_2 < I_1$; and if $L^2 > 2I_2E_k$, we label the principal axes so that $I_1 < I_2 < I_3$. If $L^2 = 2I_2E_k$, either labeling can be used. With this convention, L^2 always lies between $2I_2E_k$ and $2I_3E_k$. The two limits can be visualized by considering the loci of the angular momentum vector on the angular momentum sphere shown in Fig. 15-14. Motion with $L^2 = 2I_3E_k$ is pure rotation about the $\hat{\mathbf{P}}_3$ axis, that is, nutation-free motion. Motion with $L^2 = 2I_2E_k$, on the other hand, means that **L** lies on one of the loci passing throught the axis of intermediate symmetry, $\hat{\mathbf{P}}_2$. In the axial symmetry case, this locus is the equator of the angular momentum sphere relative to the $\hat{\mathbf{P}}_3$ axis. With the convention adopted here and the $\omega_3 > 0$ convention adopted below, then, **L** will lie on the $\hat{\mathbf{P}}_3$ side of the $L^2 = 2I_2E_k$ loci, which is the upper hemisphere in the axial symmetry case and a smaller surface in the asymmetric case.

When the body is axially symmetric, we define the *transverse moment of inertia*

$$I_T \equiv I_1 = I_2 \tag{16-57}$$

In this case, Euler's equations, Eq. (16-50a) through (16-50c) simplify to

$$I_T \frac{d\omega_1}{dt} = -(I_3 - I_T)\omega_3\omega_2 \tag{16-58a}$$

$$I_T \frac{d\omega_2}{dt} = (I_3 - I_T)\omega_3\omega_1 \tag{16-58b}$$

$$I_3 \frac{d\omega_3}{dt} = 0 \tag{16-58c}$$

Equation (16-58c) shows that ω_3 is a constant.[*] We choose the sense of the $\hat{\mathbf{P}}_3$ axis so that $\omega_3 > 0$. Differentiating Eq. (16-58a) with respect to t, multiplying by I_T, and substituting Eq. (16-58b) yields

---

[*] For *spherically symmetric* spacecraft, $I_3 = I_1 = I_2 = I_T$, and Eq. (16-58) shows immediately that ω is a constant vector in the body. This also follows from the fact that $\omega = I_3^{-1}\mathbf{L}$ in the spherical symmetry case. Because **L** is constant in inertial coordinates, ω must be constant also. Then, because $\omega \times \mathbf{L} = 0$, Eq. (16-44) shows that **L**, and hence ω, is constant in the body reference system.

$$I_T^2 \frac{d^2\omega_1}{dt^2} = -(I_3 - I_T)^2 \omega_3^2 \omega_1$$

which has the solution

$$\omega_1 = \omega_T \cos \omega_p (t - t_1) \tag{16-59a}$$

In this equation, ω_T is the maximum value of ω_1, t_1 is some time at which ω_1 attains its maximum value, and

$$\omega_p \equiv (1 - I_3/I_T) \omega_3 \tag{16-59b}$$

is the *body nutation rate* introduced in Section 15.1.* The derivation of Eq. (16-59a) is analogous to that of Eq. (16-56), but Eq. (16-59a) is exact if $I_1 = I_2$, whereas Eq. (16-56) is an approximation based on the smallness of ω_1 and ω_2 relative to ω_3. Combining Eqs. (16-58a) and (16-59a) gives

$$\omega_2 = -\omega_T \sin \omega_p (t - t_1) \tag{16-59c}$$

Equations (16-59a, c) show that $\omega_T = (\omega_1^2 + \omega_2^2)^{1/2}$, so ω_T is the magnitude of the component of the angular velocity perpendicular to the symmetry axis and is called the *transverse angular velocity*.

By using the addition formulas for the sine and cosine, we can rewrite Eqs. (16-59) in terms of the components $(\omega_{01}, \omega_{02}, \omega_{03})$ of ω_0, the initial value of ω in the body frame. Thus,

$$\omega_1 = \omega_{01} \cos \omega_p t + \omega_{02} \sin \omega_p t \tag{16-60a}$$

$$\omega_2 = \omega_{02} \cos \omega_p t - \omega_{01} \sin \omega_p t \tag{16-60b}$$

$$\omega_3 = \omega_{03} \tag{16-60c}$$

where

$$\omega_{01} = \omega_T \cos \omega_p t_1 \tag{16-61a}$$

$$\omega_{02} = \omega_T \sin \omega_p t_1 \tag{16-61b}$$

It is often useful to express ω_{03} and ω_T in terms of the magnitude of the angular momentum vector and the rotational kinetic energy. In the axial symmetry case, Eqs. (16-38) and (16-39) give

$$L^2 = I_T^2 \omega_T^2 + I_3^2 \omega_3^2 \tag{16-62a}$$

$$2E_k = I_T \omega_T^2 + I_3 \omega_3^2 \tag{16-62b}$$

so we have

$$\omega_{03} = \left[\frac{L^2 - 2I_T E_k}{I_3(I_3 - I_T)} \right]^{1/2} \tag{16-63a}$$

$$\omega_T = \left[\frac{L^2 - 2I_3 E_k}{I_T(I_T - I_3)} \right]^{1/2} \tag{16-63b}$$

---

*Equation (16-66) shows that this definition is equivalent to that of Eq. (15-7).

Equations (16-59) can be written in vector form as

$$\omega = \omega_{03}\hat{\mathbf{P}}_3 + \omega_T\left[\cos\omega_p(t-t_1)\hat{\mathbf{P}}_1 - \sin\omega_p(t-t_1)\hat{\mathbf{P}}_2\right] \tag{16-64}$$

Then the angular momentum vector is

$$\mathbf{L} = I_3\omega_{03}\hat{\mathbf{P}}_3 + I_T\omega_T\left[\cos\omega_p(t-t_1)\hat{\mathbf{P}}_1 - \sin\omega_p(t-t_1)\hat{\mathbf{P}}_2\right] \tag{16-65}$$

Comparing Eqs. (16-64) and (16-65), we obtain

$$\omega = I_T^{-1}\mathbf{L} + \omega_p\hat{\mathbf{P}}_3 \tag{16-66}$$

Thus, the *body nutation rate*, ω_p, and the *inertial nutation rate*, ω_I, defined by Eq. (16-59b) and

$$\omega_I \equiv L/I_T \tag{16-67a}$$

agree with the quantitities introduced in Section 15.1.

It is also clear that

$$\cos\theta = L_3/L = I_3\omega_{03}/L \tag{16-67b}$$

where θ is the angle between $\mathbf{L}$ and $\hat{\mathbf{P}}_3$. Thus,

$$\omega_p = \frac{I_T - I_3}{I_3}\omega_I\cos\theta \tag{16-68}$$

These equations form the basis for the discussion of nutation in the axial symmetry case given in Section 15.1.

The solutions to Euler's dynamic equations of motion in the asymmetric case, $I_1 \neq I_2$, cannot be written in terms of trigonometric functions. Instead, they involve the *Jacobian elliptic functions* [Milne-Thomson, 1965; Neville, 1951; Byrd and Friedman, 1971]. These solutions, found by Jacobi [1849], are discussed by Synge and Griffith [1959], MacMillan, [1936], Thomson [1963], and Morton, *et al.*, [1974]. The angular velocity components in body principal coordinates are given by

$$\omega_1 = \omega_{1m}\mathrm{cn}(\Phi|m) \tag{16-69a}$$

$$\omega_2 = -\omega_{2m}\mathrm{sn}(\Phi|m) \tag{16-69b}$$

$$\omega_3 = \omega_{3m}\mathrm{dn}(\Phi|m) \tag{16-69c}$$

where cn, sn, and dn are the Jacobian elliptic functions with *argument*

$$\Phi \equiv \omega_p(t-t_1) \tag{16-70}$$

and *parameter**

$$m \equiv \frac{(I_1 - I_2)(L^2 - 2I_3E_k)}{(I_3 - I_2)(L^2 - 2I_1E_k)} \tag{16-71}$$

As in the axial symmetry case, t_1 is a time at which $\omega_1 = \omega_{1m}$. The maximum values

---

*Many authors use the *modulus*, $k \equiv m^{1/2}$, rather than the parameter. We follow the notation of Milne-Thomson [1965] and Neville [1951].

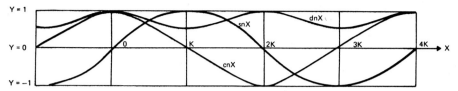

Fig. 16-3.   Graphs of the Jacobian Elliptic Functions sn($x|m$), cn($x|m$), and dn($x|m$) for $m=0.7$. The quarter-period, K, is equal to 2.07536 for this m.

of the body rate components along the three axes are

$$\omega_{1m} = \left[\frac{L^2 - 2I_3E_k}{I_1(I_1 - I_3)} \right]^{1/2} \tag{16-72a}$$

$$\omega_{2m} = \left[\frac{L^2 - 2I_3E_k}{I_2(I_2 - I_3)} \right]^{1/2} \tag{16-72b}$$

$$\omega_{3m} = \left[\frac{L^2 - 2I_1E_k}{I_3(I_3 - I_1)} \right]^{1/2} \tag{16-72c}$$

The body nutation rate is

$$\omega_p = \pm \left[\frac{(I_3 - I_2)(L^2 - 2I_1E_k)}{I_1 I_2 I_3} \right]^{1/2} = \pm \left[\frac{(I_3 - I_2)(I_3 - I_1)}{I_1 I_2} \right]^{1/2} \omega_{3m} \tag{16-73}$$

where the upper sign applies for $I_1 > I_2 > I_3$ and the lower for $I_1 < I_2 < I_3$.

The values of m given by Eq. (16-71) are always between 0 and 1. Plots of the three elliptic functions are shown in Fig. 16-3, and useful equations involving them are collected in Table 16-1. Because dn is always positive, we choose the sense of the $\hat{P}_3$ axis such that ω_3 is always positive, as in the axial symmetry case. For $m \ll 1$, the first two terms in a power series expansion in m of the Jacobian elliptic

Table 16-1.   Identities for Jacobian Elliptic Functions. In Eqs. (9), (10), and (11), the dependence on the parameter, m, has been omitted for notational convenience.

(1)	$\dfrac{d}{d\Phi}$ sn $(\Phi|m) = $ cn $(\Phi\|m)$ dn $(\Phi\|m)$	(8)	$dn^2\,(\Phi\|m) + m\,sn^2\,(\Phi\|m) = 1$
(2)	$\dfrac{d}{d\Phi}$ cn $(\Phi\|m) = -$sn $(\Phi\|m)$ dn $(\Phi\|m)$	(9)	sn $(u+v) = \dfrac{sn\ u\ cn\ v\ dn\ v + sn\ v\ cn\ u\ dn\ u}{1 - m\ sn^2\ u\ sn^2\ v}$
(3)	$\dfrac{d}{d\Phi}$ dn $(\Phi\|m) = -m$ sn $(\Phi\|m)$ cn $(\Phi\|m)$	(10)	cn $(u+v) = \dfrac{cn\ u\ cn\ v - sn\ u\ dn\ u\ sn\ v\ dn\ v}{1 - m\ sn^2\ u\ sn^2\ v}$
(4)	sn $(-\Phi\|m) = -$sn $(\Phi\|m)$	(11)	dn $(u+v) = \dfrac{dn\ u\ dn\ v - m\ sn\ u\ cn\ u\ sn\ v\ cn\ v}{1 - m\ sn^2\ u\ sn^2\ v}$
(5)	cn $(-\Phi|m) = $ cn $(\Phi\|m)$	(12)	sn $(\Phi\|m) \simeq \sin\Phi - \dfrac{1}{4}m\,(\Phi - \sin\Phi\,\cos\Phi)\,\cos\Phi$ $(m \ll 1)$
(6)	dn $(-\Phi|m) = $ dn $(\Phi\|m)$	(13)	cn $(\Phi\|m) \simeq \cos\Phi + \dfrac{1}{4}m\,(\Phi - \sin\Phi\,\cos\Phi)\,\sin\Phi$ $(m \ll 1)$
(7)	$cn^2\,(\Phi\|m) + sn^2\,(\Phi\|m) = 1$	(14)	$d_n\,(\Phi\|m) \simeq 1 - \dfrac{1}{2}m\,sin^2\,\Phi$ $(m \ll 1)$

functions given by Eqs. (12), (13), and (14) of Table 16-1 provide an analytic approximation for the rotational motion for near-axial symmetry or small nutation. These equations also show that for $m=0$, the Jacobian elliptic functions are trigonometric functions. This limit arises in the axial symmetry case, $I_1=I_2$, and also in the case of no nutation, $L^2=2I_3E_k$. In the axial symmetry limit, Eqs. (16-69) through (16-73) become equal to Eqs. (16-59) and (16-63). The $m=1$ limit is attained when $L^2=2I_2E_k$. In this limit, the Jacobian elliptic functions can be expressed as hyperbolic functions, which is in agreement with the exponential behavior for rotation about an axis of intermediate moment of inertia, for which $L^2\approx2I_2E_k$.

Equations (16-69) through (16-73) can be verified by substitution into the Euler equations of motion, with the use of Equations (1), (2), (3), (7), and (8) of Table 16-1. We can derive equations in terms of the initial body rate vector, ω_0, for the asymmetric case by using the addition laws for Jacobian elliptic functions, Eqs. (9), (10), and (11) of Table 16-1, and Eqs. (16-69) through (16-73). This gives

$$\omega_1 = \frac{\omega_{01}\operatorname{cn}\omega_p t + (\nu\omega_{02}\omega_{03}/\omega_{3m})\operatorname{sn}\omega_p t\,\operatorname{dn}\omega_p t}{1-(\mu\omega_{02}/\omega_{3m})^2\operatorname{sn}^2\omega_p t} \tag{16-74a}$$

$$\omega_2 = \frac{\omega_{02}\operatorname{cn}\omega_p t\,\operatorname{dn}\omega_p t - (\omega_{03}\omega_{01}/\nu\omega_{3m})\operatorname{sn}\omega_p t}{1-(\mu\omega_{02}/\omega_{3m})^2\operatorname{sn}^2\omega_p t} \tag{16-74b}$$

$$\omega_3 = \frac{\omega_{03}\operatorname{dn}\omega_p t + \mu^2(\omega_{01}\omega_{02}/\nu\omega_{3m})\operatorname{sn}\omega_p t\,\operatorname{cn}\omega_p t}{1-(\mu\omega_{02}/\omega_{3m})^2\operatorname{sn}^2\omega_p t} \tag{16-74c}$$

where

$$\mu \equiv \left[\frac{I_2(I_2-I_1)}{I_3(I_3-I_1)}\right]^{1/2} \tag{16-75}$$

$$\nu \equiv \left[\frac{I_2(I_2-I_3)}{I_1(I_1-I_3)}\right]^{1/2} \tag{16-76}$$

$$\omega_{01} = \omega_{1m}\operatorname{cn}\omega_p t_1 \tag{16-77a}$$

$$\omega_{02} = \omega_{2m}\operatorname{sn}\omega_p t_1 \tag{16-77b}$$

$$\omega_{03} = \omega_{3m}\operatorname{dn}\omega_p t_1 \tag{16-77c}$$

From Eqs. (16-72c), (16-38), and (16-39), we also have

$$\omega_{3m} = (\omega_{03}^2 + \mu^2\omega_{02}^2)^{1/2} \tag{16-78}$$

Equations (16-74a, b, c), are significantly more complex than Eqs. (16-60a, b, c), the analogous equations for axial symmetry. In these equations, the dependence of the Jacobian elliptic functions on the parameter has been omitted for notational convenience.

16.2.2 Torque-Free Motion—Kinematic Equations

The various forms of the kinematic equations of motion considered in Section 16.1.2 contain components of the instantaneous angular velocity vector, ω, on the right-hand side. The solutions for ω obtained above can be substituted into the kinematic equations in the torque-free case; however, this leads to rather intractable differential equations, which can be avoided by a suitable choice of coordinate system. An especially convenient inertial reference system is one in which the angular momentum vector, which is fixed in inertial space if $N \equiv 0$, lies along the third coordinate axis. Then L in body coordinates is given by

$$\begin{bmatrix} L_1 \\ L_2 \\ L_3 \end{bmatrix} = A \begin{bmatrix} 0 \\ 0 \\ L \end{bmatrix} \tag{16-79}$$

where A is the direction cosine matrix. The most convenient kinematic parameters in this case are the 3-1-3 Euler angles, so we use Eq. (12-20) for A to obtain

$$L_1 = I_1 \omega_1 = L \sin\theta \sin\psi \tag{16-80a}$$

$$L_2 = I_2 \omega_2 = L \sin\theta \cos\psi \tag{16-80b}$$

$$L_3 = I_3 \omega_3 = L \cos\theta \tag{16-80c}$$

We can choose θ to lie between 0 deg and 180 deg, so Eq. (16-80c) determines θ completely, with ω_3 given by Eq. (16-60c) or (16-69c). Note that with these conventions, θ is the *nutation angle* introduced in Section 15.1. Then, Eqs. (16-80a) and (16-80b) determine ψ completely, including the quadrant, with ω_1 given by Eq. (16-59a) or (16-69a) and ω_2 by Eq. (16-59c) or (16-69b). We cannot determine ϕ in this fashion, so we use Eq. (16-12b), which in the notation of this section is

$$\frac{d\phi}{dt} = (\omega_1 \sin\psi + \omega_2 \cos\psi)/\sin\theta \tag{16-81}$$

Using Eq. (16-80) yields the equivalent, and more useful, form

$$\frac{d\phi}{dt} = L \frac{I_1 \omega_1^2 + I_2 \omega_2^2}{I_1^2 \omega_1^2 + I_2^2 \omega_2^2} \tag{16-82}$$

In the asymmetric case, $I_1 \neq I_2$ and Eqs. (16-69a) and (16-69b) can be substituted into Eq. (16-82). Integration results in a closed-form expression for ϕ, which involves an *incomplete elliptic integral of the third kind* [Milne-Thomson, 1965; Byrd and Friedman, 1971; Morton, *et al.*, 1974]. In the axial symmetry case, on the other hand, $d\phi/dt$ is a constant, and we have

$$\frac{d\phi}{dt} = L/I_T = \omega_I \tag{16-83}$$

$$\phi = \omega_I t + \phi_0 \tag{16-84}$$

and

$$\psi = \omega_p (t - t_1) + 90° \tag{16-85}$$

where the inertial nutation rate, ω_I, was introduced in Section 15.1 and Eq. (16-67a). The initial value of ϕ in Eq. (16-84), ϕ_0, is arbitrary because the definition of the inertial reference system only specifies the location of the inertial three axis.

Because the kinematic equations of motion for the 3-1-3 Euler angles have now been solved, the direction cosine matrix can be found from Eq. (12-20). Any other set of kinematic parameters can then be evaluated by the techniques of Section 12.1, e.g., the Euler symmetric parameters from Eq. (12-14), the Gibbs vector from Eq. (12-18), or the 3-1-2 Euler angles from Eq. (12-23). The resulting parameters specify the orientation of the spacecraft body principal axes relative to an inertial frame in which the angular momentum vector is along the inertial three axis. It is frequently more convenient to specify the orientation of the spacecraft relative to some other inertial frame, such as the celestial coordinate frame. This is especially important if the resulting closed-form solution is to be used as the starting point for a variation-of-parameters analysis of the motion in the presence of torques, as described below, because the angular momentum vector is not fixed in inertial space when the torque does not vanish. Changing this reference system is straightforward if there is a convenient rule for the parameters representing the product of two successive orthogonal transformations. The most convenient product rule is Eq. (12-15) for the Euler symmetric parameters, so we will write the closed-form solution for this kinematic representation. This solution, *in the axial symmetry case*, is

$$
q(t) = \begin{bmatrix} q_4' & q_3' & -q_2' & q_1' \\ -q_3' & q_4' & q_1' & q_2' \\ q_2' & -q_1' & q_4' & q_3' \\ -q_1' & -q_2' & -q_3' & q_4' \end{bmatrix} q_0 \tag{16-86}
$$

where

$$
q_1' = u_1 \cos\alpha \sin\beta + u_2 \sin\alpha \sin\beta \tag{16-87a}
$$

$$
q_2' = u_2 \cos\alpha \sin\beta - u_1 \sin\alpha \sin\beta \tag{16-87b}
$$

$$
q_3' = u_3 \cos\alpha \sin\beta + \sin\alpha \cos\beta \tag{16-87c}
$$

$$
q_4' = \cos\alpha \cos\beta - u_3 \sin\alpha \sin\beta \tag{16-87d}
$$

$$
\alpha \equiv \tfrac{1}{2}\omega_p t \tag{16-87e}
$$

$$
\beta \equiv \tfrac{1}{2}\omega_I t \tag{16-87f}
$$

$$
\mathbf{u} \equiv \mathbf{L}_0 / |\mathbf{L}_0| = [u_1, u_2, u_3]^T \tag{16-87g}
$$

$$
\mathbf{L}_0 \equiv [L_{01}, L_{02}, L_{03}]^T \tag{16-87h}
$$

In this solution, all the constants of the motion have been reexpressed in terms of initial values of the Euler symmetric parameters and $\mathbf{L}_0$, the angular momentum vector in body principal coordinates. These initial values are arbitrary (except that the sum of squares of the Euler parameters must be unity) because the inertial reference frame can be chosen arbitrarily.

A geometrical construction, due to Poinsot, and presented in many texts (e.g.,

Goldstein [1950]; Synge and Griffith [1959]; MacMillan [1936]; Thomson [1963]; and Kaplan [1976]) pictures the rotational motion of a rigid body as the rolling of the inertia ellipsoid on an "invariable plane" normal to the angular momentum vector. In the axial symmetry case, Poinsot's construction is equivalent to the discussion in Section 15.1 of the space and body cones. In the general case, the geometrical construction is not easy to visualize, and the analytic solutions are more useful for spacecraft applications. The results for the asymmetric case are described by Morton, *et al.*, [1974].

16.2.3 Variation-of-Parameters Formulation

The solutions of the attitude dynamics equations in the torque-free case have been obtained above. The *variation-of-parameters* formulation of attitude dynamics is a method of exploiting the torque-free solutions when torques are present [Fitzpatrick, 1970; Kraige and Junkins, 1976]. Our approach follows that of Kraige and Junkins.

To introduce the basic ideas of the variation-of-parameters approach, we first consider a simple example, the translational motion of a point mass in one dimension. The equations of motion in this case are

$$\frac{dx}{dt} = v \qquad (16\text{-}88\text{a})$$

$$\frac{dv}{dt} = F(x,v,t)/m \qquad (16\text{-}88\text{b})$$

where the dependence of the force on x, v, and t is arbitrary. The solution of these equations when $F \equiv 0$ is

$$x(t) = x_0 + v_0 t \qquad (16\text{-}89\text{a})$$

$$v(t) = v_0 \qquad (16\text{-}89\text{b})$$

This is called the *forward solution* because it expresses the position and velocity, x and v, of the mass at time t in terms of its position and velocity, x_0 and v_0, at the prior time, $t = 0$. We can also write the *backward solution*:

$$x_0 = x(t) - v(t)t \qquad (16\text{-}90\text{a})$$

$$v_0 = v(t) \qquad (16\text{-}90\text{b})$$

which expresses x_0 and v_0 in terms of $x(t)$ and $v(t)$.

The central idea of the variation-of-parameters approach is to use Eqs. (16-89) to represent the motion of the mass even when a force is applied. This is possible if x_0 and v_0 are allowed to be time varying, as shown in Fig. 16-4. At each point on the trajectory of the particle, the position and velocity are the same as those of the force-free motion represented by the tangent line with intercept $x_0(t)$ and slope $v_0(t)$, i.e., with initial position and velocity $x_0(t)$ and $v_0(t)$. In this case, x_0 and v_0 are the varying parameters that would be constant in the force-free case. (It is possible to express the motion in terms of other parameters, such as the kinetic energy, but we will only consider initial conditions as the varying parameters.)

To obtain the equations of motion in the variation-of-parameters form we

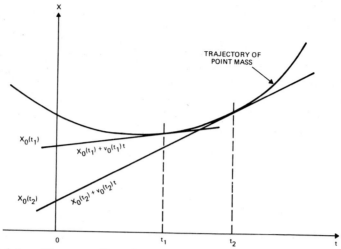

Fig. 16-4.   Variation-of-Parameters Formulation Applied to Motion of a Point Mass in One Dimension

differentiate the backward solution, Eq. (16-90):

$$\frac{dx_0}{dt} = \frac{dx}{dt} - t\frac{dv}{dt} - v \qquad (16\text{-}91a)$$

$$\frac{dv_0}{dt} = \frac{dv}{dt} \qquad (16\text{-}91b)$$

Equation (16-88) is then substituted into Eq. (16-91), yielding

$$\frac{dx_0}{dt} = v - tF(x,v,t)/m - v = -tF(x,v,t)/m \qquad (16\text{-}92a)$$

$$\frac{dv_0}{dt} = F(x,v,t)/m \qquad (16\text{-}92b)$$

Note that the right sides of these equations vanish when $F=0$ because x_0 and v_0 are constants in this case. Finally, we substitute the forward equations of motion on the right sides of Eq. (16-92) to eliminate x and v, and obtain the final equations

$$\frac{dx_0}{dt} = -tF(x_0 + v_0 t, v_0, t)/m \qquad (16\text{-}93a)$$

$$\frac{dv_0}{dt} = F(x_0 + v_0 t, v_0, t)/m \qquad (16\text{-}93b)$$

These equations must be integrated to obtain $x_0(t)$ and $v_0(t)$, and then $x(t)$ and $v(t)$ are given by Eq. (16-89).

The equations of motion in the variation-of-parameters approach, Eqs. (16-93), are generally more complicated than the original equations of motion, Eqs. (16-88). They have the advantage, however, that the right sides are smaller than the right sides of the original equations if the forces are small. Thus, larger integration steps can be taken if the equations are integrated numerically (see Section 17.1). Comparison of Eqs. (16-88a) and (16-93a) shows that the variation-of-parameters approach will be useful if

$$|F|t \ll |mv|$$

that is, if the impulse of the applied force over the time interval considered is much less than the momentum of the particle.

We now consider the attitude dynamics problem in the axial symmetry case. The parameters to be varied are the initial values of both the Euler symmetric parameters and the components of the angular momentum vector along the body principal axes. The foward solutions are Eq. (16-86) and

$$L_1(t) = L_{01}\cos 2\alpha + L_{02}\sin 2\alpha \tag{16-94a}$$

$$L_2(t) = L_{02}\cos 2\alpha - L_{01}\sin 2\alpha \tag{16-94b}$$

$$L_3(t) = L_{03} \tag{16-94c}$$

where α is given by Eq. (16-87e). These are obtained by multiplying Eqs. (16-60) by the principal moments of inertia along the three axes. The backward solutions are obtained from Eqs. (16-86) and (16-94) by interchanging L with L_0 and q with q_0, and changing the sign of t (and thus of α and β). Differentiating the backward solutions and substituting the forward equations of motion on the right-hand sides, as in the example above, yields the *variation-of-parameters equations of motion for the axial symmetry case*:

$$\frac{dq_0}{dt} = \frac{1}{2} \begin{bmatrix} 0 & \tilde{\omega}_3 & -\tilde{\omega}_2 & \tilde{\omega}_1 \\ -\tilde{\omega}_3 & 0 & \tilde{\omega}_1 & \tilde{\omega}_2 \\ \tilde{\omega}_2 & -\tilde{\omega}_1 & 0 & \tilde{\omega}_3 \\ -\tilde{\omega}_1 & -\tilde{\omega}_2 & -\tilde{\omega}_3 & 0 \end{bmatrix} q_0 \tag{16-95}$$

$$\frac{dL_{01}}{dt} = N_a + 2\left(1 - \frac{I_T}{I_3}\right) u_2 N_3 \beta \tag{16-96a}$$

$$\frac{dL_{02}}{dt} = N_b - 2\left(1 - \frac{I_T}{I_3}\right) u_1 N_3 \beta \tag{16-96b}$$

$$\frac{dL_{03}}{dt} = N_3 \tag{16-96c}$$

where β, α, and $\mathbf{u}$ are given by Eq. (16-87); and N_a, N_b, and $\tilde{\omega}$ are defined by

$$N_a \equiv N_1\cos 2\alpha - N_2\sin 2\alpha \tag{16-97a}$$

$$N_b \equiv N_2\cos 2\alpha + N_1\sin 2\alpha \tag{16-97b}$$

$$\tilde{\omega}_1 \equiv [(u_3 N_b - u_2 N_3)(1 - \cos 2\beta) - u_1 G - N_a \sin 2\beta]/L_0 \tag{16-97c}$$

$$\tilde{\omega}_2 \equiv [(u_1 N_3 - u_3 N_a)(1 - \cos 2\beta) - u_2 G - N_b \sin 2\beta]/L_0 \tag{16-97d}$$

$$\tilde{\omega}_3 \equiv [(u_2 N_a - u_1 N_b)(1 - \cos 2\beta) - u_3 G - N_3(\sin 2\beta - (1 - I_T/I_3)2\beta)]/L_0 \tag{16-97e}$$

with

$$G \equiv (u_1 N_a + u_2 N_b + u_3 N_3)(2\beta - \sin 2\beta) \tag{16-97f}$$

Note that the angle β must be expressed in radians in these equations.

The variation-of-parameters method will be most useful if $|\mathbf{N}|t \ll |\mathbf{L}|$. In this case, Eqs. (16-95) and (16-96) can be integrated with a large time step to find $q_0(t)$ and $\mathbf{L}_0(t)$. The Euler symmetric parameters, $q(t)$, representing the attitude at time t, can then be obtained from Eqs. (16-86) and (16-87); the angular momentum vector $\mathbf{L}(t)$, can be found from Eq. (16-94), if desired. Equation (16-95) has the same structure as the kinematic equation of motion for the instantaneous Euler symmetric parameters, Eq. (16-3), so any techniques used for solving the latter equation can also be used for the former. In particular, the closed-form solution of Eq. (16-4), which is discussed in Section 17.1, is applicable if the direction of $\tilde{\omega}$ does not change during the time step.

The above variation-of-parameters equations were derived from rigid body dynamics and therefore are most useful for single-spin spacecraft. Effects of reaction wheels can be added as perturbations, but the resulting equations are most useful if the deviations from the unperturbed motion are small. In particular, any spacecraft nutation (with nutation angle less than 90 deg) is modeled exactly and much more efficiently by the variation-of-parameters method than by a straightforward numerical integration of Euler's equations.

The variation-of-parameters equations for the asymmetric case, $I_1 \neq I_2$, have been studied by Kraige and Junkins [1976]. They are significantly more complicated than the equations for the axial symmetry case, largely because of the dependence of the parameter, m, of the Jacobian elliptic functions on the initial values of $\mathbf{L}$ and q.

16.3 Spacecraft Nutation

Robert M. Beard
Michael Plett

In principle, the closed-form solution (Eq. (16-69)) to Euler's equations for torque-free rigid-body motion could be used to determine spacecraft dynamic motion from telemetered attitude sensor data. In practice, this is normally impossible because initial values of the Euler angles and an accurate knowledge of spacecraft rotational kinetic energy and angular momentum magnitude are unavailable. This section discusses some simplifying assumptions which reduce the complexity of the original equations, thus permitting approximate solutions for the spacecraft motion based on attitude sensor data.

Throughout this section we assume that the spacecraft is undergoing rigid-body, torque-free motion, that the moments of inertia are known, and that the nutation angle, θ, is small. The inertial frame (X, Y, Z) is defined to have its Z axis collinear with the angular momentum vector, $\mathbf{L}$. The body frame (x, y, z) will be the principal axes. Without loss of generality, we take the body z axis to be the nominal spin axis; the 3-1-3 Euler rotation sequence (ϕ, θ, ψ) defined in Eq. (12-20) will be used to transform vectors from the inertial to the body frame.

Spacecraft nutation causes attitude sensor data which would otherwise be constant to oscillate; this oscillation may be used to determine parameters of the spacecraft dynamic motion. Virtually all types of attitude sensors are sensitive to nutation and could be used, in principle, to monitor it. For example, for a symmetric spacecraft $(I_x = I_y \equiv I_T)$, Eq. (16-60) shows that the output of gyroscopes, which measure the instantaneous angular velocity, ω, in the body frame, will oscillate at the *body nutation rate*, $\omega_p(=\dot\psi)$. Alternatively, consider an attitude sensor which measures the angle between some inertial vector, $\mathbf{S}_I$, and the body z axis. Without loss of generality, we may require the X axis of the inertial frame to be the projection of $\mathbf{S}_I$ into the plane perpendicular to $\mathbf{L}$. Then, if

$$\mathbf{S}_B \equiv \left(S_{B_x}, S_{B_y}, S_{B_z}\right)^{\mathrm{T}} = A_{313}\left(S_{I_x}, S_{I_y}, S_{I_z}\right)^{\mathrm{T}} \equiv A_{313}\mathbf{S}_I$$

is the body frame representation of $\mathbf{S}_I$, where A_{313} is the 3-1-3 transformation given in Eq. (12-20), we see that

$$S_{B_z} = S_{I_x}\sin\theta\sin\phi + S_{I_z}\cos\theta$$

For a symmetric spacecraft, as shown in Section 16.3.1, θ is a constant, and from Eq. (16-84),

$$\phi = \omega_I t + \phi_o$$

Thus, the angle measured by the sensor oscillates with the *inertial nutation rate*, ω_I $(=\dot\phi)$. Examples of such sensors include magnetometers aligned with the body z axis or star sensors which measure the coelevation of stars.

Figure 16-5 shows the simulated variation in measured Sun angle and in apparent spin rate for a symmetric spacecraft $(I_z/I_T = 1.238)$ with a Sun sensor

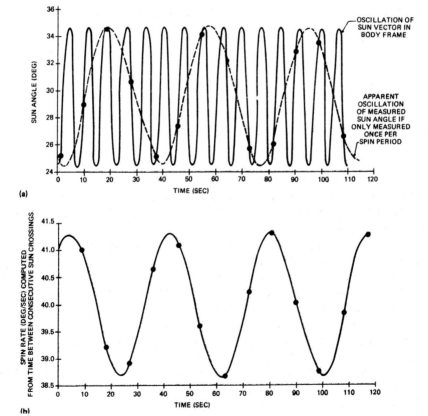

(a)

(b)

Fig. 16-5.   Effect of Nutation on Observed Sun Angles and Spin Rate for a Symmetric Spacecraft With $I_z/I_T = 1.238$, Nutation Angle of 5 Deg, and Average Spin Rate of 40 Deg/Sec

whose slit plane contains the x and z axes and having a nutation angle, θ, of 5 deg and an average measured spin period of 9 sec. We have just shown that the Sun angle oscillates at the inertial nutation rate ω_I ($=\dot{\phi}$), which for this example corresponds to a period of 7.24 sec. It can be shown in a derivation similar to that in Section 16.3.1 (but one which solves in terms of the variation of Sun crossing times) that the deviation in crossing time oscillates at the body nutation rate ω_p ($=\dot{\psi}$), or for this example, a period of 37.8 sec. Section 16.3.2 shows that for a symmetric spacecraft whose attitude sensor measures the coelevation of the inertial vector only *once* per spin period, the *measured* coelevation angles vary at a rate which is roughly approximate to the *body* nutation rate. This is shown by the dashed line in Fig. 16-5.

Section 16.3.1 derives a technique for determining the dynamic motion of a dual-spin symmetric spacecraft from body measurements of the coelevation of inertial vectors. This technique involves solving for the Euler angle rates and initial values and is suitable for a time development of the rotational dynamic motion. Section 16.3.2 derives techniques for monitoring spacecraft nutation with a digital Sun sensor. It presents approximations for the amplitude and the phase of L in the body system for a symmetric spacecraft and extends these approximations to an asymmetric spacecraft. Techniques in this section are particularly suitable for determining the amplitude and phasing of torques for active nutation damping (see Section 18.4).

16.3.1 Dynamic Motion of a Symmetric Dual-Spin Spacecraft

In this case, we assume a dual-spin spacecraft having a momentum wheel with known moments of inertia (I_{wheel}) and known constant spin rate (ω_{wheel}) relative to the body; further, we assume that the wheel rotational axis is aligned with the body z axis. The components of the total angular momentum, **L**, in the body frame are then

$$L_x = I_x\omega_x = L\sin\theta\sin\psi$$
$$L_y = I_y\omega_y = L\sin\theta\cos\psi$$
$$L_z = I_z\omega_z + h_z = L\cos\theta \tag{16-98}$$

where

$$h_z = I_{z_{wheel}}\omega_{wheel}$$

Recall from Section 16.1 that the body moments of inertia (I_x, I_y, I_z) are assumed measured with the momentum wheel "caged." Substitution of Eq. (16-98) into Eq. (16-12) (and noting that here x, y, z, are used in place of u, v, w in Section 16.1) gives

$$\dot{\theta} = L\left(\frac{1}{I_x} - \frac{1}{I_y}\right)\sin\theta\sin\psi\cos\psi$$

$$\dot{\phi} = L\left(\frac{\sin^2\psi}{I_x} + \frac{\cos^2\psi}{I_y}\right)$$

$$\dot{\psi} = L\cos\theta\left(\frac{1}{I_z} - \frac{\cos^2\psi}{I_y} - \frac{\sin^2\psi}{I_x}\right) - \frac{h_z}{I_z} \tag{16-99}$$

For the axially symmetric case, using $I_x = I_y \equiv I_T$, Eq. (16-99) becomes

$$\dot{\theta} = 0$$

$$\dot{\phi} = \frac{I_z \omega_z + h_z}{I_T \cos \theta}$$

$$\dot{\psi} = -\frac{\omega_z (I_z - I_T) + h_z}{I_T} \tag{16-100}$$

Thus, θ, $\dot{\phi}$, and $\dot{\psi}$ are each constant.

For small θ, $\cos \theta \approx 1$, and Eq. (16-11c) reduces to

$$\omega_z \approx \dot{\phi} + \dot{\psi} \equiv \tilde{\omega}$$

where $\tilde{\omega}$ is approximately the average measured spin rate (ignoring nutation*). Thus, Eq. (16-100) becomes

$$\dot{\theta} = 0$$

$$\dot{\phi} = \frac{I_z \tilde{\omega} + h_z}{I_T} \tag{16-101}$$

$$\dot{\psi} = -\frac{\tilde{\omega} (I_z - I_T) + h_z}{I_T}$$

Because the right sides of Eq. (16-101) are known, the problem of determining the time development of ϕ, θ, and ψ reduces to determining θ and the initial values of ϕ and ψ from sensor data.

Using small angle approximations for θ, the 3-1-3 transformation from the inertial to the body frame given in Eq. (12-20) reduces to

$$A_{313} \approx R \equiv \begin{bmatrix} \cos(\phi + \psi) & \sin(\phi + \psi) & \theta \sin \psi \\ -\sin(\phi + \psi) & \cos(\phi + \psi) & \theta \cos \psi \\ \theta \sin \phi & -\theta \cos \phi & 1 \end{bmatrix} \tag{16-102}$$

where θ is in radians. Thus, for each observation of a vector, $\hat{\mathbf{S}}_B$, in the body frame whose inertial position, $\hat{\mathbf{S}}_I$, is known (e.g., a star, the Sun, the magnetic field vector, the nadir vector), we have

$$\hat{\mathbf{S}}_B = \begin{bmatrix} S_x \\ S_y \\ S_z \end{bmatrix} = R \begin{bmatrix} \cos \alpha \cos \epsilon \\ \sin \alpha \cos \epsilon \\ \sin \epsilon \end{bmatrix} = R \hat{\mathbf{S}}_I \tag{16-103}$$

where α and ϵ are the azimuth and elevation, respectively, in the inertial system.

Using Eq. (16-102) and $\phi = \dot{\phi} t + \phi_0$, the third row of Eq. (16-103) becomes

$$S_z = \sin \epsilon + \theta \cos \epsilon \sin(\dot{\phi} t + \phi_0 - \alpha) \tag{16-104}$$

---

*Equation (16-103) may be used to show that if the spin rate is measured by observing the times when an inertial vector, $\mathbf{S}$, crosses a body-fixed plane containing the z axis (taken as the x-z plane without loss of generality), then, for small θ, crossings occur whenever, $\sin(\phi + \psi) - \theta(S_z/S_x)\cos \psi = 0$. Thus, for $\mathbf{S}$ sufficiently far from $\mathbf{L}$ (i.e., $\theta S_z/S_x \ll 1$), the average measured spin rate is $\dot{\phi} + \dot{\psi}$.

or

$$\frac{S_z - \sin \epsilon}{\cos \epsilon} = \theta \sin(\dot{\phi}t + \phi_0 - \alpha) \qquad (16\text{-}105)$$

and because $S_z = \sin(\epsilon_{obs})$, where ϵ_{obs} is the elevation of the observation in the body frame, we have

$$\frac{\sin(\epsilon_{obs}) - \sin \epsilon}{\cos \epsilon} = [\theta \cos \phi_0] \sin(\dot{\phi}t - \alpha) + [\theta \sin \phi_0] \cos(\dot{\phi}t - \alpha) \qquad (16\text{-}106)$$

For small ϵ the left-hand side approximately equals the elevation residual, $\epsilon_{obs} - \epsilon$, and is referred to as the *reduced elevation residual*. Because $\dot{\phi}$ is known from Eq. (16-101), this can be solved for a given set of observations with simple linear regression by recognizing that it is of the form

$$Y = C_1 X_1 + C_2 X_2 \qquad (16\text{-}107)$$

with

$$C_1 = \theta \cos \phi_0, \ C_2 = \theta \sin \phi_0$$

Then

$$\theta = \sqrt{C_1^2 + C_2^2} \qquad (16\text{-}108)$$

$$\phi_0 = \arctan(C_2/C_1) \qquad (16\text{-}109)$$

with appropriate sign checks to determine ϕ_0 on the range 0 to 360 deg.

Finally, if ζ is the observed phase angle of the body *x*-axis, we may approximate

$$\dot{\zeta} \approx \omega_z = \dot{\phi} \cos \theta + \dot{\psi} \approx \dot{\phi} + \dot{\psi} \qquad (16\text{-}110)$$

Integrating, then,

$$\zeta(t) = (\dot{\phi} + \dot{\psi})t + \phi_0 + \psi_0 \qquad (16\text{-}111)$$

or

$$\psi_0 = \zeta(t = 0) - \phi_0 \qquad (16\text{-}112)$$

and we have thus determined θ and the initial values of ϕ and ψ.

This technique was used successfully to determine the dynamic motion of the SAS-2 spacecraft from telemetered star sensor data. Figure 16-6 is a plot of the left-hand side of Eq. (16-105) (the reduced elevation residual) versus time (modulo the period of ϕ) for a selected orbit of SAS-2 star sensor data, where $I_z/I_T = 1.067$ and the average measured spin rate was $\tilde{\omega} = 1.061$ deg/sec. Fitting this data to Eq. (16-106) using the technique of Eq. (16-107) resulted in $\phi_0 = 16.73$ deg, $\theta = 0.58$ deg, $\psi_0 = 343.27$ deg, where from Eq. (16-101), $\dot{\phi} = 4.75$ deg/sec, $\dot{\psi} = -3.69$ deg/sec. For convenience the star sensor may be considered as mounted along the body *x* axis. Because the sensor field of view is small, the left-hand side of Eq. (16-105) is approximately the difference between the observed elevation in the nutating body frame and the elevation in the (nonnutating) inertial frame. The amplitude of the plot is then approximately θ. The approximate phase, ϕ_0, may be obtained from the

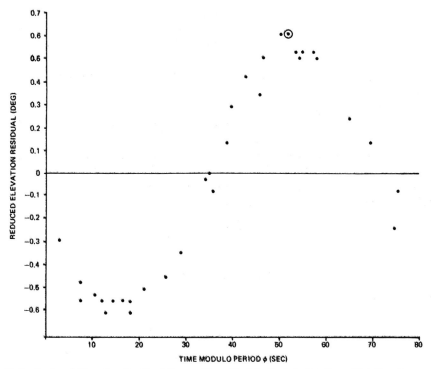

Fig. 16-6.   Reduced Elevation Residual Versus Time Modulo Period of ϕ for a Nutating Spacecraft With $\theta = 0.58$ Deg. See text for explanation.

phase of the plot and the azimuth, α, of any star sighting. For example, if the circled observation at a phase of approximately $+90$ deg occurs at 52.3 sec (modulo the period of ϕ) and is for a star whose azimuth in the $\bar{L}$ system is 175.25 deg, then $\phi_0 \approx 90° - \dot{\phi} t + \alpha \approx 16.83°$.

16.3.2 Nutation Monitoring With a Digital Sun Senor

In this subsection, we consider the problem of determining the amplitude and phase of nutation from a Sun sensor which observes the Sun approximately once per spin period. We assume that both the Sun angle and Sun sighting time are monitored by a sensor, such as the digital Sun sensor described in Section 6.1.3, whose field of view is a slit parallel to the z principal axis and perpendicular to the nominal spin plane. As illustrated in Fig. 16-5, nutation produces an oscillation in both the Sun angle and the measured spin period determined from the Sun sighting times. We wish to relate this oscillation to the nutation angle, θ, and the azimuthal orientation, ψ, of the angular momentum vector, L, in spacecraft coordinates.

Figure 16-7 shows the path on the body-fixed celestial sphere of the angular momentum unit vector, $\hat{L}$, assuming $I_x < I_y < I_z$. (This is equivalent to Fig. 15-14 with the x, y, and z axes replaced by P_3, P_2, and P_1, respectively.) Note that the nutation angle, θ, between $\hat{L}$ and the body z axis, $\hat{z}$, varies between θ_{max} and θ_{min}. A digital Sun sensor determines the measured Sun angle, $\beta_m = \arccos(\hat{z} \cdot \hat{S})$, between $\hat{z}$ and the Sun vector, $\hat{S}$, at the time that the Sun crosses the sensor slit plane. The *nominal Sun angle*, $\beta = \arccos(\hat{L} \cdot \hat{S})$, between $\hat{L}$ and $\hat{S}$ remains constant because

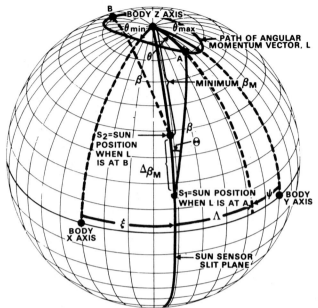

Fig. 16-7. Path of Angular Momentum Vector on Body-Fixed Celestial Sphere for a Nutating
           Spacecraft

both vectors are fixed in inertial space. Therefore, as $\hat{L}$ oscillates about $\hat{z}$, the position of the Sun at the time of a Sun angle measurement moves up and down the sensor slit plane so that β remains constant. In Fig. 16-7, when $\hat{L}$ is at A, then $\hat{S}$ is at S_1 and the measured Sun angle, β_m, is a maximum; when $\hat{L}$ is at B, $\hat{S}$ is at S_2 and β_m is a minimum.* The amplitude, $\Delta\beta_m$, of the measured Sun angle variation will be between $2\theta_{max}$ and $2\theta_{min}$ depending on the orientation of the sensor slit plane relative to the principal axes. When the Sun sensor slit plane is the $y-z$ plane, $\Delta\beta_m = 2\theta_{max}$ (assuming Sun sightings actually occur when $\theta = \theta_{max}$).

When the slit plane is the $x-z$ plane, $\Delta\beta_m$ will be a minimum. In most practical cases, this minimum value is approximately $2\theta_{min}$. However, as shown in Fig. 16-8, when $I_x \ll I_y$ and the Sun angle is sufficiently small, then the path of the angular momentum vector will be very elongated and $\Delta\beta_m$ will be greater than $2\theta_{min}$ when the slit plane is the $x-z$ plane. For Sun angles less than 90 deg, as shown in Fig. 16-8, the maximum value of β_m is $\beta + \theta_{min}$ when $\hat{L}$ is at D. However, when the maximum radius of curvature of the nutation curve is greater than β, then the minimum β_m will be $\beta - \theta_{min} - \epsilon$ and will occur at two symmetrically located points, A and B, on the nutation curve. For Sun angles greater than 90 deg, we may use Fig. 16-8, measuring β from the $-z$ axis. Thus, β_m will oscillate between $\beta + \theta_{min} + \epsilon$ and $\beta - \theta_{min}$. Note that the elevation of the Sun in the spacecraft frame oscillates at the inertial nutation rate, $\dot{\phi}$, as shown in Eq. (16-106). However, from Fig. 16-8, it is clear that the *measured* Sun angles depend on the orientation of $\hat{L}$ in the body; hence, they should oscillate at the same rate as $\hat{L}$ in the body, as will be proved later.

---

*Sun sightings occur at discrete points along the path of $\hat{L}$. However, for the geometrical arguments of Figs. 16-7 and 16-8, we assume that Sun sightings will occur for each of the possible positions of $\hat{L}$.

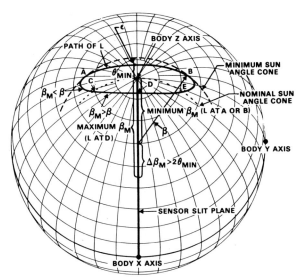

Fig. 16-8. Sun Angle Measurement Geometry for a Very Asymmetric Nutating Spacecraft. See text for explanation.

The average of the measured Sun angles will be somewhat different from the nominal Sun angle, β, as shown in Fig. 16-8. The dotted curve on Fig. 16-8 is a small circle of radius, β, centered on the sensor slit and tangent to the body z axis. At points C and E, where this curve intersects the path of $\hat{L}$, the nominal and measured Sun angles will be equal: between C and E inside the nominal Sun angle curve, $\beta_m > \beta$; outside the nominal Sun angle cone, $\beta_m < \beta$. A similar construction is possible on Fig. 16-7. It is clear from the location of points C and E that the average measured Sun angle will be less than β whenever the Sun angle is less than 90 deg. Similarly, the average β_m will be greater than β whenever the Sun angle is greater than 90 deg. Normally, this effect is small and may be masked by the granularity of the Sun angle measurement. In the extreme case in Fig. 16-8, the average β_m is about 2 deg less than β.

To obtain quantitative expressions for the nutation parameters in terms of the measured Sun angle parameters, it is convenient to define two intermediate variables. The first is R_β, the ratio of the observed Sun angle variation to the maximum nutation amplitude. Thus,

$$R_\beta \equiv \frac{\Delta\beta_m}{2\theta_{max}} < 1 \qquad (16\text{-}113)$$

For a symmetric spacecraft, the path of $\hat{L}$ in Figs. 16-7 and 16-8 is a small circle of radius θ and the maximum and minimum values of β_m are $\beta + \theta$ and $\beta - \theta$. In this case, $R_\beta = \Delta\beta_m/2\theta = 1$. For asymmetric spacecraft, R_β is a function of θ_{max}, β, the moment of inertia tensor, I, and the azimuth, ξ, of the sensor slit plane.

To determine the functional dependence of R_β on the spacecraft parameters, we define the second intermediate variable, R_θ^2, by

$$R_\theta^2 \equiv \frac{\sin^2\theta_{min}}{\sin^2\theta_{max}} = \frac{I_x(I_z - I_y)}{I_y(I_z - I_x)}, \qquad (I_x < I_y) \qquad (16\text{-}114)$$

$$\sin\theta = \frac{\sin\theta_{max}}{\sqrt{1 + \left(\dfrac{1}{R_\theta^2} - 1\right)\sin^2\psi}} \tag{16-115}$$

where θ is the nutation amplitude at phase ψ as defined in Fig. 16-7.

To derive Eqs. (16-114) and (16-115), note that θ reaches its extreme values when one component of the angular velocity reaches its maximum and the other component is zero. For the case of I_y being the principal moment of intermediate value, we have

$$\sin\theta_{max} = \frac{I_y\omega_{ym}}{L}$$

$$\sin\theta_{min} = \frac{I_x\omega_{xm}}{L}$$

Substituting for ω_{xm} and ω_{ym} (the maximum values of ω along the x and y axes, respectively) from Eq. (16-72) and taking the ratio of $\sin\theta_{min}$ to $\sin\theta_{max}$ yields Eq. (16-114). From Eq. (16-69), we have

$$L_x = I_x\omega_{xm}\text{cn}(\Phi|m) = L\sin\theta\sin\psi$$

$$L_y = -I_y\omega_{ym}\text{sn}(\Phi|m) = L\sin\theta\cos\psi$$

Substituting from Eq. (7) of Table 16-1, squaring, and adding yields

$$1 = \text{cn}^2(\Phi|m) + \text{sn}^2(\Phi|m) = \left[\frac{L^2(I_y - I_z)}{(L^2 - 2I_z E_k)I_y}\right]\sin^2\theta\left[\frac{I_y(I_x - I_z)}{I_x(I_y - I_z)}\sin^2\psi + \cos^2\psi\right]$$

where the first term in brackets is $1/\sin^2\theta_{max}$. Algebraic manipulation then yields Eq. (16-115).

From Fig. 16-7, we see that

$$\Lambda \equiv 90° - \xi - \psi \tag{16-116}$$

Note that ψ is the third Euler angle in a 3-1-3 sequence and that as ψ increases, the projection of L onto the $x - y$ plane moves clockwise. By examining the spherical triangle A-Sun-z, we obtain

$$\sin\Theta = \frac{\sin\theta\sin\Lambda}{\sin\beta} \tag{16-117}$$

and

$$\tan\left(\frac{\beta_m}{2}\right) = \tan\left(\frac{\theta + \beta}{2}\right)\frac{\cos\left(\dfrac{\Theta + \Lambda}{2}\right)}{\cos\left(\dfrac{\Theta - \Lambda}{2}\right)} \tag{16-118}$$

which gives β_m as a function of θ_{max}, β, I, and ξ. From this, $\Delta\beta_m$ can be calculated so that Eq. (16-113) can be inverted to determine θ_{max}, which is the unknown in real data. The maximum and minimum values of β_m and hence R_β are found numerically for a given θ_{max}, β, ξ, and I. Note that the dependence of R_β on I is only through the parameter R_θ^2. Values of R_β for $\theta_{max} = 2$ deg and $\beta = 90$ deg are shown in Fig. 16-9. Numerical tests indicate that R_β is insensitive to θ_{max} and β so that the curves in Fig. 16-9 are accurate to 5 to 10% when $2\theta_{max} < \beta$ and $R_\theta^2 > 0.15$. They were constructed for $I_x < I_y < I_z$ and the slit plane in the first quadrant; they may be extended to other quadrants by symmetry. These curves have the

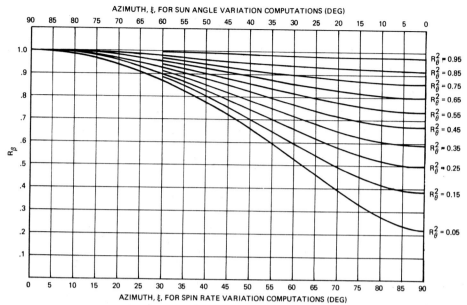

Fig. 16-9.   Ratio, R_β, of Observed Sun Angle Variation to Nutation Angle as a Function of Sun Sensor Azimuth, ξ, and R_θ^2 for $\beta = 90$ Deg, $\theta_{max} = 2$ Deg, and $I_x < I_y < I_z$. Read ξ from bottom for R_β used in computing ΔP and from the top in computing θ_{max} from the Sun angle variation. See text for explanation and example.

approximate analytic form

$$R_\beta = \left(\frac{1 + R_\theta}{2}\right) + \left(\frac{1 - R_\theta}{2}\right)\cos\left[2(90° - \xi)\right] \qquad (16\text{-}119)$$

The analytic approximation is about 5% low at $R_\theta^2 = 0.25$ and about 25% low at $R_\theta^2 = 0.05$. The oscillations of β_m with changing ψ defined by Eq. (16-118) vary from sinusoidal to nearly sawtooth depending on the spacecraft symmetry and slit location. (Note that on Fig. 16-8, points B and D are only 100 deg apart in azimuth.) The rate of the Sun angle oscillation for an asymmetric spacecraft is the asymmetric analog of Eq. (16-100) for small θ_{max}, i.e.,

$$-\dot{\psi} \approx R_I \omega_z \qquad (16\text{-}120)$$

where the intermediate variable, R_I, is defined by

$$R_I \equiv \pm \sqrt{\frac{(I_z - I_x)(I_z - I_y)}{I_x I_y}} \qquad (16\text{-}121)$$

with R_I positive for $I_x, I_y < I_z$ and negative for $I_z < I_x, I_y$.

The same numerical process that yields R_β also yields ξ_m, the azimuth of the angular momentum vector when β_m is a maximum. This information is valuable for phasing a torque to counteract the nutation and is shown in Fig. 16-10 for the same conditions as Fig. 16-9. Note that ξ_m is compressed toward 90 deg as the spacecraft becomes more asymmetrical (smaller R_θ^2). These curves are also insensitive to both θ_{max} and β. As long as $3\theta_{max} < \beta$ and $R_\theta^2 > 0.15$, the curves are accurate to about 15

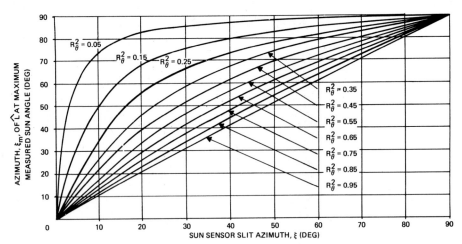

Fig. 16-10. Azimuth, ξ_m, of Angular Momentum Vector, **L**, at Which the Measured Sun Angle is a Maximum as a Function of Slit Azimuth, ξ, and R_θ^2 for $\theta_{max} = 2$ Deg and $\beta = 90$ Deg. See text for explanation and example.

deg in ξ_m. *Figures 16-9 and 16-10 are intended for estimates only; Eq. (16-118) should be solved directly when spacecraft parameters are well established.*

To illustrate nutation monitoring from Sun angle data, consider a spacecraft with relative moments of inertia $I_x = 51.5$, $I_y = 71.3$, $I_z = 90.0$ and a Sun sensor mounted 30 deg from the x axis, i.e., $\xi = 30°$. From Eq. (16-114), $R_\theta^2 = 0.35$. Assume that β_m covers the range $67.5° < \beta_m < 73.5°$. R_β is obtained from Fig. 16-9 as 0.72 and θ_{max} is then computed from

$$\theta_{max} = \frac{\Delta\beta}{2R_\beta} = \frac{6°}{2 \times 0.72} = 4.2° \tag{16-122}$$

From Fig. 16-10, the maximum value of β_m occurs at $\xi_m = 59°$, which means that **L** is at an azimuth of 59 deg at the measurement of the maximum Sun angle. A torque applied 180° out of phase with **L** at an azimuth of 239 deg would reduce the nutation amplitude. This technique was used successfully on the SSS-1 spacecraft [Flatley, 1972b].

Many Sun sensors provide the Sun crossing time so that the interval between successive crossings may be used to measure the spin period. However, this measured spin period is affected by nutation. By examining Fig. 15-8 and assuming that the Sun sensor points in the direction of P_3 to R on that figure, we see that the Sun sensor is rotating counterclockwise in inertial space at an approximately uniform rate (or, equivalently, the Sun is rotating clockwise relative to the sensor). As shown in Fig. 16-11, the angular momentum vector is rotating counterclockwise relative to the Sun sensor, where, for simplicity, we have chosen $I_x = I_y \equiv I_T < I_z$. Because I_T is less than I_z, the spacecraft is nutating more rapidly than it is rotating and $\hat{\mathbf{L}}$ is rotating faster than the Sun sensor.

It is essentially the wobbling motion of the spacecraft which is responsible for the variation in the measured spin period. Assume that in Fig. 16-11, $\hat{\mathbf{L}}$ and $\hat{\mathbf{S}}$ are at L_1 and S_1 at time t_1. At time t_2, after one *measured* spin period, the Sun has rotated 360 deg to S_2. $\hat{\mathbf{L}}$ has moved more rapidly and gone more than 360 deg to

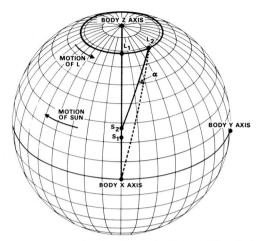

Fig. 16-11.   Body-Fixed Celestial Sphere Showing Change in Measured Spin Period From the Observation of Sun Sighting Times for a Symmetrical Nutating Spacecraft. See text for explanation.

L_2. We wish to examine the rotation of the body coordinate system relative to the inertial coordinate system defined by $\hat{\mathbf{L}}$ and $\hat{\mathbf{S}}$. At time t_1, the x axis is in the $\hat{\mathbf{L}} - \hat{\mathbf{S}}$ plane. At time t_2, it is past the $\hat{\mathbf{L}} - \hat{\mathbf{S}}$ plane by the angle α. Thus, because of the change in the orientation of the body coordinate system relative to the inertial coordinate system, it has rotated through *more than* 360 deg to pick up the Sun again and the measured spin period is greater than the true spin period. Similarly, when $\hat{\mathbf{L}}$ is to the left of center in Fig. 16-11, the measured spin period will be less than the true spin period. (Clearly, over a long term the average measured spin period must be close to the true spin period.) The spin period oscillation has the same period as $\hat{\mathbf{L}}$ in the spacecraft frame.

For a symmetric spacecraft and small nutation angles, we can express θ as a function of I, β, and the variation in the spin period, ΔP, defined by

$$\Delta P \equiv \frac{P_{max} - P_{min}}{P_{max} + P_{min}} \tag{16-123}$$

where P_{max} is the maximum measured period, and P_{min} is the minimum measured period. As shown below, the desired relation is

$$\Delta P = \frac{\theta}{180°} \left| \cot \beta \sin(180° R_I) \right| \tag{16-124}$$

Thus, given an observed variation in the spin period and the average β_m, which may be substituted for β in Eq. (16-124), we can compute the nutation angle, θ.

Equation (16-124) may be derived by extending the development of Flatley [1972a]. Let the inertial frame be as before with the Z axis collinear with $\mathbf{L}$ and the X axis along the projection of $\mathbf{S}$ into the plane perpendicular to z. Let the Sun sensor slit contain the body x and z axes. Then, Eq. (12-20) shows that the y component of $\mathbf{S}$ in the body frame, assuming θ small, is

$$S_y = \sin \beta (-\sin \psi \cos \phi - \cos \psi \cos \theta \sin \phi) + \cos \beta \cos \psi \sin \theta$$

$$\approx -\sin \beta \sin(\phi + \psi) + \theta \cos \beta \cos \psi \tag{16-125}$$

and is equal to zero at a Sun sighting. At any time, t, Eq. (16-100) gives for a single-spin symmetric spacecraft

$$\phi = \frac{I_z \omega_z t}{I_T \cos\theta} + \phi_0 \approx (R_I + 1)\omega_z t + \phi_0 \tag{16-126}$$

$$\psi = -\frac{(I_z - I_T)}{I_T}\omega_z t + \psi_0 = \psi_0 - R_I \omega_z t \tag{16-127}$$

The first Sun sighting after $t=0$ will occur at a rotation angle of approximately $2\pi - (\phi_0 + \psi_0)_{\text{mod}\,2\pi}$; and, in general, the nth Sun sighting will occur at

$$t_n = \frac{[2\pi - (\phi_0 + \psi_0)_{\text{mod}\,2\pi} + 2\pi(n-1)]}{\omega_z} + \delta t_n = \frac{-(\phi_0 + \psi_0)_{\text{mod}\,2\pi} + 2\pi n}{\omega_z} + \delta t_n \tag{16-128}$$

where δt_n is the deviation of the crossing time from that if $\theta = 0$.

Using Eqs. (16-126) and (16-127), and defining $\psi_n = \psi(t_n)$, Eq. (16-125) becomes

$$S_y = 0 \approx -\sin\beta\sin(\omega_z t_n + \phi_0 + \psi_0) + \theta\cos\beta\cos\psi_n$$

$$\approx -\omega_z \delta t_n \sin\beta + \theta\cos\beta\cos\psi_n \tag{16-129}$$

assuming $\omega_z \delta t_n$ small, or

$$\delta t_n = \frac{\theta\cos\beta\cos\psi_n}{\omega_z\sin\beta} \qquad (\beta \neq 0 \text{ or } 2\pi) \tag{16-130}$$

Thus, the period between any two consecutive Sun sightings is

$$P_n = t_n - t_{n-1} = \left[\frac{-(\phi_0 + \psi_0)_{\text{mod}\,2\pi} + 2\pi n}{\omega_z} + \delta t_n\right]$$

$$- \left[\frac{-(\phi_0 + \psi_0)_{\text{mod}\,2\pi} + 2\pi(n-1)}{\omega_z} + \delta t_{n-1}\right]$$

$$= \frac{2\pi}{\omega_z} + \frac{\theta\cos\beta}{\omega_z\sin\beta}(\cos\psi_n - \cos\psi_{n-1}) \tag{16-131}$$

Note that

$$\psi_{n-1} = \psi_0 - R_I \omega_z t_{n-1}$$

$$= \psi_n + 2\pi R_I + R_I \omega_z (\delta t_n - \delta t_{n-1})$$

$$\approx \psi_n + 2\pi R_I \tag{16-132}$$

assuming $R_I \omega_z (\delta t_n - \delta t_{n-1})$ is small. Substituting Eq. (16-132) into Eq. (16-131) and reducing yields

$$P_n = \frac{2\pi}{\omega_z} + \frac{\theta\cos\beta}{\omega_z\sin\beta} 2\sin(\pi R_I)\sin(\psi_n + \pi R_I) \tag{16-133}$$

To find the minimum and maximum P_n, we note that only the factor $\sin(\psi_n + \pi R_I)$ varies. The extrema of the sine function are ± 1 and occur at

$$\psi_n + \pi R_I = \pi/2 \pm \pi i \qquad i = 0, 1, 2, 3 \tag{16-134}$$

for $\beta \neq 0$ or π, $\omega_z \neq 0$, and $R_I \neq 0, 1, 2, 3, \ldots$. Furthermore,

$$P_n = \begin{cases} max \text{ for } \psi_n + \pi R_I = \pi/2 \pm 2\pi i \\ min \text{ for } \psi_n + \pi R_I = 3\pi/2 \pm 2\pi i \end{cases} \qquad i = 0, 1, 2, 3 \tag{16-135}$$

for $0 < \beta < \pi/2$ and $2i < R_I < 2i+1$. For $\pi/2 < \beta < \pi$ or $2i-1 < R_I < 2i$, the maximum and minimum values for P_n are reversed. Substituting Eqs. (16-133) and (16-135) into Eq. (16-123) yields Eq. (16-124).

For the case where R_I approaches an integer, Eq. (16-133) shows that P_n approaches $2\pi/\omega_z$ for all n; i.e., the spin rate variation becomes small with respect to the other approximations. The same phenomenon occurs as β approaches $\pi/2$. For either of these cases, simulations are necessary to determine the spin rate variation arising from the second-order effects neglected above.

The phase, ψ, of $\mathbf{L}$ in the spacecraft coordinate system at the time of a maximum spin period measurement is determined from Eq. (16-134) and is summarized as follows:

	$2n-1<R_I<2n$	$2n<R_I<2n+1$
$0<\beta<90°$	$270°-180R_I$	$90°-180R_I$
$90°<\beta<180°$	$90°-180R_I$	$270°-180R_I$

Note that ψ is measured clockwise from the $+y$ axis so that in terms of the azimuth, ξ (measured counterclockwise from the Sun sensor slit plane or the x axis) a maximum occurs at either $\psi=270°-180°\,R_I=90°-\xi$ or at $\xi=-180°+180°R_I$.

The variation in the Sun angle for a symmetric spacecraft may also be used to determine the phase of $\mathbf{L}$ in the body coordinate system. Applying Eq. (12-20) to the Sun vector, $\hat{\mathbf{S}}$, gives the z component of $\hat{\mathbf{S}}$ in the body frame as

$$S_z=\cos\beta_m=\sin\beta\sin\theta\sin\phi+\cos\beta\cos\theta$$
$$\approx\theta\sin\beta\sin\phi+\cos\beta\cos\theta \qquad (16\text{-}136)$$

Substituting from Eq. (16-126) for ϕ and from Eq. (16-128) for t at a Sun sighting, we have

$$\sin\phi=\sin\left[(R_I\omega_z t_n-\psi_0)+\psi_0+\phi_0+\omega_z t_n\right]$$
$$=\sin\left[-\psi+\omega_z\delta t_n\right]\approx\sin(-\psi) \qquad (16\text{-}137)$$

where Eq. (16-127) has been used to identify $-\psi$. Thus,

$$\cos\beta_m\approx-\theta\sin\beta\sin\psi+\cos\beta\cos\theta \qquad (16\text{-}138)$$

which shows that β_m varies with the period of ψ, and is a maximum when $\psi=90$ deg ($\xi=0$) and a minimum when $\psi=270$ deg ($\xi=180$ deg) for a slit in the x-z plane. This verifies the statement made earlier that the period of oscillation of the measured Sun angles for a symmetric spacecraft which measures the Sun angle only once per spin period is that of ψ.

The spin period variation observed from an asymmetric satellite depends on the orientation of the slit relative to the x and y axes. A convenient approximation based on a number of simulations with a dynamics simulator (ADSIM, described by Gray, et al., [1973] is

$$\Delta P=\frac{\theta_{max}}{180°}\cot\beta\sin(180°\,R_I)R_\beta'(\xi) \qquad (16\text{-}139)$$

where ΔP is defined by Eq. (16-123), θ_{max} is the maximum nutation angle in degrees, and R_I is defined by Eq. (16-121). R_β' has the approximate analytic form

$$R_\beta'=\left(\frac{1+R_\theta}{2}\right)+\left(\frac{1-R_\theta}{2}\right)\cos 2\xi \qquad (16\text{-}140)$$

and can be obtained more accurately either from Fig. 16-9 or by solving Eq. (16-118).

Note that R'_β has the opposite dependence on ξ from that observed with the Sun angle variation. (See Eq. (16-119).) That is, the maximum spin period variation occurs for a sensor on the x axis ($I_x < I_y$), whereas the maximum variation in Sun angle occurs for a sensor on the y axis. Simulations have shown that Eq. (16-139) holds for $\theta < 10$ deg and $\theta \ll \beta$, but that the spin rate variation does not go to zero as the Sun angle approaches 90 deg as Eq. (16-139) indicates. For a specific spacecraft, simultations are recommended to obtain a more accurate relationship between ΔP and θ. Similarly, simulations are necessary to determine the orientation of $\mathbf{L}$ at the measurement of the maximum spin period because of the complex relationship between the variations of $\dot\psi$ and the geometrical effects observed in the Sun angle variation.

To illustrate the usefulness of the spin period variation, consider the spacecraft in the previous example. For those moments of inertia $R_\theta^2 = 0.35$, $R_I = 0.442$, and (from Fig. 16-9 or Eq. (16-140)) $R'_\beta = 0.915$ for $\xi = 30$ deg. Assume $\beta_m = 50$ deg and the observed spin period ranges between 6.0288 sec and 5.9707 sec; then

$$\Delta P = 4.84 \times 10^{-3} = \frac{\theta_{max}}{180°} \cot(50°)\sin(79.56°)0.915$$

or $\theta_{max} = 1.15$ deg.

16.4 Flexible Spacecraft Dynamics

Roger M. Davis
Demosthenes Dialetis

Flexible body dynamics becomes significant when the natural frequencies of flexible spacecraft components have the same magnitude as spacecraft rigid body frequencies due to either librational motion of a gravity-gradient stabilized spacecraft (Section 18.3), nutation of a spin-stabilized spacecraft (Section 16.3), or control system response of an actively controlled spacecraft. (Section 18.3). The lowest natural frequencies of flexible components should be at least an order of magnitude greater than the rigid body frequencies before flexibility can be safely neglected.

The uncoupled lowest natural frequency, f, of a typical experiment boom with an end mass, M, extending from a compact, nonspinning central body can be estimated by the following equation, derived from linear beam theory:

$$f \approx \frac{1}{2\pi} \sqrt{\frac{3EI}{(M + 0.243\,\rho l)l^3}} \quad \text{Hz} \qquad (16\text{-}141)$$

where E is the Young's modulus of the boom structural material, I is the area moment of inertia of the boom cross section, ρ is the boom mass density per unit length, and l is the boom length. The product EI is the *bending stiffness*, which can be computed or obtained from experimental results. Typical values range from 6.0

N·m² for a very flexible antenna element such as those on the RAE, to 170 N·m²
for stiff spin axes booms on spinning spacecraft.

When the estimated natural frequencies of flexible components are close to the
rigid body frequencies, a more detailed analysis of flexibility effects is warranted.
Deformations in very flexible spacecraft will strongly influence the magnitude and
distribution of external and internal forces. Because the internal dynamics can be
highly nonlinear, a rigorous time history simulation of the spacecraft system is
required to predict attitude motions.

The need for simulation of flexible spacecraft dynamics depends on the
attitude determination accuracy required because all spacecraft are flexible at some
level. For attitude data, flexibility effects will be exhibited as either superpositions
of a high-frequency signal or as the dominant portion of the attitude motion,
depending on the flexibility of the system. Therefore, it is important to compare the
effects of flexure with the attitude requirements and with sources of error other
than flexibility. Attitude determination errors can be present in highly flexible
spacecraft because of the relative motion between the attitude sensors and experiments that require precise attitude measurements. Such systems may require
additional sensors to determine the position of the experiment relative to the prime
sensor.

Flexible components interact with the spacecraft attitude control system by
superimposing deflections and accelerations on the average measurements made by
attitude sensors and rate gyros. Consequently, the control system can give erroneous command signals that could destabilize a spacecraft. Flexibility can also
move the instantaneous center of mass and moments of inertia and thereby induce
unexpected responses to command control torques.

16.4.1 Flexibility Effects on Spacecraft Attitude Dynamics

A quantitative analysis of flexible spacecraft attitude dynamics is beyond the
scope of this section. (See, for example, the conference proceedings edited by
Meirovitch [1977]). However, we will discuss specific effects in general terms to
make the reader aware of the various phenomena that may occur. A particularly
good review and bibliography is presented by Modi [1974].

Gravity-Gradient Forces. Gravity-gradient forces are both space- and time-
dependent when acting on long flexible components such as the RAE antenna
booms. When large, the deformations cannot be treated by simple linear methods
because of the change in loading as the boom deforms. Time-dependent loading is
induced by libration of the spacecraft resulting from orbital eccentricity. Large
deformations will change the principal moments of inertia of the spacecraft system
and influence the observed attitude motions. However, axial tension due to
gravity-gradient forces can significantly increase the effective bending stiffness,
thereby raising the natural frequencies.

Solar Heating. Temperature gradients due to unequal solar heating can cause
warping of spacecraft structures. The effect on the spacecraft attitude depends on
the time history of the solar energy input, structural properties (including cross-

section geometry), thermal expansion coefficients, thermal conductivity and diffusivity, and surface properties (absorptivity and emissivity). In a fully sunlit orbit, solar heating on an Earth-pointing spacecraft can cause a bias in the attitude or induce attitude motion with a frequency equal to the orbit rate. As a satellite passes in and out of the Earth's shadow, transients can be induced by the step changes in thermal loading. The significance of these transient loadings will depend on whether the flexible spacecraft has natural frequencies that are close to multiples of the orbital frequency.

Some early spacecraft (Naval Research Laboratory's gravity-gradient satellite 164 [Goldman, 1974] and OGO IV and V [Frisch, 1969]) experienced stability problems that were attributed to solar thermal deformations. These problems have generally been overcome by designing deployable elements for minimum temperature gradients and increasing their torsional stiffness. Thermal effects can still be important, however, if the attitude is critical for experiment sensors on the end of a long boom. The attitude change, θ, at the end of a boom due to a temperature gradient can be approximated by

$$\tan\theta \approx \frac{\alpha \Delta T}{d} l$$

where α is the coefficient of thermal expansion of the boom material, ΔT is the temperature difference across the boom, d is the boom diameter, and l is the boom length. Typically, α is of order of 1.8×10^{-5} cm/(cm°K) and ΔT is in the range $0.3°K \leqslant \Delta T \leqslant 0.8°K$.

Temperature gradients in spinning spacecraft are generally not important due to the averaging effect of the spin rate. However, perturbations due to thermal lag could develop when experiment booms have a long thermal time constant and are shadowed by the spacecraft body once per spin period.

Deployment Dynamics. Coriolis forces are developed during boom development as a result of the deployed components moving relative to the body axes with a deployment velocity, $\mathbf{v}$, and the body axes themselves rotating at an angular rate $\mathbf{\Omega}$ with respect to an inertial frame. The Coriolis acceleration, $2\mathbf{\Omega} \times \mathbf{v}$, during deployment reduces the spin rate and deforms flexible appendages in a direction opposite the direction of rotation. When deployment stops, the restoring forces due to strain, centrifugal, and gravity-gradient forces will cause the flexible elements to oscillate in phase about an equilibrium position. A periodic motion will therefore be superimposed on the spin rate. The persistence of this motion will depend on the effectiveness of structural damping or boom damper devices.

Solar Pressure. Solar torques (Section 17.2) are modified in flexible spacecraft by the change in the instantaneous angle of incidence of the solar radiation due to deformations. The differential force acting on a mass element of a flexible member is proportional to the cosine of the local instantaneous angle of incidence. For very flexible spacecraft, the dynamical system is nonlinear, because the loading becomes a function of the deformation. In addition, spacecraft deformations can induce solar pressure torques due to the shift of the center of pressure from the center of mass. For most satellites at low or intermediate altitudes (up to 6500 km, solar torques due to spacecraft flexibility are negligible when compared with gravity-gradient torques. At synchronous altitude, however, solar pressure can have a significant impact on the stability of gravity-gradient stabilized spacecraft.

Aerodynamic Drag.   Deformations of flexible spacecraft modify rigid body aerodynamic torques in a manner analogous to solar pressure torques. Again, the magnitude of a differential force acting on a mass point is a function of the instantaneous angle of incidence of the air stream. Below 500 km, aerodynamic drag forces can induce significant deformation on highly flexible Earth-pointing spacecraft. The shift of the center of pressure from the center of mass due to the deformation may induce destabilizing torques that could tumble nonspinning spacecraft. Spinning spacecraft with transverse wire booms will tolerate high aerodynamic pressures if spin rates are at least 5 rpm. Aerodynamic forces will deform the wire booms; however, simulations have demonstrated that the energy absorbed during half a revolution is removed during the other half of the revolution. Spinning spacecraft in low-perigee orbits will exhibit boom oscillations but insignificant attitude perturbations due to aerodynamic drag.

System Frequencies and Modes.   Spacecraft with more than one flexible boom have system frequencies that depend on the phase relationship of boom displacements with respect to each other. The system frequencies and modes are not the same as structural bending modes because they are a combination of all flexible element modes. Antisymmetric modes will induce rotation of the central spacecraft body that will be detected by attitude sensors, as shown in Fig. 16-12. Symmetric modes do not couple to attitude motion. Hence, large symmetric element deformations cannot be sensed by attitude sensors alone. Additional position sensors are necessary when information concerning the deformed shape of an experiment boom is critical to its performance.

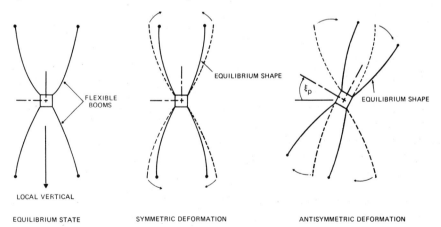

Fig. 16-12.   System Modes of a Gravity-Gradient Stabilized Spacecraft With Flexible Booms. ξ_p is the pitch angle.

Attitude Perturbations Due to Thrusting.   The dynamic response of a flexible spacecraft to thrusting can result in undesirable perturbations of the spacecraft attitude. For example, the location of the center of mass within the spacecraft may be time dependent due to deformational motion. Accordingly, the thrust from body-fixed nozzles used for orbit adjustment will induce both rotational and translational motion because the thrust vector will not pass through the instantaneous center of mass at all times. The rotational motion perturbs the attitude and changes the direction of the thrust vector.

Thrusting for attitude adjustment will cause deformation of transverse booms out of the spin plane. Repeated pulses can cause a buildup of deformations depending on the phasing of the pulses and deformations and may result in large attitude motions about the nominal rigid body orientation.

16.4.2 Modified Equations of Motion

The complexity of the flexible spacecraft equations of motion is increased by the additional degrees of freedom required for structural deformations and the coupling between translational, rotational, and deformational motion. Several methods for derivation of the equations of motion for computer simulation of flexible spacecraft are given by Likins [1970]. The generalized system mass matrix formulation is presented to illustrate the type of system equations encountered in the simulation of a flexible spacecraft.

These equations are appropriate for a rigid spacecraft with flexible solar panels and antennas and, possibly, a set of momentum wheels within the rigid structure, as shown in Fig. 16-13.

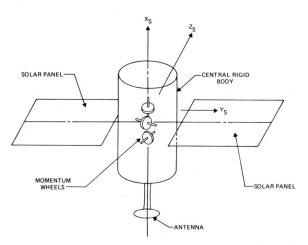

Fig. 16-13. Rigid Spacecraft with Flexible Solar Panels and Antenna and Momentum Wheels Within the Rigid Structure

To specify the spacecraft structure and motion, we define the following quantities: the mass, m, and the moment of inertia, I, of the complete spacecraft in an equilibrium configuration; the location, r, and velocity, v, of the center of mass of the complete flexible spacecraft in its rigid frame of reference; the angular rate, ω, of the rigid spacecraft and the four-component attitude quaternion, q, defined in Section 12.1; the n-vectors

$$x_2^T \equiv (\eta_1, \eta_2, \ldots, \eta_n) \tag{16-142a}$$

$$y_2^T \equiv (\dot{\eta}_1, \dot{\eta}_2, \ldots, \dot{\eta}_n) \tag{16-142b}$$

which provide the modal coordinates and velocities of the flexible spacecraft

[Goldstein, 1950]. Here, n is the number of generalized coordinates which describe the small oscillations of the spacecraft due to its flexibility. In addition, we introduce the component vectors

$$\mathbf{y}_1 \equiv \begin{pmatrix} \mathbf{v} \\ \boldsymbol{\omega} \end{pmatrix} \tag{16-143}$$

$$\mathbf{P}_1 \equiv \begin{pmatrix} \mathbf{F} \\ \mathbf{N} - \boldsymbol{\omega} \times \mathbf{L}_{tot} \end{pmatrix} \tag{16-144}$$

where $\mathbf{F}$ is the total external force and $\mathbf{N}$ is the total external torque acting on the spacecraft of total angular momentum $\mathbf{L}_{tot}$. There may also be generalized forces driving the normal modes, which will be represented by the n-vector $\mathbf{P}_2$. Finally, we introduce the $n \times n$ diagonal *damping matrix*, C_{22}; the $n \times n$ diagonal *stiffness matrix*, K_{22}; and the $(n+6) \times (n+6)$ *system mass* matrix, M, which is formed from four matrices, as follows:

$$M \equiv \begin{bmatrix} M_{11} & M_{12} \\ M_{21} & M_{22} \end{bmatrix} \tag{16-145a}$$

The 6×6 matrix M_{11} is

$$M_{11} \equiv \begin{bmatrix} m & 0 & 0 & \vdots & \\ 0 & m & 0 & \vdots & 0 \\ 0 & 0 & m & \vdots & \\ \hline & 0 & & \vdots & \overline{I} \end{bmatrix} \tag{16-145b}$$

and the $n \times n$ matrix M_{22} is diagonal and can always be reduced to the identity matrix by proper formulation of the equations of motion. The $6 \times n$ matrix M_{12} provides the interaction between the flexible modes and the rigid spacecraft and $M_{21} \equiv M_{12}^{\mathsf{T}}$. The matrices M, C_{22}, and K_{22} are obtained from a dynamic analysis and derivation of the equations of motion for a flexible spacecraft (see, for example, Heinrichs and Fee [1972]).

In terms of the quantities above, the equations of motion for a flexible spacecraft are

$$\dot{\mathbf{r}} = \mathbf{v} \tag{14-146a}$$

$$\dot{q} = \tfrac{1}{2}\Omega(\omega)q \tag{16-146b}$$

$$\dot{\mathbf{x}} = \mathbf{y}_2 \tag{16-146c}$$

$$M_{11}\dot{\mathbf{y}} + M_{12}\dot{\mathbf{y}}_2 = \mathbf{P}_1 - \dot{\mathbf{h}}_1 \tag{16-146d}$$

$$M_{21}\dot{\mathbf{y}}_1 + M_{22}\dot{\mathbf{y}}_2 = \mathbf{P}_2 - C_{22}\mathbf{y}_2 - K_{22}\mathbf{x}_2 \tag{16-146e}$$

where the 4×4 matrix $\Omega(\omega)$ is defined by Eq. (16-26). The 6-vector $\mathbf{h}_1$ refers to the moving parts of the spacecraft with respect to its rigid frame of reference. If the only moving parts are the wheels, $\mathbf{h}_1$ has the form $\mathbf{h}_1^{\mathsf{T}} = (0,0,0,\mathbf{h}^{\mathsf{T}})$, where the 3-vector $\mathbf{h}$ is the angular momentum of the wheels with respect to the rigid spacecraft. From Eq. (16-144) we see that $\mathbf{P}_1$ depends on the spacecraft total angular momentum, $\mathbf{L}_{tot}$, which is given by

$$\mathbf{L}_{tot} = I\omega + \mathbf{h} + \mathbf{L}_{flex.\ modes} \tag{16-147a}$$

where

$$L_{\text{flex. modes}} \equiv \begin{bmatrix} (M_{12})_4 \\ (M_{12})_5 \\ (M_{12})_6 \end{bmatrix} y_2 \qquad (16\text{-}147b)$$

Here, $(M_{12})_i$ is the ith row of the matrix M_{12}. Note that L_{tot} consists of three terms: the first term gives the angular momentum of the rigid spacecraft, where the moment of inertia, I, includes the mass of the wheels, the solar panels, and the antenna; the second term is the angular momentum of the wheels with respect to the spacecraft; and the third term is the angular momentum due to the flexible modes. Note that Eq. (16-146b) is identical with Eq. (16-3) and Eq. (16-147a) is an extension of Eq. (16-52) for the case of a flexible spacecraft.

Equations (16-146) and (16-147) form a complete set of equations of motion for the flexible spacecraft. In this representation, the state vector, x_y, of the flexible spacecraft is

$$x_y^T = (r^T, q^T, y_1^T, x_2^T, y_2^T) \qquad (16\text{-}148)$$

It follows from Eq. (16-146d) that in the above representation, information must be provided on the time derivative of the six-vector h_1 to solve the equations of motion. However, the available information is often the time dependence of h_1 itself. In this case, it is convenient to work in a different representation where the spacecraft state vector is

$$x_L^T = (r^T, q^T, L_1^T, x_2^T, L_2^T) \qquad (16\text{-}149)$$

The 6-vector L_1 and n-vector L_2 are

$$L_1 \equiv M_{11}y_1 + M_{12}y_2 + h_1 \qquad (16\text{-}150a)$$

$$L_2 \equiv M_{21}y_1 + M_{22}y_2 \qquad (16\text{-}150b)$$

We will show that the equations of motion for x_L do not involve the time derivative of h_1. Also, in the absence of any generalized forces, the vectors L_1 and L_2 are conserved. Thus, physical meaning can be attributed to the various vector components. In particular, the last three components of L_1 are the components of the total angular momentum of the flexible spacecraft, L_{tot} (see Eq. (16-147)). To obtain the equations of motion for L_1 and L_2, Eq. (16-150) must be solved with respect to y_1 and y_2. The result is

$$y_1 = (M_{11} - M_{12}M_{22}^{-1}M_{21})^{-1}[L_1 - h_1 - M_{12}M_{22}^{-1}L_2] \qquad (16\text{-}151a)$$

$$y_2 = (M_{22} - M_{21}M_{11}^{-1}M_{12})^{-1}[-M_{21}M_{11}^{-1}(L_1 - h_1) + L_2] \qquad (16\text{-}151b)$$

Using Eq. (16-150), Eqs. (16-146d) and (16-146e) reduce to

$$\dot{L}_1 = P_1 \qquad (16\text{-}152a)$$

$$\dot{L}_2 = P_2 - C_{22}y_2 - K_{22}x_2 \qquad (16\text{-}152b)$$

where y_2 is given by Eq. (16-151b), and the angular rate term, ω, in P_1 (see Eq.

(16-144)) is given by

$$\omega = I^{-1}\left[\mathbf{L}_{tot} - \mathbf{h} - \mathbf{L}_{flex.\ modes}\right] \qquad (16\text{-}153)$$

This follows from Eq. (16-147a). We see that the new equations of motion, namely Eqs. (16-146a), (16-146b), (16-146c), (16-152a), (16-152b), together with Eqs. (16-151b), (16-153), and (16-147b) involve only components of the state vector $\mathbf{x}_L$, and the time derivative of $\mathbf{h}_1$ is not present. Note that the equations of motion of the state vectors $\mathbf{x}_y$ and $\mathbf{x}_L$ are equivalent and should lead to identical solutions.

The equation of motion for the total angular momentum can be obtained from Eq. (16-152a) as

$$\dot{\mathbf{L}}_{tot} = \mathbf{N} - \left[I^{-1}(\mathbf{L}_{tot} - \mathbf{h} - \mathbf{L}_{flex.\ modes})\right] \times \mathbf{L}_{tot} \qquad (16\text{-}154)$$

This relation and Eq. (16-153) are extensions of Eqs. (16-55) and (16-53) for the case of a flexible spacecraft.

In Eq. (16-151b), inversion of the $n \times n$ matrix $M_{22} - M_{21}M_{11}^{-1}M_{12}$ is required. However, the useful matrix identity,

$$\left(M_{22} - M_{21}M_{11}^{-1}M_{12}\right)^{-1} = M_{22}^{-1} - M_{22}^{-1}M_{21}\left(M_{11} - M_{12}M_{22}^{-1}M_{21}\right)^{-1}M_{12}M_{22}^{-1}$$

$$(16\text{-}155)$$

may be used to reduce this to inversion of only 6×6 matrices. Moreover, substituting Eq. (16-155) into Eq. (16-151b) and noting that $M_{22} = 1$, we obtain

$$\mathbf{y}_2 = L_2 - M_{21}\left\{\left[1 - (M_{11} - M_{12}M_{21})^{-1}M_{12}M_{21}\right] \times M_{11}^{-1}(\mathbf{L}_1 - \mathbf{h}_1)\right.$$

$$\left. + (M_{11} - M_{12}M_{21})^{-1}M_{12}L_2\right\} \qquad (16\text{-}156)$$

Thus, we have replaced the multiplication of an $n \times n$ matrix by an $n \times 1$ vector involving n^2 multiplications, with multiplication of a $6 \times n$ matrix by an $n \times 1$ vector and an $n \times 6$ matrix by a 6×1 vector. This involves only $12n$ multiplications. Thus, for complex systems, Eq. (16-157) should take the place of Eq. (16-151b) in the equations of motion for the state vector $\mathbf{x}_L$.

16.4.3 Characteristics of Various Flexible Spacecraft

Flexibility effects for some past and future spacecraft are summarized in Table 16-2. The satellites are excellent examples of large flexible spacecraft [Blanchard, et al., 1968] (see Fig. 15-18). They are gravity-gradient stabilized by four 230-m-long antenna booms. The antenna booms have a double-V configuration with a nominal included angle of 60 deg. Librational motions are damped by a libration damper that is skewed a nominal 66.5 deg from the plane of the antennas. The estimated oscillation period of the RAE-1 antenna booms is 91 minutes using linear beam theory. Because the orbital period is 224 minutes, flexibility effects are obviously important. The equilibrium deformation due to gravity-gradient forces is on the order of 50 m. Hence, the linear beam theory is not adequate and detailed simulation is necessary. Axial tension due to gravity-gradient forces also increases bending frequencies. During a dynamics experiment performed on RAE-1, two distinct short-period oscillations of 19 and 6 min (corresponding to antisymmetrical

Table 16-2. Typical Spacecraft Flexibility Characteristics. "X" indicates a potentially significant mode.

SPACECRAFT	GRAVITY-GRADIENT DEFORMATION	THERMAL BENDING	AERODYNAMIC DRAG	THRUSTING PERTURBATIONS	CONTROL SYSTEMS/ STRUCTURE INTERACTION	DEPLOYMENT DYNAMICS	SPACECRAFT FLEXIBILITY CHARACTERISTICS
RAE-I & -II	X	X				X	FOUR 229-m ANTENNA BOOMS. $EI = 6.03 N \cdot m^2$, $\rho = 20.1$ g/m TWO 96-m DAMPER BOOMS. $EI = 7.17 N \cdot m^2$, $\rho = 20.2$ g/m
IMP-I						X	FOUR 61-m TRANSVERSE BOOMS. $EI = 7.17 N \cdot m^2$, OVERLAPPED, $\rho = 20.8$ g/m TWO 6.1-m SPIN AXIS BOOMS. $EI = 7.17 N \cdot m^2$, INTERLOCKED, $\rho = 20.8$ g/m
IMP-J						X	FOUR 61-m WIRE TRANSVERSE BOOMS. $EI \sim 0$, $\rho = 89$ g/m, END MASS = 3 g TWO 6.1-m SPIN AXIS BOOMS. $EI = 7.17 N \cdot m^2$, INTERLOCKED, $\rho = 20.8$ g/m
ISEE-C			X			X	FOUR 61-m WIRE TRANSVERSE BOOMS. $EI \sim 0$, $\rho = 8.52$ g/m, TIP MASS = 3 g TWO 6.1-m SPIN AXIS BOOMS
DE (HAPO)			X	X	X	X	FOUR 61-m TRANSVERSE WIRE BOOMS. $EI \sim 0$, $\rho = 8.93$ g/m, END MASS = 0.113 kg TWO 7.6-m SPIN AXIS BOOMS. $EI \sim 1200-1800$ N·m², $\rho = 82$ g/m
CTS		X				X	SOLAR PADDLES 1.3 x 7.6 m, CLOSED-LOOP CONTROL SYSTEM, MOMENTUM WHEELS, AND THRUSTERS
SEASAT						X	SOLAR PADDLES (615 m²), SYNTHETIC APERTURE RADAR ANTENNA (2 BY 10 m) REACTION WHEELS, 0.5-DEG POINTING ACCURACY
SSS-A		X					TWO 2.2-m FIBERGLASS TAPERED BOOMS. $EI \sim 25 N \cdot m^2$, $\rho \sim 320$ g/m

system modes) were superimposed on the longer period librational motion [Lawlor, et al., 1974]. By exciting antisymmetrical modes, the properties of the flexible booms could be deduced from attitude data and, hence, the antenna configuration during the steady state could be determined. The modified shape had a significant influence on the interpretation of radio astronomy scientific data.

References

1. Beard, R. M., J. E. Kronenfeld, H. Gotts, and D. Alderman, *Evaluation of the SSS-1 Star Sensor Attitude Determination*, Comp. Sc. Corp., 9101-16600-01TN, Aug. 1973.

2. Blanchard, D. L., R. M. Davis, E. A. Lawlor, and L. Beltracchi, "Design, Simulation and Flight Performance of Radio Astronomy Explorer-A Satellite," *Proceedings of the Symposium on Gravity-Gradient Attitude Stabilization*, Aerospace Corp., AD-696-694, Dec. 1968.

3. Byrd, P. F. and M. D. Friedman, *Handbook of Elliptic Integrals*. Second Edition, Berlin: Springer-Verlag, 1971.

4. Fitzpatrick, Philip M., *Principles of Celestial Mechanics*. New York: Academic Press, Inc., 1970.

5. Flatley, T., *Sun Sighting From a Spinning Spacecraft*, NASA X-732-72-139, GSFC, May 1972a.

6. ———, *Magnetic Active Nutation Damping on Explorer 45 (SSS-A)*, NASA X-732-72-140, GSFC, May 1972b.

7. Frisch, H. P., *Coupled Thermally Induced Transverse Plus Torsional Vibrations of a Thin-Walled Cylinder of Open Cross Section*, NASA X-732-69-530, GSFC, Dec. 1969.

8. Goldman, R. L., *Influence of Thermal Distortion on the Anomalous Behavior of a Gravity Gradient Satellite*, AIAA Paper No. 74-992, Aug. 1974.

9. Goldstein, Herbert, *Classical Mechanics*. Reading, MA: Addison-Wesley Publishing Company, Inc., 1950.

10. Gotts, H. S. and M. E. Plett, *Determination of Nutation Amplitude From Measured Period Variation*, Comp. Sc. Corp., 9101-16600-02TM, April 1973.

11. Gray, C. M., *et al.*, *Attitude Dynamics Data Simulator (ADSIM), version 3.1*, Comp. Sc. Corp., 3000-06000-02TR, Sept. 1973.

12. Heinrichs, Joseph A., and Joseph J. Fee, *Integrated Dynamic Analysis Simulation of Space Stations with Controllable Solar Arrays*. NASA CR-112145 Sept. 1972.

13. Jacobi, C. G. J., *Journal für Math.*, Vol. 39, p. 293, 1849.

14. Kaplan, Marshall H., *Modern Spacecraft Dynamics and Control*. New York: John Wiley & Sons, Inc., 1976.

15. Kibble, T. W. B., *Classical Mechanics*. London: McGraw-Hill, Inc., 1966.

16. Kraige, L. G. and J. L. Junkins, "Perturbation Formulations for Satellite Attitude Dynamics," *Celestial Mechanics*, Vol. 13, p. 39–64, Feb. 1976.

17. Lawlor, E. A., R. M. Davis, and D. L. Blanchard, *Engineering Parameter Determination From the Radio Astronomy Explorer (RAE-I) Satellite Attitude Data*, AIAA Paper No. 74-789, Aug. 1974.

18. Likins, P. W., *Dynamics and Control of Flexible Space Vehicles*, JPL, Jan. 1970.

19. MacMillan, William D., *Dynamics of Rigid Bodies*. New York: McGraw-Hill, Inc., 1936.

20. Meirovitch, Leonard, ed., *Dynamics and Control of Large Flexible Spacecraft, Proceedings of the AIAA Symposium, Virginia Polytechnic Institute and State University, Blacksburg, Virginia, June 13 to 15, 1977*, 1977.

21. Milne-Thomson, L. M., "Jacobian Elliptic Functions and Theta Functions," and "Elliptic Integrals," *Handbook of Mathematical Functions*, Milton Abramowitz and Irene A. Stegun, editors. New York: Dover, 1965.

22. Modi, V. J., *Attitude Dynamics With Flexible Appendages—A Brief Review*, AIAA Paper No. 74167, Feb. 1974; *J. Spacecraft*, Vol. 11, p. 743–751, 1974.

23. Morton, Harold S., Jr., John L. Junkins, and Jeffrey N. Blanton, "Analytical Solutions for Euler Parameters," *Celestial Mechanics*, Vol. 10, p. 287–301, Nov. 1974.

24. Neville, Eric Harold, *Jacobian Elliptic Functions*. Second Edition, Oxford: Oxford University Press, Inc., 1951.

25. Synge, John L. and B. Griffith, *Principles of Mechanics*. Third Edition, New York: McGraw-Hill, Inc., 1959.

26. Synge, J. L. and A. Schild, *Tensor Calculus*. Toronto: University of Toronto Press, 1964.

27. Thomson, William Tyrrell, *Introduction to Space Dynamics*. New York: John Wiley & Sons, Inc., 1963.

28. Whittaker, E. T., *A Treatise on the Analytical Dynamics of Particles and Rigid Bodies*. Fourth Edition, Cambridge: Cambridge University Press, 1937.

CHAPTER 17

ATTITUDE PREDICTION

17.1   Attitude Propagation
        General Techniques, Integration Methods
17.2   Environmental Torques
        *Gravity-Gradient Torque, Solar Radiation Torque,
        Aerodynamic Torque, Magnetic Disturbance Torque*
17.3   Modeling Internal Torques
17.4   Modeling Torques Due to Orbit Maneuvers
        *Thrust Vector Collinear With the Spin Axis, Thrust Vec-
        tor Not Collinear With Spin Axis but Nominally Passing
        Through Spacecraft Center of Mass*

To meet spacecraft attitude determination and control requirements, we must frequently predict the attitude motion for a given set of initial conditions. This requires specifying the differential equations governing the attitude motion and a method of solution. The general methods used for attitude prediction, given appropriate torque models, are discussed in Section 17.1. The necessary modeling of the environmental and internal torques is described in Sections 17.2 and 17.3. Torque modeling during orbit maneuvers is discussed in Section 17.4.

17.1 Attitude Propagation

C. B. Spence, Jr.
F. L. Markley

17.1.1 General Techniques

To model or predict the time evolution of the attitude, two basic methods are used: dynamic modeling and gyro modeling. *Dynamic modeling* consists of integrating both the dynamic and the kinematic equations of motion (see Section 16.1) using analytical or numerical models of the torque. *Gyro modeling* consists of using rate sensors or gyroscopes to replace the dynamic model such that only the kinematic equations need be integrated.

The dynamic equations of motion of a rigid spacecraft are given by Euler's equations as

$$\frac{d}{dt}\mathbf{L} = \mathbf{N}_{DIST} + \mathbf{N}_{CONTROL} - \boldsymbol{\omega} \times \mathbf{L} \tag{17-1a}$$

$$\mathbf{L} = I\boldsymbol{\omega} \tag{17-1b}$$

where I is the moment of inertia tensor and ω is the spacecraft angular velocity vector. The time derivative is taken and the vectors are resolved in a body-fixed coordinate system. The terms $\mathbf{N}_{DIST}$ and $\mathbf{N}_{CONTROL}$ are the disturbance and control torques, respectively, acting on the spacecraft. The kinematic equations can be written in differential form using the quaternion representation of the attitude (see

Section 12.1 and Appendix D) as

$$\frac{dq}{dt} = \frac{1}{2}\Omega q \qquad (17\text{-}2)$$

where

$$\Omega = \begin{bmatrix} 0 & \omega_3 & -\omega_2 & \omega_1 \\ -\omega_3 & 0 & \omega_1 & \omega_2 \\ \omega_2 & -\omega_1 & 0 & \omega_3 \\ -\omega_1 & -\omega_2 & -\omega_3 & 0 \end{bmatrix} \qquad (17\text{-}3)$$

The quaternion representation is generally prefered to the Euler angle representation because of its analytical characteristics.

As the dynamic complexity of the spacecraft increases, each degree of freedom must be represented by its appropriate equation of motion. For example, incorporating momentum or reaction wheels for attitude stability and maneuvering adds additional degrees of freedom. Momentum wheel dynamics can be included as an additional term in Euler's equations and an additional equation of motion for the wheels themselves. For this case, Eq. (17-1a) is rewritten as

$$\frac{d}{dt}(I\omega) = N_{DIST} + N_{CONTROL} - \omega \times I\omega - [\omega \times h + N_{WHEEL}] \qquad (17\text{-}4a)$$

$$L = I\omega + h \qquad (17\text{-}4b)$$

where h is the total angular momentum of the reaction wheels and N_{WHEEL} is the net torque applied to the momentum wheels, which is a function of bearing friction, wheel speed, and applied wheel motor voltage. The equation of motion of the wheels is

$$\frac{d}{dt}h = N_{WHEEL} \qquad (17\text{-}5)$$

The dynamic and kinematic equations of motion are taken as a set of coupled differential equations and integrated using one of the methods described in Section 17.1.2. The integration state vector consists of the three angular velocity body rates or angular momentum components, the attitude quaternion, and any additional degrees of freedom due to nonrigidity (wheels, movable and flexible appendages, rastering instruments, etc.) Alternatively, for spacecraft which have a set of gyros as part of their attitude determination hardware (Section 6.5), the gyro assembly performs a mechanical integration of Euler's equations (irrespective of whether the spacecraft is rigid or flexible), and consequently only the kinematic equations require numerical integration. The gyro package flown aboard a spacecraft usually consists of three or more gyros which are capable of measuring the spacecraft's angular rates. The discussion of the gyro model used to compute the spacecraft's angular velocity from the gyro measurements is described in Section 7.8.2. The attitude propagation problem is diminished for spacecraft which fly a gyro package; in many cases, however, the calibration of the gyro model (see Section 6.5) can be a significant part of the attitude determination problem.

17.1.2 Integration Methods

Once the appropriate differential equations for attitude propagation have been established, it is necessary to choose a method for solving them. Because exact closed-form solutions of the complete equations to be integrated are almost never available, an approximation method is needed. Two methods are discussed in this section: direct integration using standard methods of numerical analysis, and a method for the kinematic equations using a closed-form solution of the equations with constant body rates.

Direct Integration. The equations of motion of attitude dynamics are a set of first-order coupled differential equations of the form

$$\frac{dy}{dt} = f(t, y) \tag{17-6}$$

where f is a known vector function of the scalar t and the vector y. In this section, we will consider for simplicity the single differential equation

$$\frac{dy}{dt} = f(t, y) \tag{17-7}$$

The extension to coupled equations is straightforward, with a few exceptions that will be pointed out.

Numerical algorithms will not give the continuous solution $y(t)$, but rather a discrete set of values y_n, $n = 1, 2, \ldots$, that are approximations to $y(t)$ at the discrete times $t_n = t_0 + nh$. Values of $y(t)$ for arbitrary times can be obtained by interpolation. (For interpolation procedures, see, for example, Carnahan, et al., [1969]; Hamming [1962]; Hildebrand [1956]; Ralston [1965]; or Henrici [1964].) The parameter h is called the *step size* of the numerical integration. A minimum requirement on any algorithm is that it *converge* to the exact solution as the step size is decreased, i.e., that

$$\lim_{h \to 0} y_n = y(t_n) \tag{17-8}$$

where the number of steps, n, is increased during the limiting procedure in such a way that $nh = t_n - t_0$ remains constant.

Three important considerations in choosing an integration method are truncation error, roundoff error, and stability. *Truncation error*, or *discretization error*, is the difference between the approximate and exact solutions $y_n - y(t_n)$, assuming that the calculations in the algorithm are performed exactly. If the truncation error introduced in any step is of order h^{p+1}, the integration method is said to be of *order p*. *Roundoff error* is the additional error resulting from the finite accuracy of computer calculations due to fixed word length. An algorithm is *unstable* if errors introduced at some stage in the calculation (from truncation, roundoff, or inexact initial conditions) propagate without bound as the integration proceeds.

Truncation error is generally the limiting factor on the accuracy of numerical integration; it can be decreased by increasing the order of the method or by decreasing the step size. It is often useful to vary the step size during the integration, particularly if the characteristic frequencies of the problem change significantly; the ease with which this can be done depends on the integration method used. The computation time required is usually proportional to the number of *function evaluations*, i.e., evaluations of $f_n \equiv f(t_n, y_n)$ that are required. It is clear that decreasing the step size increases the number of function evaluations for any fixed integration algorithm.

Two families of integration methods are commonly employed. In *one-step methods*, the evaluation of y_{n+1} requires knowledge of only y_n and f_n. *Multistep methods*, on the other hand, require knowledge of *back values* y_j or f_j for some $j < n$ as well. One-step methods are relatively easy to apply, because only y_0 and f_0 are needed as initial conditions. The step size can be changed, as necessary, without any additional computations. For these reasons, one-step methods are widely used. The most common one-step methods are the *classical R-stage Runge-Kutta* methods [Lambert, 1973]

$$y_{n+1} = y_n + h\phi(t_n, y_n, h) \tag{17-9a}$$

$$\phi(t, y, h) = \sum_{r=1}^{R} c_r k_r \tag{17-9b}$$

$$\sum_{r=1}^{R} c_r = 1 \tag{17-9c}$$

$$k_1 = f(t, y) \tag{17-9d}$$

$$k_r = f\left(t + ha_r, y + h\sum_{s=1}^{r-1} b_{rs}k_s\right) \qquad r = 2, 3, \dots, R \tag{17-9e}$$

$$a_r = \sum_{s=1}^{r-1} b_{rs} \qquad r = 2, 3, \dots, R \tag{17-9f}$$

where different choices of the parameters c_r and b_{rs} (subject to the constraints of Eq. (17-9c) define different methods. The *increment function*, ϕ, is a weighted average of R evaluations of $f(t, y)$ at different points in the integration interval. Note that an R-stage method involves R function evaluations. The constants are always chosen to give the maximum order (and thus minimum truncation error) for a given R; this order is R for $R = 1, 2, 3, 4$; $R-1$ for $R = 5, 6, 7$; and $\leqslant R-2$ for $R \geqslant 8$ [Butcher, 1965]. For this reason, fourth-order four-stage Runge-Kutta methods are the most popular. It requires much tedious algebra to derive the relations among the parameters of a four-stage method that make it of order 4. This derivation leaves two free parameters, resulting in a twofold infinity of fourth-order methods [Ralston, 1965]. One popular choice is*

$$y_{n+1} = y_n + \frac{h}{6}(k_1 + 2k_2 + 2k_3 + k_4) \tag{17-10a}$$

---

*This method reduces to *Simpson's rule*

$$y_{n+1} = y_n + \frac{h}{6}\left[f(t_n) + 4f\left(t_n + \frac{1}{2}h\right) + f(t_n + h)\right]$$

if f is independent of y.

$$k_1 = f(t_n, y_n) \tag{17-10b}$$

$$k_2 = f\left(t_n + \frac{1}{2}h, y_n + \frac{1}{2}hk_1\right) \tag{17-10c}$$

$$k_3 = f\left(t_n + \frac{1}{2}h, y_n + \frac{1}{2}hk_2\right) \tag{17-10d}$$

$$k_4 = f(t_n + h, y_n + hk_3) \tag{17-10e}$$

This is the algorithm implemented in subroutine RUNGE in Section 20.3. The chief drawback of Runge-Kutta methods is the many function evaluations required per integration step.

We now turn to a discussion of multistep integration methods. A *k-step* multistep method has the form

$$y_{n+1} = h \sum_{j=0}^{k} \beta_j f_{n+1+j-k} - \sum_{j=0}^{k-1} \alpha_j y_{n+1+j-k} \tag{17-11}$$

where different choices of the parameters α_j and β_j define alternative methods. Depending on the choice of these parameters, a k-step method requires up to k back values of f_n and y_n. One drawback of these methods is that they are not self-starting; some other method, often Runge-Kutta, must be used to calculate the first k values of y_n and f_n. Another disadvantage is that step size changes are more difficult than for single-step methods; additional back values must be available if the step size is increased, and intermediate back values must be calculated by interpolation if the step size is decreased.* A third penalty is increased computer storage requirements. The chief advantage of multistep methods is that only one function evaluation is needed per integration step.

A multistep method is *explicit* if $\beta_k = 0$ and *implicit* if $\beta_k \neq 0$. Implicit methods may appear to be of dubious value, because they apparently require a knowledge of $f_{n+1} = f(t_{n+1}, y_{n+1})$ to evaluate y_{n+1}. If the original differential equation is linear, however,

$$f(t, y) = A(t)y + \phi(t) \tag{17-12}$$

an implicit method can be used directly, yielding

$$y_{n+1} = \left[1 - h\beta_k A(t_{n+1})\right]^{-1} \left[h\beta_k \phi(t_{n+1}) + h \sum_{j=0}^{k-1} \beta_j f_{n+1+j-k} - \sum_{j=0}^{k-1} \alpha_j y_{n+1+j-k}\right] \tag{17-13}$$

Such methods for linear differential equations are known as *correct-only* methods, for reasons that will be apparent shortly. If we are integrating a system of equations, A is a matrix and Eq. (17-13) requires evaluation of a matrix inverse. (The 1 in the first bracket becomes the identity matrix.)

---

* An alternative procedure is to utilize the last k values of y_n and f_n that have been evaluated, regardless of step size changes, rather than requiring the back values to be at evenly spaced points. This requires inversion of a $k \times k$ matrix at each step to find the coefficients, and the stability properties of these methods are less well understood than those of the conventional methods considered here [Yong, 1974]

Implicit methods can be made to have smaller truncation errors and better stability properties than explicit methods [Lambert, 1973], so it is desirable to have some method of finding f_{n+1} to enable an implicit method to be used for nonlinear equations. The usual procedure is to use an explicit method, known as a *predictor*, to calculate y_{n+1}. Then f_{n+1} is evaluated and an implicit method, known as a *corrector* in this application, is used to obtain a refined value of y_{n+1}, followed by a second evaluation of f_{n+1} using the new y_{n+1}. Methods in this general class are called *predictor-corrector* methods. The mode of application described above is the PECE (predict-evaluate-correct-evaluate) mode; it requires two function evaluations per step. It is also possible to apply the corrector more than once after a single use of the predictor, but analysis indicates that the PECE mode is preferable, and that decreased truncation error is better achieved by decreasing the step size than by multiple applications of the corrector [Lambert, 1973].

It can be shown that no convergent k-step method can have order greater than $k+1$ if k is odd or $k+2$ if k is even [Henrici, 1962]. However, methods of order $k+2$ have poor stability properties, so $k+1$ is the optimal order for practical k-step methods. The most commonly used k-step methods are the *Adams* methods, which are defined by choosing $\alpha_{k-1} = -1$ and $\alpha_j = 0$ for $j \neq k-1$. These have good stability properties and reduced computer storage requirements compared with other multistep methods. One explicit Adams method, the *Adams-Bashforth*, has the form

$$y_{n+1} = y_n + h \sum_{j=0}^{k-1} \beta_j f_{n+1+j-k} \qquad (17\text{-}14)$$

and is of order k. An *Adams-Moulton* method is an implicit Adams method of order $k+1$, given by

$$y_{n+1} = y_n + h \sum_{j=0}^{k} \beta_j f_{n+1+j-k} \qquad (17\text{-}15)$$

A widely used predictor-corrector pair is a p-step Adams-Bashforth predictor followed by a $(p-1)$-step Adams-Moulton corrector; both steps are of order p. One advantage of this pair is that the difference between the predicted and the corrected values of y_{n+1} gives an estimate of truncation error and can be used for step size control. This is in contrast to Runge-Kutta methods, for which step size changes are relatively easy, but estimates of truncation error are difficult to obtain. The fourth-order Adams-Bashforth-Moulton pair is given by
Predictor (explicit):

$$y_{n+1} = y_n + \frac{h}{24}(55f_n - 59f_{n-1} + 37f_{n-2} - 9f_{n-3}) \qquad (17\text{-}16a)$$

Corrector (implicit):

$$y_{n+1} = y_n + \frac{h}{24}(9f_{n+1} + 19f_n - 5f_{n-1} + f_{n-2}) \qquad (17\text{-}16b)$$

This is only an example; higher order methods are widely used and, unlike higher order Runge-Kutta methods, cost only additional storage space and not additional function evaluations. (See, for example, Hull, *et al.*, [1972] or Enright and Hull [1976].)

If values of y are needed at intermediate times and an Adams integrator is being used, it is convenient to employ an interpolation algorithm based on values of f_n rather than y_n, because the former already have to be stored for the integration routine.

In either the Runge-Kutta or the predictor-corrector calculations, some of the function evaluations may be done approximately rather than exactly (by not recalculating torques, for example) to save computational effort. These are called *pseudoevaluations*, and are represented by E*, so one often reads of a PECE* mode of a predictor-corrector.

In choosing an integration method, the factors of programming complexity, computer storage requirements, execution time, and computational accuracy must all be considered. For a specific application where the characteristic frequencies of the system are known to be nearly constant, a fixed-step method is indicated. If the step size is limited by variations in the driving terms rather than by integration error (noisy input and/or low-accuracy requirements) or if function evaluations are relatively inexpensive, a Runge-Kutta method is preferred. If, on the other hand, the integration step is set by integration error (smooth input, high accuracy), or function evaluations are expensive, a predictor-corrector method is better. Adams methods are favored in this class because they combine good stability properties with relatively low computer storage requirements and programming complexity.

If the characteristic frequencies of the system are not constant, a variable-step method should be used. A complete integration package in this class must include an algorithm for automatic step size variation, based on an estimate of local truncation error. Because predictor-corrector methods provide an automatic estimate of local truncation error, they are the preferred variable-step methods. The best general-purpose integration methods currently available are packages with variable-step and variable-order Adams-Bashforth-Moulton integrators (see Hull, *et al.*, [1972] and Enright and Hull [1976], which also include comparative tests of integration packages using a wide variety of test cases).

Approximate Closed-Form Solution for the Kinematic Equation. As described in Section 17.1.1, when a set of gyros is part of the attitude determination hardware and the method of gyro modeling is used, only the kinematic equation need be integrated for attitude propagation. Wilcox [1967] and Iwens and Farrenkopf [1971] have presented a method of processing gyro data which yields an approximate closed-form solution to the kinematic equation. If we assume that the gyro data is telemetered or sampled at a fixed rate and that the angular velocity vector in body coordinates is constant over the sampling interval, then a closed-form solution to the kinematic equation (Eq. (17-12)) is

$$q(t_{n+1}) = e^{\frac{1}{2}\Omega_n T} q(t_n) \tag{17-17}$$

where T is the sampling interval ($T = t_{n+1} - t_n$); Ω_n is Eq. (17-3) evaluated at time t_n; $q(t_n)$ is the attitude quaternion at time t_n; and $q(t_{n+1})$ is the propagated attitude quaternion at time t_{n+1}. The validity of Eq. (17-17) as a solution to the kinematic equation can be established by differentiation.

Equation (17-17) can be rewritten in a more convenient form for numerical computation by evaluating the matrix exponential using the procedures in Appendix C. (See Eq. C-79.) Specifically,

$$q(t_{n+1}) = \left[\cos\left(\frac{\omega T}{2}\right) \mathbf{1} + \frac{1}{\omega} \sin\left(\frac{\omega T}{2}\right) \Omega_n \right] q(t_n) \tag{17-18}$$

where

$$\omega \equiv \sqrt{\omega_x^2 + \omega_y^2 + \omega_z^2} \tag{17-19}$$

and **1** is the identity matrix.

For spacecraft gyros the sampling period is typically 100 to 400 ms. Because of the simplicity of Eq. (17-18), the closed-form solution has been used as the kinematic integrator for onboard computers [Fish and Chmielewski, 1977]. The gyro system is usually a rate-integrating gyro package which provides an average angular velocity vector over the sampling period (see Section 6.5). The term inside the brackets in Eq. (17-18) is then an orthogonal rotation which retains the normalization of the propagated attitude quaternion.

It remains to illustrate the general validity of the closed-form solution and the computational error. To assess the latter, the quaternion $q(t_{n+1})$ is expanded in a Taylor series about the time t_n

$$q(t_{n+1}) = q(t_n) + \frac{dq}{dt}T + \frac{1}{2}\frac{d^2q}{dt^2}T^2 + \cdots \tag{17-20}$$

By repeated use of the kinematic equation (Eq. (17-2)), the Taylor series can be rewritten as

$$q(t_{n+1}) = \left[1 + \frac{1}{2}T\Omega_n + \frac{\frac{1}{4}T^2\Omega_n^2}{2!} + \frac{\frac{1}{8}T^3\Omega_n^3}{3!} + \cdots \right] q(t_n) + \frac{1}{4}T^2\dot{\Omega}_n q(t_n)$$

$$+ \left[\frac{1}{12}\dot{\Omega}_n\Omega_n + \frac{1}{24}\Omega_n\dot{\Omega}_n \right] T^3 q(t_n) + \frac{1}{12}T^3\ddot{\Omega}_n q(t_n) + \cdots \tag{17-21}$$

The series of terms in the first bracket on the right-hand side of Eq. (17-21) is the Taylor series expansion of $\exp[\frac{1}{2}\Omega_n T]$. The remaining terms constitute the error introduced for a sampling period T in assuming a constant body rate equal to ω_n.

In general, the rates are not constant, and information from a rate-integrating gyro package can be used to form the 4×4 matrix

$$\bar{\Omega} \equiv \frac{1}{T}\int_{t_n}^{t_{n+1}} \Omega(t)dt = \Omega_n + \frac{1}{2}\dot{\Omega}_n T + \frac{1}{6}\ddot{\Omega}_n T^2 + \cdots \tag{17-22}$$

The terms in Eq. (17-21) can be rearranged to yield

$$q(t_{n+1}) = \left[1 + \frac{1}{2}T\bar{\Omega} + \frac{\frac{1}{4}T^2\bar{\Omega}^2}{2!} + \frac{\frac{1}{8}T^3\bar{\Omega}^3}{3!} + \cdots \right] q(t_n)$$

$$+ \frac{1}{48}\left[\dot{\Omega}_n\Omega_n - \Omega_n\dot{\Omega}_n \right] T^3 q(t_n) + \cdots \tag{17-23}$$

The first of the two terms on the right-hand side is the Taylor series expansion of the closed-form expression

$$q(t_{n+1}) = e^{\frac{1}{2}\bar{\Omega}T} q(t_n) \tag{17-24}$$

which differs from Eq. (17-17) in using time-averaged rate information rather than instantaneous body rates. It can also be written in the form of Eq. (17-18), with $\overline{\Omega}$ replacing Ω_n. Equation (17-23) shows that the error in this closed-form expression is of order T^3 and vanishes if $\Omega_n \dot{\Omega}_n = \dot{\Omega}_n \Omega_n$, or equivalently, if the vectors ω_n and $\dot{\omega}_n$ are parallel. Thus, the order T^3 correction to the closed-form expression using $\overline{\Omega}$ is zero if the axis of rotation is fixed, even though the rates may be time dependent.

17.2 Environmental Torques

C. B. Spence, Jr.

As described in Section 17.1, attitude prediction requires a model of the environmental disturbance torques acting on the spacecraft. To numerically integrate Euler's equations, the torque must be modeled as a function of time and the spacecraft's position and attitude. As listed in Table 1-2, the dominant sources of attitude disturbance torques are the Earth's gravitational and magnetic fields, solar radiation pressure, and aerodynamic drag.

17.2.1 Gravity-Gradient Torque

Any nonsymmetrical object of finite dimensions in orbit is subject to a gravitational torque because of the variation in the Earth's gravitational force over the object. This *gravity-gradient torque* results from the inverse square gravitational force field; there would be no gravitational torque in a uniform gravitational field. General expressions for the gravity-gradient torque on a satellite of arbitrary shape have been calculated for both spherical [Nidey, 1960; Roberson, 1961; Hultquist, 1961] and nonspherical [Roberson, 1958b] Earth models. For most applications, it is sufficient to assume a spherical mass distribution for the Earth. If more accuracy is required, this may be obtained from the general potential function for the Earth given in Section 5.2. Alternatively, the effect of the Earth's oblateness can be accounted for in the motion of the orbital plane [Hultquist, 1961; Holland and Sperling, 1969].

In this section, we assume that the spacecraft's moment-of-inertia tensor is known for some arbitrary body reference frame whose origin need not coincide with the spacecraft's center of mass and that the spacecraft is orbiting a spherical Earth. The gravitational force $d\mathbf{F}_i$ acting on a spacecraft mass element dm_i located at a position $\mathbf{R}_i$ relative to the geocenter is

$$d\mathbf{F}_i = \frac{-\mu \mathbf{R}_i dm_i}{R_i^3} \qquad (17-25)$$

where $\mu \equiv GM_\oplus$ is the Earth's gravitational constant. The torque about the spacecraft's geometric center due to a force, $d\mathbf{F}_i$, at a position, $\mathbf{r}_i$, relative to the spacecraft's geometric center (see Fig. 17-1) is

$$d\mathbf{N}_i = \mathbf{r}_i \times d\mathbf{F}_i = (\boldsymbol{\rho} + \mathbf{r}_i') \times d\mathbf{F}_i \qquad (17-26)$$

The vector $\boldsymbol{\rho}$ is measured from the geometric center to the center of mass and the vector $\mathbf{r}_i'$ is measured from the center of mass to the mass element dm_i. The

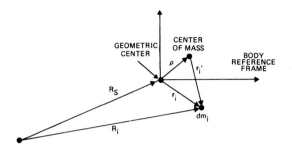

Fig. 17-1.   Coordinate System for the Calculation of Gravity-Gradient Torque

gravity-gradient torque on the entire spacecraft is obtained by integrating Eq. (17-26) to obtain

$$N_{GG} = \int r_i \times dF_i = \int (\rho + r_i') \times \frac{-\mu R_i}{R_i^3} dm_i \qquad (17\text{-}27)$$

The geocentric position vector for the ith mass element can be expressed in terms of the geocentric position vector of the origin of the body reference frame, R_s, as

$$R_i = R_s + r_i = R_s + \rho + r_i' \qquad (17\text{-}28)$$

For a practical artificial satellite $R_i = R_s + \rho + r_i' \gg \rho + r_i'$; therefore

$$R_i^{-3} = (R_i \cdot R_i)^{-\frac{3}{2}} = \left\{ R_s^2 \left[1 + \frac{2R_s \cdot (\rho + r_i')}{R_s^2} + \frac{(\rho + r_i')^2}{R_s^2} \right] \right\}^{-\frac{3}{2}} \approx R_s^{-3} \left[1 - \frac{3R_s \cdot (\rho + r_i')}{R_s^2} \right]$$
$$(17\text{-}29)$$

Substituting Eqs. (17-28) and (17-29) into Eq. (17-27) and performing some algebraic manipulation, the gravity-gradient torque may be rewritten as

$$N_{GG} = \frac{\mu M}{R_s^2} (\hat{R}_s \times \rho) + \frac{3\mu}{R_s^3} \int (r_i \times \hat{R}_s)(r_i \cdot \hat{R}_s) dm_i \qquad (17\text{-}30)$$

where $\int r_i' dm_i = 0$ by definition of the center of mass and M is the total mass of the satellite. Note that the first term is zero when the geometric center is chosen to be the center of mass. The integral in the second term may be rewritten in terms of the moments of inertia. Defining the vectors r_i and $\hat{R}_s$ along the body reference axes $(\hat{X}, \hat{Y}, \hat{Z})$, the gravity-gradient torque (assuming $\rho = 0$) can be expressed as

$$N_{GG} = \frac{3\mu}{R_s^3} \left[\hat{R}_s \times (I \cdot \hat{R}_s) \right] \qquad (17\text{-}31)$$

where I is the moment-of-inertia tensor. From Eq. (17-31), several general characteristics of the gravity-gradient torque may be deduced: (1) the torque is normal to the local vertical; (2) the torque is inversely proportional to the cube of the geocentric distance; and (3) within the approximation of Eq. (17-29), the torque vanishes for a spherically symmetric spacecraft.

Many spacecraft rotate about one of the principal axes. Because the transverse axes (the principal axes normal to the axis of rotation) are continuously changing their inertial position, it is convenient to replace Eq. (17-31) with the average torque over one spacecraft rotation period. Let the spacecraft spin about the $\hat{\mathbf{Z}}$ axis with spin rate ω. The body coordinate system at time t can be expressed in terms of an inertially fixed reference frame $\hat{\mathbf{X}}_0$, $\hat{\mathbf{Y}}_0$, and $\hat{\mathbf{Z}}_0$ at $t=0$ as

$$\hat{\mathbf{X}} = \cos\theta\,\hat{\mathbf{X}}_0 + \sin\theta\,\hat{\mathbf{Y}}_0$$

$$\hat{\mathbf{Y}} = -\sin\theta\,\hat{\mathbf{X}}_0 + \cos\theta\,\hat{\mathbf{Y}}_0 \qquad (17\text{-}32)$$

$$\hat{\mathbf{Z}} = \hat{\mathbf{Z}}_0$$

where $\theta = \omega t$. The unit vector $\hat{\mathbf{R}}_s$ can also be written as

$$\hat{\mathbf{R}}_{s1} = \hat{\mathbf{R}}^0_{s1}\cos\theta + \hat{\mathbf{R}}^0_{s2}\sin\theta$$

$$\hat{\mathbf{R}}_{s2} = -\hat{\mathbf{R}}^0_{s1}\sin\theta + \hat{\mathbf{R}}^0_{s2}\cos\theta \qquad (17\text{-}33)$$

$$\hat{\mathbf{R}}_{s3} = \hat{\mathbf{R}}^0_{s3}$$

where $\hat{\mathbf{R}}^0_{s1}$, $\hat{\mathbf{R}}^0_{s2}$, and $\hat{\mathbf{R}}^0_{s3}$ are components of $\hat{\mathbf{R}}_s$ along $\hat{\mathbf{X}}_0$, $\hat{\mathbf{Y}}_0$, and $\hat{\mathbf{Z}}_0$ at $t=0$. The instantaneous gravity-gradient torque from Eq. (17-31) is averaged over one spin period to obtain

$$\langle \mathbf{N}_{GG} \rangle_s = \frac{1}{2\pi} \int_0^{2\pi} \mathbf{N}_{GG}\,d\theta \qquad (17\text{-}34)$$

Substituting Eqs. (17-31) and (17-33) into Eq. (17-34), the spin-averaged gravity-gradient torque becomes

$$\langle \mathbf{N}_{GG_s} \rangle = \frac{3\mu}{R_s^3}\left[I_{zz} - \left(\frac{I_{xx}+I_{yy}}{2} \right) \right] (\hat{\mathbf{R}}_s \cdot \hat{\mathbf{Z}})(\hat{\mathbf{R}}_s \times \hat{\mathbf{Z}}) \qquad (17\text{-}35)$$

where the products of inertia average to zero over the spin period.

Some spacecraft consist of both an inertially fixed component and a spinning component. For example, the lower portion of the OSO-8 spacecraft spins to provide gyroscopic stability while the upper portion, which consists of solar and instrument panels, is servo controlled to keep the panels pointing toward the Sun in azimuth. For such a composite spacecraft, Hooper [1977] has shown that Eq. (17-31) can be used to calculate the gravity gradient torque along the principal body axes frame by defining an effective moment of inertia. For a composite satellite, with both spinning and inertially fixed components, the effective moments of inertia applicable to gravity-gradient torques are defined as

$$I_{xx} = \left(I_{xx_I} + \rho_I^2 M_I \right) + \left(\frac{I_{xx_S}+I_{yy_S}}{2} + \rho_S^2 M_S \right)$$

$$I_{yy} = \left(I_{yy_I} + \rho_I^2 M_I \right) + \left(\frac{I_{xx_S}+I_{yy_S}}{2} + \rho_S^2 M_S \right) \qquad (17\text{-}36)$$

$$I_{zz} = I_{zz_I} + I_{zz_S}$$

where the subscripts S and I refer to the spinning and inertially fixed components, respectively. The moments of inertia on the right-hand side of Eq. (17-36) are defined about their respective component's center of mass. The symbols M and ρ are the total component mass and the distance of the component's center of mass from the center of mass of the composite structure.

For some spacecraft, it is convenient to average the gravitational torque over an orbit to obtain the net angular momentum impulse imparted to the spacecraft. The magnitude of the time-averaged or *secular* torque is often needed for the design of attitude control systems [Hultquist, 1961; Nidey, 1961]. The time averaged value of the gravity-gradient torque $\langle \mathbf{N}_{GG} \rangle_O$ for an inertially fixed satellite is defined by integrating Eq. (17-33) over one orbit,

$$\langle \mathbf{N}_{GG} \rangle_O = \frac{1}{2\pi} \int_0^{2\pi} \mathbf{N}_{GG} \, d\mu \qquad (17\text{-}37)$$

where μ is the mean anomaly which is proportional to the elapsed time. The integration can best be carried out by changing the variable of integration from the mean anomaly to the true anomaly (see Section 3.1):

$$\langle \mathbf{N}_{GG} \rangle_O = \frac{1}{2\pi a^2 \sqrt{1 - e^2}} \int_0^{2\pi} R_s^2 \mathbf{N}_{GG} \, d\nu \qquad (17\text{-}38)$$

where e, a, and ν are the orbital eccentricity, semimajor axis, and true anomaly, respectively. Because the spacecraft is inertially fixed, the body reference axes, $\hat{\mathbf{X}}$, $\hat{\mathbf{Y}}$, $\hat{\mathbf{Z}}$ are constant and only $\mathbf{R}_s$ is a function of ν. This relation is

$$R_s \equiv |\mathbf{R}_s| = \frac{a(1 - e^2)}{1 + e \cos \nu} \qquad (17\text{-}39)$$

Choosing a coordinate system $(\hat{\mathbf{h}}, \hat{\mathbf{p}}, \hat{\mathbf{q}})$ such that $\hat{\mathbf{h}}$ is the direction of the orbit normal, $\hat{\mathbf{p}}$ is in the direction of perigee, and $\hat{\mathbf{q}} = \hat{\mathbf{h}} \times \hat{\mathbf{p}}$, the components of $\hat{\mathbf{R}}_s$ are given by

$$R_{s1} = \hat{\mathbf{X}} \cdot \hat{\mathbf{p}} \cos \nu + \hat{\mathbf{X}} \cdot \hat{\mathbf{q}} \sin \nu$$

$$R_{s2} = \hat{\mathbf{Y}} \cdot \hat{\mathbf{p}} \cos \nu + \hat{\mathbf{Y}} \cdot \hat{\mathbf{q}} \sin \nu \qquad (17\text{-}40)$$

$$R_{s3} = \hat{\mathbf{Z}} \cdot \hat{\mathbf{p}} \cos \nu + \hat{\mathbf{Z}} \cdot \hat{\mathbf{q}} \sin \nu$$

Substituting Eq. (17-40) into Eq. (17-38) and performing the integration, the average torque can be written as

$$\langle \mathbf{N}_{GG} \rangle_O = \frac{3\mu}{2a^3 \sqrt{(1 - e^2)^3}} \hat{\mathbf{X}} \Big[(I_{yy} - I_{zz})(\hat{\mathbf{Z}} \cdot \hat{\mathbf{h}})(\hat{\mathbf{Y}} \cdot \hat{\mathbf{h}}) + I_{xy}(\hat{\mathbf{X}} \cdot \hat{\mathbf{h}})(\hat{\mathbf{Z}} \cdot \hat{\mathbf{h}})$$

$$- I_{xz}(\hat{\mathbf{X}} \cdot \hat{\mathbf{h}})(\hat{\mathbf{Y}} \cdot \hat{\mathbf{h}}) + I_{yz}\big((\hat{\mathbf{Z}} \cdot \hat{\mathbf{h}})^2 - (\hat{\mathbf{Y}} \cdot \hat{\mathbf{h}})^2\big) \Big] + \hat{\mathbf{Y}} \Big[(I_{zz} - I_{xx})(\hat{\mathbf{X}} \cdot \hat{\mathbf{h}})(\hat{\mathbf{Z}} \cdot \hat{\mathbf{h}})$$

$$+ I_{xz}\big((\hat{\mathbf{X}} \cdot \hat{\mathbf{h}})^2 - (\hat{\mathbf{Z}} \cdot \hat{\mathbf{h}})^2\big) + I_{yz}(\hat{\mathbf{X}} \cdot \hat{\mathbf{h}})(\hat{\mathbf{Y}} \cdot \hat{\mathbf{h}}) - I_{xy}(\hat{\mathbf{Y}} \cdot \hat{\mathbf{h}})(\hat{\mathbf{Z}} \cdot \hat{\mathbf{h}}) \Big]$$

$$+\hat{\mathbf{Z}}\left[(I_{xx}-I_{yy})(\hat{\mathbf{X}}\cdot\hat{\mathbf{h}})(\hat{\mathbf{Y}}\cdot\hat{\mathbf{h}})+I_{xy}\left((\hat{\mathbf{Y}}\cdot\hat{\mathbf{h}})^2-(\hat{\mathbf{X}}\cdot\hat{\mathbf{h}})^2\right)\right.$$

$$\left.+I_{xz}(\hat{\mathbf{Y}}\cdot\hat{\mathbf{h}})(\hat{\mathbf{Z}}\cdot\hat{\mathbf{h}})-I_{yz}(\hat{\mathbf{X}}\cdot\hat{\mathbf{h}})(\hat{\mathbf{Z}}\cdot\hat{\mathbf{h}})\right] \qquad (17\text{-}41)$$

If $\hat{\mathbf{X}}$, $\hat{\mathbf{Y}}$, and $\hat{\mathbf{Z}}$ are principal body axes, then Eq. (17-41) reduces to

$$\langle\mathbf{N}_{GG}\rangle_0=\frac{3\mu}{2a^3\sqrt{(1-e^2)^3}}\left[\hat{\mathbf{X}}(I_{yy}-I_{zz})h_zh_y+\hat{\mathbf{Y}}(I_{zz}-I_{xx})h_xh_z+\hat{\mathbf{Z}}(I_{xx}-I_{yy})h_yh_x\right]$$

$$(17\text{-}42)$$

where h_x, h_y, and h_z are the components of the orbit normal unit vector along the principal axes.

From Eq. (17-42), we see that (1) if any principal axis is parallel to orbit normal, the secular gravity-gradient torque is zero and (2) if a principal axis is in the orbit plane, the secular gravity-gradient torque will be along that axis.

The secular gravity-gradient torque for a spin-stabilized satellite can also be calculated from Eq. (17-38). Substituting Eq. (17-35) into Eq. (17-38), the secular torque for a spinning satellite is given by

$$\langle\mathbf{N}_{GG_{spinning}}\rangle=\frac{3\mu\left[I_{zz}-\dfrac{(I_{xx}+I_{yy})}{2}\right]}{2\pi a^3\sqrt{(1-e^2)^3}}\int_0^{2\pi}(1+e\cos\nu)(\hat{\mathbf{R}}_s\cdot\hat{\mathbf{Z}})(\hat{\mathbf{R}}_s\times\hat{\mathbf{Z}})\,d\nu \quad (17\text{-}43)$$

Writing the unit vector $\hat{\mathbf{R}}_s$ in terms of the true anomaly as $\hat{\mathbf{R}}_s=\hat{\mathbf{p}}\cos\nu+\hat{\mathbf{q}}\sin\nu$ and assuming that $\hat{\mathbf{Z}}$, $\hat{\mathbf{p}}$, and $\hat{\mathbf{q}}$ are constant over one orbit, the average torque is

$$\langle\mathbf{N}_{GG_{spinning}}\rangle=\frac{3\mu\left[I_{zz}-\dfrac{(I_{xx}+I_{yy})}{2}\right]}{a^3\sqrt{(1-e^2)^3}}(\hat{\mathbf{h}}\cdot\hat{\mathbf{Z}})(\hat{\mathbf{Z}}\times\hat{\mathbf{h}}) \qquad (17\text{-}44)$$

From Eq. (17-44), we see that (1) the secular torque is perpendicular to $\hat{\mathbf{Z}}$ and therefore does not alter the magnitude of the angular momentum; (2) the gravity-gradient torque causes the spin axis to precess in a cone about the orbit normal with cone angle $\phi=\arccos(\hat{\mathbf{h}}\cdot\hat{\mathbf{Z}})$; and (3) the rate of precession of $\hat{\mathbf{Z}}$ is proportional to $\sin(2\phi)$ and therefore is a maximum at $\phi=45$ or 135 deg.

17.2.2 Solar Radiation Torque

Radiation incident on a spacecraft's surface produces a force which results in a torque about the spacecraft's center of mass. The surface is subjected to *radiation pressure* or force per unit area equal to the vector difference between the incident and reflected momentum flux. Because the solar radiation varies as the inverse square of the distance from the Sun, the solar radiation pressure is essentially altitude independent for spacecraft in Earth orbit. The major factors determining the radiation torque on a spacecraft are (1) the intensity and spectral distribution of the incident radiation, (2) the geometry of the surface and its optical properties, and (3) the orientation of the Sun vector relative to the spacecraft.

The major sources of electromagnetic radiation pressure are (1) solar illumination (Section 5.3), (2) solar radiation reflected by the Earth and its atmosphere, i.e., the Earth's albedo (Section 4.1), and (3) radiation emitted from the Earth and its atmosphere (Section 4.2). Of these sources, as shown in Table 17-1, direct solar radiation is the dominant source and is generally the only one considered. The force produced by the solar wind is also normally negligible relative to the solar radiation pressure (see Section 5.3).

Table 17-1.   Intensity of Radiation Sources for a Satellite Over the Subsolar Point Integrated Over All Wavelengths. (Data From NASA [1969b].)

ALTITUDE (km)	SOLAR RADIATION (W/m^2)	EARTH REFLECTANCE* (W/m^2)	EARTH RADIATION* (W/m^2)
500	1358	600	150
1,000	1358	500	117
2,000	1358	300	89
4,000	1358	180	62
8,000	1358	75	38
15,000	1358	30	14
30,000	1358	12	3
60,000	1358	7	2

* ASSUMING A SPHERICAL SPACECRAFT.

The mean momentum flux, P, acting on a surface normal to the Sun's radiation, is given by

$$P = \frac{F_e}{c} \qquad (17\text{-}45)$$

where F_e is the solar constant (see Section 5.3) and c is the speed of light. The solar constant is wavelength dependent and undergoes a small periodic variation for an Earth-orbiting spacecraft because of the eccentricity of the Earth's orbit about the Sun. If the momentum flux incident on the spacecraft's surface is known, Edwards and Bevans [1965] have shown that the reflected flux can be described analytically by the reflection distribution function and the directional emissivity. However, these properties of the irradiated surface are generally not known in sufficient detail to evaluate the required functions.

For most applications, the forces may be modeled adequately by assuming that incident radiation is either absorbed, reflected specularly, reflected diffusely, or some combination of these as shown in Fig. 17-2. Let P be the momentum flux incident on an elemental area dA with unit outward normal $\hat{N}$. (Each area consists of two surfaces with oppositely directed outward normal vectors.) *The differential*

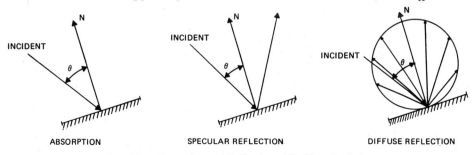

Fig. 17-2. Absorption and Reflection of Incident Radiation

radiation force (momentum transferred per unit time) due to that portion of the radiation that is completely absorbed is

$$\mathrm{df}_{absorbed} = -PC_a \cos\theta \hat{\mathbf{S}}\, \mathrm{d}A \qquad (0 \leqslant \theta \leqslant 90°) \qquad (17\text{-}46)$$

where $\hat{\mathbf{S}}$ is the unit vector from the spacecraft to the Sun, θ is the angle between $\hat{\mathbf{S}}$ and $\hat{\mathbf{N}}$, and C_a is the *absorption coefficient*. If $\cos\theta$ is negative, the surface is not illuminated and will not experience any solar force. The differential radiation force due to that portion of the radiation which is specularly reflected is

$$\mathrm{df}_{specular} = -2PC_s \cos^2\theta \hat{\mathbf{N}}\, \mathrm{d}A \qquad (0 \leqslant \theta \leqslant 90°) \qquad (17\text{-}47)$$

where the reflected radiation is in the direction $(-\hat{\mathbf{S}} + 2\hat{\mathbf{N}}\cos\theta)$. The *coefficient of specular reflection*, C_s, is the fraction of the incident radiation that is specularly reflected. For a diffuse surface, the reflected radiation is distributed over all directions with a distribution proportional to $\cos\phi$, where ϕ is the angle between the reflected radiation and $\hat{\mathbf{N}}$. The differential radiation force for diffusely reflected radiation is determined by integrating the contribution of the reflected radiation over all angles to obtain

$$\mathrm{df}_{diffuse} = PC_d\left(-\tfrac{2}{3}\cos\theta \hat{\mathbf{N}} - \cos\theta \hat{\mathbf{S}}\right)\mathrm{d}A \qquad (0 \leqslant \theta \leqslant 90°) \qquad (17\text{-}48)$$

where the *coefficient of diffuse reflection*, C_d, is the fraction of the incident radiation that is diffusely reflected. Assuming that absorption, specular reflection, and diffuse reflection all play a part (without any transmission), then the total differential radiational force is

$$\mathrm{df}_{total} = -P\int\left[(1 - C_s)\hat{\mathbf{S}} + 2\left(C_s\cos\theta + \tfrac{1}{3}C_d\right)\hat{\mathbf{N}}\right]\cos\theta\, \mathrm{d}A \qquad (17\text{-}49)$$

where $C_a + C_s + C_d = 1$. For surfaces that are not completely opaque, the incident momentum flux, P, can be modified to account for the radiation that does not impinge or interact with the surface. The differential radiation force can be written to include secondary reflections, but this is normally not a significant factor in the total radiation force [McElvain, *et al.*, 1966].

The solar radiation torque, $\mathbf{N}_{solar}$, acting on a spacecraft is given by the general expression

$$\mathbf{N}_{solar} = \int \mathbf{R} \times \mathrm{df}_{total} \qquad (17\text{-}50)$$

where $\mathbf{R}$ is the vector from the spacecraft's center of mass to the elemental area $\mathrm{d}A$. df_{total} is given by Eq. (17-49), and the integral is over the spacecraft's irradiated surface. Because of the difficulty in evaluating the radiation torque directly from Eq. (17-50) for arbitrary surfaces, the spacecraft configuration is frequently approximated by a collection of simple geometrical elements (e.g., plane, cylinder, sphere). The solar radiation force, $\mathbf{F}_i$, on each element is determined by evaluating the integral of Eq. (17-49) over the exposed surface area, that is,

$$\mathbf{F}_i = \int \mathrm{df}_{total\, i} \qquad (17\text{-}51)$$

Table (17-2) lists the solar radiation force $\mathbf{F}_i$ for some simple geometrical shapes. The torque on the spacecraft is the vector sum of the torques on the individual

Table 17-2. Solar Radiation Force for Some Simple Geometric Figures

GEOMETRIC FIGURE	FORCE
PLANE WITH SURFACE AREA A AND NORMAL $\hat{N}$: $\theta = \cos^{-1}(\hat{S}\cdot\hat{N})$	$-PA\cos\theta\left[(1-C_s)\hat{S}+2\left(C_s\cos\theta+\frac{1}{3}C_d\right)\hat{N}\right]$
SPHERE OF RADIUS r	$-P(4\pi r^2)\left(\frac{1}{4}+\frac{1}{9}C_d\right)\hat{S}$
RIGHT CIRCULAR CYLINDER OF RADIUS r, SYMMETRY AXIS $\hat{Z}$, AND HEIGHT h; ψ SUN ANGLE MEASURED FROM SYMMETRY AXIS, $A_1 = 2rh$, $A_2 = \pi r^2$	$-P\left(\left\{\left[\text{SIN }\psi\left(1+\frac{1}{3}C_s\right)+\frac{\pi}{6}C_d\right]A_1+\left(1-C_s\right)\cos\psi\,A_2\right\}\hat{S}\right.$ $+\left[\left(-\frac{4}{3}C_s\text{ SIN }\psi-\frac{\pi}{6}C_d\right)\cos\psi\,A_1+2\left(C_s\cos\psi+\frac{1}{3}C_d\right)\right.$ $\left.\left.\cos\psi\,A_2\right]\hat{Z}\right)$

elements composing the spacecraft irradiated surface, i.e.,

$$N_{solar} = \sum_{i=1}^{n} R_i \times F_i \qquad (17\text{-}52)$$

where R_i is the vector from the spacecraft center of mass to the center of pressure of the ith element. The *center of pressure* is at the intersection of the line of action of the single force which replaces the resultant radiation force and the plane passing through the center of mass of the spacecraft perpendicular to the line of action. The location of the center of pressure, r_{cp}, relative to the centroid of the geometrical sphere is given by

$$\int r \times df = r_{cp} \times F \qquad (17\text{-}53)$$

Solar radiation torques are reduced by the shadows cast by one part of the spacecraft on another. Shadowing reduces the total force and also shifts the center of pressure. The extent of shadowing is a function of the geometrical design of the spacecraft and the incident Sun angle. Examples of shadow modeling for DCSC II and AE-3 are given by Suttles and Beverly [1975] and Gottlieb *et al.*, [1974]. Although the shadow modeling for AE-3 was used to evaluate aerodynamic torque, the same method can be applied to solar radiation torque.

17.2.3 Aerodynamic Torque

The interaction of the upper atmosphere with a satellite's surface produces a torque about the center of mass. For spacecraft below approximately 400 km, the aerodynamic torque is the dominant environmental disturbance torque.

The force due to the impact of atmospheric molecules on the spacecraft surface can be modeled as an elastic impact without reflection [Beletskii, 1966]. The incident particle's energy is generally completely absorbed. The particle escapes after reaching thermal equilibrium with the surface with a thermal velocity equal to that of the surface molecules. Because this velocity is substantially less than that of the incident molecules, the impact can be modeled as if the incident particles lose their entire energy on collision. The force, df_{Aero}, on a surface element dA, with outward normal $\hat{N}$, is given by

$$df_{Aero} = -\frac{1}{2}C_D\rho V^2(\hat{N}\cdot\hat{V})\hat{V}\,dA \qquad (17\text{-}54)$$

where $\hat{\mathbf{V}}$ is the unit vector in the direction of the translational velocity, $\mathbf{V}$, of the surface element relative to the incident stream and ρ is the atmospheric density (Section 4.4). The parameter C_D is the *drag coefficient* defined in Section 3.4 and is a function of the surface structure and the *local angle of attack*, arc cos $(\hat{\mathbf{N}} \cdot \hat{\mathbf{V}})$ [Schaaf and Chambré, 1961]. For practical applications, C_D may be set to 2.0 if no measured value is available.

The aerodynamic torque $\mathbf{N}_{Aero}$, acting on the spacecraft due to the force $d\mathbf{f}_{Aero}$, is

$$\mathbf{N}_{Aero} = \int \mathbf{r}_s \times d\mathbf{f}_{Aero} \tag{17-55}$$

where $\mathbf{r}_s$ is the vector from the spacecraft's center of mass to the surface element dA. The integral is over the spacecraft surface for which $\hat{\mathbf{N}} \cdot \hat{\mathbf{V}} > 0$. Note that the translational velocity of element dA for a spacecraft spinning with angular velocity ω is

$$\mathbf{V} = \mathbf{V}_0 + \omega \times \mathbf{r}_s \tag{17-56}$$

where $\mathbf{V}_0$ is the velocity of the center of mass relative to the atmosphere. (Note that ω is relative to the rotation of the atmosphere which approximately equals the Earth's rotational rate.) Because the linear surface velocity due to the spacecraft spin is generally small compared to $\mathbf{V}_0$, second-order terms in ω can be neglected in substituting Eqs. (17-54) and (17-56) into Eq. (17-55). Thus, the total aerodynamic torque is

$$\mathbf{N}_{Aero} = \tfrac{1}{2} C_D \rho V_0^2 \int (\hat{\mathbf{N}} \cdot \hat{\mathbf{V}}_0)(\hat{\mathbf{V}}_0 \times \mathbf{r}_s) dA + \tfrac{1}{2} C_D \rho V_0 \int \left\{ \hat{\mathbf{N}} \cdot (\omega \times \mathbf{r}_s)(\hat{\mathbf{V}}_0 \times \mathbf{r}_s) \right.$$

$$\left. + (\hat{\mathbf{N}} \cdot \hat{\mathbf{V}}_0)[(\omega \times \mathbf{r}_s) \times \mathbf{r}_s] \right\} dA \tag{17-57}$$

The first term in Eq. (17-57) is the torque due to the displacement of the spacecraft's center of pressure from the center of mass. The second term is the dissipation torque due to the spacecraft spin. For a spacecraft in Earth orbit with $\omega r \ll V_0$, the second term is approximately four orders of magnitude smaller than the first and may be neglected.

The first term in Eq. (17-57) is evaluated in the same manner as the solar pressure torque. The surface area of the satellite is decomposed into simple geometric shapes and the total aerodynamic force is calculated by integrating Eq. (17-54) over the individual shapes. Table 17-3 lists the aerodynamic force for some simple geometric figures. The total torque about the center of mass of the spacecraft is the vector sum of the individual torques calculated by the cross product of the vector distance from the spacecraft's center of mass to the center of pressure of the geometric shapes and the force acting on the component.

Shadowing of one part of the spacecraft by another must also be considered in the torque evaluation. Because the aerodynamic torque increases as the spacecraft's altitude decreases, shadowing can be very important at low altitudes. The extent of shadowing is a function of the spacecraft's design and orientation relative to the velocity vector. Examples of shadowing models are given by Gottlieb, *et al.*, [1974] and Tidwell, [1970].

Table 17-3.   Aerodynamic Force for Some Simple Geometric Figures

GEOMETRIC FIGURES	FORCE
SPHERE OF RADIUS R	$-\frac{1}{2}C_D\rho V^2 \pi R^2 \hat{V}$
PLANE WITH SURFACE AREA A AND NORMAL UNIT VECTOR $\hat{N}$	$-\frac{1}{2}C_D\rho V^2 A(\hat{N}\cdot\hat{V})\hat{V}$
RIGHT CIRCULAR CYLINDER OF LENGTH L AND DIAMETER D, UNIT VECTOR $\hat{a}$ IS ALONG CYLINDER AXIS	$-\frac{1}{2}C_D\rho V^2 D L\sqrt{1-(\hat{a}\cdot\hat{V})^2}\,\hat{V}$

17.2.4 Magnetic Disturbance Torque

Magnetic disturbance torques result from the interaction between the spacecraft's residual magnetic field and the geomagnetic field. The primary sources of magnetic disturbance torques are (1) spacecraft magnetic moments, (2) eddy currents, and (3) hysteresis. Of these, the spacecraft's magnetic moment is usually the dominant source of disturbance torques. The spacecraft is usually designed of material selected to make disturbances from the other sources negligible. Bastow [1965] and Droll and Iuler [1967] provide a survey of the problems associated with minimizing the magnetic disturbances in spacecraft design and development.

The instantaneous magnetic disturbance torque, N_{mag} (in N·m), due to the spacecraft effective magnetic moment m (in A·m^2) is given by

$$N_{mag} = m \times B \tag{17-58}$$

where B is the geocentric magnetic flux density (in Wb/m^2) described in Section 5.1 and m is the sum of the individual magnetic moments caused by permanent and induced magnetism and the spacecraft-generated current loops. (See Appendix K for a discussion of magnetic units.)

The torques caused by the induced *eddy currents* and the irreversible magnetization of permeable material, or *hysteresis*, are due to the spinning motion of the spacecraft. Visti [1957] has shown that the eddy currents produce a torque which precesses the spin axis and also causes an exponential decay of the spin rate. This torque is given by

$$N_{Eddy} = k_e(\omega \times B) \times B \tag{17-59}$$

where ω is the spacecraft's angular velocity vector and k_e is a constant coefficient which depends on the spacecraft geometry and conductivity. Eddy currents are appreciable only in structural material that has a permeability nearly equal to that of free space. Table 17-4 lists values of k_e for simple geometric figures. Tidwell [1970] has outlined an alternative procedure for calculating the torque due to eddy current interaction which involves the evaluation of three different constant coefficients.

In a permeable material rotating in a magnetic field, H, energy is dissipated in the form of heat due to the frictional motion of the magnetic domains. The energy loss over one rotation period is given by

$$\Delta E_H = V \oint H \cdot dB \tag{17-60}$$

where V is the volume of the permeable material and dB is the induced magnetic induction flux in the material. The integral is over the complete path of the

Table 17-4.   Eddy Current Coefficients for Various Geometrical Figures. (Adapted From NASA [1969a].)

GEOMETRIC FIGURE	COEFFICIENT, k_e
THIN SPHERICAL SHELL OF RADIUS r THICKNESS d, AND CONDUCTIVITY σ	$\frac{2\pi}{3} r^4 \sigma d$
CIRCULAR LOOP OF RADIUS r AND CROSS-SECTIONAL AREA S LOCATED IN A PLANE CONTAINING THE SPIN AXIS	$\frac{\pi}{4} \sigma r^3 S$
THIN-WALLED CYLINDER WITH LENGTH l, RADIUS r, AND THICKNESS d	$\pi \sigma r^3 l d \left(1 - \frac{2d}{l} \text{ TANH } \frac{l}{2d}\right)$

hysteresis loop. The hysteresis effects are appreciable only in very elongated "soft" magnetic material (i.e., materials for which changes in the ambient field cause large changes in the magnetic moment). The torque due to the hysteresis is given by

$$\mathbf{N}_{Hyst} = \frac{\omega}{\omega^2} \frac{\Delta E_H}{\Delta t} \tag{17-61}$$

where Δt is the time over which the torque is being evaluated.

17.3 Modeling Internal Torques

Menachem Levitas

Internal torques are defined as torques exerted on the main body of a spacecraft by such internal moving parts as its reaction wheels, flexible booms or solar arrays, scanning or rastering instruments, tape recorder reels, liquids inside partially filled tanks, or astronauts inside a manned space station. In the absence of external torques, the total angular momentum of a spacecraft remains constant. However, internal torques can alter the system's kinetic energy and redistribute the spacecraft's angular momentum among its component parts in ways which can change its dynamic characteristics. For example, in a spinning spacecraft, angular momentum can be transferred from the nominal spin axis to another principal axis, resulting in nutation (Sections 15.2 and 16.3), uncontrolled tumbling [Thompson, 1964], or *flat spin* (spinning about a principal axis, other than the nominal spin axis; see Section 15.2 and Gebman, [1976]). These undesirable results are often best countered by attitude-stabilization systems based on other internally generated torques, such as gas jets (Section 6.8), nutation dampers (Section 18.4). reaction wheels (Sections 6.6 and 18.2), and other movable-mass stabilizing mechanisms [Childs, 1971; Childs and Hardison, 1974; Edward, 1974]. In this section we discuss three internal disturbance torques which alter the spacecraft attitude: (1) mass expulsion torques, (2) propellant slosh loads, and (3) the motion of internal hardware and astronauts. The effects of spacecraft flexibility are discussed in Section 16.4.

Mass Expulsion Torques.   Whenever mass is ejected from a spacecraft, disturbance torques result which can degrade the control system performance, lead to premature fuel depletion, or cause mission failure. Knowledge in this area has developed primarily from experience, when investigations of anomalous spacecraft behavior are traced to mass expulsion disturbance torques. An excellant summary is given by Schalkowski and Harris [1969]. Three design considerations are important in dealing with mass expulsion torques: identification of the sources and assessment of the torque magnitudes, determination of acceptable magnitudes, and control over design and development to ensure that the acceptable magnitudes will not be exceeded. Mass expulsion torques can be grouped into two major categories according to the nature of their sources:

1. *Unintentional control system torques.* These torques result from faulty design or equipment failure and include most of the mass expulsion disturbances identified to date. The most common are leakage of fuel or pressurizing agents,* thrust vector misalignment [Schalkowski and Harris, 1969], reaction forces resulting from plume impingement on the vehicle [Schalkowski and Harris, 1969; NASA, 1968; General Electric, 1964; and Victor, 1964], and anomalous thruster firing times [Schalkowski and Harris, 1969].

2. *Torques resulting from sources intended to expel mass.* These torques are natural byproducts of processes not intended to produce torque, such as dumping residual propellants [Schalkowski and Harris, 1969; MSFC, 1966], sublimation [Mobley and Fischell 1966], payload separation and ejection, and equipment jettison [Schalkowski and Harris, 1969]. Such processes occur infrequently, sometimes only once during the spacecraft lifetime. The associated disturbances cause problems only when they are overlooked or when their magnitude is underestimated.

The major problem associated with assessing the effects of mass expulsion disturbances is that of identifying the source. Once this has been done, testing or simulation may be used to determine the magnitude of the associated torque. Accurate analytic models are generally unavailable, but estimates of the upper bounds of various torques, based on test or simulation results, are usually sufficient. Due to obvious difficulties, direct measurements of mass expulsion torques are rarely made. Instead, tests are generally conducted on components to provide input data for torque calculations.

Although disturbance torques from jettisoned solids can be obtained analytically, ground testing of the ejection mechanism is normally used as a checking procedure. The separation impulse can also be computed from the photographed trajectory of the jettisoned object. Because the expelled mass is no longer regarded as part of the spacecraft, the effect of mass expulsion is to alter the "spacecraft's" total angular momentum, even though the torques are internally generated.

Propellant Slosh Loads. *Propellant sloshing* refers to free surface oscillations of a fluid in a partially filled tank resulting from translational or angular acceleration of the spacecraft caused by an attitude or orbit control system, elastic deformation of the vehicle, or an environmental disturbance. Once sloshing begins, it may persist for a long time due to the small damping effects of the tank walls unless damping devices, such as baffles, are provided. Propellant sloshing can result in attitude precession or nutation, spacecraft instability, or damage to the propellant tank.

The extent of propellant sloshing and the consequent forces on the spacecraft depend on the tank geometry, propellant properties, the effective damping, the

---

*See, for example, Schalkowski and Harris [1969], NASA Research Center Pioneer Project Office [1967], Massey [1968], Mariner-Mars 1964 Project Report MPR [1965], Dobrotin, *et al.*, [1969], Bourke, *et al.*, [1969], NASA [1968], and General Electric [1964].

height of the propellant in the tank, the acceleration field, and the perturbing motion of the tank [Langley Research Center, 1968]. The parameters which are normally adjustable include the tank structure and the damping devices. The tank geometry influences the natural sloshing frequency modes, the forced response, and the resulting pressure forces and torques acting on the tank. Baffles, as shown in Fig. 17-3, increase the effective fluid damping and thereby reduce the duration of the free oscillations and the magnitude of forced oscillations. Dynamic coupling between sloshing propellants and elastic structures may also have significant influence on the vibration frequencies and mode shapes of elastic tanks and can cause dynamic instabilities [Langley Research Center, 1968].

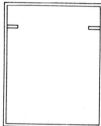

Fig. 17-3. Cross Section of a Cylindrical Tank With a Single-Ring Baffle to Dampen Propellant Sloshing

The dynamic response of vehicles to sloshing liquids is difficult to determine experimentally, especially in the case of large containers at low gravity [Dodge and Garza, 1967]. The major characteristic of low gravity is a small *Bond Number* which is proportional to the ratio of the weight of a unit depth of liquid to its surface tension. Small Bond Numbers can be simulated even at Earth gravity, but only for small containers. (Dodge and Garza, [1967] tested cylinders up to 3.3 cm in diameter.) Fortunately, the dynamic response of a vehicle can be determined analytically by representing the liquid dynamics by an equivalent mechanical system, consisting of fixed and oscillating masses connected to the tank by spring or pendulums and dashpots. This technique has been used with considerable success to derive the dynamic characteristics of sloshing liquids.* The analytical models are designed so that they have the same resultant pressure force, torque, damping, and frequency as the actual system. Procedures to determine the natural sloshing frequencies, mode shapes, and equivalent mechanical systems for axially symmetric tanks are described by Abramson, *et al.*, [1966], Lomen [1965b], Lawrence, *et al.*, [1958], Lomen [1965a], and Moiseev and Petrov [1966]. When used with similar representations for other spacecraft components, the vehicle dynamics can be calculated. When tanks become large, as in large space vehicles, the forces exerted by the propellent increase and sloshing occurs at lower frequencies which could cause serious stability problems. This can be overcome by subdividing the tanks into smaller compartments [Bauer, 1960].

Crew Motion. The effects of crew movements inside a spacecraft are difficult to predict accurately, chiefly because of the random nature of the movements.†

---

* Specific tank geometrics were studied by Dodge and Garza [1967], Abramson [1966], Lomen [1965b], Abramson, *et al.*, [1961], Bauer [1960, 1964], Rathayya [1965], Koelle [1961], and Dodge and Kana [1966].

† Although individual human motions may be random, the motions of astronauts inside a space vehicle do follow fixed statistical patterns.

Intuition and experience indicate, however, that resulting disturbances are directly proportional to the amplitude of the motion and the ratio of the human's mass to the spacecraft moment of inertia. This is illustrated in Fig. 17-4, which shows the X axis jitter rate (i.e., rate of angular deviation from the nominal direction) for Skylab (Fig. 17-5) due to the motion of the three astronauts. To provide protection against such jitter, the pointed experiment mounting package was decoupled from the main body of the spacecraft as much as possible. Equations of motion describing the dynamics of a vehicle containing an arbitrary number of moving parts (treated as point mass particles) were first developed by Roberson [1958a] and later by Grubin [1962]. Fang [1965], gives expressions for the kinetic energy and angular momentum about the variable center of mass, in terms of body-fixed coordinates. Each of the above assumes fixed masses confined to definite paths. Neither assumption, of course, is strictly valid with regard to astronauts.

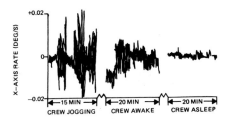

Fig. 17-4.   Jitter Caused by Crew Motion Onboard Skylab. (Adapted from Chubb, *et al.*, [1975].)

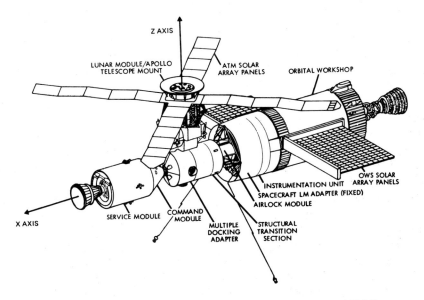

Fig. 17-5.   Skylab Spacecraft Configuration (drawing courtesy NASA)

The potential instability of spinning space stations under the influence of crew motion was studied by Thomson and Fung [1965]. They considered effects due to one or two point masses executing several types of circumferential and radial motions and concluded that an astronaut could rock a space station and cause it to tumble if the period of his motion is in the neighborhood of certain integral multiples of half the space station's spin period. The exact multiples vary with the type of motion, and the size of the neighborhood increases monotonically with the mass of the astronauts and with the amplitude of the motion. Poli [1971] concludes that when an astronaut executes a closed path motion onboard a space vehicle, the total angular momentum does not necessarily return to its original value in spacecraft coordinates—a fact which becomes clear when we observe that the astronaut can add mechanical energy to the system.

In contrast to the above deterministic works, Davidson and Armstrong [1971] investigated the effect of crew motion on spacecraft orientation from a probabilistic, random walk point of view. Recognizing that control systems consist of mass-expelling, or energy-consuming hardware, and that such hardware would be activated whenever a disturbance due to crew motion reached a certain value, the authors calculated how often stabilizing torques would be required and, hence, what the depletion rate of mass or the consumption rate of energy would be. They assumed that the crew motion followed a fixed statistical pattern and, therefore, that the use of frequency-versus-magnitude histograms of the crew's motion was legitimate. They used discrete matrix methods for limited motion and the diffusion equation in the case of large multimanned space stations. In the latter case, only the mean value and the variance of the histogram affected the outcome.

Internal Torques Produced by Moving Hardware. The motion of hardware components onboard a space vehicle is normally compensated for, such that the main body experiences no torques. In some cases, this compensation is straightforward; for example, in principle, every rotor can be balanced by an identical rotor moving in the opposite direction. In other cases, such as the Advanced Atmospheric Sounding and Imaging Radiometer (AASIR) to be flown on STORMSAT in 1982 [White, *et al.*, 1976], the motion may be complicated, requiring detailed numerical analysis to compute compensating commands to an independent torquing device, such as a magnetic coil or gas jets.

Cloutier [1975] gives a graphical technique which permits rapid evaluation of the effects of gimballed, stepping, and scanning devices on the spacecraft. Beard, *et al.*, [1974] describe how turning a tape recorder on and off affected the spin rate of SAS-2. Devices containing internal moving parts—whose primary function is to generate stabilizing torques, absorb mechanical energy, and damp nutation—are described by Childs [1971], Childs and Hardison [1974], Edward [1974], and in Section 18.4.

17.4 Modeling Torques Due to Orbit Maneuvers

Gyanendra K. Tandon

In this section we discuss the modeling of torques due to orbit maneuvers for a spin-stabilized spacecraft. The principal feature affecting the computation of this torque is the mounting configuration of the rocket used to perform the maneuver. In general, two kinds of engine mountings are used: (1) those for which the thrust vector is nominally collinear with the spin axis and (2) those for which the thrust vector is not collinear with the spin axis but nominally passes through the spacecraft center of mass.

17.4.1 Thrust Vector Collinear With the Spin Axis

This engine mounting configuration is normally used for large velocity changes such as those produced by the apogee boost motor used to change the elliptical transfer orbit into a near-circular orbit for geosynchronous satellites. This is the most desirable mounting for a spin-stabilized spacecraft because it has the following three distinct advantages over alternative mountings:

1. There is no loss of thrust due to the spacecraft spin.
2. The thrust vector always passes through the center of mass of the spacecraft if the fuel burns symmetrically and, therefore, no torque will be present.
3. The engine can be fired continuously.

However, if the thrust vector from the motor does not pass through the spacecraft center of mass, due to misalignments, then a disturbance torque will be generated which will cause the spacecraft to precess and nutate. This will affect the velocity change in two ways. First, the magnitude of the final velocity change will be reduced since a component of the thrust will be perpendicular to the new spin axis and will cancel out over a complete nutation period. Second, the resulting velocity change may be in the wrong direction, because the geometric z axis of the spacecraft will not be in the initial spin axis direction in inertial space throughout the engine firing. These errors in the magnitude and direction of the velocity vector will necessitate using more fuel for later orbital corrections and produce a corresponding reduction in the weight available for useful payload.

There are three potential angular misalignments and three offset misalignments which could lead to a torque being generated during the motor firing. Each misalignment can have both an x and a y component because the x and y axes of the spacecraft may not be equivalent. These misalignments are defined in Fig. 17-6.

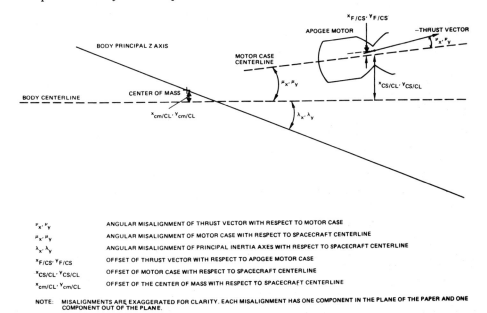

ν_x, ν_y	ANGULAR MISALIGNMENT OF THRUST VECTOR WITH RESPECT TO MOTOR CASE
μ_x, μ_y	ANGULAR MISALIGNMENT OF MOTOR CASE WITH RESPECT TO SPACECRAFT CENTERLINE
λ_x, λ_y	ANGULAR MISALIGNMENT OF PRINCIPAL INERTIA AXES WITH RESPECT TO SPACECRAFT CENTERLINE
$x_{F/CS}, y_{F/CS}$	OFFSET OF THRUST VECTOR WITH RESPECT TO APOGEE MOTOR CASE
$x_{CS/CL}, y_{CS/CL}$	OFFSET OF MOTOR CASE WITH RESPECT TO SPACECRAFT CENTERLINE
$x_{cm/CL}, y_{cm/CL}$	OFFSET OF THE CENTER OF MASS WITH RESPECT TO SPACECRAFT CENTERLINE

NOTE: MISALIGNMENTS ARE EXAGGERATED FOR CLARITY. EACH MISALIGNMENT HAS ONE COMPONENT IN THE PLANE OF THE PAPER AND ONE COMPONENT OUT OF THE PLANE.

Fig. 17-6. Definition of Misalignments for a Rocket Motor Nominally Aligned With the Spin Axis of a Spinning Spacecraft

An analytic model for the spacecraft motion during the engine firing, including the above six misalignments, can be developed with the following simplifying assumptions:

1. Rigid body dynamics are applicable.
2. The engine is a solid fuel motor and the fuel burns symmetrically about the motor case centerline.
3. The total spacecraft mass, moments of inertia, and the location of the center of mass in the spacecraft are linear functions of time during the motor firing.
4. The motor firing does not distort the spacecraft; i.e., the misalignments remain constant during the motor firing.
5. The exhaust gases carry away angular momentum equal to that of the fuel which was burned.

The last assumption is applicable to a motor which possesses a single, large, centrally mounted nozzle. A solid fuel motor of this type is generally used for large velocity changes. In this case, the exhaust gases spend so short an interval in the engine that they have no time to exchange any angular momentum with the spacecraft before being ejected and hence the spin rate of the spacecraft will not change if the alignments are correct. This is in agreement with the observed very small spin rate change during the apogee motor firing on CTS ($+0.4$ deg/s), GOES-1 ($+1.8$ deg/s), GOES-2 (-3.2 deg/s), and SIRIO ($+0.4$ deg/s) [Tandon and Smith (1976); Page (1975); Chen and McEnnan (1977)].

If the engine is different from the one discussed above, especially if it possesses more than one nozzle, an appropriate jet damping model should be used in place of assumption 5. The term *jet damping* refers to the phenomenon in which the rotation of the motor exhaust gases carries away a portion of the component of the spacecraft's angular momentum perpendicular to the nominal exhaust direction. This serves to damp the nutation induced by the motor firing. The jet damping theory is discussed by Thomson and Reiter [1965], Warner and Snyder [1968], Katz [1968], and Papis [1968]. The basic dynamics model consists of three sets of differential equations and an algebraic vector equation. These are summarized in vector form in Fig. 17-7. N_F is the portion of the torque, N, which is induced by the motor thrust, F, and N_J is the portion which models the effect of the angular momentum carried away by the exhaust gases. N_J will depend on the jet damping model used. The equation for N_J, using assumption 5, is

$$N_J = \dot{I}\omega \qquad (17\text{-}62)$$

where all vector quantities are resolved along the spacecraft principal axes. The detailed derivation of the equations in Fig. 17-7 is given by Keat and Shear [1974].

Assumptions 1 through 4, together with an approximate jet damping model in place of assumption 5, were used to simulate the performance of the CTS spacecraft during apogee motor firing by Keat and Shear [1974]. The signs of the misalignments were selected so that their effect was cumulative (i.e., the worst case for the combined effect of all of the misalignments was simulated). The results of the simulations indicated that for the nominal specified misalignments for the CTS spacecraft, the principal Z axis (the nominal spin axis) would wander up to 2 deg from its initial position in inertial space during the motor firing and this would cause a 0.5-deg error in the direction of the velocity change vector. The additional fuel needed to correct the effects of this directional error on the orbit would be

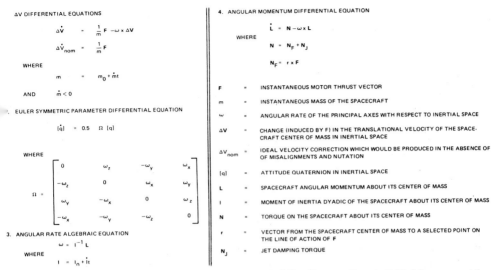

Fig. 17-7. Summary of the Dynamics Equations for Modeling Torques Due to Orbit Maneuvers. All vector quantities are resolved along spacecraft principal axes.

about 1.6 kg out of a total fuel budget of 11.2 kg for fine orbit correction maneuvers. The simulation runs with 10 times the nominal specified misalignments indicated that the effects would be 10 times larger. In view of the above results, the spacecraft hardware was aligned with extra care so that all the misalignments were within the nominal specified limits.

17.4.2 Thrust Vector Not Collinear With Spin Axis but Nominally Passing Through Spacecraft Center of Mass

This configuration is used for small velocity changes, where it is possible to tolerate some fuel wastage. In this configuration, the engine must be fired in a pulsed mode so that a net thrust in the desired direction is generated.

As the fuel is used, the spacecraft center of mass will move on the spin axis. The thrust vector will not always pass through the center of mass of the spacecraft and hence a torque will be generated which will cause the spacecraft to precess and nutate.

The effect of misalignments can be modeled similarly to that for the first configuration, the main difference being in the modeling of the angular momentum of the exhaust gases, because for small rocket engines, a liquid or gas fuel is normally used. This fuel must be moved from storage tanks to the engine before use, resulting in a change in the spacecraft moments of inertia before engine firing. In addition, the engine firing in a pulsed mode must be modeled.

References

1. Abramson, H. N., W. H. Chu, and G. E. Ramsleben, Jr., "Representation of Fuel Sloshing in Cylindrical Tanks by an Equivalent Mechanical Model," *Am. Rocket Society J.*, Vol. 31, p. 1967–1705, 1961.
2. Abramson, H. N., editor, *The Dynamic Behavior of Liquids in Moving Containers*, NASA SP-106, 1966.

3. Bastow, J. G., editor, *Proceedings of the Magnetic Workshop, March 30-April 1, 1965*, JPL Tech Memo 32-316, 1965.
4. Bauer, H. F., *Theory of the Fluid Oscillations in a Circular Cylindrical Ring Tank Partially Filled with Liquid*, NASA TN D-557, 1960.
5. ———, *Fluid Oscillations in the Containers of a Space Vehicle and Their Influence Upon Stability*, NASA TR R-187, 1964.
6. Beard, R. M., J. E. Kronenfeld, and E. Areu, *Small Astronomy Satellite-2 (SAS-2), Dynamic Attitude Determination System (DYNAD) Mathematical Models and Results from Processing*, Comp. Sc. Corp., 3000-33900-01TR, Dec. 1974.
7. Beletskii, V. V., *Motion of an Artifical Satellite About its Center of Mass*, NASA TT F-429, 1966.
8. Bourke, Roger D., Stephen R. McReynolds, and Kathryn L. Thuleen, "Translational Forces on Mariner V from the Attitude Control System," *J. Spacecraft*, Vol. 6, p. 1063-1066, 1969.
9. Butcher, J. C., "On the Attainable Order of Runge-Kutta Methods," *Math. Comp.* Vol. 19, p. 408–417, 1965.
10. Carnahan, Brice, H. A. Luther, and James O. Wilkes, *Applied Numerical Methods*, New York: John Wiley & Sons, Inc., 1969.
11. Chen, L. C. and J. J. McEnnan, *SIRIO Attitude Analysis Postlaunch Report*, Comp. Sc. Corp., CSC/TM-77/6264, Oct. 1977.
12. Childs, Dara W., "A Movable-Mass Attitude-Stabilization System for Artificial-g Space Station," *J. Spacecraft*, Vol. 8, p. 829–834, 1971.
13. ——— and Therman L. Hardison, "A Movable-Mass Attitude-Stabilization System for Cable-Connected Artificial-g Space Station," *J. Spacecraft*, Vol. 11, p. 165–172, 1974.
14. Chubb, W. B., H. F. Kennel, C. C. Rupp, and S. M. Seltzer, "Flight Performance of Skylab Attitude and Pointing Control System," *J. Spacecraft*, Vol. 12, p. 220–227, 1975.
15. Cloutier, Gerald J., "Elevation Stepping of Gimballed Devices on Rotor-stablized Spacecraft," *J. Spacecraft*, Vol. 12, p. 511–512. 1975.
16. Davidson, John R. and Robert L. Armstrong, "Effect of Crew Motion on Spacecraft Orientation," *AIAA J.*, Vol. 9, p. 232–238, 1971.
17. Dobrotin, B., E. A. Laumann, and D. Prelewicz, *Mariner Limit Cycles and Self-Disturbance Torques*, AIAA Paper No. 69–844, AIAA Guidance, Control, and Flight Mechanics Conference, Aug. 1969.
18. Dodge, Franklin T. and Daniel D. Kana, "Moment of Inertia and Damping of Liquids in Baffled Cylindrical Tanks," *J. Spacecraft*, Vol. 3, p. 153–155, 1966.
19. Dodge, F. T., and L. R. Garza, "Experimental and Theoretical Studies of Liquid Sloshing at Simulated Low Gravities," *J. Appl. Mech.*, Vol. 34, p. 555–562, 1967.
20. Droll, P. W. and E. J. Iuler, "Magnetic Properties of Selected Spacecraft Materials," *Proc., Symposium on Space Magnetic Exploration and Technology*, Engineering Report No. 9, p. 189–197, 1967.

21. Edward, Terry L. and Marshall H. Kaplan, "Automatic Spacecraft Detumbling by Internal Mass Motion," *AIAA J.*, Vol. 12, p. 496–502, 1974.

22. Edwards, D. K. and J. T. Bevans, "Radiation Stresses on Real Surfaces," *AIAA Journal*, Vol. 3, p. 522–523, 1965.

23. Enright, W. H., and T. E. Hull, "Test Results on Initial Value Methods for Non-Stiff Ordinary Differential Equations," *SIAM J. Numer. Anal.*, Vol. 13, p. 944–961, 1976.

24. Fang, Bertrand T., "Kinetic Energy and Angular Momentum About the Variable Center of Mass of a Satellite," *AIAA J.*, Vol. 3, p. 1540–1542, 1965.

25. Fish, V. R. and B. G. Chmielewski, *Flight Program Requirements Document for the High Energy Astronomy Observatory-B Attitude Control and Determination Subsystem (FPH-B)*, TRW Systems Group Doc. No. DO1137B, April 1977.

26. Gebman, Jean R. and D. Lewis Mingori, "Perturbation Solution for the Flat Spin Recovery of a Dual-Spin Spacecraft," *AIAA J.*, Vol. 14, p. 859–867, 1976.

27. General Electric Co., *Plume and Thrust Tests on Nimbus Attitude Control Nozzles*, Information Release 9461-135, May 22, 1964.

28. Gottlieb, D. M., C. M. Gray, and S. G. Hotovy, *An Approximate Shadowing Technique to Augment the Aerodynamic Torque Model in the AE-C Multi-Satellite Attitude Prediction and Control Program (MSAP/AE)*, Comp. Sc. Corp., 3000-257-01TM, Oct. 1974.

29. Grubin, C, "Dynamics of a Vehicle Containing Moving Parts," *J. of Applied Mech.*, Vol. 29, p. 486–488, 1962.

30. Hamming, R. W., *Numerical Methods for Scientists and Engineers*. New York: McGraw-Hill, Inc., 1962.

31. Henrici, P. H., *Discrete Variable Methods in Ordinary Differential Equations*. New York: John Wiley & Sons, Inc., 1962.

32. ———, *Elements of Numerical Analysis*. New York: John Wiley & Sons, Inc., 1964.

33. Hildebrand, F. B., *Introduction to Numerical Analysis*. New York: McGraw-Hill, Inc., 1956.

34. Holland, R. L. and H. J. Sperling, "A First-order Theory for the Rotational Motion of a Triaxial Rigid Body Orbiting an Oblate Primary," *Astronomical Journal*, Vol. 74, p. 490, 1969.

35. Hooper, L., Private Communication, 1977.

36. Hull, T. E., W. H. Enright, B. M. Fellen, and A. E. Sedgwick, "Comparing Numerical Methods for Ordinary Differential Equations," *SIAM J. Numer. Anal.*, Vol. 9, p. 603–637, 1972.

37. Hultquist, P. F., "Gravitational Torque Impulse on a Stabilized Satellite," *ARS Journal*, Vol. 31, p. 1506–1509, 1961.

38. Iwens, R. P. and R. Farrenkopf, *Performance Evaluation of a Precision Attitude Determination System (PADS)*, AIAA Paper No. 71-964, Guidance Control and Flight Mechanics Conference, Hofstra U., Hempstead, NY, Aug. 1971.

39. Katz, Paul, "Comments on 'A Re-Evaluation of Jet Damping'," *J. Spacecraft*, Vol. 5, p. 1246, Oct. 1968.

40. Keat, J. and M. Shear, *Apogee Motor Firing Dynamics Study for the Com-*

munications Technology Satellite, Comp. Sc. Corp, 3000-05600-08TN, May 1974.

41. Koelle, H. H., editor, *Handbook of Astronautical Engineering*. New York: McGraw-Hill Book Co., Inc., 1961.

42. Lambert. J. D., *Computational Methods in Ordinary Differential Equations*. New York: John Wiley & Sons, Inc., 1973.

43. Langley Research Center, *Propellant Slosh Loads*, NASA SP-8009, Aug. 1968.

44. Lawrence, H. R., C. J. Wang, and R. B. Reddy, "Variational Solution of Fuel Sloshing Modes," *Jet Propulsion*, Vol. 128, p. 729–736, 1958.

45. Lomen, D. O., *Liquid Propellant Sloshing in Mobile Tanks of Arbitrary Shape*, NASA CR-222, April 1965a.

46. ———, *Digital Analysis of Liquid Propellant Sloshing in Mobile Tanks with Rotational Symmetry*, NASA CR-230, May 1965b.

47. Mariner-Mars 1964 Project Report: *Mission and Spacecraft Development, vol. I: From Project Inception through Midcourse Maneuver*, JPL, Tech. Report 32740, March 1965.

48. Marshall Space Flight Center, *The Meteroid Satellite Project Pegasus*, First Summary Report, NASA TN D-3505, 1966.

49. Massey, W. A., *Pioneer VI Orientation Control System Design Survey*, Control System Laboratory, TRW, Report No. 06314-6006-R001, Rev. 1 (Contract NAS 12-110), Sept. 1968.

50. McElvain, R. J. and L. Schwartz, "Minimization of Solar Radiation Pressure Effects for Gravity-Gradient Stabilized Satellites," *J. Basic Eng.*, p. 444–451, June 1966.

51. Mobley, F. F. and R. E. Fischell, *Orbital Results from Gravity-Gradient Stabilized Satellite*, APL, Johns Hopkins U. Tech. Memo. TG-826, Oct. 1966. (Also available from NASA Ames Research Center as *Symposium on Passive Gravity Gradient Stabliazation*, p. 237, 1965.)

52. Moiseev, N. N. and A. A. Petrov, "The Calculation of Free Oscillations of a Liquid in a Motionless Container," *Advances in Applied Mech.* New York: Academic Press, Inc., Vol. 9, p. 91–154, 1966.

53. NASA, *Application Technology Satellite*, Vol. 1-8, Tech. Data Report. GSFC, NASA TM X-61130, 1968.

54. ———, *Spacecraft Magnetic Torques*, NASA SP-8018, March 1969a.

55. ———, *Spacecraft Radiation Torques*, NASA SP-8027, Oct. 1969b.

56. ———, Research Center Pioneer Project Office, *Pioneer VI Mission*, NASA Ames Research Center, Moffet Field, CA, May 22, 1967.

57. Nidey, R. A., "Gravitational Torque on a Satellite of Arbitrary Shape," *ARS Journal*, Vol. 30, p. 203–204, 1960.

58. Nidey, R. A., "Secular Gravitational Torque on a Satellite in a Circular Orbit," *ARS Journal*, Vol. 31, p. 1032, 1961.

59. Page, G., Private Communication, Nov. 1975.

60. Papis, T., "Comments on 'A Re-Evaluation of Jet Damping'," *J. Spacecraft*, Vo. 5, p. 1246–1247, Oct. 1968.

61. Poli, Corrado R., "Effect of Man's Motion on the Attitude of a Satellity," *J. Spacecraft*, Vol. 4, p. 15–20, 1971.

62. Ralston, A., *A First Course in Numerical Analysis*. New York: McGraw-Hill, Inc., 1965.

63. Rathayya, J. V., *Sloshing of Liquid in Axisymmetric Ellipsoidal Tanks*, AIAA Paper No. 65–114, Jan. 1965.

64. Roberson, R. E., "Torques on a Satellite Vehicle From Internal Moving Parts," *J. of Applied Mech.*, Vol. 25, Trans. ASME, Vol. 80, p. 196–200, 1958a.

65. ———, "Gravitational Torque on a Satellite Vehicle," *J. Franklin Institute*, Vol. 265, p. 13–22, 1958b.

66. ———, "Alternate Form of a Result by Nidey," *ARS Journal*, Vol. 31, p. 1292, 1961.

67. Schaaf, S. A. and P. L. Chambré, *Flow of Rarefied Gas*. Princeton, N.J.: Princeton University Press, 1961.

68. Schalkowski, S. and M. Harris, *Spacecraft Mass Expulsion Torques*, NASA SP-8034, Dec. 1969.

69. Suttles, T. E. and R. E. Beverly, *Model for Solar Torque Effects on DSCS II*, AAS/AIAA paper No. AAS 25-095, AAS/AIAA Astrodynamics Specialist Conference, Nassau, Bahamas, July 1975.

70. Tandon, G. K. and P. M. Smith, *Communications Technology Satellite (CTS) Postlaunch Report*, Comp. Sc. Corp., CSC/TM-76/6104, May 1976.

71. Thomson, W. T. and Y. C. Fung, "Instability of Spinning Space Stations Due to Crew Motion," *AIAA J.*, Vol. 3, p. 1082–1087, 1965.

72. Thomson, W. T. and G. S. Reiter, "Jet Damping of a Solid Rocket: Theory and Flight Results," *AIAA J.*, Vol. 3, p. 413–417, 1965.

73. Tidwell, N. W., "Modeling of Environmental Torques of a Spin-Stabilized Spacecraft in a Near-Earth Orbit," *J. Spacecraft*, Vol. 7, p. 1425–1435, 1970.

74. Victor, P. T., *Initial Stabilization Control Nozzle Plume Impingement Study*, General Electric Co., Data Memo 1:57, Oct. 1964.

75. Visti, J. P., *Theory of the Spin of a Conducting Satellite in the Magnetic Field of the Earth*. Ballistic Research Laboratories, Aberdeen, MD, Report No. 1020, July 1957.

76. Warner, G. C. and V. W. Snyder, "A Re-Evaluation of Jet Damping," *J. Spacecraft*, Vol. 5, p. 364–366, March 1968.

77. White, R. A., H. W. Robinson, and D. I. Berman, *STORMSAT Ground System Concept Study*, Comp. Sc. Corp. CSC/SD-76/6088, Nov. 1976.

78. Wilcox, J. C., "A New Algorithm for Strapped-Down Inertial Navigation," *IEEE Trans. on Aerospace and Electronic Systems*, Vol. AES-3, no. 5, p. 796–802, Sept, 1967.

79. Yong, K., *NASA Goddard Space Flight Center Sounding Rocket Division, MASS Program Documentation*, AVCO Systems Division: Seabrook, MD, Aug. 1974.

CHAPTER 18

ATTITUDE STABILIZATION

18.1 Automatic Feedback Control
18.2 Momentum and Reaction Wheels
       Momentum Bias Control Systems, Reaction Wheel Systems
18.3 Autonomous Attitude Stabilization Systems
       Inertially Referenced Spacecraft (HEAO-1), Earth-Referenced Spacecraft
18.4 Nutation and Libration Damping
       Passive Nutation Damping, Active Nutation Damping, Libration Damping

Chapters 18 and 19 describe the various techniques used for attitude control. These techniques may be divided into two categories. *Attitude stabilization*, discussed in this chapter, consists of maintaining an existing orientation. *Attitude maneuver control*, discussed in Chapter 19, consists of reorienting the spacecraft from one attitude to another. Although this is a convenient categorization for analysis, the two areas are not totally distinct. For example, we include in attitude stabilization the process of maintaining one axis toward the Earth, which implies a continuous change in the inertial orientation.

Section 18.1 introduces the principles of control theory, derived largely from electrical engineering. Section 18.2 then describes the general principles of inertial guidance and reaction wheel control. Section 18.3 provides several specific examples of attitude stabilization systems. Finally, Section 18.4 describes both active and passive methods of nutation damping.

18.1 Automatic Feedback Control

Jawaid Bashir
Gerald M. Lerner

Feedback, or *closed loop control*, is the process of sensing a system parameter to control its value—for example, using a thermostat to control the temperature of a room by regulating the operation of a furnace in response to a changing environment. Automatic feedback control is used for attitude control of many spacecraft. Using feedback control, commands to generate control torques are automatically issued to correct the spacecraft attitude whenever it has been sufficiently perturbed. Typically, the control torques are implemented by mass expulsion devices such as jets, momentum storage devices such as reaction wheels, or magnetic coils.

A *block diagram* is a convenient schematic representation of either a physical system or the set of mathematical equations characterizing its components. Figure 18-1 is a typical spacecraft attitude control system block diagram. The blocks are the *transfer elements* which represent functional relationships between the various inputs and outputs. The operations of addition and subtraction are represented by a small circle, called a *summing point*. The output of the summing point is the algebraic sum of the inputs, each with its appropriate algebraic sign. A typical input for a three-axis stabilized spacecraft is a disturbance torque and the output is an error signal indicating the deviations between the desired and the actual values. The *plant* is that part of the control system which needs to be controlled, i.e., the spacecraft dynamics. The dynamic characteristics of the plant are generally determined by the specific hardware used. The *disturbances* are external torques which affect overall system performance. They can be either deterministic or random in nature. For example, gravity-gradient and magnetic torques on the spacecraft are deterministic in the sense that they are known functions of the spacecraft position and orientation. In contrast, the torques produced by the impact of meteoroids are randomly distributed [Levinson, 1977]. The output of the system, θ_M, is measured, processed by the *feedback* loop of the control system, and compared with a reference or desired value, θ_{REF}, to obtain an *error signal*, $\theta \equiv \theta_M - \theta_{REF}$. The error signal is processed by the *controller* to generate a control torque to counter the effect of the *input* disturbance torque and thus control the *output* θ_M near θ_{REF} (or θ near zero). For convenience, we will normally assume that θ_M is a measured attitude angle (although, in practice, it is usually a time or voltage) and set $\theta_{REF} = 0$, so that $\theta = \theta_M$.

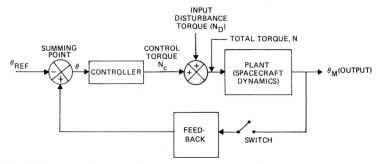

Fig. 18-1.   Block Diagram of a General Spacecraft Attitude Feedback Control System. The system output is a measured angle θ_M which is to be controlled near a reference or desired value θ_{REF}. The controller issues a torque based on the error signal, $\theta = \theta_M - \theta_{REF}$, to control the effect of the disturbance torques on the spacecraft dynamics.

If the switch in Fig. 18-1 is open, we have an *open-loop system* in which the controller response is independent of the actual output. For example, the issuance of magnetic control commands from the ground is an open-loop procedure. Conversely, if the switch is closed, we have a *closed-loop or feedback system* in which the input to the controller is modified based on information available from the actual output. For example, for the attitude control of a three-axis stabilized, Earth-oriented spacecraft, we may continuously monitor pitch and roll angles (and often rates) by attitude sensors (and gyros) and provide this information to the *controller*, which computes commands according to a control law and issues these commands to a torquing device or *actuator*. A *control law* is a principle on which the controller is designed to achieve the desired overall system performance.

The input-output relation of each element of the control system (i.e., the controller, plant, or feedback) is generally defined in terms of a *transfer function* (see Section 7.4). This idea of representing a physical system is a natural outgrowth of Laplace transform operational methods to solve linear differential equations (see Appendix F). The transfer function of each system element is defined as the ratio of the Laplace transform of its output to the Laplace transform of the input, assuming that all initial conditions are zero. Generally, the transfer function is represented as the ratio of two polynomials in s, as

$$G(s) = \frac{n(s)}{d(s)} = \frac{a_{m+1}s^m + a_m s^{m-1} + \cdots + a_1}{s^n + b_n s^{n-1} + \cdots + b_1}$$

The m values of s, for which $n(s)$ is zero, are known as the *zeros* of $G(s)$ and the n values of s, for which $d(s)$ is zero, are known as the *poles* of $G(s)$. The transfer function, $G(s)$, thus has m zeros and n poles.

The transfer function of the plant element may be obtained by taking the Laplace transform of the equation which describes the system dynamics. For example, if the equation describing the plant is

$$I\ddot{\theta} = N \tag{18-1}$$

where I is a constant, it may be transformed to obtain

$$Is^2 \mathcal{L}(\theta) = \mathcal{L}(N) \tag{18-2}$$

Thus, the transfer function, $G(s)$, for the plant described by Eq. (18-1) is

$$G(s) \equiv \frac{\mathcal{L}(\text{output})}{\mathcal{L}(\text{input})} = \frac{\mathcal{L}(\theta)}{\mathcal{L}(N)} = \frac{1}{Is^2} \qquad \text{(plant)} \tag{18-3}$$

The transfer function of the feedback element commonly describes a filtering, smoothing, or calibration of the sensed output signal; however, in this section we will assume that the measured and reference angles are compared directly and thus the feedback transfer function is unity.

The transfer function of the controller is obtained by first relating the control torque to the error signal in terms of a control law. The simplest of the control laws is *proportional control*, for which

$$N_C = -K\theta \tag{18-4}$$

where N_C is the control torque and K is the *system gain*. Proportional control is rarely used because it results in large oscillations in θ.

A common method for spacecraft attitude control is a *position-plus-rate control law* for which

$$N_C = -K_1\dot{\theta} - K_2\theta \tag{18-5}$$

Here, the control torque, N_C, is directly proportional to the error signal and its time derivative. The $K_1\dot{\theta}$ term provides damping. However, more sophisticated instruments, such as rate gyros, are needed to implement this control law. The transfer function for this controller is

$$G(s) \equiv \frac{\mathcal{L}(\theta)}{\mathcal{L}(N_C)} = \frac{-1}{K_1 s + K_2} \qquad \text{(controller)} \tag{18-6}$$

As an example, we consider the pitch control of a spacecraft with a reaction wheel with its axis along the pitch axis using a position-plus-rate control law. The equation for the pitch angle, θ, is,

$$N_D = \frac{d}{dt} L \equiv \frac{d}{dt} (I\dot{\theta} + h) = I\ddot{\theta} + \dot{h} \tag{18-7}$$

where L is the total angular momentum (wheel plus spacecraft body), h is the wheel momentum, N_D is the disturbance torque, $N_C = -\dot{h}$ is the control torque which alters the speed of the reaction wheel, and I is the moment of inertia of the spacecraft about the pitch axis. It is convenient to rewrite the position-plus-rate control law in the form

$$N_C = -\dot{h} = -K(\tau\dot{\theta} + \theta) \tag{18-8}$$

where τ is the *lead time constant* and K is the *pitch gain*. Equation (18-7) then yields

$$N_D = I\ddot{\theta} + K\tau\dot{\theta} + K\theta \tag{18-9}$$

This is a simple second-order differential equation. Using Table F-1 to take the Laplace transform of Eq. (18-5), we obtain

$$\mathcal{L}(N_D) = \left[Is^2 + K\tau s + K \right] \mathcal{L}(\theta) \tag{18-10}$$

The block diagram of this system, described by the plant of Eq. (18-3), is shown in Fig. 18-2, and the transfer function of this closed-loop system is

$$\frac{\mathcal{L}(\theta)}{\mathcal{L}(N_D)} = \frac{1}{Is^2 + K\tau s + K} \tag{18-11}$$

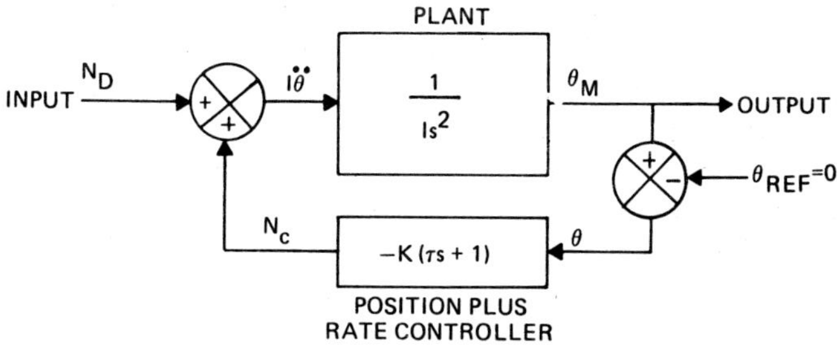

Fig. 18-2. Position-Plus-Rate Pitch Control Block Diagram. See text for explanation.

Comparing Eq. (18-9) with the second-order equation of a mass-spring-damper system (see, for example, Melsa and Shultz [1969]), we define the *natural frequency*, ω_n, and the *damping ratio*, ρ, of our system as

$$\omega_n \equiv \sqrt{\frac{K}{I}} \qquad \rho \equiv \frac{\tau}{2}\sqrt{\frac{K}{I}} \tag{18-12}$$

and rewrite Eq. (18-11) as

$$\frac{\mathcal{L}(\theta)}{\mathcal{L}(N_D)} = \frac{1/I}{s^2 + 2\rho\omega_n s + \omega_n^2} \tag{18-13}$$

We now discuss the response of this system when the input disturbance torque is a step function. Because the Laplace transform of a step function of magnitude N_0 is N_0/s (see Appendix F), Eq. (18-13) reduces to

$$\mathcal{L}(\theta) = \frac{N_0/I}{s(s^2 + 2\rho\omega_n s + \omega_n^2)} \tag{18-14}$$

This may be rewritten as the sum of partial fractions, to obtain for $\rho < 1$

$$\mathcal{L}(\theta) = \frac{N_0}{K}\left\{\frac{1}{s} - \left[\frac{s + 2\rho\omega_n}{(s + \rho\omega_n)^2 + \omega_1^2}\right]\right\} \tag{18-15}$$

where

$$\omega_1 = \omega_n\sqrt{1 - \rho^2}$$

Using the inverse Laplace transforms listed in Table F-1, we can obtain the time response of the pitch angle as

$$\theta(t) = \frac{N_0}{K}\left[1 - (1 - \rho^2)^{-1/2}\exp(-\rho\omega_n t)\sin(\omega_1 t + \psi)\right] \tag{18-16}$$

where $\psi = \arctan[(1 - \rho^2)^{1/2}/\rho] = \arccos(\rho)$.

Figure 18-3 shows a plot of the system response to a step function assuming that $N_0/K = 1$. The shape of the response curve depends on the damping ratio, ρ, and the time scale is determined by the natural frequency, ω_n. When $\rho = 0$, the system is called *undamped* and undergoes a bounded sinusoidal oscillation. As ρ

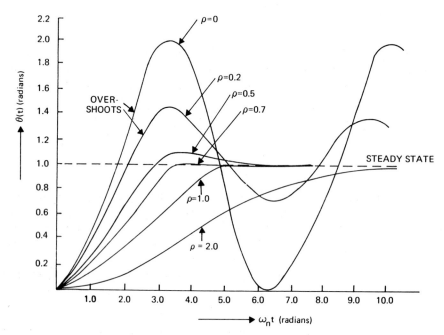

Fig. 18-3.   Time Response of a Simple Second-Order System With $N_0/K = 1$. See text for explanation.

increases, the *overshoot* and the number of oscillations decrease, and the system eventually attains a steady-state value equal to N_0/K. The gain K is chosen to achieve a specified steady-state error for an assumed magnitude of the disturbance torque, N_0. When $\sqrt{2}/2 \leqslant \rho < 1$, there is only one overshoot and no *undershoots*. When $\rho > 1$, the system is *overdamped* and acts as a simple first-order system. If $\rho = 1$, the system is *critically damped*. In many applications, overshoots are undesirable. However, if we choose a value of $\rho > 1$, the response of the system is slow; therefore, we will consider the value of $\rho = 1$ (critically damped).* For this case, Eq. (18-13) reduces to

$$\frac{\mathcal{L}(\theta)}{\mathcal{L}(N_D)} = \frac{1}{I(s+\omega_n)^2} \qquad (18\text{-}17)$$

The performance of a control system is generally expressed in terms of acceptable steady-state error for a specified disturbance. The *steady-state error* is defined as the difference between the desired output, θ_{REF}, and the actual output. The maximum steady-state error is determined using the final value theorem (see Appendix F) as

$$\theta(\infty) = \lim_{s \to 0} (s\mathcal{L}(\theta)) \qquad (18\text{-}18)$$

As an example of pitch control design, we will consider a solar radiation pressure torque of the order of 10^{-8} N·m (typical for MMS satellites). The steady-state error may be calculated using Eqs. (18-13) and (18-18) with $\mathcal{L}(N_D) = 10^{-8}s$ as

$$\theta(\infty) = \lim_{s \to 0} \left[s \frac{10^{-8}\,\text{N}\cdot\text{m}}{Is(s^2 + 2\rho\omega_n s + \omega_n^2)} \right] = \frac{10^{-8}}{I\omega_n^2} = \frac{10^{-8}}{K}\,\text{radians} \qquad (18\text{-}19)$$

where K is in N·m. Using this expression, the value of the pitch gain is chosen so that the steady-state error is within the given constraints. Having chosen the pitch gain, we can then calculate the lead time constant, τ, of the pitch control system to achieve a desired damping ratio from Eq. (18-12). The value of τ so determined should be significantly smaller than the orbital period of the satellite.

A third common control law is *bang-bang control* defined by

$$N_C = \frac{-\theta}{|\theta|} N_{max} = -N_{max}\text{sign}\,\theta \qquad (18\text{-}20)$$

where N_{max} is the maximum control torque and θ is the angular error. An example of this law is the attitude control of a spacecraft using jets to apply a constant torque in a direction to null the attitude error. The block diagram for a bang-bang control system is shown in Fig. 18-4. The control torque depends only on the sign of the difference between the desired and the actual output.

A block diagram for a *bang-bang-plus-dead zone* controller is shown in Fig. 18-5. Here the control torque is characterized by a dead zone followed by a

---

*The damping ratio $\rho = \sqrt{2}/2$ is frequently chosen because of its desirable frequency response characteristics. See Section 18.4 and DiStefano, *et al.*, [1967].

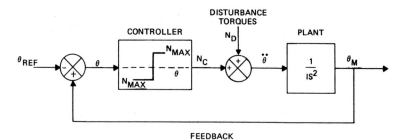

Fig. 18-4. Block Diagram for a Bang-Bang Control System

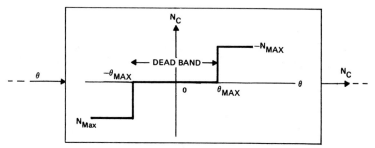

Fig. 18-5. Bang-Bang Control System With Deadband Controller

maximum torque. The law in functional form is

$$N_C = f(\theta)N_{max} \qquad (18\text{-}21)$$

where

$$f(\theta) \equiv -1 \text{ for } \theta > \theta_{max}$$

$$\equiv \quad 0 \text{ for } -\theta_{max} \leqslant \theta \leqslant \theta_{max}$$

$$\equiv \quad 1 \text{ for } \theta < -\theta_{max}$$

and θ_{max} is the half-width of the deadband.

System Stability.   The purpose of any feedback control system is to maintain a definite and known relationship between the desired output and the acutal output of the system. To achieve this goal, the system must respond to any temporary disturbances by eventually decaying to its desired or *steady-state* value. A linear system is *stable* if its output remains bounded for every bounded input. System stability can be investigated by applying a unit step function disturbance torque to the system in steady state and examining the output as time advances. If the variation of the output about the initial steady-state value approaches zero, the system is stable. If the output increases indefinitely with time, the system is *unstable*. If the output undergoes continuous bounded oscillations, the system is *marginally stable*. Finally, if the output attains a constant value other than the initial steady-state one, the system has *limited stability*.

In general, the stability of a system may be deduced by examining the poles of its closed-loop transfer function shown schematically in Fig. 18-6. An examination

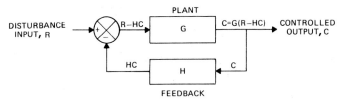

Fig. 18-6.   Schematic Block Diagram of a Closed-Loop Control System

of the figure yields

$$(R - HC)G = C$$

from which the general expression for the closed-loop transfer function is obtained as

$$\frac{C}{R} = \frac{G}{1 + GH} \qquad (18\text{-}22)$$

Comparing Figs. 18-2 and 18-6, we obtain for the previous example of the pitch control system:

$$R = \mathcal{L}(N_D)$$

$$C = \mathcal{L}(\theta)$$

$$G = 1/Is^2$$

$$H = K(\tau s + 1)$$

Hence, the closed-loop transfer function is

$$\frac{C}{R} = \frac{\mathcal{L}(\theta)}{\mathcal{L}(N_D)} = \frac{G}{1 + GH} = \frac{1}{Is^2 + K(\tau s + 1)} \qquad (18\text{-}23)$$

which agrees with Eq. (18-11). The poles of the transfer function are given by the zeros of

$$Is^2 + K(\tau s + 1) \equiv (s - \omega_+)(s - \omega_-) \qquad (18\text{-}24a)$$

$$\omega_\pm \equiv \left[-K\tau \pm \sqrt{K^2\tau^2 - 4KI} \right]/2I \qquad (18\text{-}24b)$$

The necessary and sufficient criterion for system stability is that all of the poles of the closed-loop transfer function lie in the left half of the complex s-plane; i.e., for the above example the requirement is

$$\text{Re}(\omega_\pm) < 0 \qquad (18\text{-}25)$$

where $\text{Re}(s)$ denotes the real part of s.

The general relationship between the location of the poles of the transfer function and the system stability may be determined by considering the linear differential equation

$$A\theta(t) = N(t) \qquad (18\text{-}26)$$

where the linear operator A with constant coefficients a_i is

$$A = \sum_{i=0}^{n} a_i \frac{d}{dt^i} \qquad (18\text{-}27)$$

and $N(t)$ is a disturbance torque. Substituting a trial solution of the form

$$\theta(t) = \exp(\omega t) \tag{18-28}$$

into Eq. (18-26) with $N(t) = 0$, we obtain the *characteristic equation*

$$a_n \omega^n + a_{n-1} \omega^{n-1} + \cdots + a_1 \omega + a_0 = 0 \tag{18-29}$$

The complementary solution to Eq. (18-26), i.e., $A\theta_C = 0$, is

$$\theta_C(t) = \sum_{i=1}^{n} C_i \exp(\omega_i t) \tag{18-30}$$

where the n values of ω_i are the roots or zeros of the characteristic equation. Thus, a necessary condition for system stability (i.e., for the $\lim_{t \to \infty} \theta_c(t)$ to exist) is that, for all i, $\mathrm{Re}(\omega_i) < 0$.

The question of stability has thus been reduced to investigating the characteristics of the roots of Eq. (18-29). A pure imaginary root, i.e., $\mathrm{Re}(\omega_i) = 0$, results in an undamped oscillatory component of the solution, while a root with a positive real part results in an exponentially increasing component of the solution. All of the roots of Eq. (18-29) have negative real parts if and only if the *Routh-Hurwitz criteria* [Korn and Korn, 1968] are satisfied. These are

1. $a_i > 0$ for all i $\hspace{4cm}$ (18-31a)

2. Either all the even or all the odd T_i, $i \leqslant n$, defined below, are positive.

$$T_0 = a_n \tag{18-31b}$$

$$T_1 = a_{n-1} \tag{18-31c}$$

$$T_2 = \begin{vmatrix} a_{n-1} & a_n \\ a_{n-3} & a_{n-2} \end{vmatrix} \tag{18-31d}$$

$$T_3 = \begin{vmatrix} a_{n-1} & a_n & 0 \\ a_{n-3} & a_{n-2} & a_{n-1} \\ a_{n-5} & a_{n-4} & a_{n-3} \end{vmatrix} \tag{18-31e}$$

$$T_4 = \begin{vmatrix} a_{n-1} & a_n & 0 & 0 \\ a_{n-3} & a_{n-2} & a_{n-1} & a_n \\ a_{n-5} & a_{n-4} & a_{n-3} & a_{n-2} \\ a_{n-7} & a_{n-6} & a_{n-5} & a_{n-4} \end{vmatrix} \tag{18-31f}$$

and so on.

A number of other methods of determining system stability—such as the *Nyquist criterion*, and root locus diagrams—have been developed in the last three decades (see, for example, Melsa and Shultz [1969] and Greensite [1970]). The most common of these is the *root locus diagram*, which is a plot in the complex s-plane of all possible locations of the roots of the characteristic equation of the system's *closed-loop transfer* function as the gain, K, is increased from zero to infinity.

Let GH, the *open-loop transfer function*, be represented as the ratio of two polynomials in s:

$$GH = \frac{K(s^m + a_{n-1}s^{m-1} + \cdots + a_0)}{s^n + b_{n-1}s^{n-1} + \cdots + b_0} \equiv \frac{Kn(s)}{d(s)} \equiv KB(s) \qquad (18\text{-}32)$$

where K is the system gain. Then the closed-loop transfer function is

$$\frac{C}{R} = \frac{G(s)}{1 + K\dfrac{n(s)}{d(s)}} = \frac{G(s)}{1 + KB(s)} = \frac{d(s)G(s)}{d(s) + Kn(s)} \qquad (18\text{-}33)$$

The poles of the closed-loop transfer function are the roots of $d(s) + Kn(s) = 0$. As the value of K changes, the location of these roots in the complex s-plane also changes. A root locus diagram is the locus of these roots as a function of K. The locus of a particular root is a *branch* on the root locus diagram. For $K = 0$, the roots of the characteristic equation are the roots of $d(s) = 0$, that is, the poles of the open-loop transfer function. As K increases from zero to infinity, these roots approach the roots of $n(s)$, i.e., the zeros of the open-loop transfer function. Therefore, as the value of K increases from zero to infinity, the loci of the poles of the closed-loop transfer function start at the open-loop poles and terminate at the open-loop zeros. If, for a given K, none of the roots of the characteristic equation has positive real parts, then the system is stable.

A set of general rules for constructing and interpreting root locus diagrams follows:

1. The number of loci, or branches of the root locus, is equal to the number of poles of the open-loop transfer function, $GH = KB(s)$.

2. The root loci are continuous curves. The slopes of the root loci are also continuous except for points at which either $dB(s)/ds = 0$, $K = 0$, or $B(s)$ is infinite.

3. Loci begin at poles of $B(s)$ where $K = 0$, and terminate at zeros of $B(s)$, where K is infinite.

4. If the open-loop transfer function, $KB(s)$, has p finite poles and z finite zeros, there will also be $p - z$ zeros at infinity if $p \geqslant z$.

5. For a branch of the root locus diagram to pass through a particular value of s—say, s_1—s_1 must be one of the roots of the characteristic equation $d(s_1) + Kn(s_1) = 0$ for some real value of K. The condition for which s_1 is the root of the characteristic equation is that $B(s_1)$ must have a phase angle and magnitude given by

$$|B(s_1)| = \left|\frac{1}{K}\right|$$

$$\arg B(s_1) = \begin{cases} (2l+1)\pi \text{ radians}, & K > 0 \\ 2l\pi \text{ radians}, & K < 0 \end{cases} \qquad (18\text{-}34)$$

where l is an arbitrary integer. To satisfy Eq. (18-34), the magnitude of $KB(s_1)$ must be equal to unity, and its associated phase angle must be an odd multiple of π radians. These two criteria are known as *magnitude* and *angle criteria*, respectively.

6. Branches of the root locus are symmetrical with respect to the real axis because all complex roots appear in conjugate pairs.

7. For $p > z$ and $|s| \gg 0$, branches of the root locus approach a set of asymptotic straight lines. The asymptotes to the loci at infinity meet at the centroid of $B(s)$ given by

$$s_C = -\frac{\sum\limits_{i=1}^{p} p_i - \sum\limits_{i=1}^{z} z_i}{p - z} \qquad (18\text{-}35a)$$

where p_i and z_i denote the ith pole and zero of $KB(s)$. The angles between the asymptotes and the real axis are

$$\theta_l = \begin{cases} \dfrac{(2l+1)\pi}{p-z} \text{ radians,} & K > 0 \\[2mm] \dfrac{2l\pi}{p-z} \text{ radians,} & K < 0 \end{cases} \qquad (18\text{-}35b)$$

where $l = 0, 1, 2, \ldots, p - z - 1$.

9. A *breakaway point* s_B is a point on the real axis where two or more branches arrive or depart. This point is calculated by solving the equation $dB(s)/ds = 0$ and calculating its roots.

As an example of the above rules, we construct the root locus diagram shown in Fig. 18-7 of a feedback system whose open-loop transfer function is given by

$$GH = KB(s) = \frac{K}{(s+2)(s+3)(s+4)} \qquad (18\text{-}36)$$

We first determine the poles and zeros of GH. For this function, there is no zero and there are three poles on the real axis at $s = -2$, -3, and -4, marked by crosses in the figure. Because there are no zeros, the branches are asymptotic to straight lines at infinity. The center of these asymptotes is at

$$s_C = \frac{\sum\limits_{i=1}^{p} p_i - \sum\limits_{i=1}^{z} z_i}{p - z} = \frac{-2-3-4}{3} = -3$$

The asymptotes make an angle θ with the real axis where

$$\theta = \frac{(2l+1)180°}{p-z} = \frac{(2l+1)180°}{3}, \qquad (K > 0, l = 0, 1, 2)$$

$$= 60°, 180°, \text{ and } 300°$$

and

$$\theta = \frac{360°l}{p-z} = \frac{360°l}{3}, \qquad (K < 0, l = 0, 1, 2)$$

$$= 0°, 120°, \text{ and } 240°$$

The breakway point is determined by

$$\frac{dB(s)}{d(s)} = 0$$

that is,

$$\frac{d}{ds}(s^3 + 9s^2 + 26s + 24) = 0$$

or

$$3s^2 + 18s + 26 = 0$$

with the solutions

$$s_1 = -3.58, \qquad s_2 = -2.45$$

For a system with positive gain, as K increases from $K=0$, the root at $s=(-2,0)$ moves to the left along the real axis until it reaches the breakaway point, $s_B = s_2$, where it becomes complex, moves into the second quadrant, and approaches the $60°/240°$ asymptote as $K \to \infty$. Similarly, the root at $s=(-3,0)$

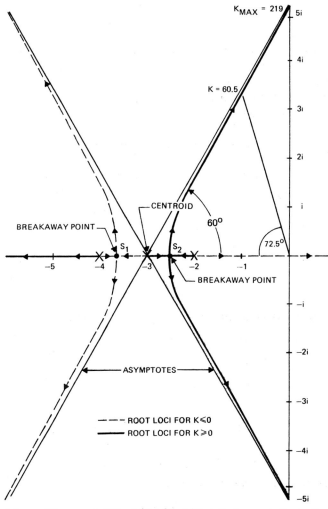

Fig. 18-7.   Root Locus Diagram of $B(s) = 1/(s+2)(s+3)(s+4)$. Arrows indicate direction for which $|k|$ increases. See text for explanation.

moves to the right along the real axis and into the third quadrant at s_2. The real root at $s=(-4,0)$ remains real, moving to the left along the axis $(0°/180°$ asymptote). For a negative gain system, as K decreases from $K=0$, the real root at $s=(-2,0)$ remains real, moving to the right along the real axis. The roots at $s=(-4,0)$ and $s=(-3,0)$ move to the breakaway point $s_B=s_1$, where they become complex and approach the $120°/300°$ asymptote as $K\to-\infty$.

The maximum value of K for which the system remains stable corresponds to a pole of the closed-loop transfer function which lies at $s=i\omega$ on the imaginary axis and is located by inspection from the root locus diagram. From Rule 5, Eq. (18-34), the gain is

$$K_{max} = \left| \frac{1}{B(s)} \right|_{s=i\omega} \tag{18-37}$$

The shape of the transient response of the system is another design criterion and is controlled by the damping ratio, ρ. The gain factor, K, required to give a specified damping ratio, is calculated from the root locus diagram by drawing a line from the origin at an angle of $\pm\theta$ with the negative real axis, where $\theta=\arccos(\rho)$. The gain at the point of intersection of this line with the root locus is the required value of K.

For the example shown in Fig. 18-7, the K_{max} is calculated for a pole located near $s=\pm3\sqrt{3}\,i$; that is,

$$K_{max} = \sqrt{(27+4)(27+9)(27+16)} = 219.1$$

The gain required for a damping factor of $\rho=0.3$ is obtained by drawing a line at an angle $\theta=\arccos(0.3)=72.5$ deg to the real axis as shown in Fig. 18-7. When the $\rho=0.3$ line is drawn, it intersects the root locus for $K>0$ at a point s_3 near where the lines $y=(x+3)\tan60°$ and $y=-x\tan72.5°$ intersect. The solution for this intersection is

$$x=\mathrm{Re}(s_3)=-1.06$$

$$y=\mathrm{Im}(s_3)=3.36$$

The gain, at s_3, is

$$|K|=\sqrt{[(x+2)^2+y^2][(x+3)^2+y^2][(x+4)^2+y^2]} = 60.5$$

By inspection of Fig. 18-7, the $\rho=0.3$ line cannot intersect the locus for $K<0$ on the left-hand side of the s-plane so that the system is stable with a damping ratio $\rho=0.3$ and $K=+60.5$.

18.2 Momentum and Reaction Wheels

Dale Headrick

As discussed in Sections 6.6 and 15.3, momentum and reaction wheels are used to provide attitude stability and control. Various wheel arrangements are used. For example, the *momentum bias* control system includes one or more momentum wheels to provide a *bias*, or nominal angular momentum different from zero. This design is often used on Earth-oriented spacecraft, such as the ITOS and AE series, to provide continuous scanning over the Earth. This design is sometimes called a *dual-spin* spacecraft to indicate that it has two parts rotating at different rates. One component may be completely despun, or rotating at a controlled rate, such as one revolution per orbit such that it maintains the same side pointing toward the Earth.

OAO and IUE are examples of an alternative arrangement in which a system of three orthogonal reaction wheels, with control signals from a set of gyroscopes, is used to provide three-axis stability and high pointing accuracy. This type of system can operate completely despun, with the reaction wheels absorbing all disturbance torques. It can also serve to reorient the spacecraft to a new target attitude by performing a series of *slew maneuvers*, or rotations about a reaction wheel axis. A hybrid configuration, flown on the Nimbus series, consists of a pitch momentum wheel with reaction wheels in the roll-yaw plane to absorb cyclic torques.

18.2.1 Momentum Bias Control Systems

In a momentum bias control system, a momentum wheel is spun up to maintain a large angular momentum relative to disturbance torques. This design is common in Earth-oriented spacecraft where the momentum wheel is along the pitch axis, nominally parallel to orbit normal. The advantages of the momentum bias design are: (1) short-term stability against disturbance torques, similar to spin stabilization; (2) roll-yaw coupling that permits yaw angle stabilization without a yaw sensor for pitch axis pointing; (3) a momentum wheel that may be used as an actuator for pitch angle control; and (4) a momentum wheel that may be used to provide scanning motion across the celestial sphere for a horizon sensor. Thus, momentum bias systems can provide three-axis control with less instrumentation than a three-axis reaction wheel system.

By incorporating horizon scanners into the momentum wheel as described in Section 6.2, roll and pitch error signals may be provided to the control system as on the ITOS and AE series. Yaw control can be achieved without a yaw sensor through the kinematics of *quarter-orbit gyroscopic coupling* as shown in Fig. 18-8.

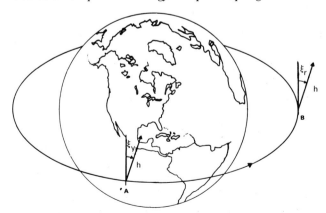

Fig. 18-8. Interchange of Yaw and Roll Attitude Components for a Momentum Wheel With Angular Momentum, **h**, Fixed in Inertial Space. The yaw error when the spacecraft is at *A* becomes a roll error when the spacecraft moves to *B*. (Compare with Fig. 2-4.)

Here, a yaw error, ξ_y, at one point in the orbit becomes a roll error, ξ_r, a quarter of an orbit later.

In a typical momentum bias system, closed-loop pitch angle control is maintained by comparing a pitch index fixed in the spacecraft body to the midscan horizon-crossing signal (see Section 6.2). Open-loop roll control is often performed using magnetic coils, as on AE or ITOS. In the AE system, the attitude is determined on the ground, and magnetic coil commands are generated to null the roll error by reorienting the pitch axis toward orbit normal. In addition to compensating for attitude disturbance torques, adjustment must also be made for the change in direction of the orbit normal due to precession of the orbit (see Section 3.4). Transferring momentum between the wheel and the spacecraft body to change the body spin rate may be used for switching between spining and nonspinning operations or for changing the pitch angle in the despun mode. The component of the total angular momentum, L_P, about the pitch axis is given by

$$L_p = I_p \omega_p + h \tag{18-38}$$

where I_p is the moment of inertia of the body of the spacecraft about the pitch axis, ω_p is the body spin rate about the pitch axis, and h is the angular momentum of the pitch wheel where the wheel momentum is oriented along the positive pitch axis. From conservation of angular momentum, the change in body rate due to a change in wheel momentum is

$$\Delta\omega_p = -\frac{\Delta h}{I_p} \tag{18-39}$$

With constant body spin-rate control, any secular disturbance torques cause a systematic increase or decrease in wheel momentum. When the wheel momentum approaches the maximum wheel capacity or minimum desired momentum, *momentum dumping* or *desaturation* must be performed using gas jets or spin-plane magnetic coils. (See Sections 19.2 and 19.3.)

An alternative design for a momentum bias control system is illustrated by the SEASAT system which uses a pair of canted scanwheels (see Section 6.2) in the pitch-yaw plane, as shown in Fig. 18-9. The scanwheels use the pitch and roll attitude error signals to maintain closed-loop three-axis attitude control. The pitch and yaw momentum components are given by

$$h_p = (h_1 + h_2)\cos\alpha$$
$$h_y = (h_1 - h_2)\sin\alpha \tag{18-40}$$

where α is the cant angle between the pitch axis and the momentum wheels. The

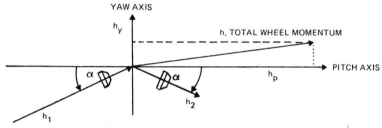

Fig. 18-9.   Canted Momentum Wheels in the Pitch-Yaw Plane. h_1 and h_2 are the wheel momenta.

momentum wheels are nominally operated at the same speed, such that $h_1 = h_2$, and the total momentum is along the pitch axis with $h_y = 0$. When the horizon scanners sense a roll angle error, a controlled yaw momentum component is generated by differentially torquing the two wheels to reduce the anticipated yaw error which will occur one-fourth of an orbit later. Because of the large moments of inertia of the SEASAT spacecraft, the scanwheel momentum is augmented by a pitch momentum wheel; a roll reaction wheel is used for roll angle control. The operation of the SEASAT control system is described in detail in Section 18.3.

18.2.2 Reaction Wheel Systems

Because the disturbance torques in high Earth orbit are very small (see Section 17.2), it is possible to use small reaction wheels to absorb them with an active control system to maintain three-axis stability. In such a system, gyroscopes are generally used to sense and feed back any body motion to the wheel torque motors on each axis. The torque motors then apply a compensating torque to each reaction wheel, which effectively absorbs the disturbance torques. Thus, the angular momentum vector changes slowly with time, and the attitude remains fixed in inertial space. When the wheels near saturation, the angular momentum is adjusted using gas jets or magnetic coils. Ideally, the attitude is controlled to the same steady-state value during desaturation, although in practice transient attitude errors are induced.

A slew, or attitude reorientation maneuver, can be executed using the set of reaction wheels to rotate the body about a commanded axis, usually one of the wheel axes, as described in Section 19.4. As shown in Fig. 18-10, the angular momentum vector remains inertially fixed, although the attitude angles change as do the angular momentum components in a body-fixed coordinate system. In the example shown, the x axis wheel might approach saturation at the final attitude just to absorb the larger momentum component. Note that in addition to estimating the attitude, it is also necessary to keep track of the wheel momenta for calculating momentum dumping commands and slew execution times.

The advantages of a three-axis stabilized reaction wheel system are: (1) capability of continuous high-accuracy pointing control, (2) large-angle slewing

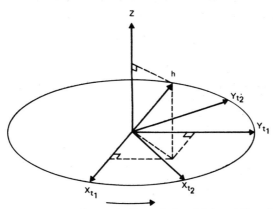

Fig. 18-10.   A Slew Rotation About the z Axis, Shown in Inertial Space. Note that the x axis wheel has to absorb additional momentum when moving to its location at t_2.

maneuvers without fuel consumption, and (3) compensation for cyclic torques without fuel consumption. This system, however, generally requires an onboard computer to implement the control laws and achieve the target attitudes.

Configurations of four reaction wheels provide control even if one wheel fails. For systems with more than three wheels operating simultaneously, a *steering law* is required to distribute the momentum between the wheels during a maneuver. For example, on the Space Telescope [Glaese, *et al.*, 1976], the total angular momentum of the four-wheel system $\mathbf{h}_{tot}$ is given by

$$\mathbf{h}_{tot} = A \left[h_1, h_2, h_3, h_4 \right]^T \qquad (18\text{-}41)$$

where h_i is the magnitude of the momentum of the ith wheel and the transformation matrix, A, depends only on the mounting angles of the wheels. The reaction wheel steering law is derived using the pseudoinverse A^R (see Appendix C) of the matrix A, where $A^R \equiv A^T(AA^T)^{-1}$. The wheel torque four-vector, $\mathbf{N}$, is given by

$$\mathbf{N} = A^R \mathbf{N}_C + k(1, -1, -1, 1)^T \qquad (18\text{-}42)$$

where $\mathbf{N}_C$ is the control torque vector in body coordinates, the vector $(1, -1, -1, 1)^T$ represents the specific wheel geometry along the diagonals of the octants with positive x, and k is an arbitrary scalar which signifies the one remaining degree of freedom. The scalar, k, can be used to achieve a desired reaction wheel momentum distribution. If k is set to zero, the steering law (Eq. (18-42)) will minimize the norm of the wheel torques. If a wheel fails or is disabled, the scalar, k, can be chosen to null the failed component of the wheel torque vector, $\mathbf{N}$, and thus avoid storing several different forms of the distribution matrix A^R.

18.3 Autonomous Attitude Stabilization Systems

Gerald M. Lerner

Section 15.3 outlined two basic techniques for attitude control. *Open-loop* control utilizes ground-based software and analysis to determine the attitude and compute and uplink commands to an onboard torquing system. Open-loop control may either maintain the spacecraft at a given orientation, which we define as *stabilization*, or maneuver the spacecraft to a new attitude, as discussed in Chapter 19. Thus, open-loop stabilization and maneuvers differ principally in the arc length separating the actual and the desired attitude. In contrast, *closed-loop* control uses attitude errors measured by sensors to automatically activate torquing devices via an onboard computer or analog electronics and thereby maintain the attitude errors within specified limits. For missions such as the planned Space Telescope, pointing requirements may be as stringent as 0.01 arc-second [Elson, 1977]. However, most current autonomous spacecraft have much more modest requirements, in the range 0.2 to 1 deg.

Closed-loop control can provide a significant improvement in both cost and accuracy over open-loop systems, which require frequent, complex, and expensive ground-based operational support to maintain a 1-deg pointing accuracy. Autonomy, however, is no panacea because of the added hardware cost, complexity, and weight and the reduced flexibility. Autonomous systems are less fault tolerant of data or environmental anomalies than are ground-based systems. They do not have available either the sophisticated ground processing software or the "common sense" judgment of the operator or analyst. Modfications to analog control systems are, of course, impossible after launch and expensive before launch, and changes to onboard software, while possible, are more difficult than comparable changes to ground-based software.

In this section, we describe the characteristics of and design considerations for several typical autonomous control systems. Two basic configurations are discussed: *inertially referenced spacecraft* which maintain a nearly fixed attitude relative to a stellar target and *Earth-referenced spacecraft* which maintain a nearly fixed attitude relative to the nadir and orbit normal. The inertial rotation rate of Earth-referenced spacecraft varies from 4 deg/minute for near-Earth satellites to 15 deg/hour for geosynchronous satellites.

18.3.1 Inertially Referenced Spacecraft (HEAO-1)

The first High Energy Astronomy Observatory (HEAO-1) operated during the early mission in a *celestial point mode* in which the body Z axis was pointed to and maintained within 1 deg of an inertial target while the spin rate about the Z axis was maintained within 10% of 0.18 deg/sec. A computed attitude reference was propagated onboard using a set of gyros (see Section 17.1) and periodically updated via ground-based command software utilizing star tracker data. The HEAO-1 control logic, implemented via an onboard computer (see Section 6.9), compares the target and observed attitude and issues a corrective thruster command when an error signal based on the attitude and attitude rate errors exceeds a preselected value.

Let q_T and q_O be the quaternions which parameterize the target and observed attitudes, respectively (in some arbitrary reference frame), and $B(q)$ be the 3×3 matrix constructed from the quaternion q. Then, as shown in Section 16.1 and Appendix D, the matrix that rotates the body axes to the target attitude is

$$B(q_E) = B(q_T)B(q_O^{-1}) \qquad (18\text{-}43a)$$

or, in quaternion notation,

$$q_O^{-1}q_T \equiv q_E = \begin{bmatrix} q_{T4} & q_{T3} & -q_{T2} & q_{T1} \\ -q_{T3} & q_{T4} & q_{T1} & q_{T2} \\ q_{T2} & -q_{T1} & q_{T4} & q_{T3} \\ -q_{T1} & -q_{T2} & -q_{T3} & q_{T4} \end{bmatrix} \begin{bmatrix} -q_{O1} \\ -q_{O2} \\ -q_{O3} \\ q_{O4} \end{bmatrix} \qquad (18\text{-}43b)$$

where $q_K = (q_{K1}, q_{K2}, q_{K3}, q_{K4})^T$; $K = E$, T, or O; and q_E is the *error quaternion*. If the observed and target quaternions are equal, then the error quaternion is $q_E = (0,0,0,1)^T$ and $B(q_E)$ is the 3×3 identity matrix.

One goal of the control laws is to minimize the projections of the observed X and Y body axes on the target Z axis; thus, we require that

$$\hat{X}_O \cdot \hat{Z}_T \cong 0$$

$$\hat{Y}_O \cdot \hat{Z}_T \cong 0. \qquad (18\text{-}44)$$

Because these projections are the 1,3 and 2,3 components of $B \equiv B(q_E)$, Eq.

(18-44) can be written as

$$\hat{X}_O \cdot \hat{Z}_T = B_{13} = 2(q_{E1}q_{E3} - q_{E2}q_{E4}) \cong 0$$

$$\hat{Y}_O \cdot \hat{Z}_T = B_{23} = 2(q_{E2}q_{E3} + q_{E1}q_{E4}) \cong 0 \qquad (18\text{-}45)$$

Let ξ_r, ξ_p, and ξ_y denote the infinitesimal rotations about the body X, Y, and Z axes required to achieve the target attitude. Then $B(\mathbf{q}_E)$ transforms vectors from the body to the target frame and is given by

$$B = \left[A_x(\xi_r) A_y(\xi_p) A_z(\xi_y) \right]^T \cong \begin{bmatrix} 1 & 0 & 0 \\ 0 & 1 & -\xi_r \\ 0 & +\xi_r & 1 \end{bmatrix} \begin{bmatrix} 1 & 0 & +\xi_p \\ 0 & 1 & 0 \\ -\xi_p & 0 & 1 \end{bmatrix} \begin{bmatrix} 1 & -\xi_y & 0 \\ +\xi_y & 1 & 0 \\ 0 & 0 & 1 \end{bmatrix}$$

or

$$B \cong \begin{bmatrix} 1 & -\xi_y & +\xi_p \\ +\xi_y & 1 & -\xi_r \\ -\xi_p & +\xi_r & 1 \end{bmatrix} \qquad (18\text{-}46)$$

where A_x, A_y, and A_z are the Euler rotation matrices (see Section 12.1) and terms of order ξ^2 are omitted in Eq. (18-46) and throughout this section. Thus, $B_{13} = +\xi_p$ and $B_{23} = -\xi_r$. The error quaternion is measured onboard the HEAO-1 spacecraft by continuously propagating a reference attitude (measured on the ground using star tracker data) with rate-integrating gyros (see Section 6.5.2). The gyros also measure the body rates, which may be compared with the desired rates of $\omega_X \approx \omega_Y \approx 0$ and $\omega_Z \approx 0.18$ deg/sec. The position and rate error for either the X or Y axis are combined as shown schematically in Fig. 18-11 to yield a desired thruster burn duration,

$$t_X = K_X(-B_{23} + \tau_X \omega_X)$$

$$t_Y = K_Y(B_{13} + \tau_Y \omega_Y)$$

where K_X and K_Y are the system gains, τ_X and τ_Y are lead time constants, and t_X and t_y are the desired thruster burn durations. Errors about the X and Y body axes are corrected independently (the small gyroscopic coupling between the X and Y axes through the angular momentum about the Z body axis is ignored).

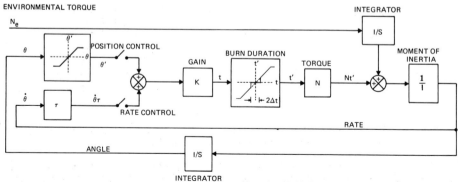

Fig. 18-11.   HEAO-1 Position Plus Rate Controller Block Diagram ($\theta = \xi_p$ or ξ_r and $\dot{\theta} = \omega_y$ or ω_x).

Either position only, rate only, or position plus rate control may be achieved by the control law shown in Fig. 18-11, where θ and $\dot{\theta}$ are the angular position and rate errors about a body axis and t is the computed thrust duration. Note that, in the figure, environmental disturbance torques, N_e, are integrated and added to the commanded control torques. To avoid excessive thruster activity, commands are issued only if the required thrust duration, t, exceeds a minimum time, Δt. For large t, the thrust interval is set equal to the sampling rate and the thruster fires continuously.

The action of the position plus rate controller is illustrated in the state space diagram in Fig. 18-12 where the ordinate is the rate error and the abscissa is the position error. If the attitude state lies within the shaded region or *deadband*, thruster commands are inhibited. However, if the attitude enters the region above or below the deadband, corrective thrusts are commanded. Above and to the right of the deadband, the angular velocity is decreased by the control law; below and to the left, it is increased. After large errors are removed, each thrust yields a minimum rate change of $\Delta \dot{\theta} = N \Delta t / I$, where N is the thruster torque, Δt is the minimum thrust duration, and I is the moment of inertia about the controlled axis. Initially, thrust commands will be issued by the controller, causing approximately a vertical trajectory* on Fig. 18-12, until the deadband boundary is reached. The

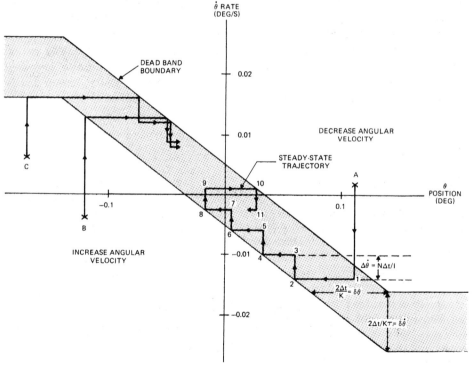

Fig. 18-12. HEAO-1 Controller State-Space Diagram. A, B, and C are various initial states. An uncontrolled trajectory is an approximately horizontal line.

*The vertical lines are actually sections of parabolas, $\theta = (\dot{\theta} - \dot{\theta}_0)^2 / 2a + \theta_0$, where the initial angle and rate are θ_0 and $\dot{\theta}_0$ and the angular acceleration during the thrust is a. For thrusters, a is large, and thus $\theta \approx \theta_0$ during the thrust.

attitude state then moves horizontally within the deadband at a constant angular velocity; when the deadband boundary is next crossed, another thrust is commanded. The trajectory, in the absence of environmental torques, for representative initial conditions A, B, and C is shown in Fig. 18-12. The path 7-8-9-10-11,..., is a *steady-state trajectory* which is approximately maintained by the control laws.

The size of the attitude deadband in state-space is determined by the system gain, K, time constant, τ, and minimum thruster duration, Δt, with the permitted position and rate dimensions given by $\delta\theta = 2\Delta t / K$ and $\delta\dot{\theta} = 2\Delta t / K\tau$, respectively, as shown in Fig. 18-12. The mean time, $\langle t \rangle$, between thrusts in steady state depends on the minimum angular rate change and is given by

$$\langle t \rangle \cong (2\Delta t / K)/(N\Delta t / I) = 2I / KN \qquad (18\text{-}47)$$

The general effect of environmental torques is to perturb the attitude, as shown in Fig. 18-12; this results in curved trajectories in the state-space diagram.

The HEAO-1 control law has the advantage of simplicity, and gains and deadbands may be selected to suit various applications. The major disadvantages are that (1) the response of the system to disturbance torques is undamped, which results in a waste of expendables as the attitude state is driven within the deadband; and (2) the attitude pointing accuracy is severely limited by the requirement for complex ground-based support to provide periodic updates to the reference attitude, q_O; typically every 12 hours. One obvious improvement on the HEAO-1 control system is to provide an autonomous capability for updating the reference attitude. Such a system, using a star tracker (see Section 6.4) as the sensing device will be used on HEAO-B [Hoffman, 1976]. In addition to providing periodic reference attitude updates, the HEAO-B control system continuously estimates the gyro drift bias (see Section 7.8) using an onboard version of the Kalman filter discussed in Section 13.5. A similar system for SMM, using a precise Sun sensor as the primary attitude reference, is described by Markley [1978].

18.3.2 Earth-Referenced Spacecraft

The two basic limitations of the HEAO-1 control system—the lack of both damping and an autonomous attitude reference—are easily overcome for Earth-referenced spacecraft. A momentum wheel provides gyroscopic rigidity and thereby permits damping. In addition, the control system may utilize either gravity-gradient torque or horizon sensors to measure absolute position errors and, consequently, does not require extensive ground support.

The spacecraft considered here rotate at one revolution per orbit in an orbit of moderate eccentricity (say, $e < 0.1$) at altitudes where atmospheric drag may be neglected. We will first reformulate Euler's equations for these spacecraft by deriving an expression for the gravity-gradient torque. Next, the general characteristics and approximations underlying the equations are described. Finally, they are applied to GEOS-3, HCMM, SEASAT, and CTS to illustrate the analytical procedures used for the evaluation of arbitrary stabilization systems.

We assume that the nominal mission attitude is as shown in Fig. 2-4, where the body Z axis (normally the payload axis) is along the nadir, and the body Y axis is along the negative orbit normal. For a circular orbit, the body X axis is along the velocity vector.* The attitude angles are defined as roll, pitch, and yaw, which are small rotational errors about the velocity vector, negative orbit normal, and nadir. (Alternatively, these may be thought of as small errors about the body X, Y, and Z axes.) The roll, pitch, and yaw angles (in radians) are denoted by ξ_r, ξ_p, and ξ_y, respectively. The transformation matrix from the orbit reference frame to the body frame is,

$$A = B^T \simeq \begin{bmatrix} 1 & \xi_y & -\xi_p \\ -\xi_y & 1 & \xi_r \\ \xi_p & -\xi_r & 1 \end{bmatrix} \qquad (18\text{-}48)$$

where B is the transformation from the body frame to the reference frame defined by Eq. (18-46). The matrix A transforms any vector, $\mathbf{V}$, from orbital coordinates $(\mathbf{V}_O)$ to body coordinates $(\mathbf{V}_B)$; that is, $A\mathbf{V}_O = \mathbf{V}_B$. The order of the three rotations in Eq. (18-48) is irrelevant because infinitesimal rotations commute.

The angular velocity vector in the body frame is (see Section 16.1) approximately

$$\omega_B = \begin{bmatrix} \dot{\xi}_r \\ \dot{\xi}_p \\ \dot{\xi}_y \end{bmatrix} + A \begin{bmatrix} 0 \\ -\omega_o \\ 0 \end{bmatrix} = \begin{bmatrix} \dot{\xi}_r - \omega_o \xi_y \\ \dot{\xi}_p - \omega_o \\ \dot{\xi}_y + \omega_o \xi_r \end{bmatrix} \qquad (18\text{-}49)$$

where $\omega_o^2 = \mu_\oplus / R^3$ is the orbital angular velocity of a spacecraft in a circular orbit of radius R, and $\mu_\oplus \equiv GM_\oplus$ is the Earth's gravitational constant. The zenith vector in body coordinates is

$$\hat{\mathbf{r}}_B = A(0,0,-1)^T = (\xi_p, -\xi_r, \doteq 1)^T \qquad (18\text{-}50)$$

and hence the gravity-gradient torque (see Section 17.2) is

$$\mathbf{N}_{GG} = 3\omega_o^2 \hat{\mathbf{r}}_B \times (I \cdot \hat{\mathbf{r}}_B) = 3\omega_o^2 (\xi_r (I_z - I_y), \xi_p (I_z - I_x), 0)^T \qquad (18\text{-}51)$$

where the moment of inertia tensor, I, is assumed diagonal with components I_x, I_y, and I_z along the body axes. Note that to first order there is no gravity-gradient torque along the yaw axis. (There is, however, a yaw-restoring torque which results from gyroscopic roll/yaw coupling, as described in the next subsection.)

With the previous definitions, Euler's equations in body coordinates for a spacecraft with internal angular momentum, h_x, h_y, and h_z along the body X, Y, and Z axes are (see Section 16.2)

$$\frac{d}{dt}\mathbf{L} + \omega \times \mathbf{L} = \sum \mathbf{N} = \mathbf{N}_E + \mathbf{N}_C + \mathbf{N}_{GG} \qquad (18\text{-}52)$$

where $\mathbf{N}_E$, $\mathbf{N}_C$, and $\mathbf{N}_{GG}$ are the environmental, external control, and gravity-gradient torques, respectively. Writing Eq. (18-52) in component form gives

$$I_x \ddot{\xi}_r + \left[4\omega_o^2(I_y - I_z) - h_y \omega_o\right]\xi_r - \left[h_y + (I_x - I_y + I_z)\omega_o\right]\dot{\xi}_y = N_{Ex} + N_{Cx} + h_z \omega_o - \dot{h}_x$$

$$(18\text{-}53a)$$

---

* For orbits of nonzero eccentricity, the velocity vector is replaced by the cross product of the negative orbit normal and the nadir vector.

$$I_y \ddot{\xi}_p + 3\omega_o^2 (I_x - I_z)\xi_p = N_{Ey} + N_{Cy} - \dot{h}_y \tag{18-53b}$$

$$I_z \ddot{\xi}_y + \left[\omega_o^2(I_y - I_x) - h_y\omega_o\right]\xi_y + \left[h_y + (I_x - I_y + I_z)\omega_o\right]\dot{\xi}_r = N_{Ez} + N_{Cz} - h_x\omega_o - \dot{h}_z \tag{18-53c}$$

where the total angular momentum is $\mathbf{L} = \mathbf{I}\cdot\boldsymbol{\omega} + \mathbf{h}$, and we have assumed that $\omega_o \gg \max(|\dot{\xi}_p|, |\dot{\xi}_r|, |\dot{\xi}_y|)$ and have neglected all second-order terms including those involving h_x and h_z in Eq. (18-53b). Equation (18-53) is central to the remainder of this section. We first describe its general characteristics and underlying approximations and then apply it to several representative spacecraft.

The pitch equation is decoupled from the roll and yaw equations which are coupled through the *bias momentum*, h_y, and the orbit rate term $(I_x - I_y + I_z)\omega_o$. Control torques, including dampers, generally increase the coupling between the roll and yaw equations but leave the pitch equation uncoupled.* For gravity-gradient stability, $I_x > I_z$ and the gravity-gradient force provides a restoring torque proportional to pitch with frequency $\sqrt{3\sigma_y}\,\omega_o$ where $\sigma_y = (I_x - I_z)/I_y$. The effect of orbital eccentricity on the pitch behavior of gravity-gradient stabilized satellites may be seen with the aid of Fig. 18-13. The rate of change of angular momentum about the pitch axis, ignoring environmental torques and with the pitch wheel speed constant, is

$$I_y(\ddot{\nu} + \ddot{\xi}_p) = N_{GG} \approx -3\omega_o^2(I_x - I_z)\xi_p \tag{18-54}$$

where ν is the true anomaly. For an orbit with small eccentricity, e, we have (see Eq. (3-11)),

$$\nu = M + 2e\sin M \tag{18-55}$$

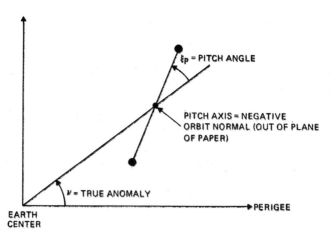

EARTH
CENTER

Fig. 18-13. Influence of Orbital Eccentricity on a Gravity-Gradient Stabilized Spacecraft. (Adapted from Pisacane, *et al.*, [1967].) The orbit is in the plane of the paper.

---

*Off-diagonal terms in the moment-of-inertia tensor lead to coupling and are usually treated as disturbance torques.

where $M = \omega_o t + M_o$ is the mean anomaly. Substitution of Eq. (18-55) into Eq. (18-54) gives

$$\ddot{\xi}_p + 3\omega_o^2 \frac{(I_x - I_z)}{I_y} \xi_p = 2\omega_o^2 e \sin M \qquad (18\text{-}56)$$

with the solution

$$\xi_p = \frac{2e}{3\sigma_y - 1} \sin M + C_1 \cos\left(\sqrt{3\sigma_y}\, \omega_o t + \phi_1\right) \qquad (18\text{-}57)$$

where C_1 and ϕ_1 are integration constants associated with the complementary solution.

The particular solution to the differential Eq. (18-57), $\xi_p = 2e \sin M/(3\sigma_y - 1)$, results in a sinusoidal steady-state error, which for GEOS-3 ($e = 0.0054$, $\sigma = 0.984$), had an amplitude of 0.3 deg. For spacecraft with $\sigma_y \approx 1/3$, there is a pitch resonance and therefore this configuration is avoided.

The coupled roll/yaw expressions from Eq. (18-53) in the absence of roll and yaw wheels, control, and environmental torques (other than gravity-gradient) are

$$I_x \ddot{\xi}_r + 4\omega_o^2 (I_y - I_z)\xi_r - \omega_o (I_x - I_y + I_z)\dot{\xi}_y = 0$$

$$I_z \ddot{\xi}_y + \omega_o^2 (I_y - I_x)\xi_y + \omega_o (I_x - I_y + I_z)\dot{\xi}_r = 0 \qquad (18\text{-}58)$$

With the notation $\sigma_x = (I_y - I_z)/I_x$ and $\sigma_z = (I_y - I_x)/I_z$, these may be rewritten in Laplace transform notation (see Appendix F) as

$$\begin{bmatrix} s^2 + 4\omega_o^2 \sigma_x & -\omega_o(1 - \sigma_x)s \\ \omega_o(1 - \sigma_z)s & s^2 + \omega_o^2 \sigma_z \end{bmatrix} \begin{bmatrix} \mathcal{L}(\xi_r) \\ \mathcal{L}(\xi_y) \end{bmatrix} = 0 \qquad (18\text{-}59)$$

with the characteristic equation

$$s^4 + \omega_o^2 (3\sigma_x + 1 + \sigma_x \sigma_z)s^2 + 4\omega_o^4 \sigma_x \sigma_z = 0 \qquad (18\text{-}60)$$

For roll/yaw stability, the roots to Eq. (18-60) must have no positive real part (see Section 18.1) and hence

$$\frac{s^2}{\omega_o^2} = \frac{-(3\sigma_x + 1 + \sigma_x \sigma_z) \pm \left[(3\sigma_x + 1 + \sigma_x \sigma_z)^2 - 16\sigma_x \sigma_z\right]^{1/2}}{2} \qquad (18\text{-}61)$$

must be real and negative.* Therefore, a necessary and sufficient condition for stability is

$$4\sqrt{\sigma_x \sigma_z} < 3\sigma_x + 1 + \sigma_x \sigma_z \qquad (18\text{-}62)$$

where

$$\sigma_x \sigma_z > 0$$

Figure 18-14 illustrates the regions of gravity-gradient stability defined by the

---

*If s_1 is a root of Eq. (18-60), then $-s_1$ is also. Thus, for both s_1 and $-s_1$ to have no positive real part, s_1 must be pure imaginary and s_1^2 is real and negative.

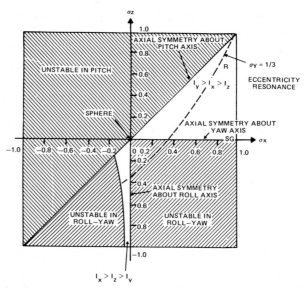

Fig. 18-14. Gravity-Gradient Stability for Various Moments of Inertia. Three-axis stability corresponds to the unshaded region (adapted from Kaplan [1976]). The letters S, G, and R denote the configurations of three gravity-gradient stabilized spacecraft—SEASAT-A, GEOS-3, and RAE-2. (See Appendix I.)

above inequalities. (As described previously, $I_x > I_z$ or $I_x\sigma_x > I_z\sigma_z$ is required for pitch stability.)

18.3.3 Examples of Earth-Referenced Spacecraft

The remainder of this section concerns the analysis of Eq. (18-53) for four representative spacecraft configurations and control systems, as described in Table 18-1. Although all four spacecraft are of the momentum bias design as discussed in Section 18.2, they are physically very different. The moments of inertia of the asymmetric HCMM spacecraft are three orders of magnitude smaller than those of the symmetric SEASAT-A spacecraft. All four spacecraft orbits are near-circular, but the CTS orbit is equatorial at synchronous altitude, whereas the others are at 500 to 850 km in polar orbits. GEOS-3, shown in Fig. 19-17, uses an extendable 6.5-m boom with a 45-kg end mass to achieve a large, gravity-gradient restoring torque which, combined with a damper, provides passive 1-deg pitch control. For

Table 18-1. Physical Characteristics of Representative Earth-Oriented Spacecraft. N/A indicates *not applicable*. See Appendix I for other spacecraft characteristics.

SPACECRAFT	MOMENTS OF INERTIA (kg·m²)			INERTIAL SPIN RATE (SEC⁻¹)	WHEEL MOMENTA (N·M·S)			ORBIT		
	I_x	I_y	I_z	(SEC⁻¹)	h_x	h_y	h_z	a(km)	e	i(deg)
GEOS-3	2157	2157	35.5	0.00103	N/A	−2.41	N/A	7221	0.0054	115
CTS	1130	92	1130	0.000073	N/A	−20.3[†]	N/A	GEOSYNCHRONOUS		
SEASAT-A	25,100	25,100	3000	0.00105	0[††]	−24.4[†]	0[††]	7153	0	108
HCMM	19.4	26.7	18.4	0.00108	N/A	−5.22[†]	N/A	6978	0	98

[†]PITCH CONTROL VIA MOMENTUM EXCHANGE.

[††]REACTION WHEELS AT NOMINAL ZERO BIAS FOR ROLL/YAW CONTROL.

SEASAT-A, the large, symmetric Agena (the final stage of the launch vehicle) remains attached to the experimental module and provides a large gravity-gradient restoring torque; however, the 0.5-deg pitch control requirement can be satisfied only by adding a pitch sensor and controlling the pitch wheel speed. For both HCMM and CTS, the effect of the gravity-gradient restoring torque is negligible, and pitch control is achieved by controlling the pitch wheel speed. Roll/yaw control, which is equivalent to maintaining the wheel angular momentum along the orbit normal, is accomplished as shown in Table 18-2. Except for GEOS-3, for which the gravity-gradient resoring torque is large, active control is achieved by using a sensed roll error to drive a torquing system. For all four spacecraft, yaw is controlled indirectly via quarter-orbit coupling with roll (see Section 18.2) because of the lack of simple, effective yaw sensors in Earth orbit.* Both SEASAT and CTS, however, use an indirect method of augmenting the yaw control, referred to as WHECON (an acronym for wheel control) by commanding a yaw torque based on the sensed roll error.

For spacecraft with magnetic coils, dampers, or residual dipoles, the geomagnetic field in the spacecraft coordinates is required. Assuming a dipole field and an orbit passing over the magnetic pole, the magnetic field in orbital coordinates (see Section 5.1 and Appendix H) is

$$\mathbf{B}_O = B_M (\cos\lambda, 0, 2\sin\lambda)^T \qquad (18\text{-}63)$$

where $B_M = 7.96 \times 10^6 / R^3$ (Teslas) as defined in Eq. (H-18), R is the geocentric distance in kilometers, and λ is a continuous measure of the latitude (i.e., λ is the

Table 18-2.   Control Systems for Spacecraft in Table 18-1

SPACECRAFT	PITCH			ROLL		
	METHOD	SENSOR	CONTROL ACCURACY (DEG)	METHOD	SENSOR	CONTROL ACCURACY (DEG)
GEOS-3	GRAVITY GRADIENT	NONE	1.0	GRAVITY GRADIENT	NONE	1.0
CTS	PITCH WHEEL SPEED	NESA[1]	0.1	JETS	NESA[1]	0.1
SEASAT-A[2]	PITCH WHEEL SPEED GRAVITY GRADIENT	SCANWHEELS[3]	0.5	ROLL REACTION WHEEL GRAVITY GRADIENT	SCANWHEELS	0.5
HCMM[2]	PITCH WHEEL SPEED	SCANWHEEL[3]	1.0	MAGNETIC TORQUING	SCANWHEEL	1.0

SPACECRAFT	YAW			MOMENTUM MAINTENANCE	DAMPING
	METHOD	SENSOR	CONTROL ACCURACY (DEG)		
GEOS-3	QUARTER-ORBIT COUPLING	NONE	1.5	NONE	MAGNETIC
CTS	WHECON[4]	NONE	0.1	JETS	JETS
SEASAT-A[2]	WHECON[4]	NONE	0.5	MAGNETIC	WHEELS
HCMM[2]	QUARTER-ORBIT COUPLING	NONE	2.0	MAGNETIC	MAGNETIC

[1]NONSPINNING EARTH SENSOR ASSEMBLY (SEE SECTION 6.2)

[2]PROPOSED SPACECRAFT.

[3]REGISTERED TRADEMARK OF ITHACO CORPORATION. (SEE SECTION 6.2)

[4]ACRONYM FOR WHEEL CONTROL (SEE SEASAT DISCUSSION IN TEXT)

---

*Direct yaw control can be achieved using a gyroscope, as in inertial guidance systems; using either onboard processing of Sun sensor or star tracker data; or using ground commands to periodically update the yaw reference.

latitude when the satellite is traveling north and 180 deg minus the latitude when the satellite is traveling south).

GEOS-3. The GEOS-3 control system is described in Table 18-2 and the spacecraft is shown in Fig. 19-17. The control torque from the magnetic eddy current damper is (see Section 18.4 and [Pettus, 1968])

$$\mathbf{N}_C = k_D \hat{\mathbf{B}} \times \frac{d}{dt} \hat{\mathbf{B}} \tag{18-64}$$

where k_D is the damper constant, the geomagnetic field is

$$\mathbf{B} = A\mathbf{B}_O = B_M A (\cos\lambda, 0, 2\sin\lambda)^{\mathrm{T}} \tag{18-65}$$

and A is defined by Eq. (18-48).

Substitution of Eqs. (18-48) and (18-65) into Eq. (18-64) gives the damper torque in body coordinates as

$$\mathbf{N}_C = -k_D \left\{ \begin{bmatrix} 0 \\ \dot{\xi}_p \\ 0 \end{bmatrix} + \frac{1}{\cos^2\lambda + 4\sin^2\lambda} \begin{bmatrix} 4\dot{\xi}_r\sin^2\lambda - \dot{\xi}_y\sin 2\lambda + 2\xi_y\omega_o \\ 2\omega_o \\ -\dot{\xi}_r\sin 2\lambda + \dot{\xi}_y\cos^2\lambda - 2\xi_r\omega_o \end{bmatrix} \right\} \tag{18-66}$$

where for a circular, polar orbit, $\dot{\lambda} = \omega_o$. Substituting Eq. (18-66) into Eq. (18-54) and using Table 18-1 with $h_x = h_z = 0$, $h_y \equiv -h$, and $I_x = I_y \equiv I$ gives the result

$$I\ddot{\xi}_r + \frac{4k_D\sin^2\lambda}{\cos^2\lambda + 4\sin^2\lambda}\dot{\xi}_r + \left(4\omega_o^2(I - I_z) + h\omega_o\right)\xi_r$$

$$+ \left(h - I_z\omega_o - \frac{k_D\sin 2\lambda}{\cos^2\lambda + 4\sin^2\lambda}\right)\dot{\xi}_y + 2k_D\omega_o\xi_y = N_{Ex}$$

$$I\ddot{\xi}_p + k_D\dot{\xi}_p + 3\omega_o^2(I - I_z)\xi_p = \frac{-2\omega_o k_D}{\cos^2\lambda + 4\sin^2\lambda} + N_{Ey}$$

$$I_z\ddot{\xi}_y + \frac{k_D\cos^2\lambda}{\cos^2\lambda + 4\sin^2\lambda}\dot{\xi}_y + h\omega_o\xi_y - \left(h - I_z\omega_o + \frac{k_D\sin 2\lambda}{\cos^2\lambda + 4\sin^2\lambda}\right)\dot{\xi}_r - 2k_D\omega_o\xi_r = N_{Ez}$$

$$\tag{18-67}$$

The pitch equation has the form of a forced harmonic oscillator which may be solved by expanding the right-hand side of the equation or the *forcing function* in a Fourier series [Repass, *et al.*, 1975],

$$f(t) \equiv \frac{-2\omega_o k_D}{1 + 3\sin^2\omega_o t} + N_{Ey}(t)$$

$$= -\omega_o k_D(1 + \tfrac{2}{3}\cos 2\omega_o t + \tfrac{2}{9}\cos 4\omega_o t + \cdots)$$

$$+ \frac{a_0}{2} + \sum_{n=1}^{\infty}(a_n\cos n\omega_o t + b_n\sin n\omega_o t) \tag{18-68}$$

where* $\lambda \equiv \omega_o t$ and a_n and b_n are the Fourier coefficients given by

$$a_n = \frac{1}{\pi} \int_0^{2\pi} N_{Ey} (\lambda / \omega_o) \cos n\lambda \, d\lambda$$

$$b_n = \frac{1}{\pi} \int_0^{2\pi} N_{Ey} (\lambda / \omega_o) \sin n\lambda \, d\lambda \qquad (18\text{-}69)$$

Taking the Laplace transform of the pitch equation and rearranging yields (see Appendix F)

$$\Xi_p(s) \equiv \mathcal{L}(\xi_p(t)) = \left\{ \mathcal{L}\left[-\omega_o k_D \left(1 + \tfrac{2}{3}\cos 2\omega_o t + \tfrac{4}{9}\cos 4\omega_o t + \cdots + \right) \right] \right.$$

$$\left. + \mathcal{L}\left[a_0/2 + \sum_{n=1}^{\infty} (a_n \cos n\omega_o t + b_n \sin n\omega_o t) \right] \right\} / \zeta(s) \qquad (18\text{-}70)$$

where the time constants and frequencies associated with the decay of pitch oscillations are given by the zeros of the characteristic equation

$$\zeta(s) = Is^2 + k_D s + 3\omega_o^2 (I - I_z) \qquad (18\text{-}71)$$

Thus, the zeros of $\zeta(s)$ are[†]

$$s_\pm = \left[-k_D \pm \sqrt{k_D^2 - 12\omega_o^2 (I - I_z)I} \right] / 2I \qquad (18\text{-}72)$$

and the time constant, τ, and oscillation frequency, f, are*

$$\tau = -1/\text{Real}(s_+) = 2I/k_D \qquad (18\text{-}73)$$

$$f = \text{Im}(s_+) = \sqrt{12\omega_o^2 (I - I_z)I - k_D^2} \, / 2I \qquad (18\text{-}74)$$

The steady-state solution, $\bar{\xi}_p(t)$, is obtained using Eq. (F-33) and the principle of superposition,

$$\bar{\xi}_p(t) = \frac{-2}{3\sigma\omega_o \tau} + \frac{a_0}{6\sigma I \omega_o^2} + \frac{a_1 \cos(\omega_o t - \varphi_1)}{I\omega_o^2 \left[(3\sigma - 1)^2 + 4/(\omega_o \tau)^2 \right]^{1/2}}$$

$$+ \frac{(a_2 - 2\omega_o k_D/3)\cos(2\omega_o t - \varphi_2)}{I\omega_o^2 \left[(3\sigma - 4)^2 + 16/(\omega_o \tau)^2 \right]^{1/2}} + \ldots, + \qquad (18\text{-}75)$$

where $\sigma = (I - I_z)/I \lesssim 1$, $\varphi_n = \arctan^{-1}[k_D n/I\omega_o (3\sigma - n^2)]$, and we assume $b_n = 0$.

The GEOS-3 design tradeoff can be seen by comparing Eqs. (18-73) and (18-75). Rapid transient response is obtained by *decreasing* $\tau = 2I/k_D$. However,

---

* Without loss of generality, we assume that the spacecraft is traveling north at the Equator at $t = 0$.
[†] The complementary solution to the pitch equation (i.e., for zero forcing terms) may be shown to be of the form

$$\xi_p(t) = \exp(-t/\tau)[A \exp(ift) + B \exp(-ift)]$$

by substitution into Eq. (18-67).

the steady-state error is reduced by *increasing* τ and I. Consequently, for GEOS-3, satisfactory steady-state performance was achieved at the cost of poor transient performance. The parameters were $I = 2157$ kg-m^2, $k_D = 0.012$ N·m·s, and thus $\tau = 4.2$ days and the root-mean-square steady-state pitch error was approximately 0.5 deg [Lerner and Coriell 1975]

The Laplace transform of the roll-yaw equations (Eq. (18-67)) may be written in matrix form as

$$
\begin{bmatrix} Is^2 + \dfrac{2k_D}{3}s + 4\omega_o^2(I - I_z) + h\omega_o & (h - I_z\omega_o)s + 2k_D\omega_o \\[2mm] -(h - I_z\omega_o)s - 2k_D\omega_o & I_z s^2 + \dfrac{k_D}{3}s + h\omega_o \end{bmatrix} \begin{bmatrix} \Xi_r(s) \\[2mm] \Xi_y(s) \end{bmatrix} \equiv M \begin{bmatrix} \Xi_r(s) \\[2mm] \Xi_y(s) \end{bmatrix}
$$

$$
= \begin{vmatrix} \mathcal{L}(N_{Ex}) \\[2mm] \mathcal{L}(N_{Ez}) \end{vmatrix} \tag{18-76}
$$

where $\Xi_r(s) \equiv \mathcal{L}(\xi_r(t))$, $\Xi_y(s) \equiv \mathcal{L}(\xi_y(t))$, and the slowly varying coefficients of the damping term have been replaced with their orbit averaged values,

$$
\left\langle \frac{2\sin\omega_o t \cos\omega_o t}{\cos^2\omega_o t + 4\sin^2\omega_o t} \right\rangle = 0
$$

$$
\left\langle \frac{4\sin^2\omega_o t}{\cos^2\omega_o t + 4\sin^2\omega_o t} \right\rangle = \frac{2}{3} \tag{18-77}
$$

$$
\left\langle \frac{\cos^2\omega_o t}{\cos^2\omega_o t + 4\sin^2\omega_o t} \right\rangle = \frac{1}{3}
$$

Equation (18-76) may be formally solved for roll and yaw to yield

$$
\begin{bmatrix} \xi_r(t) \\ \xi_y(t) \end{bmatrix} \equiv \mathcal{L}^{-1} \begin{bmatrix} \Xi_r(s) \\ \Xi_y(s) \end{bmatrix} = \mathcal{L}^{-1} \left[M^{-1} \begin{pmatrix} \mathcal{L}(N_{Ex}) \\ \mathcal{L}(N_{Ez}) \end{pmatrix} \right] \tag{18-78}
$$

The time constants and frequencies which describe the decay of transient roll and yaw oscillations are related to the zeros of the determinant of M which are given by

$$
\left[Is^2 + \frac{2k_D}{3}s + 4\omega_o^2(I - I_z) + h\omega_o \right] \left(I_z s^2 + \frac{k_D}{3}s + h\omega_o \right) + (h - I_z\omega_o)^2 s^2
$$

$$
+ 4k_D\omega_o(h - I_z\omega_o) + 4k_D^2\omega_o^2 = 0 \tag{18-79}
$$

Denoting the roots of this fourth-order equation by s_1, s_1^*, s_2, and s_2^*, the time constants and frequencies are

$$
\tau_1 = -1/\text{Re}(s_1) \approx 6.1 \text{ days}
$$

$$
\tau_2 = -1/\text{Re}(s_2) \approx 5.1 \text{ hours} \tag{18-80}
$$

$$f_1 = \mathrm{Im}(s_1) \approx 0.0016 \ \mathrm{sec}^{-1}$$

$$f_2 = \mathrm{Im}(s_2) \approx 0.012 \ \mathrm{sec}^{-1} \qquad (18\text{-}81)$$

The fact that these time constants are much greater than the orbit period (100 minutes) justifies the orbit averages of Eq. (18-77). The characteristic frequencies, f_1 and f_2, are related to the gravity-gradient restoring torque and nutation, respectively. The very long time constants associated with the decay of transient pitch and roll oscillations required the development of a procedure which minimized the pitch and roll errors after attitude acquisition (see Section 19.5).

HCMM. The HCMM control system, as outlined in Table 18-2, provides pitch control by torquing the pitch wheel based on horizon scanner pitch angle data. The pitch wheel torque command, shown schematically in Fig. 18-15, is computed from the pitch angle and rate as

$$-\dot{h}_y \equiv \dot{h} = -K_p \xi_p - K_{\dot{p}} \dot{\xi}_p \qquad (18\text{-}82)$$

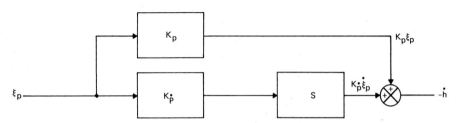

Fig. 18-15.   Position-Plus-Rate for Pitch Angle Control

The rate gain, $K_{\dot{p}} \approx 0.8 \ \mathrm{N \cdot m \cdot s}$, provides damping of the pitch response with time constant $\tau = 2I_y/K_{\dot{p}} \approx 67$ s and the position gain, $K_p \approx 0.012 \ \mathrm{N \cdot m}$, provides a restoring torque with frequency $f \approx \sqrt{K_p/I_y} \approx 0.021 \ \mathrm{s}^{-1}$. The pitch loop response to a 5-deg initial error is shown in Fig. 18-16.

Substituting Eq. (18-82) into Eq. (18-53b) and taking the Laplace transform gives

$$\left\{ I_y s^2 + K_{\dot{p}} s + \left[3\omega_o^2 (I_x - I_z) + K_p \right] \right\} \Xi_p(s) = \mathcal{L}(N_{Ey}) \qquad (18\text{-}83)$$

and the roots of the characteristic equation are

$$s_\pm = \left[-K_{\dot{p}} \pm i\sqrt{4I_y\left[K_p + 3\omega_o^2(I_x - I_z)\right] - K_{\dot{p}}^2} \ \right]/2I_y \qquad (18\text{-}84)$$

The gains, K_p and $K_{\dot{p}}$ are chosen to provide near-critical damping and minimize the overshoot, as discussed in Section 18.1. For critical damping,

$$K_{\dot{p}} = 2\sqrt{I_y\left[K_p + 3\omega_o^2(I_x - I_z)\right]} \approx 1.1 \ \mathrm{N \cdot m \cdot s} \qquad (18\text{-}85)$$

Thus, the design value, $K_{\dot{p}} = 0.8 \ \mathrm{N \cdot m \cdot s}$ corresponds to a damping ratio of $\rho \approx 1/\sqrt{2}$ (see Section 18.1), which results in one overshoot and no undershoots.

Roll and yaw control are achieved by commanding the y axis electromagnet based on magnetometer and horizon scanner roll angle data [Stickler, et al., 1976].

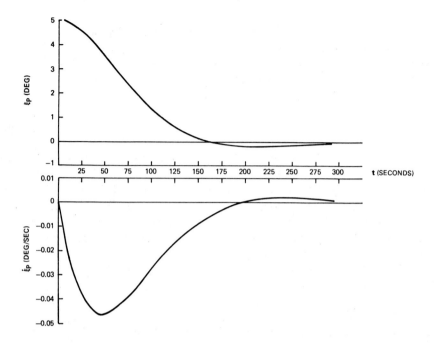

Fig. 18-16. HCMM Pitch Loop Response to a 5-Deg Error

The electromagnet strength is commanded according to the control law

$$D_y = K_N \dot{B}_y + K_P B_x \xi_r \qquad (18\text{-}86)$$

where **B** is the measured magnetic field in the body and K_P and K_N are the precession and nutation gain, respectively. Although the magnitude of D_y is limited to 10 A·m² (10,000 pole·cm) by hardware constraints, we will ignore this complication in the subsequent analysis. Substitution of Eqs. (18-48) and (18-63) into Eq. (18-86) gives the control torque

$$N_C = \mathbf{D} \times \mathbf{B} = \left\{ k_n \left[\sin\lambda (2\dot{\xi}_r + \omega_o \xi_y) - \cos\lambda (\dot{\xi}_y - 2\omega_o \xi_r) \right] + k_p \cos\lambda \xi_r \right\} \begin{bmatrix} \sin\lambda \\ 0 \\ -\cos\lambda \end{bmatrix}$$

$$(18\text{-}87)$$

where the gains, magnetic field strength, and unit conversions have been absorbed into the constants k_n and k_p. Substituting Eq. (18-87) into Eq. (18-53) and taking the Laplace transform leads to the coupled roll and yaw equations in matrix notation as

$$M(s)\, \mathcal{L} \begin{bmatrix} \xi_r \\ \xi_y \end{bmatrix} \equiv M(s) \begin{bmatrix} \Xi_r \\ \Xi_y \end{bmatrix} = \mathcal{L} \begin{bmatrix} N_{Ex} \\ N_{Ez} \end{bmatrix} \qquad (18\text{-}88a)$$

where

$$M(s) = \begin{bmatrix} I_x s^2 - 4k_n s \sin^2\lambda \\ + \left[4\omega_o^2(I_y - I_z) + h\omega_o - \sin 2\lambda(2k_n\omega_o + k_p)\right] & \left[h - (I_x - I_y + I_z)\omega_o + k_n\sin 2\lambda\right]s \\ & - 2k_n\omega_o\sin^2\lambda \\ \hline \left[(I_z + I_x - I_y)\omega_o - h + k_n\sin 2\lambda\right]s & I_z s^2 - k_n s\cos^2\lambda + (I_y - I_x)\omega_o^2 \\ + (k_p + 2\omega_o k_n)\cos^2\lambda & + h\omega_o + k_n\omega_o\sin 2\lambda/2 \end{bmatrix}$$

$$(18\text{-}88\text{b})$$

The treatment of the variable $\omega_o t = \lambda$ as a constant in Eq. (18-88) should be noted. The nonlinear differential equation may be solved without resort to this approximation by the technique of *multiple time scales* [Alfriend, 1975] which is based on the two widely differing periods that characterize the HCMM dynamics, i.e, the nutation period (20 seconds) and the orbital period (100 minutes). Here, we are concerned with a qualitative description of the HCMM dynamics as a function of the mean anomaly or subsatellite latitude.

Although many approximations were employed in obtaining Eq. (18-88) and simulations using detailed hardware and environmental models (particularly for the magnetic field) are required to evaluate control system performance, most of the characteristics of the HCMM control system are contained in the relatively simple model described by that equation. For a given latitude and control gains, the zeros of the characteristic equation

$$\det(M(s)) = 0 \qquad (18\text{-}89)$$

may be computed. In general, there are four roots to the fourth-order Eq. (18-89). In the absence of control torques, these roots are pure imaginary, $\pm i\omega_1$ and $\pm i\omega_2$, where $\omega_1 \approx h/\sqrt{I_x I_z}$ is the nutation frequency and $\omega_2 \approx \omega_o$ is the orbital frequency. With nutation control but no precession control ($k_n < 0$, $k_p = 0$), the roots are complex conjugate pairs with negative real parts and the system is damped and stable. The damping time constant associated with the nutation, τ_n, is shown as a function of latitude in Fig. 18-17(a); at the Equator ($\lambda = 0$) $\tau_n \approx -0.6/k_n$.

The Routh-Hurwitz criteria (see Section 18.1) may be applied to Eq. (18-89) to obtain the necessary conditions for stable precession control as

$$\left[4\omega_o^2(I_y - I_z) + h\omega_o - \sin 2\lambda(2k_n\omega_o + k_p)\right]\left[\omega_o^2(I_y - I_z) + h\omega_o + \tfrac{1}{2}k_n\omega_o\sin 2\lambda\right]$$
$$+ \tfrac{1}{2}k_n\omega_o\sin^2 2\lambda(2k_n\omega_o + k_p) > 0 \qquad (18\text{-}90\text{a})$$

$$-4k_n\sin^2\lambda\left[(I_y - I_x)\omega_o^2 + h\omega_o + \tfrac{1}{2}k_n\omega_o\sin 2\lambda\right]$$
$$-k_n\cos^2\lambda\left[4(I_y - I_z)\omega_o^2 + h\omega_o - (2k_n\omega_o + k_p)\sin 2\lambda\right]$$
$$-2k_n\omega_o\sin^2\lambda\left[(I_y - I_x - I_z)\omega_o + h - k_n\sin 2\lambda\right] \qquad (18\text{-}90\text{b})$$
$$-(2k_n\omega_o + k_p)\cos^2\lambda\left[(I_y - I_x - I_z)\omega_o + h + k_n\sin 2\lambda\right] > 0$$

$$k_n < 0 \qquad (18\text{-}90\text{c})$$

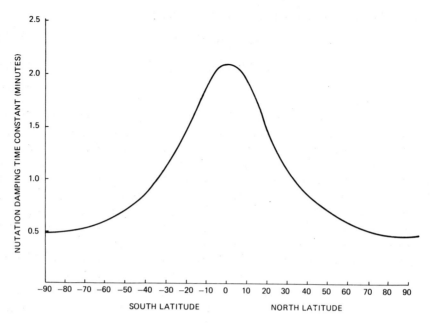

(a) NUTATION DAMPING TIME CONSTANT

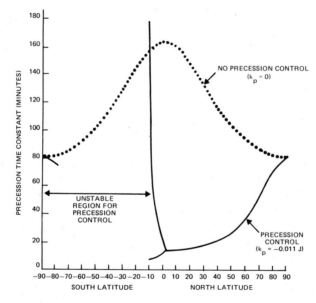

(b) PRECESSION TIME CONSTANT

Fig. 18-17.   Control System Time Constants, as a Function of Latitude (Spacecraft is Traveling North). See text for explanation.

For HCMM, the spacecraft parameters satisfy the inequalities

$$h \gg |4(I_y - I_z)\omega_o|$$
$$h \gg |(I_y - I_x)\omega_o|$$
$$|k_p| \gg |2k_n\omega_o| \qquad (18\text{-}91)$$

and Eq. (18-90) may be rewritten to good approximation as

$$h\omega_o - k_p \sin 2\lambda > 0 \tag{18-92a}$$

$$-k_n \omega_o (6 \tan^2\lambda + 1) - k_p > 0 \tag{18-92b}$$

$$k_n < 0 \tag{18-92c}$$

In Fig. 18-17(b), the dotted line shows that the mission attitude (pitch=roll =yaw=0) is a position of stable equilibrium even in the absence of active precession control, although the time constant is too long (approximately 120 minutes) to counter the effect of typical disturbance torques. The solid line shows the precession time constant for the control law defined by Eq. (18-86). For the HCMM parameters, this precession control law is ineffective in the Southern Hemisphere and unstable between 14° and 76° south latitude when the spacecraft is traveling north. (The Northern Hemisphere is the region of ineffective control system performance when the spacecraft is traveling south.)

Consequently, the HCMM control system deactivates precession control (i.e., sets $k_p = 0$) whenever $|B_z/B_x| > 1.4$. This "magnetic blanking" results in active control only within about 35 deg of the equator. Detailed parametric studies can thus be conducted to establish near-optimal gains and control laws and to obtain regions of stability by solving algebraic equations without the need for time-consuming simulation.

SEASAT. As outlined in Table 18-2, pitch control for SEASAT-A is essentially the same as for HCMM; however, roll/yaw control is achieved using roll and yaw reaction wheels with the control torques based on the horizon scanner roll error signal and wheel speeds [Beach, 1976]:

$$-\dot{h}_x = -K_r\xi_r - K_{\dot{r}}\dot{\xi}_r - \omega_o h_z$$

$$-\dot{h}_z = K_{ry}\xi_r + \omega_o h_x \tag{18-93}$$

The commanded roll reaction wheel torque, $\dot{h}_x$, based on position and rate errors, provides a roll restoring torque together with damping. The wheel-speed-dependent terms in Eq. (18-93) cancel like terms in Euler's equation and effectively remove the roll and yaw dependence upon the reaction wheel speeds (although, of course, the gyroscopic pitch momentum wheel coupling remains). The yaw wheel is torqued 180-deg out of phase with the roll error signal to provide yaw damping.

Substitution of the wheel control torque, Eq. (18-93) into Eq. (18-53) yields the roll/yaw equation in Laplace transform notation as

$$\begin{bmatrix} I_x s^2 + K_{\dot{r}}s + K_x + K_r & Hs \\ -Hs - K_{ry} & I_z s^2 + K_z \end{bmatrix} \begin{pmatrix} \Xi_r(s) \\ \Xi_y(s) \end{pmatrix} = \mathcal{L}\begin{pmatrix} N_{Ex} \\ N_{Ez} \end{pmatrix} \tag{18-94}$$

with the characteristic equation,

$$I_x I_z s^4 + K_{\dot{r}} I_z s^3 + \left[I_x K_z + I_z(K_x + K_r) + H^2 \right] s^2 + (K_{\dot{r}} K_z + K_{ry} H) s$$

$$+ K_z(K_x + K_r) = 0 \tag{18-95}$$

where $K_x = 4\omega_o^2(I - I_z) + h\omega_o = 0.011$ N·m, $K_z = h\omega_o = 0.026$ N·m, $H = h - I_z\omega_o = 21.3$ N·m·s, and $I_x = I_y \equiv I = 25100$ kg·m².

The selection of the gains, K_r, $K_{\dot{r}}$ and K_{ry}, is done most conveniently using the root locus plot shown in Fig. 18-18. With all three gains equal to zero, the two roots marked by x are pure imaginary with associated frequencies, $1.5\omega_o$ and $3.9\omega_o$, related to the nutation and the gravity-gradient torque. As the roll position gain, K_r, increases to 0.39 N·m, the roots migrate away from the origin along the imaginary s axis to the point marked by the arrowheads, which implies a faster response to a roll error. Addition of roll rate gain, $K_{\dot{r}}$, moves the roots into the third quadrant, which implies damping of both roll and yaw errors. Note, however, that the magnitude of the real part of one root remains small and the associated time constant is therefore large (74 minutes at $K_{\dot{r}} = 116$ N·m·s) and, consequently, the yaw damping is slow. The addition of a roll error to yaw torque gain, K_{ry}, substantially reduces the longer time constant and thereby improves the system's performance. The gain, $K_{ry} = 0.08$ N·m, is chosen such that the decay constants associated with the two roots are approximately equal, $\tau \approx (1.1\omega_o)^{-1} = 14$ minutes. The addition of the roll error to the yaw torque gain is fundamental to the WHECON wheel control concept which is frequently encountered in attitude control literature (see, for example, Dougherty, et al., [1968].)

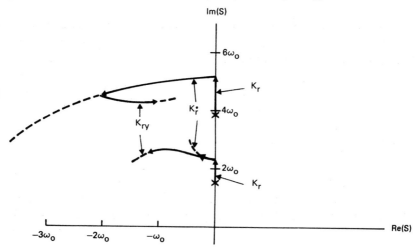

Fig. 18-18.   Root Locus Plot For SEASAT Control System. See text for explanation.

CTS.   The CTS control system, shown in Fig. 18-19, is similar to SEASAT except that gravity-gradient torques are negligible at the synchronous altitude of CTS and thrusters, offset at an angle α from the yaw axis in the roll/yaw plane, are used instead of reaction wheels to control the wheel axis attitude. The wheel angular momentum, h, is chosen such that

$$h \gg \max\left[\,|4\omega_o(I_y - I_z)|, |\omega_o(I_z + I_x - I_y)|, |\omega_o(I_y - I_x)|, |\omega_o(I_x - I_z)|\,\right] \quad (18\text{-}96)$$

and the roll/yaw equation (Eq. (18-53)) is approximately

$$\begin{bmatrix} I_x s^2 + h\omega_o & hs \\ -hs & I_z s^2 + h\omega_o \end{bmatrix}\begin{pmatrix} \Xi_r(s) \\ \Xi_y(s) \end{pmatrix} = \mathcal{L}\begin{bmatrix} N_{Ex} + N_{Cx} \\ N_{Ez} + N_{Cz} \end{bmatrix} \quad (18\text{-}97)$$

with the characteristic equation

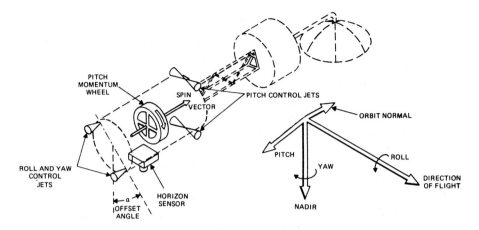

Fig. 18-19.   CTS Attitude Control System (From Dougherty, *et al.*, [1968])

$$[I_x s^2 + h\omega_o][I_z s^2 + h\omega_o] + h^2 s^2 = 0 \tag{18-98}$$

The roots of the characteristic equation are approximately $\pm i\omega_o$ and $\pm i\omega_n$, where $\omega_n = h/\sqrt{I_x I_z}$ is the nutation frequency. The control torque, N_c, is based on the roll error signal, derived from horizon sensors,

$$\mathcal{L}\begin{pmatrix} N_{Cx} \\ N_{Cz} \end{pmatrix} = -K_r(\tau s + 1)\Xi_r(s)\begin{pmatrix} \cos\alpha \\ -\sin\alpha \end{pmatrix} \tag{18-99}$$

or in the time domain

$$\begin{pmatrix} N_{Cx} \\ N_{Cz} \end{pmatrix} = -K_r(\tau\dot{\xi}_r + \xi_r)\begin{pmatrix} \cos\alpha \\ -\sin\alpha \end{pmatrix} \tag{18-100}$$

where K_r is the system gain and τ is the lead-time constant (see Section 18.1).

Substituting Eq. (18-100) into Eq. (18-97) gives the closed-loop roll/yaw equations as

$$\begin{bmatrix} I_x s^2 + K_r\cos\alpha(\tau s + 1) + h\omega_o & hs \\ -(h + K_r\tau\sin\alpha)s - K_r\sin\alpha & I_z s^2 + h\omega_o \end{bmatrix}\begin{bmatrix} \Xi_r(s) \\ \Xi_y(s) \end{bmatrix} = \mathcal{L}\begin{bmatrix} N_{Ex} \\ N_{Ez} \end{bmatrix} \tag{18-101}$$

Inverting Eq. (18-101) yields the control block diagram for the roll channel shown in Fig. (18-20). The closed-loop transfer function for this system is

$$\frac{\Xi_r(s)}{G_1(s)\mathcal{L}(N_{Ex}) + G_2(s)\mathcal{L}(N_{Ez})} = \frac{G_3(s)}{1 + K_r[\cos\alpha G_1(s) - \sin\alpha G_2(s)]H(s)G_3(s)} \tag{18-102}$$

where

$$G_1(s) \equiv I_z s^2 + h\omega_o$$

$$G_2(s) \equiv -hs$$

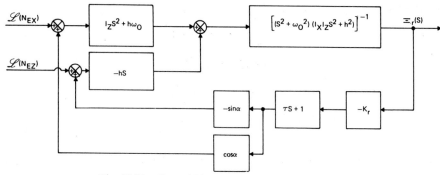

Fig. 18-20.   Control Block Diagram for CTS Roll Angle

$$G_3(s) \equiv \left[(s^2 + \omega_o^2)(I_x I_z s^2 + h^2) \right]^{-1}$$

$$H(s) \equiv \tau s + 1$$

Thus, the closed-loop poles of the system are the zeros of the characteristic equation

$$\zeta(s) = \left[G_3(s) \right]^{-1} + K_r \left[\cos\alpha G_1(s) - \sin\alpha G_2(s) \right] H(s) \qquad (18\text{-}103)$$

As the system gain is increased from $K_r = 0$ to $K_r = \infty$, the zeros of $\zeta(s)$ migrate from the zeros of $[G_3(s)]^{-1}$ to the zeros of $[\cos\alpha G_1(s) - \sin\alpha G_2(s)]H(s)$ which are located at

$$s = -1/\tau \qquad (18\text{-}104a)$$

and

$$s = \frac{-h\sin\alpha \pm \left(h^2\sin^2\alpha - 4I_z h\omega_o\cos^2\alpha \right)^{1/2}}{2I_z\cos\alpha} \qquad (18\text{-}104b)$$

The roots in Eq. (18-104b) are negative and real provided that

$$\tan\alpha \geqslant 2\sqrt{I_z\omega_o/h} \qquad (18\text{-}105)$$

For a high-gain system, the equality in Eq. (18-105) is chosen; this provides the best transient yaw response [Dougherty, et al., 1968].

The steady-state performance of the control system may be obtained by applying the final value theorem (see Appendix F) to Eq. (18-101):

$$\lim_{s \to 0} s \left\{ \left[I_x s^2 + K_r \cos\alpha(\tau s + 1) + h\omega_o \right] \Xi_r(s) + hs\Xi_y(s) \right\} = \lim_{s \to 0} s \mathcal{L}(N_{Ex})$$

$$\lim_{s \to 0} s \left\{ \left[-(h + K_r\tau\sin\alpha)s - K_r\sin\alpha \right] \Xi_r(s) + (I_z s^2 + h\omega_o)\Xi_y(s) \right\} = \lim_{s \to 0} s \mathcal{L}(N_{Ez})$$

$$(18\text{-}106)$$

which may be rewritten as

$$(K_r\cos\alpha + h\omega_o) \lim_{s \to 0} s\Xi_r(s) = \lim_{s \to 0} s \mathcal{L}(N_{Ex})$$

$$- K_r\sin\alpha \lim_{s \to 0} s\Xi_r(s) + h\omega_o \lim_{s \to 0} s\Xi_y(s) = \lim_{s \to 0} s \mathcal{L}(N_{Ez}) \qquad (18\text{-}107)$$

By the final value theorem,

$$\lim_{s \to 0} s \Xi_r(s) = \xi_r(\infty)$$

$$\lim_{s \to 0} s \Xi_y(s) = \xi_y(\infty)$$

$$\lim_{s \to 0} s \mathcal{L}(N_{Ex}) = N_{Ex}(\infty)$$

$$\lim_{s \to 0} s \mathcal{L}(N_{Ez}) = N_{Ez}(\infty) \qquad (18\text{-}108)$$

and, hence, for a high-gain system such as thrusters, where $K_r \gg h\omega_o$,

$$\xi_r(\infty) = N_{Ex}(\infty)/K_r \cos\alpha$$

$$\xi_y(\infty) = [N_{Ez}(\infty) + \tan\alpha N_{Ex}(\infty)]/h\omega_o \qquad (18\text{-}109)$$

Equation (18-109) may be used to estimate the system gain, K_r, and angular momentum, h.

18.4 Nutation and Libration Damping

Ashok K. Saxena

A spacecraft undergoes periodic motion if it is disturbed from a stable equilibrium position. For a spin-stabilized spacecraft, this periodic motion is rotational and is known as *nutation* (Section 15.1), whereas for a gravity-gradient stabilized spacecraft, it is oscillatory and is known as *libration*.

Nutation and libration occur as a result of control or environmental torques, separation from the launch vehicle, or the motion of spacecraft subsystems such as the tilting of experiment platforms or the extension of booms and arrays. Normally, an attempt is made to suppress or *damp* this motion because it affects the performance of sensors, pointed instruments, and antennas. However, Weiss, *et al.*, [1974] have shown how nutation can be advantageously used to scan the Earth. In such cases, a desired scan motion can be reversed without the use of energy by exciting controlled nutation modes.

Nutation or libration can be damped by passive or active devices. A *passive damper* is one which does not require attitude sensing, is driven by the motion itself, and dissipates energy. The frequency of the damper is intentionally kept near or equal to the rigid body frequency so that it significantly affects the motion of the spacecraft. An *active* nutation damper may be used if the initial amplitude of the motion is large, if the damping time of a passive damper is prohibitively long, or if passive damping leads to an undesirable final state (Section 15.2). In such cases, the attitude control system provides the necessary damping torques.

18.4.1 Passive Nutation Damping

As discussed in Sections 15.2 and 16.2, a rigid spacecraft can be stabilized by spinning it about the axis of maximum or minimum moment of inertia, called the *major* and *minor* axes, respectively. Nutation occurs if a spacecraft does not spin about a principal axis. Thus, the problem of *nutation damping* is that of aligning the nominal spin axis with the angular momentum vector by dissipating the excess kinetic energy associated with nutational motion. For a rigid body, this is possible only if the spin axis is the *major axis*, i.e., the principal axis having the largest moment of inertia. Table 18-3 summarizes the characteristics of several types of passive dampers. Real spacecraft always have some damping and associated energy dissipation. This may be either inherent in the system (structural damping), due to spacecraft components (fuel slosh, heat pipes), or due to the nutation damper hardware. Lord Kelvin [Chatayev, 1961] showed that a body which has been stabilized by gyroscopic means can lose its stability in the presence of dissipative forces. In 1958, Bracewell and Garriott [1958] showed that a slightly flexible spacecraft with no rotors or motors can be spin stabilized only about its major axis. During the course of publication, this result was confirmed when Explorer I, launched in February 1958, started tumbling in the first orbit because it was spinning about its minor axis. A dual-spin spacecraft with two axisymmetric rotating components can be stabilized about a minor axis in the presence of damping on one of the components [Landon and Stewart, 1964]. In this case, damping in the slower rotating component has a stabilizing effect and overcomes the destabilizing effect of damping in the faster rotating component.

Table 18-3.   Representative Passive Nutation Dampers

DAMPER	ENERGY DISSIPATION MECHANISM	CHARACTERISTICS
PENDULUM	FLUID FRICTION	STURDY, LONG LIFE
EDDY CURRENT	EDDY CURRENTS	DELICATE, HIGH-ENERGY DISSIPATION RATE, VARIABLE DAMPING CONSTANT
BALL-IN-TUBE	ROLLING AND VISCOUS FRICTION	STURDY, LONG LIFE, REMAINS TUNED FOR DIFFERENT SPIN RATES. CANNOT BE USED ON THE DESPUN PORTION OF A DUAL-SPIN SPACECRAFT
VISCOUS RING	FLUID FRICTION	SIMPLE CONSTRUCTION, LONG LIFE

The earliest duel-spin satellite (OSO) used a two-degree-of-freedom pendulum *nutation damper* which consisted of a brass ball mounted at the top of a flexible steel rod [Cloutier, 1975]. The damping was provided by immersing the ball in silicone. Currently, one-degree-of-freedeom nutation dampers, such as those described below, are preferred.

Mathematical techniques used to study passive nutation damping include the Energy Sink method used by Likins [1967] and the Routh-Hurwitz stability method [Likins, 1967]. Nutational stability has been studied using Liapunov's second method by Pringle [1969]. If the satellite has many rotating components with many dampers, the resulting equations have periodically varying coefficients and stability can be studied using Floquet analysis [Meirovitch, 1970]; this approach has been used by Johnson [1974].

Eddy Current Damper.   In an eddy current damper, the energy dissipation required for nutation damping is provided by the motion of a conducting plate relative to a magnet. The energy dissipation rate per unit weight due to the generation of eddy currents in the conductor is much greater than that of fluid dampers.

A typical pendulum eddy current damper, such as that used by the SAS series, consists of a Ni/Pt torsion wire parallel to the spin axis. The wire carries a pendulous copper vane which oscillates between the poles of an electromagnet. The drag force is proportional to the relative velocity between the vane and the electromagnet. If d and ρ are the thickness and resistivity of the vane, and B is the magnetic induction between the poles, then the damping constant, c, is given by [Haines and Leondes, 1973]

$$c = \frac{KB^2d}{\rho}$$

where K is a constant which depends on the shape and size of the poles. Eddy current dampers have the advantage of a variable damping constant, because the strength of the electromagnet can be changed by ground command. The SAS dampers could also be tuned in flight for different spin rates by changing the spring stiffness of the damper and hence its frequency of vibration.

Ball-In-Tube Damper. A ball-in tube damper, shown in Fig. 18-21, consists of a closed, curved tube in which a ball is allowed to roll freely. The damping caused by rolling friction may be augmented by viscous damping if the tube is filled with a viscous fluid. The ends of the tube may have energy-absorbing bumpers. The damper behaves like a centrifugal pendulum and its frequency of vibration is directly proportional to the spin rate of the body on which it is mounted. Hence, if such a damper is tuned initially, it remains tuned for different spin rates. These dampers are mounted on the spinning portion of a dual-spin spacecraft and on single-body spinning spacecraft. They have been used by most of the early ESRO spacecraft, including ESRO-II and -IV and HEOS-1 and -2.

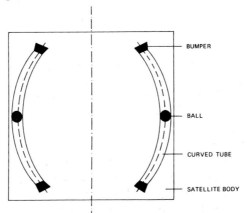

Fig. 18-21. Ball-in-Tube Nutation Damper

Viscous Ring Damper. Viscous ring dampers dissipate energy by fluid motion inside a ring. Although the study of these dampers began in 1960 [Carrier and Miles, 1960], interest in this type of damper has increased because of energy dissipation in heat pipes. A *heat pipe* is typically a fluid-filled aluminum pipe used to maintain near-isothermal conditions during the nonspinning phase of a spacecraft mission. During the spinning phase, the heat pipes act like viscous ring dampers. ATS-5, launched in August 1969, was spinning about its minor axis and was supposed to spin about its major axis after the apogee boost motor had fired and the casing had been ejected. However, the unexpectedly high energy dissipation in the heat pipe caused the spacecraft to enter a flat spin before the casing was ejected. After ejection of the casing, the spacecraft started spinning about the desired major axis, although the spin direction was opposite that desired. Viscous ring dampers may be mounted either in, or perpendicular to, the spin plane. We discuss first the performance of a damper mounted in the spin plane and later discuss a simplified model of a damper mounted in a plane perpendicular to the spin plane.

The damper flown on the ARYABHATA satellite consists of a fiberglass tube partially filled with mercury. The damping produced by viscous ring dampers depends on whether the flow through the tube is laminar or turbulent. For small nutation angles ($\leqslant 1$ deg), the fluid in a damper mounted in the spin plane is spread around the outer portion of the tube and has a free surface at the center of the tube. In such a mode, the damper performance depends on the frequency of the surface waves. For larger nutation angles, the fluid acts as a lumped mass or slug. The motion of the slug depends on the viscous drag and centripetal acceleration. If the nutation angle is small (1 deg to 10 deg), the viscous drag is large compared with the centripetal acceleration and the slug is dragged around the body with a small oscillatory motion. This is called the *spin-synchronous* mode and a small oscillation is superimposed on the nutation angle decay [Alfriend, 1974]. For larger nutation angles, the force due to centripetal acceleration exceeds the viscous drag force and the slug rotates with respect to the body at the body nutation rate. In this *nutation-synchronous* mode the slug is slightly offset from the $\mathbf{Z/L}$ plane because of the viscous drag. The transition angle between the spin-synchronous and the nutation-synchronous modes depends on the damper parameters and it is possible that the "spin-synchronous" mode may not exist if the ring is not eccentric to the spin axis.

A simple model of a viscous ring damper mounted in a plane parallel to the spin axis of a dual-spin spacecraft is shown in Fig. 18-22. The damper is modeled as a viscously coupled momentum wheel of momentum h, radius of gyration x_o, fluid mass m, and damping constant c. The angular momentum of the system is given by

$$\mathbf{L} = I\boldsymbol{\omega} - h\hat{\mathbf{k}}$$

where we have assumed that the wheel momentum is along the negative z axis. I is the moment-of-inertia tensor of the spacecraft and $\boldsymbol{\omega}$ is the spacecraft angular velocity. The model equations are

$$I_x\dot{\omega}_x = N_x + (I_z - I_x)\omega_y\omega_z + h\omega_y - I_o\dot{u}$$

$$I_y\dot{\omega}_y = N_y + (I_z - I_x)\omega_x\omega_z - h\omega_x - I_o u\omega_z$$

$$I_z\dot{\omega}_z = N_z + (I_x - I_y)\omega_x\omega_y + \dot{h} + I_o u\omega_y$$

$$I_o\dot{u} = I_o\dot{\omega}_x - c_u$$

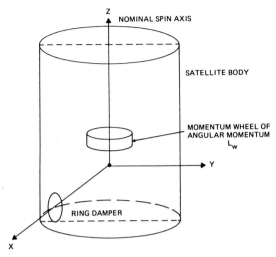

Fig. 18-22. Viscous Ring Nutation Damper in a Plane Parallel to the Spin Axis

where u is the angular velocity of the fluid relative to the spacecraft body, $I_o = mx_o^2$ is the moment of inertia of the damper, and N is the external torque. If we assume that the main body of the spacecraft is not spinning ($\omega_y = 0$) and that the damper does not significantly affect the moment of inertia of the spacecraft (mx_o^2 is small compared to the principal moments of interia), then when the damper is tuned, the optimum damping constant is

$$c_{opt} = mx_o^2\omega_p$$

where $\omega_p = h/\sqrt{I_x I_y}$ is the nutation frequency and the corresponding time constant is

$$\tau_{opt} = \frac{4I_z}{c_{opt}}$$

18.4.2 Active Nutation Damping

Active nutation damping involves the use of a sensor to measure the nutation phase and possibly its amplitude as described in Section 16.3, and an actuator to change the angular momentum of the system. Sensors commonly used are Sun sensors, horizon sensors, magnetometers, and rate gyros. The actuator may be a magnetic coil, a gas jet, a momentum/reaction wheel, or a control moment gyroscope. Active damping may be done by an *open-loop system*, in which case the actuator is activated by ground command, or by a *closed-loop system*, which requires some onboard logic circuitry between the sensors and the actuators.

Magnetic Nutation Damping. An open-loop magnetic damping method has been developed and used at Goddard Space Flight Center [Flatley, 1972] to handle circumstances in which a spacecraft may be nutating unexpectedly. This was used when SSS-1 developed unexpected nutation due to thermo-elastic flutter and when the pendulum damper of SAS-3 became stuck at its maximum amplitude.

We assume that the active nutation damping procedure is performed over a sufficiently short time so that we can neglect the change in the orientation of the geomagnetic field vector with the spacecraft's motion. When a constant current is passed through the spin axis coil, the direction of the resulting torque in inertial space is nearly constant, although it will change slightly depending on the size of the nutation cone. The transverse component L_T of the angular momentum vector rotates in space at the inertial nutation rate and in the body at the body nutation rate. If the spin axis is the major axis, the spin plane component of the geomagnetic field rotates in the opposite direction at the spin rate of the body and its direction can be measured by a magnetometer. The nutation phase is determined from Sun sensor data, as described in Section 16.3. A time and polarity for the current is selected so that the magnetic torque is opposite in direction L_T for half the inertial nutation period, after which the polarity of the current is reversed. The resulting torque will oppose L_T over the second half of the inertial nutation period. Thus, this nutation damping technique consists of (1) the proper selection of a time and polarity for the magnetic coil current and (2) reversal of it every half inertial nutation period. Thus, L_T and consequently the nutation can be reduced to zero.

An onboard control scheme is being considered for nutation damping for the HCMM spacecraft. The roll angle is observed by a horizon sensor and the nutation is sensed by the pitch axis magnetometer. The roll angle, ξ_r, can be minimized and the nutation damped simultaneously by using a pitch axis magnet control law of the form

$$M_y = K_N \dot{B}_y + K_P B_x \xi_r$$

where M_y is the coil strength, $\mathbf{B}$ is the geomagnetic field intensity, and K_N and K_P are constants. (See Section 18.3.)

Gas Jets. An open-loop gas jet nutation damping scheme was planned but not needed for the CTS spacecraft. As described in Section 1.1, CTS has 16 low-thrust engines which are mounted around a wheel at different distances from the spin axis. The nutation phase and amplitude can be determined on the ground from Sun sensor and rate gyro data (see Section 16.3). Nutation is damped by adding angular momentum equal and opposite to the transverse angular momentum and the timing is such that the spin axis is closest to the desired direction. (Section 16.3 presents an analysis of timing requirements for gas jet nutation damping.) The first of these conditions can be fulfilled by an appropriately timed single thrust. However, the second condition can be fulfilled only by changing the direction of the thrust vector in the body coordinate system. This can be done by choosing an appropriate combination of available jets. A precession maneuver from one attitude to another which includes nutation damping can ideally be accomplished by firing the jets twice, as shown in Fig. 18-23. The first impulse is such that the new angular momentum vector points halfway between the two attitudes and the second impulse of the same polarity and magnitude is fired after half a nutation period.

Dougherty, et al., [1968] have proposed an active, closed-loop gas jet nutation damping scheme known as the WHECON system. (See also Section 18.3.) The spacecraft is assumed to have a body-fixed pitch wheel and two gas jets slightly offset (about 10 deg) from the roll plane. The constant impulse jets are activated

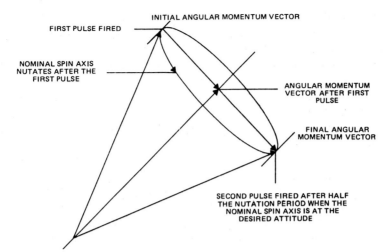

Fig. 18-23.   Gas Jet Precession Maneuver Including Nutation Damping

from roll angle data to keep the satellite within predetermined roll and yaw deadbands. The system is sensitive to changes in the nutation rate (wheel speed, moments of inertia) and to any change in the magnitude of the jet impulses. Iwens, *et al.*, [1974] have studied the stability of such a system and suggest that a time delay of about 5/8th nutation period be used between impulses. The control logic is such that when the roll deadband is encountered, a corrective pulse is fired followed by another pulse of the same polarity 5/8th nutation period later, provided that no other deadband activated impulse has been fired during this period.

Control Moment Gyroscope.   A single-axis control moment gyroscope whose gimbal can rotate about an axis perpendicular to the nominal spin axis can be used for active and semipassive nutation damping for a dual-spin satellite. In active damping, the nutation is sensed by an accelerometer whose output is used to control the gimbal angle. In the semipassive mode, the gimbal is restrained by a torsional spring and dashpot. In this case, the control moment gyroscope may be mounted on either the spinning or the nonspinning component and the stability criterion is the same as that for passive dampers. The damping time constant of such a system is much smaller than that of a passive system.

18.4.3 Libration Damping

Gravity-gradient stabilized spacecraft may librate as a result of the initial attitude acquisition process or environmental torques. For such spacecraft we are interested in aligning a principal axis to the local vertical, i.e., in *libration damping*. A gravity-gradient satellite is nominally in a stable equilibrium position when its minor axis lies along the local vertical and its major axis is perpendicular to the orbital plane.* In this section we discuss some of the methods used for passive libration damping. In the early 1950s, Roberson and Breakwell [1956] suggested

---

*Except for a small set of moments of inertia corresponding to the Delp region [DeBra and Delp, 1961].

that libration could be damped by the dissipation of energy in a flexible part of a spacecraft. Mobley and Fischell [1965] have suggested a method which utilizes eddy current rods rigidly attached to the spacecraft. Section 18.3 includes a detailed example of libration damping during the GEOS-3 attitude acquisition.

Spring Damper. The TRAAC spacecraft used a damper consisting of a spring with a small mass attached to a boom as shown in Fig. 18-24. The spring was released after boom deployment and could expand to about 12 m. As the spacecraft librated, the spring expanded and contracted, resulting in energy dissipation due to the high structural damping in the spring. This was provided by mechanically soft cadmium which covered the beryllium-copper wire of the spring. The spring was then coated with silver to prevent sublimation of the cadmium in the vacuum of space. This damper is more efficient for removing librations in the orbit plane than for librations in the plane perpendicular to the orbit.

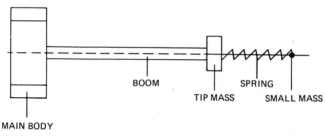

BOOM   SPRING

TIP MASS   SMALL MASS

MAIN BODY

Fig. 18-24.   Spring Libration Damper for a Gravity-Gradient Stabilized Spacecraft

Magnetically Anchored Eddy Current Damper. A damper such as that used by GEOS-3 consists of two concentric spheres which can move relative to each other and are separated by silicone oil to provide viscous damping. The inner sphere is attached to a magnet which aligns or anchors itself to the Earth's magnetic field vector. The outer sphere is attached to the spacecraft's boom and is made of pyrolitic graphite for diamagnetic centering of forces on the inner sphere and aluminum for energy dissipation through eddy currents. The damping torque N_d is

$$N_d = c\hat{\mathbf{B}} \times \left(\frac{d\hat{\mathbf{B}}}{dt} \right)$$

where c is the damping constant and $\hat{\mathbf{B}}$ is the direction of the net magnetic field in the body coordinate system. If A is the attitude matrix, ω the angular velocity of the satellite, and $\hat{\mathbf{B}}_I$ the direction of the magnetic field in inertial space, then the above expression can be rewritten in the form

$$N_d = c\left[A\left(\hat{\mathbf{B}}_I \times \hat{\mathbf{B}}_I \right) + \omega - \hat{\mathbf{B}}(\omega \cdot \hat{\mathbf{B}}) \right]$$

Because the eddy currents are generated by a strong permanent magnet, the damping is strong over a wide range of altitudes. After the spacecraft is in its equilibrium position, the magnet continues to track the Earth's magnetic field and thus creates a disturbance torque on the spacecraft. Under certain combinations of damping coefficient, orbital parameters and moments of inertia, the damper tends to move the spacecraft into a nonzero bias attitude. In such cases, a limit cycle exists because the gravity-gradient moment unloaded is equal to but out of phase with the momentum added due to the magnetic field.

Eddy Current Rods. Ferromagnetic rods which are coated with a conducting copper sheet and fixed along the principal axes of a spacecraft can be used for libration damping. As the spacecraft librates, eddy currents are generated in the rods because of the change in the geomagnetic field relative to the body coordinate system. The number of rods used does not create a proportional increase in the damping coefficient because the flux density is reduced in each rod. The instantaneous power dissipated is proportional to the square of the rate of change of the magnetic field vector along the longitudinal axis of a rod. Because the eddy currents are generated by the geomagnetic field, the damping produced is inversely proportional to the sixth power of the orbital radius and is not adequate at high altitudes.

References

1. Alfriend, Kyle T., "Partially Filled Viscous Ring Nutation Damper," *J. Spacecraft*, Vol. 11, p. 456–462, 1974.
2. ———, "Magnetic Attitude Control System for Dual-Spin Satellites," *AIAA Journal*, Vol. 13, p. 817, 1975.
3. Beach, S. W., *Linear Analysis of the SEASAT Orbital Attitude Control System*, Lockheed Missiles and Space Co., Inc., GCS/3874/5211, July 30, 1976.
4. Bracewell, R. N., and O. K. Garriott, "Rotation of Artificial Earth Satellites," *Nature*, Vol. 182, p. 760–762, Sept. 20, 1958.
5. Carrier, G. F., and J. W. Miles, "On the Annular Damper for a Freely Precessing Gyroscope," *J. of Applied Mech.*, Vol. 27, p. 237–240, 1960.
6. Chetayev, N. G., *The Stability of Motion*. New York: Pergamon Press, 1961.
7. Cloutier, G. J., *Variable Spin Rate, Two-Degrees-of-Freedom Nutation Damper Dynamics,* Paper No. AAS 75-045 AAS/AIAA Astrodynamics Specialist Conference, Nassau, Bahamas, July 28–30, 1975.
8. DeBra, D. B., and R. H. Delp, "Rigid Body Attitude Stability and Natural Frequencies in a Circular Orbit," *J. of Astonautical Sci.*, Vol. 8, p. 14–17, 1961.
9. DiStefano, Joseph J. III, Allen R. Stubberud, and Ivan J. Williams, *Schaum's Outline of Theory and Problems of Feedback and Control Systems*. New York: McGraw-Hill, Inc., 1967.
10. Dougherty, H. J., E. D. Scott, and J. J. Rodden, *Analysis and Design of WHECON-An Attitude Control Concept*, AIAA 2nd Communication Satellite Conference, San Francisco, CA., Apr. 1968.
11. Elson, Benjamin M., "Design Phase of Space Telescope Nears," *Aviation Week and Space Technology*, Vol. 107, No. 6, p. 54–59, Aug. 8, 1977.
12. Flatley, T., *Magnetic Active Nutation Damping on Explorer 45 (SSS-A)*, NASA X-732-72-140, GSFC, May 1972.
13. Glaese, J. R., H. F. Kennel, G. S. Nurre, S. M. Seltzer, and H. L. Shelton, "Low-Cost Space Telescope Pointing Control System," *J. Spacecraft*, Vol. 13, p. 400–405, 1976.
14. Greensite, Arthur L., *Elements of Modern Control Theory*. New York: Spartan Books, 1970.

15. Grell, M. G., *AEM-A Attitude Control Contingency Study*, Comp. Sc. Corp., CSC/TM-76/6203, Oct. 1976.
16. Haines, Gordon A. and Corelius T. Leondes, "Eddy Current Nutation Dampers for Dual-Spin Satellites," *J. of Astronautical Sci*, Vol. 21 p. 1, 1973.
17. Hoffman, D. P., "HEAO Attitude Control Subsystem—A Multimode Multimission Design," *Proceedings AIAA Guidance and Control Conference*, San Diego, CA., p. 89, Aug. 1976.
18. Hsu, J. C., and A. U. Meyer, *Modern Control Principles and Applications*, New York: McGraw-Hill, Inc., 1968.
19. Iwens, R. P., A. W. Fleming, and V. A. Spector, *Precision Attitude Control With a Single Body-Fixed Momentum Wheel*, AIAA Paper No. 74-894, Aug. 1974.
20. Johnson, D. A., "Effect of Nutation Dampers on the Attitude Stability of n-Body Symmetrical Spacecraft," *Gyrodynamics* (P. Y. Willems, ed.). New York: Springer-Verlag, 1974.
21. Kaplan, Marshal H., *Modern Spacecraft Dynamics and Control*. New York: John Wiley & Sons, Inc., 1976.
22. Korn, Granino A., and Theresa M. Korn, *Mathematical Handbook for Scientists and Engineers*. McGraw-Hill, Inc., New York, 1968.
23. Landon, Vernon D., and Brian Stewart, "Nutational Stability of an Axisymmetric Body Containing a Rotor," *J. Spacecraft*, Vol. 1, p. 682–684, 1964.
24. Lerner, G. M. and K. P. Coriell, *Attitude Capture Procedures for GEOS-3*, AIAA Paper No. 75-029, AIAA Astrodynamics Specialist Conference, Nassau, Bahamas, July 28–30, 1975.
25. Lerner, G. M., w. Huang, and M. D. Shuster, *Analytic Investigation of the AEM-A/HCMM Attitude Control System Performance*, AAS/AIAA Astrodynamics Conference, Jackson Hole, WY., Sept. 1977.
26. Levinson, David A., "Effects of Meteoroid Impacts on Spacecraft Attitude Motion," *J. of Aeronautical Sci.*, Vol. 25, p. 129–142, 1977.
27. Likins, Peter W., "Attitude Stability Criteria for Dual Spin Spacecraft," *J. Spacecraft*, Vol. 4, p. 1638–1643, 1967.
28. Lyons, M. G., K. L. Lebsock, and E. D. Scott, *Double Gimballed Reaction Wheel Attitude Control System for High Altitude Communication Satellites*, AIAA Paper No. 71-949, Aug. 1971.
29. Markley, F. L., *Attitude Control Algorithms for the Solar Maximum Mission*, AIAA Paper No. 78-1247, 1978 AIAA Guidance and Control Conference, Palo Alto, CA., 1978.
30. Meirovitch, L., *Methods of Analytical Dynamics*. New York: McGraw-Hill, Inc., 1970.
31. Melsa, James L. and Donald Shultz, *Linear Control Systems*. New York: McGraw-Hill, Inc., 1969.
32. Mobley, F. F., and R. E. Fischell, "Orbital Results From Gravity-Gradient Stabilized Satellites," NASA SP-107, *Symposium on Passive Gravity-Gradient Stabilization*, Ames Research Center, Moffett Field, CA., May 1965.
33. Pettus, W. W., "Performance Analyses of Two Eddy Current Damping Systems for Gravity-Gradient Stabilized Satellites," *Proceedings of the Symposium on Gravity-Gradient Attitude Stabilization*, El Segundo, CA., Dec. 1968.

34. Pisacane, Vincent L., Peter P. Pardoe, and B. Joy Hook, "Stabilization System Analysis and Performance of the GEOS-A Gravity Gradient Satellite (EXPLORER XXIX)," *J. Spacecraft*, Vol. 4, p. 1623–1630, 1967.

35. Pringle, R., Jr., "Stability of the Force Free Motions of a Dual-Spin Spacecraft," *AIAA Journal*, Vol. 7, p. 1054–1063, 1969.

36. Repass, G. D., G. M. Lerner, K. P. Coriell, and J. S. Legg, Jr., *Geodynamics Experimental Ocean Satellites (GEOS-C) Prelaunch Report*, NASA X-580-75-23, GSFC, Feb. 1975.

37. Roberson, R. E., and J. V. Breakwell, "Satellite Vehicle Structure," United States Patent 3,031,154, April 24, 1962 (filed September 20, 1956).

38. Stickler, A. Craig, and K. T. Alfriend, "Elementary Magnetic Attitude Control System," *J. Spacecraft*, Vol. 13, p. 282–287, May 1976.

39. Weiss, R., R. L. Bernstein, Sr., and A. J. Besonis, *Scan-By-Nutation, A New Spacecraft Concept*, AIAA Paper No. 74-896, Aug. 1974.

CHAPTER 19

ATTITUDE MANEUVER CONTROL

19.1 Spin Axis Magnetic Coil Maneuvers
19.2 Spin Plane Magnetic Coil Maneuvers
   Momentum and Attitude Maneuvers, Optimal Command Procedures, Representative Example of AE-5 Magnetic Maneuvers
19.3 Gas Jet Maneuvers
19.4 Inertial Guidance Maneuvers
   Single-Axis Slews, Multiple-Axis Slews
19.5 Attitude Acquisition
   Classification of Attitude Acquisition, Acquisition Maneuvers, Representative Acquisition Sequence

This chapter describes procedures for reorienting a spacecraft from one attitude to another. Sections 19.1 and 19.2 describe maneuvers using magnetic coils and Section 19.3 describes maneuvers using gas jets. Section 19.4 then describes procedures for inertial guidance maneuvers. Finally, Section 19.5 discusses the special class of *attitude acquisition maneuvers* in which the spacecraft starts in an unknown or uncontrolled attitude and ends in an attitude appropriate for mission operations. This chapter uses the general attitude control concepts introduced in Section 15.3 as well as the equations of spacecraft motion presented in Sections 16.1 and 16.2.

19.1 Spin Axis Magnetic Coil Maneuvers

B. L. Gambhir
Des R. Sood

In this section we consider precessional motion generated by a magnetic coil wound around the spin axis of a nonnutating spin-stabilized spacecraft. For such a spacecraft, the angular momentum $\mathbf{L}$ can be expressed as

$$\mathbf{L} = L\hat{\mathbf{s}} \qquad (19\text{-}1)$$

where L is the magnitude of the angular momentum and $\hat{\mathbf{s}} \equiv \hat{\boldsymbol{\omega}}$ is a unit vector along the spin axis. The magnetic moment, $\mathbf{M}$, of an electromagnet aligned with the spin axis (i.e., the spin-axis-coil) may be expressed as

$$\mathbf{M} = m_0 u\hat{\mathbf{s}} \qquad -1 \leqslant u \leqslant 1 \qquad (19\text{-}2)$$

where m_0 is the maximum attainable magnetic moment, and u is a commandable coil state parameter which is proportional to the current through the coil and is either positive or negative depending on whether the direction of current flow is counterclockwise or clockwise relative to $\hat{\mathbf{s}}$ (see Section 6.7). The magnetic dipole generated by the coil interacts with the geomagnetic field, $\mathbf{B}$, to produce a torque, $\mathbf{N}$, on the spacecraft, given by (see, for example, Jackson [1965])

$$\mathbf{N} = \mathbf{M} \times \mathbf{B} = m_0 u\hat{\mathbf{s}} \times \mathbf{B} \qquad (19\text{-}3)$$

By definition, the time rate of change of angular momentum is equal to the total impressed torque, i.e.,

$$\mathbf{N} \equiv \frac{d\mathbf{L}}{dt} \tag{19-4}$$

From Eq. (19-3), $\mathbf{N}$ is orthogonal to both $\mathbf{B}$ and $\mathbf{M}$. Because $\mathbf{M}$ is either parallel or antiparallel to $\hat{\mathbf{s}}$, the torque is also orthogonal to $\mathbf{L}$. Therefore, the magnitude of $\mathbf{L}$ remains constant, so that

$$\mathbf{N} = \frac{d\mathbf{L}}{dt} = L\frac{d\hat{\mathbf{s}}}{dt} \tag{19-5}$$

Combining Eqs. (19-3) through (19-5) gives

$$\frac{d\hat{\mathbf{s}}}{dt} = \left(\frac{m_0 u}{L}\right)\hat{\mathbf{s}} \times \mathbf{B} = \mathbf{\Omega}_p \times \hat{\mathbf{s}} \tag{19-6}$$

where

$$\mathbf{\Omega}_p \equiv -\left(\frac{m_0 u}{L}\right)\mathbf{B}$$

This is a well-known equation (see, for example, Goldstein, [1950]), describing the precession of $\hat{\mathbf{s}}$ about the magnetic field, $\mathbf{B}$, with an angular velocity, $\mathbf{\Omega}_p$, which is either parallel or antiparallel to $\mathbf{B}$ depending on the sign of the coil state parameter, u. This is illustrated in Fig. 19-1.

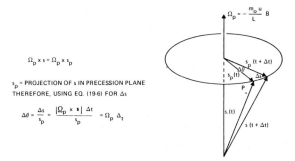

Fig. 19-1.   Spin Axis Precession Due to Interaction Between a Magnetic Dipole Aligned Along the Spin Axis and the Geomagnetic Field

It is instructive to express Eq. (19-6) in terms of time rates of change of the right ascension, α, and declination, δ, of the spin axis. In terms of these quantities, the celestial rectangular components of $\hat{\mathbf{s}}$ and $d\hat{\mathbf{s}}/dt$ can be written as

$$\hat{\mathbf{s}} = \cos\delta\cos\alpha\hat{\mathbf{x}} + \cos\delta\sin\alpha\hat{\mathbf{y}} + \sin\delta\hat{\mathbf{z}}$$

$$\frac{d\hat{\mathbf{s}}}{dt} = \cos\delta\frac{d\alpha}{dt}\hat{\mathbf{x}}_s + \frac{d\delta}{dt}\hat{\mathbf{y}}_s \tag{19-7}$$

where $\hat{\mathbf{x}}_s$ is a unit vector along $\hat{\mathbf{z}} \times \hat{\mathbf{s}}$ and is given by

$$\hat{\mathbf{x}}_s \equiv -\sin\alpha\hat{\mathbf{x}} + \cos\alpha\hat{\mathbf{y}}$$

Similarly, $\hat{\mathbf{y}}_s$ is a unit vector along $\hat{\mathbf{s}} \times \hat{\mathbf{x}}_s$ and is given by

$$\hat{\mathbf{y}}_s \equiv \hat{\mathbf{s}} \times \hat{\mathbf{x}}_s = -\sin\delta\cos\alpha\hat{\mathbf{x}} - \sin\delta\sin\alpha\hat{\mathbf{y}} + \cos\delta\hat{\mathbf{z}}$$

Taking components of Eq. (19-6) along $\hat{\mathbf{x}}_s$ and $\hat{\mathbf{y}}_s$, we obtain

$$\frac{d\alpha}{dt} = \frac{um_0}{L}\left[(B_x\cos\alpha + B_y\sin\alpha)\tan\delta - B_z\right] \tag{19-8}$$

$$\frac{d\delta}{dt} = \frac{um_0}{L}(-B_x\sin\alpha + B_y\cos\alpha) \tag{19-9}$$

Here, B_x, B_y, B_z are the celestial rectangular components of the geomagnetic field.

Figure 19-2 provides a physical interpretation of Eqs. (19-8) and (19-9). The torque component along $\hat{\mathbf{x}}_s$ rotates the equatorial projection of the spin vector around the celestial z axis and therefore is the cause of right ascension change. Similarly, the torque component along $\hat{\mathbf{y}}_s$ pulls the spin axis toward the celestial z axis, which results in declination change.

Integration of Eqs. (19-8) and (19-9) to accurately predict the total spin axis motion requires an accurate knowledge of the geomagnetic field. However, the analytical characteristics of magnetic control maneuvers may be obtained from the dipole model presented in Appendix H. We discuss spin axis maneuvers for two limiting cases of satellite orbits: equatorial and polar.

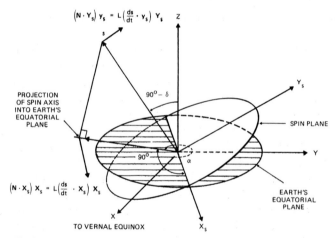

Fig. 19-2.   Resolution of Torque Into Components in the Spin Plane

For an equatorial orbit, the magnetic field components along the celestial rectangular coordinate axes are given (see Appendix H) by

$$B_{xeq} = \frac{M_G\sin\theta'_m}{2R^3}\left[\cos\alpha_m + 3\cos(2\nu' - \alpha_m)\right] \tag{19-10}$$

$$B_{yeq} = \frac{M_G\sin\theta'_m}{2R^3}\left[\sin\alpha_m + 3\sin(2\nu' - \alpha_m)\right] \tag{19-11}$$

$$B_{zeq} = -\frac{M_G\cos\theta'_m}{R^3} \tag{19-12}$$

where M_G is the strength of the geomagnetic dipole ($\cong 8.0\times10^{25}$ gauss·cm$^3 \cong 8.0\times 10^{15}$ Wb·m), R is the distance from the center of the Earth to the spacecraft, θ'_m is colatitude (≈ 168.6 deg) of the dipole, ν' is the azimuth in the orbit plane of the spacecraft position vector measured from the celestial x axis, and α_m is the right ascension of the dipole axis; $\alpha_m = \alpha_{m0} + \omega_E(t - t_0)$, where ω_E is the Earth's rotation rate and $\alpha_m = \alpha_{m0}$ at some reference time, t_0.

Note from Eqs. (19-10) through (19-12) that the equatorial magnetic field components, being proportional to $\sin\theta'_m (\approx 0.2)$, are much smaller than the component along the celestial z axis. Hence, for a satellite in an equatorial orbit, it is easier to accomplish right ascension changes than declination changes (see Eqs. (19-8) and (19-9)).

For a polar orbit, the magnetic field components are given by

$$B_{xp} = \frac{M_G}{R^3} \left\{ \frac{3}{2} \cos\theta'_m \cos\Omega \sin 2\nu' + \sin\theta'_m \left[3\cos\Omega \cos^2\nu' \cos(\alpha_m - \Omega) - \cos\alpha_m \right] \right\}$$

(19-13)

$$B_{yp} = \frac{M_G}{R^3} \left\{ \frac{3}{2} \cos\theta'_m \sin\Omega \sin 2\nu' + \sin\theta'_m \left[3\sin\Omega \cos^2\nu' \cos(\alpha_m - \Omega) - \sin\alpha_m \right] \right\}$$

(19-14)

$$B_{zp} = \frac{M_G}{R^3} \left[\frac{1}{2} \cos\theta'_m (1 - 3\cos 2\nu') + \frac{3}{2} \sin\theta'_m \sin 2\nu' \cos(\alpha_m - \Omega) \right] \quad (19\text{-}15)$$

where Ω is the right ascension of the ascending node and ν' is the azimuth in the orbital plane of the spacecraft position vector, measured from the ascending node. Note that the x and y components are of the same order of magnitude as the z component. Hence, for a satellite in a polar orbit, declination changes can be accomplished as easily as right ascension changes.

For both the equatorial and the polar orbit cases, the x and y components of the geomagnetic field are oscillatory, involving angular frequencies that are combinations of twice the orbital rate (from terms involving ν'), the Earth's rotation rate (terms involving α_m) and the rate of change of the ascending node, Ω. For near-Earth satellites, the orbital period is much smaller than the Earth's rotation period (a day), whereas the ascending node completes a cycle in several weeks or months. Therefore, the dominant oscillations in the x and y components of the geomagnetic field have twice the orbital frequency. Consequently, for a given coil state u, the time rate of change of declination (Eq. (19-9)) follows an approximate sine curve, whereas only a part of the right ascension rate is oscillatory in nature (Eq. (19-8)). The general characteristics of these results are valid even when more exact models of the geomagnetic field are used and are valuable for the development of magnetic control strategies.

Magnetic Control Strategies   The reorientation of near-Earth satellites using a spin axis coil dates from TIROS-2 in 1960. In early missions, magnetic control potentialities could not be fully exploited because of the need for ground contact to change the coil state, u (magnitude and/or polarity of the coil current). The coil was left on for a number of orbits or days, during which time, the spin axis would precess at a very slow average rate. This mode of control is referred to as *continuous torquing*.

The launch of TIROS-9, in January 1965, saw an innovation in magnetic control system design. Through an onboard timer, the coil polarity was switched four times per orbit. Therefore, this system was called *quarter-orbit magnetic attitude control*, or *QOMAC*. QOMAC takes advantage of the fact that the geomagnetic field oscillates with a period of approximately half the orbital period by switching coil polarity in consonance with the geomagnetic field. To allow some flexibility in maneuver planning, QOMAC hardware provides for control of the initial phase and polarity and the switching period. Both continuous and QOMAC torquing have been used for OSO-8. The decision as to which to use for a particular maneuver depends on whether a straight line or a square wave function best represents the desired coil state history for that maneuver.

Recent spacecraft have been equipped with *delayed command systems (DCSs)*, in which a preselected sequence of coil-state-change commands, covering an extended period of time, is loaded into an onboard memory during a station pass and is executed at the appointed times without subsequent ground contact. This provides maximum flexibility and is limited only by the size of the onboard memory. The improved precision in the timing of coil commands provides three improvements over the QOMAC method: (1) minimization of the time required to complete a maneuver, (2) minimization of the arc-length error between the desired attitude and the attitude obtainable within a specified time, and (3) minimization of the energy expended for completion of a maneuver.

The control strategies used for generating the preselected commands for the DCS load are normally based on the optimization of one or more of these three improvement criteria as dictated by mission requirements and hardware limitations. One such optimization control algorithm, developed by Werking and Woolley [1973], is adaptable to varying mission constraints and has been used successfully for SAS-3 and several spacecraft in the AE series. The governing equations of this algorithm are based on energy optimization. The time optimal conditions are obtained as a special case as shown later. Because energy expended is proportional to the square of the current, we wish to minimize

$$J = \int_{t_1}^{t_2} u^2(t) \, dt \tag{19-16}$$

subject to the end point conditions

$\hat{s}(t_1) = \hat{s}_1$, the initial attitude

$\hat{s}(t_f) = \hat{s}_2$, the final attitude and $t_f \leqslant t_2$

t_f is the time when the final attitude is attained   (19-17)

$u(t_1) = u(t_2) = 0$

and the attitude dynamics equations obtained from Eq. (19-6) with the addition of environmental disturbance torques **D**; thus,

$$\frac{d\hat{s}}{dt} = \hat{s} \times \left(\frac{m_0 u}{L} \mathbf{B} + \frac{1}{L} \mathbf{D} \right) \tag{19-18}$$

Here, the components of **D** orthogonal to $\hat{s}$ have been expressed in the form $\hat{s} \times \mathbf{D}$. The disturbance torques along $\hat{s}$ can only produce small fluctuations in the magnitude of the angular momentum vector, $\mathbf{L} = L\hat{s}$, and therefore have minimal effect on the attitude dynamics. The disturbance torques along the spin axis have been neglected in Eq. (19-18), and L is assumed to be constant.

The attitude dynamics constraint can be directly incorporated into the optimization integral by introducing three Lagrange multipliers represented by the vector λ. Thus,

$$J = \int_{t_1}^{t_2} \left\{ u^2 + \left[\hat{s} \times \left(\frac{m_0 u}{L} \mathbf{B} + \frac{1}{L} \mathbf{D} \right) - \frac{d\hat{s}}{dt} \right] \cdot \lambda \right\} dt \qquad (19-19)$$

Integrating by parts the $(d\hat{s}/dt) \cdot \lambda$ term gives

$$J = \hat{s}_1 \cdot \lambda(t_1) - \hat{s}_2 \cdot \lambda(t_2) + \int_{t_1}^{t_2} \left[u^2 + \hat{s} \times \left(\frac{m_0 u}{L} \mathbf{B} + \frac{1}{L} \mathbf{D} \right) \cdot \lambda + \hat{s} \cdot \frac{d\lambda}{dt} \right] dt \qquad (19-20)$$

The conditions under which J is at an extremum (maximum, minimum, or stationary) are obtained by requiring that the variation of the integral resulting from infinitesimal changes in the path along which the system evolves (from the state $[s_1, u(t_1)]$ to the state $[s_2, u(t_2)]$) must vanish identically; i.e., the partial derivative of the integrand with respect to u and the gradient with respect to $\hat{s}$ must be zero. Thus, for energy optimization, the following conditions must be satisfied:

$$2u + \left(\hat{s} \times \frac{m_0}{L} \mathbf{B} \right) \cdot \lambda = 0 \qquad (19-21)$$

$$\lambda \times \left(\frac{m_0 u}{L} \mathbf{B} + \frac{1}{L} \mathbf{D} \right) = \frac{d\lambda}{dt} \qquad (19-22)$$

In writing Eq. (19-22), it has been assumed that the disturbance torques are independent of the attitude over the range of the integral. This assumption is valid for maneuvers involving small arc motions, but becomes questionable for large maneuvers.

Equation (19-21) shows that the component of λ along $\hat{s}$ has no effect on the coil state u and, hence, does not influence the path of the system in the $[\hat{s}(t), u(t)]$ space. Also, Eqs. (19-18) and (19-22) correspond to precession of both λ and $\hat{s}$ about the instantaneous force field $m_0 u \mathbf{B} + \mathbf{D}$ with the instantaneous angular frequency $(m_0 u \mathbf{B} + \mathbf{D})/L$. (See the discussion following Eq. (19-6).) Therefore, λ and s maintain constant magnitudes and the angle between them remains fixed. Thus, without loss of generality, the constant angle between $\hat{s}$ and λ may be set equal to 90 deg. To underscore this fact, let us define a unit vector, $\hat{q}$, which is 90 deg ahead of λ in the plane normal to $\hat{s}$,

$$\hat{q} \equiv \hat{s} \times \hat{\lambda} \qquad (19-23)$$

Because $\hat{q}$ differs from $\hat{\lambda}$ only in a phase angle, it obeys the same dynamic equation

$$\frac{d\hat{q}}{dt} = \hat{q} \times \left(\frac{m_0 u}{L} \mathbf{B} + \frac{1}{L} \mathbf{D} \right) \qquad (19-24)$$

Equation (19-21) can now be rewritten in terms of $\hat{s}$, $\hat{q}$, and λ, using $\hat{s} \times \mathbf{B} \cdot \lambda = -\hat{s} \times \lambda \cdot \mathbf{B}$, as

$$2u = \frac{m_0 \hat{q}}{L} \cdot \mathbf{B} \lambda \qquad (19-25)$$

Thus far we have assumed that u is a continuous variable. For most missions, the coil current can be set to only a few levels and the appropriate magnitude is easily selected on the basis of the amount of arc motion desired, so that the only commands to be computed are coil-on, coil-off, and polarity selection. This corresponds to three possible values of u, namely, $u = 1$, 0, or -1. Under these conditions, exact energy optimization cannot be attained. Heuristically, the closest approach to energy optimization will be to replace Eq. (19-25) by the following set of rules. For any time, t, in the maneuver interval, set

$$u = \begin{cases} 1 & \text{if } \quad \dfrac{m_0}{L}\mathbf{B}\cdot\hat{\mathbf{q}} \geqslant \dfrac{1}{\lambda} \\[2mm] 0 & \text{if } \quad -\dfrac{1}{\lambda} < \dfrac{m_0}{L}\mathbf{B}\cdot\mathbf{q} < \dfrac{1}{\lambda} \\[2mm] -1 & \text{if } \quad \dfrac{m_0}{L}\mathbf{B}\cdot\hat{\mathbf{q}} \leqslant -\dfrac{1}{\lambda} \end{cases} \qquad (19\text{-}26)$$

Equations (19-17) and (19-18) determine the satellite attitude history and Eqs. (19-23), (19-24), and (19-26) provide the framework for obtaining the desired coil state history. However, Eq. (19-26) is not deterministic because λ is a free parameter. Also, Eq. (19-23) does not fix the initial phase (ϕ_q) of $\hat{\mathbf{q}}$ in the spin plane. Equation (19-26) shows that parameters, λ, ϕ_q have a strong influence on the coil state history. Whenever the maneuver is feasible, there will be paired sets of values of (λ, ϕ_q) which allow the desired attitude to be reached. The feasible values of (λ, ϕ_q) then parameterize the paths for which the integral J is close to being extremum. Among these paths, the one requiring the least amount of coil-on time is the *minimum energy path* and the corresponding coil commands are the *energy optimal commands*.

To obtain the minimum maneuver time commands, note that in Eq. (19-26) if $1/\lambda = 0$, then the coils will always be on in either the positive or the negative sense. As before, more than one "nearly extremum" path is possible. However, they need now be labeled by only one parameter, ϕ_q. Thus, the generation of commands for minimum time maneuvers involves only a one-parameter search.

A byproduct of the minimum time maneuvers is the determination of the final attitudes that can be obtained as the parameter ϕ_q is varied. Thereby a boundary can be generated about the initial attitude such that all attitudes within this boundary will be attainable within the prescribed time. The attitude within the boundary that is closest to the desired attitude can be determined and the corresponding minimum time commands can be generated. Thus, a single algorithm can be used for minimization of time, energy, or arc-length error. Details of the implementation of this algorithm are given by Werking and Woolley [1973].

19.2 Spin Plane Magnetic Coil Maneuvers

Mihaly G. Grell
Malcolm D. Shuster

The main function of a spin plane magnetic coil is to control the magnitude of the spacecraft total angular momentum, which we will call simply the *momentum*.

On most spacecraft, orientation and momentum control are handled separately by dipoles mounted along the spin axis for orientation control (Section 19.1) and dipoles mounted perpendicular to the spin axis for momentum control. However, orientation and momentum control can be strongly coupled in spin plane magnetic coil maneuvers. For low-inclination orbits, where the spin axis is closely aligned with the geomagnetic field vector, the spin axis coil is inefficient and the spin plane coil may be a better choice for both momentum and attitude control. Spin plane magnetic coil maneuvers can be carried out on both spinning and despun spacecraft, although the types of commands are different for each.

A spinning spacecraft is controlled by turning the magnetic coil on and off and by changing its polarity twice per rotation period at constant phase angles relative to the geomagnetic field vector. This mode of operation is called *commutation*. The phase angle of the magnetic dipole relative to an appropriate reference axis at which the polarity is changed is called the *commutation angle*, ψ. The polarity is changed at angles ψ and $\psi + \pi$. (All angles in this section are in radians.) This control mode is flexible because both momentum and attitude can be changed in any direction within the torque plane, i.e., the plane normal to the geomagnetic field.

The inertial coordinate system in which the magnetic control torque is most easily calculated is shown in Fig. 19-3. Here $\hat{s}$ is the spin axis and $\mathbf{B}$ is the geomagnetic field vector. We define an orthonormal triad $\hat{i}, \hat{j}, \hat{k}$, which is assumed to be fixed for one spin period by

$$\hat{k} \equiv \hat{B}, \qquad \hat{i} \equiv \frac{\mathbf{B} \times \hat{s}}{|\mathbf{B} \times \hat{s}|}, \qquad \hat{j} \equiv \hat{k} \times \hat{i}$$

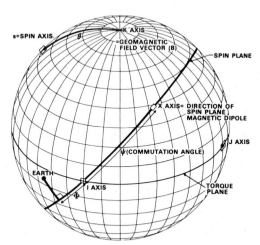

Fig. 19-3.   Coordinate Systems for Determining Magnetic Control Torques

In this system $\mathbf{L} = L\hat{s} = L(-\sin\theta\hat{j} + \cos\theta\hat{k})$. A magnetic dipole, $\mathbf{m}$, in the magnetic field, $\mathbf{B}$, produces a torque, $\mathbf{N}$, given by

$$\mathbf{N} = \mathbf{m} \times \mathbf{B} \qquad (19\text{-}27)$$

Then $\mathbf{m} = m(\cos\psi\hat{i} + \sin\psi\cos\theta\hat{j} + \sin\psi\sin\theta\hat{k})$ and, if the dipole is perpendicular to the spin axis, Eq. (19-27) can be expressed as

$$N = mB(\sin\psi\cos\theta\hat{i} - \cos\psi\hat{j}) \tag{19-28}$$

The geometry is shown in Fig. 19-3.

If the dipole polarity is reversed every half rotation, the average torque, $\langle N \rangle$, for a full spin period is

$$\langle N \rangle = \frac{1}{\pi}\int_{\psi}^{\psi+\pi}(m\times B)d\psi' = \frac{mB}{\pi}\int_{\psi}^{\psi+\pi}(\sin\psi'\cos\theta\hat{i} - \cos\psi'\hat{j})d\psi'$$

$$= \frac{2mB}{\pi}(\cos\psi\cos\theta\hat{i} + \sin\psi\hat{j})$$

and the angular momentum change, ΔL, of the spacecraft in the time interval, Δt, is

$$\Delta L = \int_0^{\Delta t}\langle N \rangle dt = \frac{2mB}{\pi}(\cos\theta\cos\psi\hat{i} + \sin\psi\hat{j})\Delta t \tag{19-29}$$

where we assume that Δt is an integral multiple of the spin period.

Equation (19-29) shows that the available angular momentum changes form an ellipse in the plane perpendicular to the geomagnetic field vector, B. We may think of both orientation and momentum maneuvers as simply changing the spacecraft's angular momentum vector, L; specifically, attitude maneuvers change the direction of L, or the spin axis, $\hat{s}$, and the momentum maneuvers change its magnitude.

19.2.1 Momentum and Attitude Maneuvers

Given the geomagnetic field and the initial target angular momentum vectors (L_0 and L_T), we wish to generate a set of commands which may be used to carry out the maneuver. The commands consist of coil-on and coil-off times and commutation angles, which are kept constant for the duration of each command. In addition, we may wish to minimize the number of commands to achieve L_T or to minimize coil-on time for energy minimization. We will assume that new commands are generated every n minutes; that each command, i, results in a change, ΔL_i; and that ϵ is the tolerance on L_T.

The goal is to determine the optimal commutation angle for each command and the times at which the coils are to be turned on and off. As a first approximation, we will pick the most favorable commutation angle at some point of each command interval and keep this angle constant. The most favorable commutation angle depends on the type of maneuver. We consider four maneuver types:

Type 1. Achieving the target momentum regardless of attitude changes
Type 2. Achieving the target attitude regardless of momentum changes
Type 3. Achieving attitude and momentum objectives simultaneously
Type 4. Achieving the target attitude without a significant change in momentum (this is a special case of type 3)

Type 1 maneuvers can be achieved by maximizing the $\hat{y}$ component of ΔL, i.e., by choosing $\psi = \pm\pi/2$ in Eq. (19-29), with the sign chosen to increase or decrease the momentum. Specifically,

$$\psi = \frac{\pi}{2}\text{sign}[(L_T - L_0)\cos\theta] \qquad \text{(Type 1)} \tag{19-30}$$

When the angle, θ, between L and B is small, type 1 maneuvers are inefficient because of small momentum and large attitude changes. To avoid inefficient operation, the magnetic coil should be turned off whenever $\theta < \theta_{\text{lim}}$. The maneuver is terminated and the coil turned off when

$$|L_T - L| \leqslant \epsilon \qquad (19\text{-}31)$$

Type 2 maneuvers are somewhat more complicated. Let $\mathbf{Q}$ be defined by

$$\mathbf{Q} = \mathbf{L} \times (\mathbf{L}_T \times \mathbf{L}) = L_T L^2 \left[\hat{\mathbf{L}}_T - (\hat{\mathbf{L}} \cdot \hat{\mathbf{L}}_T) \hat{\mathbf{L}} \right] \qquad (19\text{-}32)$$

$\mathbf{Q}$ lies in the $(\mathbf{L}, \mathbf{L}_T)$ plane and is perpendicular to $\mathbf{L}$.

For type 2 maneuvers, the best performance occurs when $\Delta \mathbf{L}$ is in the torque plane such that $\langle \mathbf{N} \rangle \cdot \mathbf{Q}$ is a maximum. Using Eq. (19-29), $\mathbf{Q} \cdot \Delta \mathbf{L}$ is a maximum when $\partial (\mathbf{Q} \cdot \Delta \mathbf{L}) / \partial \psi = 0$ and $\partial^2 (\mathbf{Q} \cdot \Delta \mathbf{L}) / \partial \psi^2 < 0$, or

$$\psi = \arctan \left[\frac{\hat{\mathbf{j}} \cdot \mathbf{Q}}{(\hat{\mathbf{i}} \cdot \mathbf{Q}) \cos \theta} \right] \qquad \text{(Type 2)} \qquad (19\text{-}33)$$

From the two solutions for ψ, we pick the one for which $\Delta \mathbf{L} \cdot \mathbf{Q} > 0$.

By introducing an efficiency angle, ζ_{lim}, the magnetic coil is turned off whenever

$$\hat{\mathbf{Q}} \cdot \Delta \hat{\mathbf{L}} < \cos \zeta_{\text{lim}} \qquad (19\text{-}34)$$

As in the type 1 maneuver, this conserves spacecraft power when the torque is small because the angle between $\mathbf{B}$ and $\mathbf{m}$ is small. The maneuver is complete when

$$\Delta \hat{\mathbf{L}} \cdot \hat{\mathbf{L}}_T \geqslant 1 - \epsilon \qquad (19\text{-}35)$$

For type 3 maneuvers we choose $\Delta \mathbf{L}$ such that $\Delta \mathbf{L} \cdot (\mathbf{L}_T - \mathbf{L})$ is maximized, from which we obtain

$$\psi = \arctan \left\{ \frac{\hat{\mathbf{j}} \cdot (\mathbf{L}_T - \mathbf{L})}{\left[\hat{\mathbf{i}} \cdot (\mathbf{L}_T - \mathbf{L}) \right] \cos \theta} \right\} \qquad \text{(Type 3)} \qquad (19\text{-}36)$$

From the two solutions for ψ, we pick the one for which $\Delta \mathbf{L} \cdot (\mathbf{L}_T - \mathbf{L}) > 0$. For efficient operation, the magnetic coil is turned off whenever

$$\Delta \hat{\mathbf{L}} \cdot \frac{\mathbf{L}_T - \mathbf{L}}{|\mathbf{L}_T - \mathbf{L}|} < \cos \zeta_{\text{lim}} \qquad (19\text{-}37)$$

The maneuver is complete when

$$|\mathbf{L}_T - \mathbf{L}| \leqslant \epsilon \qquad (19\text{-}38)$$

For type 4 maneuvers, there are two directions in the torque plane, $\psi = 0$ and $\psi = \pi$, which are perpendicular to the current angular momentum vector and, therefore, change only the orientation and not the momentum magnitude. The direction is chosen for which $\Delta \mathbf{L} \cdot (\mathbf{L}_T - \mathbf{L}) > 0$, that is,

$$\psi = \frac{\pi}{2} \left\{ 1 + \text{sign} \left[\hat{\mathbf{i}} \cdot (\mathbf{L}_T - \mathbf{L}) \cos \theta \right] \right\} \qquad \text{(Type 4)} \qquad (19\text{-}39)$$

For the efficiency angle and the convergence limit, the same parameters can be used as for type 2 maneuvers.

After the commutation angle has been determined, it must be translated into a hardware command. This can be done using the measured magnetic field directly as on SAS-3 [Gambhir and Sood, 1976] or using an Earth horizon sensor, as on the AE series [Phenneger, *et al.*, 1975]. In the latter case, the commutation angle is referenced from the nadir vector, whose orientation is sensed onboard; in this case, an extra rotation angle, Φ, is added, which is measured from the nadir vector, $\hat{\mathbf{E}}$, to the $\hat{\mathbf{i}}$ axis, as shown in Fig. 19-3.

19.2.2 Optimal Command Procedures

The above discussion is valid for fixed reference vectors. Because the magnetic reference vectors and, consequently, the optimal commutation angles are continuously changing, the commutation angles computed at any time t do not remain optimal for the duration of the command. To achieve a solution which is more nearly optimal we perform instead a discrete sequence of commands, each lasting for a specified time interval.

Let us suppose that at each time t_i, $i=1,\dots,n$, a commutation angle ψ_i is chosen and maintained until the time t_{i+1}. At some final time, t_{n+1}, the magnetic coil is turned off and the maneuver is completed. The final angular momentum $\mathbf{L}$ will be a function of these commutation angles, i.e.,

$$\mathbf{L}=\mathbf{L}(\psi) \tag{19-40}$$

where ψ denotes the n-dimensional vector $(\psi_1,\psi_2,\dots,\psi_n)^{\mathrm{T}}$. The goal of any spin plane magnetic coil maneuver is to bring $\mathbf{L}$ (or some function of $\mathbf{L}$) as close as possible to the target momentum $\mathbf{L}_T$ (or some function of $\mathbf{L}_T$). We can obtain this by requiring that the ψ_i be chosen to minimize a loss function, $F(\psi)$, or maximize a gain function, appropriate to the particular maneuver type. For type 1 maneuvers, $F(\psi)$ has the form

$$F_1(\psi)\equiv(\mathbf{L}(\psi)-\mathbf{L}_T)^2=\text{Minimum} \qquad \text{(Type 1)} \tag{19-41}$$

Here the objective is to minimize the difference between the magnitudes of the target and actual angular momenta. For type 2 maneuvers, we wish to minimize the angle between the target and the actual angular momentum vectors. Thus,

$$F_2(\psi)\equiv\hat{\mathbf{L}}(\psi)\cdot\hat{\mathbf{L}}_T=\text{Maximum} \qquad \text{(Type 2)} \tag{19-42}$$

For type 3 maneuvers, we wish to minimize the norm of the difference vector between the target and the actual angular momenta; thus,

$$F_3(\psi)\equiv(\mathbf{L}(\psi)-\mathbf{L}_T)^2=\text{Minimum} \qquad \text{(Type 3)} \tag{19-43}$$

Finally, for type 4 maneuvers, no optimization is required because the objective is met by selecting the commutation angles to be 0 or π.

Having chosen a maneuver type, the next step is to find the solution of the optimization function, F. Although various nonlinear programming techniques can be applied (see Sections 13.4 and 13.5), we will use the gradient search method to obtain an iterative solution. An initial guess $\psi^{(0)}$ is first obtained. This may be done, for example, by choosing the instantaneous solutions of Section 19.2.1 evaluated near the midpoint of each command interval. The function F decreases or increases most rapidly when ψ is varied along the gradient, $\nabla F(\psi)$, of F in ψ-space, defined by (see Appendix C):

$$\nabla F(\psi) \equiv \left(\frac{\partial F(\psi)}{\partial \psi_1}, \ldots, \frac{\partial F(\psi)}{\partial \psi_n} \right)^{\mathrm{T}} \tag{19-44}$$

Thus, subsequent iterations for ψ are chosen by

$$\psi^{(k+1)} = \psi^{(k)} + \lambda \nabla F(\psi^{(k)}) \tag{19-45}$$

where λ is a constant selected by trial and error to make the iterations converge as quickly as possible. If λ is too large, the iterations may oscillate and never come very close to the optimal solution. If λ is too small, the iterations approach the optimal solution only very slowly.

A nearly optimum value of λ may be determined by performing an iteration sequence up to some given order, computing the final loss value for each λ, and then extrapolating to the optimum λ, which minimizes the loss function. At least three trials for different values of λ will be necessary because a two-point extrapolation is a straight line and has no minimum or maximum. Once this optimum λ is found, the iteration is performed once more to determine the optimum ψ.

To calculate the gradient vector we note that $F(\psi)$ depends on ψ only through $L(\psi)$; that is,

$$F(\psi) = F(L(\psi)) \tag{19-46}$$

Hence,

$$\frac{\partial F(\psi)}{\partial \psi_i} = \nabla_L F(L) \cdot \frac{\partial L}{\partial \psi_i} \tag{19-47}$$

where ∇_L denotes the gradient with respect to the three components of L. For example, for a type 3 maneuver (Eq. (19-41)),

$$\nabla_L F_1(\psi) = 2(L(\psi) - L_T) \tag{19-48}$$

To determine $\partial L / \partial \psi_i$, we note that the final angular momentum, $L(\psi)$, may be written

$$L(\psi) = L_0 + \sum_{i=1}^{n} \Delta L_i(\psi_i) \tag{19-49}$$

where ΔL_i is the angular momentum change in the time interval from t_i to t_{i+1}. (Actually ΔL_i depends on all previous commutation angles and not on ψ_i alone.) Using Eq. (19-49), we have

$$\frac{\partial L}{\partial \psi_i} = \frac{\partial}{\partial \psi_i} \Delta L_i(\psi_i) \tag{19-50}$$

If ΔL_i and the attitude do not change greatly during this time interval, the quantity above may be determined from Eq. (19-29). Otherwise, it may be approximated from the quotient of differences as

$$\frac{\partial \mathbf{L}}{\partial \psi_i} \approx \frac{\Delta \mathbf{L}_i(\psi_i + \Delta \psi_i) - \Delta \mathbf{L}_i(\psi_i)}{\Delta \psi_i} \tag{19-51}$$

Thus, the partial derivative of $F(\psi)$ with respect to ψ_i for the various maneuver types is given by

$$\frac{\partial F_1(\psi)}{\partial \psi_i} = 2[L - L_T]\hat{\mathbf{L}} \cdot \left(\frac{\partial \mathbf{L}_i(\psi_i)}{\partial \psi_i}\right) \qquad \text{(Type 1)} \qquad (19\text{-}52)$$

$$\frac{\partial F_2(\psi)}{\partial \psi_i} = \frac{1}{L}\left[\hat{\mathbf{L}}_T - (\hat{\mathbf{L}}_T \cdot \hat{\mathbf{L}})\hat{\mathbf{L}}\right] \cdot \left(\frac{\partial \mathbf{L}_i(\psi_i)}{\partial \psi_i}\right) \qquad \text{(Type 2)} \qquad (19\text{-}53)$$

$$\frac{\partial F_3(\psi)}{\partial \psi_i} = 2[\mathbf{L} - \mathbf{L}_T] \cdot \left(\frac{\partial \mathbf{L}_i(\psi_i)}{\partial \psi_i}\right) \qquad \text{(Type 3)} \qquad (19\text{-}54)$$

The command procedure is as follows: After selecting the desired maneuver type, the instantaneous commutation angles are computed for each command interval. By using the termination criteria, we can see approximately how much time is needed to get to the target. The first approximation constitutes the initial value for the optimization scheme, which will further improve the performance of the magnetic coils by reducing the coil-on time. The command angles generated will be a minimum time set. If we want to achieve minimum energy commands, to save electric power, the parameter to optimize on for a given time interval will be the limit angle, ζ_{lim}. For a given value of ζ_{lim}, the coils will be turned off when the angle between the optimal torque vector and the desired direction is greater than ζ_{lim}. From an initial value $\zeta_{\text{lim}} = 180$ deg, we have to proceed by reducing the angle to a threshold value, under which no convergence can be obtained to the target value. The minimum energy optimization will work only if the command time given is more than the minimum time.

19.2.3 Representative Example of AE-5 Magnetic Maneuvers

The control scheme described has been used for the magnetic control of the AE-5 spacecraft launched in November 1975 [Phenneger, et al., 1975; Grell, 1977]. To maintain an orbit-normal orientation, daily commands are required to compensate for a 3-deg attitude drift due to orbital precession and atmospheric drag. The average daily change in angular momentum is 2 kg m^2/sec or about 1% of the total angular momentum. The command performance is constant for up to 15 to 20 min of command time and begins to deteriorate at about 25 min. With a total coil strength of 197.2 A m^2, 5-deg attitude maneuvers or 4 kg m^2/sec angular momentum maneuvers can be achieved in a 2-hour orbit. The combined maneuvers average half the efficiency of the pure orientation or momentum maneuvers, as shown in Fig. 19-4. For a typical set of optimized commands, the attitude favorable and momentum favorable sections of the combined maneuvers alternate during the orbit. The attitude maneuvers with no momentum change show the worst performance, but they may be necessary when mission constraints do not allow a change in the magnitude of the angular momentum.

The optimization scheme can increase the efficiency by as much as 20% relative to instantaneous commands. Generally, 20 iterations are enough to con-

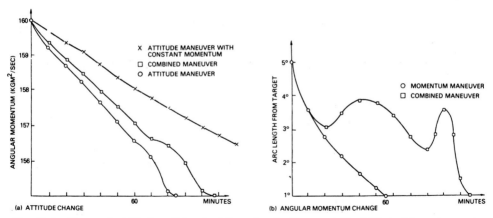

Fig. 19-4.   Changes in Attitude and Angular Momentum for Representative Attitude, Momentum, and Combined Maneuvers for the AE-5 Spacecraft

verge to the optimal solution, and the optimization gives the closest possible solution even if there is not sufficient time available to converge to the target. The initial guess for the optimization does not affect the final solution, even if the starting commutation angles are randomly selected. The most frequently used maneuver is the combined maneuver, with attitude and momentum maneuvers used as backup for various contingencies. It has also been found that for the combined maneuver, the Lagrangian optimization converges much faster (in three iterations) but diverges when insufficient time is given in which to reach the target.

19.3 Gas Jet Maneuvers

Robert S. Williams

Gas jet maneuvers can be conveniently divided into two classes according to whether the inertial direction of the torque vector is constant or changing during jet firing. Most maneuvers are of the first type and are analytically straightforward. When the direction of the applied torque is constant, the jet can be fired for as long or short a time as necessary to produce the desired change, and prediction of the spacecraft response requires only a straightforward integration of the equation for angular acceleration. The second type of maneuver is represented by precession of a spinning spacecraft using a jet fixed to the spacecraft body as described in Section 1.2.3. Here, the torque vector is approximately perpendicular to the spin axis and rotates at the spin rate. The jet must be pulsed on and off; otherwise, the torque will average to zero over a spin period. Successive pulses must be controlled by an inertial reference, commonly the Sun vector, to achieve a net cumulative motion of the spin axis in the intended direction.

The fundamental equation for gas jet control is the equation for rate of change of angular momentum:

$$\frac{d}{dt}(I\omega) = N - \dot{m}l^2\omega \qquad (19\text{-}55)$$

where I is the moment-of-inertia tensor, ω is the angular velocity, N is the applied torque, $\dot{m}$ is the rate of consumption of propellant ($\dot{m} > 0$ by definition), and l is the perpendicular distance from the spin axis to the thruster. The last term accounts for the angular momentum lost as propellant is expelled, which may be nonzero even if the thruster orientation is such that the thrust produces no net torque. Except for detailed dynamic analysis, it is normally adequate to assume that the angular velocity is parallel to one of the principal axes and that the torque is applied about this or another principal axis. The time rate of change in the moment of inertia will be included in the following examples, although as noted in Section 7.10, the effects of this term can usually be ignored.

Spin Rate Control. The spacecraft spin rate is ordinarily controlled by nozzles mounted perpendicular to and displaced from the spin axis, approximately in the plane perpendicular to the spin axis which contains the center of mass. A component of torque perpendicular to the spin axis will reduce the effective torque about the spin axis but will have no other effect, as the perpendicular torque components will average to zero over a long, continuous firing or a sequence of randomly timed pulses. In this case, Eq. (19-55) reduces to

$$N - \omega l^2 \dot{m} = \frac{\mathrm{d}}{\mathrm{d}t}(I\omega) = \dot{\omega}I - \omega d^2 \dot{m} \qquad (19\text{-}56)$$

where N is the applied torque, ω is the instantaneous spin rate, l and $\dot{m}$ are defined as in Eq. (19-55), $\dot{\omega}$ is the angular acceleration, I is the instantaneous moment of inertia about the spin axis, and d is the instantaneous *radius of gyration*, or distance from the spacecraft center of mass to the center of mass of the propellant. The equations which must be solved are then

$$\dot{\omega} = \left[N - \omega(l^2 - d^2)\dot{m} \right] / I \qquad (19\text{-}57)$$

and

$$I = d^2 m \qquad (19\text{-}58)$$

which may be used to predict the time required to produce a given change in the spin rate. The value of d will change slightly during a maneuver; this can usually be ignored if an average of initial and final values is used. When the spin rate can be directly measured, and when the spacecraft can be monitored and controlled in real time, an accurate prediction is not required because the maneuver can be extended or terminated prematurely to achieve the desired spin rate.

Momentum Unloading. Spacecraft for which all three axes must remain inertially fixed are usually controlled by a combination of gyroscopes and momentum wheels. Secular disturbance torques may change the angular momentum of the spacecraft beyond the capacity of the momentum wheels to compensate. When this happens, gas jets can be used to *dump* or *unload* excess momentum, or conversely, to *add* or *load* deficit momentum. A jet is fired to produce a torque opposite the direction of the accumulated angular momentum while the spacecraft is commanded to maintain its attitude; the result is that the momentum wheel accelerates at the rate necessary to counteract the applied torque. A detailed description of this maneuver depends on the control laws governing the momentum wheels and gyroscopes (see Section 18.2); the sole function of the jet is to introduce a "disturbance" torque of appropriate direction and magnitude.

Attitude Control With an Inertially Fixed Jet. A three-axis stabilized spacecraft, or a dual-spin spacecraft in which control thrusters are mounted on the despun portion, can be maneuvered with an inertially fixed jet. For three-axis stabilized spacecraft, maneuvers can be treated the same as spin rate control maneuvers (apart from interactions with the stabilization system) by integrating the spin rate to find the rotation angle. The inversion maneuver required for the AE spacecraft is an example of precession of a dual-spin spacecraft by continuous firing of a thruster mounted on the despun portion. The object of the maneuver is to rotate the spacecraft spin axis by 180 deg by rotating $\mathbf{L}$ about the spacecraft x axis. The thrust geometry is shown in Fig. 19-5. The equation of motion is

$$\frac{d\mathbf{L}}{dt} = L \frac{d\hat{\mathbf{L}}}{dt} = \mathbf{N} = \mathbf{r} \times \mathbf{F} = rF\left[\hat{\mathbf{L}} \times (-\hat{\mathbf{x}})\right] \qquad (19\text{-}59)$$

where $\mathbf{L}$ is the total angular momentum, $\mathbf{N}$ is the applied torque, $\mathbf{F}$ is the thrust, and $\mathbf{r}$ is the vector from the center of mass to the jet and $\mathbf{r} \cdot \mathbf{N} = 0$. Expressing the direction of the applied torque in terms of the unit vectors $\hat{\mathbf{x}}$ and $\hat{\mathbf{L}}$ allows the equation to be written in the form

$$\frac{d\hat{\mathbf{L}}}{dt} = + \frac{rF}{L} \hat{\mathbf{L}} \times (-\hat{\mathbf{x}}) \qquad (19\text{-}60)$$

which is the equation for uniform precession about the $-\hat{\mathbf{x}}$ axis at a rate $\omega = rF/L$. The AE maneuver thruster is aligned so that the thrust vector intersects the z axis as closely as possible so as to generate only precession torques. It is also assumed that the onboard control system will keep the torque vector inertially fixed. This assumption breaks down somewhat in practice, however, and correction maneuvers with the magnetic control system are usually required after the inversion maneuver.

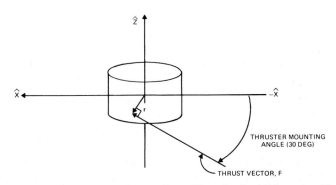

Fig. 19-5.   Thrust Geometry in Spacecraft x-z Plane for the AE Inversion Maneuver

Precession of a Spinning Spacecraft. As noted above, attitude control of a spinning spacecraft requires that the control thruster be operated in a pulsed mode, each pulse lasting a fraction of a spin period. Successive pulses are correlated to achieve a cumulative motion of the spin axis, ordinarily by using either the Sun or the Earth as an inertial reference. See Section 1.2.3 for a qualitative description of this maneuver.

Formally, Eq. (19-55) can be integrated directly to determine the motion of the spin axis resulting from a series of pulses. In practice, each pulse is assumed to produce an instantaneous change in spin rate and attitude, and successive pulses are summed to complete the maneuver. Spin rate change is computed from Eq. (19-56); the torque component along L will generally be zero unless a thruster is misaligned,* producing a spin component of the torque. The direction and magnitude of the change in attitude are computed from Eqs. 7-148 and 7-149. Normally the change in spin rate accompanying each pulse is small enough to neglect in computing the attitude change, although the cumulative change may be large enough to require that the spin rate be updated periodically in the course of simulating a single maneuver. Each pulse typically precesses the spin axis by an amount of the order of 0.1 deg. Therefore, the precession arc length is approximately determined from the fractional change in the angular momentum vector shown in Fig. 19-6:

$$\Delta\psi \approx \frac{\Delta L}{L} = \frac{N\Delta t}{I\omega} \qquad (19\text{-}61)$$

where $\Delta\psi$ is the angular change in the orientation of L; ΔL is the magnitude of the change in L, equal to the impulse N of the applied torque (from Eq. (7-149)), and assumed perpendicular to L; and $I\omega$ is the magnitude of the angular momentum about the spin axis, i.e., $L = I\omega$. The pulse centroid is calculated from Eq. (7-148). The precession torque vector equals $\mathbf{r}\times\mathbf{F}$, where $\mathbf{r}$ is the position vector to the thruster at the time of the pulse centroid, and is parallel to $\mathbf{r}\times\mathbf{L}$, because only the component of $\mathbf{F}$ parallel to L produces a precession torque.

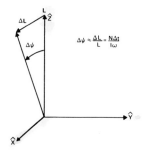

Fig. 19-6.   Single-Pulse Precession of the Momentum Vector, L

The *heading angle*, $\Theta \equiv \arccos[(\widehat{\mathbf{r}\times\mathbf{L}})\cdot(\widehat{\mathbf{S}\times\mathbf{L}})]$, where S is the Sun vector, is the rotation angle about L from the L/S plane to the L/r plane. (See Fig. 1-11.) In a *rhumb line maneuver*, the heading is fixed for the duration of the maneuver. This is nominally the case if the thruster pulse time is fixed relative to the Sun detection time, but the heading will vary if the thrust profile or the spin rate changes during the maneuver. The fact that Θ is nominally constant suggests that a convenient coordinate system for analysis is one in which the Sun vector is parallel to the z axis and L lies in the x/z plane at the start of the maneuver. In this coordinate system, the attitude *trajectory* or time history makes a constant angle, Θ, with lines of latitude as shown in Fig. 19-7.

A further simplification results if the attitude trajectory is plotted in a

---

*A thruster may be aligned deliberately to produce a torque component which will cancel the remaining terms in Eq. (19-56) so that $\dot\omega = 0$.

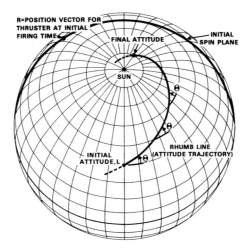

Fig. 19-7. Rhumb Line Attitude Maneuver for Constant Heading of $\Theta = 32°$ Relative to the Sun. For the manuever illustrated, $\beta_i = 50°$, $\beta_f = 10°$, and $\phi_f = 153°$. The arc length between the initial and final attitudes is 59 deg and the rhumb length is 75 deg.

*Mercator representation** with the Sun at the pole, as shown in Fig. 19-8 for the same maneuver shown in Fig. 19-7. The heading angle, Θ, can be read directly from the rhumb line joining the initial and final attitudes if the horizontal and vertical scales are equal at the equator.

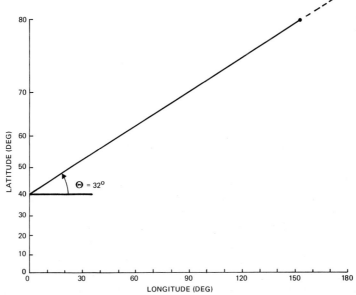

Fig. 19-8. Mercator Plot of Rhumb Line Maneuver Illustrated in Fig. 19-7. See text for explanation.

---

*The Mercator representation is a conformal mapping devised by Gerhardus Mercator, a sixteenth-century Flemish geographer. Points on the surface of the sphere are plotted on the x/y plane. Longitude is plotted directly on the x axis, but $-\log_e(\tan(\phi/2))$, where ϕ is the colatitude, is plotted on the y axis. A straight line connecting any two points on the map is a *loxodrome* or *rhumb line*, a line making a constant angle with both parallels and meridians, which made the Mercator map useful to early terrestrial navigators.

The *rhumb length*, λ, is the arc length between the initial and final attitudes along the rhumb line and is greater than or equal to the arc length of the great circle joining the two points. The values of Θ and λ for a rhumb line maneuver can be calculated from the initial and final attitudes in the coordinate system with the Sun at the pole from

$$\tan \Theta = \left[\log_e \tan(\beta_i/2) - \log_e \tan(\beta_f/2) \right] / \phi_f \qquad (19\text{-}62)$$

$$\begin{aligned} \lambda &= |\beta_f - \beta_i|/|\sin \Theta| & \beta_f \neq \beta_i \\ \lambda &= |\phi_f \sin \beta_f| & \beta_f = \beta_i \end{aligned} \qquad (19\text{-}63)$$

where β_i and β_f are the initial and final coelevation angles, respectively (measured from the Sun), of the spin axis in this coordinate system; ϕ_f is the final azimuth angle (in radians); and the initial azimuth angle is zero by definition of the coordinate system. (For a derivation, see Williams [1971].) The algebraic signs of the numerator and denominator of Eq. (19-62) correctly indicate the quadrant of Θ, which has a range of 360 deg; if ϕ_f is zero, Θ is $+90$ deg if the numerator is positive and -90 deg if the numerator is negative. Although the rhumb length λ is longer than the arc length between the initial and final attitudes, the difference is small for short arcs, rhumb lines near the equator, or rhumb lines heading nearly directly toward or away from the coordinate system pole. Where the difference is significant, a great circle maneuver can be approximated by a series of short rhumb lines.

Equations (19-62) and (19-63), and the inverse equations for β_f and ϕ_f, can be used for an initial prediction of the commands required to perform a particular maneuver and to determine the resulting trajectory. A pulse-by-pulse simulation can then be used to refine the computation based on the amount by which the simulated final attitude misses the required final attitude. A miss may occur because spin rate or pulse characteristics are known to change during a maneuver, or simply because the resolution of the control system does not allow arbitrary heading angles or arc lengths to be generated. A Mercator plot of the attitude trajectory can be used to monitor the progress of a maneuver if data can be obtained and processed in near real time. If attitude points lie on the predicted trajectory but do not progress toward the final attitude at the predicted rate, the maneuver can be lengthened or shortened and subsequent predictions adjusted proportionally. If attitude points lie off the predicted trajectory, a correction maneuver will be required, and the computation of the pulse centroid will have to be modified. If all subsequent maneuvers are performed at the same spin rate with the same pulsewidth, a constant adjustment is indicated. If maneuvers are performed at different spin rates, it may be found in some cases that the pulse timing is incorrect, in which case the centroid angular error will depend on spin rate; in other cases, the orientation of the thruster relative to the Sun sensor is incorrect, leading to a centroid angular error which is independent of spin rate.

19.4 Inertial Guidance Maneuvers

Dale Headrick

Inertial guidance maneuvers use only information obtained internally from gyroscopes or accelerometers. On IUE, for example, the telescope is commanded to move from one target to another with the maneuver execution based only on control error signals derived from rotation rates sensed by a set of gyros. (The terminal phase, however, uses data from a star tracker operated as a finder telescope.) The actuators for inertial guidance maneuvers are usually reaction wheels, although gas jets may be used instead, as on HEAO-1 (see Section 18.3). The reaction wheels may be commanded to perform a maneuver by a sequence of single-axis slews as on OAO and IUE or by simultaneously maneuvering all three axes as planned for HEAO-B.

19.4.1 Single-Axis Slews

A single-axis slew, in which the spacecraft body rotates about a fixed axis, can be executed by transferring momentum from a wheel to the body, causing the body to rotate about the wheel axis. The rotation rate is controlled with rate feedback information from gyros to provide damping to avoid overshooting the target. The required slewing time depends on the wheel capacity, the current momentum bias (even in a nominal zero-bias system), and any attitude or attitude rate limits which may be imposed. Figure 19-9 is a diagram of torque, wheel momentum, body rate, and angular position for an idealized system. At time t_0 a new angular position is commanded, and the torque motor goes full on. The wheel momentum increases linearly with time until either the maximum permissible body rate or wheel momentum is reached at time t_1. During this time, conservation of angular momentum requires the body rate, $d\theta/dt$, to decrease linearly, causing the position angle, θ, to change quadratically with time. The body rate remains constant until time t_2, when a braking torque is applied to slow the body rate to approach the target position angle with a small angular velocity. After approximately reaching the target at time t_3, the terminal phase involves the elimination of small residual errors by the stabilization control system as described in Section 18.3. When the control system, illustrated in Fig. 19-9, is implemented in discrete form (i.e., computer controlled), modifications are required from the analog position-plus-rate law to optimize the maneuver performance and minimize the computational requirements. An on-off or bang-bang type of impulsive control is preferred because it simplifies the actuator electronics by eliminating the need for continuous control and reduces the required sampling rate. Even for a discrete control system, however, the stability and optimization analysis are usually performed for the continuous case.

As an example of a bang-bang control system, consider the single-axis control loop used to place a target star in a telescope aperture slit, as shown in Fig. 19-10. The system obeys the equation $I\ddot{\theta} = N$, where the applied torque, N, is a constant. The available control is $u \equiv N/I$ with values $\pm U$ or 0 and the equations of motion are

$$\frac{d\theta}{dt} = \omega(t) \qquad (19\text{-}64a)$$

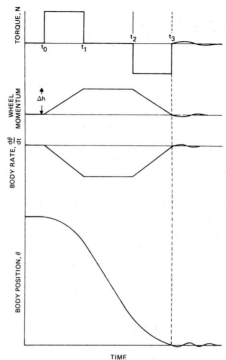

Fig. 19-9.   Example of a Single-Axis Slew Maneuver. See text for explanation.

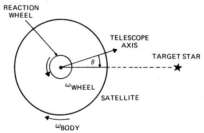

Fig. 19-10.   Reaction Wheel Attitude Control To Null the Angle, θ, Between the Telescope and the Target Star

$$\frac{d\omega}{dt} = u(t) \qquad (19\text{-}64\text{b})$$

The block diagram for a bang-bang control system is shown in Fig. 19-11. We will consider the basic bang-bang control law and several variations of it following the development of Hsu and Meyer [1968], who give a more extensive discussion of the subject with additional cases.

Position-Only Control.   As a simple case, consider *position-only feedback*, where the *rate gain* or *amplification*, a, is zero. The sampled output angle, θ, is fed back as input control signal by the function

$$u(t) = -U \operatorname{sign} \theta(t) \qquad (19\text{-}65)$$

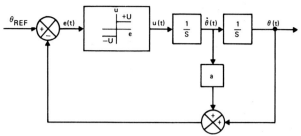

Fig. 19-11. Bang-Bang Attitude Control System. (e (t) is the control error signal.) Compare with the bang-bang system shown in Fig. 18-4 where there is no rate gain, a. If a is zero, the diagrams are equivalent.

Because usually both position and rate are of interest, the attitude behavior can be conveniently visualized in *state-space*, or *phase space*, as it is more commonly called in physics. For this single-axis control system, state-space becomes the *state plane* of position versus rate. The time history of the state parameters θ and $\dot{\theta}$ determine a trajectory in the state plane. The equation of motion is obtained by integrating Eq. (19-64) under the condition of Eq. (19-65). The state-plane trajectories of the system shown in Fig. 19-12(a) will consist of a set of parabolas given by

$$\frac{\theta^2}{2} \pm U\theta = c, \quad \text{for } u = \pm U \tag{19-66}$$

where c is a constant. They are connected about the $\dot{\theta}$ axis at the *switching line* as shown in Fig. 19-12(a). The $\dot{\theta}$ axis is called the switching line because $u = +U$ to the left of it and $u = -U$ to the right of it. Thus, the wheel torque motor control signal will change sign as the state trajectory $(\theta, \dot{\theta})$ crosses the switching line. Each state-plane trajectory is closed, with its size depending on the initial condition. Physically, the spacecraft will oscillate indefinitely about the equilibrium condition, without achieving the target attitude at the desired zero rate condition.

Position-Plus-Rate Control. Adding a term to the control error signal which is proportional to the attitude rate provides damping and has the effect of a lead network in electrical systems in predicting the state at a future time. For *position-plus-rate control* the switching function becomes

$$u(t) = -U \, \text{sign} \, (\theta + a\dot{\theta}) \tag{19-67}$$

and the switching line, instead of being the $\dot{\theta}$ axis, becomes the straight line $\theta + a\dot{\theta} = 0$, with slope $-1/a$.

The state trajectories are found from integrating Eq. (19-64) with $\dot{\omega} = -U$ for $(\theta + a\dot{\theta}) > 0$ and with $\dot{\omega} = +U$ for $(\theta + a\dot{\theta}) < 0$. The trajectories will again be families of parabolas whose curvature, or acceleration, changes sign at the switching lines as shown in Fig. 19-12(b). A system originally at A will follow the trajectory shown, reversing control at B and again at C, spiraling in toward the center. As it approaches the origin from C, however, the system trajectory crosses the switching line after shorter and shorter time intervals, causing the control relay to rapidly switch states. This condition is called chattering, and although the system will continue to move toward the origin in a damped fashion, it could lead to actuator wear. In the discrete version of this system, the relay remains on for a

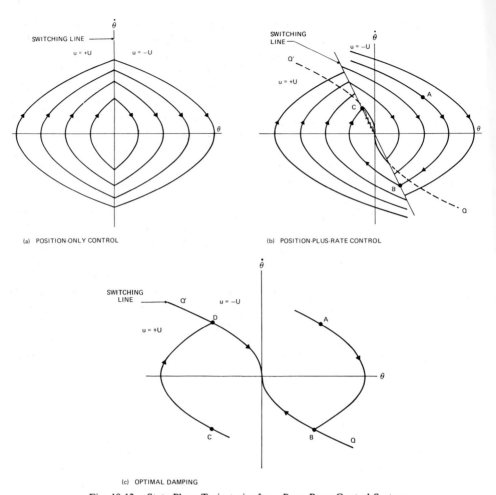

(a) POSITION-ONLY CONTROL

(b) POSITION-PLUS-RATE CONTROL

(c) OPTIMAL DAMPING

Fig. 19-12.   State Plane Trajectories for a Bang-Bang Control System

brief time after the switching line is crossed. This approach reduces actuator wear but may lead to instabilities.

Optimal Damping. We would like to design a system which will approach the origin in an efficient manner for any initial condition. From Fig. 19-12(b) it is apparent that the origin can be approached from only two directions, labeled Q and Q'. Thus, we would prefer to use as switching lines those parabolas which pass through the origin. This is algebraically described by

$$u = -U \operatorname{sign}\left[U\theta + \frac{1}{2}|\dot{\theta}|\dot{\theta} \right]$$

$$= \begin{cases} +U, & \text{for} \quad \theta < \dfrac{|\dot{\theta}|\dot{\theta}}{2U} \\[2ex] -U, & \text{for} \quad \theta > \dfrac{|\dot{\theta}|\dot{\theta}}{2U} \end{cases} \qquad (19\text{-}68)$$

As shown in Fig. 19-12(c), after the trajectory intersects the switching line, it will proceed directly toward the origin. For example, an initial state at A will intersect the switching line at B, change directions, and move along the switching line to the origin where the control is set to zero. Similarly, an initial state at C will intersect the switching line at D and proceed to the origin. This case is discussed in more detail by Hsu and Meyer [1968]. In practical applications, modifications are required to take into account any instrument or system rate limits which may exist. Also, the wheel torque (and acceleration) are neither perfectly known nor necessarily constant over the entire wheel speed range. The *minimum* expected wheel torque should be used in designing the control law to avoid relay chattering. A modified version of this optimally damped bang-bang control law has been implemented in the IUE and SMM attitude control system computers.

When a system has been shown to be stable and optimal (in some sense) when operating continuously, it does not necessarily follow that the discrete or pulsed version is optimal or controllable except in the limit of very high pulse rates. A discrete system periodically samples attitude sensors and controls the actuator. The *time step* is the time between consecutive samples and may be limited by the time constants of an analog control system or the capacity of an onboard computer. As the length of the time step is increased, control may be lost at periods corresponding to the resonant frequencies of the system. For example, if the time step is an integral multiple of the nutation period, the oscillations will not be damped. An example of this problem is given by Schmidtbauer, et al., [1973]. In such cases, further analysis, including digital simulations, will be required to analyze the system performance.

19.4.2 Multiple-Axis Slews

For an orthogonal set of reaction wheels aligned with the body axes, we can reorient the spacecraft from one celestial target to another by a sequence of single-axis slews. The required direction cosine matrix is

$$C = R_3^T(\alpha_2)R_2^T(-\delta_2)R_1^T(\beta_2)R_1(\beta_1)R_2(-\delta_1)R_3(\alpha_1) \qquad (19\text{-}69)$$

where $R_i(\theta)$ is a matrix representing the rotation about axis i through an angle θ, and the target coordinates (α, δ, β) are the right ascension and declination of the target and an azimuthal rotation angle about the target. The matrix C is uniquely specified by the initial and final target coordinates. This does not, however, completely specify the slew sequence.

A sequence of three slews is sufficient to accomplish any maneuver in the general case where the rotation angles, denoted for convenience as roll, pitch, and yaw, can take any value between 180 deg and -180 deg. There are 12 possible combinations of the form roll-pitch-yaw, where consecutive rotations about the same axis, such as roll-roll-yaw are excluded but nonconsecutive rotations about the same axis, such as roll-pitch-roll, are allowed. There are $3 \times (3-1) \times (3-1) = 12$ such combinations.

As an example, if we choose to perform a roll-pitch-yaw maneuver, we could calculate the required angles by setting (see Appendix E)

$$C = R_3(\psi)R_2(\theta)R_1(\phi)$$

$$= \begin{bmatrix} \cos\psi\cos\theta & \cos\psi\sin\theta\sin\phi + \sin\psi\cos\phi & -\cos\psi\sin\theta\cos\phi + \sin\psi\sin\phi \\ -\sin\psi\cos\theta & -\sin\psi\sin\theta\sin\phi + \cos\psi\cos\phi & \sin\psi\sin\theta\cos\phi + \cos\psi\sin\phi \\ \sin\theta & -\cos\theta\sin\phi & \cos\theta\cos\phi \end{bmatrix}$$

$$(19\text{-}70)$$

Because the C_{ij} matrix elements are known, the angles can be calculated as follows:

$$\text{Pitch: } \sin\theta = C_{31}, \quad \cos\theta = \pm\sqrt{1 - C_{31}^2}$$

$$\text{Roll: } \tan\phi = \frac{-C_{32}/\cos\theta}{C_{33}/\cos\theta} \quad\quad (19\text{-}71)$$

$$\text{Yaw: } \tan\psi = \frac{-C_{21}/\cos\theta}{C_{11}/\cos\theta}$$

This sequence is ambiguous because either sign of the radical can be taken, leading to separate roll-pitch-yaw slew sequences where the rotation angles are supplements of each other.* This same ambiguity exists for all cases, yielding a total of 24 possible slew sequences to perform a maneuver.

Because the computer operations involved are rapid, all 24 possibilities can be computed and ordered according to a suitable criterion. Typically, the minimum total path or minimum slewing time is chosen, but other criteria may be used. On OAO it was found that the performance error due to gyro misalignments grew linearly with slew angle, and the sequence was chosen which had the shortest maximum slew leg.

Although errors may arise from gyro misalignments, no error is caused if a wheel is misaligned because any undesired momentum components will be sensed by the gyros and controlled to zero. The design for the IUE spacecraft takes advantage of this feature by mounting a redundant fourth wheel along the diagonal of the orthogonal cube formed by the three primary wheels. If one of the three orthogonal wheels fails, relays will be set in the control electronics to send its commands to the skewed wheel. The rate component along the desired axis will be executed at $1/\sqrt{3}$ times the normal rate due to the projection of the commanded axis on the skewed axis, while the undesired components are temporarily absorbed by the reaction wheels on the other two axes.

A major advantage of using a sequence of single-axis slews is that the other two axes are controlled to zero, which minimizes the coupling of the axes in Euler's equations. This allows the three attitude angles to be estimated separately according to the accumulated gyro angles. Another advantage of this single-axis sequence lies in the ease of checking constraints. On IUE, for example, maneuvering is severely restricted by constraints on the position of the body axes with respect to the Sun, the Earth, and the Moon both while the spacecraft is pointed and while

---

*The ambiguity may be resolved, however, by limiting the intermediate rotation to the range 0 to 180 deg. In the above example, we would then have $\theta = \arccos(+\sqrt{1 - C_{31}^2})$. This convention is generally adopted for Euler angle sequences. See, for example Goldstein [1950] and Appendix E.

moving. Constraint checking is simplified by first transforming these inertially known vectors into the body system using the current attitude direction cosine matrix. By the geometry of the single-axis slews it is possible to perform even the dynamic constraint checks geometrically without simulating the maneuver.

An alternative to the single-axis slew sequence, called the *eigenaxis* method, [TRW, 1976] has been designed for HEAO-B, where a quaternion attitude estimate is available in the onboard processor. This method uses quaternion multiplication to compute the unique rotation axis, $\hat{e}$, and angle, ϕ, which can achieve the desired three-axis reorientation. The error quaternion, q_E, is given by

$$q_E = q^{-1}q_D \tag{19-72}$$

where q is the current attitude state and q_D is the desired attitude (see Section 12.1 and Appendix D). The eigenaxis, $\hat{e}$, is identified by expressing the four components of the error quaternion as

$$q_E = \left[\hat{e} \sin\frac{\phi}{2}, \cos\frac{\phi}{2} \right] \tag{19-73}$$

and performing the maneuver by rotating about $\hat{e}$ through the angle ϕ. In theory, this yields a minimum path maneuver, but in the actual design, large deviations are expected from the eigenaxis due to system nonlinearities such as torque motor response, torque or wheel saturation, and nonorthogonal reaction wheels. Compensation is made for these effects by continually updating the eigenaxis and recomputing the motor torques.

19.5 Attitude Acquisition

Gerald M. Lerner

Attitude acquisition consists of the series of attitude maneuvers, commands, and procedures necessary to reorient and reconfigure the spacecraft from the attitude state at separation from the launch vehicle to an attitude state suitable for the initiation of normal mission operations. The latter configuration is referred to as *mission mode*. This section describes the problems and procedures unique to attitude acquisition, including the deployment of extendable booms, antennas, and solar panels and the inflight checkout of both hardware and software.

Most missions require some period of attitude acquisition. The simplest acquisition sequences require maneuvers such as despin, deployment of solar panels, activation of the onboard sensors and experimental hardware, and a maneuver to the first mission attitude. This sequence is similar to that used by SAS-3 and most stellar-oriented missions. Slightly more complex sequences are required for geosynchronous spacecraft which employ a transfer orbit such as SMS/GOES or CTS. For these missions, a prolonged sequence of interspersed attitude and orbit maneuvers, lasting a week or more, is required to attain the proper position, orbit, and attitude. (See Section 1.1 for an example.) A sequence of maneuvers lasting 5 months was employed by RAE-2, initially to achieve a circular orbit about the Moon and finally to deploy four extendable antennas in a timed sequence to a total length of 450 m to achieve a three-axis, gravity-gradient stabilized attitude [Werking, *et al.*, 1974].

Some missions may require periods of *attitude reacquisition* to reacquire mission mode in the event of hardware or software failure or operator error. As an example, reacquisition is required for autonomous missions, such as the IUE, if the attitude error after a commanded maneuver exceeds the Fine Error Sensor field of view [Blaylock and Berg, 1976].

Clearly, specific details of attitude acquisition are very mission dependent—a function not only of attitude requirements, but of onboard hardware, ground support hardware (e.g., the availability of telemetry and command stations), ground support software, and power and thermal constraints. Despite the numerous constraints placed on this phase of a mission, some of which may be quite severe, considerable flexibility is available to the mission planner and the opportunities for innovative solutions are great. Although the implemented procedures for GEOS-3 (see Section 19.5.3) and CTS [Basset, 1976] were both specific and intricate, numerous alternatives were considered [Repass, *et al.*, 1975; Lerner, *et al.*, 1976; Kjosness, 1976] and discarded. Frequent improvements to the baseline procedures were made in the days preceding launch and probably could have continued. Although the end points of attitude acquisition sequences are fixed, the possible paths are distinctly nonunique; many must be traveled and pitfalls mapped in prelaunch planning before the best can be selected.

19.5.1 Classification of Attitude Acquisition

Attitude acquisition may be categorized by the degree of autonomy of the spacecraft hardware or, conversely, by the amount of ground support required. The spacecraft may be (1) *fully autonomous*; (2) *semiautonomous*, i.e., using a mixture of onboard and ground support; or (3) *ground controlled*. Fully autonomous attitude acquisition is accomplished either through the use of analog, preprogrammed electronics or a digital onboard computer, or OBC (see Section 6.9). Sensor data is used in a control law which is implemented via the analog electronics or OBC to command torquing devices such as electromagnets, wheels, and thrusters. For example, the German Aeronomy satellite, AEROS, used error signals from an analog Sun sensor and a magnetometer to control an electromagnet and torque the spin axis to the Sun. HCMM uses magnetometers and a wheel-mounted horizon scanner to control a magnetic torquing system to achieve a stable, three-axis Earth-pointing attitude [Stickler and Alfriend, 1974].

For semiautonomous spacecraft, a mixture of onboard and ground support is used to achieve acquisition. HEAO-1 used an onboard analog control system to place the spin axis within several degrees of the Sun and ground software using star tracker data to determine a precise three-axis attitude and calibrate the gyro-based control system. After calibration, the control system maneuvered the spacecraft to a target attitude and maintained it there using hydrazine thrusters to null the difference between the target and the gyro-propagated onboard attitude (see Section 19.4).

Ground-based attitude control may be either open-loop or closed-loop. *Closed-loop* control is similar to that provided on board: sensor data is telemetered in real time to the ground support computer; the data are processed and torquing commands are computed; and, finally, the software uplinks the requisite commands. The main advantage of closed-loop control is the flexibility and power of large ground-based computers. Continuous, rapid-response commanding capability is provided without requiring increased onboard weight and attendant complexity. The main disadvantages are the requirement for continuous uplink and downlink contact during operations and increased opportunities for hardware or software failure or operator error due to the extended communication lines. Attitude acquisition for the CTS spacecraft [Basset, 1976] used closed-loop control with a Hewlett Packard 2100A minicomputer.

Open-loop ground-based attitude control uses ground software to process and display sensor data and to compute and evaluate (often via simulation) command sequences. Analysts then select appropriate commands which are uplinked for execution on board. Open-loop control requires a time delay from 30 sec to several hours between receipt of sensor data and command execution whereas closed-loop ground-based control delays are of the order of several seconds. The advantage of open-loop control is the software simplicity and reliability afforded by relaxing the severe time constraints to permit analysts to evaluate and verify computed commands while retaining the power and flexibility of the ground-based computer facilities. The analyst can also respond to contingencies not foreseen in prelaunch analysis and rely on his judgment and experience in evaluating commands. The disadvantages are the limited control afforded by the slow response time,* and the increased possibility of operator error when many individual decisions and actions are required from computation to uplink of commands. Open-loop control has been the most widely used to date. Examples include the generation of command sequences for attitude maneuvers or maintenance for AE, SMS/GOES, and CTS and the GEOS-3 acquisition sequence described in Section 19.5.2.

19.5.2. Acquisition Maneuvers

This subsection describes attitude maneuvers that are unique to attitude acquisition. Table 19-1 (page 666) illustrates the types of maneuvers and constraints required for representative acquisition sequences. The initial state is determined largely by the configuration of the spacecraft within the last rocket stage and whether or not that stage is spin stabilized. The detailed release mechanism for spacecraft separation from the last stage and the performance of the yo-yo despin mechanism, described below, are also important. The final state includes the attitude, attitude rate, and spacecraft configuration (e.g., solar panel and antenna deployment, momentum wheel spinup, and attitude sensor and experiment turn-on).

In the event that the spacecraft cannot be commanded due to an onboard or ground support failure, intermediate attitudes between the initial and final state should be "safe harbors," *viz* capable of being maintained for prolonged periods without endangering the success of the mission. As an alternative, opportunities for easy access to safe harbors should be mapped and exploited as necessary by, for example, loading backup commands to be executed automatically on board the spacecraft at some later time.

Yo-Yo Despin. This maneuver is frequently employed for reducing the spacecraft's spin rate shortly after separation from the last stage of a booster

---

*Open-loop control employing several analysts and telephone lines was employed for RAE-1 to minimize time delays [IBM, 1968].

rocket. The last stage of rockets such as the Scout and Delta is often spin stabilized at a high angular velocity, e.g., 150 rpm, and this spin rate is maintained by the spacecraft at separation.

As illustrated in Fig. 19-13, we assume that a cylindrical spacecraft with axial moment of inertia I and radius R is rotating without nutation about its longitudinal axis with angular velocity Ω. Two equal masses, m_1 and m_2, are attached to separate cables of length l wrapped around the spacecraft perimeter opposite the

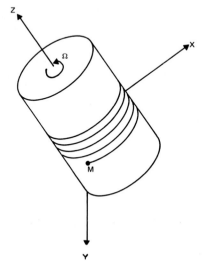

Fig. 19-13.   Mechanism of Yo-Yo Despin. Two masses are attached to cables wrapped around the spacecraft. When the masses are released, they carry away much of the spacecraft angular momentum.

direction of rotation. At time $t=0$, the masses are released and travel tangentially away from the spacecraft. As the cables unwind, they increase the moment of inertia of the system about the z axis and decrease the spacecraft's angular velocity. When completely unwound, the cables and attached masses are jettisoned, carrying off a substantial fraction of the system angular momentum. The relationship between the final spin rate, the spacecraft size and inertia, the cable length, and the yo-yo masses is derived as follows.

We define the *body coordinate* frame, $\hat{\mathbf{x}}$, $\hat{\mathbf{y}}$, and $\hat{\mathbf{z}}$, to be fixed in the spacecraft which is rotating about $\hat{\mathbf{z}}$ at angular velocity Ω relative to inertial space (see Fig. 19-14). The masses are initially at $\pm\mathbf{x}$. Similarly, the *cable frame*, $\hat{\mathbf{i}}$, $\hat{\mathbf{j}}$, and $\hat{\mathbf{k}}$, is rotating in the body such that the cables are tangent to the spacecraft perimeter along the $\pm\hat{\mathbf{i}}$ axis; i.e., the coordinates of the points where the cables are tangent to the body are fixed in the cable frame at $(\pm R, 0, 0)$.

Assuming no energy loss during the despin, we can compute $\Omega(t)$ from the conservation of energy and angular momentum. By symmetry, the angular momentum and kinetic energy of both masses are equal and it suffices to consider only m_1. The position and velocity of m_1 in the cable frame (see Fig. 19-14) are

$$\mathbf{r}_1 = R(\hat{\mathbf{i}} - \phi\hat{\mathbf{j}}) \qquad (19\text{-}74a)$$

$$\dot{\mathbf{r}}_1 = -R\dot{\phi}\hat{\mathbf{j}} \qquad (19\text{-}74b)$$

where ϕ is the angular separation between the cable and body frames in radians. The angular velocity of the cable frame, relative to inertial space, is

$$\omega = (\Omega + \dot{\phi})\hat{\mathbf{k}} \qquad (19\text{-}75)$$

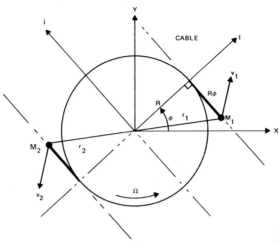

Fig. 19-14.   Yo-Yo Despin Geometry in Cable and Spacecraft Frames. The $\hat{\mathbf{z}}$ and $\hat{\mathbf{k}}$ axes are out of the plane of the figure.

From Section 16.1, the velocity, $\mathbf{v}_1$, of m_1, in inertial space, expressed in the cable frame, can be written in terms of ω, $\mathbf{r}_1$, and $\dot{\mathbf{r}}_1$ as,

$$\mathbf{v}_1 = \dot{\mathbf{r}}_1 + \omega \times \mathbf{r}_1 = R\dot{\phi}(\Omega + \dot{\phi})\hat{\mathbf{i}} + R\Omega\hat{\mathbf{j}} \qquad (19\text{-}76)$$

The angular momentum of m_1 in the cable frame is

$$\mathbf{h}_1 = \mathbf{r}_1 \times m_1\mathbf{v}_1 = m_1R^2\left[\Omega + \phi^2(\Omega + \dot{\phi})\right]\hat{\mathbf{k}} \qquad (19\text{-}77)$$

Because $\mathbf{k}$ is fixed in inertial space in the direction of total system angular momentum, we may use the conservation of angular momentum to obtain

$$\left(I + 2m_1R^2\right)\Omega_0 = I\Omega(t) + 2m_1R^2\left[\Omega(t) + \phi^2(\Omega + \dot{\phi})\right] \qquad (19\text{-}78)$$

where $\Omega_0 \equiv \Omega(t=0)$ and the total moment of inertia is the sum of the inertia of the spacecraft body (I) and the two masses $(2m_1R^2)$.

The kinetic energy of m_1 is

$$T_1 = \frac{1}{2}m_1\mathbf{v}_1 \cdot \mathbf{v}_1 = \frac{1}{2}m_1R^2\left[\phi^2(\Omega + \dot{\phi})^2 + \Omega^2\right] \qquad (19\text{-}79)$$

From the conservation of energy we obtain

$$\left(\frac{1}{2}I + m_1R^2\right)\Omega_0^2 = \frac{1}{2}I\Omega^2(t) + m_1R^2\left[\phi^2(\Omega + \dot{\phi})^2 + \Omega^2\right] \qquad (19\text{-}80)$$

Equations (19-78) and (19-80) may be solved simultaneously for ϕ and Ω to obtain

Table 19-1. Representative Attitude Acquisition Sequences

MISSION	LAUNCH DATE	ATTITUDE DETERMINATION HARDWARE	ATTITUDE CONTROL HARDWARE	INITIAL STATE	FINAL STATE	SEQUENCE DURATION	ATTITUDE CONSTRAINTS	GROUND SUPPORT	MALFUNCTIONS
RAE-2	JUNE 10, 1973	1-AXIS DIGITAL SUN SENSORS; 2-AXIS DIGITAL SUN SENSORS; PANORAMIC SCANNER	COLD GAS THRUSTER; EXTENDABLE ANTENNAS; LIBRATION DAMPER BOOM	50 RPM TRANS-LUNAR ORBIT	1 RPO, 3-AXIS, GRAVITY-GRADIENT STABI-LIZED CIRCULAR, LUNAR ORBIT, 4 ANTENNAE DE-PLOYED TO 229 M; NEGLIGIBLE AT-TITUDE RATES	5 MONTHS	POWER AND THERMAL IN TRANSLUNAR ORBIT; ANTENNA DEPLOYMENT SEQUENCED TO MINI-MIZE ATTITUDE LIBRA-TIONS & BOOM BENDING; MOON LIGHT-ING CONDITIONS SUIT-ABLE FOR PAS DATA PRIOR TO INITIAL ANTENNA DEPLOY-MENT	OPEN-LOOP, REAL-TIME GRAPHICS DISPLAYS TO MON-ITOR GAS JET MANEUVERS AND TO INITIATE ANTEN-NA DEPLOYMENT	SUN INTERFERENCE IN PAS DATA; 2-AXIS SUN SENSOR FAILURE AT LOW SPIN RATE; APPARENT FAIL-URE OF ALL ANTENNAS TO FULLY DEPLOY; COUNTER, WHICH DETER-MINED DIRECTION OF PRECESSION, SHIFTED DURING MANEUVER; DESPIN JETS IN PULSED MODE REMAINED ON BEYOND PRESET TIME INTERVAL
GEOS-3	APRIL 9, 1975	2-AXIS DIGITAL SUN SENSOR; MAGNETOM-ETER	2-AXIS ELECTRO-MAGNET; EXTEND-ABLE BOOM; PITCH AXIS MOMENTUM WHEEL; MAGNET-ICALLY ANCHORED EDDY CURRENT DAMPER	RANDOM ORIENTATION; TUMBLE RATE LESS THAN 0.5 RPM	1 RPO GRAVITY-GRADIENT-STABILIZED; MOMENTUM WHEEL AT SYNCHRONOUS RPM ALONG ORBIT NORMAL; BOOM DEPLOYED TO 6.47 M	5 DAYS (75 HOURS FOR AC-TIVE MAN-EUVERS; THE RE-MAINDER FOR PASSIVE DAMPING)	COMMANDING OVER DAYLIGHT, REAL-TIME STATIONS; ELECTROMAGNET OPERATION LIMITED BECAUSE OF POWER CONSTRAINTS; INVERTED CAPTURE (PITCH = 180°) LIMITED TO SEVER-AL ORBITS BECAUSE OF THERMAL CON-STRAINTS.	OPEN-LOOP REAL-TIME GRAPHIC DISPLAYS UTI-LIZED TO MONITOR ATTITUDE PRIOR TO COMMANDING	ANOMALOUS EXTENSION RATES FOR BOOM MECHANISM; DAMPING TIME CONSTANT 52% OF NOMINAL.
CTS (SPIN)	JANUARY 17, 1976	DIGITAL SUN SENSORS; IR EARTH SENSORS (BODY MOUNTED)	HYDRAZINE THRUSTERS	60 RPM	60 RPM, Z-AXIS ALONG NEGATIVE OR-BIT NORMAL.	2 WEEKS	THERMAL AND POWER ATTITUDE OPERA-TIONS LIMITED TO 1 HOUR NEAR PERIGEE DURING TRANSFER ORBIT BECAUSE OF EARTH SENSOR CONFIGURATION	OPEN-LOOP MANEUVERS MONITORED IN REAL TIME	ELECTRONICS CONTROLLING PRESSURE AND TEMPERATURE GAUGES ON ONE (OF TWO) HYDRAZINE TANKS FAILED; BALKY VALUES IN THRUSTER SYSTEM
(DESPIN)	JANUARY 30, 1976	2-AXIS DIGITAL SUN SENSORS; IR EARTH SENSORS (SCANNING)	HYDRAZINE THRUSTERS; PITCH AXIS MOMENTUM WHEEL; EXTEND-ABLE SOLAR ARRAYS	60 RPM; Z-AXIS ALONG NEGATIVE OR-BIT NORMAL	1 RPO, EARTH LOCKED, ON-BOARD CONTROL SYSTEM OPERATING; MOMENTUM WHEEL AT NOMINAL RPM AND ALONG ORBIT NORMAL; 6M SOLAR ARRAYS DEPLOYED	2 DAYS	YAW AXIS MUST BE ORIENTED TOWARD SUN AND SOLAR ARRAYS DEPLOYED WITH-IN 2 HOURS AFTER DESPIN; ATTITUDE MUST BE FAVOR-ABLE FOR COM-MUNICATIONS; LIMITED FOV FOR IR SCANNERS	CLOSED LOOP VIA HP2100A MINICOM-PUTER, TABULAR DISPLAYS OF RAW SENSOR AND COM-PUTED ATTITUDE DATA; STRIP CHART RECORDERS	NONE

$$\phi = \Omega_0 t \tag{19-81a}$$

$$\Omega(t) = \Omega_0 (a - \Omega_0^2 t^2)/(a + \Omega_0^2 t^2) \tag{19-81b}$$

where

$$a \equiv I/(2m_1 R^2) + 1 \tag{19-81c}$$

Thus, Eq. (19-81a) shows that the cable unwinds at a constant rate equal to the initial spacecraft angular velocity. Equation (19-81b) may be rewritten in terms of the unwound cable length, $l = R\phi = R\Omega_0 t$ as

$$\Omega(l) = \Omega_0 (aR^2 - l^2)/(aR^2 + l^2) \tag{19-82}$$

Note that $\Omega = 0$ when

$$l = \sqrt{a}\, R = \sqrt{I/2m_1 + R^2} \tag{19-83}$$

which is *independent* of the initial spin rate. For example, for the HCMM spacecraft, with $I = 18.4$ kg m^2, $R = 0.5$ m and two yo-yo masses of 2 kg each, a cable length $l = (4.6 + 0.25)^{1/2} = 2.2$ m would completely despin the spacecraft for any initial spin rate.

Momentum Transfer. *Momentum transfer* is an acquisition maneuver used for dual-spin spacecraft. Initially, the body is spin stabilized at an angular velocity Ω and the wheel is fixed in the body frame (Fig. 19-15a). Finally, the body is despun and most of the momentum is transferred to the wheel (Fig. 19-15b). For Earth-oriented missions, the desired final attitude is such that the wheel axis and orbit normal are collinear and the residual body spin rate, ω_0, is 1 revolution per orbit about the orbit normal.

The maneuver for an Earth-oriented mission is illustrated in Fig. 19-15 and described by Barba and Aubrun [1975] and Gebman and Mingori [1975]. Essentially, the maneuver involves a transfer from an initial configuration in which the body is spin stabilized with the attitude antiparallel to the orbit normal and the wheel is despun, to a final configuration in which the body is despun and the wheel is spinning with its axis antiparallel to the orbit normal.* Because the wheel axis (body y axis) is normal to the initial spin axis (body z axis), the maneuver results in the erection of the wheel axis to the orbit normal. Although the total angular momentum vector is conserved during the maneuver and the *magnitude* of the wheel angular momentum may be controlled via the wheel speed, the partitioning of the total angular momentum vector between the body and the wheel cannot be fully controlled. Consequently, the total transfer of momentum to the wheel cannot, in general, be obtained and the final state will consist of the wheel axis nutating about the conserved total angular momentum vector.

As the wheel is accelerated, the body rate about the y axis first increases, reaches a maximum, and finally decreases to near zero. The body rate about the z axis decreases rapidly and then oscillates about zero with a frequency proportional to the wheel momentum. The residual offset angle, θ, between the body y axis and the orbit normal declines from 90 deg and oscillates about a minimum residual

---

*As described in Section 18.3, we assume the body y axis is to be aligned with the pitch axis of the orbital coordinate system or the negative orbit normal.

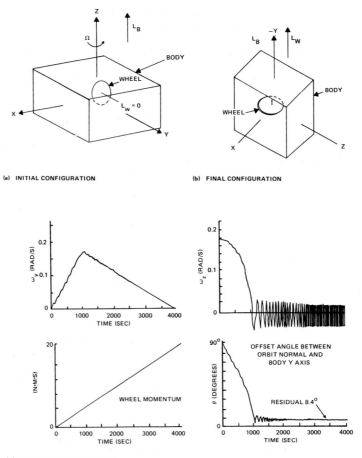

(a) INITIAL CONFIGURATION   (b) FINAL CONFIGURATION

(c) MANEUVER CHARACTERISTICS

Fig. 19-15.   Computer Simulation Results for Momentum Transfer Maneuver (from Barba and Aubrun, [1975])

offset, which is typically 5 to 10 deg. Figure 19-15c illustrates the characteristics of the maneuver.

If the transverse wheel moment of inertia, K_1, is assumed to be small compared with the body moment of inertia about the z axis, I_z, the initial state is

$$\mathbf{L}_B(0) = I_z \Omega \hat{\mathbf{n}} \qquad (19\text{-}84)$$

$$\mathbf{L}_W(0) = 0$$

and the desired final state, in an inertial frame, is

$$\mathbf{L}_B(T) = I_y \omega_0 \hat{\mathbf{n}} \qquad (19\text{-}85)$$

$$\mathbf{L}_W(T) = h \hat{\mathbf{n}}$$

where $\mathbf{L}_B$ is the angular momentum of the body, $\mathbf{L}_W$ is the angular momentum of the wheel, $\hat{\mathbf{n}}$ is a unit vector in the direction of the orbit normal, I_y is the moment of inertia of the spacecraft about the y axis, and h is the magnitude of the wheel momentum. Conservation of the total angular momentum, $\mathbf{L}_T$, during wheel accleration ensures that, in an inertial frame,

$$\mathbf{L}_T = \mathbf{L}_B(0) + \mathbf{L}_W(0) = I_z \Omega \hat{\mathbf{n}} = \mathbf{L}_B(T) + \mathbf{L}_W(T) \tag{19-86}$$

Control over wheel speed during acceleration permits the relation $\mathbf{L}_W(T) = -h\hat{\mathbf{y}}$ to be obtained in *body coordinates*, but the identity

$$L_T^2 - h^2 = \text{constant} = L_B^2(T) + 2\mathbf{L}_B(T) \cdot \mathbf{L}_W(T) \tag{19-87}$$

does not guarantee that $\mathbf{L}_B(T)$ is near the orbit normal although that desired configuration is consistent with the conservation of energy and momentum and is nearly obtained. The offset angle, θ, is approximately [Gebman and Mingori, 1975]

$$\theta = \epsilon^{1/2} \left\{ 0.939 \left[1 + \frac{1}{2} \left(\frac{I_1 - I_2}{I_2 + K_1} \right) \right]^{1/4} \right\} + \mathcal{O}(\epsilon^{3/2}) \tag{19-88}$$

where

$$\epsilon = \frac{h/T}{(I_1 + I_W - I_3) \left(\dfrac{I_1 + K_1}{I_2 + K_1} - 1 \right) \Omega^2} \tag{19-89}$$

and T is the wheel acceleration time (assuming a constant torque), $I_1 > I_2 > I_3$ are the ordered body moments of inertia, K_1 is the transverse wheel moment of inertia, and I_W is the axial wheel inertia. Equations (19-88) and (19-89) *assume* that the I_3 axis (the smallest moment-of-inertia axis) is parallel to the wheel axis (I_y in the example). Thus, reduced offset angles are achieved, for a given configuration, by reducing the wheel acceleration torque, h/T.

As an example of the application of the momentum transfer maneuver, we consider a proposed acquisition sequence for CTS [Lerner, *et al.*, 1976]. The mission mode angular momentum was 20 N·m·s (3750 rpm) and 0.01 N·m·s (1 rpo) for the wheel and body, respectively, along the positive orbit normal. The angular momentum at the start of the acquisition was $-972\hat{\mathbf{n}}$ N·m·s (60 rpm along the negative orbit normal). The proposed acquisition sequence was as follows:

1. Use gas jets to despin to -1.25 rpm to obtain a total angular momentum of 20 N·m·s $\hat{\mathbf{n}}$.
2. Accelerate the wheel until the body rate is $\sim$1 rpo, at which time the wheel speed will be near 3750 rpm.
3. Damp the resultant nutation, using thrusters as described in Section 18.4 (typical half-cone angles are 8 to 10 deg).
4. Use thrusters to precess the attitude to orbit normal, as described in Section 19.3, to achieve the final attitude (typical attitude errors are 7 to 13 deg).

Table 19-2 summarizes the results of simulated momentum transfer sequences for CTS as a function of wheel acceleration time.

Deadbeat Boom Deployment. *Deadbeat deployment* consists of either extending booms or antennas so as to minimize attitude librations after deployment or using extendable appendages to remove existing librations. The former procedure was used on RAE-1 and -2 and the latter on GEOS-3. Such maneuvers are called

Table 19-2.   Simulated Momentum Transfer Acquisition Maneuvers for CTS

WHEEL ACCELERATION TIME (SEC)	NET WHEEL TORQUE (N·M)	FINAL WHEEL SPEED (RPM)	ATTITUDE ERROR AND NUTATION			
			BEFORE DAMPING BURN (DEG)		AFTER DAMPING BURN (DEG)	
			θ_T^*	$\gamma_w^\dagger$	θ_T'	γ_w'
3850	0.005	3850	7.2	9.7	11.0	2.3
1875	0.011	3750	10.3	12.1	7.2	1.5
1000	0.020	3750	14.1	21.5	21.4	3.1

* OFFSET ANGLE BETWEEN ANGULAR MOMENTUM VECTOR AND ORBIT NORMAL.

† NUTATION HALF-CONE ANGLE, $\gamma_w = \text{ARCTAN} \left(\left[(I_x \omega_x)^2 + (I_z \omega_z)^2 \right]^{1/2} / (6 I_w S) \right)$, WHERE I_x, I_z, AND I_w ARE THE X, Z, AND WHEEL INERTIAS; ω_x AND ω_z, THE BODY RATES (DEG/S); AND S, THE WHEEL SPEED (RPM).

deadbeat, meaning no recoil, after the stroke employed by drummers, and are based on the conservation of angular momentum. Consider a spacecraft librating under the influence of gravity-gradient torques about the pitch axis with a boom fully extended along the yaw axis. At any time, the attitude state may be represented as a point in the pitch/pitch rate state-space as shown in Fig. 19-16. If an initially extended boom is retracted to an intermediate length at near zero pitch and minimum pitch rate, the decrease in inertia about the pitch axis will cause the pitch rate to increase (become less negative to conserve angular momentum) and follow the trajectory depicted by the inner circle.* If subsequently the boom is reextended at a pitch angle near zero and maximum pitch rate, the increase in inertia will reduce the pitch rate and remove the pitch librations. The proper choice of an intermediate moment of inertia, I_r, is derived as follows. Assume that the retraction and extension maneuvers are instantaneous. Conservation of angular momentum at retraction requires

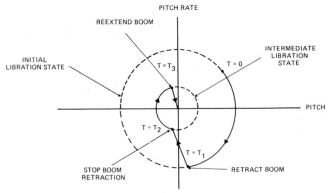

Fig. 19-16.   Deadbeat Maneuver for Removal of Pitch Libration Using Extendable Boom. The origin of the figure corresponds to pitch = pitch rate = 0, but the scale of the axes is arbitrary. Note that pitch = 0 implies an inertial rate about the pitch axis of minus 1 revolution per orbit (rpo) or 1 rpo about the positive orbit normal.

---

*We assume that the inertia change is instantaneous. For typical configurations, boom maneuvers require 1 to 10 minutes whereas libration periods are typically 1 hour ($\approx$orbital period/$\sqrt{3}$). The external torques are proportional to pitch (and near zero) when the extension and retraction occur.

$$I_e(-\dot{p}_i+\omega_o)=I_r(-\dot{x}+\omega_o) \tag{19-90}$$

where I_e is the moment of inertia about the pitch axis with the boom extended, $\dot{p}_i$ and $\dot{x}$ are the pitch rate amplitudes before and after extension, and ω_o is the orbital angular velocity. At extension, for the pitch rate to vanish, conservation of angular momentum yields

$$I_r(\dot{x}+\omega_o)=I_e\omega_o \tag{19-91}$$

Substitution of Eq. (19-91) into Eq. (19-90) yields

$$I_r=I_e(1-\dot{p}_i/2\omega_o) \tag{19-92}$$

which is the required moment of inertia after retraction to remove a libration rate amplitude $\dot{p}_i$.

The equation of motion for pitch, p, considering only gravity-gradient torque and assuming a small roll and roll rate, is (see Section 18.3)

$$\ddot{p}+\left(3\omega_o^2/2I_y\right)(I_x-I_z)\sin 2p=0 \tag{19-93}$$

where I_y, I_x, and I_z are the moments of inertia about the body pitch (y), roll (x), and yaw (z) axes and ω_o is the orbital rate. Letting $\tau \equiv \omega_p t$ where

$$\omega_p=\omega_o\left[3(I_x-I_z)/I_y\right]^{1/2} \tag{19-94}$$

Equation (19-93) can be rewritten as

$$\frac{d^2p}{d\tau^2}+\frac{1}{2}\sin 2p=0 \tag{19-95}$$

with the integral

$$\frac{dp}{dt}=\pm\omega_p\left[\frac{1}{2}(\cos 2p-\cos 2A)\right]^{1/2} \tag{19-96}$$

where A, the maximum value of pitch, is an integration constant.

For a pencil-shaped spacecraft, $I_x\approx I_y\gg I_z$, and $\omega_p\approx\sqrt{3}\,\omega_o$. The maximum rate occurs when $p=0$, so that

$$\dot{p}_i=\left(\frac{dp}{dt}\right)_{p=0}\approx\omega_o\left[\frac{3}{2}(1-\cos 2A)\right]^{1/2} \tag{19-97}$$

and Eq. (19-92) may be rewritten in terms of the libration amplitude as

$$I_r/I_e=1-\left[\frac{3}{8}(1-\cos 2A)\right]^{1/2} \tag{19-98}$$

Magnetic Stabilization. As a final acquisition maneuver type, *magnetic stabilization* is a technique in which a spacecraft axis is induced to track the Earth's magnetic field about the orbit. This is used for high inclination spacecraft to provide a reference angular momentum direction normal to the orbit plane. Consider a spacecraft in a polar orbit with an electromagnet along the yaw axis and an onboard damper. Regardless of the initial attitude and attitude rate, the interaction of the electromagnet and external field will cause the yaw axis to track the field (minimum energy configuration) and induce an average spin rate of 2 rpo about the orbit normal. Magnetic stabilization was used for GEOS-2 and proposed for GEOS-3 as the first step in the attitude acquisition because it is passive (no ground support is required) and converts a random initial state into a well-defined state suitable for subsequent acquisition maneuvers.

19.5.3 Representative Acquisition Sequence

In this section, we describe the attitude acquisition sequence employed for the *Geodynamics Experimental Ocean Satellite, GEOS-3*, launched on April 9, 1975, from Vandenberg Air Force Base, California, on a Delta 1410 rocket. Other acquisition sequences are described by Basset [1976] for CTS, Byrne, *et al.*, [1978] for HCMM, and Markley [1978] for SMM. GEOS-3 demonstrated the utility of spaceborne radar altimeters for oceanography and served as a bridge between the earlier geodetic satellites, GEOS-1 and GEOS-2, and the ocean resources program, SEASAT. The spacecraft, illustrated in Fig. 19-17, was placed in a circular orbit at an altitude of 843 km and an inclination of 115 deg to provide coverage of the North Atlantic Ocean, the area of primary experimental interest.

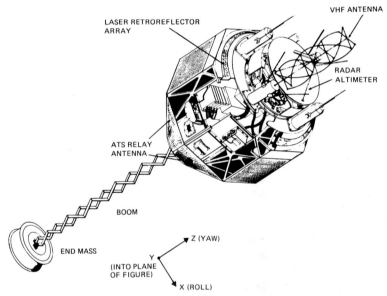

Fig. 19-17.   GEOS-3 Spacecraft

The ground-based, open loop attitude acquisition sequence for GEOS-3 was designed to achieve a gravity-gradient stabilized, three-axis attitude with the spacecraft *z* and *y* axes in the nadir and negative orbit normal directions, respectively. The GEOS-3 control hardware consisted of a 6.5-m boom extendable along the negative *z* axis; a passive, magnetically anchored eddy current damper (Section 18.4) located at the end of the boom; a *z* axis electromagnet; and a momentum wheel with its axis along the *y* axis. Attitude determination hardware consisted of two-axis digital Sun sensors and magnetometers. Pitch and roll stability in the mission mode was provided by gravity-gradient torque and yaw stability was accomplished via quarter-orbit coupling with roll through the momentum wheel (see Section 18.2).

The goal of the GEOS-3 acquisition sequence was to achieve mission mode and begin experimental operations as rapidly as possible. The mission constraints are given in Table 19-1. The initial attitude acquisition plan for the GEOS-3 was to activate the z axis electromagnet soon after spacecraft separation to achieve magnetic stabilization (Section 19.5.2), to extend the boom over a high northern latitude command station to achieve proper (pitch$\approx$0) gravity-gradient stabilization, and, finally, to accelerate the wheel to achieve yaw stabilization. This procedure was abandoned because the large amplitude librations induced by the boom and wheel maneuvers coupled with the long system damping time constants

Table 19-3.   GEOS-3 Attitude Acquisition Profile

EVENT	TIME FROM LAUNCH HR/MIN	DATE		STATION	REMARKS
		MO/DAY/YR	HHMMSS* UT		
1. LAUNCH	0	4/9/75	235801		
2. SEPARATION	1/0	4/10/75	006821	TANANARIVE	1-DEG/S TUMBLE ABOUT PITCH AXIS
3. RELEASE BOOM	3/12	4/10/75	031000	ALASKA	BOOM LENGTH 0.26 M AFTER RELEASE
4. EXTEND BOOM TO 0.72 M	4/59	4/10/75	045700	ALASKA	ACHIEVE PROPER GRAVITY-GRADIENT CAPTURE
5. DEPLOY BOOM TO 6.47 M	50/44	4/12/75	024237	ALASKA	AT YAW 81 DEG AND ROLL AMPLITUDE 2 DEG
6. ACTIVATE WHEEL	51/17	4/12/75	031503	ORRORAL	AT PITCH −11 DEG, RESIDUAL PITCH LIBRATION WAS 12 DEG
7. DEADBEAT TRIM	70/56 74/21	4/12/75 4/13/75	225400 021900	WINKFIELD WINKFIELD	RESIDUAL PITCH AMPLITUDE WAS 3 DEG

*HOUR:MINUTE:SECOND.

with the boom extended combined to require an estimated 30 days to achieve stability [Pettus, 1973].

Further prelaunch analysis yielded an improved procedure incorporating several attitude acquisition strategies:

1. Gravity-gradient capture at a boom length of approximately 1 m could be achieved by first allowing the spacecraft to despin under the influence of gravity-gradient and damper torques and subsequently extending the boom approximately 0.5 m when pitch$\approx$roll$\approx$0 [Repass, et al., 1975].

2. Pitch, but not roll, librations could be removed by a sequenced boom retraction and extension as described in Section 19.5.2; therefore, roll librations would need to be removed before boom extension.* Roll librations could be removed by the damper if gravity-gradient stabilization could be achieved before boom extension to 6.5 m because, at a boom length of 1 m, the damping time constant is only 13 hours [Davis and Yong, 1975].

3. After gravity-gradient stabilization, the acceleration of the wheel at the proper point in the pitch libration cycle, when the wheel acceleration reaction torque and gravity-gradient restoring torque cancel, could minimize subsequent librations [Pettus, 1973].

---

* When a 4- to 8-day damping time constant dominates the dynamics.

These strategies were incorporated into the acquisition sequence outlined in Table 19-3 and were implemented using a combination of passive stabilization based on gravity-gradient torque and active, open-loop, commanding using real-time graphic displays. After spacecraft separation, the attitude data indicated a slow spin, 1 deg/s, about the body y axis and the boom and damper magnet were released to permit the spacecraft to despin. Figure 19-18 illustrates the theoretical attitude behavior during the despin (i.e., events 3 to 4 in Table 19-3).

The damper magnet is driven through the Earth's magnetic field by the spacecraft's orbital motion. As described in Section 19.5.2, this induces the magnet to spin about the orbit normal at a mean inertial rate of 2 rpo. Consequently, the reaction torque on the spacecraft damps roll motion and induces the spacecraft to spin at a steady-state pitch rate of 1 rpo. Equation (19-96) describes this pitch motion, where $\omega_p = 0.087$ deg/s for a 1-m boom length. Representative attitude solutions are illustrated in Fig. 19-18. Closed trajectories about pitch 0 and 180 deg represent proper and inverted capture, respectively; open trajectories for positive

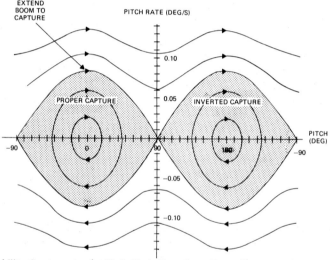

Fig. 19-18.　Stability Contours in the Pitch-Pitch Rate State Space for a 1-m Boom Length. Shaded regions indicate attitude capture.

and negative pitch rate represent backward and forward tumbling, respectively. The action of the damper causes the attitude to move to trajectories with smaller rates and, ultimately, to proper or inverted capture. Extending the boom to approximately 1.3 m at the point marked by the arrow in Fig. 19-18 increases the transverse moment of inertia by approximately 30%, halves the pitch rate, and causes the attitude to switch to a closed trajectory about pitch zero.

The real-time displays observed during actual boom extension and wheel acceleration are shown in Fig. 19-19. The attitude solutions typically lagged 20 to 40 sec behind real time and command initiation and uplink required an additional 5 to 10 sec. Figure 19-19(a) illustrates both that the boom extension command was sent after roll librations had damped and that the pitch rate decreased after extension to conserve angular momentum. The wheel acceleration command (see Fig. 19-19b) was transmitted 33 min after boom extension at a predetermined pitch angle so that the gravity-gradient and wheel acceleration reaction torques would oppose and minimize pitch librations after the maneuver.

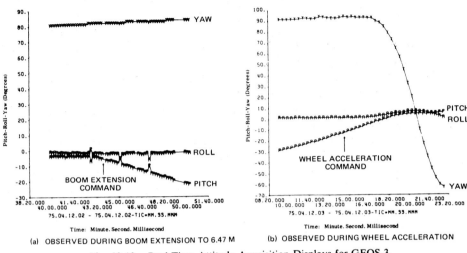

Time: Minute. Second. Millisecond

(a) OBSERVED DURING BOOM EXTENSION TO 6.47 M

Time: Minute. Second. Millisecond

(b) OBSERVED DURING WHEEL ACCELERATION

Fig. 19-19.   Real-Time Attitude Acquisition Displays for GEOS-3

Operational considerations dictated that the two commands be sent over a pair of command stations in daylight with adequate telemetry visibility before and after each command. Dynamical considerations dictated a half-hour command separation (for torque opposition) and a small yaw angle to ensure yaw capture and minimize roll-yaw librations. These constraints limited command opportunities to an average of approximately one per day. After 2 days of monitoring potential command opportunities, the maneuvers were initiated despite a yaw angle near the maximum value for satisfactory dynamics [Lerner and Coriell, 1975]. Figure 19-20 compares the observed attitude data (dots) with a postlaunch simulation (solid line) during and after the boom and wheel commands. After boom extension, a pitch libration, with an amplitude of approximately 40 deg, was induced while roll and yaw remained near their initial values of 2 deg and 80 deg, respectively. The wheel was accelerated 33 minutes later, when the wheel acceleration and gravity-gradient torques were in opposition and the 40-deg pitch libration was consequently reduced to about 15 deg. The wheel momentum coupled the roll and yaw motion, resulting in an initial 60-deg amplitude yaw oscillation and an 8-deg roll oscillation with a 9-minute period. In Fig. 19-20(b), the attitude behavior 16 hours later is shown. As described in Section 18.3, pitch, roll, and yaw oscillations have decayed with time constants of approximately 4 days, 6 days, and 5 hours, respectively. Thus, the large yaw oscillation rapidly decayed, the roll oscillation remained near the small initial value, and the pitch oscillation was subsequently reduced from 10 to 3 deg by a deadbeat boom retraction and extension as described in Section 19.5.2. Experimental operations were initiated 4 days after launch.

References

1.   Barba, P., and J. Aubrun, *Satellite Attitude Acquisition by Momentum Transfer*, Paper No. AAS 75-053, AAS/AIAA Astrodynamics Specialist Conference, Nassau, Bahamas, July 1975.

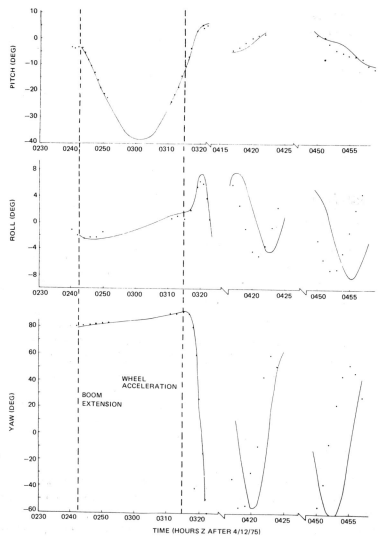

Fig. 19-20(a). Observed and Simulated Attitude Data During GEOS-3 Attitude Acquisition. Attitude data immediately after boom extension and wheel acceleration.

2. Basset, D. A., *Ground Controlled Conversion of Communications Technology Satellite (CTS) From Spinning to Three-Axis Stabilized Mode*, Paper No. AIAA 76-1928, AIAA Guidance and Control Conference, San Diego, CA, Aug. 1976.

3. Blaylock, B. T., and R. Berg, *International Ultraviolet Explorer (IUE) Attitude Recovery Software Description and Operating Guide*, Comp. Sc. Corp., CSC/SD-76/6146, June 1976.

4. Byrne, R., D. Niebur, S. Hotovy, F. Baginski, W. Nutt, M. Rubinson, G. Lerner, and R. Nankervis, *Applications Explorer Missions-A/Heat Capacity Mapping Mission (AEM-A/HCMM) Attitude Analysis and Support Plan*, NASA X-581-78-4, GSFC, March 1978.

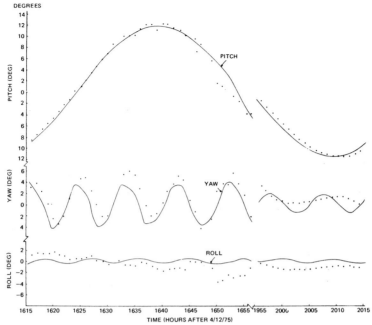

Fig. 19-20(b). Observed and Simulated Attitude Data During GEOS-3 Attitude Acquisition. Attitude data 16 hours after wheel acceleration. Roll, which was less than 5 deg before boom extension, has remained small. The yaw libration amplitude has decayed rapidly from near 90 deg to 4 deg. The pitch amplitude remains near the value reached after wheel acceleration.

5. Davenport, P. B., *Mathematical Analysis for the Orientation and Control of the OAO Satellite*, NASA TN D-1668, GSFC, 1963.

6. Davis, R., and K. Yong, *Mission Planning Study for the GEOS-C Spacecraft*, Comp. Sc. Corp., 6050-00000-02TR, Jan. 1975.

7. Gambhir, B. L., and D. R. Sood, *Spin Axis Attitude Perturbation Due to Spin/Despun Dipoles for the SAS-3 Spacecraft*, Comp. Sc. Corp., CSC/TM-76/6010, Jan. 1976.

8. Gebman, J., and D. Mingori, *Perturbation Solution for the Flat Spin Recovery of a Dual-Spin Spacecraft*, Paper No. AAS 75-044, AAS/AIAA Astrodynamics Specialist Conference, Nassau, Bahamas, July 1975.

9. Goldstein, Herbert, *Classical Mechanics*, Reading, MA: Addison-Wesley Publishing Co., Inc., 1950.

10. Grell, M. G., "Magnetic Attitude Control of the Atmosphere Explorer-E Spacecraft," *Proc. Summer Computer Simulation Conf.*, pp. 320–324, Chicago, IL, Simulation Councils, Inc., July 1977.

11. Hsu, J. C., and A. U. Meyer, *Modern Control Principles and Applications*. New York: McGraw-Hill, Inc., 1968.

12. IBM Corporation, *Radio Astronomy Explorer Attitude Determination Systems (RAEADS)*, NASA Contract NAS 5-10022, 1968.

13. Jackson, John David, *Classical Electrodynamics*. New York: John Wiley & Sons, Inc., 1965.

14. Kjosness, D. H., *CTS Attitude Acquisition Sequence Detailed Operating Procedures*, SED Systems LTD, 5143-TR-101, Jan. 1976.

15. Lerner, Gerald M., and Kathleen P. Coriell, *Attitude Capture Procedures for GEOS-C*, Paper No. AAS 75-029, AAS/AIAA Astrodynamics Specialist Conference, Nassau, Bahamas, July, 1975.

16. Lerner, G., K. Yong, J. Keat, B. Blaylock, and J. Legg, Jr., *Evaluation of the Communication Technology Satellite Attitude Acquisition Algorithms and Procedures*, Comp. Sc. Corp., CSC/TM-76/6003, Jan. 1976.

17. Markley, F. L., *Attitude Control Algorithms for the Solar Maximum Mission*, AIAA Paper no. 78-1247, AIAA Guidance and Control Conference, Palo Alto, CA, Aug. 1978.

18. NASA, *Investigation of the Dynamic Characteristics of a V-Antenna for the RAE Satellite*, NASA CR-962, 1968.

19. Pettus, W., *Optimization of Boom Deployment and Momentum Wheel Activation for GEOS-C*, Comp. Sc. Corp. Memorandum, May 1973.

20. Phenneger, M. C., M. E. Plett, M. A. Firestone, M. G. Grell, and P. V. Rigterink, *Atmospheric Explorer-D and -E Attitude Determination and Control Prelaunch Report Analysis and Operations Plan*, Comp. Sc. Corp., CSC/TR-75/6018, Oct. 1975.

21. Repass, G., G. Lerner, K. Coriell, and J. Legg, Jr., *Geodynamics Experimental Ocean Satellite-C (GEOS-C) Postlaunch Report*, NASA X-580-75-23, GSFC, Feb. 1975.

22. Schmidtbauer, B., Hans Samuelsson, and Arne Carlsson, *Satellite Attitude Control and Stabilization Using On-Board Computers*, ESRO-CR-100, July 1973.

23. Stickler, A., and K. T. Alfriend, *An Elementary Magnetic Attitude Control System*, AIAA Paper No. 74-923, AIAA Mechanics and Control of Flight Conference, Anaheim, CA, Aug. 1974.

24. TRW Systems Group, *HEAO-B Attitude Determination and Control Subsystem Preliminary Design Review*, Feb. 1976.

25. Werking, R. D., and R. D. Woolley, "Computer Simulation for Time Optimal or Energy Optimal Attitude Control of Spin-Stabilized Spacecraft," *Summer Computer Simulation Conference, Montreal, Canada, July 17–19, 1973, Proceedings*, Vol. 1, p. 448–453, Society for Computer Simulation Inc., La Jolla, CA, 1973.

26. Werking, R. D., R. Berg, K. Brokke, T. Hattox, G. Lerner, D. Stewart, and R. Williams, *Radio Astronomy Explorer-B Postlaunch Attitude Operations Analysis*, NASA X-581-74-227, GSFC, July 1974.

27. Williams, Robert S., *An Analysis of the RAE-B Attitude Control System*, Comp. Sc. Corp. 5023-06300-07TR, March 1971.

PART V

MISSION
SUPPORT

CONTENTS

PART V

MISSION SUPPORT

Chapter

20   Software System Development   681

21   Software System Structure   696

22   Discussion   714

CHAPTER 20

SOFTWARE SYSTEM DEVELOPMENT

20.1 Safeguards Appropriate for Mission Support Software
20.2 Use of Graphic Support Systems
20.3 Utility Subroutines
   Vector and Matrix Algebra Routines, Time-Conversion
   Routines, Ephemeris Routines, Plotting Routines

In practice, much of the time devoted to preparation for mission support is spent in the development of computer software systems. Although some progress has been made in the standardization of software, the variations in attitude determination and control hardware, mission requirements, and processing sophistication have meant that most spacecraft series have required largely new attitude determination and control software systems. Therefore, questions of software structure and performance are central to the practical problems of mission support. This chapter describes the general principles for the development of attitude software and the use of executive support systems and utility subroutines.

20.1 Safeguards Appropriate for Mission Support Software

Myron A. Shear

Attitude determination requirements may be divided into three categories: *real time, near real time,* and *definitive*. A real-time requirement implies that attitude must be determined within seconds of the receipt of data and is usually associated with monitoring an attitude maneuver or attitude acquisition sequence. A near-real-time requirement implies that attitude must be determined within minutes or hours of the receipt of data, usually to compute control commands to achieve or maintain a desired attitude. A definitive requirement implies that an accurate attitude history is to be generated, perhaps weeks or months after the fact, generally for use in analysis of experimental results. The most critical demands on mission support software generally arise in real-time or near-real-time support, when results must be obtained shortly after the receipt of data. Failure to obtain accurate results within the prescribed time may jeopardize the success of the mission. Therefore, software intended for use in real-time or near-real-time mission support must be designed to meet particularly high standards of reliability, flexibility, and ease of operation. In some missions, even a minor software error could lead to total mission failure; furthermore, software must be capable of handling contingencies as well as nominal mission conditions. For example, if one attitude sensor fails, the software should still be capable of supporting the mission to the extent that the remaining attitude sensors permit. In real-time and near-real-time support, there is no time to make software modifications either to correct errors or to add new capabilities. Even a minor modification to a large software system may require hours or days to implement and the system reliability would be in doubt until extensive testing had been performed. For these reasons, specific safeguards should be considered in the design, implementation, and testing of mission support software. This section describes some of the safeguards used in mission software developed for the Attitude Determination and Control Section of NASA's Goddard Space Flight Center.

The software environment for mission support programs at Goddard Space Flight Center is typically a multiprogrammed, large-scale computer with interactive graphics terminals, card readers, printers, and other peripheral devices. Mission support programs are assigned relatively high priority, and nonmission support programs are run only as resources permit. Most mission support software is designed for interactive graphics operation, primarily because of the greater flexibility provided by allowing an analyst to examine the input data and program results and change the processing options accordingly. Graphics operation, described further in Section 20.2, also allows rapid correction of user input errors. Nongraphic systems utilizing card input and printed output are normally limited to utility programs which do not directly process telemetry data and, therefore, require fewer processing options. The safeguards discussed in this section can be applied to either nongraphic or graphic systems.

Error Checking.　If a program terminates unexpectedly and must be restarted, a period of 15 minutes or more may be required to resubmit the job, schedule the required resources, and initialize the program. A delay of this magnitude is unacceptable in real-time support, and very inconvenient in near-real-time support. For this reason, mission support software must be fully protected against failures due to user errors or unexpected telemetry data. An interactive graphics program must not be allowed to terminate abnormally except for the most severe error conditions; most common errors can be corrected by the user if the program provides *appropriate* error messages. Thus, mission support software must check for all foreseeable error conditions, provide standard corrective actions whenever possible, or provide an error message which is clear enough to allow the user to diagnose and correct the problem promptly. If further diagnosis is required, the user must be able to request intermediate displays to obtain additional information.

User input errors are almost inevitable; these may be simple typographical errors, or logical errors resulting from specifying an inconsistent set of input parameters. The program should check user input parameters for validity, especially in cases in which a user input error could lead to abnormal program termination. For example, a user error which leads to overfilling an array or infinite looping may result in a program termination which is difficult to diagnose and relate to the original error.

Potential mathematical singularities should also be checked to avoid errors such as division by zero, square root of a negative argument, or inverse trigonometric functions of invalid arguments. These singularities may result from user input errors, invalid telemetry data, or spacecraft hardware noise. Most operating systems provide features to intercept such errors, apply standard fixups, print warning messages, and/or terminate the program. However, these operating system features (such as the FORTRAN monitor) are generally inadequate for an interactive graphics system. Occurrences of mathematical singularities may require a standard corrective action, a count of the errors for later display, or an execution halt to inform the user of the error with an appropriate message. If standard corrective actions are applied indiscriminately, the program may generate meaningless results with no indication of the cause of the problem.

In addition to user input, mission support software generally obtains input from telemetry data files, ephemeris files, attitude history files, and other sources. Because these files are computer generated, they are not normally as susceptible to errors as user input. However, the data in these files must still be checked for validity, especially for errors which might result in program termination. The most common errors are an ephemeris file which does not cover the time span of the data being processed; a file generated in the wrong format or out of time order due to human error or software errors in the generating program; the header record of a file which disagrees with the data on the file; I/O errors which lead to random bit changes on the file; and files containing no data records.

If, in spite of error checks, a program abnormally terminates, or *abends*, then as a last resort the graphics executive should intercept the abend and allow a rapid recovery or restart of the program. Intercepting an abend is not as satisfactory as detecting an error within the program, because diagnosis of the abend may be more difficult; however, it does permit recovery from errors detected within operating system routines in cases in which prior error detection by the user would be inconvenient or the potential for error was unforeseen.

Flexibility. Mission support software should provide enough flexibility to handle contingencies such as spacecraft hardware malfunction, telemetry errors, or mission timeline changes. It is generally not possible to foresee all contingencies, nor is it feasible to provide special capabilities for every contingency which can be foreseen. However, if the software is sufficiently flexible, it is often possible to improvise a technique for handling a contingency simply by altering the available program options. For this reason, all program parameters should be variables which can be changed via interactive graphics; parameters such as tolerances and calibration constants should not be hard-coded within the program. Data set specifications should be flexible to permit processing multiple data sets or switching between data sets without terminating the program to change job control language. Processing flow within the program should be flexible and should be controlled by user input parameters. The program should optionally provide displays of all input, output, and intermediate results in a variety of formats (plots, tables, summaries); these formats should be designed to allow for the display of nonnominal as well as nominal data. A variety of options should be provided for editing telemetry data based on criteria such as maximum and minimum tolerances, residual tests, and consistency checks. As a last resort, the program should allow the operator to manually flag individual data items or override or modify any or all of the telemetry data items.

Ease of Operation. Ease of operation is more than just a convenience in mission support software; a system with the degree of flexibility described above will normally have hundreds of user input parameters and can easily become so complex as to tax the skill of any user. Operator interaction must be minimized, both to reduce the chance of human error and to minimize delays in near-real-time operation. To minimize the need for user input, typical default values should be provided for all input parameters, for example, via FORTRAN BLOCK DATA subprograms. Those parameters which must be changed from their default values and which are known a few hours or days in advance (such as orbit parameters or calibration constants) can be specified on input cards which are read when the program is initialized (e.g., NAMELIST card input). The user then will change only those parameters which differ from their expected values. Once the user changes a parameter, the new value should be used by the program until the user changes it again.

All displays should be clear and self-explanatory to a trained user; there should be no need for the experienced user to consult a user's guide for definitions of input parameters or error codes. The most frequently altered input options should be grouped together at the beginning of displays, so that the user can skip rarely used options. All input and output should be in the units and format which are most convenient for the user; for example, if times are to be expressed in calendar format for purposes of communicating with the control center, then the program should make the necessary conversions. There should be no need for the operator to do hand calculations or consult tables because the error rate for such operations is unacceptably high. Output displays should be designed to communicate information as rapidly as possible. A plot display can often be interpreted by the user many times faster than a tabular display, but only if proper attention is given to automatic scaling and exclusion of spurious points.

Reliability. A software system is said to be *reliable* if it meets its specifications (i.e., obtains correct results) for all possible sets of input data. The error-checking features discussed above are considered part of the specifications; thus, a reliable program must generate appropriate error messages for invalid input data.

Reliability begins with program design. The design should be simple, straight-forward, and modular. If, after a detailed study, it appears that there does not exist any simple design which can meet the specifications (including execution time and core requirements), then it is advisable to consider relaxing the specifications rather than proceeding with the development of a system which may never achieve the required reliability.

Following detailed design, the program should be coded in a higher level language with features designed to minimize or eliminate bookkeeping-type errors, thereby allowing the programmer to concentrate on program logic. Some useful language features include the ability to define COMMON blocks in a single library, the ability to check calling sequences in the called routine against those in the calling routine, structured programming constructs to eliminate GO TO statements, tests for variables which are never initialized, simplified vector and matrix operations, and automatic enforcement of programming standards. Most versions of FORTRAN lack these features; however, *precompilers* can be used to add these features to FORTRAN or other programming languages. A structured FORTRAN precompiler providing many of these features is described by Chu [1977].

Enforcement of good programming standards can eliminate many common errors. The standards used will necessarily depend on the application, programming language, and computing environment. One such set of FORTRAN programming standards has been described by Berg and Shear [1976].

Table 20-1.   Methods for Avoiding Common Software Errors

ERROR	METHODS FOR AVOIDING
INTERFACE ERRORS: VALUES NOT IN EXPECTED UNITS OR COORDI-NATE SYSTEM (DEG/RAD; KM/EARTH RADII/A.U.; TIME IN SECONDS FROM REFERENCE/JULIAN DATE/CALENDAR TIME/DAY OF YEAR; COORDINATES IN MEAN OF 1950.0/TRUE OF DATE; GEOCENTRIC/SPACECRAFT-CENTERED; ETC.)	USE ONE CONSISTENT SET OF UNITS AND COOR-DINATES FOR ALL MODULE-TO-MODULE INTER-FACES. PERFORM CONVERSIONS WITHIN EACH MODULE AS NECESSARY USE ONE CONSISTENT SET OF UNITS FOR USER-TO-PROGRAM INTERFACE. IF CONVERSIONS ARE REQUIRED, DO THE CONVERSIONS AT THE POINT OF I/O E.G., SUGGESTED TIME SYSTEM: USE SECONDS FROM A FIXED REFERENCE FOR ALL MODULE TO-MODULE INTERFACES (COMMON TIME REFERENCES ARE GIVEN IN SECTION 1.4). FOR INPUT/OUTPUT, USE CALENDAR TIME FORMAT, AND PERFORM THE CONVERSION AT THE POINT WHERE I/O IS PERFORMED
VARIABLE TYPES DO NOT MATCH (SINGLE PRECISION/DOUBLE PRECISION, REAL/INTE-GER/LOGICAL); FORGOT TO DECLARE VARI-ABLE TYPE	USE FORTRAN FIRST-LETTER CONVENTIONS FOR VARIABLE NAMES. IF DOUBLE PRECISION IS RE-QUIRED, USE IT THROUGHOUT THE MODULE. EXCEPTION: LARGE ARRAYS CAN BE EXPLICITLY DECLARED SINGLE PRECISION TO SAVE CORE, SINCE A DECLARATION IS REQUIRED FOR ARRAYS IN ANY CASE
COMMON STATEMENTS DISAGREE	USE A PRECOMPILER TO COPY COMMON STATE-MENTS FROM A SINGLE LIBRARY. (NOTE THAT THIS ALSO IMPLIES THAT THE SAME NAME WILL BE USED FOR THE SAME PARAMETER IN EACH MODULE) MINIMIZE USE OF COMMON STATEMENTS BY USING CALLING SEQUENCES TO RESTRICT AND DEFINE MODULE INTERFACES
CHANGED THE VALUE OF AN INPUT PARAME-TER, THEREBY LEADING TO ERRORS ON SUB-SEQUENT CALLS	DO NOT USE THE SAME VARIABLE FOR BOTH INPUT AND OUTPUT TO A MODULE; i.e., DO NOT CHANGE INPUT PARAMETERS
LOGICAL ERROR: FAILURE OF SPECIAL CASES: COORDINATE SINGULARITIES (SPHERICAL COORDINATES AT ±90° DECLINATION; DIS-CONTINUITY AT 0/360° RIGHT ASCENSION)	USE X, Y, Z COORDINATES FOR ALL CALCULA-TIONS (SEE SECTION 2.2) CONVERT TO SPHERICAL CORDINATES FOR I/O ONLY. CHECK FOR 0/360° CROSSOVER − e.g., ADJUST RIGHT ASCENSIONS TO LIE WITHIN 180° OF A SPECIFIED NOMINAL VALUE OR FORCE ROTATION ANGLE RESIDUALS TO THE RANGE −180° TO 180°
INSUFFICIENT COMPUTATIONAL PRECISION NEAR MATHEMATICAL SINGULARITIES. (MATRIX INVERSION, INVERSE TRIGONOME-TRIC FUNCTIONS, ETC.)	USE STANDARD VECTOR/MATRIX UTILITIES WHICH HANDLE SINGULARITIES PROPERLY (SEE SECTION 20.3)
TIME CONVERSION ERRORS FOR LEAP YEARS, CROSSOVER AT END OF YEAR, CROSSOVER AT END OF DAY, ETC.	USE STANDARD UTILITY ROUTINES FOR CALEN-DAR TIME CONVERSION WHICH HANDLE ALL SPECIAL CASES (SEE SECTION 20.3). USE TIME IN SECONDS FROM A FIXED REFERENCE FOR ALL INTERNAL CALCULATIONS (SEE SECTION 1.4)
INSUFFICIENT ERROR CHECKING: FAILURE TO TEST FOR ERROR CONDITIONS; DETECTED AN ERROR CONDITION AND RE-TURNED WITHOUT SETTING ALL EXPECTED OUTPUTS	PROVIDE AN ERROR RETURN CODE FROM ALL MODULES IN WHICH AN ERROR MIGHT BE DE-TECTED. CHECK THE ERROR CODE ON RETURN AND TAKE AN APPROPRIATE ACTION (SUCH AS SETTING THE ERROR CODE FOR THE NEXT HIGHER LEVEL MODULE AND RETURNING). IF PARTIAL OUTPUT MAY BE OBTAINED FOR SOME ERROR CONDITIONS, SET ALL OUTPUTS TO A DEFAULT VALUE BEFORE BEGINNING PROCESSING IN THE MODULE
INSUFFICIENT FLEXIBILITY: UNABLE TO TURN OFF ALL PRINTOUT UNABLE TO CHANGE TOLERANCES UNABLE TO CHANGE FORTRAN UNIT NUM-BERS FOR I/O	PROVIDE USER INPUT PARAMETERS TO CONTROL THESE OPTIONS
DOCUMENTATION: INCOMPLETE OR MISLEADING	SPECIFY UNITS, COORDINATE SYSTEMS, TIME SYSTEMS, VARIABLE TYPES, DIMENSIONS LIST MODULE RESTRICTIONS, ASSUMPTIONS, DESCRIPTIONS OF RETURN CODES, AND DEFINI-TIONS OF FLAGS

Quality assurance of coding is the next step toward reliable software. All coding should be reviewed by someone other than the original author. This review serves to enforce programming standards, detect coding errors, and reduce interface problems between systems which are being developed independently.

The final step in the development of reliable software is extensive testing by an independent testing group, using realistic simulated data. Test cases should be selected to exercise all program options and to generate results which can be independently verified; this should include testing for proper handling of error conditions. Testing of filtering algorithms is described in detail in Section 14.1. Myers [1975] and Dahl, et al., [1972] suggest additional techniques for the development of reliable software. Table 20-1 lists some of the more common causes of software errors encountered in mission support programs, along with suggested methods for avoiding or minimizing these errors. The table is based primarily on experience with FORTRAN scientific applications programs.

Finally, standardization of software can significantly improve reliability and reduce software development costs. Using standard interfaces between modules and standard units (such as the fundamental SI units) for variables reduces program complexity and facilitates reuse of the module. In general, the reliability of a module tends to increase with time, as errors are detected and corrected; however, this is true only if the original specifications for the module remain fixed. Thus, if a library of standard multimission utility routines is developed and maintenance of the library is carefully controlled, these utilities can achieve a very high reliability. Specific utility routines appropriate for attitude systems are discussed in Section 20.3.

20.2 Use of Graphic Support Systems

Department Staff

In standard usage, *graphic* implies a pictorial representation. The terms *conversational* and *interactive* are used interchangeably to define a mode of processing which involves an exchange of information and control between a user at a terminal and a computer. In this section, we will use the term *graphic* to describe conversational processing with a cathode ray tube as the user's terminal.

Attitude support software systems operate in both the graphic mode and the *batch*, or *nongraphic*, mode. Some systems are designed to operate strictly nongraphically, some are designed to operate only with the direction of an operator at a display terminal, and some are designed to operate in either mode. Each processing technique provides some operational benefit.

Batch processing provides no means for interaction. Intermediate results may not be viewed and control parameters may not be modified during processing. Systems designed to operate solely in the batch mode use automated techniques, such as multiple sets of input parameters, for processing several segments of data in one job. In addition, batch systems are often programmed with logical switches which determine the level of output and the options to be employed. Because batch processing systems operate without the intervention and guidance of an operator, they normally require less core residence time than do graphic systems. This approach minimizes the use of resources and susceptibility to human error. However, the inability to dynamically modify control parameters after viewing intermediate results and to redirect program flow is a disadvantage for any system which is required to process data acquired under a wide variety of circumstances.

Graphic processing allows for modification of parameters and data and for redirection of processing by a display operator. This flexibility can be invaluable if the right parameters are available for modification. That is, the designer of a graphic system must make available to the display operator those items which may require modification. Because these parameters may be difficult to identify in advance, most systems display more parameters than are normally modified in practice.

In addition to flexibility, graphic processing techniques free the system designer from providing algorithms for all contingencies because the choice of processing options can be left to the judgment of the display operator. Although the flexibility provided by graphic processing appears to make this technique far superior to a strictly batch system, graphic processing is costly. A knowledgeable display operator must be present to operate the system, a graphic device must be allocated, and the core residence time of the system is increased because of the long idle periods while the program is awaiting operator action at a display.

The best approach to attitude software design is a system which can be executed in either mode. Such a system provides the flexibility required for nonnominal conditions but does not use the resources required by a graphic system when data conditions are nominal. The system's designer can provide both automatic recycling and contingency procedures for nongraphic runs and display parameters, data, and flow control switches for modification when a display device is available.

Early attitude support systems at NASA's Goddard Space Flight Center used the general-purpose *Graphic Subroutine Package (GSP)* to perform graphic functions [IBM, 1972]. This package supports the construction of display images and allows operator intervention. However, a knowledge of the package and a limited knowledge of the device are required to develop the graphic interface. Adding a new display image to the system generally entails developing and testing a new subroutine.

An alternative to generalized packages such as GSP is a *graphic support system* which provides fewer capabilities but is easier to use. A graphic support system offers the advantages of standard display formats and operating procedures, simple display creation, and usability without knowledge of a graphic device. The use of standard display formats means that display images can be considerably more sophisticated than would be possible with a generalized package, given the same amount of development time. The graphic support system permits the analyst to use many techniques previously available only to a few experts. In addition, an operator or analyst working with several systems can understand and use a new display even though he may be completely unfamiliar with the particular attitude system being used. This aspect is particularly important for bringing past experience to bear on current problems.

The features which make a graphic support system beneficial also tend to reduce operating efficiency. Because of the limited capabilities, which make it easy to use, the support system may not be adaptable to special-purpose requirements.

Similarly, the internal code of the support system cannot be written as efficiently as a mission-specific system because of its need to maintain generality within the limited capabilities provided. Ideally, a graphic support system provides all necessary graphic services without the complex protocol required by a generalized package.

The Graphic Executive Support System. An example of a graphic support system is the *Graphic Executive Support System, GESS*, which has been used in various forms for attitude support at Goddard Space Flight Center since 1972 [Hoover, *et al.*, 1975]. GESS provides execution sequence control, data management, error recovery, and graphic services. Execution sequence control allows the display operator to transfer control to alternative subroutines at given points in the processing flow, to move backward or forward in the program, or to skip entire subsystems. Data management includes such functions as data compression to delete bad data elements, scrolling to add recent data to an array and to discard old data, and graphic data entry under operator control. Error recovery allows a job to continue processing after an abnormal termination condition has been detected by the operating system. Graphic capabilities are the most commonly used features of GESS and consist of plot, tabular, and message displays.

A GESS plot display contains up to six functions plotted on a rectangular coordinate system and graphically displayed. A typical GESS plot is illustrated in Fig. 20-1. Each function may be plotted as discrete points, connected points, or characters. The functions to be plotted and the method for plotting each function

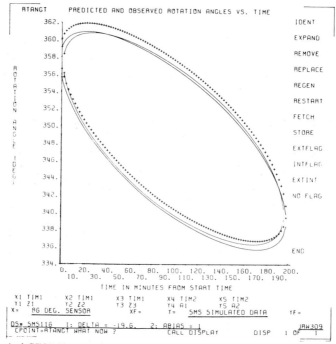

Fig. 20-1.   Typical GESS Plot. The words underlined were added by the display operator and serve to identify various characteristics of the plots.

are described in a plot display table which allows the programmer to create complex, multipurpose plots with only a few lines of code. For example, the complete code used to define the display image illustrated in Fig. 20-1 is as follows:

```
RTANGT    DISPLOT 'PREDICTED AND OBSERVED ROTATION ANGLES VS. TIME'
            'TIME IN MINUTES FROM START TIME','ROTATION ANGLE (DEG)',
            ((TIM1,-999999.),(TIM2,-999999.)),
            ((Z1,,-999999.,,,1),(Z2,,-999999.,,,1),
            (Z3,,-999999.,,,1),(A1,POINT,-999999.,,,,,2),
            (A2,POINT,-999999.,,,,,2))
```

Here, five functions are plotted: arrays Z1, Z2, and Z3 (as connected lines) versus TIM1 and A1 and A2 (as sets of discrete points) versus TIM2. " − 999999" is a flag which identifies points which are not to be plotted.

The actual data included in the functions may vary from run to run. Once the plot has been displayed, the display operator may modify the appearance of the plot or change the value of any plotted data element. The plot modifications are made available to the display operator through the option menu to the right of the plot (IDENT, EXPAND, REMOVE, FETCH, etc.) These options include identification of plotted functions, expansion of a selected area of the plot, removal of functions, data flagging, and retrieval of numeric values.

A GESS tabular display consists of control parameters displayed next to descriptive text or data arrays displayed in columns with descriptive headings. The descriptive text and headings, the formats for displaying the data, and the location of the data are described in a tabular display table similar to that used for plots. The table entries for control parameters may contain criteria against which the parameter is to be validated. For example, a parameter may be required to lie within a certain range or to match one of a list of values equated to words. In the latter case, the word is substituted for the value in the display image and the operator changes the value by entering a different word (e.g., "USE OB-LATENESS MODEL (YES, NO)   YES" for which the operator may leave the "YES" response unchanged or replace "YES" with "NO," thus changing the control option). Displayed data which do not conform to the validation criteria must be corrected by the display operator before processing can proceed.

A GESS message display consists of as many as 814 characters of text and is normally used to inform the display operator of conditions detected by the program, such as the processing status, errors encountered, or the starting time of a data block. The text of a message is defined through a subroutine calling sequence instead of a table.

All displays may be printed by the display operator on a line printer or a CalComp plotter (Fig. 20-1). All displays may also be presented strictly for information and require no modification or action by the display operator. Displays presented in this mode do not cause the system to wait for operator response.

GESS facilitates the incorporation of graphic capabilities into an attitude support software system and provides graphic support on any of several display devices. The operational and flow control restrictions have not made GESS unacceptable for any existing attitude support system, but the very ease of display creation which has made the system successful has also caused some attitude support systems to be designed for strictly graphic processing. Any graphic support system can be more effectively and efficiently used as the graphic vehicle for a system which requires graphic capability on option but which is capable of operating without it.

20.3 Utility Subroutines

Myron A. Shear

This section briefly describes several utility routines used frequently in attitude calculations. The source code for each routine described here, including internal documentation explaining input and output parameters, is available from

COSMIC
Barrow Hall
University of Georgia
Athens, GA 30601

by asking for *Program Number GSC 12421, Attitude Determination and Control Utilities*. The routines are divided into four categories: vector and matrix algebra, time conversion, ephemeris calculations, and plotting. As discussed in Section 2.2, most of the computer routines use vector components in rectangular coordinate systems; however, conversion routines between spherical and rectangular coordinates are provided.

Single- and double-precision versions of the same routine are not provided. For this and other reasons, some users may find it advisable to modify the standard routines provided. However, before developing another routine to perform any of these functions, the reader should understand the routines as they exist, because they have been extensively tested and are designed to provide a combination of accuracy, reliability, compactness, and speed.

The routines described here are written in FORTRAN IV-H for the IBM System 360. The basic algorithms should be easily implemented for any other compiler or machine, with the possible exception of the printer plot routine, which depends on character manipulation features, and the spherical grid plotting routine, which requires the use of a CalComp plotting package.

20.3.1 Vector and Matrix Algebra Routines

Routine	*Function*

Vector Routines

UNVEC	Unitizes a vector and computes its magnitude.
RADECM	Computes the right ascension and declination in degrees of a vector and the magnitude of the vector.
VEC	Converts right ascension and declination in degrees to the three components of a unit vector.
ANGLED	Computes the angle in degrees between two unit vectors. For vectors which are nearly parallel, the cross product is used for greater accuracy.

Routine	*Function*
PHASED	Computes the rotation angle, Φ, in degrees defined by three unit vectors as shown in Fig. 20-2.
CROSSP	Computes a vector cross product.
VPHASE	Computes the unit vector defined by a given arc length, θ, and rotation angle, Φ, with respect to two known unit vectors, $\hat{A}$ and $\hat{B}$, as shown in Fig. 20-2.

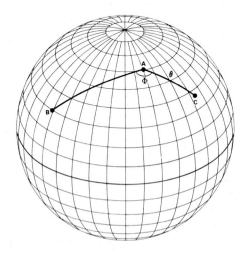

Fig. Fig. 20-2. Given **A**, **B**, θ, and Φ, Subroutine VPHAZE Computes **C**. Given **A**, **B**, **C**, Subroutine PHASED Computes Φ.

CONES8	Computes the two unit vectors defined by the intersections of two cones, where each cone is defined by a unit vector and a half-cone angle.

Matrix Algebra Routines

MATMPY	Multiplies two matrices of arbitrary dimensions.
INVERT	Inverts a matrix and/or solves a set of linear equations, using the Gauss-Jordan method with optimal pivoting.

Least-Squares Routines

POLYFT	Performs a least-squares polynomial fit to a set of data points. Execution is substantially faster than many common techniques.
DC	Given a user-supplied function and a set of data points, DC performs a standard differential correction to obtain a least-squares fit for a state vector of from 1 to 20 parameters. Requires a user-supplied routine to compute derivatives and predicted observations.

Routine	*Function*
RECUR	Given a user-supplied function, RECUR performs a standard recursive estimation to process a single observation and update the estimate of a state vector to provide a least-squares fit to a set of observations.

Integration of Differential Equations

RUNGE	General Runge-Kutta integrator, to be used with a user-supplied routine for computing derivatives.

20.3.2 Time-Conversion Routines

Routine	*Function*
JD	Converts year, month, and day to Julian date, using the algorithm of Fliegel and Van Flandern [1968].
DATE	Converts Julian date to year, month, and day. JD and DATE together provide the basis for all calendar time conversions.
TCON40	Converts time in the format YYMMDD.HHMMSS to seconds from 0 hours UT, September 1, 1957.
TCON20	Converts time in seconds from 0 hours UT, September 1, 1957, to the form: YYMMDD.HHMMSS. TCON20 and TCON40 together provide examples of how any general time conversion can be performed easily using JD and DATE.

20.3.3 Ephemeris Routines

Routine	*Function*

Analytic Ephemeris Utilities

ELEM	Converts position, velocity, and gravitational constant of the central body into classical Keplerian elements. Handles hyperbolic, parabolic, circular, or elliptical orbits.
ORBGEN	Two-body orbit generator that computes position and velocity given Keplerian elements, time from epoch, and gravitational constant of the central body. Useful for Earth or Moon orbits.
RAGREN	Computes the right ascension of the Greenwich meridian ($=$ sidereal time at Greenwich, see Appendix J) in degrees. Uses a first-order method accurate to 0.01 deg for times from 1900 to 2100 A.D.
EQUIN	Rotates coordinates from mean equator and equinox of time 1 to mean equator and equinox of time 2, using a first-order method accurate to 0.01 deg for time periods of 50 years or less.

Routine	*Function*
MAGFLD	Set of routines that compute the Earth's magnetic field vector at any desired time and position according to the International Geomagnetic Reference Field described in Appendix H.
SUN1X	Computes the position of the Sun using a rapid analytical technique accurate to 0.012-deg arc length over the period 1971 to 1981. (The epoch date of the parameters is 1900; thus we anticipate that the accuracy should remain close to this limit for times beyond 1981.)
SMPOS	Computes positions of the Sun and the Moon using an analytic technique which includes 21 perturbation terms for the Moon and 2 perturbation terms for the Sun. It is accurate to within 0.25-deg arc length for the Moon and 0.012-deg arc length for the Sun over the period 1971 to 1981. (The epoch date of the parameters is 1900; thus we anticipate that the accuracy should remain close to these limits for times beyond 1981.)
PLANET	Computes positions of all nine planets using a two-body heliocentric orbit generator. Accurate to 0.02-deg arc length for times within ±2 years of the epoch. Elements and epoch time may be updated periodically using values from the *American Ephemeris and Nautical Almanac*.

Ephemeris Utilities Which Read Data Sets

EPHEMX	General ephemeris routine for the Sun, the Moon, and spacecraft. Uses any combination of the routines GETHDR, ORBGEN, RJPLT, RO1TAP, SUNRD and SUN1X. May be used for Earth- or Moon-orbiting spacecraft.
GETHDR, GETV, DELTIM, HEMITR, INTP	Set of routines to read a standard Goddard Trajectory Determination System orbit file. (See Section 5.4 for contents of the file.)
RO1TAP, ROUND	Set of routines to read a standard Goddard Trajectory Determination System ephemeris tape. (See Section 5.4 for contents of the tape.)
SUNRD	Routine to obtain positions of the Sun, the Moon, and the first seven planets from a direct access file (SLP file, Section 5.5) containing polynomial coefficients derived from the standard Jet Propulsion Laboratory planetary ephemeris tape. The routine's accuracy is comparable to that of the Jet Propulsion Laboratory tape. (See Section 5.5.)
RJPLT	Routine to read a standard Jet Propulsion Laboratory planetary ephemeris tape to obtain positions of the Sun and the Moon. (See Section 5.5.)

20.3.4 Plotting Routines

Routine	*Function*
GRAPH, SCALE	General printer plot package. Generates a Cartesian plot of any set of data on a line printer, with scaling performed automatically. Plot covers up to 80 lines, with up to 132 characters per line. A sample plot is shown in Fig. 20-3.

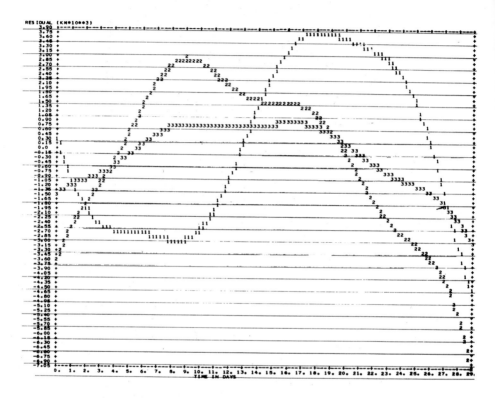

Fig. 20-3.   Sample Plot Generated by Subroutine GRAPH

| SPHCNV, SPHGRD, SPHPLT | Spherical grid CalComp plotting routines which generate a perspective drawing of the celestial sphere as seen from any orientation and plot user-specified lines, points, or other characters on the sphere. A sample plot is shown in Fig. 20-4. |

CTS T. O.. BEGINNING OF JAN. 13 LAUNCH WINDOW. AMFR (JOBCO8)

CENTER: α=100 . δ= 15   ATTITUDE: α= 11.0 . δ=-22.6

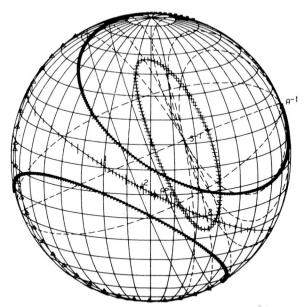

Fig. 20-4.   Sample Celestial Sphere Plot Generated by Subroutines SPHCNV, SPHGRD, and SPHPLT

References

1. Berg, R. A., and M. A. Shear, *High Energy Astronomy Observatory-A Attitude Ground Support System Design and Development Methodology*, Comp. Sc. Corp., CSC/TM-76/6134, June 1976.

2. Chu, Ben, *Structured FORTRAN Preprocessor (SFORT)*, Comp. Sc. Corp., CSC/TM-77/6256, Sept. 1977.

3. Dahl, O. J., E. W. Dijkstra, and C. A. R. Hoare, *Structured Programming*. New York: Academic Press, Inc., 1972.

4. Fliegel, Henry F., and Thomas C. Van Flandern, "A Machine Algorithm for Processing Calendar Dates," *Communications of the ACM*, Vol. 11, p. 657, 1968.

5. Hoover, J. E., T. E. Board, and A. M. Montgomery, *Graphic Executive Support System (GESS) User's Guide*, Comp. Sc. Corp., CSC/SD-75/6057, Aug. 1975.

6. International Business Machines Corporation, *IBM System/360 Operating System Graphic Subroutine Package (GSP) for FORTRAN IV, COBOL, and PL/I*, GC27-6932, Nov. 1972.

7. Myers, Glenford J., *Reliable Software Through Composite Design*. New York: Petrocelli-Charter, 1975.

CHAPTER 21

SOFTWARE SYSTEM STRUCTURE

21.1   General Structure for Attitude Software Systems
21.2   Communications Technology Satellite Attitude Support System
21.3   Star Sensor Attitude Determination Systems
       Components of a Star Sensor Attitude Determination System, Construction of Batch and Sequential Attitude Systems
21.4   Attitude Data Simulators

This chapter describes the overall structure of attitude support systems as they have been used for mission support in the Attitude Determination and Control Section at NASA's Goddard Space Flight Center. Section 21.1 gives the general framework that has proved useful in mission support. Sections 21.2 and 21.3 illustrate how this was implemented in attitude software for particular mission types. Section 21.4 briefly describes the function and operation of attitude data simulators.

21.1 General Structure for Attitude Software Systems

Myron A. Shear

The requirements for attitude support software systems vary considerably from mission to mission, depending on spacecraft and ground support hardware, data volume, telemetry format, and the mission timeline. However, certain features are common to most attitude support systems, and there is a general software structure which has proved useful for a variety of missions. This section describes that general structure and discusses the tradeoffs to be considered when modifying this general structure for special mission requirements.

The basic software requirement for most missions is to take spacecraft-telemetered data and perform ground processing to determine the attitude. These attitude results may then be used to compute control commands which are transmitted to the spacecraft. The computation of control commands is typically, though not necessarily, done with a separate software system. As discussed in the introduction to Section 20.1, attitude determination requirements may be either real time, near-real time, or definitive, depending on the time constraints. Fortunately, the same software system can often be used to satisfy all three requirements if appropriate options are provided to handle each mode of operation.

Figure 21-1 shows the general structure for a typical attitude support system used in the Attitude Determination and Control Section at Goddard Space Flight Center. This system consists of several processing subsystems, operating under the control of a driver and utilizing a graphics package (see Section 20.2) to provide interactive graphics capabilities for all subsystems.

The *telemetry processor* subsystem reads the raw telemetry data set, constructs frames of data from the telemetry stream, and converts data from binary-coded

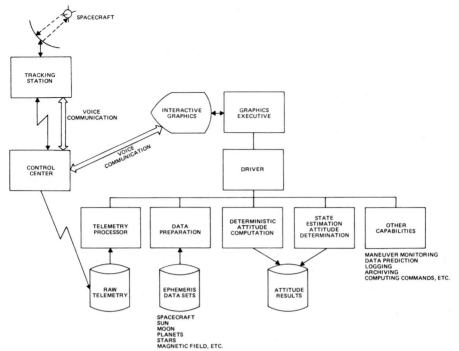

Fig. 21-1.   General Structure for an Attitude Software System. Arrows indicate direction of data flow.

values to engineering units. (Chapter 8 contains a detailed discussion of data transmission and manipulation up to and including the telemetry processor.) The *data preparation* subsystem performs data selection, editing, smoothing, calibration, and adjustment, with or without operator interactive control (see Chapter 9). Here, the data can be displayed in tabular or graphical form so the operator may examine it for anomalies. By this point in the processing, absolute times should have been attached to data items so that ephemeris data can be obtained. The required ephemerides may include the spacecraft, the Sun, the Moon, planets, stars, and the Earth's magnetic field, depending on the sensor hardware used.

The *deterministic attitude subsystem* is normally used next to obtain a rough attitude for control purposes or an initial attitude estimate for the state estimation subsystem. Deterministic methods, discussed in detail in Chapters 11 and 12, are advantageous because they can be used in the absence of an a priori attitude and in the presence of a substantial amount of spurious data. However, deterministic methods are generally limited to solving for not more than two or three parameters. Thus, sensor bias determination and calibration cannot be done in the deterministic subsystem, and the presence of these systematic errors will generally limit the accuracy of the deterministic solution. However, deterministic processing may still use a significant amount of computer time. For example, horizon sensor data may require iterative techniques to resolve attitude and central body ambiguities and to reject spurious data and terminator crossings. Similarly, star sensor data may require complex star identification procedures.

In contrast to the deterministic subsystem, the *state estimation subsystem* will generally assume that an accurate a priori attitude is available and that spurious data points have been rejected. Therefore, this subsystem can use least-squares state estimation techniques (described in Chapters 13 and 14), such as Kalman filtering, recursive estimation, or differential correction, to solve for a state vector with perhaps a dozen or more parameters, including the attitude, sensor biases, and attitude dynamics. The sensor biases determined in this subsystem may subsequently be used by the deterministic subsystem to improve the accuracy of the deterministic results for subsequent data passes.

Not all missions require both a deterministic and a state estimation subsystem; if an accurate a priori attitude is available, for example, from an onboard control system, then it may be possible to eliminate the deterministic subsystem. However, some form of attitude initialization may still be required to determine an a priori attitude immediately after launch. Conversely, if the attitude accuracy requirements do not necessitate bias determination (or if the spacecraft dynamics or mission timeline do not permit bias determination), the state estimation subsystem may be eliminated.

The "other capabilities" of Fig. 21-1 will depend on mission characteristics and may include routines to monitor maneuvers in real time, predict the availability of future data passes, compute control commands, or perform solution logging and data archiving functions. These capabilities may be provided in separate software systems or designed as subsystems invoked from the driver. The tradeoff here involves speed and ease of operation versus programming complexity and ease of maintenance. Separate utility programs are generally easier to develop and maintain because interfaces are minimized. If a separate graphics device and other computer resources are available, then the utility program can be executed concurrently with the main attitude system. However, if the main attitude system must be terminated to provide resources for the utility, then the extra time involved in terminating and reinitializing the attitude system must be considered, especially for near-real-time applications.

The general structure described above has been used successfully on many missions including CTS, GOES, SIRIO, RAE, IMP, SMS, ISEE, and IUE. The success of this structure is due primarily to its modularity and flexibility. A modular structure implies that each subsystem has a minimum number of interfaces with each other subsystem. This results in ease of development and maintenance, because subsystems can be developed concurrently and almost independently. When modifying the system for future missions, it may be possible to replace only the telemetry processor subsystem and support a spacecraft with a totally different telemetry format but similar sensor hardware. System flexibility results from the fact that subsystems can be invoked in almost any sequence, under operator control. For example, attitude processing can be repeated on the same data using different processing options without repeating telemetry processing, data preparation, and ephemeris accessing. Similarly, state estimation can be (and usually is) repeated many times, solving for a different set of parameters or changing the filtering options, without repeating the deterministic attitude processing. Thus, the system minimizes execution time for the most frequently repeated functions.

The modular system also lends itself to a simple overlay structure, allowing each subsystem to share the same core storage. The system provides interactive graphics control at each step in the processing; we have found this to be essential in attitude support systems to handle the unpredictable problems that occur in real data, requiring operator intervention to select and edit the data and ensure the quality of the attitude results.

There are several ways in which this general structure can be modified to handle special requirements. The structure in Fig. 21-1 assumes that all the subsystems are part of the same program and that core storage interfaces are used for communication between subsystems. This arrangement is used for CTS, as described in Section 21.2. However, one or more of the subsystems could be implemented as separate programs and one or more of the subsystem interfaces could be implemented via data sets. The tradeoffs here are among program complexity, computer resources, and operational timeline requirements. For example, the telemetry processor could be split off as a separate program which could then operate on a minicomputer. This would have the advantage of freeing resources on the primary computer; however, it would have the disadvantage of reducing the flexibility and ease of operation of the attitude system. If the telemetry processor incorrectly constructed frames from the telemetry stream, there is no way the attitude system could correct this error in the processed telemetry, and it would be necessary to reexecute the telemetry processor with different processing options. This could require a human interface between the primary computer and the minicomputer, which, for real-time operation, might prove impractical. Similarly, a hardware failure on the minicomputer would be just as serious as a failure of the primary computer, increasing the risk of computer failure for real-time and near-real-time requirements.

As another example, the state estimation subsystem could be a separate program, interfacing with the remainder of the attitude system via a data set containing preprocessed telemetry. This arrangement is anticipated for MAGSAT processing. In this case, the state estimation system could be run on another computer to distribute the computing load and allow the real-time requirements of the deterministic attitude system to proceed concurrently. The major advantages of a separate program are ease of maintenance and simplification of the interfaces; the major disadvantages are the increased operational difficulty involved in creating and maintaining the interface data set, the time delay involved in an extra processing step, and the reduced flexibility which results from not being able to reaccess the original telemetry data from the state estimation system.

Data set interfaces between subsystems can be used even if the subsystems are combined in a single program. A data set interface requires additional I/O processing time; there is not necessarily any reduction in core storage because generally at some point a block of data for processing must still be held in core. However, a data set interface reduces the possible interaction between subsystems and thus reduces interface problems. If the observations can be processed singly in the attitude determination subsystems, a data set interface can reduce core requirements. In this case, a data set interface is probably more convenient than the alternative of cycling between the telemetry processor and the attitude system for each observation. Data set interfaces can also provide for a more rapid restart and reduce the need for reprocessing in the event of machine failure.

The use of separate programs does not necessarily imply data set interfaces. Core storage interfaces can be used between separate programs, even operating on separate machines; however, the use of a core interface tends to increase the interdependency of the programs, thus reducing the advantages normally assoc-iated with separate programs. For this reason, and because additional system software is required to interface the programs via core storage, separate programs are normally interfaced via data sets.

For real-time maneuver monitoring, special capabilities must be provided to minimize or eliminate the need for operator interaction. In real-time operation, the telemetry processor normally reads one or a small number of data samples, and the driver automatically invokes the data preparation and deterministic attitude sub-systems to process this small set of data. Then a special maneuver monitoring subsystem is invoked to generate displays showing the actual maneuver trajectory versus the expected or desired trajectory. Computed results are also displayed to indicate whether the maneuver is within expected tolerances and to warn of any potential problems, such as violating Sun angle constraints or maneuvering outside antenna coverage. While these displays remain on the screen, program flow returns to the telemetry processor to read all the data which have been received since the previous call to this subsystem. Typically, a complete cycle through these subsys-tems will take 10 sec or less, which is well within the real-time requirements for a system which operates with a manual interface to the control center. If the real-time control requirements were much more severe than this, the manual interface would have to be eliminated and the control loop would have to be closed within the support computer (see Section 19.5.1). This would require much more sophisticated control monitoring software to make reliable control decisions without operator assistance. Fortunately, most missions are designed to make this type of ground-based, closed-loop control unnecessary.

21.2 Communications Technology Satellite Attitude Support System

Gyanendra K. Tandon

As an example of the general structure discussed in Section 21.1, we describe how that structure was implemented for the Communications Technology Satellite (CTS) Attitude Support System. This system provided adequate computational support throughout the CTS mission, even when a balky latch valve in the control system caused substantial changes in the nominal timeline. During this emergency situation, the system provided all the information needed to define an alternate timeline in real time.

The CTS Attitude Support System consists of two major programs: the CTS Attitude Determination System, CTSADS, and the CTS Maneuver Control Pro-gram, CTSMAN. In addition, the following utility programs were available for use: the CSMP/AMF Dyanmics Program (Section 17.4); simulators CTSSIM and ODAP (Section 21.4); the orbit geometry program OSAG [Shear, 1972]; and a set of standard programs for checking, archiving, and purging the data from the attitude data link, ADL, file (Section 8.1) and for checking the archived data.

The basic system structure and data flow of the CTSADS system [Nelson, *et al.*, 1975], are shown in Fig. 21-2. Graphic displays of control parameters and data are available throughout the system for controlling and monitoring program operation. Input to the system includes control parameters via NAMELIST data sets or cards for each subsystem, although these are not shown in Fig. 21-2.

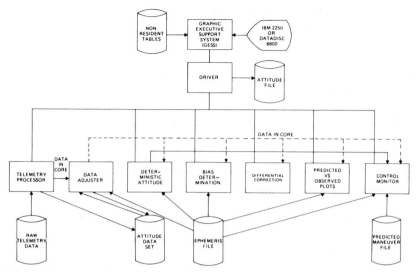

Fig. 21-2.   CTS Attitude Determination System, CTSADS, Baseline Diagram. The card input option for control parameters for each subsystem is not shown.

CTSADS uses the Graphic Executive Support System, GESS, described in Section 20.2 for operation on an IBM 2250, or a Data Disc 6600 graphics display device. GESS provides execution sequence control, data management, error recovery, and graphic services. The driver is the main control module of CTSADS, providing the interface between the GESS executive and each subsystem in CTSADS. It permits the operator to select any desired subsystem or to terminate the job. CTSADS can also be executed in a nongraphic mode for analytical purposes.

The CTSADS program consists of seven major subsystems and an attitude status file subsystem (not shown in Fig. 21-2) used for writing the current spacecraft attitude to a direct-access disk file. The seven major subsystems are as follows:

1. *Telemetry Processor.* This subsystem reads the raw telemetry data provided by the control center (see Section 8.1) on disk or tape and provides the attitude determination system with the spin rate and Sun and Earth sensor data. During attitude maneuvers it also provides the engine firing pulse counts to the control monitor subsystem. In addition, it can perform three levels of telemetry time checks (Section 8.3) and three types of data smoothing (Section 9.2).

2. *Data Adjuster.* The data adjuster selects a working set of the telemetry data and obtains the corresponding ephemeris information. In addition, it provides the operator with options for selecting a subset of the data, smoothing or adjusting data, overriding individual values, rejecting invalid data points, and selecting the ephemeris sources. The operator can examine the data both before and after adjustment, using a variety of character and plot displays.

3. *Deterministic Attitude.* The deterministic attitude subsystem computes the attitude using any combination of seven deterministic methods; each method uses a closed-form analytical technique to compute one or more attitudes from a pair of observables (see Section 11.1). These single-frame attitude solutions are then averaged to resolve the ambiguous solutions and yield a best estimate of the

average attitude over a block of data. During the attitude maneuvers, the best estimate of the average attitude for each single frame of data is determined and is passed to the control monitor for maneuver monitoring.

4. *Bias Determination*. The bias determination subsystem is used to determine attitude and sensor biases using either a least-squares differential correction or recursive estimation procedure. The program uses up to five observation models to solve for any subset of up to 20 state vector elements. Details of observation models and filtering techniques are described in Chapter 13. This subsystem and the differential correction subsystem provide two alternative programs for bias and attitude determination.

5. *Differential Correction*. The differential correction subsystem provides an alternative approach to attitude and bias determination and serves as a backup to the bias determination subsystem. It first converts the raw data (Sun angles, spin rates, and Earth times) into arc lengths and/or rotation angles. Based on these angles, the subsystem uses a least-squares state estimation algorithm (see Sections 13.4 and 13.5) to solve for a state vector which includes separate biases in each type of arc-length or rotation angle measurement or to solve for polynomial coefficients for right ascension and declination as functions of time, up to first order. The biases here are numerically convenient parameters in contrast to the physically motivated parameters of the bias determination subsystem.

6. *Predicted-Versus-Observed Plots*. This subsystem provides the operator with a visual display of the observed Earth sensor data compared with the Earth sensor data predicted using any specified set of attitude and bias parameters. These plots are used to evaluate the attitude and bias solutions obtained by the various subsystems. The plots can also be used as a backup method of attitude and bias determination, by varying the state parameters manually to obtain the best fit to the data. In addition, the predictions can be generated for arbitrary times to determine data coverage for future data passes. For examples of these plots, see Section 9.4.

7. *Control Monitor*. This subsystem monitors attitude reorientation maneuvers in real time, to determine whether they are proceeding in the right direction at the proper rate. In the monitor mode, the system automatically cycles through the telemetry processor, data adjuster, deterministic attitude, and control monitor subsystems. Ordinarily on each cycle, a single telemetry record (10 sec of data) is retrieved from the raw telemetry data file and processed through each applicable subsystem. The control monitor accumulates the results from processing each record and updates displays which show the observed attitude motion versus the predicted attitude motion as obtained from the predicted maneuver file. The predicted maneuver file is generated by the CTSMAN program described below. The control monitor can also compute new command parameters necessary to correct a maneuver if it is not proceeding as predicted.

Figure 21-3 shows the normal operating procedure for attitude determination. In CTSADS, the subsystems may be invoked in any desired order. However, the data adjuster must be executed immediately after the telemetry processor to select data for processing and to choose ephemeris options before any other subsystem can be executed. The routine steps followed for determining an attitude solution from a batch of data are delineated in Fig. 21-3.

Figure 21-4 shows the data flow during maneuver monitoring. The control monitor first reads the predicted maneuver file to obtain the predicted attitudes

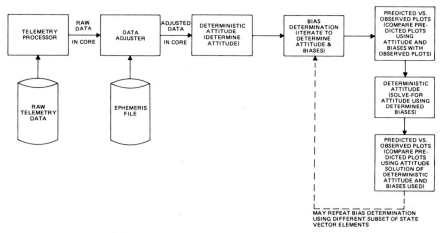

Fig. 21-3.   Normal Operating Procedure for Attitude Determination Using CTSADS

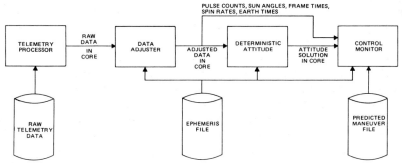

Fig. 21-4.   Data Flow During Maneuver Monitoring for CTS. The control monitor receives attitude solutions from the deterministic attitude subsystem and the rest of the data from the data adjustment subsystem.

and other parameters for the scheduled maneuver. During the actual maneuver the observed pulse counts, Sun angles, spin rates, and Earth times from the data adjuster and the single-frame attitude solutions from the deterministic attitude subsystem are passed into the control monitor for comparison with the predicted values.

Finally, the CTSMAN program [Tandon, 1975; Rochkind, 1973, 1974] computes the ground station commands necessary to perform an attitude reorientation maneuver, given an initial attitude and a desired final attitude. In addition, the program computes the full maneuver sequence and the history of maneuver-related parameters. A subset of the computed maneuver-related parameters is stored on a disk data set, called the predicted maneuver file, for use by the control monitor in monitoring the maneuver in real time.

21.3 Star Sensor Attitude Determination Systems

Lawrence Fallon, III

This section provides an overview of attitude determination systems which use star sensor data. Such systems generally use a model of the spacecraft attitude motion which is periodically updated from star sensor measurements. This model requires an initial attitude estimate usually provided by other sensor types; however, occasionally an initial three-axis estimate is calculated using star sensor data when only a single-axis estimate is available externally. The motion model used for attitude calculation may be either a simple kinematic description of a uniformly spinning spacecraft or a complicated dynamics model involving environmental and control torques. Alternatively, a system of rate or rate-integrating gyroscopes may be used to provide a mechanical substitute for a spacecraft dynamics model.

The methods which use star sensor measurements to update an attitude model may be divided into two categories: batch and sequential. In a *batch* updating system, observations made at different times are related to an epoch time using the attitude model, and collectively identified with stars whose coordinates are supplied by a star catalog. An average or least-squares attitude solution is then calculated and used to update the attitude at the epoch time. Additional model parameters are frequently included in this procedure. In a *sequential* updating system, the dynamics model is used to extrapolate the attitude to the time of each star sensor observation in succession. When the extrapolation process reaches the time of a particular observation, an attempt at star identification is made. If the identification is successful, it is used to update the attitude at the time of the measurement, and perhaps to update other parameters in the spacecraft model as well. The updated model is then extrapolated to the time of the next observation, and so on. The frequency with which the attitude reference must be updated in either system is dependent on the accuracy of the star sensor measurements, the accuracy of the motion model, and the desired accuracy of output attitudes.

21.3.1 Components of a Star Sensor Attitude Determination System

In addition to a telemetry processor and other auxiliary features, star sensor attitude determination software systems normally consist of five components, as shown in Fig. 21-5. The detailed makeup of these components depends on the type and accuracy of sensor measurements; the quality of the attitude estimates provided by other attitude hardware; the field-of-view size, orientation, and sensitivity of the sensor; the complexity and accuracy of the attitude model; and the desired accuracy of the attitude solutions.

In *star catalog acquisition*, a subcatalog is acquired from a whole sky star catalog, as described in Section 5.6. The estimated accuracy of the initial attitude, the field-of-view size, the expected motion of the sensor's optical axis, and the sensor's magnitude sensitivity dictate the size and shape of the subcatalog. For example, a spherical cap subcatalog was generated for the star tracker mounted parallel to the SAS-3 spin axis. The 13.3-deg-wide cap was selected to accommodate the rotating 8- by 8-deg field of view and approximately 2 deg of potential spin axis error. It contained 30 to 50 stars brighter than the 7.5 instrumental magnitude limit. A 12-deg wide band subcatalog generated for the SAS-3 tracker perpendicular to the spin axis contained approximately 2000 stars brighter than an instrumental magnitude of 7.5.

Data selection and correction is the most hardware-dependent of the five components. Editing, selection, correction, and calibration of the sensor data are generally done here. For example, data from the SAS-2 N-slit star scanner (Section

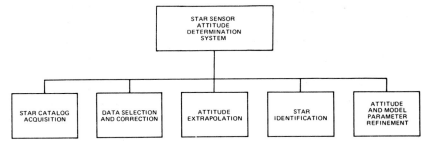

Fig. 21-5. Major Components of a Star Sensor Attitude Determination System. A telemetry processor and possibly other auxiliary components as described in Section 21.1 are also required.

7.6) consists of a series of voltages which must be examined to determine when a transit occurred, i.e., when a star crossed one of the slits. By examining the differences in time between all transits of slits 1 and 3 (see Fig. 7-24), many differences are found corresponding to θ/ω, where θ is the angular separation of slits 1 and 3 and ω is the spin rate. If the number of transit pairs separated by approximately θ/ω is plotted as a function of spin period, as shown in the spin rate histogram of Fig. 21-6, an initial estimate of the spin rate is obtained. This spin rate is then used to group the transits into triplets corresponding to the three slit crossings by the same star. As another example, SAS-3 star tracker data is examined to remove all points except those corresponding to valid star sightings by examining various telemetry flags. The star sightings are then calibrated and converted into unit vectors in the spacecraft frame, as described in Section 7.6.

Before stars can be identified, an attitude estimate must be available at the time of each sensor observation. The procedure which supplies this attitude estimate is called *attitude extrapolation*. The attitude estimate may be a three-axis attitude relative to inertial space, or to the spacecraft frame at some epoch time (for which some estimate of the inertial attitude usually is available). In either case, the

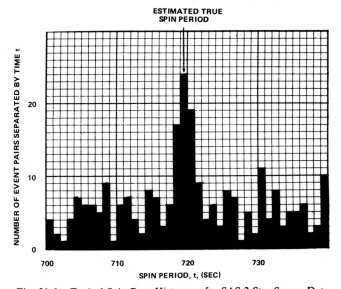

Fig. 21-6. Typical Spin Rate Histogram for SAS-2 Star Sensor Data

attitude estimate is provided by extrapolating an initial attitude using a model of the spacecraft attitude motion.

For example, after triplets of SAS-2 N-slit star scanner observations have been grouped together using the initial spin rate, a better spin rate estimate is obtained from the time differences between the slit 1 and slit 3 crossings within the same triplet. This improved spin rate is then used to calculate observed star unit vectors at some nearby epoch time. This set of observed star vectors at an epoch time is called a *snapshot*. It is assumed that within the period of interest the spacecraft is spinning uniformly and, therefore, that each star transits slits 1 and 3 of the sensor with the same time difference. Any deviation of the spacecraft from the uniform spin model may cause significant distortion in the snapshot and thus interfere with star identification.

If an accurate snapshot is needed, greater precision and sophistication in the spacecraft model is usually required. The system used for the SAS-3 star trackers modeled spacecraft motion as the simple spin and nutation of a symmetric rigid body in a torque-free environment (Section 16.3). The variation of the individual measurements of the same star with respect to time is used to estimate spin and nutation parameters, which are then used to extrapolate a relative attitude from the epoch time to the time of each measurement. This relative attitude is then used to calculate an observed star snapshot in the spacecraft frame at the epoch time. If the accuracy of the model begins to degrade because of spin rate variation, for example, the snapshot may frequently be improved by shortening the time span of data used in its creation.

The HEAO-1 system uses gyros to propagate an attitude estimate to the time of each star tracker observation (Section 17.1). This procedure is capable of providing very accurate attitude estimates for snapshot generation. Even if the initial estimate is imprecise, the relative accuracy of snapshots created in this fashion is generally greater than in either of the previously described methods.

After the star coordinates and corresponding attitude estimates are computed at a reference time, *star identification* is attempted. The star coordinates are generally in the form of a snapshot, which may contain only one star or many stars. If it contains only one star, it must be identified using a direct-match algorithm, as described in Section 7.7. If it contains more than one star, any of several pattern-matching techniques may be appropriate, depending on the size and quality of the snapshot and the accuracy of the initial attitude estimate. A typical SAS-3 star tracker snapshot with superimposed catalog stars is shown in Fig. 21-7. Observations are denoted by open circles and catalog stars by plus signs. Double circles correspond to multiple sightings of the same star.

After pattern matching, all observations but one have been identified with catalog stars. The star which caused the unidentified observation was apparently not in the catalog. Note that after matching, the catalog stars have been shifted to the right and upward about 2 deg. Details on alternative pattern-matching methods, including the direct match, are given in Section 7.7.

After observations have been identified with catalog stars, the final step is *attitude model refinement*. This involves the calculation of attitude using some averaging or optimization process, such as least-squares or Kalman filtering. This procedure may also include the optimization of other model parameters, such as environmental torque variables, angular momentum components, or gyro drift rates.

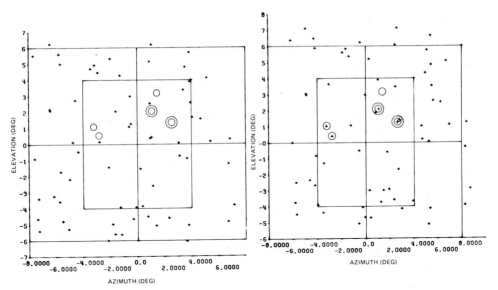

Fig. 21-7.   Snapshot of SAS-3 Spin Plane Star Tracker Observations

21.3.2 Construction of Batch and Sequential Attitude Systems

Whether the star sensor attitude determination components should be assembled to form a batch or a sequential system depends on various circumstances. Sequential systems require less computer resources because information regarding only one (or, at most, a small number) of stars must be stored at any one time. For this reason, sequential systems are usually chosen for onboard processing. Because they are frequently restricted to direct-match star identification algorithms, sequential systems are often unsuitable when more sophisticated star identification techniques are needed. Therefore, a batch system will be most appropriate when star identification may be difficult—for example, when errors in the initial attitude estimate or the attitude model are large relative to the spacing between stars visible to the sensor. If the system must operate in a variety of attitude accuracy environments, a hybrid system which incorporates both sequential and batch capabilities may be desirable.

The software support systems for the N-slit star scanner in the three SAS missions were all batch systems. The SAS-2 Star System, as described by Rigterink, et al., [1973], received an initial attitude estimate accurate to approximately 2 deg from a Sun sensor/magnetometer system. The program was then required to identify stars observed during 30- to 60-min intervals, which generally spanned several spacecraft spin periods, and to calculate attitude solutions accurate to 0.25 and 0.5 deg about the spin and lateral axes, respectively. After selection of transits, initial spin rate estimation, and association of triplets, an improved spin rate was calculated and a snapshot was created using a uniformly spinning spacecraft model. A congruent triangle distance-matching technique [Fallon, et al., 1975] was then used to identify the observations with stars in a band catalog. An attitude was then calculated for the epoch time of the snapshot using a least-squares process. A spin rate smoothing and phase angle computation procedure then allowed attitude computation for any time within the segment.

The system being used to support HEAO-1 is also a batch system. A coarse single-axis attitude estimate is provided by a Sun sensor. This estimate, a snapshot consisting of star tracker data obtained from one spacecraft rotation (approximately 30 minutes), and a band catalog are used in a phase search star identification procedure enhanced with distance-matching tests in an effort to identify specific stars. A batch least-squares program then uses the identification results to calculate an attitude which is required to be accurate to at least 1 deg. This attitude is then propagated forward in time to provide an initial estimate for a triangle-type star identification algorithm which attempts to identify 5 to 20 star tracker observations measured within a 5- to 8-minute time span. Assuming that identifications are successful, the batch least-squares program calculates a snapshot attitude solution accurate to 0.005 to 0.010 deg. This solution is then used to estimate an attitude correction which is sent to the spacecraft's onboard computer to improve its attitude reference. The onboard computer uses the gyro data to propagate its attitude reference forward in time. Because this onboard reference is required by spacecraft control procedures to maintain at least 0.25 deg (3σ) accuracy, it is updated typically 5 to 20 times a week using ground attitude solutions to counteract the effects of gyro-related errors. Comparison of attitudes propagated by the onboard computer during the periods between attitude updates with ground attitude solutions calculated at the same times as corresponding propagated attitudes yields information regarding gyro drift and misalignment parameters (Section 7.8). Refined gyro calibration parameters are then sent to the spacecraft to improve the quality of the propagated attitude reference.

The performance of the HEAO-1 reference and gyro calibration update procedure can be assessed by examining the total arc difference between ground attitude solutions and corresponding onboard propagated attitudes. Figure 21-8 shows the onboard versus ground profile for the week following September 16, 1977. During the first 5 days of this week, onboard attitude accuracy was maintained to within approximately 0.05 deg—significantly better than the 0.25-deg accuracy requirement. Note that the drift rate update sent on September 18 significantly decreased the onboard attitude error growth due to gyro-related errors. On September 21, however, a commanded scan rate change caused a rapid increase in onboard error due to the strong dependence of drift rate solutions on the scan rate. A new drift rate was estimated using data following the scan rate change and sent to the spacecraft on September 22. Propagation accuracy then returned to the 0.05-deg level. Additional details concerning the structure and performance of the HEAO-1 attitude ground support system are given by Fallon and Sturch [1977].

An example of a sequential system which uses a spacecraft dynamics generator with simple environmental torque models is given by Foudriat [1969]. An attitude reference is extrapolated to the time of each star scanner measurement by the dynamics generator and then updated using a limited-memory Kalman filter, assuming that the direct-match star identification attempts are successful. By appropriate selection of star scanner measurements for attitude refinement, attitude and model parameters may be refined well enough to permit attitude extrapolation for periods as long as 1000 sec with arc-second accuracy. Another example of a sequential system is the Space Precision Attitude Reference System (SPARS) developed by Lockheed and Honeywell for onboard attitude determination [Paulson, et al., 1969]. SPARS uses gyro data for attitude extrapolation instead of a

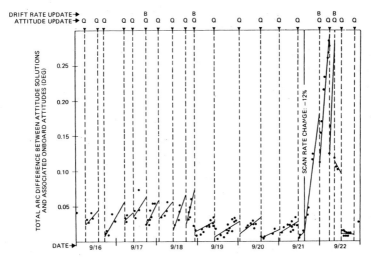

Fig. 21-8. HEAO-1 Onboard Attitude Propagation Accuracy for the Week Following September 16, 1977 [Fallon and Sturch, 1977]

spacecraft dynamics model. The direct-match identified transits are used by a Kalman filter to sequentially refine attitude and gyro drift parameters.

21.4 Attitude Data Simulators

Peter M. Smith

In the design and development of mission software, the attitude data simulator is usually the first system to be built. The simulator is used in all mission phases; therefore, it is important to understand in advance the functional and operational requirements for the entire satellite program. A summary of these requirements is presented below. Their implementation within the simulator software is then illustrated by discussion of the structure of two specific simulators. This section is concerned with mission-dependent software, which is used in conjunction with mission support software, rather than mission-independent programs such as ADSIM [Gray, *et al.*, 1973], ODAP [Joseph and Shear, 1973], and FSD [NASA, 1978] which are used primarily for prelaunch analytical studies.

Functional Requirements. Attitude data simulators are used primarily in the following application areas:
1. Development and testing of mission support attitude determination systems. Here, the simulator is used to provide data to exercise and test all capabilities of the attitude determination system.
2. Prelaunch simulation sessions in which both nominal data and contingency data situations are generated to train mission support personnel.
3. Analytical studies to aid in the planning of mission timelines and maneuver control procedures. The testing of new analytic procedures is most readily carried out using simulated data because of data control and the knowledge of all parameters which define the simulated data set. Mission requirements

will dictate the need for spacecraft dynamic modeling, based on either a slowly varying attitude responding to time-averaged external torque or a detailed model, including such effects as onboard torquing devices and flexible appendages.

4. Real-time mission support for the identification of systematic variations (such as the Pagoda effect, described in Section 9.4) and prediction of the availability and quality of future data.

For a simulator to provide adequate support in all of the above areas, its data generation capabilities should satisfy the following criteria:

1. The simulator should.generate realistic mission data (e.g., constant attitude data, maneuver data, or nutating data). The sophistication level of simulator modeling should equal or surpass that of the attitude determination system to allow the latter to be tested to the limits of its accuracy.

2. Provisions should be included for noise, random bit errors, quantization errors, and realistic sensor biases to test the performance of the attitude determination system with data that have been degraded to increasing levels of severity and to clearly identify the effect of various errors and biases.

3. For real-time support requirements, the simulator should be able to generate data both at a reduced rate with artificial delays added to simulate real-time spacecraft telemetry and at the normal high-speed rate with no delays applied.

Simulator Structure. As an example of the implementation of these functional requirements, we describe the structure of two specific simulators, CTSSIM and PLOTOC, used for the Communications Technology Satellite, launched in January 1976. CTSSIM [Smith, 1975] is an independent simulator used in the development, testing, and prelaunch phases; PLOTOC, which is capable of comparing real and simulated data, [Plett, et al., 1975; Nelson, et al., 1975] is an integrated subsystem of the attitude determination system, utilized primarily for launch support.

Figure 21-9 shows the functional baseline diagram for CTSSIM. The simulator operates under the Graphics Executive Support System, described in Section 20.2, and uses the core allocation/deallocation, graphics displays, and interactive processing services provided by this executive. Program flow through the simulator proceeds from left to right. Starting conditions for a simulation run are set via graphic, card, or data set NAMELIST input. Attitudes for a maneuver trajectory may be either internally generated or read from a tape created by an external program. Ephemeris data is read from disk data sets or tapes or is generated internally. Simulated data may be perturbed by applying noise and biases to sensor hardware parameters. Plot displays allow the user to critically examine the simulated data. The interface with the attitude determination system is either a raw or a processed telemetry data set. Data is generated on a frame-by-frame basis with a variable sampling frequency controlled by the timing routine. The simulator can also read processed telemetry data (e.g., nutating data generated by a dynamic simulator) and use it to simulate raw telemetry data.

In a typical prelaunch simulation, data covering a large time interval (approximately 6 hours), with noise and sensor biases applied, are generated in a single simulation run, carried out at the high-speed rate. The attitude determination

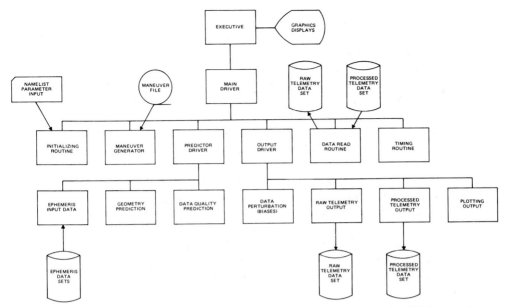

Fig. 21-9.   Functional Baseline Diagram for the Attitude Data Simulator for the CTS Spacecraft

system is then required to process these data and solve for the attitude and applied biases. In contrast, maneuver simulations are generally carried out in real time to provide launch support personnel with realistic monitoring conditions. During real-time simulations, data are generated at a rate close to that expected during mission support. Spacecraft control commands, provided by an external control system, are used as input to the simulator to generate the maneuver data; these data are then analyzed in real time. The maneuver may be allowed either to continue to completion or be stopped and retargeted, depending on the current simulated attitude. Further stopping and retargeting is carried out until the thruster has been calibrated and the maneuver is proceeding on target.

The PLOTOC simulator is a subsystem of the attitude determination system. Its prime function is to allow the operator to compare the observed infrared Earth sensor data with simulated data based on the attitude and bias solution obtained by the attitude determination system, as illustrated in Fig. 21-10. The plus signs in the figure represent observed data points for the horizon-in and horizon-out rotation angles measured by the Earth sensor. The solid lines represent the predicted rotation angles for two possible attitude solutions. The inner pair of rotation angle curves clearly represents the superior solution. Predicted-versus-observed rotation angle plots can reveal the presence of systematic variations in the observed data (e.g., the Pagoda effect) and by judiciously editing the data span, a more reliable solution may be obtained. PLOTOC can also be used to generate predicted data in advance of the observed data to provide Earth sensor coverage information.

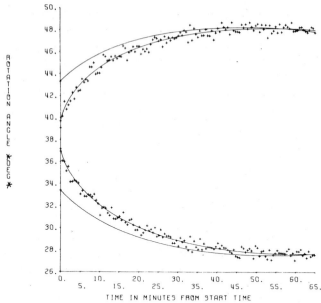

Fig. 21-10. Sample Output From Simulator Used for Comparison of Predicted and Observed Data. Plus signs represent observed data; solid lines represent predicted data for two distinct attitude solutions.

References

1. Fallon, L., D. M. Gottlieb, C. M. Gray, and S. G. Hotovy, *Generalized Star Camera Attitude Determination System Specifications*, Comp. Sc. Corp., CSC/TM-75/6129, June 1975.

2. Fallon, L., and C. R. Sturch, *Performance of Ground Attitude Determination Procedures for HEAO-1*, Flight Mechanics/Estimation Theory Symposium, GSFC, October 18–19, 1977.

3. Foudriat, E. G., "A Limited Memory Attitude Determination System Using Simplified Equations of Motion," *Proceedings of the Symposium on Spacecraft Attitude Determination, Sept. 30, Oct. 1–2, 1969*, El Segundo, CA; Air Force Report No. SAMSO-TR-69-417, Vol. I; Aerospace Corp. Report No. TR-0066(5306)-12, Vol. I, 1969.

4. Gray, C. M., *et al., Attitude Dynamics Data Simulator (ADSIM), Version 3.1*, Comp. Sc. Corp., 3000-06000-02TR. Sept. 1973.

5. Joseph, M., and M. A. Shear, *Optical Aspect Data Prediction (ODAP) System Description and Operating Guide, Version 4.0*, Comp. Sc. Corp., 3000-06000-04TM. Dec. 1973

6. NASA, *A User's Guide to the Flexible Spacecraft Dynamics Program II*, GSFC, 1978.

7. Nelson, R. W., *et al., CTS Attitude Determination System User's Guide*, GSFC Internal Document, Dec. 1975.

8. Paulson, D. C., D. B. Jackson, and C. D. Brown, "SPARS Algorithms and Simulation Results," *Proceedings of the Symposium on Spacecraft Attitude Determination, Sept. 30, Oct. 1–2, 1969*, El Segundo, CA: Air Force Report No. SAMSO-TR-69-417, Vol. I; Aerospace Corp. Report No. TR-0066(5306)-12, Vol. I, 1969.

9. Plett, M. E., *et al.*, *Multisatellite Attitude Determination/Optical Aspect Bias Determination (MSAD/OABIAS) System Description and Operating Guide*, Vol. 3, Comp. Sc. Corp., CSC/TR-75/6001, April 1975.

10. Rigterink, P. V., E. A. Brinker, R. C. Galletta, and J. S. Legg, *Small Astronomy Satellite-B (SAS-B) Star Subsystem (Verslon 7.0) System Description*, Comp. Sc. Corp., 9101-07100-05TR, March 1973.

11. Rochkind, A. B., *Synchronous Meteorological Satellite (SMS) Maneuver Control Program (SMSMAN) Task Specification*, Comp. Sc. Corp., 3101-00800-02TN, July 1973.

12. ———, *Synchronous Meteorological Satellite Maneuver Control Program (SMSMAN) User's Manual*, Comp. Sc. Corp., 3000-02800-03TM, March 1974.

13. Shear, M. A., *System Description and Operating Guide for the Orbit, Sun, and Attitude Geometry Program (OSAG)*, Comp. Sc. Corp., 5035-22300-04TR, Jan. 1972.

14. Smith, P. M., *Communications Technology Satellite Attitude Data Simulator (CTSSIM) Program Description and Operating Guide*, Comp. Sc. Corp., CSC/SD-75/6073, Dec. 1975.

15. Tandon, G. K., *Acceptance Test Plan For Communications Technology Satellite Maneuver Control Program (CTSMAN)*, Comp. Sc. Corp., CSC/TM-75/6154, Aug. 1975.

CHAPTER 22

DISCUSSION

James R. Wertz

This chapter provides a subjective discussion of the state of the art in attitude determination and control and identifies specific problems and areas in which information or further development would be valuable. The major changes that will effect future mission profiles were described in Section 1.1. The most important of these are increased spacecraft autonomy and launch via the Space Shuttle, which will substantially increase the volume of space activity and somewhat reduce costs. (Sixty shuttle launches per year are anticipated from 1984 onward.) In addition, accuracy requirements are becoming increasingly stringent, as illustrated in Fig. 22-1. Thus, the major requirements for the 1980s are to handle increased data volume and to obtain greater accuracy at reduced costs.

In contrast to orbital mechanics, the area of spacecraft attitude determination and control has developed almost entirely in two decades, so much of the fundamental work remains incomplete. For example, there is no geometrical analysis of attitude accuracy for three-axis stabilized spacecraft comparable to that discussed in Chapter 11 and, although orbit determination hardware and measurements have become at least somewhat standardized, attitude determination hardware and measurements have not; there is no clearly superior measurement type or generally accepted standard hardware or analytic techniques.

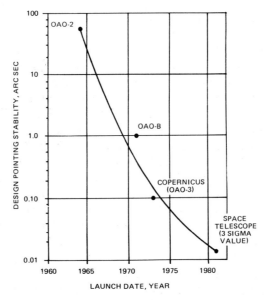

Fig. 22-1. Progression of Pointing Stability Requirements for Astronomical Telescopes in Space (From Proise [1973])

The emphasis of current research in spacecraft attitude determination and control can be approximately determined by a literature survey, as shown in Table 22-1. The categories were arbitrarily assigned in an attempt to represent the topics covered. Over half of the papers were concerned with new control systems or hardware, either proposed, planned, or recently flown. This reflects the diversity of hardware and methods in use at the present time. In contrast, only 5% of the papers dealt with any aspect of attitude determination and another 5% with evaluating disturbance torques or methods of measuring the properties of spacecraft relative to attitude stability and control.

Table 22-1. Distribution of Topics in a 1975–1976 Spacecraft Attitude Determination and Control Literature Survey Based on the NASA Scientific Technical Information Data Base. Miscellaneous additional topics and papers which could not be categorized by the title or the abstract have been omitted.

TOPIC	NUMBER OF PAPERS
DESCRIPTION, ANALYSIS OR REQUIREMENTS OF ATTITUDE CONTROL SYSTEMS PROPOSED OR PLANNED FOR FUTURE MISSIONS	45
EVALUATION OF OPERATIONAL CONTROL SYSTEMS OR HARDWARE	37
CONTROL LAWS AND CONTROL PROCEDURES	27
ATTITUDE DYNAMICS AND SYSTEM STABILITY	19
ATTITUDE DETERMINATION (ALL ASPECTS)	7
NUTATION CONTROL	5
MEASURING OR EVALUATING DISTURBANCE TORQUES	4
MEASURING SPACECRAFT PROPERTIES RELATED TO ATTITUDE STABILITY AND CONTROL	4
TOTAL	148

The Attitude Systems Operation of Computer Sciences Corporation's System Sciences Division provides attitude analysis and operations support for the Attitude Determination and Control Section of NASA's Goddard Space Flight Center. In contrast to the general literature survey, the recent analytic work of this group falls into five major areas: (1) development of more sophisticated procedures for obtaining high reliability with star sensors for missions such as HEAO or MAGSAT, (2) improvements in understanding attitude geometry, primarily for spin-stabilized spacecraft such as AE, SMS/GOES, CTS, SIRIO, IUE, and ISEE; (3) development of procedures for obtaining increased accuracy from Earth horizon sensors, both for spin-stabilized spacecraft and Earth-oriented spacecraft such as SEASAT, HCMM, DE, or MAGSAT; (4) detailed planning for attitude acquisition maneuvers for spacecraft which undergo major changes in determination and control procedures between launch and the initiation of normal mission operations, such as RAE, CTS, GEOS, or HCMM; (5) development of procedures for processing the increased volume of attitude data that is anticipated with the increased launch potential of the Space Shuttle in the 1980s; and (6) satisfying the increasingly stringent attitude determination and control requirements and ensuring the quality of computed attitude solutions.

The remainder of this section discusses developments which are necessary for the continued evolution of spacecraft attitude determination and control techniques. We have divided these developments into five categories: quality assurance, sensor design, hardware standardization, software standardization, and basic analysis.

Quality Assurance.   A major problem that must be resolved in the Shuttle era is *quality assurance*—the designing of software systems that will provide accurate attitude information without operator intervention under normal circumstances and that will recognize abnormal circumstances. The need for increased quality assurance comes from four principal changes in future attitude operations: increased autonomy with the use of onboard computers, increased accuracy requirements, increased data volume (both more spacecraft and more data per spacecraft), and the need for reduced costs. Increased data volume and reduced costs require a system with minimal operator intervention. Increased accuracy implies more sophisticated modeling procedures, more potentially adjustable attitude parameters, and more complex analysis and filtering procedures. Finally, increased autonomy implies software which executes with minimal external intervention and which is capable of recognizing abnormal data.

One solution to quality assurance and increased autonomy may lie in multiple component systems such as that described for the CTS spacecraft in Section 21.2. Here a deterministic processor (or a differential corrector with only a limited number of state vector elements) is used for normal operations either on the ground or on board the spacecraft, and a more sophisticated ground-based differential corrector is used occasionally for bias determination. The bias parameters determined in the latter system are then used as input parameters to the "normal" processor to perform routine operations. The character of the routine operations may be monitored both by internal checks and flags and by summary displays that permit an operator to gain an overview of the system operation and to examine in more detail any abnormal data segments. However, in the CTS system, even the telemetry processor and deterministic attitude component require operator intervention and interactive graphics for effective use.

A second possible solution may lie in the direction of a hybrid or *evolutionary* attitude system. Initially, such a system behaves like a very flexible data filter as described in Chapters 13, 14, 20, and 21, with operator control at essentially every stage via interactive graphics. As biases are resolved and the data quality and nature of the most common anomalies are determined, specific automatic options are chosen and the system becomes increasingly autonomous. Operator intervention is required only when anomalies are encountered, which are identified by a series of checks and flags in the data. After the most common anomalies have been identified, automatic procedures for handling these are initiated (via, for example, alternative processing parameters when specific anomalous conditions are encountered). At this point, the system is essentially fully autonomous—having been "designed" by the experience of the operator to handle the particular data characteristics specific to that mission. The system then operates routinely in a noninteractive mode* with occasional interactive runs to ensure that the system is behaving properly or to account for changing mission conditions.

Sensor Design.   One of the principal requirements for the future is a need for attitude analysis at earlier phases of mission planning and hardware design—particularly when attitude accuracy requirements approach the limit of sensor

---

* For complex ground-based systems, a low level of continuous, interactive control may be necessary to e٢ ١re the availability of data sets and hardware devices.

accuracies. Characteristic of this need for early analysis is a dichotomy between two distinct procedures for obtaining attitude measurements—the use of simple sensors with complex output versus the use of complex sensors with simple output. This distinction may be illustrated by the Sun sensors used for SIRIO and the SMS/GOES missions (Chen, et al., [1976]; Chen and Wertz [1975]). During the transfer orbit to synchronous altitude, both satellites are spin stabilized and use Sun sensors consisting of approximately V-shaped slits as shown in Fig. 22-2. The Sun angle is determined by measuring the time between Sun sightings for the two slits. The relationship between the Sun angle, β, and ratio of the time between Sun sightings, Δt, to the spin period, P (determined by the time between Sun sightings for the vertical slit), for both sensor types is shown in Fig. 22-3. By examination, it is clear that the SIRIO system is a relatively simple sensor with an output that is at least moderately complex to interpret, since it becomes very nonlinear as the Sun moves toward the spin axis. In contrast, the SMS/GOES Sun sensor has a more complex structure, but the output signal is nearly linear over the sensor's range of performance.

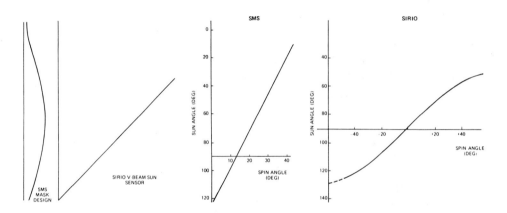

Fig. 22-2. Shape of SMS Sun Sensor Mask and Relative Orientation of Slits in the SIRIO Sun Sensor

Fig. 22-3. Sun Angle Versus Spin Angle for SMS and SIRIO Sun Sensors. The spin angle is the azimuthal rotation of the spacecraft between observations of the Sun by the two sensor components.

The relative advantages of the two sensor types do not become apparent until we ask what the measurement is to be used for and how it is to be processed. If the Sun angle is to be measured by simply attaching a scale to the sensor output with no analysis possible, as might be required in a simple display device or analog use of the data for onboard attitude control, then the SMS/GOES configuration is superior because of the linear output. However, in any attitude determination system or display device where there is software available for processing, the SIRIO design becomes distinctly superior. First, a straightforward analysis of the spherical geometry involved in the SIRIO sensor (Section 7.1) shows that

$$\tan \beta = \tan \theta / \sin(\Delta t / 2\pi P) \qquad (22\text{-}1)$$

where θ is the angle between the two linear slits in the SIRIO design. In contrast, the SMS/GOES design requires a table of calibration values which must be stored and interpolated to compute the Sun angle. In addition, the SIRIO sensor is very amenable to bias determination and in-flight calibration. If we assume only that the slits have indeed been made linear, then only three bias parameters fully characterize the relative orientation between the two slits and the spin axis (Section 7.1). In principle, these parameters can be established as elements of a state vector or determined manually from several data segments. Thus, it is at least possible to have a very accurate in-flight calibration of the SIRIO Sun sensor parameters. (In practice this may be impossible for any specific mission because of limitations in the amount or quality of available attitude data, the geometry, or mission timeline constraints.) In contrast, there is no analytic procedure for general determination of bias parameters for the SMS/GOES Sun sensor because each segment of the calibration curve would have to be corrected separately. We may carry out a bias determination procedure at any particular Sun angle, but extrapolation of the results to other Sun angles is tenuous at best.

As is clear from the above example, the complex sensor with a simple output may be preferable in systems such as limited-capacity onboard processors or control center displays where the output *must* be used directly with no algebraic manipulation, bias determination, or state estimation. However, the simple sensor with output which can be analytically modeled is preferred whenever there is software available for processing or whenever bias determination or state estimation techniques are available to increase attitude accuracies. This would include spacecraft using onboard processors for which biases can be telemetered up from the ground, as described for SMM below. Thus, high accuracy requirements would suggest the need for sensors which can be modeled analytically to make the best use of sophisticated data analysis techniques.

Hardware Standardization.   Related to the problem of sensor design is the need for software systems to provide greater reliability and reduced costs and to permit more effort to be applied to the new and unique problems which arise. The principal problem in the standardization of hardware is that missions have widely varying requirements and hardware systems have normally been designed to meet specific mission conditions at minimum cost. Therefore, the main precursor to standard sensors or standard software is the development of a hardware system with sufficient *flexibility* to meet the requirements of many missions. Two basic approaches to this problem are available: to work with combinations of existing hardware or to design new hardware with the specific intent of designing flexibility.

The use of combinations of existing hardware is a major goal of the Multi-Mission Spacecraft, MMS, series, the first of which will be the Solar Maximum Mission, SMM. As shown in Fig. 22-4, the MMS spacecraft consists of three standard modules (Power, Attitude Control, and Communications and Data Handling) in a triangular frame with space for the payload equipment at one end [GSFC, 1975]. The attitude control subsystem shown in Fig. 22-5 contains the following attitude sensing equipment: a set of three two-degree-of-freedom gyroscopes; two Ball Brothers CT 401 star trackers (described in Section 6.6); one precision digital Sun sensor; and three orthogonal magnetometers. All MMS spacecraft will have a coarse Sun sensor system, and a high-accuracy payload sensor may also be used for some missions. Attitude control will normally be provided by a set of three orthogonal, 20 N·m·s reaction wheels (possibly with a

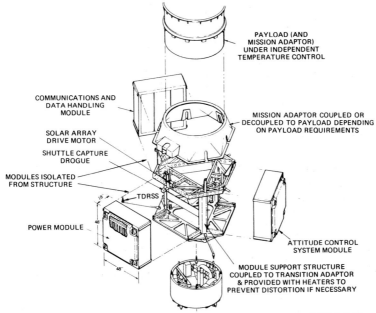

Fig. 22-4.   Multi-Mission Spacecraft (MMS) Design (From GSFC [1975])

fourth wheel). In addition, 100 A·m² electromagnets will be used for initial acquisition and momentum maintenance and will provide a minimum reorientation rate of 0.2 deg/sec for spacecraft with moments of inertia up to 400 kg·m². The MMS spacecraft includes an onboard computer described in Section 6.9 that processes all attitude data and, in conjunction with other stored information, generates control commands.

The first of the MMS missions, SMM, has an attitude accuracy requirement of 5 arc-seconds (0.001 deg) in pitch and yaw and 0.1 deg in roll. The attitude support anticipated for SMM will be based on considerable interaction between the

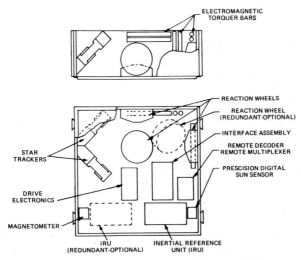

Fig. 22-5.   Attitude Control System Module Within the MMS Spacecraft

onboard processor and the ground software [Werking, 1976]. Definitive attitude will be determined by the onboard processor. For SMM, the onboard computer will spend approximately 30% of its time in attitude-related activities. Ground processing will be used for early mission support (stabilization and attitude acquisition), for calibration or bias determination (to determine sensor parameters which will then be telemetered for use in the onboard processor), and for operational control support (system monitoring, TDRSS scheduling, ephemeris and target star uplink, and maneuver planning). The SMM example is representative of the types of functions that can best be carried out by splitting data processing between the onboard computer and the more sophisticated ground-based systems, and again indicates the need for sensor systems that can be analytically modeled so that bias determination and state estimation is possible.

The MMS attitude control module is intended to support Earth-observation, solar physics, and astronomy payloads in both near-Earth and geosynchronous orbits. At present, it is not clear how widespread use of the MMS spacecraft design will become.* As with other attempts at standardization, the main problem is obtaining sufficient flexibility at low cost.

A second standard spacecraft configuration will be used for the Applications Explorer Missions, AEM, the first of which will be the Heat Capacity Mapping Mission, HCMM [Smith, 1974].† The AEM series will be relatively low-cost missions using a small spacecraft (up to 165 kg at launch) orbited initially by the Scout launch vehicle and by the Space Shuttle for later missions. The missions will generally be Earth oriented, with attitude control to approximately 1 deg in pitch and roll and 2 deg in yaw. The main components of the attitude control system are two reaction wheels incorporating horizon scanners similar to those used for Nimbus and LANDSAT, which provide both attitude determination and control. (HCMM will use only a single scanner.) A triad of magnetic torquers will be used for initial acquisition, roll control, and momentum control. Digital Sun sensors and a triaxial fluxgate magnetometer will be used for yaw angle determination. A third reaction wheel along the roll axis may be included on some missions.

An alternative approach to hardware standardization is to design it with the specific goal of extreme flexibility. For example, an attitude determination package might consist of a coarse element and a fine element, both of which provide redundancy and nearly full sky coverage. The fine element could consist of either a specific payload sensor or a combination of redundant gyroscopes (Section 6.5) and fixed head star trackers (Section 6.4), capable of accurate attitude determination in any orientation. The fine element would be used only for attitude refinement based on a good a priori estimate and would be included only on missions for which pointing accuracies of less than about 0.5 deg are required.

The coarse attitude element would be used for attitude acquisition, orbit maneuvers, and initial attitude estimates for the fine element. It would be the only attitude reference for spacecraft with accuracy requirements of about 0.5 deg or

---

* As of January 1978, MMS spacecraft were intended for use in seven missions (SMM, STORMSAT, CLIMSAT, LANDSAT-D, LFO-E, GRO, and PSCTS) and possibly for further missions in these series. See Appendix I for acronyms and additional mission details.

† As of January 1978, spacecraft in the AEM series were HCMM, SAGE, COBE, ERBS, EUVE, and STEREOSAT and possible follow-ons. See Appendix I for acronyms and additional mission details.

larger. An appropriate coarse attitude package flown on COS-B and ISEE-B consists of two pairs of slit sensors capable of triggering on both the Sun and the Earth and distinguishing between them (see Section 6.2 and Massart [1976].) A single package consisting entirely of three slit sensors has also been proposed [Wertz, 1975]. Table 22-2 compares the redundancy and coverage of the celestial sphere for the package of three slit sensors and the package of five horizon sensors and two Sun sensors flown on the SMS/GOES missions. From the table we see that the three-sensor package provides nearly full sky coverage for both the Sun and the Earth and the potential for attitude determination even with the loss of any two of the three sensors. The seven-sensor package provides substantially less sky coverage and redundancy.

Table 22-2.   Comparison of Sky Coverage and Redundancy for Sensors Flown on the Synchronous Meteoroligical Satellite Witha Possible Package of Three Slit Sensors. (Data from Chen and Wertz [1975] and Wertz [1975].)

PROPERTY	SMS	SLIT SENSOR PACKAGE
NUMBER OF SENSORS	2 120° SUN SENSORS 5 HORIZON SENSORS	1 SLIT PARALLEL TO SPIN AXIS 2 SLITS CANTED TO SPIN AXIS
SUN ANGLE MEASUREMENTS	1 OVER 100% OF SKY 2 OVER 50% OF SKY	1 OVER 99% OF SKY 3 OVER 87% OF SKY
NADIR ANGLE MEASUREMENTS (AT SYNCHRONOUS ALTI- TUDE)	NO MEASUREMENT OVER 42% OF SKY 1 OVER 58% OF SKY 2 OVER 9% OF SKY	1 OVER 100% OF SKY 2 OVER 99% OF SKY 3 OVER 93% OF SKY
EFFECT OF LOSS OF ONE CRITICAL SENSOR	MODERATELY SERIOUS: EARTH 9% COVERED; SUN 50% COVERED	NO PROBLEM: EARTH AND SUN SENSING 100% COVERED; SUN ANGLE 87% COVERED
EFFECT OF LOSS OF TWO MOST CRITICAL SENSORS	VERY CRITICAL: ALL SUN OR EARTH OBSERVATIONS LOST	SERIOUS: ALL SUN ANGLE AND MIDSCAN NADIR ANGLE MEASURE- MENTS LOST; SUN-EARTH AND EARTH-WIDTH ANGLES 87% COVERED

Software Standardizatlon.   The need for software standardization comes from increased data volume and the demand for both reduced costs and greater accuracies. By spreading the costs over multiple missions, sophisticated, expensive hardware and ground or onboard processing tools can be developed. In addition, complex systems become more reliable and more accurate after they have been used several times and system characteristics have been identified. Expertise in use and interpretation also increases when systems are used for several missions.

In the area of attitude software, several multimission support programs have been developed at Goddard Space Flight Center. At the present time, however, most of the standardization has come at the level of subroutines, system components, and relatively small utility programs. For example, the Graphics Executive Support System described in Section 20.2 has proved to be a very valuable executive for systems using interactive graphics and is used for most mission support systems.

Three multimission systems which have been used at Goddard Space Flight Center serve to illustrate the advantages and problems of standardization. The Attitude Dynamics Data Simulator, ADSIM, has been used to support over 10 missions. It simulates spacecraft dynamics and sensor data for a generalized, rigid spacecraft and has an expandable structure to accommodate additional features. However, ADSIM is relatively complex to use; it has a large set of input parameters, operates in four coordinate systems, and requires considerable analysis to set up for a particular mission. In addition, sensor simulation is time consuming (both analytically and in computer time) because it must follow a more general procedure than is permitted for a system designed to work with specific, known components which have well-established frequencies associated with them.

The Optical Aspect Bias Determination System, OABIAS, has been used to support bias determination activities for eight missions and has been modified and expanded to become the bias determination subsystem of the CTS attitude system (Section 21.2). Although OABIAS has been successful in supporting diverse missions, it is limited to spin-stabilized spacecraft using Sun sensors and horizon sensors. Thus, the system can support only a limited class of hardware and cannot be used for three-axis stabilized missions. The Multi-Satellite Attitude Prediction Program, MSAP, has supported a greater variety of missions than OABIAS because it is modified for each new spacecraft to take into account the aerodynamic drag and control laws for each specific mission. Of course this reduces the versatility and efficiency of the system because considerable analysis and programming is required for each new mission.

The generalized software systems have had only limited success, primarily because of the widely varying types of hardware and mission constraints. Generalized software capable of satisfying varied requirements will contain more options and control parameters than any one spacecraft will use. General software is more difficult to design; requires more core storage; and is generally more complex, inefficient, and expensive (although the cost may be spread over many missions). Even though there are substantial advantages to generalized software, it is unlikely to become widely used *until spacecraft hardware becomes more standardized.*

Basic Analysis. Because the fundamentals of attitude determination and control have been studied for only a few years, considerable future effort will be required to expand and systematize our knowledge of attitude-related activities. These analytic needs may be divided into five overlapping categories: (1) data evaluation and bias determination; (2) filtering theory and computer processing techniques, both ground based and onboard; (3) spacecraft dynamics; (4) environmental modeling; and (5) attitude geometry, including attitude accuracy and bias observability.

In the area of *data evaluation*, we need to determine what quality assurance procedures are appropriate for data from autonomous spacecraft; what data should be telemetered by the onboard computer to optimize the division between onboard and ground-based functions; and how the ground-based software should process and react to these data. We would like to understand how to formalize the process of identifying data anomalies as described in Chapter 9. We need to determine which bias parameters can or should be estimated and what constraints should be placed on an attitude state vector to obtain "reasonable" answers in the presence of systematic residuals. We also need improved sensor models, including the operational evaluation and refinement of the horizon sensor electronics model described in Section 7.4 and further development of star sensor instrumental magnitudes discussed in Section 5.6.

In the area of *estimation theory*, we should improve our understanding of the *behavior* of estimation algorithms, particularly when the errors are dominated by quantization. Similarly, we need to determine how best to estimate attitude parameters in the presence of large systematic residuals, because systematic errors are ordinarily the limiting factor in the attitude determination process. This understanding should lead to improved standardized procedures for processing attitude measurements with standard estimators. A better analysis of the effect of preaveraging on subsequent state estimation is necessary to provide a high-quality two-stage filter. The goal here is an estimator consisting of a preaveraging stage which reduces the data volume to a manageable level while retaining as much of the original information content as possible, and a second stage which is a complex estimator of the type described in Chapter 13 that can be iterated many times on the preaveraged data to find the best available solution. Another state estimation problem is to find the best balance between onboard and ground processing, such that the economy and strength of both methods can be fully utilized. This includes estimation techniques for onboard processors and the choice of bias, calibration, or environmental parameters to be supplied to the onboard processor by ground-based systems.

In the area of *spacecraft dynamics*, much of the current literature is related to stability and control of complex systems; this literature is the best source of information on future requirements. Additional work is also needed in the identification and modeling of internal disturbance torques, as described in Section 17.3, and in obtaining simple, approximate relations in flexible spacecraft dynamics (see Section 16.4). We should obtain a better understanding of the impact of spacecraft dynamics on attitude accuracy as more stringent requirements develop. A more complete analysis of the approximate effect of small amplitude nutation on both observed data and attitude solutions (as described in Section 16.3) would be useful.

Additional analysis is necessary in several aspects of *environmental modeling*, including more detailed models of environmental torques, as described in Section 17.2. For torque computations, models of the Earth's atmosphere which do not require frequent input of observed data would be valuable, but may be impossible. Improved models of the Earth's horizon, as described in Section 4.2, are needed; these would include variations in the infrared radiation profiles and models of the variations in the CO_2 layer which incorporate the effects of weather and horizontal temperature gradients. Particularly important here is an evaluation of *real* spacecraft attitude data obtained together with other high-accuracy attitude data, such as that from star sensors. Further development of star catalogs and star positional data, as described in Section 5.6, will become more important with the increasing use of star sensors as an accurate attitude reference.

In the area of *attitude geometry*, we have mentioned the need for a geometrical analysis of attitude accuracy for three-axis stabilized spacecraft. In addition, further analysis of bias observability over long data passes, as described in Section 14.3, is required. Convenient procedures for representing the time variations in the geomagnetic field (as sensed by the spacecraft) or the long-term position of the Sun for Earth-oriented spacecraft would be useful in mission planning and analysis. We should also obtain more information on the spherical geometry characteristics of attitude determined from multiple nearby sources, as in the case of star sensors or landmark tracking. A reliable and systematic procedure is needed to incorporate geometric error analyses into attitude solutions and uncertainty computations.

Finally, we should undertake a systematic evaluation of the relative advantages and disadvantages of various hardware types and possible attitude reference vectors. This would include particularly the use of landmark tracking data and the observability of both orbit and attitude parameters from "attitude" data taken on board the spacecraft. This could lead to an autonomous orbit/attitude system.

Summary. There are two major goals in spacecraft attitude determination and control in the Space Shuttle era: (1) reliable, flexible, and economical coarse attitude determination for attitude acquisition, orbit and attitude maneuver control, and attitude determination and control when requirements are not stringent; and (2) reliable, high-accuracy procedures for semiautonomous attitude determination and control, probably using landmark tracking for Earth-oriented spacecraft and star sensors or payload sensors for other missions. Such systems should use hardware that can be analytically modeled and onboard routine processing using calibration, bias, or environmental data supplied by sophisticated ground-based systems. Achievement of these goals will necessitate some degree of hardware standardization, which is a precursor to standardization of software and processing techniques. The support of common systems should provide an opportunity to develop a more basic understanding of attitude-related problems, rather than continually redeveloping and revising procedures to ensure the success of individualized missions.

Another major problem to be overcome is that of improved communication between the frequently discrete groups working in the area of attitude determination and control—that is, those primarily involved with hardware, stability and control theory, and ground processing and attitude support. A major goal of this book is to provide some common background so that we can reduce the communications barrier and improve our mutual understanding.

References

1. Chen, L. C., and J. R. Wertz, *Analysis of SMS-2 Attitude Sensor Behavior Including OABIAS Results*, Comp. Sc. Corp., CSC/TM-75/6003, April 1975.
2. Chen, L. C., H. L. Hooper, J. E. Keat, and J. R. Wertz, *SIRIO Attitude Determination and Control System Specifications*, Comp. Sc. Corp., CSC/TM-76/6043, Feb. 1976.
3. GSFC, *Low Cost Modular Spacecraft Description*, NASA X-700-75-140, May 1975.
4. Massart, J. A., *Preliminary Assessment of the COS-B Attitude Measurement System and Attitude Control System and the Related Ground Software*, ESA Doc. No. ESOC/OAD-WP-50, March 1976.
5. Proise, M., *Fine Pointing Performance Characteristics of the Orbiting Astronomical Observatory (OAO-3)*, AIAA Paper No. 73-869, AIAA Guidance and Control Conference, Key Biscayne, FL, Aug. 20–22, 1973.
6. Smith, Sterling R., editor, *Applications Explorer Missions (AEM) Mission Planners Handbook*, GSFC, May 1974.
7. Werking, R. D., *A Ground Based Attitude Support Plan for the Solar Maximum Mission*, presented at the Goddard Space Flight Center Flight Dynamics-/Estimation Theory Symposium, May 5–6, 1976.
8. Wertz, J. R., "The Optical Slit Sensor as Standard Sensor for Spacecraft Attitude Determination," *Flight Mechanics/Estimation Theory Symposium*, compiled by a Carmelo E. Velez, NASA X-582-75-273, p. 59–68, Aug. 1975.

PART VI

APPENDICES

CONTENTS

PART VI

APPENDICES

Appendix

A   Spherical Geometry   727

B   Construction of Global Geometry Plots   737

C   Matrix and Vector Algebra   744

D   Quaternions   758

E   Coordinate Transformations   760

F   The Laplace Transform   767

G   Spherical Harmonics   775

H   Magnetic Field Models   779

I   Spacecraft Attitude Determination and Control Systems   787

J   Time Measurement Systems   798

K   Metric Conversion Factors   807

L   Solar System Constants   814

M   Fundamental Physical Constants   826

Index   830

APPENDIX A

SPHERICAL GEOMETRY

James R. Wertz

A.1 Basic Equations
A.2 Right and Quadrantal Spherical Triangles
A.3 Oblique Spherical Triangles
A.4 Differential Spherical Trigonometry
A.5 Haversines

Finding convenient reference material in spherical geometry is difficult. This appendix provides a compilation of the most useful equations for spacecraft work. A brief discussion of the basic concepts of spherical geometry is given in Section 2.3. The references at the end of this appendix contain further discussion and proofs of most of the results presented here.

A.1 Basic Equations

Algebraic Formulas. Let P_i be a point on the unit sphere with coordinates (α_i, δ_i). The **arc-length distance**, $\theta(P_1, P_2)$, between P_1 and P_2 is given by:

$$\cos\theta(P_1, P_2) = \cos\theta(P_2, P_1)$$

$$= \sin\delta_1 \sin\delta_2 + \cos\delta_1 \cos\delta_2 \cos(\alpha_1 - \alpha_2) \qquad 0 \leqslant \theta \leqslant 180° \qquad \text{(A-1)}$$

The **rotation angle**, $\Lambda(P_1, P_2; P_3)$, from P_1 to P_2 about a third point, P_3, is cumbersome to calculate and is most easily obtained from spherical triangles (Sections A.2 and A.3) if any of the triangle components are already known. To calculate directly from coordinates, obtain as intermediaries the arc-length distances $\theta(P_i, P_j)$, between the three pairs of points. Then

$$\cos\Lambda(P_1, P_2; P_3) = \frac{\cos\theta(P_1, P_2) - \cos\theta(P_1, P_3)\cos\theta(P_2, P_3)}{\sin\theta(P_1, P_3)\sin\theta(P_2, P_3)} \qquad 0 \leqslant \Lambda \leqslant 360° \qquad \text{(A-2)}$$

with the quadrant determined by inspection.

The **equation for a small circle** of angular radius ρ and centered at (α_0, δ_0) in terms of the coordinates, (α, δ), of the points on the small circle is, from Eq. (A-1),

$$\cos\rho = \sin\delta \sin\delta_0 + \cos\delta \cos\delta_0 \cos(\alpha - \alpha_0) \qquad \text{(A-3)}$$

The **arc length**, β, along the arc **of a small circle** of angular radius ρ between two points on the circle separated by the rotation angle, Φ (Φ measured at the center of the circle) is

$$\beta = \Phi \sin\rho \qquad \text{(A-4)}$$

The **chord length**, γ, along the great circle chord of an arc **of a small circle** of angular radius ρ is given by

$$\cos\gamma = 1 - (1 - \cos\Phi)\sin^2\rho \qquad 0 \leqslant \gamma \leqslant 180° \qquad (A\text{-}5)$$

where Φ is as defined above.

The **equation for a great circle** with pole at (α_0, δ_0) is, from Eq. (A-3) with $\rho = 90°$,

$$\tan\delta = -\cot\delta_0\cos(\alpha - \alpha_0) \qquad (A\text{-}6a)$$

The inclination, i, and azimuth of the ascending node (point crossing the equator from south to north when moving along the great circle toward increasing azimuth), ϕ_0, of the great circle are

$$i = 90° - \delta_0$$

$$\phi_0 = 90° + \alpha_0 \qquad (A\text{-}6b)$$

Therefore, the equation for the great circle in terms of inclination and ascending node is

$$\tan\delta = \tan i\,\sin(\alpha - \phi_0) \qquad (A\text{-}6c)$$

The equation of a great circle through two arbitrary points is given below. Along a great circle, the arc length, the chord length, and the rotation angle, Φ, are all equal, as shown by Eqs. (A-4) and (A-5) with $\rho = 90°$.

Finally, the **direction of the cross product** between two unit vectors associated with points P_1 and P_2 on the unit sphere is the pole of the great circle passing through the two points. Find the intermediary, β_1, from

$$\cot\beta_1 = \left(\frac{\tan\delta_2}{\tan\delta_1}\right)\frac{1}{\sin(\alpha_2 - \alpha_1)} - \cot(\alpha_2 - \alpha_1) \qquad (A\text{-}7a)$$

As shown in Fig. A-1, β_1 is the azimuth of point P_1 relative to the ascending node

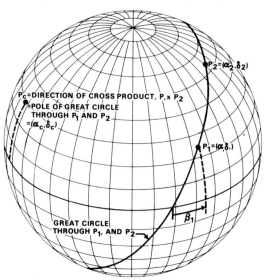

Fig. A-1. $\hat{\mathbf{P}}_c$ is the pole of the great circle passing through $\hat{\mathbf{P}}_1$ and $\hat{\mathbf{P}}_2$ and is also in the direction of the cross product $\hat{\mathbf{P}}_1 \times \hat{\mathbf{P}}_2$. β_1 is an intermediate variable used for computations.

of the great circle through P_1 and P_2. The coordinates, (α_c, δ_c), of the cross product $\hat{P}_1 \times \hat{P}_2$ are given by

$$\alpha_c = \alpha_1 - 90° - \beta_1$$

$$\tan \delta_c = \begin{cases} \dfrac{\sin \beta_1}{\tan \delta_1} & (\delta_1 \neq 0) \\[2ex] \dfrac{\sin(\alpha_2 - \alpha_1)}{\tan \delta_2} & (\delta_1 = 0) \end{cases} \tag{A-7b}$$

Combining Eqs. (A-7b) and (A-6a) gives the **equation for a great circle through points P_1 and P_2**:

$$\tan \delta = \begin{cases} \dfrac{\tan \delta_1}{\sin \beta_1} \sin(\alpha - \alpha_1 + \beta_1) & (\delta_1 \neq 0) \\[2ex] \dfrac{\tan \delta_2}{\sin(\alpha_2 - \alpha_1)} \sin(\alpha - \alpha_1 + \beta_1) & (\delta_1 = 0) \end{cases} \tag{A-8}$$

Area Formulas. All areas are measured on the curved surface of the unit sphere. For a sphere of radius R, multiply each area formula by R^2. All arc lengths are in radians and all angular areas are in steradians (sr), where

$$1\,\text{sr} \equiv \text{solid angle enclosing an area equal to the square of the radius}$$

$$= \left(\frac{180}{\pi}\right)^2 \text{deg}^2$$

The **surface area of the sphere** is

$$\Omega_s = 4\pi \tag{A-9}$$

The **area of a lune** bounded by two great circles whose inclination is Θ radians is

$$\Omega_l = 2\Theta \tag{A-10}$$

The **area of a spherical triangle** whose three rotation angles are Θ_1, Θ_2, and Θ_3 is

$$\Omega_t = \Theta_1 + \Theta_2 + \Theta_3 - \pi \tag{A-11}$$

The **area of a spherical polygon** of n sides, where Θ is the sum of its rotation angles in radians, is

$$\Omega_p = \Theta - (n-2)\pi \tag{A-12}$$

The **area of a small circle** of angular radius ρ is

$$\Omega_c = 2\pi(1 - \cos\rho) \tag{A-13}$$

The **overlap area between two small circles** of angular radii ρ and ϵ, separated by a center-to-center distance, α, is

$$\Omega_o = 2\pi - 2\cos\rho \arccos\left[\frac{\cos\epsilon - \cos\rho\cos\alpha}{\sin\rho\sin\alpha}\right]$$

$$-2\cos\epsilon \arccos\left[\frac{\cos\rho - \cos\epsilon\cos\alpha}{\sin\epsilon\sin\alpha}\right]$$

$$-2\arccos\left[\frac{\cos\alpha - \cos\epsilon\cos\rho}{\sin\epsilon\sin\rho}\right] \qquad (|\rho - \epsilon| \leqslant \alpha \leqslant \rho + \epsilon) \qquad \text{(A-14)}$$

Recall that area is measured on the *curved* surface.

A.2 Right and Quadrantal Spherical Triangles

Example of an Exact Right Spherical Triangle. For testing formulas, the isosceles right spherical triangle shown in Fig. A-2 is convenient. The sides and angles shown are exact values.

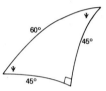

Fig. A-2.   Example of an Exact Right Spherical Triangle. $\Psi = \arcsin\sqrt{2/3} \approx 54.7356° \approx 54°44'08''$.

Napier's Rules for Right Spherical Triangles. A right spherical triangle has five variable parts, as shown in Fig. A-3. If these components and their complements (*complement* of $\Phi \equiv 90° - \Phi$) are arranged in a circle, as illustrated in Fig. A-3, then the following relationships hold between the five components in the circle:

> *The sine of any component equals the product of either*
> 1. *The tangents of the adjacent components, or*
> 2. *The cosines of the opposite components*

For example,

$$\sin\lambda = \tan\phi\tan(90° - \Phi) = \cos(90° - \Lambda)\cos(90° - \theta)$$

Quadrants for the solutions are determined as follows:
1. An oblique angle and the side opposite are in the same quadrant.
2. The hypotenuse (side θ) is less than 90 deg if and only if ϕ and λ are in the same quadrant and more than 90 deg if and only if ϕ and λ are in different quadrants.

Note: Any two components in addition to the right angle completely determine the triangle, *except* that if the known components are an angle and its opposite side, then two distinct solutions are possible.

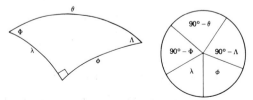

Fig. A-3.   Diagram for Napier's Rules for Right Spherical Triangles. Note that the complements are used for the three components farthest from the right angle.

The following formulas can be derived from Napier's Rules for right spherical triangles:

$$\sin\lambda = \tan\phi\cot\Phi = \sin\theta\sin\Lambda \tag{A-15}$$

$$\sin\phi = \tan\lambda\cot\Lambda = \sin\theta\sin\Phi \tag{A-16}$$

$$\cos\theta = \cot\Phi\cot\Lambda = \cos\phi\cos\lambda \tag{A-17}$$

$$\cos\Lambda = \tan\phi\cot\theta = \cos\lambda\sin\Phi \tag{A-18}$$

$$\cos\Phi = \tan\lambda\cot\theta = \cos\phi\sin\Lambda \tag{A-19}$$

Napier's Rules are discussed in Section 2.3. Proof of these rules can be found in most spherical geometry texts, such as those of Brink [1942]; Palmer, *et al.*, [1950]; or Smail [1952].

Napier's Rules for Quadrantal Spherical Triangles. A quadrantal spherical triangle is one having one side of 90 deg. If the five variable components of a quadrantal triangle are arranged in a circle, as shown in Fig. A-4, then Napier's Rules as quoted above apply to the relationships between the parts. (Note that the bottom component is *minus* the complement of Θ.) The rules for defining quadrants are modified as follows:

1. An oblique angle (other than Θ, the angle opposite the 90-deg side) and its opposite side are in the same quadrant.
2. Angle Θ (the angle opposite the 90-deg side) is *more* than 90 deg if and only if λ and ϕ are in the same quadrant and *less* than 90 deg if and only if λ and ϕ are in different quadrants.

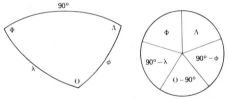

Fig. A-4. Diagram for Napier's Rules for Quadrantal Spherical Triangles. Note that complements are used in the three components farthest from the 90-deg side and the component opposite the 90-deg side is *minus* the complement of Θ.

The following formulas can be derived from Napier's Rules for quadrantal spherical triangles:

$$\sin\Lambda = \tan\Phi\cot\phi = \sin\Theta\sin\lambda \tag{A-20}$$

$$\sin\Phi = \tan\Lambda\cot\lambda = \sin\Theta\sin\phi \tag{A-21}$$

$$\cos\Theta = -\cot\phi\cot\lambda = -\cos\Phi\cos\Lambda \tag{A-22}$$

$$\cos\lambda = -\tan\Phi\cot\Theta = \cos\Lambda\sin\phi \tag{A-23}$$

$$\cos\phi = -\tan\Lambda\cot\Theta = \cos\Phi\sin\lambda \tag{A-24}$$

A.3 Oblique Spherical Triangles

Three fundamental relationships—the law of sines, the law of cosines for angles, and the law of cosines for sides—hold for all spherical triangles. These may be used to derive Napier's Rules (Section A.2) or may be derived from them. The components of a general spherical triangle as used throughout this section are shown in Fig. A-5.

Law of Sines.

$$\frac{\sin\theta}{\sin\Theta} = \frac{\sin\lambda}{\sin\Lambda} = \frac{\sin\phi}{\sin\Phi} \tag{A-25}$$

Law of Cosines for Sides.

$$\cos\lambda = \cos\theta\cos\phi + \sin\theta\sin\phi\cos\Lambda \tag{A-26}$$

Similar relationships hold for each side.

Law of Cosines for Angles.

$$\cos\Lambda = -\cos\Theta\cos\Phi + \sin\Theta\sin\Phi\cos\lambda \tag{A-27}$$

Similar relationships hold for each angle.

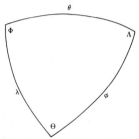

Fig. A-5.   Notation for Rules for Oblique Spherical Triangles

Half-Angle Formulas.   A spherical triangle is fully specified by either three sides or three angles. The remaining components are most conveniently expressed in terms of half angles. Specifically,

$$\sin\left(\frac{1}{2}\Lambda\right) = \sqrt{\frac{\sin(\sigma-\theta)\sin(\sigma-\phi)}{\sin\theta\sin\phi}} \tag{A-28}$$

where

$$\sigma \equiv \frac{1}{2}(\theta + \lambda + \phi)$$

and

$$\cos\left(\frac{1}{2}\lambda\right) = \sqrt{\frac{\cos(\Sigma-\Theta)\cos(\Sigma-\Phi)}{\sin\Theta\sin\Phi}} \tag{A-29}$$

where

$$\Sigma \equiv \frac{1}{2}(\Theta + \Lambda + \Phi)$$

Similar relationships may be found for the other trigonometric functions of half angles in most spherical geometry texts.

General Solution of Oblique Spherical Triangles. Table A-1 lists formulas for solving any oblique spherical triangle. In addition, in any spherical triangle, the

Table A-1. Formulas for Solving Oblique Spherical Triangles. See also Table A-2.

KNOWN	TO FIND	FORMULA	COMMENTS
θ, λ, ϕ	Λ	$\sin \frac{1}{2} \Lambda = \sqrt{\dfrac{\sin(\sigma - \theta)\sin(\sigma - \phi)}{\sin\theta \sin\phi}}$	$\sigma = \frac{1}{2}(\theta + \lambda + \phi)$ UNIQUE SOLUTION[1]
Θ, Λ, Φ	λ	$\cos \frac{1}{2} \lambda = \sqrt{\dfrac{\cos(\Sigma - \Theta)\cos(\Sigma - \Phi)}{\sin\Theta \sin\Phi}}$	$\Sigma = \frac{1}{2}(\Theta + \Lambda + \Phi)$ UNIQUE SOLUTION[1]
θ, λ, Φ	ϕ	$\cos\phi = \cos\theta \cos\lambda + \sin\theta \sin\lambda \cos\Phi$	UNIQUE SOLUTION[1]
	Λ	$\tan\Lambda = \dfrac{\sin\Gamma_1 \tan\Phi}{\sin(\theta - \Gamma_1)}$	$\tan\Gamma_1 = \tan\lambda \cos\Phi$ UNIQUE SOLUTION
θ, Λ, Φ	Θ	$\cos\Theta = \sin\Lambda \sin\Phi \cos\theta - \cos\Lambda \cos\Phi$	UNIQUE SOLUTION
	λ	$\tan\lambda = \dfrac{\tan\theta \sin\Gamma_2}{\sin(\Phi + \Gamma_2)}$	$\tan\Gamma_2 = \tan\Lambda \cos\theta$ UNIQUE SOLUTION
	ϕ	$\sin(\phi + \Gamma_3) = \dfrac{\cos\theta \sin\Gamma_3}{\cos\lambda}$	$\cot\Gamma_3 = \cos\Theta \tan\lambda$ 2 VALID SOLUTIONS
θ, λ, Θ	Λ	$\sin\Lambda = \dfrac{\sin\Theta \sin\lambda}{\sin\theta}$	2 VALID SOLUTIONS
	Φ	$\sin(\Phi + \Gamma_4) = \sin\Gamma_4 \tan\lambda \cot\theta$	$\tan\Gamma_4 = \tan\Theta \cos\lambda$ 2 VALID SOLUTIONS
	Φ	$\sin(\Phi - \Gamma_5) = \dfrac{\cos\Theta \sin\Gamma_5}{\cos\Lambda}$	$\cot\Gamma_5 = \tan\Lambda \cos\theta$ 2 VALID SOLUTIONS
θ, Θ, Λ	λ	$\sin\lambda = \dfrac{\sin\theta \sin\Lambda}{\sin\Theta}$	2 VALID SOLUTIONS
	ϕ	$\sin(\phi - \Gamma_6) = \cot\Theta \tan\Lambda \sin\Gamma_6$	$\tan\Gamma_6 = \cos\Lambda \tan\theta$ 2 VALID SOLUTIONS

[1]SECTION A-5 PRESENTS AN ALTERNATIVE FORMULA.

following rules are sufficient to determine the quadrant of any component:
1. If one side (angle) differs from 90 deg by more than another side (angle), it is in the same quadrant as its opposite angle (side).
2. Half the sum of any two sides is in the same quadrant as half the sum of the opposite angles.

A.4 Differential Spherical Trigonometry

The development here follows that of Newcomb [1960], which contains a more extended discussion of the subject.

Differential Relations Between the Parts of a Spherical Triangle.   In general, any part of a spherical triangle may be determined from three other parts. Thus, it is of interest to determine the error in any part produced by infinitesimal errors in the three given parts. This may be done by determining the partial derivatives relating any four parts of a spherical triangle from the following differentials, where the notation of Fig. A-5 is retained.
Given three angles and one side:

$$-\sin\lambda\sin\Phi\,d\theta + d\Theta + \cos\phi\,d\Lambda + \cos\lambda\,d\Phi = 0 \qquad \text{(A-30)}$$

Given three sides and one angle:

$$-d\theta + \cos\Phi\,d\lambda + \cos\Lambda\,d\phi + \sin\phi\sin\Lambda\,d\Theta = 0 \qquad \text{(A-31)}$$

Given two sides and the opposite angles:

$$\cos\theta\sin\Phi\,d\theta - \cos\phi\sin\Theta\,d\phi + \sin\theta\cos\Phi\,d\Phi - \sin\phi\cos\Theta\,d\Theta = 0 \qquad \text{(A-32)}$$

Given two sides, the included angle, and one opposite angle:

$$-\sin\Phi\,d\theta + \cos\lambda\sin\Theta\,d\phi + \sin\lambda\,d\Theta + \cos\Phi\sin\theta\,d\Lambda = 0 \qquad \text{(A-33)}$$

As an example of the determination of partial derivatives, consider a triangle in which the three independent variables are the three sides. Then, from Eq. (A-31),

$$\left.\frac{\partial\Theta}{\partial\phi}\right|_{\lambda,\theta} = -\frac{\cos\Lambda}{\sin\phi\sin\Lambda} = -\frac{\cot\Lambda}{\sin\phi}$$

Infinitesimal Triangles.   The simplest infinitesimal spherical triangle is one in which the entire triangle is small relative to the radius of the sphere. In this case, the spherical triangle may be treated as a plane triangle if the three rotation angles remain finite quantities. If one of the rotation angles is infinitesimal, the analysis presented below should be used.
Figure A-6 shows a spherical triangle in which two sides are of arbitrary, but nearly equal, length and the included rotation angle is infinitesimal. Then the

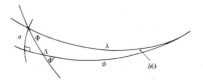

Fig. A-6.   Spherical Triangle With One Infinitesimal Angle

change in the angle by which the two sides intercept a great circle is given by

$$\delta\Phi \equiv \Phi' - \Phi = 180° - (\Lambda + \Phi)$$
$$= \delta\Theta \cos\lambda \qquad (A\text{-}34)$$

The perpendicular separation, σ, between the two long arcs is given by

$$\sigma = \delta\Theta \sin\lambda \qquad (A\text{-}35)$$

If two angles are infinitesimal (such that the third angle is nearly 180 deg), the triangle may be divided into two triangles and treated as above.

A.5 Haversines

A convenient computational tool for spherical trigonometry is the *haversine*, defined as

$$\text{haversine}\,\theta \equiv \text{hav}\,\theta \equiv \frac{1}{2}(1 - \cos\theta) \qquad (A\text{-}36)$$

The principal advantage of the haversine is that a given value of the function corresponds to only one angle over the range from 0 deg to 180 deg, in contrast to the sine function for which there is an ambiguity as to whether the angle corresponding to a given value of the sine falls in the range 0 deg to 90 deg or 90 deg to 180 deg. Given the notation of Fig. A-5, two fundamental haversine relations in any spherical triangle are as follows:

$$\text{hav}\,\lambda = \text{hav}(\theta - \phi) + \sin\theta\sin\phi\,\text{hav}\,\Lambda \qquad (A\text{-}37)$$

$$\text{hav}\,\Lambda = \frac{\sin(\sigma - \theta)\sin(\sigma - \phi)}{\sin\theta\sin\phi} \qquad (A\text{-}38)$$

where

$$\sigma \equiv \frac{1}{2}(\theta + \lambda + \phi)$$

The first three formulas from Table A-1 can be expressed in a simpler form to evaluate in terms of haversines, as shown in Table A-2. Most spherical geometry

Table A-2. Haversine Formulas for Oblique Spherical Triangles

KNOWN	TO FIND	FORMULA	COMMENTS
θ, λ, ϕ	Λ	$\text{hav}\,\Lambda = \dfrac{\text{hav}\,\lambda - \text{hav}\,(\theta - \phi)}{\sin\theta\,\sin\phi}$	
Θ, Λ, Φ	λ	$\text{hav}\,\lambda = \dfrac{-\cos\Sigma\,\cos\,(\Sigma - \Lambda)}{\sin\Theta\,\sin\Phi}$	$\Sigma = \dfrac{1}{2}\,(\Theta + \Lambda + \Phi)$
θ, λ, Φ	ϕ	$\text{hav}\,\phi = \text{hav}\,(\theta - \lambda) + \sin\theta\sin\lambda\,\text{hav}\,\Phi$	

texts (e.g., Brink [1942] or Smail [1952]) carry a further discussion of haversine formulas. The function is tabulated in Bowditch's *American Practical Navigator* [1966].

References

1. Bowditch, Nathaniel, *American Practical Navigator*. Washington, D.C.: USGPO, 1966.
2. Brink, Raymond W., *Spherical Trigonometry*. New York: Appleton-Century-Croft, Inc., 1942.
3. Newcomb, Simon, *A Compendium of Spherical Astronomy*. New York: Dover, 1960.
4. Palmer, C. I., C. W. Leigh, and S. H. Kimball, *Plane and Spherical Trigonometry*. New York: McGraw-Hill, Inc., 1950.
5. Smail, Lloyd Leroy, *Trigonometry: Plane and Spherical*, New York: McGraw-Hill, Inc., 1952.

APPENDIX B

CONSTRUCTION OF GLOBAL GEOMETRY PLOTS

James R. Wertz

Global geometry plots, as used throughout this book, are convenient for presenting results and for original work involving geometrical analysis on the celestial sphere. The main advantage of this type of plot is that the orientation of points on the surface is completely unambiguous, unlike projective drawings of vectors between three orthogonal axes. This appendix describes procedures for manually constructing these plots. Methods for obtaining computer-generated plots

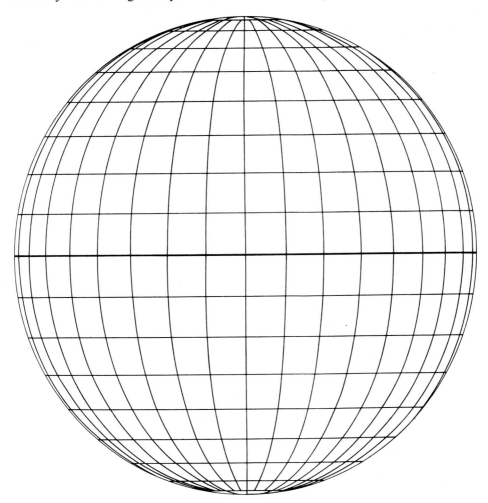

Fig. B-1. Equatorial Projection Grid Pattern

are described in Section 20.3 (subroutines SPHGRD, SPHPLT, and SPHCNV). Interpretation and terminology for the underlying coordinate system is given in Section 2.1.

For most applications related to attitude geometry, the spacecraft is thought of as being at the center of the globe. Therefore, an arrow drawn on the globe's equator from right to left would be viewed by an observer on the spacecraft as going from left to right. This geometrical reversal is illustrated in Figs. 4-3 and 4-4, which show the Earth as a globe viewed from space and as viewed on the spacecraft-centered celestial sphere. Similarly, Figs. 11-25 and 11-26 show the orbit of the Earth about the spacecraft as viewed by the spacecraft. This spacecraft-centered geometry allows a rapid interpretation of spacecraft observations and attitude-related geometry.

In this book, we use four basic globe grids showing the unit sphere from the perspective of infinitely far away (i.e., half the area of the sphere is seen on each globe) as seen by observers at 30-deg latitude intervals from the equator to the pole. The four grids are shown in Figs. B-1 through B-4 and are intended for

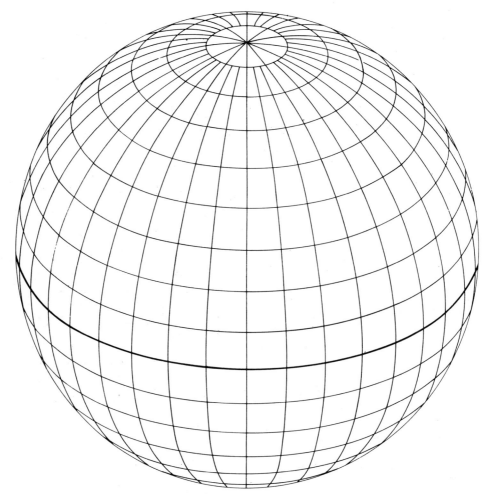

Fig. B-2.   30-Deg Inclination Grid Pattern

reproduction by interested users. (For accurate construction, the globes must be reproduced to the same size as nearly as possible. Therefore, reproductions of the different projections should be made at the same time on the same equipment.) Coordinate lines are at 10-deg intervals in latitude and longitude except within 10 deg of the poles, where the longitude intervals are 30 deg. The globe originals are handdrawn and are accurate to about 1 deg in the central regions and 2 deg near the perimeter.

The most important feature of the globes for the purpose of plot construction is that the geometry of figures constructed on the sphere does *not* depend on the underlying grid pattern. For example, if we take the globe from Fig. B-2, we may draw a small circle of 20-deg radius centered on the pole and an equilateral right spherical triangle between the pole and the equator as shown in Fig. B-5(a). (Any parallel of latitude is a small circle and the equator or any meridian of longitude is a great circle.) Having constructed the figure, we may rotate or tilt the underlying grid pattern without affecting the geometrical construction. Thus, in Fig. B-5(b) the

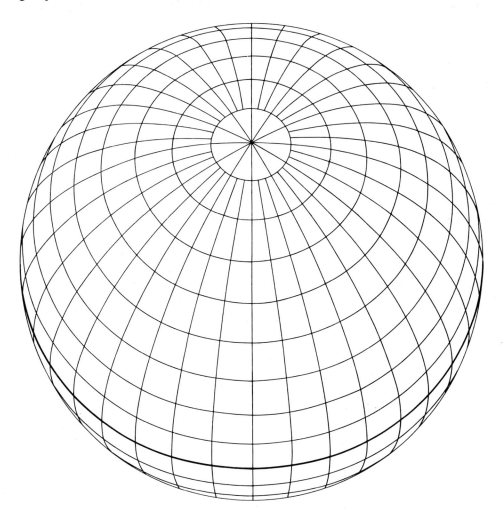

Fig. B-3.   60-Deg Inclination Grid Pattern

underlying coordinate grid has been rotated about 120 deg counterclockwise and the triangle/circle pattern has been left unchanged. Thus, in the new coordinate system (viewed by rotating the page 120 deg clockwise), we have constructed a small circle with a 20-deg angular radius centered at about 21 deg below the equator. By rotating the underlying grid an appropriate amount, we could center the small circle at any desired latitude.

Because of the symmetry of the underlying sphere, we may not only rotate the grid pattern, but also interchange the underlying grids among the four shown in Figs. B-1 through B-4. For example, Fig. B-5(c) shows the circle/triangle pattern with the underlying grid changed to the equatorial view and rotated somewhat counterclockwise. (Again, the grid may be rotated to any convenient angle.) The meridian lines on the grid pattern, along with the imaginary meridians between those that are shown, are great circles. Therefore, in Fig. B-5(c), the dashed line is a great circle passing through one vertex of the triangle and tangent to the small circle. Finally, in Fig. B-5(d), we have left the constructed figures unchanged and returned the underlying grid pattern to its original orientation from Fig. B-5(a).

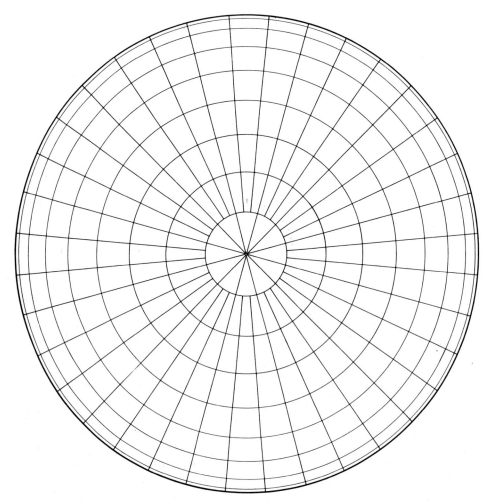

Fig. B-4. Polar Projection Grid Pattern

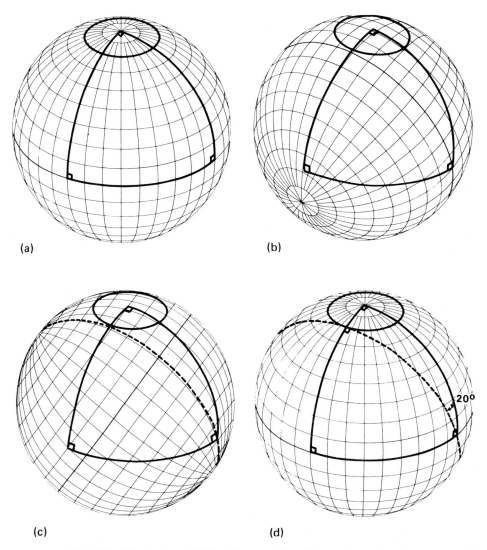

(a)

(b)

(c)

(d)

Fig. B-5. Construction of Global Geometry Plots. See text for explanation.

Thus, by using the grid pattern of Fig. B-1, we have constructed a great circle tangent to a small circle of 20-deg radius about the pole or, equivalently, at an inclination of 70 deg to the equator.

  In practice, this construction is performed by first drawing the original figure of B-5(a), and then placing it on top of the equatorial view on a light table so that both grids can be seen. After rotating the grid patterns relative to each other until the desired orientation for the dashed curve is obtained, we can trace the dashed curve directly on the grid of Fig. B-5(a). This general procedure for drawing great and small circles by superposing grids on a light table (or window) and rotating them until the desired orientation is obtained has been used to construct nearly all the globe figures in this book. Note that *whenever figures are constructed using superposed grids, the centers of the two grids must be on top of each other or, equivalently, the perimeters of the two grids must be superposed.* This principle of grid superposition can be applied to the construction of various figures, as discussed below.

Constructing Great Circles Through Two Points. Figure B-1 is the basic figure for constructing all great circles on the celestial sphere, because all possible great circles are represented by the meridian lines on the figure and the imaginary meridians between the ones drawn. To construct a great circle between any two points of any of the globes, place the globe with the two points on top of a copy of Fig. B-1. Keeping the perimeters of the two figures superposed, rotate the globes relative to each other until the two points lie over the same meridian on the underneath grid. This meridian is then the great circle defined by the two points. Note that this great circle is most precisely defined when the two points are nearly 90 deg apart and is poorly defined if the two points are nearly 0 deg or 180 deg apart.

Measuring Arc Lengths. The grid pattern in Fig. B-1 can also be used to measure the arc-length separation between any two points on the sphere. The parallels of latitude (i.e., the horizontal straight lines in Fig. B-1) are separated by 10 deg of arc along each meridian. Therefore, to determine the arc length between two points, superpose the globe with the two points over a copy of Fig. B-1 and rotate it until the meridian forming the great circle between the two points is found. The arc length is then determined by using the parallels of latitude along the meridian as a scale. For example, consider the dashed great circle in Fig. B-5(c). Because the triangle is a right equilateral triangle, the distance between any vertex and the opposite side must be 90 deg. This may be confirmed by counting the parallels of latitude along the dashed great circle. Also, the diameter of the small circle in Figs. B-5(b) and (c) may be measured along the *meridian* passing through the center of the circle. In both subfigures, the measured angular diameter is 40 deg, as required. Note that arc length must be measured along a great circle; it *cannot* be measured along parallels of latitude or other small circles.

Constructing Great Circles From General Criteria. In general, any great circle is constructed by first finding two points on it and then drawing the great circle between these points. For example, to draw a great circle at a given inclination to the equator, first pick the intercept point on the equator. Measure along the equator to the right or the left (depending on the slope desired) 90 deg and then up from the equator (along a meridian) an angle equal to the inclination. This point and the intercept point on the equator define the great circle. This method could have been used to construct the dashed great circle in Fig. B-5(c) directly without considering the tangent to the small circle.

Figure B-6 illustrates the procedure for constructing a great circle through a given point, A, perpendicular to a given great circle, AA'. Locate the point A' along the given great circle 90 deg from A by the method described above. Measure along the meridian through A' 90 deg in either direction to the point B. The great circle through A and B is perpendicular to AA'.

Constructing Small Circles. This construction has already been demonstrated by the example of Fig. B-5. The method described there may be used to construct

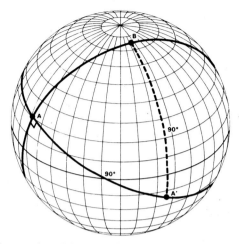

Fig. B-6. Construction of Great Circle *AB* Perpendicular to Great Circle *AA'*

small circles whose center is on the perimeter, 30 deg from the perimeter, 60 deg from the perimeter, or at the center. Small circles centered on the perimeter are straight lines on the plot and small circles centered in the middle of the plot are circles. The radius of the small circles constructed by this method is the colatitude (distance from the pole) of the parallel of latitude chosen on the underlying coordinate grid. For most purposes, one of these four sets of small circles is sufficient. They have been used for all of the constructions in this book. If it is necessary to construct a small circle at an intermediate arc distance from the perimeter, first construct a small circle of the desired radius at the desired latitude and as near the desired longitude as possible. Transform each point on this circle along a parallel of latitude a distance in longitude equal to the difference in longitude between the desired center and the constructed center. This procedure is illustrated in Fig. B-7.

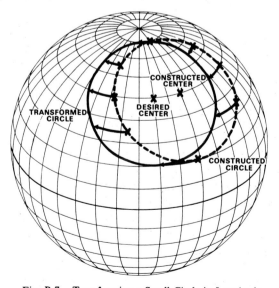

Fig. B-7. Transforming a Small Circle in Longitude

APPENDIX C

MATRIX AND VECTOR ALGEBRA

F. L. Markley

C.1 Definitions
C.2 Matrix Algebra
C.3 Trace, Determinant, and Rank
C.4 Matrix Inverses and Solutions to Simultaneous Linear Equations
C.5 Special Types of Square Matrices, Matrix Transformations
C.6 Eigenvectors and Eigenvalues
C.7 Functions of Matrices
C.8 Vector Calculus
C.9 Vectors in Three Dimensions

C.1 Definitions

A *matrix* is a rectangular array of scalar entries known as the *elements* of the matrix. In this book, the scalars are assumed to be real or complex numbers. If all the elements of a matrix are real numbers, the matrix is a *real matrix*. The matrix

$$A \equiv \begin{bmatrix} A_{11} & A_{12} & \cdots & A_{1n} \\ A_{21} & A_{22} & \cdots & A_{2n} \\ \vdots & \vdots & & \vdots \\ A_{m1} & A_{m2} & \cdots & A_{mn} \end{bmatrix} \equiv [A_{ij}] \tag{C-1}$$

has m rows and n columns, and is referred to as an $m \times n$ *matrix* or as a *matrix of order* $m \times n$. The equation $A = [A_{ij}]$ should be read as, "A is the matrix whose elements are A_{ij}." The first subscript labels the rows of the matrix and the second labels the columns.

Two matrices are equal if and only if they are of the same order and all of the corresponding elements are equal; i.e.,

$$A = B \text{ if and only if } A_{ij} = B_{ij}; i = 1, \ldots, m; j = 1, \ldots, n \tag{C-2}$$

An $n \times n$ matrix is called a *square matrix* and is usually referred to as being of order n rather than $n \times n$.

The *transpose* of a matrix is the matrix resulting from interchanging rows and columns. The transpose of A is denoted by A^T, and its elements are given by

$$A^T \equiv [(A^T)_{ij}] \equiv [A_{ji}] \tag{C-3}$$

As an example, the transpose of the matrix in Eq. (C-1) is

$$A^T = \begin{bmatrix} A_{11} & A_{21} & \cdots & A_{m1} \\ A_{12} & A_{22} & \cdots & A_{m2} \\ \vdots & \vdots & & \vdots \\ A_{1n} & A_{2n} & \cdots & A_{mn} \end{bmatrix}$$

It is clear that the transpose of an $m \times n$ matrix is an $n \times m$ matrix, and that the transpose of a square matrix is square. The transpose of the transpose of a matrix is equal to the original matrix:

$$\left(A^{\mathrm{T}}\right)^{\mathrm{T}} = A \qquad\qquad (\text{C-4})$$

The *adjoint* of a matrix, denoted by $A^\dagger$, is the matrix whose elements are the complex conjugates of the elements of the transpose of the given matrix,* i.e.,

$$A^\dagger \equiv \left[\left(A^\dagger\right)_{ij}\right] \equiv \left[A_{ji}^*\right] \qquad\qquad (\text{C-5})$$

The adjoint of the adjoint of a matrix is equal to the original matrix:

$$\left(A^\dagger\right)^\dagger = A \qquad\qquad (\text{C-6})$$

The adjoint and the transpose of a real matrix are identical.

The *main diagonal* of a square matrix is the set of elements with row and column indices equal. A *diagonal matrix* is a square matrix with nonzero elements only on the main diagonal, e.g.,

$$D = \begin{bmatrix} D_{11} & 0 & \cdots & 0 \\ 0 & D_{22} & \cdots & 0 \\ \vdots & \vdots & & \vdots \\ 0 & 0 & \cdots & D_{nn} \end{bmatrix} \qquad\qquad (\text{C-7})$$

The *identity matrix* of a given order is the diagonal matrix with all the elements on the main diagonal equal to unity. It is denoted by $\mathbf{1}$, or by $\mathbf{1}_n$ to indicate the order explicitly.

A matrix with only one column is a *column matrix*. An $n \times 1$ column matrix can be identified with a vector in n-dimensional space, and we shall indicate such matrices by boldface letters, as used for vectors.† A matrix with only one row is a *row matrix*; its transpose is a column matrix, so we denote it as the transpose of a vector. The elements of a row or column matrix will be written with only one subscript; for example,

$$\mathbf{B} \equiv \begin{bmatrix} B_1 \\ B_2 \\ \vdots \\ B_n \end{bmatrix}, \qquad \mathbf{C}^{\mathrm{T}} \equiv \left[C_1, \, C_2, \ldots, C_m\right] \qquad\qquad (\text{C-8})$$

A set of m $n \times 1$ vectors $\mathbf{B}^{(i)}$, $i = 1, 2, \ldots, m$, is *linearly independent* if and only if the only coefficients a_i, $i = 1, 2, \ldots, m$, satisfying the equation

$$\sum_{i=1}^{m} a_i \mathbf{B}^{(i)} = a_1 \mathbf{B}^{(1)} + a_2 \mathbf{B}^{(2)} + \cdots + a_m \mathbf{B}^{(m)} = \mathbf{0} \qquad\qquad (\text{C-9})$$

---

*The word adjoint is sometimes used for a different matrix in the literature.

†Strictly speaking, a vector is an abstract mathematical object, and the column matrix is a concrete realization of it, the matrix elements being the components of the vector in some coordinate system.

are $a_i = 0$, $i = 1, 2, \ldots, m$. There can never be more than n linearly independent $n \times 1$ vectors.

C.2 Matrix Algebra

Multiplication of a matrix by a scalar is accomplished by multiplying each element of the matrix by the scalar, i.e.,

$$sA \equiv \left[sA_{ij} \right] \tag{C-10}$$

Addition of two matrices is possible only if the matrices have the same order. The elements of the matrix sum are the sums of the corresponding elements of the matrix addends, i.e.,

$$A + B \equiv \left[A_{ij} + B_{ij} \right] \tag{C-11}$$

Matrix subtraction follows from the above two rules by

$$A - B \equiv A + (-1)B = \left[A_{ij} - B_{ij} \right] \tag{C-12}$$

Multiplication of two matrices is possible only if the number of columns of the matrix on the left side of the product is equal to the number of rows of the matrix on the right. If A is of order $l \times m$ and B is $m \times n$, the product AB is the $l \times n$ matrix given by

$$AB \equiv \left[(AB)_{ij} \right] \equiv \left[\sum_{k=1}^{m} A_{ik} B_{kj} \right] \tag{C-13}$$

Matrix multiplication is *associative*

$$A(BC) = (AB)C \tag{C-14}$$

and *distributive over addition*

$$A(B + C) = AB + AC \tag{C-15}$$

but is **not** *commutative*, in general,

$$AB \neq BA \tag{C-16}$$

In fact, the products AB and BA are both defined and have the same order only if A and B are square matrices, and even in this case the products are not necessarily equal. For the square matrices $A = \left[\begin{smallmatrix} 1 & 2 \\ 3 & 4 \end{smallmatrix}\right]$ and $B = \left[\begin{smallmatrix} 5 & 6 \\ 7 & 8 \end{smallmatrix}\right]$, for example, we have

$$AB = \begin{bmatrix} 19 & 22 \\ 43 & 50 \end{bmatrix} \neq BA = \begin{bmatrix} 23 & 34 \\ 31 & 46 \end{bmatrix}$$

If $AB = BA$, for two square matrices, A and B, we say that A and B *commute*. One interesting case is diagonal matrices, which always commute.

The adjoint (or transpose) of the product of two matrices is equal to the product of the adjoints (or transposes) of the two matrices taken in the opposite order:

$$(AB)^\dagger = B^\dagger A^\dagger \tag{C-17}$$

$$(AB)^T = B^T A^T \tag{C-18}$$

This result easily generalizes to products of more than two matrices.

Multiplying any matrix by the identity matrix of the appropriate order, on the left or the right, yields a product equal to the original matrix. Thus, if B is of order $m \times n$,

$$1_m B = B 1_n = B \tag{C-19}$$

The product of an $n \times m$ matrix and an m-dimensional vector (an $m \times 1$ matrix) is an n-dimensional vector; thus,

$$\mathbf{Y} \equiv A\mathbf{X} \equiv \left[\sum_{j=1}^{m} A_{ij} X_j \right] \tag{C-20}$$

A similar result holds if a row matrix is multiplied on the right by a matrix,

$$\mathbf{Y}^T = \mathbf{X}^T A^T = \left[\sum_{j=1}^{m} A_{ji} X_j \right] \tag{C-21}$$

An important special case of the above is the multiplication of a $1 \times n$ row matrix (on the left) by an $n \times 1$ column matrix (on the right) which yields a scalar,

$$s \equiv \mathbf{Y}^T \mathbf{X} \equiv \sum_{j=1}^{n} X_j Y_j \tag{C-22}$$

For real vectors, this scalar is the *inner product*, or *dot product*, or *scalar product* of the vectors $\mathbf{X}$ and $\mathbf{Y}$. For vectors with complex components, it is more convenient to define the inner product by using the adjoint of the left-hand vector rather than the transpose. Thus, in general,

$$\mathbf{Y} \cdot \mathbf{X} \equiv \mathbf{Y}^\dagger \mathbf{X} = \sum_{i=1}^{n} Y_i^* X_i \tag{C-23}$$

Note that, in general,

$$\mathbf{Y} \cdot \mathbf{X} = (\mathbf{X} \cdot \mathbf{Y})^* \tag{C-24}$$

This definition reduces to the usual definition for real vectors, for which the inner product is independent of the order in which the vectors appear. Two vectors are *orthogonal* if their inner product is zero. The inner product of a vector with itself

$$\mathbf{X} \cdot \mathbf{X} = \sum_{i=1}^{n} X_i^* X_i = \sum_{i=1}^{n} |X_i|^2 \tag{C-25}$$

is never negative and is zero if and only if all the elements of $\mathbf{X}$ are zero, i.e., if $\mathbf{X} = \mathbf{0}$. This product will be denoted by $\mathbf{X}^2$ and its positive square root by $|\mathbf{X}|$ or by X, if no confusion results. The scalar $|\mathbf{X}|$ is called the *norm* or *magnitude* of the vector, $\mathbf{X}$, and can be thought of as the length of the vector. Thus, with our definition of the inner product,

$$|\mathbf{X}| = 0 \text{ if and only if } \mathbf{X} = \mathbf{0} \tag{C-26}$$

which would not be true if we defined the inner product using the transpose rather than the adjoint, because the square of a complex number generally is not positive.

If we multiply an $n \times 1$ row matrix (on the left) by a $1 \times m$ matrix (on the right), we obtain an $n \times m$ matrix. This leads to the definition of the *outer product* of two vectors

$$\mathbf{XY}^\dagger \equiv \left[(\mathbf{XY}^\dagger)_{ij} \right] \equiv \left[X_i Y_j^* \right] \tag{C-27}$$

If the vectors are real, the adjoint of $\mathbf{Y}$ is the transpose of $\mathbf{Y}$, and the ijth element of the outer product is $X_i Y_j$.

Matrix division can be defined in terms of matrix inverses, which are discussed in Section C.4.

C.3 Trace, Determinant, and Rank

Two useful scalar quantities, the *trace* and the *determinant*, can be defined for any *square* matrix. The *rank* of a matrix is defined for any matrix.

The *trace* of an $n \times n$ matrix is the sum of the diagonal elements of the matrix

$$\operatorname{tr} A \equiv \sum_{i=1}^{n} A_{ii} \tag{C-28}$$

The trace of a product of square matrices is unchanged by a cyclic permutation of the order of the product

$$\operatorname{tr}(ABC) = \sum_{i=1}^{n} \sum_{j=1}^{n} \sum_{k=1}^{n} A_{ij} B_{jk} C_{ki} = \operatorname{tr}(CAB) \tag{C-29}$$

However, $\operatorname{tr}(ABC) \neq \operatorname{tr}(ACB)$, in general.

The *determinant* of an $n \times n$ matrix is the complex number defined by

$$\det A \equiv |A_{ij}| \equiv \sum (-1)^p A_{1p_1} A_{2p_2} \cdots A_{np_n} \tag{C-30}$$

where the set of numbers $\{p_1, p_2, \ldots, p_n\}$ is a *permutation*, or rearrangement, of $\{1, 2, \ldots, n\}$. Any permutation can be achieved by a sequence of pairwise interchanges. A permutation is uniquely *even* or *odd* if the number of interchanges required is even or odd, respectively. The exponent p in Eq. (C-30) is zero for even permutations and unity for odd ones. The sum is over all the $n!$ distinct permutations of $\{1, 2, \ldots, n\}$. It is not difficult to show that Eq. (C-30) is equivalent to

$$\det A = \sum_{j=1}^{n} (-1)^{i+j} A_{ij} M_{ij} \tag{C-31}$$

for any *fixed* $i = 1, 2, \ldots, n$, where M_{ij} is the *minor* of A_{ij}, defined as the determinant of the $(n-1) \times (n-1)$ matrix formed by omitting the ith row and jth column from A. This form provides a convenient method for evaluating determinants by successive reduction to lower orders. For example,

$$\begin{vmatrix} 1 & 2 & 3 \\ 4 & 5 & 6 \\ 7 & 8 & 9 \end{vmatrix} = 1 \times \begin{vmatrix} 5 & 6 \\ 8 & 9 \end{vmatrix} - 2 \times \begin{vmatrix} 4 & 6 \\ 7 & 9 \end{vmatrix} + 3 \times \begin{vmatrix} 4 & 5 \\ 7 & 8 \end{vmatrix}$$

$$= (5 \times 9 - 8 \times 6) - 2(4 \times 9 - 7 \times 6) + 3(4 \times 8 - 7 \times 5) = 0 \tag{C-32}$$

The determinant of the product of two square matrices is equal to the product of the determinants

$$\det(AB) = (\det A)(\det B) \tag{C-33}$$

The determinant of a scalar multiplied by an $n \times n$ matrix is given by

$$\det(sA) = s^n \det A \tag{C-34}$$

The determinants of a matrix and of its transpose are equal:

$$\det A^T = \det A \tag{C-35}$$

Thus, the determinant of the adjoint is

$$\det A^\dagger = (\det A)^* \tag{C-36}$$

The determinant of a matrix with all zeros on one side of the main diagonal is equal to the product of the diagonal elements.

The *rank* of a matrix is the order of the largest square array in that matrix, formed by deleting rows and columns, that has a nonvanishing determinant. Clearly, the rank of an $m \times n$ matrix cannot exceed the smaller of m and n. The matrices A, A^T, $A^\dagger$, $A^\dagger A$, and $AA^\dagger$ all have the same rank.

C.4 Matrix Inverses and Solutions to Simultaneous Linear Equations

Let A be an $m \times n$ matrix of rank k. An $n \times m$ matrix B is a *left inverse* of A if $BA = 1_n$. An $n \times m$ matrix C is a *right inverse* of A if $AC = 1_m$. There are four possible cases: k is less than both m and n, $k = m = n$, $k = m < n$, and $k = n < m$. If k is less than both m and n, then no left or right inverse of A exists.[*] If $k = m = n$, then A is *nonsingular* and has a unique *inverse*, A^{-1}, which is both a left and a right inverse:

$$A^{-1}A = AA^{-1} = 1 \quad (k = m = n) \tag{C-37}$$

A nonsingular matrix is a square matrix with nonzero determinant; all other matrices are *singular*. If $k = m < n$, then A has no left inverse but an infinity of right inverses, one of which is given by

$$A^R = A^\dagger (AA^\dagger)^{-1} \quad (k = m < n) \tag{C-38}$$

If $k = n < m$, then A has no right inverse but an infinity of left inverses, one of which is

$$A^L = (A^\dagger A)^{-1} A^\dagger \quad (k = n < m) \tag{C-39}$$

A^L or A^R is called the *generalized inverse* or *pseudoinverse* of A.

Consider the set of m simultaneous linear equations in n unknowns; $X_1, X_2, \ldots, X_n$;

---

[*] It is possible to define a *pseudoinverse* for a general matrix, which in this case is neither a left nor a right inverse. In the other three cases, the pseudoinverse is identical with A^{-1}, A^R, and A^L, respectively. The results on solutions of simultaneous linear equations can be generalized with this definition [Wiberg, 1971].

$$AX = Y \qquad \text{(C-40)}$$

If $k = m = n$, then $X = A^{-1}Y$ is the unique solution to the set of equations. It follows that a nonzero solution to $AX = 0$ is possible only if A is singular, i.e.,

$$AX = 0 \text{ for } X \neq 0, \text{ only if } \det A = 0 \qquad \text{(C-41)}$$

If $k = m < n$, there are more unknowns than equations, so there are an infinite number of solutions. The solution with the smallest norm, $|X|$, is

$$X = A^R Y \qquad \text{(C-42)}$$

If $k = n < m$, there are more equations than unknowns; therefore, no solution exists, in general. However, the vector X that comes closest to a solution, in the sense of minimizing $|AX - Y|$, is

$$X = A^L Y \qquad \text{(C-43)}$$

Note that although $AA^L \neq 1_m$, it is possible that $AA^L Y = Y$ for the particular Y in Eq. (C-40). In this case, Eq. (C-40) has a unique solution given by Eq. (C-43).

It is easy to see that if A is nonsingular, then A^{-1} is nonsingular also and

$$\left(A^{-1}\right)^{-1} = A \qquad \text{(C-44)}$$

Likewise, if A is nonsingular, then A^T and $A^\dagger$ are nonsingular and their inverses are given by

$$\left(A^T\right)^{-1} = \left(A^{-1}\right)^T \qquad \text{(C-45)}$$

$$\left(A^\dagger\right)^{-1} = \left(A^{-1}\right)^\dagger \qquad \text{(C-46)}$$

If two matrices, A and B, are nonsingular, their product is nonsingular also; and the inverse of the product is the product of the inverses, taken in the opposite order:

$$(AB)^{-1} = B^{-1}A^{-1} \qquad \text{(C-47)}$$

This result easily generalizes to products of more than two matrices.

Various algorithms exist for calculating matrix inverses; several are described by Carnahan, et al., [1969] and by Forsythe and Moler [1967]. An example of a subroutine for this purpose is INVERT, described in Section 20.3.

C.5 Special Types of Square Matrices, Matrix Transformations

A *symmetric* matrix is a square matrix that is equal to its transpose:

$$A^T = A, \qquad A_{ij} = A_{ji} \qquad \text{(C-48)}$$

A *skew-symmetric* or *antisymmetric* matrix is equal to the negative of its transpose:

$$A^T = -A, \qquad A_{ij} = -A_{ji} \qquad \text{(C-49)}$$

Clearly, a skew-symmetric matrix must have zeros on its main diagonal. An example of a skew symmetric matrix is Ω in Section 16.1. A *Hermitian* matrix is equal to its adjoint:

$$A^\dagger = A, \qquad A_{ij} = A_{ji}^* \tag{C-50}$$

A real symmetric matrix is a special case of a Hermitian matrix. An *orthogonal* matrix is a matrix whose transpose is equal to its inverse:

$$A^T = A^{-1}, \qquad AA^T = A^T A = 1 \tag{C-51}$$

A *unitary* matrix is a matrix whose adjoint is equal to its inverse:

$$A^\dagger = A^{-1}, \qquad AA^\dagger = A^\dagger A = 1 \tag{C-52}$$

A *real orthogonal* matrix is a special case of a unitary matrix. The product of two unitary (or orthogonal) matrices is unitary (or orthogonal). This result generalizes to products of more than two matrices. A similar result generally does **not** hold for Hermitian or symmetric matrices. A *normal* matrix is a matrix that commutes with its adjoint

$$A^\dagger A = A A^\dagger$$

Thus, both Hermitian matrices and unitary matrices are special cases of normal matrices.

By the rules for determinants of products and adjoints, it is easy to see that if A is unitary

$$|\det A|^2 = 1 \tag{C-53}$$

Thus, $\det A$ is a complex number with absolute value unity. Similarly, if A is orthogonal,

$$(\det A)^2 = 1, \qquad \det A = \pm 1 \tag{C-54}$$

An orthogonal matrix with positive determinant is a *proper* orthogonal matrix; an orthogonal matrix is *improper* if its determinant is negative.

Let $\mathbf{X}$ be an n-dimensional vector and let A be an $n \times n$ matrix. Then $A\mathbf{X}$ is another n-dimensional vector and can be thought of as the transformation of $\mathbf{X}$ by A. If $\mathbf{X}$ and $\mathbf{Y}$ are two vectors, the inner product of $A\mathbf{X}$ and $A\mathbf{Y}$ is

$$(A\mathbf{X}) \cdot (A\mathbf{Y}) = (A\mathbf{X})^\dagger (A\mathbf{Y}) = \mathbf{X}^\dagger A^\dagger A \mathbf{Y}$$

If A is unitary,

$$(A\mathbf{X}) \cdot (A\mathbf{Y}) = \mathbf{X} \cdot \mathbf{Y} \tag{C-55}$$

The dot product is unchanged if both vectors are transformed by the same unitary matrix. This result with $\mathbf{X} = \mathbf{Y}$ shows that the norm of a vector is unchanged, too, so the unitary matrix can be thought of as performing a *rotation* of the vector in n-dimensional space, thereby preserving its length. If the vectors are real, the rotations correspond to *proper real orthogonal matrices*.

The transformations of a matrix are defined analogously to the transformations of a vector, but they involve mutliplying the matrix on both the left and right sides, rather than only on the left side. Several kinds of transformations are defined. If B is nonsingular, then

$$A_S = B^{-1} A B \tag{C-56}$$

is a *similarity transformation* on A. We say that A_S is *similar* to A. A special case occurs if B is unitary. In this case we have a *unitary transformation* on A,

$$A_U = B^\dagger A B \tag{C-57}$$

A second special case occurs if B is orthogonal, in which case

$$A_O = B^T A B \tag{C-58}$$

defines an *orthogonal transformation* on A.

It follows directly from the invariance of the trace to cyclic permutations of the order of matrix products, Eq. (C-28), that

$$\operatorname{tr} A_S = \operatorname{tr} A_U = \operatorname{tr} A_O = \operatorname{tr} A \tag{C-59}$$

Also, by the rules on determinants,

$$\det A_S = \det A_U = \det A_O = \det A \tag{C-60}$$

It is easy to see that

$$A_U^\dagger = B^\dagger A^\dagger B \tag{C-61}$$

and

$$A_O^T = B^T A^T B \tag{C-62}$$

Thus, A_U is Hermitian (or unitary) if A is Hermitian (or unitary), and A_O is symmetric (or orthogonal) if A is symmetric (or orthogonal).

C.6 Eigenvectors and Eigenvalues

If A is an $n \times n$ matrix and if

$$A\mathbf{X} = \lambda \mathbf{X} \tag{C-63}$$

for some nonzero vector $\mathbf{X}$ and scalar λ, we say that $\mathbf{X}$ is an *eigenvector* of A and that λ is the corresponding *eigenvalue*. We can rewrite Eq. (C-63) as

$$(A - \lambda \mathbf{1})\mathbf{X} = \mathbf{0} \tag{C-64}$$

so we see from Eq. (C-41) that λ is an eigenvalue of A if and only if

$$\det(A - \lambda \mathbf{1}) = 0 \tag{C-65}$$

This is called the *characteristic equation* for A. It is an nth-order equation for λ and has n roots, counting multiple roots according to their multiplicity.

Because the equation $A\mathbf{X} = \lambda \mathbf{X}$ is unchanged by multiplying both sides by a scalar s, it is clear that $s\mathbf{X}$ is an eigenvector of A if $\mathbf{X}$ is. This freedom can be used to *normalize* the eigenvectors, i.e., to choose the constant so that $\mathbf{X} \cdot \mathbf{X} = 1$. From n eigenvectors of A, $\mathbf{X}^{(i)}$, $i = 1, 2, \ldots, n$, we can construct the matrix

$$P \equiv \begin{bmatrix} X_1^{(1)} & X_1^{(2)} & X_1^{(3)} & \cdots & X_1^{(n)} \\ X_2^{(1)} & X_2^{(2)} & X_2^{(3)} & \cdots & X_2^{(n)} \\ \vdots & \vdots & \vdots & & \vdots \\ X_n^{(1)} & X_n^{(2)} & X_n^{(3)} & \cdots & X_n^{(n)} \end{bmatrix} \tag{C-66}$$

Matrix multiplication and the eigenvalue equation (Eq. (C-63)) give

$$AP = \begin{bmatrix} \lambda_1 X_1^{(1)} & \lambda_2 X_1^{(2)} & \cdots & \lambda_n X_1^{(n)} \\ \lambda_1 X_2^{(1)} & \lambda_2 X_2^{(2)} & \cdots & \lambda_n X_2^{(n)} \\ \vdots & \vdots & & \vdots \\ \lambda_1 X_n^{(1)} & \lambda_2 X_n^{(2)} & \cdots & \lambda_n X_n^{(n)} \end{bmatrix} = P\Lambda \qquad (C\text{-}67)$$

where Λ is the diagonal matrix

$$\Lambda = \begin{bmatrix} \lambda_1 & 0 & \cdots & 0 \\ 0 & \lambda_2 & \cdots & 0 \\ \vdots & \vdots & & \vdots \\ 0 & 0 & \cdots & \lambda_n \end{bmatrix} \qquad (C\text{-}68)$$

The matrix P is nonsingular if and only if the n eigenvectors are linearly independent. In this case,

$$\Lambda = P^{-1}AP \qquad (C\text{-}69)$$

and we say that A is *diagonalizable* by the similarity transformation $P^{-1}AP$. If A is a normal matrix, we can choose n eigenvectors that are *orthonormal*, or simultaneously orthogonal and normalized:

$$\mathbf{X}^{(i)} \cdot \mathbf{X}^{(j)} = \delta_i^j \equiv \begin{cases} 0 & i \neq j \\ 1 & i = j \end{cases} \qquad (C\text{-}70)$$

When the eigenvectors are orthonormal, P is a unitary matrix and A is diagonalizable by the unitary transformation $\Lambda = P^\dagger AP$. Any square matrix can be brought into *Jordan canonical form* [Hoffman and Kunze, 1961] by a similarity transformation

$$J = P^{-1}AP \qquad (C\text{-}71)$$

where the matrix J has the eigenvalues of A on the main diagonal and all zeros below the main diagonal. It follows from Eqs. (C-71), (C-59), and (C-60) that the trace of A is equal to the sum of its eigenvalues, and the determinant of A is equal to the product of its eigenvalues; i.e.,

$$\text{tr}\,A = \sum_{i=1}^{n} \lambda_i \qquad (C\text{-}72)$$

$$\det A = \lambda_1 \lambda_2 \cdots \lambda_n \qquad (C\text{-}73)$$

Many algorithms exist for finding eigenvalues and eigenvectors of matrices, several of which are discussed by Carnahan, et al., [1969] and by Stewart [1973]. Using Eq. (C-61), we can see that the eigenvalues of Hermitian matrices are real numbers and the eigenvalues of unitary matrices are complex numbers with absolute value unity. Because the characteristic equation of a real matrix is a polynomial equation with real coefficients, the eigenvalues of a real matrix must either be real or must occur in complex conjugate pairs.

The case of a real orthogonal matrix deserves special attention. Because such a matrix is both real and unitary, the only possible eigenvalues are $+1$, -1, and complex conjugate pairs with absolute value unity. It follows that the determinant of a real orthogonal matrix is $(-1)^m$ where m is the multiplicity of the root $\lambda = -1$ of the characteristic equation. A proper real orthogonal matrix must have an even number of roots at $\lambda = -1$, and thus an even number for all $\lambda \neq 1$, because complex roots occur in conjugate pairs. Thus, an $n \times n$ proper real orthogonal matrix *with n odd* must have at least one eigenvector with eigenvalue $+1$. This is the basis of Euler's Theorem, discussed in Section 12.1.

It is also of interest to establish that the eigenvectors of a real symmetric matrix can be chosen to be real. The complex conjugate of the eigenvector equation, Eq. (C-63), is $A\mathbf{X}^* = \lambda\mathbf{X}^*$, because both A and λ are real. Thus, $\mathbf{X}^*$ is an eigenvector of A with the same eigenvalue as $\mathbf{X}$. Now, either $\mathbf{X} = \mathbf{X}^*$, in which case the desired result is obtained, or $\mathbf{X} \neq \mathbf{X}^*$. In the latter case, we can replace $\mathbf{X}$ and $\mathbf{X}^*$ by the linear combinations $\mathbf{X} + \mathbf{X}^*$ and $i(\mathbf{X} - \mathbf{X}^*)$, which are real eigenvectors corresponding to the eigenvalue λ. Thus, we can always find a real orthogonal matrix P to diagonalize a real symmetric matrix A by Eq. (C-69).

C.7 Functions of Matrices

Let $f(x)$ be any function of a variable x, for example, $\sin x$ or $\exp x$. We want to give a meaning to $f(M)$, where M is a square matrix. If $f(x)$ has a power series expansion about $x = 0$,

$$f(x) = \sum_{n=0}^{\infty} a_n x^n \tag{C-74}$$

then we can formally (i.e., ignoring questions of convergence) define $f(M)$ by

$$f(M) \equiv \sum_{n=0}^{\infty} a_n M^n \tag{C-75}$$

with the same coefficients a_n. It is clear that $f(M)$ is a square matrix of the same order as M. If M is a diagonalizable matrix, then by Eq. (C-69),

$$M = P\Lambda P^{-1} \tag{C-76}$$

where P is the matrix of eigenvectors defined by Eq. (C-66), and Λ is the diagonal matrix of eigenvalues. Then,

$$M^n = (P\Lambda P^{-1})^n = P\Lambda^n P^{-1} \tag{C-77}$$

and

$$f(M) = P\left(\sum_{n=0}^{\infty} a_n \Lambda^n\right) P^{-1} = P \begin{bmatrix} f(\lambda_1) & 0 & \cdots & 0 \\ 0 & f(\lambda_2) & \cdots & 0 \\ \vdots & \vdots & & \vdots \\ 0 & 0 & \cdots & f(\lambda_n) \end{bmatrix} P^{-1} \tag{C-78}$$

If M is a diagonalizable matrix, Eq. (C-78) gives an alternative definition of $f(M)$ that is valid when $f(x)$ does not have a power series expansion, and agrees with Eq. (C-75) when a power series expansion exists.

As an example, consider $\exp(\frac{1}{2}\Omega t)$, where Ω is the 4×4 matrix introduced in Section 16.1,

$$\Omega = \begin{bmatrix} 0 & \omega_3 & -\omega_2 & \omega_1 \\ -\omega_3 & 0 & \omega_1 & \omega_2 \\ \omega_2 & -\omega_1 & 0 & \omega_3 \\ -\omega_1 & -\omega_2 & -\omega_3 & 0 \end{bmatrix}$$

Matrix multiplication shows that $\Omega^2 = -(\omega_1^2 + \omega_2^2 + \omega_3^2)\mathbf{1} \equiv -\omega^2\mathbf{1}$, so it follows that

$$\Omega^{2k} = (-1)^k\omega^{2k}\mathbf{1}$$

$$\Omega^{2k+1} = (-1)^k\omega^{2k}\Omega$$

for all nonnegative k. Now,

$$\exp\left(\frac{1}{2}\Omega t\right) = \sum_{n=0}^{\infty} \frac{\left(\frac{1}{2}\Omega t\right)^n}{n!} = \sum_{k=0}^{\infty}\left[\frac{\left(\frac{1}{2}\Omega t\right)^{2k}}{(2k)!} + \frac{\left(\frac{1}{2}\Omega t\right)^{2k+1}}{(2k+1)!}\right]$$

$$= \mathbf{1}\sum_{k=0}^{\infty} \frac{(-1)^k\left(\frac{1}{2}\omega t\right)^{2k}}{(2k)!} + \Omega\omega^{-1}\sum_{k=0}^{\infty} \frac{(-1)^k\left(\frac{1}{2}\omega t\right)^{2k+1}}{(2k+1)!}$$

$$= \mathbf{1}\cos\left(\frac{1}{2}\omega t\right) + \Omega\omega^{-1}\sin\left(\frac{1}{2}\omega t\right)$$

$$= \begin{bmatrix} c & n_3 s & -n_2 s & n_1 s \\ -n_3 s & c & n_1 s & n_2 s \\ n_2 s & -n_1 s & c & n_3 s \\ -n_1 s & -n_2 s & -n_3 s & c \end{bmatrix} \tag{C-79}$$

where

$$c \equiv \cos\left(\frac{1}{2}\omega t\right)$$

$$s \equiv \sin\left(\frac{1}{2}\omega t\right)$$

$$n_i \equiv \omega_i/\omega \qquad i = 1,2,3$$

This example shows that the matrix elements of $f(M)$ are not the functions f of the matrix elements of M, in general.

C.8 Vector Calculus

Let ϕ be a scalar function of the n arguments $X_1, X_2, \ldots, X_n$. We consider the arguments to be the components of a column vector

$$\mathbf{X} \equiv [X_1, X_2, \ldots, X_n]^T$$

The n partial derivatives of ϕ with respect to the elements of $\mathbf{X}$ are the components of the *gradient* of ϕ, denoted by

$$\frac{\partial \phi}{\partial \mathbf{X}} \equiv \left[\frac{\partial \phi}{\partial X_1}, \frac{\partial \phi}{\partial X_2}, \ldots, \frac{\partial \phi}{\partial X_n} \right] \tag{C-80}$$

Note that $\partial \phi / \partial \mathbf{X}$ is considered a $1 \times n$ row matrix. If we eliminate the function ϕ from Eq. (C-80), we obtain the *gradient operator*

$$\frac{\partial}{\partial \mathbf{X}} \equiv \left[\frac{\partial}{\partial X_1}, \frac{\partial}{\partial X_2}, \ldots, \frac{\partial}{\partial X_n} \right] \tag{C-81}$$

The matrix product of the $1 \times n$ gradient operator with an $n \times 1$ vector $\mathbf{Y}$ yields a scalar, the *divergence* of $\mathbf{Y}$, which we denote by

$$\frac{\partial}{\partial \mathbf{X}} \cdot \mathbf{Y} \equiv \sum_{i=1}^{n} \frac{\partial Y_i}{\partial X_i} \tag{C-82}$$

The dot is used to emphasize the fact that the divergence is a scalar, although the usage is somewhat different from that in Eq. (C-23).

The mn partial derivatives of an m-dimensional vector $\mathbf{Y}$ with respect to $X_1, X_2, \ldots, X_n$ can be arranged in an $m \times n$ matrix denoted by

$$\frac{\partial \mathbf{Y}}{\partial \mathbf{X}} \equiv \left[\frac{\partial Y_i}{\partial X_j} \right] \tag{C-83}$$

This is like an outer product of $\mathbf{Y}$ and $\partial / \partial \mathbf{X}$; however, the analogy is not perfect because the gradient operator appears on the right in the matrix product sense and on the left in the operator sense.

C.9 Vectors in Three Dimensions

In this section, we only consider vectors with three real components. For three-component vectors, three products are defined: the dot product, the outer product, and the *cross product*. The cross product, or *vector product*, is a vector defined by

$$\mathbf{U} \times \mathbf{V} \equiv \begin{bmatrix} U_2 V_3 - U_3 V_2 \\ U_3 V_1 - U_1 V_3 \\ U_1 V_2 - U_2 V_1 \end{bmatrix} \tag{C-84}$$

The following identities are often useful:

$$\mathbf{U} \cdot \mathbf{V} \equiv U_1 V_1 + U_2 V_2 + U_3 V_3 = UV \cos \theta \tag{C-85a}$$

$$|\mathbf{U} \times \mathbf{V}| = UV \sin \theta \tag{C-85b}$$

where $\theta (0 \leqslant \theta \leqslant 180°)$ is the angle between $\mathbf{U}$ and $\mathbf{V}$. In addition,

$$\mathbf{U} \times \mathbf{V} = -\mathbf{V} \times \mathbf{U} \tag{C-86}$$

$$\mathbf{U} \cdot (\mathbf{U} \times \mathbf{V}) = 0 \tag{C-87}$$

$$\mathbf{U} \cdot (\mathbf{V} \times \mathbf{W}) = \mathbf{V} \cdot (\mathbf{W} \times \mathbf{U}) = \mathbf{W} \cdot (\mathbf{U} \times \mathbf{V}) = \begin{vmatrix} U_1 & U_2 & U_3 \\ V_1 & V_2 & V_3 \\ W_1 & W_2 & W_3 \end{vmatrix} \tag{C-88}$$

$$[\mathbf{U}\cdot(\mathbf{V}\times\mathbf{W})]^2=(\mathbf{U}\times\mathbf{V})\cdot[(\mathbf{V}\times\mathbf{W})\times(\mathbf{W}\times\mathbf{U})]$$

$$=\mathbf{U}^2\mathbf{V}^2\mathbf{W}^2-\mathbf{U}^2(\mathbf{V}\cdot\mathbf{W})^2-\mathbf{V}^2(\mathbf{U}\cdot\mathbf{W})^2-\mathbf{W}^2(\mathbf{U}\cdot\mathbf{V})^2$$

$$+2(\mathbf{U}\cdot\mathbf{V})(\mathbf{V}\cdot\mathbf{W})(\mathbf{W}\cdot\mathbf{U}) \tag{C-89}$$

$$\mathbf{U}\times(\mathbf{V}\times\mathbf{W})=\mathbf{V}(\mathbf{U}\cdot\mathbf{W})-\mathbf{W}(\mathbf{U}\cdot\mathbf{V}) \tag{C-90}$$

$$\mathbf{0}=\mathbf{U}\times(\mathbf{V}\times\mathbf{W})+\mathbf{V}\times(\mathbf{W}\times\mathbf{U})+\mathbf{W}\times(\mathbf{U}\times\mathbf{V}) \tag{C-91}$$

$$(\mathbf{U}\times\mathbf{V})\cdot(\mathbf{W}\times\mathbf{X})=(\mathbf{U}\cdot\mathbf{W})(\mathbf{V}\cdot\mathbf{X})-(\mathbf{U}\cdot\mathbf{X})(\mathbf{V}\cdot\mathbf{W}) \tag{C-92}$$

The following identity provides a means of writing the vector $\mathbf{W}$ in terms of $\mathbf{U}$, $\mathbf{V}$, and $\mathbf{U}\times\mathbf{V}$, if $\mathbf{U}\times\mathbf{V}\neq\mathbf{0}$:

$$[(\mathbf{U}\times\mathbf{V})\cdot(\mathbf{U}\times\mathbf{V})]\mathbf{W}=[(\mathbf{V}\times\mathbf{U})\cdot(\mathbf{V}\times\mathbf{W})]\mathbf{U}+[(\mathbf{U}\times\mathbf{V})\cdot(\mathbf{U}\times\mathbf{W})]\mathbf{V}$$

$$+[\mathbf{W}\cdot(\mathbf{U}\times\mathbf{V})]\mathbf{U}\times\mathbf{V} \tag{C-93}$$

If A is a real orthogonal matrix,

$$(A\mathbf{U})\times(A\mathbf{V})=\pm A(\mathbf{U}\times\mathbf{V}) \tag{C-94}$$

where the positive sign holds if A is proper, and the negative sign if A is improper.

The tangent of the *rotation angle* from $\mathbf{V}$ to $\mathbf{W}$ about $\mathbf{U}$ (the angle of the rotation in the positive sense about $\mathbf{U}$ that takes $\mathbf{V}\times\mathbf{U}$ into a vector parallel to $\mathbf{W}\times\mathbf{U}$) is

$$\tan\Theta=\frac{|\mathbf{U}|\mathbf{U}\cdot(\mathbf{V}\times\mathbf{W})}{\mathbf{U}^2(\mathbf{V}\cdot\mathbf{W})-(\mathbf{U}\cdot\mathbf{V})(\mathbf{U}\cdot\mathbf{W})} \tag{C-95}$$

The quadrant of Θ is given by the fact that the numerator is a positive constant multiplied by $\sin\Theta$, and the denominator is the same positive constant multiplied by $\cos\Theta$. If $\mathbf{U}$, $\mathbf{V}$, and $\mathbf{W}$ are unit vectors, Θ is the same as the rotation angle on the celestial sphere defined in Appendix A. Equation (C-95) is derived in Section 7.3. (See Eqs. (7-57).)

References

1. Bellman, R. E., *Introduction to Matrix Algebra*. New York: McGraw-Hill, Inc., 1960.
2. Carnahan, B., H. A. Luther and J. O. Wilkes, *Applied Numerical Methods*. New York: John Wiley & Sons, Inc., 1969.
3. Forsythe, George E., and C. Moler, *Computer Solution of Linear Algebraic Systems*. Englewood Cliffs, NJ: Prentice-Hall, Inc., 1967.
4. Halmos, P. R., ed., *Finite-Dimension Vector Spaces*, Second Edition. Princeton, NJ: D. Van Nostrand Company, Inc., 1958.
5. Hoffman, Kenneth and Ray Kunze, *Linear Algebra*. Englewood Cliffs, NJ: Prentice-Hall, Inc., 1961.
6. Noble, Ben, *Applied Linear Algebra*. Englewood Cliffs, NJ: Prentice-Hall, Inc., 1969.
7. Stewart, G. W., *Introduction to Matrix Computations*. New York: Academic Press, Inc., 1973.
8. Wilberg, Donald M., State Space and Linear Systems, *Schaum's Outline Series*, New York: McGraw-Hill, Inc., 1971.

APPENDIX D

QUATERNIONS

Lawrence Fallon, III

The quaternion representation of rigid body rotations leads to convenient kinematical expressions involving the Euler symmetric parameters (Sections 12.1 and 16.1). Some important properties of quaternions are summarized in this appendix following the formulation of Hamilton [1866] and Whittaker [1961].

Let the four parameters (q_1, q_2, q_3, q_4) form the components of the *quaternion*, *q*, as follows:

$$q \equiv q_4 + iq_1 + jq_2 + kq_3 \tag{D-1}$$

where i, j, and k are the hyperimaginary numbers satisfying the conditions

$$i^2 = j^2 = k^2 = -1$$
$$ij = -ji = k$$
$$jk = -kj = i$$
$$ki = -ik = j \tag{D-2}$$

The conjugate or inverse of *q* is defined as

$$q^* \equiv q_4 - iq_1 - jq_2 - kq_3 \tag{D-3}$$

The quantity, q_4, is the *real* or *scalar part* of the quaternion and $iq_1 + jq_2 + kq_3$ is the *imaginary* or *vector part*.

A vector in three-dimensional space, **U**, having components U_1, U_2, U_3 is expressed in quaternion notation as a quaternion with a scalar part of zero,

$$\mathbf{U} = iU_1 + jU_2 + kU_3 \tag{D-4}$$

If the vector **q** corresponds to the vector part of *q* (i.e., $\mathbf{q} = iq_1 + jq_2 + kq_3$), then an alternative representation of *q* is

$$q = (q_4, \mathbf{q}) \tag{D-5}$$

Quaternion multiplication is performed in the same manner as the multiplication of complex numbers or algebraic polynomials, except that the order of operations must be taken into account because Eq. (D-2) is not commutative. As an example, consider the product of two quaternions

$$q'' = qq' = (q_4 + iq_1 + jq_2 + kq_3)(q'_4 + iq'_1 + jq'_2 + kq'_3) \tag{D-6}$$

Using Eq. (D-2), this reduces to

$$q'' = qq' = (-q_1 q'_1 - q_2 q'_2 - q_3 q'_3 + q_4 q'_4)$$
$$+ i(q_1 q'_4 + q_2 q'_3 - q_3 q'_2 + q_4 q'_1)$$
$$+ j(-q_1 q'_3 + q_2 q'_4 + q_3 q'_1 + q_4 q'_2)$$
$$+ k(q_1 q'_2 - q_2 q'_1 + q_3 q'_4 + q_4 q'_3) \tag{D-7}$$

If $q' = (q_4', \mathbf{q}')$, then Eq. (D-7) can alternatively be expressed as

$$q'' = qq' = (q_4 q_4' - \mathbf{q} \cdot \mathbf{q}', q_4 \mathbf{q}' + q_4' \mathbf{q} + \mathbf{q} \times \mathbf{q}') \tag{D-8}$$

The *length* or *norm* of q is defined as

$$|q| \equiv \sqrt{qq^*} = \sqrt{q^*q} = \sqrt{q_1^2 + q_2^2 + q_3^2 + q_4^2} \tag{D-9}$$

If a set of four Euler symmetric parameters corresponding to the rigid body rotation defined by the transformation matrix, A (Section 12.1), are the components of the quaternion, q, then q is a representation of the rigid body rotation. If q' corresponds to the rotation matrix A', then the rotation described by the product $A'A$ is equivalent to the rotation described by qq'. (Note the inverse order of quaternion multiplication as compared with matrix multiplication.)

The transformation of a vector $\mathbf{U}$, corresponding to multiplication by the matrix A,

$$\mathbf{U}' = A\mathbf{U} \tag{D-10}$$

is effected in quaternion algebra by the operation

$$\mathbf{U}' = q^*\mathbf{U}q \tag{D-11}$$

See Section 12.1 for additional properties of quaternions used to represent rigid body rotations.

For computational purposes, it is convenient to express quaternion multiplication in matrix form. Specifically, let the components of q form a four-vector as follows:

$$q = \begin{bmatrix} q_1 \\ q_2 \\ q_3 \\ q_4 \end{bmatrix} \tag{D-12}$$

This procedure is analogous to expressing the complex number $c = a + ib$ in the form of the two-vector,

$$\mathbf{c} = \begin{bmatrix} a \\ b \end{bmatrix}$$

In matrix form, Eq. (D-7) then becomes

$$\begin{bmatrix} q_1'' \\ q_2'' \\ q_3'' \\ q_4'' \end{bmatrix} = \begin{bmatrix} q_4' & q_3' & -q_2' & q_1' \\ -q_3' & q_4' & q_1' & q_2' \\ q_2' & -q_1' & q_4' & q_3' \\ -q_1' & -q_2' & -q_3' & q_4' \end{bmatrix} \begin{bmatrix} q_1 \\ q_2 \\ q_3 \\ q_4 \end{bmatrix} \tag{D-13}$$

Given the quaternion components corresponding to two successive rotations, Eq. (D-13) conveniently gives the quaternion components corresponding to the total rotation.

References

1.   Hamilton, Sir W. R., *Elements of Quaternions*. London: Longmans, Green and Co., 1866.
2.   Whittaker, E. T., *A Treatise on the Analytical Dynamics of Particles and Rigid Bodies*. Cambridge: Cambridge University Press, 1961.

APPENDIX E

COORDINATE TRANSFORMATIONS

Gyanendra K. Tandon

E.1   Cartesian, Spherical, and Cylindrical Coordinates
E.2   Transformations Between Cartesian Coordinates
E.3   Transformations Between Spherical Coordinates

E.1 Cartesian, Spherical, and Cylindrical Coordinates

The components of a vector, **r**, in cartesian, spherical, and cylindrical coordinates are shown in Fig. E-1 and listed below.

Cartesian	(x,y,z)
Spherical	(r,θ,ϕ)
Cylindrical	(ρ,ϕ,z)

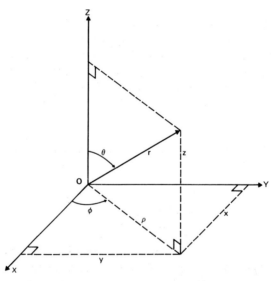

Fig. E-1. Components of a Vector, **r**, in Cartesian (x,y,z), Spherical (r,θ,ϕ), and Cylindrical (ρ,ϕ,z) Coordinates

The declination, δ, used in celestial coordinates is measured from the equatorial plane (x-y plane) and is related to θ by the equation

$$\delta \equiv 90° - \theta \qquad \text{(E-1)}$$

The components in cartesian, spherical, and cylindrical coordinates are related by the following equations:

$$x = r\sin\theta\cos\phi \qquad\qquad = \rho\cos\phi \qquad\qquad \text{(E-2a)}$$
$$y = r\sin\theta\sin\phi \qquad\qquad = \rho\sin\phi \qquad\qquad \text{(E-2b)}$$

$$z = r \cos \theta \qquad\qquad = z \qquad\qquad\text{(E-2c)}$$

$$r = (x^2 + y^2 + z^2)^{1/2} \qquad = (\rho^2 + z^2)^{1/2} \qquad\qquad\text{(E-2d)}$$

$$\theta = \arccos\left\{ z/(x^2 + y^2 + z^2)^{1/2} \right\} \quad = \arctan(\rho/z) \quad 0 \leqslant \theta \leqslant 180° \qquad\text{(E-2e)}$$

$$\phi = \arctan(y/x) \qquad\qquad = \phi \qquad\qquad 0 \leqslant \phi < 360° \qquad\text{(E-2f)}$$

$$\rho = (x^2 + y^2)^{1/2} \qquad\qquad = r \sin \theta \qquad\qquad\text{(E-2g)}$$

The correct quadrant for ϕ in Eq. (E-2f) is obtained from the relative signs of x and y.

E.2 Transformations Between Cartesian Coordinates

If $\mathbf{r}$ and $\mathbf{r}'$ are the cartesian representations of a vector in two different cartesian coordinate systems, then they are related by

$$\mathbf{r}' = A\mathbf{r} + \mathbf{a} \qquad\qquad\text{(E-3)}$$

where $\mathbf{a}$ represents the translation of the origin of the unprimed system in the primed system and the matrix A represents a rotation. For most attitude work, $\mathbf{a} = \mathbf{0}$.

The transformation matrix A (called the *attitude matrix* or *direction cosine matrix* in this book) can be obtained by forming the matrix product of matrices for successive rotations about the three coordinate axes as described in Section 12.1. The elements of matrix A are direction cosines of the primed axes in the unprimed system and satisfy the orthogonality condition. Because A is an orthogonal matrix, its inverse transformation matrix will be its transposed matrix; symbolically,

$$A^{-1} = A^{\mathrm{T}} \qquad\qquad\text{(E-4)}$$

For many applications, the definition of the direction cosine matrix in terms of the orthogonal coordinate unit vectors in the two coordinate systems,

$$A \equiv \begin{bmatrix} \hat{\mathbf{x}}' \cdot \hat{\mathbf{x}} & \hat{\mathbf{x}}' \cdot \hat{\mathbf{y}} & \hat{\mathbf{x}}' \cdot \hat{\mathbf{z}} \\ \hat{\mathbf{y}}' \cdot \hat{\mathbf{x}} & \hat{\mathbf{y}}' \cdot \hat{\mathbf{y}} & \hat{\mathbf{y}}' \cdot \hat{\mathbf{z}} \\ \hat{\mathbf{z}}' \cdot \hat{\mathbf{x}} & \hat{\mathbf{z}}' \cdot \hat{\mathbf{y}} & \hat{\mathbf{z}}' \cdot \hat{\mathbf{z}} \end{bmatrix} \qquad\qquad\text{(E-5a)}$$

is useful for computations. As an example, let the primed coordinate system have its coordinate axes aligned with the spacecraft-to-Earth vector $\mathbf{R}$, the component of $\mathbf{V}$ perpendicular to $\mathbf{R}$, and the orbit normal vector $\mathbf{R} \times \mathbf{V}/|\mathbf{R} \times \mathbf{V}|$, where $\mathbf{V}$ is the spacecraft velocity:

$$\hat{\mathbf{x}}' \equiv \mathbf{R}/|\mathbf{R}|$$

$$\hat{\mathbf{y}}' \equiv (\mathbf{R} \times \mathbf{V}) \times \mathbf{R}/|(\mathbf{R} \times \mathbf{V}) \times \mathbf{R}|$$

$$\hat{\mathbf{z}}' \equiv (\mathbf{R} \times \mathbf{V})/|\mathbf{R} \times \mathbf{V}| \qquad\qquad\text{(E-5b)}$$

Then, substituting Eq. (E-5b) into Eq. (E-5a) gives an expression for A which does not require the evaluation of trigonometric functions.

Euler's Theorem. Euler's theorem states that any finite rotation of a rigid body can be expressed as a rotation through some angle about some fixed axis. Therefore, the most general transformation matrix A is a rotation by some angle,

Φ, about some fixed axis, $\hat{e}$. The axis $\hat{e}$ is unaffected by the rotation and, therefore, must have the same components in both the primed and the unprimed systems. Denoting the components of $\hat{e}$ by e_1, e_2, and e_3, the matrix A is

$$A = \begin{bmatrix} \cos\Phi + e_1^2(1-\cos\Phi) & e_1e_2(1-\cos\Phi)+e_3\sin\Phi & e_1e_3(1-\cos\Phi)-e_2\sin\Phi \\ e_1e_2(1-\cos\Phi)-e_3\sin\Phi & \cos\Phi+e_2^2(1-\cos\Phi) & e_2e_3(1-\cos\Phi)+e_1\sin\Phi \\ e_1e_3(1-\cos\Phi)+e_2\sin\Phi & e_2e_3(1-\cos\Phi)-e_1\sin\Phi & \cos\Phi+e_3^2(1-\cos\Phi) \end{bmatrix} \quad \text{(E-6)}$$

In this case, the inverse transformation matrix may be obtained by using Eq. (E-4) or by replacing Φ by $-\Phi$, in Eq. (E-6), that is, a rotation by the same amount in the opposite direction about the axis $\hat{e}$.

Euler Symmetric Parameters. The Euler symmetric parameters, q_1 through q_4, used to represent finite rotations are defined by the following equations:

$$q_1 \equiv e_1\sin\frac{\Phi}{2}$$

$$q_2 \equiv e_2\sin\frac{\Phi}{2}$$

$$q_3 \equiv e_3\sin\frac{\Phi}{2}$$

$$q_4 \equiv \cos\frac{\Phi}{2} \quad \text{(E-7a)}$$

Clearly,

$$q_1^2 + q_2^2 + q_3^2 + q_4^2 = 1 \quad \text{(E-7b)}$$

The transformation matrix A in terms of Euler symmetric parameters is

$$A = \begin{bmatrix} q_1^2-q_2^2-q_3^2+q_4^2 & 2(q_1q_2+q_3q_4) & 2(q_1q_3-q_2q_4) \\ 2(q_1q_2-q_3q_4) & -q_1^2+q_2^2-q_3^2+q_4^2 & 2(q_2q_3+q_1q_4) \\ 2(q_1q_3+q_2q_4) & 2(q_2q_3-q_1q_4) & -q_1^2-q_2^2+q_3^2+q_4^2 \end{bmatrix} \quad \text{(E-8)}$$

The inverse transformation matrix in this case may be obtained by use of Eq. (E-4), or by replacing q_1, q_2, and q_3 by $-q_1, -q_2$, and $-q_3$, respectively, in Eq. (E-8) and leaving q_4 unaltered.

The Euler symmetric parameters may be regarded as components of a quaternion, q, defined by

$$q = q_4 + iq_1 + jq_2 + kq_3 \quad \text{(E-9)}$$

where i, j, and k are as defined in Appendix D. The multiplication rule for successive rotations represented by Euler symmetric parameters is given in Appendix D. The Euler symmetric parameters in terms of the 3-1-3 Euler angle rotation ϕ, θ, ψ (defined below) are as follows:

$$q_1 = \sin(\theta/2)\cos((\phi-\psi)/2)$$

$$q_2 = \sin(\theta/2)\sin((\phi-\psi)/2)$$

$$q_3 = \cos(\theta/2)\sin((\phi+\psi)/2)$$

$$q_4 = \cos(\theta/2)\cos((\phi+\psi)/2) \quad \text{(E-10)}$$

Gibbs Vector. The Gibbs vector (components g_1, g_2, and g_3) representation (see Section 12.1) for finite rotations is defined by

$$g_1 \equiv q_1/q_4 = e_1 \tan(\Phi/2)$$

$$g_2 \equiv q_2/q_4 = e_2 \tan(\Phi/2)$$

$$g_3 \equiv q_3/q_4 = e_3 \tan(\Phi/2) \tag{E-11}$$

The transformation matrix A in terms of the Gibbs vector representation is as follows:

$$A = \frac{1}{1+g_1^2+g_2^2+g_3^2} \begin{bmatrix} 1+g_1^2-g_2^2-g_3^2 & 2(g_1 g_2+g_3) & 2(g_1 g_3-g_2) \\ 2(g_1 g_2-g_3) & 1-g_1^2+g_2^2-g_3^2 & 2(g_2 g_3+g_1) \\ 2(g_1 g_3+g_2) & 2(g_2 g_3-g_1) & 1-g_1^2-g_2^2+g_3^2 \end{bmatrix} \tag{E-12}$$

The inverse of A can be obtained in this case by the method of Eq. (E-4), or by replacing g_i by $-g_i$ in Eq. (E-12).

Euler Angle Rotation. The Euler angle rotation (ϕ, θ, ψ) is defined by successive rotations by angles ϕ, θ, and ψ, respectively, about coordinate axes i, j, k (Section 12.1). The i-j-k Euler angle rotation means that the first rotation by angle ϕ is about the i axis, the second rotation by angle θ is about the j axis, and the third rotation by angle ψ is about the k axis. There are 12 distinct representations for the Euler angle rotation which divide equally into two types:

TYPE 1. In this case, the rotations take place successively about each of the three coordinate axes. This type has a singularity at $\theta = \pm 90$ deg, because for these values of θ, the ϕ and ψ rotations have a similar effect.

TYPE 2. In this case, the first and third rotations take place about the same axis and the second rotation takes place about one of the other two axes. This type has a singularity at $\theta = 0$ deg and 180 deg, because for these values of θ, the ϕ and ψ rotations have a similar effect.

Table E-1 gives the transformation matrix, A, for all of the 12 Euler angle representations. The 3-1-3 Euler angle representation is the one most commonly used in the literature. The Euler angles ϕ, θ, and ψ can be easily obtained from the elements of matrix A. A typical example from each type is given below.

TYPE 1: 3-1-2 Euler Angle Rotation

$$\phi = \arctan(-A_{21}/A_{22}) \qquad 0 \leqslant \phi < 360° \tag{E-13a}$$

$$\theta = \arcsin(A_{23}) \qquad -90° \leqslant \theta \leqslant 90° \tag{E-13b}$$

$$\psi = \arctan(-A_{13}/A_{33}) \qquad 0 \leqslant \psi < 360° \tag{E-13c}$$

The correct quadrants for ϕ and ψ are obtained from the relative signs of the elements of A in Eqs. (E-13a) and (E-13c), respectively.

TYPE 2: 3-1-3 Euler Angle Rotation

$$\phi = \arctan(A_{31}/-A_{32}) \qquad 0 \leqslant \phi < 360° \tag{E-14a}$$

$$\theta = \arccos(A_{33}) \qquad 0 \leqslant \theta \leqslant 180° \tag{E-14b}$$

$$\psi = \arctan(A_{13}/A_{23}) \qquad 0 \leqslant \psi < 360° \tag{E-14c}$$

Table E-1. The Attitude Matrix, A, for the 12 Possible Euler Angle Representations ($S \equiv$ sine, $C \equiv$ cosine, $1 \equiv x$ axis, $2 \equiv y$ axis, $3 \equiv z$ axis)

TYPE-1

EULER ANGLE REPRESENTATION	MATRIX A
1-2-3	$\begin{bmatrix} C\psi C\theta & C\psi S\theta S\phi + S\psi C\phi & -C\psi S\theta C\phi + S\psi S\phi \\ -S\psi C\theta & C\psi C\phi - S\psi S\theta S\phi & S\psi S\theta C\phi + C\psi S\phi \\ S\theta & -C\theta S\phi & C\theta C\phi \end{bmatrix}$
1-3-2	$\begin{bmatrix} C\psi C\theta & C\psi S\theta C\phi + S\psi S\phi & C\psi S\theta S\phi - S\psi C\phi \\ -S\theta & C\theta C\phi & C\theta S\phi \\ S\psi C\theta & S\psi S\theta C\phi - C\psi S\phi & S\psi S\theta S\phi + C\psi C\phi \end{bmatrix}$
2-3-1	$\begin{bmatrix} C\theta C\phi & S\theta & -C\theta S\phi \\ -C\psi S\theta C\phi + S\psi S\phi & C\psi C\theta & C\psi S\theta S\phi + S\psi C\phi \\ S\psi S\theta C\phi + C\psi S\phi & -S\psi C\theta & C\psi C\phi - S\psi S\theta S\phi \end{bmatrix}$
2-1-3	$\begin{bmatrix} C\psi C\phi + S\psi S\theta S\phi & S\psi C\theta & -C\psi S\phi + S\psi S\theta C\phi \\ C\psi S\theta S\phi - S\psi C\phi & C\psi C\theta & S\psi S\phi + C\psi S\theta C\phi \\ C\theta S\phi & -S\theta & C\theta C\phi \end{bmatrix}$
3-1-2	$\begin{bmatrix} C\psi C\phi - S\psi S\theta S\phi & C\psi S\phi + S\psi S\theta C\phi & -S\psi C\theta \\ -C\theta S\phi & C\theta C\phi & S\theta \\ S\psi C\phi + C\psi S\theta S\phi & S\psi S\phi - C\psi S\theta C\phi & C\psi C\theta \end{bmatrix}$
3-2-1	$\begin{bmatrix} C\theta C\phi & C\theta S\phi & -S\theta \\ -C\psi S\phi + S\psi S\theta C\phi & C\psi C\phi + S\psi S\theta S\phi & S\psi C\theta \\ C\psi S\theta C\phi + S\psi S\phi & C\psi S\theta S\phi - S\psi C\phi & C\psi C\theta \end{bmatrix}$

TYPE-2

EULER ANGLE REPRESENTATION	MATRIX A
1-2-1	$\begin{bmatrix} C\theta & S\theta S\phi & -S\theta C\phi \\ S\psi S\theta & C\psi C\phi - S\psi C\theta S\phi & C\psi S\phi + S\psi C\theta C\phi \\ C\psi S\theta & -S\psi C\phi - C\psi C\theta S\phi & -S\psi S\phi + C\psi C\theta C\phi \end{bmatrix}$
1-3-1	$\begin{bmatrix} C\theta & S\theta C\phi & S\theta S\phi \\ -C\psi S\theta & C\psi C\theta C\phi - S\psi S\phi & C\psi C\theta S\phi + S\psi C\phi \\ S\psi S\theta & -S\psi C\theta C\phi - C\psi S\phi & -S\psi C\theta S\phi + C\psi C\phi \end{bmatrix}$
2-1-2	$\begin{bmatrix} C\psi C\phi - S\psi C\theta S\phi & S\psi S\theta & -C\psi S\phi - S\psi C\theta C\phi \\ S\theta S\phi & C\theta & S\theta C\phi \\ S\psi C\phi + C\psi C\theta S\phi & -C\psi S\theta & -S\psi S\phi + C\psi C\theta C\phi \end{bmatrix}$
2-3-2	$\begin{bmatrix} C\psi C\theta C\phi - S\psi S\phi & C\psi S\theta & -C\psi C\theta S\phi - S\psi C\phi \\ -S\theta C\phi & C\theta & S\theta S\phi \\ S\psi C\theta C\phi + C\psi S\phi & S\psi S\theta & -S\psi C\theta S\phi + C\psi C\phi \end{bmatrix}$
3-1-3	$\begin{bmatrix} C\psi C\phi - S\psi C\theta S\phi & C\psi S\phi + S\psi C\theta C\phi & S\psi S\theta \\ -S\psi C\phi - C\psi C\theta S\phi & -S\psi S\phi + C\psi C\theta C\phi & C\psi S\theta \\ S\theta S\phi & -S\theta C\phi & C\theta \end{bmatrix}$
3-2-3	$\begin{bmatrix} C\psi C\theta C\phi - S\psi S\phi & C\psi C\theta S\phi + S\psi C\phi & -C\psi S\theta \\ -S\psi C\theta C\phi - C\psi S\phi & -S\psi C\theta S\phi + C\psi C\phi & S\psi S\theta \\ S\theta C\phi & S\theta S\phi & C\theta \end{bmatrix}$

The correct quadrants for ϕ and ψ are obtained from the relative signs of the elements of A in Eqs. (E-14a) and (E-14c), respectively.

Kinematic Equations of Motion.   For convenience, the kinematic equations of motion (Section 16.1) for the 12 possible Euler angle representations are given in Table E-2. The kinematic equations of motion for other representations of the attitude matrix can be found in Section 16.1.

Table E-2. Kinematic Equations of Motion for the 12 Possible Euler Angle Representations ($1 \equiv x$ axis, $2 \equiv y$ axis, $3 \equiv z$ axis; ω_1, ω_2, ω_3 are components of the angular velocity along the body x, y, z axes.)

AXIS SEQUENCE		INDEX VALUES			KINEMATIC EQUATIONS OF MOTION
		I	J	K	
TYPE 1	1 - 2 - 3	1	2	3	$\dot{\phi} = (\omega_I \cos\psi - \omega_J \sin\psi)\sec\theta$
	2 - 3 - 1	2	3	1	$\dot{\theta} = \omega_J \cos\psi + \omega_I \sin\psi$
	3 - 1 - 2	3	1	2	$\dot{\psi} = \omega_K - (\omega_I \cos\psi - \omega_J \sin\psi)\tan\theta$
	1 - 3 - 2	1	3	2	$\dot{\phi} = (\omega_I \cos\psi + \omega_J \sin\psi)\sec\theta$
	3 - 2 - 1	3	2	1	$\dot{\theta} = \omega_J \cos\psi - \omega_I \sin\psi$
	2 - 1 - 3	2	1	3	$\dot{\psi} = \omega_K + (\omega_I \cos\psi + \omega_J \sin\psi)\tan\theta$
TYPE 2	1 - 2 - 1	1	2	3	$\dot{\phi} = (\omega_K \cos\psi + \omega_J \sin\psi)\csc\theta$
	2 - 3 - 2	2	3	1	$\dot{\theta} = \omega_J \cos\psi - \omega_K \sin\psi$
	3 - 1 - 3	3	1	2	$\dot{\psi} = \omega_I - (\omega_K \cos\psi + \omega_J \sin\psi)\cot\theta$
	1 - 3 - 1	1	3	2	$\dot{\phi} = - (\omega_K \cos\psi - \omega_J \sin\psi)\csc\theta$
	3 - 2 - 3	3	2	1	$\dot{\theta} = \omega_J \cos\psi + \omega_K \sin\psi$
	2 - 1 - 2	2	1	3	$\dot{\psi} = \omega_I + (\omega_K \cos\psi - \omega_J \sin\psi)\cot\theta$

E.3 Transformations Between Spherical Coordinates

Figure E-2 illustrates the spherical coordinate system on a sphere of unit radius defined by the north pole, N, and the azimuthal reference direction, R, in the equatorial plane. The coordinates of point P are (ϕ, θ). A new coordinate system is defined by the north pole, N', at (ϕ_0, θ_0) in the old coordinate system. The new azimuthal reference is at an angle, ϕ_0' relative to the NN' great circle. The coordinates (ϕ', θ') of P in the new system are given by:

$$\cos\theta' = \cos\theta_0 \cos\theta + \sin\theta_0 \sin\theta \cos(\phi - \phi_0)$$

$$\sin(\phi' - \phi_0') = \sin(\phi - \phi_0)\sin\theta/\sin\theta' \tag{E-15}$$

where θ and θ' are both defined over the range 0 to 180 deg, and $(\phi - \phi_0)$ and $(\phi' - \phi'_0)$ are both in the range 0 to 180 deg or in the range 180 to 360 deg. Simplified forms of Eqs. (E-15) in two special cases are as follows:

Case 1: $\phi = \phi'_0 = 90°$

$$\cos\theta' = \cos\theta_0\cos\theta + \sin\theta_0\sin\theta\sin\phi_0$$

$$\cos\phi' = -\cos\phi_0\sin\theta/\sin\theta' \qquad \text{(E-16)}$$

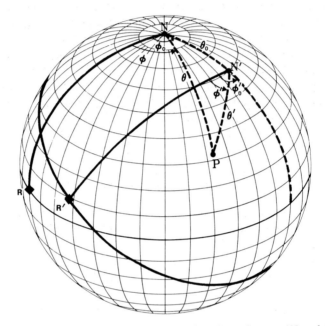

Fig. E-2.   Transformation Between Spherical Coordinate Systems NR and $N'R'$

Case 2: $\phi = \phi'_0 = 0$

$$\cos\theta' = \cos\theta_0\cos\theta + \sin\theta_0\sin\theta\cos\phi_0$$

$$\sin\phi' = -\sin\phi_0\sin\theta/\sin\theta' \qquad \text{(E-17)}$$

The most common spherical inertial coordinates for attitude analysis are the celestial coordinates (α, δ) defined in Section 2.2. The right ascension, α, and the declination, δ, are related to ϕ and θ by

$$\alpha = \phi$$

$$\delta = 90° - \theta \qquad \text{(E-18)}$$

APPENDIX F

THE LAPLACE TRANSFORM

Gerald M. Lerner

Laplace transformation is a technique used to relate time- and frequency-dependent linear systems. A *linear system* is a collection of electronic components (e.g., resistors, capacitors, inductors) or physical components (e.g., masses, springs, oscillators) arranged so that the system output is a linear function of system input. The input and output of an electronic system are commonly voltages, whereas the input to an attitude control system is a sensed angular error and the output is a restoring torque. Most systems are linear only for a restricted range of input.

Laplace transformation is widely used to solve problems in electrical engineering or control theory (e.g., attitude control) that may be reduced to linear differential equations with constant coefficients. The *Laplace transform* of a real function, $f(t)$, defined for real $t > 0$ is

$$\mathcal{L}(f(t)) \equiv F(s) = \int_{0^+}^{\infty} f(t)\exp(-st)dt \tag{F-1}$$

where 0^+ indicates that the lower limit of the integral is evaluated by taking the limit as $t \to 0$ from above. The argument of the Laplace transform, $F(s)$, is complex,

$$s \equiv \sigma + i\omega$$

where $i \equiv \sqrt{-1}$. For most physical applications, t and ω denote time and frequency, respectively, and σ is related to the decay time.

The *inverse Laplace transform* is

$$\mathcal{L}^{-1}(F(s)) \equiv f(t) = \frac{1}{2\pi i} \int_{C-i\infty}^{C+i\infty} F(s)\exp(st)ds \tag{F-2}$$

where the real constant C is chosen such that $F(s)$ exists for all $\mathrm{Re}(s) > C$, that is, to the right of any singularity.

Properties of the Laplace Transform and the Inverse Laplace Transform*. The Laplace transform and its inverse are *linear operators*, thus

$$\mathcal{L}(af(t) + bg(t)) = a\mathcal{L}(f(t)) + b\mathcal{L}(g(t))$$
$$\equiv aF(s) + bG(s) \tag{F-3}$$
$$\mathcal{L}^{-1}(aF(s) + bG(s)) = a\mathcal{L}^{-1}(F(s)) + b\mathcal{L}^{-1}(G(s))$$
$$= af(t) + bg(t) \tag{F-4}$$

where a and b are complex constants.

---

* For further details, see DiStefano, *et al.*, [1967]

The *initial value theorem* relates the initial value of $f(t)$, $f(0^+)$, to the Laplace transform,

$$f(0^+) = \lim_{s \to \infty} sF(s) \tag{F-5}$$

and the *final value theorem*, which is widely used to determine the steady-state response of a system, relates the final value of $f(t)$, $f(\infty)$, to the Laplace transform,*

$$f(\infty) = \lim_{s \to 0} sF(s) \tag{F-6}$$

The Laplace and inverse Laplace transformations may be scaled in either the time domain (*time scaling*) by

$$\mathcal{L}(f(t/a)) = aF(as) \tag{F-7}$$

or the frequency domain (*frequency scaling*) by

$$\mathcal{L}^{-1}(F(as)) = f(t/a)/a \tag{F-8}$$

The Laplace transform of the *time-delayed* function, $f(t - t_0)$, is

$$\mathcal{L}(f(t - t_0)) = \exp(-st_0)F(s) \tag{F-9}$$

where $f(t - t_0) = 0$ for $t \leqslant t_0$. The inverse Laplace transform of the frequency shifted function, $F(s - s_0)$, is

$$\mathcal{L}^{-1}(F(s - s_0)) = \exp(s_0 t)f(t) \tag{F-10}$$

Laplace transforms of exponentially damped, modulated, and scaled functions are

$$\mathcal{L}(\exp(-at)f(t)) = F(s + a) \tag{F-11a}$$

$$\mathcal{L}(\sin \omega t f(t)) = [F(s - i\omega) - F(s + i\omega)]/2i \tag{F-11b}$$

$$\mathcal{L}(\cos \omega t f(t)) = [F(s - i\omega) + F(s + i\omega)]/2 \tag{F-11c}$$

$$\mathcal{L}(t^n f(t)) = (-1)^n \frac{d^n}{ds^n} F(s) \tag{F-11d}$$

$$\mathcal{L}(f(t)/t) = \int_s^\infty F(u)du \tag{F-11e}$$

The Laplace transform of the product of two functions may be expressed as the *complex convolution integral*,

$$\mathcal{L}(f(t)g(t)) = \frac{1}{2\pi i} \int_{C - i\infty}^{C + i\infty} F(\omega)G(s - \omega)d\omega \tag{F-12}$$

Multiplying the Laplace transform of a function by s is analogous to differentiating the original function; thus,

---

*The final value theorem is valid provided that $sF(s)$ is analytic on the imaginary axis and in the right half of the s-plane; i.e., it applies only to stable systems.

$$\mathcal{L}^{-1}(sF(s)) = \frac{df}{dt} + f(0)\delta_D(t) \tag{F-13}$$

Dividing the Laplace transform of a function by s is analogous to integrating the original function; thus,

$$\mathcal{L}^{-1}(F(s)/s) = \int_0^t f(u)\,du \tag{F-14}$$

The inverse Laplace transform of a product may be expressed as the *convolution integral*

$$\mathcal{L}^{-1}(F(s)G(s)) = \int_{0+}^t f(t)g(t-\tau)\,d\tau = \int_{0+}^t g(t)f(t-\tau)\,d\tau \tag{F-15}$$

which may be inverted to yield

$$F(s)G(s) = \mathcal{L}\int_{0+}^t f(t)g(t-\tau)\,d\tau = \mathcal{L}\int_{0+}^t g(t)f(t-\tau)\,d\tau \tag{F-16}$$

A short list of Laplace transforms is given in Table F-1; detailed tables are given by Abramowitz and Stegun [1968], Korn and Korn [1968], Churchill [1958], and Erdélyi, *et al.*, [1954].

Table F-1.   Laplace Transforms

g(t)	G(S)	g(t)	G(S)
$\dfrac{df}{dt}$	$SF(S) - f(0^+)$	$u(t-a)$[†]	$\exp(-as)/s$
$\dfrac{df^n}{dt^n}$	$S^n F(S) - S^{n-1} f(0^+)$ $- S^{n-2}\dfrac{df}{dt}\Big\|_{0^+} - \cdots \dfrac{df^{n-1}}{dt^{n-1}}\Big\|_{0^+}$	t	$1/s^2$
		t^n	$n!/S^{n+1}$
$\displaystyle\int_0^t f(\tau)\,d\tau$	$F(S)/S$	t^a	$\Gamma(a+1)/S^{n+1}$
		$\exp(-at)$	$1/(S+a)$
		$t^n \exp(-at)$	$1/(S+a)^{n+1}$
$t^n f(t)$	$(-)^n \dfrac{d^n F(S)}{dS^n}$	$\sin \omega t$	$\omega/(S^2+\omega^2)$
		$\cos \omega t$	$S/(S^2+\omega^2)$
$f(t)/t$	$\displaystyle\int_S^\infty F(u)\,du$	$\exp(-at)\sin \omega t$	$\omega/[(S+a)^2+\omega^2]$
		$\exp(-at)\cos \omega t$	$(S+a)/[(S+a)^2+\omega^2]$
$f(t/a)$	$a F(aS)$	$[\exp(-at)-\exp(-bt)]/[a-b]$	$1/[(S+a)(S+b)]$
$f(t-t_o)$	$\exp(-t_o S)F(S)$	$[a\exp(-at)-b\exp(-bt)]/[b-a]$	$S/[(S+a)(S+b)]$
$\exp(ts_o)f(t)$	$F(S-S_o)$	$\sinh(at)$	$a/(S^2-a^2)$
		$\cosh(at)$	$S/(S^2-a^2)$
$\delta_D(t-a)$[*]	$\exp(-aS)$		

NOTE: $F(S) = \displaystyle\int_{0^+}^\infty f(t)\exp(-st)\,dt$; $G(S) = \displaystyle\int_{0^+}^\infty g(t)\exp(-st)\,dt$

n DENOTES A POSITIVE INTEGER; a AND b DENOTE POSITIVE REAL NUMBERS.

*δ_D IS THE DIRAC DELTA FUNCTION.

†u IS HEAVISIDE'S UNIT STEP FUNCTION WHICH IS DEFINED BY u = 0 FOR t < a, u = 1 FOR t ⩾ a.

Solution of Linear Differential Equations. Linear differential equations with constant coefficients may be solved by taking the Laplace transform of each term of the differential equation, thereby reducing a differential equation in t to an algebraic equation in s. The solution may then be transformed back to the time domain by taking the inverse Laplace transform. This procedure simplifies the analysis of the response of complex physical systems to frequency-dependent stimuli, such as the response of an onboard control system to periodic disturbance torques.

The solution to the linear differential equation

$$\sum_{i=0}^{n} a_i \frac{d^i y}{dt^i} = x(t) \tag{F-17}$$

with $a_n = 1$ and forcing function $x(t)$ is given by

$$y(t) = \mathcal{L}^{-1}\left[\frac{X(s)}{\mathfrak{z}(s)}\right] + \mathcal{L}^{-1}\left[\frac{\displaystyle\sum_{i=1}^{n}\sum_{k=0}^{i-1} a_i s^{i-1-k} y_0^{(k)}}{\mathfrak{z}(s)}\right] \tag{F-18}$$

where $X(s) \equiv \mathcal{L}(x(t))$, $\mathfrak{z}(s) \equiv \displaystyle\sum_{i=0}^{n} a_i s^i$ is the *characteristic polynomial* of Eq. (F-17) and

$$y_0^{(k)} \equiv \frac{d^k y}{dt^k}\bigg|_{t \to 0^+}$$

are the initial conditions.

Any physically reasonable forcing function, including impulses, steps, and ramps, may be conveniently transformed (see Table F-1). The analysis of the algebraic transformed equation is generally much easier than the original differential equation. For example, the steady-state solution, $f(\infty)$, of a differential equation is obtained from the Laplace transform by using the final value theorem, Eq. (F-6).

The first term on the right-hand side of Eq. (F-18) is the *forced response* of the system due to the forcing function and the second term is the *free response* of the system due to the initial conditions. The forced response, $\mathcal{L}^{-1}(X(s)/\mathfrak{z}(s))$, consists of two parts: *transient* and *steady state*.

Solving the differential equation (Eq. (F-17)) is equivalent to finding the inverse Laplace transform of the algebraic functions of s in Eq. (F-18). One technique involves expressing rational functions of the form

$$R(s) = \sum_{i=0}^{m} b_i s^i \bigg/ \sum_{i=0}^{n} a_i s^i \equiv N(s)/\mathfrak{z}(s) \tag{F-19}$$

as a sum of partial fractions ($n \geqslant m$) using the fundamental theorem of algebra. The characteristic polynomial, $\mathfrak{z}(s)$, may be factored as

$$\mathfrak{z}(s) = \prod_{i=1}^{r} (s + p_i)^{m_i} \tag{F-20}$$

where $-p_i$ is the ith zero of $\zeta(s)$ with multiplicity m_i and

$$\sum_{i=1}^{r} m_i = n \tag{F-21}$$

The partial fraction expansion is

$$R(s) = b_n + \sum_{i=1}^{r} \sum_{k=1}^{m_i} \frac{C_{ik}}{(s+p_i)^k} \tag{F-22}$$

where

$$C_{ik} \equiv \frac{1}{(m_i - k)!} \frac{d^{m_i - k}}{ds^{m_i - k}} \left[(s+p_i)^{m_i} R(s) \right]_{s = -p_i}$$

and $b_n = 0$ unless $m = n$. The coefficients C_{i1} are the *residues* of $R(s)$ at the *poles* $-p_i$. If no roots are repeated, Eq. (F-22) may be rewritten as

$$R(s) = b_n + \sum_{i=1}^{n} \frac{C_{i1}}{s+p_i} \tag{F-23}$$

where

$$C_{i1} \equiv (s+p_i)R(s)|_{s = -p_i}$$

The zeros of $\zeta(s)$ may be determined using various numerical methods [DiStefano, et al., 1967].

The inverse Laplace transform of expressions in the form of Eq. (F-23) may be obtained directly from Table F-1. Other techniques for computing inverse Laplace transforms include series expansions and differential equations [Spiegel, 1965].

Example: Forced Harmonic Oscillator. The equation describing a 1-degree-of-freedom gyroscope (Sections 6.5 and 7.8) is

$$\frac{d^2\theta}{dt^2} + \frac{D}{I_G} \frac{d\theta}{dt} + \frac{K\theta}{I_G} = \frac{L}{I_G} \omega(t) \tag{F-24}$$

where I_G is the moment of inertia of the gyroscope about the output axis, D is the viscous damping coefficient about the output axis, K is the restoring spring constant about the output axis, L is the angular momentum of the rotor, and $\omega(t)$ is the angular velocity about the input axis which is to be measured (see Fig. 6-45).

We assume that the input angular velocity is sinusoidal* with amplitude, A, and frequency, ω_e; i.e.,

$$\omega(t) = A \cos \omega_e t \tag{F-25}$$

The solution of Eq. (F-24) is given by (F-18) as

$$\theta(t) = \mathcal{L}^{-1} \left[\frac{X(s)}{\zeta(s)} \right] + \mathcal{L}^{-1} \left[\frac{D\theta_0/I_G + s\theta_0 + \dot{\theta}_0}{\zeta(s)} \right] \tag{F-26}$$

---

*This is not as severe a restriction as it might seem because any physically reasonable $\omega(t)$ may be expanded in a Fourier series. The result for a general $\omega(t)$ is then obtained by linear superposition.

where

$$\zeta(s)=s^2+Ds/I_G+K/I_G$$

$$X(s)=\mathcal{L}(AL\cos\omega_e t/I_G)=ALs\left[(s^2+\omega_e^2)I_G\right]^{-1}$$

$$\theta_0\equiv\theta|_{t=0}$$

$$\dot{\theta}_0\equiv\frac{d\theta}{dt}\bigg|_{t=0}$$ (F-27)

The characteristic polynomial, $\zeta(s)$, may be factored as

$$\zeta(s)=(s+p_1)(s+p_2)$$

where

$$p_1\equiv\left(D+i\sqrt{4KI_G-D^2}\right)\Big/(2I_G)$$

$$p_2\equiv\left(D-i\sqrt{4KI_G-D^2}\right)\Big/(2I_G)$$ (F-28)

and we assume $4KI_G-D^2>0$.
    Substitution of Eq. (F-28) into Eq. (F-26) yields

$$\theta(t)=\mathcal{L}^{-1}\left[\frac{ALs/I_G}{(s^2+\omega_e^2)(s+p_1)(s+p_2)}\right]+\mathcal{L}^{-1}\left[\frac{D\theta_0/I_G+s\theta_0+\dot{\theta}_0}{(s+p_1)(s+p_2)}\right]$$ (F-29)

The second term on the right-hand side of Eq. (F-29) is given in Table F-1 as

$$\mathcal{L}^{-1}\left[\frac{D\theta_0/I_G+s\theta_0+\dot{\theta}_0}{(s+p_1)(s+p_2)}\right]=\left(\frac{1}{p_1-p_2}\right)\{(D\theta_0/I_G+\dot{\theta}_0)[\exp(-p_1t)-\exp(-p_2t)]$$

$$-\theta_0[p_1\exp(-p_1t)-p_2\exp(-p_2t)]\}$$ (F-30)

The first term on the right-hand-side of Eq. (F-29) may be expanded in partial fractions as

$$\mathcal{L}^{-1}\left[\frac{s}{(s^2+\omega_e^2)(s+p_1)(s+p_2)}\right]=\frac{\exp(-i\omega_e t)}{2(p_1-i\omega_e)(p_2-i\omega_e)}+\frac{\exp(i\omega_e t)}{2(p_1+i\omega_e)(p_2+i\omega_e)}$$

$$+\frac{1}{(p_1-p_2)}\left\{\frac{p_1\exp(-p_1t)}{p_1^2+\omega_e^2}-\frac{p_2\exp(-p_2t)}{p_2^2+\omega_e^2}\right\}$$ (F-31)

Equation (F-30) and the last term on the right-hand-side of Eq. (F-31) are the transient response of the system to the initial conditions and the forcing function. The transient response decays with a time constant and frequency given by

$$[\mathrm{Re}(p_1)]^{-1}=\tau_0=2I_G/D$$

$$\mathrm{Im}(p_1)=\omega_1=\sqrt{4KI_G-D^2}\Big/2I_G$$ (F-32)

For $t\to\infty$, the *steady-state solution*, the first two terms on the right-hand side of Eq. (F-31) dominate $\theta(t)$. These two terms may be rewritten as

$$\lim_{t\gg\tau_0}\theta(t)=\frac{I_GAL}{K^2+\omega_e^2(D^2-2KI_G)+I_G^2\omega_e^4}\left[\left(\frac{K}{I_G}-\omega_e^2\right)\cos\omega_e t+\frac{D\omega_e}{I_G}\sin\omega_e t\right]$$

$$=\frac{AL\cos(\omega_e t-\phi)}{\left[I_G^2\left(\frac{K}{I_G}-\omega_e^2\right)^2+D^2\omega_e^2\right]^{1/2}}\tag{F-33}$$

where

$$\tan\phi=\frac{D\omega_e}{K-I_G\omega_e^2}$$

Several features of gyroscope design are evident from these equations:

1. The output response of the system to a constant or low-frequency input, $\omega_e\approx0$, is linearly related to the input for $t\gg\tau_0$; for example,

$$\lim_{\omega_e\to0}\theta(t)=A(L/K)\qquad(t\gg\tau_0)$$

2. The viscous damping constant, D, must be sufficiently high so that τ_0 is small compared with the gyro sampling period. However, if the damping is too high, the system output becomes frequency dependent and lags the input.

3. Systems with negligible damping, $D\approx0$, *resonate* at input frequencies near the *characteristic frequency* of the system,

$$\omega_0=\sqrt{K/I_G}$$

Integral Equations. *Integral equations* have the general form

$$y(t)=f(t)+\int_{u_1}^{u_2}k(u,t)y(u)du\tag{F-34}$$

where $k(u,t)$ is the *kernel* of the equation. The limits of the integral may be either constants or functions of time. If u_1 and u_2 are constants, Eq. (F-34) is called a *Fredholm* equation, whereas if u_1 is a constant and $u_2=t$, then Eq. (F-34) is called a *Volterra* equation of the second kind [Korn and Korn, 1968].

If the functional form of the kernel may be expressed as

$$k(u,t)=k(u-t)\tag{F-35}$$

then the Volterra equation

$$y(t)=f(t)+\int_0^t k(u-t)y(u)du\tag{F-36}$$

may be solved by Laplace transform methods. Taking the Laplace transform of Eq. (F-36) and rearranging, we obtain

$$Y(s)=F(s)/(1-K(s))\tag{F-37}$$

where $Y(s) \equiv \mathcal{L}(y(t))$, $F(s) \equiv \mathcal{L}(f(t))$, and $K(s) \equiv \mathcal{L}(k(t))$, which may be solved for $y(t)$ by taking the inverse Laplace transform.

References

1. Abramowitz, Milton, and Irene A. Stegun, *Handbook of Mathematical Functions*. Washington, D.C., National Bureau of Standards, 1970.
2. Churchill, R. V., *Operational Mathematics*, Second Edition. New York: McGraw-Hill, Inc., 1958.
3. DiStefano, Joseph J., III, Allen R. Stubberud, and Ivan J. Williams, *Feedback and Control Systems*, Schaum's Outline Series. New York: McGraw Hill, Inc., 1967.
4. Erdélyi, A., *et al.*, *Tables of Integral Transforms*. New York: McGraw Hill, Inc., 1954.
5. Korn, Granino A., and Theresa M. Korn, *Mathematical Handbook for Scientists and Engineers*. New York: McGraw Hill, Inc., 1968.
6. Spiegel, Murray R., *Laplace Transform*, Schaum's Outline Series. New York: Schaum Publishing Co., 1965.

APPENDIX G

SPHERICAL HARMONICS

John Aiello

Laplace's Equation, $\nabla^2 U = 0$, can be written in the spherical coordinate system of Section 2.3 as:

$$\frac{\partial^2 U}{\partial r^2} + \frac{2}{r}\frac{\partial U}{\partial r} + \frac{1}{r^2}\frac{\partial^2 U}{\partial \theta^2} + \frac{\cot\theta}{r^2}\frac{\partial U}{\partial \theta} + \frac{1}{r^2\sin^2\theta}\frac{\partial^2 U}{\partial \phi^2} = 0 \qquad \text{(G-1)}$$

If a trial substitution of $U(r,\theta,\phi) = R(r)Y(\theta,\phi)$ is made, the following equations are obtained through a separation of variables:

$$r^2\frac{d^2 R(r)}{dr^2} + 2r\frac{dR(r)}{dr} - n(n+1)R(r) = 0 \qquad \text{(G-2)}$$

$$\frac{\partial^2 Y(\theta,\phi)}{\partial \theta^2} + \cot\theta\frac{\partial Y(\theta,\phi)}{\partial \theta} + \frac{1}{\sin^2\theta}\frac{\partial^2 Y(\theta,\phi)}{\partial \phi^2} + n(n+1)Y(\theta,\phi) = 0 \qquad \text{(G-3)}$$

where $n(n+1)$ has been chosen as the separation constant. Solutions to Eq. (G-2) are of the form

$$R(r) = Ar^n + Br^{-(n+1)} \qquad \text{(G-4)}$$

Thus, solutions to Laplace's Equation (Eq. (G-1)) are of the form

$$U = \left[Ar^n + Br^{-(n+1)}\right]Y(\theta,\phi), \qquad n = 0,1,2,\ldots, \qquad \text{(G-5)}$$

These functions are referred to as *solid spherical harmonics*, and the $Y(\theta,\phi)$ are known as *surface spherical harmonics*. We wish to define U over a domain both interior and exterior to a spherical surface of radius r, and to have U continuous everywhere in the domain and to assume prescribed values $U_0(\theta,\phi)$ on the surface. Under these conditions, Eq. (G-5) with $B=0$ gives the form of U for the interior region of the sphere and with $A=0$ represents its form in the exterior region.

To determine the surface spherical harmonics, the trial substitution

$$Y(\theta,\phi) = P(\cos\theta)\Phi(\phi) \qquad \text{(G-6)}$$

is made in Eq. (G-3).

Multiplying by $\sin^2\theta/P\Phi$ and choosing a separation constant of m^2 yields

$$\frac{d^2 P(\cos\theta)}{d\theta^2} + \cot\theta\frac{dP(\cos\theta)}{d\theta} + \left[n(n+1) - \frac{m^2}{\sin^2\theta}\right]P(\cos\theta) = 0 \qquad \text{(G-7)}$$

$$\frac{d^2\Phi(\phi)}{d\phi^2} + m^2\Phi(\phi) = 0 \qquad \text{(G-8)}$$

The solutions to Eq. (G-8) are readily found to be

$$\Phi(\phi) = C \cos m\phi + S \sin m\phi \tag{G-9}$$

in which m must be an integer, because $\Phi(\phi)$ is required to be a single valued function. Equation (G-7) can be rewritten substituting $x = \cos\theta$ as,

$$\frac{d}{dx}\left[(1-x^2)\frac{dP}{dx}\right] + \left[n(n+1) - \frac{m^2}{1-x^2}\right]P = 0 \tag{G-10}$$

which is the generalized Legendre equation [Jackson, 1962]. For $m = 0$, the solutions to Eq. (G-10) are called *Legendre polynomials* and may be computed from either Rodrigues' formula

$$P_n(x) = \frac{1}{2^n n!}\left(\frac{d}{dx}\right)^n (x^2 - 1)^n \tag{G-11}$$

or from a recurrence relation convenient for computer use [Arfken, 1970].

$$P_{n+1}(x) = 2xP_n(x) - P_{n-1}(x) - [xP_n(x) - P_{n-1}(x)]/(n+1) \tag{G-12}$$

Rodrigues' formula can be verified by direct substitution into Eq. (G-10), and the recurrence relation can be verified by mathematical induction. When $m \neq 0$, solutions to Eq. (G-10) are known as *associated Legendre functions* (of degree, n, and order, m), and may be computed by [Yevtushenko, et al., 1969]

$$P_{nm}(x) = \frac{(1-x^2)^{m/2}}{2^n n!} \frac{d^{n+m}(x^2-1)^n}{dx^{n+m}} \tag{G-13}$$

or by [Heiskanen and Moritz, 1967]

$$P_{nm}(x) = 2^{-n}(1-x^2)^{m/2}\sum_{k=0}^{l}(-1)^k\frac{(2n-2k)!}{k!(n-k)!(n-m-2k)!}x^{n-m-2k} \tag{G-14}$$

where l is either $(n-m)/2$ or $(n-m-1)/2$, whichever is an integer. Table G-1 lists the associated Legendre functions up to degree and order 4 in terms of $\cos\theta$ [Fitzpatrick, 1970].*

Table G-1. Explicit Forms of Associated Legendre Functions Through Degree $n=4$ and Order $m=4$

n \ m	0	1	2	3	4
0	1				
1	$\cos\theta$	$\sin\theta$			
2	$\frac{3}{2}\cos^2\theta - \frac{1}{2}$	$3\sin\theta\cos\theta$	$3\sin^2\theta$		
3	$\frac{5}{2}\left(\cos^3\theta - \frac{3}{5}\cos\theta\right)$	$\frac{15}{2}\sin\theta\left(\cos^2\theta - \frac{1}{5}\right)$	$15\sin^2\theta\cos\theta$	$15\sin^3\theta$	
4	$\frac{35}{8}\left(\cos^4\theta - \frac{6}{7}\cos^2\theta + \frac{3}{35}\right)$	$\frac{35}{2}\sin\theta\left(\cos^3\theta - \frac{3}{7}\cos\theta\right)$	$\frac{105}{2}\sin^2\theta\left(\cos^2\theta - \frac{1}{7}\right)$	$105\sin^3\theta\cos\theta$	$105\sin^4\theta$

---

*Because Eq. (G-10) is a homogeneous equation in P, it does not define the normalization of P. Equations (G-11) and (G-13) define the conventional *Neumann normalization*, but other normalizations are used (see Appendix H or Chapman and Bartels [1940]).

Using Eq. (G-13), the functions P_{nm} can be shown to be orthogonal; that is,

$$\int_{-1}^{+1} P_{pm}(x) P_{qm}(x) dx = \frac{2}{2q+1} \frac{(q+m)!}{(q-m)!} \delta_p^q \qquad (G\text{-}15)$$

where δ_p^q is the Kronecker delta.

It is now possible to write the complete solution to Laplace's equation as

$$U(r,\theta,\phi) = \sum_{n=0}^{\infty} \left(\frac{a}{r}\right)^{n+1} \sum_{m=0}^{n} \left[C_{nm}\cos m\phi + S_{nm}\sin m\phi \right] P_{nm}(\cos\theta) \qquad (G\text{-}16)$$

describing the potential exterior to a spherical surface of radius a. Customarily, Eq. (G-16) is written in the form

$$U(r,\theta,\phi) = \sum_{n=0}^{\infty} \left(\frac{a}{r}\right)^{n+1} J_n P_{n0}(\cos\theta)$$

$$+ \sum_{n=1}^{\infty} \sum_{m=1}^{n} \left(\frac{a}{r}\right)^{n+1} \left[C_{nm}\cos m\phi + S_{nm}\sin m\phi \right] P_{nm}(\cos\theta) \qquad (G\text{-}17)$$

where $J_n \equiv C_{n0}$. Terms for which $m=0$ are called *zonal* harmonics and the J_n are *zonal harmonic coefficients*. Nonzero m terms are called *tesseral* harmonics or, for the particular case of $n=m$, *sectoral* harmonics.

Visualizing the different harmonics geometrically makes the origin of the names clear. The zonal harmonics, for example, are polynomials in $\cos\theta$ of degree n, with n zeros, meaning a sign change occurs n times on the sphere ($0° \leqslant \theta \leqslant 180°$), and the sign changes are independent of ϕ. Figure G-1 shows the "zones" (analogous to the temperate and tropical zones on the Earth) for the case of

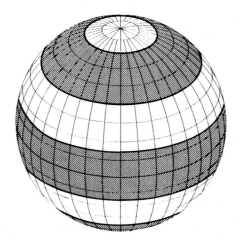

Fig. G-1.   Zones for $P_6(\cos\theta)$ Spherical Harmonics

$P_6(\cos\theta)$. The tesseral and sectoral harmonics have $n-m$ zeros for $0° < \theta < 180°$, and $2m$ zeros for $0° \leqslant \phi \leqslant 360°$. Figure G-2, the representation of $P_{63}(\cos\theta)\cos 3\phi$, illustrates the division of the sphere into alternating positive and negative *tesserae*. The word "tessera" is Latin for tiles, such as would be used in a mosaic. When $n = m$, the tesseral pattern reduces to the "sector" pattern in Fig. G-3.

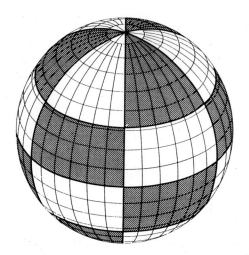

Fig. G-2. $P_{63}(\cos\theta)\cos 3\phi$ Showing Alternating Positive and Negative Tesseral Harmonics

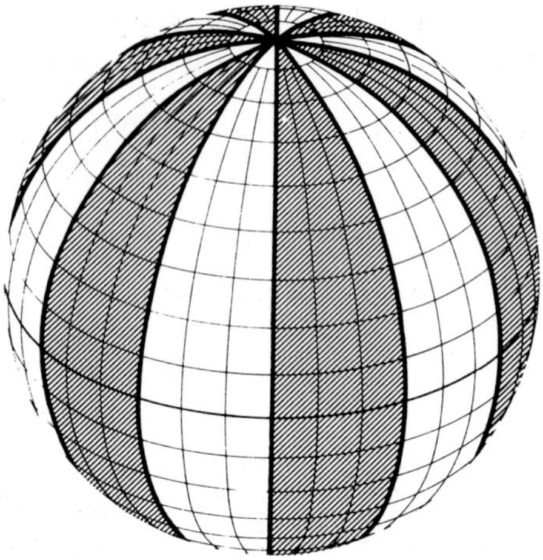

Fig. G-3. $P_{66}(\cos\theta)\cos 6\phi$ Showing Tesseral Pattern Reduced to Sectoral Pattern

For a more detailed discussion of spherical harmonics, see Hobson [1931].

References

1. Arfken, G., *Mathematical Methods for Physicists*. New York: Academic Press, Inc., 1970.
2. Chapman, Sydney, and Julius Bartels, *Geomagnetism*. Oxford: Clarendon Press, pp. 609–611, 1940.
3. Fitzpatrick, P. M., *Principles of Celestial Mechanics*. New York: Academic Press, Inc., 1970.
4. Hieskanen, W., and H. Moritz, *Physical Geodesy*. San Francisco: W. H. Freeman, 1967.
5. Hobson, E. W., *The Theory of Spherical and Ellipsoidal Harmonics*. New York: Chelsea Publishing Co., 1931.
6. Jackson, John David, *Classical Electrodynamics*. New York: John Wiley & Sons, Inc., 1962.
7. Yevtushenko, G., et al., *Motion of Artificial Satellites in the Earth's Gravitational Field*, NASA, TTF-539, June 1969.

APPENDIX H

MAGNETIC FIELD MODELS

Michael Plett

Spherical Harmonic Model. This appendix presents some computational aspects of geomagnetic field models. A more qualitative description of the field characteristics is given in Section 5.1. As discussed there, the predominant portion of the Earth's magnetic field, **B**, can be represented as the gradient of a scalar potential function, V, i.e.,

$$\mathbf{B} = -\nabla V \tag{H-1}$$

V can be conveniently represented by a series of spherical harmonics,

$$V(r,\theta,\phi) = a \sum_{n=1}^{k} \left(\frac{a}{r}\right)^{n+1} \sum_{m=0}^{n} (g_n^m \cos m\phi + h_n^m \sin m\phi) P_n^m(\theta) \tag{H-2}$$

where a is the equatorial radius of the Earth (6371.2 km adopted for the International Geomagnetic Reference Field, IGRF); g_n^m and h_n^m are Gaussian coefficients (named in honor of Karl Gauss); and r, θ, and ϕ are the geocentric distance, coelevation, and East longitude from Greenwich which define any point in space.

The Gaussian coefficients are determined empirically by a least-squares fit to measurements of the field. A set of these coefficients constitutes a *model* of the field. The coefficients for the IGRF (Section 5.1; [Leaton, 1976]), are given in Table H-1. The first-order time derivatives of the coefficients, called the *secular*

Table H-1.   IGRF Coefficients for Epoch 1975. Terms indicated by a dash (–) are undefined.

n	m	g(nT)	h(nT)	ġ(nT/yr)	ḣ(nT/yr)	n	m	g(nT)	h(nT)	ġ(nT/yr)	ḣ(nT/yr)
1	0	−30186	––	25.6	—	6	2	15	102	2.0	−0.1
1	1	−2036	5735	10.0	−10.2	6	3	−210	88	2.8	−0.2
2	0	−1898	—	−24.9	---	6	4	−1	−43	0.0	−1.3
2	1	2997	−2124	0.7	−3.0	6	5	−8	−9	0.9	0.7
2	2	1551	−37	4.3	−18.9	6	6	−114	−4	−0.1	1.7
3	0	1299	—	−3.8	—	7	0	66	—	0.0	—
3	1	−2144	−361	−10.4	6.9	7	1	−57	−68	0.0	−1.4
3	2	1296	249	−4.1	2.5	7	2	−7	−24	0.0	−0.1
3	3	805	−253	−4.2	−5.0	7	3	7	−4	0.6	0.3
4	0	951	—	−0.2	—	7	4	−22	11	0.9	0.3
4	1	807	148	−2.0	5.0	7	5	−9	27	0.3	−0.7
4	2	462	−264	−3.9	0.8	7	6	11	−17	0.3	0.1
4	3	−393	37	−2.1	1.7	7	7	−8	−14	−0.5	0.8
4	4	235	−307	−3.1	−1.0	8	0	11	—	0.2	—
5	0	−204	—	0.3	—	8	1	13	4	0.3	−0.2
5	1	368	39	−0.7	1.2	8	2	3	−15	0.0	−0.4
5	2	275	142	1.1	2.3	8	3	−12	2	0.2	−0.2
5	3	−20	−147	−1.6	−2.0	8	4	−4	−19	−0.4	−0.3
5	4	−161	−99	−0.5	1.3	8	5	6	1	−0.3	0.4
5	5	−38	74	1.0	1.1	8	6	−2	18	0.6	−0.3
6	0	46	—	0.2	—	8	7	9	−6	−0.3	−0.6
6	1	57	−23	0.5	−0.5	8	8	1	−19	−0.1	0.3

terms, are also given in Table H-1. With these coefficients and a definition of the associated Legendre functions, P_n^m, it is possible to calculate the magnetic field at any point in space via Eqs. (H-1) and (H-2).

The coeffients of the IGRF assume that the P_n^m are *Schmidt normalized* [Chapman and Bartels, 1940], i.e.,

$$\int_0^\pi [P_n^m(\theta)]^2 \sin\theta \, d\theta = \frac{2(2 - \delta_m^0)}{2n + 1} \tag{H-3}$$

where the Kronecker delta, $\delta_j^i = 1$ if $i = j$ and 0 otherwise. This normalization, which is nearly independent of m, is chosen so that the relative strength of terms of the same degree (n) but different order (m) can be gauged by simply comparing the respective Gaussian coefficients. For Schmidt normalization, the $P_n^m(\theta)$ have the form

$$P_n^m(\theta) \equiv \left\{ \left[\frac{(2 - \delta_m^0)(n - m)!}{(n + m)!} \right]^{1/2} \frac{(2n - 1)!!}{(n - m)!} \right\} \sin^m\theta$$

$$\times \left\{ \cos^{n-m}\theta - \frac{(n - m)(n - m - 1)}{2(2n - 1)} \cos^{n-m-2}\theta \right.$$

$$\left. + \frac{(n - m)(n - m - 1)(n - m - 2)(n - m - 3)}{2 \cdot 4(2n - 1)(2n - 3)} \cos^{n-m-4}\theta - \cdots \right\} \tag{H-4}$$

where $(2n - 1)!! \equiv 1 \cdot 3 \cdot 5 \cdots (2n - 1)$. The square root term in Eq. (H-4) is the only difference between the Schmidt normalization and the common Neumann normalization described in Appendix G. The computation time required for the field models can be significantly reduced by calculating the terms in Eq. (H-4) recursively, i.e., expressing the nth term as a function of the $(n - 1)$th term. The first step is to convert the coefficients in Table H-1 from Schmidt to Gauss normalization, which saves about 7% in computation time [Trombka and Cain, 1974]. The Gauss functions, $P^{n,m}$, are related to the Schmidt functions, P_n^m, by

$$P_n^m = S_{n,m} P^{n,m} \tag{H-5a}$$

where

$$S_{n,m} \equiv \left[\frac{(2 - \delta_m^0)(n - m)!}{(n + m)!} \right]^{1/2} \frac{(2n - 1)!!}{(n - m)!} \tag{H-5b}$$

The factors $S_{n,m}$ are best combined with the Gaussian coefficients because they are independent of r, θ, ϕ and so must be calculated only once during a computer run. Thus, we define

$$g^{n,m} \equiv S_{n,m} g_n^m$$

$$h^{n,m} \equiv S_{n,m} h_n^m \tag{H-6}$$

Using mathematical induction, it is possible to derive the following recursion relations for $S_{n,m}$:

APPENDIX H

MAGNETIC FIELD MODELS

Michael Plett

Spherical Harmonic Model. This appendix presents some computational aspects of geomagnetic field models. A more qualitative description of the field characteristics is given in Section 5.1. As discussed there, the predominant portion of the Earth's magnetic field, **B**, can be represented as the gradient of a scalar potential function, V, i.e.,

$$\mathbf{B} = -\nabla V \qquad \text{(H-1)}$$

V can be conveniently represented by a series of spherical harmonics,

$$V(r,\theta,\phi) = a \sum_{n=1}^{k} \left(\frac{a}{r}\right)^{n+1} \sum_{m=0}^{n} (g_n^m \cos m\phi + h_n^m \sin m\phi) P_n^m(\theta) \qquad \text{(H-2)}$$

where a is the equatorial radius of the Earth (6371.2 km adopted for the International Geomagnetic Reference Field, IGRF); g_n^m and h_n^m are Gaussian coefficients (named in honor of Karl Gauss); and r, θ, and ϕ are the geocentric distance, coelevation, and East longitude from Greenwich which define any point in space.

The Gaussian coefficients are determined empirically by a least-squares fit to measurements of the field. A set of these coefficients constitutes a *model* of the field. The coefficients for the IGRF (Section 5.1; [Leaton, 1976]), are given in Table H-1. The first-order time derivatives of the coefficients, called the *secular*

Table H-1. IGRF Coefficients for Epoch 1975. Terms indicated by a dash (–) are undefined.

n	m	g(nT)	h(nT)	ġ(nT/yr)	ḣ(nT/yr)	n	m	g(nT)	h(nT)	ġ(nT/yr)	ḣ(nT/yr)
1	0	−30186	−−	25.6	—	6	2	15	102	2.0	−0.1
1	1	−2036	5735	10.0	−10.2	6	3	−210	88	2.8	−0.2
2	0	−1898	—	−24.9	−−	6	4	−1	−43	0.0	−1.3
2	1	2997	−2124	0.7	−3.0	6	5	−8	−9	0.9	0.7
2	2	1551	−37	4.3	−18.9	6	6	−114	−4	−0.1	1.7
3	0	1299	—	−3.8	—	7	0	66	—	0.0	—
3	1	−2144	−361	−10.4	6.9	7	1	−57	−68	0.0	−1.4
3	2	1296	249	−4.1	2.5	7	2	−7	−24	0.0	−0.1
3	3	805	−253	−4.2	−5.0	7	3	7	−4	0.6	0.3
4	0	951	—	−0.2	—	7	4	−22	11	0.9	0.3
4	1	807	148	−2.0	5.0	7	5	−9	27	0.3	−0.7
4	2	462	−264	−3.9	0.8	7	6	11	−17	0.3	0.1
4	3	−393	37	−2.1	1.7	7	7	−8	−14	−0.5	0.8
4	4	235	−307	−3.1	−1.0	8	0	11	—	0.2	—
5	0	−204	—	0.3	—	8	1	13	4	0.3	−0.2
5	1	368	39	−0.7	1.2	8	2	3	−15	0.0	−0.4
5	2	275	142	1.1	2.3	8	3	−12	2	0.2	−0.2
5	3	−20	−147	−1.6	−2.0	8	4	−4	−19	−0.4	−0.3
5	4	−161	−99	−0.5	1.3	8	5	6	1	−0.3	0.4
5	5	−38	74	1.0	1.1	8	6	−2	18	0.6	−0.3
6	0	46	—	0.2	—	8	7	9	−6	−0.3	−0.6
6	1	57	−23	0.5	−0.5	8	8	1	−19	−0.1	0.3

terms, are also given in Table H-1. With these coefficients and a definition of the associated Legendre functions, P_n^m, it is possible to calculate the magnetic field at any point in space via Eqs. (H-1) and (H-2).

The coeffients of the IGRF assume that the P_n^m are *Schmidt normalized* [Chapman and Bartels, 1940], i.e.,

$$\int_0^\pi [P_n^m(\theta)]^2 \sin\theta \; d\theta = \frac{2(2-\delta_m^0)}{2n+1} \tag{H-3}$$

where the Kronecker delta, $\delta_j^i = 1$ if $i = j$ and 0 otherwise. This normalization, which is nearly independent of m, is chosen so that the relative strength of terms of the same degree (n) but different order (m) can be gauged by simply comparing the respective Gaussian coefficients. For Schmidt normalization, the $P_n^m(\theta)$ have the form

$$P_n^m(\theta) \equiv \left\{ \left[\frac{(2-\delta_m^0)(n-m)!}{(n+m)!} \right]^{1/2} \frac{(2n-1)!!}{(n-m)!} \right\} \sin^m\theta$$

$$\times \left\{ \cos^{n-m}\theta - \frac{(n-m)(n-m-1)}{2(2n-1)} \cos^{n-m-2}\theta \right.$$

$$\left. + \frac{(n-m)(n-m-1)(n-m-2)(n-m-3)}{2\cdot 4(2n-1)(2n-3)} \cos^{n-m-4}\theta - \cdots \right\} \tag{H-4}$$

where $(2n-1)!! \equiv 1\cdot 3\cdot 5 \cdots (2n-1)$. The square root term in Eq. (H-4) is the only difference between the Schmidt normalization and the common Neumann normalization described in Appendix G. The computation time required for the field models can be significantly reduced by calculating the terms in Eq. (H-4) recursively, i.e., expressing the nth term as a function of the $(n-1)$th term. The first step is to convert the coefficients in Table H-1 from Schmidt to Gauss normalization, which saves about 7% in computation time [Trombka and Cain, 1974]. The Gauss functions, $P^{n,m}$, are related to the Schmidt functions, P_n^m, by

$$P_n^m = S_{n,m} P^{n,m} \tag{H-5a}$$

where

$$S_{n,m} \equiv \left[\frac{(2-\delta_m^0)(n-m)!}{(n+m)!} \right]^{1/2} \frac{(2n-1)!!}{(n-m)!} \tag{H-5b}$$

The factors $S_{n,m}$ are best combined with the Gaussian coefficients because they are independent of r, θ, ϕ and so must be calculated only once during a computer run. Thus, we define

$$g^{n,m} \equiv S_{n,m} g_n^m$$

$$h^{n,m} \equiv S_{n,m} h_n^m \tag{H-6}$$

Using mathematical induction, it is possible to derive the following recursion relations for $S_{n,m}$:

$$S_{0,0} = 1$$

$$S_{n,0} = S_{n-1,0}\left[\frac{2n-1}{n}\right] \qquad\qquad n \geqslant 1 \qquad\qquad \text{(H-7)}$$

$$S_{n,m} = S_{n,m-1}\sqrt{\frac{(n-m+1)(\delta_m^1+1)}{n+m}} \qquad m \geqslant 1$$

The $P^{n,m}$ can be similarly obtained from the following recursion relations:

$$P^{0,0} = 1$$

$$P^{n,n} = \sin\theta P^{n-1,n-1} \qquad\qquad\qquad\qquad\qquad \text{(H-8)}$$

$$P^{n,m} = \cos\theta P^{n-1,m} - K^{n,m}P^{n-2,m}$$

where

$$K^{n,m} \equiv \frac{(n-1)^2 - m^2}{(2n-1)(2n-3)} \qquad n > 1 \qquad\qquad \text{(H-9)}$$

$$K^{n,m} \equiv 0 \qquad\qquad\qquad n = 1$$

Because the gradient in Eq. (H-1) will lead to partial derivatives of the $P^{n,m}$, we need

$$\frac{\partial P^{0,0}}{\partial\theta} = 0$$

$$\frac{\partial P^{n,n}}{\partial\theta} = (\sin\theta)\frac{\partial P^{n-1,n-1}}{\partial\theta} + (\cos\theta)P^{n-1,n-1} \qquad n \geqslant 1 \quad \text{(H-10)}$$

$$\frac{\partial P^{n,m}}{\partial\theta} = (\cos\theta)\frac{\partial P^{n-1,m}}{\partial\theta} - (\sin\theta)P^{n-1,m} - K^{n,m}\frac{\partial P^{n-2,m}}{\partial\theta}$$

Also note that

$$\cos m\phi = \cos((m-1)\phi + \phi)$$

$$= \cos((m-1)\phi)\cos\phi - \sin\phi\sin((m-1)\phi) \qquad \text{(H-11)}$$

A similar recursion relation can be derived for $\sin m\phi$. The computational advantage of Eq. (H-11) is that it greatly reduces the number of times that sine and cosine functions must be calculated.

Given the coefficients $g^{n,m}$ and $h^{n,m}$ and recursion relations in Eqs. (H-7) through (H-11), the field **B** is calculated from Eqs. (H-1) and (H-2). Specifically,

$$B_r = \frac{-\partial V}{\partial r} = \sum_{n=1}^{k}\left(\frac{a}{r}\right)^{n+2}(n+1)\sum_{m=0}^{n}(g^{n,m}\cos m\phi + h^{n,m}\sin m\phi)P^{n,m}(\theta)$$

$$B_\theta = \frac{-1}{r}\frac{\partial V}{\partial\theta} = -\sum_{n=1}^{k}\left(\frac{a}{r}\right)^{n+2}\sum_{m=0}^{n}(g^{n,m}\cos m\phi + h^{n,m}\sin m\phi)\frac{\partial P^{n,m}(\theta)}{\partial\theta}$$

$$B_\phi = \frac{-1}{r\sin\theta}\frac{\partial V}{\partial\phi} = \frac{-1}{\sin\theta}\sum_{n=1}^{k}\left(\frac{a}{r}\right)^{n+2}\sum_{m=0}^{n}m(-g^{n,m}\sin m\phi + h^{n,m}\cos m\phi)P^{n,m}(\theta)$$

$$\text{(H-12)}$$

Here, B_r is the radial component (outward positive) of the field, B_θ is the coelevation component (South positive), and B_ϕ is the azimuthal component (East positive). (See Fig. 2-5, Section 2.3.) The magnetic field literature, however, normally refers to three components X, Y, Z, consisting of North, East, and nadir relative to an oblate Earth. These components are obtained from Eq. (H-12) by

$$X(\text{“}North\text{”}) = -B_\theta\cos\epsilon - B_r\sin\epsilon$$
$$Y(\text{“}East\text{”}) = B_\phi \tag{H-13}$$
$$Z(\text{“}Vertical\text{”} \text{ inward positive}) = B_\theta\sin\epsilon - B_r\cos\epsilon$$

where $\epsilon \equiv \lambda - \delta < 0.2°$, λ is the geodetic latitude, and $\delta \equiv 90° - \theta$ is the declination. The correction terms in $\sin\epsilon$ are of the order of 100 nT or less [Trombka and Cain, 1974].

The geocentric inertial components used in satellite work are

$$B_x = (B_r\cos\delta + B_\theta\sin\delta)\cos\alpha - B_\phi\sin\alpha$$
$$B_y = (B_r\cos\delta + B_\theta\sin\delta)\sin\alpha + B_\phi\cos\alpha$$
$$B_z = (B_r\sin\delta - B_\theta\cos\delta) \tag{H-14}$$

Note that $\mathbf{B}$ is still a function of longitude, ϕ, which is related to the right ascension, α, by:

$$\phi = \alpha - \alpha_G \tag{H-15}$$

where α_G is the right ascension of the Greenwich meridian or the sidereal time at Greenwich (Appendix J).

Dipole Model. Equations (H-6) through (H-14) are sufficient to generate efficient computer code. However, for analytic purposes, it is convenient to obtain a *dipole model* by expanding the field model to first degree ($n = 1$) and all orders ($m = 0, 1$). Eq. (H-2) then becomes

$$V(r,\theta,\phi) = \frac{a^3}{r^2}\left[g_1^0 P_1^0(\theta) + (g_1^1\cos\phi + h_1^1\sin\phi)P_1^1(\theta) \right]$$

$$= \frac{1}{r^2}\left(g_1^0 a^3\cos\theta + g_1^1 a^3\cos\phi\sin\theta + h_1^1 a^3\sin\phi\sin\theta \right) \tag{H-16}$$

The $\cos\theta$ term is just the potential due to a dipole of strength $g_1^0 a^3$ aligned with the polar axis. (See, for example, Jackson [1965].) Similarly, the $\sin\theta$ terms are dipoles aligned with the x and y axes. Relying on the principle of linear superposition, these three terms are just the Cartesian components of the dipole component of the Earth's magnetic field. From Table H-1, we find that for 1978,

$$g_1^0 = -30109 \text{ nT}$$

$$g_1^1 = -2006 \text{ nT}$$

$$h_1^1 = 5704 \text{ nT} \tag{H-17}$$

Therefore, the total dipole strength is

$$a^3 H_0 = a^3\left[g_1^{0^2} + g_1^{1^2} + h_1^{1^2} \right]^{1/2} = 7.943 \times 10^{15} \text{ Wb·m} \tag{H-18}$$

The coelevation of the dipole is

$$\theta'_m = \arccos\left(\frac{g_1^0}{H_0}\right) = 168.6° \tag{H-19}$$

The East longitude of the dipole is

$$\phi'_m = \arctan\left(\frac{h_1^1}{g_1^1}\right) = 109.3° \tag{H-20}$$

Thus, the first-order terrestrial magnetic field is due to a dipole with *northern* magnetization pointed toward the southern hemisphere such that the northern end of any dipole free to rotate in the field points roughly toward the north celestial pole. The end of the Earth's dipole in the northern hemisphere is at 78.6° N, 289.3° E and is customarily referred to as the "North" magnetic pole. Frequently, dipole models in the literature use the coordinates of the North magnetic pole and compensate with a minus sign in the dipole equation.

The above calculations were performed for 1978 by adding the secular terms to the Gaussian coefficients of epoch 1975. The location of the dipole in 1975 can be similarly calculated and compared with the 1980 dipole. That comparison yields a 0.45% decrease in dipole strength between 1975 and 1980 and a 0.071-deg drift northward and a 0.056-deg (arc) drift westward for a total motion of 0.09-deg arc.

The dipole field in local tangent coordinates is given by

$$B_r = 2\left(\frac{a}{r}\right)^3\left[g_1^0\cos\theta + (g_1^1\cos\phi + h_1^1\sin\phi)\sin\theta\right]$$

$$B_\theta = \left(\frac{a}{r}\right)^3\left[g_1^0\sin\theta - (g_1^1\cos\phi + h_1^1\sin\phi)\cos\theta\right] \tag{H-21}$$

$$B_\phi = \left(\frac{a}{r}\right)^3\left[g_1^1\sin\phi - h_1^1\cos\phi\right]$$

The field could be converted to geocentric inertial coordinates using Eq. (H-14), but the exercise is arduous and not particularly instructive. However, we may take advantage of the dipole nature of the dominant term in the field model to approximate the magnetic field of the Earth as due to a vector dipole, **m**, whose magnitude and direction are given by Eqs. (H-18) through (H-20). Thus,

$$\mathbf{B}(\mathbf{R}) = \frac{a^3 H_0}{R^3}\left[3\,(\hat{\mathbf{m}}\cdot\hat{\mathbf{R}})\hat{\mathbf{R}} - \hat{\mathbf{m}}\right] \tag{H-22}$$

where **R** is the position vector of the point at which the field is desired. Because this is a vector equation, the components of **B** may be evaluated in any convenient coordinate system. As an example, the field in geocentric inertial components can be obtained from the dipole unit vector,

$$\hat{\mathbf{m}} = \begin{Bmatrix} \sin\theta'_m\cos\alpha_m \\ \sin\theta'_m\sin\alpha_m \\ \cos\theta'_m \end{Bmatrix} \tag{H-23}$$

$$\alpha_m = \alpha_{G0} + \frac{d\alpha_G}{dt}t + \phi'_m$$

where α_{G0} is the right ascension of the Greenwich meridian* at some reference time ($\alpha_{G0} = 98.8279°$ at 0^h UT, December 31, 1979), $d\alpha_G/dt$ is the average rotation rate of the Earth (360.9856469 deg/day), t is the time since reference, and (θ'_m, ϕ'_m) $= (168.6°, 109.3°)$ in 1978.

Then

$$\hat{\mathbf{m}} \cdot \hat{\mathbf{R}} = R_x \sin\theta'_m \cos\alpha_m + R_y \sin\theta'_m \sin\alpha_m + R_z \cos\theta'_m \qquad \text{(H-24)}$$

where R_x, R_y, and R_z are the geocentric inertial direction cosines of $\mathbf{R}$. The field components are

$$B_x = \frac{a^3 H_0}{R^3} \left[3(\hat{\mathbf{m}} \cdot \hat{\mathbf{R}}) R_x - \sin\theta'_m \cos\alpha_m \right]$$

$$B_y = \frac{a^3 H_0}{R^3} \left[3(\hat{\mathbf{m}} \cdot \hat{\mathbf{R}}) R_y - \sin\theta'_m \sin\alpha_m \right] \qquad \text{(H-25)}$$

$$B_z = \frac{a^3 H_0}{R^3} \left[3(\hat{\mathbf{m}} \cdot \hat{\mathbf{R}}) R_z - \cos\theta'_m \right]$$

These equations are useful for analytic computations and for checking computer calculations. For example, if $\mathbf{R}$ is in the Earth's equatorial plane, then $R_z = 0$ and

$$B_z = \frac{a^3 H_0}{R^3} (-\cos 168.6°) \qquad \text{(H-26)}$$

which is positive, i.e., *north*. Because the direction of the field line is customarily defined as that indicated by a compass needle, Eq. (H-22) is self-consistent.

For analytical work, the most useful coordinate system is the 1, *b*, *n* orbit plane system (Section 2.2), in which $\mathbf{R}$ has the particularly simple representation

$$R_1 = R(v')\cos v'$$

$$R_b = R(v')\sin v' \qquad \text{(H-27)}$$

$$R_n = 0$$

where v' is the true anomaly measured from the ascending node. Vectors are transformed into the l,b,n system from the geocentric inertial system by first rotating about the inertial z axis through Ω, the right ascension of the ascending node, followed by a rotation about the ascending node by the angle i, the orbital inclination. Using this transformation, the unit magnetic dipole is

$$m_l = \sin\theta'_m \cos(\Omega - \alpha_m)$$

$$m_b = -\sin\theta'_m \cos i \sin(\Omega - \alpha_m) + \cos\theta'_m \sin i \qquad \text{(H-28)}$$

$$m_n = \sin\theta'_m \sin i \sin(\Omega - \alpha_m) + \cos\theta'_m \cos i$$

---

*This technique of computing α_G is good to about 0.005° for 1 year on either side of the reference date. At times more distant from the reference date, a new α_{G0} can be computed as described in Appendix J. Note that α_{G0} is equal to the Greenwich sidereal time at the reference time of 0^h UT, December 31, 1978.

where Ω is the right ascension of the ascending node and i is the inclination of the orbit.

Substituting Eqs. (H-27) and (H-28) into Eq. (H-22) yields the magnetic field in the l,b,n system. Although the equations are moderately complex, they can still be useful. Due to the simple form for **R**, especially for circular orbits, it is possible to analytically integrate the torque due to a spacecraft dipole moment as has been done for ITOS [Kikkawa, 1971].

A circular equatorial orbit is particularly simple because $i = \Omega = 0$ and, therefore,

$$\hat{\mathbf{m}} \cdot \hat{\mathbf{R}} = \sin\theta'_m(\cos\alpha_m\cos\nu' + \sin\alpha_m\sin\nu') \qquad \text{(H-29)}$$

Substituting into Eq. (H-22) and simplifying yields

$$B_l = \frac{a^3 H_0}{2R^3}\sin\theta'_m\left[3\cos(2\nu' - \alpha_m) + \cos\alpha_m\right]$$

$$B_b = \frac{a^3 H_0}{2R^3}\sin\theta'_m\left[3\sin(2\nu' - \alpha_m) + \sin\alpha_m\right] \qquad \text{(H-30)}$$

$$B_n = -\frac{a^3 H_0}{R^3}\cos\theta'_m$$

As in Eq. (H-26), the minus sign in the orbit normal component, B_n assures the northward direction of the field lines.

The torque resulting from a spacecraft magnetic dipole interacting with B_n is in the orbit plane, or, in this case, the equatorial plane. This torque causes precession around the orbit normal, or, for $i = 0$, right ascension motion. Torque out of the orbit plane is caused by the ascending node component B_l and the component B_b. For $i = 0$, out of plane is the same as declination motion. Thus, for an equatorial orbit, the ratio of declination motion to right ascension motion is at most on the order of

$$\left|\frac{B_{l,b}}{B_n}\right| \cong \left|\frac{2\sin\theta'_m}{\cos\theta'_m}\right| = 0.4 \qquad \text{(H-31)}$$

Consequently, for a satellite in an equatorial or low-inclination orbit, the right ascension is the easier to control.

Note that the declination terms B_l and B_b in Eq. (H-30) oscillate with a frequency of twice the orbital period. That is, the direction of the magnetic field in the orbit plane system rotates through 720 deg during the orbit.

Thus, B_l and B_b change signs four times during the orbit. Declination motion then can be obtained in a certain direction by switching the polarity of the magnetic control coil four times or every quarter orbit. This is the basis for QOMAC control theory. If the satellite has a residual magnetic dipole, the B_n term will cause a secular drift in right ascension and the B_l, B_b terms will cause an oscillation in declination at twice the orbital period and a diurnal oscillation in declination due to the rotation of the Earth.

References

1. Leaton, B. R., "International Geomagnetic Reference Field 1975," *Trans., Amer. Geophysical Union (E $\oplus$ S)*, Vol. 57, p. 120, 1976.
2. Chapman, Sydney and Julius Bartels, *Geomagnetism*. Oxford, Clarendon Press, 1940.
3. Trombka, B. T. and J. C. Cain, *Computation of the IGRF I. Spherical Expansions*, NASA X-922-74-303, GSFC, Aug. 1974.
4. Jackson, John David, *Classical Electrodynamics*. New York: John Wiley & Sons, Inc., 1965.
5. Kikkawa, Shigetaka, *Dynamic Attitude Analysis Study*, Comp. Sc. Corp., CSC 5023-10000-01TR, Jan. 1971.

APPENDIX I

SPACECRAFT ATTITUDE DETERMINATION AND CONTROL SYSTEMS

Ashok K. Saxena

I.1   Spacecraft Listed by Stabilization Method
I.2   Spacecraft Listed by Attitude Determination Accuracy Requirements
I.3   Spacecraft Listed by Type of Control Hardware
I.4   Spacecraft Listed by Type Of Attitude Sensors

This appendix summarizes spacecraft attitude systems and serves as a *guide to mission specific attitude determination and control literature*. The main table is an alphabetical listing of satellites by acronym with pertinent data as available. Subsequent sections list these spacecraft by stabilization method, accuracy requirements, control system, and sensor system. For example, if you are interested in gravity-gradient stabilization in low-Earth orbit, Section I.1 lists DODGE, GEOS-3, and RAE-2 as gravity-gradient stabilized. The main table lists GEOS-3 as the only one of these in low-Earth orbit. Normal automated literature search procedures may then be used to obtain available literature citations for GEOS-3.

The material in this appendix has been collected from literature searches, the *TRW Space Log*, and Joseph and Plett [1974]. Design values are quoted for upcoming spacecraft, which are denoted by an asterisk after the spacecraft acronym. The superscript "b" is used for body and "w" is used for wheel.

I.1 Spacecraft Listed by Stabilization Method

Missions with mutiple phases (e.g., RAE-2) are listed in all appropriate categories.

Single Spin
> AEROS-1,2, ALOUETTE-1,2, ARIEL-III, ARYABHATA, ATS-3, CTS, DE-A*, ESRO-IV, GOES-1, HEAO-1,C, HEOS-1, IMP-6,7,8, ISEE-1,B,C*, ISIS-I, II, ISS, IUE, LES-5,7, RAE-2, SIRIO, SM-3, SMS-1,2, SKYNET (U.K.1,2), SSS-1

Dual Spin
> ANS, ATS-6, DODGE, FLTSATCOM, HEAO-B*, IUE, NIMBUS-5,6, OAO-2,3, OGO-1, SAGE*, SEASAT*, SMM*, ST, SYMPHONIE

Momentum Wheels
> ANS, ATS-6, DODGE, FLTSTCOM, HEAO-B*, IUE, NIMBUS-5,6, OAO-2, 3, OGO-1, SAGE*, SEASAT*, SMM*, ST, SYMPHONIE

Gravity Gradient
> DODGE, GEOS-3, RAE-2

Magnetic
> AZUR-I, HCMM*

Gas Jets
> HEAO-1,C*

ACRONYM	NAME	PRINCIPAL COUNTRY	MASS (KG)	LAUNCH DATE AND VEHICLE	ORBIT PARAMETERS			
					INCLINA- TION (DEG)	PERIGEE[1] HEIGHT (KM)	APOGEE[1] HEIGHT (KM)	PERIOD (MINUTES)
AE-3	ATMOSPHERE EXPLORER-3	USA	454 PLUS PROPEL- LANT	DEC. 16, 1973, BY DELTA	68.1	150	4,300	130
AE-5	ATMOSPHERE EXPLORER-5	USA	454 PLUS PROPEL- LANT	NOV. 19, 1975, BY DELTA	19	159	1,618	102.7
AEM	GENERAL NAME FOR APPLICATIONS EX- PLORER MISSION; e.g., HCMM*, SAGE*	USA						
AEROS-1	AERONOMY SATELLITE-1	GERMANY	127	DEC. 16, 1972, BY SCOUT	97.2	250	800	97.5
						SUN SYNCHRONOUS		
AEROS-2	AERONOMY SATELLITE-2	GERMANY	127	JULY 16, 1974, BY SCOUT	96.8	227	889	95.8
						SUN SYNCHRONOUS		
ALOUETTE-1		CANADA	145.6	SEPT. 29, 1962, BY THOR-AGENA B	80.5	996	1,031	105.4
ALOUETTE-2		CANADA	146.1	NOV. 28, 1965, BY THOR-AGENA B	80.5	507	2,892	121.4
ANS	NETHERLANDS ASTRONOMICAL SATELLITE	NETHERLANDS	130	AUG. 30, 1974, BY SCOUT	98.0	260	1,100	98.2
ARIEL-III		U.K.		MAY 5, 1967, BY SCOUT	80	500	~ 600	95.6
ARYABHATA		INDIA	360	APR. 19, 1975	50.4	564	623	96.4
ATS-3	APPLICATIONS TECHNOLOGY SATELLITE-3	USA	361	NOV. 5, 1967, BY ATLAS- AGENA D	0.4	35,772		1,436
						GEOSYNCHRONOUS		
ATS-6	APPLICATIONS TECHNOLOGY SATELLITE-6	USA	1,360	MAY 30, 1974, BY TITAN-III C	1.3	35,759	35,820	1,436
						GEOSYNCHRONOUS		
AZUR-1, ALSO CALLED GRS-A	GERMAN RESEARCH SATELLITE	GERMANY	71.3	NOV. 8, 1969, BY SCOUT	102.9	239	1,955	121.8
CTS	COMMUNICATIONS TECHNOLOGY SATELLITE	CANADA		JAN. 17, 1976, BY DELTA	0.1	35,786		1,436
						GEOSYNCHRONOUS		
DE-A*	DYNAMICS EXPLORER-A	USA	258	1981, BY DELTA	90	275	23,918	417
DE-B*	DYNAMICS EXPLORER-B	USA	305	1981, BY DELTA	90	275	1,200	100
DODGE	DEPARTMENT OF DEFENSE GRAVITY EXPERIMENT	USA	195.2	JULY 1, 1967, BY TITAN-III C	7.2	33,243	33,602	1,317
ERTS	EARTH RESOURCES TECHNOLOGY SATEL- LITE RENAMED LANDSAT							
ESRO-IV	EUROPEAN SPACE RESEARCH ORGANIZATION-IV	ESRO	120	NOV. 21, 1972, BY SCOUT	91.06	254	1,143	99

[1]PERIGEE AND APOGEE HEIGHT ARE MEASURED FROM THE SURFACE OF THE EARTH.

STABILIZATION TECHNIQUE (SPIN RATE IN RPM)	ATTITUDE CONTROL HARDWARE	POINTING ACCURACY	ATTITUDE SENSORS	MISSION OBJECTIVES
DUAL SPIN ω_b = 0.01–16 ω_w = 214.4	MAGNETIC Z AXIS COIL MAGNETIC SPIN COILS MOMENTUM WHEEL GAS JETS	± 1°	BODY-MOUNTED IR HORIZON SCANNERS WHEEL-MOUNTED IR HORIZON SCANNERS DIGITAL SUN SENSORS	ATMOSPHERE PROBE WITH SOLAR ABSORPTION EXPERIMENTS
DUAL SPIN ω_b = 0.01–16 ω_w = 214.4	MAGNETIC Z AXIS COIL MAGNETIC SPIN COILS MOMENTUM WHEEL GAS JETS	± 1°	BODY-MOUNTED IR HORIZON SCANNERS WHEEL-MOUNTED IR HORIZON SCANNERS DIGITAL SUN SENSORS	ATMOSPHERE PROBE
SPIN ω_b = 10	MANUAL AND AUTOMATIC MAGNETIC COIL FOR PRECESSION AUTOMATIC SPIN CORRECTION DURING DAYLIGHT	± 0.2° FOR SOLAR ASPECT ANGLE ± 3° FOR SOLAR AZIMUTH	TWO IR SENSORS ANALOG FINE SUN SENSOR DIGITAL SOLAR ASPECT SENSOR THREE-AXIS MAGNETOMETER	AERONOMY, MEASURES PHYSICAL PROPERTIES OF UPPER ATMOSPHERE IN CORRELATION WITH EXTREME ULTRAVIOLET RADIATION
SPIN ω_b = 10	MANUAL AND AUTOMATIC MAGNETIC COIL FOR PRECESSION AUTOMATIC SPIN CORRECTION DURING DAYLIGHT	± 0.2° FOR SOLAR ASPECT ANGLE ± 3° FOR SOLAR AZIMUTH	TWO IR SENSORS ANALOG FINE SUN SENSOR DIGITAL SOLAR ASPECT SENSOR THREE-AXIS MAGNETOMETER	AERONOMY, MEASURES PHYSICAL PROPERTIES OF UPPER ATMOSPHERE IN CORRELATION WITH EXTREME ULTRAVIOLET RADIATION
SPIN ω_b = 1.43				MEASUREMENT OF EARTH'S IONOSPHERE BY RADIO SOUNDING
SPIN ω_b = 2.25				IONOSPHERIC RESEARCH
MOMENTUM WHEELS ONBOARD COMPUTER	THREE ORTHOGONAL MOMENTUM WHEELS X, Y, AND Z MAGNETIC COILS	± 0.3° FOR SCAN AND SLOW SCAN ± 1° FOR STAR, OFFSET, AND X-RAY	STAR SENSOR HORIZON SENSOR SOLAR SENSOR MAGNETOMETER	ASTRONOMY
SPIN		± 2°	SOLAR ASPECT SENSORS	MEASURE VERTICAL DISTRIBUTION OF MOLECULAR OXYGEN, MAP RADIO FREQUENCY NOISE IN GALAXY
SPIN ω_b = 10–90	NITROGEN JETS FOR SPIN	± 1°	DIGITAL SOLAR SENSOR TRIAXIAL FLUGATE MAGNETOMETER	X-RAY ASTRONOMY SOLAR PHYSICS AERONOMY
SPIN ω_b = 100	HYDRAZINE JETS	± 1°	STAR SENSOR V-SLIT SUN SENSOR SCANNING RADIOMETER	COMMUNICATIONS METEOROLOGY
MOMENTUM WHEELS	THREE MOMENTUM WHEELS GAS JETS	± 0.1°	IR EARTH SENSORS DIGITAL SUN SENSORS POLARIS TRACKER INERTIAL REFERENCE GYROSCOPES INTERFEROMETER	ERECT A LARGE ANTENNA STRUCTURE PROVIDE GOOD TV SIGNAL TO LOW-COST GROUND RECEIVERS DEMONSTRATE USER-ORIENTED APPLICATION EXPERIMENTS
PASSIVE MAGNETIC	TWO FIXED MAGNETS SUCH THAT THE RESULTING DIPOLE MOMENT POINTS IN THE DIRECTION OF THE Z AXIS WITH AN ACCURACY OF 8 MINUTES	ANGLE BETWEEN THE GEOMAGNETIC FIELD VECTOR AND THE Z-AXIS WAS LESS THAN 12°	SUN SENSOR TWO-AXIS MAGNETOMETER	MEASUREMENT OF RADIATION AND PARTICLES IN THE POLAR ZONE AND IN THE VAN ALLEN BELT
SPIN IN TRANSFER AND DRIFT ORBITS ω_b = 60 MOMENTUM WHEEL ω_w = 3750	HYDRAZINE CATALYTIC THRUSTERS MOMENTUM WHEEL	± 1° TRANSFER AND DRIFT ORBIT ± 0.1° PITCH AND ROLL } MISSION ORBIT ± 1.1° YAW	TWO NONSPINNING EARTH SENSORS TWO IR HORIZON SCANNERS NONSPINNING SUN SENSOR SPINNING DIGITAL SUN SENSOR	COMMUNICATIONS
SPIN ω_b = 10	HYDRAZINE THRUSTERS	± 1°	V-HORIZON SCANNERS SUN SENSORS	INVESTIGATE ELECTRODYNAMIC EFFECTS IN UPPER ATMOSPHERE
DUAL SPIN ω_b = 1 REV/ORBIT	PITCH MOMENTUM WHEEL MAGNETIC COILS HYDRAZINE THRUSTERS	± 1°	HORIZON SCANNERS SUN SENSORS	INVESTIGATE ELECTRODYNAMIC EFFECTS IN UPPER ATMOSPHERE
GRAVITY-GRADIENT AUGMENTED BY CONSTANT-SPEED ROTOR AND MAGNETIC DIPOLE FOR YAW	MOMENTUM WHEEL	± 16°	SUN SENSORS TV CAMERAS	DEMONSTRATE GRAVITY-GRADIENT STABILIZATION NEAR GEOSYNCHRONOUS ALTITUDE
SPIN ω_b = 65	MAGNETIC COILS	ATTITUDE DETERMINATION ± 2° ATTITUDE CONTROL ± 5°	IR HORIZON SCANNER DIGITAL SUN SENSOR TRIAXIAL FLUX GATE MAGNETOMETER	STUDY NEAR-EARTH MAGNETOSPHERE AND IONOSPHERE

ACRONYM	NAME	PRINCIPAL COUNTRY	MASS (KG)	LAUNCH DATE AND VEHICLE	ORBIT PARAMETERS			
					INCLINA-TION (DEG)	PERIGEE[1] HEIGHT (KM)	APOGEE[1] HEIGHT (KM)	PERIOD (MINUTES)
EXPLORER	GENERAL NAME GIVEN TO SATELLITES; e.g., EXPLORER-45 IS SAME AS SSS-1							
FLTSATCOM	FLEET SATELLITE COMMUNICATION SYSTEM	USA	1,861	ATLAS-CENTAUR		GEOSYNCHRONOUS		1,436
GEOS-3	GEODYNAMICS EXPERIMENTAL OCEAN SATELLITE-3	USA	340	APR. 9, 1975, BY DELTA	114	835	846	102
GOES-1	GEOSTATIONARY OPERATIONAL ENVIRONMENTAL SATELLITE-1	USA	295	OCT. 16, 1975, BY DELTA	1.0	35,770	35,796 GEOSYNCHRONOUS	1,435.9
HCMM*	HEAT CAPACITY MAPPING MISSION	USA	~ 95	MAY 1978, BY SCOUT	98	600	600	97
HEAO-1	HIGH ENERGY ASTRONOMY OBSERVATORY-1	USA	3,150	AUG. 12, 1977, BY ATLAS-CENTAUR	22.75	417	434	93
HEAO-B*	HIGH ENERGY ASTRONOMY OBSERVATORY-B	USA	3,150	NOV. 1978, BY ATLAS-CENTAUR	22.75	~ 435	~ 435	~ 95
HEAO-C*	HIGH ENERGY ASTRONOMY OBSERVATORY-C	USA	3,150	LATE 1979, BY ATLAS-CENTAUR	22.75	~ 460	~ 460	~ 95
HEOS-1	HIGHLY ECCENTRIC ORBIT SATELLITE	ESRO	106	DEC. 5, 1968, BY DELTA	28.28	424	223,428	6,792
IMP-6	INTERPLANETARY MONITORING PLATFORM-6	USA	286	MAR. 13, 1971, BY THOR-DELTA	28.8	146	122,146	5,628
IMP-7	INTERPLANETARY MONITORING PLATFORM-7	USA	376	SEPT. 23, 1972, BY DELTA	17.2	197,391	235,389	17,439
IMP-8	INTERPLANETARY MONITORING PLATFORM-8	USA	371	OCT. 26, 1973, BY DELTA	28.2	196,041	235,754	17,362
ISEE-1	INTERNATIONAL SUN-EARTH EXPLORER-1	USA	315	OCT. 22, 1977, BY DELTA	28.4	288	134,000	2.29 DAYS
ISEE-B	INTERNATONAL SUN-EARTH EXPLORER-B	ESA	165	LAUNCHED SIMULTANE-OUSLY WITH ISEE-A ON THE SAME LAUNCH VEHICLE	28.4	288	134,000	2.29 DAYS
ISEE-C*	INTERNATIONAL SUN-EARTH EXPLORER-C	USA	469	JULY 1978, BY DELTA	IN ECLIPTIC PLANE[2]	PERIHELION AT 0.973 AU[2]	APOHELION AT 1.007 AU[2]	1 YEAR
ISIS-I	INTERNATIONAL SATELLITE FOR IONOSPHERIC STUDIES-1	CANADA	238	JAN. 30, 1969, BY DELTA	88.43	574	3,522	128.3
ISIS-II	INTERNATIONAL SATELLITE FOR IONOSPHERIC STUDIES-2	CANADA	264	APR. 1, 1971, BY THOR-DELTA	88.16	1,356	1,423	113
ISS	IONOSPHERE SOUNDING SATELLITE		135		~ 70	~ 1,000	~ 1,000	
ITOS-H = NOAA-5	IMPROVED TIROS OPERATIONAL SATELLITE-8 = NATIONAL OCEANIC AND ATMOSPHERIC ADMINISTRATION-5	USA	340	JULY 29, 1976, BY DELTA	102	1,507	1,522	116.2

[1]PERIGEE AND APOGEE HEIGHT ARE MEASURED FROM THE SURFACE OF THE EARTH.

STABILIZATION TECHNIQUE (SPIN RATE IN RPM)	ATTITUDE CONTROL HARDWARE	POINTING ACCURACY	ATTITUDE SENSORS	MISSION OBJECTIVES
MOMENTUM WHEELS	THREE MOMENTUM WHEELS AT 3000 ± 300 RPM, HYDRAZINE THRUSTERS	± 0.2°	TWO SPINNING EARTH SENSORS TWO SPINNING SUN SENSORS ONE NONSPINNING EARTH SENSOR TWO NONSPINNING SUN SENSORS	WORLDWIDE COMMUNICATION SYSTEM FOR THE NAVY SUPPORT AIR FORCE COMMAND AND CONTROL
GRAVITY GRADIENT DUAL SPIN ω_w = 2000	PITCH WHEEL REACTION BOOM Z AXIS MAGNETIC COIL	± 1.5° YAW ± 1.2° POINTING	THREE DIGITAL SUN SENSORS THREE-AXIS MAGNETOMETER	SOLID EARTH PHYSICS AND OCEANOGRAPHY
SPIN ω_b = 90	HYDRAZINE JETS	± 1.0° (TRANSFER ORBIT) ± 0.1° (MISSION ORBIT)	FIVE IR HORIZON SCANNERS TWO DUAL-SLIT SUN SENSORS	OPERATIONAL METEOROLOGICAL SATELLITE
THREE-AXIS ONBOARD CONTROL, DUAL SPIN ω_b = 1 REV/ORBIT ω_w = 1940	ONE MOMENTUM WHEEL THREE MAGNETIC COILS	ATTITUDE DETERMINATION ± 0.7° IN ROLL, ± 0.5° IN PITCH, ± 2° IN YAW ATTITUDE CONTROL ± 1° IN ROLL AND PITCH ± 2° IN YAW	ONE IR SCANNER THREE DIGITAL SUN SENSORS TRIAXIAL FLUX GATE MAGNETOMETER	MAP HEAT CAPACITY OF THE EARTH
ONBOARD COMPUTER SINGLE-AXIS SPIN, THREE-AXIS STABILIZED	HYDRAZINE JETS	± 1°	COARSE SUN SENSORS FINE SUN SENSORS TWO 8° x 8° FOV STAR TRACKERS FOUR SDOF GYROSCOPES	ALL SKY X- AND GAMMA-RAY SURVEY
ONBOARD COMPUTER THREE-AXIS STABILIZED	MOMENTUM WHEELS MAGNETIC COILS HYDRAZINE JETS	ONBOARD: ± 1° ARC MINUTE; GROUND ATTITUDE DETERMINATION: ± 0.1 ARC MINUTE	COARSE SUN SENSORS FINE SUN SENSORS THREE 2° x 2° FOV STAR TRACKERS SIX SDOF GYROSCOPES	X-RAY TELESCOPE TO DETERMINE INTENSITY, SPECTRA, POSITION, AND TIME VARIATIONS OF X-RAY SOURCES
ONBOARD COMPUTER SINGLE-AXIS SPIN-STABILIZED	HYDRAZINE JETS	ATTITUDE CONTROL ± 1° ATTITUDE DETERMINATION ± 0.05°	COARSE SUN SENSORS FINE SUN SENSORS TWO 8° x 8° FOV STAR TRACKERS FOUR SDOF GYROSCOPES	ALL SKY GAMMA-RAY SURVEY
SPIN ω_b = 10	NITROGEN JETS		TWO EARTH ALBEDO SENSORS TWO SOLAR ASPECT SENSORS SOLAR GATE SENSOR	STUDY INTERPLANETARY RADIATION, SOLAR WIND, AND MAGNETIC FIELDS OUTSIDE THE MAGNETOSPHERE
SPIN ω_b = 5	FREON JETS	± 1°	OPTICAL HORIZON SCANNER DIGITAL SUN SENSOR	STUDY INTERPLANETARY PARTICLES AND ELECTROMAGNETIC RADIATION
SPIN ω_b = 46	FREON JETS	± 1°	OPTICAL TELESCOPE DIGITAL SUN SENSOR	STUDY SOLAR PLASMA, SOLAR WIND, SOLAR AND COSMIC RADIATION, ELECTROMAGNETIC FIELD VARIATIONS, EARTH'S MAGNETIC TRAIL
SPIN ω_b = 12–68	FREON JETS	± 0.5°	OPTICAL TELESCOPE DIGITAL SUN SENSOR	STUDY SOLAR PLASMA, SOLAR WIND, SOLAR AND GALACTIC COSMIC RADIATION, ELECTROMAGNETIC FIELD VARIATIONS, AND INTERPLANETARY MAGNETIC FIELD
SPIN ω_b = 20	COLD GAS JETS	± 1°	PANORAMIC ATTITUDE SENSOR SOLAR ASPECT SENSOR	STUDY MAGNETOSPHERE, INTERPLANETARY SPACE AND INTERACTIONS BETWEEN THEM
SPIN ω_b = 20	COLD GAS JETS	± 3°	V-SLIT COMBINED SUN/EARTH ALBEDO SENSOR	STUDY MAGNETOSPHERE, INTERPLANETARY SPACE AND INTERACTIONS BETWEEN THEM
SPIN ω_b = 20	HYDRAZINE JETS	± 1°	PANORAMIC ATTITUDE SENSOR SOLAR ASPECT SENSOR	STUDY MAGNETOSPHERE, INTERPLANETARY SPACE AND INTERACTIONS BETWEEN THEM
SPIN ω_b = 2.938	MAGNETIC COIL WHEN COMMUTATED WITH SUNLINE OR GEOMAGNETIC FIELD VECTOR GIVES ATTITUDE OR SPIN RATE CHANGE			MEASUREMENT OF EARTH'S IONOSPHERE BY RADIO SOUNDING
SPIN ω_b = 2.9	MAGNETIC COIL FOR ATTITUDE AND SPIN RATE CHANGE			MEASUREMENT OF EARTH'S IONOSPHERE BY RADIO SOUNDING
SPIN ω_b = 10		± 1.5°	IR SENSOR DIGITAL SUN SENSOR MAGNETOMETER	WORLD MAP OF ELECTRON DENSITY OF THE ATMOSPHERE
DUAL SPIN ω_b = 1 REV/ORBIT ω_w = 140	MAGNETIC COILS	SUFFICIENT TO MAINTAIN CONSTRAINTS	IR HORIZON SENSORS SOLAR ASPECT SENSORS	METEOROLOGICAL SATELLITE

ACRONYM	NAME	PRINCIPAL COUNTRY	MASS (KG)	LAUNCH DATE AND VEHICLE	ORBIT PARAMETERS			
					INCLINA-TION (DEG)	PERIGEE[1] HEIGHT (KM)	APOGEE[1] HEIGHT (KM)	PERIOD (MINUTES)
ITOS-I*, J*	IMPROVED TIROS OPERATIONAL SATELLITE-I, J	USA		DELTA	102	~ 1,459	~ 1,466	115.2
IUE*	INTERNATIONAL ULTRAVIOLET EXPLORER	USA	430	JAN. 26, 1978, BY AUG-MENTED THRUST DELTA	28.7	25,706	45,876	1,436
					ELLIPTICAL GEOSYNCHRONOUS			
LANDSAT-1	ORIGINALLY CALLED ERTS	USA	959	JULY 23, 1972, BY DELTA	99	915	917	103.3
LANDSAT-2		USA	959	JUNE 22, 1975, BY DELTA	99	898	916	103.1
LANDSAT-C*		USA		THIRD QUARTER 1977, BY DELTA	99	916	916	103.3
LANDSAT-D		USA		~ 1981	98	700	700	99
LES-5	LINCOLN EX-PERIMENTAL SATELLITE-5	USA	102	JULY 1, 1967, BY TITAN-III C (MULTIPLE PAY-LOAD)	7.2	33,300	33,600	1,319
LES-7	LINCOLN EX-PERIMENTAL SATELLITE-7	USA	~ 300		~ 0	NEAR GEOSTATIONARY		
MAGSAT*	GEOMAGNETIC FIELD SATELLITE	USA	~ 165	1979–1980	97	325	550	96
MMS	MULTIMISSION MODULAR SPACE-CRAFT SERIES (e.g., SMM*)							
NIMBUS-5		USA	772	DEC. 11, 1972, BY DELTA	99.95	1,089	1,102	107.25
NIMBUS-6		USA	829	JUNE 12, 1975, BY DELTA	99.9	1,100	1,113	107.4
OAO-2	ORBITING ASTRO-NOMICAL OBSER-VATORY-2	USA	1,998	DEC. 7, 1968, BY ATLAS-CENTAUR	35	764	774	100.3
OAO-3	ORBITING ASTRO-NOMICAL OBSER-VATORY-3, ALSO COPERNICUS	USA	2,204	AUG. 21, 1972, BY ATLAS-CENTAUR	35	738	750	99.7
OGO-1	ORBITING GEO-PHYSICAL OBSERVA-TORY	USA	487	SEPT. 4, 1964, BY ATLAS-AGENA B	31.1	260	150,000	64 HOURS
OSO-7	ORBITING SOLAR OBSERVATORY-7	USA	635	SEPT. 29, 1971, BY DELTA	33	323	571	93.4
OSO-8	ORBITING SOLAR OBSERVATORY-8	USA	1,090	JUNE 21, 1975, BY DELTA	32.95	545.9	562.4	95.74
RAE-2	RADIO ASTRONOMY EXPLORER-2	USA	205 (ORBIT) 328 (LAUNCH)	JUNE 10, 1973, BY AUGMENTED THRUST LONG-TANK DELTA		1,100	1,100	225
					SELENOCENTRIC ORBIT			
SAGE*	STRATOSPHERIC AEROSOL AND GAS EXPERIMENT, ALSO KNOWN AS AEM-B*	USA	~ 105	FEB. 1979	~ 50°	~ 600	~ 600	~ 97
SAS-2	SMALL ASTRONOMY SATELLITE-2	USA	185	NOV. 15, 1972, BY SCOUT	2	443	632	96

[1]PERIGEE AND APOGEE HEIGHT ARE MEASURED FROM THE SURFACE OF THE EARTH.

[2]ORBIT ABOUT SUN-EARTH LAGRANGIAN POINT L_1.

STABILIZATION TECHNIQUE (SPIN RATE IN RPM)	ATTITUDE CONTROL HARDWARE	POINTING ACCURACY	ATTITUDE SENSORS	MISSION OBJECTIVES
DUAL SPIN ω_b = 1 REV/ORBIT ω_w = 140	MAGNETIC COILS	SUFFICIENT TO MAINTAIN CONSTRAINTS	IR HORIZON SENSORS SOLAR ASPECT SENSORS	METEOROLOGICAL SATELLITES
SPIN ABOUT AXIS OF MINIMUM MOMENT OF INERTIA; THREE-AXIS STABILIZED IN MISSION NODE	HYDRAZINE JETS REACTION WHEELS	1 ARC-SECOND	PANORAMIC ATTITUDE SENSOR SPIN MODE SUN SENSOR FINE SUN SENSOR ANALOG SUN SENSOR GYROSCOPES FINE ERROR SENSOR	ULTRAVIOLET SPECTROGRAPHY OF STELLAR SOURCES
DUAL SPIN		ATTITUDE CONTROL ± 0.7º ATTITUDE DETERMINATION ± 0.07º	HORIZON SENSORS	EARTH'S RESOURCES
DUAL SPIN		ATTITUDE CONTROL ± 0.7º ATTITUDE DETERMINATION ± 0.07º	HORIZON SENSORS	EARTH'S RESOURCES
DUAL SPIN		ATTITUDE CONTROL ± 0.7º ATTITUDE DETERMINATION ± 0.07º	HORIZON SENSORS	EARTH'S RESOURCES
	MMS PACKAGE (SEE SMM)	± 0.01º	TWO STAR TRACKERS GYROSCOPES THEMATIC MAPPER FOR LANDMARK TRACKING	EARTH SURVEY
SPIN ω_b = 10	TWO ORTHOGONAL MAGNETIC COILS	± 2º	EARTH SENSOR WITH 3 FAN BEAM FIELD OF VIEW FOUR SOLAR GATE SENSORS	EXPERIMENTAL SATELLITE
SPIN ω_w = 1100	GIMBALED FLYWHEEL PLASMA JETS	± 0.1º	IR EARTH SENSORS	EXPERIMENTAL SATELLITE
DUAL SPIN	MAGNETIC COILS	ATTITUDE DETERMINATION: 20 ARC-SEC	TWO STAR CAMERAS FINE SUN SENSOR GYROSCOPES ATTITUDE TRANSFER SYSTEM	ACCURATE MAPPING OF THE EARTH'S MAGNETIC FIELD STUDY OF THE EARTH'S CRUST, MANTLE, AND CORE
MOMENTUM WHEELS	FOUR MOMENTUM WHEELS (TWO ON ROLL AXIS)	± 1º	HORIZON SENSORS	METEOROLOGICAL
MOMENTUM WHEELS	FOUR MOMENTUM WHEELS (TWO ON ROLL AXIS)	± 1º	HORIZON SENSORS	METEOROLOGICAL
MOMENTUM WHEELS	THREE MOMENTUM WHEELS	± 0.02º	STAR TRACKER RATE INTEGRATING GYROSCOPES	ULTRAVIOLET ASTRONOMY
MOMENTUM WHEELS	THREE MOMENTUM WHEELS	± 0.1 ARC-SEC = 3×10^{-5} DEG	FOUR GIMBALED STAR TRACKERS RATE-INTEGRATING GYROSCOPES	ULTRAVIOLET ASTRONOMY
MOMENTUM WHEELS	THREE MOMENTUM WHEELS AIR JETS	± 2º	TWO HORIZON TRACKERS YAW AND SOLAR ARRAY SUN SENSORS PITCH RATE GYROSCOPES	GEOPHYSICAL
DUAL SPIN ω_{SAIL} = 0 ω_w = 30	Z AXIS MAGNETIC COIL NITROGEN JETS	± 0.1º	V-SLIT STAR SENSOR V-SLIT SUN SENSOR MAGNETOMETER	MEASURE SOLAR AND COSMIC X-RAYS, GAMMA RAYS, ULTRAVIOLET RADIATION, AND OTHER ASPECTS OF SOLAR ACTIVITY
DUAL SPIN ω_{SAIL} = 0 ω_w = 6	Z AXIS MAGNETIC COIL NITROGEN JETS	± 0.1º	V-SLIT STAR SENSOR V-SLIT SUN SENSOR MAGNETOMETER	SOLAR AND COSMIC X-RAYS, MEASURE SOLAR ULTRAVIOLET LINE PROFILES, AERONOMY
SPIN THREE-AXIS GRAVITY GRADIENT	FREON JETS FOR SPIN AND PRECESSION	ATTITUDE DETERMINATION: ± 1º IN SPIN ± 3º IN GRAVITY-GRADIENT CONTROL ATTITUDE CONTROL: ± 10º PITCH, ROLL ± 20º YAW	OPTICAL HORIZON SCANNER WITH VARIABLE MOUNTING ANGLE DIGITAL SUN SENSOR	STUDY RF SOURCES
MOMENTUM WHEELS ω_b = 1 REV/ORBIT	TWO MOMENTUM WHEELS THREE MAGNETIC COILS	ATTITUDE DETERMINATION: ± 0.5º PITCH ± 0.7º ROLL ± 2º YAW ATTITUDE CONTROL: ± 1º IN PITCH, ROLL ± 2º YAW	TWO IR SENSORS FIVE DIGITAL SUN SENSORS MAGNETOMETERS	STUDY THE OZONE LAYER
DUAL SPIN ω_b = 0.5 TO 1º/SEC ω_w = 2,000	MAGNETIC COILS FOR SPIN AND PRECESSION	± 1º	N-SLIT STAR SENSOR DIGITAL SUN SENSOR MAGNETOMETERS	GAMMA-RAY ASTRONOMY

| ACRONYM | NAME | PRINCIPAL COUNTRY | MASS (KG) | LAUNCH DATE AND VEHICLE | ORBIT PARAMETERS | | | |
					INCLINA-TION (DEG)	PERIGEE[1] HEIGHT (KM)	APOGEE[1] HEIGHT (KM)	PERIOD (MINUTES)
SAS-3	SMALL ASTRONOMY SATELLITE-3	USA	189	MAY 7, 1975, BY SCOUT	2.9	503	511	94.8
SEASAT*		USA	2,315	JUNE 1978, BY ATLAS-AGENA	108	790	790	100
SIRIO	ITALIAN INDUSTRIAL OPERATIONS RE-SEARCH SATELLITE	ITALY	190	AUG. 25, 1977, BY DELTA	0.3	35,800 GEOSYNCHRONOUS	35,800	1,436
SKYNET (UK−1)		U.K.	129	NOV. 22, 1969, BY DELTA	2.4	34,700 GEOSYNCHRONOUS	36,680	1431.0
SM-3	SAN MARCO-3	ITALY	362	APRIL 24, 1971, BY SCOUT	3	138	449	93.8
SMS-1, -2	SYNCHRONOUS METEOROLOGICAL SATELLITE-1, -2	USA	243	MAY 17, 1974; FEB. 6, 1975, BY DELTA	1.8 0.4	35,785 35,482 GEOSYNCHRONOUS	35,788 36,103	1,436.1 1,436.5
SMM*	SOLAR MAXIMUM MISSION (FIRST MULTI-MISSION SATELLITE)	USA	2,087	1979, BY DELTA	28.5 OR 33	~ 560	~ 560	~ 96
SSS-1	SMALL SCIENTIFIC SATELLITE-1	USA	52	NOV. 15, 1971, BY SCOUT	3	222	28,876	517
ST	SPACE TELESCOPE	USA		~ 1983, BY SHUTTLE	28°	500	500	95
SYMPHONIE		FRANCE/ GERMANY	221	DEC. 19, 1974, BY DELTA	0.5	35,017 GEOSYNCHRONOUS	35,852	1,418
TIROS-IX		USA	135	JAN. 22, 1965, BY DELTA	96.4	800	3,000	119

[1]PERIGEE AND APOGEE HEIGHT ARE MEASURED FROM THE SURFACE OF THE EARTH.

STABILIZATION TECHNIQUE (SPIN RATE IN RPM)	ATTITUDE CONTROL HARDWARE	POINTING ACCURACY	ATTITUDE SENSORS	MISSION OBJECTIVES
DUAL SPIN ω_b = 0–5 REV/ORBIT	MAGNETIC COILS FOR SPIN AND PRECESSION	± 1'	TWO STAR CAMERAS N-SLIT STAR SENSOR IR HORIZON SCANNER SPINNING AND NONSPINNING SUN SENSOR MAGNETOMETER	X-RAY ASTRONOMY
DUAL SPIN ω_b = 1 REV/ORBIT ω_w = 2300	MOMENTUM WHEEL REACTION WHEEL THREE MAGNETIC COILS	ATTITUDE DETERMINATION: ± 0.2° IN PITCH, ROLL, AND YAW ATTITUDE CONTROL: ± 0.5° IN PITCH, ROLL, AND YAW	TWO IR HORIZON SENSORS FOUR DIGITAL SUN SENSORS MAGNETOMETERS	OCEAN PHYSICS
SPIN ω_b = 90	HYDRAZINE JETS	± 1°	IR SLIT HORIZON SENSOR IR HORIZON TELESCOPE PLANAR AND V BEAM SUN SENSORS	SUPER-HIGH-FREQUENCY COMMUNICATIONS 12–18 GIGAHERTZ
SPIN	GAS JETS		EARTH HORIZON SENSOR DIGITAL SUN SENSOR	COMMUNICATIONS
SPIN	MAGNETIC COIL	± 1°	SUN SENSOR MAGNETOMETER TRIAD	OBTAIN DATA ON ATMOSPHERIC DENSITY AND MOLECULAR TEMPERATURE; DETERMINE NITROGEN CONCENTRATION
SPIN ω_b = 90	HYDRAZINE JETS	± 1.0° (TRANSFER ORBIT) ± 0.1° (MISSION ORBIT)	FIVE IR HORIZON SCANNERS TWO OVAL-SLIT SUN SENSORS	METEOROLOGY
MOMENTUM WHEELS ONBOARD COMPUTER	FOUR MOMENTUM WHEELS SIX MAGNETIC COILS	± 0.1° ROLL ABOUT SUNLINE; ± 5 ARC-SEC IN PITCH AND YAW	THREE TWO-AXIS GYROSCOPES REDUNDANT FINE POINTING SUN SENSORS REDUNDANT TRIAXIAL MAGNETOMETERS TWO FIXED-HEAD STAR TRACKERS COARSE SUN SENSOR	SOLAR ASTROPHYSICS
SPIN ω_b = 7	MAGNETIC COIL ACTIVATED AT PERIGEE ONLY	± 1°	STAR SCANNER OPTICAL HORIZON SENSOR DIGITAL SUN SENSOR	STUDY MAGNETOSPHERE
ONBOARD COMPUTER USING FINE ERROR SENSOR	MOMENTUM WHEELS MAGNETIC TORQUING	POINTING ACCURACY: ± 0.1 ARC-SEC STABILITY: ± 0.01 ARC-SEC	TWO STAR TRACKERS SUN SENSORS FINE ERROR SENSOR	MAJOR ASTRONOMICAL TELESCOPE
MOMENTUM WHEELS	MOMENTUM WHEELS GAS JETS	± 1°	EARTH HORIZON SENSOR DIGITAL SUN SENSOR	EXPERIMENTAL COMMUNICATIONS SATELLITE
SPIN	MAGNETIC COILS	± 1°	CONICAL SCAN HORIZON SENSORS	METEOROLOGICAL

I.2 Spacecraft Listed by Attitude Determination Accuracy Requirements

⩽ 1′	⩽ 0.1°	⩽ 0.25°	⩽ 0.5°
ANS	ATS-6	AEROS-1,2	HCMM*
HEAO-B*	CTS (mission)	FLTSATCOM	IMP-8
IUE	GOES-1 (mission)	ISEE-C*	
LANDSAT-D*	HEAO-1,C*	SEASAT*	
MAGSAT*	LANDSAT-1,2,C*		
OAO-3	LES-7		
SAS-3	OAO-2		
SMM*	OSO-7,8		
ST*	SMS-1,-2 (mission)		

⩽ 1°		⩽ 1.5°	⩽ 2°
AE-3,5	RAE-2	GEOS-3	ARIEL-III
ARYABHATA	SAS-2	ISS	LES-5
ATS-3	SIRIO		OGO-1
CTS (acquisition)	SM-3		
DE-A*,B*	SMS-1,-2 (transfer)		
GOES-1 (transfer)	SSS-1		
IMP-6,7	SYMPHONIE		
ISEE-1,B	TIROS-IX		
NIMBUS-5,6			

I.3 Spacecraft Listed by Type of Control Hardware

Magnetic

AE-3,-5, AEROS-1,2, ANS, AZUR-1 (passive), DE-B*, ESRO-IV, GEOS-3 (acquisition), HCMM*, HEAO-B*, ISIS-I,II, ITOS-8,I*,J*, LANDSAT-D*, LES-5, MAGSAT, OSO-7,8, SAS-2,3, SM-3, SSS-1, ST*, TIROS-IX

Jets

AE-3,5, ARYABHATA, ATS-3,6, CTS, DE-A*,-B*, FLTSATCOM, GOES-1, HEAO-1,B*,C*, HEOS-1, IMP-6,7,8, ISEE-1,B,C*, IUE, LES-7(plasma), OGO-1, OSO-7,8, RAE-2 (acquisition), SIRIO, SKYNET (U.K.1,2), SYMPHONIE

Momentum Wheel

AE-3,5, ANS, ATS-6, GEOS-3 (mission), FLTSATCOM, HCMM*, HEAO-B*, IUE, LANDSAT-D*, MAGSAT*, NIMBUS-5,6, OAO-2,-3, OGO-1, SAGE*, SEASAT*, SMM*, ST*, SYMPHONIE

Gimbaled Flywheel

LES-7

Reaction Boom

GEOS-3

I.4 Spacecraft Listed by Type of Attitude Sensors

Star Sensors

Star Scanner

ANS, ATS-3, OSO-7,8, SAS-2,3, SSS-1

Fixed-Head Star Trackers

HEAO-1,B*,C*, LANDSAT-D*, MAGSAT*, SAS-3, SMM*, ST*

Gimbaled Star Trackers

ATS-6: OAO-2,3

Horizon Scanners

Optical

DODGE, IMP-6,7,8, ISEE-1,C*, IUE, RAE-2, SSS-1

Infrared

AE-3,-5, AEROS-1,-2, ATS-6, CTS, ESRO-IV, FLTSATCOM, ISS, GOES-1, HCMM*, ITOS-8,I*,J*, LES-7, MAGSAT*, SAGE*, SAS-3, SEASAT*, SMS-1,-2, SIRIO

Sun Sensors

Analog Sun Sensor

AEROS-1,2, ATS-3,6, GOES-1, IUE, SIRIO, SMM*, SMS-1,-2

One-Axis Digital Sun Sensors

AE-3,5, AEROS-1, ARYABHATA, ATS-6, CTS, ESRO-IV, IMP-8,9,10, ISEE-1,C*, ISS, ITOS-8,I*,J*, IUE, RAE-2, SAS-2,3, SKYNET (U.K.1,2), SM-3, SSS-1, SYMPHONIE

Two-Axis Digital Sun Sensors

ATS-6, CTS, GEOS-3, HCMM*, HEAO-1,B*,C*, IUE, MAGSAT*, RAE-2, SAGE*, SAS-3, SEASAT*, SMM*, ST*

Magnetometers

AE-3-5, AEROS-1, ANS, ARYABHATA, GEOS-3, HCMM*, ISS, MAG-SAT*, OAO-2,-3, OSO-7,8, SAGE*, SAS-2,3, SEASAT*, SM-3, SMM*

Gyroscopes

ATS-6, HEAO-1,B*,C*, IUE, LANDSAT-D*, MAGSAT*, OAO-2,-3, SMM*

References

1. Joseph, M. and M. Plett, *Sensor Standardization Study Task Report*, Comp. Sc. Corp., 3000-19300-01TN, May 1974.
2. TRW Systems Group, Public Relations Staff, *TRW Space Log*, Redondo Beach, CA (Annual Report).

APPENDIX J

TIME MEASUREMENT SYSTEMS

Conrad R. Sturch

International Atomic Time, TAI, which is provided by atomic clocks, is the basis for the two time systems used in spacecraft time measurements. *Ephemeris Time, ET*, which is used in the preparation of ephemerides, is a uniform or "smoothly flowing" time and is related to *TAI* by

$$ET = TAI + 32.18 \text{ sec}$$

In contrast to *ET, Coordinated Universal Time, UTC*, uses the *TAI* second as the fundamental unit, but introduces 1-second steps occasionally to make *UTC* follow the nonuniform rotation of the Earth. *UTC* is necessary for terrestrial navigation and surveying for which the rotational position of the Earth at a given instant is critical. It is this time which is broadcast internationally and is used for tagging spacecraft data and for all civil timekeeping. Finally, *sidereal time* is a direct measure of the rotational orientation of the Earth relative to the "fixed" stars and, therefore, is used to estimate the position of a spacecraft relative to points on the Earth's surface. The characteristics of the various time systems are summarized in Table J-1.

Any periodic phenomenon may be used as a measure of time. The motion of the Earth, the Moon, and the Sun relative to the fixed stars has traditionally been used for this purpose. However, the need for increasingly accurate time measurements has resulted in the development of several alternative time systems. The requirement for high accuracy comes from the cumulative effect of timing errors. An accuracy of 1 sec/day (1 part in 10^5) would appear satisfactory for most scientific or technical purposes. However, an error of this magnitude in an ephemeris of the Earth causes an error of 10,800 km in the position of the Earth

Table J-1. Time Systems

KIND OF TIME	DEFINED BY	FUNDAMENTAL UNIT	REGULARITY	USE
SIDEREAL	EARTH'S ROTATION RELATIVE TO STARS	SIDEREAL DAY, 1 ROTATION OF EARTH	IRREGULAR	ASTRONOMICAL OBSERVA- TIONS; DETERMINING UT AND ROTATIONAL ORIENTA- TION OF EARTH
SOLAR:				
APPARENT	EARTH'S ROTATION RELATIVE TO TRUE SUN	SUCCESSIVE TRANSITS OF SUN	IRREGULAR AND ANNUAL VARIA- TIONS	SUNDIALS
MEAN	EARTH'S ROTATION RELATIVE TO FICTITIOUS MEAN SUN	MEAN SOLAR DAY	IRREGULAR	—
UNIVERSAL UT0	OBSERVED UT	MEAN SOLAR DAY	IRREGULAR	STUDY OF EARTH'S WANDERING POLE
UT1	CORRECTED UT0	MEAN SOLAR DAY	IRREGULAR	SHOWS SEASONAL VARIATION OF EARTH'S ROTATION
UT2	CORRECTED UT1	MEAN SOLAR DAY	IRREGULAR	BASIC ROTATION OF EARTH
UTC = GMT = Z	ATOMIC SECOND AND LEAP SECONDS TO APPROXIMATE UT1	MEAN SOLAR DAY	UNIFORM EXCEPT FOR LEAP SEC- ONDS	CIVIL TIMEKEEPING; TER- RESTRIAL NAVIGATION AND SURVEYING; BROADCAST TIME SIGNALS
EPHEMERIS, ET	FRACTION OF TROPICAL YEAR 1900	EPHEMERIS SECOND	UNIFORM	EPHEMERIDES
ATOMIC, TAI	FREQUENCY OF 133 Ce RADIA- TION	ATOMIC SECOND = EPHEMERIS SECOND	UNIFORM	BASIS OF ET AND UTC

after only 1 year, several orders of magnitude worse than what is acceptable for many unsophisticated measurements. Thus, the generation of accurate ephemerides requires a precise time measurement system.

The diurnal motion of celestial objects is the most obvious timekeeper. Until the Middle Ages "seasonal hours," one-twelfth of daylight or nightime periods, was used. Of course, this unit varies both with the season and with the observer's latitude. A more uniform unit of time is the *apparent solar day*, defined as the interval between two successive passages of the Sun across the observer's meridian. As discussed below, this interval varies throughout the year due to variations in the Earth's orbital speed and the inclination of the ecliptic. The Earth's orbital motion does not affect the *sidereal day*, the interval between two succesive meridian passages of a fixed star. However, irregularities in the rotation of the Earth cause both periodic and secular variations in the lengths of the sidereal and solar days.

The annual motions of celestial objects provide a measurement of time which is independent of the irregular variations in the rotation of the Earth. The *tropical year*, upon which our calendar is based, is defined as the interval of time from one vernal equinox to the next. The *ephemeris second* is defined as $1/31556925.9747$ of the tropical year for 1900. Because of the precession of the equinoxes (Section 2.2.2), the tropical year is about 20 minutes shorter than the orbital period of the Earth relative to the fixed stars. This latter period is known as the *sidereal year*. Due to secular variations in the orbit and rate of precession of the Earth, the lengths of both types of year (in units of s_e, the ephemeris second) vary to first order according to the relations:

$$Tropical\ year = 31556925.9747 - .530\,T$$

$$Sidereal\ year = 31558149.540 + .010\,T$$

where T is the time in units of Julian Centuries of 36525 days from 1900.0 [Newcomb, 1898].

The first satisfactory alternative to celestial observations for the measurement of time was the pendulum clock. The period of a pendulum is a function of the effective acceleration of gravity, which varies with geography and the position of the Sun and the Moon. The resonance frequency of quartz crystals has recently been employed in clocks; this frequency depends on the dimensions and cut of the crystal and its age, temperature, and ambient pressure. Atomic clocks are based on the frequency of microwave emission from certain atoms. An accuracy of 10^{-14} (fractional standard deviation) may be achieved with atomic clocks; corresponding accuracies for quartz and pendulum clocks are 2×10^{-13} and 10^{-6}, respectively. For an extended discussion of time systems, see Woolard and Clemence [1966], the *Explanatory Supplement to the Astronomical Ephemeris and the American Ephemeris and Nautical Almanac* [H.M. Nautical Almanac Office, 1961] and Müller and Jappel [1977].

Solar Time. The *celestial meridian* is the great circle passing through the celestial poles and the observer's zenith. As shown in Fig. J-1, the *hour angle, HA*, is the azimuthal orientation of an object measured westward from the celestial meridian. As the Earth rotates eastward, a celestial object appears to move westward and its *HA* increases with time. It takes 24 hours for an object to move completely around the celestial sphere or 1 hour to move 15 deg in *HA*; thus, 1 deg of *HA* corresponds to 4 minutes of time.

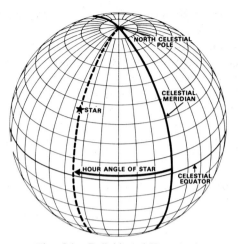

Fig. J-1. Definition of Hour Angle

The *apparent solar time* is equal to the local *HA* of the Sun, expressed in hours, plus 12 hours. Apparent solar time can be measured with a simple sundial constructed by driving a long nail perpendicularly through a flat piece of wood. If the nail is then pointed toward the celestial pole, the plane of the wood is parallel to the equatorial plane, and the shadow of the nail cast by the Sun onto the wood is a measure of the *HA*.

Due to the Earth's orbital motion, the Sun appears to move eastward along the ecliptic throughout the year. Because the Earth travels in an elliptical orbit, it moves faster when near the Sun and slower when it is more distant; therefore, the length of the solar day varies. Even if the Earth were in a circular orbit with a constant speed, the azimuthal component of the Sun's motion (parallel to the celestial equator) would vary due to the inclination of the ecliptic relative to the equator. To illustrate this, consider a satellite in a nearly polar orbit, as shown in Fig. J-2. The satellite changes azimuth slowly while near the equator and rapidly while near the poles. Although the variation in the length of the day due to the

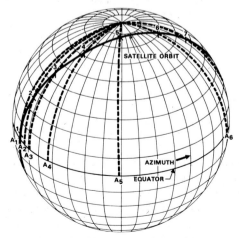

Fig. J-2. Variation in Azimuthal Rate for a Satellite Moving Uniformly in its Orbit. *A*1, *A*2, ..., *A*5 are azimuthal projections of the orbital points 1, 2, ... 5 and are equally spaced in time.

eccentricity and inclination of the Earth's orbit is small, the cumulative variation reaches a maximum of 16 minutes in November.

To provide more uniform time than the real Sun, a fictitious *mean Sun*, which moves along the equator at a constant rate equal to the average annual rate of the Sun, has been introduced. *Mean solar time* is defined by the *HA* of the mean Sun. The difference between the mean and apparent solar times is called the *equation of time*.

Standard Time. Mean solar time is impractical for communication and transportation because it varies continuously with longitude. Therefore, the world has been divided into 24 time zones of approximately 15 deg each. Normally, these zones are centered on standard meridians which are multiples of 15 deg in longitude. The uniform time throughout each zone is referred to as *Standard Time*, and usually differs by an integral number of hours from the mean solar time at 0 deg longitude, or *Universal Time*, as discussed below. Table J-2 lists the standard meridians for time zones in the continental United States. The apparent solar time is converted to Standard Time by adding the equation of time for the date and subtracting the algebraic difference (expressed in units of time) between the observer's longitude and the standard meridian.

Table J-2. Standard Time

TIME ZONE	STANDARD MERIDIAN (DEG, EAST LONG.)	UT MINUS STANDARD TIME (HOURS)	UT MINUS DAYLIGHT TIME (HOURS)
EASTERN	285	5	4
CENTRAL	270	6	5
MOUNTAIN	255	7	6
PACIFIC	240	8	7

Greenwich Mean Time, Universal Time. The 0-deg longitude line is referred to as the *Greenwich meridian* because it is defined by the former site of the Royal Greenwich Observatory. *Greenwich Mean Time, GMT,* is the mean solar time at 0 deg longitude; that is, *GMT* is the *HA* of the mean Sun observed at Greenwich (called the *GHA*) in hours plus 12 hours, modulo 24. Greenwich Mean Time is also called *Universal Time, UT,* and, in spaceflight operations, *Zulu,* or *Z*.

Uncorrected *UT* or *UT0* (read "UT Zero") is found from observations of stars, as explained in the discussion of sidereal time below. *UT0* time as determined by different observatories is not the same, however, due to changes in the longitudes of the observatories caused by the wandering of the geographic pole. Therefore, *UT0* is corrected for this effect to give *UT1*, which is then a measure of the actual angular rotation of the Earth. The Earth's rotation is subject to periodic seasonal variations, apparently caused by changes in, for example, the amount of ice in the polar regions. When *UT1* is corrected by periodic terms representing these seasonal effects, the result is *UT2*. Even *UT2* is not a uniform measure of time. Evidence from ancient eclipse records and other sources shows that the Earth's rate of rotation is slowing; also, unpredictable irregularities in the rotation rate are observed.

Before 1972, the broadcast time signals were kept within 0.1 sec of *UT2*. Since January 1, 1972, however, time services have broadcast *Coordinated Universal*

Time, UTC. A second of *UTC* is equal to a second of International Atomic Time, but *UTC* is kept within 0.90 sec of *UT1* by the introduction of 1-sec steps, usually at the end of June and December.

Ephemeris Time. The irregularities in the Earth's rotation cannot be predicted; however, gravitational theories have been formulated for the orbital motions of the Earth, the Moon, and the planets. In particular, Simon Newcomb's *Tables of the Sun*, [1898], published at the end of the 19th Century, gives the position of the Sun for regular time intervals. These intervals define a uniform time called *Ephemeris Time, ET*. In theory, Ephemeris Time is determined from observations of the Sun. In practice, observations of the Moon are used because the Sun moves slowly and its position is difficult to observe. One method is to record the *UT* of a lunar occultation of a star; the tabulated value of *ET* for the observed lunar position, corrected for effects such as parallax, is noted and the difference

$$\Delta T = ET - UT$$

is determined. A table of approximate ΔT values, both in the past and extrapolated into the future, is provided in *The American Ephemeris and Nautical Almanac* [U.S. Naval Observatory, 1973]. Ephemeris time at any instant is given by

$$ET = \Delta T + UT$$

International Atomic Time. The cesium nuclide, 133 Ce, has a single outer electron with a spin vector that can be either parallel or antiparallel to that of the nucleus. The flip from one orientation to the other, a hyperfine transition, is accompanied by the absorption or emission of microwave radiation of a given frequency. In an atomic clock, the number of these transitions is maximized in a resonator by the introduction of microwave radiation from an oscillator tuned to the same frequency. The cycles of the oscillator are counted to give a unit of time. In 1967, the 13th General Conference on Weights and Measures established the *Systeme Internationale* (SI) *second* as the duration of 9 192 631 770 periods of the radiation from the above transition in 133 Ce. This unit is the basis of *International Atomic Time, TAI*, and was chosen to make the SI second equal to the ephemeris second. The reference epoch for *TAI* is January 1, 1958, when $0^h0^m0^s$ *TAI* equaled $0^h0^m0^s$ *UT2*. For most purposes, ephemeris time may be considered to be equal to *TAI* plus 32.18 sec, the value of ΔT for January 1, 1958.

Sidereal Time. *Sidereal time, ST*, is based on the rotation of the Earth relative to the stars and is defined as the *HA* of the vernal equinox, ♈. The *local sidereal time, LST*, is defined as the local *HA* of ♈, *LHA* ♈, and the *sidereal time at Greenwich, GST*, is defined as the Greenwich *HA* of ♈, *GHA* ♈. Sidereal time may also be determined from the *HA* and *right ascension, RA*, of any star. The *RA* of a star is the azimuthal component of the star's position measured eastward from ♈ (see Section 2.2.2). From Fig. J-3 we see that

$$LST = LHA\ ♈ = LHA^* + RA^*, \text{modulo 24} \tag{J-1}$$

where LHA^* and RA^* are the *HA* and *RA* (both converted to time) of the star. In the example in Fig. J-3, LHA^* is 135 deg or 9 hours, RA^* is 90 deg or 6 hours, and the *LST* is 15 hours. Similarly,

$$GST = GHA\ ♈ = GHA^* + RA^*, \text{modulo 24} \tag{J-2}$$

where *GHA** is the *GHA* of the star (converted to time). In Fig. J-3, *GHA** is 45 deg or 3 hours; thus, *GST* is 9 hours. Note that the *sidereal time at Greenwich is equal to the right ascension of the Greenwich meridian*. The difference between *LST*

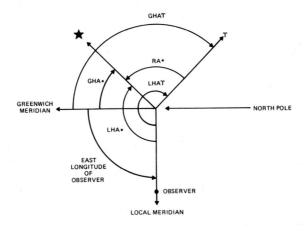

Fig.   J-3.   Sidereal Time. (View looking down on the Earth's North Pole.)

and *GST* (6 hours in this example) corresponds to the observer's East longitude (90 deg in this example). In general,

$$LST = GST + EL/15 \qquad (J-3)$$

where *EL* is the observer's East longitude in degrees. From the definition of mean solar time, it follows that *GMT* or *UT* equals the *GHA* of the fictitious mean Sun plus 12 hours, or

$$UT = 12 \text{ hours} + GST - R_u$$

where R_u is the right ascension of the mean Sun. For a given *UT* of any calendar date,

$$GST = R_u - 12 \text{ hours} + UT$$

$$= 6^h 38^m 45^s.836 + 8640184^s.542 \, T + 0^s .0929 \, T^2 + UT \qquad (J-4)$$

where *T* is the number of Julian centuries of 36,525 days which have elapsed since noon (GMT) on January 0, 1900 [Newcomb, 1898]. The corresponding equation for *GST* expressed in degrees is

$$GST = 99°.6910 + 36000°.7689 \, T + 0°.0004 \, T^2 + UT \qquad (J-5)$$

where *UT* is in degrees and *T* is in Julian centuries. Julian dates, or *JD* (Section 1.4), are convenient for determining *T* in Eqs. (J-4) and (J-5). The *JD* for Greenwich mean noon on January 0, 1900 (i.e., January 0.5, 1900), is 2 415 020.0. *JD*'s for any date in the last quarter of this century may be obtained by adding the day number of the year to the *JD* for January 0.0 UT of that year listed in Table J-3. For example, to find the *GST* for 3^h UT, July 4, 1976:

Day number of July 4.125 ($=3^h$ UT July 4), 1976	186.125
$+ JD$ for January 0.0, 1976	$+2\ 442\ 777.500$
$= JD$ for July 4.125, 1976	$=2\ 442\ 963.625$
$- JD$ for January 0.5, 1900	$-2\ 415\ 020.000$
$= T$ in days	27 943.625
$\div 36{,}525 = T$ in Julian centuries	0.765054757
$8640184.542\ T + 0.0929\ T^2$	6 610 214.340 sec
	$= 76^d 12^h 10^m 14^s\!.340$
$+$ first term Eq. (J-4)	6 38 45.836
$+\ UT$	3 0 0.000
GST	$21^h 49^m 0^s\!.176$

Due primarily to the varying distances of the Sun and the Moon, a small amplitude oscillation, known as *astronomical nutation*, is superimposed on the precession of the equinoxes. Sidereal time corrected for this effect is called *mean sidereal time*. *GST* in Eqs. (J-4) and (J-5) is mean sidereal time. The maximum difference between mean and apparent sidereal time is only about 1 sec.

Table J-3.   Julian Date at the Beginning of Each Year From 1975 to 2000

YEAR	JD FOR JAN 0.0 UT	YEAR	JD FOR JAN 0.0 UT	YEAR	JD FOR JAN 0.0 UT
	2 400 000+		2 400 000+		2 400 000+
1975	42 412.5	1984	45 699.5	1993	48 987.5
1976	42 777.5	1985	46 065.5	1994	49 352.5
1977	43 143.5	1986	46 430.5	1995	49 717.5
1978	43 508.5	1987	46 795.5	1996	50 082.5
1979	43 873.5	1988	47 160.5	1997	50 448.5
1980	44 238.5	1989	47 526.5	1998	50 813.5
1981	44 604 5	1990	47 891.5	1999	51 178.5
1982	44 969.5	1991	48 256.5	2000	51 543.5
1983	45 334.5	1992	48 621.5		

Because of the orbital motion of the Earth, a solar day is longer than a sidereal day. As illustrated in Fig. J-4, the fixed stars are sufficiently far away that lines connecting one of them to the Earth are essentially parallel. Because the Earth's orbital period is approximately 360 days, angle A is approximately 1 deg. A *sidereal day* is defined as one complete rotation of the Earth, 360 deg, relative to the stars. The Earth has to rotate $360 + A$ deg to complete a solar day. The ratio of the mean solar day to the mean sidereal day is 1.00273 79093; the mean sidereal day equals 23 hours, 56 min, 4.09054 sec of mean solar time and the mean solar day equals 24 hours, 3 min, 55.55536 sec of mean sidereal time [U.S. Naval Observatory, 1973]. Note that the "76 days" in the above example indicates the excess number of sidereal days, one for each year, that had occurred since the beginning of the century.

Sidereal time and mean solar time are affected proportionally by variations in the Earth's rotation. Although the irregular fluctuations in the Earth's rotation cannot be predicted, the general deceleration can be seen in Fig. J-5. The lengths of

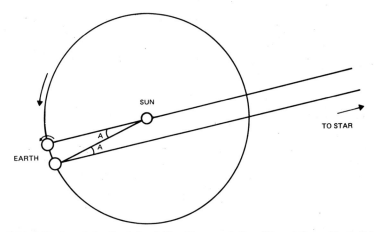

Fig. J-4. Orbital Motion of the Earth for 1 Day (Exaggerated) as Viewed From North Ecliptic Pole

the two types of day in s_e are given approximately by:

$$Sidereal\ day = 86164.09055 + 0.0015\,T$$
$$Mean\ solar\ day = 86400 + 0.0015\,T$$

where T is in Julian centuries from 1900.0 [Allen, 1973]. These terms are the dashed line in Fig. J-5.

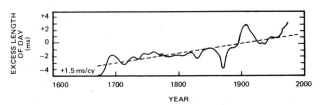

Fig. J-5. Excess Length of the Day Compared With the Day Near 1900. Note the very irregular fluctuations about the mean slope of 1.5 ms/century [Morrison, 1973].

Using Sidereal Time to Compute the Longitude of the Subsatellite Point.   To determine the direction of geographic points on the Earth as seen from a spacecraft, it is necessary to know both the spacecraft ephemeris and the longitude of the subsatellite point. For any UT, Eq. (J-4) can be used to determine the Greenwich sidereal time, GST, which in turn can be used to determine the East longitude, EL_{spc}, of the subsatellite point for any satellite for which the right ascension of its position in geocentric coordinates is known. From Eq. (J-3), we have

$$EL_{spc} = RA_{spc} - GST \text{ (in degrees)}$$

where RA_{spc} is the right ascension in degrees of the spacecraft at time GST. Because UTC is accurate to about 1 sec, the accuracy of the resulting longitude will be about 0.005 deg if the spacecraft ephemeris is known precisely.

References

1. Allen, C. W., *Astrophysical Quantities*, Third Edition. London: The Athlone Press, 1973.

2. H. M. Nautical Almanac Office, *Explanatory Supplement to the Astronomical Ephemeris and the American Ephemeris and Nautical Almanac*. London: Her Majesty's Stationery Office, 1961.

3. Morrison, L. V., "Rotation of the Earth from AD 1663-1972 and the Constancy of G," *Nature*, Vol. 241, p. 519–520, 1973.

4. Müller, Edith A. and Arndst Jappel, Editors, *International Astronomical Union, Proceedings of the Sixteenth General Assembly, Grenoble, 1976*. Dordrecht, Holland: D. Reidel Publishing Co., 1977.

5. Newcomb, Simon, *Astronomical Papers Prepared for the Use of the American Ephemeris and Nautical Almanac*, Bureau of Equipment, U.S. Department of the Navy, Washington, DC, 1898.

6. U.S. Naval Observatory, *The American Ephemeris and Nautical Almanac*. Washington, DC: U.S. G. P. O., 1973.

7. Woolard, Edgar W., and Gerald M. Clemence, *Spherical Astronomy*. New York: Academic Press, 1966.

APPENDIX K

METRIC CONVERSION FACTORS

The metric system of units, officially known as the *International System of Units*, or *SI*, is used throughout this book, with the single exception that angular measurements are usually expressed in degrees rather than the SI unit of radians. By international agreement, the fundamental SI units of length, mass, and time are defined as follows (see, for example, NBS Special Publication 330 [NBS, 1974]):

The *metre* is the length equal to 1 650 763.73 wavelengths in vacuum of the radiation corresponding to the transition between the levels $2p_{10}$ and $5d_5$ of the krypton-86 atom.

The *kilogram* is the mass of the international prototype of the kilogram (a specific platinum-iridium cylinder stored at Sevres, France).

The *second* is the duration of 9 192 631 770 periods of the radiation corresponding to the transition between two hyperfine levels of the ground state of the cesium-133 atom.

Additional base units in the SI system are the *ampere* for electric current, the *kelvin* for thermodynamic temperature, the *mole* for amount of substance, and the *candela* for luminous intensity. Mechtly [1973] provides an excellent summary of SI units for scientific and technical use.

The names of multiples and submultiples of SI units are formed by application of the following prefixes:

Factor by which unit is multiplied	Prefix	Symbol
10^{12}	tera	T
10^9	giga	G
10^6	mega	M
10^3	kilo	k
10^2	hecto	h
10	deka	da
10^{-1}	deci	d
10^{-2}	centi	c
10^{-3}	milli	m
10^{-6}	micro	μ
10^{-9}	nano	n
10^{-12}	pico	p
10^{-15}	femto	f
10^{-18}	atto	a

For each quantity listed below, the SI unit and its abbreviation are given in parentheses. For convenience in computer use, most conversion factors are given to the greatest available accuracy. Note that some conversions are exact definitions and some (speed of light, astronomical unit) depend on the value of physical constants. All notes are on the last page of the list.

To convert from	To	Multiply by	Notes
Mass (kilogram, kg)			
Atomic unit (electron)	kg	$9.109\ 6 \times 10^{-31}$	(1)
Atomic mass unit, amu	kg	$1.660\ 53 \times 10^{-27}$	(1)
Ounce mass (avoirdupois)	kg	$2.834\ 952\ 312\ 5 \times 10^{-2}$	E
Pound mass, lbm (avoirdupois)	kg	$4.535\ 923\ 7 \times 10^{-1}$	E
Slug	kg	$1.459\ 390\ 294 \times 10^{1}$	
Short ton (2000 pound)	kg	$9.071\ 847\ 4 \times 10^{2}$	E
Metric ton	kg	1.0×10^{3}	E
Solar mass	kg	1.989×10^{30}	(1)
Length (metre, m)			
Angstrom	m	1.0×10^{-10}	E
Micron	m	1.0×10^{-6}	E
Mil (10^{-3} inch)	m	2.54×10^{-5}	E
Inch	m	2.54×10^{-2}	E
Foot	m	3.048×10^{-1}	E
Statute mile (U.S.)	m	$1.609\ 344 \times 10^{3}$	E
Nautical mile (U.S.)	m	1.852×10^{3}	E
Earth equatorial radius	m	$6.378\ 140 \times 10^{6}$	(3)
Vanguard unit	m	$6.378\ 166 \times 10^{6}$	(4)
Solar radius	m	$6.959\ 9 \times 10^{8}$	(1)
Astronomical unit, AU	m	$1.495\ 978\ 70 \times 10^{11}$	(5)
Light year (tropical year)	m	$9.460\ 530 \times 10^{15}$	(1)
Parsec (distance for which stellar parallax is 1 arc-sec.)	m	$3.085\ 678 \times 10^{16}$	(1)
Time (second, s)			(7)
Sidereal day, d_* (ref. $= \Upsilon$)	s	$8.616\ 409\ 18 \times 10^{4}$	(1)*
		$= 23^{\text{h}}\ 56^{\text{m}}\ 4.0918^{\text{s}}$	(1)*
Ephemeris day, d_e	s	8.64×10^{4}	E
Ephemeris day, d_e	d_*	$1.002\ 737\ 89$	(1)*
Vanguard unit	s	$8.068\ 124\ 2 \times 10^{2}$	(4)
Keplerian period of a satellite in low-Earth orbit	min	$1.658\ 669 \times 10^{-4} \times a^{3/2}$ a in km	(6)
Keplerian period of a satellite of the Sun	d_e	$3.652\ 569 \times 10^{2} \times a^{3/2}$ a in AU	(6)
Tropical year (ref. $= \Upsilon$)	s	$3.155\ 692\ 555\ 1 \times 10^{7}$	(7)*
Tropical year (ref. $= \Upsilon$)	d_e	$3.652\ 421\ 938\ 8 \times 10^{2}$	(7)*
Sidereal year (ref. $=$ fixed stars)	s_e	$3.155\ 314\ 954\ 8 \times 10^{7}$	(7)*
Sidereal year (ref. $=$ fixed stars)	d_e	$3.652\ 563\ 605\ 1 \times 10^{2}$	(7)*

* Epoch 1980.

To convert from	To	Multiply by	Notes
Calendar year (365 days)	s_e	$3.153\ 6 \times 10^7$	E
Julian century	d	$3.652\ 5 \times 10^4$	E
Gregorian calendar century	d	$3.652\ 425 \times 10^4$	E

Velocity (metre/second, m/s)

Foot/minute, ft/min	m/s	5.08×10^{-3}	E
Inch/second, ips	m/s	2.54×10^{-2}	E
Kilometre/hour, km/hr	m/s	$(3.6)^{-1} = 0.277777\ldots$	E
Foot/second, fps or ft/sec	m/s	3.048×10^{-1}	E
Miles/hour, mph	m/s	$4.470\ 4 \times 10^{-1}$	E
Knot	m/s	$5.144\ 444\ 444 \times 10^{-1}$	
Miles/minute	m/s	$2.682\ 24 \times 10^1$	E
Miles/second	m/s	$1.609\ 344 \times 10^3$	E
Astronomical unit/sidereal year	m/s	$4.740\ 388\ 554 \times 10^3$	
Vanguard unit	m/s	$7.905\ 389 \times 10^3$	(4)
Velocity of light, c	m/s	$2.997\ 925 \times 10^8$	(1)

Acceleration (metre/second^2, m/s^2)

Gal (galileo)	m/s^2	1.0×10^{-2}	E
Inch/second^2	m/s^2	2.54×10^{-2}	E
Foot/second^2	m/s^2	3.048×10^{-1}	E
Free fall (standard), g	m/s^2	$9.806\ 65$	E
Vanguard unit	m/s^2	$9.798\ 299$	(4)

Force (Newton $\equiv$ kilogram $\cdot$ metre/second^2, N $\equiv$ kg $\cdot$ m/s^2)

Dyne	N	1.0×10^{-5}	E
Poundal	N	$1.382\ 549\ 543\ 76 \times 10^{-1}$	E
Ounce force (avoirdupois)	N	$2.780\ 138\ 5 \times 10^{-1}$	(8)
Pound force (avoirdupois), lbf $\equiv$ slug $\cdot$ foot/second^2	N	$4.448\ 221\ 615\ 260\ 5$	E

Pressure (Pascal $\equiv$ Newton/metre^2 $\equiv$ kilogram $\cdot$ metre^{-1} $\cdot$ second^{-2}, $Pa \equiv N/m^2 \equiv kg \cdot m^{-1} \cdot s^{-2}$)

Dyne/centimetre^2	Pa	1.0×10^{-1}	E
lbf/foot^2	Pa	$4.788\ 025\ 8$	(8)
Torr (0° C)	Pa	$1.333\ 22 \times 10^2$	(8)
Centimetre of Mercury (0° C)	Pa	$1.333\ 22 \times 10^3$	(8)
Inch of Mercury (32° F)	Pa	$3.386\ 389 \times 10^3$	(8)
lbf/inch^2, psi	Pa	$6.894\ 757\ 2 \times 10^3$	(8)
Bar	Pa	1.0×10^5	E
Atmosphere	Pa	$1.013\ 25 \times 10^5$	E

To convert from	To	Multiply by	Notes

Energy or **Torque** (Joule$\equiv$Newton$\cdot$metre$\equiv$kilogram$\cdot$metre^2/second^2,   (2)
$\quad$J$\equiv$N$\cdot$m$\equiv$kg$\cdot$m^2/s^2)

Electron volt, eV	J	$1.602\ 191\ 7 \times 10^{-19}$	(8)
Mass-energy of 1 amu	J	$1.492\ 41 \times 10^{-10}$	(1)
Erg$\equiv$gram$\cdot$centimetres^2/second^2			
$\quad$ = pole$\cdot$centimetre$\cdot$oersted	J	1.0×10^{-7}	E
Ounce inch	J	$7.061\ 551\ 6 \times 10^{-3}$	(8)
Foot poundal	J	$4.214\ 011\ 0 \times 10^{-2}$	(8)
Foot lbf = slug$\cdot$foot^2/second^2	J	$1.355\ 817\ 9$	(8)
Calorie (mean)	J	$4.190\ 02$	(8)
British thermal unit, BTU (mean)	J	$1.055\ 87 \times 10^3$	(8)
Kilocalorie (mean)	J	$4.190\ 02 \times 10^3$	(8)
Kilowatt hour	J	3.6×10^6	E
Ton equivalent of TNT	J	4.20×10^9	(8)

Power (Watt$\equiv$Joule/second$\equiv$kilogram$\cdot$metre^2/second^3, W$\equiv$J/s$\equiv$kg$\cdot$m^2/s^3)

Foot lbf/second	W	$1.355\ 817\ 9$	(8)
Horsepower (550 ft lbf/s)	W	$7.456\ 998\ 7 \times 10^2$	(8)
Horsepower (electrical)	W	7.46×10^2	E
Solar luminosity	W	3.826×10^{26}	(1)

Moment of Inertia (kilogram$\cdot$metre^2, kg$\cdot$m^2)

Gram$\cdot$centimetre^2	kg$\cdot$m^2	1.0×10^{-7}	E
lbm$\cdot$inch^2	kg$\cdot$m^2	$2.926\ 397 \times 10^{-4}$	
lbm$\cdot$foot^2	kg$\cdot$m^2	$4.214\ 011 \times 10^{-2}$	
Slug$\cdot$inch^2	kg$\cdot$m^2	$9.415\ 402 \times 10^{-3}$	
Inch$\cdot$lbf$\cdot$seconds^2	kg$\cdot$m^2	$1.129\ 848 \times 10^{-1}$	
Slug$\cdot$foot^2 = ft$\cdot$lbf$\cdot$seconds^2	kg$\cdot$m^2	$1.355\ 818$	

Angular Measure (radian, rad). Degree (abbreviated deg) is the basic unit used
$\quad\quad\quad\quad$ in this book.

Degree	rad	$\pi/180$	E
		$\approx 1.745\ 329\ 251\ 994\ 329\ 577 \times 10^{-2}$	
Radian	deg	$180/\pi$	E
		$\approx 5.729\ 577\ 951\ 308\ 232\ 088 \times 10^1$	

Solid Angle (steradian, sr)

Degree^2, deg^2	sr	$(\pi/180)^2$	E
		$\approx 3.046\ 174\ 197\ 867\ 085\ 993 \times 10^{-4}$	
Steradian	deg^2	$(180/\pi)^2$	E
		$\approx 3.282\ 806\ 350\ 011\ 743\ 794 \times 10^3$	

To convert from	*To*	*Multiply by*	*Notes*

Angular Velocity (radian/second, rad/s). Degrees/second is the basic unit used in this book.

Degrees/second, deg/s	rad/s	$\pi/180$	E
		$\approx 1.745\ 329\ 251\ 994\ 329\ 577 \times 10^{-2}$	
Revolutions/minute, rpm	rad/s	$\pi/30$	E
		$\approx 1.047\ 197\ 551\ 196\ 597\ 746 \times 10^{-1}$	
Revolutions/second, rev/s	rad/s	2π	E
		$\approx 6.283\ 185\ 307\ 179\ 586\ 477$	
Revolutions/minute, rpm	deg/s	6.0	E
Radians/second, rad/s	deg/s	$180/\pi$	E
		$\approx 5.729\ 577\ 951\ 308\ 232\ 088 \times 10^{1}$	
Revolutions/second, rev/s	deg/s	3.6×10^{2}	E

Angular Momentum (kilogram·metre²/second, kg·m²/s)

Gram·centimetre²/second,			
g·cm²/s	kg·m²/s	1.0×10^{-7}	E
lbm·inch²/second	kg·m²/s	$2.926\ 397 \times 10^{-4}$	
Slug·inch²/second	kg·m²/s	$9.415\ 402 \times 10^{-3}$	
lbm·foot²/second	kg·m²/s	$4.214\ 011 \times 10^{-2}$	
Inch·lbf·second	kg·m²/s	$1.129\ 848 \times 10^{-1}$	
Slug·foot²/second = ft·lbf·second	kg·m²/s	1.355 818	

Magnetic Flux (Weber ≡ Volt·second ≡ kilogram·metre²·Ampere⁻¹·second⁻², Wb ≡ V·s ≡ kg·m²·A⁻¹·s⁻²)

Maxwell (EMU)	Wb	1.0×10^{-8}	E

B, Magnetic Induction (commonly called "magnetic field", (9) Telsa ≡ Weber/metre² ≡ kilogram·Ampere⁻¹·second⁻², T ≡ Wb/m² ≡ kg·A⁻¹·s⁻²)

Gamma (EMU)	T	1.0×10^{-9}	E, (9)
Gauss (EMU)	T	1.0×10^{-4}	E, (9)

H, Magnetic Field Strength (ampere turn/metre, A/m) (9)

Oersted (EMU)	A/m	$(1/4\pi) \times 10^{3}$	E, (9)
		$\approx 7.957\ 747\ 154\ 594\ 766\ 788 \times 10^{1}$	

Magnetic Moment (ampere·turn·metre² ≡ Joule/Telsa, A·m² ≡ J/T)

Abampere·centimetre² (EMU)	A·m²	1.0×10^{-3}	E, (9)
Ampere·centimetre² (Practical)	A·m²	1.0×10^{-4}	E

To convert from	*To*	*Multiply by*	*Notes*

Magnetic Dipole Moment (Weber·metre≡kilogram·metre·Ampere^{-1}·second^{-2},
Wb·m≡kg·m·A^{-1}·s^{-2})

Pole·centimetre (EMU)	Wb·m	$4\pi \times 10^{-10}$	E, (9)
		$\approx 1.256\ 637\ 061\ 435\ 917\ 295 \times 10^{-9}$	
Gauss·centimetre^3 (Practical)	Wb·m	1.0×10^{-10}	E

Temperature (Kelvin, K)

Celsius, C	K	$t_K = t_C + 273.15$	E
Fahrenheit, F	K	$t_K = (5/9)(t_F + 459.67)$	E
Fahrenheit, F	C	$t_C = (5/9)(t_F - 32.0)$	E

Notes:

E (*Exact*) indicates that the conversion given is exact by definition of the non-SI unit or that it is obtained from other exact conversions.

(1) Values are those of Allen [1973].

(2) In common usage "Joule" is used for energy and "Newton-metre" for torque.

(3) Value is that adopted by the International Astronomical Union in 1976 [Müller and Jappel, 1977]. Reported values of the equatorial radius of the Earth differ by about 20 m. It is therefore recommended that this unit not be used except in internal calculations, where it is given a single defined value.

(4) Values are those adopted in subroutine RO1TAP, described in Section 20.3. Vanguard units should be avoided if possible because of differences in the definitions of the units involved. The Vanguard unit of length is equatorial radius of the Earth; the Vanguard unit of time is the time for an Earth satellite to move 1 radian if the semimajor axis is 1 Vanguard unit.

(5) Value is that adopted by the International Astronomical Union in 1976 [Müller and Jappel, 1977].

(6) Value is calculated from mass parameters adopted by the International Astronomical Union in 1976 [Müller and Jappel, 1977]. Actual period will differ due to various perturbation effects. (See Section 3.4.)

(7) For high-precision work, consult Appendix J on time measurement systems. The conversions for the length of the year are derived from values, given by Newcomb [1898], which define the unit of ephemeris time. The most convenient method for determining the time interval between events separated by several days or more is to use the Julian Date. See Section 1.4 and subroutine JD in Section 20.3 for a convenient algorithm for determining the Julian date and Appendix J for a table of Julian Dates.

(8) Values are those of Mechtly [1973].

(9) Care should be taken in transforming magnetic units, because the dimensionality of magnetic quantities (**B**, **H**, etc.) depends on the system of units. Most of the conversions given here are between SI and EMU (electromagnetic). The following equations hold in both sets of units:

$N = m \times B = d \times H$

$B = \mu H$

$m = I A$ for a current loop in a plane

$d = \mu m$

with the following definitions

$N \equiv$ torque

$B \equiv$ magnetic induction (commonly called "magnetic field")

$H \equiv$ magnetic field strength or magnetic intensity

$m \equiv$ magnetic moment

$I \equiv$ current in loop

$A \equiv$ vector normal to the plane of the current loop (in the direction of the angular velocity vector of the current about the center of the loop) with magnitude equal to the area of the loop

$d \equiv$ magnetic dipole moment

$\mu \equiv$ magnetic permeability

The permeability of vacuum, μ_0, has the following values, by definition:

$$\mu_0 \equiv 1 \text{ (dimensionless)} \quad \text{EMU}$$
$$\mu_0 \equiv 4\pi \times 10^{-7} \text{ N/A}^2 \quad \text{SI}$$

Therefore, in electromagnetic units in vacuum, magnetic induction and magnetic field strength are equivalent and the magnetic moment and magnetic dipole moment are equivalent. For practical purposes of magnetostatics, space is a vacuum but the spacecraft itself may have $\mu \neq \mu_0$.

References

1. Allen, C. W., *Astrophysical Quantities*, Third Edition. London: The Athlone Press, 1973.
2. Mechtly, E. A., *The International System of Units: Physical Constants and Conversion Factors, Second Revision*, Washington, DC, NASA SP-7012, 1973.
3. Müller, Edith A. and Arndst Jappel, editors, *International Astronomical Union, Proceedings of the Sixteenth General Assembly, Grenoble, 1976*. Dordrecht, Holland: D. Reidel Publishing Co., 1977.
4. National Bureau of Standards, U.S. Department of Commerce, *The International System of Units (SI)*, NBS Special Publication 330 (1974 edition), 1974.
5. Newcomb, Simon, *Astronomical Papers Prepared for the Use of the American Ephemeris and Nautical Almanac*. Washington, DC: Bureau of Equipment, Navy Department, 1898.

APPENDIX L

SOLAR SYSTEM CONSTANTS

James R. Wertz

L.1   Planets and Natural Satellites
L.2   The Sun
L.3   The Earth
L.4   The Moon
L.5   Potential Spacecraft Orbits

The mass, size, and gravitational parameters are those adopted by the International Astronomical Union (IAU) in 1976 [Muller and Jappel, 1977]. The geocentric and geographical coordinate system conversions are based on the method adopted by the *American Ephermeris and Nautical Almanac* [H.M. Nautical Almanac Office, 1961] using the updated value for the Earth's flattening adopted by the IAU [Muller and Jappel, 1977]. The properties of artificial satellites, both in orbit about solar system objects and in transfer orbits between objects, are calculated from the parameters given and are based on Keplerian orbits with no perturbative corrections. The properties of the Earth's upper atmosphere are from the 1972 COSPAR International Reference Atmosphere [1972], and the U.S. Standard Atmosphere [1976]. Additional constants are from Allen [1973], which is an excellent source of additional astronomical information. See Chapter 3 for definitions of orbital quantities and planetary magnitudes.

L.1 Planets and Natural Satellites

Table L-1 lists the orbital properties of the major planets. Because of orbital perturbations, the data here are not tabulated with the full precision normally used for ephemerides. If greater accuracy is needed, consult the *American Ephemeris and Nautical Almanac* for current osculating elements or Section 5.4 for epemerides for computer use. Quoted data are from Allen [1973].

Table L-2 lists the physical properties of the Moon and planets. Additional data on the Earth and Moon are given in Sections L.3 and L.4. Properties of the natural satellites of the planets are given in Table L-3.

Table L-1.   Planetary Orbits

PLANET	SEMIMAJOR AXIS (AU)	SEMIMAJOR AXIS (10^6 KM)	SIDEREAL PERIOD (TROPICAL YEARS)	SIDEREAL PERIOD (DAYS)[1]	SYNODIC PERIOD (DAYS)[1]	MEAN DAILY MOTION[2] (DEG)	MEAN ORBITAL VELOCITY (KM/SEC)	ECCENTRICITY 1970
MERCURY	0.387099	57.9	0.24085	87.969	115.88	4.092339	47.89	0.205628
VENUS	0.723332	108.2	0.61521	224.701	583.92	1.602131	35.03	0.006787
EARTH	1.000000	149.6	1.00004	365.256	—	0.985609	29.79	0.016722
MARS	1.523691	227.9	1.88089	686.980	779.94	0.524033	24.13	0.093377
JUPITER	5.202803	778.3	11.86223	4332.589	398.88	0.083091	13.06	0.04845
SATURN	9.53884	1427.0	29.4577	10759.22	378.09	0.033460	9.64	0.05565
URANUS	19.1819	2869.6	84.0139	30685.4	369.66	0.011732	6.81	0.04724
NEPTUNE	30.0578	4496.6	164.793	60189	367.49	0.005981	5.43	0.00858
PLUTO	39.44	5900	247.7	90465	366.73	0.003979	4.74	0.250

PLANET	INCLINATION TO ECLIPTIC, i 1970 (DEG)	LONGITUDE OF THE ASCENDING NODE, 1900 Ω (DEG)	$\Delta\Omega$ (DEG/CENTURY)	MEAN LONGITUDE OF PERIHELION,[3] $\tilde{\omega}$, 1900 (DEG)	$\Delta\tilde{\omega}$ (DEG/CENTURY)	PLANETARY LONGITUDE, L, JANUARY 0.5, 1970 (DEG)	DATE OF PERIHELION PASSAGE 1970 OR EARLIER
MERCURY	7.0042	47.1458	+1.1853	75.8983	+1.5544	47.9825	DEC. 25, 1970
VENUS	3.3944	75.7797	+0.8997	130.1527	+1.3917	265.4144	MAY 21, 1970
EARTH	—	—	—	101.2197	+1.7167	99.7422	JAN. 1, 1970
MARS	1.8500	48.7863	+0.7711	334.2183	+1.8406	12.6752	OCT. 21, 1969
JUPITER	1.3047	99.4416	+1.0108	12.7208	+1.6106	203.4197	SEPT. 26, 1963
SATURN	2.4894	112.7888	+0.8728	91.0972	+1.9583	43.0055	SEPT. 8, 1944
URANUS	0.7730	73.4783	+0.4989	171.53	+1.5	184.2902	MAY 20, 1966
NEPTUNE	1.7727	130.6811	+1.0983	46.67	+1.4	238.9233	SEPT. 2, 1876
PLUTO	17.17	109.73	—	223	—	181.65[4]	OCT. 24, 1741

[1] ONE DAY = 86,400 SI SECONDS.

[2] MEAN DAILY MOTION IS THE MEAN CHANGE IN TRUE ANOMALY AS VIEWED FROM THE SUN.

[3] THE LONGITUDE OF PERIHELION, $\tilde{\omega}$, IS MEASURED FROM THE VERNAL EQUINOX; THAT IS, $\tilde{\omega} = \Omega + \omega$, WHERE Ω IS MEASURED ALONG THE ECLIPTIC FROM THE VERNAL EQUINOX EASTWARD TO THE ASCENDING NODE AND ω IS THE ARGUMENT OF PERIHELION MEASURED FROM THE ASCENDING NODE ALONG THE ORBIT IN THE DIRECTION OF THE PLANET'S MOTION TO PERIHELION.

[4] AT EPOCH SEPTEMBER 23.0, 1960.

Table L-2.   Physical Properties of the Moon and Planets

PLANET	EQUATORIAL RADIUS (km)	ELLIPTICITY, $(R_e-R_p)/R_e$	PLANETARY GRAVITATIONAL CONSTANT, GM[1] $(10^{12} m^3/s^2)$	MASS[1] $(10^{24} kg)$	MEAN DENSITY (gm/cm^3)	INCLINATION OF EQUATOR TO ORBIT (DEG)	MAJOR ATMOSPHERIC COMPONENTS (IN ORDER OF ABUNDANCE)
THE MOON	1738.2	0.00054	4.902786	0.073483	3.341	6.68	NONE
MERCURY	2,439	0.0	22.03208	0.33022	5.4	< 28	NONE
VENUS	6,052	0.0	324.8588	4.8690	5.2	3	CO_2, N_2, O_2, H_2O
EARTH	6378.140	0.00335281	403.5033	6.0477	5.518	~ 23.44	N_2, O_2, Ar, H_2O, CO_2
MARS	3397.2	0.009	42.82829	0.64191	3.95	23.98	CO_2, Ar, CO, H_2O
JUPITER	71,398	0.063	126,712.0	1,899.2	1.34	3.08	$H_2, He, H_2O, CH_4, NH_3$
SATURN	60,000	0.098	37,934.0	568.56	0.70	26.73	H_2, CH_4, NH_3[5]
URANUS	25,400	0.06	5803.2	86.978	1.58	97.92	H_2, CH_4
NEPTUNE	24,300	0.021	6871.3	102.99	2.30	28.80	H_2, CH_4
PLUTO	2,500	—	40	0.7	—	—	UNKNOWN

PLANET	BOND ALBEDO[6] A	VISUAL MAGNITUDE AT UNIT DISTANCE,[7] V (1,0)	MEAN VISUAL MAGNITUDE AT OPPOSITION[8] V_0	COLOR INDEX,[10] B-V	SMALL ANGLE PHASE VARIATION,[11] A_1	SUBSOLAR TEMPERATURE (°K)	DARK SIDE TEMPERATURE (°K)	SIDEREAL, EQUATORIAL ROTATION PERIOD
THE MOON	0.067	+0.23	−12.73	+0.91	+0.026	—	104	27.321661 DAYS
MERCURY	0.056	−0.36	− 0.2[7]	0.91	0.027	600	100	59 DAYS
VENUS	0.72	−4.34	− 4.22[7]	0.79	0.013	240	240	244.3 DAYS[2]
EARTH	0.39	−3.9	—	0.2	—	295	280	23.93447 HOURS
MARS	0.16	−1.51	− 2.02	1.37	0.016	250	—	24.62294 HOURS
JUPITER	0.70	−9.25	− 2.6	0.8	0.014	120	..	9.8417 HOURS[3]
SATURN	0.75	−9.0[8]	+ 0.7[9]	1.0	0.044	90	—	10.23 HOURS[4]
URANUS	0.90	−7.15	+ 5.5	0.55	0.001	65	—	10.82 HOURS
NEPTUNE	0.82	−6.90	+ 7.9	0.45	0.001	50	—	15.80 HOURS
PLUTO	0.145	−1.0	+14.9	0.79	—	—	—	6.4 DAYS

[1] MASSES AND GRAVITATIONAL CONSTANTS INCLUDE PLANET PLUS ATMOSPHERE PLUS SATELLITES (i.e., "EARTH" VALUE IS EARTH PLUS MOON); ACTUAL MASSES ARE LIMITED BY THE KNOWLEDGE OF THE UNIVERSAL GRAVITATIONAL CONSTANT, G (TAKEN AS $6.672 \times 10^{-11} m^3 kg^{-1} s^{-2}$). THEREFORE, FOR PRECISION WORK, THE PLANETARY GRAVITATIONAL CONSTANT $\equiv$ G X ($M_{planet} + M_{sat}$) SHOULD BE USED. VALUES GIVEN ARE DERIVED FROM THOSE ADOPTED BY THE IAU (MULLER AND SAPPEL, 1977).

[2] RETROGRADE.

[3] ROTATION PERIOD ~ 9.928 HOURS AT HIGH LATITUDES.

[4] ROTATION PERIOD ~ 10.63 HOURS AT HIGH LATITUDES.

[5] TITAN, SATURN'S LARGEST SATELLITE, HAS AN ATMOSPHERE WHOSE MAJOR COMPONENT IS CH_4.

[6] RATIO OF TOTAL REFLECTED LIGHT TO TOTAL INCIDENT LIGHT.

[7] V (1, 0) IS THE VISUAL MAGNITUDE WHEN THE OBSERVER IS DIRECTLY BETWEEN THE SUN AND THE PLANET AND THE PRODUCT OF THE SUN-PLANET DISTANCE (IN AU) AND OBSERVER-PLANET DISTANCE (IN AU) IS 1. SEE SECTION 3.5 FOR FORMULAE FOR OTHER VIEWING ANGLES AND DISTANCES.

[8] AS VIEWED FROM THE EARTH; MAGNITUDES FOR MERCURY AND VENUS ARE AT GREATEST ELONGATION.

[9] WITH RING SYSTEM VIEWED EDGE-ON.

[10] BLUE MAGNITUDE MINUS VISUAL MAGNITUDE. SEE SECTION 5.6.

[11] FOR SMALL PHASE ANGLES, ξ, IN DEGREES, THE VARIATION OF VISUAL MAGNITUDE WITH PHASE IS APPROXIMATELY $V (\xi) \approx V (0) + A_1 \xi$. SEE SECTION 3.5. FOR SATURN, V DEPENDS STRONGLY ON THE ORIENTATION OF THE RING SYSTEM.

Table L-3.   Natural Satellites of the Planets

PLANET	SATELLITE	ORBIT SEMIMAJOR AXIS (10^3 KM)	MAXIMUM SEPARATION FROM PRIMARY[2] (DEG)	SIDEREAL PERIOD (DAYS)	ORBIT INCLINATION[3] (DEG)	ORBIT ECCENTRICITY	RADIUS (KM)	SATELLITE-TO-PLANET MASS RATIO	ESTIMATED MASS (10^{21} KG)	VISUAL MAGNITUDE AT OPPOSITION[2]	ESCAPE VELOCITY (KM/SEC)
EARTH	MOON[4]	384		27.321661	23	0.055	1738	0.01230002	73.5	−12.7	2.3735
MARS	1 PHOBOS	9	0.0069	0.318910	1	0.021	7			+11.5	~0.02
	2 DEIMOS	23	0.0172	1.262441	2	0.003	4			+12.6	~0.01
JUPITER[5]	1 IO	422	0.0383	1.769138	0	0.000	1810	4.70×10^{-5}	89.3	+ 4.9	2.57
	2 EUROPA	671	0.0611	3.551181	1	0.000	1480	2.56×10^{-5}	48.6	+ 5.3	2.09
	3 GANYMEDE	1070	0.0975	7.154553	0	0.001	2600	7.84×10^{-5}	149	+ 4.6	2.76
	4 CALLISTO	1883	0.1716	16.689018	0	0.007	2360	5.6×10^{-5}	106	+ 5.6	2.45
	5 AMALTHEA	181	0.0163	0.418178	0	0.003	80			+13	~0.1
	6 HIMALIA	11476	1.0455	250.566	28	0.158	50			+14.2	~0.1
	7 ELARA	11737	1.0694	259.65	26	0.207	12			+17	~0.03
	8 PASIPHAE	23500	2.15	739	147	0.40	10			+18	~0.02
	9 SINOPE	23600	2.17	758	156	0.275	9			+18.6	~0.02
	10 LYSITHEA	11700	1.0694	259.22	29	0.12	8			+18.8	~0.01
	11 CARME	22600	2.05	692	163	0.207	9			+18.6	~0.02
	12 ANANKE	21200	1.93	630	147	0.169	8			+18.7	~0.01
SATURN	1 MIMAS	186	0.0083	0.942422	2	0.020	270		0.04	+12.2	0.13
	2 ENCELADUS	238	0.0105	1.370218	0	0.004	300		0.08	+11.8	0.18
	3 TETHYS	295	0.0133	1.887802	1	0.000	500		0.64	+10.5	0.34
	4 DIONE	377	0.0169	2.736916	0	0.002	480		1.1	+10.6	0.35
	5 RHEA	527	0.0236	4.417503	0	0.001	650		2.3	+ 9.9	0.51
	6 TITAN	1222	0.0547	15.945449	0	0.029	2440	2.41×10^{-4}	137	+ 8.3	2.74
	7 HYPERION	1483	0.0663	21.276657	1	0.104	220		1.1	+14	0.21
	8 IAPETUS	3560	0.1597	79.33084	15	0.028	550		1.1	+10.7	0.68
	9 PHOEBE	12950	0.5808	550.33	150	0.163	120			+15	~0.1
	10 JANUS	159	0.0072	0.7490	0	0.0	150			+14	~0.3
URANUS	1 AERIEL	192	0.0039	2.52038	0	0.003	350		1.3	+14.3	0.46
	2 UMBRIEL	267	0.0056	4.14418	0	0.004	250		0.5	+15.1	0.30
	3 TITANIA	438	0.0092	8.70588	0	0.002	500		4.3	+13.9	0.75
	4 OBERON	586	0.0122	13.46326	0	0.001	450		2.6	+14.1	0.60
	5 MIRANDA	130	0.0028	1.414	0	0.00	120		0.1	+16.8	0.18
NEPTUNE	1 TRITON	355	0.0047	5.87654	160	0.00	1900	2×10^{-3}	206	+13.6	3.7
	2 NEREID	5562	0.0733	359.88	28	0.75	120			+19.1	~0.2

[1] SATELLITE/PLANET MASS RATIO IS GIVEN ONLY FOR THOSE SATELLITES FOR WHICH DYNAMIC ESTIMATES ARE AVAILABLE AND ARE THOSE ADOPTED BY THE IAU IN 1976 [MULLER AND JAPPEL, 1977].

[2] VIEWED FROM EARTH.

[3] RELATIVE TO PRIMARY EQUATORIAL PLANE. GREATER THAN 90 DEG INDICATES RETROGRADE MOTION

[4] SEE TABLES L-8 AND L-9.

[5] FOR AN INTERESTING DISCUSSION OF THE NOMENCLATURE OF JOVIAN SATELLITES, SEE OWEN [1976]

L.2 The Sun

Table L-4 lists the principal physical properties of the Sun. See Section 5.3 for properties of solar radiation and the solar wind.

Table L-4.   Physical Properties of the Sun

PROPERTY	VALUE
RADIUS OF THE PHOTOSPHERE (VISIBLE SURFACE)	6.9599×10^5 km
ANGULAR DIAMETER OF PHOTOSPHERE AT 1 AU	0.53313 Deg
MASS	1.989×10^{30} kg
MEAN DENSITY	1.409 gm/cm^3
TOTAL RADIATION EMITTED	3.826×10^{26} J/sec
TOTAL RADIATON PER UNIT AREA AT 1 AU	1358 J s^{-1} m^{-2}
ESCAPE VELOCITY FROM THE SURFACE	617.7 km/sec
POLAR MAGNETIC FIELD AT SUNSPOT MINIMUM	(1 to 2) $\times 10^{-4}$ T
APPARENT VISUAL MAGNITUDE AT 1 AU	−26.74
ABSOLUTE VISUAL MAGNITUDE (MAGNITUDE AT DISTANCE OF 10 PARSECS)	+4.83
COLOR INDEX, B−V (SEE SECTION 5.6)	+0.65
SPECTRAL TYPE	G2 V
EFFECTIVE TEMPERATURE	5770° K
VELOCITY OF THE SUN RELATIVE TO NEARBY STARS*	15.4 km/sec TOWARD $\alpha = 268$ Deg, $\delta = +26$ Deg
INCLINATION OF THE EQUATOR TO THE ECLIPTIC	7.25 Deg
LONGITUDE OF THE ASCENDING NODE OF THE EQUATOR IN 1980 PLUS RATE OF CHANGE IN LONGITUDE	75.48 Deg + 0.014 Deg/Year
SIDEREAL ROTATION RATE OF THE SUNSPOT ZONE, AS A FUNCTION OF LATITUDE, L ($|L| \lesssim 40^\circ$)	$(14.44^\circ - 3.0^\circ \sin^2 L)$ per Day
ADOPTED PERIOD OF SIDEREAL ROTATION (L = 17°)	25.38 Days
CORRESPONDING SYNODIC ROTATION PERIOD (RELATIVE TO THE EARTH)	27.275 Days
MEAN SUNSPOT PERIOD	11.04 Years
DATES OF FORMER MAXIMA	1957.9, 1968.9
MEAN TIME FROM MAXIMUM TO SUBSEQUENT MINIMUM	6.2 Years

*THE QUANTITY LISTED IS THE MODE OF THE VELOCITY DISTRIBUTION. THE MEAN OF THE DISTRIBUTION, WHICH IS MORE STRONGLY INFLUENCED BY HIGH VELOCITY STARS, IS 19.5 KM/SEC TOWARD $\alpha = 271^\circ$ $\delta = +30^\circ$. FOR FURTHER DETAILS, SEE MIHALAS [1968].

L.3 The Earth

The principal physical properties of the Earth are listed in Table L-5. For general characteristics, see also Tables L-1 through L-3. See Appendix J for a discussion of the length of day and year and the nonuniform rotation of the Earth. See Chapter 4 for an extended discussion of Earth models.

Table L-5. Physical Properties of the Earth

PROPERTY	VALUE
EQUATORIAL RADIUS, a	6378.140 km
FLATTENING FACTOR (ELLIPTICITY) $\frac{a-c}{a} \equiv f$	0.00335281 = 1/298.257
POLAR RADIUS[†], c	6356.755 km
MEAN RADIUS[†] $(a^2 c)^{1/3}$	6371.00 km
ECCENTRICITY[†] $\left(\sqrt{a^2-c^2}\right)/a$	0.0818192
SURFACE AREA	5.10066×10^8 km^2
VOLUME	1.08321×10^{12} km^3
ELLIPTICITY OF THE EQUATOR $(a_{MAX} - a_{MIN})/a_{MEAN}$	$\sim 1.6 \times 10^{-5}$ $(a_{max} - a_{min} \approx 100$ m$)$
LONGITUDE OF MAXIMA, a_{MAX}	20° W and 160° E
RATIO OF THE MASS OF THE SUN TO THE MASS OF THE EARTH	332,946.0
GEOCENTRIC GRAVITATIONAL CONSTANT, $GM_E \equiv \mu_E$	3.986005×10^{14} m^3 s^{-2}
MASS OF THE EARTH[‡]	5.9742×10^{24} kg
MEAN DENSITY	5.515 gm/cm^3
GRAVITATIONAL FIELD CONSTANTS $\begin{cases} J2 \\ J3 \\ J4 \end{cases}$	$+1082.63 \times 10^{-6}$ -2.54×10^{-6} -1.61×10^{-6}
MEAN DISTANCE OF EARTH CENTER FROM EARTH-MOON BARYCENTER	4671 km
AVERAGE LENGTHENING OF THE DAY (SEE FIGURE J-5)	0.0015 sec/Century
GENERAL PRECESSION IN LONGITUDE (i.e., PRECESSION OF THE EQUINOXES) PER JULIAN CENTURY, AT EPOCH 2000	1.39697128 Deg/Century
RATE OF CHANGE OF PRECESSION	$+6.181 \times 10^{-4}$ Deg/Century^2
OBLIQUITY OF THE ECLIPTIC, AT EPOCH 2000	23.4392911 Deg
RATE OF CHANGE OF THE OBLIQUITY (T IN JULIAN CENTURIES)	$(-1.30125 \times 10^{-2}$ T -1.64×10^{-6} T^2 $+5.0 \times 10^{-7}$T$^3)$ Deg
AMPLITUDE OF EARTH'S NUTATION	2.5586×10^{-3} Deg
LENGTH OF SIDEREAL DAY, EPOCH 1980[§]	86,164.0918 sec = 23 hr 56 min 4.0918 sec
LENGTH OF SIDEREAL YEAR, EPOCH 1980[§]	3.1588149548×10^7 sec = 365.25636051 Days[‖]
LENGTH OF TROPICAL YEAR (REF $=\Upsilon$) EPOCH 1980[§]	3.156925551×10^7 sec = 365.24219388 Days[‖]
LENGTH OF ANOMALISTIC YEAR (PERIHELION TO PERIHELION), EPOCH 1980[§]	3.1558433222×10^7 sec = 365.25964377 Days[‖]

[†] BASED ON THE ADOPTED VALUES OF f AND a.

[‡] ASSUMING G = 6.672×10^{-11} m^3 kg^{-1} s^{-2}; THE VALUE OF GM_E IS MORE ACCURATELY KNOWN.

[§] SEE APPENDIX J FOR FORMULAE AND DISCUSSION.

[‖] ONE DAY $\equiv$ 86,400 SI SECONDS.

Table L-6 summarizes the properties of the upper atmosphere of the Earth. The mean profiles between 25 and 500 km are from the COSPAR International Reference Atmosphere, CIRA 72 [1972]. Between 500 and 1000 km, the CIRA 72 profile for $T_\infty = 1000K$ was used to indicate the densities to be expected. The maximum and minimum values of the density between 100 and 500 km were extracted from the explanatory material in CIRA 72 and indicate the variation in densities which can be obtained with the models. Sea level temperature and density are from the *U.S. Standard Atmosphere* [1976].

Geocentric and Geodetic Coordinates on the Earth. The *geocentric latitude*, ϕ', of a point, *P*, on the surface of the Earth is the angle at the Earth's center between *P* and the equatorial plane. The *geodetic* or *geographic latitude*, ϕ, is the angle between the normal to an arbitrarily defined reference ellipsoid (chosen as a close approximation to the oblate Earth) and the equatorial plane. *Astronomical latitude* and *longitude* are defined relative to the *local vertical*, or the normal to the equipotential surface of the Earth. Thus, *astronomical latitude* is defined as the angle between the local vertical and the Earth's equatorial plane. Maximum values

Table L-6.   The Upper Atmosphere of the Earth

ALTITUDE (KM)	MEAN KINETIC TEMPERATURE ($^{\circ}$K)	DENSITY (kg/m^3)			SCALE HEIGHT (KM)
		MINIMUM	MEAN	MAXIMUM	
0	288.2		$1.225 \times 10^{+0}$		8.44
25	221.7		3.899×10^{-2}		6.49
30	230.7		1.774×10^{-2}		6.75
35	241.5		8.279×10^{-3}		7.07
40	255.3		3.972×10^{-3}		7.47
45	267.7		1.995×10^{-3}		7.83
50	271.6		1.057×10^{-3}		7.95
55	263.9		5.821×10^{-4}		7.73
60	249.3		3.206×10^{-4}		7.29
65	232.7		1.718×10^{-4}		6.81
70	216.2		8.770×10^{-5}		6.33
75	205.0		4.178×10^{-5}		6.00
80	195.0		1.905×10^{-5}		5.70
85	185.1		8.337×10^{-6}		5.41
90	183.8		3.396×10^{-6}		5.38
95	190.3		1.343×10^{-6}		5.74
100	203.5	3.0×10^{-7}	5.297×10^{-7}	7.4×10^{-7}	6.15
110	265.5	6.0×10^{-8}	9.661×10^{-8}	3.0×10^{-7}	8.06
120	334.5	1.0×10^{-8}	2.438×10^{-8}	6.0×10^{-8}	11.6
130	445.4	4.5×10^{-9}	8.484×10^{-9}	1.6×10^{-8}	16.1
140	549.0	2.0×10^{-9}	3.845×10^{-9}	6.0×10^{-9}	20.6
150	635.2	1.2×10^{-9}	2.070×10^{-9}	3.5×10^{-9}	24.6
160	703.1	6.5×10^{-10}	1.244×10^{-9}	2.0×10^{-9}	26.3
180	781.2	2.4×10^{-10}	5.464×10^{-10}	9.0×10^{-10}	33.2
200	859.3	1.0×10^{-10}	2.789×10^{-10}	3.2×10^{-10}	38.5
250	940.2	4.0×10^{-11}	7.248×10^{-11}	1.6×10^{-10}	46.9
300	972.8	1.6×10^{-11}	2.418×10^{-11}	8.8×10^{-11}	52.5
350	986.5	2.0×10^{-12}	9.158×10^{-12}	6.0×10^{-11}	56.4
400	992.6	3.7×10^{-13}	3.725×10^{-12}	5.0×10^{-11}	59.4
450	995.7	9.0×10^{-14}	1.585×10^{-12}	3.8×10^{-11}	62.2
500	997.3	1.3×10^{-14}	6.967×10^{-13}	3.0×10^{-11}	65.8
600	1000.0		1.454×10^{-13}		79
700	1000.0		3.614×10^{-14}		109
800	1000.0		1.170×10^{-14}		164
900	1000.0		5.245×10^{-15}		225
1000	1000.0		3.019×10^{-15}		268

Table L-7.   Coefficients for Determining Geocentric Rectangular Coordinates from Geodetic
Coordinates on the Surface of the Earth. Based on $f = 1/298,257$.

ϕ (DEG)	S	C	ϕ (DEG)	S	C
± 0	0.993306	1.000000	± 50	0.995262	1.001970
5	0.993331	1.000025	55	0.995544	1.002254
10	0.993406	1.000101	60	0.995809	1.002520
15	0.993528	1.000224	65	0.996048	1.002761
20	0.993695	1.000392	70	0.996255	1.002969
25	0.993900	1.000598	75	0.996422	1.003138
30	0.994138	1.000838	80	0.996546	1.003262
35	0.994401	1.001103	85	0.996622	1.003338
40	0.994682	1.001386	± 90	0.996647	1.003364
± 45	0.994972	1.001678			

of the *deviation of the vertical,* or the angle between the local vertical and the normal to a reference ellipsoid, are about 1 minute of arc. Maximum variations in the height between the reference ellipsoid and *mean sea level* (also called the *geoid* or *equipotential surface*) are about 100 m, as illustrated in Fig. 5-8.

The coordinate transformations given here are intended for use near the Earth's surface to correct for an observer's height above sea level and are valid only for altitudes much less than the radius of the Earth. For satellite altitudes, the coordinates will depend on the definition of the subsatellite point or the method by which geodetic coordinates are extended to high altitudes. For a discussion of geodetic coordinates at satellite altitudes, see Hedman [1970] or Hedgley [1976].

Geodetic and geocentric latitude are related by [H.M. Nautical Almanac Office, 1961]:

$$\tan\phi = \tan\phi'/(1-f)^2 \approx 1.006740 \tan\phi'$$

$$\phi - \phi' = \left(f + \tfrac{1}{2}f^2\right)\sin 2\phi - \left(\tfrac{1}{2}f^2 + \tfrac{1}{2}f^3\right)\sin 4\phi + \tfrac{1}{2}f^3\sin 6\phi$$

$$\approx 0.19242°\sin 2\phi - 0.000323°\sin 4\phi$$

where $f \approx 1/298.257$ is the flattening factor of the Earth as adopted by the IAU in 1976 [Muller and Jappel, 1977].

Let h be the height of P above the reference ellipsoid in metres; let $R_\oplus$ be the equatorial radius of the Earth in metres; and let d be the distance from P to the center of the Earth in units of $R_\oplus$. Then d and h are related by

$$h = R_\oplus\left[d - (1-f)/\sqrt{1 - f(2-f)\cos^2\phi'}\,\right]$$

$$d = \frac{h}{R_\oplus} + 1 - \left(\frac{1}{2}\right)f - \left(\frac{5}{16}\right)f^2 + \left(\frac{1}{2}\right)f\cos 2\phi - \left(\frac{5}{16}\right)f^2\cos 4\phi + \mathcal{O}(f^3)$$

$$\approx (1.5679 \times 10^{-7})h + 0.998327 + 0.001676\cos 2\phi - 0.000004\cos 4\phi$$

To convert geographic or geodetic coordinates to geocentric rectangular coordinates in units of $R_\oplus$, use the following:

$$d\sin\phi' = (S + h \times 1.5679 \times 10^{-7})\sin\phi$$

$$d\cos\phi' = (C + h \times 1.5679 \times 10^{-7})\cos\phi$$

$$\tan\phi' = (0.993305 + h \times 1.1 \times 10^{-9})\tan\phi$$

Table L-8.   Physical and Orbital Properties of the Moon

PROPERTY	VALUE
MEAN DISTANCE FROM EARTH	384401 ± 1 km
EXTREME RANGE	356400 to 406700 km
ECCENTRICITY OF ORBIT	0.0549
INCLINATION OF ORBIT TO ECLIPTIC (OSCILLATING ± 0.15 DEG WITH PERIOD OF 173 DAYS)	5.1453 Deg
SIDEREAL PERIOD (RELATIVE TO FIXED STARS) WHERE T IS IN CENTURIES FROM 1900.0	$27.32166140 + T \times 1.6 \times 10^{-7}$ Ephemeris Days
SYNODICAL MONTH (NEW MOON TO NEW MOON)	$29.5305882 + T \times 1.6 \times 10^{-7}$ Ephemeris Days
TROPICAL MONTH (EQUINOX TO EQUINOX)	$27.32158214 + T \times 1.3 \times 10^{-7}$ Ephemeris Days
ANOMALISTIC MONTH (PERIGEE TO PERIGEE)	$27.5545505 - T \times 4 \times 10^{-7}$ Days
NODICAL MONTH (NODE TO NODE)	27.212220 Days
NUTATION PERIOD = PERIOD OF ROTATION OF THE NODE (ROTROGRADE)	18.61 Tropical Years
PERIOD OF ROTATION OF PERIGEE (DIRECT)	8.85 Years
OPTICAL LIBRATION IN LONGITUDE (SELENOCENTRIC DISPLACEMENT)	± 7.6 Deg
OPTICAL LIBRATION IN LATITUDE (SELENOCENTRIC DISPLACEMENT)	± 6.7 Deg
SURFACE AREA NEVER VISIBLE FROM EARTH	41%
INCLINATION OF EQUATOR TO ECLIPTIC TO ORBIT	1.542 Deg 6.68 Deg
RADII: a TOWARD EARTH, b ALONG ORBIT, c TOWARD POLE	
MEAN RADIUS (b+c)/2 a—c a—b b—c	1738.2 km 1.09 km 0.31 km 0.78 km
MEAN ANGULAR DIAMETER AT MEAN DISTANCE FROM EARTH	0.5182 Deg
RATIO OF MASS OF MOON TO MASS OF EARTH	0.01230002
MASS OF THE MOON*	7.3483×10^{22} kg
MEAN DENSITY	3.341 g cm^{-3}
SURFACE GRAVITY	162.2 cm s^{-2}
SURFACE ESCAPE VELOCITY	2.38 km/s

*ASSUMING G = 6.672×10^{-11} m^3 kg^{-1} s^{-2}; MASS RATIOS ARE MORE ACCURATE.

where:

$$C \equiv \left[\cos^2\phi + (1-f)^2 \sin^2\phi \right]^{-1/2}$$

$$S \equiv (1-f)^2 C$$

values of S and C are given in Table L-7. In terms of S and C, the distance to the center of the Earth for $h = 0$ is

$$d^2 = \tfrac{1}{2}(S^2 + C^2) + \frac{1}{2}(C^2 - S^2)\cos 2\phi$$

L.4 The Moon

The physical and orbital properties of the Moon are summarized in Table L-8. Additional general characteristics are given in Tables L-2 and L-3. To determine the Moon's visual magnitude, $V(R,\xi)$, at any distance and phase, let R be the observer-Moon distance in AU and ξ be the phase angle at the Moon between the Sun and the observer. Then

$$V(R,\xi) = 0.23 + 5\log_{10}R - 2.5\log_{10}P(\xi)$$

where the phase law, $P(\xi)$, for the Moon is given in Table L-9 [Allen, 1973]. For additional details and a sample computation, see Section 3.5. Note that the visual magnitude of the Moon at opposition (i.e., full Moon) at the mean distance of the Moon from the Earth is -12.73.

Table L-9. Phase Law and Visual Magnitude of the Moon

ξ (DEG)	P (ξ)	V (R, ξ) − V (R, 0)	ξ (DEG)	P (ξ)	V (R, ξ) − V (R, 0)
0	1.000	0.00	80	0.127	2.24
5	0.929	0.08	90	0.089	2.63
10	0.809	0.23	100	0.061	3.04
20	0.625	0.51	110	0.041	3.48
30	0.483	0.79	120	0.027	3.93
40	0.377	1.06	130	0.017	4.44
50	0.288	1.35	140	0.009	5.07
60	0.225	1.62	150	0.004	5.9
70	0.172	1.91	160	0.001	7.5

L.5 Potential Spacecraft Orbits

Table L-10 lists the transfer time and velocity required for Hohmann transfer orbits between the planets. The values cited are for minimum energy transfer orbits between the mean distances of the planets from the Sun. The upper number is the one-way transfer time in days; the lower number is the velocity change required to go from the orbital velocity of the planet of origin to the transfer orbit in km/sec. See Section 3.3 for relevant formulae. Finally, Table L-11 gives the velocity of escape, circular velocity, and synchronous altitude and velocity for potential artificial satellites of the Moon and planets.

Table L-10. Hohmann Transfer Orbit Properties. (See text for explanation.)

DESTINATION	ORIGIN								
	MERCURY	VENUS	EARTH	MARS	JUPITER	SATURN	URANUS	NEPTUNE	PLUTO
MERCURY	–	75.6 5.8	105 7.5	171 8.8	853 8.2	2,020 6.9	5,590 5.4	10,800 4.6	16,300 4.1
VENUS	75.6 6.7	–	146 2.5	217 4.8	931 6.6	2,120 6.0	5,730 5.0	11,000 4.3	16,400 3.8
EARTH	105 9.6	146 2.7	–	259 2.6	997 5.6	2,210 5.4	5,850 4.7	11,200 4.1	16,700 3.7
MARS	171 12.6	217 5.8	259 2.9	–	1,130 4.3	2,380 4.6	6,080 4.2	11,500 3.7	16,900 3.4
JUPITER	853 17.4	931 11.4	997 8.8	1,130 5.9	–	3,650 1.5	7,780 2.4	13,500 3.7	19,300 2.5
SATURN	2,020 18.5	2,120 12.7	2,210 10.3	2,380 7.6	3,650 1.8	–	9,940 1.3	16,100 1.7	22,100 1.8
URANUS	5,590 19.1	5,730 13.6	5,850 11.3	6,080 8.7	7,780 3.3	9,940 1.5	–	22,300 0.6	29,000 0.9
NEPTUNE	10,800 19.4	11,000 13.9	11,200 11.7	11,500 9.2	13,500 4.0	16,100 2.2	22,300 0.7	–	37,400 0.3
PLUTO	16,300 19.5	16,400 14.1	16,700 11.8	16,900 9.4	19,300 4.3	22,100 2.6	29,000 1.1	37,400 0.4	–
THE MOON			5.0 3.2						

Table L-11.   Parameters for Potential Artificial Satellites of the Moon and Planets

PLANET	VELOCITY OF ESCAPE (KM/SEC)	VELOCITY IN CIRCULAR ORBIT AT THE SURFACE (KM/SEC)	SYNCHRONOUS ORBIT	
			ALTITUDE ABOVE SURFACE (KM)	VELOCITY (KM/SEC)
THE MOON	2.376	1.679	86,710	0.235
MERCURY	4.263	3.014	241,400	0.301
VENUS	10.346	7.316	1,536,000	0.459[1]
EARTH	11.180	7.905	35,786	3.075
MARS	5.023	3.552	17,033	1.448
JUPITER	59.62	42.16	87,820[2]	28.22
SATURN	35.53	25.12	49,150[2]	18.63
URANUS	21.77	15.39	36,130	9.78
NEPTUNE	23.40	16.55	57,480	9.12
PLUTO	5.4	3.8	68,000	0.81

NOTE: THE VELOCITY OF ESCAPE FROM THE SOLAR SYSTEM AT THE DISTANCE OF THE EARTH'S ORBIT
        IS 29.785 KM/SEC.

[1]RETROGRADE.

[2]FOR EQUATORIAL ROTATION; THE PLANET'S ROTATION IS SLOWER AT HIGHER LATITUDES.

References

1. Allen, C. W., *Astrophysical Quantities*, Third Edition. London: The Athlone Press, 1973.

2. COSPAR, *COSPAR International Reference Atmosphere*. Berlin: Akademie-Verlag, 1972.

3. Hedgley, David R., Jr., *An Exact Transformation from Geocentric to Geodetic Coordinates for Nonzero Altitudes*, NASA TR R-458, Flight Research Center, March 1976.

4. Hedman, Edward L., Jr., "A High Accuracy Relationship Between Geocentric Cartesian Coordinates and Geodetic Latitude and Altitude," *J. Spacecraft*, Vol. 7, p. 993–995, 1970.

5. H.M. Nautical Almanac Office, *Explanatory Supplement to the Astronomical Ephemeris and the American Ephemeris and Nautical Almanac*. London: Her Majesty's Stationery Office, 1961.

6. Mihalas, Dimitri, *Galactic Astronomy*. San Francisco: H. W. Freeman and Company, 1968.

7. Muller, Edith A. and Arndst Jappel, Editors, *International Astronomical Union, Proceedings of the Sixteenth General Assembly, Grenoble 1976*. Dordrecht, Holland: D. Reidel Publishing Co., 1977.

8. Owen, Tobias, "Jovian Satellite Nomenclature," *Icarus*, Vol. 29, p. 159–163, 1976.

9. *U.S. Standard Atmosphere*, Washington, U.S.G.P.O., 1976.

APPENDIX M

FUNDAMENTAL PHYSICAL CONSTANTS

The physical constants are those compiled by Cohen and Taylor [1973a, 1973b] under the auspices of the CODATA (Committee on Data for Science and Technology of the International Council of Scientific Unions) Task Group on Fundamental Constants and officially adopted by CODATA. The astronomical constants are those compiled by Commission 4 of the International Astronomical Union and adopted at the 1976 IAU meeting in Grenoble [Muller and Jappel, 1977]. The uncertainties are the 1σ standard deviation expressed in parts per million (ppm). Additional constants are listed in Appendix K (Conversion Factors), Appendix L (Solar System Constants), Allen [1973], and Rossini [1974].

DIMENSIONLESS NUMBERS (Note 1):

$$\pi = 3.141\ 592\ 653\ 589\ 793\ 238\ 462\ 643\ \ldots$$

$$e = 2.718\ 281\ 828\ 459\ 045\ 235\ 360\ 287\ \ldots$$

PHYSICAL CONSTANTS:

Quantity	Symbol	Value	Units	Uncertainty (ppm)
Elementary Charge	e	$1.6021892 \times 10^{-19}$	C	2.9
Planck Constant	h	6.626176×10^{-34}	J$\cdot$s	5.4
Permeability of Vacuum	μ_0	$4\pi \times 10^{-7}$	N/A^2	Exact
Fine Structure Constant $(\mu_0 c e^2/2h)$	α	7.2973506×10^{-3}	None	0.82
Avogadro Constant	N_A	6.022045×10^{23}	mol^{-1}	5.1
Atomic Mass Unit	u	$1.6605655 \times 10^{-27}$	kg	5.1
Electron Rest Mass	m_e	9.109534×10^{-31}	kg	5.1
Proton Rest Mass	m_p	$1.6726485 \times 10^{-27}$	kg	5.1
Neutron Rest Mass	m_n	$1.6749543 \times 10^{-27}$	kg	5.1
Muon Rest Mass	m_μ	1.883566×10^{-28}	kg	5.6
Electron Magnetic Moment	μ_e	9.284832×10^{-24}	J/T	3.9
Proton Magnetic Moment	μ_p	$1.4106171 \times 10^{-26}$	J/T	3.9
Muon Magnetic Moment	μ_μ	4.490474×10^{-26}	J/T	3.9
Rydberg Constant $(\mu_0^2 c^3 m_e e^4/8h^3)$	R_∞	1.097373177×10^7	m^{-1}	0.075
Bohr Radius $(\alpha/4\pi R_\infty)$	a_0	$5.2917706 \times 10^{-11}$	m	0.82
Classical Electron Radius $(\alpha^3/4\pi R_\infty)$	r_e	$2.8179380 \times 10^{-15}$	m	2.5

Quantity	Symbol	Value	Units	Uncertainty (ppm)
Compton Wavelength of the electron $(h/m_e c)$	λ_c	$2.4263089 \times 10^{-12}$	m	1.6
Molecular volume of ideal gas at S.T.P.	V_m	2.241383×10^{-2}	m^3/mol	31
Boltzman constant	k	1.380662×10^{-23}	J/K	32

ASTRONOMICAL CONSTANTS:

Quantity	Symbol	Value	Units	Uncertainty (ppm)
Speed of Light in Vacuum	c	2.99792458×10^8	m/s	0.004
Gaussian Gravitational Constant	k	$1.720209895 \times 10^{-2}$	rad/day	Note (2)
Earth Equatorial Radius	$R_\oplus$	6.378140×10^6	m	0.78
Earth Dynamical Form Factor	J_2	1.08263×10^{-3}	None	9.2
Earth Flattening Factor	f	3.35281×10^{-3}	None	6.0
$1/f$		2.98257×10^2	None	6.0
Earth Gravitation Constant	$GM_\oplus$	3.986005×10^{14}	m^3/s^2	0.75
Moon Gravitation Constant	GM_M	4.902794×10^{12}	m^3/s^2	3.6
Sun Gravitation Constant	$GM_\odot$	$1.32712438 \times 10^{20}$	m^3/s^2	0.038
Gravitational Constant	G	6.6720×10^{-11}	$m^3/(kg \cdot s^2)$	615. (3)
Mass of the Moon	M_M	7.3483×10^{22}	kg.	615. (3)
Mass of the Sun	$M_\odot$	1.9891×10^{30}	kg	615. (3)
Mass of the Earth	$M_\oplus$	5.9742×10^{24}	kg	615. (3)
Ratio of the Mass of the Moon to the Mass of the Earth	$M_M/M_\oplus$	1.230002×10^{-2}	None	3.6
Obliquity of the Ecliptic at Epoch 2000	ϵ	$23°26'21''.448$ $= 2.34392911 \times 10^1$	deg	1.2
General Precession in Longitude per Julian Ephemeris Century, at Epoch 2000	P	1.39697128	$\left(\dfrac{deg}{century}\right)$	30
Constant of Nutation, at Epoch 2000	N	2.55858×10^{-3}	deg	600
Astronomical Unit	AU	$1.49597870 \times 10^{11}$	m	0.013
Solar Parallax	$\pi_\odot$	2.442819×10^{-3}	deg	0.80

Ratio of the mass of the Sun to those of the planetary systems (planetary system masses include both atmosphere and satellites):

Mercury	6.023600×10^6		580
Venus	4.085235×10^5		5.2
Earth + Moon	3.289005×10^5		1.5

Quantity	Symbol	Value	Units	Uncertainty (ppm)
Mars		3.098710×10^6		26
Jupiter		1.047355×10^3		24
Saturn		3.4985×10^3		430
Uranus		2.2869×10^4		9900
Neptune		1.9314×10^4		3900
Pluto		3×10^6		$\binom{-0.3}{+4} \times 10^6$

Notes:

(1) Values from Abramowitz and Stegun [1970].

(2) The Gaussian gravitational constant has the given value by definition and serves to define the other astronomical constants.

(3) The gravitational constant enters theories of orbital dynamics only through the product GM. This product is well known for the various objects in the solar system. G itself, and consequently the masses in kilograms of the Sun and planets, is not as well known. Therefore, accurate analyses should use directly the product $GM_\odot$ and the ratios of the masses of the planets and the Sun.

EARTH SATELLITE PARAMETERS

ALTITUDE, h (km)	ANGULAR RADIUS OF THE EARTH (deg)	PERIOD (min)	VELOCITY (km/sec)	REQUIRED ENERGY (MJ/kg)
0	90.00	84.49	7.905	31.14
100	79.92	86.48	7.844	31.62
200	75.84	88.49	7.784	32.09
300	72.76	90.52	7.726	32.54
400	70.22	92.56	7.669	32.98
500	68.02	94.62	7.613	33.41
600	66.07	96.69	7.558	33.83
800	62.69	100.87	7.452	34.62
1,000	59.82	105.12	7.350	35.37
2,000	49.58	127.20	6.898	38.60
3,000	42.85	150.64	6.519	41.14
4,000	37.92	175.36	6.197	43.18
5,000	34.09	201.31	5.919	44.87
10,000	22.92	347.66	4.933	50.22
20,000	13.99	710.60	3.887	54.83
35,786 (SYNCHRONOUS)	8.70	1436.07 (1 SIDEREAL DAY)	3.075	57.66
∞	0.0	∞	0.0	62.39

In the above table, the angular radius, period, and energy required are valid for elliptical orbits of arbitrary eccentricity; however, the velocity is correct only for circular orbits. For noncircular orbits, h should be interpreted as the instantaneous altitude when determining the angular radius of the Earth and as the mean altitude when determining the period and required energy. The mean altitude is $h_m \equiv (P + A)/2$, where P and A are the perigee and apogee altitudes, respectively.

References

1. Abramowitz, Milton and Irene A. Stegun, *Handbook of Mathematical Functions with Formulas, Graphs and Mathematical Tables*. Washington, DC: National Bureau of Standards, 1970.
2. Allen, C. W., *Astrophysical Quantities*, Third Edition. London: The Athlone Press, 1973.
3. Cohen, E. Richard and B. N. Taylor, *J. Phys. Chem. Ref. Data*, vol. 2, p. 663, 1973a
4. ———, CODATA Bulletin No. 11, Dec. 1973b.
5. Müller, Edith A. and Arndst Jappel, Editors, *International Astronomical Union, Proceedings of the Sixteenth General Assembly, Grenoble 1976*. Dordrecht, Holland: D. Reidel Publishing Co., 1977.
6. Rossini, Frederick D., *Fundamental Measures and Constants for Science and Technology*. Cleveland, Ohio: CRC Press, 1974.

INDEX

—A—

Aberration 144
Abnormal termination in mission support
   software, handling of 682–683
Absorption bands, in Earth's atmosphere 91–92
Absorption of radiation, torque due to 572
AC two-phase induction motor, torque
   profile for 270–271
Acceleration, units and conversion factors 809
Accelerometers, as attitude determination
   reference 17
Acquisition of signal (See In-triggering)
Acquisition phase, of space mission 3–8
Active attitude control 18, 503, 506–509
Actuator (control system component) 502, 589
Adams integrators 563–564
Adams-Bashforth integrators 563–564
Adams-Bashforth-Moulton integrators 563–564
Adams-Moulton integrator 563
Adaptive filter, for state estimation 467
Adcole Corporation, Sun sensors 157, 161–166
Adjoint, of a matrix 745
ADL (See Attitude data link)
Advanced range instrumentation aircraft
   284, 287
AE (Atmospheric Explorer)—
   Application of block averaging to
      attitude solutions 371
   Attitude determination accuracy 397
   Attitude sensor bias evaluation 402
   Attitude system of 788–789
   Data collection for bias determination 475
   Data curve fitting for 318
   Data sample from AE-3 313
   Earth-width data 233–234
   Horizon sensors 177
   Magnetic coil control system 509
   Momentum wheel 202–203, 601, 602
   Optimal magnetic maneuvers 640
   Orbit generator accuracy 138
   Pitch angle measurement 360
   Shadow modeling for 573
   Spin axis magnetic coils 205
   Spin plane magnetic control 646, 648–649
   Stabilization method 503
   Sun sensor 157
   Telemetry data errors 311
   Use of body-mounted horizon sensor 173
   Use of carbon dioxide band horizon
      sensor 92
   Use of open loop control 663
   Yaw inversion maneuver 651
AEM (Applications Explorer Mission)—
   (See also HCMM (AEM-A); SAGE
     (AEM-B))
   Attitude system of 720, 788–789

Sun sensor 157
Aerodynamic drag, effect on flexible
   spacecraft 551
   Effect on orbit 63–65
Aerodynamic stabilization 19
Aerodynamic torque 17, 573–575
AEROS (German Aeronomy Satellite)—
   Attitude acquisition 662
   Attitude system of 788–789
   Stabilization method 503
   Sun sensor 157
Agena (upper stage of launch vehicle) 53
Akademik Sergey Korolev (Soviet tracking
   ship) 290
Albedo 83
   Of the Moon and planets, table 816
   Torque due to 571
   Types of 79
Albedo sensor (See also Horizon sensor) 83
Algebraic method of three-axis attitude
   determination 421, 424–426
Alignment, Math. model for
   slit sensors 219–221
   Math. model for two-axis sensors 221–223
Alouette (Canadian Ionospheric Research
   Satellite)—
   Attitude system of 788–789
Altitude of Earth satellites, table of period and
   Earth size vs. 828
Ambiguous attitude solutions, methods for
   resolution of 371–373
American Ephemeris and Nautical
   Almanac 36, 139–140
Analog Sun sensors 156–159
Analog-to-digital conversion (A/D) 279, 298
   For magnetometers 250
Analysis of variance 429
Angle of attack (in aerodynamic torque) 574
ANGLED (subroutine) 690
Angstrom 808
Angular measure, units and conversion
   factors 810
Angular momentum 516–521
   Orbital 40
   Storage in flywheel (See Momentum wheels)
   Units and conversion factors 811
Angular momentum axis, of spacecraft 487
Angular momentum control, with spin
   plane magnetic coils 644, 646–649
Angular momentum sphere 499–501
Angular separation, between two points 727
Angular separation match, for star
   identification 259, 262–263
Angular velocity, math. model for
   gyroscope measurement of 267–268
   Units and conversion factors 811

Angular velocity vector, for Earth-referenced spacecraft 609
Annular eclipse 72, 76
Anomalistic period, for Earth satellite 67
Anomalistic year 48
Anomaly (orbit parameter) 44–46
ANS (Netherlands Astronomy Satellite)—
Attitude system of 788–789
Antennas, Examples of tracking and command 286–289
Telemetry and command 284–288
Antipoint, on celestial sphere 22
Antisolar point 22
Antisymmetric matrix 750
Aphelion 44
Apofocal distance 44
Apofocus 44
Apogee 44
Apogee height 44
Maneuvers to change 59
Apollo 53
Visual magnitude of during trip to Moon 79
Apollo 15 subsatellite, nutation of 495–497
Apolune 44
Apparent solar time 798, 800
Apsides 44
Rotation of (*See Perigee, rotation of*)
Arc, of great circle 23
Arc length, measurement of on global geometry plot 742
Arc length distance, between two points 727
Arc length measurement 23, 346–349
Notation for 23
Singularity conditions 406–407
Use in deterministic attitude solutions 364–365, 368–370
Arc length uncertainty, of single-axis attitude solutions 372, 374–383
Arc minute 24
Arc second 24
Area on celestial sphere, formulas for 729–730
Areal velocity 46, 47
Argument of perigee 45–46
Argument of perihelion 49
Ariel (U.K. Satellite)—
Attitude system of 788–789
Aryabhata (Indian satellite)—
Attitude system of 788–789
Nutation damper 628
Ascending node 44
Maneuver to change 59
Aspects of the planets (*See Planetary configurations*)
Asteroids, Lagrange point orbits 55
Astrodynamics 487
Astronauts, disturbance torques due to motion of 576, 578–580
Astronomical constants 827–828
Astronomical Ephemeris 36, 139
Astronomical latitude and longitude 820
Astronomical symbols 50
Astronomical Unit 41, 808

Atmosphere (unit of pressure) 809
Atmosphere of Earth (*See Earth, atmosphere of*)
Atomic clocks 799, 802
ATS (Applications Technology Satellite)—
Attitude system of 788–789
Horizon sensors 175
Kalman filter state & covariance propagation 465–467
Momentum wheel 202
Momentum wheel control system 508
Nutation damping 628
Orbit generator accuracy 138
Polaris tracker 189
Stabilization method 503
Sun sensor 157
Telemetry data errors 311
Attitude
Definition 1
Introduction to 1–21
Introduction to analysis 343–361
Parameterizations of 410–420
—Table of relative advantages 412
Transmission of results 292
Attitude acquisition (*See also Acquisition phase*) 502, 661–667
Classifications of 662–663
CTS example 672
GEOS-3 example 672–677
Attitude acquisition maneuvers 663–672
Attitude control 2, 588–678
Areas of current work 715
Closed loop and open loop 662–663
Example of 14–16
Example of function of 7
Functions of 2
Introduction to 502–509
Needed analytic projects 722–723
Table of methods 19, 503
Transmission of commands 279, 292
Attitude control hardware (*See also specific item, e.g., Gas jets*) 201–213
Earth-oriented spacecraft (table) 613
Introduction to 502–509
Listed by spacecraft & sensor type 796
Mathematical models 270–275
Attitude control strategy 16
Attitude control system 502
Attitude Data Link (ADL) 292
Attitude data simulators 709–712
Examples of structure 710–712
Functional requirements 709–710
Attitude determination (*See also specific method, e.g., Earth width / Sun angle*) 2, 343–484
Areas of current work 715
Block averaging 370–373
Definitive vs. real time 681
Deterministic methods 362–435
—Advantages of 437–438
—Solution behavior 402–408

Deterministic vs. state estimation   436–438
Example of   10–14
Example of function of   7
Functions of   2
Hardware (*See Attitude sensors*)
Introduction to analysis   343–361
Methods of   16–18
Mission support software   681–713
Needed analytic projects   722–723
Proc. for elim. data anomalies   334–339
Reference sources, table of   17
Reference vector   10
—Attitude accuracy limits   393–397
Single axis   362–409
Spinning spacecraft   363–370
Star sensor methods   703–709
Three-axis   420–428
Uncertainties   345–346
—Expressions for   375–376, 381–382, 384
Attitude determination accuracy   397
Analytic solutions for   373–402
Direct calculation of   373, 429–435
Effect of Earth oblateness   105
Estimating systematic error   476
Estimating systematic uncertainty   402–408
Evolution of over time   714
For continuous measurements   376–382
For correlated uncertainties
                              382–383, 431–432
For nonorthogonal measurements   377–382
For orthogonal measurements   376–377
For quantized measurements   374–376, 431
For uncorrelated uncertainties
                              374–382, 431–432
Geometrical limitations on   389–402
Identification of singularities   406–407
—Table   407
Limited by systematic error   476
Sample computation (IUE)   384–386
Single axis   373–409
Single frame, summary table   384
Spacecraft requirements, list of   796
Three axis   429–434
Attitude disturbance torques (*See
    Disturbance torques*)
Attitude dynamics   487–587
Flexible spacecraft   548–556
Introduction to   487, 498–502
Mathematical models   521–523
Model of, in state vector formulation   440
—For attitude propagation   558–559
Attitude error ellipse (*See Error ellipse*)
Attitude error parallelogram (*See Error
    parallelogram*)
Attitude extrapolation, for star sensor
    attitude determination   705–706
Attitude geometry (*See also Global
    geometry plots*)
As limitation on attitude accuracy   22–35
As limitation on attitude accuracy   389–402
For single-axis attitude solutions   362–402
Unresolved analytic problems   723

Attitude kinematics   510–521
Approximate closed-form solutions
    for   564–566
Equations for attitude propagation   558–559
Introduction to   487–497
Model of, in state vector formulation   440
Attitude maneuver control   2, 502, 636–678
Attitude maneuver control program for
    CTS spacecraft   700, 703
Attitude maneuver monitoring, in CTS
    attitude system   702–703
Software for   700
Attitude matrix (*See Direction cosine matrix*)
Attitude measurements (*See also Arc
    length measurement; Compound attitude
    measurement; Rotation angle measurement*)
Equivalence of   344
Intersection of loci   344–346
Types of   346
Attitude measurement density   345–346
As limit on attitude
    accuracy   392–393, 394–396
Expressions for   384, 386–388
For rotation angles   352
Attitude measurement uncertainty (*See
    Attitude determination accuracy;
    Uncertainty*)
Attitude motion (*See also Attitude
    kinematics; Attitude Dynamics*)
Math. model for gyro measurement of   267
Attitude perturbations, due to flexible
    spacecraft   548–556
Attitude prediction (attitude
    propagation)   2, 558–587
Accuracy of for HEAO-1   708–709
Attitude sensor electronics   242–249
Attitude Sensor Unit (combined Earth/Sun
    sensor)   155–201, 178–179
Attitude sensors (*See also item sensed,
    e.g., Horizon sensors*)   10
Distinction between hardware & math.
    models   217
Listed by spacecraft & sensor type   797
Mathematical models   217–270
Need for standardization   718–721
Representative telemetry data errors   311
Simple vs. complex designs   716–718
Use in attitude control loop   502
Attitude stabilization (*See also
    Stabilization methods, e.g., magnetic
    stabilization*)   2–3, 502, 588–635
Spacecraft listed by method   787
Table of methods   19
Attitude stabilization systems   604–625
AU (*See Astronomical unit*)
Automatic control of spacecraft (*See also
    Onboard control; Inertial guidance*)
Automatic threshold adjust (for digital Sun
    sensor)   163–165
Autumnal equinox   27

Symbol for 50
Averaged attitude solution, estimating
   reliability of 373
Averaging of attitude solutions 370–373
Axial symmetry 524
Azimuth angle, attitude component 360
   Component of a spherical coordinate
     system 25
   Determination of, for spin-stabilized
     spacecraft 366
Azimuth biases 239–242
   In horizon sensors 235
Azur (German Research Satellite)—
   Attitude system of 788–789

—B—

Baffles, in fuel tanks 578
Ball-in-tube nutation damper 626, 627
Ballistic coefficient 64
Ballistic trajectory 52
Bang-bang control law 593
   Example of 655–658
Bang-bang-plus-dead-zone control law 593
Bar (unit of pressure) 809
Barycenter 38
Batch estimator 437, 439, 448
   For star data 704, 707–709
Batch least-squares estimator 448–456
   Advantages & disadvantages 456
   Convergence 455–456
   Example of 454–455, 456–459
Batch mode of program operation 686
Bays, magnetic 123
BC/CD/CPD Number (star catalogs) 143
Bending stiffness, of spacecraft booms 548–549
Bessel, Friedrich Wilhelm 45
Betelgeuse (star), angular diameter of 167
Bias determination (*See also Differential
   correction; estimation theory*)
   Application of scalar checking to 329–330
   Choice of observation models for 447
   Choice of state vector elements,
     example 440–441
   Geometrical conditions for 478–483
   Need for 407–408
   Need for "simple" sensors for 717–718
   Operational procedures for 473–476
Bias momentum, dual-spin spacecraft 610
Biases (*See individual item; e.g.,
   Magnetometer biases*)
   Effect on deterministic solution behavior 404
   Types of 477
Binary codes 295–298
Bipropellant gas jet 206
Block (of attitude solutions) 370
Block averaging 370–373
Block diagram, for control system 588
Bode's law 49

Body cone 491–492
Body-fixed coordinates (*See Spacecraft-
   fixed coordinates*)
Body-mounted horizon sensor 169, 173
   Mathematical models 231–237
Body nutation rate 490, 525–526, 535
Bolometer, as energy detector for horizon
   sensor 171, 178
   Misalignment of in wheel-mounted
     horizon sensor 236–237
Bond albedo 79
Bond number 578
Bonner Durchmusterung (star catalog) 143
Boom deployment, deadbeat maneuver 669–671
Boresight (Sun sensor) 165
Brahe, Tycho 36
Branch, of root locus diagram 597
Brazilian anomaly in geomagnetic field 115
Breakaway point, of a root locus diagram 598
Brightness of planets and satellites 77–80
Brightness of spacecraft, sample calculation 79
Brouwer method, of general perturbations 137
Butterworth filter 324–327

—C—

Calendar time conversion subroutines 692
Canopus (star), as attitude reference 189
Cape Canaveral, Florida (*See Eastern Test
   Range*)
Cape Photographic Durchmusterung (star
   catalog) 143
Carbon dioxide, absorption bands 91–92
   In Earth's atmosphere 108
   Use of absorption band for horizon
     sensing 92
Cartesian coordinate transformations 761–765
Cartesian coordinates 760
Cartesian plot subroutines (graph, scale) 694
Catalog of Bright Stars 143, 146
Causal (in time-dependent linear systems) 243
Cayley-Klein parameters 414
Cayley, Arthur 416
Celestial coordinate systems (*See
   Spherical coordinate systems*)
Celestial coordinates (right ascension,
   declination) 26–28
   Transformation subroutines (RADECM,
     VEC) 690
Celestial equator 26
Celestial mechanics 1
Celestial meridian 799
Celestial poles 26
Celestial sphere 22–24
   Plots of (*See Global geometry plots*)
Center of mass, of a Keplerian orbit
   (barycenter) 38
Center of pressure 573
Central Limit Theorm 431

Centroid of torque   274
Characteristic equation, for a matrix   752
Characteristic frequency, of a linear system 773
Characteristic polynomial, of a differential
  equation   770
Charge coupled device star tracker   189
Chebyshev polynomials (curve fitting)   318–322
Chi-squared function   318
Chord length, of a small circle   727–728
Circular velocity   42
Classical elements, of an orbit (See
  Keplerian elements)
CLIMSAT (climatology satellite)—
  Use of MMS spacecraft   720
Clock angle (component of sensor
  orientation)   422
Clocks   799
  Atomic   799, 802
  Ground based   299–302
  Spacecraft   298–299
  Sundials   800
Closed-loop control   604, 662
Closed-loop control system   502, 588–600
Closed-loop poles, of root locus diagram   597
Closed-loop rate gyroscope   198–199
Closed-loop transfer function, in control
  systems   595–596
CMG (See Control moment gyroscope)
Coasting phase, of space mission   53
COBE, use of AEM spacecraft   720
Coding of data for transmission   295–298
Cold gas jets   206, 209–210
Colored noise   269
Column matrix   745
Commands, transmission of   279, 292
Committee on Space Research (COSPAR),
  U.N. Committee   52
Commutation   293–294
Commutation angle, in magnetic control
  system   643
Commutation mode, in magnetic control
  system   643
Commutator   293
Commutator channel   293–294
Component uncertainty (of attitude)   375
Compound attitude measurements   357–359
  In single-axis attitude solutions   370
Computer environment, Goddard Space
  Flight Center   682
Computer programs (See Software)
Computers, onboard spacecraft   210–213
Cone angle   23
Cone angle measurement (See Arc length
  measurement)
Cone intersections, analytic procedure for
  determining   364
  Attitude solutions using   368–370
  Subroutine for (CONES8)   691
CONES8 (subroutine)   691
  Use of in attitude solutions   364–370

Conic sections   38–40
Coning (spacecraft attitude motion)   489
Conjunction   49
  Inferior vs. superior   49
Conservation of angular momentum, related
  to Kepler's Second Law   40
Constants, general   826–829
  Solar system   814–825
  Unit conversion factors   807–813
Consumables   18
Control (See Attitude control)
Control hardware (See specific item, e.g.,
  Gas jets)
Control law (See also Attitude control) 502, 589
  Implementation via onboard computer   210
Control moment gyroscopes   196, 200–201
  Nutation damping with   631
Control torque vs. disturbance torque   498
Control torques (See attitude control)
Controller, in control system   589
Convergence, in a batch least-squares
  estimator   455–456
  In a Kalman filter   467–468
  In estimation theory   451
  In integration procedures   560
Conversational software system   686
Conversion, of telemetry data to
  engineering units   304, 306–307
Conversion factors, for SI (metric) units
                                       807–813
Convolution integral   243, 768–769
Convolutional encoding   282
Coordinate systems   24–31
  Notation   xi
  Parallax   30
  Table of inertial   28
  Transformations   760–766
Coordinated Universal Time (UTC)
                                    798, 801–802
  Attached to data   298
Copernicus spacecraft (OAO-3) (See OAO)
Cordoba Durchmusterung (Star Catalog)   143
Core catalog (star catalogs)   147
Cores, magnetic   205
Correction of invalid data   296–297, 307
Correlation, among measurement types   478
  Among observations & noise in state
    estimation   444–445
  In estimation theory   452–453
  Limited attitude accuracy due to
                                   390–391, 394
  Of a single measurement type at
    different times   478–482
  Of measurement uncertainties
                            374, 378–379, 382–383
Correlation angle   346, 353–357
  Expressions for   384, 388–389
  Figure summarizing relationships   391
  For a single measurement type at
    different times   478–482

Correlation coefficient   378, 429, 452
COS (European Astronomy satellite)—
  Slit horizon sensor/Sun sensor
                          169, 178–179, 721
Cosine detector (Sun sensor)   156–159
Cosines, law of (in spherical triangles)
                          33–34, 731–732
COSPAR (*See Committee on Space
  Research*)
**COSPAR International Reference
  Atmosphere**   110
Covariance   429
Covariance analysis   429–434
  In state estimation   452–453, 461, 465
Covariance matrix (*See Error covariance
  matrix*)   430–434
CO₂ (*See Carbon dioxide; Infrared radiation*)
Crescent (illumination phase)   331
Crew, disturbance torques due to motion
  of   578–580
Critical angle prism (Sun sensor)   159–160
Critically damped control system   593
Cross product (*See Vector product*)
CROSSP (subroutine)   691
Crosstalk, in magnetometers   250
CTS (Communications Technology Satellite)—
  Application of block averaging to
    attitude solutions   371
  Attitude acquisition
              661–663, 666, 669–670, 672
  Attitude data simulator   710–711
  Attitude determination   10–12
  —Accuracy of   397
  —During attitude acquisition   422–424
  Attitude software structure   698
  Attitude support system   700–703
  Attitude system of   788–789
  Control system description 612–613, 622–625
  Correlation among measurement types
                                     480–484
  Data collection for bias determination   474
  Determination of bias on Earth
    angular radius   483
  Determination of sensor mounting angle
    bias   483
  Deterministic attitude subsystem   370
  Effect of flexibility on attitude dynamics 556
  Gas jet control system   506
  Horizon sensors   169, 175–176
  Mission profile   4–8
  Modeling torque due to orbit maneuvers 582
  Multiple component software   716
  Nutation damping   630
  Spacecraft   6
  Spin rate change due to orbit maneuvers 582
  Stabilization method   503
  Sun sensor   157
  Telemetry data errors   311
  Use of body-mounted horizon sensor   173
  Use of carbon dioxide band horizon

    sensor   92
  Use of open-loop control   663
Curie point   115
Curve fitting   317, 318–322
Cusp   90
Cusp region, geomagnetic field   120
Cylindrical coordinates   760

—D—

D'Alembert, Jean   38
Damping (*See also Nutation damping;
  Libration damping*)
  Of a control system   591–593
  Of inertial control systems   658–659
Damping matrix (for flexible spacecraft)   553
Dark angle   88
Data—
  Acquisition and transmission process
                                     278–298
  Correction if invalid   296–297, 307
  Generation and handling of, onboard
    spacecraft   278–283
  Handling invalid data   315
  Processing at receiving stations   303–304
  Time tagging   298–304
  Transmission from receiving station to
    attitude computer   292
Data adjuster, of CTS attitude system   701
Data anomalies, procedures for identifying
                                     334–339
Data averaging, for single-frame attitude
  solutions   370–373
Data conversion, in telemetry processor
                              304, 306–307
Data dropout   310
Data errors   310–311
  Checking for in mission support
    software   682–683
  Table of representative examples   311
Data filters (*See also State estimators*)
Data filters, Butterworth vs. least-
  squares quadratic vs. averaging   325–327
  Definition   437
  Use for data smoothing   317, 322–327
Data flagging (*See Flags*)
Data handling, in telemetry processor   304–308
Data preparation subsystems, in attitude
  software systems   697
**Data selection requiring attitude
  information**   334–339
Data smoothing (*See Smoothing*)
Data transmission (*See Telemetry*)
Data validation (*See Validation*)
Data weighting (*See also Attitude
  determination accuracy*)   370–373
DATE (subroutine)   692
Date, conversion subroutines for Julian

dates (JD, DATE)   692
Day, apparent solar   799
  Mean solar   805
  Sidereal   799, 805
DC (subroutine)   691
DE (Dynamics Explorer satellite)—
  Attitude system of   788–789
  Effect of flexibility on attitude dynamics 556
  Use of carbon dioxide band horizon
    sensor   92
Deadband, in attitude control   607
Deadbeat boom deployment   669–671
Decay, of Earth satellite orbits   64
Declination (*See also Celestial coordinates*) 28
Deep Space Network, Jet Propulsion
    Laboratory   284
**Definitive attitude determination
    requirements**   681
Definitive orbit   62
Delayed command system   640
Delta functions   xii
Delta Launch Vehicle   3–5
DELTIM (subroutine)   693
Density of attitude loci (*See Measurement
    density*)
Density of Earth's atmosphere   107
Descending node   44
Design of mission support software 681–713
Determinant, of a square matrix   748–749
Deterministic attitude (*See Attitude
    determination, deterministic methods*)
Deterministic subsystem, in attitude
    software systems   697
  Of CTS attitude system   701
Development of mission support software
                                       681–695
Deviation of the vertical   821
Diagonal matrix   745
Diagonalization, of a matrix   752–754
Differential correction (*See State
    estimation; Data filters; Least squares*)
Differential correction subroutine (DC) 691
Differential correction subsystem, of CTS
    attitude system   702
Differential equations, solution using
    Laplace transform   770–771
Differential spherical trigonometry 734–735
Diffuse reflection, torque due to   572
Digital codes   295–298
Digital processors (*See Onboard computers*)
Digital sun sensor   156, 161–166
  Spinning or one-axis, mathematical
    model   223–224
  Two-axis, mathematical model   224–227
Dihedral angle   23
Dihedral angle measurement (*See Rotation
    angle measurement*)
Dipole model, of geomagnetic field (*See
    Geomagnetic field, dipole model*)
Dipole moment, magnetic   204

Dirac delta function   xii
Direct match, for star identification
                                   259, 260–262
Direct orbit   53
Direction cosine matrix (attitude matrix)
                                       411, 424
  Kinematic equations of motion for   512
  Parameterizations   410–420
  Summary of properties   761–762
  Table of as function of Euler angles   764
Directrix   40
Discretization error, in integration
    procedures   560
Disturbance torques (*See also Specific
    torques, e.g., Aerodynamic*)   502
  As control system inputs   589
  Distinguished from control torques   498
  Due to engine misalignments   580–583
  Due to orbit maneuvers   580–583
  Environmental   566–576
  —Frequencies of   318
  Internal   576–580
  Mathematical models of   558–587
  Table of   17
  Treatment of for attitude propagation   558
Divergence—
  In Kalman filters   467–468
  Of a vector function   756
DODGE (Department of Defense Gravity
    Experiment Satellite)—
  Attitude system of   788–789
Dot product, of vectors   747
Double stars, in star catalogs   145–146
Downlink   278
Draconic month   52
Drag, on spacecraft orbits   64
Drag coefficient   64, 574
Drift—
  In gyroscopes   200
  Of spacecraft attitude   498
Drift rate ramp, in gyroscopes   200
DSCS (Defense Satellite Communications
    System)—
  Shadow modeling for   573
Dual-flake horizon sensors   171
Dual-scanner single-axis attitude solutions 368
Dual-spin spacecraft   202–203, 601
  Attitude acquisition via momentum
    transfer   667–669
  List of   787
  Nutation of   536–539
  Pitch control   617
  Stabilization by   503–505
Dumbbell, rotation of   487–488
Duty Cycle—
  Horizon sensor output   172
  Of reaction wheel command   270
Dynamic equations of motion   521–523
  For flexible spacecraft   552–555
  Torque free   524–529

Dynamics (*See Attitude dynamics*)
Definition 487

—E—

e (Base of natural logarithms), value 826
Earth (*See also Nadir*)
  Albedo of 83
  Appearance from space 83–106
  —Due to oblateness 99–103
  Appearance of horizon at 14 to 16
    microns 92–98
  As attitude determination reference
    source 17
  Atmosphere of 106–110
  —Composition 108
  —Effect of during eclipses 77
  —Horizontal temperature gradients 94–96
  —Models 109–110
  —Structure 106–109
  —Table of properties 820
  —Variation in structure 109
  —Vertical temperature cross section 96
  Bias in sensed angular radius 235
  —Procedure to measure 483
  Dark angle 88
  Geocentric and geodetic coordinates
    820–822
  Geometrical distortion of surface as seen
    from space 87
  Gravitational field models 123–129
  Horizon of (*See Horizon*)
  Illumination of as seen by nearby
    spacecraft 334
  Inertial rotational position of (GST) 802–805
  —Irregularities in 805
  Infrared appearance of 90–98
  Infrared radiation from 82, 90–98
  Magnetic field (*See Geomagnetic field*)
  Modeling procedures 82–110
  Models of surface shape, table 98
  Oblateness—
  —Effect on orbit 65–69
  —Modeling 98–106
  —Term in gravitational potential 124
  Orbit of (*See also Ecliptic*) 44, 48–51
  Path of conical scan on 87, 178
  Position of relative to Sun, Moon, and
    planets 138–142
  Properties of, table 819–821
  Radiation balance, table 82
  Radiation from 82–98
  Shadow cone of 75
  Shape of 99–103
  Symbol for 50
  Terminator
  —Identification 331–334
  —Modeling 86–90

  Thermodynamic equilibrium of 82
  Torque due to reflected sunlight from 571
  Visual appearance of 83–87
Earth-in (*See In-triggering*)
Earth-Moon Lagrange points 55
Earth-out (*See Out-triggering*)
Earth referenced spacecraft 605, 608–625
  Table of characteristics 612
Earth satellites, period vs. altitude table 828
Earth sensor (*See also Horizon sensor*)
  Combined Earth/Sun sensor 178–179
  Visible vs. infrared 83, 169
Earth-width measurement 172, 347–349
  Density of 348
  Error due to unmodeled oblateness 105
  Pagoda effect in 336–339
Earth-width model, for horizon sensors
    231, 233–234
Earth-width/Sun angle single-axis attitude
  solutions 368
  Singularities in 406–407
Eastern Test Range launch site 3, 5
Eccentric anomaly 45–46
Eccentricity 38, 46
Echo I satellite 65
Eclipse 72, 75–77
  Conditions for 75–77
  Of the Sun (*See Solar eclipse*)
Ecliptic 44
  As reference for solar system orbits 48
  Obliquity of 48
  Relative to celestial coordinates 27
Ecliptic coordinates 28
  Use in three-axis attitude system 425
Eddy current libration damper 632
Eddy current nutation damper 626–627
Eddy current nutation damping 614
Eddy current rods, use of for libration
  damping 633
Eddy currents, torque due to 575
Effective torque, for gas jet 274
Eigenaxis inertial guidance maneuvers 661
Eigenvalue 752
Eigenvector 752
Eigenvectors and Eigenvalues—
  Interpret. of for attitude matrix 411–412
  Of moment-of-inertia tensor 519
Electric thrusters, for attitude control 19
Electromagnetic units 811–813
Electronic noise (gyroscopes), mathe-
  matical model 269
Electronics modeling, attitude sensors 242–249
ELEM (subroutine) 692
  Basis of 60–62
Elements—
  Of an orbit (*See also Orbit elements,*
    Keplerian orbits) 46
  Of a matrix 744
Elevation—
  Component in local tangent coordinates 30
  Component in spherical coordinates 25

Ellipse   38–40
Elliptical orbit, table of properties   47
Ellipticity, of the Earth   99
Elongation, of a planet   49–50
Encoder   280
Energy—
  Dissipation, effect of on rotation   499–501
  Required for spacecraft launch   54
  Units and conversion factors   810
Energy ellipsoid   499–501
  Discovery of   521
Energy optimal magnetic maneuver   642, 648
Environmental torques (See Disturbance
  torques)
EPHEM file (See Ephemeris file)
Ephemeris   36
  Algebraic approx. for Sun, Moon, and
    planets   140–143
  Spacecraft   133–134
  Sun, Moon, and planets   138–142
Ephemeris file   133
  Format   135
  Subroutines for reading   693
Ephemeris subroutines—
  Analytic-for spacecraft, Sun, Moon,
    planets   692–693
  Definitive (i.e., data set)   693
  General-purpose Sun, Moon, spacecraft
    (EPHEMX)   693
Ephemeris Time (ET)   798, 802
EPHEMX (subroutine)   693
Epoch (orbit parameter)   46
Epoch, of celestial coordinates   27
Equation of the center   140
Equation of time   801
Equator (type of coordinate system)   28
Equator, of a spherical coordinate system   24
Equatorial electrojet   123
EQUIN (subroutine)   692
Equinoxes   48
  Precession of   27
  —Sub. for updating coord. (EQUIN)   692
ERBS, use of AEM spacecraft   720
Error-correcting codes   296–297
Error covariance matrix   452
  Analysis for attitude determination
                                        378–379, 429–434
Error ellipse   376–381, 434
Error parallelogram   345–346, 374–376
Error signal, in control system   502–503
Errors in software, avoidance of   682–683, 685
ERTS (Earth Resources Technology
    Satellite) (See also Landsat)—
  Attitude system of   788–789
  Horizon sensors   176–178
Escape velocity (See Velocity of escape)
ESRO (European Space Research
    Organization Satellites)—
  Attitude system of   788–789
  Nutation damping   627
Estimation theory (See State estimation)

Estimation theory techniques   447–470
Euler angles   417–420
  Formulas   763–765
  Kinematic equations of motion for
                                          513–514, 765
Euler axis, of rotation   413
Euler rotation angle   413
Euler symmetric parameters (See also
    Quaternions)   414–416, 583, 758–759, 762
Euler's equations   522, 558
  General form for Earth-referenced space-
    craft   609–610
  Mechanical integration via gyroscopes   559
  Solutions for torque-free motion   524–528
Euler's theorm   412, 487–488, 761–762
  Basis of   754
EUVE, use of AEM spacecraft   720
Evaluation of state estimators   471–473
Even parity   295
Exosphere   106
Exospheric temperature   106
Explorer (general U.S. satellite name)   790
Explorer I—
  Instability of rotation   501
  Nutation   626

                        —F—

Fading memory (in Kalman filters)   463
Fall time (gas jet)   273
False sightings, in star sensors   192
Feedback control systems   502, 588–600
Filters (See Data filters; State estimators)
Final value theorm   768
Fine Sun sensor   166
  Mathematical models   227–230
  Reticle pattern and photocell output   228
First point of aries (See Vernal equinox)
  Origin of term   27
Fixed-head star trackers   186, 189–190, 193–195
  List of spacecraft using   797
Flags—
  Associated with telemetry data   313–315
  Internal vs. external   315
  Set by telemetry processor   307
Flake (component of horizon sensor)   171
Flat spin   501, 576
Flattening, of the Earth   99
Flexible spacecraft dynamics   18, 548–556
  Effects on equations of motion   552–555
Flight path angle   61
Float torque derivative noise (gyroscopes),
    mathematical model   269
Float torque noise (gyroscopes), mathe-
    matical model   269
FLTSATCOM (Fleet Satellite Communica-
    ti·n System)—
  Attitude system of   790–791
Fluxgate magnetometer   182–184
  Mathematical model   249–254

Flyby trajectory 60
Flywheel 201
Focus, of an ellipse 38
Force, Units and conversion factors 809
Forced attitude solutions 370
Forced response, of a linear system 770
Forcing function, of differential equation 770
Fourier series, use to solve linear
    differential equations 614
Fourier transform 243
Frame, of data 293
Frame synchronization signal 293
Framing of telemetry data 293
Fredholm equation 773
Free response, of a linear system 770
Freon, as gas jet fuel 210
Fresnel reflection 156
Friction modeling, for reaction wheels 271
Fuel—
    Budget for gas jets 207
    Loss due to engine misalignments 582–583
    Tanks, torques in 577–578
    Used for gas jets 207

—G—

Gain matrix, in sequential estimators 461
Gal (unit of acceleration) 809
Galactic coordinates 28
Gamma (unit of magnetic induction) 811
Gas jets—
    Attitude control systems using 503, 506
    Attitude maneuver analysis for 649–654
    Disturbance torques due to propellant
        slosh 577–578
    Effects of thrusting on flexible space-
        craft 551–552
    Example of use in attitude control 14–16
    Hardware description 206–210
    List of spacecraft using 796
    Mathematical models 272–275
    Nutation damping with 630–631
    Use for attitude control 19
    Use for attitude stabilization 622–625
Gauss (unit of magnetic induction) 811
Gauss, Karl 113, 779
Gauss' Equation 45
Gauss-Newton least-squares procedure 448, 455
Gaussian coefficients, Geomagnetic field 117
    Table of 779
Gaussian measurement errors 430–431
    Probabilities associated with 381
    Sample computation 434
GCI (See Geocentric Inertial Coordinates)
Gemini program, use of horizon sensors
    167, 168, 180
General perturbations, method of (orbit
    analysis) 139
Geocentric coordinates, conversions with
    geodetic 820–822

Geocentric Inertial Coordinates 29
Geodetic coordinates, on the Earth's
    surface 820–822
Geoid 98, 99, 125
Geoid height 99
    Map of 125
Geomagnetic field (See also Magnetic,
    magnetometers, etc.) 113–123, 779–786
    Accuracy of models 118–119
    Analytic approximations for 782–785
    As attitude determination reference
        source 17
    Dipole model 782–785
    —For Earth-referenced spacecraft 613
    —Rectangular components 784
    —Spherical components 783
    Diurnal variation 122–123
    General description 113–120
    Index of geomagnetic activity 122
    Magnetic storms 121
    Mathematical models 779–785
    Models 117–123
    Perturbations of 120–123
    Secular drift 113
    Solar perturbations 120–123
    Spherical harmonic model 779–782
    Subroutine for (MAGFLD) 693
Geometric albedo 79
Geometric method of three-axis attitude
    determination 421–424
Geometrical axes of spacecraft 487
Geometrical limitations of attitude
    accuracy 389–402
    Applications 397–402
Geometry—
    Attitude 22–35
    Effect of changes on information
        content of measurements 478–482
    Spherical (See Global geometry plots;
        Spherical geometry)
GEOS (Geodynamics Experimental
    Ocean Satellite)—
    Attitude acquisition 662, 666, 672–677
    Attitude system of 790–791
    Control system description 612–617
    Data records 304
    Data sample 312, 313, 314
    Deadbeat boom deployment 669
    Fitting magnetometer data 321
    Libration damping 632
    Magnetic stabilization 672
    Momentum wheel 202–203
    Spacecraft 672
    Stabilization method 503
    State vector for bias determination 441
    Sun sensor 157
    —Data correction 330
    Use of open-loop control 663
GESS (See Graphics Executive Support
    System)
GETHDR (subroutine) 693

Use of 133
GETV (subroutine) 693
Gibbous (illumination phase) 331
Gibbs vector 416, 763
  Kinematic equations of motion for 512–513
Gibbs, J. Willard 416
Gimbal (gyroscope support) 196
Gimbal rotation axis (gyroscope) 196
Gimbaled star trackers 186, 187–189
  List of spacecraft using 797
Global geometry plots
  Construction of 737–743
  Explanation of 22–26
  For attitude determination 397–399
  Spacecraft orbit on 398
  Subroutines for (SPHGRD, SPHCNV,
    SPHPLT) 694–695
Global Positioning System 8–10
GMT (See Universal Time)
Goddard Space Flight Center—
  Attitude Data Link 292
  Computer environment 682
  Information Processing Division
      292, 299–303
  Network Operations Control Center 284
  Role in CTS mission 7
  Role in receiving & relaying data 284
  SCAMA (Switching, Conferencing, etc.) 291
Goddard Trajectory Determination
    System (GTDS) 133–134
  subroutines for 693
GOES (Geostationary Operational
    Environmental Satellite)—
  Application of block averaging to
    attitude solutions 371
  Attitude acquisition 661
  Attitude determination accuracy 397
  Attitude software structure 698
  Attitude system of 790–791
  Correlation among measurement
    types 480–484
  Data collection for bias determination 474
  Determination of sensor mounting angle
    bias 483
  Fitting attitude solutions 322–323
  Orbit generator accuracy 138
  Spin rate change due to orbit maneuvers 582
  Sun sensor analysis 717–718
  Telemetry data errors 311
  Use of carbon dioxide band horizon
    sensor 92
  Use of open-loop control 663
  Use of body-mounted horizon sensor 173
Goodness-of-fit function 318
GPS (See Global Positioning System)
Gradient, of a scalar function 756
Gradient operator 756
Gradient search, method of differential
  correction 455
GRAPH (subroutine) 694
Graphic software systems 686–690

Graphic Subroutine Package (GSP) 687
Graphic support systems 686–690
Graphics (See also Interactive graphics)
Graphics Executive Support System
  (GESS) 688–690
Graphing subroutines 694–695
Gravitation, Newton's development of
  laws 36–38
Gravitational constant, accuracy of 41
Gravitational constants, Earth, Moon, and
  Sun 827
Gravitational field models 123–129
Gravitational potential 123–129
Gravity assist trajectory 60
Gravity-gradient attitude control 614–617
Gravity-gradient capture sequence 672–677
Gravity-gradient stabilization 19, 503, 505–506
  Conditions for 611–612
  List of spacecraft using 787
Gravity-gradient tensor 128–129
Gravity-gradient torque 17
  Effect on flexible spacecraft 549, 551
  For dual-spin spacecraft, math.
    model 568–570
  For Earth-referenced spacecraft 609
  Mathematical model 566–570
Gray Code—
  Algorithm for conversion to binary 306–307
  Conversion table 164
  Output vs. Sun angle for sensors using 165
  Reason for use 163–164
  Reticle pattern for 164
Great circle 22
  Construction of on global geometry plot 742
  Equations for 728–729
  Properties of 32
Greatest elongation 50
Greenwich Hour Angle 802–803
Greenwich Mean Time (GMT) (See also
  Universal Time) 19, 801
Greenwich meridian 801
  Subroutine for right ascension of
    (RAGREN) 692
Greenwich Sidereal Time (GST) 802–804
Gregorian calendar century 809
GRO, use of MMS spacecraft 720
GSFC (See Goddard Space Flight Center)
GTDS (See Goddard Trajectory
  Determination System)
Gyroscopes—
  Accuracy of attitude propagation with
    for HEAO-1 708–709
  As attitude determination reference 17
  Attitude propagation with 564–566, 558–559
  Biases 198, 200
  Effect of misalignments on slew
    maneuvers 660
  Hardware description 196–201
  Mathematical models 266–270, 558, 559
  Measurements from (rate and rate
    integrating) 266–270

Modeling noise effects   268–270
Solution of differential equation   771–773
Spacecraft using   797
Gyrotorquer (*See Control moment gyroscope*)

—**H**—

Half-angle formulas, for spherical triangles 732
Hamming Code   296–297
Harmonic oscillator—
    Equation for forced   614
    Solution for forced   771–773
Haversines   735–736
    Advantages óver normal trig functions 735
HCMM (Heat Capacity Mapping Mission)
    (*See also AEM*)—
    Attitude acquisition   662, 672
    Attitude system   720, 790–791
    Control system description   612–613, 617–621
    Momentum wheel   202
    Nutation damping   630
    Scanwheels horizon sensor   176–178
    Use of carbon dioxide band horizon
        sensor   92
    Yo-yo despin   667
HD number (star catalogs)   143
Heading (orbit parameter)   61
Heading angle (gas jet precession) 652
HEAO (High Energy Astronomy
    Observatory)—
    Attitude acquisition   662
    Attitude system of   790–791
    Control system description   605–608
    Fixed-head star trácker   190, 193–195
    Gyroscopes measurements   266–270
    Image dissector tube star sensor
        measurements   256–258
    Inertial guidance maneuvers   655, 661
    Inertial reference assembly   197
    Instrumental magnitude for star camera 258
    Large data volume   308
    Momentum wheel   202–203
    Onboard computer   211, 212–213
    Star catalog for   147
    Star tracker attitude determination
                                     706, 708–709
    Two-axis Sun sensor   158
    Use of q method for attitude
        determination   427
Heat pipe   627
Heat sink, use in horizon sensors 171
Height, used for distances measured from
    the Earth's surface   43
Heliocentric coordinates   29
Helmholz coil, for magnetometer testing 250
HEMITR (subroutine)   693
Henry Draper star catalog   143
HEOS (Highly Eccentric Orbit Satellite)
    Attitude system of   790–791

Nutation damping   627
Hermitian matrix   750
Hohmann transfer orbit (*See also Plane
    change orbit maneuvers*)   56–59
    Between the planets, table   824
Horizon—
    Appearance of at 14 to 16 microns 92–98
    Definition dependent on sensor   167
    Identification of   331–334
    Of an oblate Earth   99
Horizon crossing vector (*See also Earth
    width; Sun sensor / horizon sensor
    rotation angle*)
    Computation of   238–239
    For oblate Earth   103–105
    In single-axis attitude solutions 370
Horizon plane, for an oblate Earth 101
Horizon sensors—
    Analysis of representative poor geometry
        for   397–399
    Biases   234–237
    Components of   169–172
    Data validation   329
    Example of use in attitude
        determination   10–12
    Geometry of   231
    Hardware   166–180
    List of spacecraft using   797
    Mathematical models   230–242
    Mathematical models of electronics 244–249
    Model of azimuth biases   239–242
    Optical system of   170
    Output   171–172
    Pagoda effect bias at small Earth
        widths   336–339
    Path of scan on the Earth   87
    Representative output   172
    Representative spectral response 170
    Representative telemetry data errors 311
    Rotation angle (from Sun sensor)
        models   237–242
    Slit horizon sensor/Sun sensor   178–179
    Use for single-axis attitude   362–409
    Use for three-axis attitude   426
    Visible vs. infrared   83, 169
Horizon spheroid, for an oblate Earth 101
Horizon/Sun rotation angle (*See Sun
    sensor / horizon sensor rotation angle*)
Hot gas jets   206, 207–209
Hour angle   799
Hour angle of Greenwich Meridian,
    subroutine for (RAGREN)   692
Hour Angle, Greenwich   802–803
Hour, measure of right ascension 28
HR number (star catalogs)   143
Huygens, Christian   38
Hydrazine—
    As gas jet fuel   207–209
    Thrust characteristics for   273–274
Hydrogen peroxide, as gas jet fuel 207
Hyperbola   38–40

Hyperbolic anomaly 47
Hyperbolic orbit, table of properties 47
Hyperbolic velocity 42
Hypergolic fuel 53
Hysteresis, torque due to 575

—I—

Identity matrix 745
IGRF (*See International Geomagnetic Reference Field*)
Illumination during partial eclipse 76
Illumination of planet, as seen by nearby spacecraft 334
Illumination of sphere, as function of phase, distance 78–79, 89
Image dissector, in star tracker 189
Image dissector tube star measurements, mathematical model 256–259
IMP (Interplanetary Monitoring Platform)—
  Attitude software structure 698
  Attitude system of 790–791
  Effect of flexibility on attitude dynamics 556
  Telemetry data errors 311
  Use of body-mounted horizon sensor 173
  Use of convolutional encoding 282
Improper orthogonal matrix 751
Impulse (of force) 206
Impulse (of torque) 274
Impulse response function 242
Inclination, (orbit parameter) 44, 46
  Maneuver to change 59
  Of Earth satellite orbit 53
In-crossing (*See In-triggering*)
Induction magnetometer 181–184
Inertia wheel 201
Inertia, moment of (*See Moment of inertia*)
Inertial coordinate systems 26–28
  Table of 28
Inertial guidance 16
Inertial guidance maneuvers 655–661
Inertial nutation rate 490, 526, 535
Inertial reference assembly, for HEAO-1 197
Inertial spin rate 490
Inertially referenced spacecraft 605–608
Inferior conjunction 49–50
Inferior planet 49
Infinitesimal spherical triangles 734
Information Processing Division (IPD) 292, 299–303
Infrared horizon sensors (*See Horizon sensors*)
Infrared radiation, from Earth 82, 90–98
Initial Value Theorm 768
Injection 53
  Determination of orbit elements from 60–62
Inner product, of vectors 747
In-plane orbit maneuvers 56–59
Input axis (gyroscope) 196
Instantaneous rotation axis, of

spacecraft 487–488
Instrumental star magnitudes 258
Integral equations, solution using Laplace transform 773–774
Integration methods 560–564
  Choice of 564
  Errors in 560
  Subroutine (RUNGE) 692
Interactive graphics, use in mission support software 682, 686–690
Interactive software system 686
Interfaces, data set vs. core storage 699
Internal torques on spacecraft 576–580
International Astronomical Union, 1976 adopted astronomical constants 827–828
International Atomic Time (TAI) 798, 802
International designation, of spacecraft 52
International Geomagnetic Reference Field 118
  Coefficients of 779
  Subroutine for (MAGFLD) 693
International System of Units 807–813
  Prefixes 807
International Telecommunication Union 283
Interplanetary environment 130–132
Interplanetary flight—
  Sample calculation 58–59
  Table of orbit properties 824–825
Interplanetary probe, distinguished from satellite 52
Inter-Range Instrumentation Group 282
Intersecting cones, attitude solutions 12
Intersection, of attitude measurement loci 345–346
Interstellar probe 52, 60
INTP (subroutine) 693
In-triggering (Earth-in) 11, 172, 358
Inverse Laplace transform 767
Inverse, of a matrix 749–750
INVERT (subroutine) 691
Ion jets 206
Ion thrusters, for attitude control 19
IPD (*See Information Processing Division*)
IR (*See Infrared*)
ISEE (International Sun Earth Explorer)—
  Application of block averaging to attitude solutions 371
  Attitude determination accuracy 397
  Attitude software structure 698
  Attitude system of 790–791
  Earth and Moon coverage for ISEE-C 402
  Effect of flexibility on attitude dynamics 556
  Gas jet control system 210
  Panoramic scanner 169, 173–175
  Slit horizon sensor/Sun sensor 169, 178–179, 721
  Stabilization method 503, 504
  Use of convolutional encoding 282
ISIS (International Satellite for Ionospheric Studies)—
  Attitude system of 790–791
ISS (Ionosphere Sounding Satellite)—

Attitude system of   790–791
ITOS (Improved Tiros Operational
  Satellite)—
  Attitude system of   790–793
  Momentum wheel   202, 601, 602
IUE (International Ultraviolet Explorer)—
  Application of block averaging to
    attitude solutions   371
  Attitude acquisition   662
  Attitude determination accuracy   397
  Attitude software structure   698
  Attitude system of   792–793
  Computation of attitude determination
    accuracy   384–386
  Gas jet control system   207–209
  Inertial guidance maneuvers   655, 659, 660
  Momentum wheel   203
  Onboard computer   210
  Panoramic scanner   169, 173–175
  Reaction wheels   270, 272, 601
  Star tracker   190
  Sun sensor   157, 166
  Thruster characteristics   273
  Use of convolutional encoding   282

—J—

Jacchia atmosphere models   110
Jacobian elliptic functions   526–528
JD (*See Julian Day*)
JD (subroutine)   692
JDS (*See Julian Day for Space*)
Jet damping   582
Jet Propulsion Laboratory, Deep Space
  Network   284
Jets (*See Gas jets*)
Jordan canonical form, for a square matrix   753
Joseph Algorithm (in sequential estimators)   461
JPL (*See Jet Propulsion Laboratory*)
JPL ephemeris tapes   140
  Subroutine for (RJPLT)   693
JSC (*See Lyndon B. Johnson Space Center*)
Julian century   809
Julian Day   20
  Conversion subroutines for (JD, DATE)   692
  Table   804
Julian Day for Space   20
Julian period (basis of Julian Day)   20
Jupiter, effectiveness for gravity assist
  trajectory   60
J_2 perturbations   67–69
J_2 term, in gravitational potential   126–127

—K—

Kalman filter   448, 462–469
  For star data   708–709
  Propagation of state & error
    covariance matrix   464–467

Kapustin Yar (Soviet launch site)   4
Kepler, Johannes   36–38
Kepler's equation   45
  Numerical solutions of   46, 134, 140
Kepler's First Law   37–40
Kepler's Laws   37–42
Kepler's Second Law   37–40
Kepler's Third Law   37, 41–42
Keplerian orbit   35–37
  Table of properties   47
Keplerian orbit elements   46
  As function of position & velocity,
    subroutine for   692
  As function of injection conditions   60–62
  Table of   46
Kernal, of integral equation   773
Kilogram, definition   807
Kinematic equations of motion   511–514
  Euler angle representations   765
  Torque free   529–531
Kinematics (*See Attitude kinematics*)
Kinetic energy—
  Of rotational motion   517–519
  Orbital (*See also Vis viva equation*)   38
Knudsen number   108
Kosmonavt Vladimir Komarov (Soviet
  tracking ship)   290
Kronecker delta   xii

—L—

Lagrange, Joseph   55
Lagrange point orbits   55
Lagrange points   55
Lambert sphere   79
  Reflected intensity from   85
Landmark tracking, potential use for
  attitude determination   724
Landsat (*See also ERTS; Earth Survey*
  Satellite) —
  Attitude system of   792–793
  Horizon sensors   180
  Landsat-D, use of MMS spacecraft   720
  Momentum wheel   202
Laplace transform   767–774
  Application to control systems   590
  Example of use to solve linear differential
    equations   615–616
  Table   769
Laplace's equation, in spherical coordinates   775
Latitude, geocentric vs. geodetic vs.
  astronomical   820–822
Latitude component, of a spherical
  coordinate system   25
Launch
  Added velocity required for prograde
    orbit   53
  Energy required for   54
Launch phase, of space mission   3–7
Launch sites

Soviet   3
U.S.   2–3
Launch vehicles
   Reignitable upper stage   53
   U.S.   3
Launch window, attitude accuracy
   constraints   399–401
Law of Cosines   33–34
Law of Sines   33
LBN coordinate system (orbit defined) 28
Lead time constant, in control systems 591
Least-squares estimator (*See also Batch*
   least-squares estimator)   437
   Analytic basis   447–470
   Example of   454–455, 456–459
Least-squares quadratic filter   322–327
Least-squares subroutines   691–692
Legendre functions, & Legendre
   polynomials   776
Legrange, Joseph   45
Leibnitz, Goffried   38
Length, units and conversion factors 808
LES (Lincoln Experimental Satellite)—
   Attitude system of   792–793
LFO (Landsat Follow-on Satellite)—
   Use of MMS spacecraft   720
Libration   625
Libration damping   631–633
   Use of deadbeat maneuver for   669–671
Libration points (three-body orbit)   55
Lifetime, of Earth satellites   64
Lift (aerodynamic)   63
Light year   808
Lighting conditions, on spacecraft   71–80
Limit checking   314
Limited stability, of a control system 594
Line of apsides   44
Line of nodes   44
Linear independence, of a set of vectors 745
Linear operators   767
Linear system   242
   Relationship to Laplace transformation 767
Lit horizon   85
Local horizontal coordinates   30
Local tangent coordinates   30
Locator (horizon sensor signal
   processing)   94, 172
Locus—
   In defining attitude measurements   344
   Intersection of in attitude
      measurements   344–346
Long Range Navigation-C
   (LORAN-C)   299–300
Longitude component, of a spherical
   coordinate system   25
Longitude of subsatellite point   805
Longitude of the ascending node   49
LORAN-C (*See Long Range Navigation-C*)
Loss function (in estimation theory) 449–451
Loss of signal (*See Out-triggering*)
Loxodrome (rhumb line)   653

Lubrication, of spacecraft wheel bearings 202
Lunar Orbiter (spacecraft), attitude
   reference system   189
Lunar parallax   142–143
Lyndon B. Johnson Space Center   285

—M—

MAGFLD (subroutine)   693
Magnetic attitude control (*See also*
   Magnetic stabilization)   18–19, 617–621
   Hardware for   204–205
   Maneuver strategy   639–642, 644–649
   Maneuvers   636–649
Magnetic coil control systems 503, 508–509
Magnetic coils   204–205
Magnetic dipole moment   204
   Units and conversion factors   812, 813
Magnetic disturbance torque   575–576
Magnetic equator   114
Magnetic field—
   Interplanetary   130–132
   Of the Earth (*See Geomagnetic field*)
Magnetic field strength, units and
   conversion factors   811, 813
Magnetic flux, units and conversion factors 811
Magnetic induction field   251
Magnetic induction, units and conversion
   factors   811, 813
Magnetic materials   205
Magnetic moment   204
   Of a current distribution   252
   Units and conversion factors   811, 813
Magnetic nutation damping   614, 629–630
Magnetic observatories   122
Magnetic permeability   813
Magnetic precession   636–649
Magnetic stabilization   671–672
   List of spacecraft using   787
Magnetic storm   121
Magnetic systems of units   811–813
Magnetic torque   17
Magnetic torquing—
   Continuous   639
   List of spacecraft using   796
   Quarter orbit (QUOMAC)   640
Magnetometers   180–184
   Bias determination   329–330
   Biases   251–254
   Data—
   —Curve fitting for   321
   —Residual errors in   328
   —Validation of   328
   Example of use in attitude
      determination   13–14
   List of spacecraft using   797
   Mathematical models   249–254
   On Apollo 15 subsatellite   495–496
Magnetopause   106, 120

Magnetosheath	120
Magnetosphere	106, 120
Magnetotail	120
Magnitude (scale of brightness)	77
Instrumental	258
Moon and planets, table of	816–817, 823
Of planets, satellites, and spacecraft	77–80
Sample calculation for spacecraft	79
Stellar	144–145
MAGSAT (Magnetic Satellite)—	
Attitude software structure	699
Attitude system of	792–793
Evaluation of horizon sensor data	471
Fixed-head star tracker	193–195
Need for accurate Sun sensor	166
Scanwheels horizon sensor	176–178
Star catalog for	147
Use of carbon dioxide band horizon sensor	92
Major frame	293–294
Major principal axis	500, 625–626
Maneuver control (See Attitude maneuver control)	
Marginal stability, of a control system	594
Mariner (spacecraft), attitude reference system	189
Mariner 10	60
Marquardt's Algorithm	455–456
Mars—	
Hohmann transfer orbit to	58–59
Oppositions of	51
Mask detector (Sun sensor)	158
Mass expulsion system (See Gas jets)	
Mass expulsion torques	576–577
Mass, units and conversion factors	808
Master catalog (star catalogs)	147
Master frame (telemetry)	293–294
Master station (time signals)	300
MATMPY (subroutine)	691
Matrix	744
Matrix algebra	744–757
Subroutines	691
Matrix functions	754–755
Matrix inversion	749–750
Subroutine (INVERT)	691
Matrix multiplication subroutine (MATMPY)	691
Matrix notation	x
Matrix Riccati equation	465
Matrix transformations	751–752
Mean (of a random variable)	429
Mean angular motion	47, 67
for Earth satellite	67
Mean anomaly	45–46
Rate of change of	67
Mean distance, in an elliptical orbit	38
Mean free path, in atmosphere	108–109
Mean of date coordinates	28
Mean orbital elements	46
Mean sea level (See also Geoid; geoid height)	98, 99
Mean Solar Time	798, 801, 805
Mean Sun	801
Measurement, as used in state estimation	444
Measurement covariance matrix	461
Measurement density (See Attitude measurement density)	
Measurement uncertainty (See Attitude determination accuracy; Uncertainty)	
Mercator, Gerhardus	653
Mercator representation	653
Mercury (planet), relativistic rotation of perhelion	63
Mercury program, use of horizon sensors	167, 168
Meridian	24
Mesopause	106
Mesosphere	107
Message vector (in a Hamming code)	296
Metre, definition	807
Metric conversion factors	807–813
Micrometeorites, torque due to	17
Minicomputers, use in attitude software system	699
Minor, of a matrix	748
Minor frame	293
Minor frame counter (minor frame ID)	293–294
Minor principal axis	625–626
Misalignment, of rocket engine, torque due to	580–583
Mission Control Center	285
Mission mode, of space flight	661
Mission operations phase, of space mission	3–8
Mission orbit	53
Mission profile—	
Future changes in	8–12
Representative	3–12
Mission support (See also Software)	681–713
Example of role of attitude determination & control	3–8
Requirements during Space Shuttle era	716–722, 724
Software	681–713
MMS (Multimission Modular Spacecraft) (See also SMM-MMS-A)	
MMS series spacecraft—	
Attitude system	718–720
Computer used on	210
Momentum wheels	203
Sun sensor	166–167
MOD coordinates (See Mean of date coordinates)	
Modified Julian Day	21
Modulus, of Jacobian elliptic functions	526
Molniya (Soviet communications satellite)	290
Moment of inertia	
Estimate of for Apollo 15 subsatellite	497
Of spacecraft	489
Transverse	524
Units and conversion factors	810
Moment-of-inertia tensor	518–520

Momentum bias	201
Momentum bias control system	600–603
Design of	203
Momentum dumping	602
Need for	203
Using gas jets	650
Momentum transfer maneuvers	667–669
Momentum wheels (*See also Dual-spin spacecraft; Reaction wheels*)	
As part of horizon sensor system	176–178
Attitude control systems using	503, 507–508
Dynamic equations of motion for spacecraft with	522–523
Effect on spacecraft nutation	536–539
Euler's equations for	559
Hardware description	201–203
List of spacecraft using	787, 796
Torque model	656
Use in attitude acquisition sequence	667–669
Use in attitude stabilization	600–603
Use in inertial guidance maneuvers	655–661
Monopropellant gas jet	206
Month—	
Types of	52
—Numerical values for	822
Moon—	
Analytic ephemeris subroutine (SMPOS)	693
Dark angle	88
Definitive ephemeris subroutines (SUNRD, RJPLT)	693
Effect on acceleration of Earth satellite	127
Effect on Earth satellite orbit	63, 70–71
Identification of from RAE-2 data	335
Lagrange points with Earth	55
Located horizon dependent on temperature	168
Magnitude and phase law of, table	823
Multipurpose ephemeris subroutine (EPHEMX)	693
Numerical values of different types	822
Orbit of	139, 141–143
Parallax of	142–143
Properties of	822–823
Properties of orbit	51–52
Shadow cone of	75
Types of	52
Moon-centered coordinates (*See Seleno-centric coordinates*)	
Moving arc filter	322
Moving edge tracker (Earth sensor)	179–180
Multimission software	686, 721–722
Multiplexor	280
Multistep integrators	561

—N—

Nadir	22, 83
Nadir angle	12, 23
Error due to unmodeled oblateness	105
Nadir angle measurement	344

Density of	348
Information content	482
Nadir angle/Sun angle measurement, information content	480–482
Nadir angle/Sun angle single-axis attitude solutions	368
Singularities of	406–407
Nadir cone	12
Use in attitude determination	365
Nadir vector	12
Nadir vector projection model, for horizon sensors	232–233
Napier, John	34
Napier's Rules	34
For quadrantal spherical triangles	731
For right spherical triangles	730, 731
NASA Communications Network	291
NASA Monograph atmosphere model	110
NASA Standard Spacecraft Computers (NSSC-1 & NSSC-2)	210–212
NASCOM (*See NASA Communications Network*)	
National Bureau of Standards, timekeeping system	299
Natural frequency, of control system	591
NAVSTAR (*See Global positioning system*)	
Near-real-time attitude determination requirements	681
Network Operations Control Center	284
Neumann normalization, of Legendre polynomials	776
Neutral sheet, geomagnetic field	121
Newcomb, Simon	802
Newton, Issac	36–38
Nimbus (meteorology satellite)—	
Attitude system of	792–793
Digital Sun sensor	157, 162
Horizon sensors	176–178
Momentum and reaction wheels	601
Momentum wheel	202
State vector for bias determination	441–442
Telemetry data errors	311
Yaw reaction wheel	270
Nineteen Fifty (1950) Coordinates	27
Node, of an orbit	44
Motion of (*See Regression of nodes*)	
Nodical month	52
Noise, White vs. colored	269
Noise correlation, as criteria for observation model selection	444–445
Nongravitational forces, effect on orbit	63–65
Nonintersecting loci, use of in attitude solutions	370
Nonreturn to zero level pulse generation	281
Nonreturn to zero mark pulse generation	281
Nonspherical mass distribution, effect on orbit	63, 65–69
Nonspinning Earth Sensor Assembly	169, 175–176

Normal matrix   751
Notation—
  For attitude angles   349
  Use in this book   x–xii
NSSC (*See NASA Standard Spacecraft*
  Computers)
Null (attitude geometry parameter)   350
  Use of in evaluating correlation angles   391
Null (optical center line of a sensor)   188
Nutation—
  Astronomical   804
  Description from spacecraft frame
            494, 499–501
  Effect on sensor data   534–548
  Effect on spin period measurement   544–548
  Effect on Sun angle data   535–536
  Equation for, in spacecraft frame   522, 525
  Example of Apollo 15 subsatellite   495–497
  Measurement of   534–548
  Monitoring via Sun angle data   539–548
  Physical description   490–494
Nyquist criterion, for system stability   596

—O—

OAO (Orbiting Astronomical
  Observatory)—
  Attitude system of   792–793
  Evolution of accuracy requirements   714
  Inertial guidance maneuvers   655, 660
  Momentum wheel   202
  Onboard computer   210
  Reaction wheels   601
  Stabilization method   503
  Star trackers   188
  Sun sensor   157
Oblate spacecraft   491
Oblateness of Earth (*See Earth, oblateness*)
Oblique spherical triangle—
  Equations for components of   731–734
  Table of general solutions   733, 735
Obliquity of the ecliptic   48
Observability—
  In least-squares estimators   450
  Of state vector elements   443
Observation (as used in state estimation)
            443–444
Observation models—
  As used in state estimation   444
  Construction of   443–447
  Criteria for selecting   443–447
  Testing of   473
Observation model vector   439, 449
Observation residual vector   449
Observation vector   439
OCC (*See Operations Control Center*)
Occultation   72–75
Odd parity   295
Oersted (unit of magnetic field strength)   811

OGO (Orbiting Geophysical Observa-
  tory)—
  Attitude system of   792–793
  Deformation due to solar heating   550
  Horizon sensor   168, 180
  Rubidium vapor magnetometer   184
On-off control law (*See Bang bang control
  law*)
Onboard computers   210–213
  Interaction with ground-support
    facilities   719–720
  Use in attitude acquisition   662
  Use of for HEAO attitude propagation   708
Onboard processing   8–9
One-step integrators   561
One's complement arithmetic   297
Open-loop control system   502, 589, 604, 663
Open-loop transfer function, in control
  system   596–598
Open loop zeros, of root locus diagram   597
Operational procedures—
  For identifying data anomalies   334–336
  For use of state estimators   473–476
Operations Control Center   285, 292, 299
  Role in CTS mission   7
Operations phase, of space mission (*See
  Mission operations*)
Opposition (planetary configuration)   49
  Of Mars, table of   51
Optical double stars   146
Optical pumping   184
Optimal attitude determination methods (*See
  also State estimation*)   426–428
ORBGEN (subroutine)   141, 692
  Basis of   134–135
Orbit decay   64
Orbit defined coordinate systems   28–29
  Use for three-axis attitude   425
Orbit determination   1, 132
Orbit elements—
  As function of injection conditions   60–62
  Subroutine for determining from position
    & velocity   692
Orbit file format   133–134
Orbit generators—
  Keplerian (ORBGEN)   692
  Numerical   134–138
  —Accuracy and applications of   137–138
Orbit maneuvers   56–60
  Torques due to   580–583
Orbit normal   8
Orbit perturbations   62–71
Orbit vs. trajectory   53
Orbital motion of spacecraft, attitude
  correction for   365–366
Orbits (*See also Ephemeris subroutines*)
  Apparent shape when viewed obliquely   73
  Definitive spacecraft   133–134
  Earth satellites   828
  Example of types in typical mission   6–8
  Lunar and planetary   138–142

Numbering of 53
Planetary 48–52
Potential artificial satellites of planets 825
Solar system 815, 817, 824–825
Subroutine for elements from position
    and velocity (ELEM) 692
Table of equations 47
Two-body generator (ORBGEN) 692
Orthogonality, of vectors 747
Orthogonal matrix 751
Orthogonal transformation 752
Osculating orbital elements 47
OSO (Orbiting Solar Observatory)
Attitude system of 792–793
Dual-spin stabilization 505
Gas jet control system 209
Gravity-gradient torque on 568
Magnetic torquing on OSO-8 640
Momentum wheel 203
Nutation damping 626
Orbit generator accuracy 138
Spin axis magnetic coils 205
Stabilization method 503
Telemetry data errors 311
V-slit star scanner 187, 190–192
Out-of-plane orbit maneuvers 56
Out-crossing (*See Out-triggering*)
Out-triggering (Earth-out) 11, 172, 359
Outer product, of vectors 748
Output axis (gyroscope) 196
Overdamped control system 593
Overshoot, in control systems 593
Ozone, absorption bands 91–92

—P—

Pagoda effect 336–339
Mathematical model 244–249
Panoramic Scanner
(attitude sensor) 169, 173–175
Data from 335
Mathematical models 231–242
Model of biases relative to Sun sensor 241
Parabola 38–40
Parabolic anomaly 47
Parabolic orbit, table of properties 47
Parabolic velocity 42
Parallax 30–31
Lunar 142–143
Parallel (component of a spherical
    coordinate system) 24
Parallel telemetry formats 279
Parameter, of Jacobian elliptic functions 526
Parity code 295
Parking orbit 5–6, 53
Parsec (unit of distance) 808
Partial derivatives—
Numerical vs. analytical evaluation 432–433
Procedure for testing correctness of 473

Partial eclipse 72, 76
Passive attitude control 503–506
Passive attitude stabilization 18–19
Passive nutation damper 625
PECE, integration method 563–564
Pendulum nutation damper 626
Penumbra 72
Penumbral eclipse 72
Periastron 42
Pericyanthiane 42
Perifocal distance 42
Perifocus 42
Perigee 42
Rotation of 66
—Numerical formula for 68–69
Perigee height 43
Maneuvers to change 59
Perihelion 42
Perilune 42
Period (orbit parameter)—
In an elliptical orbit 41
Of Earth or Sun satellite as function of
    semimajor axis 808
Of Earth satellite as function of altitude,
    table 828
Of Earth satellite, numerical values 54, 828
Permalloy (use in magnetic coils) 205
Permeability of vacuum 813
Permendur (use in magnetic coils) 205
Perturbations—
In solar, lunar, and planetary
    ephemerides 139
Of orbits (*See Orbit perturbations*)
Phase angle (*See also Azimuth angle*)
Of solar illumination 78
Phase law—
Of Moon, table 823
Of planets and satellites 78–79
Phase match, for star identification
259, 264–265
PHASED (subroutine) 691
Use in attitude computations 366
Photochemical reactions, in Earth's
    atmosphere 108
Photodiode, as energy detector for horizon
    sensor 170, 178
Physical constants 826–829
Pi (π), value 826
Pioneer, nuclear power supply 156
Pioneer 10, 11 60
Pitch angle 360
Pitch axis 29
Pitch control, for dual-spin spacecraft 617
Pitch gain, in control systems 591
Plane change, orbit maneuvers 56, 59
PLANET (subroutine) 693
Analytic basis 141–142
Planetary configurations (planetary aspects) 49
Symbols for 50
Planetary index, of geomagnetic activity 122
Planets—

Analytic ephemeris subroutine for
(PLANET) 693
Definitive ephemeris subroutine
(SUNRD) 693
Illumination of as seen by nearby
spacecraft 334
Magnitudes of 77–80
Orbits of 48–52, 138–142
—Table 815
Properties of 814–817
Spheres of influence on satellite orbits 71
Symbols for 50
Plant, in control system 589
Plesetsk (Soviet launch site) 4
PLOTOC (attitude data simulator) 711–712
Plots—
Computer generated (*See Graphic
support systems*)
Of celestial sphere (*See Global geometry
plots*)
Plotting subroutines 694–695
Plume (gas jet exhaust) 208
Poinsot's construction, for rigid body
rotation 530–531
Polar electrojet 123
Polaris (pole star) 27
As attitude reference 189
Pole centimetre (unit of magnetic dipole
moment) 812
Pole, of a spherical coordinate system 24
Poles, of control system transfer function 590
POLYFT (subroutine) 691
Polygon match, for star identification 263
Polynomial fit, subroutine for (POLYFT) 691
Poor geometry regions, for attitude
determination 389–402
Position-only control system 656–657
Position-plus-rate control system 590, 657–658
Use of by HCMM 617
Use of by HEAO-A 606–608
**Postprocessing of attitude results vs. data
preprocessing** 318
Poundal (unit of force) 809
Power, units and conversion factors 810
Poynting-Robertson effect 64
Poynting vector 156
Preaveraging (to reduce data volume) 317
Precession (attitude motion) 14, 498
Precession of the equinoxes 27, 48
Subroutine for updating coordinate
(EQUIN) 692
Precomplier, use in software systems 684
Predicted vs. observed plots, for CTS
attitude system 702
Predictor-corrector integrators 563–564
Preprocessing of attitude data (*See also
Smoothing; Validation; Telemetry
processor*) 310–334
Contrasted with postprocessing of
results 318

Effect on statistics 317
Pressure, units and conversion factors 809
Primary (one of two objects in an orbit) 38
Prime meridian 25
Principal axes 519
Discovery of 521
Of spacecraft 488–489
—Stability of rotation about 523
Principal moments of inertia 489, 519
Printer plot subroutines (GRAPH, SCALE) 694
Probability density (of attitude) 375, 376
Process (in state estimation) 438
Product of inertia 519
Prograde orbit 53
Programmable telemetry format 295
Programming (*See Software*)
Programming standards 684–686
Project Operations Control Center 285
Project Scanner 92–96
Prolate spacecraft 491
Propagation (*See Attitude propagation*)
Propellant (*See also Fuel*)—
Disturbance torques due to slosh 577–578
Proper motion 144
Proper orthogonal matrix 751
Proportional control system 590
Proton precession magnetometer 184
PSCTS, use of MMS spacecraft 720
Pseudoevaluation (in numerical
integration) 564
Pseudoinverse, of a matrix 749
Pseudoinverse state estimator 468–469
Pulse amplitude modulation 281
Pulse code modulation 280–281
Pulse duration modulation 281
Pyroelectric detectors, for Earth sensing 171

—Q—

Q method, of three-axis attitude
determination 421, 426–428
Quadrantal spherical triangle 34
Equations for components of 731
Quadrature 50
Quality assurance—
Of attitude solutions, need for 716
Of software 686
Quality flag, on attitude data 314
Quantized measurements 374
Attitude determination accuracy 374–376
Variance of 431
Quantum magnetometer 181, 184
Quarter orbit coupling 601
**Quarter Orbit Magnetic Attitude
Control** (QUOMAC) 640
Basis for 785
Quaternions (*See also Euler symmetric
parameters*)
Algebra of 758–759

Components of   758
Kinematic equations of motion for   511–512
Norm of   759
Representation of attitude by   414
Use in control laws   605
Use for attitude propagation
                                         558–559, 564–566
Quicklook displays   305
QUOMAC (*See Quarter Orbit Magnetic
    Attitude Control*)

—R—

RADECM (subroutine)   690
Radiance profiles, of Earth in infrared   95
Radiation balance, for the Earth   82
Radiation pressure, torque due to   570–573
Radiometric balance Earth sensor   180
Radius of gyration, of gas jet propellant   650
RAE (Radio Astronomy Explorer)—
    Application of block averaging to
        attitude data   371
    Attitude acquisition   661, 666
    Attitude software structure   698
    Attitude system of   792–793
    Data from panoramic scanner   335
    Deadbeat boom deployment   669
    Effect of flexible booms   549, 555, 556
    Gas jet control system   210
    Gravity-gradient stabilization   505–506
    High-altitude attitude magnetometers   181
    Panoramic scanner   169, 173–174
    Stabilization method   503
    Sun sensor   157
    Telemetry data errors   311
    Thruster characteristics   273
    Time-tagging of playback data   302–303
    Use of convolutional encoding   282
RAGREN (subroutine)   692
Random error, in attitude measurements   402
Range and range rate, orbit measurements   132
Rank, of a matrix   749
Rate gyroscopes   196, 197–198
Rate-integrating gyroscopes   196, 199–200
Reaction wheels—
    Hardware description   201–203
    Mathematical models   270–272
    Use for attitude control   19
    Use in attitude stabilization   603–604
    Use in inertial guidance maneuvers   655–661
**Real-time attitude determination
    requirements**   681
Receiving stations   291–292
    Data processing at   303–304
Rectangular coordinate system,
    advantages relative to spherical   25–26
RECUR (subroutine)   692
    Application of on AE spacecraft data   318
Recursive estimator   437, 439, 448

Subroutine (RECUR)   692
Recursive least-squares estimator   459–462
    Advantages & disadvantages   461–462
Reference meridian   25
Reference orbit   62
Reference point, in spherical coordinate
    system   25
Reference spheroid, in Earth models   99
Reference vectors (*See Attitude determina-
    tion, Reference sources, see also indi-
    vidual reference vectors, e.g., Sun, mag.
    field*)
Reflected binary code (Gray code)   295
Reflection, specular vs. diffuse   571
Regression of nodes   66
    Numerical formula for   68
Relativistic effects on orbit   37, 63
Residual, rms   453
Residual editing   320
Residual magnetic dipole   252
Retrograde orbit   53
Return-to-zero pulse generation   281
Revolution vs. rotation   53
Rhumb length   654
Rhumb line   653
Rhumb line attitude maneuver   652–654
Rigid spacecraft motion   523–524
Right ascension (*See also Celestial
    coordinates*)   28
    Related to time   802–803
**Right ascension of the Greenwich
    Meridian** (Greenwich Sidereal Time)   803
    Subroutine for (RAGREN)   692
Right ascension of the ascending node   44, 46
    Maneuver to change   59
    Motion of (*See Regression of nodes*)
Right spherical triangle   34
    Equations for components   730–731
    Example of exact triangle   730
Rise time (gas jet)   273
RJPLT (subroutine)   140, 693
Roberts atmosphere model   110
Rocket engine misalignments, torque
    due to   580–583
Roll axis   29
Roll, Pitch, Yaw coordinate system   29
Root locus diagram   596–600
    Use for selection of control gains   622
Rosman STDN tracking station   284–289
Rotating coordinate frames, rate of
    change of vectors in   514–515
Rotation—
    Distinction from nutation and coning   489
    Distinction from revolution   53
Rotation angle   23
    Formula for   727
    Formula for, in vector notation   757
    Subroutines for (PHASED, VPHAZE)   691
Rotation angle measurement   23, 346, 349–352
    Density   352
    Notation for   23

Singularity conditions   406–407
Use in deterministic attitude solutions
                                  364, 365, 369
Rotation angle models (Sun sensor/horizon
    sensor)   237–242
Rotation axis of spacecraft   487–488
Rotational motion (*See Attitude dynamics,
    attitude kinematics*)
ROUND (subroutine)   693
Round-off error, in integration procedures 560
Routh-Hurwitz criteria—
    Example of use   619
    For nutation damper study   626
    For system stability   596
Row matrix   745
Royal Greenwich Observatory   20
RO1TAP (subroutine)   693
    Use of   134
RPY coordinates (*See Roll, Pitch, Yaw
    coordinates*)
RUNGE (subroutine)   692
    Analytic basis for   562
Runge Kutta, integration method   561–562
    Subroutine for (RUNGE)   692

                  —S—

S-band data transmission   282
S-band telemetry subbands   282
SAGE (Stratospheric Aerosol Gas Experi-
    ment satellite) (*See also AEM*)—
    Attitude system of   792–793
    Scanwheels horizon sensor   176–178
    Use of AEM spacecraft   720
San Marco Platform launch site   4
SAO number (star catalogs)   143
SAS (Small Astronomy Satellite)—
    Analysis of dynamic motion   538
    Attitude acquisition   661
    Attitude system of   792–795
    Constant current source on SAS-3   206
    Data records   304
    Disturbance torques   580
    Earth-width data   233–234
    Evaluation of horizon sensor data   471
    Fixed head star tracker   193–195
    Image dissector tube star sensor
        measurements   256–258
    Instrumental magnitude for star camera 258
    Large data volume   308
    Launch of SAS-1 (Uhuru)   4
    Momentum wheel   202–203
    N-slit star scanner analysis   705, 706, 707
    Nutation damping   627, 629
    Observation model for bias
        determination   445–446
    Optimal magnetic maneuvers   640
    Programmable telemetry formats   295
    Scanwheels horizon sensor   176–178

Spin axis magnetic coils   205
Spin plane magnetic control   646
Stabilization method   503
Star catalog for   147
Star identification for   263
Star scanner   187
Star tracker analysis   706–707
State vector for bias determination   440–441
Sun sensor   157
Telemetry data errors   311
Telemetry processor   305
Satellite—
    Defined   38, 52
    Distinguished from interplanetary probe 52
    Local mean time of subsatellite point 68
    Longitude of subsatellite point   805
    Magnitude of (brightness measurement)
                                        77–80
    Names (*See Spacecraft, names*)
    Orbit of   52–62
    —Lifetime   64
    —Period around Earth or Sun vs.
        semimajor axis   808
    —Period vs. altitude above Earth, table 828
    —Potential artificial satellites of planets 825
    —Properties   825
    —Utility subroutines for   692–693
Satellite Automatic Tracking Antennas 284
Satellite Command Antenna   284, 289
Satellite tracking stations (*See Tracking
    stations*)
Satellites, natural, table of properties 817
Saturation limit, of momentum wheel   508
Saturn V launch vehicle   3
Scalar checking, for data validation 328–334
Scalar product, of vectors   747
SCALE (subroutine)   694
Scale height, of atmosphere   108
Scaliger, Joseph   20
SCAMA (*See Switching, Conferencing,
    and Monitoring Arrangement*)
Scanning mechanism, employed by horizon
    sensor   169
Scanwheels horizon sensor   169, 176–178
Schmidt normalization, for spherical
    harmonics   780
Score (numerical measure of star
    identification)   260
Scout launch vehicle   3
Search coil magnetometer   181
Search pattern (in fixed head star trackers) 189
SEASAT (ocean studies satellite)—
    Anticipated horizon radiance variations 98
    Attitude system of   794–795
    Canted momentum wheels   602–603
    Clock   299
    Control system description 612–613, 621–622
    Effect of flexibility on attitude dynamics 556
    Locator used on horizon sensor   172
    Momentum wheel   202
    Scanwheels horizon sensor   170, 176–178

Sun sensor   157, 166
—Coverage of   402
Use of carbon dioxide band horizon
    sensor   92
Second (ephemeris)   799
Second (SI unit)   802
    Definition   807
Secondary (one of two objects in an orbit)   38
Sectoral harmonic coefficients,
    explanation of   777–778
Secular drift, geomagnetic field   113
Secular terms, geomagnetic field model   779
Selenocentric coordinates   29
Self-correcting codes   296–297
Semiconjugate axis, of hyperbola   40
Semimajor axis   38, 46
Semiminor axis   38
Semitransverse axis, of hyperbola   40
Sensor electronics, mathematical models
                                               242–249
Sensors (See Attitude sensors; see also
    Object sensed, e.g., Horizon sensors)
Sequential estimator   437, 448
    For star data   704, 707–709
Serial telemetry formats   279
Shadow bar Sun sensor   159
Shadow cone   72
    Of Earth and Moon   75
Shadowing—
    Effect on aerodynamic torque   574
    Effect on radiation torque   573
Short period variations, in orbital
    elements   65
Shuttle (See Space shuttle)
SI (metric) units   807–813
    Prefixes   807
Sidereal day   804–805
Sidereal month   52
Sidereal period   50
    Distinguished from solar period for
        Earth satellite   55
Sidereal time   798, 802–805
    Subroutine for (RAGREN)   692
Sidereal year   48
Sifting (to reduce data volume)   317
Signal conditioner   279
Similarity transformation   752
Simpson's Rule, for integration   561
Simulators, for attitude data   709–712
Simultaneous linear equations, solution of
                                               749–750
Sines, law of (in spherical triangles)
                                          33, 731–732
Single-axis attitude   343–346
Single-axis attitude determination—
    Accuracy   373–409
    Methods   362–409
Single-degree-of-freedom gyroscope   196
Single-spin spacecraft   503–504
Singular matrix   749
Singularities, in attitude solutions   403

Singularity conditions in attitude
    determination   406–407
    Table of   407
Sinter, Use in thermistor flake   171
SIRIO (Italian experimental
    communications satellite)—
    Application of block averaging to atti-
        tude solutions   371
    Attitude accuracy constraints on launch
        window   399–401
    Attitude determination accuracy   397
    Attitude software structure   698
    Attitude system of   794–795
    Correlation among measurement types   480
    Slit horizon sensor/Sun sensor   178–179
    Spin rate change due to orbit maneuvers   582
    Sun sensor analysis   717–718
    Use of body-mounted horizon sensor   173
    Use of carbon dioxide band horizon
        sensor   92
Skew-symmetric matrix   750
Skylab—
    Attitude control system   197, 201
    Disturbance torques due to crew motion   579
    Spacecraft configuration   579
SKYMAP, star catalog   147
SKYNET (U.K. Communications
    satellite)— Attitude system of   794–795
Skywave (in radio broadcasts)   301
Slave station (time signals)   300
Slew maneuvers   601, 655–661
Slit horizon/Sun sensor   178–179
    As possible standard coarse sensor   721
    Math. model for misalignment   219
Slit sensors, Analysis of alternative
    designs   717–718
Slit star sensor, mathematical model   254–256
SLP ephemeris files   140
    Subroutine for reading (SUNRD)   693
Slug (unit of mass)   808
SM (San Marco satellite)
    Attitude system of   794–795
Small circle (spherical geometry)   22, 32
    Area formulas   729–730
    Construction of on global geometry plot
                                          739, 742–743
    Equations for   727
Smithsonian Astrophysical Observatory,
    star catalog   143–144, 146–147
SMM (Solar Maximum Mission) (See
    also MMS)—
    Attitude acquisition   672
    Attitude control law for   659
    Attitude system   718–720, 794–795
    Control system   608
    Data collection for bias determination   475
    Fine Sun sensor   166–167
    Inertial reference assembly   187
    Onboard computer   210
Smoothing, of attitude data and
    results   315–327

Applications of 316
Guidelines for 317–318
SMPOS (subroutine) 693
Analytic basis 141–142
SMS (Synchronous Meteorological
Satellite)—
Application of block averaging to
attitude solutions 371
Attitude acquisition 661
Attitude determination accuracy 397–399
Attitude software structure 698
Attitude solutions from 373
Attitude system of 794–795
Behavior of single-frame solutions 403–405
Correlation among measurement types
478, 480–482
Data collection for bias determination 474
Earth-width data 233–234
Horizon sensor electronics
modeling 244–249
Launch of 5
Pagoda effect 336–339
Sensor package characteristics 721
State vector for bias determination 440
Sun sensor analysis 717–718
Telemetry data errors 311
Use of body-mounted horizon sensor 173
Use of carbon dioxide band horizon
sensor 92
Use of open-loop control 663
View of Earth by 84, 91
Snapshot, of star sensor data 706
Snell's law 223
Software—
Avoidance of errors in 682–683, 685
Development of 681–713
Example of attitude support software
structure 700–703
For multimission support 686, 721–722
General structure for attitude
support 696–700
Goddard Space Flight Center
environment 682
Safeguards for mission support 681–686
Standardization of 686, 721–722
Systems, general structure for 696–700
Test procedures for state estimators 471–473
Utility subroutines 690–695
Solar eclipse 72
Solar heating, effect on flexible spacecraft
549, 550
Solar mass ratio, for planets 827–828
Solar parallax 31
Solar radiation—
Flux 130
Pressure, effect on orbit 64–65
—Effect on flexible spacecraft 550–551
Stabilization 19
Torque 17, 570–573
Solar sail 64
Solar System—

Orbits 48–52
Properties 814–825
Solar time 798, 799–801
Solar wind 120, 129–132
Sectors 131–132
Solid angle 23
Formulas 729–730
Units and conversion factors 810
Solid spherical harmonics 775
Sounding rocket 52
South Atlantic Anomaly (*See Brazilian
Anomaly*)
Soviet Space Program 3
Launch sites 3
Tracking and data acquisition 290
Space cone 491–492
Space Mission, profile of—
Future changes in 8–12
Representative 3–12
Space navigation 1
**Space Precision Attitude Reference
System (SPARS)** 708
Space shuttle 3, 8–9
Effect on attitude determination and
control 714, 724
Orbit ephemeris 134
Payload mass as a function of altitude 9
Star tracker for 190
Thrust 53
Space Telescope—
Attitude system of 794–795
Onboard computer 211
Pointing accuracy 714
Reaction wheels 604
Stability requirements 604
Spacecraft—
Data generation and handling
onboard 278–283
Effects of flexibility on dynamics 548–556
Gyroscope measurement of angular
velocity 267–268
Magnitude of when viewed from a
distance 79
Names and international designations 52
Stabilization and control, methods of 18–19
Stabilization, methods of (*See also
Attitude stabilization*) 3
Spacecraft attitude control (*See Attitude
control*)
**Spacecraft attitude determination and
control systems** 787–797
Spacecraft attitude dynamics (*See
Attitude dynamics*)
Spacecraft attitude motion—
Example of Apollo 15 subsatellite 495–497
Introduction to 487–502
Spacecraft axes, alternative systems 487–489
Spacecraft-centered celestial sphere 22–24
Spacecraft-centered coordinates 26–29
Spacecraft clocks 298–299
Spacecraft ephemerides—

Definitive subroutines 693
Two-body orbit generator (ORBGEN) 692
Spacecraft fixed coordinates 26
Spacecraft orbits 52–62, 132–138
As function of injection conditions 60–62
Multipurpose ephemeris subroutine
(EPHEMX) 693
Spacecraft stability (*See also Attitude dy-
namics; Disturbance torques;
Nutation; Flexible spacecraft dynamics*) 523
Apollo 15 subsatellite 495–497
With respect to libration (GEOS
example) 674
Spaceflight Tracking and Data Network
(STDN) 283–290
Time-tagging by 299–301
Spacelab, onboard computer 211
SPARS (*See Space Precision Attitude
Reference System*)
Special perturbations, method of (orbit
analysis) 139
Specular reflection 84
Torque due to 572
SPHCNV (subroutine) 694
Sphere, illumination of as function of phase,
distance 78–79, 89
Sphere of influence—
For spacecraft orbits 69–71
Table of for planets 69–71
Spherical coordinate systems 24–31, 760
Advantages relative to rectangular 25–26
Properties of 24–26
Transformations between 765–766
Spherical excess 32
Spherical geometry 31–35
Construction of global geometry plots
737–743
Equations for 727–736
Spherical harmonics 775–778
Expansion of gravitational potential in 124
Recursion relations 781
Representation of the geomagnetic
field 779–782
Schmidt normalization for 780
Spherical plot subroutines (SPHCNV,
SPHGRD, SPHPLT) 694–695
Spherical plots (*See Global geometry plots*)
Spherical triangle 23
Infinitesimal 734
Notation for 33
Properties of 32–35
Spherical trigonometry 33–35
Differential 734–735
Example of 34–35
Table of general solutions 733, 735
SPHGRD (subroutine) 694
SPHPLT (subroutine) 694
Spin-axis drift, during rocket engine firing 582
Spin-axis precession, magnetic 636–649
Spin rate—
Change during rocket engine firing 582

Control using gas jets 650
Effect of nutation on measurement
of 544–548
Spin stabilization 19, 503–504
Spin-stabilized spacecraft 3
List of 787
Split-to-index time, horizon sensors 172
Math. model (split-angle model) 231, 234
Square root filter 467
SSS (Small Scientific Satellite)
Attitude system of 794–795
Effect of flexibility on attitude dynamics 556
Nutation damping 544, 629
Telemetry data errors 311
ST (*See Space Telescope*)
Stability—
Of control systems 594–600
Of rotation about a principal axis 523
Of spacecraft (*See Spacecraft stability*)
Stabilization (*See Attitude stabilization;
Nutation damping; Libration damping*)
Standard deviation 429
Standard notation x–xii
For attitude angles 23, 349
Standard symbols xi–xii, 50
Standard Time 801
Standardization—
Of attitude hardware 718–721
Of attitude software 686, 721–722
Star azimuth 259
Star camera 193
Star catalog, acquisition of by an attitude
system 704
Star catalogs 143–150
Star longitudes 149–150
Star scanners 186, 187, 190–192
List of spacecraft using 797
Star sensors—
Attitude determination methods for 703–709
Characteristics vs. accuracy require-
ments 190
Data selection and correction 704–705
Example of use for nutation
monitoring 538–539
Hardware 184–195
List of spacecraft using 797
Mathematical model of intensity
response 258–259
Mathematical models 254–259
Overview of attitude determination
with 703–709
Representative telemetry data errors 311
Star trackers—
Fixed-head 186, 189–190, 193–195
Gimbaled 186, 187–189
List of spacecraft using 797
Stars—
Angular diameter of 167
As attitude determination reference
source 17
Densities 145

Distribution, mathematical model of   263
Identification   706
Identification techniques   259–266
Instrumental magnitudes of   258
Magnitudes   144–145
Position modeling   143–150
State (in differential correction or "state"
  estimation)   439
State estimation   436–483
  Advantages relative to deterministic 437–438
  Need for   407–408
  Operational limits on accuracy   476
  Subroutine for (DC)   691
  Use in CTS attitude system   702
State estimators—
  Analytic basis   447–470
  Operational use of   471–484
  Prelaunch evaluation   471–473
  Subsystem in attitude software
    system   697–698
  Unresolved analytic problems   722–723
State noise covariance matrix   465
State plane trajectory   657
  Examples of   658
  For HEAO-1   607
State space   657
State transition matrix   450
State vector   436, 438–443
  Choice of elements to be solved for 476–483
  Construction of   438–443
  Need for   407–408
  Subroutine for updating (RECUR)   692
State vector elements—
  Choice of   439–443
  Observability of   443
  —Limitations   476–483
State weight matrix   449
STDN (*See Spaceflight Tracking and Data
  Network*)
Steady state (in Kalman filters)   467
Steady-state error, in control systems 593
Steady-state system response   770
Steady-state trajectory   608
Steepest descent, method of (differential
  correction technique)   455
Steering law   604
Stellar parallax   31
Stepsize, in numerical integration 560
STEREOSAT, use of AEM spacecraft   720
Stiction   272
Stiffness matrix (for flexible spacecraft) 553
STORMSAT—
  Disturbance torques   580
  Use of MMS spacecraft   720
Strapdown torque rebalanced gyroscope 199
Stratopause   107
Stratosphere   107
Subcommutated data   293–294
Subsatellite, Apollo 15, nutation of 495–497
Subsatellite point (*See also Nadir*) 22
  Alternative definitions   83
  Local mean time of   68

Longitude of   805
Subsolar point   84
Summing point, in control system   589
Sun— (*See also Solar*)
  Analytic ephemeris subroutines
    (SUN1X, SMPOS)   693
  Approach by spacecraft   60
  As attitude determination reference
    source   17
  Definitive ephemeris subroutines
    (SUNRD, RJPLT)   693
  Effect on acceleration of Earth satellite 127
  Effect on geomagnetic activity   120–123
  Energy flux from   130
  Expression for mean motion   141
  Interference with horizon sensors   169
  Multipurpose ephemeris subroutine
    (EPHEMX)   693
  Properties of   818
  Solar wind   120, 129–132
  Symbol for   50
Sun angle measurement   11, 23, 344
  Density of   347
  For nutating spacecraft   539–548
  Information content   480
**Sun angle/nadir angle single-axis attitude
  solutions**   368
  Singularities in   406–407
Sun cone   12
  Use in attitude determination   364
Sun presence detector   156, 159–161
Sun sensor   11, 13
  Analysis of SIRIO vs. SMS design   717–718
  Calculation of coverage of celestial
    sphere   34–35
  Calibration constants   230
  Combination Sun/Earth horizon
    sensor   178–179
  Data validation   329–330
  Example of use in attitude control   14–16
  Example of use in attitude
    determination   10–14
  Field of view   225–226
  Hardware   155–166
  List of spacecraft by Sun sensor type 797
  Mathematical models   218–230
  Model of azimuth biases relative to
    horizon sensor   239–242
  Nutation monitoring with   539–548
  Simple vs. complex   716–718
  Two axis, accuracy analysis   355–357
  Use for single-axis attitude   362–409
  Use for three-axis attitude   426
**Sun sensor/horizon sensor rotation angle
  measurement**   357–359
  Information content   480–482
  Models   237–242
  Use in attitude determination   364–365, 369
Sun shade (for star sensors)   186
Sun-synchronous orbit   68
Sundials   800

SUNRD (subroutine)   140, 693
SUN1X (subroutine)   693
  Analytic basis   141
Supercommutated data   293–294
Superior conjunction   49–50
Superior planet   49
Surface spherical harmonics   775
Surveyor (spacecraft), attitude reference
  system   189
Switching, Conferencing, and Monitoring
  Arrangement (SCAMA)   291
Switching line (component of control
  system)   657
Symbols, astronomical   50
Symmetric mass distribution, principal
  axis of   489
Symmetric matrix   750
Symmetric spacecraft   524
SYMPHONIE (French/German
  communications satellite)—
  Attitude system of   794–795
  Horizon sensors   180
Synch speed, of reaction wheel   270
Synchronization pattern, quality flag for   314
Synchronization signal, role in telemetry   293
Synchronization word, in NASCOM data
  format   291
Synchronous satellite   55
  Of various planets, table   825
Syndrome vector   296
Synodic month   52
Synodic period   50
System gain—
  Of a control system   590
  Role in root locus diagram   597–600
  Selection of for attitude control   622, 624
System mass matrix (for flexible space-
  craft)   553
Systematic error, in attitude measurements   402
  Effect on deterministic solution behavior   404
Syzygy   49

—T—

Tachometers, for measuring wheel speed   202
Tangent height   93
Tangent plane coordinates (*See Local
  tangent coordinates*)
TCON20 (subroutine)   692
TCON40 (subroutine)   692
TDRSS (*See Tracking and Data Relay
  Satellite System*)
Teflon—
  Use in gas jets   206
  Use in momentum wheel bearings   202
Telemetry   293–298
  Generation and transmission of   278–298
  Time tagging   298–304
Telemetry antennas   284–288

Telemetry data errors (*See also Data
  errors*)   310–311
  Table of representative examples   311
Telemetry formats   279, 293–295
Telemetry On-Line Processing System
  (TELOPS)   292
Telemetry processor   304–308
  In attitude software systems   696–697
  Of CTS attitude system   701
Telemetry word   293
Tell-tale (data flag)   313–315
TELOPS (Telemetry On-Line Processing
  System)   292
Temperature—
  Of Earth's atmosphere   107
  Units and conversion factors   812
Tensor   519
Terminator   84, 86–90
  Identification of   331–334
Tesseral harmonic coefficients, explana-
  tion of   777–778
Testing—
  Of attitude software   686
  Of state estimators   471–473
Thermal radiation (*See Infrared radiation*)
  Definition of   83
Thermistor, as energy detector for horizon
  sensor   171, 178
Thermopile, as energy detector for horizon
  sensor   171, 180
Thermosphere   107
Third-body interactions, effect on orbit
                                              63, 69–71
Three-axis attitude   343, 359–361
Three axis attitude determination—
  Accuracy   429–434
  Example of least-squares estimator   456–459
  Methods   410–434, 420–428
Three-axis stabilized spacecraft   3
Thrust profile (of gas jet)   207, 272–275
Thrusters (*See Gas jets*)
Time—
  Local mean time of subsatellite point   68
  Measurement and broadcast facilities
                                              299–302
  Units and conversion factors   808–809
Time checking, of telemetry data   307–308
Time measurement systems   18–21, 798–806
  Conversion subroutines for   692
  Table of   798
Time optimal magnetic maneuver   642, 648
Time tagging   278
  Near-real-time data   302
  Playback data   302
  Representative telemetry data errors   311
  Telemetry data   298–304
TIROS (meteorology satellite)—
  Attitude system of   794–795
  First Use of Quarter Orbit Magnetic
    Attitude Control   639

Titius-Bode law (*See Bode's Law*)
TOD (*See True of date*)
Torque (*See also Disturbance torques; Attitude control*)
    Average of gas jet   274
    Due to magnetic moment   813
    Effect of, on spacecraft motion   498–502
    Internal vs. external   521
    Units and conversion factors   810
Torque-free motion, of spacecraft   487–497
Torque-free solutions, for attitude motion   524–531
Torr (unit of pressure)   809
Total eclipse   72, 76
TRAAC, libration damping   632
Trace, of a matrix   748
Track pattern (in fixed-head star trackers)   189
Tracking and Data Relay Satellite System (TDRSS)   8–10, 287–290
Tracking stations   283–290
    Location of   284–285
    Timing systems   299–302
Trajectory, of gas jet precession   652
Trajectory, of spacecraft   53
Transfer elements, of a control system   588
Transfer function   244
    Mathematical model of horizon sensor electronics   244–248
    Of horizon sensor electronics   172
    Use in control systems   589–590
    Use of to evaluate stability   591–593
Transfer orbit (*See also Hohmann transfer orbit*)   5–6, 53
Transfer time, in Hohmann transfer orbit   58–59
    Between the planets (table)   824
Transformations between coordinate systems   760–766
Transit   71–75
Transit time, in slit star scanner   254
Transmission, of data and commands (*See also Telemetry*)   278–292
Transpose, of a matrix   744
Transverse angular velocity   525
Transverse moment of inertia   524
Trapezoid model, of gas jet profiles   275
Trigonometry, spherical (*See Spherical trigonometry*)
Trojan asteroids   55
Tropical year   48
Tropopause   106
True anomaly   45–46
True of date coordinates   27–28
Truncation error, in integration procedures   560
Tumbling, of spacecraft due to crew motion   579–580
Turbopause   108
Turn angle, of hyperbola   40, 60–61
Two-axis Sun sensor (*See Sun sensor*)
Two-degree-of-freedom gyroscope   196

Two thousand (2000) coordinates   27
Two's complement arithmetic   297
Tyuratan (Soviet launch site)   4

—U—

UBV magnitudes   144–145
UDU filter   467
Uhuru (*See SAS*)
Umbra   72
Uncertainty in attitude measurements   345–346
    Correlated   374, 378–379, 382–383
    Due to systematic errors   402–408, 476
    Expressions for   375–376, 381–382, 384
    Uncorrelated   374–382
Uncertainty of state estimator solutions   476
Undershoot, in control systems   593
Unitary matrix   751
Unitary transformation   752
Units and conversion factors   807–813
Universal Time (UT) = GMT = Z   19, 798, 801–802
    Attached to data   298
Unpacking, of telemetry data   304–306
UNVEC (subroutine)   690
Uplink   278
Upper stage rocket vehicles   53
U.S. Coast Guard, time keeping system   299
U.S. Naval Observatory, time keeping system   299–300
U.S. Standard Atmosphere   110
UT (*See Universal Time*)
UTC (*See Universal Time; Coordinated Universal Time*)
Utility subroutines   690–695

—V—

V-brush, Sun sensor code   163
V-slit Sun sensor (V beam Sun Sensor)   161
    Mathematical model   218–221
Validation—
    Data flags and sensor identification   313–314
    Of attitude data   315–327
    Of telemetry data   307, 312–334
    Requiring attitude information   334–339
Vandenberg Air Force Base, Calif. (*See Western Test Range*)
Vanguard (tracking ship)   284, 287
Vanguard units   808–809
Variance   429
Variation of parameters, formulation of attitude dynamics   531–534
VEC (subroutine)   690
Vector algebra   744–757
    Subroutines   690–691
Vector calculus   755–756
Vector identities, in three dimensions   756–757

Vector magnetometer, mathematical
   model   250–254
Vector multiplication, inner and outer
   products   747–748
Vector notation   x, xii
Vector product (cross product)   756
   Equation for direction of   728
   Subroutine for (CROSSP)   691
Velocity—
   In a Keplerian orbit   54, 828
   Units and conversion factors   809
Velocity of escape   42
   Planets and satellites, table 817, 825
Vernal equinox   27, 48
   Symbol for   50
VHF data transmission   282–283
Vidicon, in star sensors   189
Viking (Mars mission), launch dates
   relative to opposition   51, 57–58
Vis viva **controversy**   38
Vis viva **equation**   38, 42
   Origin of   38
Viscous ring nutation damper 626, 627–629
Visibility of satellites and spacecraft 80
Visibility of spacecraft   71–80
Visible light sensor (*See also Horizon sensor*) 83
Visual magnitude, for spacecraft, sample
   calculation   79
Volterra equation   773
VOP (*See Variation of parameters*)
VPHAZE (subroutine)   691

—W—

Wallops Island, launch site   4
Water, absorption bands   91–92
Weighting (in estimation theory) 449–450

Weighting data   370–37
Weightlessness   41–4
Western Test Range (launch site)
WHECON, wheel control system 613, 62
   Active nutation damping with 630–631
Wheel-mounted horizon sensor (*See*
   also Scanwheels)   169
   Mathematical models   234, 236
   Table of characteristics   176–178
White noise   269
White Sands (launch site)   4
World Warning Agency   52
WWV time signals   299
WWVH time signals   299

—Y—

Yaw axis   29
Year—
   Sidereal   799
   Tropical   799
   Types of   48
Yo-yo despin maneuvers   663–669

—Z—

Z (time unit) (*See Universal Time*)
Zenith   22
Zenith angle   85
Zero crossing magnetometer   250
Zeros, of control system transfer function 590
Zonal harmonic coefficients—
   Explanation of   777
   In gravitational potential 124, 127
Zulu Time   801

ASTROPHYSICS AND SPACE SCIENCE LIBRARY

Edited by

J. E. Blamont, R. L. F. Boyd, L. Goldberg, C. de Jager, Z. Kopal, G. H. Ludwig, R. Lüst,
B. M. McCormac, H. E. Newell, L. I. Sedov, Z. Švestka, and W. de Graaff

1. C. de Jager (ed.), *The Solar Spectrum, Proceedings of the Symposium held at the University of Utrecht, 26–31 August, 1963.* 1965, XIV + 417 pp.
2. J. Orthner and H. Maseland (eds.), *Introduction to Solar Terrestrial Relations, Proceedings of the Summer School in Space Physics held in Alpbach, Austria, July 15–August 10, 1963 and Organized by the European Preparatory Commission for Space Research.* 1965, IX + 506 pp.
3. C. C. Chang and S. S. Huang (eds.), *Proceedings of the Plasma Space Science Symposium, held at the Catholic University of America, Washington, D.C., June 11–14, 1963.* 1965, IX + 377 pp.
4. Zdeněk Kopal, *An Introduction to the Study of the Moon.* 1966, XII + 464 pp.
5. B. M. McCormac (ed.), *Radiation Trapped in the Earth's Magnetic Field. Proceedings of the Advanced Study Institute, held at the Chr. Michelsen Institute, Bergen, Norway, August 16–September 3, 1965.* 1966, XII + 901 pp.
6. A. B. Underhill, *The Early Type Stars.* 1966, XII + 282 pp.
7. Jean Kovalevsky, *Introduction to Celestial Mechanics.* 1967, VIII + 427 pp.
8. Zdeněk Kopal and Constantine L. Goudas (eds.), *Measure of the Moon. Proceedings of the 2nd International Conference on Selenodesy and Lunar Topography, held in the University of Manchester, England, May 30–June 4, 1966.* 1967, XVIII + 479 pp.
9. J. G. Emming (ed.), *Electromagnetic Radiation in Space. Proceedings of the 3rd ESRO Summer School in Space Physics, held in Alpbach, Austria, from 19 July to 13 August, 1965.* 1968, VIII + 307 pp.
10. R. L. Carovillano, John F. McClay, and Henry R. Radoski (eds.), *Physics of the Magnetosphere, Based upon the Proceedings of the Conference held at Boston College, June 19–28, 1967.* 1968, X + 686 pp.
11. Syun-Ichi Akasofu, *Polar and Magnetospheric Substorms.* 1968, XVIII + 280 pp.
12. Peter M. Millman (ed.), *Meteorite Research. Proceedings of a Symposium on Meteorite Research, held in Vienna, Austria, 7–13 August, 1968.* 1969, XV + 941 pp.
13. Margherita Hack (ed.), *Mass Loss from Stars. Proceedings of the 2nd Trieste Colloquium on Astrophysics, 12–17 September, 1968.* 1969, XII + 345 pp.
14. N. D'Angelo (ed.), *Low-Frequency Waves and Irregularities in the Ionosphere. Proceedings of the 2nd ESRIN-ESLAB Symposium, held in Frascati, Italy, 23–27 September, 1968.* 1969, VII + 218 pp.
15. G. A. Partel (ed.), *Space Engineering. Proceedings of the 2nd International Conference on Space Engineering, held at the Fondazione Giorgio Cini, Isola di San Giorgio, Venice, Italy, May 7–10, 1969.* 1970, XI + 728 pp.
16. S. Fred Singer (ed.), *Manned Laboratories in Space. Second International Orbital Laboratory Symposium.* 1969, XIII + 133 pp.
17. B. M. McCormac (ed.), *Particles and Fields in the Magnetosphere. Symposium Organized by the Summer Advanced Study Institute, held at the University of California, Santa Barbara, Calif., August 4–15, 1969.* 1970, XI + 450 pp.
18. Jean-Claude Pecker, *Experimental Astronomy.* 1970, X + 105 pp.
19. V. Manno and D. E. Page (eds.), *Intercorrelated Satellite Observations related to Solar Events. Proceedings of the 3rd ESLAB/ESRIN Symposium held in Noordwijk, The Netherlands, September 16–19, 1969.* 1970, XVI + 627 pp.
20. L. Mansinha, D. E. Smylie, and A. E. Beck, *Earthquake Displacement Fields and the Rotation of the Earth, A NATO Advanced Study Institute Conference Organized by the Department of Geophysics, University of Western Ontario, London, Canada, June 22–28, 1969.* 1970, XI + 308 pp.
21. Jean-Claude Pecker, *Space Observatories.* 1970, XI + 120 pp.
22. L. N. Mavridis (ed.), *Structure and Evolution of the Galaxy. Proceedings of the NATO Advanced Study Institute, held in Athens, September 8–19, 1969.* 1971, VII + 312 pp.

23. A. Muller (ed.), *The Magellanic Clouds. A European Southern Observatory Presentation: Principal Prospects, Current Observational and Theoretical Approaches, and Prospects for Future Research, Based on the Symposium on the Magellanic Clouds, held in Santiago de Chile, March 1969, on the Occasion of the Dedication of the European Southern Observatory.* 1971, XII + 189 pp.

24. B. M. McCormac (ed.), *The Radiating Atmosphere. Proceedings of a Symposium Organized by the Summer Advanced Study Institute, held at Queen's University, Kingston, Ontario, August 3–14, 1970.* 1971, XI + 455 pp.

25. G. Fiocco (ed.), *Mesospheric Models and Related Experiments. Proceedings of the 4th ESRIN-ESLAB Symposium, held at Frascati, Italy, July 6–10, 1970.* 1971, VIII + 298 pp.

26. I. Atanasijević, *Selected Exercises in Galactic Astronomy.* 1971, XII + 144 pp.

27. C. J. Macris (ed.), *Physics of the Solar Corona. Proceedings of the NATO Advanced Study Institute on Physics of the Solar Corona, held at Cavouri-Vouliagmeni, Athens, Greece, 6–17 September 1970.* 1971, XII + 345 pp.

28. F. Delobeau, *The Environment of the Earth.* 1971, IX + 113 pp.

29. E. R. Dyer (general ed.), *Solar-Terrestrial Physics/1970. Proceedings of the International Symposium on Solar-Terrestrial Physics, held in Leningrad, U.S.S.R., 12–19 May 1970.* 1972, VIII + 938 pp.

30. V. Manno and J. Ring (eds.), *Infrared Detection Techniques for Space Research. Proceedings of the 5th ESLAB-ESRIN Symposium, held in Noordwijk, The Netherlands, June 8–11, 1971.* 1972, XII + 344 pp.

31. M. Lecar (ed.), *Gravitational N-Body Problem. Proceedings of IAU Colloquium No. 10, held in Cambridge, England, August 12–15, 1970.* 1972, XI + 441 pp.

32. B. M. McCormac (ed.), *Earth's Magnetospheric Processes. Proceedings of a Symposium Organized by the Summer Advanced Study Institute and Ninth ESRO Summer School, held in Cortina, Italy, August 30–September 10, 1971.* 1972, VIII + 417 pp.

33. Antonin Rükl, *Maps of Lunar Hemispheres.* 1972, V + 24 pp.

34. V. Kourganoff, *Introduction to the Physics of Stellar Interiors.* 1973, XI + 115 pp.

35. B. M. McCormac (ed.), *Physics and Chemistry of Upper Atmospheres. Proceedings of a Symposium Organized by the Summer Advanced Study Institute, held at the University of Orléans, France, July 31–August 11, 1972.* 1973, VIII + 389 pp.

36. J. D. Fernie (ed.), *Variable Stars in Globular Clusters and in Related Systems. Proceedings of the IAU Colloquium No. 21, held at the University of Toronto, Toronto, Canada, August 29–31, 1972.* 1973, IX + 234 pp.

37. R. J. L. Grard (ed.), *Photon and Particle Interaction with Surfaces in Space. Proceedings of the 6th ESLAB Symposium, held at Noordwijk, The Netherlands, 26–29 September, 1972.* 1973, XV + 577 pp.

38. Werner Israel (ed.), *Relativity, Astrophysics and Cosmology. Proceedings of the Summer School, held 14–26 August, 1972, at the BANFF Centre, BANFF, Alberta, Canada.* 1973, IX + 323 pp.

39. B. D. Tapley and V. Szebehely (eds.), *Recent Advances in Dynamical Astronomy. Proceedings of the NATO Advanced Study Institute in Dynamical Astronomy, held in Cortina d'Ampezzo, Italy, August 9–12, 1972.* 1973, XIII + 468 pp.

40. A. G. W. Cameron (ed.), *Cosmochemistry. Proceedings of the Symposium on Cosmochemistry, held at the Smithsonian Astrophysical Observatory, Cambridge, Mass., August 14–16, 1972.* 1973, X + 173 pp.

41. M. Golay, *Introduction to Astronomical Photometry.* 1974, IX + 364 pp.

42. D. E. Page (ed.), *Correlated Interplanetary and Magnetospheric Observations. Proceedings of the 7th ESLAB Symposium, held at Saulgau, W. Germany, 22–25 May, 1973.* 1974, XIV + 662 pp.

43. Riccardo Giacconi and Herbert Gursky (eds.), *X-Ray Astronomy.* 1974, X + 450 pp.

44. B. M. McCormac (ed.), *Magnetospheric Physics. Proceedings of the Advanced Summer Institute, held in Sheffield, U.K., August 1973.* 1974, VII + 399 pp.

45. C. B. Cosmovici (ed.), *Supernovae and Supernova Remnants. Proceedings of the International Conference on Supernovae, held in Lecce, Italy, May 7–11, 1973.* 1974, XVII + 387 pp.

46. A. P. Mitra, *Ionospheric Effects of Solar Flares.* 1974, XI + 294 pp.

47. S.-I. Akasofu, *Physics of Magnetospheric Substorms.* 1977, XVIII + 599 pp.

48. H. Gursky and R. Ruffini (eds.), *Neutron Stars, Black Holes and Binary X-Ray Sources.* 1975, XII + 441 pp.

49. Z. Švestka and P. Simon (eds.), *Catalog of Solar Particle Events 1955–1969. Prepared under the Auspices of Working Group 2 of the Inter-Union Commission on Solar-Terrestrial Physics.* 1975, IX + 428 pp.

50. Zdeněk Kopal and Robert W. Carder, *Mapping of the Moon.* 1974, VIII + 237 pp.

51. B. M. McCormac (ed.), *Atmospheres of Earth and the Planets. Proceedings of the Summer Advanced Study Institute, held at the University of Liège, Belgium, July 29–August 8, 1974.* 1975, VII + 454 pp.

52. V. Formisano (ed.), *The Magnetospheres of the Earth and Jupiter. Proceedings of the Neil Brice Memorial Symposium, held in Frascati, May 28–June 1, 1974.* 1975, XI + 485 pp.

53. R. Grant Athay, *The Solar Chromosphere and Corona: Quiet Sun.* 1976, XI + 504 pp.

54. C. de Jager and H. Nieuwenhuijzen (eds.), *Image Processing Techniques in Astronomy. Proceedings of a Conference, held in Utrecht on March 25–27, 1975.* XI + 418 pp.

55. N. C. Wickramasinghe and D. J. Morgan (eds.), *Solid State Astrophysics. Proceedings of a Symposium, held at the University College, Cardiff, Wales, 9–12 July 1974.* 1976, XII + 314 pp.

56. John Meaburn, *Detection and Spectrometry of Faint Light.* 1976, IX + 270 pp.

57. K. Knott and B. Battrick (eds.), *The Scientific Satellite Programme during the International Magnetospheric Study. Proceedings of the 10th ESLAB Symposium, held at Vienna, Austria, 10–13 June 1975.* 1976, XV + 464 pp.

58. B. M. McCormac (ed.), *Magnetospheric Particles and Fields. Proceedings of the Summer Advanced Study School, held in Graz, Austria, August 4–15, 1975.* 1976, VII + 331 pp.

59. B. S. P. Shen and M. Merker (eds.), *Spallation Nuclear Reactions and Their Applications.* 1976, VIII + 235 pp.

60. Walter S. Fitch (ed.), *Multiple Periodic Variable Stars. Proceedings of the International Astronomical Union Colloquium No. 29, held at Budapest, Hungary, 1–5 September 1976.* 1976, XIV + 348 pp.

61. J. J. Burger, A. Pedersen, and B. Battrick (eds.), *Atmospheric Physics from Spacelab. Proceedings of the 11th ESLAB Symposium, Organized by the Space Science Department of the European Space Agency, held at Frascati, Italy, 11–14 May 1976.* 1976, XX + 409 pp.

62. J. Derral Mulholland (ed.), *Scientific Applications of Lunar Laser Ranging. Proceedings of a Symposium held in Austin, Tex., U.S.A., 8–10 June, 1976.* 1977, XVII + 302 pp.

63. Giovanni G. Fazio (ed.), *Infrared and Submillimeter Astronomy. Proceedings of a Symposium held in Philadelphia, Penn., U.S.A., 8–10 June, 1976.* 1977, X + 226 pp.

64. C. Jaschek and G. A. Wilkins (eds.), *Compilation, Critical Evaluation and Distribution of Stellar Data. Proceedings of the International Astronomical Union Colloquium No. 35, held at Strasbourg, France, 19–21 August, 1976.* 1977, XIV + 316 pp.

65. M. Friedjung (ed.), *Novae and Related Stars. Proceedings of an International Conference held by the Institut d'Astrophysique, Paris, France, 7–9 September, 1976.* 1977, XIV + 228 pp.

66. David N. Schramm (ed.), *Supernovae. Proceedings of a Special IAU-Session on Supernovae held in Grenoble, France, 1 September, 1976.* 1977, X + 192 pp.

67. Jean Audouze (ed.), *CNO Isotopes in Astrophysics. Proceedings of a Special IAU Session held in Grenoble, France, 30 August, 1976.* 1977, XIII + 195 pp.

68. Z. Kopal, *Dynamics of Close Binary Systems,* XIII + 510 pp.

69. A. Bruzek and C. J. Durrant (eds.), *Illustrated Glossary for Solar and Solar-Terrestrial Physics.* 1977, XVIII + 204 pp.

70. H. van Woerden (ed.), *Topics in Interstellar Matter.* 1977, VIII + 295 pp.

71. M. A. Shea, D. F. Smart, and T. S. Wu (eds.), *Study of Travelling Interplanetary Phenomena.* 1977, XII + 439 pp.

72. V. Szebehely (ed.), *Dynamics of Planets and Satellites and Theories of Their Motion. Proceedings of IAU Colloquium No. 41, held in Cambridge, England, 17–19 August 1976.* 1978, XII + 375 pp.

73. James R. Wertz (ed.), *Spacecraft Attitude Determination and Control.* 1978, XVI + 858 pp.

74. Peter J. Palmadesso and K. Papadopoulos (eds.), *Wave Instabilities in Space Plasmas. Proceedings of a Symposium Organized Within the XIX URSI General Assembly held in Helsinki, Finland, July 31–August 8, 1978.* 1979, VII + 309 pp.

75. Bengt E. Westerlund (ed.), *Stars and Star Systems. Proceedings of the Fourth European Regional Meeting in Astronomy held in Uppsala, Sweden, 7–12 August, 1978.* 1979, XVIII + 264 pp.

76. Cornelis van Schooneveld (ed.), *Image Formation from Coherence Functions in Astronomy. Proceedings of IAU Colloquium No. 49 on the Formation of Images from Spatial Coherence Functions in Astronomy, held at Groningen, The Netherlands, 10–12 August 1978.* 1979, XII + 338 pp.

77. Zdeněk Kopal, *Language of the Stars. A Discourse on the Theory of the Light Changes of Eclipsing Variables.* 1979, VIII + 280 pp.

78. S.-I. Akasofu (ed.), *Dynamics of the Magnetosphere. Proceedings of the A.G.U. Chapman Conference 'Magnetospheric Substorms and Related Plasma Processes' held at Los Alamos Scientific Laboratory, N.M., U.S.A., October 9–13, 1978.* 1980, XII + 658 pp.

79. Paul S. Wesson, *Gravity, Particles, and Astrophysics. A Review of Modern Theories of Gravity and G-variability, and their Relation to Elementary Particle Physics and Astrophysics.* 1980, VIII + 188 pp.

80. Peter A. Shaver (ed.), *Radio Recombination Lines. Proceedings of a Workshop held in Ottawa, Ontario, Canada, August 24–25, 1979.* 1980, X + 284 pp.